ROUTLEDGE HANDBOOK OF FOREST ECOLOGY

The *Routledge Handbook of Forest Ecology* is an essential resource covering all aspects of forest ecology from a global perspective.

This new edition has been fully revised and updated throughout to reflect the profound and unprecedented changes in both forests and climates since the publication of the first edition in 2015. The handbook reflects key developments in the field of forest dynamics and large-scale processes, as well as the changes that are now manifesting in different types of forests across the globe as a result of climate change. It covers both natural and managed forests, from boreal, temperate, sub-tropical and tropical regions of the world. In this second edition, the breadth of the handbook has been expanded with new chapters on mountain forests, monodominance, pathogens and invertebrate pests and amphibians and reptiles in forest ecosystems. Original author teams are complemented by the addition of new authors to offer fresh perspectives, and the second edition places greater emphasis on the applicability of each topic at a global level. The handbook is divided into seven parts:

- Part I: The forest
- Part II: Forest dynamics
- Part III: Forest flora and fauna
- Part IV: Energy and nutrients
- Part V: Forest conservation and management
- Part VI: Forest and climate change
- Part VII: Human ecology

The *Routledge Handbook of Forest Ecology* is an essential reference text for a wide range of students and scholars of ecology, environmental science, forestry, geography and natural resource management.

Kelvin S.-H. Peh is an associate professor of conservation science in the Faculty of Environmental and Life Sciences, University of Southampton, UK.

Richard T. Corlett is an Emeritus professor at the Xishuangbanna Tropical Botanical Garden, Chinese Academy of Sciences, Yunnan, China, and an honorary research fellow at the Royal Botanic Gardens, Kew, UK. He was previously a professor at the National University of Singapore and the University of Hong Kong, China.

Yves Bergeron is an Emeritus professor of forest ecology and management at Université du Québec en Abitibi-Témiscamingue and Université du Québec à Montréal, Canada.

ROUTLEDGE HANDBOOK OF FOREST ECOLOGY

Second Edition

Edited by Kelvin S.-H. Peh, Richard T. Corlett and Yves Bergeron

Designed cover image: Valery Rybakou / Getty Images

Second edition published 2025
by Routledge
4 Park Square, Milton Park, Abingdon, Oxon, OX14 4RN

and by Routledge
605 Third Avenue, New York, NY 10158

Routledge is an imprint of the Taylor & Francis Group, an informa business

First edition published by Routledge 2015

British Library Cataloguing-in-Publication Data
A catalogue record for this book is available from the British Library

ISBN: 978-1-032-34838-4 (hbk)
ISBN: 978-1-032-34842-1 (pbk)
ISBN: 978-1-003-32407-2 (ebk)

DOI: 10.4324/9781003324072

Typeset in Galliard
by Newgen Publishing UK

Access the Support Material: https://doi.org/10.6084/m9.figshare.26097829.v1

CONTENTS

List of contributors *x*
List of figures *xvii*
List of tables *xxiv*

1 Introduction 1
Kelvin S.-H. Peh, Yves Bergeron and Richard T. Corlett

PART I
The forest **5**

2 Boreal forests 7
Jean-Pierre Saucier, Ken Baldwin, Pavel Krestov and Torre Jorgenson

3 Temperate forests 29
Lee E. Frelich and Rebecca A. Montgomery

4 Subtropical forests 49
Richard T. Corlett and Alice C. Hughes

5 Tropical forests 61
Richard T. Corlett

6 Temperate and boreal mountain forests 74
Christopher Carcaillet and Peter Z. Fulé

PART II
Forest dynamics **93**

7 Insect disturbances in forest ecosystems 95
Daniel Kneeshaw, Brian R. Sturtevant, Barry Cooke, Timothy Work, Deepa Pureswaran, Louis DeGrandpré and David A. MacLean

8 Fire in forest ecosystems 119
David F. Greene, Madeleine A. Lopez and Sean T. Michaletz

9 Ecological effects of strong winds on forests 133
Stephen M. Turton and Mohammed Alamgir

10 Forest succession and gap dynamics 148
Rebecca A. Montgomery and Lee E. Frelich

11 Tree genetic diversity and gene flow in forest ecosystems 164
Francine Tremblay

12 Monodominance in tropical lowland forests 188
Kelvin S.-H. Peh

PART III
Forest flora and fauna **209**

13 The ecology of lianas and their increasing influence in tropical forests 211
Stefan A. Schnitzer

14 Vascular epiphytes in forest ecosystems 230
Amanda Taylor

15 Insects in forest ecosystems 241
Andrea Battisti

16 Bryophytes in forest ecosystems: diversity, function and management 255
Nicole J. Fenton, Kristoffer Hylander, Emma Pharo and Charles E. Zartman

17 Lichens in forest ecosystems 268
Per-Anders Esseen, Göran Thor and Darwyn Coxson

18 Mammals in forest ecosystems 283
Richard T. Corlett and Alice C. Hughes

19 Ecology and conservation of forest birds 299
Malcolm C. K. Soh, Ding Li Yong, Richard T. Corlett and Kelvin S.-H. Peh

20 Amphibians and reptiles of forest ecosystems 314
David Bickford, Adrian Garda, Seshadri K.S., Umilaela Arifin, Hiral Naik, Hanyeh Ghaffari and Barbod Safaei-Mahroo

21 Global patterns of biodiversity in forests 327
Christine B. Schmitt and João de Deus Vidal Jr.

PART IV
Energy and nutrients **341**

22 Mycorrhizal symbiosis in forest ecosystems 343
Leho Tedersoo

23 Biogeochemical cycling 356
David Paré, Daniel Markewitz and Håkan Wallander

24 Forest hydrology 372
André St-Hilaire

25 Primary production and allocation in forest ecosystems 380
Frank Berninger and Kelvin S.-H. Peh

PART V
Forest conservation and management **397**

26 Natural regeneration after harvesting 399
Nelson Thiffault, Lluís Coll, Douglass F. Jacobs, Marie Ange Ngo Bieng and Eliott Maurent

27 Tropical deforestation, forest degradation, and the role of REDD+ in climate change mitigation 412
John A. Parrotta

28 Restoration of forest ecosystems 426
John A. Stanturf

29 Forest fragments and fragmentation 445
Michael D. Pashkevich, Badrul Azhar, Damayanti Buchori, Robert J. Fletcher, Jr., Jake L. Snaddon and Edgar C. Turner

30 The ecology of logged forests 459
Paul Woodcock, Panu Halme and David P. Edwards

31 Pollution in forests 475
Mikhail V. Kozlov and Elena L. Zvereva

32 Biological invasions in forests and forest plantations 491
Marcel Rejmánek

33 Pathogens and invertebrate pests in North American forest ecosystems 516
Sandy M. Smith and Louis Bernier

PART VI
Forest and climate change 533

34 Fire and climate: using the past to predict the future 535
Marion Lestienne, Justin Waito, Chéïma Barhoumi, Laurent Bremond, Martin P. Girardin, Jacques C. Tardif, Adam A. Ali, Hermann Behling, Julia Unkelbach and Christelle Hély

35 The ecological consequences of droughts in forests 552
Richard T. Corlett

36 From wood formation to tree ring: How tree growth responds to climate change 564
Roberto Silvestro, Minhui He, Jian-Guo Huang, Hubert Morin and Sergio Rossi

37 Plant movements in response to rapid climate change 579
Richard T. Corlett

38 Forest carbon budgets and climate change 589
Yadvinder Malhi, Tina Christmann, Xiongjie Deng, Huanyuan Zhang-Zheng, Sam Moore and Terhi Riutta

PART VII
Human ecology **611**

39 Multiple roles of non-timber forest products in ecologies, economies and livelihoods 613
Charlie M. Shackleton

40 Agriculture in the forest: Ecology and rationale of shifting cultivation 625
Olivier Ducourtieux

41 Indigenous forest knowledge 642
Hugo Asselin

42 Wild meat hunting in tropical forests 654
Julia E. Fa and Stephan M. Funk

Index *677*

CONTRIBUTORS

Kelvin S.-H. Peh, Associate Professor, School of Biological Sciences, University of Southampton, Southampton, UK.

Yves Bergeron, Emeritus Professor, Institut de recherche sur les forêts Université du Québecen Abitibi- Témiscamingue and Département des sciences biologiques, Université du Québec à Montréal, Canada.

Richard T. Corlett, Emeritus Professor, Center for Integrative Conservation and Yunnan Key Laboratory for the Conservation of Tropical Rainforests and Asian Elephants, Xishuangbanna Tropical Botanical Garden, Chinese Academy of Sciences; Honorary Research Fellow, Royal Botanic Gardens Kew.

Jean-Pierre Saucier, Forester and ecologist, Nemoralis Écologie- Foresterie, Québec, Canada.

Ken Baldwin, Forest Ecologist, retired from Natural Resources Canada, Canadian Forest Service, Canada.

Pavel Krestov, Biogeographer and Vegetation Ecologist; Director, Botanical Garden-Institute of the Far Eastern Branch, Russian Academy of Sciences, Vladivostok, Russia.

Torre Jorgenson, Landscape Ecologist, Alaska Ecoscience, USA.

Lee E. Frelich, Director, Center for Forest Ecology, University of Minnesota, St. Paul, USA.

Rebecca A. Montgomery, Professor, Department of Forest Resources, University of Minnesota, St. Paul, USA.

Alice C. Hughes, Associate Professor, School of Biological Sciences, University of Hong Kong, Hong Kong.

Christopher Carcaillet, Professor, École Pratique des Hautes Études, Paris Sciences & Lettres University (EPHE- PSL), France; University of Lyon, Université Claude Bernard Lyon 1, Villeurbanne, France; Churchill College, University of Cambridge, UK.

Peter Z. Fulé, Regents' Professor, School of Forestry, Northern Arizona University, Flagstaff, AZ, USA.

Daniel Kneeshaw, Professor, Centre d'étude de la forêt et Département des sciences biologiques Université du Québec à Montréal, Canada.

Brian R. Sturtevant, Research Ecologist, Northern Research Station, United States Department of Agriculture Forest Service, USA.

Barry Cooke, Research Scientist, Northern Forestry Centre, Natural Resources Canada, Canada.

Timothy Work, Entomologist, Département des sciences biologiques, Université du Québec à Montréal, Canada.

Deepa Pureswaran, Research Scientist, Laurentian Forestry Centre, Natural Resources Canada, Government of Canada.

Louis DeGrandpré, Research Scientist, Laurentian Forestry Centre, Natural Resources Canada, Government of Canada.

David A. MacLean, Professor, Faculty of Forestry and Environmental Management, University of New Brunswick, Canada.

David F. Greene, Professor, Department of Forestry, Fire, and Rangeland Management, Cal Poly Humboldt, Arcata, USA.

Madeleine A. Lopez, Doctoral Researcher, Department of Geography, Planning and Environment, Concordia University, Montréal, Canada.

Sean T. Michaletz, Assistant Professor, Biodiversity Research Centre, Vancouver BC, Canada.

Stephen M. Turton, Adjunct Professor of Environmental Geography, Division of Research, Central Queensland University, Rockhampton, Australia.

Mohammed Alamgir, Adjunct Senior Research Fellow, Centre for Tropical Environmental and Sustainability Science, James Cook University, Cairns, Australia.

Lee E. Frelich, Director, Center for Forest Ecology, University of Minnesota, St. Paul, USA.

Francine Tremblay, Retired Professor, Forest Research Institute, Université du Québec en Abitibi Témiscamingue; Canada.

Stefan A. Schnitzer, Mellon Distinguished Professor, Marquette University, Milwaukee, WI, USA; and Smithsonian Tropical Research Institute, , Balboa, Republic of Panama.

Amanda Taylor, Postdoctoral Researcher, Biodiversity, Macroecology and Biogeography, Faculty of Forest Sciences and Forest Ecology, University of Göttingen, Göttingen, Germany.

Andrea Battisti, Professor, University of Padova, Padova, Italy.

Nicole J. Fenton, Professor, Forest Research Institute, Université du Québec en Abitibi-Témiscamingue, Canada.

Kristoffer Hylander, Professor, Department of Ecology, Environment and Plant Sciences, Stockholm University, Sweden.

Emma Pharo, Associate Professor, School of Land and Food, University of Tasmania, Australia.

Charles E. Zartman, Researcher, Department of Biodiversity, National Institute of Amazonian Research, Brazil.

Per-Anders Esseen, Professor Emeritus, Department of Ecology and Environmental Science, Umeå University, Umeå, Sweden.

Göran Thor, Professor, Department of Ecology, Swedish University of Agricultural Sciences,, Uppsala, Sweden.

Darwyn Coxson, Professor, University of Northern British Columbia, Ecosystem Science and Management Program, Prince George, BC, Canada.

Malcolm C. K. Soh, Principal Researcher, National Parks Board (NParks), Singapore

Ding Li Yong, Regional Flyway Coordinator, BirdLife International Asia, Singapore

David Bickford, Chair of the International Herpetological Committee, World Congress of Herpetology, USA.

Adrian Garda, Associte Professor, DBEZ- Centro de Biociências, Campus Universitário, Lagoa Nova Universidade Federal do Rio Grande do Norte, Natal, Brazil.

Seshadri K.S., Fellow in Residence, Ashoka Trust for Research in Ecology and the Environment, Bangalore, India.

Umilaela Arifin, Postdoctoral Researcher, Centre for Taxonomy and Morphology (ztm), Leibniz Institute for the Analysis of Biodiversity Change, Hamburg, Germany.

Hiral Naik, Doctoral Researcher, School of Animal, Plant and Environmental Sciences, University of the Witwatersrand, Johannesburg, South Africa.

Hanyeh Ghaffari, Assistant Professor, Department of Environmental Sciences, Faculty of Natural Resources, University of Kurdistan, Sanandaj, Iran.

Barbod Safaei-Mahroo, Chairman and Cofounder, Institute Pars Herpetologists Institute, Tehran, Iran.

Christine B. Schmitt, Professor of Physical Geography with a focus on Human–Environment Research, Geography Section, University of Passau, Germany.

João de Deus Vidal Jr., Postdoctoral researcher, Geography Section, University of Passau, Germany.

Leho Tedersoo, Professor, Institute of Ecology and Earth Sciences, University of Tartu, Estonia.

David Paré, Research Scientist, Laurentian Forestry Centre, Canadian Forest Service, Natural Resources Canada, Québec, Canada.

Daniel Markewitz, Professor, Warnell School of Forestry and Natural Resources, University of Georgia, Athens, GA.

Håkan Wallander, Professor, Department of Biology, Lund University, Sweden.

André St-Hilaire, Professeur- chercheur, Institut National de la Recherche Scientifique, Centre Eau Terre Environnement, Québec, Canada.

Frank Berninger, Professor, Department of Environmental and Biological Sciences, University of Eastern Finland, Joensuu, Finland.

Nelson Thiffault, Research Scientist, Natural Resources Canada, Québec, Canada.

Lluís Coll, Professor, Department of Agricultural and Forest Sciences and Engineering, University of Lleida, Spain.

Douglass F. Jacobs, Fred M van Eck Professor of Forest Biology, Purdue University, West Lafayette, USA.

Marie Ange Ngo Bieng, CIRAD, Université de Montpellier, UR Forêts & Sociétés, Montpellier, 34398, France, CATIE, Centro Agronómico Tropical de

Eliott Maurent, CIRAD, Université de Montpellier, UR Forêts & Sociétés, Montpellier, France.

John A. Parrotta, National Program Leader, International Science Issues, USDA Forest Service, Research & Development, Washington, DC, USA.

John A. Stanturf, Visiting Professor, Institute Forestry and Rural Engineering, Estonian University of Life Sciences, Tartu, Estonia.

Michael D. Pashkevich, Marshall Sherfield Fellow, Department of Zoology, Universityof Cambridge, Cambridge, UK; Research Scientist, Natural Resources

Badrul Azhar, Associate Professor, Department of Forest Science and Biodiversity, Faculty of Forestry and Environment, Universiti Putra Malaysia.

Damayanti Buchori, Professor, Center for Transdisciplinary and Sustainability Sciences, IPB University, Bogor; and Department of Plant Protection, Faculty of Agriculture, IPB University, Bogor.

Robert J. Fletcher, Jr., Professor, Department of Wildlife Ecology and Conservation, University of Florida; Department of Zoology, University of Cambridge, Cambridge, UK.

Jake L. Snaddon, Lecturer, School of Geography and Environmental Science, University of Southampton and University of Belize Environmental Research Institute.

Edgar C. Turner, Professor, Department of Zoology, University of Cambridge, Cambridge, UK.

Paul Woodcock, Biodiversity Evidence Specialist, Joint Nature Conservation Committee, Peterborough, UK.

Panu Halme, Senior Lecturer, Department of Biological and Environmental Science, University of Jyväskylä, Finland.

David P. Edwards, Professor, Department of Plant Sciences and Conservation Research Institute, University of Cambridge, Cambridge, UK.

Mikhail V. Kozlov, Adjunct Professor, Department of Biology, University of Turku, Finland.

Elena L. Zvereva, Adjunct Professor, Department of Biology, University of Turku, Finland.

Marcel Rejmánek, Professor Emeritus, Department of Evolution and Ecology, University of California, Davis, Davis, CA, USA.

Sandy M. Smith, Professor, Institute of Forestry & Conservation, John H Daniels Faculty of Architecture, Landscape & Design, University of Toronto, Toronto, Ontario, Canada.

Louis Bernier, Emeritus Professor, Université Laval, Centre d'étude de la forêt (CEF) / Centre for Forest Research (CFR) & Institut de biologie intégrative et des systèmes (IBIS), Québec, Canada.

Marion Lestienne, Teaching Assistant, ISEM, University of Montpellier, Montpellier, France.

Justin Waito, Researcher, Centre for Forest Interdisciplinary Research (C- FIR), University of Winnipeg, Winnipeg, Manitoba, Canada.

Chéïma Barhoumi, Teaching Assistant, ISEM, University of Montpellier, Montpellier, France.

Laurent Bremond, Researcher, ISEM, University of Montpellier, Montpellier, France.

Martin P. Girardin, Researcher, Canadian Forest Service, Natural Resources Canada, Quebec, Quebec, Canada.

Jacques C. Tardif, Researcher, Centre for Forest Interdisciplinary Research (C-FIR), University of Winnipeg, Winnipeg, Manitoba, Canada.

Adam A. Ali, Researcher, ISEM, University of Montpellier, Montpellier, France.

Hermann Behling, Researcher, Department of Palynology and Climate Dynamics, Albrechtvon-Haller- Institute for Plant Sciences, Göttingen, Germany.

Julia Unkelbach, Researcher, Department of Palynology and Climate Dynamics, Albrechtvon- Haller- Institute for Plant Sciences, Göttingen, Germany.

Christelle Hély, Researcher, ISEM, EPHE, Université PSL, University of Montpellier, Montpellier, France.

Roberto Silvestro, Postdoctoral Researcher, Laboratoire sur les écosystèmes terrestres boréaux, Département des Sciences Fondamentales, Université du Québec à Chicoutimi, Chicoutimi (QC) G7H2B1, Canada.

Minhui He, Postdoctoral Researcher, Laboratoire sur les écosystèmes terrestres boréaux, Département des Sciences Fondamentales, Université du Québec à Chicoutimi, Chicoutimi, Canada.

Jian-Guo Huang, Professor, MOE Key Laboratory of Biosystems Homeostasis and Protection, College of Life Sciences, Zhejiang University, Hangzhou, China.

Hubert Morin, Professor, Laboratoire sur les écosystèmes terrestres boréaux, Département des Sciences Fondamentales, Université du Québec à Chicoutimi, Chicoutimi (QC), Canada.

Sergio Rossi, Professor, Laboratoire sur les écosystèmes terrestres boréaux, Département des Sciences Fondamentales, Université du Québec à Chicoutimi, Chicoutimi, Canada.

Yadvinder Malhi, Professor, Environmental Change Institute, School of Geography and theEnvironment, University of Oxford, Oxford, UK.

Tina Christmann, Doctoral Researcher, Environmental Change Institute, School of Geography and the Environment, University of Oxford, Oxford, UK; Lecturer in Environmental Science, University of Southampton, Southampton, UK.

Xiongjie Deng, Doctoral Researcher, Environmental Change Institute, School of Geography and the Environment, University of Oxford, Oxford, UK.

Huanyuan Zhang- Zheng, Doctoral Researcher, Environmental Change Institute, School of Geography and the Environment, University of Oxford, Oxford, UK.

Sam Moore, Postdoctoral Researcher, Environmental Change Institute, School of Geography and the Environment, University of Oxford, Oxford, UK.

Terhi Riutta, Postdoctoral Researcher, Environmental Change Institute, School of Geography and the Environment, University of Oxford, Oxford, UK.

Charlie M. Shackleton, Professor, Department of Environmental Science, Rhodes University, Makhanda, South Africa.

Olivier Ducourtieux, Associate Professor, Comparative Agriculture Unit/ UMR PRODIG, AgroParisTech, Paris-Saclay University, France.

Hugo Asselin, Professor, School of Indigenous Studies, Université du Québec en Abitibi-Témiscamingue, Canada.

Julia E. Fa, Senior Research Associate, Center for International Forestry Research (CIFOR), CIFOR Headquarters, Bogor, Indonesia and Professor of Biodiversity and Human Development, Department of Natural Sciences, School of Science and the Environment, Manchester Metropolitan University, Manchester, UK.

Stephan M. Funk, Nature Heritage, Director, Jersey, Channel Islands, and Lemu Earth Spa, Chief Science Officer, Santiago, Chile.

FIGURES

2.1	Extent of boreal forests and woodlands around the globe, with floristic subdivisions and non-forest boreal vegetation zones	8
2.2	Distribution of permafrost zones in the Northern Hemisphere	22
3.1	Location of the world's temperate forests	30
3.2	Northern hemisphere forests	31
3.3	Nothofagus forest in New Zealand	32
3.4	Eucalyptus forest in southeastern Australia	33
3.5	Dacrycarpus forest in New Zealand	33
4.1	Potential forest cover between 23.4° and 32.0°, under the assumption that all areas with >700 mm annual rainfall and <4,000 m elevation are capable of supporting forest	50
6.1	Examples of boreal and temperate mountain forests, ordered from the North to the South	75
6.2	Distribution of mountain forests per continent or per vegetation type in area (Mha) and in proportion (%)	76
6.3	Main environmental characteristics that organize mountain forest ecosystems, distinct from the driving characteristics of plain forests	78
6.4	Mountain forest landscape naturally disturbed by wildfires and avalanche/rockfall, resulting in a mosaic of communities along slope and valley gradients	83
6.5	Animals of mountain forests	85
7.1	Area disturbed by insects (mountain pine beetle, forest tent caterpillar and spruce budworm combined), by fire and by harvesting from 2005 to 2020 in Canada.	96
7.2	A schematic representation of different outbreak patterns in time	98
7.3	(a) Spatially, outbreak intensity is distributed according to a latitudinal belt-shaped gradient pattern, shown for spruce budworm defoliation in Quebec and Ontario, Canada, from 1938 to 1992. (b) Cyclic eruptions occur every few decades, with pulses of defoliation occurring every three to seven years, which are sustained through a slow decline in host forest condition	99

7.4 Factors influencing insect populations 100
7.5 Forest dynamics following different outbreak severities 107
8.1 Diagram illustrating the phases and products of combustion for a forest fuel element 120
8.2 Fire shape is the outcome of fire spread rates in all directions from the ignition point (black points) 121
8.3 Mechanisms of heat transfer from a fire to plant root, stem and crown surfaces 123
8.4 Recruitment density across a 1000 m-wide burned patch at the 2008 Klamath Complex Fire in northern California 126
8.5 The limitation of small germinants to the thinnest post-fire organic substrates: an example from the Monday fire in the boreal forest of Saskatchewan 128
9.1 Structural, functional, successional and biodiversity effects of strong winds on forests ecosystems 135
9.2 Graph shows principles of intermediate disturbance hypothesis 139
10.1 Stages of stand development 149
10.2 The intermediate disturbance hypothesis 155
10.3 Three alternate successional pathways for forest development, showing the relative levels of structural complexity exhibited in each seral stage 157
11.1 Clinal variation in quantitative traits 176
12.1 Types of monodominance in tropical lowland rain forests 192
12.2 Model of possible mechanisms (grey boxes) and their consequences (white boxes) leading to classical (persistent) monodominance in tree species growing in environmental conditions similar to adjacent high-diversity forests 202
13.1 The change in liana density with increasing latitude from the southern edge of the tropics to the northern hemisphere 212
13.2 The response of trees to liana removal in 25 separate ecological experiments conducted in tropical and subtropical forests 214
13.3 The change in liana and tree density with mean annual precipitation in lowland tropical forests 218
13.4 Mean bootstrapped relative growth and annualized relative growth for lianas (first column, N = 648 individuals / 54 species) and trees (second column, *N* = 1,117 individuals/128 species) from 2011 to 2016 on Gigante Peninsula in central Panama 220
14.1 Global patterns of epiphyte species richness per 10,000 km^2 231
14.2 The relationship between absolute latitude and (A) the total number of vascular epiphytes (black points) and terrestrial plants (grey points) 233
14.3 Examples of vascular epiphyte assemblages of seed plants (A) and pteridophytes (D), epiphyte–animal interactions (B), and (C) the advantage of a tank growth form to live in extreme environments, such as tree canopies or — this in this case — a telephone wire 234
14.4 Common morphological traits of vascular epiphytes in relation to microclimate gradients within the canopy 237
15.1 Effects of latitude and elevation on the distribution of organisms in the Northern Hemisphere 242

15.2 Distribution of the number of species among classes of the animal kingdom represented by the relative size of a component of each group 243
15.3 Section of a larch trunk from the Italian Central Alps with the indication of the growth rings periodically reduced, as shown by the graph 244
16.1 Species accumulation curves for bryophytes in scattered 0.02 ha plots (in most cases 10 x 20 m) in different biomes and forest environments across the world 257
16.2 Summary of effects of bryophyte mats in various positions (over mineral soil, on branches or on leaves) on water (solid lines) and nutrients (dashed lines) fluxes 260
17.1 The crustose lichen *Calicium tigillare* on wood in boreal forest, Norway (top left). The foliose lichens *Lobaria pulmonaria* and *L. scrobiculata* (dark grey, in centre) epiphytic on *Picea glauca* in sub-boreal forest in northern British Columbia, Canada (top middle). The fruticose, pendulous lichens *Alectoria sarmentosa* (light coloured) and *Bryoria* spp. (dark coloured) on *Abies lasiocarpa* in subalpine forest, British Columbia, Canada (top right). The foliicolous lichen *Eremothecella calamicola* on a leaf of *Arenga engleri* in subtropical forest, Iriomote Island, South Japan (middle left). The foliose lichen *Pseudocyphellaria coronata* on *Dacrycarpus dacrydioides* in south temperate New Zealand rainforest (middle right). The foliose lichen *Anzia japonica* on bark of *Abies sachalinensis* in hemiboreal forest, Hokkaido, Japan (bottom left). The fruticose lichen *Cladonia sulphurina* growing in boreal forest floor, Norway (bottom right). Photographs (top left) and (bottom right) by Einar Timdal, (top middle) and (middle right) by Darwyn Coxson, (top right) by Per-Anders Esseen, (middle left) by Kento Miyazawa, and (bottom left) by Yoshihito Ohmura 270
17.2 Relationship between biomass of epiphytic macrolichens per branch and tree age (A), and between number of invertebrates per branch and lichen biomass (B) in lower canopy of *Picea abies* forests in northern Sweden 274
17.3 Contrast between heavily grazed (left) and ungrazed forest floor (enclosure from 1949) with reindeer lichens in dry, boreal *Pinus sylvestris* forest in Sweden 275
17.4 A comparison of the vertical distribution of canopy epiphyte functional groups in old-growth (>400 years old) coastal (redrawn from McCune, 1993) and interior (redrawn from Benson and Coxson 2002) wet temperate rainforests in western North America 276
19.1 The proportion of vertebrate-dispersed angiosperm families dispersed by birds is higher than by mammals (here, primates and bats) 302
19.2 Oriental honey buzzards' migratory routes in autumn and spring (Sugasawa and Higuchi, 2019) 305
19.3 Spatial distribution of colour diversity in passerines 306
19.4 Snared birds in Southeast Asian markets 307
19.5 Warmer temperatures are driving montane species upslope on the Cerro de Pantiacolla in southern Peru 308
21.1 Global centres (shown in black) of bird species endemism (a) and richness (b) 329

21.2 Forest and non-forest areas for the 14 WWF biomes 332
21.3 Ranked species-abundance distributions for 262 woody species in the Ethiopian moist montane forests (logarithmic scale) 334
22.1 Structural differences among four major types of mycorrhiza 344
22.2 Relative placement of mycorrhizal fungi in the parasitism–mutualism–saprotrophy continuum (above dashed line) 345
22.3 Simplified structure of metanetworks in forest ecosystems emphasizing on the relationships of mycorrhizal fungi and plants with other major organism groups aboveground (top) and belowground (bottom) 349
23.1 Generalized model of biogeochemical cycling of elements in ecosystems demonstrating a balance of inputs, outputs and change in storage 360
23.2 Patterns of soil N accumulation in primary successional sequences 361
23.3 Conceptualized patterns of changes with time in rates of nutrient fluxes along primary and secondary successions 363
23.4 Patterns of soil N and P availability throughout a primary chronosequence over a millennial timescale from Hawaii and a secondary chronosequence over a decadal timescale from the Brazilian Amazon 365
23.5 Ectomycorrhizal fungi (EM) effects on C and N cycling in forest soil 366
25.1 Productivity of a boreal, a temperate, and a humid tropical ecosystem (all in g C m^{-2} yr^{-1}) 382
25.2 Rates of photosynthesis of Scots pine at different light levels 384
25.3 Rates of photosynthesis of different coniferous (closed symbols) and deciduous (open symbols) species in Wisconsin as a function of the leaf nitrogen concentration 386
25.4 Shoot fraction for pendolous birch (*Betula pendula*) (closed circles), Scots pine (*Pinus sylvestris*) (triangles), Contorta pine (*Pinus contorta*) (open circles), and Norway spruce (*Picea abies*) (crosses) as a function of relative nitrogen supply 389
25.5 Biomass fraction of root (RMF), stem (SMF), leaf (LMF), and reproductive (ReMF) as affected by drought in different plant forms 391
26.1 Conceptual representation of regeneration methods within a gradient of severity, clearing size, and frequency 400
26.2 Main factors affecting natural regeneration following harvesting 402
26.3 (a) Natural seedling of balsam fir established in the understory of a mature stand in northern Québec (Canada). (Photograph by N. Thiffault.) (b) Harvest trail from a selection cut in a sugar maple-dominated stand of Maine (USA). (Photograph by N. Thiffault.) (c) Experimental group selection cuttings of 0.2 and 0.4 ha applied in a Supra-Mediterranean Scots pine stand near Solsona, Lleida (Spain). (Photograph by Santiago Martín Alcón.) (d) Damage to soil in a trail used for harvesting in a tropical stand in Costa Rica. (Photograph by E. Maurent.) 404
27.1 The major carbon fluxes in forest ecosystems 417
27.2 Economic and social impacts of REDD+ management actions on different stakeholders within a landscape 422
28.1 Forests in Germany degraded by severe drought and bark beetle infestations 427
28.2 Degraded forest reserve in Ghana restored using the taungya system 429

28.3 Natural regeneration development in an exclosure 430
28.4 Hybrid poplar nurse crop over direct seeded *Fagus sylvatica* on private land in western Denmark 434
28.5 Mountain afforestation can require physical structures to control runoff and terracing for planting spots 437
29.1 Forest fragmentation studies over time, showing that the number of studies has increased dramatically over the last 30 years 446
29.2 Location of some of the most influential large-scale experiments focused on forest fragmentation 447
29.3 Identifying tractable strategies that restore degraded and fragmented forests is needed to conserve forests worldwide 453
30.1 A five-year-old clear-cut boreal forest in Finland, where energy-wood was harvested, has very little leftover deadwood 462
30.2 Examples of recently (<8 years) relogged forest in Borneo and showing variation in vegetation structure in areas more heavily logged (a) compared to less heavily logged (b) 463
30.3 A boreal clear-cut in Finland 9 years after harvesting 468
31.1 (a) Industrial barren located 4 km south of the nickel–copper smelter in Monchegorsk, north-west Russia. This barren area evolved from dense Norway spruce forest. Note the dead trees, the absence of field layer vegetation, and extensive soil erosion. (b) Modifications of the crown structure of Norway spruce (foreground) and Scots pine (background) in this barren site 478
31.2 A meta-analysis on the effects of industrial pollution on different characteristics and different trophic levels of biota 480
31.3 Relationships between the duration of pollution impacts and the magnitudes of the pollution effects on individual fitness, abundance, and diversity of organisms 481
31.4 A meta-analysis on the effects of industrial pollution on organisms in different biomes 485
31.5 Relationships between mean midsummer temperatures in the polluter's locality and the magnitudes of the observed effects of pollution on organisms from four trophic levels 486
32.1 The global donor–acceptor network of invasive alien trees and shrubs 492
32.2 *Ehrharta erecta*, a highly invasive perennial grass native to South Africa, forms a dense cover in the Douglas fir (*Pseudotsuga menziesii*) forests in the Point Reys National Seashore, California 497
32.3 Introduced eastern fox squirrel (*Sciurus niger*) eating drupes of introduced Chinese pistache (*Pistacia chinensis*) in California 509
33.1 Biological cycle of Dutch elm disease (DED), an example of fungal pathogen–insect vector association that has resulted in two devastating pandemics 519
34.1 Examples of annual area burned time series for various countries 536

34.2	(a) Time-since-fire (TSF) map for the Duck Mountain Provincial Forest, Manitoba, Canada, as of 2002. The TSF date (i.e. approximate year of burn) for each of the sites was either provided by the oldest trees in the overstory cohort, by sampling of burned material left on the ground (b), and/or by fire scar date (c)	540
34.3	Summary of the main analytical steps required for reconstructing past fire frequency from sedimentary charcoal records	541
34.4	(a) General Mediterranean vegetation dynamics and comparison with (b) modeled fire hazard (characterized by the Monthly Drought Code and the Fire Season Length (see Lestienne et al. (2020a) for details) and (c) reconstructed fire frequency	544
36.1	Transverse section of a weekly sampled microcore, observed at 400× magnification, for counting the developing tracheids, and classified as (CA) cambium, (EC) enlarging cells, (WTC) wall-thickening and lignifying cells and (MC) mature cells of the previous year	566
36.2	Number of cambial, enlarging, wall thickening and lignifying, and mature cells observed in 159 balsam firs during 2018	568
36.3	Duration of xylogenesis and amount of cell production versus the May–September temperatures and the snow-free period recorded in five study sites in the boreal forest of Quebec, Canada	570
36.4	Standardized major axis (SMA) regressions among timings of onset (light grey dots) and ending (black dots) of xylem phenology, and the total number of cells in the tree ring at the end of the growing season in 159 balsam firs	570
36.5	Location of the tree-ring sampling sites according to the International Tree-Ring Data Bank (ITRDB), the most comprehensive repository of tree growth	572
38.1	Example of vegetation and soil C pools (Mg C ha^{-1}) and C fluxes (Mg C ha^{-1} year–1) in mature, closed canopy forests in tropical, temperate and boreal zones	592
38.2	Terrestrial laser scanning (TLS), unmanned aerial vehicle (UAV), plane and satellite remote sensing technologies are part of a remote sensing system able to quantify carbon fluxes (gross primary production (GPP), autotrophic respiration (AR), net primary production (NPP), ecosystem respiration (RECO), net ecosystem production (NEP) and net biome production (NBP)) and stocks (aboveground biomass (AGB) and soil organic carbon (SOC))	594
38.3	General scheme of how disturbances, forest regrowth and atmospheric changes at a forest stand level impact local carbon flux	601
40.1	Rotation of a long fallow phase and a short cultivation phase in a shifting cultivation	629
40.2	Secondary succession in a shifting cultivation fallow	630
40.3	Shifting cultivation crisis with accelerating rotation	632
40.4	Factors of the decline in tropical forest cover	636

41.1 Two white pines (*Pinus strobus*) towering above the surrounding canopy and used as landmarks by the members of the Kitcisakik Algonquin community 646
41.2 Beaver (*Castor canadensis*) dam preventing stream water flow through a culvert 647
42.1 Examples of animal types consumed as wild meat in different parts of the world 655

TABLES

2.1 Area of boreal forests and woodlands by country and proportion of the country in closed forest or exploitable closed forest 9
2.2 Climatic values for bioclimates within the boreal macrobioclimate 10
2.3 Dominant boreal tree or shrub/herb species by continent and floristic subdivisions of the boreal zone 11
9.1 Tree and forest stand attributes and likely effects of strong winds 136
11.1 Impacts of evolutionary forces on population genetic diversity 166
11.2 Examples of observed long-distance pollen and seed dispersal in trees (>3 km for pollen and >1 km for seeds) 169
11.3 Examples of genetic diversity (microsatellite markers, allozyme loci and ESTPs of ncDNA3) studies in populations of tree species 171
12.1 Diversity of monodominant species in tropical lowland forest 189
15.1 Major feeding guilds of main insect orders 245
18.1 Major ecological roles played by the mammalian orders in forests 284
18.2 The recent global distributions of mammalian orders in forests 285
21.1 Biodiversity criteria and indicators 328
21.2 Number of species of mammals, birds, amphibians, reptiles and conifers in the IUCN forest habitats 331
23.1 Nutrient fluxes in a tropical rain forest, a deciduous temperate forest and a boreal forest 359
24.1 Combined impact of canopy and wind on snow storage, according to Dickerson-Lange et al. (2021) 373
25.1 Definition of different productivity terms 383
27.1 Relevance of management interventions to the five REDD+ activities 419
32.1 Some prominent examples of non-native plant species invading forests and forest plantations 493
32.2 Some prominent examples of non-native earthworm species invading forests and forest plantations 499
32.3 Some prominent examples of non-native insect species invading forests and forest plantations 502

32.4 Numbers of non-native forest insect species established in the continental United States 504
32.5 Some prominent examples of non-native bird species invading forests and forest plantations 507
32.6 Some prominent examples of non-native mammal species invading forests and forest plantations 508
38.1 Forest area and forest C stocks and fluxes in tropical, temperate and boreal zones 596
39.1 Illustrative examples of the range of NTFP contribution to rural household income (cash and non-cash combined) 617
42.1 Estimated sustainability and decline in population densities of mammals due to hunting 666

1
INTRODUCTION

Kelvin S.-H. Peh, Yves Bergeron and Richard T. Corlett

Almost a decade since the first edition of this handbook was published in 2015, our world is a different place. Changes in both forests and climates since then have been unprecedented and profound. On the global forest front, we are losing a large extent of primary forests annually, particularly in the tropics. In a recent study, nearly half of the more than 70,000 animal species for which data was available—many associated with forests—had decreased in their numbers. As greenhouse gas emissions continue to grow, the planet has now warmed by 1.1°C above pre-industrial levels, and the carbon dioxide concentration in the atmosphere exceeded 420 ppm in 2023. Forest ecosystems, together with all living things they harbour, are now affected by climate change more frequently and severely, through changes in precipitation patterns, temperature regimes, fire dynamics and extreme weather events.

Yet, our remaining battered forests are still stupendous. Forest ecosystems, which have been evolving over hundreds of millions of years, remain immensely biodiverse, supporting the great majority of terrestrial biodiversity. These ecosystems remain ecologically complex and support the existence of humans. However, humans also cause widespread deforestation and degradation, valuing clearing of forests for agricultural expansion to enable food and biofuel production. Finding common ground between these conflicting human needs calls for a deeper understanding of forest ecology and a greater appreciation of the benefits provided by forests. We hope that this new edition of the handbook will continue to contribute to these aims.

In addition to deforestation and climate change, our forests also face numerous interacting and severe anthropogenic threats. For example, invasive alien species outcompete and displace native forest flora and fauna. Unsustainable logging, fuelwood harvesting, and wild meat hunting deplete forest resources. Pollution and eutrophication from fertilisers and industrial emissions alter tree growth and disrupt nutrient cycles. Together with deforestation and climate change, these threats exponentially worsen the outcomes through synergistic positive feedback effects. We are still observing how the forest ecosystems cope. We stand by our original mission to provide reliable information about forests, their dynamics, biodiversity, responses to human disturbances and climate change, and the applications of ecology in addressing these global challenges, accessible to a wider audience.

This edition of the handbook serves as a state-of-the-art summary of our knowledge of forest ecology. Many chapters from the previous edition have been updated by the original team of authors who are experts in their respective field. We have also added new chapters to

DOI: 10.4324/9781003324072-1

further broaden the range of subjects subsumed in the realm of forest ecology. Some chapters have been rewritten by new expert authors to give fresh perspectives. For this edition, there is a noticeable improvement: we have achieved a truly global coverage to the greatest extent possible for most chapters, with just a few focussing on regions where the topic is most pertinent or has received the most attention. Our aim has been to offer readers a unique perspective by examining each topic across diverse geographical areas or, where relevant.

This edition, like the previous one, is divided into seven parts. The first section, 'The Forest' defines major forest biomes by latitudinal belts, including boreal forests, temperate forests, subtropical forests, and tropical forests. This part also contains a new chapter on mountainous forests. These chapters offer an introductory overview of the definitions, scopes, and forest types within each biome. Part I serves as a primer on different forest ecosystems to understand other themes, although it does not cover managed forests, which are discussed in other sections.

Part II, 'Forest Dynamics', delves into the impacts of various disturbances—insects, fires, and strong winds—on forests. The common thread here is that forests often display resistance and resilience to disturbances, with the ability to recover. However, disturbances can lead to rapid and irreversible system collapse if a critical threshold is exceeded. Therefore, it is vital to understand the factors behind these tipping points, the conditions under which they occur, and how to prevent them. This part covers chapters discussing how biological factors, from genes to interacting species communities, shape and transform forests. It also includes a new chapter examining the phenomenon of a single tree species dominating the canopy in tropical lowland forests.

Part III, 'Forest Flora and Fauna', highlights forest biodiversity and its role in supporting ecological processes. It explores the ecology of key functional groups and taxa, including lianas, vascular epiphytes, insects, bryophytes, lichens, mammals, birds, amphibians, and reptiles. The latter two taxa are included in a new chapter. This section then aptly concludes with a chapter assessing global forest biodiversity and our knowledge of species diversity. However, this part does not cover microbial pathogens and insect pests, as they are discussed in another section.

Part IV, 'Energy and Nutrients', explores the intricacies of the vital ecological functions that govern how energy and nutrients are acquired, used, and recovered. All ecological processes work in a precise manner. Any processes that are slowed down or disrupted by disturbance could be further hampered by negative feedback loops. For example, a reduced nutrient cycling rate greatly diminishes net primary productivity, reducing plant nutrient availability and worsening nutrient limitations. These chapters emphasise the fragility of forest systems, offering an ecological perspective on the intricate engineering of entire forest ecosystems, which serve as a foundational support for all life.

Part V, 'Forest Conservation and Management', examines the ecology of anthropogenic threats and conservation approaches that inform policy and management. These topics play a central role in forest restoration or pose significant challenges to the conservation of biodiversity and function. A new chapter on pathogens and invertebrate pests in North American forest ecosystems highlights how addressing the interplay of multiple challenges will become increasingly important for effective forest management in the future.

Part VI, 'Forest and Climate Change', focusses on the effects of various climate change phenomena such as fires and droughts. While we address the threats of climate change individually, the final chapter of this section explores how 21st century atmospheric changes, disturbances and processes of regrowth and recovery are affecting the carbon balance of each major forest biome.

As forests and people have become increasingly intertwined and indigenous people and local communities are asserting their forest rights, the last section, 'Human Ecology', explores how people directly use forests and the broader implications for maintaining the ability of forests to provide ecosystem services. These chapters discuss significant services offered by forests, including non-timber forest products, wild meat, and land for shifting cultivation. Collectively, this part highlights the link between human well-being and the overall health of our global forests.

This handbook represents a delicate balance between the need to disseminate extensive knowledge while ensuring that the narrative remains coherent and engaging. Our primary intent with this work is to create an authoritative text that not only caters to experts in the field but is also accessible and appealing to the layperson seeking an introduction to forest ecology. Furthermore, it serves as a valuable resource for postgraduate students, providing them with a comprehensive collection of the latest research to facilitate further exploration and discussion in the realm of forest ecology.

Beyond these educational objectives, our aim is to shed light on the ecological principles underpinning current approaches and trends in forest management, which can be invaluable to conservation practitioners. In this endeavour, we express our gratitude to each contributor who has generously shared their cutting-edge knowledge, enriching the contents of this handbook. Finally, we extend our heartfelt thanks to a dedicated team of anonymous reviewers who invested their time and effort in reviewing the contents. Their insightful feedback and thoughtful recommendations have been essential to improve the quality and depth of this handbook.

PART I

THE FOREST

2
BOREAL FORESTS

Jean-Pierre Saucier, Ken Baldwin, Pavel Krestov and Torre Jorgenson

The boreal biome is one of the largest forested biomes on Earth and forms a circumpolar belt of forests and woodlands between the treeless arctic zone and the temperate zone. It is typically characterized by a cold continental climate with relatively short, mild summers and long, cold, snowy winters. Much of the region was covered by glaciers during the Late Pleistocene so in a substantial portion of the region the forests have only developed within the Holocene. Almost half of the biome occurs in the zone of permafrost (Brown et al. 1997; Helbig et al. 2018). Boreal forests and woodlands represent approximately 27 per cent of the world's forested area (FAO & UNEP 2020). Boreal forests are dominated by a relatively few, primarily coniferous, genera (*Picea, Abies, Larix, Pinus*) that are adapted to cold temperatures, low-nutrient conditions, and recurrent disturbance types. Short growing seasons and cold, acidic mineral soils under conifer canopies result in extensive feathermoss carpets on upland sites. *Sphagnum* mosses occupy landscape positions with permanently high water tables, resulting in acidic organic soils that often develop into peatlands. On both upland and lowland sites, decomposition rates are slow and nutrient cycling is typically restricted to the upper soil layers where oxygen and increased temperatures support microbial and fungal metabolism. Understory vegetation of boreal forests is also dominated by a relatively few botanical families, especially the Ericaceae (the heath or blueberry family), which are adapted to cold, nutrient impoverished habitat conditions. With proximity to the oceans, winters become milder and summers cooler; snow covers the ground for longer periods and the growing season is shortened. In such oceanic boreal climates, conifers are generally absent and are replaced by birch (*Betula* spp.), alder (*Alnus* spp.) or ericaceous shrublands that can tolerate such harsh conditions. We include these non-forested regions to provide a comprehensive description of the boreal biome.

Extent of the boreal zone

The boreal zone is the northernmost forest zone forming a large belt between latitudes 45°N and 72°N (Rivas-Martínez et al. 2011) (Figure 2.1). It covers approximately 12.1 million km^2 (Kuusela 1992) and represents 8.4 per cent of the surface of the earth. Boreal forests are also called taiga.

DOI: 10.4324/9781003324072-3

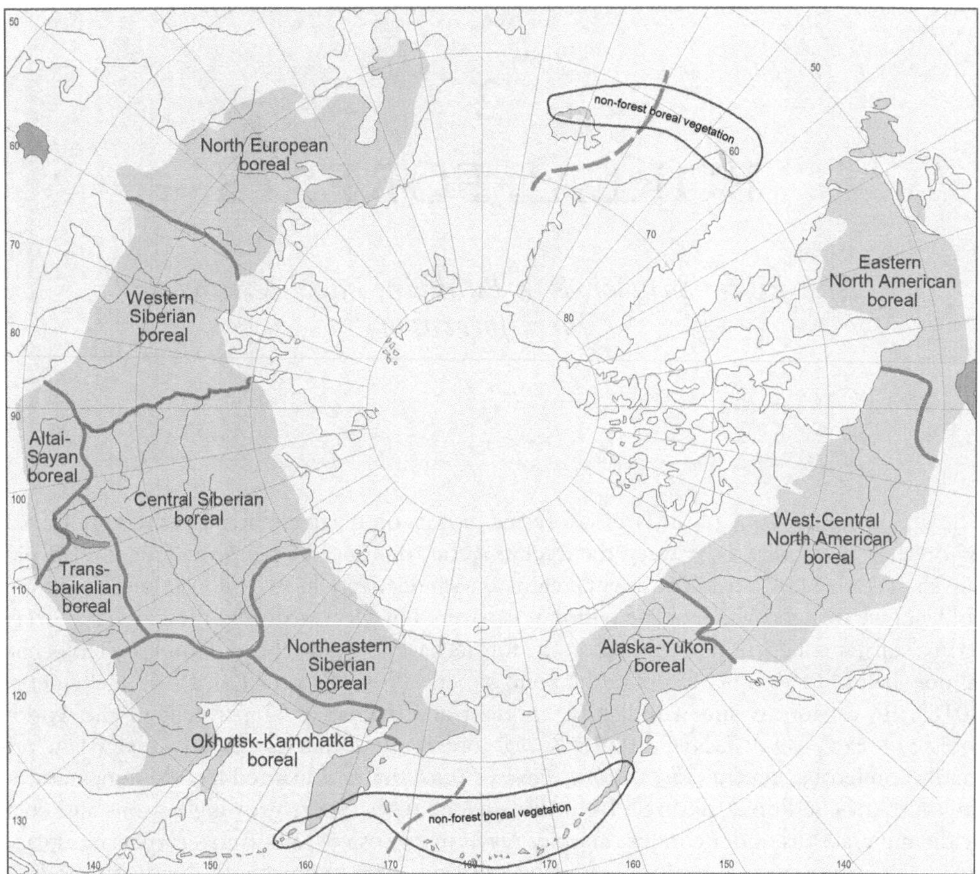

Figure 2.1 Extent of boreal forests and woodlands around the globe, with floristic subdivisions and non-forest boreal vegetation zones.

The boreal zone covers large areas of North America, Europe and Russia. In North America, boreal forests extend from Alaska (USA) through Canada to Labrador and the island of Newfoundland. In Europe, boreal forests occupy most of the areas of Sweden and Finland, a lesser part of Norway, Russian Karelia and east to the Ural Mountains. In Asia, the boreal zone stretches from the Ural Mountains eastward through Siberia and the Far East to the Pacific coast. Non-forested boreal vegetation also covers some islands in the Atlantic (Iceland, Faroe Islands and the southernmost part of Greenland) and in the Pacific (Kuril and Aleutian Island chains). Closed forest covers 76 per cent of the boreal zone, while commercially exploitable closed forest occupies about 53 per cent of the total area (Table 2.1).

The northern limit between the boreal zone and the arctic zone is the continental treeline (i.e., the extent of the habitat in which trees are capable of growing). Moving northward from closed forests through open forests and woodlands, climatic and/or site conditions eventually become too harsh (cold, dry, windy, infertile) to support tree growth. Trees are replaced by low or prostrate shrubs, especially dwarf ericaceous, birch and willow species, together with herbs, mosses and lichens as sparse woodlands gradually change to arctic tundra.

Table 2.1 Area of boreal forests and woodlands by country and proportion of the country in closed forest or exploitable closed forest

Area of forest and other wooded land within the boreal zone (million km²)							
	Russia	*Alaska*	*Canada*	*Norway*	*Sweden*	*Finland*	*Total*
Area of boreal forests and woodlands (million km²)	7.90	0.46	3.27	0.07	0.21	0.23	12.14
Proportion of the total	65.1%	3.8%	26.9%	0.6%	1.8%	1.9%	100.0%
Proportion of the boreal zone covered by closed forest or exploitable closed forest							
Closed forest	85.2%	10.9%	60.6%	84.3%	86.0%	85.9%	75.8%
Exploitable closed forest	57.0%	10.9%	44.0%	72.9%	75.2%	83.7%	52.6%

Source: Adapted from Kuusela (1992).

Between the boreal and arctic zones, there is a transition called hemiarctic, or subarctic. This broad ecotone is characterized by a landscape matrix of woodlands and treeless barrens with rare stands of trees growing in sheltered locations. Accumulated snow usually protects the trees during the coldest months. These tree stands are often embedded in krummholz vegetation. We included the hemiarctic as part of the boreal zone, since its flora retains species (especially the tree species) that are at the northern limit of their ranges.

The southern boundary of the boreal zone is less obvious as it is based on vegetation physiognomy, species composition and ecosystem dominance on the landscape. This boundary represents a gradual shift of forest composition where, moving southward, thermophilous species replace the boreal species (Brandt 2009). It is usually marked by a change from a coniferous-dominated landscape, associated with a cold climate, to a mixedwood forest landscape with a milder climate. In certain areas, with very dry climates the boreal forest is replaced by steppe at its southern margin.

Between the boreal zone and the temperate zone, the transition is called hemiboreal (Hämet-Ahti 1981; Tuhkanen 1984, Ermakov 2003). In the hemiboreal, boreal species form mixed forest types with species that are less cold-tolerant. Some authors classify the hemiboreal as a part of the boreal zone (Hämet-Ahti 1981). Considering that temperate species found in this transition zone are at the northern limit of their ranges and extend far south into the temperate zone, we consider that hemiboreal vegetation fits better into the temperate zone (Saucier 2008, Brandt 2009).

Conifer forests outside the boreal zone

In the temperate zone, some climatically or edaphically extreme habitats support vegetation communities that share characteristics similar to those of the boreal zone.

The colder climate at higher elevations, associated with more severe exposure to wind and weather events, often with shallow or eluviated soils, favours species that show similar adaptations to those of boreal species. Therefore, forest types similar to boreal ones (i.e., dominated by *Picea* spp., *Abies* spp., *Pinus* spp.) are often found at high elevations of mountainous regions in the temperate zone. See Chapter 6 on mountain forest. Also, bogs and poor fens occurring in temperate regions contain boreal species that are characteristic of cold, acidic, nutrient-poor peatlands.

Table 2.2 Climatic values for bioclimates within the boreal macrobioclimate

Macrobioclimate	*Bioclimate*	*Mean annual temperature (°C)*	*Continentality index[1] (°C)*	*Annual ombrothermic index[2]*
Boreal	Boreal hyperoceanic	< 6.0	< 11	> 3.6
	Boreal oceanic	≤ 5.3	11–21	> 3.6
	Boreal subcontinental	≤ 4.8	21–28	> 3.6
	Boreal continental	≤ 3.8	28–46	> 3.6
	Boreal hypercontinental	< 0.0	> 46	–
	Boreal xeric	≤ 3.8	< 46	> 3.6

Source: From Rivas-Martínez et al. (2011).

1 The continentality index represents the annual thermic interval (mean monthly temperature of the hottest month minus the mean monthly temperature of the coldest month).

2 The annual ombrothermic index provides an indication of the amount of water available for evapotranspiration (total precipitation of the months with mean monthly temperature > 0°C divided by the sum of mean monthly temperatures > 0°C times 10).

Climate of the boreal zone

The boreal zone is generally characterized by a cold climate with relatively short summers and long, snowy winters. To distinguish boreal from other ecological zones, FAO and UNEP (2020) retained the criteria proposed by Köppen and modified by Trewartha based on the number of months with a mean monthly temperature over 10°C. The boreal zone has up to three months with a mean monthly temperature over 10°C, while the temperate zone has four to eight, and the arctic zone (or polar) has none. Winter (mean monthly temperature < 0°C) usually lasts from five to seven months. The length of the growing season (mean daily temperature > 5°C) ranges from 80 to 150 days but can be as low as 50 to 70 days at the continental treeline and along oceanic coasts, marking the transition between forests and non-forest vegetation.

The boreal zone cannot be easily defined by mean annual temperature because temperature and precipitation act as compensating climatic factors. In continental climates, winters are so cold that low temperatures affect the survival of tree species, while summers are quite hot, sometimes resulting in a water deficit for plants. Conversely, in oceanic climates, winters are not as cold but summer temperatures stay quite low, usually with very high humidity; thus, resulting in a very short growing season. Rivas-Martínez et al. (2011) use three climatic indices to define six bioclimates in the boreal macrobioclimate (Table 2.2). In this system, depending on continentality, the mean annual temperature of boreal sub-zones can range from less than 0°C up to 6°C (Rivas-Martínez et al. 2011). Conifer tree species characterize all boreal bioclimates except boreal hyperoceanic. In this bioclimate, vegetation is dominated by broadleaved woodlands, krummholz, ericaceous dwarf shrubs or herbaceous meadows.

Vegetation of boreal forests

Main floristic subdivisions of the boreal zone

Boreal forests cover northern latitudes around the globe and overall have very similar floristic composition in their different regions. Boreal forests are characterized by coniferous

Table 2.3 Dominant boreal tree or shrub/herb species by continent and floristic subdivisions of the boreal zone

North American floristic subdivisions		
Alaska–Yukon	*West–Central North American*	*Eastern North American*
Picea glauca (White spruce)	*P. glauca* (White spruce)	*P. glauca* (White spruce)
Picea mariana (Black spruce)	*P. mariana* (Black spruce)	*P. mariana* (Black spruce)
Larix laricina (Tamarack)	*L. laricina* (Tamarack)	*L. laricina* (Tamarack)
Betula neoalaskana (Alaska paper birch)	*Pinus contorta* (Lodgepole pine)	*P. banksiana* (Jack pine)
	Pinus banksiana (Jack pine)	*A. balsamea* (Balsam fir)
Betula kenaica (Kenai birch)	*Abies lasiocarpa* (Subalpine fir)	*B. papyrifera* (Paper birch)
Populus tremuloides (Quaking aspen)	*Abies balsamea* (Balsam fir)	*B. pubescens* (Downy birch)
	Betula papyrifera (Paper birch)	*P. tremuloides* (Quaking aspen)
Populus balsamifera (Balsam poplar)	*Betula nana* (Dwarf birch)	*P. balsamifera* (Balsam poplar)
	P. tremuloides (Quaking aspen)	
Populus trichocarpa (Black cottonwood)	*P. balsamifera* (Balsam poplar)	
European and Northern Maritime (non-forested) floristic subdivisions		
North European	*North Pacific Maritime*	*North Atlantic Maritime*
Picea abies (Norway spruce)	*Alnus viridis* (Alder)	*Kalmia angustifolia* (Sheep laurel)
Pinus sylvestris (Scots pine)	*Salix glauca* (Gray-leaved willow)	*E. nigrum* (Black crowberry)
Betula pendula (Silver birch)	*Empetrum nigrum* (Black crowberry)	*Vaccinium vitis-idaea* (Mountain cranberry)
B. pubescens (Downy birch)	*Phyllodoce aleutica* (Aleutian heather)	
	Vaccinium uliginosum (Bog bilberry)	*Rhododendron groenlandicum* (Common Labrador tea)
B. nana (Dwarf birch)	*Calamagrostis nutkaensis* (Nootka reedgrass)	
Populus tremula (Eurasian aspen)	*Dryopteris expansa* (Spiny wood fern)	*Calluna vulgaris* (Common heather)
	Athyrium filix-femina (Lady fern)	*Erica cinerea* (Bell heather)
Asian floristic subdivisions		
Western Siberian	*Altai–Sayan*	*Central Siberian*
Picea obovata (Siberian spruce)	*P. obovata* (Siberian spruce)	*Larix gmelinii* (Dahurian larch)
Abies sibirica (Siberian fir)	*A. sibirica* (Siberian fir)	*P. obovata* (Siberian spruce)
Larix sibirica (Siberian larch)	*L. sibirica* (Siberian larch)	*A. sibirica* (Siberian fir)
Pinus sibirica (Siberian pine)	*P. sibirica* (Siberian pine)	*L. sibirica* (Siberian larch)
P. sylvestris (Scots pine)	*P. sylvestris* (Scots pine)	*Larix cajanderi* (Cajander's larch)
B. pendula (Silver birch)	*B. pendula* (Silver birch)	*P. sibirica* (Siberian pine)
P. tremula (Eurasian aspen)	*Betula platyphylla* (Asian white birch)	*P. sylvestris* (Scots pine)
	P. tremula (Eurasian aspen)	*B. platyphylla* (Asian white birch)
		P. tremula (Eurasian aspen)

(*Continued*)

Table 2.3 (Continued)

Transbaikalian	*Northeastern Siberian*	*Okhotsk–Kamchatka*
L. gmelinii (Dahurian larch)	*L. cajanderi* (Cajander's larch)	*Picea jezoensis* (Jezo spruce)
P. sylvestris (Scots pine)	*Pinus pumila* (Siberian dwarf pine)	*Abies nephrolepis* (Manchurian fir)
P. sibirica (Siberian pine)	*P. sylvestris* (Scots pine)	*Abies sachalinensis* (Sakhalin fir)
P. obovata (Siberian spruce)	*B. platyphylla* (Asian white birch)	*L. cajanderi* (Cajander's larch)
A. sibirica (Siberian fir)	*P. tremula* (Eurasian aspen)	*Betula ermanii* (Erman's birch)
L. sibirica (Siberian larch)		*B. platyphylla* (Asian white birch)
B. pendula (Silver birch)		*P. pumila* (Siberian dwarf pine)
P. tremula (Eurasian aspen)		*P. tremula* (Eurasian aspen)

tree species in the genera *Picea, Abies, Larix* and *Pinus*, often in association with broadleaved species in the genera *Betula, Populus, Alnus,* and *Salix*. Understory floristic composition, dominated by tall, low or dwarf shrubs, and mosses and lichens, also shows global similarity due to the shared biogeographic history of this biome. Because of these general similarities, Takhtajan (1986) combined the entire boreal zone into a single circumboreal floristic region subdivided into 15 floristic provinces. The variation in floristic composition between these provinces represent differences in regional climate and other environmental factors within the boreal zone, as well as differences in the history of glaciation, climatic events, and related evolutionary pathways for boreal species.

For this chapter, we propose 12 broad floristic subdivisions of the boreal zone (Figure 2.1): North America is divided into three floristic subdivisions (Alaska–Yukon, West–Central and Eastern); Europe contains one floristic subdivision (North European); Asia is divided into six floristic subdivisions (Western Siberian, Altai–Sayan, Central Siberian, Transbaikalian, Northeastern Siberian and Okhotsk–Kamchatka); and there are two non-forested maritime subdivisions (Atlantic and Pacific). For Eurasia, our subdivisions are based on Takhtajan's floristic provinces, for North America they are based on floristic regionalizations developed for the Canadian National Vegetation Classification (CNVC 2014), while the maritime subdivisions are adapted from Takhtajan. Each of these floristic subdivisions has its own set of characteristic tree species, shrub/herb species and forest types, although some species are present in more than one subdivision (Table 2.3).

North American floristic subdivisions

Alaska–Yukon boreal

The Alaska–Yukon boreal floristic subdivision of North America extends from western Alaska, where forest grades into arctic tundra and boreal oceanic shrublands and heaths, to the North American Cordillera in Yukon Territory (Jorgenson and Meidinger 2015). Latitudinally, it extends from the southern Brooks Range to the northern side of the Alaskan coastal mountains. The region is dominated by mixedwood forests that range from early to late successional stages due to the prevalence of fire disturbance. While open to closed *Picea glauca* forests represent the potential natural vegetation on zonal sites, *P. glauca, Betula neoalaskana* and *Populus tremuloides* are often intermixed in a patchy mosaic of differing stand ages, with all three species typically present in every successional stage (Viereck et al. 1992). The Alaska–Yukon boreal is differentiated from the West–Central North American boreal mainly by the presence of *B. neoalaskana*, the absence of *Pinus contorta* var. *latifolia* and *Abies lasiocarpa*, and extensive areas of unglaciated terrain that has allowed more vegetation and soil development over time. Forests are typically found up to 900 m in elevation.

The Alaska–Yukon boreal comprises diverse forest types related to climate, topography, soil moisture, nutrients and permafrost, and disturbance by fire and thermokarst (Chapin et al. 2006, Jorgenson and Meidinger 2015). Along a climatic gradient, the region has been subdivided into a cold northern sector of mainly coniferous woodlands, a mesothermal central sector with mixed forests, and a warm southern sector with fern-rich mixed forests. In the northern sector, deciduous trees form only a minor component of the vegetation due to colder temperatures and nearly continuous permafrost. The understory is dominated by low shrubs and a ground cover of feathermosses and lichens. The central sector has more permafrost-free areas in the uplands, where vegetation succession after fire has several stages: herb, shrub and sapling, deciduous forest (*B. neoalaskana, P. tremuloides*) after 30–130 years, and coniferous forest (*P. glauca*) after 100–250 years. Permafrost-affected soils on north-facing slopes and lowlands support coniferous woodlands (*Picea mariana, Larix laricina*), which are highly susceptible to fire. Thermokarst is prevalent in permafrost-rich lowlands and creates a variety of non-forested bog and fen ecosystems (Jorgenson et al. 2013). In warmer, southern areas, where the climate is subcontinental and permafrost is lacking, *Betula kenaica* is common in some areas. On floodplains, *Populus balsamifera* is common in the north-central sectors, while *Populus trichocarpa* is common in the southern sector.

West–Central North American boreal

The West–Central boreal floristic subdivision of North America includes boreal forests of central Canada (northwestern Ontario, Manitoba, Saskatchewan, Northwest Territories, and eastern Alberta), northern British Columbia and southern Yukon Territory, as well as the Cordilleran foothill forests of western Alberta. Sub-humid, continental climates prevail throughout the region, which extends from approximately longitude 87°W in northwestern Ontario to approximately longitude 136°W in northwestern British Columbia/southwestern Yukon, where it extends northwards to approximately 62°N of latitude. Characteristic forest cover comprises closed-crown conifer forests, although more open forests or even woodlands are common in colder, northern areas, when site-level ecological factors become limiting or when a short interval between disturbances has occurred.

Conifer forests of this subdivision are characterized by spruce and pine—feathermoss types. *P. mariana* dominates on nutrient-poor upland soils and on forested wetlands; *P. glauca* is the dominant species on more fertile upland and riparian sites. Both species range northwards to the continental treeline, with stand physiognomy changing into open spruce–lichen woodlands in northern areas. West of the Cordilleran foothills of western Alberta, *P. contorta* var. *latifolia* is very abundant on the landscape, responding to the intensity and frequency of forest fire. East of the foothills, *Pinus banksiana* occupies the same ecological niche as *P. contorta*.

With the short fire return interval in this dry, continental climate, the early seral broadleaved species *P. tremuloides* and *Betula papyrifera* are common and abundant on the landscape. *P. tremuloides* is very common in southern portions of the region, where it often forms mixedwood stands with *P. glauca*. At the southern margin of the boreal forest in Alberta, Saskatchewan and Manitoba, these species form parkland conditions as the boreal formation transitions to grasslands on the Great Plains of North America. *B. papyrifera* ranges further north than aspen, intergrading with *B. neoalaskana* in the northwestern part of this region. Approaching elevational and continental treeline, shrub birch species (e.g., *Betula nana*) become prevalent as the abundance of trees on the landscape declines.

Abies spp. are rare on the landscape because of the high-frequency fire regime. At low elevations in the Cordillera and in British Columbia and southern Yukon, *A. lasiocarpa* occurs occasionally on fire-protected sites, but its abundance increases at higher elevations where it is an important subalpine species. Similarly, east of the Cordillera, *A. balsamea* occurs occasionally where fire return intervals are longer.

Eastern North American boreal

The Eastern boreal floristic subdivision of North America extends from northwestern Ontario (approximate longitude 87°W), through Québec to the Atlantic Ocean in Newfoundland and Labrador (longitude 53°W). Characteristic forest cover consists of coniferous closed-crown forests, but more open forests or even woodlands are frequent when the climate is colder or when ecological factors become more limiting. In southern portions of this region, broadleaved and mixedwood forests dominate the landscape.

The most characteristic and widely distributed forest type is the *P. mariana*—feathermoss forest. In these forests, *P. mariana* dominates the canopy, with *A. balsamea*, *P. banksiana* and sometimes *P. glauca* as variable companion species. Closer to the northern limit of the boreal, the canopy opens to form a *P. mariana*—lichen woodland. *P. banksiana* is most frequent in the more continental parts of this region where the fire cycle is relatively short (around 100 years).

A. balsamea is associated with oceanic influences and milder climates; it is most common in the humid areas of eastern Québec and Newfoundland and Labrador. Here, it mixes with *P. mariana* after long periods without fire. In these cases, the shade-tolerant fir can regenerate under canopy cover and survive with low light availability, waiting for the canopy to open after a catastrophic event or gap-phase replacement of senescent trees. It is helped periodically by insect epidemics (e.g., spruce budworm [*Choristoneura fumiferana*] or hemlock looper [*Lambdina fiscellaria*]) that create these gaps and release the understory regeneration. Under these conditions, *P. mariana* often maintains itself by vegetative layering. *P. glauca* is much more frequent in hyperoceanic and oceanic bioclimates.

In southern areas of the eastern boreal, *A. balsamea* mixes with the shade-intolerant broadleaved species, *B. papyrifera* and *P. tremuloides*, mostly as boreal mixedwood forests. The *A. balsamea–B. papyrifera* forest type is characteristic of the southern latitudes of the boreal zone in eastern Québec and Newfoundland and Labrador. In the more continental parts of this subdivision (Ontario and western Québec), mixedwood forests are dominated by *P. tremuloides*.

The understory of conifer forests is typically dominated by mosses and, in oceanic areas, liverworts. The shrub layer is represented by heath species. On colder or less fertile sites, often associated with coarse-textured and dry soils, the tree canopy is more opened and lichens cover the ground. On richer sites or warmer parts of the eastern boreal, the herb and broadleaved shrub layers are well developed to the detriment of the moss layer.

European floristic subdivision

North European boreal

The North European boreal floristic subdivision extends from Norway in the west (approximate longitude 5°E) to the Ural Mountains in the east (approximate longitude 60°E). Northern European boreal forests have simple stand structures, characterized by *Picea abies*, with species-poor floristics.

The typical and most widespread zonal forest condition is dominated by *P. abies* forming a tall overstory with closed-canopy cover. *Betula pendula* and *Pinus sylvestris* may occasionally be present in the tree layer. These species rapidly increase their presence after disturbance and gap formation in the spruce canopy. The shrub layer is not abundant in these forests. The herb and dwarf shrub layers are usually well developed.

P. sylvestris is more abundant in the canopy on nutrient-poor sites and, on well-drained sites with sandy nutrient-poor soils, it becomes a fully dominant species in stands with a dense lichen cover and scattered ericaceous species in the understory. Pine also occupies paludified habitats with thick organic soils and a water table at the ground surface for several months per year.

Due to the strong climatic gradient between the Atlantic coast and interior regions, forest vegetation changes with distance from the ocean (Hytteborn et al. 2005). With proximity to the Atlantic coast, dominance of spruce decreases and the proportion of *Betula pubescens* increases. In the northeastern part of Norway, spruce disappears and there is a narrow belt of pure *B. pubescens* forests with crooked trunks and a species-rich herb layer composed of tall forbs. In eastern parts of this subdivision, the climate becomes more continental and *Larix sibirica* appears.

Asian floristic subdivisions

The boreal zone of northern Asia extends from the Ural Mountains in the west (approximate longitude 60°E) to the Bering Strait (approximate longitude 170°W) in the east. It covers an enormous climatic gradient from the hypercontinental regions in eastern Yakutia, where winter temperatures fall well below -60°C and summer temperatures reach +35°C, to the Pacific coast of Kamchatka, in an oceanic climate of mild winters, cool summers and a very short growing season. About half of this area is in the zone of continuous permafrost. Compared to other parts of the boreal zone, the Asian portion is the most diverse (Nakamura and Krestov 2005).

Western Siberian boreal

The Western Siberian boreal floristic subdivision extends from the Ural Mountains in the west (approximate longitude 60°E) to the Yenisei River in the east (approximate longitude 87°30'E). Floristically, this subdivision is very similar to the North European boreal subdivision, except for the replacement of the dominant conifer species by *Picea obovata, Abies sibirica, Pinus sibirica* and *L. sibirica. P. sylvestris* occurs with increasing presence towards the south. *B. pendula, B. pubescens* and *Populus tremula* are common broadleaved species (Ermakov and Morozova 2011). Understory composition in a typical stand of *P. obovata* dominated forest is similar to that of a *P. abies* stand in the North European subdivision. Spruce forests with *Vaccinium myrtillus* are characteristic of zonal sites.

At the landscape level, *P. obovata* dominated forests occupy habitats with well-developed soils in areas with moderate soil moisture and air humidity. *A. sibirica* is a humidity-sensitive species that occurs in habitats that receive more precipitation and have long periods of foggy weather. *P. sibirica* occupies large areas with nutrient-poor well-drained soils, where coarse-textured glacial materials form shallow stony soils around rock outcrops. *L. sibirica*, a fire-resistant and gap-dependent species, increases its presence on sites affected by low-intensity fires.

Altai–Sayan boreal

Southern Siberia is a region where the boreal zone transitions with temperate forest and steppe regions. On the southwestern edge of the Central Siberian boreal, this floristic subdivision is influenced by the vegetation of adjacent mountainous areas, including the major mountain systems from the Altai in the west to the Hamar-Daban in the east.

Zonal vegetation on the low elevation arid plains is grass steppe, however in the mountains, mixedwood (*B. pendula, P. tremula, P. sylvestris*) and conifer forests (*P. sibirica, A. sibirica*) prevail. These mountain systems are affected by humid Atlantic air masses and receive over 1,500 mm of annual precipitation. The relatively high temperatures and high humidity of the temperate zone in central Eurasia have led to the formation of unique forests characterized by boreal dominants and well-developed layers of shrubs and tall herbs. These forests have a complete set of boreal species but differ from other boreal forests by containing a very high proportion of temperate species in the understory (Nazimova et al. 2014).

Central Siberian boreal

The Central Siberian boreal floristic subdivision occupies the area between the Yenisei River in the west (approximate longitude 87°30'E) and the Verkhoyansk Range and the eastern border of the Aldan River Basin in the east (approximate longitude 128°E). The climate is hypercontinental to xeric, with severe winters with no or shallow snow cover and hot summers; the entire area lies within the zone of permafrost. The combination of cold soils and harsh climatic conditions does not favour the shade-tolerant conifer species of the *Pinus, Picea*, and *Abies* genera. Instead, *Larix* forests prevail.

Most of the area is occupied by *Larix gmelinii*, a tree species adapted to the most extreme growing conditions in the Northern Hemisphere. It can tolerate winter temperatures lower than −70°C, strong paludification, very low summer precipitation and air humidity, and very cold soils with shallow active layers above permafrost. In conditions of low precipitation and summer-long droughts, permafrost is the only source of moisture for trees. *L. gmelinii* is a long-lived species, up to 500 years in northern regions.

Larix normally forms pure even-aged stands with a single-stratum canopy, varying in cover, without any other tree species. Across the whole province, *Larix* stands alternate with *P. sylvestris* stands on sandy soils. Both species are shade-intolerant and adapted to a wide range of ecological conditions. *P. sylvestris* saplings appear to be more fire tolerant than *Larix* saplings, and pine replaces larch when surface fires are frequent. The well-developed shrub layer in *Larix* stands is composed of both circumboreal shrubs and species restricted to the eastern part of boreal Asia.

Transbaikalian boreal

This floristic subdivision lies mostly south of Lake Baikal and extends from the Northern Baikal Plateau in the west (approximate longitude 100°E) to the upper part of the Amur River in the east (approximate longitude 124°E). The southern boundary of the Transbaikalian boreal is in northeastern Mongolia. The climate is hypercontinental to xeric and, although continuous permafrost is absent, soils are characterized by severe freezing. The landscape is semi-forested, moisture being the major limiting factor for forest distribution. Forests mainly occupy the northern aspects of mountain slopes, while southern aspects and valleys are covered by steppe vegetation. Most Transbaikalian boreal forests are composed of *L. gmelinii*, except on sandy sites, which are occupied by *P. sylvestris*. At higher elevations, *P. sibirica* can form sparse stands.

Typical *L. gmelinii* forests occur within the forest–steppe ecotone of the Transbaikalian boreal. Their open canopy is dominated by *L. gmelinii* and *Betula platyphylla* in different proportions, with occasional occurrence of *P. tremula*.

Northeastern Siberian boreal

The Northeastern Siberian boreal floristic subdivision extends from the Verkhoyansk Range in the west (approximate longitude 128°E) to the coasts of the Bering and Okhotsk seas in the east, excluding the Kamchatka Peninsula. The main vegetation condition of this floristic subdivision is open *Larix cajanderi* forest and woodland, gradually changing to *Pinus pumila* thickets towards the east.

L. cajanderi forests and woodlands are humidity-dependent and occur in conditions of moderately oceanic climates along the coasts of the Arctic and north Pacific oceans as well as in the massif mountain systems of northeast Asia (Krestov et al. 2009). Forest stands are characterized by (1) a very open canopy of larch; (2) lack of shade-tolerant species characteristic of coniferous forests; (3) presence of species characteristic of larch-dominated mires; and (4) high constancy of species that are widely distributed in the subarctic/subalpine zones of northern Asia. These forests/woodlands occur across a wide range of habitats, influenced by shallow seasonally thawed soil over permafrost and by a less than three months growing season.

With proximity to the coast of the East Siberian Sea, the role of *P. pumila* in the *Larix* forests increases. North of latitude 60°N, within 1,000 km of the coast, it becomes the dominant species on the landscape. The main features of these forests include: (1) dominance by *P. pumila*, which forms a closed canopy; (2) presence of a complex of subarctic/subalpine species; and (3) presence of other tree species, usually *L. cajanderi.*

Okhotsk–Kamchatka boreal

The Okhotsk–Kamchatka boreal floristic subdivision includes the Kamchatka Peninsula, with the adjacent Commander Islands and Koraginskii Island, the Kuril Archipelago, Sakhalin Island, and the mainland area adjacent to the Okhotsk Sea and the Sea of Japan. The vegetation of this subdivision maintains features typical of boreal vegetation, but with forests dominated by Far Eastern tree species that are different from the rest of the Eurasian boreal zone. The major dominants are *Abies sachalinensis* var. *gracilis*, *A. nephrolepis*, *A. sachalinensis* and *Picea jezoensis*. Increasing oceanicity in the coastal areas causes dramatic changes in forest structure as shade-tolerant conifer species disappear and *Betula ermanii* becomes the dominant canopy species (Krestov 2003). This forest type changes to *Alnus viridis* ssp. *fruticosa* communities in the coastal areas of Kamchatka and on the northern islands of the Kuril Archipelago, and then to treeless tall forb vegetation on the Commander Islands.

Three major forest conditions related to climatic gradients of temperature and continentality can be distinguished over the range of *P. jezoensis*: (1) low diversity pure spruce forests of simple structure in the northern part of the range and in Kamchatka; (2) more complex, higher diversity mixed spruce–fir forests with *A. nephrolepis* along with broadleaved tree species, such as *Acer mono, Betula costata, Fraxinus mandshurica* or *Tilia amurensis* on the mainland south of latitude 54°N, plus the Shantar Islands; and (3) mixed spruce–fir forests with *A. sachalinensis*, often in association with *Sorbus commixta* or *Acer ukurunduense*, on the islands of Sakhalin, Iturup and Hokkaido (Krestov and Nakamura 2002).

Forests of *B. ermanii* characterize the oceanic and subcontinental regions of northeast Asia. This species forms vegetation belts in mountains influenced by oceanic air masses and is most developed in Kamchatka, where it occurs over a wide range of ecologically different sites, except in wetlands or on permafrost. *B. ermanii* distribution depends, among other factors, on snow depth and timing of snowmelt (i.e., length of snow-free period). The recorded maximum age for this species is 500 years. *A. viridis* ssp. *fruticosa*, *P. pumila* and *Sorbus sambucifolia* may form a tall shrub layer in *B. ermanii* stands. Because of the well-developed herb layer, there is usually no moss–lichen layer.

Northern Maritime floristic subdivisions (non-forested)

North Pacific Maritime boreal

The non-forested North Pacific Maritime boreal floristic subdivision, including southwest Alaska, Aleutian Islands, Commander Islands, southeastern Kamchatka, and the northern Kuril Islands, have a hyperoceanic to oceanic climate and a distinctive vegetation including species of both Asian and North American origin (Qian et al. 2003, Talbot et al. 2010). Vegetation mainly consists of dwarf shrub heath with patches of bogs, herbaceous meadows, and shrublands along streams. In the Aleutians, vegetation is dominated by dwarf shrubs and forbs (interspersed with meadows comprising tall herbs and graminoids). In western portions of this area, stunted forms of shrubs dominate.

North Atlantic Maritime boreal

In south-central Iceland, the Faroe Islands, southernmost Greenland and easternmost Canada, where a hyperoceanic boreal climate creates a short growing season, boreal vegetation is represented by treeless communities (Daniëls and de Molenaar 2011). Vegetation includes a gradient of North American to European species. In Newfoundland, Canada, maritime heathland vegetation varies by environmental conditions ranging from sparse cover in harsh environments to stunted coniferous forest patches in sheltered valleys (Baldwin et al. 2020). Mesic sites are dominated by low/dwarf shrubs while mosses are uncommon.

In Iceland, due to the maritime climate, volcanic activity, presence of glaciers and depletions of woodlands by human land use, vegetation is characterized by low-growing species associated with heaths, grasslands, wetlands and rocky barrens. Woodlands occur in small remnant patches in restricted valleys.

Survival mechanisms of boreal species

Boreal species exhibit several anatomical, physiological and reproductive strategies that facilitate survival in the harsh environmental conditions of the boreal zone.

Species adaptation to cold temperatures

In order to survive boreal winters, vegetation is adapted to cold temperatures and short growing seasons. Adaptation takes different forms: physiological ability to survive extremely low temperatures; protection of buds, seeds and evergreen leaves against desiccation and freezing; and short reproductive life cycles to successfully mature seeds in an annual growing season.

Some plants, including many coniferous species, show tolerance to extreme cold temperatures during the dormant period (i.e., hardiness). Plants use three strategies to survive freezing (Sutinen et al. 2001; Brandt 2009). The first, called supercooling, is a process by which cellular water can remain liquid at temperatures as low as –40°C. Species that limit their adaptation to supercooling are usually more frequent in the transition between boreal and temperate zones (also called hemiboreal).

A second strategy is to tolerate extracellular freezing by moving liquid outside of cell walls, thus avoiding severe damage to the internal cell structures. The third mechanism is extra-organ freezing, where water is moved out of fragile organs like buds and shoots to create a form of icy insulation when it freezes around the organs (Sutinen et al. 2001). Tree species that are endemic to the boreal, including those within the *Abies, Larix, Pinus, Picea, Betula* and *Populus* genera use all these strategies to withstand temperatures as low as –80°C (Brandt 2009). Plant hardiness also includes an ability to recover from frost damage when the warmer season returns.

Species adaptation to natural disturbance

Boreal species are well adapted to recurring natural disturbances such as fire, insect infestation and extreme weather events (windthrow, ice storms, heavy snow load) that trigger renewal of the boreal forest (Chapin et al. 2006, Shorohova et al. 2011).

Adaptation to fire can take several forms. Some species, like *P. sylvestris* and *L. sibirica*, develop thick bark that can resist light to moderate ground fires by protecting the cambium and keeping the individual tree alive. Some temperate pine species that occur in the hemiboreal transition use this strategy.

Another strategy is protection of seeds against fire. Some species, like *P. banksiana* and *P. contorta*, for instance, have serotinous cones (cones that are protected by a waxy coating) that require intense heat, such as that provided by fire, to release their seeds (Rudolph and Laidly 1990). After a forest fire, even if the individual tree does not survive, the scales of the cone open and seeds are dispersed to sites that have been cleared of competing understory species. Fire also has the effect of reducing, if not removing altogether, the soil organic layer, thus preparing ideal seedbeds for germination and seedling establishment. Some spruce species, such as *P. mariana*, have semi-serotinous cones (Viereck and Johnston 1990) that confer a post-fire advantage over species such as *A. balsamea, P. glauca* and *Thuja occidentalis* that do not have this adaptation.

Seed dispersal by wind after fire can be an important strategy; the seeds of *Populus* spp. and *Salix* spp. are extremely light and can be dispersed over large distances. The slightly heavier seeds of *Betula* spp. and *Alnus* spp., as well as those of conifers, are also winged and can be wind-dispersed over short to moderate distances. However, seeds released in winter can travel long distances over snow crusts and ice. These shade-intolerant species also have the ability to outgrow competition with fast juvenile height growth in full light conditions. The hardwood species can also regenerate from stump sprouting or root suckering after fire or other catastrophic disturbance has killed the parent tree.

Insect outbreaks can have tremendous impacts on the forested landscape. When an outbreak lasts several years in forests with a high component of species that are targeted by the insect, severe mortality can result in a stand replacement or conversion event. Some shade-tolerant boreal species survive such events by establishing a bank of seedlings and saplings that persist in the understory, ready to replace dead trees as canopy gaps open. For example, in eastern North America, *A. balsamea* is sensitive to spruce budworm (*C. fumiferana*) outbreaks. The larvae of this insect feed on new foliage of the fir trees, killing the annual growth and leaving only the old foliage, which is less efficient for photosynthesis. Because fir needles are retained on the tree for less than six years, trees become severely weakened if the infestation lasts more than three years. With a severe outbreak, high mortality rates occur after three to five years of damage by the insect, gradually opening the canopy. This shade-tolerant species produces yearly seed crops that can germinate and establish a seedling bank in the understory (Morin et al. 2009). The canopy gaps release the seedlings, effectively re-establishing a fir forest. With stands containing more than one species of different susceptibility, the result can be a shift in forest composition induced by the insect outbreak. *P. glauca* and *P. mariana*, which are common co-associates of *A. balsamea*, are less sensitive to this insect and may survive more years of defoliation. When hardwoods such as *B. papyrifera* or *P. tremuloides* are mixed with the fir, they can assume dominance over the coniferous species in a protracted outbreak.

Wind is a natural disturbance agent with varying effects on individual trees or tree stands. Continual exposure to wind often shapes the crowns of individual trees or stands. Moderate wind events remove foliage, damage branches or parts of stems and break or uproot single or groups of trees to create canopy gaps. Severe windthrow events create large gaps at the landscape scale and result in the replacement of entire stands. The local consequences of windthrow are multiple. There is an immediate loss of standing biomass, and a consequent addition of downed woody debris. Canopy gaps increase light transmission to the understory. Uprooting creates new microsites, as soil layers are intermixed and microtopography is changed (Ruel et al. 2023). For conifers that survive a severe wind event, crown swaying and disturbance of the root system can result in reduced ability to take up moisture (Janda et al. 2021, Korznikov et al. 2022). Consequently, within three to five years after exposure to strong winds, trees may become desiccated and more vulnerable to insect infestation. The frequency of extreme wind, topographic exposure, soil conditions (texture, humidity, stoniness) and fertility are all factors that affect the severity and extent of wind damage to a forest (Mitchell, 2013). In South and East Asia for example, tropical cyclones are known to be the most important agents of natural forest canopy disturbance in tropical and temperate forests. More recently, due to global climatic transformations, tropical cyclones have become more active in Northeast Asia and have caused disturbances in the region's boreal coniferous forests (Altman et al. 2018, Korznikov et al. 2022).

Species adaptation to soil conditions

Paludification and permafrost development can severely alter soil conditions and alter forest productivity and composition. Paludification is the accumulation of organic matter over time caused either by rising water tables or increasing soil moisture accompanied by *Sphagnum* colonization (Lavoie et al. 2005). Forest decline related to paludification is prevalent in Siberia (Russia) (Peregon et al. 2007), Canada (Kuhry et al. 1993; Lavoie et al. 2005) and Alaska (Jorgenson et al. 2013). Edaphic paludification occurs in wet topographic positions, where a high water table promotes the growth of peat-forming plants (e.g., *Carex* and *Sphagnum* species). Successional paludification can occur on well-drained soils, where peat-forming mosses accumulate at a faster rate than humus decomposition during forest succession between fire events, independent of site topography or drainage (Simard et al. 2007). Paludification reduces soil temperature, decomposition rates, microbial activity and nutrient availability, which together lead to lower site productivity with time after disturbance (Swanson et al. 2000). *Sphagnum* growth and expansion, in particular, is a major contributor to acidification, peat accumulation and a decrease in productivity (Klinger 1990). Paludification, however, can also provide a strong soil sink for carbon.

Perennially frozen ground (permafrost) is a characteristic of arctic, subarctic and high mountain regions that is fundamental to geomorphic processes and ecological development in tundra and boreal forests. In northeast Yakutia (Eastern Siberia), the layer of permafrost can be 500 m deep, but the effect of permafrost on vegetation depends on the depth of the seasonally thawed surface active layer. Permafrost-affected regions occupy about 23 per cent of exposed lands in the Northern Hemisphere and it is abundant in the northern portions of the boreal biome (Figure 2.2; Brown et al. 1997). Generally, permafrost is considered a limiting factor for trees: cold soils and delayed thawing of the active horizon in spring cannot be tolerated by many species.

Peatlands exhibit a general trend towards being a carbon sink rather than a source near the southern limit of their distribution (Peregon et al. 2007), and in permafrost-affected areas, impeded drainage by permafrost water impoundment in thermokarst features contributes to *Sphagnum* growth and to soil carbon accumulation (Jorgenson et al. 2013). Permafrost serves as one of the largest reservoirs of carbon and can be considered a factor in long-term boreal ecosystem dynamics (Schuur and Mack 2018). Because permafrost properties are temperature dependent, degradation of permafrost in response to climate change will have large consequences for natural ecosystems, human infrastructure (Shur and Jorgenson 2007) and future climate. Upon thawing, everything about the ecosystem changes, including surface hydrology and soil drainage, habitat changes for vegetation and wildlife, and emissions of greenhouse gases (Jorgenson et al. 2013).

Dynamic patterns of boreal forests

Multiple dynamic patterns can be observed in boreal forests. Bergeron et al. (2014) give a detailed description of the stand dynamics of North American boreal mixedwoods, while Shorohova el al. (2011) present the variability of dynamics in the circumboreal zone. Here, we provide an overview of boreal forest dynamics according to four potential pathways: (1) auto-cyclic stand replacement after major disturbance; (2) succession of species in the absence of major disturbance; (3) mixture maintenance through cyclic, but less severe, disturbance;

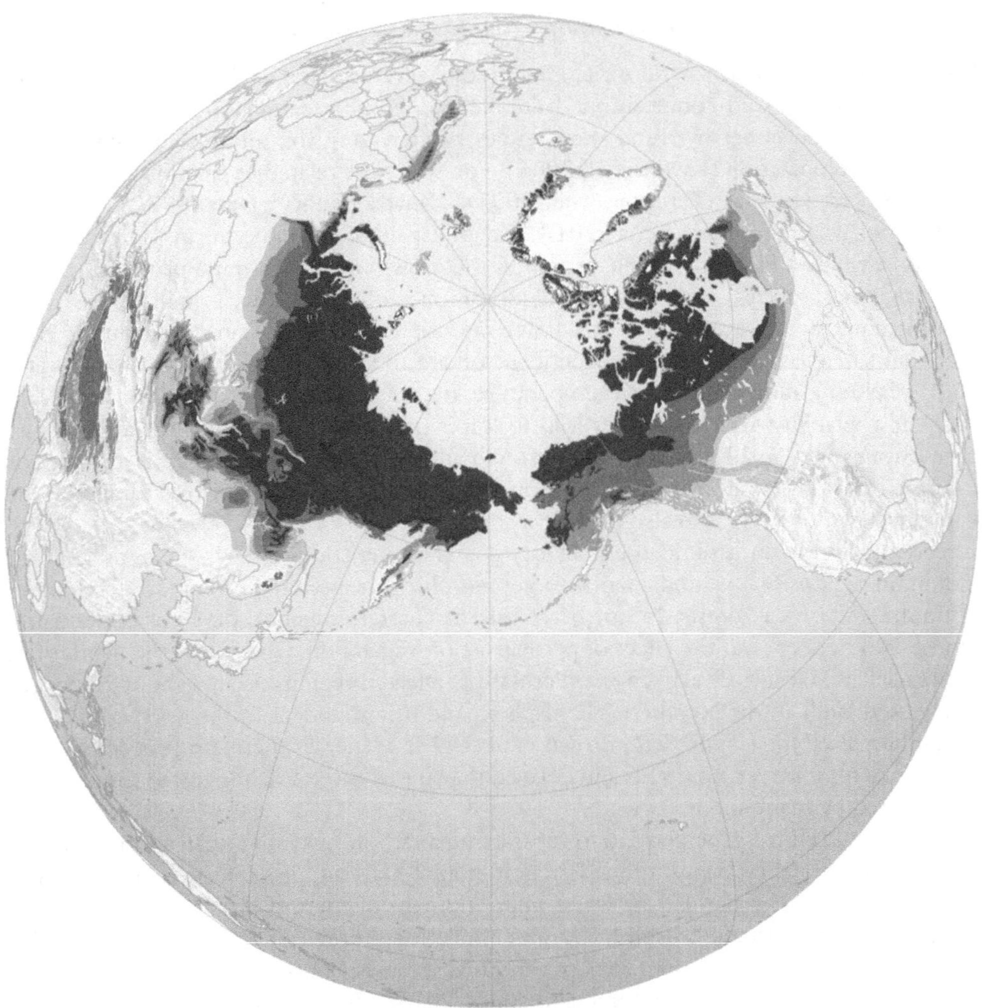

Figure 2.2 Distribution of permafrost zones in the Northern Hemisphere (NASA Earth Observatory map by Joshua Stevens, using data from the National Snow and Ice Data Center; https://earthobservatory.nasa.gov/images/87794/picturing-arctic-permafrost).

and (4) replacement of closed forest by open woodland following a failure of natural regeneration after a major disturbance.

Major disturbances, such as fire, insect outbreak, windthrow, river channel changes, flooding, avalanches, and landslides, trigger the replacement of entire forest stands. Many boreal tree species, however, are adapted to such events and are able to replace themselves in the new stand through varying strategies. Some species, such as *P. banksiana* and *P. mariana*, have serotinous or semi-serotinous cones that release their seeds after fire when seedling establishment and early growth is optimal, while others, such as *A. balsamea* and *P. sibirica* establish a bank of shade-tolerant seedlings in the understory that can sprout after disturbance. Resprouting from surviving roots is common for some species, such as *P. tremuloides* and *B. papyrifera* and *B. pendula*. Finally, light, wind-blown seeds are effective at colonizing disturbed areas for some species like aspen and birch. Recolonization can fail under certain circumstances, however, such as when fire occurs in a young stand before the trees have reached sexual maturity. This can result in a very open stand of the same species (i.e., a woodland) or a shift in species composition, usually towards shade-intolerant species with the ability to regenerate vegetatively or to disperse seeds from distant seed sources.

Succession of species involves the gradual transition over a long period of time from a stand dominated by shade-intolerant species to one dominated by shade-tolerant species. This process takes place when no major disturbance occurs for a period exceeding the lifespan of the shade-intolerant species that initiated the stand after the last major disturbance. Over time, shade-tolerant species establish in the understory. These individuals can take advantage of openings created by the death of a single or a small group of trees to grow into the canopy (Bouchard et al. 2006). This process is known as gap-phase replacement. In the continuing absence of major disturbance, the species mix of the stand will evolve until no shade-intolerant trees persist. An example of this dynamic pathway in the Eastern North American boreal is the progressive transition from stands dominated by *B. papyrifera* (a shade-intolerant species) to mixedwoods of *Betula* and *A. balsamea* (a highly shade-tolerant species) and finally to *Abies*-dominated stands that are able to auto-cycle as described above. Similarly, in the West–Central North American boreal *P. tremuloides* stands transition to *P. glauca*-dominated stands. Severe (i.e., stand-replacing) disturbance during the succession process returns the system to an early seral stage dominated by shade-intolerant species (e.g., *Betula* or *Populus* in our examples) from which the stand could evolve into a different pathway.

Mixtures of shade-intolerant and shade-tolerant tree species can establish and maintain themselves in the presence of frequent, low-intensity disturbances. Depending on the size of the gaps created in the canopy, different species can be favored. Large openings with more light reaching the forest floor are suitable for species that are less shade-tolerant. The relative proportions of species in the mixture can change in response to disturbance events. The mixture of *P. tremuloides* and *P. glauca* found in the West–Central North American boreal is an example of this process (Bergeron et al. 2014).

In case of a major or too severe disturbance in a stand too young for the trees to have reached sexual maturity, or that is isolated from seed sources, a shift in density is often observed and could lead to an open woodland. Because of these different pathways, and the low tree species diversity in the boreal, landscape diversity in stand structure is often important. Stand structures are varied because the time since disturbance and regeneration success are also variable at the landscape level.

Threats to the boreal zone

The boreal zone has historically been sparsely populated because of climatic and soil limitations to agriculture, although indigenous peoples developed cultures that utilized seasonally abundant fauna and flora for subsistence. In recent history, the boreal zone has been widely exploited for industrial forestry, agriculture, mining, oil extraction and hydroelectric activities and has undergone increasing urban expansion. Yet, global boreal forests are still largely ecologically intact and account for 44% of the world's intact forests (Potapov et al. 2008). Boreal forests are important for providing timber and non-timber products, habitat and forage for animals, as well as supporting ecosystem services, such as water purification, climate regulation and carbon storage (Chapin et al. 2006). Forest resilience in the boreal zone generally ensures regeneration of the forest after natural or anthropogenic disturbance, but the conditions of secondary forests are not similar to those of the primeval forest. Because logging usually targets dense and mature stands, it gradually replaces those stands with younger ones, with less dead wood and a different proportion of the main tree species. There are cumulative impacts of forest harvesting and other human activities over time and space, and those are augmenting the impacts of natural disturbances, not replacing them. One effect is the fragmentation and reduction of old-growth forests on the landscape (Leduc et al. 2009).

Human use of boreal forests, in the form of tree harvesting or land use for mining, farming or urban development, has effects on the resident animal species. These effects could be detrimental or beneficial depending on the species. For example, in eastern Canada, fragmentation of the forested landscape by access roads, logging and increased road traffic have been identified as a threat to the woodland caribou (*Rangifer tarandus*) (Bouchard and Garet 2014). In contrast, moose populations in northwestern Québec have been favoured by logging in mosaic patterns, with recently logged patches providing browse while adjacent patches of mature forest provide shelter (Potvin et al. 2005). In eastern Asia, the Amur tiger (*Panthera tigris altaica*) and Amur leopard (*Panthera pardus orientalis*) after being affected by habitat reduction and fragmentation find their feeding territory reduced and abundance of a number of key prey species lowered.

Efforts have been made in many northern countries to create and maintain protected areas to preserve key habitat, natural dynamic processes and ecosystem services. These protected areas may harbor examples of old-growth boreal forest and may serve as refuges for some endangered species. Complementary to forest protection are efforts to develop forest management practices that emulate landscape-scale patterns that are created by natural disturbances. An important goal of ecosystem-based management is to mitigate the impact of current management practices on natural ecosystems (Gauthier et al. 2009). With this management approach, the focus is not only on what to harvest but also on wood and forest patches to be left in place in order to ensure that essential ecosystem services continue to be provided (Leduc et al. 2009).

Climate change effects on boreal forests

The boreal biome is expected to be sensitive to climatic changes that can alter natural disturbance regimes, forest composition, permafrost stability and global carbon cycling (Chapin et al. 2006). The effects of global warming will probably be greater in the northern regions than in southern areas, while changes in precipitation patterns can also be expected. Many studies suggest that, given the ecological niches of boreal species, the ranges of these species will likely shift (Périé et al. 2014) and the known limits of the boreal zone will change. For example, species distribution modeling for *P. jezoensis* in northeast Asia predicts a considerable northward shift of suitable conditions for this species (Korznikov et al. 2023). Overall, the southern boundaries with the temperate zone and the northern boundary with the arctic are expected to move northward, as species shift their ranges in response to increasing temperatures and changing soil conditions.

The global climate system responds to changes in boreal forests through changing water and energy exchange, exchange of greenhouse gases and delivery of water to the Arctic

Ocean, but there is uncertainty whether the changes will enhance or mitigate global warming (Chapin et al. 2006). Tree canopy cover and snow cover strongly affect albedo and energy exchange. Increasing disturbance from fires, insect outbreaks and altered land cover contributes to these changes by affecting the energy exchange of forest canopies, the combustion of vegetation and surface organic layers and the emission of large quantities of carbon dioxide and methane (Chapin et al. 2006).

Soil organic carbon has accumulated over long time periods in permafrost soils in the boreal ecosystems, and the thawing and loss of this carbon has the potential to have large effects on the global climate system (Schuur and Mack 2018). Thawing of permafrost and subsequent decomposition of its organic matter could release significant amounts of greenhouse gases to the atmosphere, especially methane with its greater warming potential compared to carbon dioxide. Permafrost soils account for 70 per cent of the soil organic carbon in the northern circumpolar regions (Tarnocai et al. 2009). However, the net effects of permafrost thaw on soil carbon dynamics are presently uncertain (Jorgenson et al. 2013).

Finally, the boreal forest plays a significant role in the delivery of water to the Arctic Ocean because it dominates the landmasses of arctic watersheds (Serreze et al. 2006). Changes in runoff from permafrost thaw and reduced snow cover within the boreal forest could alter the water budget of the Arctic Ocean and change sea ice dynamics, thus affecting feedbacks to the climate system.

Conclusion

Ecosystems of the boreal zone, despite their floristic and structural simplicity, provide important habitat for many species of animals. The largest animals—wolves, moose, caribou, bears, beaver, porcupine, lynx and foxes—are widespread but can be represented in different regions by closely related taxa. Boreal ecosystems provide summer breeding range to many species of birds, most of which migrate south for the winter. Grazing ungulates can have an effect on forest regeneration, as they selectively choose their summer food and because they rely on bark, young twigs and buds in the winter (Kielland et al. 2006). Moose, particularly, affect the composition of early successional stands, with consequences for later forest structure. The same is true for rodents that feed on seeds of certain species, such as pines. Insects also consume large quantities of foliage and seeds of boreal vegetation. The extensive lichen-rich woodlands in Northern Eurasia and North America support large populations of caribou and reindeer. In Eurasia, reindeer are extensively raised by local people for multiple uses, including transportation and food. Wolves are the quintessential large predator of boreal forests, while black and brown bears serve as both predators (mainly moose and caribou calves) and important herbivores (mainly berries).

The boreal biome is one of the largest global ecosystems. Plant species of the circumboreal zone share several mechanisms of adaptation to the cold climate and to natural disturbance regimes. Boreal forests have evolved over long time periods, responding to alternating ice ages and warmer climatic periods. Due to the prevalence of permafrost-affected soils, large soil carbon reservoirs and forest structure that affects the global energy balance, changes in forest patterns and processes are likely to have large consequences for the global climate system.

Acknowledgements

We thank two anonymous reviewers (for the first edition) and Sylvie Gauthier whose comments contributed to the improvement of the manuscript.

References

Altman, J., Ukhvatkina, O.N., Omelko, A.M., Macek, M., Plener, T., Pejcha, V., Cerny, T., Petrik, P., Srutek, M., Song J. S., Zhmerenetsky, A. A.,Vozmishcheva, A. S., Krestov, P. V., Petrenk,o T. Y., Treidte, K. and Dolezal, J. (2018) 'Poleward migration of the destructive effects of tropical cyclones during the 20th century', *PNAS*, vol 115, pp. 11543–11548.

Baldwin, K, et al. (2020) 'Vegetation Zones of Canada: A Biogeoclimatic Perspective', Sault Ste. Marie, ON, Canada: Natural Resources Canada. Information Report GLC-X-25.

Bergeron, Y., Chen, H. Y., Kenkel, N. C., Leduc, A. L. and MacDonald, S. E. (2014) 'Boreal mixedwood stand dynamics: ecological processes underlying multiple pathways', *The Forestry Chronicle*, vol 90, pp. 202–213.

Bouchard, M. and Garet, J. (2014) 'A framework to optimize the restoration and retention of large mature forest tracts in managed boreal landscapes', *Ecological Applications*, vol 24, pp. 1689–1704.

Bouchard, M., Kneeshaw, D. and Bergeron, Y. (2006) 'Forest dynamics after successive spruce budworm outbreaks in mixedwood forests', *Ecology*, vol 87, pp. 2319–2329.

Brandt, J. P. (2009), 'The extent of the North American boreal zone', *Environmental Reviews*, vol 17, pp. 101–161.

Brown, J, Ferrians, O.J., Heginbottom, J. A., Jr. and Melnikov, E. S. (1997), 'Circum-Arctic map of permafrost and ground ice conditions', Reston, VI: U.S. Geological Survey. CP-45.

Canadian National Vegetation Classification project (CNVC) (2014) http://cnvc-cnvc.ca/, accessed 10 January 2014, and personal communication from Ken Baldwin.

Chapin III, F.S., Oswood, M. W., Van Cleve, K., Viereck, L. A. and Verbyla, D. L. (2006) *Alaska's Changing Boreal Forest*, Oxford University Press, New York.

Daniëls, F. J. A. and de Molenaar, J. G. (2011) 'Flora and vegetation of Tasiilaq, formerly Angmagssalik, Southeast Greenland — a comparison of data between around 1900 and 2007', *Ambio*, vol 40, pp. 650–659.

Ermakov, N. (2003) *Diversity of Boreal Vegetation: Hemiboreal Forests, Classification and Ordination*, Nauka, Novosibirsk.

Ermakov, N. and Morozova, O. (2011) 'Syntaxonomical survey of boreal oligotrophic pine forests in northern Europe and Western Siberia', *Applied Vegetation Science*, vol 14, pp. 524–536.

FAO and UNEP (2020) 'The State of the World's Forests 2020. Forests, Biodiversity and People', Rome.

Gauthier, S., Vaillancourt, M. A., Leduc, A., De Grandpré, L., Kneeshaw, D. and others, eds., (2009) *Ecosystem Management in the Boreal Forest*, Presses de l'Université du Québec. Québec, Canada.

Hämet-Ahti, L. (1981) 'The boreal zone and its biotic subdivision', *Fennia*, vol 159, pp. 69–75.

Helbig, M., Pappas, C. and Sonnentag, O. (2018) 'Permafrost thaw and wildfire: Equally important drivers of boreal tree cover changes in the Taiga Plains, Canada', *Geophysical Research Letters*, vol 43, pp. 1598–1606.

Hytteborn, H., Maslov, A. A.,Nazimova, D. I. and Rysin, L. P. (2005) 'Boreal forests of Eurasia', in F. Andersson (ed.) *Coniferous Forests* (Ecosystems of the World 6). Elsevier, Amsterdam.

Janda, P., Ukhvatkina, O. N., Vozmishcheva, A. S., Omelko, A. M., Doležal, J., Krestov, P. V., Zhmerenetsky, A.A., Song, J.S., and Altman, J. (2021). 'Tree canopy accession strategy changes along the latitudinal gradient of temperate Northeast Asia', *Global Ecology and Biogeography*, vol 30, No. 3, pp. 738–748.

Jorgenson, M. T., Harden, J., Kanevskiy, M., O'Donnel, J., Wickland, K. and others, (2013) 'Reorganization of vegetation, hydrology and soil carbon after permafrost degradation across heterogeneous boreal landscapes', *Environmental Research Letters*, vol 8, 035017.

Jorgenson, M. T. and Meidinger, D. (2015) 'The Alaska-Yukon Region of the Circumboreal Vegetation Map', Conservation of Arctic Flora and Fauna, CAFF Strategies Series, Akureyri, Iceland.

Kielland, K., Bryant, J. P. and Ruess, R. W. (2006) 'Mammalian herbivory, ecosystem engineering, and ecological cascades in Alaska boreal forests', in F. S. Chapin, M. W. Oswood, K. Van Cleve, L. A. Viereck and D. L. Verbyla (eds.) *Alaska's Changing Boreal Forests*, Oxford University Press, Oxford.
Klinger, L. F. (1990) 'Global patterns in community succession: 1. Bryophytes and forest decline', *Memoirs of the Torrey Botanical Club*, vol 24, pp. 1–50.
Korznikov, K., Kislov, D., Doležal, J. and Altman, J. (2023). 'Poleward migration of tropical cyclones induced severe disturbance of boreal forest above 50°'. *Science of the Total Environment*, vol 890, 164376. doi: 10.1016/j.scitotenv.2023.164376.
Korznikov, K., Kislov D., Doležal J., Petrenko T. and Altman J. (2022) 'Tropical Cyclones Moving into Boreal Forests: Relationships between Disturbance Areas and Environmental Drivers', *Science of The Total Environment*, vol 844, 156931.
Krestov, P. V. (2003) 'Forest vegetation of Easternmost Russia (Russian Far East)', in J. Kolbek, M. Šrůtek and E.O. Box (eds.), *Forest Vegetation of Northeast Asia*, Kluwer Academic Publishers, Dordrecht.
Krestov, P. V., Ermakov, N. B., Osipov, S. V. and Nakamura, Y. (2009) 'Classification and phytogeography of larch forests of northeast Asia', *Folia Geobotanica*, vol 44, pp. 323–363.
Krestov, P. V. and Nakamura, Y. (2002) 'Phytosociological study of the *P. jezoensis* forests of the Far East', *Folia Geobotanica*, vol 37, pp. 441–474.
Kuhry, P., Nicholson, B., Gignac, L. D., Vitt, D. H. and Bayley, S. E. (1993) 'Development of Sphagnum-dominated peatlands in boreal continental Canada', *Canadian J. of Botany*, vol 71, pp. 10–22.
Kuusela, K (1992) 'The boreal forests: an overview', *Unasylva*, vol 43, pp. 3–13.
Lavoie, M., Paré, D., Fenton, N., Groot, A. and Taylor, K. (2005) 'Paludification and management of forested peatlands in Canada: a literature review', *Environmental Reviews*, vol 13, pp. 21–50.
Leduc, A., Gauthier, S., Vaillancourt, M. A., Bergeron, Y., De Grandpré L. and others, (2009) 'Perspectives' in S. Gauthier, M.A. Vaillancourt, A. Leduc, L. De Grandpré, D. Kneeshaw, H. Morin, P. Drapeau and Y. Bergeron (eds.) *Ecosystem Management in the Boreal Forest*, Presses de l'Université du Québec. Québec, Canada.
Mitchell, S. J. (2013) 'Wind as a natural disturbance agent in forests: a synthesis', *Forestry*, vol 86, pp. 147–157.
Morin, H., Laprise, D., Simard, A.-A. and Amouch, S., (2009) 'Spruce budworm outbreak regimes in eastern North America' in. S. Gauthier, M.-A. Vaillancourt, A. Leduc, L. De Grandpré, D. Kneeshaw, H. Morin, P. Drapeau and Y. Bergeron (eds.) *Ecosystem M. in the Boreal Forest*, Presses de l'Université du Québec, Québec, Canada.
Nakamura, Y. and Krestov, P. V. (2005) 'Coniferous forests of the temperate zone of Asia', in F. Andersson (ed.) *Coniferous Forests* (Ecosystems of the World, 6). Elsevier Academic Press, New York.
Nazimova, D.I., Danilina, D. M. and Stepanov, N. V. (2014) 'Rain-barrier forest ecosystems of the Sayan Mountains', *Botanica Pacifica*, vol 3, pp. 39–47.
Peregon, A., Uchida, M. and Shibata, Y. (2007) 'Sphagnum peatland development at their southern climatic range in West Siberia: trends and peat accumulation patterns', *Environmental Research Letters*, vol 2, 045014.
Périé, C., De Blois, S., Lambert, M.-C. and Casajus, N. (2014) 'Effets anticipés des changements climatiques sur l'habitat des espèces arborescentes au Québec', Mémoire de recherche forestière n° 173. Gouvernement du Québec, Ministère des Ressources naturelles, Direction de la recherche forestière.
Potapov, P., Yaroshenko, A., Turubanova, S., Dubinin, M., Laestadius, L. and others, (2008) 'Mapping the world's intact forest landscapes by remote sensing', *Ecology and Society*, vol 13, p. 51.
Potvin, F., Breton, L. and Courtois, R. (2005) 'Response of beaver, moose, and snowshoe hare to clear-cutting in a Quebec boreal forest: a reassessment 10 years after cut', *Canadian J. of Forest Research*, vol 35, pp. 151–160.
Qian, H, Krestov, P., Fu P. Y., Wang, Q. L, Song, J. S., Chourmouzis, C. (2003) 'Phytogeography of Northeast Asia. Forest vegetation of northeast Asia', in J. Kolbeck (ed.) *Forest V. of Northeast Asia*, Kluwer Academic Publishers, Dordrecht.
Rivas-Martínez, S., Rivas Sáenz, S. and Penas, A. (2011) 'Worldwide bioclimatic classification system', *Global Geobotany*, vol 1, pp. 1–634 + maps.

Rudolph, T. D. and Laidly, P. R. (1990) 'Jack pine (*P. banksiana* Lamb.)', in R. M. Burns and B. H. Honkala, (eds.) *Silvics of North America*, Volume 1, Conifers, U.S.D.A. Forest Service Agriculture Handbook 654, Washington D.C.

Ruel, J. C., Wermelinger, B., Gauthier, S., Burton, P.J., Waldron, K. and Shorohova, E. (2023). 'Selected examples of interactions between natural disturbances', in M. M. Girona et al. (eds.), Boreal Forests in the Face of Climate Change, 2023, Advances in Global Change Research 74, Springer, Switzerland, pp. 123–141.

Saucier, J.-P. (2008) 'Defining the boreal in the ecological land classification for Québec', in S. S. Talbot (ed.) Proceedings of the Fourth International Conservation of Arctic Flora and Fauna (CAFF) Flora Group Workshop, 15–18 May 2007, Tórshavn, 'Faroe Islands', CAFF Technical Report No. 15. Akureyri, Iceland.

Schuur E. A. and Mack, M. C. (2018) 'Ecological response to permafrost thaw and consequences for local and global ecosystem services', *Annual Review of Ecology, Evolution, and Systematics*, vol 49, pp. 279–301.

Serreze, M. C., Barrett, A. P., Slater, A. G., Woodgate, R. A., Aagaard, K. and others (2006) 'The large-scale freshwater cycle of the Arctic',. *J. of Geophysical Research*, vol 111, C11010.

Shorohova, E., Kneeshaw, D., Kuuluvainen, T. and Gauthier, S. (2011) 'Variability and dynamics of old-growth forests in the circumboreal zone: implications for conservation, restoration and management', *Silva Fennica*, vol. 45, p. 72.

Shur, Y.L. and Jorgenson, M. T. (2007) 'Patterns of permafrost formation and degradation in relation to climate and ecosystems', *Permafrost and Periglacial Processes*, vol 18, pp. 7–19.

Simard, M., Lecomte, N., Bergeron, Y., Bernier, P. Y. and Paré, D. (2007) 'Forest productivity decline caused by successional paludification of boreal soils', *Ecological Applications*, vol 17, pp. 1619–1637.

Sutinen, M.-L., Arora, R., Wisniewski, M., Ashworth, E., Strimbeck, R. and Palta, J. (2001) 'Mechanisms of frost survival and freeze-damage in nature', in F. J. Bigras and S.J. Colombo (eds.) *Conifer Cold Hardiness*, Tree Physiology, Kluwer Academic Publishers, Dordrecht, vol 1, pp. 89–120.

Swanson, D. K., Lacelle, B. and Tarnocai, C. (2000) 'Temperature and the boreal-subarctic maximum in soil organic carbon', *Géographie Physique et Quaternaire*, vol 54, pp. 157–167.

Takhtajan, A., (1986) *Floristic Regions of the World*, University of California Press, Berkeley, CA.

Talbot, S. S., Schofield, W. B., Talbot, S. L., Daniëls, F. J. A., (2010) 'Vegetation of eastern Unalaska Island, Aleutian Islands, Alaska', *Botany*, vol 88, pp. 366–388.

Tarnocai, C., Canadell, J. G., Schuur, E. A. G., Kuhry, P., Mazhitova, G. and others, (2009) 'Soil organic carbon pools in the northern circumpolar permafrost region', *Global Biogeochemical Cycles*, vol 23, GB2023.

Tuhkanen, S. (1984). 'Circumboreal system of climatic-phytogeographical regions'. *Acta Botanica Fennica*, vol 127, pp. 1–50.

Viereck, L.A., Dyrness, C. T., Batten, A. R. and Wenzlick, K. J. (1992) 'The Alaska Vegetation Classification', Pacific Northwest Research Station, U.S. Forest Service, Portland, OR. Gen. Tech. Rep. PNW- GTR-286.

Viereck, L. A. and Johnston, W. F. (1990) '*P. mariana* (Mill.) B.S.P. black spruce', in R. M. Burns and B. H. Honkala (eds.) *Silvics of North America*, Volume 1, Conifers, U.S.D.A. Forest Service Agriculture Handbook 654, Washington D. C.

3
TEMPERATE FORESTS

Lee E. Frelich and Rebecca A. Montgomery

Introduction

Biogeography

Temperate forests occur at latitudes between 30 and 60 degrees in the northern and southern hemispheres and comprise about 16% of all forests on the earth (Figure 3.1; Whittaker, 1975). Biogeography includes four regions in the northern hemisphere—eastern North America (Northeastern United States and adjacent Canada), western North America (Pacific northwest United States and adjacent Canada), Europe (British Isles across northern Europe, parts of southeastern Europe and the Caucasus), northeastern Asia (China, Japan, Korean Peninsula), and two regions in the southern hemisphere—southern South America (Chile and Argentina) and Australasian (southeastern mainland Australia, Tasmania and New Zealand).

Northern hemisphere forests (Figure 3.2) are very large in spatial extent and are found between boreal forest to the north, and subtropical forests, grasslands and savannas, or Mediterranean vegetation to the south. Many of the tree genera in temperate forests occur across North America, Europe and Asia: *Acer*, *Alnus*, *Betula*, *Carpinus*, *Carya*, *Castanea*, *Celtis*, *Fagus*, *Fraxinus*, *Larix*, *Pinus*, *Populus*, *Prunus*, *Quercus*, *Picea*, *Tilia*, and *Ulmus*. Other genera—*Liriodendron*, *Pseudotsuga* and *Tsuga*—occur in North America and Asia. Some tree genera are widespread across temperate and boreal biomes (e.g., some *Betula*, *Pinus*, *Picea* and *Populus* species), others occur in the temperate zone (e.g., *Quercus*, *Tilia* and some *Acer* species), while some occupy the transition zone plus limited portions of the southern boreal and northern temperate zones, and cannot be conveniently classified as temperate or boreal species (e.g. *Fraxinus nigra* (black ash), *Pinus resinosa* (red pine) and *Picea rubens* (red spruce) in North America).

Some tree communities have similar composition at the generic level across eastern North America, Europe and northeastern Asia. For example, *Fagus*, *Acer* and *Quercus* occur together on mesic sites: *A. pictum susp. mono* (painted maple), *F. crenata* (Japanese beech) and *Q. mongolica* (Mongolian oak) in Japan, *A. saccharum* (sugar maple), *F. grandifolia* (American beech) and *Q. rubra* (red oak) in eastern North America, and *A. platanoides*

DOI: 10.4324/9781003324072-4

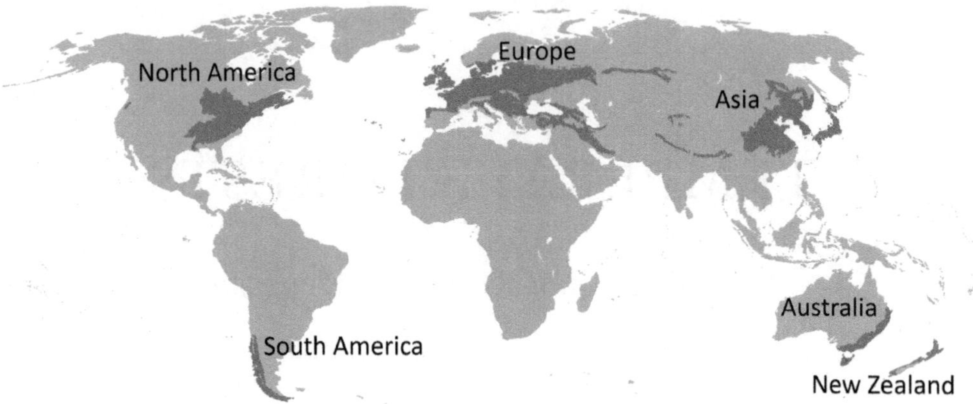

Figure 3.1 Location of the world's temperate forests. Modified from: Terpsichores (Own work) [CC BY-SA 3.0 (http://creativecommons.org/licenses/by-sa/3.0)], via Wikimedia Commons.

(Norway maple), *F. sylvatica* (European beech) and *Q. robur* (English oak) in northern Europe (Runkle, 1982; Yamamoto et al., 1995; Svenning and Skov, 2005).

Western North American temperate rainforests include mainly the coniferous species *Picea sitchensis* (Sitka spruce), *Tsuga heterophylla* (western hemlock), *Thuja plicata* (western red cedar), *Pseudotsuga menziesii* var. *menziesii* (coast Douglas-fir) and the deciduous broadleaf *Acer macrophyllum* (bigleaf maple).

Southern hemisphere temperate forests (Figures 3.3–3.5) differ substantially in composition from the northern temperate forest, with different genera of trees listed below. Although broadleaved and coniferous temperate rainforests occur in oceanic coastal climates, closed canopy dry-to-mesic temperate forests also occur in drier areas such as eastern Tasmania, as well as moving inland from the coasts of mainland Australia and western coast of southern South America, where they occur between the coastal rainforests and interior woodlands with Mediterranean climates, grasslands and savannas. Colder montane vegetation of shorter stature also occurs at higher elevations in South America and New Zealand—but unlike northern temperate forests, there are no true boreal forests in a poleward position from the southern temperate forests—poleward areas are open ocean.

Nothofagus species occur in the regions of temperate forest in southern hemisphere mentioned above (Figure 3.3), but there is regional variation in which *Nothofagus* species are present, their level of dominance and the other tree species that occur alongside them also vary dramatically. There are distinctive tree floras in southern South America, mainland Australia and Tasmania, and New Zealand (DellaSala 2011). Valdivian temperate forests and Magellanic subpolar forests occur in southern Chile and Argentina. The Valdivian ecoregion has deciduous forests in the north with *N. alpina* (rauli beech) and *N. obliqua* (roble), further south, laurel-leaved forests with broadleaved evergreens—*Laureliopsis philippiana* (laurel), *Aextoxicon punctatum* (olivillo), *Eucryphia cordifolia* (ulmo) and *Weinmannia trichosperma* (tineo), Patagonian Andean forests at higher elevations and dominated by evergreen conifers, *Araucaria araucana* (monkey puzzle) and *Fitzroya cupressoides* (alerce), and northern Patagonian forests, with broadleaf evergreens such as *N. dombeyi* (coihue) and *N. nitida* (coihue de Chiloé) and coniferous *Pilgerodendron uviferum* (Guaitecas cypress).

Figure 3.2 Northern hemisphere forests. (a) Old-growth mesic forest with yellow birch, sugar maple and hemlock in Michigan, USA. Photograph by Lee Frelich. (b) Dry-mesic red oak forest in Minnesota, USA. Photograph by David Hansen, University of Minnesota.

The world's furthest south forest is the Magellanic subpolar forest. It includes rainforest with *N. betuloides* (Magellan's beech), *Drimys winteri* (canelo) and *Pilgerodendron uviferum*, and somewhat drier forests of mixed evergreen *N. betuloides* and deciduous *N. pumilio* (lenga beech) trees.

Moving to Australia, *N. cunninghamii* (myrtle beech), *N. moorei* (Antarctic beech) and *Atherosperma moschatum* (sassafras) occur in southeast Australia, along with a number of *Eucalyptus* species (Figure 3.4). *E. baxteri* (brown stringybark), *E. viminalis* (manna gum), *E. obliqua* (messmate stringybark) and *E. cypellocarpa* (mountain gum) occur in wet

Figure 3.3 *Nothofagus* forest in New Zealand. Photograph by George Weiblen.

sclerophyll forests—although *Eucalyptus* are generally pyrophilic species, these species are able to reproduce in forests where fires have longer return intervals than in dry climates of interior Australia.

On the island of Tasmania rainforests with *N. cunninghamii* (myrtle beech) and *Atherosperma moschatum* (sassafras) share dominance with a wide variety of species, including *Anodopetalum biglandulosum* (horizontal scrub), *Athrotaxis selaginoides* (King Billy pine), *Eucryphia* species (leatherwood), *Lagarostrobos franklinii* (huon pine) and *Phyllocladus aspleniifolius* (celery top pine). Drier forests with a variety of *Eucalyptus* species, are also found on the eastern side of the island.

Evergreen rainforests of New Zealand occur in coastal areas, with variation in species composition by elevation. At low elevation, the Podocarps *Dacrydium cupressinum* (rimu) and *Prumnopitys taxifolia* (mataī) are emergent above broadleaved evergreen canopies of *Pterophylla racemosa* (kamahi) and *Beilschmiedia tawa* (tawa) while at middle elevations other species of podocarps become common: *Prumnopitys ferruginea* (miro), *Dacrycarpus dacrydioides* (kahikatea; Figure 3.5) and *Podocarpus totara* (totara) along with *Libocedrus bidwillii* (Kaikawaka), and at higher elevation (above 500 m), forests are dominated by beech—*Nothofagus menziesii* (silver beech), *N. fusca* (red beech) and *N. truncata* (hard beech).

Climate

Temperate forests have seasonal climates, with warm and cool to cold seasons. Mean temperatures < 10°C last for at least three to four months, including some temperatures below freezing. Temperate forests have enough temperature difference to produce distinct annual rings in most tree species. Northern hemisphere temperate forests in general experience

Figure 3.4 *Eucalyptus* forest in southeastern Australia. Photograph by George Weiblen.

Figure 3.5 *Dacrycarpus* forest in New Zealand. Photograph by George Weiblen.

much colder winters and more continental climates (seasonal temperature ranges of 10–30°C) than southern hemisphere temperate forests, which have oceanic climates due to the much smaller land masses in temperate latitudes of the southern hemisphere, causing relatively cool summers and warm winters (seasonal temperature ranges of 5–10°C). A number of temperate forest ecoregions have rainforest climates, including the Pacific Northwest and parts of eastern North America, areas near the Black Sea, Caspian Sea, parts of Japan, China and the Korean Peninsula, and the southern tip of South America, southeastern Australia and New Zealand.

Northern temperate forests have strong seasonality with growing season lengths ranging from about four months at the temperate-boreal ecotone in the north (long, cold winters and cool summers) to about eight months at the temperate-subtropical forest ecotone (short cool winters and long hot summers) in the south (Peel et al., 2007). Precipitation regimes vary from wet winter/dry summer in coniferous temperate rainforests such as the Pacific Northwest United States to wet throughout the year (*Tsuga* forests in the Southeastern United States), to wet summer regimes with snow during the dormant season in most of the cold parts of the temperate forest, to droughty summers with erratic rainfall in places where temperate forest and savanna/grassland ecotones occur, such as in the continental interiors of North America and Eurasia and Australia (Whittaker, 1975). In the colder temperate climates, snow plays an important role by limiting the depth to which the soil freezes, which would otherwise cause extensive root damage in mature trees and mortality to seedlings and herbaceous plants. Some colder regions, such as Upper Michigan in the United States, and northern Japan, receive extreme amounts of snow due to lake-effect or ocean-effect snowfall generated by cold winds blowing over large water bodies before hitting land. These 'snow forests' have unique species composition—for example, *Acer saccharum* (sugar maple) is the dominant tree species in such forests in the Midwestern United States, since its main competitor, *Tsuga canadensis* (hemlock)—is damaged by the load of snow landing on the branches, and restricted to areas with lower snowfall.

The transition from temperate to boreal forest depends on either low winter minimum temperatures or cool summers. The deep supercooling point of the cells in the cambium under the bark of most temperate tree species ranges from –40°C to –45°C, and if winter temperatures are commonly that cold, then many of the dominant temperate tree species in genera such as *Quercus* and *Acer* cannot survive (George et al. 1974). In other areas near large bodies of water (e.g., Lake Superior in North America), winter temperatures are nowhere near cold enough to limit temperate tree species, but short, cool summers tip the competitive balance to boreal species, creating shoreline belts of boreal forest that transition to temperate forest inland where summers are warmer. The ecotone between northern temperate and boreal forests is broad, with species from the two biomes commonly found growing together over a 500-km-wide region (Fisichelli et al., 2014). In North America, temperate *Betula alleghaniensis* (yellow birch), *Acer saccharum* (sugar maple) and *Fagus grandifolia* (American beech) are mixed with boreal *Abies balsamea* (balsam fir) and *Picea glauca* (white spruce), and in Asia, temperate *Quercus mongolica* (Mongolian oak) and *A. pictum susp. mono* (painted maple) mix with boreal *Pinus koraiensis* (Korean pine) and *P. sylvestris* var. *mongolica* (Scots pine), while in Europe, temperate *A. platanoides* (Norway maple), *F. sylvatica* (European beech) and *Q. robur* (English oak) mix with boreal *B. pendula* (silver birch) and *Picea abies* (Norway spruce). In the southern hemisphere, there are a variety of forests of short stature or shrublands at elevations higher than those where the temperate forests exist, for example, *Nothofagus antarctica* (antarctic beech) krummholz and *Olearia colensoi* (leatherwood) shrublands at high elevations in Chile and New Zealand, respectively.

Soil and site conditions

The northern portions of the temperate forest in North America and Europe as well as parts of southern Chile, New Zealand and Tasmania were glaciated and have relatively young soils (about 9,000–17,000 years old) on top of glacial deposits or rocky surfaces sculpted by the glaciers. Older soils with a wide range of substrates, fertility and topographical relief occur elsewhere. Soil age and associated changes in soil nutrient availability, especially increases in N availability early in soil development (e.g., first 10,000 years post-glaciation) and loss of P over long time periods (>100,000 years post-glaciation) can influence development of forest types in northern and southern hemispheres (Peltzer et al., 2010).

Gradients from wet to mesic and dry soils in complex topography are common in all temperate regions. In New Zealand, *Dacrydium cupressinum* (rimu), *Dacrycarpus dacrydioides* (kahikatea) and *Libocedrus bidwillii* (Kaikawaka) regenerated abundantly after disturbance on poorly to very poorly drained sites, while *Pterophylla racemosa* (kamahi) occurred mostly on well-drained sites, and *Nothofagus menziesii* (silver beech) regenerated across the wet-to-dry sites (Veblen et al. 2016). In the northern hemisphere, mesic sites with loamy soils are common and usually dominated by *Acer*, *Fagus*, *Tilia* and *Tsuga*. Dry-mesic to dry sites with soils that are sandy or shallow to bedrock are usually characterized by *Pinus* and *Quercus* species. Swamps, peatlands and riparian areas are dominated by *Acer*, *Alnus*, *Betula*, *Fraxinus*, *Populus* and *Salix* species.

Importance to society

Temperate forest ecosystems provide globally significant ecosystem services—they cover an area of 10.4 million km^2 (about 6% of all terrestrial ecosystems) and store about 21% of carbon (139 PgC) in plants and 11.2% of carbon (262 PgC) in soils (Sabine et al., 2004).

High-density human settlement and land clearing for timber and growing space for agriculture have taken place in much of the temperate forest over the last few millennia. This clearance occurred much earlier in Asia and Europe, but also in North America and the southern hemisphere over the last two centuries, associated with the development of the industrial revolution in Europe during the 1800s and its expansion to the rest of the world. For several centuries, temperate forests were also essential wood sources for naval construction, for example *Pinus sylvestris* (Scots pine) in Europe, *P. strobus* (white pine) in North America and *Lagarostrobos franklinii* (huon pine) in Tasmania (Cadman, 2020). Thus, the temperate forests have played a major role in shaping human society as it now exists. The now recovering temperate forests constitute a large and growing source of carbon storage, wildlife habitat, endemic species and watershed protection. Many important products come from the temperate forest, including sources of non-timber forest products (wild plants and fungi used as food, hunting animals for food, maple syrup [N. America] from *Acer saccharum*, and birch sap for traditional beverages and in folk medicine as an ingredient for different antiseptic and anti-inflammatory treatments [Central and northeastern Europe]), wood products (hardwood used for furniture, flooring and construction lumber), and paper products (many species used for pulp).

The forests also contribute greatly to the tourism industry through attractions such as remnant old-growth forests, spring wildflowers, brilliant fall colors and habitat for migrant bird species, songbirds and other wildlife species commonly viewed by tourists. A few examples of superlative tree species include *Eucalpytus regnans* (mountain ash) that occurs in places with relatively wet climates in southeast mainland Australia and Tasmania, and has the distinction of being the tallest flowering plant, with some specimens reportedly up to 114 m tall, *Fitzroya cupressoides* (alerce) in Chile, can reach huge sizes and ages as much as 3,600 years, *Pseudotsuga menziesii* (Douglas-fir) and *Picea sitchensis* (Sitka spruce) in western North America, which can reach heights of 100 m, and *Abies nordmannia* (Nordman fir) in southeastern Europe, which can reach heights of about 80 m.

Ecological processes

Nutrient cycling and decay processes

Soil structure in temperate forests usually has an organic horizon at the top of the soil profile composed of leaf litter in various stages of decay, from fresh leaves at the top to fragmented leaves in the middle and black humus at the bottom. The rate of decay is important for determining how fast nutrients are released from the organic horizon and is determined by temperature and soil moisture, which in turn is determined by climate and soil texture. Organic horizon thickness is greater in colder climates and drier climates within the temperate zone, where decomposition of leaf litter is slower, in forests dominated by species with high C:N ratios and low nitrogen content of leaf litter (*Fagus*, *Picea*, *Pinus*, *Quercus*), in forests on sandy soils, and forests with low earthworm abundance. Cold growing season temperatures in temperate and boreal forests depress organic matter decomposition and may suppress nutrient movements in soil and their uptake by roots. Young, postglacial soils tend to be more N- than P-limited (Reich and Oleksyn, 2004).

Changes in forestry practices that alter tree species composition, along with urbanization and other influences of human activity such as atmospheric deposition, may alter cycles of ecologically important nutrients, including nitrogen, sulfur, hydrogen ions and base cations (Ca^{2+}, Mn^{2+}, Na^{+}, and K^{+}; Hedin et al., 1995). Temperate old-growth forest ecosystems are particularly vulnerable to increases in atmospheric N deposition, which have occurred in the northern hemisphere. In response to increased N supply (both from N deposition and faster litter mineralization enhanced by increasing temperatures), the pattern of N cycling may change and be accompanied by increased selection favoring nitrophilous plant species and a decrease in ground species richness (Tamm, 1991).

Examples of large-scale changes in nutrient cycling and soil function caused by human activity have occurred in Europe and the southern hemisphere temperate forests during the late 19th and 20th centuries. In Europe, Coniferous trees including native *Picea abies* (Norway spruce) and *Pinus sylvestris* (Scots pine), and North American *Picea sitchensis* (Sitka spruce) and *Pseudotsuga menziesii* (Douglas-fir) species were widely planted to reverse 18th- and 19th-century deforestation and increase productivity, often replacing native deciduous species. Creation of large-scale coniferous monocultures contributed to further acidification of already acidic soils (65% of European topsoils have pH ≤4.5, Augusto et al., 2002). In Chile, Australia and New Zealand, extensive *Pinus radiata* (Monterey pine) plantations have contributed to lower water and nutrient retention capacities, carbon sequestration of soils and loss of functional diversity of soil invertebrates (Cifuentes-Croquevielle et al., 2020).

Disturbance and succession

Disturbance regimes in the temperate forest include large, high-severity disturbances that create landscapes with coarse-scale, mostly even-aged stands of early successional species, gap dynamics regimes with fine-grained patches of mostly late-successional tree species and multi-aged stands (also discussed below under the section 'Canopy structure and gap dynamics') and frequent fire regimes that maintain early or mid-successional forests and woodlands (Frelich, 2002).

Major disturbances that kill or level the forest canopy include logging, wind, fire, ice storms, insect infestations, landslides and volcanoes (Frelich, 2002; Veblen et al., 2016). Landslides and volcanic disturbances have been termed tectonic disturbances, and along with severe windstorms, have caused formation of large, landscape-scale, even-aged stands of early successional *Nothofagus* species in Chile and New Zealand. These stands often date to known disturbance events from a few decades to 400 years ago, and thus these tree species can be classified as long-lived, early successional species. These high-severity disturbances are the norm on some of the landscapes in this biogeographic zone, creating an irregular mosaic of even-aged forests with little evidence of succession, while on other landscapes sheltered from these disturbance types, multi-aged forests composed of late-successional species occur (Veblen et al., 2016). This dichotomy between landscapes with coarse-scale, even-aged stands of early successional trees maintained by disturbances, which occur before the early successional cohort, is replaced by late-successional species, and landscapes with multi-aged forests composed of late-successional species where most disturbances are single to multiple tree, gap-forming events, also exists in North America and Asia, although in those regions large-scale, high-intensity disturbances tend to be fires (with long-lived *Pinus* and *Quercus* species composing the even-aged stands).

Whether stand-leveling disturbance initiates an episode of succession—defined as a directional change in species composition over time, usually establishment of a dominant cohort of early successional, pioneer trees species followed by its gradual replacement by a suite of later-successional species—is context-dependent (Heinselman, 1981; Finegan, 1984). For example, canopy-leveling windstorms or logging of the large trees may leave a carpet of late-successional advanced regeneration intact on the forest floor, leading to initiation of an even-aged stand followed by a stand development sequence from even-aged to multi-aged but not necessarily a change in species composition. This has been shown to occur in North American *Acer saccharum* (sugar maple) and *Tsuga canadensis* (hemlock) forests. On the other hand, high-severity disturbances such as windthrow followed by high-intensity fire or landslide could easily wipe out the advanced regeneration, leading to establishment of shade-intolerant, early successional species, which in the absence of further severe disturbance can succeed to shade-tolerant, late-successional species over the subsequent one to two centuries. Illustrative case studies include *Betula papyrifera* (paper birch) and *Populus tremuloides* (quaking aspen) after logging followed by high-intensity fires (during European settlement in the late 1800s) succeeding to *Acer saccharum* and *Tsuga canadensis* in North America and *Betula grossa* (Japanese cherry birch) succeeding to *Tsuga sieboldii* (southern Japanese hemlock) and *Fagus crenata* (Japanese beech) in Japan (Yoshida and Ohsawa, 1996; Frelich, 2002).

The third disturbance regime mentioned above is that of frequent low-intensity fire, commonly occurring in *Quercus* and *Pinus* savannas/woodlands in the northern hemisphere and *Eucalyptus* woodlands in Australia, and often associated indigenous burning to maintain the desired vegetation type. These forests are also multi-aged (at varied spatial scales from fine-grained to coarse-grained) and have little succession over time unless the disturbance regime is disrupted, in this case by fire exclusion, by removal of indigenous cultural practices from the landscape and fire suppression. The tree species are adapted to frequent fire due to thick bark that insulates the base of the trees from being girdled (*Quercus* and *Pinus*), and/or the ability to sprout if top-killed by fire (*Betula, Populus, Quercus, Eucalyptus*), and ability to germinate in post-fire conditions with partial to full sun and lack of leaf litter. Once established, these trees commonly live a century or more and survive many subsequent fires. These forests occupy areas with dry-to-mesic climates in the transition zones from temperate forests to grasslands or Mediterranean vegetation, as well as excessively well-drained sites on sand plains and equatorward rocky hillsides within wetter climates.

It is interesting to note that few of the tree species other than *Eucalpytus* in the southern hemisphere are adapted to fire, but some species (especially shrubs and bamboo) have better preadaptations to fire than others, and thus anthropogenic fires in temperate rainforest areas of South America and New Zealand can create alternate states with shrubby vegetation, which are more flammable than forest, and have relatively poor conditions for regeneration of *Nothofagus* (Kitzberger et al., 2016).

Canopy structure and gap dynamics

Large tracts of temperate forest occur on wet to mesic sites where stand-leveling disturbances are rare, and gaps from single to several trees dying constitute the main type of disturbance (e.g., Payette et al., 1990). The dynamics of these treefall gaps, canopy turnover rates, canopy tree residence times, canopy structure and recruitment of new trees into gaps have been studied at numerous sites in North and South America, Tasmania, mainland Australia and New Zealand. Such forests are usually multi-aged, and dominated by late-successional, shade-tolerant species, with some mid-tolerant species, and an occasional specimen of early successional species.

Fluctuations in relative abundance of the late-successional species occur over time due to influences on seedling success such as preferential deer grazing on seedlings and neighborhood effects (structure and chemical properties of the leaf litter, sprouting, Brisson et al., 1994). The late-successional species generally exist as suppressed seedlings (or advanced regeneration) that will record a release from suppression if the tree above dies; this fact has been used to reconstruct the treefall history of forests going back as much as several centuries (Lorimer, 1980, Payette et al., 1990). In contrast, the seedling bank of mid-tolerant tree species is rather small due to rapid turnover as seedlings die due to low light levels, but these species are adapted to enter new gaps by seed dispersal followed by rapid height growth that may surpass that of shade-tolerant species. The average tree residence times in the canopy vary from ≈100 to 200 years in temperate forests, with some trees living two to three times the average lifespan (Runkle, 1982).

Gap formation creates microsites for seedling germination including relatively dry and wet mineral soil on mounds and in pits created by uprooted trees, above-ground light gaps and below ground 'gaps' in root occupancy of the soil created by fallen trees (Beatty and Stone, 1986). These gaps can have not only more light but higher soil and air temperatures and more water (due to lack of use by the fallen tree), leading to more nutrient availability in the first few years as the fine roots of the fallen tree decay in the relatively warm, wet environment. Tree species mid-tolerant of shade often take advantage of these gaps, which therefore help to maintain richness of the local species over time. In addition, treefalls create coarse woody debris, which, when it reaches an advanced state of decay, can also be a microsite for germination of some tree species that cannot tolerate thick leaf litter or the dense herbaceous vegetation that may cover the forest floor. Decaying logs also are small-scale hot spots of biodiversity due to the large number of species of mosses, fungi, insects and amphibians that use them as habitat due to their relatively high water content (Harmon et al., 1986).

Plant species interactions

Interactions of note in temperate forests include the occurrence of mosaics formed by patches of evergreen conifer and broadleaf deciduous species, shade-tolerant and intolerant species interactions, the presence of numerous understory species that take advantage of a brief period of combined warmth and light in spring, and heterotrophic species that don't require sunlight to fuel carbon gain (e.g. saprophytic and parasitic plants).

Evergreen-deciduous mosaics of *Tsuga* or *Picea* with *Acer*, *Betula* and *Tilia* (mesic sites), or with *Pinus* and *Quercus* (drier sites) occur in the parts of the temperate forests with cooler climates, including the northern United States, northern Europe and Asia. These mosaics may be created by slight differences in the soil environment magnified by neighborhood effects of the trees themselves, through shading, leaf-litter chemistry, seed rain, interactions with disturbance and in some cases, sprouting so that the deciduous and evergreen species each favor their own reproduction and/or disfavor seedlings of the opposite group under their own canopies. Such mosaics maintain patches of different tree communities on the landscape and thus allow for coexistence of more species than from landscape heterogeneity alone (Frelich et al., 1993).

Shade tolerance or avoiding shade is necessary for any plant species to reproduce in these forests given dense canopies and forest floor light levels ranging from 1% to 10% of full sunlight. Plant species have evolved a variety of strategies. Spring ephemerals are species that grow rapidly as soon as winter ends, taking advantage of the sunlight available on the forest floor prior to canopy leaf out. Included are species in the genera *Allium* (wild leek), *Cardamine* (toothwort), *Dicentra* (Dutchman's breeches), *Erythronium* (trout lily) and *Mertensia* (bluebell), which complete their above-ground life cycle of growth, flowering, seed set and die back to a dormant phase within four to six weeks. Some of these genera have vicarious species (related species that live in similar environments with similar ecological niche) among the major occurrences of temperate forest. For example, *Erythronium americanum* (yellow trout lily) in the eastern United States, *E. dens-canis* (dog's tooth violet) in central and southern Europe, and *E. japonicum* (Asian fawn lily) in Japan.

Other early blooming species are commonly misclassified as spring ephemerals, but they keep their leaves for much of the summer—these include geophytes such as *Trillium*, and herbs such as *Arisaema* (jack-in-the-pulpit), *Sanguinaria* (bloodroot) and *Viola* (violets), which can use energy stored in a bulb or rhizomes to produce leaves early on in full sun, but continue to photosynthesize in deep shade after canopy leaf out. Saprophytic plants, such as *Corallorhiza* spp. (coral root orchids) and *Monotropa uniflora* (Indian pipe) in North America and Asia, and *Thismia rodwayi* (Fairy lantern) in Australia and New Zealand, get their energy from fungi, which are decomposing organic matter, and therefore don't have chlorophyll or require sunlight.

A number of plant species take advantage of tree fall gaps, where higher than average light levels last a few decades, to grow up into the canopy (e.g., *Betula*, *Quercus*) or to complete their life cycle and create long-lived buried seeds that await the next gap a century or more later—for example, *Prunus pensylvanica* (pin cherry), *Rubus idaeus* (red raspberry) and *Geranium bicknellii* (Bicknell's geranium). Another group of species has short life cycles within gaps but produce plumed seeds that float long distances through the air to 'find' new gaps (e.g. *Epilobium angustifolium*, fireweed). Shade-tolerant shrubs in forest understories can create complex spatial dynamics for tree regeneration. Understory bamboo (*Sasa* spp. in Asian forests and *Chusquea* spp. in Chilean temperate forests) can exclude seedling establishment, even for shade-tolerant tree species. For example, beech (*Fagus crenata*) and maple (*Acer mono*) seedlings in Japanese forests are restricted to small patches where bamboo is absent (Yamamoto et al., 1995).

Plant–animal interactions

Many symbiotic and antagonistic plant–animal interactions occur in temperate forests, involving seed dispersal, pollination, defoliation, insect herbivory and preferential grazing by large mammals (deer, moose) that can alter successional trajectories. Although many species of temperate trees have wind-dispersed seeds, animal dispersal (zoochory) also occurs, and in such cases, the relationship between the tree species and wildlife species is often symbiotic. For example, masting of *Quercus* and *Fagus* tree species in North America, Europe and Asia, and *Nothofagus* trees in the southern hemisphere; masting refers to producing a very large crop of seeds every several years, as a way to satiate seed consumers. During mast years, seeds are left to germinate after the local wildlife species have been satiated. *Nothofagus* trees in New Zealand produce mast crops about every five years, which leads to increased breeding activity of birds, for example *Cyanoramphus auriceps* (yellow-crowned parakeet) and *Nestor meridionalis* (New Zealand kākā), and populations of mammals such as *Mustela ermine* (stoat) and *Trichosurus vulpecula* (brushtail possum). In North America, it is noted that having a mixture of several nut-producing *Carya*, *Juglans* and *Quercus* tree species, which mast in different years, provides more stable food sources for wildlife species than forests with single masting tree species.

One well-known case of tree seed dispersal is for acorns of *Quercus rubra* (northern red oak) that are cached by various animal species in the fall for later use during winter. *Sciurus carolinensis* (gray squirrels) bury acorns in the soil and are known to remember the locations of several thousand acorns, but during mast years, they often bury more than they retrieve, leaving many to germinate, and at the same time hiding those acorns from consumption by white-tailed deer (*Odocoileus virginianus*), black bear (*Ursus americanus*), turkey (*Meleagris gallopavo*) and assorted boring beetles. Blue jays (*Cyanocitta cristata*) can carry acorns and commonly fly several hundred meters to 2 km from the parent tree, scatter hoarding thousands of acorns across the landscape (Johnson and Webb, 1989). In Europe, acorns of several *Quercus* species are dispersed by *Garrrulus glandarius* (Eurasian jay). Analysis of the seed dispersal service performed by the Eurasian jay in the Stockholm National Urban Park, Sweden, has shown that depending upon seeding or planting technique chosen, the cost of replacement per pair of jays though human means in the Park is US$2,100 (seeding) to US$9,400 (planting) per hectare, respectively (Hougner et al., 2006). Such estimates provide a good example of the value of management strategies that secure critical breeding and foraging habitats of seed dispersal animals.

Seed dispersal by ants is known as myrmecochory and is common among temperate forest understory plant species (Gomez and Espadaler, 1998). The spring wildflower genera *Trillium* (North America and Asia) and *Viola* (North America, Europe and Asia) are examples with a number of species dispersed by ants, although other vectors also exist. The seeds of these species have a fat-rich elaiosome attached to each seed, which the ant can detach for consumption, commonly after moving the seed one or two meters.

A large number of bee species occur in temperate forests (often 50 or more species in one stand), which pollinate most of the forest understory species with brightly colored flowers and limited pollen production, and a few of the tree genera such as *Acer* (some species also wind-dispersed), *Prunus* and *Tilia*. Folivorous insects are common, with many species that cause ongoing low-to-moderate levels of defoliation, which interact with plant defense compounds in leaves such as phenols and terpenoids. Other insect species have periodic outbreaks every decade or longer, during which the forest may be almost totally defoliated. For example, deciduous forests in eastern North America can be defoliated by *Heterocampa guttivitta* (saddled prominent moth) caterpillars during droughts that prevent trees from defending themselves against insects, and on a more regular 10–12 years cycle by *Malacosoma americanum* (forest tent caterpillar; Horsley et al. 2002). These defoliation events usually cause scattered tree mortality, especially to older trees with other health problems, but seldom cause extensive tree death at the landscape scale. Much more serious are growth losses and tree mortality after outbreaks of *Diprion pini* (common pine sawfly). During a severe outbreak in Finland, approximately 500,000 ha was defoliated by *D. pini* resulting in high tree mortality and growth losses in the following year and later. Mortality rates in Europe after an outbreak are typically 4–24% and it can take 10–15 years for radial growth to recover (Lyytikäinen-Saarenmaa et al., 2003).

Grazing animals, especially ungulates like deer, elk and moose, can regulate plant community composition and direct succession through their plant species preferences. Many ungulates in the temperate zone consume woody plants during winter and herbaceous plants during summer. If the deer to plant ratio is high, they can regulate the balance between relatively palatable and unpalatable plant species. The strength of these influences on composition is in turn embedded in a trophic cascade. For example, *Canis lupus* (gray wolf) in Wisconsin, United States creates a patchy distribution in deer density across the landscape due to predation and deer avoidance of wolf pack territories. This in turn influences lushness, composition and species richness of herbaceous plants (Callan et al., 2013).

Soil animals such as earthworms, beetles and many other taxonomic groups (whether native or invasive) also influence plant community composition by altering the structure of the organic horizon, moving seeds to different layers within the soils, consuming seeds and regulating water and nutrient cycles within the soils. These are considerable ecological cascades within the soil. For example, earthworms can change leaf-litter structure to favor or disfavor certain fungal species that in turn may be involved in symbiotic relationships with tree roots as mycorrhizas that help trees absorb nutrients, or with symbiotic relationships with seed germination (e.g., orchid seeds); furthermore changes in leaf litter on the forest floor can affect which species of trees and understory plants germinate successfully, with cascading effects on plant community composition (Frelich et al., 2019).

Conservation issues

Land use conversion, fragmentation and harvesting

Although indigenous peoples lived in temperate forests for millennia, the magnitude of their impacts via hunter-gatherer activities—use of wood products, burning and clearance for agriculture—left the ecological legacies of forests largely intact, compared to more industrial activities such as widespread land conversion to agriculture, plantations, mining and large cities. The temperate zone has had high human populations and correspondingly large industrial-type impacts on forests for one to two millennia in Europe and Asia, and for the last two to four centuries in North America and the southern hemisphere. Therefore, much of the forest has been converted to croplands, grazing lands for cattle and sheep and urban areas and almost all land that remains forested has been logged at least once. In Europe and Asia, most forests have been logged multiple times and planted regeneration is used on a widespread basis to supplement natural regeneration after harvesting, while in North America and the southern hemisphere, most forests have been logged once or twice, and natural regeneration has been more common, although the incidence of plantations of exotic tree species has recently increased in the southern hemisphere temperate zone. In recent decades, the wildland–urban interface, an area with widely scattered isolated houses, has encroached on large tracts of forested land, creating numerous foci for the introduction of invasive species, and fragmenting the landscape, favoring certain plant and animal species that do well along the forest edges.

A very small amount of primary temperate forest (never logged) and intact temperate forest (areas with minimal influence of human economic activities at least 50,000 ha in size and at least 10 km in width) remain as compared to tropical and boreal forest biomes. Less than 1% of the original forest in eastern North America, China and Europe remains in primary/intact condition. However, higher percentages occur in more remote temperate forest areas where industrial activity came later on, during the mid-to-late 20th century, when conservation was a higher priority, including parts of the Pacific Northwest in North America, Tasmania, New Zealand, Chile and Argentina (DellaSala et al., 2011). Intact forest remnants are important because they serve as templates for restoration of secondary forests and as a baseline for the occurrence of ecosystem processes as compared to forests that are harvested. However, in many regions with thousands of years of industrial human influence, no natural templates for restoration exist. In such cases, a multi-disciplinary synthesis of historic records, silvicultural and paleoecological evidence may be needed to develop management techniques that mimic disturbances and other conditions needed to maintain the diversity of native tree species and smaller species dependent on them, and maintenance of certain cultural features of the landscape may also become a priority (Lindbladh et al., 2007).

Fragmentation manifested as small woodlands of a few to a few hundred hectares surrounded by agricultural lands or cities is common in the temperate forest biome (Wilcove et al., 1986). This has led to conservation problems including: (1) facilitation of invasive plant species; (2) over populations of edge-loving native wildlife species such as deer and certain birds like the cowbirds (*Molothrus* spp.), which then parasitize nests of songbirds; and (3) potential inbreeding over time and loss of populations of many species of native plants and animals. Island biogeography theory predicts that it is not possible to maintain as many species on a small fraction of the extent of forest that existed prior to deforestation and fragmentation. The term 'extinction debt' has been used to describe the situation where many species with long life spans still exist within fragmented landscapes but are predicted to eventually go extinct (Tilman et al. 1994), and unfortunately, extinction debt is high in many temperate forest landscapes.

Overgrazing by wildlife and domestic livestock

Overgrazing is a common problem for maintenance of biodiversity and regeneration of trees in the temperate zone (Côté et al., 2004). In the southern hemisphere, cattle grazing in Chile and Argentina can reduce or eliminate regeneration of *Araucaria araucana* (monkey puzzle tree), making it one of several important factors in the conservation of this endangered tree species (Zamorano-Elgueta et al., 2012). *Cervus elaphus* (red deer) have been introduced into temperate forests of South America, Australia and New Zealand, adding to the negative impacts of other stress factors on forest regeneration. In New Zealand, multiple species of non-native deer have been introduced, posing problems for the regeneration of tree species, even those not preferred by red deer—such as *Nothofagus* species—which are eaten by other deer species such as *Cervus nippon* (sika deer) that persist on poor-quality forage (Coomes et al., 2003). An additional problem in New Zealand is browsing of foliage and dieback in mature trees, where *Trichosorus vulpecula* (brushtail possum, a native of Australia) was introduced (Kirkpatrick and DellaSala 2011).

In North America, *Odocoileus virginianus* (white-tailed deer) prefer seedlings of *Thuja occidentalis* (northern white cedar), *Betula alleghaniensis* (yellow birch), *Quercus rubra* (red oak) and *Tsuga canadensis* (hemlock) during winter, leading to widespread difficulties in regeneration of those species. This limits the ability of forest managers to direct succession toward desired tree species and to maintain species richness. During the growing season, deer prefer many herbaceous species in the families *Liliaceae* and *Orchidaceae*. Deer can drive succession to unpalatable trees such as *Fagus grandifolia* (American beech) and understory species including ferns and sedges (*Carex* spp.) that are not favored by deer or that can tolerate heavy grazing. Effects of deer overabundance in different parts of the temperate forest can be idiosyncratic. *Cervus nippon* (sika deer) in Japan can prevent *Fagus* recruitment (rather than favoring it as in North America) but can also have similar effects on herbaceous vegetation as in North America by favoring graminoids and ferns in the herbaceous vegetation layer (Takatsuki 2009). Lowering deer densities through hunting has been controversial since members of the public often equate more deer with a healthy ecosystem and the effectiveness of hunting may become limiting in an increasingly urbanized world where relatively few people learn to hunt. Livestock such as cows can also cause similar problems if allowed to graze freely within woodlands without sufficient rotation to different areas throughout the growing season. Overgrazing may be a problem for national parks where plants as well as grazer populations are protected from human exploitation. For example, in the Bialowieza Forest National Park in Poland where *Bison bonasus* (European bison) has been successfully restored after extinction in the wild at the beginning of 20th century, a recent study has shown that consumption of trees and shrubs by bison increased with decreasing access to supplementary fodder, ranging from 16% in intensively fed bison to 65% in non-fed bison using forest habitats. Bison browsed mainly on *Carpinus, Corylus* and *Betula*, tree species of relatively low economic importance in the region, so that impacts on forestry may be relatively small. However, more investigation of bison impacts is needed to develop management plans to meet other objectives, such as maintaining plant diversity within the park and reducing damage to agriculture surrounding the park (Kowalczyk et al., 2011).

Invasive species

Intercontinental movements of temperate plant and animal species have occurred over the last few centuries, with accelerating rates of new introductions in recent decades corresponding to the larger human population and commercial trade in plants and animals (Kalusova et al., 2013). Freed from competing species adapted to their presence, from leaf and seed-eating insects, from large herbivores that prefer them over other species, and from diseases in their native habitat, invasive plants often become much more abundant on their new continent than their home continent, filling forest understories, reducing the abundance of native tree seedlings and necessitating expensive management actions to remove them and/or to restore native species. *Rhamnus cathartica* (common buckthorn), *Lonicera tatarica* (tatarian honeysuckle) and *Alliaria petiolata* (garlic mustard) have moved from Eurasia to North America, and *Prunus serotina* (black cherry), *Robinia pseudoacacia* (black locust) and *Quercus rubra* (red oak) have moved from North America to Europe, and leguminous shrubs and herbs from Europe such as *Cytisus scoparius* (Scotch broom) and lupine (*Lupinus spp.*) now grow in South American temperate forests.

Invasive animals include *Sciurus carolinensis*, the North American gray squirrel, invading European forests where they displace the native *Sciurus vulgaris* (red squirrel), and *Nyctereutes procyonoides* (raccoon dog) from Asia is invading European forests (Genovesi et al., 2012). *Oryctolagus cuniculus* (rabbit) and *Mustela vison* (mink) have invaded Valdivian rainforests of South America, while *Vulpes vulpes* (fox) and *Sus scrofa* (wild pig) have invaded Australian forests, and a number of European and Asian deer species have been introduced into forests of the southern hemisphere.

Earthworms as invasive species can have profound impacts on ecosystems around the world (Hendrix et al., 2008). A recent body of research shows that European earthworms in North America change soil structure, eliminating the organic horizon and increasing bulk density, making soils drier, depleting nutrients and changing seedbed conditions. Ecological cascade effects include reduced tree growth, reduced native plant species richness, favoring a different suite of plant species and facilitating invasive plant species that were coevolved with the worms on their home continent (Larson et al., 2010; Frelich et al., 2019).

Diseases and insect pests of trees moving from one continent to another have caused huge aesthetic, economic and habitat losses, wiping out tree species that are foundational to ecosystem function (Parry and Teale, 2011). North America has been the recipient of many devastating tree pandemics (Roy et al., 2014) including *Cryphonectria parasitica* (chestnut blight), *Ophiostoma ulmi* (Dutch elm disease), *Agrilus planipennis* (emerald ash borer), *Adelges piceae* (balsam woolly adelgid) and *Adelges tsugae* (hemlock woolly adelgid). In combination, these pests and diseases have the capacity to greatly reduce the species richness of the tree canopy in eastern North American forests, with cascading ecological impacts on nutrient and light regimes, as well as physical structure of the forest habitat.

Climate change

Climates that will support temperate forests are projected to shift polewards or upwards in elevation over the next several decades as the climate warms. In most of North America, Europe and Asia, there is plenty of room for movement of temperate species into the current boreal biome, although seed sources for temperate species and ability of plants and smaller animal species (especially plants and their pollinators moving together) to keep up with the velocity of climate change could be challenging. However, in the southern hemisphere (and some places in the northern hemisphere), there is limited or no land available in a poleward direction, but areas with higher elevations are available. The advantages are that the distance to move is not great, but the disadvantage is that less and less area is available as 'sky islands' in mountainous regions become smaller and more isolated with increasing elevation, leading to concerns regarding the availability of sufficient land for the dynamic equilibrium of disturbances and representation of all successional stages needed to provide habitat for all species in a given ecosystem.

The temperate forest climate zone in central North America and Europe is projected to shift to the north 200–700 km by the end of the 21st century, depending on local and global magnitude of warming (Galatowitsch et al., 2009; Dyderski et al., 2018). Two types of progression are likely to occur along the southern margin, where temperate forests will likely be out of equilibrium with the future climates: warm-dry scenarios, and warm-wet scenarios. Warm-dry scenarios will likely lead to increases of oak, pine and grass species at the expense of mesic forest species as the environmental niche of mesic species shrinks due to drought stress and mortality. The second, warm-wet scenarios, would be much less stressful for most tree species and would allow for gradual replacement of existing species with species from further south. Invasions of savanna and grasslands into mesic forest, and subtropical forest into temperate forest, are likely within several hundred kilometers of the southern margin of the northern temperate forest biome. At the same time, at the northern margin, temperate forest is likely to invade the southern boreal forest, at first forming a wider mixed temperate-boreal ecotone, as temperate species are freed from limitation of extreme winter cold and/ or short summers, and later on decline of boreal tree species as warm temperature thresholds for those species are crossed.

A number of factors are likely to interact with climate change, including invasive species, deer grazing, insects, windstorms, fires and fragmentation (Frelich and Reich, 2010). Invasive plant species, as a group, are generally tolerant of a wide variety of climates and disturbances, have abundant seed production and long-distance dispersal, therefore giving them an advantage in a rapidly changing forest community. Deer preference for grazing on certain species of tree seedlings could either oppose climate-induced change by consuming seedlings of tree species from further south, or exacerbate climate-induced change, by consuming seedlings of existing tree species, hastening their demise in a warming climate. In first decade of the 21st century, several consecutive years of severe drought and heat stress that occurred in most European countries caused deterioration of *Picea abies* (Norway spruce) stands, predisposing them to catastrophic bark beetle infestation (Przybył et al., 2008). Therefore, this episode of drought and insect damage is a harbinger of future impacts of a warming climate. A warmer climate with higher evaporation and more erratic precipitation is likely to lead to more fires over much of the temperate forest biome, which will potentially increase the proportion of early successional forests on the landscape and alter the dynamics of tree species migration. In large swaths of the temperate zone with human-induced (forest remnants in agricultural landscapes) or natural (sky islands in mountainous regions) fragmentation, movement of species in response to climate change will be hindered; there are a large number of species with very limited dispersal distances and long establishment times, for example the previously mentioned myrmecochorus plant species. Forest managers and the public will have to decide whether to employ the practice of assisted migration for such cases (Buma and Wessman, 2013).

Forest scientists and managers of forests in commercial and natural area settings face many challenges in the temperate forest biome. In a biome where climate change, overgrazing, invasive species and fragmentation are pervasive, creative research, comparison of remnant natural forests with commercial forests and development of adaptive management techniques will be paramount in order to allow a continued existence of productive forests capable of maintaining ecological function and native species diversity.

References

Augusto, L., Ranger, J., Binkley, D. and Rothe, A. (2002) 'Impact of several common tree species of European temperate forests on soil fertility', *Annals of Forest Science*, vol 59, pp. 233–253.

Beatty, S.W. and Stone, E.L. (1986) 'The variety of soil microsites created by tree falls', *Canadian Journal of Forest Research*, vol 16, pp. 539–548.

Brisson, J., Bergeron, Y., Bouchard, A. and Leduc, A. (1994) 'Beech-maple dynamics in an old-growth forest in southern Quebec, Canada', *Ecoscience*, vol 1, pp. 40–46.

Buma, B, and Wessman, C.A. (2013) 'Forest resilience, climate change, and opportunities for adaptation: A specific case of a general problem', *Forest Ecology and Management*, vol 306, pp. 216–225.

Cadman, S.T. (2020) 'Tasmanian temperate rainforests', *Imperiled: The Encyclopedia of Conservation*, Elsevier, Amsterdam.

Callan, R., Nebbelink, N.P., Rooney, T.P., Wiedenhoeft, J.E., and Wydeven, A.P. (2013) 'Recolonizating wolves trigger a trophic cascade in Wisconsin (USA)', *Journal of Ecology*, vol 101, pp. 837–845.

Cifuentes-Croquevielle, C., Stanton, D.E., and Armesto, J.J. (2020) 'Soil invertebrate diversity loss and functional changes in temperate forest soils replaced by exotic pine plantations', *Scientific Reports*, vol 10, 7762.

Coomes, D.A., Allen, R.B., Forsyth, D.M., and Lee, W.G. (2003) 'Factors preventing the recovery of New Zealand forests following control of invasive deer', *Conservation Biology*, vol 17, pp. 450–459.

Côté, S.D., Rooney, T.P., Tremblay, J-P, Dussault, C.D., and Waller, D.M. (2004) 'Ecological impacts of deer overabundance', *Annual Review of Ecology and Systematics*, vol 35, pp. 113–147.

DellaSala D.A. (2011) *Temperate and Boreal Rainforests of the World: Ecology and Conservation*, Island Press, Washington, D.C.

Dyderski, M.K., Paz, S., Frelich, L.E. and Jagodzinski, A.M. (2018) 'How much does climate change threaten European forest tree species distributions?', *Global Change Biology*, vol. 24, pp. 1150–1163.

Finegan, B. (1984) 'Forest succession', *Nature*, vol. 312, pp. 109–114.

Fisichelli, N.A., Frelich, L.E. and Reich, P.B. (2014) 'Temperate tree expansion into adjacent boreal forest patches facilitated by warmer temperatures', *Ecography*, vol 37, pp. 152–161.

Frelich, L.E. (2002) *Forest Dynamics and Disturbance Regimes*, Cambridge University Press, Cambridge.

Frelich, L.E., Blossey, B., Cameron, E.K., Davalos, A., Eisenahuer, N., Fahey, T., Ferlian, O., Groffman, P., Larson, E., Loss, S., Maerz, J., Nuzzo, V., Reich, P.B., Yoo, K. (2019) 'Side swiped: ecological cascades emanating from earthworm invasion', *Frontiers in Ecology and the Environment*, vol 17, pp. 502–510.

Frelich, L.E., Calcote, R.R., Davis, M.B. and Pastor, J. (1993) 'Patch formation and maintenance in an old growth hemlock-hardwood forest', *Ecology*, vol 74, pp. 513–527.

Frelich, L.E. and Reich, P.B. (2010) 'Will environmental changes reinforce the impact of global warming on the prairie-forest border of central North America?', *Frontiers in Ecology and Environment*, vol 8, pp. 371–378.

Galatowitsch, S., Frelich, L.E. and Phillips-Mao, L. (2009) 'Regional climate change adaptation strategies for biodiversity conservation in a midcontinental region of North America', *Biological Conservation*, vol 142, pp. 2012–2022.

Genovesi, P., Carnevali, L., Alonzi, A, and Scalera, R. (2012) 'Alien animals in Europe: updated numbers and trends, and assessment of the effects on biodiversity', *Integrative Zoology*, vol 7, pp. 247–253.

George, M.F., Burke, M.J., Pellett H.M. and Johnson, A.G. (1974) 'Low temperature exotherms and woody plant distribution', *Hortscience*, vol 9, pp. 519–522.

Gomez, C. and Espadaler, X. (1998) 'Myrmecochorous dispersal distances: a world survey', *Journal of Biogeography*, vol 25, pp. 573–580.

Harmon, M.E., Franklin, J.F., Swamson, F.J., Sollins, P, Gregory, S.V., Lattin, J.D., Anderson, H.H., Cline, S.P., Aumen, N.G., Sedell, J.R., Lienkaemper, G.W., Cromack Jr., K. and Cummins, K.W. (1986) 'Ecology of coarse woody debris in temperate ecosystems', *Advances in Ecological Research*, vol 15, pp. 133–302.

Hedin, L.O., Armesto, J.J. and Johnson, A.H. (1995) 'Patterns of nutrient loss from unpolluted, oldgrowth temperate forests: Evaluation of biogeochemical theory', *Ecology*, vol 76, pp. 493–509.

Heinselman, M.L. (1981) 'Fire and succession in the conifer forests of northern North America', in D.C. West, D.C., H.H. Shugart H.H., and D.B. Botkin (eds), *Forest Succession, Concepts and Application,* Springer-Verlag, New York.

Hendrix, P.F., Callaham, M.A., Jr., Drake, J.M., Huang, C-Y., James, S.W., Snyder, B.A., and Zhang, W. (2008) 'Pandora's box contained bait: the global problem of introduced earthworms', *Annual Reviews of Ecology and Systematics*, vol 39, pp. 593–613.

Horsley, S.B., Long, R.P., Bailey, S.W., Hallett, R.A. and Wargo, P.M. (2002) 'Health of eastern North American sugar maple forests and factors affecting decline', *Northern Journal of Applied Forestry*, vol 19, pp. 34–44.

Hougner, C., Colding, J. and Söderqvist, T. (2006) 'Economic valuation of a seed dispersal service in the Stockholm National Urban Park, Sweden', *Ecological Economics*, vol 59, pp. 364–374.

Johnson, W.C. and Webb III, T. (1989) 'The role of blue jays (*Cyanocitta cristata* L.) in the post-glacial dispersal of fagaceous trees in eastern North America', *Journal of Biogeography*, vol 16, pp. 561–571.

Kalusova, V., Chytry, M., Kartesz, J.T., Nishino, M and Pysek, P. (2013) 'Where do they come from and where do they go? European natural habitats as donors of invasive alien plants globally', *Diversity and Distributions*, vol 19, pp. 199–214.

Kirkpatrick, J.B. and DellaSala, D.A. (2011) 'Temperate rainforests of Australasia', in: D.A. DellaSala (ed), *Temperate and Boreal Rainforests of the World: Ecology and Conservation*, Island Press, Washington, D.C.

Kitzberger, T., Perry, G.L.W., Paritsis, J., Gowda, J.H., Tepley, A.J., Holz, A., and Veblen, T.T. (2016) 'Fire–vegetation feedbacks and alternative states: common mechanisms of temperate forest vulnerability to fire in southern South America and New Zealand', *New Zealand Journal of Botany*, vol 54, pp. 247–272.

Kowalczyk, R., Taberlet, P., Coissac, E., Valentini, A., Miquel, C., Kaminski, T and Wojcik, J.M. (2011) 'Influence of management practices on large herbivore diet—Case of European bison in Bialowieza Primeval Forest (Poland)', *Forest Ecology and Management*, vol 261, pp. 821–828.

Larson, E.R., Kipfmueller, K.F., Hale, C.M., Frelich, L.E., and Reich, P.B. (2010) 'Tree rings detect earthworm invasions and their effects in northern hardwood forests', *Biological Invasions*, vol 12, pp. 1053–1066.

Lindbladh, M., Brunet, J., Hannon, G., Niklasson, M., Eliasson, P., Eriksson, G.R., and Ekstrand, A. (2007) 'Forest history as a basis for ecosystem restoration- A multidisciplinary case study in a south Swedish temperate landscape', *Restoration Ecology*, vol 15, pp. 284–295.

Lorimer, C.G. (1980) 'Age structure and disturbance history of a southern Appalachian virgin forest', *Ecology*, vol 61, pp. 1169–1184.

Lyytikäinen-Saarenmaa P., Niemelä, P. and Annila, E. (2003) 'Growth responses and mortality of Scots Pine (*Pinus sylvestris* L.) after a Pine Sawfly outbreak' *Proccedings: IUFRO Kanazawa 2003 "Forest Insect Population Dynamics and Host Influences"*, pp. 81–85.

Parry, D. and Teale, S.A. (2011) 'Alien invasions: the effects of introduced species on forest structure and function', in J.D. Castello and S.A. Teale (eds), *Forest Health, An Integrated Perspective,* Cambridge University Press, Cambridge.

Payette, S., Filion, L. and Delawaide, A. (1990) 'Disturbance regime of a cold temperate forest as deduced from tree-ring patterns: the Tantaré Ecological Reserve, Quebec', *Canadian Journal of Forest Research*, vol 20, pp. 1228–1241.

Peel, M.C., Finlayson, B.L. and McMahon, T.A. (2007) 'Updated world map of the Köppen-Geiger climate classification', *Hydrology and Earth System Sciences*, vol 11, pp. 1633–1644.

Peltzer, D.A., Wardle, D.A., Allison, V.J., Baisden, W.T. and others (2010) 'Understanding ecosystem retrogression', *Ecological Monographs*, vol 80, pp. 509–529.

Przybył, K., Karolewski, P., Oleksyn, J., Łabędzki, A. and Reich, P.B. (2008) 'Fungal diversity of Norway spruce litter: effects of site conditions and premature leaf fall caused by bark beetle outbreak', *Microbial Ecology*, vol 56, pp. 332–340.

Reich, P.B. and Oleksyn, J. (2004) 'Global patterns of plant leaf N and P in relation to temperature and latitude', *Proceedings of the National Academy of Sciences of the United States of America*, vol 101, pp. 11001–11006.

Roy, B.A., Alexander, H.M., Davidson, J., Campbell, F.T., Burdon, J.J., Sniezko, R. and Brasier, C. (2014) 'Increasing forest loss worldwide from invasive pests requires new trade regulations', *Frontiers in Ecology and the Environment*, vol 12, pp. 457–465.

Runkle, J.R. (1982) 'Patterns of disturbance on some old-growth mesic forests of eastern North America', *Ecology*, vol 63, pp. 1533–1546.

Sabine, C.L., Heimann, M., Artaxo, P., Bakker, D.C.E., Chen, C.T.A., Field, C.B., Gruber, N., LeQuéré, C. Prinn, R.G., Richey, J.E., Lankao, P.R., Sathaye, J.A. and Valentini, R. (2004) 'Current status and past trends of the global carbon cycle', in C.B. Field and M.R. Raupach (eds), *The Global Carbon Cycle: Integrating Humans, Climate, and the Natural World, SCOPE 62*, Island Press, Washington, D.C.

Svenning, J-C., Skov, F. (2005) 'The relative roles of environment and history as controls of tree species composition and richness in Europe', *Journal of Biogeography*, vol 32, pp. 1019–1033.

Takatsuki, S. (2009) 'Effects of sika deer on vegetation in Japan: A review', *Biological Conservation*, vol 142, pp. 1922–1929.

Tamm, C.O. (1991) *Nitrogen in Terrestrial Ecosystems, Questions of Productivity, Vegetational Changes, and Ecosystem Stability*, Springer-Verlag, Berlin.

Tilman, D., May, R.M., Lehman, C.L. and Nowak, M.A. (1994) 'Habitat destruction and the extinction debt', *Nature*, vol 371, pp. 65–66.

Veblen, T.T., González, M.E., Stewart, G.H., Kitzberger, T., and Brunet, J. (2016) 'Tectonic ecology of the temperate forests of South America and New Zealand', *New Zealand Journal of Botany*, vol 54, pp. 223–246.

Whittaker, R.H. (1975) *Communities and Ecosystems*. 2nd edition, Macmillan, New York.

Wilcove, D.S., McLellan, C.H. and Dobson, A.P. (1986) 'Habitat fragmentation in the temperate zone', in M. Soulé (ed), *Conservation Biology, the Science of Scarcity and Diversity*. Sinauer Associates, Sutherland, MA.

Yamamoto, S., Nishimura, N., and Matsui, K. (1995) 'Natural disturbance and tree species coexistence in an old-growth beech - Dwarf bamboo forest, southwestern Japan', *Journal of Vegetation Science*, vol 6, pp. 875–886.

Yoshida, N., and Ohsawa, M. (1996) 'Differentiation and maintenance of topo-community patterns with reference to regeneration dynamics in mixed cooo temperate forests in the Chichibu Mountains, central Japan', *Ecological Research*, vol 11, pp. 351–362.

Zamorano-Elgueta, C., Cayuela, L, González-Espinosa, M., Lara, A., and Parra-Vázquez, M.R. (2012) 'Impacts of cattle on the South American temperate forests: Challenges for the conservation of the endangered monkey puzzle tree (*Araucaria araucana*) in Chile', *Biological Conservation*, vol 152, pp. 110–118.

4
SUBTROPICAL FORESTS

Richard T. Corlett and Alice C. Hughes

Introduction

There is no widely agreed definition of what constitutes the 'subtropics', but in recent ecological literature, the term has most often been applied to the two latitudinal belts between the tropics (±23.4°) and 30°–34° north and south of the equator (Corlett, 2013). We use 23.4°–32° in this chapter (Figure 4.1). Although this choice is debatable, particularly at the outer (32°) limit, applying a uniform standard makes it possible to compare forests in the same latitudinal belt in different parts of the world, as well to make comparisons within regions between subtropical forests and the forests of the better-studied tropical (i.e., <23.4° from the equator) and temperate (i.e., >32°) zones. Moreover, defining the subtropics by latitude alone avoids the confusion between latitudinal and altitudinal gradients that occurs in some of the literature.

Much of the subtropical belt, as defined above, is too dry (mean annual rainfall < 600–700 mm) for forest because of the descending branches of the Hadley circulation (the subtropical high), but there are, or were, extensive forests on the southeastern side of all the continents. Subtropical forest makes up around 6% of all forests globally, with particularly high coverage in China and adjacent countries (59% of the total current forest area in the subtropics based on the analysis of Global Forest Watch data for 2021 with a threshold of 50% tree coverage), and in northern Argentina and southeastern Brazil (22%), with smaller areas in northern Mexico and the southeast United States (14%), eastern Australia (3%), and southern Africa (3%), and very small areas in southeastern Madagascar and the Canary Islands. These extant areas of subtropical forest represent only part of their potential extent; around 27% if forest is assumed to have covered all land areas < 4000 m a.s.l. and with a mean annual rainfall > 700 mm (Figure 4.1). Deforestation is continuing in most regions and much of the extant forest is secondary or degraded.

In comparison with temperate forest regions, the major common feature of the climates of these subtropical forests, apart from adequate rainfall, is the occurrence of lowland temperatures that are suitable for plant growth over most or all of the year. On the other hand, in contrast with tropical forests, there are at least mild winter frosts in most areas in most years, as well as more extreme cold events at longer intervals (Smith and Sheridan, 2020; Osland et al., 2021). In coastal regions, except in South America, subtropical forests

DOI: 10.4324/9781003324072-5

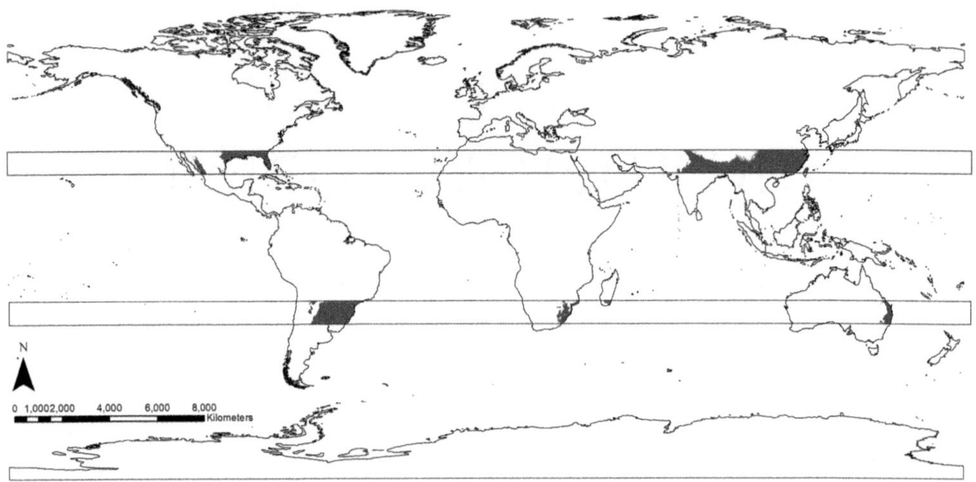

Figure 4.1 Potential forest cover between 23.4° and 32.0°, under the assumption that all areas with >700 mm annual rainfall and <4,000 m elevation are capable of supporting forest. Map produced in ArcMap 10.8 using climate and elevation data from WorldClim 2.1 (https://www.worldclim.org/).

are vulnerable to cyclones, with repeated occurrences reducing stature and biomass (e.g., McEwan et al., 2011). As in tropical forests, the soils under subtropical forests are mostly old and highly weathered, and multiple types of evidence support a predominant role for phosphorous limitation (Cui et al., 2022).

Subtropical forests have been intensively studied in China, where there are many large (15–50 ha) forest dynamics plots, numerous smaller plots, several canopy cranes, and a large-scale biodiversity and ecosystem function experiment (Mi et al., 2021). They have also received considerable attention in South America, but studies elsewhere have been largely floristic. The results of these studies are scattered across multiple journals since there is no subtropical equivalent of the international journals that focus on the tropics and are in multiple languages. Moreover, inconsistent nomenclature makes it hard to retrieve these papers electronically. The overview presented here is thus inevitably incomplete, but we hope it will encourage more such 'pan-subtropical' comparisons.

History and biogeography of the subtropics

The subtropics of the northern and southern hemisphere are separated by the 5,185-km width of the tropics. The forests of the northern subtropics, in Asia and North America, are floristically quite similar to each other, reflecting the availability of apparently frost-free land routes across the North Atlantic until the end of the Eocene, c. 34 million ago (Deng et al., 2018), and of routes for more cold-tolerant taxa until the early Miocene, c. 18–16 million years ago (Jiang et al., 2019). Subtropical fossil floras from central Tibet, North America, and Europe in the middle Eocene have many taxa in common at the genus or family level (Su et al., 2020). In contrast, the subtropical forests in the southern hemisphere have less in common with each other than they do with their nearest northern counterparts, with which they are linked by a near-continuous mountain chain in the Americas and montane stepping-stones across the tropical lowlands in other regions. Although these southern forests are all on fragments of the ancient southern supercontinent of Gondwana, the last terrestrial connections between most of them were in the early Cretaceous, before most modern groups of plants originated. However, Australia and South America were connected through Antarctica until the early Eocene, around 50 million years ago, and some elements of this 'paleo-Antarctic rainforest'—including *Araucaria*, *Podocarpus*, *Nothofagus*, and *Castanopsis* (Wilf et al., 2019)—occur in subtropical forests today.

In East Asia, there was a broad arid belt occupying subtropical latitudes during the Paleocene and early Eocene, with the Asian monsoon bringing seasonal rainfall only south of 20–22°N (Wu et al, 2023), apart from a brief period during the Paleocene-Eocene Thermal Maximum (c. 56 m years ago), when humid broadleaved forests existed at 30°N (Xie *et al.*, 2022). From the late-middle Eocene (c. 41 m years ago) onward, the monsoon expanded northward, bringing summer rains first to the southern subtropics and then intermittently at first and permanently from the late Oligocene (c. 26 million years ago) to the northern subtropics and into the temperate zone. Subtropical forest floras generally similar to those of today have existed continuously since the late Eocene in East Asia, although phylogenetic studies suggest that many characteristic groups of modern forests originated and diversified later, in the Miocene (Xiao et al., 2022). This continuity, along with the large area occupied by subtropical forests in East Asia until very recently, the persistent connections with tropical and temperate forests, and the tectonic activity of the region as a result of the ongoing collision between India and Eurasia, probably account for the exceptional diversity of East Asian subtropical forest floras today.

Subtropical forests were diverse and extensive in North Africa in the early Miocene and forests similar to those still extant in subtropical East Asia extended to at least 51°N in northern Central Europe during the Miocene Climatic Optimum, 16.9–14.7 million years ago (Kunzmann et al., 2022). However, the shrinkage of the Tethys Sea during the late Miocene drastically weakened the African summer monsoon, creating conditions at subtropical latitudes that were too arid for forest (Zhang et al., 2014), while cooling climates in the late Miocene and Pliocene progressively eliminated the warm-wet 'subtropical' elements from European floras (Steinthorsdottir et al., 2021). A relic of the North African Miocene forests persisted in the subtropical Canary Islands, which were buffered against excessive drying, but these isolated forests have been impoverished by their small total area, the difficulties of dispersal over marine barriers, and changes in climate (del Arco Aguilar et al., 2010, Fernández-Palacios et al., 2011).

In eastern Australia, the rapid northward movement of the continent since the Eocene has both progressively reduced the extent of subtropical rainforests and brought the region closer to the extensive forests and floras of Southeast Asia. As a result, there is a north–south gradient in the relative contribution of recent immigrants from the Asian tropics—which arrived mostly in the last 10–15 million years—and ancient Gondwanan lineages to humid forest floras, with immigrants enriching but not replacing the indigenous clades in humid tropical and subtropical forests (Weston and Jordan, 2015). In southern Africa, the patchy fossil record suggests forests at subtropical latitudes were progressively restricted by expanding grasslands from the mid-Miocene (Steinthorsdottir et al., 2021). In South America, subtropical forests west of the Andes were replaced by arid vegetation from the mid-Miocene onward as the Andes rose and the cold Humboldt Current was established (Le Roux, 2012), while non-forest vegetation expanded at the expense of subtropical forest east of the Andes (Steinthorsdottir et al., 2021).

Northern subtropics

In the northern subtropics, forests are most extensive by far in East Asia. There is a much smaller area in North America and none in North Africa, although a relic of the Miocene forests of North Africa and southern Europe persists in the Canary Islands. The most characteristic common feature of these forests is the abundance and diversity of Fagaceae (oaks and other genera) and, particularly following disturbance and at higher altitudes, pines (*Pinus*). Fagaceae are most diverse at both the species and genus level in subtropical Asia. Pines attain their greatest diversity in the subtropical forests of the Sierra Madre Occidental of Mexico (Cord et al., 2014), which is also a secondary centre of diversity for oaks (*Quercus*). Other genera shared between East Asia and North America include *Abies*, *Alnus*, *Ilex*, *Magnolia*, and *Picea*.

Asian subtropics

Subtropical forests were once very extensive in eastern China and the southern islands of Japan. They used to cover 25% of China's total land area, but this zone is highly suitable for agriculture and has a high human population density. Most of the original forest has now been cleared and the remnants are mostly badly degraded, except at high altitude (Corlett, 2019). The total forest area is now increasing in the Chinese subtropics, but most of this is young secondary forests and monoculture plantations, often of exotic species.

Under the influence of the East Asian monsoon, the subtropical forests of East Asia receive more rainfall (900–2,000 mm) than most other forests at these latitudes and the winter dry season is ameliorated by cooler weather that reduces evapotranspiration. As a result, these forests are composed largely of broadleaved evergreen trees, although the deciduous component increases northward. The Fagaceae and Lauraceae are usually the dominant tree families, and common genera include *Ilex* (Aquifoliaceae), *Elaeocarpus* (Elaeocarpaceae), *Castanopsis*, *Quercus* (sect. *Cyclobalanopsis*), and *Lithocarpus* (Fagaceae), *Distylium* (Hamamelidaceae), *Beilschmiedia*, *Cinnamomum*, *Cryptocarya*, *Lindera*, *Machilus*, *Neolitsea*, and *Phoebe* (Lauraceae), *Magnolia* and *Michelia* (Magnoliaceae), *Symplocos* (Symplocaceae), and *Schima* (Theaceae) (Corlett, 2019). In cold, dry areas of northern Yunnan, there is a distinctive evergreen sclerophyllous forest, dominated by hard-leaved *Quercus* species (Chen *et al.*, 2021). Young secondary forests are often dominated by pines (e.g., *Pinus massoniana*, *P. yunnanensis*) or sometimes by winter-deciduous species (e.g., *Alnus nepalensis*). Similar broadleaved evergreen forests extend west across northern Myanmar and along the foothills of the Himalayas (Corlett, 2019). At higher altitudes, these give way to forests dominated by conifers (*Abies*, *Picea*, *Pinus*, *Tsuga* etc.) and broad-leaved deciduous species.

In eastern China, the subtropical lowlands have been almost entirely deforested, but in Myanmar, northern India, Bhutan, and Nepal, they support mostly dry-season deciduous or semi-deciduous forests with a distinctly tropical flora. The semi-deciduous dipterocarp, *Shorea robusta* (sal), dominates large areas, and in both Northeast India and northern Myanmar, lowland evergreen rainforest with dipterocarps occurs at 27–28°N (Proctor *et al.*, 1998). These forests experience annual minimum temperatures below 10°C and most suffer at least occasional frosts, but this region is protected by the Himalaya from the cold air outbreaks from the north that bring low temperature extremes to East Asia.

The oceanic Ogasawara (Bonin) Islands of Japan, 1,300 km east of the Ryukyus, were covered until recently in a subtropical evergreen broadleaved forest with a flora largely of East Asian origin but lacking the poorly dispersed Fagaceae (Corlett, 2019).

North American subtropics

The largest forest area in the North American subtropics is on the Sierra Madre Occidental, a mountain range in northwestern Mexico that runs parallel to the Pacific coast and attains a maximum altitude of 3,300 m. The forests here are largely dominated by oaks (*Quercus* spp.), pines (*Pinus* spp.), or both, with pine dominance increasing with altitude (González-Elizondo et al., 2012). This mountain range (including the southern, tropical, parts) supports 54 species of *Quercus* and 24 species of *Pinus.* Relatively minor forest types include mixed coniferous forest at higher altitudes, dominated by *Abies*, *Pseudotsuga*, and/or *Picea*, along with oaks and pines. There are also small areas of montane cloud forest with *Magnolia*, *Styrax*, *Cedrela*, *Ilex*, *Tilia*, oaks, and various Lauraceae. Similar vegetation extends into the United States in southern Arizona and New Mexico on 'sky islands' rising above the semi-arid lowlands.

The Sierra Madre Oriental, in eastern Mexico, also supports diverse pine-oak forests in the more humid areas. At lower altitudes, on the dry northeastern coastal plains, in areas with accessible groundwater, there are forests with a tree layer dominated by genera of Nearctic affinities (*Alnus*, *Carya*, *Platanus*, *Quercus*, *Salix*, and *Ulmus*) (Encina-Domínguez et al., 2011). Oak- and pine-dominated forests also occur in the southeastern United States, where most have been severely disturbed, and on some of the islands of the Bahamas. In the northern Bahamas, the arrival of human settlers around 1,200 years ago resulted in the replacement of a fire-excluding forest of palms and hardwoods by a fire-maintained pine forest (Fall et al., 2021).

Canary Islands

Remnants of subtropical broadleaved forests persist on several of the Canary Islands. Similar forests occur on the island of Madeira (32°N), on the northern margin of the subtropics as defined here, and even further north, in the Azores (37–40°N), where oceanic buffering has produced frost-free lowland climates. The Fagaceae were represented in the Canary Islands by an unidentified species of *Quercus* in the Holocene forests of Tenerife, but this declined to extinction after human settlement. The surviving broadleaved evergreen forests are dominated by the family Lauraceae (del Arco Aguilar *et al.*, 2010). There are also extensive pine forests at higher altitudes and in drier areas, and small areas of sclerophyllous woodland at lower altitudes.

Southern subtropics

The absence of Fagaceae and native pines distinguishes the forests of the southern subtropics from those of the north, although pines are now widely planted and have become invasive in some areas. Gondwanan elements (e.g., Cunoniaceae, Proteaceae, and Podocarpaceae) are usually present, but most other species in these forests appear to have tropical (or northern subtropical) rather than 'southern' origins.

Madagascar

The small subtropical southeastern section of Madagascar has steep altitudinal and rainfall gradients, resulting in a wide range of forest habitats, including coastal forests on sandy soils, lowland and montane humid forests, and dry spiny forests, as well as a transition between rainforest and spiny forest (Goodman 1999; Helme and Rakotomalaza 1999; Rakotomalaza and Messmer 1999). The flora of these forests is essentially tropical, which may reflect the absence of land to the south, but elevational zones are lower than is typical in the tropics to the north.

Southern Africa

Southern Africa is mostly too dry for forests, but there are a few thousand square kilometres of forest patches scattered through the wetter areas inland from the east coast (Mucina et al., 2022). These forests have been divided into two basic types, the interior Afromontane forests, mostly above 1,000 m elevation, and the coastal lowland forests, which can then be divided further on the basis of floristics and habitats. However, there is considerable overlap in the floras of the various types, except at the extremes. Giant emergent Podocarpaceae with shade-tolerant seedlings are characteristic of old-growth Afromontane forests, but angiosperms dominate after human disturbance (Adie et al., 2013). In the northern subtropics of South Africa and Mozambique, a distinctive 'sand forest' with a more tropical flora grows on deep sands.

South America

East of the Andes, subtropical forests occur across a wide range of environments, from the cool, moist Andean montane forests, through the warm, dry Chaco lowlands, then seasonal semi-deciduous forests, to the everwet Atlantic coastal rainforests. The Atlantic forests stretch along the eastern coast of Brazil for more than 4,000 km, from 4°S to 32°S. At subtropical latitudes, these forests extend inland as far as eastern Paraguay and northern Argentina. Floristically, they form a continuum, with rainforests near the coast and seasonal semi-deciduous forests inland (Oliveira-Filho et al., 2015). Both forest types also show a strong altitudinal zonation. Inland, there used to be a large area of mixed forests, at 500–1,800 m a.s.l., with the relatively frost-tolerant Gondwanan conifer *Araucaria angustifolia* forming the upper canopy over a diverse broadleaved middle and lower storey, but only an estimated 3% of this forest remains. Tree diversity is highest in the rainforest and declines with both altitude and latitude, with low-temperature extremes probably the most important factor in excluding species. Unlike the subtropical forests of Asia, the tree flora of the subtropical Atlantic forests consists largely of tropical species approaching their southern limits, although there are some subtropical endemics and a few Gondwanan genera, including *Araucaria*, *Podocarpus*, *Weinmannia* (Cunoniaceae), and *Drimys* (Winteraceae) (Oliveira-Filho et al., 2015).

Most of the surviving natural vegetation in the dry Chaco region would not meet most definitions of forest, but there are also areas where soil, topography, and rainfall (>

500–600 mm) and the absence of burning and grazing allow the formation of a more or less closed woody canopy. In comparison with tropical forests with similar dry season length, these subtropical forests are lower, structurally simpler, smaller-leaved, and floristically less diverse (Sarmiento, 1972).

The montane forests of the Andean mountain range extend from Venezuela to Argentina. At subtropical latitudes, in northwestern Argentina, these forests grow for 700 km along the eastern slopes of the mountains, between around 300 and 3,000 m a.s.l. (Entrocassi et al., 2020). At lower elevations, they transition to the dry Chaco region and at high elevations they give way to montane grasslands and shrublands. Climate and vegetation vary strongly with elevation. At low elevations, the flora is essentially tropical, with many trees deciduous in the dry season, while the upper montane forests contains both winter-deciduous trees of north-temperate ancestry (e.g., *Alnus*) and species of Gondwanan ancestry (e.g., *Podocarpus*).

West of the Andes, in the southern subtropics of Chile, small patches of evergreen forest occur on coastal mountains facing the Pacific, where they depend on fog subsidies in areas receiving only 350 mm of rainfall (Squeo et al., 2016). Easter Island, 3,700 km west of South America, supported a forest dominated by an extinct palm species until deforestation following human settlement (Rull, 2020).

Australia

Most of Australia is too dry for forest, particularly in the subtropical zone, but a broad belt along the east coast receives enough rainfall (> c. 400 mm) to support eucalypt (*Eucalyptus* spp.) woodland and the narrow strip nearest the coast (> 800–1,000 mm rainfall) supports largely evergreen rainforest patches in a matrix of tall open eucalypt forest, with a more or less dense understory. Rainforest was much more extensive at subtropical latitudes in Australia in the early Paleogene until the late Eocene but has become severely restricted subsequently. The richest rainforest floras are found in areas where there is evidence for continuity of a moist climate (Weber et al., 2014). These areas shelter both ancient Gondwanan lineages and species derived from Asian lineages that colonized Australia from the Miocene onwards. East of Australia, distinctive subtropical forests occur on Lord Howe Island, Norfolk Island, and the Kermadec Islands. These are all volcanic islands, with frost-free oceanic climates without a pronounced dry season. All three have areas of palm-dominated forest and Norfolk Island has forests dominated by an endemic Gondwanan conifer, *Araucaria heterophylla*.

Carbon economy

There has been a lot of recent research on the forest carbon cycle in subtropical China and some in South America and Australia. In comparison with the tropics, subtropical climates have cool winters, with both the frequency and intensity of frosts increasing with latitude and elevation. These seasonal low temperatures are expected to reduce photosynthesis, and this has been confirmed at three, largely evergreen, subtropical forest sites in China (Zhang et al., 2016). This seasonal reduction is no doubt even greater in forests with a substantial deciduous component in the canopy. In contrast to temperature, water limitation does not seem to be a problem in humid subtropical forests in either China or South America (Zhang et al., 2016). In both areas, the dry and cold seasons coincide, so evapotranspiration is considerably reduced in winter. This absence of seasonal water limitation is unlikely to extend, however, to the driest subtropical forest types.

Eddy covariance data from East Asia shows that, although gross primary productivity (GPP)—the rate of carbon fixation by photosynthesis—declines from the tropics to the subtropics, the net ecosystem exchange (NEE)—the balance between photosynthesis and total ecosystem respiration (plants, animals, and microbes)—increases with latitude, before declining again in the temperate zone (Zhang et al., 2016). This unexpected finding is explained by the high sensitivity of ecosystem respiration to temperature, so that although carbon fixation declines in the cold season, carbon losses through respiration decline even more. There is also a strong negative relationship with forest age and a strong positive relationship with wet nitrogen deposition (from industry and agriculture), suggesting that both recovery from past disturbance and nitrogen fertilization contribute to the high carbon uptake. The result is that, at least in China and presumably in the moist subtropics elsewhere, subtropical forests are among the world's largest carbon sinks on a per-area basis. These data also suggest, however, that this sink may be vulnerable to global warming, as respiration increases faster than photosynthesis. The total ecosystem carbon stock (above and below-ground biomass, litter, and soil carbon) in 667 subtropical forest plots across China was also influenced negatively by mean annual temperature, again suggesting a vulnerability to warming, but tree species diversity had a positive influence and may help buffer carbon stocks (Yan et al., 2022).

Subtropical faunas

Although there is a growing literature on subtropical plant communities, there is much less on the animals. Subtropical vertebrate faunas are generally subsets of those nearer the equator, although in East Asia, the diversity of babblers and pheasants is highest in the subtropics. The tropical aspect of most subtropical faunas presumably reflects the ability of many vertebrates to avoid short periods of extreme cold behaviourally or, for mammals and birds, by thermoregulation, so that their poleward limits are more likely to be set by a seasonal gap in food supply than by climate directly (Corlett, 2019). Teng et al. (2023) argue that the extirpation of large native herbivores—particularly Asian elephants and rhinoceroses—from most subtropical forests in East Asia over the last millennium may have fundamentally changed their structure and functioning, and the same may be true of subtropical forests elsewhere. Many tropical vertebrates reach their poleward limits within the subtropics, so diversity declines with increasing latitude. There is less data for subtropical invertebrates, but diversity also declines with latitude in many groups.

Discussion

The subtropics have often been viewed by biologists as merely a transition between the tropics and the temperate zone, although there has been little agreement on where exactly this transition occurs (Corlett, 2013). Are subtropical forests sufficiently distinct from tropical and temperate forests to be worth distinguishing as a separate ecological entity? A phylogenetic classification of global tropical and subtropical forests based on tree plot data identified a 'subtropical' floristic region, which clustered subtropical and tropical montane sites in America and Asia, confirming both the floristic links between Asia and the Americas and the distinctiveness of these forests from the nearest tropical lowland forests (Slik et al., 2018).

Although the latitudes of the Tropics of Cancer and Capricorn—the equatorward boundaries of the subtropics—are ecologically arbitrary, they are usually close to the 'frost line', which seems to be a general barrier to tropical plants because of the devastating impact of ice-crystal formation inside plant cells (Corlett, 2019). In China, the poleward limit of the subtropics also coincides, more or less, with another eco-physiological boundary, where the dominant broadleaved evergreen trees give way to winter-deciduous species. This limit seems to coincide with an absolute minimum temperature of around −15°C (corresponding to a January mean of around 0°C). However, it is not clear from the available literature if a similar boundary occurs in other parts of the world.

In China, where there is a good paleoecological record, forests that would be considered subtropical today have expanded north in warmer periods and retreated south in colder ones, to be replaced by broad-leaved deciduous forests or more open vegetation types (Corlett, 2019). This suggests that it is the climate—warm, wet, with mild winters—rather than the latitude that makes a 'subtropical forest'. However, while mean temperature isotherms have shifted north and south with global climate change, temperature seasonality is largely a function of latitude, so thermal regimes have not simply shifted back and forth.

Subtropical forests are almost certainly less extensive today than at any time in the last 20–30 million years. Long-term continental drying since the mid-Miocene had already greatly limited their extent, except in monsoon Asia, and human impacts have further reduced their area in those regions where suitable climates still exist. Their limited global extent has, in turn, reinforced the assumption that there are only two major forest biotas, tropical and temperate, and that forests that do not fit into this binary classification, including tropical montane and subtropical forests, consist of mixtures of tropical and temperate lineages. Viewed from a Miocene perspective, when subtropical and 'subtropical-like' forests were at their maximum extent, a trinary classification makes more sense and, indeed, is standard in East Asia. These are human labels, however, and reality is undoubtedly more complex.

Subtropical forests have received less attention from conservationists than those in the tropics despite their importance for both biodiversity and carbon sequestration, although much of the remaining forest area is included within Conservation International's 36 global biodiversity hotspots. In East Asia, breeding-bird diversity reaches a maximum at around 25°N, rather than near the equator, with the larger land area available at subtropical latitudes than on the tropical peninsula and islands further south a plausible explanation for this pattern (Ding et al., 2006). Their relatively small extent reduces the global significance of subtropical forests for carbon sequestration, but, as shown above, some have both a high biomass and a high sink capacity on a per-area basis, so their protection and restoration have both biodiversity and carbon benefits.

Simple models of future vegetation distribution under anthropogenic climate change suggest that subtropical forests will shift poleward as the temperature rises. Even in the absence of the now ubiquitous forest fragmentation, however, tree species migration rates would probably be too slow to track the current rate of temperature change (*see* Chapter 37, *Plant Movements in Response to Climate Change*). In any case, real-world responses are likely to be more complex than a simple poleward shift, reflecting both the complexities of climate change and the many additional anthropogenic impacts that these forests are now subject to. Long-term monitoring of permanent sample plots is our best hope of understanding these changes.

References

Adie, H., Rushworth, I. and Lawes, M. J. (2013) 'Pervasive, long-lasting impact of historical logging on composition, diversity and above ground carbon stocks in Afrotemperate forest', *Forest Ecology and Management*, vol 310, pp. 887–895.

Chen, L., Deng, W., Su, T., Li, S. and Zhou, Z. (2021) 'Late Eocene sclerophyllous oak from Markam Basin, Tibet, and its biogeographic implications', *Science China Earth Sciences*, vol 64, pp. 1969–1981.

Cord, A. F., Klein, D., Gernandt, D. S., Pérez de la Rosa, J. A. and Dech, S. (2014) 'Remote sensing data can improve predictions of species richness by stacked species distribution models: a case study for Mexican pines', *Journal of Biogeography*, vol 41, pp. 736–748.

Corlett, R. T. (2013) 'Where are the subtropics?', *Biotropica*, vol 45, pp. 273–275.

Corlett, R. T. (2019) *The Ecology of Tropical East Asia*, 3rd edition. Oxford University Press, Oxford.

Cui, E., Lu, R., Xu, X., Sun, H., Qiao, Y., Ping, J., Qiu, S., Lin, Y., Bao, J., Yong, Y., Zheng, Z., Yan, W. and Xia, J. (2022) 'Soil phosphorus drives plant trait variations in a mature subtropical forest', *Global Change Biology*, vol 28, pp. 3310–3320.

del Arco Aguilar, M., González-González, R., Garzón-Machado, V. and Pizarro-Hernández, B. (2010) 'Actual and potential natural vegetation on the Canary Islands and its conservation status', *Biodiversity and Conservation*, vol 19, pp. 3089–3140.

Deng, M., Jiang, X.-L., Hipp, A. L., Manos, P. S. and Hahn, M. (2018) 'Phylogeny and biogeography of East Asian evergreen oaks (*Quercus* section *Cyclobalanopsis*; Fagaceae): Insights into the Cenozoic history of evergreen broad-leaved forests in subtropical Asia', *Molecular Phylogenetics and Evolution*, vol 119, pp. 170–181.

Ding, T. S., Yuan, H. W., Geng, S., Koh, C. N. and Lee, P. F. (2006) 'Macro-scale bird species richness patterns of the East Asian mainland and islands: energy, area and isolation', *Journal of Biogeography*, vol 33, pp. 683–693.

Encina-Domínguez, J. A., Rocha, E. M., Meave, J. A. and Zárate-Lupercio, A. (2011) 'Community structure and floristic composition of *Quercus fusiformis* and *Carya illinoinensis* forests of the Northeastern Coastal Plain, Coahuila, Mexico', *Revista Mexicana de Biodiversidad*, vol 82, pp. 607–622.

Entrocassi, G. S., Gavilán, R. G. and Sánchez-Mata, D. (2020) *Subtropical Mountain Forests of Las Yungas: Vegetation and Bioclimate*. Springer, Basel, Switzerland.

Fall, P. L., van Hengstum, P. J., Lavold-Foote, L., Donnelly, J. P., Albury, N. A. and Tamalavage, A. E. (2021) 'Human arrival and landscape dynamics in the northern Bahamas', *Proceedings of the National Academy of Sciences of the USA*, vol 118, e2015764118.

Fernández-Palacios, J. M., de Nascimento, L., Ottto, R., Delgado, J. D., García-del-Ray, E., Arévalo, J. R. and Whittaker, R. J. (2011) 'A reconstruction of Palaeo-Macaronesia, with particular reference to the long-term biogeography of the Atlantic island laurel forests', *Journal of Biogeography*, vol 38, pp. 226–246.

González-Elizondo, M. S., González-Elizondo, M., Tena-Flores, J. A., Ruacho-González, L. and López-Enríquez, I. L. (2012) 'Vegetación de la Sierra Madre Occidental, México: una síntesis', *Acta Botánica Mexicana*, vol 100, pp. 351–403.

Goodman, S. M. (1999) 'Description of the Réserve Naturelle Intégrale d'Andohahela, Madagascar, and the 1995 biological inventory of the reserve', *Fieldiana Zoology N.S.*, vol 94, pp. 1–7.

Helme, N. A. and Rakotomalaza, P. J. (1999) 'An overview of the botanical communities of the Réserve Naturelle Intégrale d'Andohahela, Madagascar', *Fieldiana Zoology N.S.*, vol 94, pp. 11–24.

Jiang, D., Klaus, S., Zhang, Y.-P., Hillis, D. M. and Li, J.-T. (2019) 'Asymmetric biotic interchange across the Bering land bridge between Eurasia and North America', *National Science Review*, vol 6, pp. 739–745.

Kunzmann, L., Li, S.-F., Huang, J., Utescher, T., Su, T. and Zhou, Z.-K. (2022) 'Assessment of phytogeographical reference regions for Cenozoic vegetation: a case study on the Miocene flora of Wiesa (Germany), *Fossil Imprint*, vol 78, pp. 1–43.

Le Roux, J. P. (2012) 'A review of tertiary climate changes in southern South America and the Antarctic Peninsula. Part 2: continental conditions', *Sedimentary Geology*, vol 247–248, pp. 21–38.

McEwan, R. W., Lin, Y.-C., Sun, I. F. *et al.* (2011) 'Topographic and biotic regulation of aboveground carbon storage in subtropical broad-leaved forests of Taiwan', *Forest Ecology and Management*, vol 262, pp. 1817–1825.

Mi, X., Feng, G., Hu, Y., Zhang, J., Chen, L., Corlett, R. T., Hughes, A. C., Pimm, P., Schmid, B., Shi, S., Svenning, J.-C. and Ma, K. (2021) 'The global significance of biodiversity science in China: an overview', *National Science Review*, vol 8, nwab032.

Mucina, L. Lötter, M. C., Rutherford, M. C., van Niekerk, A., Macintyre, P. D., Tsakalos, J. L., Timberlake, J., Adams, J. B., Riddin, T. and Mccarthy, L. K. (2022) 'Forest biomes of Southern Africa', *New Zealand Journal of Botany*, vol 60, pp. 377–428.

Oliviera-Filho, A. T., Budke, J. C., Jarenkow, J. A., Eisenlohr, P. V. and Neves, D. R. M. (2015) 'Delving into the variations in tree species composition and richness across South American subtropical Atlantic and Pampean forests', *Journal of Plant Ecology*, vol 8, pp. 242–260.

Osland, M. J., Stevens, P. W., Lamont, M. M., Brusca, R. C., Hart, K. M., Waddle, J. H., Langtimm, C. A., Williams, C. M., Keim, B. D., Terando, A. J., Reyier, E. A., Marshall, K. E., Loik, M. E., Boucek, R. E., Lewis, A. B. and Seminoff, J. A. (2021) 'Tropicalization of temperate ecosystems in North America: The northward range expansion of tropical organisms in response to warming winter temperatures', *Global Change Biology*, vol 27, pp. 3009–3034.

Proctor, J., Haridasan, K. and Smith, G. W. (1998) 'How far north does lowland evergreen tropical rain forest go?', *Global Ecology and Biogeography Letters*, vol 7, pp. 141–146.

Rakotomalaza, P. J. and Messmer, N. (1999) 'Structure and floristic composition of the vegetation in the Réserve Naturelle Intégrale d'Andohahela, Madagascar', *Fieldiana Zoology N.S.*, vol 94, pp. 51–96.

Rull, V. (2020) 'The deforestation of Easter Island', *Biological Reviews*, vol 95, pp. 124–141.

Sarmiento, G. (1972) 'Ecological and floristic convergences between seasonal plant formations of tropical and subtropical South America', *Journal of Ecology*, vol 60, pp. 367–410.

Slik, J. W. et al. (+186 authors) (2018) 'Phylogenetic classification of the world's tropical forests', *Proceedings of the National Academy of Sciences of the USA*, vol 115, 1837–1842.

Smith, E. T. and Sheridan, S. C. (2020) 'Where do cold air outbreaks occur, and how have they changed over time?', *Geophysical Research Letters*, vol 47, e2020GL086983.

Squeo, F. A., Loayza, A. P., López, R. P. and Gutiérrez, J. R. (2016) 'Vegetation of Bosque Fray Jorge National Park and its surrounding matrix in the Coastal Desert of north-central Chile', *Journal of Arid Environments*, vol 126, pp. 12–22.

Steinthorsdottir, M., Coxall, H. K., de Boer, A. M., Huber, M., Barbolini, N., Bradshaw, C. D., Burls, N. J., Feakins, S. J., Gasson, E., Hendericks, J., Holbourn, A. E., Kiel, S., Kohn, M. J., Knorr, G., Kürschner, W. M., Lear, C. H., Liebrand, D., Lunt, D. J., Mörs, T., Pearson, P. N., Pound, M. J., Stoll, H. and Strömberg, C. A. E. (2021) 'The Miocene: the future of the past', *Paleoceanography and Paleoclimatology*, vol 36, e2020PA004037.

Su, T. et al. (+ 32 authors) (2020) 'A Middle Eocene lowland humid subtropical "Shangri-La" ecosystem in central Tibet', *Proceedings of the National Academy of Sciences of the USA*, vol 117, pp. 32989–32995.

Teng, S. N., Svenning, J.-C. and Xu, C. (2023) 'Large mammals and trees in eastern monsoonal China: anthropogenic losses since the Late Pleistocene and restoration prospects in the Anthropocene', *Biological Reviews*, vol 98, pp. 1607–1632.

Weber, L. C., VanDerWal, J., Schmidt, S., McDonald, W. J. F. and Shoo, L. P. (2014) 'Patterns of rain forest plant endemism in subtropical Australia relate to stable mesic refugia and species dispersal limitations', *Journal of Biogeography*, vol 41, pp. 222–238.

Weston, P. H. and Jordan, G. J. (2015) 'Evolutionary biogeography of the Australian flora in the Cenozoic Era', In: *Australian Vegetation* (ed. D.A. Keith), Cambridge University Press, Cambridge, pp. 40–62.

Wilf, P., Nixon, K. C., Gandolfo, M. A. and Cúneo, R. (2019) 'Eocene Fagaceae from Patagonia and Gondwanan legacy in Asian rainforests', *Science*, vol 364, eaaw5139.

Wu, F., Fang, X., Yang, Y., Dupont-Nivet, G., Nie, J., Fluteau, F., Zhang, T. and Han, W. (2023) 'Reorganization of the Asian climate in relation to Tibetan Plateau uplift', *Nature Reviews Earth & Environment*, vol 3, pp. 684–700.

Xiao, T.-W., Yan, H.-F. and Ge, X.-J. (2022) 'Plastid phylogenomics of tribe *Perseeae* (Lauraceae) yields insights into the evolution of East Asian subtropical evergreen broad-leaved forests, *BMC Plant Biology*, vol 22, p. 32.

Xie, Y., Wu, F. and Fang, X. (2022) 'A transient south subtropical forest ecosystem in central China driven by rapid global warming during the Paleocene-Eocene thermal maximum', *Gondwana Research*, vol 101, pp. 192–202.

Yan, G. Y., Bongers, F. J., Trogisch, S., Li, Y., Chen, G. K., Yan, H. R., Deng, X. L., Ma, K. P. and Liu, X. J. (2022) 'Climate and mycorrhizae mediate the relationship of tree species diversity and carbon stocks in subtropical forests', *Journal of Ecology*, vol 110, pp. 2462–2474.

Zhang, Y.-J., Cristiano, P. M., Zhang, Y.-F., Campanello, P. I., Tan, Z.H., Zhang, Y.-P., Cao, K. F. and Goldstein, G. (2016) 'Carbon economy of subtropical forests', In: *Tropical Tree Physiology* (eds. G. Goldstein & L.S. Santiago), Springer, Basel, Switzerland.

Zhang, Z., Ramstein, G., Schuster, M., Li, C., Contoux, C. and Yan, Q. (2014) 'Aridification of the Sahara Desert caused by Tethys Sea shrinkage during the Late Miocene', *Nature*, vol 513, pp. 401–404.

5
TROPICAL FORESTS

Richard T. Corlett

Introduction

This chapter provides an introduction to the forests growing between the Tropics of Cancer and Capricorn, 23.4° north and south of the equator, respectively. Although these poleward boundaries are ecologically arbitrary, they are widely used and avoid having to choose among multiple alternative ecological definitions of the tropics. In this book, the forests immediately to the north and south of the tropical belt are covered in the chapter on subtropical forests. These differ from their tropical equivalents largely because of the impact of winter cold, including lowland frosts in most areas, and the consequences of this weather for forest structure and floristics (*see* Chapter 4, *Subtropical forests*).

Tropical forests cover only 18% of the Earth's total land area but support 63% of all mammal species, 73% or all birds, and 76% of amphibians (Pillay et al., 2022). They also support a large proportion of the Earth's flora (Corlett, 2016) and invertebrate fauna (e.g. Kass et al., 2022). Moreover, tropical forests account for more than half of the total carbon stored in global terrestrial biomass and make major contributions to both gross carbon emissions to the atmosphere (from clearance, degradation, and fires) and gross removals (from growth) (*see* Chapter 38, *Forest carbon budgets and climate change*). Tropical forests are also more efficient water pumps than grasslands or crops because of their deeper roots, larger leaf areas, higher surface roughness, and lower albedos, and deforestation has been linked to reduced rainfall at a regional scale (Smith et al., 2023). There is also evidence for teleconnections with climate change beyond the tropics (Wei et al., 2023). The magnitude of the carbon and water fluxes through tropical forests therefore means that small changes can have global consequences.

Tropical forests provide support—in some cases, critical support—for the livelihoods and well-being of tens of millions of people. In 2012, an estimated billion people in tropical countries lived within 5 km of forest, including 24% of Brazil's and 35% of Indonesia's populations (Newton et al., 2020). These 'forest-proximate' people range from indigenous rainforest hunter-gatherers, some of whom show evidence of genetic adaptations to this challenging environment extending back into the Late Pleistocene (Deng et al., 2022; Padilla-Iglesias et al., 2022), through agriculturalists who use forest products, including wood, bushmeat

DOI: 10.4324/9781003324072-6

(*see* Chapter 42, ***Wild meat hunting in tropical forests***) and numerous species of non-timber plants (*see* Chapter 39, ***Multiple roles of non-timber forest products in ecologies, economies and livelihoods***) to supplement their livelihoods and may benefit from forest environmental services (Newton et al., 2016), to the inhabitants of new suburban housing developments with urban jobs and lifestyles for whom forest proximity is coincidental. The harvest of timber, and to a smaller extent other forest products, also supports people involved in processing, manufacturing, and trade living far from the nearest forest, and many other people depend on forested watersheds for their water supply, consume forest products, and/or participate in forest-based recreation and tourism.

Tropical forest environments

Within the tropics, as defined here, climate varies across a multidimensional continuum, with rainfall and temperature as the ecologically most important axes. Until recently, forests occupied all but the driest (<800–1,200 mm/yr) and coldest (mean annual temperature < c. 6–7°C) parts of this continuum. Drier non-forested areas are occupied by savanna (trees over a more or less continuous layer of grasses), grassland, or desert, while colder areas above the altitudinal treeline are occupied by alpine shrublands and grasslands. On the wetness axis, forest and savanna can both occur over a range of rainfalls, depending on soil, fire, and grazing (Charles-Dominique et al., 2018). In contrast, the altitudinal treeline in the tropics is predicted well by temperature, irrespective of soil and rainfall, although the mechanism responsible for this is unclear (Körner, 2021).

Within the forested areas, the physiognomy (i.e., external appearance), structure (canopy height and closure, stratification, etc.), and floristic composition of the forest also form a continuum, although non-climatic factors, including soil properties, fire, and the biology of the dominant trees, may create sharp boundaries on a landscape and regional scale. Moreover, neither rainfall nor temperature are simple variables. The distributions of tree species and forest types in the tropics appear to depend more on dry-season length and intensity than on total rainfall and are also influenced by other factors that affect water availability, including dry-season temperatures and soil water-holding capacity (Muller-Landau et al., 2021). Most variation in temperature within the tropics is associated with elevation and is thus confounded with a predictable decline in atmospheric pressure and less predictable changes in rainfall, cloudiness, and other climatic factors, as well as geomorphology and soils.

Soil characteristics are influenced to a varying extent by climate but also have an independent influence on forest structure and composition, particularly at local and regional scales. Soils are diverse and complex entities and the characteristics used in soil classifications are aimed at identifying agricultural potential, so their predictive value for natural vegetation varies. Most tropical soils are highly weathered, acidic, and low in nutrients—Ultisols and Oxisols in the United States Department of Agriculture (USDA) classification (USDA, 2006), Acrisols, Ferralsols, and Alisols in the World Reference Base (WRB) system (FAO, 2015). There is evidence that the availability of phosphorous, in particular, may limit forest productivity on such soils, although not all studies support this finding (Wright, 2022).

Younger, more fertile soils occur in volcanic areas, on steep slopes, and on recent fluvial deposits, and a range of other soil types of intermediate fertility also occupy substantial areas. Two extreme soil types with very low agricultural potential have striking effects on forest structure and composition, described below. Histosols—peats—form in areas with near-permanent waterlogging, while Podzols (WRB)/Spodosols (USDA) occur mostly on sand or sandstone. Some other regionally rare geologies also give rise to distinct soils, although the distinctiveness of the associated forests varies (Corlett and Tomlinson, 2020). Limestone outcrops—karsts—create alkaline islands in a 'sea' of acidic soils, with consequences for the availability of both nutrients and toxic cations, while soils on ultramafic rocks may have toxic levels of nickel, cobalt, and magnesium, as well as nutrient deficiencies.

Human impacts of many types are now a major additional influence on forest structure and floristics in many tropical forests. A recent global assessment of forest canopy structure using satellite lidar and spectroradiometric data found that, although climate was the most important driver of forest structure globally, human impacts were second in importance, followed by soil, topography, and fire (Li et al., 2023). In the tropics, human factors dominated large areas in peripheral parts of Amazonia and Central Africa, and more widely in Madagascar and Asia.

Biogeography

Even when physical environments are matched as closely as possible, forests on different continents are often visibly different (Corlett and Primack, 2011). No forest plant species are shared across the tropics, and few occur in more than one region. Trees in tropical Asia are taller for the same diameter than those in other tropical regions and the above-ground biomass of tropical forests is higher in Africa and Asia than the Neotropics (Muller-Landau et al, 2021). In contrast, Neotropical forests are most diverse floristically and African forests least. Although some of these differences may be explained by differences in the proportions of different climate types or geologies, others appear to reflect historical accidents in what lineages reached where, coupled with the lack of frost-free land routes between the major biogeographical regions—the Neotropics, Africa, Madagascar, Asia, and Australia/New Guinea—in the last 30–40 million years. Tropical forests everywhere consist of a similar set of Late Cretaceous tree lineages but differ considerably in which more recent lineages dominate community composition (Slik et al., 2018). The most striking difference is the dominance of the family Dipterocarpaceae in the canopy of most tropical Asian forests which, in turn, has had consequences for many aspects of the ecology of these forests (Corlett and Primack, 2011; Ghazoul, 2016).

Classifying tropical forests

There is no single agreed pantropical classification of tropical forests and many different characteristics have been used as a basis for classification, including phenology, structure, altitude, climate, soils, and floristics (Corlett, 2015). Global classifications usually recognize two or three major types of lowland forests on the basis of tree phenology: tropical lowland rainforest, tropical deciduous (or dry) forest, and sometimes an intermediate semi-evergreen (or semi-deciduous) type. The term 'tropical moist forest' is often used to encompass both evergreen and semi-evergreen forests. Additional lowland forest types associated with extreme soil types or with inundation by fresh, brackish, or seawater, are also widely distinguished, but terminologies are sometimes difficult to match across the tropics. Most classifications also divide forests by altitude, recognizing two (lowland and montane) or more zones (three or more of lowland, premontane or submontane, lower montane, upper montane, and subalpine). The International Union for Conservation of Nature's (IUCN) Global Ecosystem Typology uses ecosystem functions and ecological processes to differentiate four 'ecosystem functional groups' within the tropical or subtropical forest biome: lowland rainforests, dry forests and thickets, montane rainforests, and heath forests (Keith et al., 2022).

Disturbance history, where known or easily detected, is also usually recognized in classifications, with primary or old growth or mature forests distinguished from secondary or second-growth forests that have developed on areas recently cleared of forest. Confusingly, the term 'secondary' is also often applied to selectively logged forests, although the process of recovery and the resulting forests are very different from those on cleared sites (Corlett, 1994).

Where enough data is available, floristic classifications—usually based on large trees only—are useful at local and regional levels and can predict the other characteristics. Floristic data from sample plots can be extrapolated using environmental variables (Yang et al., 2023), but anthropogenic impacts reduce the predictability of species assemblages. One situation where a floristic classification always makes sense is with monodominant forests, where a single species forms 60% or more of the canopy (*see* Chapter 12, *Monodominance in tropical lowland forests*).

Earlier forest classifications often had an economic motivation, so they are based on extractable timber trees and/or suitability of the land for commercial crops. The recent emphasis on tropical forests as stores of carbon and reservoirs of biodiversity, however, calls for new types of classification. A focus on carbon stores and fluxes, and other ecosystem services, requires a classification based on ecological functions rather than static attributes. Recent studies have shown that it is possible to identify and map functionally distinct forest types from satellite-based remote sensing, although ground-based studies will still be needed to assess the mechanisms and processes involved (Ordway et al., 2022). For biodiversity conservation, we need a classification that reflects patterns in both plant and animal diversity, although in practice this will have to use indicator groups, such as trees and birds, for which good data is most often available. Carbon and biodiversity have been mapped together on a global scale, using landcover and stacked vertebrates range maps, respectively, as proxies (Soto-Navarro et al., 2020), and this approach could be extended and refined using more taxa.

Major tropical forest types

Tropical rainforests

Tropical rainforests are found in lowland areas with a reliable, year-round surplus of rainfall over evapotranspiration (Keith et al., 2022). A dry season is absent or brief (< 3 months) or mitigated by accessible groundwater. Soils are not regularly inundated or waterlogged, and although typically acid- and nutrient-poor, are not extreme. Tropical rainforests have a closed canopy and complex, multilayered structure, a high primary productivity, and support a higher taxonomic and functional diversity of plants than other tropical vegetation types. The plants, in turn, support exceptional diversities of vertebrates and of many groups of invertebrates.

Tropical dry forests

Lowland areas with a substantial seasonal water deficit (dry season >3 months) support a range of forest types with drought-deciduous trees in the canopy. In general, the longer and drier the dry season, the greater the proportion of deciduous trees. In this chapter, only closed-canopy forests are included in this category since vegetation with an open tree canopy over a more or less continuous grass layer is better called savanna and treated as a non-forest habitat. In general, tropical dry forests support a lower tree biomass and less diverse flora and fauna than tropical rainforests in the same region, but patterns of diversity within the dry forest biome are complex and appear to be controlled by historical biogeography more than present-day climate (e.g. DRYFLOR, 2016). Across the tropics, dry forests occupied many of the sites most suitable for human settlements, so they have been extensively cleared and modified over millennia.

Forested wetlands

This is an extremely diverse category that includes both forests growing in areas subject to periodic or continuous flooding and others growing in soil that is saturated with water throughout the year, although rarely or never flooded. Mangrove forests occupy the upper half of the intertidal zone on muddy shores in the tropics, where they are flooded daily by seawater. Fewer than 70 species of angiosperms globally can grow under these conditions, so mangroves are floristically and structurally much simpler than other major tropical forest types. Areas subject to tidal flooding by brackish water are also often dominated by one or a few specialist species, while the smaller areas flooded daily by freshwater backed up by the tides have a more diverse but less distinctive flora. Much larger areas are flooded seasonally by freshwater, particularly in the Congo and Amazon Basins. These vary in the frequency, depth, and duration of flooding, and in the chemistry of the water and sediments (Junk et al., 2011), but there is no standard pantropical terminology.

Forested tropical peatlands—peat swamp forests—are largely confined to areas with year-round rainfall, or only a short dry season, and low-lying topography (Page et al., 2022). In these conditions, the rate of addition of organic matter to the soil surface may exceed the rate of decomposition, leading to the accumulation of partly decomposed organic matter as peat. If this continues long enough, the peat surface is raised above the surrounding topography, so water and nutrients are received only from the atmosphere, in rain and dust, leading to highly acidic, nutrient-poor conditions. Peat may also form in areas subject to flooding, but the extent of such peatlands is unclear.

The vegetation of tropical peatlands depends on peat depth and whether or not it is subject to flooding, with forests on the deepest peats typically consisting of a high density but low diversity of relatively small trees. These 'peatland pole forests' are the most carbon-dense ecosystems in Amazonia, with most of the carbon in the peat (Coronado et al., 2021). Peat swamps dominated by palms are common in Africa and the Neotropics, although not all palm swamps form peat. Tropical peatlands occupy at least 440,000 km^2 of the tropics, with the largest areas in Borneo, Sumatra, New Guinea, the Congo Basin, and the Amazon Basin (Page et al., 2022). Despite their global importance as a carbon store (*see* Chapter 38, ***Forest carbon budgets and climate change***), they are still poorly mapped, incompletely understood, and increasingly threatened in some areas.

Forests in areas with a shallow water table (< 5 m from the surface), but that don't flood, are not normally considered as wetlands, but they cover vast areas of the tropics, including a large proportion of the Amazon region, and are distinctive in structure, function, and species composition (Costa et al., 2023). These forests are understudied, but shallow water tables constrain rooting depths while increasing resilience to at least moderate droughts.

Forests on other extreme soil types

Forests on regionally rare geological substrates often have sharp boundaries with adjacent forests on more typical geologies but may also intergrade gradually. What counts as 'extreme' is fairly arbitrary, but the heath forests (or white sand forests) on deep, sandy, acidic, nutrient-poor soils (podzols, spodosols) are distinguished across the tropics because of their distinctive appearance and unsuitability for agriculture (Adeney et al., 2016; Keith et al., 2022). The vegetation on these substrates, which may also be prone to droughts and/or waterlogging, varies from open habitats without trees to tall evergreen forest, but the most characteristic is a relatively low (< 20 m), relatively open forest dominated by a few species of slender-trunked trees, often with small, leathery leaves. Forests on shallow soils over ultramafic rocks are also often sharply separated, both structurally and floristically, from adjacent forests on different rock types, but forests on deeper, well-weathered soils over the same rocks are often not distinct (van der Ent et al., 2018). Forests on limestone tend to be highly heterogeneous, with a mix of generalist and specialist tree species. Distinct forest types are also found locally and regionally on a range of other geological substrates, but none of these are as widespread as the three types mentioned above (Corlett and Tomlinson, 2020).

Tropical montane forests

Forest structure, physiognomy, and floristics all change with elevation above sea level. In general, with increasing elevation and declining temperature, trees become shorter, leaves and crowns become smaller, and cold-intolerant taxa progressively drop out and are replaced by a smaller number of cold-tolerant taxa. Meanwhile, soils become more acid and soil organic matter increases. However, there are many variations on this pattern, reflecting changes in water availability, cloudiness, soil properties, and other physical and biotic variables (de la Cruz-Amo et al., 2020). Changes in vegetation with altitude may be gradual but are sometimes more or less abrupt.

Secondary and structurally altered forests

It is useful to distinguish forests that have regenerated on sites that have lost their previous forest cover—secondary or second-growth forests—from forests that are recovering from a disturbance that did not remove the forest. Forests in this latter category are most often known as degraded or disturbed, but the term 'structurally altered' avoids the negative connotations of these words, which are often not justified. The most important difference between these two categories is that the floras of structurally altered forests mostly persist through the disturbance on-site, while the floras of secondary forests must arrive from elsewhere and are therefore usually dispersal limited. Soil structure and nutrient capital are also more likely to survive more or less intact in structurally altered than secondary forests. Secondary forests can arise from natural disturbances, such as landslides and river-channel migration, but the majority of secondary forests in the tropics today are the result of abandonment of land cleared for pasture or crops. Similarly, natural windthrow and other disturbances can change forest structure, but most structurally altered forests today are the results of selective logging, firewood collection and other forms of resource extraction, or fires (*see* Chapter 30, *The ecology of logged forests*, and Chapter 27, *Tropical deforestation, forest degradation, and the role of REDD+ in climate change mitigation*).

Rates of recovery of old-growth forest attributes during secondary succession after agricultural or pasture use vary between attributes, being fastest for soil properties (< 10 years) and for plant functioning (< 25 years), intermediate for structure and species diversity (25–60 years), and slowest for biomass and species composition (>120 years) (Poorter et al., 2021; Chazdon et al., 2022). Prolonged or intensive cultivation and isolation from forest seed sources can considerably slow recovery, however. Recovery of most forest attributes after logging is much quicker, although, at high logging intensities, up to half of above-ground biomass can be lost and post-logging recovery of carbon stocks can take several decades. Even heavily logged forests, however, are diverse and vibrant ecosystems with amplified energy flows (Malhi et al., 2022), demonstrating that the common description of such forests as 'degraded' is misleading. A recent study estimated that above-ground carbon accumulation by secondary and structurally altered humid tropical forests between 1984 and 2018 was enough to offset around a quarter of emissions from deforestation over the same period (Heinrich et al., 2023). However, for logged forests, Mills et al. (2023) show that large carbon losses from soil and dead wood continue for at least a decade after logging and more than outweigh the gains from increasing above-ground biomass, suggesting that this estimate of the total carbon offset is probably too high.

Plot-based studies

It is no exaggeration to say that most of what we know about the ecology of tropical forests comes from studies of long-term forest plots in which trees are mapped, measured, and identified, and then re-censused at intervals in order to record growth, mortality, and recruitment. More than 11,000 such plots currently exist across the tropics (Davies et al., 2021). Many thousands of additional plots have been mapped, measured, and identified, but not re-censused. Many plots are parts of networks with standardised protocols to facilitate comparisons between sites (e.g. RAINFOR, AfriTRON, and T-FORCES, which share the same field and analytical protocols, and ForestGEO). Small plots—often 1 ha, with all trees ≥10 cm diameter at breast height (dbh, 1.3 m) measured—can be established relatively rapidly (c. 136 skilled person days in the field, herbarium, and lab) and have been used to study the relationships between forest structure, biomass, biodiversity, and floristic composition and environmental variables, and to measure changes over time (ForestPlots.net et al., 2021). Most of these studies have leveraged the broad geographical and ecological spread of these plots. For example, inventory data from 243 plots distributed across the Brazilian Amazon shows that the highest diversities of trees are in the wetter northwest, but the highest diversities of very large trees (dbh > 70 cm) are in the northeast, where wind speeds and lightning frequency are lower (de Lima et al., 2023).

Much larger plots (15–50 ha), with every stem ≥1 cm diameter measured every 5 years, are a massive effort to establish—typically two or more years' work by a large team—but the large size ensures rare species are adequately represented and the inclusion of small stems (where mortality is usually highest) is important for understanding the mechanisms and processes responsible for species coexistence and diversity (Davies et al., 2021). There are fewer of these tropical 'forest dynamics plots'—32 in the ForestGEO network in 2023—and most studies using them have focussed on a single plot or compared plots in the same region.

A major advance in our understanding of tropical forests from the first large forest plots was the discovery that species assemblages at the plot scale are highly structured spatially, in relation to topography, soil nutrients, moisture, and disturbance history, while the role of biotic factors has become clearer from longer term studies (Davies et al., 2021). In addition to the basic measurements made in all plots, many plots in the ForestGEO network include seed traps and seedling plots in order to fill the gap between adult trees and 1-cm dbh saplings. These have revealed the role of seed dispersal limitation in creating the clumped distributions for many tree species. This clumping, in turn, increases intraspecific competition and the impacts of pathogens and herbivores. Interactions with natural enemies can result in conspecific negative density dependence for growth and survival (i.e. growth and survival are lower in plants surrounded by a high density of conspecifics), potentially allowing more species to coexist. In the 50-ha plot on Barro Colorado Island in Panama, analysis of 20 years of seedling census data showed that the impact of conspecific seedling density in the first year of growth was more negative in wetter years, leading to an increase in seedling diversity, which persisted for at least 15 years (Lebrija-Trejos et al., 2023). Many plots have also implemented more frequent measurements of mortality and growth (using dendrometer bands) so that short-term responses to droughts and storms can be assessed.

While studies in the few large permanent plots have made major contributions to our understanding of the mechanisms responsible for tree species diversity at the local scale, analyses of the much larger set of Amazonian forest inventory plots of all sizes have identified an opposite pattern—the 'hyperdominance' of a few hundred species that account for >50% of all individuals in each size class (Draper et al., 2021). While this suggests that it may be possible to understand the functioning of tropical forests without being overwhelmed by their diversity, there is evidence that at least some of these hyperdominants are species complexes, so dominance is less and diversity even greater than the raw plot data suggest (Bacon et al., 2022).

Plot-based studies in the tropics have a bias to wetter, old-growth, lowland forests, reflecting the importance of these forests for both biodiversity and carbon, and to areas accessible from major local and international research institutions. Dry and montane forests are underrepresented, as are secondary and logged forests, although together these account for a majority of the current forest cover. Existing protocols for small plots need to be adjusted for the smaller trees in these forests (e.g. Moonlight et al., 2021), while expansion of the large-plot network in the tropics is constrained by funding. There are also geographical gaps in plot coverage in remote areas of the Amazon, the central Congo Basin, Myanmar, and New Guinea (ForestPlots.net et al., 2021).

Other field-based approaches to tropical forest ecology

A major limitation of plot-based studies is that long-term studies of demographic processes within a plot are not compatible with major construction work (e.g. cranes or towers) or large-scale manipulative experiments within or adjacent to the plot. Canopy cranes can provide total access to the forest canopy over a limited area (0.8–2 ha) and, although very expensive and thus very few in number, provide data (e.g. on tree ecophysiology, epiphytes, canopy fauna, and ecological interactions) that cannot be collected from the ground or by tree climbing (Nakamura et al., 2017). Eddy flux covariance towers measure fluxes of carbon, water, and energy between the forest and atmosphere that cannot be directly measured any other way (e.g. Fu et al., 2018). Experiments on a wide variety of spatial and temporal scales have provided crucial—and sometimes unexpected—insights into the mechanisms behind key ecosystem processes. Examples include controlled fertilization experiments (e.g. Wright, 2022), litter manipulation experiments (Sayer et al., 2020), liana removal experiments

(Estrada-Villegas et al., 2022), soil warming experiments (Nottingham et al., 2020), and throughfall exclusion experiments to simulate drought (e.g. Bittencourt et al., 2020). The AmazonFACE programme will soon expose an old-growth forest in Brazil to future CO_2 concentrations using free-air carbon dioxide enrichment techniques (https://amazonface.unicamp.br/en/). We still lack a whole ecosystem warming experiment in a tropical forest, although there have now been many in situ warming experiments on leaves or branches, and one on a tropical forest understorey (Carter et al., 2020).

Scaling up, from plots to planet

Long-term forests plots include several thousand hectares of forest in the tropics, but this is a tiny and biased fraction of the total remaining forest area. The same limitations apply to all other field-based studies in tropical forests. Sustainable exploitation, efficient conservation planning, and the construction and calibration of the Earth System Models used to predict future climates all require the extrapolation of our field-based knowledge of tropical forests to much larger areas. This can potentially be achieved by the integration of information from plots and canopy cranes, eddy flux covariance studies, and field experiments with the data from airborne remote sensing and the continuous, wall-to-wall coverage from existing and planned satellite-based sensors (Chave et al., 2019; Davies et al., 2021; ForestPlots.net et al., 2021; Ordway et al., 2022). There are currently gaps in both the ground data and the capabilities of satellite sensors, but the additional investment needed is not very large by 'big science' standards and would massively increase both our understanding of tropical forests and their roles in the Earth System, and how these will change over the coming decades. While it may make sense to concentrate major ground investments on a few 'supersites' with existing data, ongoing funding, and local commitment (Chave et al., 2019), the thousands of smaller plots and plotless studies—existing and future—will continue to have an essential role in ensuring that the diversity of tropical forests is fully represented.

Deforestation

Despite the importance of tropical forests for biodiversity, carbon, and livelihoods, and despite international agreements intended to slow or halt it, deforestation continues (*see* Chapter 27, *Tropical deforestation, forest degradation, and the role of REDD+ in climate change mitigation*). Most sustained forest loss is associated with the expansion of agriculture—pastures and cropland— although a substantial fraction of the total area cleared does not result in agricultural production and appears to be a result of land speculation, lack of economic viability, or fires spreading from adjacent agriculture (Pendrill et al., 2022). Pasture expansion accounts for around half the deforestation that does result in production, mostly in South America. Soy, also concentrated in South America, and oil palm, largely in Southeast Asia, together account for at least 20%. Rubber, cocoa, coffee, rice, maize, and cassava account for much of the rest. Although many of these are important export crops, domestic demand in producer countries accounts for most deforestation, except in Southeast Asia. Tropical montane forests, previously protected by remoteness, high elevations, and steep slopes, have suffered from accelerating deforestation in the 21st century, particularly in Southeast Asia (He et al., 2023). Tropical forests have also become increasingly fragmented over the last 20 years, particularly in the southeastern Amazon, the Congo Basin, and Indochina (Ma et al., 2023). Logging often subsidizes agricultural expansion, and while urbanization, mining, and hydropower dams are relatively minor direct drivers on a pantropical scale, they are locally devastating. Roads and other infrastructure financed by commercial agriculture and industry facilitate deforestation by other agents.

Summary

This chapter introduces the diversity of tropical forests and their environments, and some of the ongoing research. Additional aspects of tropical forest biology are covered in many other chapters in this book, including the effects of strong winds (*see* Chapter 9, *Ecological effects of strong winds on forests*), the roles of lianas (*see* Chapter 13, *The ecology of lianas and their increasing influence in tropical forests*), the occurrence of monodominant forests (Chapter 12, *Monodominance in tropical lowland forests*), vascular epiphytes (*see* Chapter 14, *Vascular epiphytes in forest ecosystems*), lichens (*see* Chapter 17, *Lichens in forest ecosystems*), mammals (*see* Chapter 18, *Mammals in forest ecosystems*), birds (*see* Chapter 19, *Ecology and conservation of forest birds*), amphibians and reptiles (*see* Chapter 20, *Amphibians and reptiles in forest ecosystems*), forest carbon budgets (*see* Chapter 38, *Forest carbon budgets and climate change*), and the impacts of fires (*see* Chapter 34, *Fire and climate: using the past to predict the future*), logging (*see* Chapter 30, *The ecology of logged forests*), hunting (*see* Chapter 42, *Wild meat hunting in tropical forests*), and climate change (*see* Chapter 35, Ecological consequences of droughts in forests; Chapter 37, *Plant movements in response to rapid climate change*).

As these chapters show, research in tropical forests still suffers from large geographical gaps and pervasive taxonomic biases, reflecting largely ease of access from research institutions and the expertise available in them (e.g., Carvalho et al., 2023). Expeditions can fill gaps in sampling, but sustained research requires well-supported local institutions. Tropical forest ecology has a long tradition of 'helicopter research', with little or no involvement of local people and often, as a result, little or no relevance to them. This is now changing, but we still have a long way to go.

References

Adeney, J.M., Christensen, N.L., Vincentini, A. and Cohn-Haft, M. (2016) 'White-sand Ecosystems in Amazonia', *Biotropica*, vol 48, pp. 7–23.

Bacon, C.D., Hill, A., ter Steege, H., Antonelli, A. and Damasco, G. (2022) 'The impact of species complexes on tree abundance patterns in Amazonia', *American Journal of Botany*, vol 109, pp. 1525–1528.

Bittencourt, P.R., Oliveira, R.S., da Costa, A.C., Giles, A.L., Coughlin, I., Costa, P.B. et al. (2020) 'Amazonia trees have limited capacity to acclimate plant hydraulic properties in response to long-term drought'. *Global Change Biology*, vol 26, pp. 3569–3584.

Carter, K.R., Wood, T.E., Reed, S.C., Schwartz, E.C., Reinsel, M.B., Yang, X. and Cavaleri, M.A. (2020) 'Photosynthetic and respiratory acclimation of understory shrubs in response to *in situ* experimental warming of a wet tropical forest', *Frontiers in Forests and Global Change*, vol 3, 576320.

Carvalho, R.L. et al. (2023) 'Pervasive gaps in Amazonian ecological research', *Current Biology*, vol 33, pp. 3495–3504.

Charles-Dominique, T., Midgley, G.F., Tomlinson, K.W. and Bond, W.J. (2018) 'Steal the light: shade vs fire adapted vegetation in forest–savanna mosaics', *New Phytologist*, vol 218, pp. 1419–1429.

Chave, J. et al. (2019) 'Ground data are essential for biomass remote sensing missions', *Surveys in Geophysics*, vol 40, pp. 863–880.

Chazdon, R.L., Norden, N., Colwell, R.K. and Chao A. (2022) 'Monitoring recovery of tree diversity during tropical forest restoration: lessons from long-term trajectories of natural regeneration', *Philosophical Transactions of the Royal Society B*, vol 378, 20210069.

Corlett, R.T. (1994) 'What is secondary forest?', *Journal of Tropical Ecology*, vol 10, pp. 445–447.
Corlett, R.T. (2015) Classifying tropical forests. In: *Tropical Forestry Handbook* (eds. M. Köhl & L. Pancel), Springer, Berlin, Germany, pp. 1–9.
Corlett, R.T. (2016) 'Plant diversity in a changing world: status, trends and conservation needs', *Plant Diversity*, vol 38, pp. 10–16.
Corlett, R.T. and Primack, R.B. (2011) *Tropical Rainforests: An Ecological and Biogeographical Comparison*, second edition. Wiley-Blackwell, Oxford.
Corlett, R.T. and Tomlinson, K.W. (2020) 'Climate change and edaphic specialists: Irresistible force meets immovable object?', *Trends in Ecology & Evolution*, vol 35, pp. 367–376.
Coronado, E.N.H. et al. (2021) 'Intensive field sampling increases the known extent of carbon-rich Amazonian peatland pole forests', *Environmental Research Letters*, vol 16, 074048.
Costa, F.R.C., Schietti, J., Stark, S.C. and Smith, M.N. (2023) 'The other side of tropical forest drought: do shallow water table regions of Amazonia act as large-scale hydrological refugia from drought?', *New Phytologist*, vol 237, pp. 714–733.
Davies, S.J. et al. (2021) 'ForestGEO: understanding forest diversity and dynamics through a global observatory network', *Biological Conservation*, vol 253, 108907.
de la Cruz-Amo, L., Bañares-de-Dios, G., Cala, V., Granzow-de la Cerda, I., Espinosa, C.I., Ledo, A., Salinas, N., Macía, M.J. and Cayuela, L. (2020) 'Trade-offs among aboveground, belowground, and soil organic carbon stocks along altitudinal gradients in Andean tropical montane forests', *Frontiers in Plant Science*, vol 11, p. 106.
de Lima, R.B. et al. (2023) 'Giants of the Amazon: How does environmental variation drive diversity patterns of large trees?', *Global Change Biology*, vol 29, pp. 4861–4879
Deng, L., Pan, Y., Wang, Y., Chen, H., Yuan. K., Chen, S., Lu, D., Lu, Y., Mokhtar, S.S., Rahman, T.A., Hoh, B.-P. and Shu, S. (2022) 'Genetic connections and convergent evolution of tropical indigenous peoples in Asia', *Molecular Biology and Evolution*, vol 39, msab361.
Draper, F.C. et al. (2021) 'Amazon tree dominance across forest strata', *Nature Ecology & Evolution*, vol 5, pp 757–767.
DRYFLOR (2016) 'Plant diversity patterns in neotropical dry forests and their conservation implications', *Science*, vol 353, pp. 1383–1387.
Estrada-Villegas, S., Narvaez, S.S.P., Sanchez, A. and Schnitzer, S.A. (2022) 'Lianas significantly reduce tree performance and biomass accumulation across tropical forests: a global meta-analysis', *Frontiers in Forests and Global Change*, vol 4, 812066.
FAO (2015) *World reference base for soil resources 2014, update 2015.* IUSS Working Group WRB, Food and Agriculture Organization of the United Nations, Rome.
ForestPlots.net et al. (2021) 'Taking the pulse of Earth's tropical forests using networks of highly distributed plots', *Biological Conservation*, vol 260, 108849.
Fu, Z., Gerken, T., Bromley, G., Araújo, A., Bonal, D., Burban, B., Ficklin, D., Fuentes, J.D., Goulden, M., Hirano, T., Kosugi, Y., Liddell, M., Nicolini, G., Niu, S., Roupsard, O., Stefani, P., Mi, C., Tofte, Z., Xiao, J., Valentini, R., Wolf, S. and Stoy, P.C. (2018) 'The surface-atmosphere exchange of carbon dioxide in tropical rainforests: Sensitivity to environmental drivers and flux measurement methodology', *Agricultural and Forest Meteorology*, vol 263, pp. 292–307.
Ghazoul, J. (2016) *Dipterocarp Biology, Ecology, and Conservation.* Oxford University Press, Oxford.
He, X., Ziegler, A.D., Elsen, P.R., Feng, Y., Baker, J.C.A., Liang, S., Holden, J., Spracklen, D.V. and Zeng, Z. (2023) 'Accelerating global mountain forest loss threatens biodiversity hotspots', *One Earth*, vol 6, pp. 303–315.
Heinrich, V.H.A., Vancutsem, C., Dalagnol, R., Rosan, T.M., Fawcett, D., Silva-Junior, C.H.L, Cassol, H.L.G., Achard, F., Jucker, T., Silva, C.A., House, J., Sitch, S., Hales, T.C. and Aragão, L.E.O.C. (2023) 'The carbon sink of secondary and degraded humid tropical forests', *Nature*, vol 615, pp. 436–442.
Junk, W.J., Piedade, M.T.F., Schöngart, J., Cohn-Haft, M., Adeney, J.M. and Wittmann, F. (2011) 'A classification of major naturally-occurring Amazonian lowland wetlands', *Wetlands*, vol 31, pp. 623–640.
Kass, J.M. et al. (2022) 'The global distribution of known and undiscovered ant biodiversity', *Science Advances*, vol 8, eabp9908.
Keith, D.A. et al. (2022) 'A function-based typology for Earth's ecosystems', *Nature*, vol 610, pp. 513–518.

Körner, C. (2021) 'The cold range limit of trees', *Trends in Ecology & Evolution*, vol 36, pp. 979–989.

Lebrija-Trejos, E., Hernández, A. and Wright, S.J. (2023) 'Effects of moisture and density-dependent interactions on tropical tree diversity', *Nature*, vol 615, pp. 100–104.

Li, W., Guo, W.Y., Pasgaard, M., Niu, Z., Wang, L., Chen, F., Qin, Y. and Svenning, J.C. (2023) 'Human fingerprint on structural density of forests globally,' *Nature Sustainability*, vol 6, pp. 368–379.

Ma, J., Li, J., Wu, W. and Liu, J. (2023) 'Global forest fragmentation change from 2000 to 2020', *Nature Communications*, vol 14, 3752.

Malhi, Y., Riutta, T., Wearn, O.R., Deere, N.J., Mitchell, S.L., Bernard, H., Majalap, N., Nilus, R., Davies, Z.G., Ewers, R.M. and Struebig, M.J. (2022) 'Logged tropical forests have amplified and diverse ecosystem energetics', *Nature*, vol 612, pp. 707–713.

Mills, M.B., Malhi, Y., Ewers, R.M., Kho, L.K., Teh, Y.A., Both, S., Burslem, F.R.P., Majalap, N., Nilus, R., Huasco, W.H., Cruz, R., Pillco, M.M., Turner, E.C., Reynolds, G. and Riutta, T. (2023) 'Tropical forests post-logging are a persistent net carbon source to the atmosphere', *PNAS*, vol 120, e2214462120.

Moonlight, P.W. et al. (2021) 'Expanding tropical forest monitoring into dry forests: The DRYFLOR protocol for permanent plots', *Plants, People, Planet*, vol 3, pp. 295–300.

Muller-Landau, H.C., Cushman, K.C., Arroyo, E.E., Cano, I.M., Anderson-Teixeira, K.J. and Backiel, B. (2021) 'Patterns and mechanisms of spatial variation in tropical forest productivity, woody residence time, and biomass', *New Phytologist*, vol 229, pp. 3065–3087.

Nakamura, A., Kitching, R.L., Cao, M., Creedy, T.J., Fayle, T.M., Freiberg, M., et al. (2017) 'Forests and their canopies: achievements and horizons in canopy science', *Trends in Ecology and Evolution*, vol 32, pp. 438–451.

Newton, P., Kinzer, A.T., Miller, D.C., Oldekop, J.A. and Agrawal, A. (2020) 'The number and spatial distribution of forest-proximate people globally', *One Earth*, vol 3, pp. 363–370.

Nottingham, A.T., Meir, P., Velasquez, E. and Turner, B.L. (2020) 'Soil carbon loss by experimental warming in a tropical forest', *Nature*, vol 484, pp. 234–237.

Ordway, E.M, Asner, G.P, Burslem, D.F.R.P., Lewis, S.L., Nilus, R., Martin, R.E., O'Brien, M.J., Phillips, O.L., Qie, L., Vaughn, N.R. and Moorcroft, P.R. (2022) 'Mapping tropical forest functional variation at satellite remote sensing resolutions depends on key traits', *Communications Earth & Environment*, vol 3, 247.

Padilla-Iglesias, C., Atmore, L., Olivero, J., Lupo, K., Manica, A., Isaza, E.A., Vinicius, L. and Migliano, A.B. (2022) 'Population interconnectivity over the past 120,000 years explains distribution and diversity of Central African hunter-gatherers', *Proceedings of the National Academy of Sciences*, vol 119, e2113936119.

Page, S., Mishra, S., Agus, F., Anshari, G., Dargie, G., Evers, S., Jauhiainen, J., Jaya, A., Jovani-Sancho, A.J., Laurén, A., Sjögersten, S., Suspense, I.A., Wijedasa, L.S. and Evans, C.D. (2022) 'Anthropogenic impacts on lowland tropical peatland biogeochemistry', *Nature Reviews Earth & Environment*, vol 3, pp. 426–443.

Pendrill, F., Gardner, T.A., Meyfroidt, P., Persson, U.M. et al. (2022) 'Disentangling the numbers behind agriculture-driven tropical deforestation', *Science*, vol 377, eabm9267.

Pillay, R., Venter, M., Aragon-Osejo, J., González-del-Pliego, P., Hansen, A.J., Watson, J.E.M and Venter, O. (2022) 'Tropical forests are home to over half of the world's vertebrate species', *Frontiers in Ecology and the Environment*, vol 20, pp. 10–15.

Poorter, L. et al. (2021) 'Multidimensional tropical forest recovery', *Science*, vol 374, pp. 1370–1376.

Sayer, E.J., Rodtassana, C., Sheldrake, M., Bréchet, L.M., Ashford, O.S., Lopez-Sangil, L., Kerdraon-Byrne, D., Castro, B., Turner, B.L., Wright, S.J. and Tanner, E.V.J. (2020) 'Revisiting nutrient cycling by litterfall—Insights from 15 years of litter manipulation in old-growth lowland tropical forest', *Advances in Ecological Research*, Vol 62, 173–223.

Slik, J.W.F., Franklin, J., Arroyo-Rodríguez, V., et al. (2018) 'Phylogenetic classification of the world's tropical forests', *Proceedings of the National Academy of Sciences of the USA*, vol 115, pp. 1837–1842.

Smith, C., Baker, J.C.A. and Spracklen, D.V. (2023) 'Tropical deforestation causes large reductions in observed precipitation', *Nature*, vol 615, pp. 270–275.

Soto-Navarro, C. et al. (2020) 'Mapping co-benefits for carbon storage and biodiversity to inform conservation policy and action', *Philosophical Transactions of the Royal Society B*, vol 375, 20190128.

USDA (2006) *Keys to Soil Taxonomy,* 10th edition. United States Department of Agriculture, Washington, D.C.

van der Ent, A., Cardace, D., Tibbett, M. and Echevarria, G. (2018) 'Ecological implications of pedogenesis and geochemistry of ultramafic soils in Kinabalu Park (Malaysia)', *Catena*, vol 160, pp. 154–169.

Wei, S., Wang, X. and Xie, Q. (2023) 'Strengthening effect of Maritime Continent deforestation on the precipitation decline over southern China during late winter and early spring', *Climate Dynamics*, vol 60, pp.1173–1185.

Wright, S.J. (2022) 'Nutrients limit carbon capture by tropical forests', *Nature*, vol 608, pp. 476–477.

Yang, Q., Bader, M.Y., Feng, G., Li, J., Zhang, D. and Long, W. (2023) 'Mapping species assemblages of tropical forests at different hierarchical levels based on multivariate regression trees', *Forest Ecosystems*, vol 10, 100120.

6
TEMPERATE AND BOREAL MOUNTAIN FORESTS

Christopher Carcaillet and Peter Z. Fulé

Introduction

Mountain forests are present in all continents under most climates, except for polar regions (Figure 6.1). Mountain areas cover 27% of the Earth's surface, and 40% of their surfaces are covered by forests. In total, mountain forests represent 23%, i.e., 938 million hectares (Mha) of global forests (FAO 2022). Because most mountain regions are partly covered by forests, they are crucial for the global economy (timber, fuelwood, etc.) and support social and cultural functions, including for religion or spiritism, hunting and gathering, tourism, or recreation. Mountain forests need to be conserved for their socio-economical role, and their ecological services (water and soil protection, carbon cycle, etc.). Additionally, mountain forests are home to great biodiversity and endemism for plants and animals (Rivers et al. 2019), while microorganisms are still largely unknown. In mountain forests, specifically, the elevation and slope are important drivers: these two factors superimpose on other drivers common in flatter, lower elevations that determine forest biodiversity and productivity.

The present chapter is devoted to temperate and boreal mountain forests, including Mediterranean ecosystems which are warm temperate ecosystems. In terms of geography, temperate and boreal mountain forests are mainly located in the northern hemisphere (Europe, North America, Asia, and northwestern North Africa), although some are located in southern South America, southeastern Australia, New Zealand, and southern Austral Africa (Figure 6.2). Most temperate and cold mountain forests are in Europe and Russia (46% of global mountain forests), North and Central America (16%), and Asia (16%). Southern temperate mountain forests, that is, Oceania, South America, and Africa account for only 4% of global mountain forests. In Russia and Europe, which is the continent hosting the most of mountain forests (Figure 6.2), they are distributed such that 33 Mha (i.e., 12% of the total European) of warm temperate, mostly Mediterranean type, 61 Mha of temperate (22%), and 184 Mha boreal (66%) (Muhtoo 2002).

Temperate and boreal mountain forests are distributed along slopes. The physical processes (temperature, precipitation, solar irradiance and spectrum, season length) change with the increasing elevation, the slope angle and aspect, and the size and geographical structure of mountains. However, because physical processes directly act on the biota along the elevation for instance by controlling the composition of communities and their primary productivity,

 DOI: 10.4324/9781003324072-7

Figure 6.1 Examples of boreal and temperate mountain forests, ordered from the North to the South. (A) Scots pine (*Pinus sylvestris*) forests with domestic reindeers (*Rangifer tarandus*), boreal mountain, Sweden (500 m a.s.l.). (B) Cold temperate forest with eastern hemlock (*Tsuga canadensis*), sugar maple (*Acer saccharum*), red maple (*A. rubrum*), balsam fir (*Abies balsamea*), yellow birch (*Betula alleghaniensis*), white pine (*Pinus strobus* in the northern Appalachian Mountain, Quebec, eastern Canada (550 m a.s.l.). (C) Rain mountain temperate forest composed of Pacific silver fir (*Abies amabilis*), western hemlock (*Tsuga heterophylla*), Douglas fir (*Pseudotsuga menziesii*) and Sitka spruce (*Picea sitchensis*) in British Columbia, western Canada (900 m a.s.l.). (D) Mediterranean subhumid mountain forest of black pine (*Pinus nigra*) and European birch (*Fagus sylvatica*) in Corsica, France (900 m a.s.l.). (E) Aerial view (Image ©2022 Airbus) of warm-dry temperate forests dominated by Brant's oak (*Quercus brantii*); the light tree-cover (dark pixels) is well illustrated and the visible ground (light matrix) typical of steppe forests of the Zagros Mountains, Iran (1700–2000 m a.s.l.). (F) degraded woodland, few years after fire, with *Leucadendron* sp. trees and fynbos understory, Du Toitskloof mountain, Western Cape, South Africa (800 m a.s.l.). Photography: ©Christopher Carcaillet.

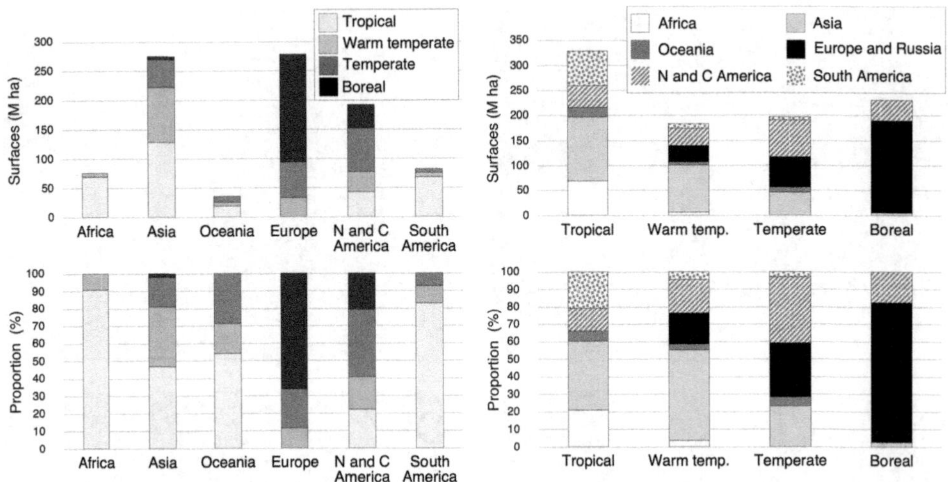

Figure 6.2 Distribution of mountain forests per continent (right side) or per vegetation type in area (Mha left side) and in proportion (%) (Sources: Muhto 2002 and FAO 2022; note forests from the Russian Federation are included in 'Europe' in FAO statistics.

secondary biotic mechanisms that alter the biota (disturbances, biotic interactions) are also modified along these slope gradients, thus producing feedbacks specific to mountain ecosystems. For instance, snow avalanches carve the vegetation landscape vertically along slopes that are organized in horizontal vegetation belts, thus allowing transfer of organisms and matter between belts, and changing the pattern of spread of disturbances through modifications of biomass load and fuel quality (Suffling 1993).

The following text summarizes, in the first section, the environmental drivers that are specific to mountain forests. Then, the second section synthesises their main biotic pattern and processes. Finally, the last section develops the societal issues of mountain forests notably in terms of land uses or their abandonment, ecosystem services and conservation.

Macro-ecology of forest mountains

To sum up the following section, environmental patterns (temperature, precipitation, partial pressure of gases, radiation, light) and their gradients are complex macroecological drivers in mountains, with specific interactions driven largely by the elevation and varying according to latitudes and the proximity to seas at macro-scales (biogeographical pattern). At mesoscale, angle of the slopes and aspect control biotic landscape patterns. At microscale, soil conditions and topography are important drivers. All these main features interact, generating specific mountain landscape gradients that directly control the physiology of plants and animals, the distribution of species, the species composition of communities, the pattern and functioning of food webs, the biogeochemical cycles, the genetic diversity and phylogeny of organisms of mountain forests, and the disturbance regimes. Mountain forest landscapes are complex, with strong gradients from small- to macro-scales. These physiographical parameters constitute the basis of the diversity and productivity of mountain forests worldwide.

Elevation, the chief driver

Elevation is the main environmental driver that controls the composition, structure, and function of mountain ecosystems. Globally, temperatures decrease with increasing elevation (Figure 6.3A). The decrease of about 0.65°C/100 m^{-1} is a global rule whatever the geographic location of mountains. However, this global climatic constant varies with seasons: in spring, lower elevations warm-up more rapidly than the higher, which can affect biomass productivity (Figure 6.3A) Other factors besides elevation act on the temperatures, notably slope and aspect.

Elevation also regulates patterns: some increase with elevation (precipitation and moisture, solar irradiance), others decrease (temperature, available oxygen and carbon dioxide), and some change their proportion (light spectrum). All these features control life distribution and functionalities. For instance, temperature is the chief process acting on the distribution of upper treelines, forest limits, and vegetation belts around the world (Körner et al. 2011) through physiological processes controlled by temperature and mitigated by water, carbon, and soil nutrients (Tranquillini 1979). For Körner (1998), the upper limits of upright trees are determined by the critical mean air temperatures of the growing season between 5.5°C and 7.5°C in the air but also in the soil.

Solar irradiance: influence of slope aspect and latitude

Solar irradiance contributes to photosynthesis and to ground temperatures. Solar irradiance is a complex process in mountainous landscapes but varies in strength and complexity when the slope is steep, and with the aspect, which can vary a lot at small scale with rugged topography. For example, for a middle latitude (45°) mountain on a 22° slope (=25%) exposed to the sun at the summer solstice, the sunlight reaches the vegetation at an angle perpendicular to the slope (90°=100%) thus providing the maximum of potential radiation (Figure 6.3B). Conversely, on the opposite slope (to the north in the northern hemisphere and to the south in the southern hemisphere), at the summer solstice, the sunlight reaches to the slope with an angle of 45° (Figure 6.3B). At this latitude, a sun-facing slope captures above 5000 $Mj.m^{-2}$ per year, while the shaded slope captures less than 3500 $Mj.m^{-2}$ per year (Piedallu and Gégout, 2007).

The flux of solar radiation theoretically increases with the elevation due to less atmospheric turbidity and less cloudiness (Körner 2003). However, the effect of cloudiness varies with the geographic location of mountains, depending on their exposure to humid airmasses (more cloudiness, less radiation). This means that mountain slopes exposed to humid air masses supplying clouds present less ground heating than slopes protected from these air masses, everything else being equal (Piedallu and Gégout, 2007).

Tree cover mitigates the solar irradiance reaching the ground. A tree cover with very high density of foliage (or "leaf area index") intercepts the light that will be used for photosynthesis without heating the ground. The leaf area index depends on the tree density but also on density of leaves per tree, and on leaf form and size that are species-dependent (Michalet et al. 2023). For instance, pine (*Pinus*) or larch (*Larix*), two common tree genera in mountains, have very low leaf area indices compared to beech (*Fagus*) or maple (*Acer*) two other common tree genera in mountains. Everything else being equal (solar irradiance, tree density), an understory of pine or larch will be warmer than an understory of oak or beech. In the latter, because the understory is colder, the evaporation will be lower and the

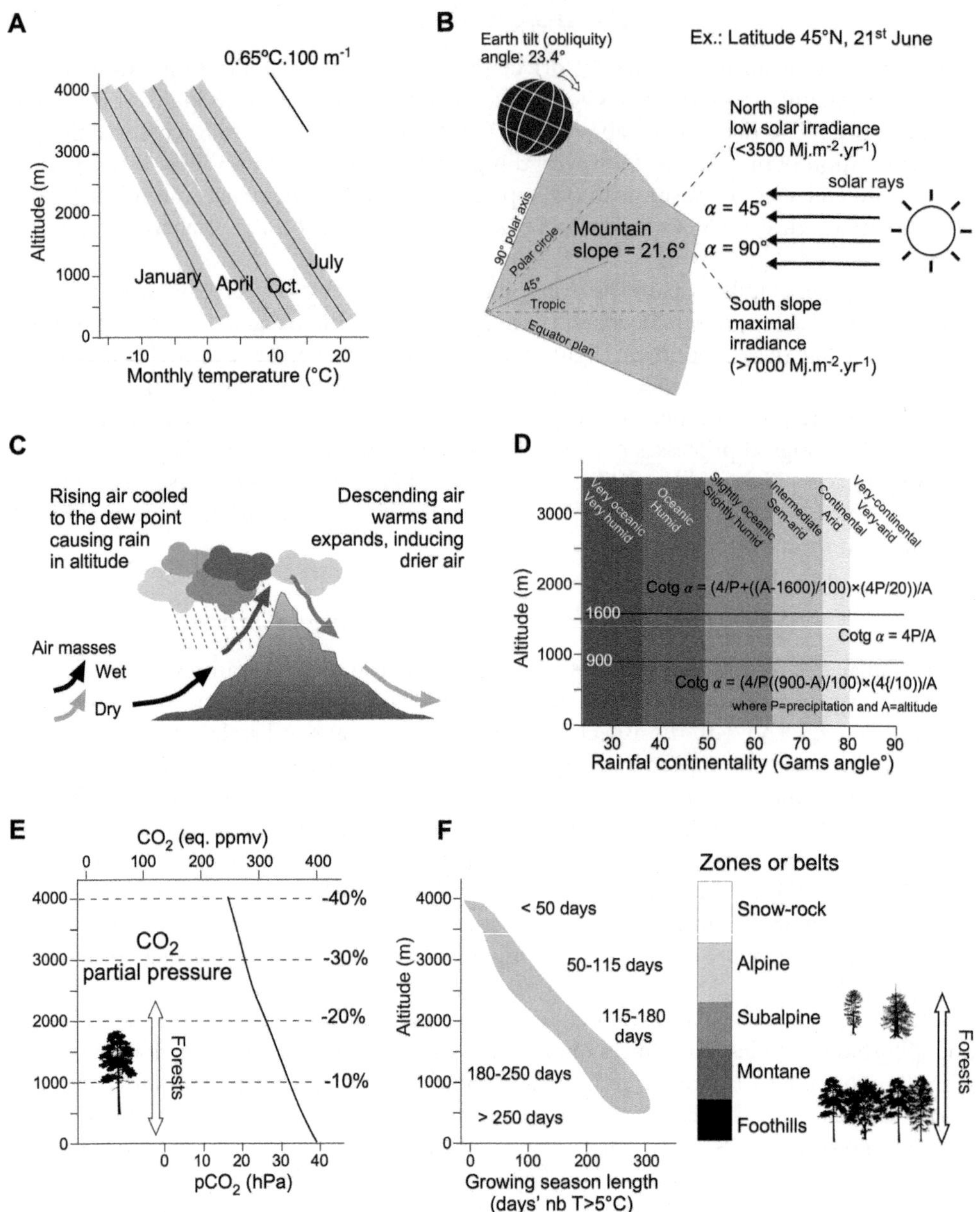

Figure 6.3 Main environmental characteristics that organize mountain forest ecosystems, distinct from the driving characteristics of plain forests. (A) Temperatures decrease with elevation; the global average of 0.65°C per 100 m varies according to the seasons as illustrated by the graph taken from the weather stations in European Alps (mid-latitude) showing for example that the low elevations warm up more quickly in spring than the high ones. (B) Independently of the air temperature, the ground temperature depends on the solar irradiance, which varies during the year and the latitude, culminating with the summer

solstice, that is, June 21 in the northern hemisphere and September 21 in the southern hemisphere; however, in the mountains thanks to the sun-facing slope, the solar irradiance increases and, conversely, decreases on shaded slope; this phenomenon explains the variability in the growing season length according to the slopes exposed or not to the sun. (C) Increase in precipitation with elevation and rain-shadow effect both structure the water balance between slopes. (D) This combined mechanism results in a pattern of precipitation continentality well assessed by the Gams index (Michalet et al. 2021), which measures the angle shaped by precipitation values plotted as a function of elevation (cf. equations), a pattern which is mitigated by the rain shadow effect; this results in dry to arid areas in valleys protected from oceanic air masses by high mountain massifs; the higher and wider the peripheral mountain, the drier the internal valleys. (E) Decrease in partial pressure of carbon dioxide (pCO_2, hPa) along the altitudinal gradient and its equivalent concentration (ppmv); this pCO_2 depends on the atmospheric pressure and the CO_2 concentration, which has been set here at 400 ppmv (global value for the year 2014). The pattern is the same for the partial pressure of oxygen (pO_2) essential for cellular respiration in all organisms but with different basal and constant O_2 concentration (21%), while CO_2 is essential for photosynthesis with basal values that vary with time. (F) The length of the growing season (a mid-latitude and temperate example) varies with air temperature as a function of elevation (A), solar irradiance as a function of latitude and slope degrees (B), and precipitation balance (C, D); low-elevation areas with sandy and shallow soils on sun-exposed slopes warm-up earlier in spring and cool-down later in autumn, giving about 2 months (60–70 days) difference between slopes, but with higher water deficit due to higher evapotranspiration, especially if forests on shaded slopes growth on deep, clay-organic soils. All these characteristics shape the altitudinal gradient of vegetation belts, species distribution and pattern of productivity.

moisture index higher than in a pine or larch forest, at equal precipitation, slope, and soil. The feedbacks of land cover to climatic drivers are important to explain macro- or micro-scale differences in mountains ecosystems.

Finally, the composition of the solar irradiance changes with the elevation, notably with more ultraviolet (UV) rays in higher mountains than in the lower ones. The UV is an energetic wavelength in the light spectrum, which cannot be used by plants. The UV radiation is intercepted by the ozone molecules that are more concentrated in higher elevations than in the lower ones, resulting a UV increase by 4% per 300 m in elevation. However, as an adaptation to higher flux of UV radiation, plants augment the thickness of their leaf epidermis, notably the cuticles. It has been suggested that plants could acclimate well to UV, notably by increasing the concentration of flavonoids and other polyphenols, which are responsible for UV absorbance (Körner 2003). However, if UV would not be a functional parameter controlling the species composition of communities due to the plasticity of species and their acclimation potential, UV would mostly be an evolutionary process through the potential mutagenic effects, which might partially explain the high plant diversity of mountains.

Precipitation linked to elevation and rain-shadow effect

The main general pattern of precipitation is an increase with increasing elevation. This is mainly due to the condensation of air moisture in the form of clouds under lower temperatures (Figure 6.3C). The top of mountains is therefore much cloudier compared to the lower elevations. This phenomenon of cloudiness occurs generally in the late day when moisture of valleys has been evaporated by hot temperatures and solar irradiance of the mid-day, explaining why precipitation often occurs late in the day during the growing season.

The asymmetrical windward-to-leeward gradient of mountains controls the precipitation, with a shelter effect called "rain-shadow effect" (Figure 6.3C). In large mountain ranges — for example, American cordillera, Andes, Alps, Caucasus, Zagros, or Himalaya—the inner valleys are sheltered from all rain provenances whatever the dominant wind. This pattern generally favours the development of moist and eventually rainforests on wind-exposed slopes, and dry continental-type ecosystems, sometimes arid in wind-sheltered areas. It results in a continentality gradient in all the mountains, which combines both elevation and distance to the main wet airmasses. Several climatic indices have been proposed during the early 20th century to illustrate this continentality gradient, which is extremely selective for mountain ecosystems, for instance, the Gams index that is a proxy of moisture (Michalet et al. 2021; Figure 6.3D).

If temperature is the chief global environmental mechanism linked to elevation and latitude, precipitation is thus an important attenuator or amplifier of temperature gradient. Indeed, water plays a key role on forest biodiversity, productivity, and recycling of organic matter. The water budget, which depends on precipitation and temperature, controls basic mechanisms of photosynthesis and decomposition by microorganisms. Then, if temperature is not the main limiting factor such as in warm regions, water availability generally regulates patterns such as the tree distribution or the biomass productivity. For instance, dry mountain regions present lower biomass, net primary productivity, and woody cover. At very high elevation, the relationship of precipitation with temperature is reversed, with low temperatures controlling the decomposition of organic matter and primary productivity.

Partial pressure of carbon dioxide (CO_2) and oxygen (O_2)

CO_2 is a nutrient for plants, needed with water for photosynthesis, and O_2 is needed for respiration (all organisms). The partial pressures of CO_2 and O_2 are mechanically linked to the atmospheric pressure, all of which monotonically decrease with elevation (Figure 6.3E). Therefore, mountain forests experience different atmospheric environments than lowland forests, with direct consequences for available CO_2 and O_2. For example, trees experience an atmospheric pressure of 1013 hPa at sea level, which decreases to 726 hPa (–29%) at 2,600 m in elevation (a typical upper tree limit in mid-latitude mountains) and decreases further to 555 hPa at 4,000 m (e.g., the Himalayas or the Mexican Cordillera) where the pressure is about 45% lower than at sea level. These partial pressures of CO_2 and O_2 are structural features of forest ecosystems by selecting species within communities for primary productivity (CO_2-related photosynthesis) or herbivores, carnivores, and decomposers (through O_2). At high elevation, trees, shrubs, and herbs must be able to photosynthesise with a low partial pressure of CO_2, and respiration must take place with less available O_2, excluding non-adapted species.

Interestingly, if the atmospheric pressure is a constant, the CO_2 concentration varies through time, notably by increasing since the industrial era. Körner (2003) further calculated that a plant at 2,600 m elevation, under the CO_2 concentration of 2000 AD, experienced a partial pressure of CO_2 that would be equivalent to the CO_2 value at sea level 200 years earlier in 1800 AD, that is, at the beginning of the industrial era before the anthropogenic global increase in CO_2. The global greenhouse gas increase changes the photosynthetic experience of mountain forest.

Season length

The length of the growing season of plants is inversely related to elevation (Figure 6.3F), mediated by temperature, precipitation, slope aspect and angle, latitude, and water vapor pressure deficit (i.e., difference between the amount of actual air moisture and the amount of moisture that air can hold at saturation), which is linked to the continentality gradient and elevation. The length of the season decreases with elevation but also on shaded slopes compared to slopes exposed to the sun. As soon as snowpack has disappeared, the balance between temperature and water vapor pressure deficit controls the length, start, and end of the growing season. For example, in warm temperate areas with summer drought but wet spring like in the Mediterranean, the season starts early in spring, possibly at the end of winter but ends quickly as soon as summer drought becomes the limiting factor hampering the primary productivity and decomposition of soil organic matter. However, these regions can have wet autumns whose precipitation offsets the drop in temperature before winter, thus potentially allowing a second growing season as well illustrated in the tree rings. The growing season length is a synthesis of all above drivers, most of them are dependent on elevation (Figure 6.3F).

Soil and site conditions

Soil is originally not linked to elevation. Soils chiefly depend on parent material (bedrocks, moraine deposits). However, the tectonic processes that compress the earth´s crust result in a straightening of stacked and stratified deep rocks that eventually emerge from the ground, creating elevated topography. Therefore, soils at short distance can come from very distinct bedrocks, creating a fine-scale mosaic of soil qualities in terms of granulometry and geochemistry. This mosaic has no equivalent of heterogeneity in plains. The granulometry controls the soil water capacity, with deep clay soils efficient to conserve water for a long time, while sandy soils dry out quickly. The geochemistry controls for the pH and nutrients.

However, elevation plays a secondary role by changing the thermodynamics of biogeochemical processes due to lower temperature and changed water balance, which reduces the weathering of bedrocks and soil particles with elevation. Further, at higher elevation, the interactions are reduced between soil microorganisms, organic matter, and geochemical elements, thus creating a complex soil landscape pattern linked first to geology, and, second, to elevation. This landscape mosaic controls for the distribution of plants and microorganisms, all species actively driven by soil pH, nutrients, and water capacity.

Biodiversity and ecological processes

Community and ecosystem structure

Mountain temperate and boreal forests are a critical link in supporting the remarkable, but threatened, diversity of life on Earth (Rivers et al. 2019; Figure 6.1). Biodiversity is most often considered at the level of individual species, but aspects of diversity are important within species (genetic, population) and among species at community and landscape level. Communities are commonly used to organize taxa that have certain attributes in common, such as the plant community or invertebrate community. More specific sub-community classifications, that is, the guild, such as bark-gleaning birds (e.g., nuthatches, woodpeckers), are useful for managers who are interested in improving habitat for a group that shares a common resource.

Forest communities dominated by co-occurring tree species, such as forests of pine-oak (*Pinus-Quercus*), spruce-fir-poplar (*Picea-Abies-Populus*), monkey-puzzle tree-southern beech (*Araucaria-Nothofagus*) or several eucalyptus species, are often addressed as a unit, although it is important to note that species ranges may increasingly vary individually as climate warms or dryness, possibly breaking apart former community groups (Weiskopf et al. 2020). Trees are crucial in forest notably through their dominant photosynthetic function over all other plants. Photosynthesis is the process used by plant to produce carbon-based molecules synthesised from atmospheric CO_2 and water (H_2O) and using solar energy. Since trees are the largest terrestrial plants in mountain forest, most of the biomass is produced by trees.

Ecosystem structure—the sizes, locations, and types of life forms—varies over landscapes and forms the habitat of plants and animals, the fuel for wildfires, the arrangement of stored carbon, and numerous other ecosystem services (see Section "Ecosystem services and conservation issues"). The vertical arrangement of structure goes from the shortest elements such as shrubs and young, suppressed trees, up through intermediate trees to the dominant and co-dominant trees of the overstory. Among mountain shrubs, the group of Ericaceae (*Erica*, *Calluna*, *Rhododendron*, *Vaccinium*, etc.) and close taxa (*Empetrum*) are very important in frequency, cover, and eventually biomass, and crucial in terms of mountain forest functioning, thanks to biochemical mechanisms that reduce organic matter recycling and productivity (Fanin et al. 2022), forming a keystone plant group. Horizontal patterns of forest structure include the distribution of stands of different species and sizes over the varying topography of a mountain or watershed, varying with disturbance type and regimes.

Disturbances and succession

In ecological terms, the word "disturbance" refers to any discrete event in time and space that disrupts ecological systems and results in a total or partial destruction of biomass (White and Pickett 1985). Important mountain forest disturbances include abiotic (non-living) factors such as fire, wind, drought, flood, and avalanches, as well as biotic factors such as insects, browsing/grazing animals, and humans (cutting, husbandry, litter removal, etc.) (Figure 6.4). Patterns of disturbance over long periods are called disturbance regimes, with characteristic attributes such as frequency, intensity (energy release), severity (ecological effects), seasonality, and spatial distribution. Species have evolved adaptations to many disturbances, such as the ability of some species to release seeds after fire, called serotiny (Pausas 2015). Following a disturbance, new plant, animal, and microorganism communities take advantage of areas where older organisms were killed or reduced in biomass. Therefore, biotic and abiotic dynamics occur, eventually characterizing a succession.

In terms of vegetation, two types of successions can be considered. First the "relay floristic" model (Clements 1916), where the first pioneers or early-successions species are often herbaceous species that are short-lived but grow fast, with high reproduction traits. Then woody shrubs and light-loving trees become established, growing until the open patch created by disturbance is re-occupied with crowns and roots. Later, trees that tolerate shade establish in the understory, growing slowly until a competitor overhead dies and creates a gap into which it can climb. Eventually an old-growth forest state may be reached as large trees reach the end of their lifetimes, a relatively stable condition where tree growth and mortality

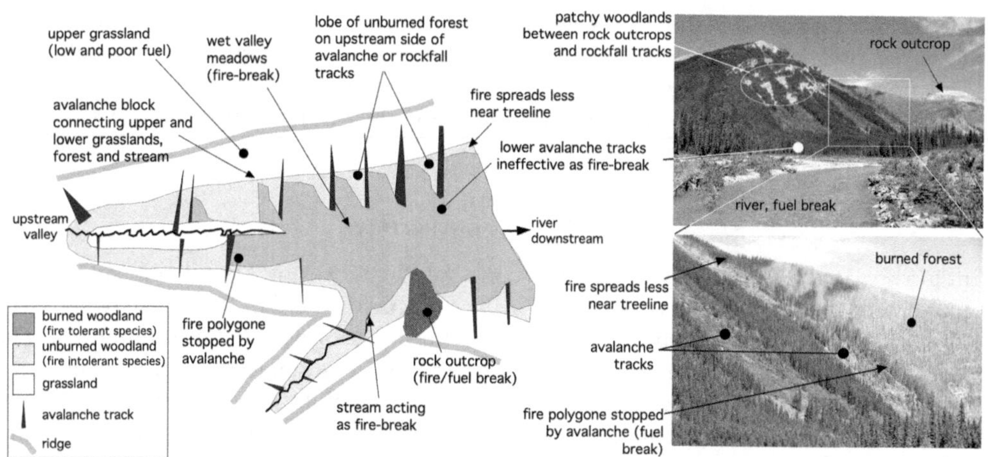

Figure 6.4 Mountain forest landscape naturally disturbed by wildfires and avalanche/rockfall, resulting in a mosaic of communities along slope and valley gradients (inspired from Suffling 1993). (Right) Example in the Canadian Rockies, Kootenay National Park, British Columbia; [right, up] the landscape presents a mountain stream with a meander torrential flow showing disturbed forest vegetation on the more or less temporary banks; the river flows at the foot of a slope covered by montane (bottom) and subalpine (upper) forests disturbed by wildfire, snow avalanches and rockfalls; [right, low] the detailed photo shows the interplay between the wildfire polygon and an avalanche corridor; avalanche corridor behaves as a fuel break controlling the fire spread, producing patches of different fire intensities; the avalanche corridor sustain low-flammable species (deciduous broadleaves trees, herbs, dwarf shrubs) while forests are mostly composed of high-flammable species (needleleaved trees: *Abies lasiocarpa, Picea glauca, P. engelmannii, Pinus contorta, P. albicaulis*); it results in a patchy complex natural forest landscape; not visible: patches of dead trees disturbed by insects. Photography: ©Christopher Carcaillet.

are balanced (Gundersen et al. 2021). Second, the "initial floristic composition" model (Egler 1954) assumes that all, or the majority, of species present during the succession are in place at the beginning of the recovery phase, and the reestablishment of the initial plant community is a rapid phenomenon. In mountain forest ecosystems, most dynamics liked to natural disturbances or low-severity tree-harvesting follow the Egler's model, while dynamics following agricultural abandonment or severe perturbations are generally characterized by the Clements' model.

The succession of different communities after disturbance can be fairly predictable, allowing foresters to plan tree harvests followed by natural regeneration, or to estimate the time since a severe fire from the age of the trees. However, successional sequences often vary in less predictable ways due to mountain physiographical patterns (elevation, slope aspect, soil, etc.), and disturbance interactions.

Litter, humus, and nutrients

Mountain forest soils are covered with an organic layer derived from the vegetation above: tree leaves and needles, fallen fruits and cones, twigs, branches, and tree trunks. The freshly fallen litter decays as the chemical constituents are consumed by bacteria and fungi, forming a nutrient-rich humus layer through which rainwater leaches organic matter into the mineral soil below. Soil organic matter plays a key role as a binder for mineral matter, resulting in stable organo-mineral aggregates. Nutrients essential for plant growth, such as nitrogen and phosphorus, are released as a result of decomposition in chemical forms that are soluble in water and can be directly taken up by plant roots.

Plant–fungi interactions

Fungi associated with plant roots, called mycorrhizal ("fungus-root") fungi, play vital roles in mountain forests. The thin, hairlike fungal hyphae extend outward from the fine roots of the trees, creating a network for absorbing water and inorganic-form nutrients from the soil. The trees benefit from these resources, but in return, the fungi are sustained by sugars produced in the tree leaves and transported down to the roots. This symbiotic or mutually beneficial relationship can make a huge difference, such that tree nurseries take care to inoculate seedlings with mycorrhizal fungi in the greenhouse before out-planting. Most of fungi associated with mountain tree roots are ectomycorrhizas, meaning that their hyphae are intercellular without penetrating the plant cells. Other fungal species, called endophytes or arbuscular mycorrhizal fungi (AMF), live within the cells of tree organs, notably needles and leaves. The ecological role of endophytes associated with trees remains a puzzle for researchers. However, it appears that their community composition varies between close mountain forests in response to complex relief (Taudière et al. 2018). AMF are also specialized to understory herbs and forbs. Last, the keystone understorey plant group, the Ericaceae, is associated to another group of fungi, the ericoid mycorrhiza.

Other important fungi species are pathogenic, attacking especially the roots and lower trunks of trees. Spores of fungi in the genera *Armillaria* and *Heterobasidion* can infect freshly cut stumps and then spread to surrounding living trees through root contact, resulting in expanding patches of dead and dying trees (Wingfield 2004).

Many fungi grow on dead trees, decaying the wood, or in the soil where they help decompose organic material. The fruiting bodies of soil-living fungi, mushrooms, and toadstools are delicious and economically important forest products, although knowledge of edible versus poisonous species is needed.

Plant–animal interactions

Trees and other plants comprise the habitat for mountain forest animals (Figure 6.5A–F). The following paragraphs outline how forests provide food, shelter, access to water, and space for life activities and reproduction, but can also result in damage, regulating plant dynamics and dispersion, selecting species, and ecosystem functioning.

Among the most important plant–animal interaction in mountain forests is that of plant–insect interactions, including pollination, seed dispersal, or herbivory, for instance, herbivory or defoliation. Some insects consume the tree leaves or needles, temporally reducing during a growing season the photosynthetic activity of trees (e.g., larch budmoth in Euro-Asian forests or *Zeiraphera diniana*; Baltensweiler and Fischlin 1988), while others can kill trees by totally removing their foliage (e.g., Douglas-fir tussock moth in North American Rockies or *Hemerocampa pseudotsugata*; Clancy 2002). Some of the smallest insects are bark beetles

Figure 6.5 Animals of mountain forests. (A) Browser: European elk (*Alces alces*) in downy birch (*Betula pubescens*) subalpine forest near the upper treeline (950 m a.s.l.), Dovrefjell Mountain, Norway. (B) Omnivore: Franklin's grouse (*Canachites franklinii*) in white spruce (*Picea glauca*)-subalpine fir (*Abies lasiocarpa*) montane forest (1200 m a.s.l.), Athabaska Valley, Alberta, Canada. (C) Frugivore: Yellow-pine chipmunk (*Tamias amoenus*) eating berries of *Shepherdia canadensis* in a spruce (*Picea engelmanii*)-pine (*Pinus albicaulis*) subalpine forest (1900 m a.s.l.), Kooteney Valley, British Columbia, Canada. (D) Omnivore: Black bear (*Ursus americana*) in lodgepole pine (*Pinus contorta*) montane forest (2000 m a.s.l.), Lamar Valley, Wyoming, USA. (E) Insectivore: Corsican fire salamander (*Salamandra corsica*) in black pine (*Pinus nigra*) montane forest (1350 m a.s.l.), Corsica, France. (F) Nectar feeder: Cape sugarbird (*Promerops cafer*) in *Erica* montane fynbos (1400 m a.s.l.), Swartberg mountain, South Africa. Photography: ©Christopher Carcaillet.

(*Scolotyneae* group) (e.g., the mountain pine beetle *Dendroctonus ponderosae* common in the North America Cordillera). They tunnel into the inner bark to lay eggs where the larvae can feed and grow to maturity, often girdling the tree. Outbreaks of bark beetles are among the largest disturbance factors in mountain forests, interplaying with global changes (Raffa et al. 2008), with major consequences for ecosystem services (Morris et al. 2018).

Soil-dwelling insects are necessary for soil health but can also cause undesired damage. Insects, along with seeds and other plant parts, form the base of food webs upon which birds, amphibians, reptiles, and many mammals depend. Aquatic invertebrates in forest streams also depend upon plant inputs such as litter as well as the cooling effect of tree shade. Larger predatory animals such as wolves or hawks feed on the herbivores.

Animals also carry out critical functions for plants, including seed dispersal. Birds such as the pinyon jay (*Gymnorhinus cyanocephalus*) carry pine seeds for notable distances. Despite being able to relocate their caches due to their impressive spatial memories, some seeds are not eaten and germinate (Bednekoff and Balda 1996). The same mechanism happens with other mountain Corvidae, for example, the nutcrackers: *Nucifraga caryocatactes* associated to the cembra pines (*Pinus cembra*, *P. siberica*, *P. pumila*) in the Alps, the Carpathian, and the Siberian mountains to Kamchatka; *N. multipunctata* in *Pinus geradiana* or *Picea smithiana* forests in the western Himalaya; and *N. columbiana* in the North American Cordillera where *Pinus albicaulis* and *P. flexilis* growth.

Pollination is another essential task done by insects, birds (Figure 6.5F), and bats. In many cases, pollinators have highly specialized relationships that resulted from co-evolution with certain plant species.

Key habitat elements in forests include snags (dead standing trees) with decaying wood that can be excavated by some birds such as woodpeckers, and then occupied for generations by other bird species. Another important role is played by downed woody debris, large logs that protect sheltered runways for small mammals, for example, rodents such as mice and voles (Muridae), or insectivores such as shrews (Soricinae). Forests provide essential foods for many species, especially in the form of fruits and seeds from trees, shrubs, and grass (Figure 6.5C). For instance, the long-tailed Aberts squirrels (*Sciurus aberti*) rely in many ways on ponderosa pine trees in North American forests. They build nests in trees and jump from canopy to canopy, avoiding predators. For food, they eat pine seeds, chew the inner bark of twigs, and dig edible fungi that are symbiotic with the pines (Yarborough et al. 2015). Koalas in Australia have a similarly strong dependence on Eucalyptus species for foraging and shelter, including in mountain forests (Johnson et al. 2018).

Finally, ungulates can be abundant in all temperate and boreal mountain forests up to and sometimes beyond the tree line, especially deer, which are the main group of large mammals in mountain forests: for example, *Cervus*, *Odocoileus*, *Alces* (Figure 6.5A), *Rangifer* (Figure 6.1A), *Capreolus*, *Dama*, etc. These species regulate tree regeneration through herbivory and girdling (bark removal), and growth through browsing of twigs. They can induce significant changes in the tree community and dynamics by suppressing palatable species or through trampling, and indirectly selecting other species that can grow and eventually compose the overstory. Sometimes, in certain regions, they create forest gaps without trees when the animals gather periodically, especially in winter, a process that generally occurs when predators (wolve, lynx, bear, etc.) are absent (Figure 6.5D).

Ecosystem services and conservation issues

Importance to society, ecosystem services

Natural and healthy mountain forests are crucial for societies, providing nature-based services and offering many and varied benefits to humankind through what we call "ecosystem services" (www.millenniumassessment.org/en/index.html). It is estimated that 12% of the global human population is living in mountains (Price, 2003), most of them living in temperate, warm temperate, and tropical areas. They directly depend on mountain forests, while other human population living in lowlands indirectly benefits from these forests.

Mountain forests provide many provisioning, regulating, and cultural ecosystem services. They potentially supply societies with raw materials (construction or furniture wood), biomass-energy (firewood), food for animals (pasture), and humans (berries, mushrooms), genetic resources for forestry or horticulture, etc. Mountain forests regulate carbon through sequestration in biomass, which helps stabilize global climate, contribute to air and water purification, control biodiversity, stabilize soils, and limit landslides and erosion. In terms of cultural services, they are areas for aesthetic and recreational experiences (ecotourism, sport, outdoor activities), providing spiritual resources and conditions for nature-based education, nurturing artistic creation, and containing many conservation areas.

Ancient anthropogenic removal of forest

Mountains have been the home of human colonisation since antiquity, to profit from forests, but also from natural grasslands to feed livestock. Asian mountains have been areas of wildlife domestication, notably about 11,000 years ago for sheep (wild ancestor: *Ovis gmelini*) and goats (wild ancestor: *Capra aegarus*) both in the Zagros Mountains in Iran, and maybe in eastern Anatolia in Turkey (Naderi et al. 2008; Portanier et al. 2022). Although these species prefer tree-less habitats, prehistoric farmers moved them in forest belts. Humans cleared forests, often using burning, to create and later enlarge mountain grassland in forest belts for grazing and hay-making areas necessary to feed goats and sheep during winters. In Europe, evidence of first farmers using mountain forests for livestock grazing dates from the Neolithic, about 7,000 years before present, a date that is later than in the Middle East, but contemporaneous of central Asia. Further north, in Scandinavian and Siberian mountains, humans domesticated reindeer in different areas, probably 2,000 years ago or later (Røed et al. 2008). In mid-season and in winter, reindeer move from alpine tundra to conifer forests, grazing lichens.

Tree overharvesting, forest overgrazing and deforestation, and other human-based damage

While the North American mountain forests were not intensively used for domestic livestock before the European colonisation since 17th century, all European and Asian mountain forests have suffered from pastoralism since millennia. Further, mountain people exploited trees for fuel and for building, sometimes forest litter or tree resin as well (Gimmi et al. 2008), which contributed to deplete forest density and eventually to change plant composition over millennia (Blarquez et al. 2012).

The multiple forest uses have necessarily reduced the forest covers and sometimes to severely fragment them. There are valleys in the Alps where there are no more trees resulting from centuries of deforestation, which in turn induced severe soil erosion and floods. To face deforestation, for instance, France started to engage between 1860 and 1882, a major programme of tree seedling and plantation of hundreds of thousands of hectares of mountain lands with introduced *Pinus nigra* from Serbia (Austrian black pine), and *Larix decidua* (European larch) of unknown provenance, resulting a huge transformation of mountain forest during the 20th century.

The overharvesting of mountain forests was so severe in areas that species have been threatened with the risk of extinction. A well-known case is *Cedrus libani* (Lebanon cedar) growing in the Near East, today mainly in Turkish mountains. It covered larger areas during antiquity but was abundantly used for building. Today, this species is threatened with extinction in Lebanon and has been extirpated from Syria. In Chile, *Araucaria araucana* (monkey puzzle tree), which grows in Andean forests, is threatened by intensive and deregulated forestry as well as too frequent and sever wildfires. In the Blue Mountain, eastern Australia, *Wollemia nobilis* (Wollemia pine), strictly protected, is threatened by mega wildfires fostered by anthropogenic climatic change. The same situation exists for *Sequoiadendron giganteum* (giant sequoia) in Californian mountains.

Another mechanism associated with the anthropogenic shrinkage and fragmentation of forest cover is the lowering of most upper forest limits in Europe by around 500 m in elevation or even more, depending on the specific mountains. This mechanism occurs during the last millennia in Europe and Asia mostly due to human land uses. All these mountain forest alterations have altered plant composition, ecosystem productivity, food webs, and other physical aspects like the snow cover or the hydric budget.

Globally, mountain forests continue to be deforested or overused, notably in Asia and North Africa, due to substantial increases in human population and economies more dependent on the uses of natural resources.

Land use abandonment and new usages

In Europe, mountain residents begun an exodus during the late 19th century, resulting in a massive land-use abandonment. In these areas, mountains are still the home of people, but less and less areas of exploitation of natural resources. This means that some European temperate forests escaped the global trends of deforestation and forest land uses. In European and North American mountains, the main trajectory is regeneration (natural spontaneous process) and reforestation (tree planting) of low productive and remote lands (Conti and Fagarazzi 2005). This pattern varies greatly between regions at micro-scale.

A consequence of this land-use abandonment is the expansion of forests that is slower in the highest elevation than in the lowest but also the expansion of heathland of Ericaceae covering large areas in European mountain forests today (Tasser et al. 2002), sometimes hindering the tree expansion due to biochemical process they induce or competition. Interestingly, paleoecology has not found evidence of such Ericaceae cover in the past. Was Ericaceae a rare plant group that expanded and benefits today from land-use abandonment, or is paleoecology failing regardless of the methods used to reconstruct past vegetations? This remains an open question that has potential aftermath for forest conservation and productivity.

Climate change threats to mountain forests

Mountain forests are impacted by the current and ongoing climatic change; the threats are several (Beniston 2003). They concern the forest productivity that is changing according to the sensitivity of species to cope with the magnitude and trend of climatic change. While a warming trend is common nearly everywhere, precipitation does not present a unique trend, with areas that will become drier while other will receive more precipitation. Further, the mean annual values mask other changes of parameters concerning the seasonality change, that is, the magnitude between summer and winter parameters, or the start and the end of the growing season and associated processes (e.g., fire, avalanches). Some mountains could be drier in winter, but wetter in summer—or vice versa—potentially resulting in no change in mean annual values but with huge changes in forest functioning.

With climate change, extreme values can be reinforced, especially in terms of precipitation or wind. Intense and long-lasting rainfall can occur despite a drier climate. These intense and lasting precipitations are generally linked to the mountains thanks to topographic relief, which causes the condensation of the air at lower temperatures; moreover, the topography can concentrate the waters as in a funnel, inducing high-energy runoff in the lower valleys proportional to the size of the catchment area and the height of the abrupt precipitation. This mechanism induces intense floods, with erosion sometimes destroying the valley forests. No mountain region escapes it. Fortunately, mountain forests play a key role in mitigating these extreme processes, reducing their magnitude and severity.

Extremely low rainfall for a long period of time or extremely high temperatures cause large and severe forest fires in mountain forests. Since the 1980s, megafires have been more frequently recorded in eastern Australia, western North America, in the Mediterranean mountains, and in the boreal mountain forests like in Scandinavia.

Wind is becoming a major issue linked to climate change, which fosters more frequent intense storms. In mountain areas, for example in the Alps or the Balkans, forests have been affected by unusually severe storms since the 1980s. This has effects on the forest economy, but also on tree composition, insect infestations, and forestry practices. Monocultures are more affected by storms than mixed forests; for instance, forests of European spruce (*Picea abies*), whose trees are easily uprooted or broken during intense storms.

Introduced and invasive species and genotypes

Non-native tree species and genotypes have been introduced through planting programmes since the 19th century and more recently under the paradigm of assisted migration. Assisted migration consists of species or populations translocations from warmer and drier areas to facilitate adaptation of forest productivity through a direct management response to global warming, sometimes by the substitution of natural species by exotics (Vitt et al. 2010). Examples of introduction include the western American *Picea engelmannii* (Engelmann spruce) in the Swedish mountains, the western American coast *Pseudotsuga menziesii* (Douglas fir), and *Picea sitchensis* (Sitka spruce) introduced in the western European mountains (Scotland, France). Example of assisted migration is the translocation into southern Europe of *Cedrus atlantica* (Atlantic cedar) from Morocco and Algeria. The Atlantic cedar and the Douglas-fir have perfectly acclimatized, sometimes regenerating naturally, creating new ecosystems and ending up excluding natural species. Under these conditions, some species can become invasive. Among the environmental issues associated with these introductions are severe soil acidification with a monospecific planting of Sitka spruce, or nitrification and acidification with monospecific translocation of Douglas-fir (Zeller et al. 2019).

Another example is the millions of hectares of black pine (*Pinus nigra*) plantations carried out in southern Europe mountains, largely from Serbian populations, notably to protect arid or unproductive soil from erosion. Today, these million hectares of artificial forests represent a threat due to the fire-prone character of black pine forests combined with the climatic change (Michalet et al. 2023).

Sometimes, planted species were natural, but their population were from other regions. For instance, European larch (*Larix decidua*) planted in the western Alps since the 19th century were from provenances in the eastern Alps. This mechanism could result in genetic degradation, with potential effect considering the ongoing climatic changes if this new population is less adapted to future climate, while original population could be better adapted thanks to tens of millennia of local adaptation (Housset et al. 2021).

Conclusion

Mountain forests are distinguished from lowland forests by the topographic relief and notably the elevation, which constitute the main drivers of the idiosyncratic microclimates, to which the forests have acclimatized and adapted. The diversity of bedrock creates soil mosaic, which, together with the microclimate, create complex ecological patterns. Each region of the world has its own set of species thanks to millions of years of evolutionary and biogeographical processes, making each mountain range quite singular. However, all mountain forests show a similar ecological functioning, varying according to geography and the set of species, which concerns the food web, the vertical forest structure, the disturbances, and the interactions between plants and fauna or plants with micro-organisms. Finally, mountain forests differ greatly by land-use history, with Asian and European forests being the most transformed for millennia, especially warm temperate forests. The least anthropized mountain forests are the boreal forests in North America, and certain cold forests of South America mountains, although much remains to be learned about these regions. Climate and land-use changes interact, altering disturbance regimes and ultimately the future composition and productivity of forests.

References

Baltensweiler, W., Fischlin, A. (1988). The Larch Budmoth in the Alps. In: Berryman, A.A. (eds), *Dynamics of Forest Insect Populations. Population Ecology*. Springer, Boston, MA. https://doi.org/10.1007/978-1-4899-0789-9_17

Bednekoff, P.A., Balda, R.P. (1996) Social caching and observational spatial memory in Pinyon Jays. *The Southwestern Naturalist* 40:180–184.

Beniston, M. (2003). Climatic change in mountain regions: A review of possible impacts. *Climatic Change* 59: 5–31.

Blarquez O., Carcaillet C., Elzein T.M., Roiron P. (2012) Needle accumulation rate model-based reconstruction of palaeo-tree biomass in the western subalpine Alps. *The Holocene* 22, 579–587. DOI: 10.1177/0959683611427333

Clancy, K.M. (2002). Mechanisms of resistance in trees to defoliators. In: Wagner, M.R., Clancy, K.M., Lieutier, F., Paine, T.D. (eds), *Mechanisms and Deployment of Resistance in Trees to Insects*. Springer, Dordrecht. https://doi.org/10.1007/0-306-47596-0_3

Clements, F.E. (1916) Plant succession. In: *An Analysis of the Development of Vegetation*, Publication 242, Carnegie Institute, Washington, DC.

Conti, G., Fagarazzi, L., 2005. Forest expansion in mountain ecosystems: "environmentalist's dream" or societal nightmare? *Planum—The European Journal of Planning* 11, 1–20.

Egler, F.E. (1954) Vegetation science concepts. I. Initial floristic composition—a factor in old-field vegetation development. *Vegetatio* 4, 412–418.

Fanin N., Clemmensen K.E., Lindahl B.D., Farrell M., Nilsson M.C., Gundale M.J., Kardol P., Wardle D.A. (2022) Ericoid shrubs shape fungal communities and suppress organic matter decomposition in boreal forests. *New Phytologist* 236, 684–697 doi: 10.1111/nph.18353

FAO. (2022) *State of the World's Forests 2022*. FAO, Rome.

Gimmi, U., Burgi, M., Stuber, M., 2008. Reconstructing anthropogenic disturbance regimes in forest ecosystems: a case study from the Swiss Rhone Valley. *Ecosystems* 11, 113–124.

Gundersen P, Thybring EE, Nord-Larsen T, et al. (2021) Old-growth forest carbon sinks overestimated. *Nature* 591, E21–E23. https://doi.org/10.1038/s41586-021-03266-z

Housset J.M., Toth E.G., Tremblay F., Motta R., Girardin M.P., Bergeron Y., Carcaillet C. (2021) Tree-rings, genetics and the environment: complex interactions at the rear edge. *Dendrochronologia* 69, 125863, DOI: 10.1016/j.dendro.2021.125863

Johnson RN, O'Meally D, Chen Z, et al. (2018) Adaptation and conservation insights from the koala genome. *Nature Genetics* 50, 1102–1111. https://doi.org/10.1038/s41588-018-0153-5

Körner C. (1998) A re-assessment of high elevation treeline positions and their explanation. *Oecologia* 115, 445–459

Körner C. (2003) *Alpine Plant Life: Functional Plant Ecology of High Mountain Ecosystems*, 2nd edition, Springer, Berlin, 344p.

Körner C., Paulsen J., Spehn E.M. (2011) A definition of mountains and their bioclimatic belts for global comparisons of biodiversity data. *Alpine Botany* 121, 73–78.

Michalet R., Carcaillet C., Delerue F., Domec J.-C., Lenoir J. (2023) Assisted migration in a warmer and drier climate: less climate buffering capacity, less facilitation and more fires at temperate latitudes? *Oikos*. https://doi.org/10.1111/oik.10248

Michalet R., Choler P., Callaway R.M., Whitham T.G. (2021) Rainfall continentality, via the winter Gams angle, provides a new dimension biogeographical distributions in the western United States. *Global Ecology and Biogeography* 30, 384–397. DOI: 10.1111/geb.13223.

Morris JL, Cottrell S, Fettig CJ, et al. (2018) Bark beetles as agents of change in social-ecological systems. *Frontiers in Ecology and the Environment* 16, S34–S43. https://doi.org/10.1002/fee.1754

Muthoo M. (2002) Mountain environment and development. *Unasylva* 208, Y3549E/y3549e00. DOI: https://www.fao.org/3/Y3549E/y3549e00.htm

Naderi S., Rezaei H.R., Pompanon F, Blum M.G.B., Negrini R., Naghash H.R., Balkız ö, Mashkour M., Gaggiotti O.E., Ajmone-Marsan P., Kencef A., Vigne J.D., and Taberlet P. (2008) The goat domestication process inferred from large-scale mitochondrial DNA analysis of wild and domestic individuals. *PNAS*, 105, 17659–17664.

Pausas JG (2015) Evolutionary fire ecology: lessons learned from pines. *Trends in Plant Science* 20:318–324. https://doi.org/10.1016/j.tplants.2015.03.001

Piedallu C, Gégout J.C. (2007) Multiscale computation of solar radiation for predictive vegetation modelling. *Annals of Forest Science* 64, 899–909.

Portanier E., Chevret P., Gélin P., Benedetti P., Sanchis F., Barbanera F., Kaerle C., Queney G., Bourgoin G., Devillard S., Garel M. (2022) New insights into the past and recent evolutionary history of the Corsican mouflon (*Ovis gmelini musimon*) to inform its conservation. *Conservation Genetics*, 23, 91–107. https://doi.org/10.1007/s10592-021-01399-2

Price M.F. (2003) Why mountain forests are important? *The Forestry Chronicle* 79, 219–222.

Raffa K.F., Aukema B.H., Bentz B.J., Carroll A.L., Hicke J.A., Turner M.G., Romme W.H. (2008) Cross-scale drivers of natural disturbances prone to anthropogenic amplification: the dynamics of bark beetle eruptions. *BioScience* 58:501–517. Doi: 10.1641/B580607

Rivers M.C., Beech E., Bazos I., Bogunić F., Buira A., Caković D., Carapeto A., Carta A., Cornier B., Fenu G., Fernandes F., Fraga P., Garcia Murillo P.J., Lepší M., Matevski V., Medina F.M., Menezes de Sequeira M., Meyer N., Mikoláš V., Montagnani C., Monteiro-Henriques T., Naranjo Suárez J., Orsenigo S., Petrova A., Reyes-Betancort J.A., Rich T., Salvesen P.H., Santana López I., Scholz S.,

Sennikov A., Shuka L., Silva L.F., Thomas P., Troia A., Villar J.L., Allen D.J. (2019) *European Red List of Trees*, IUCN, Cambridge, UK and Brussels, Belgium. viii + 60pp.
Røed K.H., Flagstad Ø., Nieminen M., HolandØ., Dwyer M.J., Røv N., Vila C. (2008) Genetic analyses reveal independent domestication origins of Eurasian reindeer. *Proceedings of the Royal Society B*, 275, 1849–1855. doi:10.1098/rspb.2008.0332
Suffling R. (1993) Induction of vertical zones in sub-alpine valley forests by avalanche-formed fuel breaks. *Landscape Ecology*, 8, 127–138.
Tasser E., Tappeiner U. (2002) Impact of land use changes on mountain vegetation. *Applied Vegetation Science* 5, 173–184.
Taudière A, Bellanger J.-M., Carcaillet C., Hugot L., Kjellberg F., Lecanda A., Lesne A., Moreau P.-A., Scharmann K., Leidel S., Richard F. (2018) Diversity of foliar endophytic ascomycetes in endemic Corsican pine forests. *Fungal Ecology* 36, 128–140. http://doi.org/10.1016/j.funeco.2018.07.008
Tranquillini W (1979) *Physiological Ecology of the Alpine Timber-line. Tree Existence at High Altitudes with Special References to the European Alps.* Ecological Studies 31, Springer, Berlin, Germany.
Vitt P., Havens K., Kramer A.T., Sollen-Berger D., Yates E. (2010) Assisted migration of plants: Changes in latitudes, changes in attitudes. *Biological Conservation* 143, 18–27. https://doi.org/10.1016/j.biocon.2009.08.015
Weiskopf S.R., Rubenstein M.A., Crozier L.G., et al. (2020) Climate change effects on biodiversity, ecosystems, ecosystem services, and natural resource management in the United States. *Science of the Total Environment* 733, 137782. https://doi.org/10.1016/j.scitotenv.2020.137782
White P., Pickett S. (1985) *The Ecology of Natural Disturbance and Patch Dynamics.* Academic Press, New York.
Wingfield M.J. (2004) Disease affecting exotic plantation species. In: *Encyclopedia of Forest Sciences*, Academic Press, New York.
Yarborough R.F., Gist J.A., Loberger C.D., Rosenstock S.S. (2015) Habitat use by Abert's squirrels (*Sciurus aberti*) in managed forests. *The Southwestern Naturalist* 60, 166–170. https://doi.org/10.1894/JKF-49.1
Zeller, B., Legout, A., Bienaimé, S., Gratia, S., Santenoise P., Bonnaud P. (2019) Douglas fir stimulates nitrification in French forest soils. *Scientific Reports* 9, 10687. https://doi-org.inee.bib.cnrs.fr/10.1038/s41598-019-47042-6

PART II

FOREST DYNAMICS

7

INSECT DISTURBANCES IN FOREST ECOSYSTEMS

Daniel Kneeshaw, Brian R. Sturtevant, Barry Cooke, Timothy Work, Deepa Pureswaran, Louis DeGrandpré and David A. MacLean

Tree-feeding insects are ubiquitous in forest ecosystems. While relatively few species cause widespread mortality, insect outbreaks are important ecological disturbances that, in some cases, can have devastating economic effects. Insect outbreaks have been considered in the context of forest disturbances for decades (Mattson and Addy 1975). However, the interest of forest managers and policymakers often waxes and wanes in cycles that follow the population dynamics of the insects themselves, despite the fact that the area affected by insect outbreaks in Canada, the United States and Europe is greater than that disturbed by fire or harvesting (FAO 2014; Kneeshaw et al. 2011) (Figure 7.1 shows patterns for Canada).

Species with extreme population fluctuations that vary over several orders of magnitude are capable of disrupting forest ecosystem functions (Wallner 1987) and generally fall into two main taxonomic groups: aggressive tree-killing scolytid bark beetles that feed within the inner bark and exhibit eruptive population dynamics (Raffa et al. 1993); and *Lepidopteran* and *Hymenopteran* species whose larvae defoliate tree leaves and needles and whose populations follow high-amplitude oscillations (Cooke et al. 2007). The spatial extent of outbreaks (i.e., spatial synchrony) combined with the degree of damage caused by the outbreak (i.e., severity) determines the degree to which humans define insects as "pests" (Liebhold et al. 2012). Depending on jurisdictional policies, large-scale tree mortality can temporarily flood regional and even continental wood markets as forest industries attempt to salvage trees before they deteriorate and lose their value (e.g., Walton 2012). Long-term legacies of severe insect outbreaks may have even greater economic impacts as extensive mortality leads to gaps in forest age–class structure and reductions in annual allowable cuts (Abbott et al. 2009).

Insect disturbances differ from other natural disturbances such as wind or fire in that dietary and other biological constraints limit the taxonomic range of tree species they can damage. Monophagous forest insects are restricted to a single tree species or genus by their specific adaptations to particular secondary metabolites, polyphagous species are more tolerant of a wider range of secondary metabolites permitting feeding on a wider (but not unlimited) taxonomic range, and oligophagous species fall between these two extremes (Jactel and Brockerhoff 2007).

DOI: 10.4324/9781003324072-9

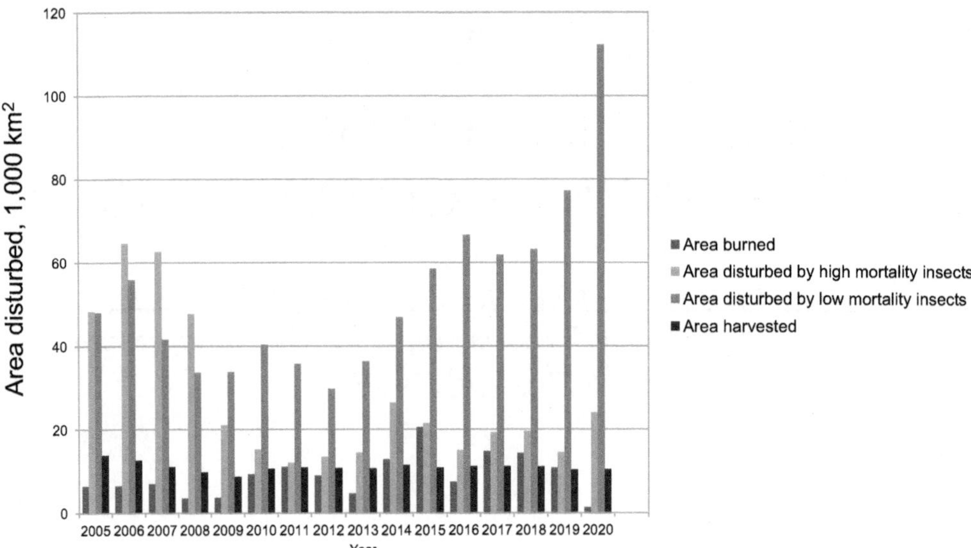

Figure 7.1 Area disturbed by insects (mountain pine beetle, forest tent caterpillar and spruce budworm combined), by fire and by harvesting from 2005 to 2020 in Canada. Data from NRCanada (https://natural-resources.canada.ca/our-natural-resources/forests/state-canadas-forests-report/disturbance-canadas-forests/16502). Insect effects are divided into both low-severity characterized primarily by growth loss and low mortality and high severity with high tree mortality events.

There is evidence that outbreaks in North America and Europe (1) have increased in severity and duration; (2) have become more synchronous; (3) have expanded their geographic range; and (4) are affecting new host species as outbreak ranges expand due to changes in climate and forest structure (see sections below). For example, mountain pine beetle (*Dendroctonus ponderosae*) in Western North America has been more severe and expanded its range due at least in part to climate warming (Samaraju et al. 2021). In Europe, range expansion and tree mortality due to the European spruce bark beetle (*Ips typographus*) have been linked to climate change-induced drought (Trubbin et al 2022)_ Outbreak range shift in the spruce budworm has also been linked to climate change (Régnière et al. 2012). However, there is also evidence of climate extremes such as heat waves or thermal shocks having lethal effects on insects or less palatable host tissues or more abundant parasitoids exerting negative population pressures (Jactel et al. 2019; Pureswaran et al. 2018). Different life-cycle stages during an insect development can also be more strongly or weakly affected by temperature extremes (Harvey et al. 2020). Although attention is often drawn to cases where outbreaks seem to be increasing due to climate change, physiological processes can both promote and restrict insect population growth and thus a case-by-case analyses is required. Furthermore, since changes in forest structure can be an equally or more important driver of outbreak severity (Cooke et al. 2024), disentangling processes is critical to understand what degree factors such as climate change or forest structure are driving changes.

Differences among biomes

Tropical forests, with high tree diversity and reduced concentration of individual tree species, are thought to inhibit widespread insect outbreaks (Janzen 1970). Insect outbreaks causing widespread damage and mortality in tropical regions are most commonly observed in stands dominated by one or a few tree species. Extensive mortality due to insect outbreak is not commonly observed in unmanaged tropical forests but more common within simple plantations consisting of few or a single tree species. For example, outbreaks of the coffee berry borer, *Hypothenemus hampei* (Ferrari), in coffee plantations are well known (Damon 2000). Outbreaks may also be localized or occur only on given species (Sutton et al. 2021). Indeed, the ecological dogma that "there are no outbreaks in tropical forests" (Elton 1958) has been challenged (Wolda 1987; Dyer et al. 2012; Nair 2012). Research shows that the population cycles of herbivorous insects in natural tropical forests vary from low to high population densities (Nair 2012) even if widespread mortality is not observed. Dyer et al. (2012) summarized examples of insect damage within natural tropical forests but acknowledged challenges limiting our understanding of tropical outbreaks. These include a general lack of understanding of tropical insect natural history, difficulties in monitoring outbreak activity (for example, monitoring insect populations with multiple generations per year), and the absence of annual tree rings for traditional historical outbreak reconstructions using tree-ring analysis.

On a global scale, North American forests experience some of the largest and most severe insect outbreaks (Hlasny et al. 2021; Shorohova et al. 2012), and these outbreaks tend to occur in boreal and montane regions dominated by natural monocultures such as spruces, firs, pines and poplar species (Kneeshaw et al. 2011). Two extreme examples for North America are the mountain pine beetle, *Dendroctonus ponderosae* (Hopkins) and spruce budworm, *Choristoneura fumiferana* (Clemens). Each species has destroyed, within a single decade, millions of hectares of cordilleran and boreal forest of relatively low tree-species diversity. Such extreme examples have not been observed in comparable forests of Eurasia. For example, in 2005, 17.3 million hectares (Mha) were affected by forest insects in Canada and 5.6 Mha in the United States whereas elsewhere in the world insects affected 3.2 Mha in China, 1.7 Mha in Russia, and 1.3 Mha in Romania (FAO 2014). Although it is tempting to speculate that higher levels of insect depredation in high-latitude forests are causally associated with lower tree-species diversity, the Eurasian boreal serves as a cautionary example of the risk of over-generalization. Tree-species diversity is not the only variable that tends to be correlated with latitude and elevation. Other factors to consider include faunal diversity, especially natural enemy diversity, and climate. It is an open question why twentieth-century post-glacial boreal and cordilleran forests in the eastern and western hemisphere seem to differ so greatly in the extent and intensity of insect disturbance.

Forest–insect pest patterns and processes—from the regulation of population fluctuations to the rise and fall of outbreaks

Forest pests are most often found in endemic or low population densities. Outbreaks occur infrequently as populations transition synchronously from endemic to epidemic levels across large spatial extents, from stands of trees to entire regions. According to Berryman (1987), outbreak species of forest insects fall into six classes based on the pattern of outbreak development in space and time. These six classes are defined by two axes relating to the fine-scale

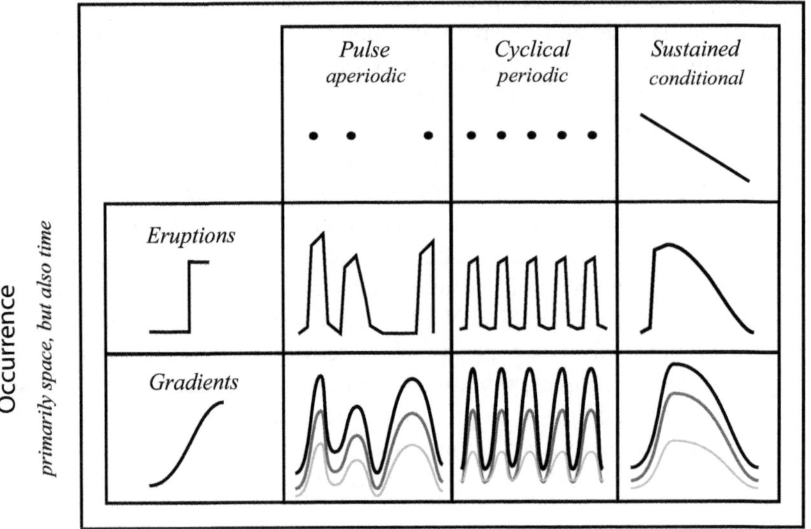

Figure 7.2 A schematic representation of different outbreak patterns in time. In the upper panel, occurrence represents the two main occurring outbreak patterns (gradient vs eruptive), which are then further broken down by their recurrence into pulsed, cyclical or sustained patterns. In the lower panel, we show how all of these patterns can be observed during an outbreak in time and space for a given outbreaking insect (Berryman 1987).

pattern of occurrence (either "eruptive" or "gradient") and the larger-scale pattern of recurrence (either "pulse", or "periodic", or "sustained"), in both time and space (Figure 7.2). "Eruptions" are defined as populations that occur abruptly in space and time, with sharp, clear transitions between endemic and epidemic states. In contrast, "gradients" refer to populations that vary more continuously in density through time and space; fluctuations are not abrupt and there are no clear transitions between discrete endemic and epidemic states. Having described these basic patterns of occurrence, Berryman described the patterns of recurrence in time and space as either "pulse", "cyclical" or "sustained", based on the regularity and duration of episodes (or patches) of high abundance (1987).

Thirty years later, we find this pattern classification scheme still has value, although range-wide analysis of many decades of new data from many pest species leads us to observe that the spatial and temporal scale of analysis, and the type of data used in that analysis (e.g., point samples of insect population density versus polygons of defoliation in discrete damage classes) will influence one's perceptions and conclusions. In short, scale matters (Figure 7.3a). Moreover, the ability to infer deterministic patterns of dynamics in space and time depends to what degree those dynamics are subject to random variations in ecological conditions (Figure 7.3b). The more stochastic the population process, the longer the system needs to be observed under a wider range of conditions in order to apply Berryman's classification scheme (1987) to a given data set (Figure 7.3).

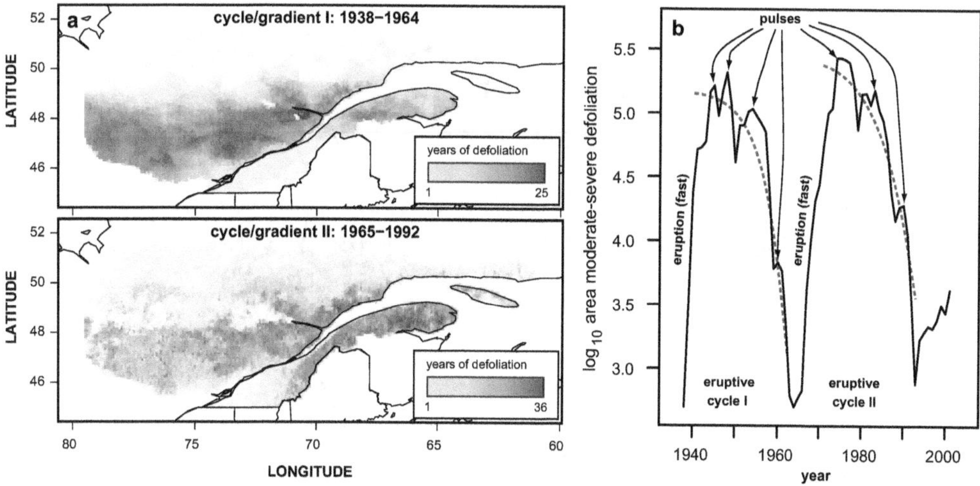

Figure 7.3 (a) Spatially, outbreak intensity is distributed according to a latitudinal belt-shaped gradient pattern, shown for spruce budworm defoliation in Quebec and Ontario, Canada, from 1938 to 1992. (b) Cyclic eruptions occur every few decades, with pulses of defoliation occurring every three to seven years, which are sustained through a slow decline in host forest condition. The dashed line indicates the pattern of decay associated with slow decline in forest condition, followed by a more rapid collapse due to faster predator–prey cycling.

This example illustrates how a single insect species, if it is observed long enough over a large enough spatial extent, may exhibit all five behaviours (eruptive, gradient, pulse, cyclic, sustained) conceptualized in Berryman's scheme (1987; Figure 7.2)

Although different forest–insect pest systems exhibit different outbreak patterns, there are some universal principles at play that serve as a common basis to understand the wide array of behaviours. All outbreaking insects are influenced by changes in forest conditions, natural enemies and climatic conditions (dF/dt, dP/dt and dC/dt; Figure 7.4). The understanding of the influence of these factors on rates of herbivore population change (dH/dt) has evolved to our current understanding of reciprocal feedbacks between natural enemies, pest herbivores and forests, where all three components are simultaneously influenced by climate. Our ability to understand and predict the outcomes of these interactions is further complicated by nonlinear population processes and feedbacks that interact at different spatio-temporal scales (Scott et al. 2021). For example, insect population turnover operates on timescales ranging from weeks to years and is therefore fast relative to tree biomass accumulation. Generally, complex dynamics are the result of the fast processes layered on top of the slow processes (Holling 2001). The same scale-crossing phenomena can also occur in space (Peters et al. 2004), for example, if the spread of outbreaks is enabled by landscape abundance of host species (Raffa et al. 2008). Cross-scale interactions can include both amplifying interactions via positive feedbacks that intensify the process (such as the Allee effect, which leads to abrupt population release or collapse), and attenuating interactions via negative feedbacks that diffuse the process (e.g., local plant diversity effect on herbivore population

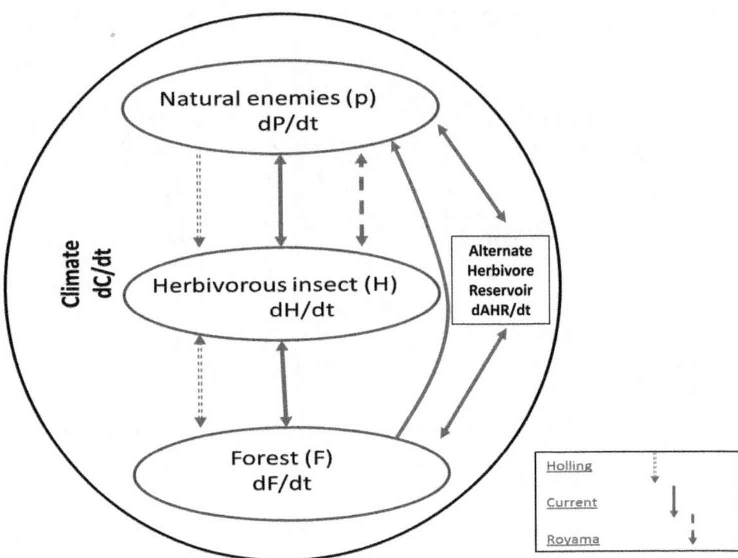

Figure 7.4 Factors influencing insect populations. Arrows indicate factors considered by different authors and the current viewpoint. The dynamic nature of the system is represented by changes in climate (dC/dt), changes in forest conditions (dF/dt), changes in natural enemies (dP/dt), changes in alternative herbivore resource (dAHR/dt) and their combined effect on outbreaking herbivorous insects (dH/dt). The current vision considers interactions between the target outbreaking insect, climate, natural enemies and the fact that these natural enemies often need alternate insect food sources which can also act as competitors with the outbreaking insect, and forest composition and structure which affects insect herbivores and thus natural enemies (Holling 1992).

growth) (Allen 2007). Feedbacks can occur within or across spatial scales, but when they occur across scales our ability to forecast the future dynamics of the system is complicated by strong interactions among the fast and slow variables. These nonlinear interactions across scales make predictions about the effects of climate change even more difficult (Boulanger et al. 2016, Scotte et al. 2021).

Despite these difficulties, there are also emergent properties of insect outbreak dynamics that are surprisingly reliable. Population dynamics of defoliators versus bark beetles behave somewhat differently due to the relative strength of host plant defences that must be overcome within each respective system. Defoliators consume a relatively renewable plant resource (foliage) while bark beetles consume an essentially non-renewable plant resource (phloem) with far more serious consequences for plant fitness on a biomass equivalent basis that are reflected by weak versus strong plant defences, respectively (Mattson et al. 1991). The mass-attack strategy that is the hallmark of tree-killing bark beetle systems (described below) is therefore unnecessary for the exponential growth of defoliator populations. The population dynamics of each respective system is therefore considered separately here. Nonetheless, we note that congregation is beneficial for avoiding mating failure (Regniere et al. 2013) and may help to explain the formation of epicentres (Bouchard & Auger 2014). There are thus many aspects of the population dynamics of these two groups of outbreaking insects where differences between the two groups blur.

Dynamics of defoliators

Defoliating larvae consume foliage and reduce the photosynthetic capacity of the tree, resulting in growth loss and eventually tree mortality. Plant secondary metabolites such as terpenes and phenols function as feeding deterrents or inhibit digestion of leaf material to influence feeding rate and development of defoliator larvae and thus the level of defoliation experienced by trees but may not have much influence during outbreak conditions (Bauce et al. 2006; Kumbasli et al. 2011). Adaptation to specific secondary metabolites and other species or genus-specific traits such as foliar phenology underlies host-specificity within defoliating species. Stand age, tree size and other characteristics of individual trees influence the severity of outbreaks. For example, spruce budworm damage is greater in mature rather than immature stands (Hennigar et al. 2008). The closely related jack pine budworm is dependent on pollen cones for its earliest growth stages, where severe defoliation can restrict pollen cone production the following year (Nealis and Lomic 1994). Similarly, there is evidence that in some deciduous forests, trees can increase the production of secondary metabolites to defend against elevated herbivory (Palo and Thomas 1984).

Among the most damaging defoliators of North American forests are spruce budworm, forest tent caterpillar (*Malacosoma disstria* Hbn.), jack pine budworm (*Choristoneura pinus pinus* Free.), spongy moth (*Lymantria dispar* L.) and hemlock looper (*Lambdina fiscellaria fiscellaria* Gn.). Spongy moth is an introduced species from Europe where it also causes large-scale damage along with the nun moth (*Lymantria monacha* L.), Siberian silk moth (*Dendrolimus sibiricus* [Chetverikov]), larch budmoth (*Zeiraphera diniana*), autumnal moth (*Epirrita autumnata*), pine processionary moth (*Thaumetopoea pityocampa*) and others. For example, an outbreak of the nun moth in the mid-1800s defoliated spruce (*Picea abies*) forests for a decade in western Russia and eastern Prussia (modern-day Poland) (Bejer 1988). Many of these species share a common trait in that they are "early season" defoliators that time their primary feeding stage with the spring flush of new foliage that is both poorly defended and nutrient-rich. However, as with all attempts at generalizing there are also outbreaking species with late-season phenologies, for example, winter moth (*Operophtera brumata*) and autumnal moth (Vindstad et al. 2021), and more complex semivoltine life histories, for example, Siberian silk moth (Kirichenko et al. 2011).

Defoliator outbreaks have been considered to be characteristically periodic or quasi-periodic in time and spatially synchronized (Myers and Cory 2013). However, recent work is challenging this notion and demonstrating more asynchronous behaviours at small scales (Berguet et al. 2021, Nenzen et al. 2018) with strong evidence of non-stationarity for species like the forest tent caterpillar (Cooke and Roland 2023).

It has been suggested that the primary oscillation in populations is generally caused by a delayed density-dependent reaction of natural enemies to their defoliator prey/host populations, leading to a classical predator–prey cycle (Royama 2005). Vertebrate predators such as songbirds and rodents can functionally respond to increasing defoliator populations but have limited numerical response and are consequently most influential during endemic population phases. For example, birds consumed 84 per cent of spruce budworm larvae and pupae at low population levels, 22 per cent during intermediate levels, but had negligible impacts on epidemic populations (Jennings and Crawford 1985). However, Régnière et al. (2021) suggest that birds contribute to reducing populations of spruce budworm in the declining phase of an outbreak.

There is more support for the top-down influence of parasitoids, which have more rapid numerical response and capacity to control outbreak populations (Roland 2005). Parasitoids (primarily wasps [*Hymenoptera*] and flies [*Diptera*]) deposit eggs on, in, or near specific immature life stages of the defoliator (i.e., eggs, larvae, or pupae) that later hatch and consume the defoliator host. The parasitoid community affecting the spruce budworm is dominated by generalist species with comparatively slow numerical response (Eveleigh et al. 2007), resulting in a characteristically long cycle with long duration (Régnière and Nealis 2007). In contrast, hemlock looper and forest tent caterpillar, parasitoid communities are dominated by specialists which have rapid numerical response, resulting in more frequent outbreaks of shorter duration (Hébert et al. 2001). *Lepidoptera* are also attacked by viral diseases. For example, forest tent caterpillar larvae are colonial and thus the rates of lateral transmission are very high in larval aggregations (Roland and Kaupp 1995).

It has been suggested that during climate change, insect pests may move into ranges where parasitoids are absent or in reduced numbers (Stireman et al. 2005). There is some specific support in the case of the spruce budworm for reduced abundance of specific parasitoids like *Apanteles fumiferana* in northern forests suggesting that parasitoid control may be reduced with pest shifts due to climate change (Legault et al. 2018).

Cycle length has additional consequences for the impact of a given outbreak, as it is the combination of defoliation intensity (i.e., the percent foliage damaged) and duration (i.e., consecutive years of defoliation) that determines tree damage (MacLean 1980; Cooke et al. 2012). For those pest systems that cause widespread mortality of tree hosts, forest collapse may contribute to the outbreak decline (Régnière and Nealis 2007; Cooke et al. 2007). Locally oscillating populations subjected to small, random but regionally correlated perturbations, typically by weather, can lead to the synchronization of those oscillations across space—a phenomenon known as the "Moran effect" (Royama 1984). Random weather correlates with variations in survival, these random perturbations accumulate over time, and if the random perturbations are spatially correlated, the cycle phases will tend to converge (Cooke et al. 2007). The Moran effect has been suggested as an important factor contributing to the regional outbreak synchrony for a wide range of *Lepidopteran* species (Ranta et al. 1997; Myers and Cory 2013). Adult dispersal and parasitoid dispersal are processes by which independently oscillating populations can become synchronized (Régnière and Lysyk 1995; Kaitala and Ranta 1998; Williams and Liebhold 2000). Imperfect synchronization can manifest as "epicentres" spreading to adjacent areas (Berryman 1987), or as "travelling waves" where outbreaks spread across large geographic areas (e.g., Bjørnstad et al. 2002; Rober et al. 2020, Tenow et al. 2013). Recent evidence from the forest tent caterpillar system shows that the forest environment, that is, the abundance and configuration of host trees, influences outbreak synchrony (Cooke and Roland 2023).

Defoliator pest species have been broadly characterized as either grazers, which attack the current year's foliage, or wasteful feeders, which attack either foliage from previous years or destroy a disproportionate quantity of foliage, which is not consumed (Cooke et al. 2007). Grazers such as spruce budworm can repeatedly remove new-year foliage over five to seven years before a tree dies (MacLean and Ostaff 1989). On hardwood species such as aspen (*Populus* spp.), trees produce a second flush of leaves following early season defoliation. However, continued defoliation for more than three years can deplete metabolic reserves enough to cause widespread tree mortality (Cooke et al. 2012). Wasteful feeding by species such as hemlock looper and Siberian silk moth can cause tree death more rapidly than grazers as they attack both current and previous years' foliage during a single year (MacLean and Ebert 1999; Iqbal et al. 2011, Kirichenko and Baranchikov 2007). Differences in feeding behaviour can lead to characteristic spatial patterns of tree mortality on the landscape (Cooke et al. 2007). Wasteful-feeding defoliators generally cause moderate to severe mortality that is aggregated locally as outbreaks collapse before they can spread across landscapes. In contrast, moderate defoliation by grazing defoliators is generally more diffuse in pattern and extensive in spatial scale.

Dynamics of bark beetles

Several species of economically important bark beetles have been recently active in western North America (Raffa et al. 2008) while the southern pine beetle (*Dendroctonus frontalis* Zimmerman) has been a long-term issue in pine-dominated systems of the south-eastern United States (Coulson and Klepzig 2011). Across Europe, bark beetles affected 8.2 Mha or 3.0 per cent of the forest area in 2005, which was second in Europe only to wind and storm damage (FAO 2014).

Bark beetles develop from eggs laid in galleries within the phloem of trees under the bark (i.e., subcortical environment). Larvae hatching from these eggs feed on nutrient-rich phloem, resulting in tunnels under the bark. The subcortical environment offers some protection from thermal extremes, desiccation (whereas to survive in external environments defoliators had to adapt through feeding shelters, motility, colonial behaviour, and increased water use efficiencies), and natural enemies, relative to the exposed conditions experienced by defoliator larvae feeding on foliage. Extensive feeding can girdle the tree, thereby preventing the transport of nutrients (Wood 1982). Consequently, the subcortical environment is among the most intensely defended tissues within a tree.

Genus-specific defences of conifers attacked by bark beetles include the production of resins that physically eject or isolate burrowing insects, the ability of wound periderm to suberize and isolate damaged phloem, and the production of toxic phenolic compounds (Franceschi et al. 2005). To overcome host defences, bark beetles rely on a mass attack where large numbers of attacking beetles combined with symbiotic fungal associates overwhelm host defences. Bark beetles can detoxify host terpenes through oxidation and convert them to aggregation pheromones, which further encourage the mass attack (Wood 1982; Seybold and Tittiger 2003). This adaptation to resin-based defences in conifers is thought to account for the affiliation of tree-killing bark beetles with coniferous tree species (Seybold et al. 2006).

During the endemic stage, tree-killing bark beetles persist by attacking weakened trees with poor defences. A certain threshold population density is required to successfully mass attack and kill healthy trees. However, once that threshold is reached, the reproductive capacity of tree-killing bark beetles increases exponentially before levelling or dropping off at extreme population densities (Boone et al. 2011). Outbreaks occur as key population thresholds are surpassed due to triggering events such as widespread wind disturbance or drought that weaken large populations of otherwise healthy host trees (Raffa et al. 2008). Threshold temperatures may also heavily influence both generation time and overwintering success—each of which may have dramatic effects on bark beetle outbreak dynamics (Powell and Bentz 2009; see "Response to climate change" section below).

The natural enemy complex affecting bark beetles includes predatory beetles and woodpeckers, and parasitic flies and wasps (Wermelinger 2004; Safranyik and Carroll 2006;

Coulson and Klepzig 2011). Predation and parasitism impacts are far more challenging to measure within the subcortical environment by comparison with free-ranging defoliators. Clerid beetles have been shown to have a significant influence on southern pine beetle population dynamics (Reeve 1997). There is an absence of quantitative evidence for analogous effects in other bark beetle systems, including mountain pine beetle, where the bottom-up effects of host plant defences are thought to outweigh the top-down influence of predation and parasitism. For example, Safranyik et al. (1999), in a model of mountain pine beetle population dynamics, include the effects of predation as a weak fixed effect. A key limitation in all studies of bark beetle population dynamics is the context-dependent nature of predation and parasitism. Unfortunately, there is a heavy bias towards the study of post-eruptive populations and a paucity of studies targeted outside the focal epicentre, on the endemic populations in the surrounding landscape.

Transitions between discrete endemic and epidemic states are abrupt within tree-killing bark beetle systems. These state transitions are accompanied by both behavioural shifts in host preferences and attack behaviour (Safranyik and Carroll 2006) and a loss of spatial patterning, as focal epicentres coalesce into relatively homogeneous, landscape-scale patterns of mortality. The rate of transition from endemic to eruptive behaviour depends on thermal parameters governing the number of generations per year. Southern pine beetle, occupying the temperate southern United States (and also areas of Mexico and Central America), may have several generations per year (i.e., multi-voltine), so spot growth proceeds relatively quickly. Mountain pine beetle, in contrast, usually has one generation per year (i.e., univoltine), and sometimes has fewer than one generation per year (i.e., partial voltinism), so spot growth occurs across years rather than within the year.

Severe outbreaks of bark beetle tend to collapse as a result of exhaustion of suitable host trees, although they may be truncated prematurely by extremely unfavourable weather (Raffa et al. 2008). The spatial pattern of bark beetle outbreak—either clustered or continuous—depends on whether the outbreak is allowed to develop to maturity. Mountain pine beetle outbreaks often terminate before host trees are completely exhausted. As the forest is thinned, the outbreak tends to fizzle, and it is even thought that there is a critical host patch size below which outbreaks are not possible (Heavilin and Powell 2008). This illustrates that bark beetles both influence, and are influenced by, host forest structure—the definition of reciprocal feedback.

Invasive species

The previous two sections illustrate how the interplay between bottom-up (i.e., tree defence) and top-down (i.e., natural enemies) forces result in consequent spatio-temporal patterns of insect damage within native forest–pest systems. The introduction of nonindigenous insects via global trade represents an increasingly prevalent threat to forests because of the potential for unrestricted population growth due to the lack of natural enemies and/or host resistance (Work et al. 2005). It is important to understand that accidental introductions are quite common, but only a select few species establish in new areas (Gandhi and Herms 2010. Nealis et al. 2016). Yet the arrival of species such as the emerald ash borer (*Agrilus planipennis*) in North America has had and will continue to have catastrophic effects on entire genera of trees. Understanding factors that permit establishment and spread of newly arrived forest insects has therefore become a critical issue for both forest health and biodiversity worldwide (Gandhi and Herms 2010).

Two brief case studies illustrate the processes and issues underlying outbreak behaviour as species invade new systems. The emerald ash borer is a wood-boring beetle of the family Buprestidae. While little is known about its life history within its native range in China, it causes only minor damage to associated ash (*Fraxinus* spp.) trees there. Emerald ash borer was first detected in North America in the interior port city of Detroit, where it became apparent that North American ash species have virtually no resistance to its attack. The species readily attacks and kills ash trees of any size down to 2 cm in diameter. While the species has limited dispersal capability, its spread has been massively exacerbated by human transport of colonized wood. While a few introduced predators hold promise for biocontrol, the emerald ash borer exemplifies the worst-case scenario (as also witnessed in the pathogens that rendered functionally extinct the American chestnut and decimated the American Elm) of unrestricted population growth that threatens the existence of entire tree genera at biome to continental scales.

European spongy moth (*Lymantria dispar*) was introduced to North America as a potential silk producer, but subsequently escaped captivity and became established in hardwood forests of Massachusetts at the turn of the nineteenth century. This species is a polyphagous defoliator with a particular affinity for oak species (*Quercus* spp.). Many natural enemies have been introduced for the biocontrol of spongy moth, with unfortunate unintended consequences for some native *Lepidopteran* species. Defoliation damage by the spongy moth, while extensive, is far less virulent in its impacts. It has been suggested that outbreaks may be worse following the management interventions as they may maintain populations of the insect (Petrovsky and MacKay 2010). Following invasion, its behaviour is often similar to that in its native range, that is, periodic, regionally synchronized outbreaks causing moderate damage, although naturalized populations do take time to fully equilibrate and surprises still occur (Liebhold et al. 2022).

The impacts of introduced insect species therefore range from benign to catastrophic. The very real risk of the worst-case scenario requires extreme vigilance to avoid accidental introductions, requiring close cooperation across jurisdictional boundaries around the globe. If exotic species become established and invasive, quick response strategies can limit large-scale tree loss, including effective and early identification of infested trees, the eradication of isolated populations, slowing the spread (e.g., by reducing or eliminating human transport) and diversifying forests (i.e., urban forests).

Insect disturbance effects on forest dynamics

As forest characteristics affect outbreak severity, the resultant tree mortality from outbreaks also alters forest structure and ecosystem processes. Tree death creates canopy gaps of various sizes and increases the volume of snags and coarse woody debris (Kneeshaw and Bergeron 1998; Taylor and MacLean 2007; Coulson and Klepzig 2011, Przepióra et al. 2020). As stressed throughout this chapter, insect disturbance differs from most other types of disturbances in that it is host-specific—and this host-specificity has implications for resulting patterns of growth reduction, tree mortality, and forest response to those processes through time. In some cases, succession is accelerated as shade-intolerant overstory tree species are selectively killed during outbreaks and composition shifts towards more shade-tolerant species. For example, multiple years of trembling aspen defoliation by the forest tent caterpillar has led to a shift from shade-intolerant aspen to more shade-tolerant balsam fir (Moulinier et al. 2013). In contrast, spruce budworm outbreaks in balsam fir—tolerant hardwood stands can shift the stand towards yellow birch or sugar maple dominance (Amos-Bink and MacLean 2010, 2016). Shade-tolerant species can persist following overstory mortality when advance regeneration is abundant. However, non-host species may be recruited if advance regeneration is limited (Reyes et al. 2010).

The balsam fir–spruce budworm system is a classic example, in which the insect was labelled a "super silviculturist" by Baskerville (1975a) as overstory mortality resulted in the release of an understory seedling bank of the host species balsam fir (Spence and MacLean 2012) that would then be ready to be "harvested" by the insect in a subsequent outbreak. Similar re-establishment of shade-tolerant species has been observed in central Europe after severe bark beetle (*Ips typographus*) outbreaks where Norway spruce forest regenerates well after the overstory mortality of spruce (Jonasova and Prach 2004; Svoboda et al. 2010). In natural forests, however, these dynamics are a function both of forest structure and outbreak severity (Figure 7.5). In mixedwood stands, mortality of mature balsam fir trees caused by the spruce budworm has been identified as providing opportunities for companion species to recruit (Bouchard et al. 2006) or for equally shade-tolerant species such as eastern white cedar to increase in dominance (Kneeshaw and Bergeron 1998) (Figure 7.5). Similar dynamics are also observed following mountain pine beetle outbreaks where mortality of pine in mixed stands can accelerate succession to more shade-tolerant companion species.

In contrast, in the spruce budworm system balsam fir is a shade-tolerant species and deciduous species, such as paper birch (*Betula papyrifera*) and aspen (*Populus tremuloides*), where present, may also increase in abundance in response to outbreaks opening the canopy (Taylor and Chen 2011). This could lead to an increase in site productivity given their positive effect on soil nutrient availability (Légaré et al. 2005). The presence of hardwood species can reduce the level of defoliation of host species during spruce budworm outbreaks (Su et al. 1996; Zhang et al. 2018), possibly due to greater herbivore and natural enemy diversity and to early-instar dispersal losses (Zhang et al. 2020).

Following insect outbreaks, spikes in nitrogen due to the accumulation of insect frass and insect mortality have been observed (Paré and Bergeron 1996) creating a fertilizing effect with the greatest effect occurring in nitrogen-limited ecosystems. Forests defoliated by spongy moth have been shown to release nitrogen loads into the watershed (Townsend et al. 2004). It has also been shown that nutrient enrichment of soils, due to frass, dead larva, etc. positively feedback into the severity of spruce budworm outbreaks (DeGrandpré et al. 2022). On poor-quality sites canopy opening resulting from insect outbreak could lead to further losses in productivity by permitting ericaceous species (e.g., *Kalmia* spp.) to increase; such ericaceous species subsequently inhibit the regeneration of spruce and other tree species and are associated with reduced nutrient availability and slow tree growth (Mallik 2003).

Forest management effects: positive and negative

Insect populations are constrained by food availability, which is affected by forest management through changes to tree-species composition (i.e., hosts to non-hosts), to size or age–class structure (size and age/vigour affect insect growth and damage) and configuration (fragmentation and isolation affect dispersal). In the following, we will use examples from the spruce budworm and mountain pine beetle systems as case study examples of the potential effects of forest management.

Forest management practices that increase host species will lead to more severe (greater mortality) outbreaks and these may be occurring over larger areas than in the past (Blais

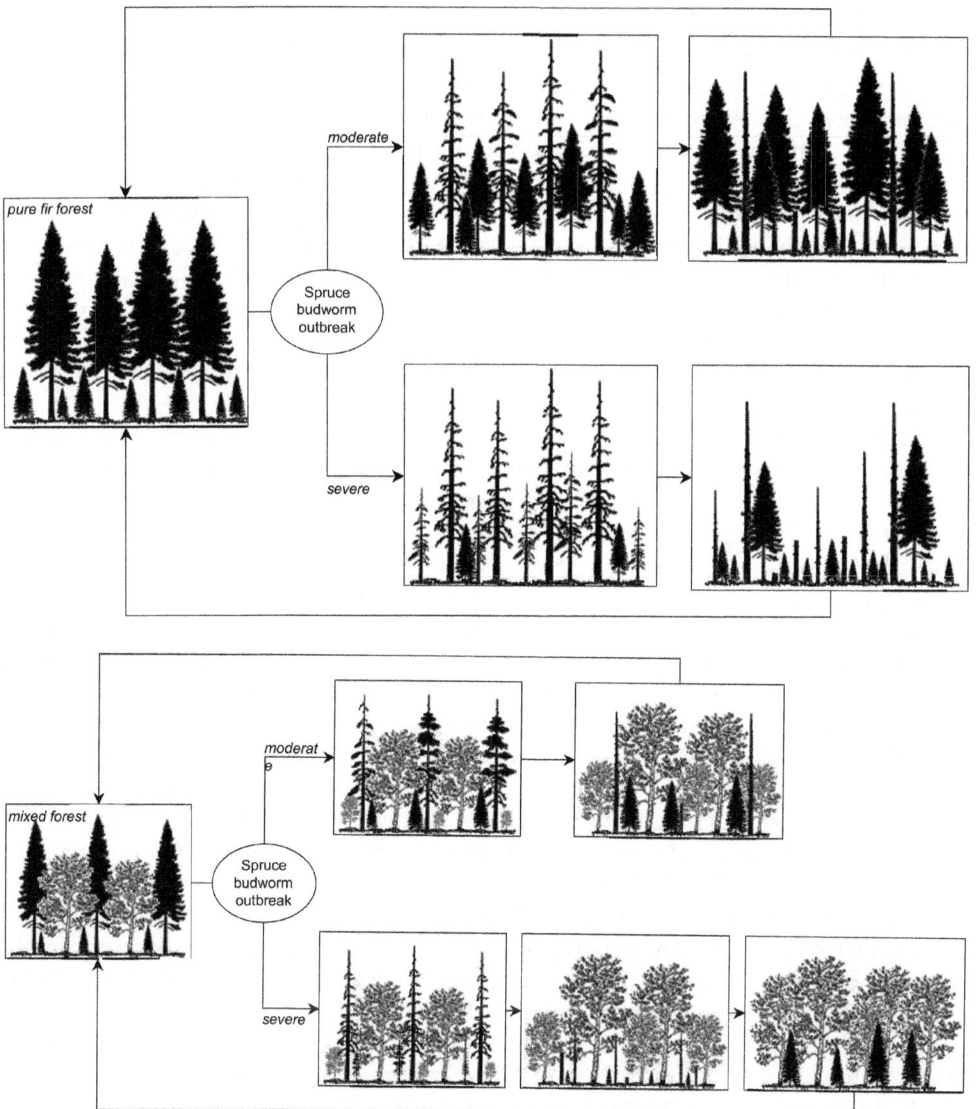

Figure 7.5 Forest dynamics following different outbreak severities. In panel A, a moderate outbreak in a host-dominated forest releases the understory, which then recruits to the overstory, following a severe outbreak the advance regeneration is also reduced and stand development to a fully stocked host stand takes much longer. In panel B, the initial stand conditions are a mixedwood stand. A moderate outbreak can lead to the maintenance of a mixed stand although species proportions may change. After a severe outbreak, stand composition can be shifted to dominance of companion species. Over time, shade-tolerant conifer host species may re-establish.

1983; Raffa et al. 2008). Among the most pervasive effects of forest management on insect outbreaks is fire suppression, which can favour host species and older stands, both of which increase forest vulnerability (Sturtevant et al. 2012). Forest management and fire suppression have promoted the availability of large contiguous stands of lodgepole pine, a major host species for the mountain pine beetle, and this was identified as a cause of the extent of the current outbreak (Raffa et al. 2008). Stand-level treatments by themselves have little if any influence on defoliator dynamics but changes in forest composition affect damage severity (Muzika and Liebhold 2000; Weso³owski and Rowiński 2006). Pure stands of mature host trees lead to greater and widespread mortality while forests mixed with hosts and phylogenetically distant tree species tend to have reduced herbivory (Jactel and Brockerhoff 2007). For example, higher content of non-host hardwoods can reduce spruce budworm, reducing defoliation and mortality in host fir and spruce stands (Bergeron et al. 1995; Su et al. 1996; Zhang et al. 2018).

Age and size class are directly modified by forest management. For example, Hennigar et al. (2008) report that mature stands of host species experience greater mortality than immature stands of the same species. The mountain pine beetle, on the other hand, attacks only trees greater than 20–25 cm in diameter. Harvesting and the establishment of plantations can lead to younger forests that are typically less vulnerable, but not invulnerable to species such as the spruce budworm.

Landscape studies have demonstrated that human-caused modifications to forested landscapes affect outbreak characteristics. For example, spruce budworm outbreaks within unmanaged forest landscapes were found (1) to be more synchronous; (2) to have more trees per site affected; and (3) to be less frequent than within commercially managed forests (Robert et al. 2012, 2018). In aspen-dominated forests, outbreaks of the forest tent caterpillar were more severe and lasted longer in fragmented landscapes (Robert et al. 2020); whereas in Europe the winter moth was found to cause less defoliation during outbreaks in management-fragmented landscapes (Wesołowski and Rowiński 2006). Host connectivity may influence outbreak synchrony and mitigate forest damage by affecting defoliator dispersal (Ims et al. 2008), movement of natural enemies from adjacent habitats (Roland and Taylor 1997), or both. Although guidelines have been proposed to reduce risk and loss of timber in severe outbreaks (e.g. reducing the connectivity of the most vulnerable host species and size classes), the mechanisms and the efficacy of the proposed treatments are still questioned although evidence from combined host outbreaking insect systems are providing insights. Large blocks of host trees favour high-amplitude cycling although too many hosts may lead to a loss of cycle stability whereas when host habitat is low the spatial dynamics of eruptive spread dominates over the cycling dynamics in time (Cooke et al 2024) A landscape-scale risk analyses also show that outbreak proofing a forest landscape would require modifying forest conditions on scales that seem implausible (McNie et al. 2023) and yet decades of allowing host build-up can lead to predictable surprises (Kneeshaw et al. 2021)

Some authors suggest that greater tree diversity offers a protective effect (Su et al. 1996; Zhang et al. 2018; MacLean and Clark 2021), but there is still controversy on the strength of any effect (Miller and Rusnock 1993; Koricheva et al. 2006). An explanation for the lack of a consistent relationship is that diversity effects are strongest when mixtures are composed of phylogenetically distant trees (Jactel and Brockerhoff 2007). Even in the absence of protective effects, tree mixtures of non-host tree species will reduce the potential for widespread mortality compared to contiguous stands of host species (Baskerville 1975b). However, increases in non-host species may not always be desirable if these are not commercially valuable (Needham et al. 1999). In such a case, foresters are losing potentially valuable timber fibre to protect against uncertain losses. In other cases, increasing the proportion of non-host species for one insect may simply mean increasing the proportion of a species that is vulnerable to another insect pest (Robert 2014). For example, increasing the proportion of aspen and birch in a forest as non-host species to the spruce budworm may increase vulnerability to forest tent caterpillar outbreak. Trade-offs that occur due to manipulating forest composition need to be carefully evaluated, which can be done using stand growth models and decision support systems for forest insects (e.g., MacLean et al. 2001; Hennigar et al. 2011; Iqbal et al. 2012).

Response to climate change

Weather and climate affect survival and development of individual life stages and thus can contribute to incipient outbreaks or to their collapse—in both defoliator and bark beetle systems. In general, warm conditions favour rapid development of larvae whereas cold temperatures can limit the northern distribution of insects if thermal units are insufficient to complete a generation. Increased access to alternative susceptible hosts at higher altitudes and milder winter temperatures have contributed to the recent unprecedented mountain pine beetle outbreak in western North America (Bentz et al. 2010). In particular, temperature-related transitions in generation rates influence the invasibility of many species around the northern hemisphere (Ghandi and Hofstetter 2021). For species such as spruce budworm, warm autumn temperatures delay diapause and increase mortality as larvae exhaust metabolic resources needed to overwinter (Régnière et al. 2012). In the early stages of a spruce budworm outbreak, previous year outbreak conditions and spring climate predict spruce budworm population changes in the following year (Li et al. 2020). Climate governs at least the broadest outlines of species distributions (Sexton et al. 2009). Our greatest certainty is that following climate change, current patterns in disturbances will not remain the same (Dale et al. 2001; Haynes et al. 2014). In terms of insect outbreaks, climate change will influence outbreak dynamics by modifying insect reproduction, survival and ranges (Candau and Fleming 2011, Pureswaran et al. 2018), and host species distribution (Collier et al. 2020). Along a 700-km latitudinal gradient in eastern Canada, balsam fir regeneration density was negatively affected by increased mean annual temperature in interaction with height class, while height and lateral branch growth rates were generally positively affected by increased summer precipitation (Collier et al. 2024). As described below, changes can also be expected between hosts, insect pests and controlling parasitoids and predators.

Some authors have suggested that insect outbreaks will increase in frequency and severity as climate warms (Fleming et al. 2002; Schelhaas et al. 2003). Range shifts are also predicted with outbreaks extending beyond traditional limits as climate becomes more or less hospitable (Régnière et al. 2012, Lesk et al. 2017, Rozenberg et al. 2020). These predictions are already supported by observations of outbreak range expansions and phenological matching of pests and tree hosts in some northern forest ecosystems (Jepsen et al. 2011). These range shifts can expose previously unattacked (or rarely attacked) tree species in the same family or genus to novel herbivory, which may be exacerbated by a decoupling of forest pests and their parasitoids (Stireman et al. 2005), thus leading to greater severity outbreaks in the new range. Individualist responses of different species to climate change have been observed in Europe, with some species' outbreaks increasing in severity and others having outbreak population cycles collapse (Haynes et al. 2014). Other research suggests that outbreaks may decrease in amplitude and severity if heat stress or drought affects the insect pest directly (Pureswaran et al. 2018)

Tree hosts are also affected by climate and may be more vulnerable to insect attack if stressed by non-optimal conditions (Mattson and Haack 1987; Zhang et al. 2014). For example, drought creates stress and lowers tree defences, making them less resistant to herbivorous insect attacks (the plant-stress hypothesis) (Larsson 1989). Drought stress has been implicated as a major driver underlying widespread bark beetle outbreaks by multiple species in western North American forests (Raffa et al. 2008, Gandi and Hofstetter 2021). The evidence for the impact of drought on defoliators is, however, mitigated (Kolb et al. 2016).

In contrast, higher mean temperatures may be of greater benefit to the natural enemy complex by increasing the development rate, fecundity and search rate of multiple parasitoid species (Gray 2008). However, the control response of parasitoids is equivocal as Stireman et al. (2005) suggested that parasitoids may be disadvantaged by slower dispersal during range expansion. Greater knowledge is also needed about how temperature and precipitation at different times of the year affect insect pests and their parasites. Although the effects of climate on major insect pests have been studied to some extent, very little is known about climatic controls on parasitoid species.

Conclusion

Although insects occur throughout the world, those that cause economically damaging large-scale outbreaks occur primarily in temperate and boreal forests with relatively few tree species. The outbreaking insects can be separated into two large groups: defoliators and bark beetles. For both groups, it is the larval stages that cause damage and mortality and outbreaking species are characterized by populations with large oscillations. Combinations of weather, forest condition (i.e., host abundance, age and size class distribution and landscape configuration) and natural enemies control the population dynamics of outbreaking insects. These factors all work together to influence the rise and fall of outbreaks. In the case of invasive exotic species, insect outbreaks may occur in the new ranges due to the absence of natural enemies and reduced host resistance.

It is our contention that to reduce the impacts of forest outbreaks, greater focus should be placed on forest pests not only during the epidemic phases during which management responses are primarily reactive, but also during endemic phases when landscape-scale forest management decisions will have a large effect on subsequent outbreak severity. Often for political and economic reasons, decisions are made that increase conditions for future outbreaks (Kneeshaw et al. 2021). These authors argue that we have the scientific knowledge to reduce future impacts, but that decision making is too often based on short-term considerations. Modelling Daniel approaches based on simple rules that link outbreak severity with forest conditions are being used more frequently (MacLean et al. 2001; Hennigar et al. 2011). These decision support systems are useful during outbreaks to target interventions (pesticide application, salvage and/ or harvest rescheduling) so as to minimize both short-term and long-term losses and for preventive planning (forest restructuring) during endemic stages. During these stages, landscapes and stands can be managed to reduce host connectivity, reduce host composition and maximize stand and age–class diversity as pest management strategies to reduce the impact of future outbreaks. And during incipient outbreak periods, an 'early intervention strategy' of intensive monitoring of overwintering larval populations, combined with biological insecticide treatment of 'hotspots' before defoliation is evident, has been shown to prevent a spruce budworm outbreak (Johns et al. 2019; Liu et al. 2019; MacLean et al. 2019).

Changes in forest landscapes and climate change are currently leading to, or are projected to lead to, changes in the extent and distribution of forest types and insect populations. Changes in host, predator and parasite relationships may either aggravate or attenuate outbreaks such that some insect pests will be of greater concern and others will cause less damage, and locations of outbreaks may differ. Combined, these nonlinear and cross-scale interactions (Holling 1992; Peters et al. 2004) conspire to increase long-term uncertainty in future forest–insect dynamics. Indeed, there is already evidence from around the world of changes in outbreak characteristics that have been linked to global changes (Dale et al 2001; Schelhaas et al. 2003).

Although insect outbreaks can cause widespread economic impact by causing millions of hectares of damage, outbreaks are also a natural process. As such they can be creative forces of re-organization leading to nutrient pulses, shifts in tree composition and forest structure from stand to landscape scales.

References

Abbott, B., Stennes, B. and van Kooten, G.C. 2009. Mountain pine beetle, global markets, and the British Columbia forest economy. *Canadian Journal of Forest Research* 39: 1313–1321.

Allen, C.D. 2007. Interactions across spatial scales among forest dieback, fire, and erosion in northern New Mexico landscapes. *Ecosystems* 10: 797–808.

Amos-Binks, L.J., and MacLean, D.A. 2016. The influence of natural disturbances on developmental patterns in Acadian mixedwood forests from 1946–2008. *Dendrochronologia* 37: 9–16.

Amos-Binks, L.J., MacLean, D.A., Wilson, J.S. and Wagner, R.G. 2010. Temporal changes in species composition of mixedwood stands in northwest New Brunswick: 1946–2008. *Canadian Journal of Forest Research* 40: 1–12.

Baskerville, G. 1975a. Spruce budworm: super silviculturist. *The Forestry Chronicle* 51: 138–140.

Baskerville, G. 1975b. Spruce budworm: the answer is forest management: or is it? *The Forestry Chronicle* 51: 157–160.

Bauce, E., Kumbasli, M., Van Fankenhuyzen, K. and Carisey, N. 2006. Interactions among white spruce tannins, Bacillus thuringiensis subsp. kurstaki, and spruce budworm (Lepidoptera: Tortricidae), on larval survival, growth and development. *Journal of Economic Entomology* 99: 2038–2047.

Bejer, B. 1988. The nun moth in European spruce forests. In: *Dynamics of Forest Insect Populations*, Springer, New York, pp. 211–231.

Bentz, B.J., Régnière, J., Fettig, C.J., Hansen, E.M., Hayes, J.L. and others. 2010. Climate change and bark beetles of the western United States and Canada: direct and indirect effects. *BioScience* 60: 602–613.

Bergeron, Y., Leduc, A., Morin, H. and Joyal, C. 1995. Balsam fir mortality following the last spruce budworm outbreak in northwestern Quebec. *Canadian Journal of Forest Research* 25: 1375–1384.

Berguet, C., Martin, M., Arseneault, D. and Morin, H., 2021. Spatiotemporal dynamics of 20th-century spruce budworm outbreaks in eastern Canada: Three distinct patterns of outbreak severity. *Frontiers in Ecology and Evolution* 8: 544088.

Berryman, A.A. 1987. Equilibrium or non-equilibrium: Is that the question? *Bulletin Ecological Society of America* 68: 500–502.

Bjørnstad, O.N., Peltonen, M., Liebold, A.M. and Baltensweiler, W. 2002. Larch bud moth population dynamics and forest defoliation in the European Alps. *Science* 298: 1020–1023.

Boone, C.K., Aukema, B.H., Bohlmann, J., Carroll, A.L. and Raffa, K.F. 2011. Efficacy of tree defense physiology varies with bark beetle population density: a basis for positive feedback in eruptive species. *Canadian Journal of Forest Research* 41: 1174–1188.

Bouchard, M. and Auger, I. 2014. Influence of environmental factors and spatio-temporal covariates during the initial development of a spruce budworm outbreak. *Landscape Ecology* 29: 111–1126.

Bouchard, M., Kneeshaw, D. and Bergeron, Y. 2006. Forest dynamics after successive spruce budworm outbreaks in mixedwood forests. *Ecology* 87: 2319–2329.

Boulanger, Y., Gray, D.R., Cooke, B.J. and De Grandpré, L. 2016. Model-specification uncertainty in future forest pest outbreak. *Global Change Biology* 22(4): 1595–1607.

Candau, J. and Fleming, R. 2011. Forecasting the response of spruce budworm defoliation to climate change in Ontario. *Canadian Journal of Forest Research* 41: 1948–1960.

Collier, J., MacLean, D.A., D'Orangeville, L. and Taylor, A.R. 2022. A review of climate change effects on the regeneration dynamics of balsam fir. *The Forestry Chronicle* 98: 54–65.

Collier, J., MacLean, D.A., Taylor, A.R., and D'Orangeville, L. 2024. Warming adversely affects density but not growth of balsam fir regeneration across a climatic gradient in the Acadian rest Region of eastern Canada. *Canadian Journal of Forest Research* 54: 134–146 (accepted July 2023).

Cooke, B.J. 2024. Forest tent caterpillar (Lepidoptera: Lasiocampidae) across Canada, 1938–2001: I. Periodic outbreaks; episodic impacts. *The Canadian Entomologist* 156, e8: 1–16.

Cooke, B.J., MacQuarrie, C.J. and Lorenzetti, F. 2012. The dynamics of forest tent caterpillar outbreaks across east-central Canada. *Ecography* 35: 422–435.

Cooke, B.J., Nealis, V.G. and Régnière, J. 2007. Insect defoliators as periodic disturbances in northern forest ecosystems. In: *Plant Disturbance Ecology: The Process and the Response.* Edited by E.A. Johnson and K. Miyanishi. Elsevier Academic Press, Burlington, MA, pp. 487–525.

Cooke, B.J., Robert, L.E., Sturtevant, B.R., Kneeshaw, D. and Thapa, B., 2024. Confronting the cycle synchronisation paradigm of defoliator outbreaks in space and time—Evidence from two systems in a mixed-species forest landscape. *Journal of Ecology*, 112(1:152–173.

Cooke, B.J. and Roland, J., 2023. Variable synchrony in insect outbreak cycling across a forest landscape gradient: multi-scale evidence from trembling aspen in Alberta. *Canadian Journal of Forest Research*, 53(11): 839–854.

Coulson, R.N. and Klepzig, K.D. 2011. Southern Pine Beetle II. Gen. Tech. Rep. SRS-140. U.S. Department of Agriculture Forest Service, Southern Research Station, Asheville, NC, 512 pp.

Dale, V.H., Joyce, L.A., McNulty, S., Neilson, R.P., Ayres, M.P. and others. 2001. Climate change and forest disturbances. *BioScience* 51: 723–734.

Damon, A. 2000. A review of the biology and control of the coffee berry borer, Hypothenemu hampei (Coleoptera: Scolytidae). *Bulletin of Entomological Research* 90: 453–465.

DeGrandpré, L., Marchand, M., Kneeshaw, D.D., Paré, D., Boucher, D., Bourassa, S., Gervais, D., Simard, M., Griffin, J.M. and Pureswaran, D.S., 2022. Defoliation-induced changes in foliage quality may trigger broad-scale insect outbreaks. *Communications Biology*, 5(1): 463.

Dyer, L.A., Carson, W.P. and Leigh, E.G. Jr. 2012. Insect outbreaks in tropical forests: patterns, mechanisms and consequences. In: *Insect Outbreaks Revisited.* Edited by P. Barbosa, D.K. Letourneau and A.A. Agrawal.Wiley Blackwell, Hoboken NJ, pp. 219–245.

Elton, C.S. 1958. *The Ecology of Invasions by Animals and Plants.* Methuen, London.

Esper, J., Buntgen, U., Frank, D.C., Nievergelt, D. and Liebhold, A. 2007. 1200 years of regular outbreaks in alpine insects. *Proceedings of the Royal Society B* 274: 671–679.

Eveleigh, E.S., McCann, K.S., McCarthy, P.C., Pollock, S.J., Lucarotti, C.J. and others. 2007. Fluctuations in density of an outbreak species drive diversity cascades in food webs. *Proceedings of the National Academy of Sciences* 104: 16976–16981.

FAO (Food and Agriculture Organization). 2014. Disturbance affecting forest health and vitality (1 000 ha) by FRA categories, year, country. Forest Health, CountrySTAT, Food and Agriculture Organization of the United Nations. Rome, Italy. Available online at http://countrystat.org/home.aspx?c=FOR&tr=5 [viewed 26 March 2014].

Fleming, R.A., Candau, J.N. and McAlpine, R.S. 2002. Landscape-scale analysis of interactions between insect defoliation and forest fire in central Canada. *Climatic Change* 55: 251–272.

Franceschi, V.R., Krokene, P., Christiansen, E. and Krekling, T. 2005. Anatomical and chemical defenses of conifer bark against bark beetles and other pests. *New Phytologist* 167: 353–376.

Gandhi, K.J.K. and Herms, D.A. 2010. Direct and indirect effects of alien insect herbivores on ecological processes and interactions in forests of eastern North America. *Biological Invasions* 12: 389–405.

Gandhi, K.J.K. and Hofstetter, R.W. eds., 2021. *Bark Beetle Management, Ecology, and Climate Change.* Academic Press, London.

Gray, D.R. 2008. The relationship between climate and outbreak characteristics of the spruce budworm in eastern Canada. *Climatic Change* 87: 361–383.

Harvey, J.A., Heinen, R., Gols, R. and Thakur, M.P., 2020. Climate change-mediated temperature extremes and insects: From outbreaks to breakdowns. *Global Change Biology* 26(12): 6685–6701.

Haynes, K.J., Allstadt, A.J. and Klimetzek, D. 2014. Forest defoliator outbreaks under climate change: effects on the frequency and severity of outbreaks of five pine insect pests. *Global Change Biology* 20: 2004–2018.

Heavilin, J. and Powell, J. 2008. A novel method of fitting spatio-temporal models to data with applications to the dynamics of mountain pine beetles. *Natural Resource Modeling* 21: 489–524.

Hébert, C., Berthiaume, R., Dupont, A. and Auger, M. 2001. Population collapses in a forecasted outbreak of Lambdina fiscellaria (Lepidoptera: Geometridae) caused by spring egg parasitism by Telenomus spp. (Hymenoptera: Scelionidae). *Environmental Entomology* 30: 37–43.

Hennigar, C.R., MacLean, D.A., Quiring, D.T. and Kershaw, J.A. Jr. 2008. Differences in spruce budworm defoliation among balsam fir and white, red, and black spruce. *Forest Science* 54: 158–166.

Hennigar, C.R., Wilson, J.S., MacLean, D.A. and Wagner, R.G. 2011. Applying a spruce budworm decision support system to Maine: projecting spruce-fir volume impacts under alternative management and outbreak scenarios. *Journal of Forestry* 109: 332–342.

Hlásny, T., Zimová, S. and Bentz, B., 2021. Scientific response to intensifying bark beetle outbreaks in Europe and North America. *Forest Ecology and Management* 499, 119599.

Holling, C.S. 1992. Cross-scale morphology, geometry, and dynamics of ecosystems. *Ecological Monographs* 62: 447–502.

Holling, C.S. 2001. Understanding the complexity of economic, ecological, and social systems. *Ecosystems* 4: 390–405.

Ims, R.A., Henden, J.A. and Killengreen, S.T. 2008. Collapsing population cycles. *Trends in Ecology and Evolution* 23: 79–86.

Iqbal, J., Hennigar, C.R. and MacLean, D.A. 2012. Modeling insecticide protection versus forest management approaches to reducing balsam fir sawfly and hemlock looper damage. *Forest Ecology and Management* 265: 150–160.

Iqbal, J., MacLean, D.A. and Kershaw, J.A. 2011. Impacts of hemlock looper defoliation on growth and survival of balsam fir, black spruce and white birch in Newfoundland, Canada. *Forest Ecology and Management* 261: 1106–1114.

Jactel, H. and Brockerhoff, E.G. 2007. Tree diversity reduces herbivory by forest insects. *Ecology Letters* 10: 835–848.

Jactel, H., Koricheva, J. and Castagneyrol, B., 2019. Responses of forest insect pests to climate change: not so simple. *Current Opinion in Insect Science* 35: 103–108.

Janzen, D.H. 1970. Herbivores and the number of tree species in tropical forests. *The American Naturalist* 104: 510–528.

Jennings, D.T. and Crawford, H.S. 1985. Predators of the spruce budworm. *U.S. Department of Agriculture, Agriculture Handbook No. 644.* Department of Agriculture, Washington, DC.

Jepsen, J.U., Kapari, L., Hagen, S.B., Schott, T., Vindstad, O.P.L. and others. 2011. Rapid northwards expansion of a forest insect pest attributed to spring phenology matching with sub-Arctic birch. *Global Change Biology* 17: 2071–2083.

Johns, R., Bowden, J., Carleton, D., Cooke, B., Edwards, S., Emilson, E., James, P., Kneeshaw, D., MacLean, D., Martel, V., Moise, E., Mott, G. C. Norfolk, E. Owens, D. Pureswaran, D. Quiring, J. Régnière, Richard, B., and Stastny, M. 2019. A conceptual framework for the spruce budworm Early Intervention Strategy: can outbreaks be stopped? *Forests* 10, 910.

Jonasova, M. and Prach, K. 2004. Central-European mountain spruce (*Picea abies* (L.) Karst.) forests: regeneration of tree species after a bark beetle outbreak. *Ecological Engineering* 23: 15–27.

Kaitala, V. and Ranta, E. 1998. Traveling wave dynamics and self-organisation in spatio-temporally structured populations. *Ecology Letters* 1: 186–192.

Kirichenko, N., & Baranchikov, Y. N. (2007). Appropriateness of needles of different conifer species for the feeding and growth of larvae from two populations of the Siberian moth. *Russian Journal of Ecology* 38(3): 198–203. https://doi.org/10.1134/S1067413607030083

Kirichenko, N., Flament, J., Baranchikov, Y. et al. 2011.Larval performances and life cycle completion of the Siberian moth, Dendrolimus sibiricus (Lepidoptera: Lasiocampidae), on potential host plants in Europe: a laboratory study on potted trees. *European Journal of Forest Research* 130: 1067–1074 https://doi.org/10.1007/s10342-011-0495-3

Kneeshaw, D.D. and Bergeron, Y. 1998. Canopy gap characteristics and tree replacement in the south-eastern boreal forest. *Ecology* 79: 783–794.

Kneeshaw, D.D., Bergeron, Y. and Kuuluvainen, T. 2011. Forest ecosystem dynamics across the circumboreal forest. In: *Handbook of Biogeography*. Edited by A.C. Millington, M.A. Blumler, G. MacDonald and U. Shickhoff. Sage, Washington, DC, Chapter 14.

Kneeshaw, D.D., Sturtevant, B.R., DeGrandpé, L., Doblas-Miranda, E., James, P.M.A., Tardif, D. and Burton, P.J. (2021). The vision of managing for pest-resistant landscapes: Realistic or utopic? *Current Forestry Reports* 7(2), 97–113. https://doi.org/10.1007/s40725-021-00140-z

Kolb, T.E., Fettig, C.J., Ayres, M.P., Bentz, B.J., Hicke, J.A., Mathiasen, R., Stewart, J.E. and Weed, A.S., 2016. Observed and anticipated impacts of drought on forest insects and diseases in the United States. *Forest Ecology and Management* 380: 321–334.

Koricheva, J., Vehvilainen, H., Riihimaki, J., Ruohomaki, K., Kaitaniemi, P. and Ranta, H. 2006. Diversification of tree stands as a means to manage pests and diseases in boreal forests: myth or reality? *Canadian Journal of Forest Research* 36: 324–336.

Kumbasli, M., Bauce, É., Rochefort, S. and Crépin, M. 2011. Effects of tree age and stand thinning related variations in balsam fir secondary compounds on spruce budworm Choristoneura fumiferana development, growth and food utilization. *Agricultural and Forest Entomology* 13: 131–141.

Larsson, S. 1989. Stressful times for the plant stress: insect performance hypothesis. *Oikos* 277–283.

Légaré, S., Bergeron, Y. and Paré, D. 2005. Effect of aspen (Populus tremuloides) as a companion species on the growth of black spruce (Picea mariana) in the southwestern boreal forest of Quebec. *Forest Ecology and Management* 201: 211–222.

Legault, S. and James, P.M., 2018. Parasitism rates of spruce budworm larvae: testing the enemy hypothesis along a gradient of forest diversity measured at different spatial scales. *Environmental Entomology*, 47(5): 1083–1095.

Lesk, C., Coffel, E., D'Amato, A. et al. 2017. Threats to North American forests from southern pine beetle with warming winters. *Nature Climate Change* 7: 713–717. https://doi.org/10.1038/nclimate3375

Li, M., MacLean, D.A., Hennigar, C.R., and Ogilvie, J. 2020. Previous year outbreak conditions and spring climate predict spruce budworm population changes in the following year. *Forest Ecology and Management* 458: 117737.

Liebhold, A.M., Brockerhoff, E.G., Garrett, L.J., Parke, J.L. and Britton, K.O. 2012. Live plant imports: the major pathway for forest insect and pathogen invasions of the US. *Frontiers in Ecology and the Environment* 10: 135–143.

Liebhold, A.M., Hajek, A.E., Walter, J.A. et al. 2022. Historical change in the outbreak dynamics of an invading forest insect. *Biological Invasions* 24: 879–889. https://doi.org/10.1007/s10530-021-02682-6

Liu, E.Y, Lantz, V., MacLean, D.A., and Hennigar, C. 2019. Economics of early intervention to suppress a potential spruce budworm outbreak on Crown land in New Brunswick, Canada. *Forests* 10: 481.

MacLean, D.A. 1980. Vulnerability of fir spruce stands during uncontrolled spruce budworm outbreaks: a review and discussion. *The Forestry Chronicle* 56: 213–221.

MacLean, D.A., Amirault, P., Amos-Binks, L., Carleton, D., Hennigar, C., Johns, R., and Régnière, J. 2019. Positive results of an early intervention strategy to suppress a spruce budworm outbreak after five years of trials. *Forests* 10: 448.

MacLean, D.A., and Clark, K.L. 2021. Mixedwood management positively affects forest health during insect infestations in Eastern North America. *Canadian Journal of Forest Research* 51: 910–920.

MacLean, D.A. and Ebert, P. 1999. The impact of hemlock looper (Lambdina fiscellaria fiscellaria (Guen.)) on balsam fir and spruce in New Brunswick, Canada. *Forest Ecology and Management* 120: 77–87.

MacLean, D.A., Erdle, T.A., MacKinnon, W.E., Porter, K.B., Beaton, K.P. and others. 2001. The spruce budworm decision support system: forest protection planning to sustain long-term wood supplies. *Canadian Journal of Forest Research* 31: 1742–1757.

MacLean, D.A. and Ostaff, D.P. 1989. Patterns of balsam fir mortality caused by an uncontrolled spruce budworm outbreak. *Canadian Journal of Forest Research* 19: 1087–1095.

Mallik, A.U. 2003. Conifer regeneration problems in boreal and temperate forests with ericaceous understory: role of disturbance, seedbed limitation, and keystone species change. *Critical Reviews in Plant Sciences* 22: 341–366.
Mattson, W.J. and Addy, N.D. 1975. Phytophagous insects as regulators of forest primary production. *Science* 190: 515–522.
Mattson, W.J. and Haack, R.A. 1987. The role of drought in outbreaks of plant-eating insects. *Bioscience* 37: 110–118.
Mattson, W.J., Herms, D.A., Witter, J.A. and Allen, D.C. 1991. Woody plant grazing systems: North American outbreak folivores and their host plants. In: *Forest Insect Guilds: Patterns of Interactions with Host Trees.* Edited by Y.N. Baranchikov, W.J. Mattson, F.P. Haine and T.L. Payne. U.S. Department of Agriculture, Forest Service, General Technical Report NE-153, pp. 53–85.
Miller, A. and Rusnock, P. 1993 The rise and fall of the silvicultural hypothesis in spruce budworm (Choristoneura fumiferana) management in Eastern Canada. *Forest Ecology and Management* 61: 171–189.
Moulinier, J., Lorenzetti, F. and Bergeron, Y. 2013. Effects of a forest tent caterpillar outbreak on the dynamics of mixedwood boreal forests of eastern Canada. *Ecoscience* 20: 182–193.
Muzika, R.M. and Liebhold, A.M. 2000. A critique of silvicultural approaches to managing defoliating insects in North America. *Agricultural and Forest Entomology* 2: 97–105.
Myers, J.H. and Cory, J.S. 2013. Population cycles in forest Lepidoptera revisited. *Annual Review of Ecology, Evolution, and Systematics* 44: 565–592.
Nair, K.S.S. 2012. *Tropical Forest Insect Pests: Ecology, Impact and Management*, Cambridge University Press, Cambridge.
Nealis, V.G. and Lomic, P.V., 1994. Host-plant influence on the population ecology of the jack pine budworm, Choristoneura pinus (Lepidoptera: Tortricidae). *Ecological Entomology* 19: 367–373.
Needham, T., Kershaw, J., MacLean, D.A., and Su, Q. 1999. Effects of mixed stand management to reduce impacts of spruce budworm defoliation on balsam fir stand-level growth and yield. *Northern Journal of Applied Forestry* 16: 19–24.
Nenzén, H.K., Peres-Neto, P. and Gravel, D., 2018. More than Moran: coupling statistical and simulation models to understand how defoliation spread and weather variation drive insect outbreak dynamics. *Canadian Journal of Forest Research* 48: 255–264.
Novotny, V., Basset, Y., Miller, S.E., Weiblen, G.D., Bremer, B. and others. 2002. Low host specificity of herbivorous insects in a tropical forest. *Nature* 416: 841–844.
Palo, A. and Thomas, R. 1984. Distribution of birch (Betula spp.), willow (Salix spp.), and poplar (Populus spp.) secondary metabolites and their potential role as chemical defense against herbivores. *Journal of Chemical Ecology* 10: 499–520.
Paré, D. and Bergeron, Y. 1996. Effect of colonizing tree species on soil nutrient availability in a clay soil of the boreal mixedwood. *Canadian Journal of Forest Research* 26: 1022–1031.
Parry, D., Spence, J.R. and Volney, W.J.A. 1998. Budbreak phenology and natural enemies mediate survival of first-instar forest tent caterpillar (Lepidoptera: Lasiocampidae). *Environmental Entomology* 27: 1368–1374.
Peters, D.P., Pielke, R.A., Bestelmeyer, B.T., Allen, C.D., Munson-McGee, S. and Havstad, K.M. 2004. Cross-scale interactions, nonlinearities, and forecasting catastrophic events. *Proceedings of the National Academy of Sciences of the United States of America* 101: 15130–15135.
Petrovskii, S.V. and McKay, K., 2010. Biological invasion and biological control: a case study of the gypsy moth spread. *Aspects of Applied Biology* 104: 37–48.
Powell, J.A. and Bentz, B.J. 2009. Connecting phenological predictions with population growth rates for mountain pine beetle, an outbreak insect. *Landscape Ecology* 24: 657–672.
Przepióra, F., Loch, J. and Ciach, M., 2020. Bark beetle infestation spots as biodiversity hotspots: Canopy gaps resulting from insect outbreaks enhance the species richness, diversity and abundance of birds breeding in coniferous forests. *Forest Ecology and Management* 473: 118280.
Pureswaran, D.S., Roques, A. and Battisti, A., 2018. Forest insects and climate change. *Current Forestry Reports* 4: 35–50.
Raffa, K., Aukema, B., Bentz, B., Carroll, A., Hicke, J. and others. 2008. Cross-scale drivers of natural disturbances prone to anthropogenic amplification: the dynamics of bark beetle eruptions. *Bioscience* 58: 501–517.

Raffa, K.F., Phillips, T.W. and Salom, S.M. 1993. Strategies and mechanisms of host colonization by bark beetles. In: *Beetle-Pathogen Interactions in Conifer Forests.* Edited by T.D. Schowalter and G.M. Filip. Academic Press, San Diego, CA, pp. 103–128.

Ranta, E., Kaitala, V., Lindström, J. and Helle, E. 1997. The Moran effect and synchrony in population dynamics. *Oikos* 78: 136–142.

Reeve, J.D. 1997. Predation and bark beetle dynamics. *Oecologia* 112: 48–54.

Régnière, J., Delisle, J., Pureswaran, D.S. and Trudel, R., 2013. Mate-finding allee effect in spruce budworm population dynamics. *Entomologia Experimentalis et Applicata* 146(1): 112–122.

Régnière, J. and Lysyk, T.J. 1995. Population dynamics of the spruce budworm, Choristoneura fumiferana. In: *Forest Insect Pests in Canada.* Edited by J.A. Armstrong and W.G.H. Ive. Canadian Forest Service, Ottawa, Ontario, Canada, pp. 95–105.

Régnière, J. and Nealis, V.G. 2007. Ecological mechanisms of population change during outbreaks of the spruce budworm. *Ecological Entomology* 32: 461–477.

Régnière, J., St-Amant, R. and Duval, P. 2012. Predicting insect distributions under climate change from physiological responses: spruce budworm as an example. *Biological Invasions* 14: 1571–1586.

Régnière, J., Venier, L. and Welsh, D., 2021. Avian predation in a declining outbreak population of the spruce budworm, Choristoneura fumiferana (Lepidoptera: Tortricidae). *Insects* 12(8): 720.

Reyes, G.P., Kneeshaw, D., DeGrandpre, L. and Leduc, A. 2010. Changes in woody vegetation abundance and diversity after natural disturbances causing different levels of mortality. *Journal of Vegetation Science* 21: 406–417.

Robert, L. 2014. Influence à l'échelle du paysage des legs associés à l'aménagement forestier sur les épidémies d'insectes. Thesis UQAM, Montreal, Canada.

Robert, L., Kneeshaw, D. and Sturtevant, B.R. 2012. Effects of forest management legacies on spruce budworm (Choristoneura fumiferana) outbreaks. *Canadian Journal of Forest Research* 42: 463–475.

Robert, L., Sturtevant, B.R., Cooke, B.J., James, P.M., Fortin, M.J., Townsend, P.A., Wolter, P.T. and Kneeshaw, D., 2018. Landscape host abundance and configuration regulate periodic outbreak behavior in spruce budworm Choristoneura fumiferana. *Ecography* 41: 1556–1571.

Robert, L., Sturtevant, B. R., Kneeshaw, D., James, P. M. A., Fortin, M., Wolter, P. T., Townsend, P. A., & Cooke, B. J. (2020). Forest landscape structure influences the cyclic-eruptive spatial dynamics of forest tent caterpillar outbreaks. *Ecosphere* 11(8): e03096. https://doi.org/10.1002/ecs2.3096

Roland, J. 2005. Are the "seeds" of spatial variation in cyclic dynamics apparent in spatially-replicated short time-series? An example from the forest tent caterpillar. *Annales Zoologici Fennici* 42: 397–407.

Roland, J. and Kaupp, W.J. 1995. Reduced transmission of forest tent caterpillar (Lepidoptera: Lasiocampidae) nuclear polyhedrosis virus at the forest edge. *Environmental Entomology* 24: 1175–1178.

Roland, J. and Taylor, P.D. 1997. Insect parasitoid species respond to forest structure at different spatial scales. *Nature* 386: 710–713.

Royama, T. 1984. Population dynamics of the spruce budworm Choristoneura fumiferana. *Ecological Monographs* 54: 429–462.

Royama, T. 2005. Moran effect on nonlinear population processes. *Ecological Monographs* 75: 277–293.

Rozenberg, P., Pâques, L., Huard, F., & Roques, A. (2020). Direct and indirect analysis of the elevational shift of larch budmoth outbreaks along an elevation gradient. *Frontiers in Forests and Global Change*, 3. https://doi.org/10.3389/ffgc.2020.00086.

Safranyik, L., Barclay, H., Thomson, A.J, and Riel, W.G. 1999. *A Population Dynamics Model for the Mountain Pine Beetle, Dendroctonus ponderosae Hopk.* (Coleoptera: Scolytidae). Canadian Forest Service, Information Report BC-X-386 Pacific Forestry Centre. 35 pp.

Safranyik, L. and Carroll, A.L. 2006. The biology and epidemiology of the mountain pine beetle in lodgepole pine forests. In: *The Mountain Pine Beetle: A Synthesis of Biology, Management and Impacts on Lodgepole Pine.* Edited by L. Safranyik and W.R. Wilson. Natural Resources Canada, Canadian Forest Service, Pacific Forestry Centre, Victoria, BC, Canada, pp. 3–66.

Sambaraju, K.R. and Goodsman, D.W., 2021. Mountain pine beetle: an example of a climate-driven eruptive insect impacting conifer forest ecosystems. *CABI Reviews* 16, No. 018.

Schelhaas, M.J., Nabuurs, G.J. and Schuck, A. 2003. Natural disturbances in the European forests in the 19th and 20th centuries. *Global Change Biology* 9: 1620–1633.

Scott, E.R., Wei, J.P., Li, X., Han, W.Y. and Orians, C.M., 2021. Differing non-linear, lagged effects of temperature and precipitation on an insect herbivore and its host plant. *Ecological Entomology* 46(4): 866–876.

Sexton, J.P., McIntyre, P.J., Angert, A.L. and Rice, K.J. 2009. Evolution and ecology of species range limits. *Annual Review of Ecology, Evolution, and Systematics* 40: 415–436.

Seybold, S.J., Huber, D.P.W., Lee, J.C., Graves, A.D. and Bohlmann, J. 2006. Pine monoterpenes and pine bark beetles: a marriage of convenience for defense and chemical communication. *Phytochemistry Reviews* 5: 143–178.

Seybold, S.J. and Tittiger, C. 2003. Biochemistry and molecular biology of de novo isoprenoid pheromone production in the scolytidae. *Annual Reviews Entomology* 48: 425–453.

Shorohova, E., Kneeshaw, D.D., Kuuluvainen, T. and Gauthier, S. 2012. A comparison of old-growth forest dynamics across the circum-boreal forest zone. *Silva Fennica* 45: 785–806.

Spence, C.E. and MacLean, D.A. 2012. Regeneration and stand development following a spruce budworm outbreak, spruce budworm-inspired harvest, and salvage harvest. *Canadian Journal of Forest Research* 42: 1759–1770.

Stireman, J.O., Dyer, L.A., Janzen, D.H., Singer, M.S., Lill, J.T. and others. 2005. Climatic unpredictability and parasitism of caterpillars: implications of global warming. *Proceedings of the National Academy of Sciences of the United States of America* 102: 17384–17387.

Sturtevant, B.R., Miranda, B.R., Shinneman, D.J., Gustafson, E.J. and Wolter, P.T. 2012. Comparing modern and presettlement forest dynamics of a subboreal wilderness—Does spruce budworm enhance fire risk? *Ecological Applications* 22: 1278–1296.

Su, Q., MacLean, D.A. and Needham, T.D. 1996. The influence of hardwood content on balsam fir defoliation by spruce budworm. *Canadian Journal of Forest Research* 26: 1620–1628.

Sutton, S., Pasquini, S.C., Swanson, T. and Carson, W.P., 2021. On the occurrence of a highly localized outbreak of a saturniid in lowland east Ecuador: a case study and literature review. *Neotropical Biodiversity* 7(1): 39–44.

Svoboda, M., Fraver, S., Janda, P., Bače, R. and Zenáhlíková, J. 2010. Natural development and regeneration of a Central European montane spruce forest. *Forest Ecology and Management* 260: 707–714.

Taylor, A.R. and Chen, H.Y.H. 2011. Multiple successional pathways of boreal forest stands in central Canada. *Ecography* 34: 208–219.

Taylor, S.L. and MacLean, D.A. 2007. Deadwood dynamics in declining balsam fir and spruce stands in New Brunswick, Canada. *Canadian Journal of Forest Research* 37: 750–762.

Tenow, O., Nilssen, A.C., Bylund, H., Pettersson, R., Battisti, A. and others. 2013. Geometrid outbreak waves travel across Europe. *Journal of Animal Ecology* 82: 84–95.

Townsend, P.A., Eshleman, K.N. and Welcker, C. 2004. Remote sensing of gypsy moth defoliation to assess variations in stream nitrogen concentrations. *Ecological Applications* 14: 504–516.

Trubin, A., Mezei, P., Zabihi, K., Surový, P. and Jakuš, R., 2022. Northernmost European spruce bark beetle Ips typographus outbreak: Modelling tree mortality using remote sensing and climate data. *Forest Ecology and Management* 505: 119829.

Vindstad, O.P.L., Jepsen, J.U., Molvig, H. and Ims, R.A., 2022. A pioneering pest: the winter moth (Operophtera brumata) is expanding its outbreak range into Low Arctic shrub tundra. *Arctic Science* 8(2): 450–470.

Wallner, W.E. 1987. Factors affecting insect population dynamics: differences between outbreak and non-outbreak species. *Annual Review of Entomology* 32: 317–340.

Walton, A. 2012. Update of the infestation projection based on the Provincial Aerial Overview Surveys of Forest Health conducted from 1999 through 2011 and the BCMPB model (year 9). B.C. Forest Service Report. http://www.for.gov.bc.ca/ftp/hre/external/!publish/web/bcmpb/year9/BCMPB. v9.BeetleProjection.Update.pdf

Wermelinger, B. 2004. Ecology and management of the spruce bark beetle Ips typographus—a review of recent research. *Forest Ecology and Management* 202: 67–82.

Werner, R.A., Holsten, E.H., Matsuoka, S.M. and Burnside, R.E. 2006. Spruce beetles and forest ecosystems in south-central Alaska: a review of 30 years of research. *Forest Ecology and Management* 227: 195–206.

Weso³owski, T. and Rowiński, P. 2006. Tree defoliation by winter moth Operophtera brumata L. during an outbreak affected by structure of forest landscape. *Forest Ecology and Management* 221: 299–305.

Williams, D.W. and Liebhold, A.M. 2000. Spatial synchrony of spruce budworm outbreaks in Eastern North America. *Ecology* 81: 2753–2766.
Wolda, H. 1987. Altitude, habitat and tropical insect diversity. *Biological Journal of the Linnean Society* 30: 313–323.
Wood, D.L. 1982. The role of pheromones, kairomones and allomones in the host selection and colonization of bark beetles. *Annual Review of Entomology* 27: 411–446.
Work, T.T., McCullough, D.G., Cavey, J.F. and Komsa, R. 2005. Arrival rate of nonindigenous insect species into the United States through foreign trade. *Biological Invasions* 7: 323–332.
Zhang, B., MacLean, D.A., Johns, R.C., and Eveleigh, E.S. 2018. Effects of hardwood content on balsam fir defoliation during the building phase of a spruce budworm outbreak. *Forests* 9: 530
Zhang, B., MacLean, D.A., Johns, R.C., Eveleigh, E.S., and Edwards, S. 2020. Hardwood-softwood composition influences early-instar larval dispersal mortality during a spruce budworm outbreak. *Forest, Ecology and Management* 463: 118035.
Zhang, X., Lei, Y., Ma, Z., Kneeshaw, D.D. and Peng, C. 2014. Insect-induced tree mortality of boreal forests in eastern Canada under a changing climate. *Ecology and Evolution* 4: 2384–2394.

8

FIRE IN FOREST ECOSYSTEMS

David F. Greene, Madeleine A. Lopez and Sean T. Michaletz

Introduction

Natural disturbance not only kills plants and animals but also makes resources available for subsequent cohorts. Consequently, it plays a decisive role in births and deaths of individuals and is the major arbiter of woody plant demography. In this chapter, we review the recent advances in our understanding of the processes linking climate, fire behavior, and fire effects on plants and animals. More specifically, we examine the ecological patterns that emerge from a coupling of fire behavior, physiology, and demographic processes. For example, fire effects on vegetation can be understood by considering the mechanisms governing fire behavior, heat transfer into plants, heat injury of plant tissues, and the physiology linking tissue injuries to whole-plant growth, mortality, and reproduction (Michaletz and Johnson 2007; Michaletz et al. 2012; Kleynhans et al. 2021).

Fire is a combustion process in which rates of gaseous fuel production via pyrolysis (endothermic) are controlled by positive feedback of heat from the combustion process itself (exothermic; Quintiere 2006). We can distinguish between two types of combustion in forest fires: flaming combustion and smoldering combustion (Figure 8.1). These two types of combustion are distinguished by the rate of gaseous fuel supply; flaming combustion occurs when pyrolysis rates are high enough to support combustion in the air adjacent to the fuel elements, and smoldering combustion occurs when pyrolysis rates are low and/or limited by buildup of ash and char so that combustion only occurs on the surface of fuel elements. The differences between these two types of combustion can be illustrated by comparing a candle (flaming combustion) and incense (smoldering combustion). Although the behaviors of flaming and smoldering combustion are quite different, both have important ecological effects in fire-prone areas.

Characterizing fire

Four conditions are necessary for forest fire initiation and spread (reviewed in Macias Fauria et al. 2011). First, there must be sufficient fuel to allow propagation of the burning front. Forest fuels are generally classified into three strata: (1) ground fuels comprising below-ground organic soil (duff) and plant roots near the soil surface, (2) surface fuels consisting

DOI: 10.4324/9781003324072-10

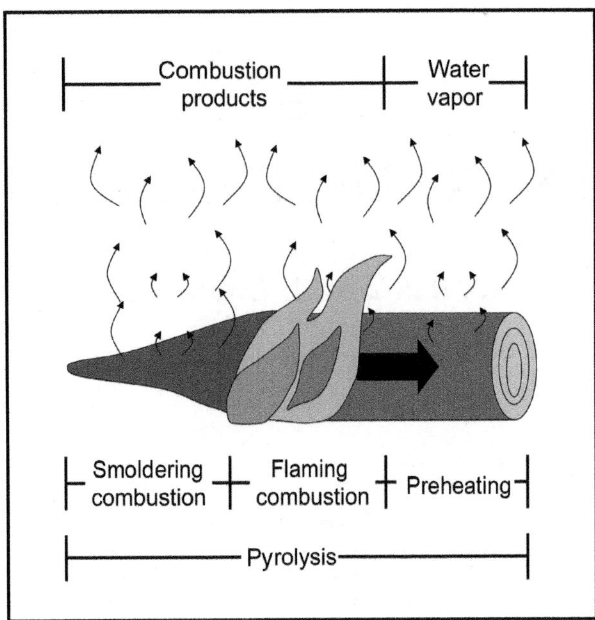

Figure 8.1 Diagram illustrating the phases and products of combustion for a forest fuel element. Fire spread progress from left to right with passage of the flaming combustion front followed by smoldering combustion.

of litter and small plants on the forest floor, and (3) crown fuels located within the forest canopy. In all strata, the flaming fire front consumes only fine fuels less than approximately 6 mm in diameter. Second, there must be a period of at least 10–21 days with low relative humidity and little to no precipitation. This can dry the fuels sufficiently to sustain combustion because fine fuels have a large surface area to volume ratio and thus respond rapidly to changes in moisture conditions (Nelson 2001; Dickman et al. 2023). In boreal and temperate areas, the drying period often results from persistent positive mid-tropospheric height anomalies that prevent precipitation by blocking zonal flow and encouraging meridional flow of warm, dry air. In the tropics, fires are more common during dry El Niño events (Burton et al 2020). Third, there must be an ignition source, which is invariably lightning or human activity. Fourth, strong winds are necessary for creating high rates of spread and, consequently, large fire sizes (Figure 8.2). This is especially apparent in some regions adjacent to mountain ranges such as the California chaparral or the Eucalyptus forests of southeast Australia, where hot, dry downslope winds (e.g., katabatic winds and föhn winds) enhance fuel drying and fire spread (Keeley and Fotheringham 2001; Sharples et al. 2010).

Given these four conditions, it is clear why fire is less of an important disturbance in certain forested biomes. In temperate deciduous forests, high relative humidity and frequent convective storms during the growing season preclude fuel drying. In other biomes, it is less clear why fire is unimportant. Tropical deciduous forest fires occur when there is a long period of drought during the low-sun season, which is when the relative humidity is also low. Nonetheless, natural fires are extremely rare in this biome, possibly due to high water content in live fuels of many tree species (Hayden and Greene 2008; Dickman et al.

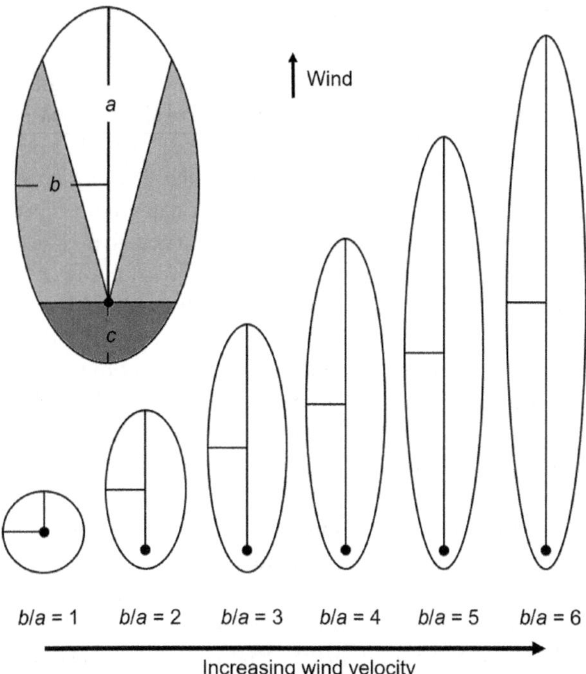

Figure 8.2 Fire shape is the outcome of fire spread rates in all directions from the ignition point (black points). Fire shapes can be idealized as ellipses characterized by the length-to-breadth ratio b/a, where a and b are the rates of spread for head and flank fires, respectively. Wind generally enhances spread rates in the direction of the wind (a) and suppresses spread rates perpendicular to (b) and against the wind (c), resulting in fire shapes with larger length-to-breadth ratios. Dark gray represents the area burned by a back fire, light gray the area burned by flank fire, and white the area burned by head fire. After Van Wagner (1969) and Finney (2001).

2023). The moisture content among living plants can span from 0 to 300 per cent, with higher retention of moisture working to inhibit consumption of fuels. Finally, there are some forested biomes, for example, many tropical evergreen forests, where fire is presently quite common, mainly because of human activities associated with land clearance (Krawchuk et al. 2009). Additionally, small patches of forest are still burned in the wet tropics by indigenous groups. As these tropical regions experience changes in rainfall patterns in combination with increased deforestation for agricultural purposes, drier biomass becomes available to promote fire spread, and the possibility of greater fire frequency and severity increases. For example, the Amazon rainforest experienced a record number of fires in 2019, with an estimated total of 906,000 hectares burned (Lizundia-Liola et al. 2020). That said, lightning-started fires remain quite rare. Here we focus on forests where lightning ignitions play a large role in the total area burned. Transitional vegetation types such as savannas are beyond the scope of this chapter.

We distinguish between three classes of forest fires based on the fuels involved (ground, surface, or crown), the type of combustion (flaming or smoldering), and the fire intensity (the rate of heat production per unit length of the fireline): ground fires, surface fires, and crown fires. Ground fires involve smoldering combustion of duff and litter and generally occur after passage of the flaming fire front. Ground fires can persist for long periods in the absence of significant precipitation—sometimes up to months or years. Surface fires involve flaming combustion of litter and live fine fuels of herbs and shrubs located on the forest floor. Surface fires generally have relatively low frontal intensities, as higher intensities lead to ignition of crown fuels (Cruz et al. 2006). Crown fires involve flaming combustion of fine fuels located in the forest canopy. Crown fires can be further classified as passive, active, or independent based on the degree to which ignition of unburned fuel relies on heat fluxes from the surface fire below (Michaletz and Johnson 2007). As with many other disturbance types, fire is a low-frequency, high-magnitude event. Although small fires may be more frequent, it is the rare, large fires (ranging from a few hundred to a few hundred thousand ha) that constitute most of the area burned in most fire-prone forest biomes. In Canada, for instance, about 3 per cent of the fires accounted for 97 per cent of the total area burned between 1959 and 1999 (Stocks et al. 2002). Since large fires burn most of the area, they have the greatest influence on forest structure and dynamics and are consequently of primary interest from an ecological perspective.

Burns are invariably elongated, reflecting the importance of wind speed and direction on fire shape (Figure 8.2). Fires also have characteristically large perimeter-to-area ratios (Anderson 1983). Furthermore, small, unburned areas (residual stands) are common within large fires (Kolden et al 2012). Including the edges of these residual stands, Greene (unpublished) has shown for large fires in Saskatchewan that the median distance from any edge to a randomly chosen point in the fire is only about 150 m, and this unexpectedly small distance is due to the remarkable invagination of burn perimeters. While this still represents a serious constraint for seed dispersal from the burn edge, the situation is not as bleak as one might initially imagine for say a 50,000 ha fire containing large severe patches.

Fire effects on plants and animals

The heat produced by combustion can cause physical injuries to plant roots, stems, or crowns that can affect post-fire growth, mortality, and/or reproduction (Figure 8.3; Michaletz and Johnson 2007; Michaletz 2018; O'Brien et al. 2018; Bär et al. 2019; Kleynhans et al. 2021). This occurs as a result of heat transfer from the fire to key tissues such as meristems, vascular tissues, and embryos. Three heat transfer mechanisms link the heat produced by combustion to heat fluxes on plant surfaces: conduction, radiation, and convection. Root surfaces can be heated via conduction through soil, while stem and crown surfaces can be heated via radiation and convection from the flame and buoyant plume. Heat fluxes on plant surfaces can then drive heat conduction from the surface toward the interior of the plant. Injuries can have immediate effects on reproduction and mortality via necrosis of key plant parts like seeds or meristems, or can have delayed effects resulting from an interaction of injuries such as leaf and partial cambium necrosis (Michaletz and Johnson 2006, 2007; Michaletz et al. 2012; Bär et al. 2019; Kleynhans et al. 2021). Stem injury is governed by rates of heat conduction through the stem to tissues such as the vascular cambium, phloem, and xylem (Michaletz and Johnson 2007; Michaletz et al. 2012; Michaletz 2018). The rates of conduction vary with the depth of the tissue in question and the thermal conductivity surface heating drives heat conduction into the plant, resulting in injuries to several key tissues and organs of the stem. Stem moisture content is a key variable influencing the rates of stem heating, because water is a latent heat sink that limits stem temperature increases to a maximum of 100°C.

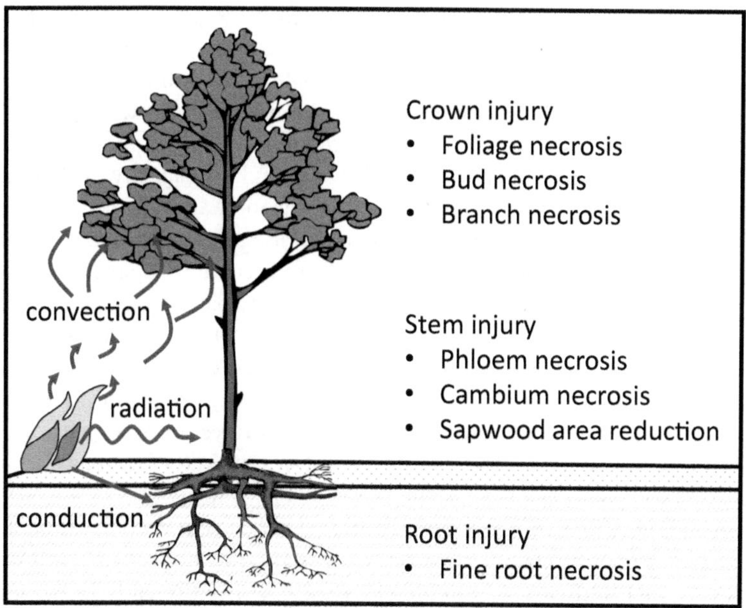

Figure 8.3 Mechanisms of heat transfer from a fire to plant root, stem and crown surfaces.

The retained moisture also reduces flame heights and temperatures, as the internal water must be volatilized from the fuel (Dickman et al. 2023). As the water vapor surrounds the fuel particle, less oxygen is available to support combustion. For woody plants, bark provides additional resistance to conduction and helps insulate the underlying phloem, cambium, and xylem tissues. As larger stems have thicker bark, they are more insulated than smaller stems. However, stem size also influences air-flow patterns, with larger diameter stems producing a turbulent leeward wake that can increase flame residence times and leeward convection rates. This often causes uneven heating around the stem circumference, leading to increased injury on the leeward side (e.g. fire scars resulting from phloem and vascular cambium necrosis). In cases where phloem and cambium necrosis occur around the entire stem circumference (girdling), carbon translocation to roots is prevented and root growth must rely upon stored carbon reserves. When these reserves are depleted, fine root production will cease and the entire plant will die as a result of water stress (Balfour and Midgley 2006; Michaletz et al. 2012; Michaletz 2018). Stem heating can also reduce the hydraulic conductivity of sapwood via enhanced air seed cavitation and conduit wall deformation of xylem (Michaletz et al. 2012; Michaletz 2018; Bär et al. 2018; Bär et al. 2019), which results in more rapid mortality than girdling alone. Experiments with an African Acacia showed that stem heating caused rapid plant mortality that was associated with reduced sapwood area (Balfour and Midgley 2006).

Injuries to plant crown components (branches, buds, foliage, and seed-housing structures) occur via convective and radiative fluxes from flame and plume. In low-intensity surface fires, convection is the dominant heat transfer process and height of crown injury varies with the two-thirds power of fireline intensity (Van Wagner 1973; Michaletz and Johnson 2006). In high-intensity crown fires, canopy fuels are consumed and injuries can be present at all heights. Small crown components such as leaves and buds are "thermally thin", meaning their rates of heating are controlled primarily by surface resistances to radiation and convection and not internal resistance to conduction (Michaletz and Johnson 2006); consequently, thermally thin objects generally do not have internal temperature gradients. The crown component width, surface area, shape, orientation, and degree of shielding by foliage control surface heat fluxes, and the mass, water content, and specific heat capacity determine how much energy is required to cause a temperature increase (Bison et al. 2022). For example, species with relatively large buds such as ponderosa pine and longleaf pine are far less susceptible to bud necrosis than are species with relatively small buds such as sugar maple and American beech. The seasonal phenology of crown functional traits plays a central role in the impact of fire on crown injuries (Bison et al. 2022). For larger crown components like branches and seed-housing structures, internal temperature gradients do exist and conduction within the component becomes important. In these cases, size plays an important role in helping to insulate key tissues against heat necrosis, as conduction rates decrease with distance. Thus, bark can help insulate key tissues like cambium and xylem. Likewise, many cones and fruits can insulate embryo tissues against heat necrosis in fires (Splawinski et al. 2019). This has been widely recognized in serotinous species that store seeds in aerial seed banks, but it can also be important for non-serotinous species provided the fire occurs during a temporal window (late summer and early autumn) when seeds are germinable but not yet abscised (Lopez 2022). For instance, it has been estimated that about 10 per cent of *Picea glauca* seeds within closed cones can survive high-intensity crown fires when the fire event occurs between approximately 600 and 1,200 degree days (Michaletz et al. 2012). Finally, extreme vapor pressure deficits in fire plumes might also cause cavitation of xylem conduits in plant crowns that could lead to branch and leaf death (Kavanagh et al. 2010).

Despite our growing knowledge of how plant tissues are injured during fires, we still have a limited understanding of how they interact to control whole-plant function (Michaletz and Johnson 2007; Michaletz et al. 2012; Michaletz 2018; Bär et al. 2019). It is generally accepted that stem girdling or death of all vegetative buds will cause mortality, but it is less clear how multiple injuries to multiple tissues interact to affect post-fire growth and mortality. This is of interest as some studies have used air photos of recent burns to argue that the proportion of tree mortality is extraordinarily variable within fires. However, interactions of multiple injuries can often lead to delayed mortality meaning that many plants suffering partial crown scorch will die within a few years after the initial imagery.

Necrosis of root meristems may occur as a result of heat conduction through soil (Michaletz and Johnson 2007). Most root meristems are small and do not have internal temperature gradients, so the primary resistance to heating is provided by the overlying soil. Consequently, the key variables controlling rates of root heating are meristem depth, soil structure, and soil composition (water, organic, and mineral content; Hillel 1998). Root heating decreases with depth and soil water content and increases with heating duration, soil surface heat flux, and soil thermal conductivity. Meristem necrosis most commonly occurs as a result of smoldering combustion, as the period of flaming combustion is generally too short to cause lethal temperatures at appreciable depths (Hartford and Frandsen 1992). For a given soil type, the susceptibility of different sizes or species of plants can be predicted by comparing their vertical root distributions; plants with buds distributed to greater depths will be less susceptible to root meristem necrosis (Michaletz and Johnson 2007). Generally, at least some perennating tissues of most individual or clonal shrubs and perennial herbs survive a forest fire except in the upper few centimeters of organic material (which invariably smolders) and in those small patches (lateral scale of less than a meter or two) where the smoldering continues down to or very near the mineral soil. It is presumed that the hyphae of fungi likewise survive in the same parts of the organic layer and upper mineral soil. Mosses tend to be killed outright by smoldering although it has been noted that species with characteristically long setae (e.g. *Sphagnum*) can at times survive a fire (Greene et al 2006). Most moss species in coniferous forests ("feathermosses") tend to be clustered under tree crowns—precisely where smoldering tends to consume organic material down to mineral soil.

We have discussed a number of plant traits that control injury of key tissues in forest fires. These traits are fire adaptive if they confer a fitness advantage to the plant, although traits that are fire adaptive in one fire regime are not necessarily adaptive in another (Keeley et al. 2011). For example, in fire regimes typified by frequent, low-intensity surface fires, useful traits include thick bark, elevated crowns, and large buds. These traits are adaptive in these fire regimes because they permit survival and reproduction for this suite of fire behavior characteristics. On the other hand, in fire regimes typified by high-intensity, stand-replacing fires, traits such as thin bark, low crowns, and small buds are adaptive because the biomass not allocated to structure can instead be invested in other areas that promote fitness such as seeds or root carbon reserves (resprouting). Finally, we stress that although fire adaptive traits confer a fitness advantage to the plant, they did not necessarily arise in response to fire as a selective agent (i.e., they are not necessarily adaptations; Keeley et al. 2011).

Large animals generally avoid fire. Most vertebrates—even toads—have the instinctive response to move away from smoke, as such direct mortality from the fire is quite low. At the Yellowstone fires in 1988, half a million ha were burned and yet the mortality rate as measured immediately after the burns was about 0.5 percent for large mammal species such as elk, deer, and bison (Romme et al 2011). Undoubtedly, the old and sick were well-represented among these casualties.

It is more difficult for small animals to avoid fire, because their net speed is typically lower than the fire spread rate. However, this does not necessarily mean that all small animals are killed by fire as some fish, amphibians, and soil invertebrates can be relatively unaffected by the direct effects of flames or smoke because they are insulated by their surrounding soil or water media. In most other cases, post-fire mortality rates can be high and population densities can be strongly reduced. Ground-nesting bird populations can be dramatically reduced by fire, and one well-known example is the rare *Tympanuchus cupido cupido* (heath hen), which was driven to the brink of extinction by a wildfire. Small mammals such as deer mice and voles are not killed by heating but rather by asphyxiation, and thus generally experience high mortality rates from direct fire effects. Nonetheless, small mammal mortality rates are generally not measured directly but instead estimated indirectly via comparisons of population densities in burned and unburned areas. However, this approach provides a rather poor estimate of mortality because (1) surveys do not typically occur immediately after the fire, so recolonization is already underway and (2) the distances from the burn sampling plots to the nearest unburned edges are not defined, which confounds density estimates. In consequence, we lack direct measurements of small mammal mortality rates as well as estimates of the pace of post-fire recolonization.

Recolonization of fires

The recolonization of burns by plants can occur via resprouting or recruitment from seeds. Resprouting is rare among conifers but quite common among angiosperms. For angiosperm trees, the typical source of new shoots is from dormant buds around the root collar near the soil surface. These shoots are often initially numerous but are reduced through self-thinning so that only one or two remain at maturity. A very minor number of angiosperm trees (but many shrubs and herbs) can also produce suckers from dormant buds along shallow roots or stolons. This latter form of asexual reproduction is far more useful than basal sprouts because it permits the clone to extend laterally. Some *Populus* (aspen) clones are among the largest organisms on earth. Likewise, much of the hyphal network of fungal clones survives in the duff and coarse woody debris. Many species, for example *Morchella* (morels) and *Geopyxis* (pixie cups) produce copious numbers of mushrooms immediately following wildfire but almost exclusively where the organic soil (duff) has been substantially reduced by smoldering combustion. It is unclear why fungal fruiting bodies are mainly produced where hyphae have been pruned.

Sexual spread via seeds occurs even with the species relying primarily on asexual recruitment to persist in situ. However, given the seed dispersal constraint, dispersal by seed is far less reliable than asexual reproduction for restocking the interiors of large severely burned patches. Consider the case of *Pseudostuga menziesii* (Douglas fir) in Figure 8.4. For an extensive array of conspecific seed sources (an area source) with a full display of leaves, deposited seed densities 200 m from the burn edge are reduced by more than 90 per cent from those at the burn edge (Greene and Johnson 1996).

Species with an aerial seedbank (i.e., delayed seed abscission leading to persistent seed crops) tend to produce a similar number of seeds each year. By contrast, species without aerial seedbanks tend to instead exhibit masting, which results in tremendous inter-annual variation in crop size. It has been demonstrated from examination of tree ages that a mast year occurring one or two years after a large fire will subsequently produce the largest post-fire cohort (e.g., Peters et al 2005). Most woody species in fire-prone landscapes have wind-dispersed seeds and thus do not rely on animal dispersal. While the larger animals that can carry seeds beyond wind-dispersal distances are found in recent fires because of the

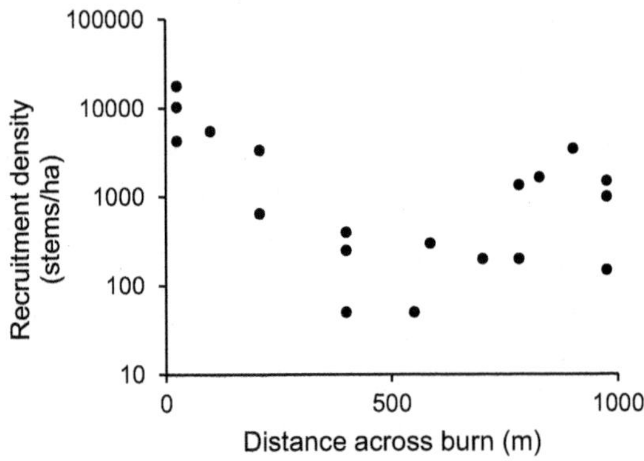

Figure 8.4 Recruitment density across a 1000 m-wide burned patch at the 2008 Klamath Complex Fire in northern California. There were living seed sources at the transect ends.

abundance of browse, they deposit seeds in infrequent clumps as they defecate or regurgitate. By contrast, while the wind cannot routinely disperse seeds more than a few hundred meters, it can ensure that seeds are more evenly distributed across an area. Given the large size of burns, an equitable spread of seeds is far more valuable than a scattering of seed-filled feces.

Germination success depends greatly on the seedbeds created by fire as the seeds, and therefore the germinants arising from them, tend to be small. The wind-dispersed woody plants of fire-prone landscapes necessarily have small seeds because the rate of fall (and thus the distance traveled) is allometrically constrained by the ratio of seed mass to the area (including lift-producing wings or drag-producing fibers). For example, species in the fire-prone boreal forest have no seeds that weigh more than about 10 mg. In turn, small seeds produce small germinants; given the isometric scaling of seed length, volume, and mass, germinant length generally scales as one-third of the power of seed mass.

There are three main reasons why the transition from deposited seed to established seedling is so improbable. First, as discussed above, granivory takes a large fraction of the seeds. On average about 50 per cent of the seeds are eaten according to a review by Greene and Johnson (1998), although variation in predation among studies ranged from 0 to 100 per cent. However, initially in a large severe fire, it is highly likely that the abundance of small mammals is low in the interior of the burn (Charron and Greene 2002). This is due to small mammal populations having been decimated by smoke inhalation, and presumably (we know of no studies for burned forests detailing re-invasion rates of small predators) granivore density remains low in the interior of a large severe patch for a few years. By contrast, seed-eating birds have no serious initial dispersal constraint but typically occur in such low numbers that they are not the major source of pre-germination seed loss.

Second, a few rainless days cause drying of the poorly decomposed upper organic soil layer (duff), which can lead to desiccation and death of the germinant unless its radicle reaches the underlying mineral soil or humus layers that can remain wet via capillary flow from below.

Third, while the turgor pressure of a hydrating germinant is more than sufficient to push up dried leaves, deep litter layers can prevent photosynthetic tissues from reaching sufficient light, which can lead to germinant death. For these reasons, small seed establishment requires a substrate with little to no organic layer (Figure 8.5). In the dry and wet tropics, there is essentially no duff due to the highly active metabolism of the decomposition community, presumably the result of warm and wet soils, at least during the rainy season in the dry tropics. Moreover, hot and dry conditions in the dry tropics limit vegetative growth, which hinders organic layer accumulation. In short, at lower latitudes essentially all the ground surface represents clement seedbeds.

Where duff is present, fires can create optimal seedbeds for small seeds via smoldering combustion of the duff. After the flames have burned a thin layer of the duff, smoldering combustion will subsequently reduce organic material down to the humus or, in some cases, the mineral soil (Miyanishi and Johnson 2002). Prime seedbeds occur at the scale of a meter or two in a matrix of barely combusted organics, and germinants are effectively found only on these better microsites.

While varying tremendously within or among forest fires, the average proportion of optimal seedbeds (arbitrarily defined as <3 cm depth following fire) is around 40 per cent for a forest type with deep pre-fire duff, such as the boreal forest (Greene et al 2007; Figure 8.5). As for the dry forests of Australia, or the southwestern United States, it is close to 100 per

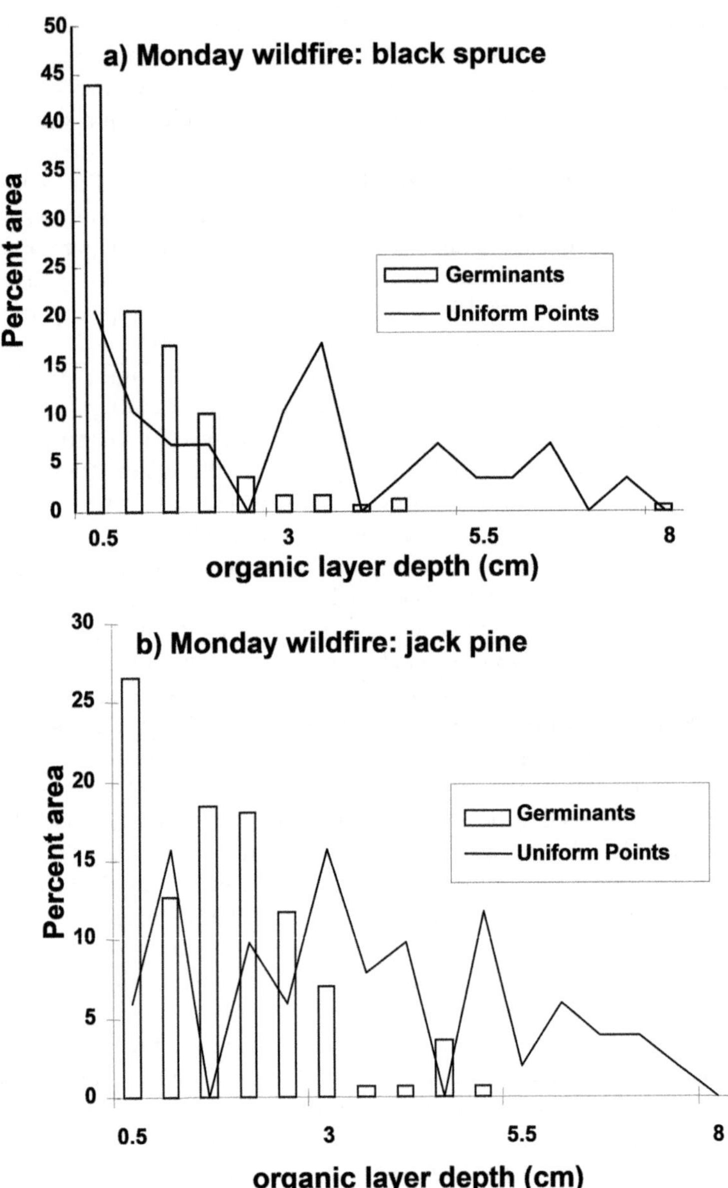

Figure 8.5 The limitation of small germinants to the thinnest post-fire organic substrates: an example from the Monday fire in the boreal forest of Saskatchewan. Germination success declines inversely with organic depth, and neither black spruce (*Picea mariana*) nor jack pine (*Pinus banksiana*) has many germinants at depths >3 cm. "Uniform points" refers to the actual percentage of the various organic layer depth classes; thus, the survivorship from the seed to the germinant stage is far higher on thin organic layers.

cent. In contrast, good seedbeds comprise only around 1–5 per cent coverage in a typical undisturbed closed canopy forest.

Moss and fungal spores are even smaller than the smallest tree seeds—only a few hundred microns in diameter—and consequently are even more dependent on substrate clemency than are small seeds. The first arriving mosses, such as *Ceratodon purpureus*, tend to be tolerant of desiccation in open areas. Given that spores are produced in large numbers and spread widely by the wind, one can use the bright green setae of, for example, new *Polytrichum mosses* as "phytometers" to measure seedbed quality (moisture availability) for conifer and angiosperm seeds. The immediate post-fire substrate in a boreal forest often becomes a mosaic of mushrooms, mosses, and small-seeded, sexually recruited species on the well-combusted surfaces, with asexually responding plants on the adjacent much less combusted substrates. Such a mosaic occurs at the scale of a few meters. Moisture availability in the first year after germination is much less of a problem in the wet tropics. In the dry tropics, germination is cued to the onset of the rainy season, and thus again the role of water availability becomes less important in determining post-fire recruitment density.

Optimal seedbeds do not last long following fire. As the herbaceous vegetation accrues, it deposits litter (much more rapidly than slower-growing woody plants) that begins to present a mechanical constraint to small seed germination. It has been argued, based on the few available studies, that this window of opportunity for small-seeded species is about four years (Charron and Greene 2002). Of course, the width of this window depends on herbaceous and shrub plant density and their growth rates.

The foregoing argument referred to biomes such as the circumboreal or mid-latitude higher elevation forests where pre-fire organic layers are thick. In areas where low severity fires are (or were) the predominant fire type (e.g., savannas and dry forests in Mediterranean climates), pre-fire organic layers are typically much thinner, and thus smoldering combustion is not as crucial a process. In such forests often lacking a closed canopy, the main inhibitor of germination is lack of water in the upper soil horizon (due to the paucity of rainfall) rather than evaporation from a porous, thick post-fire duff layer.

There are a limited number of plant strategies for coping with recurrent, large fires. The first, as already mentioned, is to possess a capacity for asexual reproduction, generally via apical meristems on stems or roots (e.g., axillary or adventitious buds) that are insulated against heating by bark or soil. The second is aerial seedbanks (serotiny). A compelling trait associated with aerial seedbanks is that these are not masting species, and thus at the time of fire a population is assured a large amount of available seeds with no dramatic dispersal constraint given that burned stems can provide seeds, even in the middle of a severe patch. This requires that the seeds reside within a structure that provides sufficient resistance to heat transfer. The resistance allows some fraction of the seeds to survive. Persistence of serotinous species on a landscape is dependent on a fire return interval that is substantially longer than the juvenile period required to produce the first reproductive propagules. This may be the main limit for expansion of any serotinous tree species abutting the parkland transition from forest to grassland, a biome invariably characterized by short return intervals (Brown and Johnstone 2012). During mast years, non-serotinous species can, just like serotinous species, produce such dense clusters of seed-bearing organs that the outer structures diminish the heat transfer sufficiently to permit the survival of seeds within the interior of the cluster. For example, seeds within small *Kunzea* (Myrtaceae) capsules of Australia can survive longer periods of heating when the capsules are clustered than when they are isolated (cf. Pounden et al. 2014).

A third strategy for post-fire colonization is to diffuse seeds from surviving wind-dispersed plants at the fire edge or within residual stands, but this should be viewed as a lack of a more refined adaptation. Asexual reproduction and dispersal from aerial seedbanks are far more effective mechanisms for repopulating a large fire than is solely dispersing from an edge (Greene and Johnson, 1996). Despite the well-invaginated, elongate shape of big fires, only about 5 per cent of wind-dispersed seeds of trees at the edge will travel more than 200 m from a surviving edge (Figure 8.4). The percentage decreases with distance from the burn edge.

Many herbaceous and shrub species mix these strategies. For example, the modestly shade-tolerant fireweed (*Chamerion angustifolia*) persists in ever-dwindling numbers as a forest stand ages. Immediately after a fire, these few individuals then perennate from surviving organs within the forest floor. Fueled by the high light environment following fire, fireweeds produce a first-summer crop of small seeds with a fibrous drag-producing appendage that is well-dispersed by wind. With a very low rate of fall (around 0.08 m s^{-1}), the dispersal distance from an area source of fireweed is remarkably high, and post-fire slopes are soon carpeted with fireweed for the first few years before they are overtopped by woody plants.

Another connection to masting is exhibited by several Australian shrub species that resprout after fire and then produce a very large seed crop from carbohydrates stored in the roots. Experiments have shown that the mast crop produced after burning in these "fire-stimulated" species is indeed triggered by burning of the plant (Lamont and Downes 2011).

Many animal species are reliant on large fires and, in some cases, help promote fire spread. Like fireweed, they readily colonize recent fires, quickly increase their populations, and then begin a slow decline in abundance until the next large fire (Saint-Germain and Greene 2009). Such species include many saprophagous beetle species such as *Bupestris* or *Monochamus* that can sense wood smoke even from a great distance and move upwind toward the source. In turn, bird species specialized on saprophagous insects, most famously woodpeckers, increase their populations suddenly in the aftermath of a fire. In Australia, some raptors have even been documented to intentionally spread fire via burning sticks during fire events to push prey out for forage (Bonta et al., 2017).

Before we leave the topic of recolonization by plants and animals, it is important to mention the very recent trend toward rapid salvage. As shown by a number of authors, having a second disturbance following immediately after a fire is a stress that most fire-adapted species cannot tolerate. As such, most plant species arriving primarily by seed are adversely affected (e.g., Leverkus et al. (2014) in Mediterranean forests), as are saprophagous species and their predators (Saint-Germain and Greene 2009).

References

Anderson, H. E. 1983. Predicting wind-driven wild land fire size and shape. USDA Forest Service Research Paper INT-305.

Balfour, D. A. and Midgley, J. J. (2006) 'Fire induced stem death in an African acacia is not caused by canopy scorching', *Austral Ecology*, vol 31, pp. 892–896.

Bär, A., Michaletz, S.T. and Mayr, S. 2019. Fire effects on tree physiology. *New Phytologist*, vol 223, pp. 1728–1741.

Bär, A., Nardini, A. and Mayr, S. (2018) 'Post-fire effects in xylem hydraulics of *Picea abies*, *Pinus sylvestris* and *Fagus sylvatica*', *New Phytologist*, vol 217, pp. 1484–1493.

Bison, N.N., Partelli Feltrin, R. and Michaletz, S. T. (2022) 'Trait phenology and fire seasonality co-drive seasonal variation in fire effects on tree crowns', *New Phytologist*, vol 234, pp. 1654–1663.

Bonta, M., Gosford, R., Eussen, D., Ferguson, N., Loveless, E. and Witwer, M. (2017) 'Intentional fire-spreading by "Firehawk" raptors in Northern Australia', *Journal of Ethnobiology*, vol 37, pp. 700–718.

Brown, C. D. and Johnstone, J. F. (2012) 'Once burned, twice shy: Repeat fires reduce seed availability and alter substrate constraints on *Picea mariana* regeneration', *Forest Ecology and Management*, vol 266, pp. 34–41.

Burton, C., Betts, R. A., Jones, C. D., Feldpausch, T. R., Cardoso, M., Anderson, L. O. (2020). El Niño driven changes in global fire 2015/16, *Frontiers in Earth Science*, vol 8, Art 199. DOI:10.3389/feart.2020.00199

Charron, I. and Greene, D. F. (2002) 'Post-fire seedbeds and tree establishment in the southern mixedwood boreal forest', *Canadian Journal of Forest Research*, vol 32, pp. 1607–1615.

Cruz, M. G., Butler, B. W., Alexander, M. E., Forthofer, J. M. and Wakimoto, R. H. Wakimoto. (2006) 'Predicting the ignition of crown fuels above a spreading surface fire. Part I: Model idealization', *International Journal of Wildland Fire*, vol 15, pp. 47–60.

Dickman, L.T., Jonko, A., Linn, R., Altintas, I., Atchley, A.L., Bär, A., Collins, A. D., Dupuy, J.-l., Gallagher, M. R., Hiers, J. K., Hoffman, C. M., Hood, S. M., Hurteau, M. D., Jolly, M., Josephson, A., Loudermilk, E. L., Ma, W., Michaletz, S. T., Nolan, R. H., O'Brien, J., Parsons, R., Partelli Feltrin, R., Pimont, F., Resco de Dios, V., Restaino, J., Sartor, K. A., Schultz-Fellenz, E., Serbin, S., Shuman, J., Sieg, C. H., Skowronski, N. S., Weise, D. R., Wright, M., Xu, C., Yebra, M. and Younes, N. (2023) 'Integrating plant physiology into simulation of fire behavior and effects', *New Phytologist*, vol 238, pp. 952–970.

Finney, M. A. (2001) 'Design of regular landscape fuel treatment patterns for modifying fire growth and behavior', *Forest Science*, vol 47, pp. 219–228.

Greene, D. F., Gauthier, S., Noel, J., Rousseau, M. and Bergeron, Y. (2006) 'A field experiment to determine the effect of post-fire salvage on seedbeds and tree regeneration', *Frontiers in Ecology and the Environment*, vol 4, pp. 69–74.

Greene, D. F. and Johnson, E. A. (1996) 'Wind dispersal of seeds from a forest into a clearing', *Ecology*, vol 77, pp. 595–609.

Greene, D. F. and Johnson, E. A. (1998) 'Seed mass and juvenile survivorship of trees in clearings and shelterwoods', *Canadian Journal of Forest Research*, vol 28, pp. 1307–1316.

Greene, D. F., Macdonald, S. E., Haeussler, S., Domenicano, S., Noël J. et al. (2007) 'The reduction of organic layer depth by wildfire in the North American boreal forest and its effect on tree recruitment by seed', *Canadian Journal of Forest Research*, vol 37, pp. 1012–1023.

Hartford, R. A. and Frandsen, W. H. (1992) 'When it's hot, it's hot etc. or maybe it's not! (Surface flaming may not portend extensive soil heating)', *International Journal of Wildland Fire*, vol 2, pp. 139–144.

Hayden, B. and Greene, D. F. (2008) 'The ecology of tropical dry forests', in: K. Del Claro, P. S. Oliveira,V. Rico- Gray, A. A. Almeida Barbosa, A. Bonet, and others (eds), *International Commission on Tropical Biology and Natural Resources, Encyclopedia of Life Support Systems (EOLSS)*, Developed under the Auspices of the UNESCO, EOLSS Publishers, Oxford, www.eolss.net

Hillel, D. (1998) *Environmental Soil Physics*, Academic Press, New York.

Kavanagh, K. L., Dickinson, M. B. and Bova, A. S. (2010) 'A way forward for fire-cased tree mortality prediction: Modeling a physiological consequence of fire', *Fire Ecology*, vol 6, pp. 80–94.

Keeley, J. E. and Fotheringham, C. J. (2001) 'Historic fire regime in southern California shrublands' *Conservation Biology*, vol 15, pp. 1536–1548.

Keeley, J. E., Pausas, J. G., Rundel, P. W., Bond, W. J. and Bradstock R. A. (2011) 'Fire as an evolutionary pressure shaping plant traits', *Trends in Plant Science*, vol 16, pp. 406–411.

Kleynhans, E. J., Atchley, A. and Michaletz, S. T. (2021) 'Modelling fire effects on plants: From organs to ecosystems', in E. A., Johnson and K. Miyanishi (eds). *Plant Disturbance Ecology: The Process and the Response*, 2nd edition, Academic Press, New York.

Kolden, C. A., Lutz, J. A., Key, C. H., Kane, J. T. and van Wagtendonk, L. W. (2012) 'Mapped versus actual burned area within wildfire perimeters: Characterizing the unburned', *Forest ecology and Management*, vol 286, pp. 38–47.

Krawchuk, M. A., Moritz, M. A., Parisien, M.-A., Van Dorn, J. and Hayhoe, K. (2009) 'Global pyrogeography: The current and future distribution of wildfire', *PLos One*, vol 4, e5102.

Lamont, B. B. and Downes, K. S. (2011) 'Fire-stimulated flowering among resprouters and geophytes in Australia and South Africa', *Plant Ecology*, vol 212, pp. 2111–2125.

Leverkus, A. B., Lorite, J., Navarro, F. B., Sanchez-Canete, E. P. and Castro, J. (2014) 'Post-fire salvage logging alters species composition and reduces cover, richness, and diversity in Mediterranean plant communities', *Journal of Environmental Management*, vol 133, pp. 323–331.

Lizundia-Loiola, J., Pettinari, M. L. and Chuvieco, E. (2020) 'Temporal anomalies in burned area trends: Satellite estimations of the Amazonian 2019 fire crisis', *Remote Sensing*, vol 12, p. 151.

Lopez, M. A. (2022) 'Investigating seed maturation and mortality: a mechanism for post-fire regeneration in non-serotinous conifers', Cal Poly Humboldt theses and projects 536. https://digitalcommons.humboldt.edu/etd/536

Macias Fauria, M., Michaletz, S. T. and Johnson, E. A. (2011) 'Predicting climate change effects on wildfires requires linking processes across scales', *WIRES Climate Change*, vol 2, pp. 99–112.

Michaletz, S. T. (2018) 'Evaluating the kinetic basis of plant growth from organs to ecosystems', *New Phytologist*, vol 219, pp. 37–44.

Michaletz, S. T. and Johnson, E. A. (2006) 'A heat transfer model of crown scorch in forest fires', *Canadian Journal of Forest Research*, vol 36, pp. 2839–2851.

Michaletz, S. T. and Johnson, E. A. (2007) 'How forest fires kill trees: a review of the fundamental biophysical processes', *Scandinavian Journal of Forest Research*, vol 22, pp. 500–515.

Michaletz, S. T., Johnson, E. A. and Tyree, M. T. (2012) 'Moving beyond the cambium necrosis hypothesis of post-fire tree mortality: Cavitation and deformation of xylem in forest fires', *New Phytologist*, vol 194, pp. 254–263.

Miyanishi, K. and Johnson, E. A. (2002) 'Process and patterns of duff consumption in the mixedwood boreal forest', *Canadian Journal of Forest Research*, vol. 32, pp. 1285–1295.

Nelson, R. M. Jr. (2001) 'Water relations of forest fuels', in E. A., Johnson and K. Miyanishi (eds), *Forest Fires: Behavior and Ecological Effects*. Academic Press, New York.

O'Brien, J. J., Hiers, J. K., Varner, J. M., Hoffman, C. M., Dickinson, M. B., Michaletz, S. T., Loudermilk, L. L., Butler, B. W. (2018) 'Advances in mechanistic approaches to quantifying biophysical fire effects', *Current Forestry Reports*, vol 4, pp. 161–177.

Peters, V. S., MacDonald, S. E. and Dale, M. R. T. (2005) 'The interaction between masting and fire is key to white spruce regeneration', *Ecology*, vol 86, pp. 1744–1750.

Pounden, E., Greene, D. F. and Michaletz, S. (2014) 'Non-serotinous woody plants behave as aerial seed bank species when a fire late in the seed maturation period coincides with a mast seeding year', *Ecology and Evolution*, vol 4, pp. 3830–3840.

Quintiere, J. G. (2006) *Fundamentals of Fire Phenomena*, John Wiley and Sons, New York.

Romme, W. H., Boyce, M. S., Gresswell, R., Merrill, E. H., Minshall, G. W. et al. (2011) 'Twenty years after the 1988 Yellowstone fires: lessons about disturbance and ecosystems', *Ecosystems*, vol 14, pp. 1196–1215.

Saint-Germain, M. and Greene, D. F. (2009) 'Salvage logging in the boreal and cordilleran forests of Canada: integrating industrial and ecological concerns in management plans', *Forestry Chronicle*, vol 85, pp. 120–134.

Sharples, J. J., Mills, G. A., McRae, R. H. D. and Weber, R. O. (2010) 'Foehn-like winds and elevated fire danger conditions in southeastern Australia', *Journal of Applied Meteorology and Climatology*, vol 49, pp. 1067–1095.

Splawinski TB, Greene DF, Michaletz ST, Gauthier S, Houle D, Bergeron Y. (2019) 'Position of cones within cone clusters determines seed survival in black spruce during wildfire', *Canadian Journal of Forest Research*, vol 49, pp. 121–127.

Stocks, B. J., Mason, J. A., Todd, J. B., Bosch, E. M., Wotton, B. M. et al. (2002) 'Large forest fires in Canada, 1959–1997', *Journal of Geophysical Research: Atmospheres*, vol 108, 8149.

Van Wagner, C. E. (1969) 'A simple fire growth model', *Forestry Chronicles*, vol 45, pp. 103–104.

Van Wagner, C. E. (1973) 'Height of crown scorch in forest fires', *Canadian Journal of Forest Research*, vol 3, pp. 373–378.

9

ECOLOGICAL EFFECTS OF STRONG WINDS ON FORESTS

Stephen M. Turton and Mohammed Alamgir

Introduction

Forest ecosystem dynamics are highly dependent on physical disturbances (like strong wind events) that reshape ecosystem structure and composition, modulate ecosystem functioning, and reset and accelerate succession (Franklin et al., 2002; Turner, 2010; Mitchell, 2013; Thom et al., 2013). Strong winds—typically those sustained above gale force or 61 km/h—are among the most important exogenic disturbance agents affecting forest ecosystems across the world, at a range of spatiotemporal scales (Proctor et al., 2001; Zhao et al., 2006; Lugo, 2008; Turton, 2008; Wang and Xu, 2009; Yoshida et al., 2011; Turton, 2019).

Tropical cyclones—also known as hurricanes and typhoons—affect wet and dry tropical forest regions adjacent to eight tropical ocean basins around the world: (1) northwest Pacific; (2) north Indian; (3) southwest Indian; (4) southeast Indian; (5) southwest Pacific; (6) northeast Pacific; (7) north Atlantic/Caribbean; and (8) south Atlantic (Turton, 2019). Forests between about 5–7° north and south of the equator rarely experience tropical cyclones due to the weak Coriolis effect near the equator. Nonetheless, there are many anecdotal reports of severe damage to forests over several square kilometres outside the typhoon belt in Southeast Asia (e.g. Whitmore and Burslem, 1998), and there is evidence that these events are common enough to have an influence on the structure of many of the forests in the region (e.g. Proctor et al., 2001; Baker et al., 2005). Forest blowdowns, associated with supercell thunderstorms, are also a common phenomenon in equatorial Amazonian forests in South America (Espírito-Santo et al., 2014).

Tropical cyclones also affect the forests of subtropical and temperate regions of both hemispheres as they often transition into intense extra-tropical cyclones as they move into higher latitudes (Everham III and Brokaw, 1996; Simonson, 2020). Winter windstorms, produced by intense mid-latitude depressions (extra-tropical lows), also affect the temperate deciduous forests of western and central Europe, northeast Asia, eastern and northwest parts of North America and many temperate forest areas in the Southern Hemisphere, while tornadoes regularly affect the temperate forests of North America, and occasionally those in Europe, Asia and elsewhere (Fischer et al., 2013; Fortuin et al., 2023).

DOI: 10.4324/9781003324072-11

Strong winds affect forests at spatial scales ranging from a few hundred square metres (e.g. tree crown damage from a single severe thunderstorm) to thousands of square kilometres (e.g. landscape-scale damage from a large synoptic-scale intense low-pressure weather system). Other events produce severe winds and forest damage at more intermediate spatial scales (e.g. a tornado). Winds affect forests at temporal scales ranging from a few minutes (e.g. a single microburst from a severe thunderstorm) to several days (e.g. a slow-moving severe tropical cyclone or intense extra-tropical cyclone). Other events produce strong winds that impact on forests at scales of a few hours (e.g. a slow-moving supercell thunderstorm). Wind damage to forests is due to both horizontal and vertical wind gusts, the latter often associated with downdrafts (micro-bursts) from severe supercells or large thunderstorms.

It is well-established that strong winds play an important role in shaping and moderating ecological processes in forests—from the tropics to the high latitudes. In this chapter, we consider the ecological effects of strong winds on the main forest biomes of the world—tropical, temperate, and boreal—at a range of scales. We first consider the effects of strong winds on forest structure, forest function, forest succession and biodiversity. We conclude with an evaluation of the likely effects of anthropogenic climate change on extreme wind events and the possible consequences for forest ecosystems.

Scales of forest disturbances

Wind is a universal phenomenon and 'average' winds rarely damage forests. However, there are some forests that experience 'chronic' wind stress from persistent winds of lower speeds that have a dramatic effect on forest structure (Mitchell, 2013; Moore et al., 2018). The Nanjenshan forest plot in southern Taiwan provides a good example of a lowland forest heavily influenced by prevailing monsoon winds (Chao et al., 2010; Ku et al., 2022).

Strong winds—associated with intense weather systems—may affect forests across a range of spatial scales. In temperate and boreal forests, the disturbance regimes are relatively large in North American forests, small in European forests and intermediate to large in East Asian forests (Fischer et al., 2013; Gregow et al., 2020). Severe wind events are responsible for more than 50 per cent of forest damage from catastrophic events by volume in Europe, according to Gardiner et al. (2011). A severe tropical cyclone (hurricane/typhoon) may affect thousands of square kilometres of forest area and can alter ecosystems significantly. By comparison, tornadoes usually affect small forest areas but cause the most severe localized damage in affected areas (Fischer et al., 2013).

Impacts of strong winds on forest ecosystems

Strong winds have both highly visible and cryptic impacts across all forest biomes (Everham III and Brokaw, 1996; Lugo, 2008; Webb et al., 2014; Gardiner et al., 2016; Zimmerman et al. 2020; Jackson et al., 2021; Quine et al., 2021). Winds affect forests at levels ranging from the smallest functional unit to the largest structural unit. In Figure 9.1, we highlight the effects on various components of a forest ecosystem affected by a severe wind event. The figure also represents the changing dimensions of ecological effects from the short to long term that typically occur in forests in the aftermath (or recovery phase) of a strong wind event. The immediate ecological effects are closely related to changes in forest structure, whereas the long-term effects usually affect ecosystem functioning, forest succession and

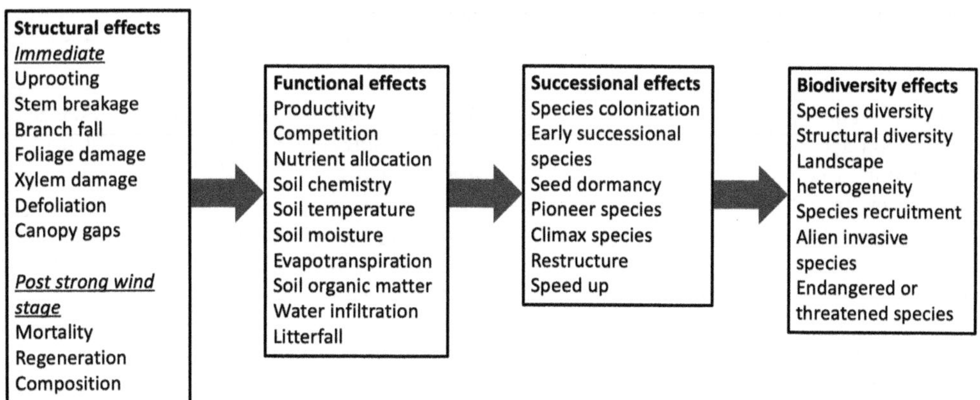

Figure 9.1 Structural, functional, successional and biodiversity effects of strong winds on forests ecosystems.

Source: Adapted from Seidl and Blennow (2012).

biodiversity. Strong winds also affect crucial ecosystem services provided by forests, including carbon storage (Delphin et al., 2013).

Forest ecosystem structure

The main structural elements of forest ecosystems are trees. Strong winds substantially affect trees in any forest biome due to their prominence and density compared with other terrestrial biomes, for example, grasslands. The scale and intensities of wind effects vary from individual tree- to stand- to landscape levels, depending on wind intensity and the spatial extent of the impacting weather system. Individual tree-level effects include: (1) breakage of branches, foliage and main stem; (2) defoliation; (3) shedding of bark; (4) root damage; and (5) uprooting of the entire tree. Stand-level effects include: (1) creation of gaps in the canopy by uprooting a large tree or breakage of several canopy trees; (2) mortality of seedlings and saplings due to sudden exposure to unfavourable environmental factors such as more sunlight, higher temperatures, and lower humidity; and (3) sometimes landslides. At the landscape level, strong winds may create landscape heterogeneity (Foster and Orwig, 2006; Mitchell, 2013; Turton, 2019; Fortuin et al., 2023). Some forest areas may be badly damaged while others remain relatively intact.

Table 9.1 summarizes the main tree and forest attributes in relation to strong wind events. Trees along the forest edges are more susceptible to snapping whereas interior forest trees—away from edges—are more susceptible to uprooting (Yoshida et al., 2011; Turton, 2019). The regular exposure to wind makes edge trees more wind resistant, as such trees often snap before experiencing root damage. Generally, damage is greater for the larger canopy trees compared with understory trees (Lugo, 2008). Apart from wind velocity, tree size is also a determining factor in tree damage due to strong winds (Webb, 1988; Zhao et al., 2006). Generally, larger canopy trees (Webb, 1988; Lugo, 2008) and intermediate trees (Dyer and Baird, 1997) are more susceptible to damage than smaller trees (Webb, 1988). Alternatively, more established root systems might make larger trees more resistant to wind damage and less vulnerable to uprooting; consequently, there is no disproportionate damage in larger

Table 9.1 Tree and forest stand attributes and likely effects of strong winds

Attributes	*Likely effects*
Higher wood density trees	More uprooting, less snapping, more resistance
Lower wood density trees	Less uprooting, more snapping, less resistance
Slow-growing trees	More resistance
Fast-growing trees	Less resistance
Trees on forest edge	More snapping, less uprooting
Interior trees	More uprooting, less snapping
Larger trees (dbh* and height)	Less resistance**
Smaller trees (dbh and height)	More resistance
Trees with shallow root system	Less resistance
Trees with deep root system	More resistance
Canopy trees	More damage
Understory trees	Less damage
Denser forest stand	More tolerant*

* Diameter at breast height (1.3 m).
** Opposite arguments are also available in literature.

trees (Dyer and Baird, 1997). However, more stem rot and root rot may result in large old (senescing) trees being more vulnerable to wind damage (Yoshida et al., 2011).

In the mangrove forests of Florida, larger diameter trees suffered from greater mortality and more snapping after Hurricane Andrew than smaller diameter trees (Baldwin et al., 1995). Responses to high-energy storms also differ among mangrove species. For Caribbean mangroves, Imbert (2018) found the red mangrove, *Rhizophora mangle*, was significantly less resistant to hurricane damage than was the black mangrove, *Avicennia germinans*. In tropical forests, trees with higher wood density are more susceptible to uprooting than snapping, whereas trees with lower wood density are more susceptible to snapping and post-disturbance mortality due to pathogens and insect attack (Lugo, 2008).

The structural effects of strong winds on forest ecosystems are direct, immediate, and highly visible. In the short term, visible damage occurs to the structural elements of the forest, but in the long term, structural damage substantially affects the species composition of the forest and its vertical structure, stand dynamics, landscape heterogeneity and ecosystem functioning (Figure 9.1). Severe structural damage may turn a carbon 'sink' forest into a carbon 'source' forest for a period, only to return to a carbon sink as regrowth ensues.

Fragmented forests (often within an agricultural matrix) are more vulnerable to strong winds than areas of contiguous forests (Laurance and Curran, 2008; Metcalfe et al., 2014; Turton, 2019). This is due to their abrupt artificial margins and open and often exposed surrounding landscapes that enhance wind speeds near the edges of forest remnants. However, forest structure, measured up to six months after Tropical Cyclone Larry in north-east Australia, did not differ between small (<40 ha) forest remnants and larger intact forest areas; the severity of effects in both sizes was largely determined by distance from the path of the cyclone across the landscape (Catterall et al., 2008). Hence, both local and regional contexts are relevant when evaluating the vulnerability of fragmented forest landscapes to strong wind events (Turton, 2019).

Forest ecosystem functioning and nutrient cycling

The effect of strong winds on forest ecosystem functioning is a critical element because disruption to any ecosystem functioning processes may have profound long-term effects on forests (Figure 9.1). After strong wind events, residual trees in forests often suffer from root and xylem damage that ultimately reduces the overall productivity of affected forests in the following three to five years (Zhao et al., 2006). In the long term, the recovering forest might be more productive. This might be due to more light availability (Pascarella, 1997), reduced biotic competition, and increased supply of nutrients and rainwater in the residual trees. For example, Zhao et al. (2006) found that after 12 years of hurricane damage in the bottomland hardwood forests of South Carolina in the United States that stem density—both in the tree and sapling layers—exceeded pre-hurricane levels.

Uprooting of trees due to strong winds mixes the upper soil layer that is advantageous for site quality and plant growth, but uprooting creates pits and mounds in the local topography that may increase pest infestation at the site (Dyer and Baird, 1997). Partially damaged trees tend to become more susceptible to insect and fungal attack (Lugo, 2008).

If strong winds destroy the habitat of micro- and macro-fauna in a forest, they typically suffer for food and shelter, resulting in the death of cryptic micro-fauna that are very important for nutrient cycling, while visible macro-fauna are very important for the maintenance of food webs in any forest ecosystem. Strong winds may also have an impact on soil temperature, moisture, and soil compaction. Increased soil compaction after strong wind events reduces water infiltration capacity and increases surface run-off and erosion of sediment into catchments in the subsequent years after the event. Furthermore, river flows often increase in wind-disturbed forest catchments due to decreased evapotranspiration from the denuded forest cover (Foster and Orwig, 2006).

Strong wind disturbance produces high canopy litterfall in forests that eventually decomposes and adds nutrients to the soil, and ultimately has a strong influence on the primary productivity of the forest (Huang et al., 2023). For example, in the subtropical humid forests of Puerto Rico—where hurricanes are somewhat infrequent—the sudden increase of nutrients in the soil from wind-induced litterfall significantly altered the nutrient patterns and forest responses in the understory (Lodge et al., 1991). However, the impact may be very different in forest regions where tropical cyclones are more frequent, such as parts of Southeast and East Asia (Lin et al., 2003). Tropical forest ecosystems are very efficient at decomposing litterfall and making the micronutrients available for trees within six months of a strong wind event (Ostertag et al., 2003). In forests where decomposition rates are very slow, like temperate and boreal forests, litterfall—due to strong winds—may prevent germination on the forest floor by creating a 'space' between seeds and mineral soils. Such effects extend beyond temperate and boreal forests. For example, Shiels et al. (2010) reported the death of seedlings from the deposition of a layer of canopy-derived debris on the ground after a hurricane in Puerto Rico. Hence, strong winds have profound effects on ecosystem functioning and nutrient cycling, but this varies among forest biomes.

Forest ecosystem succession

Forest ecosystem dynamics substantially depend on natural disturbances like strong winds that play an important role in forest succession (Thom et al., 2013; Xi et al., 2019). They may redirect forest succession (Figure 9.1) and speed up forest succession, depending on the context (Zhao et al., 2006). During tropical cyclones, the upper canopy layer trees in forests are more damaged than intermediate canopy layer trees. So, tropical cyclones tend to accelerate the growth of intermediate-sized trees, which are usually later successional species in forests (Webb and Scanga, 2001). Tropical cyclones also create opportunities for the regeneration of early successional species on the forest floor by creating canopy gaps and allowing more sunlight to reach the understory (Batista and Platt, 2003). They also promote growth of shade-tolerant species within the seedling and sapling layer (Zhao et al., 2006). Larger trees in forests (usually shade-intolerant) generally suffer more from tropical cyclone damage and post-disturbance mortality. Therefore, strong winds may facilitate colonization of shade-intolerant species in some forests (Zhao et al., 2006). Additionally, strong winds sometimes break the seed dormancy of successional species, which may have been inactive in the soil over many years. This process helps the regeneration of dormant species at the site as well as restructuring forest succession (Pascarella, 1997).

Forest biodiversity

Strong winds may have profound impacts on forest biodiversity at a range of scales (Figure 9.1). Immediately after forest disturbance, species diversity usually increases due to increased nutrients, sunlight, and space availability, which together enhance regeneration and recruitment of new species (Lugo, 2008; Xi et al., 2019). Zhao et al. (2006) demonstrated that hurricane disturbances restructure species composition and enrich species diversity via this process. Fischer et al. (2013) reported that strong wind disturbance may kill large trees but may create necessary conditions for new tree cohorts, thereby increasing the species diversity in disturbed forests. Strong wind disturbances increase the abundance of climbers, vines, and herbaceous plants (Lugo, 2008). These plants may suppress the growth of young trees and recruits, which may be a limiting factor for increasing species diversity after wind disturbances.

Strong winds create structural diversity in forests by promoting growth from suppressed understory saplings and seedlings (Zhao et al., 2006). Foster and Terborgh (1998) discovered in the Amazon Forest that selective killing of large trees due to wind storms contributed to structural heterogeneity. Similar tree mortality patterns have occurred in temperate forests after severe wind damage from tornadoes (Glitzenstein and Harcombe, 1988).

Over longer time periods, severe wind disturbances may create landscape diversity (heterogeneity) in forests (Dyer and Baird, 1997). This diversity may arise from landslides or from differential recovery rates of different forest patches after severe meteorological or other disturbances, for example, fire, dieback due to disease. Forest fragmentation provides an opportunity for recruitment of new species into forest ecosystems, including non-native invasive species (Murphy and Metcalfe, 2016). Bellingham et al. (2005) found that strong wind disturbances accelerated the invasion of non-native plant species in tropical forests in Jamaica.

Tropical cyclones (hurricanes/typhoons), tornadoes and even windstorms in European forests may uproot large trees or break crowns of large trees. These large trees act as a seed bank for natural regeneration in forests and provide important microhabitat for fauna and specialized flora, for example, epiphytes and vines. So, if there is no intermediate tree species in the seed bank, the forest suffers due to a lack of enough regeneration (Yoshida et al., 2011), which ultimately affects the whole forested landscape. Animal diversity may also suffer following wind disturbance across a fragmented forest landscape due to a loss of food resources for vulnerable species (Turton, 2019).

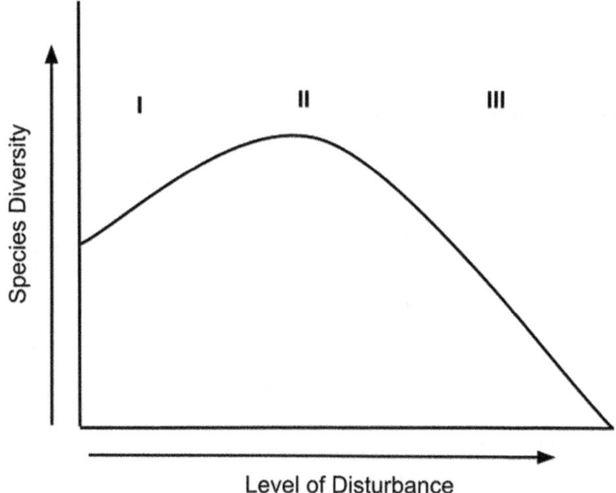

Figure 9.2 Graph shows principles of intermediate disturbance hypothesis: (I) at low levels of ecological disturbance species richness decreases as competitive exclusion increases, (III) at high levels of disturbance species richness is decreased due an increase in species movement, (II) at intermediate levels of disturbance, diversity is maximized because species that thrive at both early and late successional stages can coexist (after Connell, 1978).

Source: Creative Commons.

Natural disturbances—like strong winds—are very important factors in maintaining high biodiversity in forests (Zhao et al., 2006). The influence of strong winds on biodiversity is mainly dependent on the scale of disturbance, both in terms of intensity and frequency (Xi et al., 2019). A rigorous debate exists about the scale of disturbance required to maintain the highest level of biodiversity in forests. Here we describe (Figure 9.2) one of the very early and most popular theories—the 'intermediate disturbance theory' developed by Connell (1978). This theory argues that ecosystems that experience small- and large-scale disturbances typically have lower levels of biodiversity compared with those that have intermediate levels of disturbance, which typically maintain the highest biodiversity. If there is no or low disturbance, there is more competition and ultimately low biodiversity. If there is a high level of disturbance, there is more mortality and disturbed forests have less time for recruitment and recovery, and consequently fewer species may survive (Connell 1978). Post-disturbance mortality also has an influence on biodiversity after a very large-scale disturbance. For example, strong winds may be catastrophic for animal diversity, particularly if the affected forests are habitat for endangered and threatened species because they may also die out during the post-disturbance stage.

Forest recovery (resilience)

Compared with other ecosystems (e.g., grasslands and wetlands), forest ecosystems usually require a long time for recovery after a strong wind event due to their relatively complex, slow-growing and long-life characteristics. For example, it may take several decades after severe tropical cyclones for a highly damaged forest to return to its original state (Metcalfe et al., 2008, Murphy et al., 2014). Forest recovery after strong wind events depends on the scale and intensity of the disturbance regimes, composition and structure of the forest, site quality, forest resistance and resilience to strong winds and the adaptive capacity of the residual trees to the changing environment, notably the microclimate (Turton, 2019).

More resilient forests have more capacity to recover in the short term after disturbances, while more complex forests are more resistant to strong wind than less complex forests (Lugo et al., 1983; Zhao et al., 2006). Higher biodiversity is an indicator of a complex forest. In tropical forests, recovery after cyclones is relatively rapid due to higher production rates associated with humid tropical environments (Metcalfe et al., 2008). However, Tanner and Bellingham (2006) reported for forests in Jamaica that higher elevation forests—with lower species diversity—were more resistant to strong winds than higher diversity, lower elevation forests. Following wind disturbance, temperate and boreal forests recover more slowly than tropical forests due to their lower rates of primary productivity. Forests containing trees with sprouting ability may recover more rapidly than forests with trees lacking this function and are more dependent on other means of regeneration, for example, seeding and sapling regrowth. In the early stage of recovery invasive species are abundant in disturbed forest areas. These invasive species limit the opportunity for regeneration and recruitment of many native species. They may delay forest recovery and change the species composition, particularly in large open areas and within riparian vegetation along the river edges. However, in contiguous forest areas, weed invasions tend to be ephemeral following severe wind disturbance (Murphy and Metcalfe, 2016).

Notable strong wind events with forest disturbance regimes

Strong wind disturbances regularly affect forest biomes across the world. In European forest ecosystems, strong wind is the most important disturbance factor in terms of the volume of timber damage (Schelhaas et al., 2003). In the forests of the eastern United States, hurricanes are also one of the most important disturbance factors. For example, in terms of total area affected, hurricanes are the most important disturbance factor in the temperate New England forests in the United States (Foster and Orwig, 2006). Tropical cyclones are also one the most important disturbance factors for tropical forests in northeast Australia (Turton, 2008, 2019). Box 9.1 provides examples of significant wind disturbance events for three contrasting forest biomes.

Factors that determine effects of strong winds on forest ecosystems

Forest damage from strong winds largely depends on wind speed—or its velocity—as the damaging forces of wind are directly proportional to the fourth power of the velocity of the wind. Apart from the physical dimensions of wind, the degree of damage also depends on several endogenic factors within forests. So, a strong wind event with equivalent force affects different forest types at varying degrees depending on a range of endogenic factors (Mitchell, 2013; Xi, 2019):

- Forest structure and species composition
- Topographic conditions
- Soil characteristics.

Box 9.1 Examples of significant wind disturbance events for three contrasting forest biomes

It was found that in European forests, the average structural damage due to strong winds was 1.8 million cubic metres from 1950 to 2000. For example, the storm Gudrun, from 8 to 9 January 2005 affected the spruce forests of Norway and Sweden and was the worst storm on record for Sweden. It caused an estimated economic damage of €2.4 billion, damaged over 75 million cubic metres of wood and resulted in significant growth reduction of residual trees (Schelhaas et al., 2003; Nilsson et al., 2004; Seidl and Blennow, 2012).

A severe hurricane in 1938 affected the forest areas of central Massachusetts in the United States and was the most destructive storm in the last 175 years. It damaged 3 billion board feet of timber, increased river flow and caused major ecosystem changes and damage to 80 per cent of canopy trees (Patric, 1974; Cooper-Ellis et al., 1999; Foster and Orwig, 2006).

Severe Tropical Cyclone Yasi affected the equivalent of about 9,000 km^2 of forest (roughly 75 per cent of the total forest in the Wet Tropics bioregion of Australia), and produced a mortality of about 300 million trees representing about 65 Tg of carbon—equivalent to a large fraction of the net carbon sink from Australian forest land in 2014 (Negron-Juarez et al., 2014).

Mangrove forests and coastal forests are more vulnerable to strong winds due to their proximity to the ocean where wind speeds are higher than adjacent land areas. Mangroves are possibly also vulnerable because they have shallow roots to avoid anoxic soils (Turton, 2019). Wang and Xu (2009) found that in Lower Pearl River Valley (United States) coastal wetland forests were more susceptible to strong wind damage than inland mixed forests and evergreen forests. The damage within a forest stand from strong winds also varies with tree size (Webb, 1989; Xi et al., 2008).

Different tree species have different degrees of resistance to strong winds (Paz et al., 2018). Trees with higher wood densities are more wind resistant than those with lower wood densities. Higher wood density provides trees with more internal strength in the face of forces from strong winds. Slow-growing trees are more resistant to wind damage than fast-growing trees. Slow-growing trees may become wind resistant due to more wind exposure and disturbance during their longer life spans.

Topographic conditions have an influence on the intensity and scale of damage on forests due to strong winds (Foster and Boose, 1992). The primary topographic factors are slope, aspect and elevation. Forests on windward slopes may experience more damage than forests on leeward slopes by their exposure to winds (Foster and Boose, 1992). However, because wind speeds usually peak at ridges, leeward slopes sometimes experience severe downslope turbulence due to gravity waves (Finnigan and Brunet, 1995; Turton, 2008, 2019). Interestingly, even though winds increase with elevation above sea level, forests at lower elevations generally receive more damage from strong winds than forests at higher elevations, mainly because upland forests are often lower in stature with aerodynamically smooth canopies (Reilly et al., 2002).

Soil characteristics have a profound impact on root development of trees (Nicoll et al., 2006), which provide the anchoring strength against the effects of strong winds. Trees with shallow root systems are less resistant to strong winds than trees with deep root systems.

Trees grown in poor drainage and seasonal waterlogging conditions usually develop shallow root systems (Mayer, 1989; Ray and Nicoll, 1998), resulting in their low resistance to strong winds; consequently, they often suffer from greater damage after a strong wind event.

Likely effects of climate change on winds and forests

Damage from strong winds may be increasing in some forest biomes because of increased intensity of winds due to global climate change (Intergovernmental Panel on Climate Change, 2021). Extra-tropical cyclones and tropical cyclones have likely shifted poleward since the 1980s, and these latitudinal shifts in the average paths of cyclonic storms are likely to continue with future global warming (Turton, 2023). This will bring tropical and extra-tropical cyclones in increasing contact with exposed forests that have previously been less affected by strong wind events.

Wind damage in boreal and temperate forests in Europe is likely to interact with other biological and climatic drivers, including bark beetle infestations, root rot, lack of winter snow cover and hence exposure of tree seedlings to frost (Machado Nunes Romeiro et al., 2022). One global study showed that future changes in forest disturbance are likely to be most pronounced in coniferous forests and the boreal biomes (Seidl et al., 2017).

Damage to tropical forests (especially mangrove forests) of Southeast Asia is increasing due to more intense tropical cyclones in that region (Alamgir and Turton, 2013). Studies show that the intensity of strong winds associated with tropical cyclones will increase significantly over the 21st century under projected climate change although there is more uncertainty of cyclone frequency in the future, except that is likely to vary significantly among tropical ocean basins (Knutson et al., 2010; Turton, 2023). There has also been a discernible global slowdown in the forward motion of tropical cyclones, that similarly shows differences among tropical ocean basin (Kossin, 2018).

Given that climate change is highly likely to increase the risk of more extreme wind events in the future across all forest biomes (Intergovernmental Panel on Climate Change, 2021), it might be prudent to consider the following questions:

- Will the observed poleward shift in the range of tropical and extra-tropical cyclones continue under global warming, bringing them into contact with forest ecosystems that currently do not experience severe cyclones or wind storms?
- Will forest biodiversity decline in forested landscapes because of more extreme wind events as suggested by the 'intermediate disturbance theory' (Figure 9.2)?
- How will the slowdown of forward motion of tropical cyclones interact with forest ecosystem processes, notably longer intervals of prolonged high winds during storm events?
- Will forest structure change over time towards a generally lower stature?
- Will a shift in the severity spectrum for severe wind events across all forest biomes tend to favour species more resistant to strong winds, including less desirable non-native species?
- How will contiguous forest areas fare in comparison with exposed forest remnants contained in agricultural and peri-urban matrices?
- How might projected increases in wind intensity act synergistically with other ongoing changes in forest ecosystems, such as those driving habitat loss and fragmentation, loss of biodiversity, spread of pathogens and associated desiccation and increased fire risk?

Synergies between climate change, more extreme weather events, forest habitat loss and fragmentation, and increased risk for the spread of pathogens pose a significant threat to the world's forests. We must plan for the management of our forest ecosystems with knowledge that they may not recover to their original state in the aftermath of severe wind events. Instead, we should prepare ourselves for witnessing "trajectories of responses" that may culminate—over time—in forest ecosystems that are structurally and floristically very different to their pre-disturbance states.

Conclusions

- Strong winds—at a range of scales—are among the most important exogenic disturbance agents affecting forest ecosystems across the world. While extreme wind events are important to all forest biomes, we also need to consider the influences of 'chronic' high wind regimes on some forests.
- Strong winds may affect forests at spatial scales ranging from a few hundred square metres to thousands of square kilometres, while other events may produce severe winds and damage to forests at more intermediate spatial scales. Winds may affect forests at temporal scales ranging from a few minutes to several days, while other events may produce strong winds that impact forests at scales of a few hours.
- Strong winds play an important role in shaping and moderating ecological processes in forests from the tropics to the high latitudes; notably, they affect forest structure, forest function, forest succession and biodiversity at a range of scales.
- Climate change is directly and indirectly affecting all the forest biomes on Earth. Among the climate change drivers includes an increase in average wind speeds and extreme wind events. Synergies between background global heating, more extreme weather events, forest habitat loss and fragmentation, and spread of forest pathogens mean we must accept that our forest ecosystems may not recover to their original state in the aftermath of severe wind events.

References

Alamgir, M. and Turton, S. M. (2013) 'Climate change and organic carbon storage in Bangladesh', in N. Tutja and S. S. Gill (eds), *Climate Change and Plant Abiotic Stress Tolerance*, Wiley-Blackwell, Oxford.

Baker, P. J., Bunyavejchewin, S., Oliver, C. D. and Ashton, P. S. (2005) 'Disturbance history and historical stand dynamics of a seasonal tropical forest in western Thailand', *Ecological Monographs*, vol 75, pp. 317–343.

Baldwin, A. H., Platt, W. J., Gathen, K. L., Lessmann, J. M. and Rauch, T. J. (1995) 'Hurricane damage and regeneration in fringe mangrove forests of southeast Florida, USA', *Journal of Coastal Research*, vol 21, pp. 169–183.

Batista, W. B. and Platt, W. J. (2003) 'Tree population responses to hurricane disturbance: syndromes in southeastern USA old growth forest', *Journal of Ecology*, vol 91, pp. 197–212.

Bellingham, P. J., Tanner, E. V. J. and Healy, J. R. (2005) 'Hurricane disturbance accelerates invasion by the alien tree Pittosporum undulatum in Jamaican montane rain forests', *Journal of Vegetation Science*, vol 16, pp. 675–684.

Catterall, C. P., McKenna, S., Kanowski, J. and Piper, S. D. (2008) 'Do cyclones and forest fragmentation have synergistic effects? A before-after study of rainforest vegetation on the Atherton Tableland, Australia', *Austral Ecology*, vol 33, pp. 471–484.

Chao, W.-C., Song, G.-Z. M., Chao, K.-J., Liao, C.–C, Fan, S.–W. and others (2010) 'Lowland rainforests in southern Taiwan and Lanyu, at the northern border of Paleotropics and under the influence of monsoon wind', *Plant Ecology*, vol 210, pp. 1–17.

Connell, J. H. (1978) 'Diversity in tropical rain forests and coral reef', *Science*, vol 199, no 4335, pp. 1302–1310.

Cooper-Ellis, S. D., Foster, R., Carlton, G. and Lezberg, A. (1999) 'Response of forest ecosystems to catastrophic wind: evaluating vegetation recovery on an experimental hurricane', *Ecology*, vol 80, pp. 2683–2696.

Delphin, S., Escobedo, F. J., Abd-Elrahman, A. and Cropper, W. (2013) 'Mapping potential carbon and timber losses from hurricanes using a decision tree and ecosystem services driver model', *Journal of Environmental Management*, vol 129, pp. 599–607.

Dyer, J. M. and Baird, P. R. (1997) 'Wind disturbance in remnant forest stands along the prairie-forest ecotone, Minnesota, USA', *Plant Ecology*, vol. 29, pp. 121–134.

Espírito-Santo, F. B., Gloor, M., Keller, M. et al. (2014) 'Size and frequency of natural forest disturbances and the Amazon forest carbon balance', *Nature Communincations*, vol. 5, 3434.

Everham III, E. M. and Brokaw, N. V. L. (1996) 'Forest damage and recovery from catastrophic wind', *Botanical Reviews*, vol. 62, no 2, pp. 113–185.

Finnigan, J. J. and Brunet, Y. (1995) 'Turbulent airflow in forests on flat and hilly terrain', in M. P. Coutts and J. Grace (eds), *Wind and Trees*, Cambridge University Press, Cambridge.

Fischer, A., Marshall, P. and Camp, A. (2013) 'Disturbances in deciduous temperate forest ecosystems of the northern hemisphere: their effects on both recent and future forest development', *Biodiversity Conservation*, vol. 22, pp. 1863–1893.

Fortuin, C. C., Montes, C. R., Vogt, J. T. and Gandhi, K. J. K. (2023) 'Stand and tree characteristics influence damage severity after a catastrophic hurricane disturbance', *Forest Ecology and Management*, vol. 532, 120844.

Foster, D. R. and Boose, E. R. (1992) 'Patterns of forest damage resulting from catastrophic wind in Central New England, USA', *Journal of Ecology*, vol. 80, pp. 79–98.

Foster, D. R. and Orwig, D. A. (2006) 'Preemptive and salvage harvesting of New England forests: when doing nothing is a viable alternative', *Conservation Biology*, vol. 20, pp. 959–970.

Foster, M. S. and Terborgh, J. (1998) 'Impact of a rare storm event on an Amazonian forest', *Biotropica*, vol. 30, pp. 470–474.

Franklin, J. F., Spies, T. A., Pelt, R., Van Carey, A. B., Thornburgh, D. A. and others (2002) 'Disturbances and structural development of natural forest ecosystems with silvicultural implications, using Douglas- fir forests as an example', *Forest Ecology and Management*, vol. 155, pp. 399–423.

Gardiner, B., Berry, P. and Moulia, B. (2016) 'Review: Wind impacts on plant growth, mechanics and damage', *Plant Science*, vol. 245, pp. 94–118.

Gardiner, B., Blennow, K., Carnus, J-M., Fleischer, P., Ingemarson, F. and others (2011) *Destructive Storms in European Forests: Past and Forthcoming Impacts*, European Forest Institute, Final Report to European Commission.

Glitzenstein, J. S. and Harcombe, P. (1988) 'Effects of December 1983 tornado on forest vegetation of the Big Thicket, South-east Texas, USA', *Forest Ecology and Management*, vol. 25, pp. 269–290.

Gregow, H., Rantanen, M., Laurila, T. K. and Mäkelä, A. (2020) 'Review on winds, extratropical cyclones and their impacts in Northern Europe and Finland', *Finish Meteorological Institute Report* 2020:3.

Huang, C., Liu, C., Hu, T., Chung, H. and Wang, J. (2023) 'Variation of seasonal litterfall in subtropical montane cloud forests to typhoon severity and environmental factors', *Biotropica*, vol. 55, pp. 132–144.

Imbert, D. (2018) 'Hurricane disturbance and forest dynamics in east Caribbean mangroves', *Ecosphere*, vol. 9, e02231.

Intergovernmental Panel on Climate Change (2021) 'Summary for Policymakers', in V.P. Masson-Delmotte, A. Pirani, S.L. Connors, C. Péan, S. Berger, N. Caud, Y. Chen, L. Goldfarb, M.I. Gomis, M. Huang, K. Leitzell, E. Lonnoy, J.B.R. Matthews, T.K. Maycock, T. Waterfield, O. Yelekçi, R. Yu, and B. Zhou (eds), *Climate Change 2021: The Physical Science Basis. Contribution of Working Group I to the Sixth Assessment Report of the Intergovernmental Panel on Climate Change*, Cambridge University Press, Cambridge.

Jackson, T. D., Sethi, S., Dellwik, E., Angelou, N., Bunce, A., van Emmerik, T., et al. (2021) 'The motion of trees in the wind: a data synthesis', *Biogeosciences*, vol. 18, pp. 4059–4072.

Knutson, T. R., McBride, J. L., Chan, J., Emanuel, K., Holland, G. and others (2010) 'Tropical cyclones and climate change', *Nature Geoscience*, vol. 3, pp. 157–163.

Kossin, J. P. (2018) 'A global slowdown of tropical cyclone translation speed', *Nature*, vol. 558, pp. 104–107.

Ku, C., Chao, K., Song, G., Lin, H., Fan, S. and Chao, W. (2022) 'How the strength of monsoon winds shape forest dynamics', *Diversity*, vol. 14, 169.

Laurance, W. F. and Curran, T. J. (2008) 'Impacts of wind disturbance on fragmented tropical forests: an international review', *Austral Ecology*, vol. 33, pp. 399–408.

Lin, K., Hamburg, S. P., Tang, S., Hsia, Y. and Lin, T. (2003) 'Typhoon effects on litterfall in a subtropical forest', *Canadian Journal of Forest Research*, vol. 33, pp. 2184–2192.

Lodge, D. J., Scatena, F. N., Asbury, C. E. and Sánchez, M. J. (1991) 'Fine litterfall and related nutrient inputs resulting from hurricane Hugo in subtropical wet and lower montane rain forests of Puerto Rico' *Biotropica*, vol 23, no. 4a, pp. 336–342.

Lugo, A. E. (2008) 'Visible and invisible effects of hurricanes on forest ecosystems: an international review', *Austral Ecology*, vol. 33, pp. 368–398.

Lugo, A. E., Applefield, M., Pool, D. J. and McDonald, R. B. (1983) 'The impact of Hurricane David on the forests of Dominica', *Canadian Journal of Forest Research*, vol. 13, pp. 201–211.

Machado Nunes Romeiro, J., Eid, T., Antón-Fernández, C., Kangas, A. and Trømborg, E. (2022) 'Natural disturbances risks in European boreal and temperate forests and their links to climate change — a review of modelling approaches', *Forest Ecology and Management*, vol. 509, 120071.

Mayer, H. (1989) 'Windthrow', *Philosophical Transactions of the Royal Society of London Series B, Biological Sciences*, vol. 324, pp. 267–281.

Metcalfe, D. J., Bradford, M. G. and Ford, A. J. (2008) 'Cyclone damage to tropical rain forests: species- and community-level impacts', *Austral Ecology*, vol. 33, pp. 432–441.

Metcalfe, D. J., O'Malley, T., Lawson, T. J. and Ford, A. J. (2014) *Mapping Littoral rainforest and coastal vine thickets of eastern Australia in the Wet Tropics: Mission Beach Pilot Study*, Report to the National Environmental Research Program, Reef and Rainforest Research Centre Limited, Cairns, Queensland, Australia.

Mitchell, S. J. (2013) 'Wind as a natural disturbance agent in forests: a synthesis', *Forestry: An International Journal of Forest Research*, vol. 86, pp. 147–157.

Moore, J., Gardiner, B. and Sellier, D. (2018) 'Tree mechanics and wind loading', in A. Geitmann and J. Gril (eds), *Plant Biomechanics*, Springer, Cham.

Murphy, H. T. and Metcalfe., D. J. (2016) The perfect storm: weed invasion and intense and intense storms in tropical forests. *Austral Ecology*, vol. 41, pp. 846–874.

Murphy, H. T., Metcalfe, D. J., Bradford, M. G. and Ford, A. J. (2014) 'Community divergence in a tropical forest following a severe tropical cyclone', *Austral Ecology*, vol. 39, pp. 696–709.

Negron-Juarez, R. I., Chambers, J. Q., Hurt, G. C., Annane, B., Cocke, S., Powell, M. Stott, M., Goosem, S., Metcalfe, D. J. and S. S. Saatchi, S. S. (2014) 'Remote sensing assessment of forest disturbance across complex mountainous terrain: the pattern and severity of impacts of Tropical Cyclone Yasi on Australian rainforests', *Remote Sensing of Environment*, vol. 6, pp. 5633–5649.

Nicoll, B. C., Gardiner, B. A., Rayner, B. and Peace, A. J. (2006) 'Anchorage of coniferous trees in relation to species, soil type, and rooting depth', *Canadian Journal of Forest Research*, vol. 36, pp. 1871–1883.

Nilsson, C., Stjernquist, I., Bärring, L., Schlyter, P., Jonsson, A. M. and Samuelsson, H. (2004) 'Recorded storm damage in Swedish forests 1901–2000', *Forest Ecology and Management*, vol. 199, no 1, pp. 165–173.

Ostertag, R., Scatena, F. N. and Silver, W. L. (2003) 'Forest floor decomposition following hurricane litter inputs in several Puerto Rican forests', *Ecosystems*, vol. 6, pp. 261–273.

Pascarella, J. B. (1997) 'Hurricane disturbance and the regeneration of Lysiloma latisiliquum (Fabaceae): a tropical tree in south Florida', *Forest Ecology and Management*, vol. 92, pp. 97–106.

Patric, J. H. (1974) 'River flow increases in central New England after the hurricane of 1938', *Journal of Forestry*, vol. 72, pp. 21–25.

Paz, H., Vega-Ramos, F. and Arreola-Villa, F. (2018) 'Understanding hurricane resistance and resilience in tropical dry forest trees: A functional traits approach', *Forest Ecology and Management*, vol. 426, pp. 115–122.

Proctor, J., Brearley, F. Q., Dunlop, H., Proctor, K., Supramono and Taylor, D. (2001) 'Local wind damage in Barito Ulu, Central Kalimantan: a rare but essential event in a lowland dipterocarp forest?', *Journal of Tropical Ecology*, vol. 17, pp. 473–475.

Quine, C. P., Gardiner, B. A. and Moore, J. (2021) 'Wind disturbance in forests: The process of wind created gaps, tree overturning, and stem breakage,' in E. A. Johnson, K. and Miyanishi (eds) *Plant Disturbance Ecology*, second edition, Academic Press, New York.

Ray, D. and Nicoll, B. C. (1998) 'The effect of soil water-table depth on root-plate development and stability of Sitka spruce', *Forestry*, vol. 71, pp. 169–182.

Reilly, J., Mayer, M. and Harnisch, J. (2002) 'The Kyoto Protocol and non-CO_2 greenhouse gases and carbon sinks', *Environmental Modeling and Assessment*, vol 7, pp. 217–229.

Schelhaas, M-J., Nabuurs, G-J. and Schuck, A. (2003) 'Natural disturbances in the European forests in the 19th and 20th centuries' *Global Change Biology*, vol 9, no. 11, pp. 1620–1633.

Seidl, R. and Blennow, K. (2012) 'Pervasive growth reduction in Norway spruce forests following wind disturbance', *PLos One*, vol. 7, no. 3, e33301.

Seidl, R., Thom, D., Kautz, M. et al. (2017) 'Forest disturbances under climate change', *Nature Climate Change*, vol. 7, pp. 395–402.

Shiels, A. B., Zimmerman, J. K., García-Montiel, D. C., Jonckheere, I., Holm, J. and others (2010) Plant responses to simulated hurricane impacts in a subtropical wet forest, Puerto Rico, *Journal of Ecology*, vol. 98, pp. 659–673.

Simonson, J. (2020) 'Extratropical cyclones and associated climate impacts in the northeastern United States', PhD thesis, The University of Maine, USA. *Electronic Theses and Dissertations*, 3224. https://digitalcommons.library.umaine.edu/etd/3224

Tanner, E. V. J. and Bellingham, P. J. (2006) 'Less diverse forest is more resistant to hurricane disturbance: evidence from montane rain forests in Jamaica' *Journal of Ecology*, vol. 94, pp. 1003–1010.

Thom, D., Seidl, R., Steyrer, G., Krehan, H. and Formayer, H. (2013) 'Slow and fast drivers of the natural disturbance regime in Central European forest ecosystems', *Forest Ecology and Management*, vol. 307, pp. 293–302.

Turner, M. G. (2010) 'Disturbance and landscape dynamics in a changing world', *Ecology*, vol. 91, pp. 2833–2849.

Turton, S. M. (2008) 'Landscape-scale impacts of Cyclone Larry on the forests of northeast Australia including comparisons with previous cyclones impacting the region between 1858 and 2006', *Austral Ecology*, vol. 33, e02571.

Turton, S. M. (2019) 'Reef-to-ridge ecological perspectives of high-energy storm events in northeast Australia', *Ecosphere*, vol. 10, pp. 1–20.

Turton, S. M. (2023) *Surviving the Climate Crisis: Australian Perspectives and Solutions*, CRC Press, Boca Raton, FL.

Wang, F. and Xu, Y. J. (2009) 'Hurricane Katrina-induced forest damage in relation to ecological factors at landscape scale', *Environmental Monitoring and Assessment*, vol. 156, pp. 491–507.

Webb, E. L., Webb, R. C., Tualaulelei, A. and Carrasco, L. R. (2014) 'Factors affecting tropical tree damage and survival after catastrophic wind disturbance', *Biotropica*, vol. 46, pp. 32–41.

Webb, S. L. (1988) 'Windstorm damage and microsite colonization in two Minnesota forests', *Canadian Journal of Forest Research*, vol. 18, pp. 1186–1195.

Webb, S. L. (1989) 'Contrasting windstorm consequences in two forests, Itasca State Park, Minnesota', *Ecology*, vol. 70, pp. 1167–1180.

Webb, S. L. and Scanga, S. E. (2001) 'Windstorm disturbance without patch dynamics: 12 years of change in Minnesota'. *Forest Ecology*, vol. 82, pp. 893–897.

Whitmore, T. C. and Burslem, D. F. R. P. (1998) 'Major disturbances in tropical rain forests', in D. M. Newbery, H. H. T. Prins and N. Brown (eds), *Dynamics of Tropical Communities*, Blackwell Science, Oxford.

Xi, W. and Peet, R. K. (2008) 'Hurricane effects on the Piedmont forests: patterns and implications', *Ecological Restoration*, vol. 26, pp. 295–298.

Xi, W., Peet, R. K., Lee, M. T. et al. (2019) 'Hurricane disturbances, tree diversity, and succession in North Carolina Piedmont forests, USA', *Journal of Forestry Research*, vol. 30, pp. 219–231.

Yoshida, T., Noguchi, M., Uemura, S., Yanaba, S., Miya, H. and Hiura, T. (2011) 'Tree mortality in a natural mixed forest affected by stand fragmentation and by a strong typhoon in northern Japan', *Journal of Forest Research*, vol. 16, pp. 215–222.

Zhao, D., Allen, B. and Sharitz, R. R. (2006) 'Twelve-year response of old-growth southeastern bottomland hardwood forests to disturbance from Hurricane Hugo', *Canadian Journal of Forest Research*, vol. 36, pp. 3136–3147.

Zimmerman, J. K., Willig, M. R. and Hernández-Delgado, E. A. (2020) Resistance, resilience, and vulnerability of social-ecological systems to hurricanes in Puerto Rico', *Ecosphere*, vol. 11, e03159.

10

FOREST SUCCESSION AND GAP DYNAMICS

Rebecca A. Montgomery and Lee E. Frelich

Overview

Succession is a foundational concept in forest ecology. While succession has been defined variously, the most common definition is the sequence of communities or species that successively replace each other through time after a disturbance. The term is also used to describe the process of change itself. Many include change in forest and community structure and system-level properties such as diversity and productivity in their definition. Succession can be divided into two major types: (i) primary succession, change that occurs on a previously unvegetated site, and (ii) secondary succession, change that occurs after disturbance to an existing ecosystem. In secondary succession, biological legacies left by past vegetation remain and influence succession. This chapter focuses on secondary succession, as it is the most common and widespread in forested ecosystems. We use a broad definition of succession discussing compositional, structural and system-level change. We also discuss management strategies for forested ecosystems that are based on principles of ecological succession and challenges associated with global change.

A brief history of the concept

The study of ecological succession was advanced early in ecological circles through the work of Henry Chandler Cowles. Cowles studied the patterns and processes that transform sand dunes into hardwood forests along the shores of Lake Michigan in the United States (Cowles 1899). Frederick Clements created a general classification of causes that is still applied today (Clements 1916). His scheme involved (1) *nudation*, the removal of vegetation by disturbance; (2) *migration*, the arrival of organisms at the open site; (3) *ecesis*, the establishment of organisms at the open site; (4) *competition*,[1] the interaction of organisms at the site; (5) *reaction*, the alteration of the site by the organisms. The result of these five was *stabilization* in the form of the climax. For Clements, the climax was the major unit of vegetation and the permanent and final stage of succession (Clements 1916, 1936). The nature of the climax was intimately intertwined with the climate. He considered the climax an expression of climate and a super-organism with its own particular trajectory of growth and development. The growth and development or succession terminated in the climax and more specifically in the

 DOI: 10.4324/9781003324072-12

mono-climax, which could maintain itself indefinitely if not disturbed. Although the conception of succession as a process where species successively replace each other through time has endured, the concept of a mono-climax, the primary role of climate as mechanism and the climax as a super-organism proved too restrictive for a number of early ecologists (Cooper 1913; Gleason 1927). One of the prominent critics was Henry A. Gleason who asserted that succession is an "extraordinarily mobile phenomenon" that may not be repeatable or predictable (Gleason 1927). He recognized multiple causes that could lead succession down multiple pathways. In other words, Gleason recognized succession as a complex process without a fixed endpoint. Moreover, he argued that any predictions of successional trajectories must recognize that multiple forces operate and that some may act in opposing directions. Gleason espoused an individualistic view of communities that was later championed by Whittaker (cf. Whittaker 1957) in which species respond to their abiotic and biotic environment in an individualistic way and the patterns in succession reflect the ebb and flow of populations of species through time and space.

Compositional and structural change

The process of succession involves changes in both the species composition and the physical structure of a forest through time. Forest development describes the change in forest structure through time. In general, all forested ecosystems go through a predictable sequence of changes in structure as they develop following disturbances that kill or level the forest canopy (Figure 10.1). Stages include two generally even-aged stages—stand initiation or establishment phase and the stem exclusion or thinning phase; a third stage involving transition to an uneven-age canopy or understory reinitiation, and a final multi-aged stage, sometimes referred to as old-growth or steady-state (Oliver 1981; Frelich 2002; but see Franklin et al. 2002 for critique). These changes may or may not be associated with compositional change. It is important to note that structural change and compositional change can occur together or separately.

The stand initiation or establishment phase is defined as the period of time immediately after disturbance that new stems are establishing and filling available growing space. This is

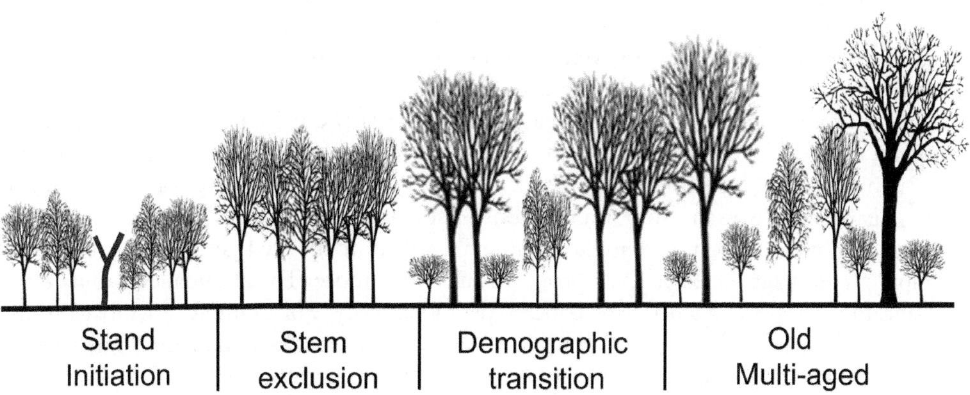

Figure 10.1 Stages of stand development.
Source: From Frelich (2002).

a time of very high density of small stems. These stems establish and grow until all growing space is occupied and one or more resources becomes limiting. At this point, new stems are not established. The stem exclusion phase is one of intense competition among established individuals during which one or several species become dominant. Shade-intolerant species that do not become dominant die while those tolerant of shade can persist in the understory. Over time, as trees grow larger and use more resources, there is growing space for fewer stems and those less vigorous or slower growing die while more vigorous stems expand their crowns into the space left behind. Eventually, neighboring trees cannot fill space opened by death of canopy trees and existing suppressed or newly established stems begin to grow in gaps made by canopy tree death. This is the transition or understory reinitiation phase. The characteristics of the multi-aged stage of development vary within and among forest types and include rare steady-state stands with a constant rate of canopy turnover in the form of small canopy gaps, as well as more common stands where a variety of partial disturbances creates a constantly changing mix of gap sizes, age class structures and mixtures of tree species life histories.

At the landscape scale, the frequency of major disturbances that kill or level the forest canopy regulates the proportion of stands among the stages of succession and development, a concept proposed by Watt (1947). Boreal forests with high-intensity fires at intervals less than tree lifespan may be composed mostly of stands in initiation and stem exclusion stages with rare multi-aged stands, whereas in regions where stand-leveling disturbance is rare (e.g. mesic cold-temperate forests or tropical rainforest), the distribution of stands among stages of development is the opposite, with a high proportion of multi-aged or old-growth stands.

Stand-level structural changes during forest development and succession include changing diameter distributions, stem density, maximum tree height and stand basal area that follow general patterns across global forested ecosystems (Guariguata and Ostertag 2001; Frelich 2002; Powers *et al.* 2009). During forest succession, stands increase in overall height and basal area, eventually reaching an asymptotic maximum height and basal area. The time to reach asymptotic height and basal area differs with forest types, climate, edaphic factors and disturbance history. Bormann and Likens (1979) proposed a similar model for biomass accumulation post-disturbance. Density changes appear more variable but generally show a more peaked pattern, rising and then declining, though the degree of decline varies among forests. Generally, the range of diameters present in a stand increases and becomes more heterogeneous as stands move from even- to multi-aged composition and stands become more diverse in vertical and horizontal structure.

An important aspect of post-disturbance structure is the existence of legacy structures (Franklin et al. 2002). Legacy structures include living organisms that survive the disturbance such as mature trees and dead materials such as standing dead snags or downed wood. These legacy structures are important to post-disturbance successional processes providing seed sources of new colonists, habitat for animals, substrate for seed germination, bud banks for rapid recolonization through resprouting and much more. The abundance and nature of biological legacies varies with disturbance type and severity. For example, a windstorm may selectively kill species that are not wind firm leaving other species undamaged and releasing suppressed individuals in the understory (Zimmerman et al. 1994, Rich et al. 2007).

Compositional changes during succession in many ways have defined the concept. Often but not always the early-successional community is dominated by fast-growing, relatively short-lived species. As a relatively even-aged early-successional canopy thins, the species that reinitiate the understory are often but not always more shade-tolerant species than in the canopy. These species may have been present from the point of disturbance or they may establish as resource conditions change. These two models were originally described as initial floristics and relay floristics, respectively (Egler 1954). Relay floristics describes a process by which species at any given stage either facilitate the establishment of the next or inhibit colonization, leading to new individuals and species establishing and dominating sequentially. In contrast, initial floristics describes a process by which most species establish early after disturbance but due to life history characteristics, species dominate at different points along the successional sequence.

The propensity for species with particular life history characteristics and functional traits to be associated with different successional stages has led to the widespread use of the terms early-, mid- and late-successional to describe species. In forests, there is often the dichotomy between shade-intolerant pioneers and shade-tolerant late-successional species (Swaine and Whitmore 1988). Early-successional tree species tend to produce many, small seeds that can lie dormant for many years. These tend to be wind dispersed (e.g., *Populus*) in the temperate zone and bird (e.g. *Macaranga*) and bat (e.g. *Cecropia*) dispersed in the tropics. They have fast growth rates; suffer considerable herbivory; are short-lived; and have high rates of photosynthesis, nutrient-rich rapidly decomposing leaves and low wood density. In contrast, shade-tolerant late-successional species have larger seeds that either lack or show short periods of dormancy. They are slow-growing, possess defenses again structural damage and herbivory, are long-lived and tall and have relatively low rates of photosynthesis, lower leaf nutrient levels, slower decomposition rates and high wood density. Of course, this is a simplified view of the diversity of forest plant strategies. In reality, trees and other forest plants fall along a continuum of strategies that involves a mix of dispersal traits, growth rates, demographic strategies and positions in the vertical and horizontal structure of the forest. Several studies have found that in addition to the fast–slow axis of life history trade-offs described above, there exists a stature-recruitment trade-off (Chazdon et al. 2010, Ruger et al. 2018). Long-lived pioneers recruit in high light but have long life spans and are tall. Short-lived breeders recruit prolifically in low-light but have slow growth, low survival and are of short stature. Combining aspects of both trade-offs can explain interspecific variation in recruitment and mortality during secondary succession. For example in tropical secondary forests in central Panama, area-based photosynthetic rates, maximum height and seed dry mass were related to changing recruitment and mortality through secondary succession with functional traits explaining 70 per cent in interspecific variation in recruitment and 79 per cent of variation in mortality (Lai et al. 2021).

Mechanisms

A general theory of ecological succession has failed to emerge despite many efforts. Such theory rests on laying out causes or mechanisms of succession, which are manifold. In attempts to develop theory, Connell and Slatyer (1977) laid out successional pathways that invoke the mechanisms of facilitation, tolerance and inhibition. The Clementsian view of succession asserted the primacy of facilitation by early colonists who modify the environment in such a way that later colonists can establish. Evidence in forests for this mechanism comes from succession on old fields in the temperate zone and abandoned pastures in the tropics. Tolerance has been interpreted as both an active and passive mechanism (Pickett et al. 1987) that involves either the ability to endure low resources and slowly grow in stature eventually shading the early dominants (active) or life cycle complementarity (passive) where two species with different growth rates establish simultaneously and the slow grower eventually succeeds the fast grower. In contrast, inhibition invokes structural or competitive dominance by early colonists that prevents other species from establishing or becoming dominant. Evidence for inhibition in forests is poor. Moreover, inhibition is also intertwined with tolerance: late-successional species accumulate because they can tolerate low resources and then just wait for a disturbance to release them from suppression. This is demonstrated in forests that feature small-scale disturbances such as gaps (Uhl 1988; Abe et al. 1995). For some Connell and Slatyer's 'models' were too rigid and reductionist and a general hierarchical framework of causes promised a more broadly applicable approach to creating a predictive framework and a general theory of ecological succession (Finegan 1984, Pickett et al. 1987).

The general hierarchical framework proposed by Pickett et al. (1987) remains one of the best syntheses of the causes and mechanisms of succession. It continues to inspire thinking about the importance and future directions of successional research (Meiners et al. 2015). The framework starts with the question 'What causes succession?' and answers with the following three responses: open sites are available, species are differentially available at the open site, and species differ in the capacity to deal with one site or another (Pickett et al. 1987). These then are the three core sources of variation or differentials: site conditions and history, species availability and species performances (Meiners et al. 2015). Successional drivers are then organized within these broad classes. The second question or level of hierarchy delves into drivers asking, "What interactions, processes or conditions contribute to the general causes of succession?" These might include dispersal, resource availability, ecophysiology, competition, herbivory and disease. Finally, a third question asks, "What site-specific factors or behaviors determine outcome of interactions?" This question invokes interactions of the organism and site-specific features. We illustrate the framework with the following example. Open sites are made available by a crown destroying fire. Post-disturbance stands are initiated from several sources: mobile seeds that come from outside the stand or from reproductive individuals that survived the fire (legacy trees); seeds in the seed bank; and individuals that survived and resprouted. The nature of the regenerating forest in part depends on the relative importance of these different sources of colonists (differential species availability), which may depend on disturbance severity (a site-specific factor) and on the interaction of species with characteristics of the site (differential species performance as a result of post-disturbance resource availability).

Gap dynamics

Gaps are openings in the forest created by small-scale disturbance created by mortality of one or multiple trees due to density-independent mechanisms such as windthrow or insect damage (Brokaw 1982). Gaps alter resource availability and microclimate, increasing light, soil moisture and nutrients and the extremes of temperature and humidity, especially near the forest floor. Gap dynamics have received the most attention in forests where they dominate late-successional processes, such as mesic temperate and tropical forests (Runkle 1981, 1982; Denslow 1987). However, there is growing evidence that under certain disturbance regimes, gaps play a role in boreal forest dynamics in North America and Eurasia (Kneeshaw and Bergeron 1998; Shorohova et al. 2009; Anyomi et al. 2022).

Gaps play several roles in forest dynamics depending in part on gap size and other conditions at time of gap formation. Gaps provide sites for the colonization and establishment of shade-intolerant species that require high light for seed germination and growth. Usually, the establishment of these species requires large gaps created by more than a single tree. These species are maintained within the forest ecosystem by the existence of regeneration sites in large tree fall gaps. In this role, gaps create patches of early-successional forest embedded in late-successional forests (Watt 1947). Gaps also release suppressed shade-tolerant trees that have formed the advanced regeneration layer. There is good evidence that small-scale disturbances associated with gaps in late-successional forests are sufficient to maintain the landscape species composition in a relatively steady state (Runkle 1981).

Gaps are considered one of the mechanisms that maintain diversity in forested ecosystems (Denslow et al. 1987; Brokaw and Busing 2000); however, how gaps promote coexistence remains poorly characterized (Gravel et al. 2010). An early articulation of mechanism was the gap-partitioning hypothesis, which posits that species colonize gaps of different sizes and different areas of gaps depending on their resource requirements (the regeneration niche; Grubb 1977). In a large gap, shade-intolerant species dominate the center, mid-tolerant species dominate parts of the gap closer to the edges and shade-tolerant species dominate the edges (Denslow et al. 1987; Schnitzer and Carson 2000). A similar scheme applies to gaps that range in size from small, single-tree falls to very large gaps (Runkle et al. 1982). This hypothesis applies not only to the tree component of forests but also the forest understory layer of shrubs and herbs. However, gap colonization or capture can also be a function of stochastic processes. For example, gap capture may be dependent on the species that happen to be in the advance regeneration layer at time of canopy disturbance (Brokaw and Busing 2000). There is evidence for a significant role of stochastic processes in the tree replacement process in systems with high numbers of shade-tolerant trees (Hubbell et al. 1999; Gravel et al. 2012).

Niche-based models have attempted to resolve conflicting perspectives. By incorporating size-structured growth, multiple physiological trade-offs, multiple axes of differentiation and both evolutionary and ecological assembly, Falster et al. (2017) successfully modeled the coexistence of multi-species in forests. Moreover, they found that near neutrality (e.g. Hubbell et al. 1999) developed in low light from the niche-based model. Similarly, Detto et al. (2022) modeled the coexistence of many species by combining species-specific allocation trade-offs (e.g., growth versus fecundity), asymmetric light competition and disturbance. They found that coexistence emerged through partitioning of light through time as well as space.

Boreal forests, whose disturbance regimes are dominated by fire, have generally not been considered ecosystems where gaps play a role. However, recent work in North America and Eurasia suggests that small disturbances can influence forest dynamics between larger high-intensity fires. Such canopy gap processes are important in certain regions of the boreal biome where intervals between fires are greater than 200 years, such as eastern Canada and parts of Scandinavia (Kneeshaw and Bergeron 1998; Shorohova et al. 2009; Anyomi et al. 2022).

Greater appreciation of heterogeneity and complexity in forest ecosystems coupled with changing management goals has led to a reconsideration of traditional silvicultural systems that often assume and prescribe homogeneity with the goal of timber production (Coates and Burton 1997). In particular, applying lessons from ecological research on gap dynamics to achieve ecosystem management goals through silviculture has become common in systems involving partial harvest of the ecosystem (Coates and Burton 1997). Ecosystem management refers to an approach that delivers goods and services while sustaining forest function, diversity and structure in perpetuity. Increasingly, it has been adopted in forestry to achieve multiple goals that include not only timber production, but also biodiversity conservation, wildlife habitat and sustaining ecosystem functions (e.g. nutrient cycling, productivity, water quality). In this framework, if the goal of a management intervention is to increase the life history diversity of the forest or to increase understory biodiversity while still producing timber, then a range of harvested patch sizes might be used to simulate a range of gap sizes and concomitant effects on resource availability and tree regeneration.

Diversity and forest succession

Although patterns of change in composition through succession have been widely described and are a hallmark of the process, system-scale measures such as diversity have yielded conflicting patterns. Three major hypotheses regarding change in diversity through succession have emerged, all with some support. The classic treatment of patterns of species diversity through succession is that of low-species diversity in early succession and increasing diversity through time due to increasing structural complexity or higher rates of immigration than extinction in successional stands. For example, in the forests of subtropical China, species richness increases through successional time and changes are largely due to continuous immigration of new species that enrich the sites (Brunlheide et al. 2011). Alternately, diversity could decrease if initial floristic composition dominates, with species being eliminated through time as succession proceeds (Egler 1954). Species diversity could also decline through succession as forest density declines due to the positive relationship between density and richness. A third model suggests that diversity patterns tend to shift with forest development stages. Diversity is initially relatively high as species resprout or recolonize available sites after disturbance. Diversity of trees drops as one or several species attains canopy dominance, creating a low-resource environment in the understory that can also lead to declines in ground-layer diversity. Diversity increases as the forest becomes more vertical and horizontally stratified, eventually reaching a mid-successional peak. However, given the general importance of small disturbances in later successional forests, gap formation may increase and maintain diversity (Runkle 1982)—no mid-successional peak. However, if small-scale disturbances are not prevalent, diversity can decline if one or several species are able to dominate and exclude others. This pattern of diversity through succession was key to the formulation of the intermediate disturbance hypothesis described first for grasslands and later applied to tropical rain forests (Connell 1978). It posits that most communities are kept in a non-equilibrium state such that most patches are in highly diverse mid-successional stages. Frequent disturbance or large disturbance moves communities toward dominance by a few early-successional species, while rare or small and low severity disturbance allows dominance by those species that can competitively exclude other members (Connell 1978; Figure 10.2). One possibility for these different models may be due to the difficulty in actually sampling a full successional sequence and hence any one study only sees part of the story (Howard and Lee 2003). It is also possible that no single model describes changes in species diversity because the causes of succession may differ among forests in different locations, or in the same forest over time.

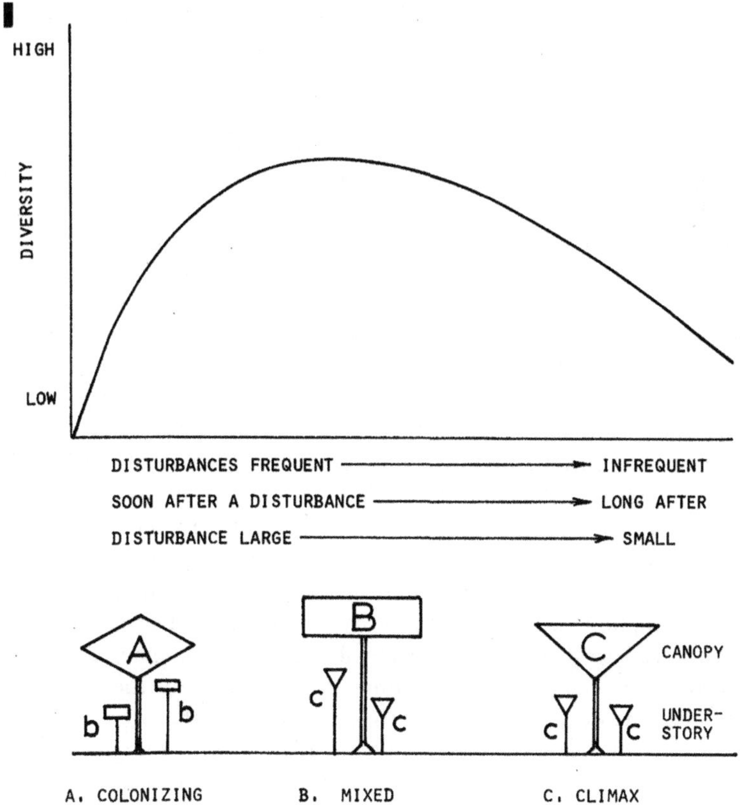

Figure 10.2 The intermediate disturbance hypothesis.
Source: Adapted from Connell et al. (1978).

Stochasticity, complexity and multiple pathways

The preceding sections have presented general patterns of change associated with succession. Often, those patterns were accompanied by caveats such as 'but not always'. In essence, for every pattern presented in this chapter, there is undoubtedly an exception illustrating the complex nature of succession and the myriad pathways of change that can occur due to variation in the dominant causes in each place-based successional sequence. Myriad pathways also result, in part, from the action of stochastic processes and disturbance interactions.

A key stochastic process influencing forest succession is recruitment limitation, when species fail to recruit into sites most favorable for their growth (Hubbell et al. 1999). After major or small-scale disturbances, new colonists often are a function of chance, related to the species producing seeds at the time of disturbance or the species already present, rather than species that are most competitive on the site. There is ample evidence that recruitment limitation plays an important role in forest and gap dynamics (cf. Brokaw and Busing 2000).

Another related process, dispersal limitation, is critical for predicting succession, especially in the tropics. In general, classical theories of succession ignore seed dispersers. The source-disperser limitation framework links niche-based theories of succession with dispersal limitation (Dent and Estrada-Villegas 2021), positing that the interaction between seed sources and dispersers regulates seed movement during succession. This movement sets the stage for recruitment at regeneration sites. Established seedlings are then subject to niche-based processes that affect growth and survival and forest succession.

To illustrate stochasticity, complexity and multiple pathways, we use two examples. In the first, we explore how complexity usually associated with later successional stages can occur much earlier in sites where early and late-successional species establish together after disturbance and in which establishment is sparse (Donato et al. 2012). The 'precocious succession model' describes a system where structural complexity develops early in a natural successional sequence rather than only during late-successional states (Figure 10.3). In this pathway, after major disturbance, the tree establishment phase is protracted due to large patch sizes, long distances to seed sources, unfavorable environment conditions for seed establishment, and early competition from other vegetation (e.g. shrubs). An open community is established early and the low density and/or competition from shrubs stratifies tree establishment and can create an early interspecific competitive exclusion phase or forgo canopy closure all together (Donato et al. 2012). In such young stands, structural aspects that imbue old-growth stands with complexity are present early including clumped, widely spaced trees; vertical heterogeneity in the canopy and among tree crowns; coexistence of under-, mid- and overstory; and facilitation of shade-tolerant species (Donato *et al.* 2012).

Our second example involves interactions among species traits, herbivory and multiple disturbances that lead to divergent successional pathways that maintain white pine dominance versus transition stands to maple dominance in forests in the Lakes States, United States (Frelich 2002, Tester et al. 1997). In the pine forests of the Lakes States, fire and wind are dominant disturbance agents. In mature white pine stands, young cohorts of pine must establish in the understory and grow large enough to survive the next surface fire. They can fail to do so for many reasons, including deer browse, drought, poor soils or taller maple or spruce saplings in the understory. If they fail to grow large enough before the occurrence of surface fires, then they, along with maple and spruce saplings, will be killed by surface fire. The longer it takes to grow large enough the more likely a surface fire will occur that kills them. Moreover, longer times for saplings to grow large enough also increase the likelihood that the adult white pine will be killed by a windstorm, leaving no seed sources for white pine to recolonize. In this compound disturbance regime, white pine stands transition to hardwoods. In contrast, if the white pine saplings grow large enough, they survive surface fire while maples are killed. Moreover, if canopy individuals are killed by windstorm, subcanopy individuals of white pine are released to become the future canopy and seed source.

Both examples include disturbance interactions. Disturbance interactions can alter the pathways and timing of succession by affecting the resistance and resilience of forest ecosystems. They are critical to predicting succession through enhanced understanding of the interconnections of disturbance drivers, the mechanisms of resistance and recovery, and the role of legacies (Buma 2015). Disturbance interactions can be conceptualized as those that affect resistance to disturbance (linked disturbance) or alter the resilience or recovery from disturbance (compound disturbance; Buma 2015). Linked disturbances occur when one disturbance alters the resistance of the system by changing the likelihood, spatial extent, severity or intensity of another disturbance. Compound disturbances arise when elements of multiple disturbances combine to create a new or novel disturbance that changes rates and extent of ecosystem recovery.

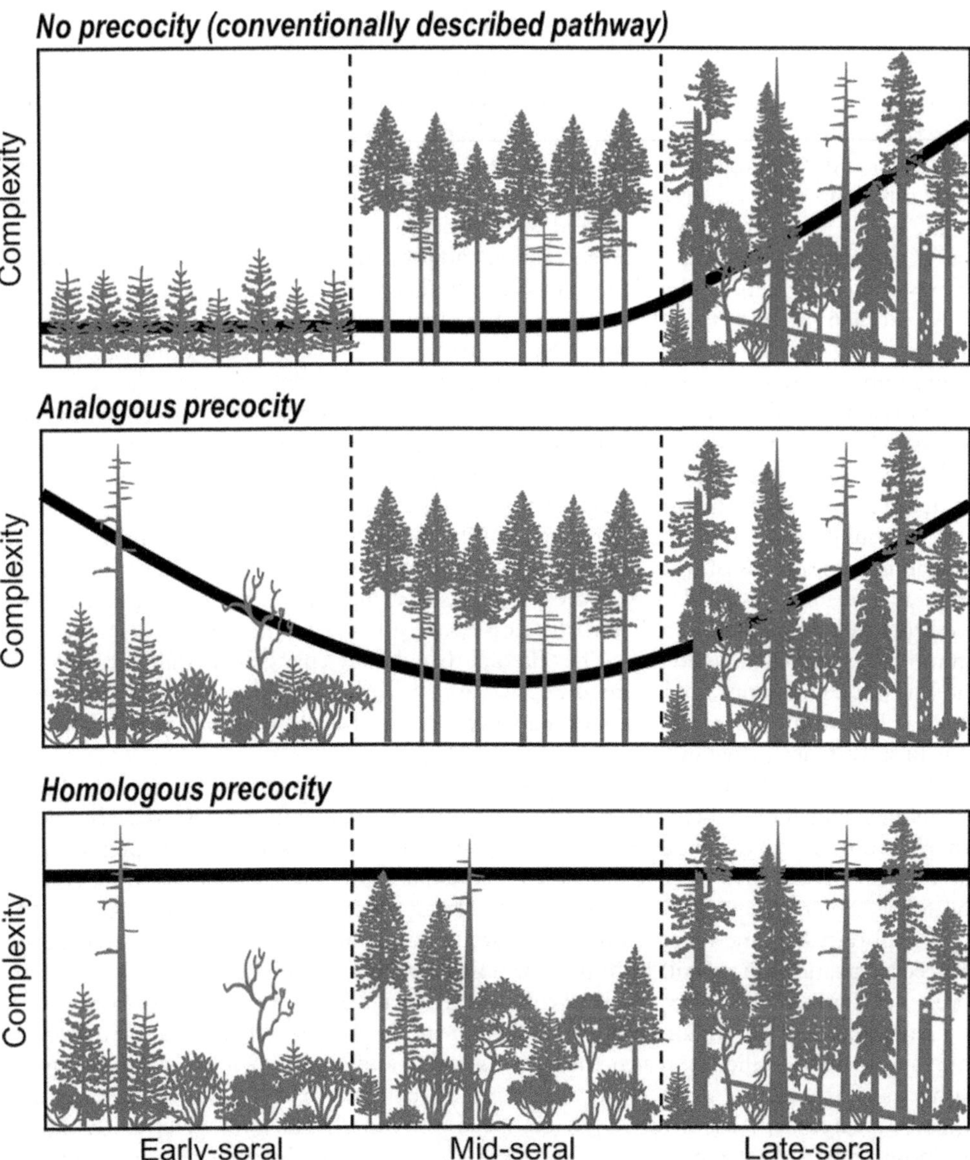

Figure 10.3 Three alternate successional pathways for forest development, showing the relative levels of structural complexity exhibited in each seral stage. In the conventional successional model, both early- and mid-seral conditions are dominated by a relatively even-aged tree cohort, and structural complexity does not arise until the latest stage of development. In the case of analogous precocity, early-successional stands exhibit structural complexity in some ways similar to that in old stands, but canopy closure results in reduced complexity during mid-succession. In the case of homologous precocity, the lack of a tree canopy-closure phase results in a continuity of complexity throughout forest development.

Source: From Donato et al. (2012).

Succession and global change

Global change, such as elevated CO_2, warming temperatures, altered disturbance and browsing regimes and global movement of species has and will continue to impact forest ecosystems, often in ways detrimental to the cultural, ecological and economic roles they play. To understand the consequences of global change requires prediction of forest dynamics into the future. This must be attempted despite uncertainty over exact environmental conditions or the nature of biotic interactions and disturbance regimes. Predicting future dynamics is one of the hallmarks of research on succession and thus succession provides a foundation for understanding impacts of global change on forested ecosystems.

An excellent example is provided by work in boreal forests of central Alaska, United States (Johnstone et al. 2010). Many boreal North American ecosystems are dominated by high-severity fire regimes and can show repeatable successional cycles that are characterized by re-establishment of dominant pre-fire vegetation. The majority of the current landscape is occupied by black spruce whose serotinous seeds help ensure its post-fire dominance. In sites with deciduous dominance, successional shifts to conifers occur at time intervals considerably longer than average disturbance cycles. Thus, at present, patches of black spruce and patches of deciduous species tend to be self-replacing. What might happen to successional trajectories and landscape patterns of spruce and deciduous forests in the future? Research after a severe fire year in central Alaska, United States, suggests that high fire severity and frequent fires, predicted by climate change, will favor successional pathways to deciduous species especially on moderate to well-drained sites. They predict that climate change may shift the balance at the landscape level from predominantly spruce forests toward deciduous-dominated forests. Such a shift would have regional consequences for local climate, forest flammability and human subsistence activities (Johnstone et al. 2010).

More frequent and intense disturbances in the future will alter forest dynamics and result in younger and more open forests. This contrasts with current and historical conditions where forest change occurs in pulses within periods of relative stability. The period immediately after disturbance has been called the reorganization phase, a short window in which the forest either renews itself and follows a similar trajectory or shifts to another trajectory. As forests become more disturbed, understanding patterns and processes that govern the reorganization phase will become more important as this phase sets the stage for the forest development for many years (Seidl and Turner 2022). The response pathways of the reorganization phase can be organized along the dimensions of structure and composition and include resilience, restructuring, reassembly and replacement. Resilience response pathways are those that don't result in changes in structure or composition. Restructuring involves changes in structure but not composition, while reassembly involves changes in composition but not structure. Finally, replacement response pathways have changes in both. To understand and predict forest changes in a changing climate, identifying the processes that influence these pathways is critical (Seidl and Turner 2022).

Another pattern of change that will undoubtedly alter forest succession is the increased prevalence of climate-driven tree mortality and dieback, especially from hotter droughts (Hammond et al. 2022). Climate change is expected to increase temperature, atmospheric drought and soil water deficits. An increase in temperature during the growing season increases evapotranspiration. This is expected to increase drought stress for forests in relatively dry climates adjacent to deserts and grasslands; however, even in wet climates, increased evapotranspiration in a warmer climate can outpace precipitation or projected increases in precipitation. This can cause increased drought-related mortality in places like the Amazon rainforest (Phillips et al. 2009). Water savings associated with growing at elevated CO_2 may reduce some drought stress due to climate warming. A global fingerprint of tree mortality associated with hotter droughts has been found in most biomes. Elevated tree mortality was associated with multiple climate measures of drought and heat stress (Hammond et al. 2022). Since species differ in drought and heat tolerance, differences in species performance may lead to shifting dominance patterns associated with climate change-induced droughts (Cavin et al. 2013). Secondary effects of drought and heat stress will also raise tree mortality rates—these include increased likelihood of fire, insect herbivores and pathogens (Ayers and Lombardero 2000).

Arrested succession has been noted in some systems where ground-layer and understory vegetation retards the establishment and growth of trees. Royo and Carson (2006) reviewed the studies of species that form dense and persistent understory layers. They found that such layers were often the result of anthropogenic factors. Specifically, recalcitrant understory layers appear to result when there is a high level of canopy disturbance, when there are high levels of browsing by large vertebrate herbivores, and when fire regimes are altered. Recalcitrant understory layers are now common globally (Royo and Carson 2022). The factors appear causing recalcitrance are unlikely to change in the future and suggest that arrested succession may become more common in the future. One such factor, deer herbivory, has been widely studied in the eastern deciduous biome of North America. Deer herbivory can alter gap recruitment such that browse-sensitive species are unable to recruit (Van der Molen & Webster 2021). This leads to depauperate understories dominated by browse-tolerant species. The exclusion of deer cannot reverse these losses (Royo and Carson 2022) and the mounting stressors on these forests threaten to further erode diversity and function.

Lastly, climate change and the global movement of species may alter successional pathways by introducing novel biotic interactions into forested ecosystems. Climate change could alter the relative competitive ability of species at range limits through alleviation of cold-limitation of species at their cold-edge of the range. To detect range shifts, most studies have focused on the regeneration layer, looking at composition, relative abundances and performance of the juveniles with the implicit assumption that these individuals are expected to become members of the canopy in the future. Such studies have shown evidence for enhanced recruitment of currently cold-limited species at range edges and declines in recruitment of species at their warm range edge (e.g. Fisichelli et al. 2014). Shifts in recruitment suggest a high potential for turnover in species composition in forests in regions where species with different ranges overlap (e.g. ecotones).

Moreover, the global movement of species has resulted in an increase in invasive plants and animals. These species can alter successional pathways by eliminating tree species or altering the environment so that species from one or more successional stages are no longer successful. Hemlock woolly adelgid (*Adelges tsugae*), an insect pest from Asia, has eliminated eastern hemlock (*Tsuga canadensis*), a dominant late-successional tree species, from forests in parts of the eastern United States (Ford et al. 2012). European earthworms invading the Great Lakes Region of the United States have changed the seedbed conditions so that *Acer saccharum*, a late-successional dominant over large areas, cannot compete successfully at the seed germination and seedling growth stages of its life cycle (Hale et al. 2006). It remains to be seen what new successional pathways will develop in response to changes wrought by these species.

Summary

We close with some philosophical musings. While succession represents a foundational concept in forest ecology and management with a long history of scholarship and application, has its importance been overemphasized? The industrial revolution fueled increasing demand for forest products that spurred European colonization and settlement of North America and expansion of industrial activities in Europe and elsewhere. This led to widespread clearing of late-successional forest types and removal of burning regimes by Indigenous peoples that formerly held the forests at a desired early-successional stage. This caused succession to occur across vast tracts of land in the temperate zone during the 1900s, the time when the science of ecology developed, giving ecologists a misleading impression of succession as being the dominant process across the landscape. In contrast, under natural conditions, episodes of succession occurred only occasionally when rare disturbance combinations break the legacy lock of late-successional communities (Johnstone et al. 2010) or when the chance absence of fire allowed late-successional species to move into fire-dependent systems. That said, successional theory has an important role to play. It guides the design of silviculture prescriptions and forest management plans and as forests face growing challenges associated with global change, studies of forest succession can guide our understanding of the potential trajectory of forests into the future. Despite common threads of cause and consequence in forest succession across the globe, it is clear that complexity is the rule rather than the norm. Moreover, as an inherently place- and time-based subject, there will always be more to learn.

Note

1 Clements called it competition but other interactions are generally recognized today.

References

Abe, S., Masaki, T. and Nakashizuka, T. (1995) 'Factors influencing sapling composition in canopy gaps of a temperate deciduous forest' *Vegetatio*, vol 120, no 1, pp. 21–31.

Anyomi, K. A., Neary, B., Chen, J., and Mayor, S. J. (2022) 'A critical review of successional dynamics in boreal forests of North America' *Environmental Reviews* vol. 30, pp. 563–594.

Ayres, M.P., and Lombardero, M.J. (2000) 'Assessing the consequences of global change for forest disturbances for herbivores and pathogens'. *The Total Science of the Environment,* vol 262, pp. 263–286.

Bormann, F.H. and Likens, G.E. (1979) *Pattern and Process in a Forested Ecosystem.* Springer-Verlag, New York.

Brokaw, N.V.L. (1982) 'The definition of treefall gap and its effects on measures of forest dynamics'. *Biotropica*, vol. 14, pp. 158–160.

Brokaw, N.V.L. and Busing R.T. (2000) 'Niche versus chance and tree diversity in forest gaps'. *Trends in Ecology and Evolution,* vol 15, pp. 183–188.

Bruelheide H., Bohnke M., Both S., Fang T., Assmann T., et al. (2011) 'Community assembly during secondary forest succession in a Chinese subtropical forest'. *Ecological Monographs,* vol 81, pp. 25–41.

Buma, B., (2015) 'Disturbance interactions: characterization, prediction, and the potential for cascading effects'. *Ecosphere,* vol 6, pp. 1–15.

Cavin, L., Mountford, E.P., Peterken, G.F. and Jump, A.S. (2013) 'Extreme drought alters competitive dominance within and between tree species in a mixed forest stand'. *Functional Ecology*, vol 27, no 6, pp. 1424–1435.

Chazdon, R.L., Finegan, B., Capers, R.S., Salgado-Negret, B., Casanoves, F., Boukili, V. and Norden, N. (2010) 'Composition and dynamics of functional groups of trees during tropical forest succession in Northeastern Costa Rica'. *Biotropica,* vol 42, no 1, pp. 31–40.

Clements, F.E. (1916) 'Plant succession: An analysis of the development of vegetation'. Carnegie Institute of Washington Publication 242.

Clements, F.E. (1936) 'Nature and structure of the climax'. *Journal of Ecology*, vol. 24, pp. 252–284.

Coates, K.D. and Burton, P.J. (1997) 'A gap-based approach for development of silvicultural systems to address ecosystem management objectives'. *Forest Ecology and Management*, vol 99, pp. 337–354.

Connell, J.H. (1978) 'Diversity in tropical rain forests and coral reefs'. *Science*, vol 199, pp. 1302–1310.

Connell, J.H. and Slatyer, R.O. (1977) 'Mechanisms of succession in natural communities and their role in community stability and organization'. *American Naturalist*, vol 111, pp. 1119–1144.

Cooper, W.S. (1913) 'The climax forests of Isle Royale, Lake Superior and its development'. *Botanical Gazette*, 55: I 1-44, II 115-140, III 189–235.

Denslow, J.S. (1987) 'Tropical forest gaps and tree species diversity'. *Annual Review of Ecology and Systematics*, vol 18, pp. 431–451.

Dent, D. H., and Estrada-Villegas, S. (2021) 'Uniting niche differentiation and dispersal limitation predicts tropical forest succession'. *Trends in Ecology & Evolution*, vol 36, pp. 700–708.

Detto, M., Levine, J. M., and Pacala, S. W. (2022) 'Maintenance of high diversity in mechanistic forest dynamics models of competition for light'. *Ecological Monographs*, vol 92. e1500.

Donato, D.C., Campbell, J.L. and Franklin, J.F. (2012) 'Multiple successional pathways and precocity in forest development: can some forests be born complex'. *Journal of Vegetation Science*, vol 23, pp. 576–584.

Egler, F. E. (1954) 'Vegetation science concepts. I. Initial floristic composition, a factor in old field vegetational development'. *Vegetatio*, vol 4, pp. 412–417.

Falster, D. S., Brännström, Å., Westoby, M., and Dieckmann, U. (2017) 'Multitrait successional forest dynamics enable diverse competitive coexistence'. *Proceedings of the National Academy of Sciences*, vol 114, E2719–E2728.

Finegan, B. (1984) 'Forest succession'. *Nature*, vol 312, pp. 109–114.

Fisichelli, N.A., Frelich, L.E. and Reich, P.B. (2014) 'Temperate tree expansion into adjacent boreal forest patches facilitated by warmer temperatures'. *Ecography,* vol. 37, pp. 52–161.

Ford, C.R., Elliot, K.J., Clinton, B.D., Kloeppel, B.D. and Vose, J.M. (2012) 'Forest dynamics following eastern hemlock mortality in the southern Appalachians'. *Oikos*, vol 121, pp. 523–536.

Franklin, J.F., Spies, T.A., Van Pelt, R., Carey, A.B., Thornburgh, D.A., Berg, D.R., Lindenmayer, D.B., Harmon, M.E., Keeton, W.S., Shaw, D.C., Bible, K. and Chen, J. (2002) 'Disturbances and structural development of natural forest ecosystems with sivicultural implications, using Douglas-fir forests as an example. *Forest Ecology and Management*, vol 155, pp. 399–423.

Frelich, L.E. (2002) *Forest Dynamics Disturbance Regimes: Studies from Temperate Evergreen–Deciduous Forests*, Cambridge University Press, Cambridge.

Gleason, H. A. (1927) 'Further views on the succession concept'. *Ecology*, vol 8, pp. 299–326.

Gravel, D., Canham, C.D., Beaudet, M. and Messier, C. (2010) 'Shade tolerance, canopy gaps and mechanisms of coexistence of forest trees'. *Oikos*, vol 119, pp. 475–484.

Grubb, P.J. (1977) 'The maintenance of species-richness in plant communities: the importance of the regeneration niche'. *Biological Review,* vol 52, pp. 107–145.

Guariguata, M. R. and Ostertag, R. (2001) 'Neotropical secondary forest succession: changes in structural and functional characteristics'. *Forest Ecology and Management*, vol 148, pp. 185–206.

Hale, C.M., Frelich, L.E. and Reich, P.B. (2006) 'Changes in cold-temperate forest understory plant communities in response to invasion by European earthworms. '*Ecology*, vol 87, pp. 1637–1649.

Hammond, W. M., Williams, A. P., Abatzoglou, J. T., Adams, H. D., Klein, T., López, R., Sáenz-Romero, C., Hartmann, H., Breshears, D. D. and Allen, C. D. (2022) 'Global field observations of tree die-off reveal hotter-drought fingerprint for Earth's forests'. *Nature Communications*, vol 13. 1761.

Howard, L. F. and Lee T. D. (2003) 'Temporal patterns of vascular plant diversity in southeastern New Hampshire forests'. *Forest Ecology and Management*, vol 185, pp. 5–20.

Hubbell, S.P., Foster, R.B., O'Brien, S.T., Harms, K.E., Condit, R., Wechsler, B., Wright, S.J. and Loo de Lao, S. (1999) 'Light-gap disturbances, recruitment limitation, and tree diversity in a neotropical forest' *Science*, vol 283, pp. 554–557.

Johnstone J.F., Hollingsworth T.N., Chapin F.S. III and Mack M.C. (2010) 'Changes in fire regime break the legacy lock on successional trajectories in Alaskan boreal forest'. *Global Change Biology*, vol 16, no 4, pp. 1281–1295.

Kneeshaw, D.D. and Bergeron Y. (1998) 'Canopy gap characteristics and tree replacement in the southeastern boreal forest'. *Ecology*, vol 79, pp. 783–794.

Lai, H. R., Craven, D., Hall, J. S., Hui, F. K. C., and Breugel, M. (2021) 'Successional syndromes of saplings in tropical secondary forests emerge from environment-dependent trait–demography relationships'. *Ecology Letters*, vol 24, pp. 1776–1787.

Meiners, S. J., Cadotte, M. W., Fridley, J. D., Pickett, S. T. A. and Walker, L. R. (2015) 'Is successional research nearing its climax? New approaches for understanding dynamic communities'. *Functional Ecology*, vol 29, pp. 154–164.

Oliver, C.D. (1981) 'Forest development in North America following major disturbances'. *Forest Ecology and Management*, vol 3, pp. 153–168.

Phillips, O.L., Aragão, L.E.O.C., Lewis, S.L., Fisher, J.B., Lloyd, J., López-González, G., *et al.* (2009) 'Drought sensitivity of the Amazon rainforest' *Science*, vol 323, pp. 1344–1347.

Pickett, S.T.A., Collins, S.L. and Armesto, J.J. (1987) 'Models, mechanisms and pathways of succession'. *Botanical Review*, vol 53, no 3, pp. 335–371.

Powers, J.S., Becknell, J.M., Irving, J. and Perez-Aviles, D. (2009) 'Diversity and structure of regenerating tropical dry forests in Costa Rica: geographic patterns and environmental drivers'. *Forest Ecology and Management*, vol 258, pp. 959–970.

Rich, R.L., Frelich, L.E. and Reich, P.B. (2007) 'Wind-throw mortality in the southern boreal forest: effects of species, diameter and stand age' *Journal of Ecology*, vol 95, pp. 1261–1273.

Royo, A.A. and Carson, W.P. (2006) 'On the formation of dense understory layers in forests worldwide: consequences and implications for forest dynamics, biodiversity, and succession'. *Canadian Journal of Forest Research*, vol 36, pp. 1345–1362.

Royo, A.A. and Carson, W.P. (2022). 'Stasis in forest regeneration following deer exclusion and understory gap creation: A 10-year experiment'. *Ecological Applications*, vol 32, e2569.

Rüger, N., Comita, L. S., Condit, R., Purves, D., Rosenbaum, B., Visser, M. D., Wright, S. J. and Wirth, C. (2018) 'Beyond the fast–slow continuum: demographic dimensions structuring a tropical tree community'. *Ecology Letters*, vol 21, pp. 1075–1084.

Runkle, J.R. (1981) 'Gap regeneration in some old-growth forests of the eastern United States'. *Ecology*, vol 62, pp. 1041–1051.

Runkle, J.R. (1982) 'Patterns of disturbance in some old-growth mesic forests eastern North America'. *Ecology*, vol 63, pp. 1533–1546.

Schnitzer, S.A. and Carson, W.P. (2000) 'Have we forgotten the forest because of the trees?' *Trends in Ecology and Evolution*, vol 15, pp. 375–376.

Seidl, R., and Turner, M. G. (2022) 'Post-disturbance reorganization of forest ecosystems in a changing world'. *Proceedings of the National Academy of Sciences*, vol 119. e2202190119.

Shorohova, E., Kuuluvainen, T., Kangur, A. and Jõgiste, K. (2009) 'Natural stand dynamics, disturbance regimes and successional dynamics in the Eurasian boreal forests: a review with special reference to Russian studies' *Annals of Forest Science*, vol 66, no. 201, pp. 1–20.

Swaine, M.D., and Whitmore, T.C. (1988) 'On the definition of ecological species groups in tropical rain forests'. *Vegetatio*, vol 75, pp.81–86.

Tester, J., Starfield, A. and Frelich, L.E. (1997) 'Modeling for ecosystem management in Minnesota pine forests'. *Biological Conservation*, vol 80, pp. 313–324.

Uhl, C., Clark, K. and Maquirno, P. (1988) 'Vegetation dynamics in Amazonian treefall gaps'. *Ecology*, vol 69, pp. 751–763.

Van der Molen, M. S., and Webster, C. R. (2021) 'Influence of deer herbivory on regeneration dynamics and gap capture in experimental gaps, 18 years post-harvest'. *Forest Ecology and Management*, vol 501, 119675.

Watt, A. S. (1947) 'Pattern and process in the plant community' *Journal of Ecology*, vol 35, pp. 1–22.
Whittaker, R. G. (1957) 'Recent evolution of ecological concepts in relation to the eastern forests of North America'. *American Journal of Botany*, vol 44, pp. 197–206.
Zimmerman, J. K., Everham III, E. M., Waide, R. B., Lodge, D. J., Taylor, C. M., and Brokaw, N. V. (1994) 'Responses of tree species to hurricane winds in subtropical wet forest in Puerto Rico: implications for tropical tree life histories'. *Journal of Ecology*, vol 82, pp. 911–922.

11

TREE GENETIC DIVERSITY AND GENE FLOW IN FOREST ECOSYSTEMS

Francine Tremblay

Introduction

Genetic diversity provides all living organisms with the potential to adapt and evolve in response to a changing environment. Forest tree species are known for their high level of genetic diversity (Hamrick, 1979). This source of variation is expressed in a large tree-to-tree variation called phenotypic variation, which is a generated by a combination of the effects of the genotype and of the growing environment. In other words, "*it is the tree that you see*". This source of variation has been exploited by foresters to implement tree genetic improvement programs all over the world (Mullin et al. 2011) The selected material is widely used in forest plantations as it helps increase forest productivity, timber supply and the provision of several ecological services to the community.

Trees pose significant challenges for geneticists. Their biological attributes, long juvenile phase, life-span and predominant outcrossing mating system all impose constraints on experimental genetic studies. Nevertheless, understanding the genetics of trees is essential for gaining insights into the evolution, conservation and sustainability of forest diversity worldwide. In this chapter, our goal is to synthesize the biological and ecological factors that impact the genetic diversity of forest tree species in boreal and tropical ecosystems. We discuss the main life-history traits of trees, specifically, reproduction, dispersal potential and demography, with respect to genetic dynamics of tree populations. Furthermore, we examine the relative importance of both adaptive and neutral levels of genetic diversity in tree populations. The discussion will address the forest tree's evolutionary response to climate changes in relation to these traits.

We focused our attention on boreal and tropical ecosystems because they represent two contrasting cases to illustrate the potential interaction of biological constraints, environmental conditions, and human influence on the genetic diversity of trees. The boreal zone is characterized by the presence of a small number of tree species (generally less than six) in any one stand, with large, monotype stands being quite common. In moist tropical forests, typically, there is a high number of tree species but a low population density (Finkeldey and Hattemer 2006). In tropical Asia, for instance, one hectare may contain anywhere from 150 to 250 tree species, with approximately 26–36 per cent of all tree species represented by a

 DOI: 10.4324/9781003324072-13

single tree. At a global scale, the boreal forest in North America experiences minimal human impact from urban and agricultural developments, while land-use pressures remain high in the tropics. Pereira et al. (2010) identify the current conversion of tropical forests to agricultural systems as a major contributor to global biodiversity loss.

Basic concepts of population genetics

Genetic variation in an individual tree is measured by the presence of distinct alleles at a given locus, located on a pair (diploid organism) of homologous chromosomes. Therefore, when one considers a group of individuals belonging to the same species, the difference between genomes (sets of genes) within this group is designated as intraspecific genetic diversity. The term intraspecific diversity reflects the fact that the individuals from one species are not all genetically identical. The level of genetic diversity can be reflected in the variation in phenotypic traits in trees growing in similar ecological conditions. However, it may be more subtle when variation in non-coding DNA (or genes) is unexpressed in the phenotype.

Population genetics

The aim of population genetics is to understand and to quantify how genetic variability is transmitted from one generation to the next and to examine how this variability evolves over generations. The level, structure and partitioning of genetic diversity within and between populations can be characterized using different tools (e.g., neutral markers) and indices of diversity. The Hardy–Weinberg model is commonly applied in population genetic studies. It is based on the principle that genotype frequencies can be predicted from gene/allele frequencies which remained constant over generations in random mating populations. It is easy to use, but the underlying assumptions (no mutation, selection, or migration) for its use are rarely met simultaneously.

What are the sources of genetic variation and population differentiation?

Evolution refers to the change of the genetic (allelic or genotypic) structure of a population at one or several gene loci. Two types of genetic markers, neutral genetic markers and adaptive genetic markers, are used in population genetics to study genetic variation within and between populations. Neutral genetic markers are subject to random fluctuation of allele frequencies over generations and are not influenced by natural selection. They are useful for inferring population history, migration, and demographic changes. Adaptive genetic markers, on the other hand, are genetic variations that influence fitness, enabling individuals with specific genetic traits to have a better chance of survival and reproduction. These markers are subjected to natural selection and are used to investigate local adaptations. The evolutionary forces and their impacts shaping the genetic variation patterns within and among populations are briefly summarized below (Table 11.1).

Random genetic drift definition and consequences

Genetic drift occurs when the frequency of alleles (different versions of a gene) in a gene pool changes randomly over time due to chance events rather than natural selection. The

Table 11.1 Impacts of evolutionary forces on population genetic diversity

	Selective forces	
	Drift/Mutation/Migration *Random*	*Selection* *Directional*
Differentiation among populations	+/ +/ −	+
Variation within populations	−/ +/ +	-
Heterozygosity	− / +/ +	+/ −

From Musch et al. (2004).

gradual deviation from the expected frequency of alleles under random mating leads to a departure from the Hardy–Weinberg equilibrium. Genetic drift is expected to take place in finite populations that undergo a significant reduction in their original size, or in newly established populations that originate from a small number of individuals. When genetic drift persists for multiple generations, it can result in the random fixation of one allele and the loss of the other. The anticipated duration for the loss or fixation of an allele relies on the original population size. Genetic drift is intensified by inbreeding, or the crossing of related individuals, leading to a higher frequency of homozygotes, and decreased genetic variation in the population. Additionally, genetic drift is impacted by associative mating, where only a portion of the reproductive population contributes to a larger proportion of the new generations.

Some evidence shows that large effective population sizes combined with life-history traits such as long-distance pollination and seed dispersal (migration) prevent many tree species from genetic isolation and its consequences (drift, inbreeding). Nevertheless, there are exceptions. *Pinus rubra* (red pine), for instance, is considered one of the most genetically depauperate conifer species in North America despite its large geographic range and life-history traits (Mosseler et al. 1992). This pattern may be attributed to a significant reduction in population size (genetic bottleneck) that occurred during the Holocene's glacial episodes.

Migration

Migration is the transfer of gene variants from one population to another through pollen, seeds, propagules, or plant parts and the source of gene flow between populations. At the intra-population level, migration increases the genetic diversity and effective population size by bringing new alleles. However, at the same time, it limits the genetic divergence between populations and opposes local adaptation by preventing the change in the frequency of more adaptive alleles from one population to another. A small level of migration can have a significant impact on intra-population diversity and prevent marked divergence between populations, even those subjected to genetic drift. Migration also gives birth to new populations when a seed produces a viable individual in a site where the species was not present before. This phenomenon is more frequently observed at the front or the rear edge of a species' range. One side effect of migration is that it can be difficult to clearly establish the exact size of a given population and delineate its boundary.

Mutation

Mutation is the source of new alleles and produces changes that may be heritable. The rate of spontaneous mutation is generally low, on the order of 10^{-4} to 10^{-6} mutations per gene per generation. In conifers, the rate of nucleotide substitution is even lower, estimated at 0.68×10^{-9} synonymous substitutions per site per year. Organisms with longer generation times are associated with lower rates of mutation (calibrated to the year) and molecular evolution. The mitotic rate hypothesis has established a relationship between plant size and the rate of genome copying. According to Lanfear et al. (2013), genetic changes that occur during cell division in plant shoots could potentially be passed on to future generations. In the long term, the rate of cell division and genome copying slows down in taller plants. As a result, somatic mutations accumulate more rapidly in fast-growing but short-lived plant species that have faster cell cycles than in relatively slow-growing but long-lived species such as trees. Mutations are therefore very important on an evolutionary time scale but are unlikely to be important on an ecological time scale (e.g. climate change). If a mutation has no effect on the function of the modified gene, it is considered selectively neutral. The rate of fixation of non-neutral mutations generally occurs at low frequencies because many non-synonymous changes are deleterious and are removed from the population. Therefore, their consequences are almost negligible on our time scale. In certain cases, somatic mutations can affect the ability of a tree to adapt to changes in its environment over its lifetime. In the mosaic tree *Eucalyptus melliodora*, small changes in DNA within a single tree (on the order of ten nucleotide differences between two branches) lead to differential upregulation of a secondary metabolism from one branch to another and to differential susceptibility (Padovan et al. 2013). Whether these mutations are heritable and transmitted to the next generation remains an open question.

Selection

Not all trees in a population contribute equally to the next generation. An individual's fitness depends on its viability, probability of reaching reproductive age, and fertility. Natural selection operates on the phenotypes interacting with the environment, altering the population's average value for traits that are subject to selection. When populations within a species occupy similar environments, natural selection will impede divergence as the filter for adaptation remains consistent across all populations.

Natural selection can affect phenotypic traits governed by either monogenic (one gene) or commonly polygenic (two or more genes) factors. Although infrequent among tree species, the resistance of five-needle white pine species to the introduced disease, white pine blister rust (*Cronartium ribicola*), is an example of a trait determined by a single gene (Kinloch 1992). The disease resistance is determined by a dominant allele of a single major gene called major gene resistance (MGR) that initiates a hypersensitive response (HR) in needles (Schoettle et al. 2014). This allele's presence in either heterozygous or homozygous genotypes (i.e., one or two copies of the resistance allele) leads to resistance to this disease.

Genetic diversity in the population

More than 90 per cent of the overall neutral marker genetic diversity of tree species is found within populations exhibiting low genetic differentiation, which is commonly observed, according to Hamrick (2004). In comparison to annuals and herbaceous perennials, trees have a mean differentiation among populations (G_{ST} based on allozymes) estimated at 8.9 per cent, compared to 35.5 and 25.6 per cent, respectively (Hamrick 2004). The high level of gene flow between populations is the prevailing explanation for within-population diversity. Wind-dispersed pollen is common in boreal and temperate tree species, which are primarily outcrossing. Kremer et al. (2012) reviewed that wind-dispersed viable pollen can travel up to 600 km. However, pollination distances that result in successful mating—as determined by genetic parentage analysis—range between 3 and 100 km (Table 11.2). In tropical regions, animal-dispersed pollen is prevalent and can be effective, with dispersal reaching tens of kilometers. The mean distances of seed dispersal mediated by animals range from 100 m to 1 km, while the range for pollen movement is 100 m to 14 km.

The life cycle of trees is involved in the partitioning of genetic diversity among and within populations. Their extended juvenile phase is particularly significant in this process (Austerlitz et al. 2000). Tree stands typically consist of individuals from various age or size groups with overlapping generations. Following a colonization event and before the first cohort of trees reach the reproductive age (during the first few decades), there is a continuous inflow of new migrants, or seed flow, and no in-stand reproduction. Hence, in a newly established population, a considerable portion of space is already occupied by juveniles from seeds that arrived years earlier. Lesser et al. (2013) demonstrated a rapid accumulation of alleles in populations of ponderosa pine (*Pinus ponderosa*) after initial colonization. They also found that contemporary levels of genetic diversity are established early in the development of tree populations. Allele accumulation reached saturation at population sizes of around 100 individuals. High levels of gene flow in the early stages of population growth can lead to a rapid accumulation of alleles, resulting in relatively homogenous genetic patterns among populations.

Boreal tree species

Numerous studies have recorded the distribution of genetic diversity among and between populations of boreal species (Table 11.3 provides some examples). Most of these studies focused on economically important conifers that are common and possess average F_{ST} values below 0.1 (boreal species with a continuous range have a mean value of approximately 0.050). In both black spruce (*Picea mariana*) and white spruce (*Picea glauca*), the genetic variation among populations is no more than 1 per cent (Gamache et al. 2003; Jaramillo-Correa et al. 2001). Trembling aspen (*Populus tremuloides*) maintains high levels of neutral genetic variation and clonal diversity both locally (Wyman et al. 2003) and at a continental scale (Latutrie et al. 2019). High levels of genetic diversity are also observed among fragmented populations of coniferous species in the northern edge (Gamache et al. 2003; Xu et al. 2012).

Tropical tree species

The Baker–Fedorov hypothesis (BFH) predicted that genetic drift would facilitate speciation over limited spatial scales, leading to the typical community structure pattern in tropical forests. This hypothesis was rooted in the notion that tropical trees are highly inbred due to restricted gene flow. However, several decades of research have invalidated this hypothesis, at least for canopy trees. Many tropical tree species exhibit self-incompatibility with high rates of outcrossing and long-distance pollen dispersal (Bawa 1990). Despite this, tropical

Table 11.2 Examples of observed long-distance pollen and seed dispersal in trees (> 3 km for pollen and > 1 km for seeds)

Species	*Dispersal system*					*Dispersal distance*		
	Propagule	*Vector*	*Type*[1]	*Location*	*Method*	*Maximum Range*	*Proportion≥ threshold*[2]	*References*
Pinus banksiana and *Picea glauca*	Pollen	Wind	Potential	Canada	Aerobiologic analysis	3000 km		Campbell et al. (1999)
Pinus sylvestris	Pollen	Wind	Potential	Northern Europe	Aerobiologic analysis and phenological analysis	600 km		Varis et al. (2009)
Populus trichocarpa	Pollen	Wind	Effective	Western North America	Genetic Parentage analysis		5% > ~ 5–10 km	Slavov et al. (2009)
Ficus spp.	Pollen	Wind	Potential	Central America	Genetic parental reconstruction	14 km (isolated mother trees)		Nason et al. (1998)
Ficus sycomorus	Pollen	Insects	Effective	Namibia	Genetic Parentage analysis	165 km		Ahmed et al. (2009)
Swietenia humilis	Pollen	Insects	Effective	Central America	Genetic paternity analysis		40–80% ‡ 4 km (in small fragments)	White et al. (2002)
Xylopia hypolampra and seven other species	Seeds	Birds	Potential	Cameroon	Empirically based simulations of vector movements and seed passage time	6.9 km		Holbrook & Smith (2000)

(*Continued*)

Table 11.2 (Continued)

Species	*Dispersal system*					*Dispersal distance*		
	Propagule	*Vector*	*Type*[1]	*Location*	*Method*	*Maximum Range*	*Proportion≥ threshold*[2]	*References*
Tamarindus indica	Seed	Elephant	Potential	Myanmar (Burma)	Empirically based simulations of vector movements and seed passage time	5.4 km	50% > 1.2 km	Campos-Arceiz et al. (2008)
Duroia duckei and two other species	Seed	Fish	Potential	Peru	Empirically based simulations of vector movements and seed passage time	5.5 km	5% > 1.7 km	Anderson et al. (2011)
Aucoumea klaineana	Seed Pollen	Wind	Potential	Gabon Central Africa	Parentage and sex assignment analyses	118 ± 62 m[1]		Born et al. (2011b)
Swietenia humilis	Pollen	Bees, moths, thrips	Potential	Costa Rica	Between forest fragments		up to 4.5km	White et al. (2002)
Ficus sp.	Pollen	Fig wasps	Potential	Panama			5.8–14.2km	Nason et al. (1998)

Note: The table is arranged first by propagule type (pollen or seed), then by vector type (wind, insects) and dispersal type (potential, viable or effective).

1 Dispersal types are respectively: (1) potential, or the distance dispersed by a propagule in any condition; (2) effective, or the pollen that gives rise to seeds, or seeds that established yielding seedlings, saplings, or young adult plants.

2 The proportion (percentage) of propagules dispersed to equal or greater distances than the specified threshold. The threshold was defined by the authors of each study, often arbitrarily or according to features of the study landscape and/or populations (adapted from Kremer et al. 2012).

Table 11.3 Examples of genetic diversity (microsatellite markers[1], allozyme loci[2] and ESTPs of ncDNA[3]) studies in populations of tree species

Species	*A*	A_R	H_0	H_E	F_{IS}	F_{ST} / G_{ST}	*Reference*
Protium subserratum[1]	4.8 (2.8–6.0)	—	0.43 (0.05–0.66)	0.53 (0.36–0.67)	0.07 (0.01–0.16)	(0.02-0.46)	Misiewicz and Fine (2014)
Vriesea gigantea[1]	2.83 (1.18–3.49)	—	0.43 (0.03–0.61)	0.57 (0.06–0.72)	0.27 (0.05–0.48)	0.21	Palma-Silva et al. (2009)
Shorea xanthophylla	8.07 (5.0–12.3)	—	0.61 (0.40–0.79)	0.66 (0.46–0.8)	0.06 (-0.13–0.31)	—	Kettle et al. (2009)
Parashorea tomentella[1]	10.67 (4.92–17.52)	—	0.57 (0.34–0.77)	0.60 (0.34–0.86)	0.05 (-0.008–0.15)	—	Kettle et al. (2009)
Dipterocarpus grandiflorus[1]	13.67 (9.63–16.66)	—	0.63 (0.55-0.73)	0.68 (0.55–0.76)	0.06 (-0.03-0.26)	—	Kettle et al. (2009)
Aucoumea klaineana[1]		—	0.45 (0.34–0.53)	0.50 (0.44-0.59)		(0.03-0.16)	Born et al. (2011b)
Shorea javanica[1]		—	0.42	0.47	0.38 (-0.05-0.64)	0.064	Rachmat et al. ((2012)
Swietenia macrophylla[1]	9.5 (7.6-10.7)	—	0.75 (0.68-0.81)	0.78 (0.75–0.81)	0.03 (–0.004-0.10)	0.09	Lemes et al. (2003)
Carapa guianensis[1]		—	0.66	0.64	(–0.02-0.14)	(0.001-0.08)	Dayanandan et al. (1999)
Pinus banksiana[2]	2.3-2.5		0.18-0.16	0.16-0.15		0.014-0.11	Gauthier et al. (1992)
Picea glauca[2]	3.03	2.14	0.34	0.34	0.002		O'Connell et al. (2006)
Picea mariana[3]	2.37 (0.05)	0.15 (0.02)	0.22 (0.01)	0.21 (0.01)		0.01	Gamache et al. (2003)
Populus tremuloides[1]	—	3.43 (2.81–3.95)	0.64 (0.53-0.80)	0.65 (0.53-071)	0.013 (-0.242–0.143)		Latutrie et al. (2019)
P. tremuloides[1]	7.44 (6.25-8.2)		0.55 (0.47–0.70)	0.725 (0.69–0.76)	0.201 (-0.054–0.325)	0.03	Wyman et al. (2003)
L. laricina[2]	1.7		0.14	0.15		0.03	Ying and Morgenstern (1991)

(*Continued*)

Table 11.3 (Continued)

Species	*A*	A_R	H_0	H_E	F_{IS}	F_{ST} / G_{ST}	*Reference*
P. strobus[1]	9.43 (9.23–9.62)		0.52 (0.50–0.53)	0.60 (0.59–0.61)	—	—	Rajora et al. (2000)*
P. strobus[1]	—	6.7	0.47	0.48	0.01	—	Marquardt and Epperson (2004)*
T. occidentalis[1]	7.3 (5.67–9.33)	6.8 (5.16–8.51)	0.60 (0.49–0.66)	0.61 (0.49–0.67)	0.01 (-0.06–0.10)	0.07	Pandey and Rajora (2012)
T. occidentalis[1]	7.8 (5.0–10.0)	5.9 (4.6–6.9)	0.73 (0.46–0.88)	0.77 (0.71–0.84)	0.14	0.06	Xu et al. (2012)

[1] Microsatellites or single sequence repeats (SSRs) are short DNA fragments of usually only two or three base pairs in length which are repeated several times in a particular location of the DNA.

[2] Allozyme loci are the number of forms of the same enzyme.

[3] Expressed sequence tag polymorphisms (ESTPs) of ncDNA.

* Study used logging or non-natural forest; we only reported populations from old-growth and natural forests. *A*, mean number of alleles per locus; A_R, mean allelic richness; H_0, mean observed heterozygosity; H_E, mean expected heterozygosity; F_{IS}, inbreeding coefficient, F_{ST}, mean pairwise F_{ST}, G_{ST}, mean pairwise G_{ST}. Range values are given in parentheses (adapted from Graignic 2014).

tree species tend to have a greater genetic differentiation between populations compared to boreal species. Typical values of F_{ST} range from 0.034 to 0.17 for populations separated by hundreds of meters to kilometers (Hamrick 1994; Dick et al. 2008) (Table 11.3). Species with abiotic modes of seed dispersal, such as barochorous and anemochorous, display significantly higher differentiation among populations on average (G_{ST} = 0.138) compared to zoochorous species (G_{ST} = 0.050) according to Finkeldey and Hattemer (2006). This suggests that low tree density has little effect on gene exchange, which is likely due to the presence of effective pollinators or animal dispersers with extensive home ranges capable of spanning significant distances between individuals.

Spatially explicit analyses of genetic variation for tree species that occur at low density in tropical forests may provide more detailed information than the traditional approach of dividing overall genetic variation among and within populations (Duminil et al. 2016). However, identifying trees within a densely populated tropical forest that belong to the same population is often challenging. The non-random spatial patterns of gene dispersal imposed by animal pollinators and seed dispersers can lead to highly structured genetic diversity within populations of animal-dispersed species (Jordano and Godoy 2002; Schroeder et al. 2014; Vieira et al. 2012). However, Cristobal-Pérez et al. (2020) found low fine-scale genetic structure (FSGS) and no differences among size classes in *Spondias purpurea*, a tropical dioecious tree. The low FSGS observed in *S. purpurea* is likely due to long-distance seed dispersal and early density-dependent factors that have influenced the fine-scale genetic patterns.

Phylogeographic pattern of variation

The genetic structure of tree species in all continents has been shaped by the characteristics of the cool and dry glacial periods. As a result of these episodes, forest ranges in North America, Europe, and the tropics have been severely restricted. Research into the geographic patterns of genetic diversity, known as phylogeography, is conducted using various markers. Cytoplasmic DNA is a typical marker used to trace lineages that have spread across continents following glacial retreat. In conifers, mitochondrial DNA (mtDNA) is inherited from the mother and chloroplastic DNA (cpDNA) is inherited from the father. However, there are two exceptions to this rule, *Taxodiaceae* and *Cupressaceae*, where mtDNA is strictly inherited from the father. In flowering plants, both mtDNA and cpDNA are inherited from the mother. Unlike nuclear DNA, (ncDNA) cytoplasmic DNAs are highly conserved.

Boreal tree species

During the late Quaternary and previous glaciations, the boreal region was covered in ice, displacing organisms south of their current limits. As the climate began to warm and the glaciers receded, many species migrated northward. Research on the transcontinental boreal distribution of white spruce (*P. glauca*) has identified the presence of three glacial refugia, namely the Beringian, Mississippian, and East Appalachia regions (de Lafontaine et al. 2010). A study on the distribution of mitotypes (mtDNA) in black spruce (*P. mariana*) revealed the presence of four well-defined groups: two found only in the west and east parts of Canada, one extensive group widespread throughout its range, and a fourth group exclusively observed in the northeast region (Jaramillo-Correa et al. 2001). A range-wide analysis of balsam poplar (*Populus balsamifera*) populations in North America identified three geographically distinct groups situated in the northern, central, and eastern areas of the species' range (Keller et al. 2010). These findings indicate recent population growth and a significant expansion from the center toward the north and east. A phylogeographic analysis of range expansion in trembling aspen (*P. tremuloides*) in the northwestern part of the species' range revealed a low level of population differentiation. No evidence was found to indicate that either Beringia or the "ice-free corridor" served as refugia (Latutrie et al. 2016).

Tropical tree species

To date, the majority of phylogeographic research has focused on tree species found in temperate and boreal regions, and as such, investigations of tropical tree species are not as common despite their significance. The impact of Pleistocene climate on tropical trees is still not well understood, partly due to limited palynological records. During glacial episodes, the climate in the tropics was dry and cool, while during interglacial episodes, it was like current conditions. Climatic changes have undoubtedly caused the forest to experience contraction and expansion, as well as transitions between various types of forest communities. The biogeographic processes induced by the climatic changes of the Pleistocene era have impacted both the genetic and species diversities of the forest. The study of nuclear DNA and cpDNA haplotypes in Central Africa shows the existence of Pleistocene Forest refuges for the Bush Mango, *Irvingia gabonensis*, and a north–south phylogeographic disjunction in *Greenwayodendron suaveolens* (Lowe et al. 2010). In Central Africa, Born et al. (2011a) demonstrated the existence of four genetic units in *Aucoumea klaineana* Pierre (*Burseraceae*), an African pioneer tropical rain tree species. These genetic units resulted from the expansion of subdivided source populations, a pattern that is congruent with the hypothesis of population fragmentation during the Late Glacial Maximum (LGM). Genetic signatures indicate past population fragmentation and demographic changes in two African forest legume trees that are widely distributed within the Guineo-Congolian rainforest of Africa (Demenou et al. 2020). Both species exhibit similar phylogeographic structures, which may result from the stability of forest refuges and comparable forest contraction/expansion events during successive glaciations.

A phylogeographic pattern of variation has been identified in South America. Lemes et al. (2010) conducted an analysis on the natural populations of mahogany (*Swietenia macrophylla* King (*Meliaceae)*) ranging from Mexico to Panama, in Central America, and across 2,100 km of the southern arc of the Amazon Basin. They observed a strong break between Central and South American mahogany populations. Furthermore, the researchers reported high levels of population differentiation in the Amazon Basin, while relatively low differentiation was found across Central America. A comparative phylogeographical analysis demonstrated a congruent signal of population size decline in trees within the Panamanian rainforest (Turchetto-Zolet et al. 2013). In contrast, Honorio Coronado et al. (2019) found no evidence of a shared phylogeographic break that would align with either geological paleoarches or putative Pleistocene refugia in five Amazonian rainforest tree species. According to the authors, "Amazonian tree species seem to have originated and occupied environments individualistically rather than as congruent components of a coherent Amazonian flora."

Geographic variation in quantitative traits

Neutral markers are not reliable indicators of the variation levels in adaptive traits, which affect tree fitness. To investigate the plastic responses of tree populations to environmental conditions, multiple-site provenance trials have been established. These trials used common garden experiments to test different seed sources from defined geographical regions (provenances) in locations with varying environmental conditions. The differentiation for quantitative traits, such as growth, phenology, and survival, often exhibited continuous variation along an environmental gradient based on altitude, latitude, and longitude (Morgenstern, 1996). This clinal variation is typically assumed to be influenced by natural selection. The average Q_{ST} estimate or the proportion of quantitative genetic variation resulting from inter-population differences in large northern areas is ten times greater than the average F_{ST} (mean Q_{ST} = 0.463; mean F_{ST} = 0.044). The Q_{ST} estimates for diameter and height growth vary from negligible (0) to almost 1, while estimates for traits associated with bud set timing and cold hardiness, which are crucial for survival and growth exceed 0.5 (Alberto et al. 2013). Leites and Garzón (2023) conducted a literature review utilizing provenance trial data that mostly consists of forest tree species possessing commercial value from temperate and boreal biomes. Their analysis provides valuable insights into how local climate affects intraspecific genetic adaptation. Clines in fitness-related traits were associated with temperature variables in 57 per cent of the species, with precipitation in 14% of the species, and with both in 25 per cent of the species.

Boreal tree species

A given tree genotype can result in various functional trait phenotypes, which depend on the environment. This occurs either through multiple biochemical pathways associated with high genetic diversity (i.e., heterozygosity) or natural selection, which favors alleles tolerant to broad environmental differences (Morgenstern 1996). Boreal tree species commonly show high genetic diversity with clinal variation in quantitative traits (Figure 11.1a). For instance, the bud flush is initiated by the cumulation of cold and heat sums exceeding a certain temperature threshold. Such temperature thresholds and critical temperature sums, determined genetically, differ between species and, to some extent, among populations of the same species. The photoperiod mainly regulates bud set while temperatures and drought modulate it. In a warming climate, spring phenology can potentially advance without significant genetic changes if the chilling requirement is met. However, a genetic change in photoperiodic responses is more likely required for a change in bud set date during the fall.

This clinal variation in phenological timing (bud set, bud flush), growth (height), and physiological characteristics may occur over short or long distances and is often documented in boreal species (see Kilsz et al. 2023 for a review). For instance, variation in Douglas fir (*Pseudotsuga menziesii*) is associated with the elevation of the seed source (Rehfeldt 1983). In this case, genetic variation is organized into multiple local specialist populations adapted to a narrow range of environments. Guo et al. (2021) reported a clinal pattern of variation in bud phenology among black spruce (*P. mariana*) provenances in common garden experiments, reflecting an adaptation to colder environments and shorter growing seasons in northern provenances. Therefore, although the evergreen-black spruce is predicted to

A)

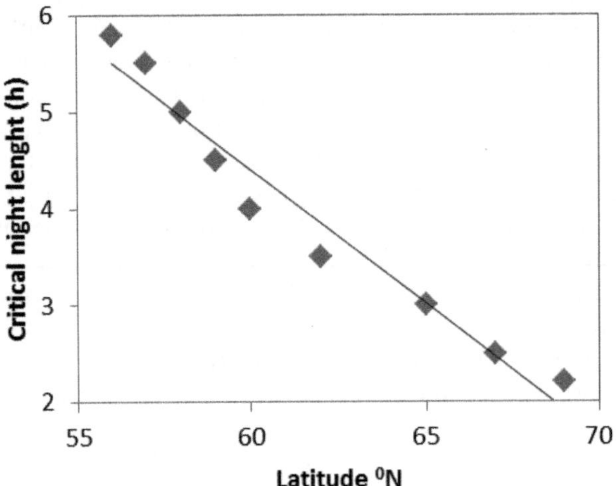

B)

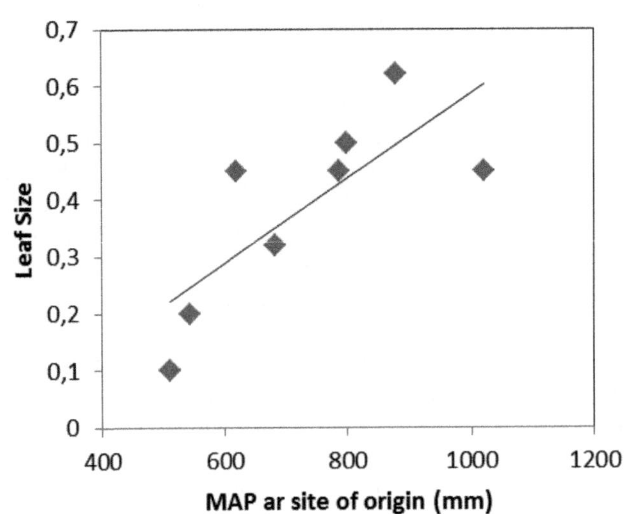

Figure 11.1 Clinal variation in quantitative traits: (a) relationship between critical night length (*h*) and timing of growth cessation in *P. abies* populations from different latitudes; (b) relationships between trait plasticity (relative trait range (RTR)) of nine *E. tricarpa* provenances and the mean annual precipitation (MAP) at their site of origin, plasticity of leaf size showed a positive linear correlation with MAP. The figures are redrawn from data of (a) Savolainen et al. (2007).and (b) McClean et al. (2014).

experience a longer photosynthesis period under warmer conditions, the duration of bud development may remain unchanged.

Tropical tree species

Clinal variation in tropical regions often corresponds to the response to drought, as shown in Figure 11.1b. In West Africa's Sahel region, there exists a transition area between the relatively humid savannah woodlands in the south and the Sahara Desert in the north. The first progeny test of *Prosopis africana* indicates that growth and survival are linked to rainfall gradients (Weber et al. 2008). Provenances from drier areas of the region demonstrated superior growth and survival compared to those from wetter regions when tested in a relatively dry site in Niger. *Eucalyptus tricarpa*, a species found across a climatic gradient in southeastern Australia, a region that is experiencing increasing aridity, has been shown to possess local adaptation to drought (McLean et al. 2014). Provenances from drier regions exhibit greater plasticity for several functional traits and display varying growth rates in common gardens. Specialization across edaphically diverse environments can also contribute to population-level divergence, even in the absence of geographic isolation. Recently, Barton et al. (2020) reported clinal variation across a rainfall gradient in *Metrosideros polymorpha* (*Myrtaceae*), a tropical foundation tree species. The study reveals that trees with high dispersal abilities can exhibit considerable local adaptation in drought tolerance. This implies that intraspecific variation may also be crucial for other foundational tropical tree species. In their research, Misiewicz and Fine (2014) found that Mesoamerican mahogany *S. macrophylla* King (Big-leaf mahogany) displays significant morphological variation, which is related to soil type. They reported greater genetic differentiation and reduced migration rates among adjacent populations residing on distinct soil types as compared to populations on the same soil type but residing at geographically distant locations.

Impacts of forest harvesting

Boreal forest landscapes undergo succession and disturbance. In North America, these landscapes are characterized by a limited number of tree species spread over a wide area. These species regenerate after natural disturbances, including fire, insects, and windthrow, or anthropogenic disturbances such as logging. Clear-cutting is the primary harvesting technique applied in boreal forests, but partial cut, selection cut, and shelterwood methods are also employed. Little population differentiation emerged between managed and wild lodgepole pine (*Pinus contorta*) stands (MacDonald et al. 2001). Buchert et al. (1997) found negative genetic diversity effects for white pine (*Pinus strobus*) after 75 per cent of trees (seed trees cut) were removed. However, most studies have shown that clear-cutting followed by natural or artificial regeneration does not have a negative genetic effect on several species of conifers. Fageria and Rajora (2013) found similar genetic structure in white spruce (*P. glauca*) between un-harvested controls, pre-harvest old growth, and post-harvest natural regeneration. The populations in the landscape contribute to the preservation of genetic diversity in harvested stands through gene flow.

Conserving genetic diversity in trees within a tropical ecosystem poses significant challenges due to the concurrent deforestation, habitat destruction, and population fragmentation associated with harvesting in this region. The most frequently used harvesting approach in tropical forests is partial cutting, also known as selective cut, with cutting limits set to minimum diameters. Deforestation may alter population density, but more importantly, it can impact the abundance and behavior of pollinator and seed disperser communities (Dick 2010). Certain research reports indicate notable differences in outcrossing rates between unmodified and logged plots, as well as a significant rise in self-fertilization rates within fragmented forest patches. For instance, in Costa Rica, studies have reported lower outcrossing rates in disrupted environments with minimal populations of reproductive *Symphonia globulifera* (Aldrich and Hamrick 1998) and *Pachira quinata* (Fuchs et al. 2003). It should be noted that the disturbance levels and population densities influenced the genetic effects in both cases. In contrast, Cloutier et al. (2007) observed no effect of selective logging on *Carapa guianensis* populations in Brazil regarding gene diversity, inbreeding, pollen dispersal, and spatial genetic structure. However, selective logging led to a reduction in genetic diversity of *Hymenaea courbaril* in the same forest area in Brazil despite the maintenance of outcrossing and gene flow (Carneiro et al. 2011). Meanwhile, the study by Wagar et al. (2021) highlights how past deforestation and other human disturbances have impacted the genetic diversity of *Manilkara multifida* populations in southern Bahia, Brazil. They reported a significant reduction in genetic variability between generations of adult and juvenile populations and concluded that anthropogenic activities have caused a loss of alleles that may be necessary to maintain the evolutionary fitness of these populations. Model simulations further supported the concept that species-level genetic information is crucial for predicting the long-term effects of selective logging on tree species in the Brazilian Amazon tree species and derive sustainable management scenarios for tropical forests (Vinson et al. 2014).

Impact of climate change

The fate of natural populations of trees in the context of global change is unpredictable. In the Northern Hemisphere, most models foresee a northward shift in the current range of many tree species. The ability of trees to cope with the anticipated rapid onset of change from a genetic point of view is uncertain (for a review, see Alberto et al. 2013). Trees that exhibit a high degree of phenotypical plasticity are most likely to persist and survive when exposed to climatic variation. However, according to current evidence, there is an increase in tree mortality in many populations, while forest health is declining (Clark 2007; Peng et al. 2011). For instance, the primary causes of forest dieback at the southern edge of the boreal forests have been identified as direct climate impacts of heat and drought (Allen et al. 2010). Furthermore, over the last two decades, common boreal tree species have demonstrated increased mortality (Peng et al. 2011). It is challenging to separate the distinct impact of present-day global change from the historical context (human activities) in study sites (Clark, 2007).

The adaptability of trees may be limited by their long generation times and relatively slow mutation rate. On the other hand, forest trees possess a high level of genetic diversity within populations, which, theoretically, provide the basic materials for adaptation to new climate conditions. Trees also often have very large effective population sizes, which makes selection more effective (relative to drift). Long-distance gene flow and heterogeneous selection contribute to maintaining genetic variability within populations. Therefore, gene flow increases the genetic variance of a population available for selection (Hamrick and Nason 2000) and may provide pre-adapted genotypes that facilitate adaptation (Kremer et al. 2012). Numerous observations have shown that genes can move, through seeds and pollen, over larger spatial scales than the habitat shifts predicted by climate change models within one generation. Kremer and Le Corre (2012) proposed that adaptation may occur without a detectable change in allele frequency. Their simulations showed that allelic associations were rapidly modified under moderate to strong divergent selection. The response elapsed during less than 10–20 generations. They suspect that the build-up of allelic associations is enhanced by the very high genetic diversity residing within populations, which is sustained by a high gene flow.

Epigenetic effects may contribute to tree adaptation to new environmental conditions. Maternal environment factors, such as temperature and photoperiod, can have a significant impact on phenotypic traits during the reproductive phase, thereby influencing progeny performance. Such effects have been observed in boreal species like Norway spruce (*Picea abies*) (Solvin 2020) and scots pine (*Pinus sylvestris*) (Bose et al. 2020). On average, these effects are 20 per cent smaller than the true additive genetic variation for quantitative traits. This finding may hold significant implications for interpreting provenance trial data as it may explain some of the phenotypic variation among populations, which are frequently misinterpreted as genetic variation. Raj et al. (2011) demonstrated that genetically identical poplar clones exhibit varied stomatal behavior in drought conditions, depending on their maternal environment. The study conducted on *Pinus pinea*, a genetically depauperated species, revealed significant cytosine methylation levels in the genome, as well as variations in methylation between the trees (Saez-Laguna et al. 2014). This implies the potential role of cytosine methylation in the regulation of gene expression and variation in phenotypic traits without detectable genetic variation. Such epigenetic effects allow for immediate adaptation to new climatic conditions and provide sufficient time for an evolutionary response. However, it remains unclear if these effects will be passed on to the progeny.

Different forest management strategies have been proposed to reduce the potential negative impact of climate change on forest ecosystems. To maintain or restore current ecosystems, resistance and resilience strategies may be applied to highly valued resources (Millar et al. 2007). Another approach is to facilitate the transitions to new ecosystems that may require the human-assisted migration of species. However, the most cited approach consists of a better match of genotypes of commercial species for future climates in reforestation programs. In other words, assisted migration (moving seed source to a new climate) is restricted to the movement of populations within a species range (Klisz et al. 2023). For these purposes, data from the provenance trials are used and combined with statistical models (response functions, transfer models) to predict a species' response to climate change (Pedlar, McKenney, and Lu, 2021). Seed zone maps are produced with the assumption that environmental variation and genetic variation in adaptive traits are largely correlated. However, the extent to which genetic variation reflects environmental conditions experienced by tree populations varies among species.

These species-specific seed zones will be an indication of whether species have an innate tendency to be adaptive generalists or specialists. However, as only 15–20 per cent of the variation between populations (or 2–3 per cent of the total variation) may be linked to climate, it raises the question of whether assisted migration strategies will need to be implemented to address climate change. Using tree-ring signatures of cold damage from common garden trials, Montwé et al. (2018) reported that transferring lodgepole pine (*P. contorta*) genotypes, which are adapted to southern and warm climates and are more productive, to northward climates resulted in significant growth loss and a permanent rank change following a spring frost event. Mura et al. (2022) also emphasized the significance of intraspecific variability in the bud phenology timings and the risk of frost damage on developing buds in provenances of black spruce (*P. mariana*). In certain instances, spring frosts can be so severe that local sources are equal in vulnerability to displaced ones (Benomar et al. 2022). These findings emphasize the significance of choosing appropriately acclimatized plant populations when performing seed transfers. Additionally, the intentional movement of provenances or species does not come without criticism. The debate is largely focused on the assessment of risks and benefits. Uncertainty about future climate conditions and possible impairment of ecological function and structure are major constraints to its implementation (Aitken and Whitlock 2013).

To help mitigate the uncertainties associated with climate change, managed forests could incorporate composite provenancing. These composites should include material that is locally sourced, that is proximate in origin, and that is eco-geographically matched. In addition, a smaller proportion of material (ranging from 10 to 30 per cent based on inferred gene flow dynamics) should originate from more distant sources. This approach minimizes the potential risks of assuming that populations can adapt, versus the risk of missing distinct gene pools in populations.

Future perspectives

The advent of rapid and inexpensive next-generation sequencing (NGS) technology has opened up opportunities to apply new approaches to the study of tree genomes. The increased availability of sequencing data is aiding the comprehension of standing variation for quantitative traits in trees. Poplar (*Populus trichocarpa*) is the pioneering tree for which the genome was sequenced (Tuskan et al. 2006), establishing this species as the exemplary research system for long-lived woody perennials. The relative genome size of poplar is smaller than that of conifers, with genome sizes in conifers being among the largest (typically 20–30 gigabases of pairs (Gb)) among all organisms. This makes genome-wide analyses particularly challenging for gymnosperms. Draft genomes of Norway spruce (*P. abies*) (Nystedt et al. 2013) and white spruce (*Picea glauca*) (Birol et al. 2013) have been published. Additionally, complete genome sequences are available for woody perennial species such as *Eucalyptus* (Myburg et al. 2014), *Papaya* (Ming et al. 2008), and *Theobroma cacao* (Argout et al. 2011). The increasing number of sequences available supports the development of powerful genomic tools. A vast body of research literature exists on this topic, which cannot be fully covered here (for further reading, see Neale and Kremer 2011; Groover and Jansson 2014; Isabel, Holliday, and Aitken 2020). Expressed sequence tag collections (ESTs) have been developed to help select candidate genes or detect single nucleotide polymorphisms (SNPs) that may be candidates for selection. A total of 17 SNPs were linked to serotiny in maritime pine (*Pinus pinaster*) populations growing in a fire-prone environment (Budde et al. 2014). This study illustrates the significance of identifying genes that are specifically linked with ecologically relevant traits that are under strong selection. SNPs with associations to bud set and cold hardiness phenotypes were identified by Holliday et al. (2010) in the widespread conifer Sitka spruce (*Picea sitchensis*). Numerous SNPs correlated with a minimum of one climate variable are also associated with phenotypic traits. Recently, the integration of population genomics, gene expression patterns, and phenotypic traits has facilitated a comprehensive understanding of the molecular mechanisms underlying environmental adaptation in trees. In this approach, genetic associations identify candidate genes associated with adaptive traits, and transcriptomic analysis reveals how these genes are regulated under specific environmental conditions (Depardieu et al. 2021). Genomic selection (also known as genomic breeding) is a promising strategy now being used in forest tree breeding programs. Genomic selection relies on genome-wide distributed markers and phenotypic data to efficiently select for desired traits (Beaulieu et al. 2020; Laverdière et al. 2022). It has the potential to speed up the breeding process because the data (genotypic/phenotypic) collected on individuals in genetic tests can be used to identify superior trees at a young age, thus reducing the time required to make selection decisions. Together these approaches help to increase our understanding of traits that could be used to model capacity for adaptation responses as well as provide tools for molecular breeding strategies.

References

Ahmed, S., Compton, S.G., Butlin, R.K., Gilmartin, P.M. (2009) Wind-borne insects mediate directional pollen transfer between desert fig trees 160 kilometers apart, *Proceedings of the National Academy of Sciences USA*, vol. 106, pp. 20342–20347.

Aitken, S.N., Whitlock, M.C. (2013) Assisted gene flow to facilitate local adaptation to climate change, *Annual Review of Ecology Evolution and Systematics*, vol.44, pp. 367–388.

Alberto, F.J., Aitkens, S.N., Alía, R., Gonzαlez-Martínez, S.C., Hänninenk, H. and others (2013) Potential for evolutionary responses to climate change—evidence from tree populations, *Global Change Biology*, vol. 19, pp. 1645–1661. doi: 10.1111/gcb.12181

Aldrich, P.R., Hamrick, J.L. (1998) Reproductive dominance of pasture trees in a fragmented tropical forest mosaic, *Science*, vol. 281, pp.103–105.

Allen, C.D., Macalady, A.K., Chenchouni, H., Bachelet, D., McDowell, N. and others (2010) A global overview of drought and heat-induced tree mortality reveals emerging climate change risks for forests, *Forest Ecology and Management*, vol.259, pp. 660–684.

Anderson, J.T., Nuttle, T., Saldanã Rojas, J.S., Pendergast, T.H., Flecker, A.S. (2011) Extremely long-distance seed dispersal by an overfished Amazonian frugivore, *Proceedings of the Royal Society B: Biological sciences*, vol. 278, pp. 3329–3335.

Argout, X., Salse, J., Aury, J.M., Guiltinan, M.J., Droc, G. and others (2011) The genome of *Theobroma cacao*. *Nature Genetics*, vol. 43, pp. 101–108. doi: 10.1038/ng.736

Austerlitz, F., Mariette, S., Machon, N., Gouyon, P.H., Godelle, B. (2000) Effects of colonization processes on genetic diversity: differences between annual plants and tree species, *Genetics*, vol. 154, pp.1309–1321.

Barton, K.E., Jones C., Edwards, K.F., Shiels A., Knight T. (2020) Local adaptation constrains drought tolerance in a tropical foundation tree, *Journal of Ecology*, vol.108, pp.1540–1552.

Bawa, K.S. (1990) Plant-pollinator interactions in tropical rainforests, *Annual Review of Ecology Evolution and Systematics*, vol. 21, pp. 399–422.

Beaulieu, J., Nadeau, S., Ding, C., Celedon, J.M., Azaiez, A. and others (2020) Genomic selection for resistance to spruce budworm in white spruce and relationships with growth and wood quality traits, *Evolutionary Applications*, vol. 13, pp. 2704–2722. doi: 10.1111/eva.13076

Benomar, L., Bousquet, J., Perron, M., Beaulieu, J., Lamara, M. (2022) Tree Maladaptation Under Mid-Latitude Early Spring Warming and Late Cold Spell: Implications for Assisted Migration, *Frontiers in Plant Sciences*, vol. 13, 920852. doi:10.3389/fpls.2022.920852

Birol, I., Raymond, A., Jackman, S.D., Pleasance, S., Coope, R. and others (2013) Assembling the 20 Gb white spruce (*Picea glauca*) genome from whole-genome shotgun sequencing data, *Bioinformatics*, vol. 29, pp. 1492–1497.

Born, C., Alvarez, N., McKey, D., Ossari, S., Wickings, E.J. and others (2011a) Insights into the biogeographical history of the Lower Guinea Forest Domain: evidence for the role of refugia in the intraspecific differentiation of *Aucoumea klaineana*, *Molecular Ecology*, vol. 20, 131–142.

Born, C., Kjellberg, F., Chevallier, M.H., Vignes, H., Dikangadissi, J.-T. and others (2011b) Colonization processes and the maintenance of genetic diversity: insights from a pioneer rainforest tree, *Aucoumea klaineana Proceedings of the Royal Society B: Biological Sciences*, vol. 275, pp. 2171–2179.

Bose A.K., Moser, B., Rigling, A., Lehmann, M.M., Milcu, A. and others (2020) Memory of environmental conditions across generations affects the acclimation potential of Scots pine, *Plant Cell Environment*, vol. 43, pp. 288–299.

Buchert, G.P., Rajora, O.P., Hood, J.V., Dancik, B.P. (1997) Effects of harvesting on genetic diversity in old-growth eastern White Pine in Ontario, Canada, *Conservation Biology*, vol. 11, pp. 747–758.

Budde, K.D., Heuertz, M., Hernandez-Serrano, A., Pausas, J.G., Vendramin, G.G. and others (2014) In situ genetic association for serotiny, a fire-related trait, in Mediterranean maritime Pine (*Pinus pinaster*), *New Phytologist*, vol. 201, pp. 230–241.

Campbell, I.D., McDonald, K., Flannigan, M.D., Kringayark, J. (1999) Long distance transport of pollen into the Arctic, *Nature*, vol. 399, pp. 29–30.

Campos-Arceiz, A., Larrinaga, A.R., Weerasinghe, U.R., Takatsuki, S., Pastorini, J. and others (2008) Behavior rather than diet mediates seasonal differences in seed dispersal by Asian elephants, *Ecology*, vol. 89, pp. 2684–2691.

Carneiro, F.S., Lacerda, A.E.B., Lemes, M.R., Gribel, R., Kanashiro, M. and others (2011) Effects of selective logging on the mating system and pollen dispersal of *Hymenaea courbaril* L. (*Leguminosae*) in Eastern Brazilian Amazon as revealed by microsatellites markers. *Forest Ecology and Management*, vol. 262, pp. 1758–1765.

Clark, D. (2007) Detecting tropical forests' responses to global climatic and atmospheric change: current challenges and a way forward, *Biotropica*, vol. 39, pp. 4–10.

Cloutier, D., Kanashiro, M., Ciampi, A.Y., Schoen, D.J. (2007) Impact of selective logging on inbreeding and gene dispersal in an Amazonian tree population of *Carapa guianensis* Aubl., *Molecular Ecology*, vol.16, pp. 797–809.

Cirstobal-Pérez, E.J., Fuchs, E.J., Olivares-Pinto, U., Quesada, M. (2020) Janzen-Connell effects shape gene flow patterns and realized fitness in the tropical dioecious tree *Spondias purpurea* (*Anacardiaceae*), *Scientific Reports*, vol. 10, 4584. 10.1038/s41598-020-61394-4

Dayanandan, S., Dole, J., Bawa, K., Kessel, R. (1999) Population structure delineated with microsatellite markers in fragmented populations of a tropical tree, *Carapa Guianensis* (*Meliaceae*), *Molecular Ecology*, vol. 8, pp. 1585–1592.

de Lafontaine, G., Turgeon, J., Payette, S. (2010) Phylogeography of White Spruce (*Picea glauca*) in eastern North America reveals contrasting ecological trajectories, *Journal of Biogeography*, vol. 37, pp. 741–751.

Demenou, B.B., Migliore, J., Heuerz, M., Monthe, F.K., Ojeda, D.II., and others (2020) Plastome phylogeography in two African rain forest legume trees reveals that Dahomey Gap populations originate from the Cameroon volcanic line, *Molecular Phylogeography and Evolution*, vol. 150, 106854.

Depardieu, C., Gérardi, S., Nadeau, S., Paren, J.G., Mackay, J. and others (2021) Connecting tree-ring phenotypes, genetic associations, and transcriptomics to decipher the genomic architecture of drought adaptation in a widespread conifer, *Molecular Ecology*, vol. 30, 3898–3917.

Dick, C.W. (2010) Phylogeography and Population Structure of Tropical Trees, *Tropical Plant Biology*. vol. 3, pp. 1–3.

Dick, C.W., Hardy, O.J., Jones, F.A., Petit, R.J. (2008) Spatial scales of pollen and seed-mediated gene flow in tropical rain forest trees, *Tropical Plant Biology*, vol.1, pp. 20–33.

Duminil, J. Mendene Abessolo, D.T., Ndlade Bourobou, D.T., Doucet, J.L., Loo, J., Hardy, O.J. (2016) High selfing rate, limited pollen dispersal and inbreeding depression in the emblematic African rain forest tree *Baillonella toxisperma* — Management implications, *Forest Ecology and Management,* vol. 379, pp. 20–29.

Fageria, M.S., Rajora, O.M. (2013) Effects of harvesting of increasing intensities on genetic diversity and population structure of White Spruce, *Evolutionary Applications*, vol. 6, pp. 778–794.

Finkeldey, R., Hattemer, H.H. (2006) Tropical Forest Genetics. Springer-Verlag Berlin Heidelberg New York.

Fuchs, E.J., Lobo, J.A., Quesada, M. (2003) Effects of Forest Fragmentation and Flowering Phenology on the Reproductive Success and Mating Patterns of the Tropical Dry Forest Tree *Pachira quinate*, *Conservation Biology*, vol.17, pp. 149–157.

Gamache, I., Jaramillo-Correa, J.P., Payette, S., Bousquet, J. (2003) Diverging patterns of mitochondrial and nuclear DNA diversity in subarctic black Spruce: imprint of a founder effect associated with postglacial colonization, *Molecular Ecology*, vol. 12, pp. 891–901.

Gauthier, S., Simon, J.P., Bergeron, Y. (1992) Genetic structure and variability in jack pine populations: effects of insularity, *Canadian Journal of Forest Research*, vol. 22, 1958–1965.

Graignic, N. (2014) Impact de la fragmentation sur la capacité reproductrice et la diversité génétique de l'érable à sucre (*Acer saccharum* Marshall) au Québec, PhD thesis, Institut de recherche sur les forêts, UQAT, Rouyn-Noranda, Québec.

Groover, A., Jansson, S. (2014) Comparative and Evolutionary Genomics of Forest Trees Challenges and Opportunities for the World's Forests in the 21st Century Forestry Sciences, Springer-Verlag Berlin Heidelberg New York.

Guo, X., Klisz, M., Puchalka, R., Silvestro, R., Faubert, P., and others (2021) Common-garden experiment reveals clinal trends of bud phenology in black spruce populations from a latitudinal gradient in the boreal forest, *Journal of Ecology*, vol. 110, pp. 1043–1053.

Hamrick, J.L. (1994) Genetic diversity and conservation in tropical forests. Proceedings of the International Symposium on Genetic Conservation and Production of Tropical Forest Tree Seed, ASEAN-Canada Forest Tree Seed Centre Project, Muak-Lek, Saraburi, Thailand, pp. 1–9.

Hamrick, J.L. (2004) Response of forest trees to global environmental changes, *Forest Ecology and Management*, vol. 197, pp. 323–335.

Hamrick, J.L., Nason, D. (2000) Gene flow in forest trees, CSIRO Publishing, Collingwood, Australia

Holbrook, K.M., Smith, T.B. (2000) Seed dispersal and movement patterns in two species of *Ceratogym nahornbills* in a West African tropical lowland forest, *Oecologia,* vol 125, pp. 249–257.

Holliday, J.A., Ritland, K., Aitken, S.N. (2010) Widespread, ecologically relevant genetic markers developed from association mapping of climate-related traits in Sitka spruce (*Picea sitchensis*), *New Phytologist*, vol. 188, 501–514.

Honorio Coronado, E.N., Dexter, K.G., Hart, M.L., Phillips, O.L., Pennington, RT. (2019) Comparative phylogeography of five widespread tree species: Insights into the history of western Amazonia, *Ecology and Evolution*, vol.9, pp. 7333–7345.

Isabel, N., Holliday, J. A., Aitken, S. N. (2020) Forest genomics: advancing climate adaptation, forest health, productivity, and conservation. *Evolutionary Applications*, vol. 13, pp. 3–10.

Jaramillo-Correa, J.P., Beaulieu, J., Bousquet, J. (2001) Contrasting evolutionary forces driving population structure at expressed sequence tag polymorphisms, allozymes and quantitative traits in White Spruce, *Molecular Ecology*, vol. 10, pp. 2729–2740.

Jordano, P., Godoy, J.A. (2002) Frugivore-generated seed shadows: a landscape view of demographic and genetic effects, CAB International, Wallingford, UK.

Keller, S.R., Olson, M.S., Silim, S., Schroeder, W.R., Tiffin, P. (2010) Genomic diversity, population structure, and migration following range expansion in the Balsam Poplar, *Populus balsamifera*, *Molecular Ecology*, vol. 19, pp. 1212–1226.

Kettle, C.J., Hollingsworth, P.M., Burslem, D.F.R.P., Maycock, C.R., Khoo, E., Ghazoul, J. (2010) Determinants of fine-scale spatial genetic structure in three co-occurring rainforest canopy trees in Borneo, *Perspectives in Plant Ecology,* Evolution and Systematics, vol. 13, pp. 47–56.

Kinloch, B.B. Jr (1992) Distribution and frequency of a gene for resistance to white pine blister rust in natural populations of sugar pine, *Canadian Journal of Botany*, vol. 70, pp. 1319–1323.

Klisz, M., Chakraborty, D., Cvjerković. B., Grabner, M., Lintunen A., and others (2023) Functional Traits of Boreal Species and Adaptation to Local Conditions. Boreal Forests in the Face of Climate Changes, vol 74, Springer, Cham. https://doi.org/10.1007/978-3-031-15988-6_12

Kremer, A., Le Corre, V. (2012) Decoupling of differentiation between traits and their underlying genes in response to divergent selection, *Heredity*, vol. 108 pp. 375–385.

Kremer, A., Ronce, O., Robledo-Arnuncio, J.J., Guillaume, F., Bohrer, G. and others (2012) Long-distance gene flow and adaptation of forest trees to rapid climate change, *Ecology Letters*, vol. 15, pp. 378–392.

Lanfear R., Ho, S.M.Y., Davies, T.J., Moles, A.T., Aarssen, L. and others (2013) Taller plants have lower rate of evolution, *Nature Communications*, vol.4, pp. 1–7.

Latutrie, M., Bergeron, Y., Tremblay, F. (2016) Fine-scale assessment of genetic diversity of trembling aspen in northwestern North America, *BMC Evolutionary Biology*, vol. 16, p. 231. https://doi.org/10.1186/s12862-016-0810-1

Latutrie, M., Tóth, E.G. Bergeron, Y., Tremblay, F. (2019) Novel insights into the genetic diversity and clonal structure of natural trembling aspen (*Populus tremuloides* Michx.) populations: A transcontinental study, *Journal of Biogeography*, vol. 46, pp. 1124–1137.

Laverdière, J.-P., Lenz, P., Nadeau, S., Depardieu, C., Isabel, N. and others (2022) Breeding for adaptation to climate change: genomic selection for drought response in a white spruce multi-site polycross test, *Evolutionary Applications*, vol. 15, pp. 383–402. doi:10.111/eva.13348

Leites, L., Garzón, M.B. (2023) Forest tree species adaptation to climate across biomes: Building on the legacy of ecological genetics to anticipate responses to climate change, *Global Change Biology*, vol. 29, pp. 4711–4730.

Lemes, M.R., Girbel, R., Proctor, J., Gratapaglia, D. (2003) Population genetic structure of mahogany (*Swietenia macrophylla* King, *Meliaceae*) across the Brazilian Amazon, based on variation at microsatellite loci: implications for conservation, *Molecular Ecology*, vol. 12, pp. 2875–2883.

Lemes, M.R., Dick, C.W., Navarro, C., Lowe, A.J., Cavers, S., Gribel, R. (2010) Chloroplast DNA Microsatellites Reveal Contrasting Phylogeographic Structure in Mahogany (*Swietenia macrophylla* King, *Meliaceae*) from Amazonia and Central America, *Tropical Plant Biology*, vol. 3, pp. 40–49.

Lesser, M.R., Parchman, T.L., Jacksons, S.T. (2013) Development of genetic diversity, differentiation and structure over 500 years in four ponderosa pine populations, *Molecular Ecology*, vol. 22 pp. 2640–2652.

Lowe, A.J., Harris, D., Dormontt, E., Dawson, I.K. (2010) Testing Putative African Tropical Forest Refugia Using Chloroplast and Nuclear DNA Phylogeography, *Tropical Plant Biology*, vol. 3, pp. 50–58.

MacDonald, S.E., Thomas, B.R., Cherniawsky, D.M., Pudy, B.G. (2001) Managing genetic resources of lodgepole Pine in west central. Alberta: patterns of isozyme variation in natural populations and effects of forest management, *Forest Ecology and Management*, vol.152, pp. 45–58.

McLean, E.H., Prober, S.M., Stock, W.D., Steane, D.A., Potts, B.M., and others (2014) Plasticity of functional traits varies clinally along a rainfall gradient in *Eucalyptus trica*rpa, *Plant, Cell and Environment*, vol. 37, pp. 1440–1451.

Marquardt, P.E., Epperson, B.K. (2004) Spatial and population genetic structure of microsatellites in white pine, *Molecular Ecology*, vol. 13, pp. 3305–3315.

Millar, C.I., Stephenson, N.L., Stephens, S.L. (2007) Climate change and forests of the future: Managing in the face of uncertainty, *Ecological Applications*, vol. 17, pp. 2145–2151.

Ming, R., Hou, S., Feng, Y. (2008) The draft genome of the transgenic tropical fruit tree papaya (*Carica papaya* Linnaeus), *Nature*, vol. 452, pp. 991–996.

Misiewicz, T.M., Fine, P.V.A. (2014) Evidence for ecological divergence across a mosaic of soil types in an Amazonian tropical tree: *Protium subserratum* (*Burseraceae*), *Molecular Ecology*, vol. 23, pp. 2543–2558.

Montwé, D., Isaac-Renton, M., Hamann, A., Spiecker, H. (2018) Cold adaptation recorded in tree rings highlights risks associated with climate change and assisted migration, *Nature Communications*, vol. 9. doi: 10.1038/s41467-018-04039-5

Morgenstern, E.K. (1996) Geographic Variation in Forest Trees: Genetic Basis and Application of Knowledge in Silviculture. University of British Columbia Press, Vancouver, BC, Canada.

Mosseler, A., Egger, K.N., Hughes, G.A. (1992) Low levels of genetic diversity in Red Pine confirmed by random amplified polymorphic DNA markers, *Canadian Journal of Forest Research*, vol. 22, pp. 1332–1337.

Mullin, T.J, Andersson, B., Bastien, J.C, Beaulieu, J., Burdon, R.D., and others (2011) Economic importance, breeding objectives and achievements, Chapter 2. In: Plomion C, Bousquet J, Kole C (eds), Genetics, Genomics and Breeding of Conifers. CRC Press and Science Publishers, New York, pp. 40–127.

Mura, C., Buttò, V., Silvestro, R., Deslauriers, A., Charrier, G. and others (2022) The early bud gets the cold: Diverging spring phenology drives exposure to late frost in a *Picea mariana* [(Mill.) BSP] common garden, *Physiologia Plantarum*, vol. 174, e13798. doi: 10.1111/ppl.13798

Musch, B., Valadon, A., Oddou-Mutatorio, S. (2004) A propos de génétique des populations. Rendez-vous techniques de l'ONF, hors-série n° 1, Diversité génétique des arbres forestiers, pp. 6–15.

Myburg, A., Grattapaglia, D., Tuskan, G.A., Hellsten, U., Hayes, R.D. and others (2014) The genome of Eucalyptus grandis, *Nature*, vol. 510, pp. 356–362.

Nason, J.D., Herre, E.A., Hamrick, J.L. (1998) The breeding structure of a tropical keystone plant species, *Nature*, vol 391, pp. 685–687.

Neale, D.B., Kremer, A. (2011) Forest tree genomics: growing resources and applications, *Nature Reviews Genetics*, vol.12, pp.111–122.

Nystedt, B., Street, N., Wetterbom, A., Zuccolo, A., Lin, Y.-C. and others (2013) The Norway Spruce genome sequence and conifer genome evolution, *Nature*, vol. 497, pp. 579–584.

O'Connell, L.M., Mosseler, A., Rajora, O.P. (2006) Impacts of forest fragmentation on the mating system and genetic diversity of white spruce (*Picea glauca*) at the landscape level, *Heredity*, vol. 97, 418-426.

Padovan, A., Kesze, A., Foley, W.J., Külheim, C. (2013) Differences in gene expression within a striking phenotypic mosaic *Eucalyptus* tree that varies in susceptibility to herbivory, *BMC Plant Biology*, vol 13, p. 29. https://doi.org/10.1186/1471-2229-13-29

Palma-Silva, C., Lexer, C., Paggi, G.M., Barbara, T., Bered, F., Bodanese-Zanettini, M.H. (2009) Range-wide patterns of nuclear and chloroplast DNA diversity in *Vriesea gigantea* (*Bromeliaceae*), a neotropical forest species, *Heredity*, vol.103, pp. 503–512.

Pandey, M., Rajora, O.P. (2012b) Genetic diversity and differentiation of core vs. peripheral populations of eastern white cedar, *Thuja occidentalis* (*Cupressaceae*), *American Journal of Botany*, vol. 99, pp. 690–699.

Pedlar, J.H., Mckenney, D.W., Lu, P. (2021) Critical seed transfer distances for selected tree species in eastern North America. *Journal of Ecology*; vol. 109, pp. 2271–2283.

Peng, C., Ma, Z., Lei, X., Zhu, H., Chen, W. and others (2011) A drought-induced pervasive increase in tree mortality across Canada's boreal forests, *Nature Climate Change*, vol.1, pp. 467–471.

Pereira, H.M., Leadley, P.W., Proenca, V., Alkemade, R., Scharlemann, J.P. and others (2010) Scenarios for global biodiversity in the 21st century, *Science*, vol. 330, pp. 1496–1501.

Rachmat, H.H., Kamiya, K., Harada, K. (2012) Genetic diversity, population structure and conservation implication of the endemic Sumatran lowland dipterocarp tree species (*Shorea javanica*), *International Journal of Biodiversity and Conservation*, vol. 4, pp. 573–583.

Raj, S., Bräutigama, K., Hamanishi, E.T., Wilkins, O., Thomas, B.R. and others (2011) Clone history shapes *Populus* drought responses, *Proceedings of the National Academy of Sciences*, vol. 108, pp. 12521–12526.

Rajora, O.P., Rahman, M.H., Buchert, G.P., Dancik, B.P. (2000) Microsatellite DNA analysis of genetic effects of harvesting in old-growth eastern white pine (*Pinus strobus*) in Ontario, Canada, *Molecular Ecology*, vol. 9, pp. 339–348.

Rehfeldt, G.E. (1983) Ecological adaptations in Douglas-Fir (*Pseudotsuga menziesii* var. *Glauca*) populations. III. Central Idaho, *Canadian Journal of Forest Research*, vol. 13, pp. 626–632.

Saez-Laguna, E., Guvera, M.A., Diaz, L.M., Sánchez-Gómez, D., Collada, C. and others (2014) Epigenetic variability in the genetically uniform forest tree species *Pinus pinea* L *PLoS One*, vol. 9, e103145. doi:10.1371/journal.pone.0103145

Savolainen, O., Pyhäjärvi, T., Knürr, T. (2007) Gene flow and local adaptation in trees, *Annual Review of Ecology, Evolution and Systematics*, vol. 38, pp. 595–619.

Schroeder, J. W., Trana, H. T., Dick, C. W. (2014) Fine scale spatial genetic structure in *Pouteria reticulata* (Engl.) *Eyma* (*Sapotaceae*), a dioecious, vertebrate dispersed tropical rain forest tree species. *Global Ecology and Conservation*, vol. 1, pp. 43–49.

Schoettle, A.W., Sniezko, R.A., Kegley, A., Burns, K.S. (2014) White Pine Blister Rust Resistance in Limber Pine: Evidence for a Major Gene, *Phytopathology*, vol. 104, pp.163–173.

Slavov, G.T., Leornardi, S., Burczyck, J., Adams, W.T., Strauss, S.H., Di Fazio, S.P. (2009) Extensive pollen flow in two ecologically contrasting populations of *Populus trichocarpa*, *Molecular Ecology*, vol. 18, pp. 357–373.

Solvin, T.M. (2020) Quantification of genetic and epigenetic variation in phenology and growth in Norway spruce. Philosophiae Doctor Thesis 2020:39. Norwegian University of Life Science.

Turchetto-Zolet, A., Pinheiro, F., Salgueiro, F., Palma-Silva, C. (2013) Phylogeographical patterns shed light on evolutionary process in South America, *Molecular Ecology*, vol. 22, pp. 1193–1213.

Tuskan, G.A., DiFazio, S., Jansson, S., Bohlmann, J., Grigoriev, I. and others (2006) The genome of Black Cottonwood, *Populus trichocarpa* (Torr. and Gray), *Science*, vol. 313, pp. 1596–1604.

Varis, S., Pakkanen, A., Galofre, A., Pulkkinen, P. (2009) The extent of south north pollen transfer in Finnish Scots pine. *Silva Fennica*, vol. 43, pp. 717–726.

Vieira, F. A., Fajardo, C. G., Souza, A. M., Reis, C. A. F, Carvalho, D. (2012) Fine-scale genetic dynamics of a dominant neotropical tree in the threatened Brazilian Atlantic Rainforest. *Tree Genetics and Genomes*, vol. 8, 1191–1201.

Vinson, C.C., Kanashiro, M., Sebbenn, A.M., Williams, T.C.R., Harris, S.A. and others (2014) Long-term impacts of selective logging on two Amazonian tree species with contrasting ecological and reproductive characteristics: inferences from Eco-gene model simulations, *Heredity*, vol. 115, pp. 130–139.

Waqar, Z., Moraes, R.C.S., Benchimol, M., Morante-Filho, J.C., Mariano-Neto, E., Gaiotto, F.A. (2021) Gene Flow and Genetic Structure Reveal Reduced Diversity between Generations of a Tropical Tree, *Manilkara multifida* Penn., in Atlantic Forest Fragments, *Genes*, vol.12, 2025. doi: 10.3390/genes12122025

Weber, J.C., Larwanou, M., Abasse, T.G., Kalinganire, A. (2008) Growth and survival of *Prosopis africana* provenances tested in Niger and related to rainfall gradients in the West African Sahel, *Forest Ecology and Management*, vol. 256, pp. 585–592.

White, G.M., Boshier, D.H., Powell, W. (2002) Increased pollen flow counteracts fragmentation in a tropical dry forest: an example from *Swietenia humilis Zuccarini, Proceedings of the National Academy of Sciences* USA, vol. 99, pp. 2038–2042.

Wyman, J., Bruneau, A., Tremblay, F. (2003) Microsatellite analysis of genetic diversity in four populations of *Populus tremuloïdes* in Quebec, *Canadian Journal of Botany*, vol. 81, pp. 350–367

Ying, L., Morgenstern, E.K. (1991) The population structure of *Larix laricina* in New Brunswick, Canada, *Silvae Genetica*, vol 40, pp. 5–8.

Xu, H., Tremblay, F., Bergeron, Y., Paul, V., Chen, C. (2012) Genetic consequences of fragmentation in the "arbor vitae ", the Eastern White Cedar (*Thuja occidentalis* L.), towards the northern limit of its distribution range, *Ecology and Evolution*, vol. 2, pp. 2506–2520.

Glossary

Adaptive divergence: the differentiation among the mean phenotypes of populations subject to different selective pressures.

Adaptive trait: a phenotypic trait that enhances the fitness of an individual in a particular environment. Examples for trees are the time of flushing and bud set, and physiological traits determining water use efficiency.

Allele: One of the different forms of a gene that can exist at a single locus.

Allelic richness (AR): The total number of alleles in a population.

Bud set: formation of a terminal bud at the end of the vegetative period, the timing of which is heritable and determines cold tolerance in boreal and temperate trees.

Epigenetic: all inheritable changes in gene expression that are produced without any changes in the DNA sequence. Differences in gene expression affect the functioning of traits coded by these genes and produce different phenotypes.

Expected heterozygosity (H_e): the most frequently used measure of genetic diversity within populations. It is based on the assumption that alleles are randomly combined to genotypes.

Fixation index (F_{st} or G_{st}): the proportion of the genetic variance among populations relative to the total variance.

Hardy–Weinberg equilibrium: A principle stating that both allele and genotype frequencies in a randomly mating population remain constant and remain in this equilibrium across generations unless a disturbing influence is introduced.

Heterosis: the higher fitness of progeny obtained through crosses between populations rather than within the same population.

Inbreeding coefficient (F_{is}): the mean reduction in heterozygosity of an individual due to non-random mating within a subpopulation, that is, a measure of the extent of genetic inbreeding within populations that range from −1.0 (all individuals heterozygous) to +1.0 (no observed heterozygotes)

Inbreeding depression: reduced fitness of inbred individuals

Migration load: the contribution of immigrant genes to genetic load (see genetic load)

Observed heterozygosity (H_o): the proportion of all heterozygotes from all individuals (homozygotes and heterozygotes) investigated in a population.

Parentage analysis: the probabilistic determination of the parents of an individual, frequently using genetic markers.

Phenotypic cline: a continuous change of a phenotypic trait along an environmental and/or geographical gradient.

Phenotypic plasticity: the capacity of a genotype to produce distinct phenotypes.

Provenance: the original geographic source of a population or group of individuals (used also to refer to such a population or group).

Provenance test: a common garden experiment, in one or more locations, where the genetic variation of different provenances is evaluated (see provenance).

Mean quantitative genetic differentiation (Q_{ST}) assesses the proportion of total genetic variation in a quantitative trait that is due to differences among populations.

12
MONODOMINANCE IN TROPICAL LOWLAND FORESTS

Kelvin S.-H. Peh

Introduction

Tropical lowland forests exhibit remarkable richness in plant diversity, not only when viewed from a regional perspective but also within single localities (Hubbel, 2004). However, it is imperative to dispel the general notion that all tropical forests uniformly exhibit high alpha-diversity. On the contrary, extensive swathes of tropical forests are characterized by the dominance of a solitary species in terms of the proportion of total canopy trees. Early literature has documented numerous instances of this phenomenon, known as monodominance, occurring across all major tropical regions worldwide (Table 12.1). It is worth noting that in several of these monodominant forests, the dominant species can constitute over 80 per cent of the total canopy trees (Hart et al.,1989). Nevertheless, the conventional criterion for defining monodominance requires that at least 60 per cent of canopy-level trees belong to the same species (Torti et al., 2001). This different aspect of tropical forest diversity challenges conventional perceptions and highlights the complexity inherent in the structure and composition of these ecosystems.

Monodominance in tropical lowland forests has garnered significant attention among tropical ecologists due to its relevance to the fundamental goals of ecological research, which include understanding species diversity and distribution patterns. The phenomenon of monodominant forests, where a single species dominates the canopy, challenges the typical expectation of high species richness in tropical lowland forests. This concept was initially observed by Van Zon (1915), who noted the prevalence of *Drynobalanops aromatica* as a monodominant species, often juxtaposed with mixed-species-rich forests. Pioneering studies by Hart et al. (1989) and Connell and Lowman (1989) ventured into explaining the co-occurrence of these contrasting forest types. Hart et al. (1989) postulated that monodominant forests may arise from prolonged periods of minimal large-scale disturbances. Connell and Lowman (1989), on the other hand, demonstrated that as one canopy species becomes more dominant within their study plots, overall canopy tree diversity decreases. However, the total number of species within monodominant forests remains relatively constant compared to their mixed-species counterparts. Their research suggested that monodominance could result from an ectomycorrhizal (EM) association among the dominant tree species. This

DOI: 10.4324/9781003324072-14

Table 12.1 Diversity of monodominant species in tropical lowland forest

Type	*Distribution*	*Family*	*Name*	*Reference*
Classical	Africa	Burseraceae	*Aucoumea klaineana* Pierre	Maisels (2004)
	Africa	Fabaceae	*Cynometra alexandri* C.H.Wright	Hart et al. (1989)
	Africa	Fabaceae	*Gilbertiodendron dewevrei* (De Wild.) J.Léonard	Conway (1992)
	Africa	Fabaceae	*Microberlinia bisulcata* A.Chev.	Newbery et al. (2004)
	Africa	Fabaceae	*Talbotiella gentii* Hutchinson & Greenway	Richards (1996)
	Africa	Fabaceae	*Tetraberlinia tubmaniana* J.Léonard	Connell and Lowman (1989)
	Asia	Dipterocarpaceae	*Dryobalanops aromatica* C.F.Gaertn.	Ithoh (1995)
	Asia	Dipterocarpaceae	*Parashorea chinensis* H. Wang	Van der Velden et al. (2014)
	Asia	Dipterocarpaceae	*Parashorea malaanonan* Merr.	Richards (1996)
	Asia	Dipterocarpaceae	*Shorea curtisii* Dyer ex King	Grubb et al. (1994)
	Asia	Lauraceae	*Eusideroxylon zwageri* Teijsm. & Binn.	Richards (1996)
	Neotropics	Apocynaceae	*Aspidosperma excelsum* Benth.	Richards (1996)
	Neotropics	Burseraceae	*Dacryodes excelsa* Vahl	Richards (1996)
	Neotropics	Euphorbiaceae	*Celaenodendron mexicanum* Standl.	Martijena (1998)
	Neotropics	Fabaceae	*Dicymbe corymbosa* Spruce ex Benth.	Henkel et al. (2005)
	Neotropics	Fabaceae	*Mora gonggrijpii (Kleinh.)* Sandwith	Connell and Lowman (1989)
	Neotropics	Fabaceae	*Mora oleifera* Ducke	Holdridge et al. (1971)
	Neotropics	Fabaceae	*Pentaclethra macroloba* Kuntze	Connell and Lowman (1989)
	Neotropics	Fagaceae	*Quercus oleoides* Schltdl. & Cham.	Boucher (1981)
	Neotropics	Moraceae	*Brosimum rubescens* Taub.	Marimon et al. (2001)

(*Continued*)

Table 12.1 (Continued)

Type	*Distribution*	*Family*	*Name*	*Reference*
Low-nutrient	Asia	Theaceae	*Adinandra dumosa* S. Vidal	Sim et al. (1992)
	Neotropics	Fabaceae	*Dimorphandra conjugata* Sandwith	Richards (1996)
	Neotropics	Fabaceae	*Dimorphandra hohenkerkii* Sprague & Sandwith	Richards (1996)
	Neotropics	Fabaceae	*Eperua falcata* Aubl.	Forget (1989)
	Neotropics	Fabaceae	*Eperua leucantha* Benth.	Richard (1996)
	Neotropics	Fabaceae	*Eperua obtusata* Cowan	Coomes and Grubb (1996)
	Neotropics	Fabaceae	*Peltogyne gracilipes* Ducke	Villela and Proctor (2002)
	Neotropics	Sapotaceae	*Manilkara bidentata* (A.DC.) A.Chev.	Richards (1996)
Successional	Africa	Fabaceae	*Guibourtia demeusei* (Harms) J.Léonard	Hughes and Hughes (1993)
	Africa	Moraceae	*Musanga cecropioides* R.Br.	Connell and Lowman (1989)
	Africa	Rhamnaceae	*Maesopsis eminii* Engl.	Eggeling (1947)
	Africa	Rubiaceae	*Mitragyna stipulosa* (DC.) Kuntze	Hughes and Hughes (1992)
	Africa, Asia	Euphorbiaceae	*Macaranga spp.*	Richards (1996)
	Africa, Neotropics	Bombacaceae	*Ochroma spp.*	Richards (1996)
	Asia	Dipterocarpaceae	*Shorea albida* Symington ex A.V.Thomas	Anderson (1961, 1964)
	Asia	Dipterocarpaceae	*Shorea parvifolia* Dyer	Whitmore (1984)
	Asia	Euphorbiaceae	*Mallotus spp.*	Richards (1996)
	Neotropics	Asteraceae	*Moquiniastrum polymorphum* (Less.) G. Sancho	Figueiredo et al. (2022)
	Neotropics	Cecropiaceae	*Cecropia latiloba* Miq.	Parolin et al. (2002)
	Neotropics	Cecropiaceae	*Cecropia mexicana* Hemsl.	Richards (1996)
	Neotropics	Cecropiaceae	*Cecropia sciadophylla* Mart.	Mesquita et al. (1998)
	Neotropics	Euphorbiaceae	*Alchornea castaneifolia* (Humb. & Bonpl. Ex Willd.) A.Juss.	Daly and Mitchell (2000)
	Neotropics	Fabaceae	*Macrolobium acaciifolium* Benth.	Schöngart et al. (2005)

Table 12.1 (Continued)

Type	*Distribution*	*Family*	*Name*	*Reference*
	Neotropics	Fabaceae	*Senna reticulata* (Willd.) H.S.Irwin & Barneby	Parolin et al. (2002)
	Neotropics	Salicaceae	*Salix humboldtiana* Willd.	Parolin et al. (2002)
	Neotropics	Sapindaceae	*Allophylus edulis* Niederl.	Richards (1996)
	Oceania	Apocynaceae	*Cerberiopsis candelabra* var. *candelabra*	Read et al. (2018)
	Oceania	Casuarinaceae	*Casuarina aff. cunninghamiana*	Whitmore (1984)
	Oceania	Casuarinaceae	*Casuarina papuana* S. Moore	Connell and Lowman (1989)
	Oceania	Cunoniaceae	*Codia mackeeana*	Read et al. (2018)
	Oceania	Datiscaeae	*Octomeles sumatranus* Miquel	Richards (1996)
	Oceania	Dipterocarpaceae	*Anisoptera thurifera* Blume	Whitmore (1984)
	Oceania	Dipterocarpaceae	*Anisoptera polyandra* Blume	Connell and Lowman (1989)
	Oceania	Myrtaceae	*Arillastrum gummiferum*	Read et al. (2018)
	Oceania	Myrtaceae	*Eucalyptus deglupta* Blume	Richards (1996)
	Oceania	Myrtaceae	*Metrosideros polymorpha* J.R.Forst. Ex Hook.f.	Mueller-Dombois (2000)
	Oceania	Nothofagaceae	*Nothofagus aequilateralis* (Baum.-Bodenh.) Steenis	Read et al. (2006)
	Oceania	Nothofagaceae	*Nothofagus balanse*	Read et al. (2018)
	Oceania	Nothofagaceae	*Nothofagus codonandra* (Baill.) Steenis	Read et al. (2006)
	Oceania	Nothofagaceae	*Nothofagus discoidea*	Read et al. (2018)
	Pantropical	Ulmaceae	*Trema spp.*	Richards (1996)
	Tropical Australia	Myrtaceae	*Backhousia bancroftii* F.M.Bailey	Connell and Lowman (1989)
Water-logged	Africa	Rubiaceae	*Mitragyna stipulosa* (DC.) Kuntze	Richards (1996)
	Neotropics	Fabaceae	*Mora excelsa* Benth.	Beard (1946)

(*Continued*)

Table 12.1 (Continued)

Type	*Distribution*	*Family*	*Name*	*Reference*
	Neotropics	Fabaceae	*Prioria copaifera* Griseb.	Holdridge et al. (1971)
	Neotropics	Fabaceae	*Pterocarpus officinalis* Jacq.	Janzen (1978)

Note: This list is not exhaustive.

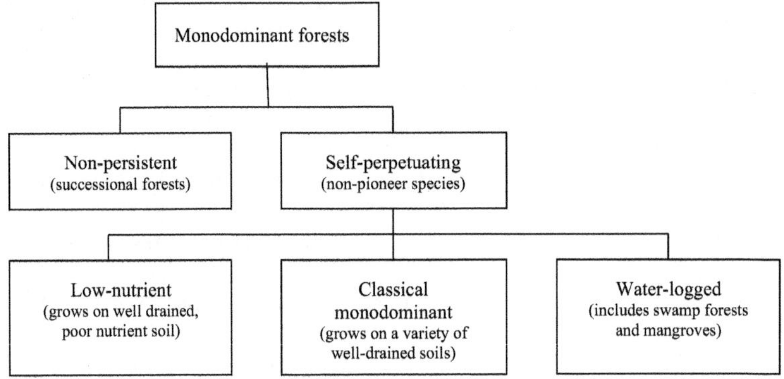

Figure 12.1 Types of monodominance in tropical lowland rain forests.

association may confer various advantages, such as enhanced nutrient acquisition and protection from pathogens, thereby facilitating monodominance.

Building upon the works of Connell and Lowman (1989) and Hart et al. (1989), an increasing number of observational studies have examined the ecological hypotheses presented to elucidate the enigmatic nature of monodominant forests (e.g., Torti et al., 2001). These investigations have aimed to provide a comprehensive understanding of the factors contributing to the prevalence of monodominance in tropical lowland forests. While numerous life-history traits of the dominant species have been hypothesized to enhance their monodominance, it is essential to note the absence of a consistent consensus regarding the mechanisms at play and the empirical evidence supporting each hypothesis.

In this context, this chapter critically evaluates these theoretical mechanisms considering the available evidence derived from both experimental and observational studies. The objective is to reconcile and synthesize existing knowledge, aiming to bridge the gap between theoretical conjectures and empirical observations. To further our understanding of these intricate ecological systems, a probabilistic framework is proposed. This framework seeks to provide a more robust and comprehensive perspective on the underlying dynamics driving monodominance in tropical lowland forests.

Diversity of monodominance in tropical lowland forests

Monodominant forests exhibit a distinct dichotomy, classifying them into two fundamental categories: non-persistent and persistent, as illustrated in Figure 12.1. Within the realm of persistent monodominant forests, a nuanced subdivision emerges, delineating them into three principal types: water-logged forests, low-nutrient forests, and classical monodominant forests. This chapter primarily focuses on the classical monodominant forests, which are defined as those flourishing under conditions where high-diversity forests are usually found. Nevertheless, it is essential to provide a comprehensive overview of monodominance within tropical lowland forests, encompassing the entire spectrum of variations within this intriguing ecological phenomenon.

Successional forests

The concept of non-persistent dominance within monodominant forest ecosystems is inherently characterized by its inability to self-perpetuate over prolonged periods beneath the primary dominant canopy. In essence, non-persistent dominance manifests for only a single generation or, at most, a few consecutive generations (Connell and Lowman, 1989). However, certain monodominant pioneer species, exemplified by *Metrosideros polymorpha* (Mueller-Dombois, 2000), exhibit a relatively longer temporal persistence, surpassing the typical single-generation duration. Other examples include long-lived, relatively shade-intolerant species such as *Nothofagus* spp., *Arillastrum gummiferum*, *Cerberiopsis candelabra*, and *Codia mackeeana* from New Caledonia (Read et al., 2018). These exceptional cases represent a transitional phase where non-persistence eventually gives way to a regressive decline. Such examples are still classified as non-persistent, highlighting the dynamic and complex nature of monodominance within tropical lowland forests.

The phenomenon of short-term dominance is often associated with secondary forest succession (Hart et al., 1989). This form of dominance typically arises during the regenerative process following anthropogenic clearance or large-scale natural disturbances such as windstorms (Nelson et al., 1994). While the extent of non-persistent monodominant forest cover remains absent in the literature, it is imperative to recognize the expanding area of secondary forest succession in tropical regions. Notably, secondary forest areas were estimated to increase at a considerable rate of 1.06×10^6 ha annually (Wright, 2005). Within these secondary forests, dominant species can comprise a substantial portion, ranging from 68 to 89 per cent of the total number of canopy trees (Connell and Lowman, 1989).

Non-persistent monodominant forests, which hold significant spatial and ecological importance, are notably prevalent in the Neotropics. A notable example of such forests can be found in seasonal floodplain ecosystems, which comprise approximately 3–4 per cent of the Amazon basin. Remarkably, a substantial 30 per cent of these floodplain forests are characterized by monodominance (Butler, 2024). Within these ecosystems, pioneer species like *Ceropia* spp., *Salix humboldtiana*, and *Senna reticulata* play a prominent role in establishing monodominant forests on newly formed sites, primarily shaped by the sedimentation processes driven by river dynamics, representing a form of primary succession (Parolin et al., 2002).

Floodplain forests also feature prominently in the African landscape, notably in the "cuvette centrale" region, where the Congo Basin presents a low-lying and flat terrain. In this context, monodominant species exhibit distinct preferences based on soil type and moisture levels. For muddy soils with slow-flowing water, *Mitragyna stipulosa* emerges as the dominant species, while on drier sandy soils, *Guibourtia demeusei* takes precedence (Hughes and Hughes, 1992).

Species with a suite of life-history traits that confer upon them the ability to effectively colonize an area following perturbations are often associated with the formation of non-persistent dominant stands. This is primarily due to the inherent limitations of these traits, which typically do not support self-regeneration under the shade of their own canopy (Connell and Lowman, 1989). Among the most conspicuous life-history traits exhibited by these tree species are those facilitating extensive establishment. These include traits such as tolerance to full sunlight, resistance to desiccation, higher nutrient-uptake and nutrient-use efficiency, rapid growth rates, favourable seed dispersal mechanisms, specific mycorrhizal associations, and an innate capacity to alter soil chemistry, all of which are identified as key factors in the ability of these species to establish dominance within their ecological niches (Connell and Lowman, 1989; Read et al., 2018). Additionally, abiotic factors, including edaphic conditions, have been observed to exert a significant influence on the composition of successional communities, further shaping the dynamics of single dominance (Sim et al., 1992).

In the realm of successional forests, those categorized as non-persistent, certain species can sustain monodominance over more than one generation, in the context of reduced successional dynamics resulting from nutrient-poor conditions. This phenomenon is particularly salient in the nutrient-deprived "igapó" forests of the Amazon basin, characterized by recurrent and prolonged inundation due to floods. However, the mechanisms underpinning monodominance among some other successional species are notably more complex, prompting critical questions about the role of their life-history traits in achieving and sustaining this dominance. For example, the association between long lifespan and slow growth exhibited by species like *Macrolobium acaciifolium* is of profound ecological significance in maintaining monodominant stands within unfavourable sites (Schöngart et al., 2005). The longevity of these tree species implies extended intervals between regeneration opportunities, typically occurring only in exceptional years, often triggered by catastrophic events such as droughts (Wittmann and Junk, 2003). In this context, the combination of long-lived individuals thriving in adverse conditions and the periodic regeneration opportunities arising from climate anomalies, often associated with the El Niño phenomenon leading to severe floods or droughts, emerges as a pivotal mechanism driving successional monodominance in floodplains (Schöngart et al., 2005). This is particularly pertinent in regions where only a limited number of tree species possess the capacity to attain significant tree ages (Loehle, 1988). Thus, the interplay between life-history traits, ecological conditions, and climate anomalies assumes a paramount role in elucidating the complexities of successional monodominance in these dynamic floodplain ecosystems.

Water-logged forests

Persistent monodominant old-growth forests encompass several well-known types, with water-logged forests, including mangroves and swamp forests, occupying a prominent position within this category (Figure 12.1). These ecosystems can be exemplified by species such as *Mora excelsa*, which can comprise a substantial proportion, ranging from 63 to 84 per cent of the total number of canopy trees (Connell and Lowman, 1989). The enduring water-logged conditions characteristic of these forests are believed to exert a suppressive effect on diversity by conferring competitive advantages upon high-water tolerant species.

A defining feature of mangroves is the occurrence of pure stands of dominant species, often in zones aligned parallel to the shoreline. The specific species dominating a particular zone within mangroves is intricately linked to factors such as the extent of tidal flooding, salinity gradients, and soil texture (Richards, 1996). Certain mangrove species possess remarkable anatomical adaptations, including adventitious roots and aerenchyma, which enable them to surmount challenges associated with high sedimentation rates, consequently facilitating their existence in monospecific stands (Richards, 1996). Although the global estimate for monodominant swamp forest coverage remains unknown, the extent of mangrove cover alone in 2020 is estimated to be approximately 14.7×10^6 ha (United Nations Environment Programme, 2023).

Low-nutrient forests

Persistent monodominant stands extend to another distinctive category, represented by forests thriving on podzolized sandy soil or other coarse-grained substrates which are poor in nutrients, known as Amazonian caatinga in South America and health forests in Southeast Asia, hereafter referred to as low-nutrient forests (Figure 12.1). Remarkably, there is a notable absence of global estimates for the extent of low-nutrient forest coverage within the existing literature. Nevertheless, the prevalence of soils with exceedingly low fertility, such as spodosols and psamments, covers a substantial expanse of 109×10^6 ha in regions occupied by tropical forests (Vitousek and Sanford, 1986).

Within low-nutrient forests, a monodominant species, *Eperua falcata*, can constitute a significant proportion, encompassing at least 67 per cent, of the total number of canopy trees within specific locales like Moraballi in Guyana (Connell and Lowman, 1989). However, the Amazonian caatinga in Venezuela is distinctively characterized by the dominance of *Eperua obtusata* (Coomes and Grubb, 1996). These observations show the diverse manifestations of monodominance within low-nutrient forests.

Low-nutrient forests, while often adjacent to mixed forests, exhibit striking differences in plant species richness and composition, distinguishing them significantly from their more diversified counterparts (Richards, 1996). These forests are characterized by coarse-textured substrates with limited water-holding capacity, frequently experiencing sporadic drought events of varying intensity.

Brünig's (1974) drought hypothesis has been a significant framework for understanding low-nutrient forest dynamics. This hypothesis postulates that the poor water-retaining capacity of the soils is the primary driver of monodominance. It predicts that dominant species, characterized by traits such as small-sized leaves and light-coloured, shiny leaf surfaces, mitigate water loss during drought periods by directly or indirectly reducing evapotranspiration, thereby reducing the impact of water stress on the vegetation. However, this hypothesis faces substantial challenges, particularly from instances where low-nutrient forests in Borneo and South America do not encounter unusually dry conditions (Richards, 1996).

Consequently, attention has shifted significantly towards the nutrient deficiency hypothesis (Medina et al., 1990). This hypothesis contends that the poor nutrient content in the soils plays a pivotal role in shaping monodominance. In contrast to the drought hypothesis, the nutrient deficiency hypothesis underscores the significance of ectotrophic mycorrhizas, integral to the root systems of dominant trees in low-nutrient forests, as they facilitate nutrient absorption from litter to roots. This assertion is underpinned by the generally low base-exchange capacity and reduced availability of soil nitrogen (N) in low-nutrient forest soils due to slow organic matter decomposition (Medina et al. 1990). However, both the drought and nutrient deficiency hypotheses lack direct experimental support, relying predominantly on the adaptive physiognomy and floristic features of low-nutrient forest vegetation. Nonetheless, Coomes and Grubb (1998) have provided experimental evidence showcasing the capacity of certain low-nutrient forest species to thrive in poor soil conditions.

An alternative perspective posits that the dominance of a single species in lowland tropical forests may be attributed to the toxicity of relatively high concentrations of specific soil elements (e.g., Marimon et al., 2001; Villela and Proctor, 2002). This hypothesis suggests that certain species possess a unique tolerance to soil nutrient concentrations that would be toxic to most other plants, facilitating their dominance. However, such claims remain largely speculative. Currently, there is a lack of experimental evidence demonstrating that the capacity to tolerate these potentially toxic soil nutrient concentrations directly accounts for monodominance (Nascimento et al., 2007).

Classical monodominant forests

In addition to water-logged and low-nutrient forests, there exists another category of persistent monodominant forests, referred to as classical monodominant forests (Figure 12.1; Peh et al., 2011a). These forests thrive in environments seemingly similar to their adjacent high-diversity forests (i.e., mixed forests), without being primarily attributed to significant edaphic disparities or recent disturbances. Among the best-studied examples of classical monodominant forests is one dominated by *Gilbertiodendron dewevrei*, forming extensive stands on the central African plateau (Torti et al., 2001). Surprisingly, the fundamental biology of these forests remains relatively poorly understood despite their prevalence. While the global extent of classical monodominant forest cover is unknown, *G. dewevrei* is presumed to cover a larger area compared to other single-dominant species (Richards,1996). Notably, classical monodominant forests often coexist alongside mixed forests, marked by distinct boundaries, and can achieve total dominance of up to 100 per cent in terms of the number of canopy trees (Connell and Lowman, 1989). The overall species richness between the two forest types typically remains comparable, with the primary distinguishing factor being the presence or absence of classical monodominant species within these ecosystems (Connell and Lowman, 1989; van der Velden et al., 2014). However, *G. dewevrei* forest exhibited lower sample-controlled species richness, species density, and population density when considering stems with a diameter at breast height (dbh) of 10 cm or more (Djuikouo et al., 2014; Hall et al., 2019; Peh et al., 2014). There is evidence that a *G. dewevrei* forest is expanding into the mixed forests (Glick et al., 2021, 2023). Examples of classical monodominant species in Southeast Asia and the Neotropics include *Shorea albida* (Anderson, 1961), and *Peltogyne gracilipes* (Villela and Proctor, 2002) (Table 12.1).

Explaining classical monodominance

At least 21 tree species representing eight families are considered potentially capable of forming classical monodominant stands (Table 12.1). Intriguingly, the existence of classical monodominant stands in environments that outwardly resemble nearby high-diversity forests remains enigmatic yet to be fully unravelled. A prevalent assumption regarding classical monodominant forests is that they have undergone minimal or no significant disturbance over extended periods (Connell and Lowman, 1989; Hart et al. 1989). Conversely, an alternative viewpoint posits that a high frequency of small-area disturbances could lead to monodominance. In this scenario, these small gaps are readily colonized by previously suppressed offspring of dominant species (*sensu* Connell's intermediate disturbance theory; Newbery et al. 2004).

Torti et al. (2001) put forth a comprehensive set of life-history traits deemed essential for gaining recruitment advantages over other species, thereby facilitating the attainment of classical monodominance. These proposed plant traits include a high canopy density that casts deep shade, effectively outcompeting light-demanding species; slow leaf litter decomposition that results in the accumulation of leaf materials, acting as a physical barrier to prevent the establishment of small-seeded species; shade-tolerant saplings, enabling survival and growth within the shade created by parent trees; ballistic dispersal, promoting gregarious habits for replacing individuals of other species; and large seeds equipped with sufficient reserves to traverse deep litter layers, facilitating survival under low-light conditions.

This chapter reviews the extent to which each of these proposed mechanisms aligns with existing empirical evidence. Each existing hypothesis of prospective mechanisms to explain classical monodominance can be defined as either an evolutionary or an ecological explanation (see Peh et al. 2011a).

Research papers published since 1989 on classical monodominance predominantly concentrate on a small number of species, notably *G. dewevrei* and *Celaenodendron mexicanum*. In the process of assessing each potential explanation for classical monodominance, it is crucial to avoid interpreting discrepancies between a specific mechanism and the existing empirical evidence as a wholesale refutation of the mechanism's relevance. Rather, the existence of counterexamples may signify that a particular mechanism, in isolation, may not be adequate to account for all instances of classical monodominance in tropical lowland forests under non-extreme environmental conditions.

Mechanism 1: lack of exogenous disturbance

The infrequent occurrence of exogenous disturbances, resulting in reduced rates of gap formation and a subsequent low abundance of shade-intolerant species, can create a physical environment conducive to the establishment of certain monodominant species (Connell and Lowman, 1989; Hart et al., 1989). This notion aligns with research findings that in *Gilbertiodendron* forests, there were notably fewer and smaller canopy gaps compared to adjacent mixed forests, suggesting that treefall disturbances were less frequent in the monodominant forest (Hart et al. 1989). Moreover, the combination of slow growth rates and limited seed dispersal capabilities exhibited by many species that achieve monodominance implies that these communities have not been subjected to major disturbance events and are not simply older secondary forests (Connell and Lowman,1989). In fact, *G. dewevrei* forests existed over the past 2,700 years mostly with no evidence of fire—only minimal fire occurred over the last 300 years—lending support for the low-disturbance hypothesis (Tovar et al., 2017).

It is, however, worth noting that the structure and dynamics of a substantial cluster of *Microberlinia bisulcata* (a climax species but with light-demanding requirement) in central Africa hint at the possible role of periods characterized by multiple disturbance events (Newbury et al., 2004). This highlights the complexity of monodominance and its relationship with disturbance regimes.

Mechanism 2: shade tolerance under closed canopy

Several studies have provided circumstantial evidence regarding the relationship between seedling survival under the forest canopy and monodominance (e.g., Marimon et al., 2012; Hall et al., 2019). Hart (1995) reported that *G. dewevrei* seeds experienced higher mortality in the forest understorey but surviving *G. dewevrei* seedlings displayed greater persistence in the understorey. This persistence resulted in higher seedling density with increasing size and age when compared to *Brachystegia laurentii* and *J. seretii* from the mixed forest. However, given that *G. dewevrei, B. laurentii, and J. seretii* can all tolerate low-light environments and germinate and establish in the shade of their respective parent trees (Hart et al., 1989; Hart, 1995), it is unlikely that *G. dewevrei*'s deep canopy and shade-tolerant characteristics are the sole features contributing to greater seedling recruitment in *G. dewevrei* stands. Additionally, the understorey of a monodominant forest does not consistently exhibit lower light levels than that of the adjacent mixed forest, although this is likely the case for *G. dewevrei* stands (Vierling and Wessman, 2000).

The high plasticity of leaf functional traits may be an important factor contributing to the long-term persistence of some classical monodominant species, in the context of shaded environment. For example, *Brosimum rubescens* in monodominant stands exhibits leaf traits such as a higher specific leaf area, larger stomata size, greater maximum stomatal pore opening, and longer petioles compared to the same species in mixed forests (Araújo et al., 2022). These leaf functional adaptations allow individuals to maximize light capture in low-light environment, enhancing their acquisition and use of resources within monodominant stands (Araujo et al., 2022). However, integrating high plasticity in leaf functional traits as an explanation for classical monodominance excludes species such as *G. dewerei* that occurs exclusively in monodominant stands.

Mechanism 3: slow decomposition rate

The distribution of forest types is intricately intertwined with soil conditions, encompassing factors such as soil nutrient cycling, nutrient availability, the accumulation and release of nutrients, soil types, and drainage characteristics (Richards, 1996). In the context of monodominant forests, both exogenous soil conditions existing beneath these ecosystems and the inherent traits of monodominant species, representing endogenous conditions, can contribute to the phenomenon of slow litter decomposition (Torti et al., 2001). Torti et al. (2001) illustrates that the litter decomposition rate in monodominant *G. dewevrei* forests is notably slower, approximately two to three times, in comparison to mixed forests within central Africa (also see Peh et al. 2012), offering a suggestive link between slow decomposition and monodominance.

The reduced litter decomposition rate in single-species dominant stands could result in a gradual release of nutrients in these forests. There is empirical evidence substantiating significant differences in soil properties between classical monodominant forests and their adjacent mixed forests (e.g., *G. dewevrei*, Hall et al., 2019; *Parashorea chinensis*, van der Velden et al., 2014). In contrast, soil surveys conducted in the Congo Basin have revealed no disparities in soil pH, carbon (C), nitrogen (N), phosphorus (P), calcium (Ca), magnesium (Mg), and potassium (K) between *G. dewevrei* forests and their mixed counterparts (Conway,1992; Hart, 1985, Hart et al., 1989; Lokonda et al., 2018; Peh et al., 2011b).

Also, soil water-holding capacity was similar between a *B. rubescens* monodominant forest and adjacent mixed forest in Neotropics (Marimon-Junior et al., 2020). These studies underscore the overall similarity in soil properties between monodominant and mixed forests. Lastly, Elias et al. (2018) provided evidence that *B. rubescens* is not subject to any edaphic restrictions for establishment.

Mechanism 4: large seed in deep leaf litter

A complementary feature accompanying the slow rate of litter decomposition in monodominant forests is the presence of a larger litter standing crop. It has been postulated that the substantial depth of leaf litter may impede seed germination and establishment, serving as a physical barrier to the soil. Additionally, it reduces light availability, alters soil temperatures, and releases chemical inhibitors during decomposition (Torti et al., 2001). Consequently, species with larger seeds may hold a competitive advantage over smaller-seeded species, potentially being less constrained by the presence of the deep litter layer (see Kazmierczak et al., 2016).

However, the explanation based solely on an average large seed size falls short for two primary reasons. First, several large-seeded forest species that coexist in mixed forests adjacent to monodominant forests do not give rise to monodominant stands themselves. Examples from Central Africa, including *Anonidium mannii*, *Mammea africana*, and *Uapaca paluosa*, and from South America, such as *Aldina latifolia*, *Swartzia polyphylla*, and *Vatairea guianensis*. Second, some monodominant species may also exhibit small seeds, exemplified by *Dryanobalanops aromatica*. In addition, the phenomenon of classical monodominance in Amazonian forests is not significantly linked to seed mass (ter Steege et al., 2019).

Martijena (1998), employing both greenhouse and field experiments, demonstrated that the germination percentage and seedling survival of a mixed forest species, *Caesalpinia eriostachys*, with an average dried seed weight of 0.22 g, remained unaltered in the presence of the thick litter layer characteristic of monodominant *C. mexicanum*, whose seeds had an average dried weight of 0.64 g (Liu et al., 2008). These findings suggest that non-dominant species may, to some extent, possess the capacity to withstand the adverse effects of the deep litter present in classical monodominant stands.

Mechanism 5: ectomycorrhizal association

The mycorrhizal hypothesis (Connell and Lowman,1989) revolves around the notion that the presence of ectomycorrhizal (EM) associations represents a pivotal characteristic of monodominant species. A substantial number of tropical single-dominant tree species were indeed associated with EM (Connell and Lowman, 1989; Fukami et al., 2017). It was even observed in seedlings of the classical monodominant *Dicymbe corymbose* do not rely strictly on adult trees for their mycobionts and can have unique assemblages of EM fungi over their development (Delevich et al., 2021). These EM associations could confer several advantages, including the more efficient exploitation of larger soil volumes, the direct decomposition of leaf litter, and the uptake of organic nitrogen (Connell and Lowman, 1989; Ebenye et al., 2017). Additionally, EM associations may enable host trees to mast fruit (see Mechanism 4), facilitated by the supply of essential phosphorus (P) from storage fungal hyphae (Turner, 2001).

However, there is no unequivocal evidence supporting the hypothesis that classical monodominance exclusively arises from an association with EM. The few occurrences of classical monodominance in Amazonian forests are not significantly liked to EM (ter Steege et al., 2019). Another case in point is the observation that the classical monodominant species *C. mexicanum*, occurring in Mexico, is instead associated with vesicular-arbuscular mycorrhizae rather than EM (Martijena,1998). This implies that an EM association is not invariably a prerequisite for monodominance to manifest. Furthermore, there are various non-EM-associated tree species, such as *Cynometra alexandri*, which, under certain circumstances, can achieve monodominance (Torti et al., 1997). Conversely, even when EM-associated tree species are present in mixed forests neighbouring monodominant stands, their competitive advantage over other species across extensive areas remains inconspicuous, as exemplified by *Julbernardia seretii*, for which little evidence of dominance exists (Hart, 1995). Additionally, the presence of some monodominant stands poses a challenge to the purported beneficial mechanisms attributed to EM associations, as the soils on which they thrive are not necessarily nutrient-poorer than nearby areas that do not exhibit single-species dominance, as discussed in Mechanism 1.

Mechanism 6: masting leading to predation satiation

Many monodominant tree species exhibit the characteristic of mast fruiting, a phenomenon marked by synchronous supra-annual flowering leading to synchronized supra-annual fruiting (e.g. Henkel and Mayor, 2019). Several authors have posited that mast fruiting, along with irregular periodicity in reproduction, plays a pivotal role in enabling several tree species to achieve monodominance (e.g., Torti, 2001). This is because the quantity of fruit available at a particular location and time exceeds the consumption capacity of the local seed predator population, thereby increasing the survival rates of offspring (Hart, 1995; Torti et al., 2001). Turner (2001) also suggests that irregular seed production may help maintain small populations of seed predators, ensuring that only a few individuals are present to consume the abundant fallen seeds during episodes of mast fruiting. Supporting this, Boucher (1981) revealed an inverse relationship between seed survivorship in monodominant tropical lowland oak, *Quercus oleoides*, and seed predator density within masting areas. In accordance with the predator satiation hypothesis, Henkel and Mayor (2019) found that seed predation of classical monodominant *Dicymbe corymbose* was proportionally higher during non-masting years.

Conversely, the escape hypothesis, as coined by Howe and Smallwood (1982), counters the predator satiation hypothesis, predicting high seed mortality for mast fruiting because clumps of fruits are easier for seed predators to locate. While some seeds may escape predation as predators are not perfect locators, evidence seems to favour the escape hypothesis. For example, low-seed survivorship was observed near a parent tree, as nearly all the seeds were destroyed (Clark and Clark, 1984). Moreover, findings focused on stands of *G. dewevrei* only partially support the predator satiation hypothesis (Hart, 1995). In masting areas of *G. dewevrei* forests where seed densities were high, mammalian predators were satiated, but specialized insect predators were not. Insect seed predators, with their shorter regeneration times, can rapidly build up large populations to exploit the available food resources (Turner, 2001). Furthermore, insect predators are highly specialized and capable of tracking the masting areas of potential host trees, infesting a high number of seeds. Hart (1995) therefore suggested that mast fruiting does not directly determine the higher canopy dominance of *G. dewevrei*. However, research has not supported the escape hypothesis in monodominant *Mora gonggrijpii* (see Hammond and Brown, 1998) and *B. rubescens* (Marimon et al., 2016), concerning insect seed predation. Consequently, further research is essential to elucidate why the effects of seed predators on masting seed survivorship vary across studies.

Mechanism 7: poor seed dispersal leading to gregarious habit

Ballistic seed dispersal may lead to clumped distribution, eventually driving a slow transformation of mixed forests into monodominant stands (Kazmierczak et al., 2016). However, the gregarious habit of *G. dewevrei* under parental trees due to poor seed dispersal does not appear to be uniquely advantageous, as claimed by Torti et al. (2001). Like *G. dewevrei*, *Julbernardia seretii* in mixed forests also possesses ballistic seed dispersal, with seeds germinating within a week of explosive pod dehiscence (Hart, 1995). However, *J. seretii* was not observed to dominate the canopy. While this trait may be necessary for achieving monodominance, it alone is insufficient to explain the phenomenon.

Mechanism 8: escape from herbivory via leaf defence

The leaf defence hypothesis, closely related to Mechanism 4, posits that low leaf damage on dominant saplings enhances their survivorship, facilitating monodominance through a strong investment in leaf defence mechanisms. These defences primarily involve the presence of high phenolic compounds and tough fibre content in leaves (Turner 2001). However, the only test of the leaf defence hypothesis in the context of monodominance comes from an observational study in *G. dewevrei*-dominated forests conducted by Gross et al. (2000). Their survey compared the rate of leaf damage on *G. dewevrei* with seven other tree species in both monodominant and adjacent mixed forests. Surprisingly, *G. dewevrei* exhibited higher leaf damage levels than the other species, and it did not show a higher phenolic content compared to its counterparts. While *G. dewevrei* did possess higher fibre content, this attribute did not correlate with herbivore damage. This study suggests that monodominance establishment and maintenance were not reliant on the avoidance of herbivory and pathogen damage.

An interesting case of a dominant tree species suffering intense herbivory by defoliating insects was reported by Maisels (2004). In this study, a monodominant stand of *Aucoumea klaineana* in Gabon was observed to be defoliated by a lepidopteran species. While the leaf defence hypothesis might mitigate damage from certain generalist herbivores, there is limited evidence to support its status as the primary mechanism upon which species depend to attain or maintain monodominance.

Linking the hypotheses: a probabilistic framework

There has been a concerted effort to apply ecological and evolutionary theories to address the existence of monodominance in forest ecosystems. The collective body of evidence has led to the perspective that specific life-history traits could provide a competitive advantage to certain species, enabling them to outperform others under conditions of minimal or no disturbance. Consequently, monodominance is believed to result from a combination of "endogenous" characteristics of a particular species (its inherent traits and implications for its resource acquisition strategies; see Hall et al. 2019) and "exogenous" characteristics of the forest system (the absence of externally imposed large-scale disturbance). Nevertheless, both theoretical frameworks and empirical studies have revealed inconsistencies, and there is no single, definitive trait that universally leads to monodominance.

Considering these concepts, a probabilistic conceptual framework is proposed by Peh et al. (2011a), where a suite of potential positive feedback mechanisms may collectively contribute to a higher probability of monodominance (Figure 12.2). This means that specific traits

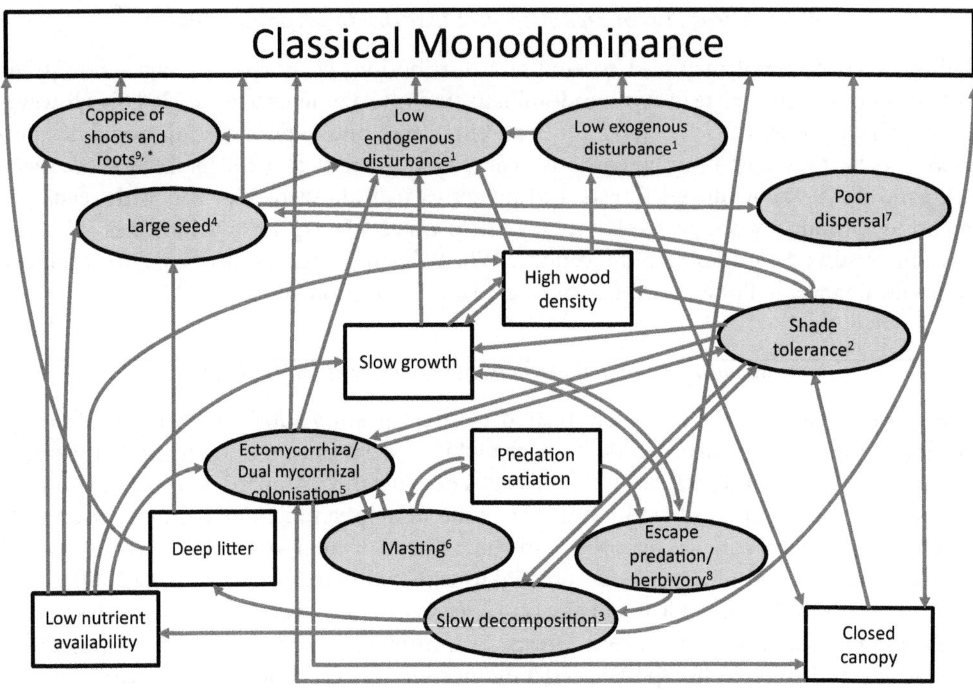

Hypothesis/ Mechanism	*References*	
	Support	*Lack of support*
1	Hart et al. (1989); Tovar et al. (2017)	Newbery et al. (2004)
2	Torti et al. (2001)	Hart et al. (1989)
3	Torti et al. (2001)	Conway (1992); Hart (1985); Peh et al. (2011b)
4	Hart et al. (1989); Torti et al. (2001)	Hart (1995)
5	Connell and Lowman (1989); McGuire (2007)	Torti et al. (2019)
6	Boucher (1981); Henkel and Mayor, (2019)	Hart (1995)
7	Torti et al. (2001)	Green and Newbery (2002); Newbery et al. (2004)
8	Ribeiro and Brown (2006)	Gross et al. (2000); Maisels (2004)
9	Woolley et al. (2008); ter Steege et al. (2019)	

Figure 12.2 Model of possible mechanisms (grey boxes) and their consequences (white boxes) leading to classical (persistent) monodominance in tree species growing in environmental conditions similar to adjacent high-diversity forests. Arrows show that one factor beneficially alters the environment to favour another in the direction of the arrow. The inconsistent results from experiments and observations relating to each published mechanism are given in the box below. The diagram shows that there are potentially multiple routes to monodominance, via differing positive feedback loops, thus apparently contradictory results may be reconciled. *This is the product from the growth of apical meristems outside the normal temporal sequence.

Source: Reproduced from Peh et al. (2011a).

exhibited by potentially dominant species can create additional conditions that favour the establishment of more individuals of the same species, ultimately leading to a monodominant system. Identified feedback mechanisms include shade tolerance and the production of large seeds, which can be favoured by environmental factors such as deep shade beneath the forest canopy and a substantial leaf litter layer, respectively. Additionally, factors like ectomycorrhizal associations, which enhance nutrient competition in low-nutrient environments, and predation satiation, potentially stemming from masting events, can further boost the probability of monodominance.

This probabilistic framework links all potential mechanisms into a potentially self-reinforcing cycle, which can help reconcile contradictory observational and experimental findings. However, external factors such as the infrequency of natural disturbances also play a crucial role in determining the likelihood of establishing monodominant forest stands. The absence of substantial disturbance, coupled with the favourable microhabitat created by the species itself, appears to be a necessary combination for a single species to outcompete all others across extensive areas. Nevertheless, all classical monodominant species are not expected exhibit all the traits in Figure 12.2. It is highly probable that there are various pathways to classical monodominance, which can explain seemingly conflicting findings in experiments and observational studies that analyse data on a mechanism-by-mechanism basis.

References

Anderson, J. A. R. (1961) 'The destruction of *Shorea albida* forest by an unidentified insect', *Empirical Forestry Review*, vol. 40, pp. 19–29.

Arújo, I., Morandi, P. S., Müller, A. O., Mariano, L. H., Alvarez, F., da Silva, I. V., Marimon Junior, B. H. and Marimon, B. S. (2022) 'Leaf functional traits and monodominance in Southern Amazonia tropical forests', *Plant Ecology*, vol 223, pp. 185–200.

Beard, J. S. (1946) 'The Mora forest of Trinidad, British West Indices', *Journal of Ecology*, vol. 33, pp. 173–192.

Boucher, D. H. (1981) 'Seed predation by mammals and forest dominance by *Quercus oleides*, a tropical lowland oak', *Oecologia*, vol. 49, pp. 409–414.

Brünig, E. F. (1974) 'Ecological studies in the Kerangas forests of Sarawak and Brunei', Borneo Literature Bureau for Sarawak Forest Department, Kuching.

Butler, R. A. (2024) The Amazon Rainforest: The world's largest rainforest. https://worldrainforests.com/amazon/

Clark D. A. and Clark D. B. (1984) 'Spacing dynamics of a tropical rain forest tree: evaluation of the Janzen-Connell model', *American Naturalist*, vol. 124, pp. 769–788.

Connell, J. H. and Lowman, M. D. (1989) 'Low-diversity tropical rain forests: some possible mechanisms for their existence', *American Naturalist*, vol. 134, pp. 88–119.

Conway, D. (1992) 'A comparison of soil parameters in monodominant and mixed forest in the Ituri Forest Reserve, Zaire' Tropical Environmental Science Honours Project. University of Aberdeen, Aberdeen.

Coomes, D. A. and Grubb, P. J. (1996) 'Amazonian caatinga and related communities at La Esmeralda, Venezuela: forest structure, physiognomy and floristics, and control by soil factors', *Vegetatio*, vol. 122, pp. 167–191.

Coomes, D. A. and Grubb, P. J. (1998) 'A comparison of 12 tree species of Amazonian caatinga using growth rates in gaps and understorey, and allometric relationships', *Functional Ecology*, vol. 12, pp. 426–435.

Daly, D. C. and Mitchell, J. D. (2000) 'Lowland vegetation of tropical South America—an overview'. In: Lentz, D. (ed.) *Imperfect Balance: Landscape Transformation in the Pre-Columbian Americas*, pp. 391–454. Columbia University Press, New York.

Delevich, C. A., Koch, R. A., Aime, M. C. and Henkel, T. W. (2021) 'Ectomycorrhizal fungal community assembly on seedlings of a Neotropical monodominant tree', *Biotropica*, vol. 53, pp. 1486–1497.

Djuikouo, M. N. K., Peh, K. S.-H., Nguembou, C. K., Doucet, J. L., Lewis, S. L., Sonké, B. (2014) 'Stand structure and species co-occurrence in mixed and monodominant Central African tropical forests', *Journal of Tropical Ecology*, vol. 30, pp. 447–455.

Ebenye, H. C. M., Taudière, A., Niang, N., Ndiaye, C., Sauve, M., Awana, N. O., Verbeken, M., De Kesel, A., Séne, S., Diédhiou, A. G., Sarda, V., Sadio, O., Cissoko, M., Ndoye, I., Selosse, M.-A. and Bâ, A. M. (2017) 'Ectomycorrhizal fungi are shared between seedlings and adults in a monodominant *Gilbertiodendron dewevrei* rain forest in Cameroon', *Biotropica*, vol. 49, pp. 256–267.

Eggeling, W. J. (1947) 'Observations on the ecology of Budongo rain forest, Uganda', *Journal of Ecology*, vol. 34, pp. 20–87.

Elias, F., Marimon, B. S., Marimon-Junior, B. H., Budke, j. C., Esquivel-Muelbert, A., Morandi, p. S., Reis, S. M. and Phillips, O. L. (2018) 'Idiosyncratic soil-tree species associations and their relationships with drought in a monodominant Amazon forest', *Acta Oecologica*, vol 97, pp. 127–136.

Figueiredo, P. H. A., Sanchez-Tapia, A., de Siqueira, M. F. and Sansevero, J. B. B. (2022) 'Linking regeneration niche to monodominance in biodiverse tropical forest landscapes', *Journal of Vegetation Science*, vol. 33, e13128.

Forget P.-M. (1989) 'La régénération naturelle d'une espèce autochore de la forêt Guyanaise: *Eperua falcate* Aublet (Casalpiniaceae)' Biotropica, vol. 21, pp. 115–121.

Fukami, T., Nkajima, M., Fortunel, C., Fine, P. V. A., Baraloto, C., Russo, S. E. and Peay, K. G. (2017) 'Geographical variation in community divergence: insight from tropical forest mondominance by ectomycorrhizal trees', *The American Naturalist*, vol 190, pp. S105–S122.

Glick, H. B., Umunay, P. M., Makana, J.-R., Tomlin, C. D., Reuning-Scherer, J. D. and Gregoire, T. G. (2021) 'Developmental dynamics of *Gilbertiodendron dewevrei* (Fabaceae) drive forest structure and biomass in the eastern Congo Basin', *Forest*, vol. 12, 738.

Glick, H. B., Umunay, P. M., Makana, J.-R., Tomlin, C. D., Reuning-Scherer, J. D. and Gregoire, T. G. (2023) 'The spatial propagation and increasing dominance of *Gilbertiodendron dewevrei* (Fabaceae) in the eastern Congo Basin', *PLoS One*, vol. 18, e0275519.

Green, J. J. and Newbery, D. M. (2002) 'Reproductive investment and seedling survival of the mast-fruiting rain forest tree, *Microberlinia bisulcata* A. Chev.', *Plant Ecology*, vol. 162, pp. 169–183.

Gross, N. D., Torti, S. D., Feener Jr., D. H. and Coley, P. D. (2000) 'Monodominance in an African rain forest: Is reduced herbivory important?', *Biotropica*, vol. 32, pp. 430–439.

Grubb, P. J., Turner, I. M. and Burslem, D. F. R. P. (1994) 'Mineral nutrient status of coastal hill dipterocarp forest and adinandra belukar in Singapore: analysis of soil, leaves and litter', *Journal of Tropical Ecology*, vol. 10, pp. 559–577.

Hall, J. S., Harris, D. J., Saltonstall, K., de Paul Medjibe, V.m Ashton, M. S. and Turner, B. L. (2019) 'Resource acquisition strategies facilitate *Gilbertiodendron dewevrei* monodominance in African lowland forests', *Journal of Ecology*, vol. 108, pp. 433–448.

Hammond, D. S. and Brown, V. K. (1998) 'Disturbance, phenology and life-history characteristics: factors influencing distance/density-dependent attack on tropical seeds and seedlings', in D. M. Newbery, H. H. T. Prins, and N. D. Brown (eds), *Dynamics of Tropical Communities*, Blackwell Science, Oxford.

Hart, T. B. (1985) 'The ecology of a single-species-dominant forest and of a mixed forest in Zaire, Africa' Ph.D. Thesis. Michigan State University.

Hart, T. B. (1995) 'Seed, seedling and sub-canopy survival in monodominant and mixed forests of the Ituri Forest, Africa' *Journal of Tropical Ecology*, vol. 11, pp. 443–459.

Hart, T. B., Hart, J. A. and Murphy, P. G. (1989) 'Monodominant and species-rich forests in the humid tropics: causes for their co-occurrence' *American Naturalist*, vol. 133, pp. 613–633.

Henkel, T. W. and Mayor, J. R. (2019) 'Implications of a long-term mast seeding cycle for climatic entrainment, seedling establishment and persistent monodominance in a Neotropical, ectomycorrhizal canopy tree', *Ecological Research*, vol 34, pp. 472–484.

Henkel, T. W., Mayor, J. R. and Woolley, L. P. (2005) 'Masting fruiting and seedling survival of the ectomycorrhizal, monodominant *Dicymbe corymbosa* (Caesalpiniaceae) in Guyana', *New Phytologist*, vol. 167, pp. 543–556.

Holdridge, L. R., Grenke, W. C., Hatheway, W. H., Liang, T. and Tosi, J. A. (1971) *Forest Environments in Tropical Life Zones*, Pergamon Press, Oxford.

Howe, H. F. and Smallwood, J. (1982) 'Ecology of seed dispersal', *Annual Review of Ecology and Systematics*, vol. 13, pp. 201–228.

Hubbell, S. P. (2004) 'Two decades of research on the BCI Forest dynamics plot' In: Losos, E.C. and Leigh Jr., E.G. (eds), *Tropical Forest Diversity and Dynamism: findings from a Large-scale Plot Network*, pp. 8–30. University of Chicago, Chicago.

Hughes, R. H. and Hughes, J. S. (1992) *A Directory of African Wetlands*, IUCN, UNEP and WCMC

Ithoh, A. (1995) 'Effects of forest floor environment on germination and seedling establishment of two Bornean rainforest emergent species' *Journal of Tropical Ecology*, vol. 11, pp. 517–527.

Janzen, D. H. (1978) 'Description of a *Pterocarpus officinale* (Leguminosae) monoculture in Corcovada National Park, Costa Rica' *Brenesia*, vol. 14–15, pp. 305–309.

Kazmierczak, M., Backmann, P., Fedriani, J. M., Fischer, R., Hrtmnn, A. K., Huth, A., May, F., Muller, M. S., Taubert, F., Grimm, V. and Groeneveld, J. (2016) 'Monodominance in tropical forests: modelling reveals emerging clusters and phase transitions', *Journal of Royal Society Interface*, vol. 13, 20160123.

Liu, K., Eastwood, R.J., Flynn, S., Turner, R.M. and Stuppy, W.H. (2008) Seed Information Database, http://data.kew.org/sid/

Loehle, C. (1988) 'Tree life-history strategies—the role of defenses', *Canadian Journal of Fores tResearch*, vol. 18, pp. 209–222.

Lokonda, M., Freycon, V., Gourlet-Fleury, S. and Kombele, F. (2018) 'Are soils under monodominant *Gilbertiodendron dewevrei* and underadjacent mixed forests similar? A case study in the Democratic Republic of Congo', *Journal of Tropical Ecology*, vol. 34, 176–185.

Maisels, F. (2004) 'Defoliation of a monodominant rain-forest tree by a noctuid moth in Gabon', *Journal of Tropical Ecology*, vol. 20, pp. 239–241.

Marimon, B. S., Felfili, J. M., Fagg, C. W., Marimon-Junior, B. H., Umetsu, R. K., Oliveira-Santos, C., Morandi, P. S., Lima, H. S. and Terra Nascimento, A. R. (2012) 'Monodominance in a forest of *Brosimum rubescens* Taub. (Moraceae): structure and dynamics of natural regeneration', *Acta Oecologica*, 43, pp. 134–139.

Marimon, B. S., Felfili, J. M. and Haridasan, M. (2001) 'Studies in monodominant forests in eastern Mato Grosso, Brazil: I. A forest of *Brosimum rubescens* Taub.' *Edinburgh Journal of Botany*, vol. 58, pp. 123–137.

Marimon, B. S., Felfili, J. M., Marimon Jr, B. H., Fagg, C. W., Da Silveira Anacleto, T. C., Umetsu, R. K., Lenza, E., Batista, J. D. and Rossete, A. N. (2016) 'Leaf herbivory and monodominance in a Cerraado-Amazonia transitional forest, Mato Grosso, Brazil', *Plant Biosystems*, vol 150, pp. 124–130.

Marimon-Junior, B. H., Du Vall Hay, J., Oliveras, I., Jancoski, H., Umetsu, R. K., Feldpausch, T. R., Galbraith, D. R., Gloor, E. U., Phillips, O. L. and Marimon, B. S. (2020) 'Soil water-holding capacity and monodominance in Southern Amazon tropical forests', *Plant Soil*, vol. 450, pp. 65–79.

Martijena, N. E. (1998) 'Soil properties and seedling establishment in soils from monodominant and high-diversity stands of the tropical deciduous forests of Mexico', *Journal of Biogeography*, vol. 25, pp. 707–719.

McGuire, K. L. (2007) 'Common ectomycorrhizal networks may maintain monodominance in a tropical rain forest' *Ecology*, vol. 88, pp. 567–574.

Medina, E., Garcia, V. and Cuevas, E. (1990) 'Sclerophyll and oligotrophic environments: relationships between leaf structure, mineral nutrient content and drought resistance in tropical rain forests of the Upper Rio Negro' *Biotropica*, vol. 22, pp. 51–64.

Mesquita, R., De C. G., Workman, S. W. and Neely, C. L. (1998) 'Slow litter decomposition in a Ceropia-dominated secondary forest of central Amazonia' *Soil Biology and Biochemistry*, vol. 30, pp. 167–175.

Mueller-Dombois, D. (2000) 'Rain forest establishment and succession in Hawaiian Islands' *Landscape and Urban Planning*, vol. 51, pp. 147–157.

Nascimento, M. T., Barbosa, R. I., Villela, D. M. and Proctor, J. (2007) 'Above-ground biomass changes over an 11-year period in an Amazon monodominant forest and two other lowland forests', *Plant Ecology*, vol. 192, pp. 181–191.

Nelson, B. W, Kapos, V., Adams, J. B., Oliveira, W. J., Braun, O. P. G. and do Amaral, I. L. (1994) 'Forest disturbance by large blowdowns in the Brazilian Amazon' *Ecology*, vol. 75, pp. 853–858.

Newbery, D. M., van der Burgt, X. M. and Moravie, M.-A. (2004) 'Structure and inferred dynamics of a large glove of *Microberlinia bisulcata* trees in central African rain forest: the possible role of periods of multiple disturbance events', *Journal of Tropical Ecology*, vol. 20, pp. 131–143.

Parolin, P., Oliverira, A. C., Piedade M. T. F., Wittmann, F. and Junk, W. J. (2002) 'Pioneer trees in Amazonian floodplains: three key species form monospecific stands in different habitats', *Folia Geobotanica*, vol. 37, pp. 225–238.

Peh, K. S.-H., Lewis, S. L., and Lloyd, J. (2011a) 'Mechanisms of monodominance in diverse tropical tree-dominated systems', *Journal of Ecology*, vol 99, pp. 891–898.

Peh, K. S.-H., Sonké, B., Lloyd, J., Quesada, C. A., and Lewis, S. L. (2011b) 'Soil does not explain monodominance in a Central African tropical forest', *PLoS One*, vol. 6, e16996.

Peh, K. S.-H., Sonké, B., Séné, O, Djuikouo, M. N. K., Nguembou, C. K., Taedoumg, H., Begne, S. K. and Lewis, S. L. (2014) 'Mixed-forest species establishment in a monodominant forest in central Africa: implications for tropical forest invasibility, *PLoS One*, vol. 9, e97585.

Peh, K. S.-H., Sonké, B., Taedoumg, H., Séné, O., Lloyd, J. and Lewis, S. L. (2012) 'Investigating diversity dependence of tropical forest litter decomposition: experiments and observations from Central Africa', *Journal of Vegetation Science*, vol. 23, pp. 223–235.

Read, J., Jaffre, T., Ferris, J. M., McCoy, S. and Hope, G. S. (2006) 'Does soil determine the boundaries of monodominant rain forest with adjacent mixed rain forest and maquis on ultramafic soils in New Caledonia?' *Journal of Biogeography*, vol. 33, pp. 1055–1066.

Read, J., McCoy, S., Jaffre, T. and Logan, M. (2018) 'Nutrient-uptake and -use efficiency in seedlings of rain-forest trees in New Caledonia: monodominants vs. subordinates and episodic vs. continuous regenerators', *Journal of Tropical Ecology*, vol. 34, pp. 277–292.

Ribeiro, S. P. and Brown, V. K. (2006) 'Prevalence of monodominant vigorous tree populations in the tropics: herbivory pressure on Tabebuia species in very different habitats', *Journal of Ecology*, vol. 94, pp. 932–941.

Richards, P. W. (1996) *The Tropical Rain Forest*, 2nd ed. Cambridge University Press, Cambridge.

Schöngart, J., Piedade, M. T. F., Wittmann, F., Junk, W. J. and Worbes, M. (2005) 'Wood growth patterns of *Macrolobium acaciifolium* (Benth.) Benth. (Fabaceae) in Amazonian black-water and white-water floodplain forests', *Oecologia*, vol. 145, pp. 454–461.

Sim, J. W. S., Tan, H. T. W. and Turner, I. M. (1992) 'Adinandra belukar: an anthrogenic heath forest in Singapore' *Vegetatio*, vol. 102, pp. 125–137.

ter Steege, H., Henkel, T. W., Helal, N., Marimon, B. S., Marimon-Junior, B. H., et al. (2019) 'Rarity of monodominance in hyperdiverse Amazionian forests', *Scientific Reports*, vol. 9, 13822.

Torti, S. D., Coley, P. D. and Janos, D. P. (1997) 'Vesicular-arbuscular mycorrhizae in two tropical monodominant tree species', *Journal of Tropical Ecology*, vol. 13, pp. 623–629.

Torti, S. D., Coley, P. D. and Kursar, T. A. (2001) 'Causes and consequences of monodominance in tropical lowland forests', *American Naturalist*, vol. 157, pp. 141–153.

Tovar, C., Harris, D. J., Breman, E., Brncic, T. and Willis, K. J. (2019) 'Tropical monodominance forest resilience to climate change in Central Africa: A *Gilbertiodendron dewevrei* forest pollen record over the past 2,700 years', *Journal of Vegetation Science*, vol. 30, pp. 575–586.

Turner, I. M. (2001) *The Ecology of Trees in the Tropical Rain Forest*, Cambridge University Press, Cambridge.

United Nations Environment Programme (2023). *Decades of Mangrove Forest Change: What Does it Mean for Nature, People and the Climate?*, UNEP, Nairobi.

Van der Velden, N., Slik, J. W. F., Hu, Y.-H., Lan, G., Lin, L., Deng, X. and Poorter, L. (2014) 'Monodominance of *Parashorea chinensis* on fertile soils on a Chinese tropical rain forest', *Journal of Tropical Ecology*, vol. 30, pp. 311–322.

van Zon, P. (1915) 'Mededeelingen omtrent den kamferboom (*Dryobalanops aromatica*)', *Tectona*, vol. 8, pp. 319–324.

Vierling, L. A. and Wessman, C. A. (2000) 'Photosynthetically active radiation heterogeneity within a monodominant Congolese rain forest canopy' *Agricultural and Forest Meteorology*, vol. 103, pp. 265–278.

Villela, D. M. and Proctor, J. (2002) 'Leaf litter decomposition and monodominance in the Peltogyne forest of Maracá Island, Brazil' *Biotropica*, vol. 34, pp. 334–347.

Vitousek, P. M. and Sanford Jr., R. L. (1986) 'Nutrient cycling in moist tropical forest' *Annual Review of Ecology and Systematics*, vol. 17, pp. 137–167.

Whitmore, T. C. (1984) *Tropical Rain Forests of the Far East*, 2nd ed. Clarendon Press, Oxford

Wittmann, F. and Junk, W. J. (2003) 'Sapling communities in Amazonian white-water forests', *Journal of Biogeography*, vol. 30, pp. 1533–1544.

Woolley, L. P., Henkel, T. W. and Sillett, S. C. (2008) 'Reiteration in the monodominant tropical tree *Dicymbe corymbosa* (Caesalpiniaceae) and its potential adaptive significance', *Biotropical*, vol. 40, pp. 32–43.

Wright, S. J. (2005) 'Tropical forests in a changing environment', *Trends in Ecology and Evolution*, vol. 20, pp. 553–560.

PART III

FOREST FLORA AND FAUNA

13

THE ECOLOGY OF LIANAS AND THEIR INCREASING INFLUENCE IN TROPICAL FORESTS

Stefan A. Schnitzer

Introduction

Lianas are woody climbing plants that use the architecture of other plants, typically trees, to ascend to the top of the forest canopy (Schnitzer and Bongers, 2002). Lianas are found in forests worldwide—in both the temperate and tropical zones—however, they peak in diversity, stem density, and biomass in lowland tropical forests (Figure 13.1; see also DeWalt et al., 2010, 2015). In many tropical forests, lianas comprise more than one-third of the woody species (lianas shrubs, and trees; e.g., Ibarra-Manríquez et al., 2015; Parthasarathy et al., 2015; Thomas et al., 2015; Schnitzer et al., 2021). Lianas are not from a single, specialized taxonomic group; instead, they belong to many different plant families and orders—nearly 40 per cent of the dicot plant families and more than 75 per cent of dicot orders have at least one climbing species (including herbaceous climbers; Gianoli, 2015). The characteristics of lianas are readily identifiable; they have specialized organs for climbing (e.g., tendrils, hooks, twining stems, twining branches, or adhesive roots), thin stems relative to the trees that they climb, and a large ratio of sun-exposed leaf area to stem area, which maximizes their efficiency in converting sunlight into carbohydrates (Putz, 1984; Schnitzer and Bongers, 2002; Wyka et al., 2013). Similar to trees (and unlike hemi-epiphytes and many epiphytes), lianas remain rooted in the ground throughout their lifetime.

Lianas have both positive and negative effects on tropical forest diversity and functioning. Their extremely high species diversity in tropical forest commonly contributes many hundreds of species (e.g., Schnitzer et al., 2015a and citations within), thus adding to forest diversity. This high liana diversity includes species that produce a wide variety of resources that are essential for animal species: nectar, pollen, fruits, sap, leaves, and edible young twigs (Arroyo-Rodriguez et al., 2015; Yanoviak, 2015). The nutrition that lianas provide may be especially important during the dry season when many animal species switch their diets to liana fruits or young leaves (Dunn et al., 2012; Arroyo-Rodriguez et al., 2015). Lianas also connect multiple tree crowns, converting the forest canopy into a complex structural environment that may maintain animal diversity and provide shelter and habitat for many animal species (e.g., Michel et al., 2015; Yanoviak, 2015, Adams et al., 2017, 2019; Schnitzer et al., 2020a).

DOI: 10.4324/9781003324072-16

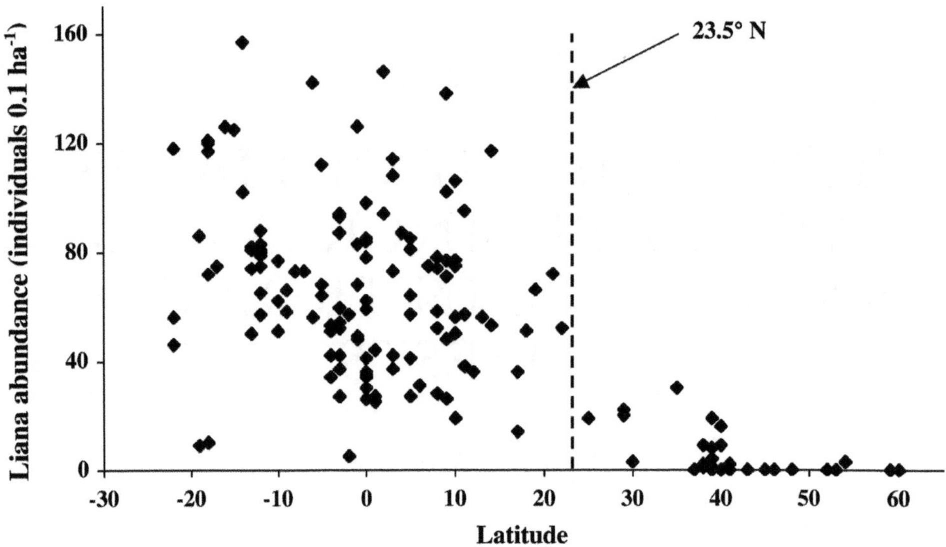

Figure 13.1 The change in liana density with increasing latitude from the southern edge of the tropics to the northern hemisphere. Data were collected by A.H. Gentry in 10 small (2 × 50 m) transects (0.1 ha total) in each forest, and the minimum diameter cutoff was ≥ 2.5 cm diameter (published in Phillips and Miller, 2002). The dashed line denotes the northern edge of the tropics. The data demonstrate a sharp decrease in liana density outside of the tropics. Based on Schnitzer (2005).

Lianas also can have a negative influence on tropical forests; they reduce canopy tree growth, regeneration, and reproduction (Toledo-Aceves, 2015; Kainer et al., 2014; Garcia-Leon et al., 2018; Estrada-Villegas et al., 2022), thus altering ecosystem functioning and potentially tree community composition. By reducing canopy tree reproduction and fecundity, lianas limit per-capita tree seed production and thus may amplify the effects of dispersal limitation. In addition, lianas recruit rapidly and in high densities following disturbance (e.g., in treefall gaps, forest edges, and in young regenerating forests; DeWalt et al., 2000; Schnitzer et al., 2000, 2021; Barry et al., 2015), where they can blanket the existing vegetation and reduce the recruitment and diversity of regenerating tree species (Putz, 1984; Schnitzer and Carson, 2010). Intense competition from lianas in both intact forest and following disturbance reduces the amount of carbon that trees sequester and store (Durán and Gianoli, 2013; Ledo et al., 2016; Estrada-Villegas et al., 2020); however, because lianas do not store as much carbon as they displace in trees, they reduce net tropical forest carbon uptake and storage capacity (van der Heijden et al., 2013, 2015a; Schnitzer et al., 2014). Lianas may also deplete soil water and nutrients, further affecting tree communities and ecosystem functioning (Reid et al., 2015; Toledo-Aceves, 2015), although more research is necessary to accurately quantify the effects of lianas on below-ground resources.

Until recently, our knowledge of liana ecology has lagged far behind that of trees. Over the past three decades, however, the study of liana ecology has grown exponentially, revealing many important contributions of lianas to forest ecology (Schnitzer et al., 2015b). In this chapter, I review the state of knowledge about the ecology of lianas and their contribution to forest ecosystems. I divide the chapter into four main sections: (1) a brief introduction to liana ecology (the current section). (2) Ecological effects of lianas on forest ecosystems. In this section, I review the state of knowledge of the detrimental effects of lianas on tree performance and diversity, as well as the mechanisms by which lianas alter tree community composition and forest carbon accumulation. I also review some of the positive contributions of lianas to forest plant and animal diversity, factors that are often overlooked or misunderstood. (3) Global distribution of lianas and the factors that control their distribution. Determining the factors that control the distribution of organisms is one of the fundamental goals of ecology and determining where lianas are most abundant (and why) is key to understanding the forest types in which lianas are likely to have the greatest influence. (4) The pattern of increasing liana abundance in tropical forests, which may be one of the most significant structural changes occurring in old-growth tropical forests, one with profound potential implications for tropical forest diversity and functioning.

Ecological effects of lianas in forest ecosystems

Negative effects of lianas on trees

Many dozens of studies in tropical forests around the world have demonstrated that lianas have a strong negative effect on trees (reviewed by Schnitzer and Bongers, 2002; Toledo-Aceves, 2015; Marshall et al., 2017; Estrada-Villegas and Schnitzer, 2018). While it remains unclear whether the negative relationship between lianas and trees is best described as competition or parasitism (Stewart and Schnitzer, 2017), empirical studies have shown unequivocally that lianas reduce tree performance, including tree recruitment, growth, and fecundity (e.g., Grauel and Putz, 2004; Ingwell et al. 2010; Schnitzer and Carson, 2010; Kainer et al., 2014, Garcia-Leon et al., 2018, Schnitzer, 2024). For example, using a meta-analytic approach with 25 ecological liana removal experiments that were conducted in the American, African, and Asian tropics, Estrada-Villegas et al. (2022) found that lianas reduced tree reproduction, recruitment, growth, leaf water potential, and forest leaf area index (Figure 13.2). Lianas also reduced tree sap flow velocity; however, this effect was marginally significant likely due to the low sample size (Figure 13.2). Surprisingly, this meta-analysis revealed that lianas did not increase tree mortality, although this finding may have been influenced by the relatively small scale of liana removal experiments and the difficulty in detecting changes in large tree mortality over small spatial and temporal scales (McDowell et al., 2019; Gora et al., 2021).

Lianas substantially reduce tropical tree reproductive output, increasing tree dispersal limitation for tropical trees. In a Brazilian forest, Kainer et al. (2014) showed that Brazil nut production 10 years following liana removal was 77 per cent higher than for Brazil nut trees that hosted lianas. In a liana removal study in a Panamanian forest, Garcia-Leon et al. (2018) reported that 150 per cent more canopy trees produced fruits in forest plots five years following liana removal compared to trees in control plots where lianas had not been removed. Understory trees showed a similar pattern, with 101 per cent more individuals producing fruits where lianas had been removed. Thus, lianas reduce both canopy and understory tree fruit production, which may limit the ability of trees to disperse in tropical forests (dispersal limitation).

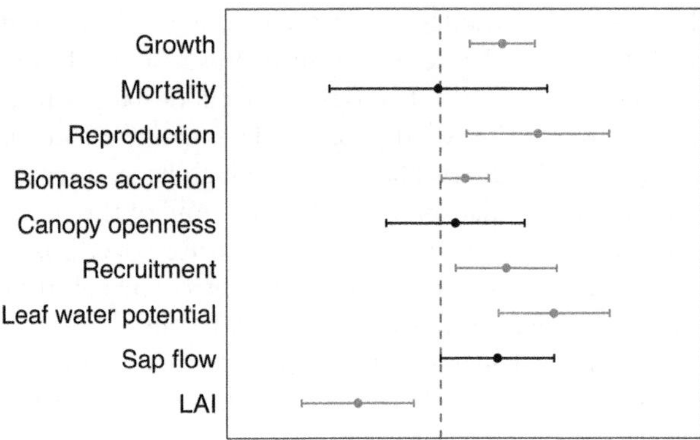

Figure 13.2 The response of trees to liana removal in 25 separate ecological experiments conducted in tropical and subtropical forests. Values are standardized effect sizes (Hedges' *g*) and their 95% confidence intervals estimated by three-level meta-analytic models. LAI denotes leaf area index. Values greater than zero indicate positive effects of liana removal.

Source: From Estrada-Villegas et al. (2022).

One way that lianas reduce tree performance is by limiting their host tree's access to light. Lianas tend to deploy a large canopy of leaves directly over those of their host tree (Avalos and Mulkey, 1999; Rodriguez-Ronderos et al., 2016), resulting in trees with underdeveloped or deformed canopies (Moorthy et al., 2018). Lianas may also reduce tree performance by limiting belowground resource availability. Soil moisture is limiting during the dry season, and lianas may further reduce available soil moisture, thus decreasing tree performance (Schnitzer et al., 2005; Tobin et al., 2012; Toledo-Aceves, 2015; Alvarez-Cansino, 2015). Lianas remain physiologically active and grow more than trees during seasonal droughts (Smith-Martin et al., 2019; Schnitzer and van der Heijden, 2019), which may allow them to deplete dry season soil moisture. Lianas also cause mechanical stress to their hosts by adding considerable weight to the host tree crown, thus forcing trees to increase stem diameter at the expense of height (Schnitzer et al., 2005, Ingwell et al., 2010). The relatively high level of nutrients in liana leaves suggests that lianas may also compete with trees for nutrients (Asner and Martin, 2015); however, there is little evidence that lianas themselves are nutrient limited (Schnitzer et al., 2020a) or that they are responsible for tree nutrient limitations. The effects of lianas on soil nutrient availability remain poorly understood.

Lianas also limit the ability of tree species to colonize tropical treefall gaps, thus decreasing tree density and diversity following gap formation (Schnitzer et al., 2000). Treefall gaps had once been thought to benefit shade-tolerant trees by providing a heterogeneous and resource-rich environment in which trees can partition resources and thus coexist (e.g., Denslow, 1987); however, gaps also promote the density and diversity of lianas (e.g., Putz, 1984; Schnitzer et al., 2021), which compete intensely with trees in gaps. Using an eight-year liana removal experiment in treefall gaps in central Panama, Schnitzer and Carson (2010) found that shade-tolerant tree species diversity increased over 60 per cent faster in treefall gaps where lianas were removed compared to control gaps with lianas. Thus, lianas appear to suppress shade-tolerant tree colonization, survival, density, and diversity. By limiting both tree seed production and the ability of trees to colonize treefall gaps, lianas may increase dispersal and colonization limitations and thus preclude the ability of trees to partition resources as predicted by gap/disturbance theory (Schnitzer and Carson, 2010).

Lianas and forest carbon dynamics

Tropical forests are critical ecosystems for global carbon dynamics, storing more than half of the earth's aboveground terrestrial forest carbon (Saatchi et al., 2011; Xu et al., 2021). In most tropical forests, trees store more than 90 per cent of this carbon. By reducing tree growth, lianas can dramatically reduce rates of carbon uptake in tropical forests (Durán and Gianoli, 2013; van der Heijden et al., 2013). Forests with high liana densities tend to have far lower biomass than forest areas with low liana densities, suggesting that lianas reduce whole-forest carbon accumulation and storage (e.g., Durán and Gianoli, 2013). For example, in an observational study in the Peruvian Amazon, lianas appeared to reduce annual tree biomass increase by around 10 per cent; however, annual liana biomass increase compensated for only around 29 per cent of the biomass displaced in trees (van der Heijden and Phillips, 2009). In a liana-removal experiment in treefall gaps on Gigante Peninsula in central Panama, Schnitzer et al. (2014) reported that lianas reduced tree biomass accumulation in gaps by nearly 300 per cent over an eight-year period—a 35 per cent annual reduction in biomass accumulation. However, since lianas themselves have low wood volume, they accumulate and store relatively little carbon, and lianas were able to compensate for only 24 per cent of the biomass accumulation that they displaced in trees (Schnitzer et al., 2014). In a liana-removal study in large permanent sampling plots (also on Gigante Peninsula; Schnitzer, 2024), van der Heijden et al. (2015) found that lianas reduced net aboveground carbon uptake by 76 per cent per year, mostly by reducing tree growth. Lianas also increased the amount of carbon stored in leaves at the expense of stems. Thus lianas alter how carbon is stored in tropical forests, storing less carbon in recalcitrant structures (wood) and more carbon in fast-turnover organs (leaves) and thus recycling carbon more rapidly (van der Heijden et al., 2015).

The finding that lianas can have a net negative effect on tropical forest carbon uptake is surprising. Most theories assume that competition among plants is a zero-sum-game in terms of carbon uptake. That is, one plant's carbon loss is another's carbon gain. This assumption, however, breaks down with competition among contrasting plant growth forms that inherently store different amounts of carbon, and the effect of lianas on trees results in a net carbon loss for tropical forests (Schnitzer, 2018). Considering that tropical forests store more than one-half of all aboveground terrestrial forest carbon (Saatchi et al., 2011; Xu et al., 2021), with most of this carbon located in the trunks of living trees, by reducing tropical tree carbon accumulation and storage, lianas alter the global carbon budget. Furthermore, the ongoing increase in liana abundance throughout the neotropics (see below) will likely amplify the negative effects of lianas on global carbon dynamics.

The negative effects of lianas on tree carbon uptake and storage appear to occur in tropical forests across the globe and across seasons (wet and dry seasons). For example, Estrada-Villegas et al. (2022) found that lianas in tropical forests in Asia, Africa, and the Americas substantially reduced tree biomass accretion (Figure 13.2; see also Durán and Gianoli, 2013; van der Heijden et al., 2013, 2015; Ledo et al., 2016). The strong negative effect of lianas on tree biomass accumulation was similar in both wet and dry seasons (in central Panama; van der Heijden et al., 2019). Thus, lianas may have a strong negative effect on tree biomass accretion in forests throughout the year and throughout the lowland tropics.

Positive contributions of lianas in forest ecosystems

Lianas also have positive effects on forest ecosystems. First, lianas are exceptionally diverse in terms of the number of species, as well as diversity at higher taxonomic (genus and family) levels. The scandent growth form arises in 38.7 per cent (162) of all angiosperm plant families (Gianoli, 2015). On the Barro Colorado Island 50-ha plot in central Panama, lianas contributed 175 species from 98 genera and 41 families (Schnitzer et al., 2012, 2021). By comparison, there were 321 tree species; thus, lianas constituted 35 per cent of the woody plant species and 28 per cent of the woody stem density (Schnitzer and Carson, 2024).

High liana diversity supports many animal species, both vertebrates and invertebrates. Many mammal species consume liana leaves, flowers, and fruits, which constitute an important component of their overall diet (Emmons and Gentry, 1983; Arroyo-Rodriguez, 2015). Primates have been observed to switch their diet to include far more lianas during the dry season compared to the wet season (Dunn et al., 2012). Lianas may be particularly important during seasonal drought because they tend to be more physiologically active during the dry season than co-occurring trees (Schnitzer, 2018; Schnitzer and van der Heijden, 2019). Thus, lianas provide critical resources for many vertebrate species.

Lianas also increase habitat heterogeneity and complexity, which may benefit animal species diversity and community stability (Huffaker, 1958). In the understory, lianas can form dense tangles of looping stems, which provide structure, forage, and a refuge from predators for many bird and small vertebrate species (e.g., Lambert and Halsey, 2015). By weaving through the forest canopy, lianas connect the crowns of many trees, providing critical inter-crown pathways for arboreal animals to traverse from tree to tree (e.g., ants, sloths, rodents, monkeys; Montgomery and Sunquist, 1978; Emmons and Gentry, 1983; Ødegaard, 2000; Arroyo-Rodriguez, 2015; Yanoviak, 2015). For example, Adams et al. (2019) found that removing lianas from large forest patches in central Panama decreased ant diversity (see also Adams et al., 2017). However, adding climbing ropes to forest patches where lianas had been removed allowed ant species diversity to recover, demonstrating that lianas have a strong positive effect on ant diversity (and likely other animal species) by connecting canopy trees and increasing habitat complexity (*sensu* Huffaker, 1958).

Lianas are important for bird diversity, providing habitat, lekking sites, and abundant resources (Michel et al., 2015). In a large-scale liana removal experiment in central Panama (Schnitzer 2024), bird diversity decreased precipitously following liana removal, with a 78 and 77 per cent decrease in bird abundance and diversity compared to control plots eight months following liana removal (Schnitzer et al., 2020b). Twenty months after liana removal, bird abundance and diversity were 40 and 51 per cent lower than in the control plots. The effect of liana removal was particularly damaging for insectivorous birds, which was the largest class of birds, and their abundance and diversity was 77 and 76 per cent lower, respectively, following liana removal than in the control plots 20 months following liana removal (Schnitzer et al., 2020b).

Lianas also provide services and resources to humans, including food (e.g., legumes, passion fruit, cucumbers, melons, gourds), stimulants (e.g., guarana from *Paullinia cupana*), medicines (e.g., curare from *Strychnos toxifera* and *Chondrodendron tomentosum*), as well as the raw materials for construction, fishing nets, hammocks, baskets, and a variety of crafts (Davis, 1983, Phillips, 1991, Guadagnin and Gravato, 2013). Lianas are also utilized as a source of intoxicants, providing the raw ingredients for the production of wine and cognac from grapes (*Vitis* spp.), beer from hops (*Humulus lupulus*), and hayauasca (from

Banisteriopsis caapi). Thus, lianas have many positive effects on forest biodiversity and wildlife, including humans.

The global distribution of lianas

Considering the strong influence that lianas have in forests worldwide, determining the distribution of lianas is critical for determining the forests where lianas will have the most influence. The abundance and diversity of lianas vary considerably within and among forests, with different mechanisms influencing liana abundance at local, regional, altitudinal, and latitudinal scales (Schnitzer, 2005). However, a single unified mechanism may explain how liana distribution is regulated at all scales (Schnitzer, 2018).

Local (within-forest) liana distribution

There is now compelling evidence to support the hypothesis that disturbance maintains liana diversity and controls the local (within-forest) distribution of tropical lianas. Lianas are known to be highly responsive to such disturbances as treefall gaps and forest edges in both temperate and tropical forests (e.g., Schnitzer et al., 2000, 2021; Londré and Schnitzer, 2006; Campbell et al., 2018). For example, in the 50-ha forest dynamics plot on Barro Colorado Island, Panama (BCI), the local distribution of more than 50 per cent of the liana species was positively related to disturbance, primarily treefall gaps resulting from tree mortality (Dalling et al., 2012; Schnitzer et al., 2012, 2021). Disturbance combined with positive density dependence (a measure of liana spatial clumping) explained the local distribution of 75 per cent of the liana species on the BCI 50-ha plot (Ledo and Schnitzer, 2014).

The rapid response of lianas to disturbance appears to be driven primarily by the ability of lianas to colonize via the clonal reproduction following disturbance. More than 50 per cent of the liana species produced significantly more clonal stems in disturbed (treefall gap) areas than in intact, undisturbed forest (Ledo and Schnitzer, 2014; see also Schnitzer et al., 2021). Furthermore, liana diversity was higher in treefall gaps than non-gaps even when controlling for stem density (Schnitzer and Carson, 2001). Ledo and Schnitzer (2014) contrasted the role of disturbance with two other putative diversity maintenance mechanisms (negative density dependence and niche partitioning) and found that disturbance was the major driver of liana diversity and distribution on Barro Colorado Island, Panama.

Pan-tropical liana distribution

Among lowland tropical forests, liana density and diversity increase with decreasing mean annual precipitation, peaking in seasonally dry areas, where mean annual rainfall is relatively low and the number of dry months is high (Schnitzer, 2005; DeWalt et al., 2010; Parolari et al., 2020). This pattern for liana density contrasts that of trees, which increase with increasing precipitation in tropical forests (Figure 13.3; Schnitzer, 2005). The unique pattern of decreasing liana density with increasing rainfall was first demonstrated with datasets collected in forests around the world by A.H. Gentry (published in Phillips and Miller, 2002), in which he surveyed lianas (≥ 2.5 cm diameter) in 10 small (2 × 50 m) transects (0.1 ha total) in each forest (Schnitzer, 2005). The sites included: Africa (*n*=8), Asia (*n*=4), Central America (*n*=9), and South America (*n*=45). The Gentry data, even considering the very low level of within forest sampling (and thus high variation among forests), revealed a clear decrease in liana density with increasing mean annual rainfall (Figure 13.3; Schnitzer, 2005).

DeWalt et al. (2010, 2015) used the Global Liana Database to confirm the pattern of decreasing liana density with increasing mean annual precipitation using 29 relatively well-sampled forest plots around the world: Africa (*n*=3), Asia (*n*=9), Mexico (*n*=2), Central America (*n*=4), and South America (*n*=11). The authors confirmed that liana abundance was negatively correlated with mean annual precipitation and also found that liana abundance was positively correlated with increasing dry-season length. Parolari et al. (2020) also confirmed the pattern of increasing liana density (and diversity) with decreasing rainfall and increasing seasonality along the strong rainfall gradient that crossed the isthmus of Panama. In Ghana, liana diversity was also higher in seasonal forests; lianas varied from 30 per cent of the vascular plant species in the wetter forests (2,000 mm annual rainfall) to 43 per cent of the vascular plant species in the drier forests (1,000 mm annual rainfall; Swaine and Grace, 2007). Consequently, there is compelling evidence that liana abundance and diversity peak in highly seasonal and relatively dry tropical forests and decrease with increasing precipitation and the loss of seasonality.

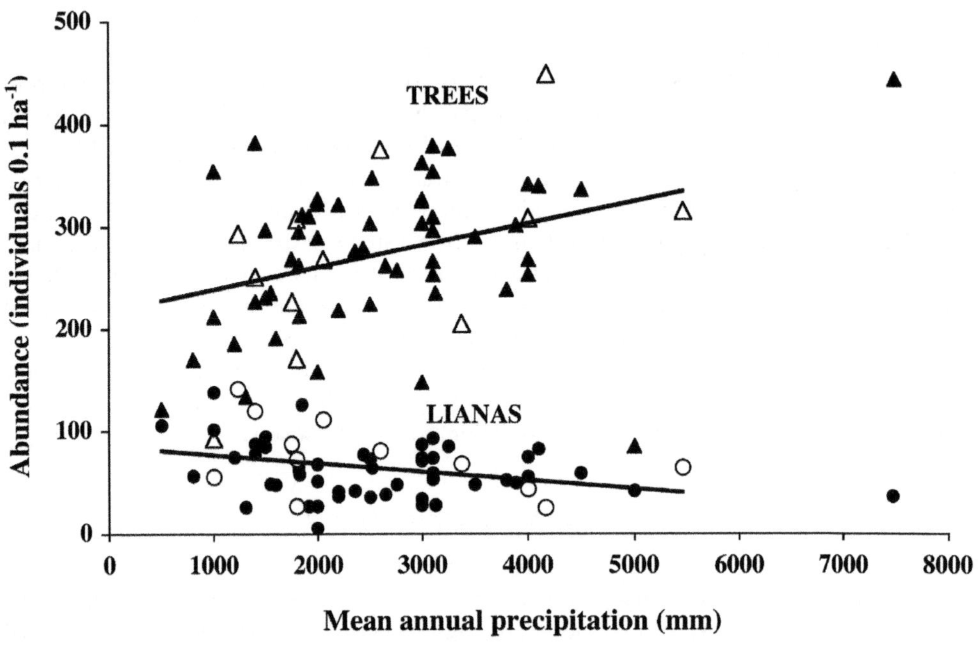

Figure 13.3 The change in liana and tree density with mean annual precipitation in lowland tropical forests. Data were collected by A.H. Gentry in 10 small (2 × 50 m) transects (0.1 ha total) in each forest, and the minimum diameter cutoff was ≥ 2.5 cm diameter (published in Phillips and Miller, 2002). Sites are from Africa (*n*=8), Asia (*n*=4), Central America (*n*=9), and South America (*n*=45). Triangles represent trees; circles represent lianas. Closed symbols are neotropical sites (Central and South America) and open symbols are old-world sites (Africa and Asia).

Source: From Schnitzer (2005).

Latitudinal liana distribution

Lianas reach their highest abundance and diversity in the tropics, and liana abundance and diversity are sharply lower outside of the tropics (Figure 13.1; Schnitzer, 2005). Lianas typically constitute 10 per cent or less of the woody plant flora in temperate forests (Durigon et al. 2014; Ladwig and Meiners, 2015; Schnitzler et al., 2016), which is substantially lower than the 35–40 per cent of liana diversity in the tropics (Schnitzer et al., 2015a and references therein). Liana abundance and diversity can be highly variable in temperate forests—even those that seem to share similar rainfall and seasonality (Londré and Schnitzer, 2006; Schnitzler et al., 2016). However, lianas in temperate forests (like in the tropics) tend to be more abundant along forest edges and in heavily disturbed forests, and they are less prevalent in intact, undisturbed areas (Londré and Schnitzer, 2006). When lianas are in high abundance, they likely have a large effect on forest communities and functioning, and thus determining the processes that control liana distributions in temperate forests remains an important question that has only recently been explored (Ladwig and Meiners, 2015; Schnitzler et al., 2016).

A unified mechanistic explanation for the distribution of lianas at multiple scales

The peak in tropical liana abundance with high seasonality and rainfall in the tropics, and the decrease in liana density and diversity with increasing latitude (and altitude), are patterns that may be explained by the unique ecological, anatomical, and physiological traits of lianas (Schnitzer, 2018). Lianas tend to have thin stems with a relatively large mass of leaves that are typically deployed at the very top of the forest canopy (Avalos and Mulkey, 1999; Rodriguez-Ronderos et al., 2016). The ratio of leaf area to conductive stem area is typically higher for lianas than trees (Wyka et al., 2013). Thin-stemmed lianas can supply adequate amounts of water to their leaves using extraordinarily large vessel elements, which provide the capacity for highly efficient water movement (e.g., Jiménez-Castillo and Lusk, 2013; Angyalossy et al., 2015; Zhang et al., 2023).

Theoretically, however, plants with such large and efficient vascular systems would generally avoid dry areas, because large vessels make them vulnerable to cavitation and embolism (e.g., Sperry, 1987). When the plant stem water column is exposed to high negative internal pressures (derived from the plant losing much more water at the leaf surface than it can pull up from the soil via its roots), the water column could break, resulting in a permanent loss of water transport capacity in some vessels. The challenge of maintaining a healthy internal water balance during dry conditions is compounded by the fact that lianas deploy most of their leaves at the top of the forest canopy, which can be a particularly arid and stressful place to grow, especially during the dry season, because of the high temperatures and low relative humidity, resulting in high vapor pressure deficit (VPD). High VPD causes plants to lose water rapidly because of the large gradient of water inside the leaf to that of the surrounding air.

It seems paradoxical for lianas to reach peak abundance in highly seasonal tropical forests that experience an extended dry season when the large vessel elements of lianas have the potential to embolize during seasonal drought (Schnitzer, 2005). However, lianas may excel in seasonal tropical forests because they can take advantage of high dry-season light availability while increasing their water use efficiency to minimize water loss during the dry season. If so, lianas would not only persist in seasonal tropical forests but they would actually grow better than co-occurring trees (Schnitzer, 2005, 2018).

Recent empirical evidence indicates that lianas have such a seasonal growth advantage (Schnitzer, 2018). In central Panama, Schnitzer and van der Heijden (2019) measured dry-season and wet-season diameter growth for 648 canopy lianas and 1,117 canopy trees over a five-year period (2013–2018). They found that lianas had the highest diameter growth and biomass gain during the dry season (Figure 13.4). In fact, lianas were able to realize 50 per cent of their annual growth during the four-month January–May dry season. By contrast, tree growth occurred mainly during the wet season, when 75 per cent of tree growth occurred (Schnitzer and van der Heijden, 2019; van der Heijden et al., 2019). Over decades of high dry season growth, this seasonal growth advantage may allow lianas to increase in

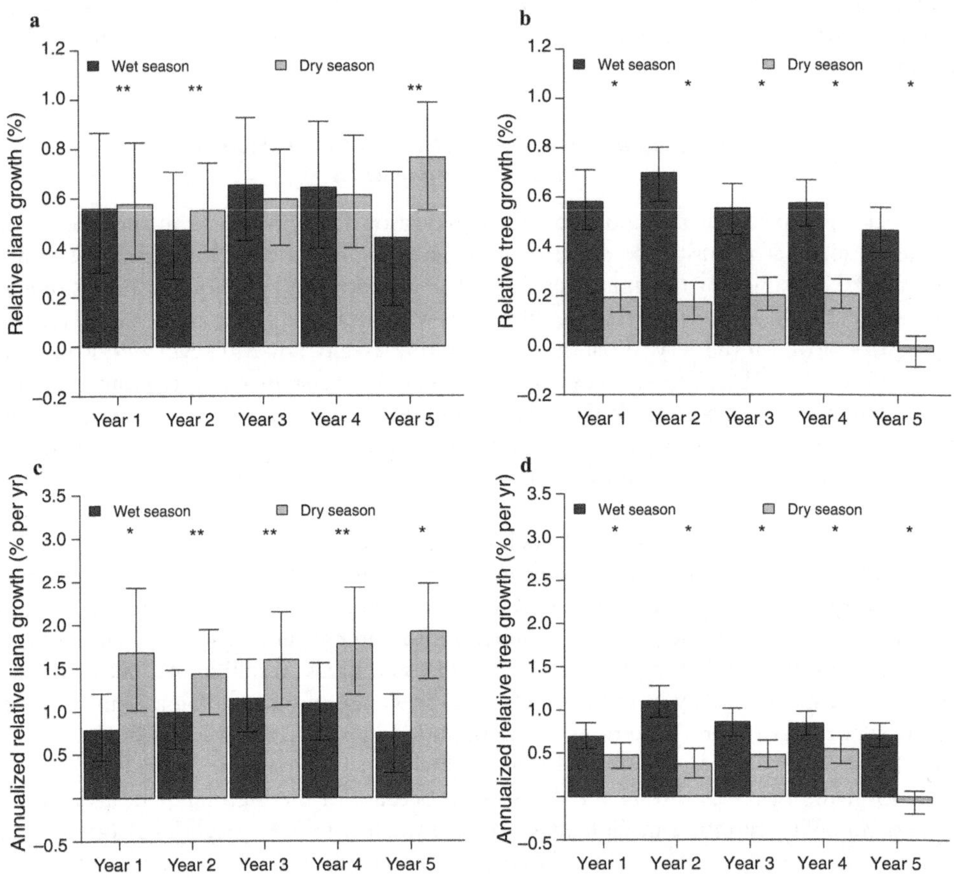

Figure 13.4 Mean bootstrapped relative growth and annualized relative growth for lianas (first column, N = 648 individuals / 54 species) and trees (second column, N = 1,117 individuals / 128 species) from 2011 to 2016 on Gigante Peninsula in central Panama. Error bars represent 95% confidence intervals based on 100,000 bootstrap iterations; * indicates $P < 0.05$, ** indicates $0.05 \le P \le 0.10$.

Source: From Schnitzer and van der Heijden (2019).

abundance relative to trees in seasonal forests compared to non-seasonal wet forests, thus explaining their pan-tropical distribution (Schnitzer, 2005, 2018).

Lianas may use two main strategies to grow during the dry season. First, lianas may fix most of their carbon early in the day by rapidly reaching their maximum photosynthetic rate, when light is plentiful, but VPD is still relatively low (fast and furious hypothesis; Schnitzer, 2018). As the day warms and VPD rises in the late morning, however, lianas close their stomata and reduce their xylem tension by taking up water from the soil the rest of the day and night. The capability of lianas to rapidly move large amounts of water may allow them to quickly reach and maintain maximum levels of photosynthesis. Indeed, lianas appear to have a suite of functional traits that are consistent with the fast and furious hypothesis (Sezen et al., 2022; Mello, 2023).

Second, lianas may be able to adjust their tolerance to drought during the dry season, which allows them to increase their water use efficiency. The increase in water use efficiency would allow lianas to take advantage of increased dry-season light. In a study of 247 tree and 47 liana species in French Guiana, Maréchaux et al. (2017) found that lianas consistently adjusted their turgor loss (wilting) point during the dry season, which theoretically would enable them to become more drought-tolerant when water availability is low. By contrast, in the same study, trees did not adjust their turgor loss point. Mello (2023) used a common garden experiment in central Panama and found a similar result; lianas significantly reduced their turgor loss point during the dry season, whereas trees did not. This dry season adjustment in turgor loss point allowed lianas to maintain sufficiently high leaf hydraulic status (see also Smith-Martin et al., 2019), which, in turn, allowed them to maintain a high dry season photosynthetic rate. By contrast, trees had lower maximum photosynthetic rates during the dry season than the wet season, apparently downregulating photosynthesis when water availability was low and the threat of water loss was high (Mello, 2023).

At the local, within forest scale, high liana abundance and diversity in treefall gaps may also be explained by their ability to grow in hot, dry, and sunny (high VPD) environments. When a tree falls in a tropical forest, lianas recruit into the resulting treefall gap as seedlings and as advanced regeneration (established plants in the forest understory). However, many lianas are also pulled into the gap along with the falling tree, where they survive the treefall and proliferate rapidly (Putz, 1984; Schnitzer et al., 2000, 2021; Schnitzer and Carson, 2001, 2010). The established adult canopy lianas that are pulled into the gaps are adapted for the hot, dry conditions found at the top of the canopy and have fully established root systems. These "naturally transplanted" lianas can proliferate immediately in the newly established gap, which provide hot and dry conditions that are similar to that of the forest canopy. By contrast, trees that were present in the dark understory prior to gap creation must replace their leaves with new leaves that can tolerate high light. These trees may also have to allocate additional resources to roots, which had not been necessary for survival in the shaded understory. Therefore, the ability of lianas to grow in the arid environment of the forest canopy, survive treefalls, and produce copious numbers of clonal stems in high VPD treefall gaps may give them an advantage in gaps that is unavailable to trees.

The explanation for the precipitous decrease in liana abundance (and species richness) with increasing latitude (and altitude) is also based on liana anatomy, particularly their highly conductive vascular system (Sperry et al., 1987; Schnitzer, 2005; Zheng et al., 2023). Large vessel elements and relatively thin stems that lack insulation make lianas particularly vulnerable to prolonged freezing conditions. When exposed to freezing temperatures, large liana vessels could rupture when water in the vessels turns to ice and expands. Or, vessels could suffer freeze–thaw embolism, wherein air bubbles form in the vessels when ice is converted to water as temperatures change from below to above freezing (Sperry et al., 1987). Both scenarios could render large portions of the vascular system inoperable and ultimately kill the plant (Jiménez-Castillo and Lusk, 2013).

Likewise, the decrease in lianas with altitude in the tropics may also be driven by the lack of resistance to cold temperatures (Jiménez-Castillo and Lusk, 2013), combined with copious rainfall, and the relatively low light, cloudy/misty environment—all of which may be unfavorable to lianas. Thus, the striking decrease in liana abundance and diversity outside of the tropics (and with increasing altitude within the tropics) appears to be driven by trade-offs of fast growth, slender stems, and a highly efficient vascular system, all of which are beneficial where the climate is warm but detrimental where the climate is seasonally cold and wet (Schnitzer, 2005, 2018).

Increasing liana abundance in neotropical forests: patterns, causes, and consequences

Liana density, biomass, and productivity appear to be increasing in neotropical forests and there are now more than 20 published studies documenting this increase (Phillips, 2002; Schnitzer and Bongers, 2011; Schnitzer et al., 2020a; Abiem et al., 2023). The pattern of increasing liana abundance has been reported in many neotropical forests, including the Bolivian Amazon, Brazilian Amazon, Costa Rica, French Guiana, Panama, and in Puerto Rico. Increasing liana abundance seems to be consistent across tropical forest types, and liana increases have been reported in mature wet, moist, and dry forests (Schnitzer, 2015).

For example, in various forests in Northwest South America, Amazonia, and Central America, Phillips et al. (2002) found that both small and large lianas had increased significantly in stem density and basal area, both in absolute values and relative to trees. The authors also found that the most rapid increase occurred in the last decade of the study, thus suggesting that liana increases will likely continue into the future. In central Amazonia, liana stem density (stems ≥ 2 cm in diameter) increased an average of 1 per cent per year from 1999 until 2012 (Laurance et al., 2014). In this same forest, liana seedling density increased 500 per cent from 1993 to 1999, while tree and herbaceous plant seedling recruitment decreased over this same period (Benítez-Malvido and Martínez-Ramos, 2003). In French Guiana, the density of large (≥ 10 cm diameter) lianas increased 1.8 per cent, while large tree abundance decreased 4.6 per cent from 1992 until 2002 (Chave et al., 2008).

On Barro Colorado Island (BCI), Panama, liana productivity (measured as leaf litter production) increased by 57 per cent from 1986 until 2002 (Wright et al., 2004), liana flower production increased more than 125 per cent faster than that of trees (Wright and Calderon, 2006), the percentage of trees with liana infestation increased from 32 per cent in 1967–1968 to 47 per cent in 1979, and then to nearly 75 per cent in 2007 (Ingwell et al., 2010). Using a longitudinal study from 2007 to 2017 of more than 117,000 rooted liana stems in the BCI 50-ha plot, Schnitzer et al. (2021) reported that liana density had increased by nearly 30 per cent and basal area by 12 per cent, representing nearly a 3 per cent annual liana density increase. In Costa Rica, liana increases have been reported in both dry and wet forests (York et al., 2013; Becknell et al., 2022). For example, in a wet forest at La Selva Biological Station in Costa Rica, Yorke et al. (2013) found a 2.9 per cent annual increase in mean liana basal area and a 2.0 per cent annual increase in mean liana density (stems ≥ 1 cm diameter) in six old-growth forest plots over an eight-year period from 1999 to 2007. While most of the liana density increases have been found in neotropical forests, lianas may also be increasing in African (Abiem et al., 2023) and Asian (Khadanga et al., 2015) forests. However, the change in lianas in African and Asian forests is not yet resolved; several studies have found either no signal of increases or even liana decreases (Bongers et al., 2020; Wright et al., 2015). Currently, there are too few long-term datasets from tropical Africa or Asia to adequately test the pattern of increasing liana abundance in these regions.

The main hypotheses proposed to explain increasing liana abundance in neotropical forests include increases in forest disturbance (including both natural disturbance and changes in land use and fragmentation), greater intensity of seasonal drought and lower annual rainfall, elevated atmospheric CO_2, elevated nutrient deposition, and increases in hunting (Schnitzer and Bongers, 2011; Schnitzer et al., 2011, 2020a). To date, only a couple of these putative mechanisms have been tested rigorously and linked to ongoing global changes. For example, Schnitzer et al. (2021) found that most of the nearly 30 per cent increase in liana density from 2007 until 2017 on BCI in Panama occurred in disturbed areas of the forest (treefall gaps). Further, they found that clonal stem production in gaps explained most of the increase pattern (Schnitzer et al., 2021). Tree mortality in forests is increasing globally (McDowell et al., 2019, 2020); thus, empirical evidence is consistent with the hypothesis that disturbance is a cause of increasing lianas.

Elevated atmospheric CO_2 and nutrient deposition may also contribute to liana increases by enabling lianas to grow more efficiently and rapidly. However, recent empirical tests did not support either of these hypotheses. For example, Marvin et al. (2016) used an open-chamber CO_2 addition experiment and found that both lianas and tree seedlings responded positively and similarly to elevated CO_2 compared to the ambient CO_2 plots. A test of the elevated nutrient deposition hypothesis also was not supported in a recent experimental test. Specifically, Schnitzer et al. (2020b) sampled lianas over a five-year period in a 20-year nutrient addition experiment in central Panama and found that neither liana density nor basal area was higher in plots that received 15 years of nutrient addition compared to control plots. Therefore, there is currently no evidence for the elevated CO_2 or nutrient deposition hypotheses; however, there is some evidence for the hypothesis that increasing disturbance is responsible for increasing tropical forest liana density and basal area. The increase in seasonal drought may also explain increasing liana density; however, this hypothesis has not yet been rigorously tested.

Summary

Lianas are now recognized as a key component of forests worldwide, particularly in highly seasonal tropical areas, where lianas peak in abundance. Lianas compete intensely with trees, decreasing tree recruitment, growth, fecundity, and diversity. Lianas also significantly reduce tree biomass accumulation; however, lianas themselves have thin stems and low wood volume, and thus they cannot compensate for the biomass that they displace in trees. Furthermore, the effects of lianas on forest community and ecosystem dynamics are becoming more pronounced in neotropical forests, where liana abundance is increasing relative to trees. Not all the effects of lianas are negative. Lianas contribute significantly to vascular plant diversity, forest structure, and complexity, and lianas are an important source of food, shelter, and habitat for animals. The structure and complexity that lianas provide may be particularly important; by linking tree canopies together, lianas provide aerial pathways for many animals to traverse the forest canopy without having to descend to the forest floor. In summary, lianas have both positive and negative effects in forests worldwide, and regardless of whether you love them, hate them, or have never thought about them, lianas are an important component of the ecology of forest ecosystems.

Acknowledgments

I thank G.G.O. Dossa and two anonymous reviewers for their helpful comments on this manuscript. Thanks also to Richard Corlett for inviting me to write this chapter. This work was made possible by financial support from US National Science Foundation grants DEB 06-13666, DEB 08-45071, DEB 10-19436, IOS 15-58093, DEB 18-22473, and DEB 20-01799.

References

Abiem, I., Kenfack D. and Chapman, H. M. (2023) 'Assessing the impact of abiotic and biotic factors in an African montane forest', *Frontiers in Forests and Global Change*, vol. 6, 1108257.

Adams, B. J., Schnitzer, S. A. and Yanoviak, S. P. (2017) 'Trees as islands: canopy ant species richness increases with the size of liana-free trees in a Neotropical forest', *Ecography*, vol. 4, pp. 1067–1075.

Adams, B. J., Schnitzer, S. A. and Yanoviak, S. P. (2019) 'Connectivity explains local ant community structure in a Neotropical forest canopy: a large-scale experimental approach', *Ecology*, vol. 100, e02673.

Alvarez-Cansino, L., Schnitzer, S. A., Reid, J. and Powers, J. S. (2015) 'Liana competition with tropical trees varies with seasonally, but not with tree species identity', *Ecology*, vol. 96, pp. 39–45.

Angyalossy, V, Pace, M. R. and Lima, A. C. (2015) 'Liana anatomy: A broad perspective on structural evolution of the vascular system', in S. A. Schnitzer, F. Bongers, R. J. Burnham and F. E. Putz (eds) *The Ecology of Lianas*, Wiley-Blackwell, Oxford.

Arroyo-Rodriguez, V., Asensio, N., Dunn, J. C., Cristóbal-Azkarate, J. and Gonzalez-Zamora, A. (2015) 'The use of lianas by primates: more than a food resource', in S. A. Schnitzer, F. Bongers, R. J. Burnham and F. E. Putz (eds) *The Ecology of Lianas*, Wiley-Blackwell, Oxford.

Asner, G.P. and Martin, R. E. (2015) Canopy chemistry expresses the life-history strategies of lianas and trees', in S. A. Schnitzer, F. Bongers, R. J. Burnham, F. E. Putz (eds) *The Ecology of Lianas*, Wiley-Blackwell, Oxford.

Avalos, G. and Mulkey, S. S. (1999) 'Seasonal changes in liana cover in the upper canopy of a neotropical dry forest', *Biotropica*, vol. 31, pp. 186–192.

Barry, K.E., S.A. Schnitzer, M. van Bruegal, J.S. Hall. 2015. 'Rapid liana colonization and community development along a secondary forest chronosequence', *Biotropica*, vol. 47, pp. 672–680.

Becknell, J. M., G. Vargas, G., Wright, L. A., Woods, N. F., Medvigy, D., Powers, J. S. (2022) 'Increasing liana abundance and associated reductions in tree growth in secondary dry seasonal forest', *Frontiers in Forests and Global Change*, vol. 5, 838357.

Benítez-Malvido J. and Martínez-Ramos, M. (2003) 'Impact of forest fragmentation on understory plant species richness in Amazonia', *Conservation Biology*, vol. 17, pp. 389–400.

Bongers, F., Ewango, C. E. N., van der Sande, M.T. and Poorter, L. (2020) 'Liana species decline in Congo basin contrasts with global patterns', *Ecology*, vol. 10, e03004.

Campbell, M. J., Edwards, W., Magrach, A., Alamgir, M., Porolak, G., Mohandass, D. and Laurance, W. F. (2018) 'Edge disturbance drives liana abundance increase and alteration of liana–host tree interactions in tropical forest fragments', *Ecology and Evolution*, vol. 8, pp. 4237–4251.

Chave J., Olivier, J., Bongers, F., Châtelet, P., Forget, P. M., van der Meer, P., Norden, N., Riéra, B. and Charles-Dominique, P. (2008) 'Aboveground biomass and productivity in a rain forest of eastern South America', *Journal of Tropical Ecology*, vol. 24, pp. 355–366.

Dalling, J.W., Schnitzer, S.A., Baldeck, C., Harms, K.E., John, R., Mangan, S.A., Lobo, E., Yavitt, J.B. and Hubbell, S.P. (2012) 'Resource-based habitat associations in a neotropical liana community', *Journal of Ecology*, vol. 100, pp. 1174–1182.

Davis, C.W. (1983) 'The enthnobotany of Chamairo: *Mussatia hyacinthia*', *Journal of Ethnopharmacology*, vol. 9, pp. 225–236.

Denslow, J. S. (1987) 'Tropical rainforest gaps and tree species diversity', *Annual Review of Ecology, Evolution, and Systematics*, vol. 18, pp. 431–451.

DeWalt, S. J., Schnitzer, S. A., Alves, L. F., Bongers, F., Burnham, R. J., Cai, Z. Q., Carson, W. p., Chave, J., Chuyong, G., Costa, F. R. C., Ewango, C. E. N., Gallagher, R., Gerwing, J. J., Gortaire, E., Hart, T., Ibarra-Manríquez, G., Ickes, K., Kenfack, D., Letcher, S., Macía, M. J., Makana, J. R., Malizia, A., Martínez-Ramos, M., Mascaro, J., Muthumperumal, C., Muthuramkumar, S., Nogueira, A., Parren, M. P. E., Parthasarathy, N., Pérez-Salicrup, D. R., Putz, F. E., Romero-Saltos, H., Sridhar Reddy, M., Sainge, M. N., Thomas, D. and van Melis, J. (2015) 'Biogeographical patterns of liana abundance and diversity', in S. A. Schnitzer, F. Bongers, R. J. Burnham, F. E. Putz (eds) *The Ecology of Lianas*, Wiley-Blackwell, Oxford.

DeWalt S. J., Schnitzer, S. A., Chave, J., Bongers, F., Burnham, R. J., Cai, Z. Q., Chuyong, G., Clark, D. B., Ewango, C. E. N., Gerwing, J. J., Gortaire, E., Hart, T., Ibarra-Manríquez, G., Ickes, K., Kenfack, D., Macía, M. J., Makana, J. R., Mascaro, J., Martínez-Ramos, M., Moses, S., Muller-Landau, H. C., Parren, M. P. E., Parthasarathy, N., Pérez-Salicrup, D. R., Putz, F. E., Romero-Saltos, H. and Thomas, D. (2010) 'Annual rainfall and seasonality predict pan-tropical patterns of liana density and basal area', *Biotropica*, vol. 42, pp. 309–317.

DeWalt, S. J., Schnitzer, S.A. and Denslow, J. S. (2000) 'Density and diversity of lianas along a chronosequence in a central Panamanian tropical forest', *Journal of Tropical Ecology*, vol. 16, pp. 1–19.

Dunn, J., Asensio, N., Arroyo-Rodriguez, V., Schnitzer, S. A. and Cristobal-Azkarate, J. (2012) 'The ranging costs of a fallback food: liana consumption supplements diet but increases foraging effort in howler monkeys', *Biotropica*, vol. 44, pp. 705–714.

Durán, S. M. and Gianoli, E. (2013) 'Carbon stocks in tropical forests decrease with liana density', *Biology Letters*, vol. 9, 20130301.

Durigon, J., Luitti, S. T. S. and Gianoli, E. (2014) 'Distribution and traits of climbing plants in subtropical and temperate South America', *Journal of Vegetation Science*, vol. 25, pp. 1484–1492.

Emmons, L. H. and Gentry, A. H. (1983) 'Tropical forest structure and the distribution of gliding and prehensile-tailed vertebrates', *The American Naturalist*, vol. 121, pp. 513–524.

Estrada-Villegas, S., Hall, J. S., van Breugel, M. and Schnitzer, S. A. (2020) 'Lianas reduce biomass accumulation in early successional tropical forests', *Ecology*, vol. 101, e02989.

Estrada-Villegas, S., Pedraza-Narvaez, S. S., Sánchez-Andrade, A., Schnitzer, S. A. (2022) 'Lianas significantly reduce tree performance and biomass accumulation across tropical forests: a global meta-analysis', *Frontiers in Forests and Global Change*, vol. 4, https://doi.org/10.3389/ffgc.2021.812066

Estrada-Villegas, S. and Schnitzer, S. A. (2018) 'A comprehensive synthesis of liana removal experiments in tropical forests', *Biotropica*, vol. 50, pp. 729–739.

Garcia-Leon, M.M., Martinez-Izquierdo, L., Mello, F. N. A., Powers, J. S. and Schnitzer, S. A. (2018) 'Lianas reduce community-level canopy tree reproduction in a Panamanian forest', *Journal of Ecology*, vol. 106, pp. 737–745.

Gianoli, E. 2015. 'Evolutionary implications of the climbing habit in plants', pp. 239–250, in S.A. Schnitzer, F. Bongers, R.J. Burnham, F.E. Putz (eds), *The Ecology of Lianas*, Wiley-Blackwell, Oxford.

Gora, E.M., Blitzer, P.M., Burchfield, J.C., Guitierez, C. and Yanoviak, S.P. (2021) 'The contributions of lightning to biomass turnover, gap formation and plant mortality in a tropical forest', *Ecology*, vol. 102, e03541.

Grauel W.T., F.E. Putz. (2004) 'Effects of lianas on growth and regeneration of Prioria copaifera in Darien, Panama', *Forest Ecology and Management*, vol. 190, pp. 99–108.

Guadagnin, D.L. and Gravato, I.C. (2013) 'Ethnobotany, availability, and use of lianas by the Kaingang people in suburban forests in southern Brazil', *Economic Botany*, vol. 67, pp. 350–362.

Huffaker, C.B. 1958. 'Experimental studies on predation: dispersion factors and predator-prey oscillations', *Hilgardia*, vol. 27, pp. 795–835.

Ibarra-Manríquez, G., P. Carrillo-Reyes, F.J. Rendón-Sandoval, G. Cornejo-Tenorio (2015) 'Diversity and distribution of lianas in Mexico', pp. 91-103 in S.A. Schnitzer, F. Bongers, R.J. Burnham, F.E. Putz (eds) *The Ecology of Lianas*, Wiley-Blackwell, Oxford.

Ingwell, L.L., Wright, S.J., Becklund, K.K., Hubbell, S.P. and Schnitzer, S.A. (2010) 'The impact of lianas on 10 years of tree growth and mortality on Barro Colorado Island, Panama', *Journal of Ecology* vol. 98, pp. 879–887.

Jiménez-Castillo M. and Lusk, C. H. (2013) 'Comparative vascular anatomy and function of woody plants in a temperate rainforest: lianas suffer higher levels of freeze–thaw embolism than associated trees', *Functional Ecology*, vol. 27, pp. 403–412.

Kainer, K. A., Wadt, L. H. O. and Staudhammer, C. L. (2014) 'Testing a silvicultural recommendation: Brazil nut responses 10 years after liana cutting', *Journal of Applied Ecology*, vol. 51, pp. 655–663.

Khadanga, S. S., Muthumperumal, C. and Parthasarathy. N. (2015) 'Changes in liana diversity over a decade in Indian tropical dry evergreen forests', in S. K. Tripathi (ed) *Biodiversity in Tropical Ecosystems 2015*, Today and Tomorrow's Printers and Publishers, New Delhi.

Ladwig, L. and Meiners, S. (2015) 'The role of lianas in temperate tree communities', in S. A. Schnitzer, F. Bongers, R. J. Burnham, F. E. Putz (eds) *The Ecology of Lianas*, Wiley-Blackwell, Oxford.

Lambert, T. D. and Halsey, M. K. (2015) 'Relationship between lianas and arboreal mammals: examining the Emmons–Gentry hypothesis', in S. A. Schnitzer, F. Bongers, R. J. Burnham, F. E. Putz (eds) *The Ecology of Lianas*, Wiley-Blackwell, Oxford.

Laurance, W. F., Andrade, A. S., Magrach, A., Camargo, J. L. C., Valsko, J. J., Campbell, M., Fearnside, P. M., Edwards, W., Lovejoy, T. E. and Laurance, S. G. (2014) 'Long-term changes in liana abundance and forest dynamics in undisturbed amazonian forests', *Ecology*, vol. 95, pp. 1604–1611.

Ledo, A., Illian, J. B., Schnitzer, S. A., Wright, S. J., Dalling, J. W. and Burslem, D. (2016) 'Prediction of fine-scale distribution of aboveground biomass in a tropical moist forest', *Journal of Ecology*, vol. 104, pp. 1819–1828.

Ledo, A. and Schnitzer, S. A. (2014) 'Disturbance, not negative density dependence or habitat specialization maintains liana diversity in a tropical forest', *Ecology*, vol. 95, pp. 3008–3017.

Londré R. A. and Schnitzer S. A. (2006). 'The distribution of lianas and their change in abundance in temperate forests over the past 45 years', *Ecology*, vol. 87, pp. 2973–2978.

Maréchaux, I., Bartlett, M. K., Iribar, A., Sack, L., Chave, J. (2017) 'Stronger seasonal adjustment in leaf turgor loss point in lianas than trees in an Amazonian forest', *Biology Letters*, vol. 13, 20160819.

Marshall, A.R., Coates, M. A., Archer, J., Kivambe, E., Mnendendo, H., Mtoka, S., Mwakisoma, R., de Figueiredo, R. J. R. L. and Njilima, F. M. (2017) 'Liana cutting for restoring tropical forests: a rare palaeotropical trial', *African Journal of Ecology*, vol. 55, pp. 282–297.

Marvin, D. C., Asner, G. P., Schnitzer, S. A. (2016) 'Liana canopy cover mapped throughout a tropical forest with high-fidelity imaging spectroscopy', *Remote Sensing of Environment*, vol. 176, pp. 98–106.

McDowell, N., Allen, C.D., Anderson-Teixeira, K., Aukema, B.H., Bond-Lamberty, B., Chini, L., Clark, J.S., Dietze, M. et al. (2020). 'Pervasive shifts in forest dynamics in a changing world', *Science*, vol. 368, eaaz9463.

McDowell, N., Allen, C. D., Anderson-Teixeira, K., Brando, P., Brienen, R., Chambers, J., Christoffersen, B. et al. (2019). 'Drivers and mechanisms of tree mortality in moist tropical forests', *New Phytologist*, vol. 219, pp. 851–869.

Mello, F.N.A. (2023) *A trait-based approach to explain plant coexistence in tropical forests.* Doctoral disseration, Marquette University, Milwaukee, Wisconsin.

Michel, N.L., Robinson, W.D. and Sherry, T.W. (2015) 'Liana–bird relationships: a review', pp. 362–397 in S.A. Schnitzer, F. Bongers, R.J. Burnham, F.E. Putz (eds) *The Ecology of Lianas*, Wiley-Blackwell, Oxford.

Montgomery, G.G. and Sunquist, M.E. (1978) 'Habitat selection and use by two-toed and threetoed sloths', pp. 329–359, in G.G. Montgomery (ed) *The Ecology of Arboreal Folivores*, Smithsonian Institution Press, Washington, DC.

Moorthy, S., Calders, K., di Porcia e Brugnera, M., Schnitzer, S.A. and Verbeeck, H. (2018). 'Terrestrial laser scanning to detect liana impact on forest structure', *Remote Sensing*, vol. 10, p. 810.

Ødegaard, F. (2000) 'The relative importance of trees versus lianas as hosts for phytophagous beetles in tropical forests', *Journal of Biogeography*, vol. 27, pp. 283–296.

Parolari, A.J., Paul, K., Griffing, A., Condit, R., Perez, R., Aguilar, S. and Schnitzer, S.A. (2020) 'Liana abundance and diversity increase with rainfall seasonality along a precipitation gradient in Panama', *Ecography*, vol. 43, pp. 25–33.

Parthasarathy, N., Muthuramkumar, S., Muthumperumal, C., Vivek, P., Ayyappan, N. and Reddy, M.S. (2015) 'Diversity and distribution of lianas in Mexico', pp. 36–49 in S.A. Schnitzer, F. Bongers, R.J. Burnham and F.E. Putz (eds) *The Ecology of Lianas*, Wiley-Blackwell, Oxford.

Phillips, O.L. (1991) 'The enthnobotany and economic botany of tropical vines', in F. E. Putz and H. A. Mooney (eds) *The Biology of Vines*, Cambridge University Press, Cambridge.

Phillips, O. L. and Miller, J. S. (2002) *Global Patterns of Plant Diversity: Alwyn H. Gentry's Forest Transect Data Set*, Missouri Botanical Garden, St. Louis.

Phillips, O.L., Vasquez Martinez, R., Arroyo, L., Baker, T.R., Killeen, T., Lewis, S.L., Malhi, Y., Mendoza, A.M., Neill, D., Vargas, P.N., Alexiades, M., Ceron, C., Di Fiore, A., Erwin, T., Jardim, A., Palacios, W., Saldias, M. and Vincenti, B. (2002) 'Increasing dominance of large lianas in Amazonian forests', *Nature*, vol. 418, pp. 770–774.

Putz, F. E. (1984) 'The natural history of lianas on Barro Colorado Island, Panama', *Ecology*, vol. 65, pp. 1713–1724.

Reid, J. P., Schnitzer, S. A. and Powers, J. S. (2015) 'Soil moisture variation after liana removal in a seasonally moist, lowland tropical forest', *PLoS One*, vol. 10, e0141891.

Rodriguez-Ronderos, M. E., Bohrer, G., Sanchez-Azofeifa, A., Powers, J. S. and Schnitzer, S. A. 2016. 'Contribution of lianas to plant area index and structure in a Panamanian forest', *Ecology*, vol. 97, pp. 3271–3277.

Saatchi, S. S., Harris, N. L., Brown, S., Lefsky, M., Mitchard, E. T. A., Salas, W., Zutta, B. R., Buermann, W., Lewis, S. L., Hagen, S. Petrova, S., White, L., Silman, M. and Morel, A. (2011) 'Benchmark map of forest carbon stocks in tropical regions across three continents', *Proceedings of the National Academies of Science*, vol. 108, pp. 9899–9904.

Schnitzer, S. A. (2005) 'A mechanistic explanation for global patterns of liana abundance and distribution', *The American Naturalist*, vol. 166, pp. 262–276.

Schnitzer, S. A. (2015) 'Increasing lianas abundance in neotropical forests: causes and consequences', in S. A. Schnitzer, F. Bongers, R. J. Burnham and F. E. Putz (eds) *The Ecology of Lianas*, Wiley-Blackwell, Oxford.

Schnitzer, S. A. (2018) 'Testing ecological theory with lianas. Tansley Review', *New Phytologist*, vol. 220, pp. 366–380.

Schnitzer, S. A. (2024) 'The gigante liana removal experiment', in H. C. Muller-Landau and S. J. Wright (eds) *The First 100 Years of Research on Barro Colorado: Plant and Ecosystem Science*, Smithsonian Institution Scholarly Press, Washington, DC.

Schnitzer, S. A. and Bongers, F. (2002) 'The ecology of lianas and their role in forests', *Trends in Ecology and Evolution*,. vol. 17, pp. 223–230.

Schnitzer, S. A. and Bongers, F. (2011) 'Increasing liana abundance and biomass in tropical forests: emerging patterns and putative mechanisms', *Ecology Letters*, vol. 14, pp. 397–406.

Schnitzer S. A., Bongers, F., Burnham, R. J. Putz F. E. (eds) (2015a) *Ecology of Lianas*, Wiley-Blackwell, Oxford.

Schnitzer, S. A., Bongers, F. and Wright, S. J. (2011) 'Community and ecosystem ramifications of increasing lianas in neotropical forests', *Plant Signaling and Behavior*, vol. 6, pp. 598–600.

Schnitzer, S. A. and Carson, W. P. (2001) 'Treefall gaps and the maintenance of species diversity in a tropical forest', *Ecology*, vol. 82, pp. 913--919

Schnitzer, S.A. and Carson., W.P. (2010). 'Lianas suppress tree regeneration and diversity in treefall gaps',. *Ecology Letters*, vol. 13, pp. 849–857.

Schnitzer, S. A. and Carson, W. P. (2024) 'Ecology of lianas on Barro Colorado nature monument', in H. C. Muller-Landau and S. J. Wright (eds) *The First 100 Years of Research on Barro Colorado: Plant and Ecosystem Science*, Smithsonian Institution Scholarly Press, Washington, DC.

Schnitzer, S. A., Dalling, J. W. and Carson, W. P. (2000) 'The impact of lianas on tree regeneration in tropical forest canopy gaps: Evidence for an alternative pathway of gap-phase regeneration', *Journal of Ecology*, vol. 88, pp. 655–666.

Schnitzer, S. A., DeFilippis, D., Visser, M., Estrada-Villegas, S., Rivera-Camaña, R., Bernal, B., Peréz, S., Valdéz, A., Valdéz, S., Aguilar, A., Broadbent, E. N., Almeyda Zambrano, A. M., Dalling, J. W., Hubbell, S. P. and Garcia-Leon, M. M. (2021). 'Local canopy disturbance as an explanation for long-term increases in liana abundance', *Ecology Letters*, vol. 24, pp. 2635–2647.

Schnitzer, S. A., Estrada-Villegas, S., Wright, S. J. (2020b) 'The response of lianas to 20 years of nutrient addition in a Panamanian forest', *Ecology*, vol. 101, e03190.

Schnitzer S.A., Kuzee, M. and Bongers, F. (2005). Disentangling above- and below-ground competition between lianas and trees in a tropical forest. *Journal of Ecology*, vol. 93, pp. 1115–1125.

Schnitzer, S. A., Mangan, S. A., Dalling, J. W., Baldeck, C. A., Hubbell, S. P., Ledo, A., Muller-Landau, H., Tobin, M. F., Aguilar, S., Brassfield, D., Hernandez, A., Lao, S., Perez, R., Valdes, O., and Yorke, S. R. (2012) 'Liana abundance, diversity, and distribution on Barro Colorado Island, Panama', *PLoS One*, vol. 7, e52114.

Schnitzer, S. A., Michel, N. L., Powers, J. S. and Robinson, W. D. 2020a. 'Lianas increase insectivorous bird abundance and diversity in a tropical forest', *Ecology*, vol. 101, e03176.

Schnitzer, S. A., Putz, F. E., Bongers, F. and Kroening, K. (2015b) 'The past, present, and potential future of liana ecology', in S. A. Schnitzer, F. Bongers, R. J. Burnham and F. E. Putz (eds). *The Ecology of Lianas*, Wiley-Blackwell, Oxford.

Schnitzer, S. A. and van der Heijden, G. M. F. (2019) 'Lianas have a seasonal growth advantage over co-occurring trees', *Ecology*, vol. 100, e02655.

Schnitzer, S. A., van der Heijden, G. M. F., Mascaro, J. and Carson, W. P. (2014) 'Lianas in gaps reduce carbon accumulation in a tropical forest', *Ecology*, vol. 95, pp. 3008–3017.

Schnitzler, A., Amigo, J., Hale, B. and Schnitzler, C. (2016) 'Patterns of climber distribution in temperate forests of the Americas', *Journal of Plant Ecology*, vol. 9, pp. 724–733.

Sezen, U. U., Worthy, S. J., Umaña, M. N., Davies, S. J., McMahon, S. M. and Swenson, N. G. (2022). 'Comparative transcriptomics of tropical woody plants supports fast and furious strategy along the leaf economics spectrum in lianas', *Biology Open*, vol. 11, bio059184.

Smith-Martin, C.M., Bastos, C. L., Lopez, O. R., Powers, J. S. and Schnitzer, S. A. (2019). 'Effects of dry-season irrigation on leaf physiology and biomass allocation in tropical lianas and trees', *Ecology*, vol. 100, e02827.

Sperry, J. S., Holbrook N. M., Zimmerman M. H. and Tyree M. T. (1987). 'Spring filling of xylem vessels in wild grapevine', *Plant Physiology*, vol. 83, pp. 414–417.

Stewart, T. E. and Schnitzer, S. A. (2017) 'Blurred lines between competition and parasitism', *Biotropica*, vol. 49, pp. 433–438.

Swaine, M. D. and Grace, J. (2007) 'Lianas may be favoured by low rainfall: evidence from Ghana', *Plant Ecology*, vol. 192, pp. 271–276.

Thomas, D., Burnham, R. J., Chuyong, G., Kenfack, D. and Nsangy Sainge, M. (2015) 'Liana abundance and diversity in Cameroon's Korup National Park', in: S. A. Schnitzer, F. Bongers, R. J. Burnham and F. E. Putz (eds) *The Ecology of Lianas*, Wiley-Blackwell, Oxford.

Tobin M. F., Wright A. J., Mangan S. A. and Schnitzer S. A. (2012) 'Lianas have a greater competitive effect than trees of similar biomass on tropical canopy trees', *Ecosphere*, vol. 3, p. 20.

Toledo-Aceves, T. (2015) 'Above and belowground competition between lianas and trees', in S. A. Schnitzer, F. Bongers, R. J. Burnham and F. E. Putz (eds) *The Ecology of Lianas*, Wiley-Blackwell, Oxford.

van der Heijden, G. M. F and Phillips, O.L. (2009) 'Liana infestation impacts tree growth in a lowland tropical moist forest', *Biogeosciences*, vol. 6, pp. 2217–2226.

van der Heijden, G. M. F., Powers, J. S. and Schnitzer, S. A. (2015) 'Lianas reduce carbon accumulation in tropical forests', *Proceedings of the National Academy of Sciences*, vol. 112, pp. 13267–13271.

van der Heijden, G. M. F., Powers, J. S. and Schnitzer, S. A. (2019) 'No seasonal differences in liana effect on forest-level tree biomass growth in a liana removal experiment in Panama', *Journal of Ecology*, vol. 107, pp. 1890–1900.

van der Heijden, G. M. F., Schnitzer, S. A., Powers, J. S. and Phillips, O. L. (2013) 'Liana impacts on carbon cycling, storage and sequestration in tropical forests', *Biotropica*, vol. 45, pp. 682–692.

Wright, S. J. and Calderon, O. (2006) 'Seasonal, El Nino and longer term changes in flower and seed production in a moist tropical forest', *Ecology Letters*, vol. 9, pp. 35–44.

Wright S. J., Calderón O., Hernandéz A. and Paton S. (2004) 'Are lianas increasing in importance in tropical forests? A 17-year record from Panamá', *Ecology*, vol. 85, pp. 484–489.

Wright, S. J., Sun, I.-F., Pickering, M., Fletcher, C. D. and Chen, Y.-Y. 2015. 'Long-term changes in liana loads and tree dynamics in a Malaysian forest', *Ecology*, vol. 96, pp. 2748–2757.

Wyka, T. P., Oleksyn, J., Karolewski, P. and Schnitzer, S. A. (2013) 'Phenotypic correlates of the lianescent growth form—a review', *Annals of Botany*, vol. 112, pp. 1667–1681.

Xu, L., Saatchi, S., Lang, Y., Yu, Y., Pongratz, J., Bloom, A. A., Bowman, K., Worden, J., Liu, J., Yin, Y., Domke, G., McRoberts, R. E., Woodall, C., Nabuurs, G.-J., De-Miguel, S., Keller, M., Harris,

N., Maxwell, S. and Schimel, D. (2021) 'Changes in global terrestrial live biomass over the 21st century', *Scientific Advances*, vol. 7, eabe9829.

Yanoviak, S.P. (2015). 'Effects of lianas on canopy arthropod community structure', pp. 345–361 in S.A. Schnitzer, F. Bongers, R.J. Burnham, and F.E. Putz) (eds) *The Ecology of Lianas*, Wiley-Blackwell, Oxford.

Yorke, S.R., Schnitzer, S. A., Mascaro, J., Letcher, S. and Carson, W. P. (2013) 'Increasing liana abundance and biomass in a tropical forest: the contribution of long-distance clonal colonization', *Biotropica*, vol. 45, pp. 317–324.

Zhang, K.-Y., Yang, D., Zhang, Y.-B., Liu, Q., Wang, Y.-S., Ke, Y., Xiao, Y., Wang, Q., Dossa, G. G. O., Schnitzer, S. A. and Zhang, J.-L. (2023) 'Vessel dimorphism and wood traits in lianas and trees among three contrasting environments', *American Journal of Botany*, vol. 110, e16154.

14
VASCULAR EPIPHYTES IN FOREST ECOSYSTEMS

Amanda Taylor

Introduction

Forest canopies are often described as biodiversity hotspots, being home to a vast array of plant and animal species adapted to life in the treetops. One of the most remarkable components of forest canopy ecosystems are vascular epiphytes, defined here as non-parasitic plants that grow on the branches and stems of host trees (termed *phorophytes*). While epiphytes account for just 9.6 per cent of the world's vascular flora (Zotz et al., 2021), a Neotropical montane cloud forest may reveal a rich variety of epiphytes that comprise over half of the species diversity. In tropical forest ecosystems, epiphytes provide various ecosystem services from microclimate regulation and rainfall partitioning (Mendieta-Leiva et al., 2020; Stuntz et al., 2002), to nutrient acquisition and cycling (Coxson and Nadkarni, 1995), to promoting diversity by providing food, water, and habitat for canopy-dwelling fauna (e.g., spiders, Méndez Castro et al., 2018).

The earliest known existence of epiphytes can be traced back to the Carboniferous period 359–299 Ma when lycophytes, ferns and now-extinct fern relatives dominated the landscape (Benzing, 2004). However, the first major radiation events of terrestrial plants into the 'epiphyte niche' did not occur until the Cretaceous period 145–66 Ma, coinciding with the rapid ascent of angiosperms, also known as flowering plants (Watkins and Cardelús, 2012). The diversification of angiosperms generated novel ecological opportunities, facilitating evolution as species adapted to new canopy substrates and microclimates in a process analogous to how species colonise and diversify into vacant niches on recently formed oceanic islands. Despite this, most modern-day epiphyte lineages diversified comparatively recently, either in parallel with mountain uplift (e.g., the Andes ~20–15 Ma, Givnish et al., 2014; Pérez-Escobar et al., 2017), or in relation to past climate shifts such as occurred in the Miocene ~23–5 Ma, which led to an expansion of tropical forest and thus new habitat for epiphytes to colonise and diversify into (Andriananjamanantsoa et al., 2016).

To date, epiphytism has evolved independently in 79 plant families (Zotz et al., 2021), although it is important to note that the taxonomic distribution of this life form is highly variable. Among the angiosperms, epiphytism is nowhere more strikingly expressed than in the Orchidaceae family, of which 75 per cent are epiphytes (Zotz et al., 2021). Other

 DOI: 10.4324/9781003324072-17

taxonomic groups that commonly adopt the epiphytic habit include pteridophytes (ferns and lycophytes, 25 per cent epiphytic), Bromeliaceae (59 per cent), and Araceae (20 per cent, Zotz et al., 2021), where the development of epiphyte-associated traits has been linked to rapid diversification events observed within these groups (Crayn et al., 2004; Givnish et al., 2015). In contrast, epiphytism is poorly represented among gymnosperms, Asteraceae, and Poaceae, with less than 1 per cent of their species being epiphytic, while the large, cosmopolitan Brassicaceae, Fabaceae, and Euphorbiaceae families have no epiphytic representatives (Zotz et al., 2021). The uneven distribution of epiphytes among plant families likely reflects a complex interplay between ecological, environmental, and evolutionary factors, making the study of epiphytes hugely important to understand the evolution of plant species and trait adaptations.

Global patterns and drivers of epiphyte distributions

Just as epiphytes are taxonomically unevenly distributed throughout the plant kingdom, their spatial distributions are also highly heterogeneous among biogeographical regions (Figure 14.1). In this section, I will outline our current understanding of the global distribution of vascular epiphytes, including the ecological and evolutionary processes that have shaped their distribution patterns over time.

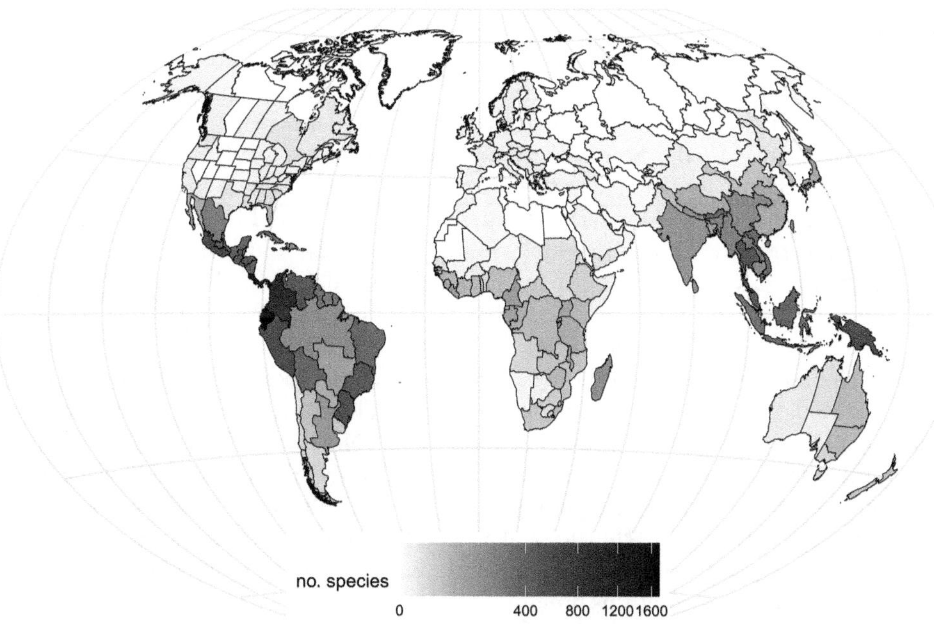

Figure 14.1 Global patterns of epiphyte species richness per 10,000 km². White regions indicate no species are present, while increasingly darker shades correlate with increasing species richness. Regions are delineated following the Taxonomic Database Working Group (TDWG) scheme (Brummitt et al., 2001), which use standardised botanical units rather than political units. Data are from Taylor et al. (2022) and are freely available for download from the Dryad depository https://doi.org/10.5061/dryad.kwh70rz46.

Arguably the most striking global distribution pattern observed among epiphytes is the significant decrease in species richness and abundance as one moves away from the tropics to the poles (Benzing, 1990). While a decrease in diversity with increasing latitude is not unusual, epiphytes decrease three times more rapidly than terrestrial species, and thus play an important role in driving the latitudinal diversity gradient for plants (Taylor et al., 2022). Why epiphytes are so tightly associated with the tropics has prompted several decades of research, with water availability being widely accepted as the primary physiological constraint on epiphyte distributions (Zotz, 2016). Due to their lack of direct contact with the ground and thus soil water, epiphytes are highly dependent on precipitation in the form of rain, fog, or humidity to meet their water requirements. Even in wet forests, water supply is often inconsistent, and epiphytes have developed specialised adaptations that maximise water uptake, storage, and use-efficiency. Tank-forming bromeliads, for example, have evolved specialised leaves that form a central reservoir for the collection and storage of water, while pseudobulb-forming orchids possess modified, bulbous stems or 'pseudobulbs', which reduce water stress during intermittent periods of drought (Yang et al., 2016). At the other end of the spectrum are atmospheric bromeliads (e.g., *Tillandsia*), which can thrive in otherwise inhospitable environments due to their ability to absorb water directly from the atmosphere (Benzing, 2000). These water-storing traits, while beneficial in a canopy environment, are highly vulnerable to frost damage in cold environments where the freezing of water damages plant cells. Most epiphyte lineages are therefore restricted to the tropics as they lack specialised functional traits to persist in both arid regions (e.g., Mediterranean climate, deserts) and regions with frequent frost, both of which are associated with higher latitudes (Taylor et al., 2022).

The lower incidence of frost and increased precipitation in the Southern Hemisphere may therefore account for a second notable biogeographic trend in epiphyte distribution—the more pronounced decrease in species richness and abundance with increasing latitude in the northern relative to the Southern Hemisphere (Benzing, 1990). For example, while Georgia (USA, 32.65° north) has just 17 epiphyte species, New South Wales (Australia, 32.18° south), which sits at a similar latitude to the south, has 113 species. Similarly, the southern island of New Zealand (44.02° south, 69 species) and southern Chile (48.21° south, 39 species) are more diverse than equivalent latitudes in the Northern Hemisphere (e.g., Yugoslavia 44.16° north, four species; Newfoundland 48.72° north, two species, Taylor et al., 2022). This extreme disparity in epiphyte diversity between Northern and Southern Hemispheres is likely not only a function of contemporary climate conditions but also related to past fluctuations in climate. During the Pleistocene epoch, which lasted from about 2.6 million to 11,700 years ago, much of the Northern Hemisphere experienced severe glaciations (Raymo, 1994), which drastically altered the landscape and led to the extinction of many plant species, including epiphytes. In contrast, the Southern Hemisphere did not experience as extreme climate oscillations (Kaplan et al., 2010), leading to the persistence of tropical, sub-tropical, and temperate rainforests that likely acted as refugia for many epiphyte lineages during this time.

Although strongly associated with the tropics, the diversity and composition of epiphytes is not uniform across the major tropical realms. Among these, the Neotropics have an exceptional level of epiphyte diversity, accounting for 63 per cent of the world's epiphyte flora, with the Indomalayan, Australasian, and Afrotropical realms reporting significantly less—18, 17, and 6 per cent, respectively (Taylor et al., 2022). This discrepancy is likely attributed to the repeated radiations of angiosperms that have occurred to a greater extent in the Neotropics than in any other tropical realm (Hagen et al., 2021). For example, several large epiphytic angiosperm families like Bromeliaceae, Araceae, and Ericaceae form significant constituents of the Neotropical epiphyte flora, yet do not appear to have diversified elsewhere (Taylor et al., 2022). Moreover, the Neotropics have experienced fewer effects from Pleistocene dry periods, which significantly impacted the least diverse Afrotropical flora (Tryon, 1986), likely contributing to its lower epiphyte diversity.

Epiphyte species composition patterns, like diversity, also vary considerably around the world. One evident pattern is that seed plants and pteridophytes (ferns and lycophytes) respond differently to the environmental gradients correlating with latitude, with seed plant diversity decreasing more rapidly with increasing latitude than pteridophytes (Figure 14.2B, C). For instance, in southern Chile, epiphytism is observed in less than 1 per cent of all seed plants, whereas approximately 34 per cent of pteridophytes in the region are epiphytic. Similarly, pteridophytes in New Zealand and southern Australia also display a significant degree of epiphytism, with 28 and 14 per cent of pteridophytes growing epiphytically, respectively (Taylor et al., 2022). Pteridophytes have several adaptations that may allow them to survive colder and drier environments compared to most epiphytic seed plants with tropical origins. For one, at least 10 per cent of all pteridophytes are classified as xerophytic epiphytes (Tryon, 1964), meaning that they are specialised to survive in arid environments, although this number is likely underestimated (Hietz, 2010). The 'trash basket' growth form may also contribute to reducing drought stress in epiphytic pteridophytes by trapping canopy debris. As well as displaying a degree of drought tolerance, experimental evidence suggests that some epiphytic pteridophytes may withstand below freezing temperatures of –20°C to –40°C (Sato, 1982), which might explain their higher prevalence in temperate forests and high elevations compared to angiosperms.

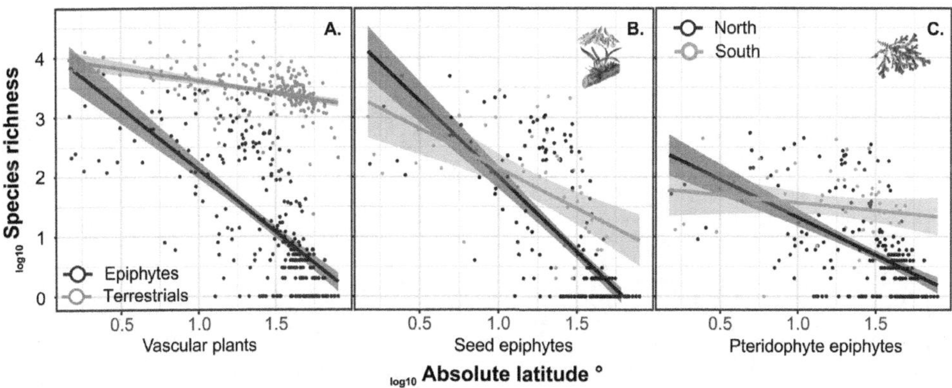

Figure 14.2 The relationship between absolute latitude and (A) the total number of vascular epiphytes (black points) and terrestrial plants (grey points). Absolute latitude is then divided among hemispheres in panels (B) seed epiphytes and (C) pteridophyte epiphytes, where black points represent the epiphyte-latitudinal gradient in the Northern Hemisphere, and grey points represent the epiphyte-latitudinal gradient in the southern-hemisphere. All shaded lines are 95 per cent confidence bands. Both axes are log10 transformed. Data are from Taylor et al. (2022) and are freely available for download from the Dryad depository https://doi.org/10.5061/dryad.kwh70rz46.

The greater role of epiphytes in forest ecosystems and threats to the epiphyte habit

While epiphytes make a substantial contribution to most global centres of plant diversity, they also play an important role in promoting the structure, functioning, and biodiversity within the forest ecosystems they occur.

One key feature of forest canopies is the vertical gradient of light intensity, temperature, and relative humidity that results from the gradual interception of light as it travels down through the canopy layers. This gradient creates a mosaic of within-tree microclimate conditions, with increasingly brighter, warmer, and drier conditions towards the tree crown and darker, cooler, and more humid conditions towards the tree base. A second major environmental axis runs horizontally from the trunk to the outer branches, where light intensity and temperature increase with increasing horizonal distance from the trunk, while humidity decreases. Depending on their specific light, moisture, and nutrient requirements, epiphytes partition and distribute themselves along these vertical and horizontal planes in a process called 'niche partitioning' (Benzing, 1990). Niche partitioning promotes species coexistence by way of natural selection, where co-occurring competitors evolve to utilise resources (e.g., water, light) in different ways. Thus, in a tropical montane cloud forest with ample moisture availability, a single tree may yield hundreds of epiphyte individuals seemingly growing on top of one another (Figure 14.3).

The resulting 'epiphytic matter' (Zotz, 2016), which consists not only of living and dead epiphyte vegetation, but also canopy soil and associated soil invertebrates and fungi, can weigh up to an estimated 44,000 kg ha^{-1} (Hofstede et al., 1993), although there is a huge variation among studies and forest types (see Gotsch et al., 2016; Zotz, 2016 for comparisons). Epiphytes therefore play an important role in forest hydrology and biogeochemical cycles through the interception and storage of water and decomposing organic material (Benzing,

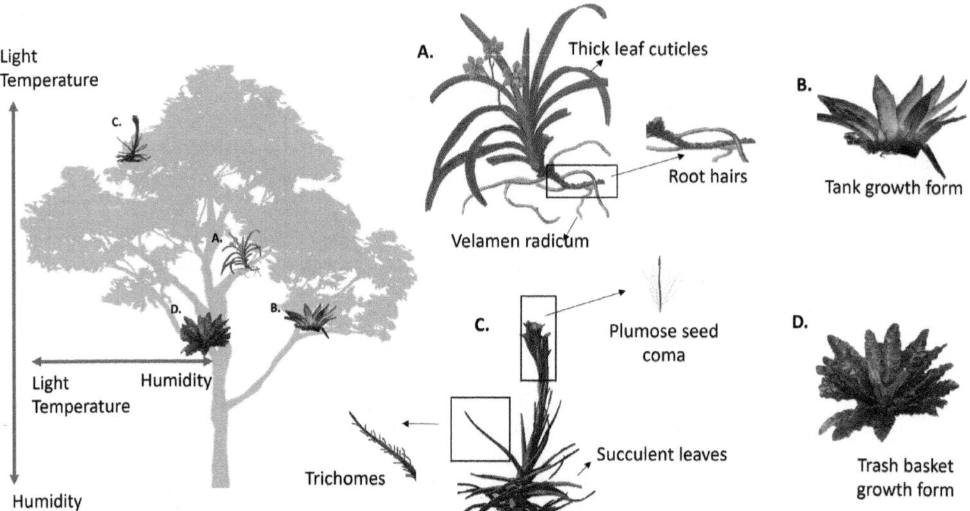

Figure 14.3 Examples of vascular epiphyte assemblages of seed plants (A) and pteridophytes (D), epiphyte–animal interactions (B), and (C) the advantage of a tank growth form to live in extreme environments, such as tree canopies or—this in this case—a telephone wire.

1990; Mendieta-Leiva et al., 2020; Stuntz et al., 2002), in regions where they are abundant. It is important to note, however, that the specific contribution by ***vascular epiphytes*** to these processes, particularly forest hydrology, is still poorly understood compared to ***non-vascular*** epiphytes (Zotz, 2016). Current evidence suggests that while non-vascular epiphytes (i.e., mosses, liverworts, hornworts) contribute significantly to global rainfall interception (Porada et al., 2018), vascular epiphytes likely play a comparatively lesser role in forest hydrological processes (Zotz, 2016). Even in large tank bromeliads, stored water evaporates after just a few days (Zotz and Thomas, 1999).

There is little doubt, however, about the importance of vascular epiphytes to the general ecology of forest ecosystems. Above all, epiphytes increase the structural complexity of forest canopies, creating a new habitat for canopy-dwelling fauna and acting as a natural buffer against the otherwise arid and hot conditions typical of many canopy ecosystems (Stuntz et al., 2002). In this context, epiphytes can be viewed as habitat island oases for canopy-dwelling fauna, who take refuge in the cooler, moist microclimates (Figure 3). Some of these plant-animal interactions are highly specialised, such as between ant-garden ants and epiphytes, a mutualism where ants benefit from the provision of food and habitat while epiphytes benefit by the provision of nutrients and protection against herbivores. Seed dispersal mutualisms are also common among fleshy-fruited epiphytes (e.g., Araceae) and birds, and to a lesser extent arboreal mammals (Vieira and Izar, 1999), thus highlighting the essential role of epiphytes in promoting biodiversity within the forest ecosystems they occur through the facilitation of species interactions.

Despite providing these crucial ecosystem services and being important in maintaining the biodiversity of forest ecosystems, how sensitive epiphytes are to anthropogenic changes is still poorly understood. Given that epiphytes are more strongly coupled with atmospheric conditions than terrestrial plants (Taylor et al., 2022, 2023), it is often assumed that they will also be more adversely affected by climate change (Zotz and Bader, 2009). However, contrasting findings have been reported, and the responses of epiphytes to anthropogenic changes likely depend on a variety of factors. For example, studies comparing epiphyte resilience between cloud forest and drier lowland forest communities report a greater resilience of epiphytes growing in the lowlands (Gotsch et al., 2018), which suggests that drought tolerance is a major factor driving epiphyte resilience to anthropogenic changes (Wolf, 2005). This also suggests that epiphytes endemic to cloud forests, which are disappearing at an alarming rate (Karger et al., 2021) are most at risk to anthropogenic changes in climate, while deforestation and habitat destruction are likely the biggest threat to epiphytes in lowland tropical forests (Zotz and Bader, 2009). Nevertheless, it is essential to conserve and protect vascular epiphytes as a critical component of forest ecosystems in which they occur, ensuring their continued provision of critical ecosystem services for the long-term sustainability of these diverse and complex systems.

Adaptations of epiphytes to life in the canopy

Due to their lack of direct access to ground soil water, epiphytes obtain water and nutrients from canopy debris, canopy soil, and atmospheric precipitation (e.g., fog, rain) and have evolved a variety of traits to overcome challenges related to water limitation and general life in the canopy. Arguably the most important traits for any epiphyte relate to dispersal and establishment, as this correlates with long-term viability (Einzmann and Zotz, 2017). Most epiphytes have small, wind-dispersed propagules (e.g., ferns, orchids, most bromeliads), which is advantageous for dispersing long distances through the canopy and settling into tree bark crevasses (Zotz, 2016). Although a high capacity for long-distance dispersal has been experimentally shown for some epiphytic orchids and bromeliads (Einzmann and Zotz, 2017), other studies report substantial dispersal limitation in these groups (Cascante-Marín et al., 2009; McCormick and Jacquemyn, 2014). In bromeliads, this may be due to the hairy appendage attached to their seeds (plumose seed coma) that increases air resistance, thus preventing them from dispersing far from parent plants (Garcia-Franco and Rico-Gray, 1988). Although plumose seed comas may limit dispersal distance, they promote establishment success through supporting anchorage to phorophytes at the seed and small seedling stage (Benzing, 2000), just as root hairs and glue adhesion promote anchorage in other species at later life stages (Tay et al., 2023).

Following successful dispersal and establishment, epiphytes must overcome constraints related to an intermittent water and nutrient supply. Most epiphytes have a specialised root tissue (velamen radicum) that acts as a spongy medium, which can rapidly absorb water and nutrients (Zotz, 2016). Likewise, trichomes, which are specialised hair-like structures or scales common among bromeliads, may aid to capture water from the surrounding environment while simultaneously reducing evaporation (Benzing, 2000). Some epiphytes exhibit 'tank' growth forms, which can trap organic matter and store water, which differ from 'trash basket' epiphytes that capture organic material but little moisture (Figure 14.4). Another important trait associated with epiphytism is succulence, whether that be succulent leaves, stems, or modified stems called pseudobulbs (Yang et al., 2016). Non-succulent taxa might instead have thick cuticles to reduce water loss (Yang et al., 2016). Finally, crassulacean acid metabolism (CAM) photosynthesis, which allows for a more economical use of water, is also reportedly common among epiphytes (Crayn et al., 2004).

Successional patterns of epiphytes are tightly linked to these functional strategies and the changes in microclimate conditions that occur as phorophytes age, which plays a pivotal role in shaping the composition and diversity patterns of epiphyte communities within forest ecosystems (Woods, 2017). Young phorophytes, for example, typically lack morphological and physiological traits that are key prerequisites for the colonisation of most epiphytes, including underdeveloped bark roughness, branching architecture, and canopy soil. Only a subset of epiphytes can establish under these conditions, such as those with creeping rhizomes (e.g., *Pyrrosia eleagnifolia*) or rhizomatous roots (e.g., *Dendrobium cunninghamii*), which can form dense mats around host tree trunks and branches. These early colonists facilitate the establishment of later colonising species by providing substrate for dispersing propagules to attach to, creating more favourable microclimate conditions, and through the development of canopy soil. Thus, as phorophytes age and increase in size, epiphytes with higher nutrient and water requirements can establish, leading to a predictable increase in epiphyte species richness with increasing tree size (Taylor and Burns, 2015).

Variation in tree age explains to some extent heterogeneity in epiphyte distributions within a forest stand; however, it does not explain why neighbouring trees of the same age can exhibit markedly different epiphyte communities. While this alludes to epiphytes being, at least to some extent, host tree-specific, empirical evidence suggests that epiphytes have preferences for certain tree characteristics, rather than host species (Wagner et al., 2015). For example, epiphytes are less likely to establish on trees with flaky or very smooth bark,

Figure 14.4 Common morphological traits of vascular epiphytes in relation to microclimate gradients within the canopy. Panels A–C photographed by Fabian Mühlberger and panel C by Fabian Brambach.

or trees with thin, simple branching architecture (López-Villalobos et al., 2008). The water-holding capacity of tree bark is also important and to a lesser degree chemical properties (Callaway et al., 2002). Lastly, variation in leaf density among tree species significantly shapes the microclimate conditions within individual trees, adding another layer of complexity to the heterogeneous composition and distribution patterns of epiphytes within forest stands.

Scale dependencies in epiphyte research

Although the patterns and drivers described above showcase epiphyte diversity across large spatial scales, it is important to consider the dependency of these relationships with scale. Scale dependencies refer to different patterns or relationships, which manifest at varying scales, necessitating careful considerations when interpreting biodiversity syntheses (Peterson et al., 1998). In general, larger areas typically support a greater number of species than smaller ones, a relationship that has been extensively studied and documented in ecology (Connor and McCoy, 1979). Epiphytes, however, show an inverse species–area relationship where diversity tends to increase with diminishing spatial scale. For example, while 39 per cent of Ecuador's flora grows epiphytically (Taylor et al., 2022), within the smaller Zamora-Chinchipe province, the proportion of epiphytes exceeds that of all other growth forms, including their host trees (Taylor et al., 2023). Thus, we can expect at consecutively smaller spatial scales (and in the wet tropics), that epiphytes become increasingly important contributors to plant diversity. Different drivers may also influence epiphyte diversity patterns across different spatial scales. At large spatial scales, recent analyses show that epiphyte diversity is driven by a combination of contemporary and past climate conditions, topography, and historical biogeography (Taylor et al., 2022, 2023), which are difficult to capture at smaller spatial scales. At these local scales, microclimate gradients and biotic interactions, such as host tree traits and facilitation by earlier-establishing vascular epiphytes and bryophytes, are likely to play a more pivotal role. Indeed, although epiphytes do not follow the general species–area relationship at large spatial scales, several studies have shown that epiphyte diversity increases with phorophyte size (e.g., Taylor and Burns, 2015), demonstrating the importance of multi-scale comparisons to fully elucidate how epiphyte communities are structured across all spatial dimensions.

References

Andriananjamanantsoa, H. N., Engberg, S., Louise Jr, E. E. and Brouillet, L. (2016) 'Diversification of *Angraecum* (Orchidaceae, Vandeae) in Madagascar: Revised phylogeny reveals species accumulation through time rather than rapid radiation', *PLoS One*, vol. 11, e0163194.

Benzing, D. H. (1990) *Vascular Epiphytes: General Biology and Related Biota* (1st ed.), Cambridge University Press, New York.

Benzing, D.H. (2000) *Bromeliaceae: Profile of an Adaptive Radiation*, Cambridge University Press, Cambridge.

Benzing, D. H. (2004) 'Vascular epiphytes', in M. D. Lowman and H. B. Rinker (eds) *Forest Canopies* (2nd ed.), Academic Press, Cambridge, MA.

Brummitt, R., Pando, F., Hollis, S. and Brummitt, N. (2001) 'World geographical scheme for recording plant distributions Vol. 951' Hunt Institute for Botanical Documentation, Carnegie Mellon University, Pittsburgh, PA: International working group on taxonomic databases for plant sciences (TDWG).

Callaway, R. M., Reinhart, K. O., Moore, G. W., Moore, D. J. and Pennings, S. C. (2002) 'Epiphyte host preferences and host traits: Mechanisms for species-specific interactions', *Oecologia*, vol. 132, pp. 221–230.

Cascante-Marín, A., von Meijenfeldt, N., de Leeuw, H. M., Wolf, J. H., Oostermeijer, J. G. B. and den Nijs, J. C. (2009) 'Dispersal limitation in epiphytic bromeliad communities in a Costa Rican fragmented montane landscape', *Journal of Tropical Ecology*, vol. 25, pp. 63–73.

Connor, E. F. and McCoy, E. D. (1979) 'The statistics and biology of the species-area relationship', *The American Naturalist*, vol. 113, pp. 791–833.

Coxson, D. S. and Nadkarni, N. M. (1995) 'Ecological roles of epiphytes in nutrient cycles of forest ecosystems', in M. D. Lowman and H. B. Rinker (eds) *Forest Canopies*, Academic Press, San Diego, CA.

Crayn, D. M., Winter, K. and Smith, J. A. C. (2004) 'Multiple origins of crassulacean acid metabolism and the epiphytic habit in the Neotropical family Bromeliaceae', *Proceedings of the National Academy of Sciences*, vol. 101, pp. 3703–3708.

Einzmann, H. J. and Zotz, G. (2017) 'Dispersal and establishment of vascular epiphytes in human-modified landscapes', *AoB Plants*, vol. 9, plx052.

Garcia-Franco, J. G. and Rico-Gray, V. (1988) 'Experiments on seed dispersal and deposition patterns of epiphytes. The case of *Tillandsia deppeana* Steudel (Bromeliaceae)', *Phytologia*, vol. 65, pp. 73–78.

Givnish, T. J., Barfuss, M. H., Van Ee, B., Riina, R., Schulte, K., Horres, R., Gonsiska, P. A., Jabaily, R. S., Crayn, D. M. and Smith, J. A. C. (2014) 'Adaptive radiation, correlated and contingent evolution, and net species diversification in Bromeliaceae', *Molecular Phylogenetics and Evolution*, vol. 71, pp. 55–78.

Givnish, T. J., Spalink, D., Ames, M., Lyon, S. P., Hunter, S. J., Zuluaga, A., Iles, W. J., Clements, M. A., Arroyo, M. T. and Leebens-Mack, J. (2015) 'Orchid phylogenomics and multiple drivers of their extraordinary diversification', *Proceedings of the Royal Society B: Biological Sciences*, vol. 282, 20151553.

Gotsch, S. G., Dawson, T. E. and Draguljić, D. (2018) 'Variation in the resilience of cloud forest vascular epiphytes to severe drought', *New Phytologist*, vol. 219, pp. 900–913.

Gotsch, S. G., Nadkarni, N. and Amici, A. (2016) 'The functional roles of epiphytes and arboreal soils in tropical montane cloud forests', *Journal of Tropical Ecology*, vol. 32, pp. 455–468.

Hagen, O., Skeels, A., Onstein, R. E., Jetz, W. and Pellissier, L. (2021) 'Earth history events shaped the evolution of uneven biodiversity across tropical moist forests', *Proceedings of the National Academy of Sciences of the United States of America*, vol. 118, e2026347118.

Hietz, P. (2010) 'Fern adaptations to xeric environments', in K. Mehltreter, L. R. Walker, and J. M. Sharpe (eds) *Fern Ecology*, Cambridge University Press, New York.

Hofstede, R. G. M., Wolf, J. H. D. and Benzing, D. H. (1993) 'Epiphytic biomass and nutrient status of a Colombian upper montane rain forest', *Selbyana*, vol. 14, pp. 37–45.

Kaplan, M. R., Schaefer, J. M., Denton, G. H., Barrell, D. J. A., Chinn, T. J. H., Putnam, A. E., Andersen, B. G., Finkel, R. C., Schwartz, R. and Doughty, A. M. (2010) 'Glacier retreat in New Zealand during the Younger Dryas stadial', *Nature*, vol. 467, pp. 194–197.

Karger, D. N., Kessler, M., Lehnert, M. and Jetz, W. (2021) 'Limited protection and ongoing loss of tropical cloud forest biodiversity and ecosystems worldwide', *Nature Ecology & Evolution*, vol. 5, pp. 854–862.

López-Villalobos, A., Flores-Palacios, A. and Ortiz-Pulido, R. (2008) 'The relationship between bark peeling rate and the distribution and mortality of two epiphyte species', *Plant Ecology*, vol. 198, pp. 265–274.

McCormick, M. K. and Jacquemyn, H. (2014) 'What constrains the distribution of orchid populations?', *New Phytologist*, vol. 202, pp. 392–400.

Méndez Castro, F., Bader, M., Mendieta-Leiva, G. and Rao, D. (2018) 'Islands in the trees: A biogeographic exploration of epiphyte-dwelling spiders', *Journal of Biogeography*, vol. 45, pp. 2262–2271.

Mendieta-Leiva, G., Porada, P. and Bader, M. Y. (2020) 'Interactions of epiphytes with precipitation partitioning', in I. Van Stan John T., E. Gutmann, and J. Friesen (eds), *Precipitation Partitioning by Vegetation: A Global Synthesis*. Springer International Publishing, Cham.

Pérez-Escobar, O. A., Chomicki, G., Condamine, F. L., Karremans, A. P., Bogarín, D., Matzke, N. J., Silvestro, D. and Antonelli, A. (2017) 'Recent origin and rapid speciation of Neotropical orchids in the world's richest plant biodiversity hotspot', *New Phytologist*, vol. 215, pp. 891–905.

Peterson, G., Allen, C. R. and Holling, C. S. (1998) 'Ecological resilience, biodiversity, and scale', *Ecosystems*, vol. 1, pp. 6–18.

Porada, P., Van Stan, J. T. and Kleidon, A. (2018) 'Significant contribution of non-vascular vegetation to global rainfall interception', *Nature Geoscience*, vol. 11, pp. 563–567.

Raymo, M. E. (1994) 'The initiation of Northern Hemisphere glaciation', *Annual Review of Earth and Planetary Sciences*, vol. 22, pp. 353–383.

Sato, T. (1982) 'Phenology and wintering capacity of sporophytes and gametophytes of ferns native to northern Japan', *Oecologia*, vol. 55, pp. 53–61.

Stuntz, S., Simon, U. and Zotz, G. (2002) 'Rainforest air-conditioning: The moderating influence of epiphytes on the microclimate in tropical tree crowns', *International Journal of Biometeorology*, vol. 46, pp. 53–59.

Tay, J. Y., Zotz, G. and Einzmann, H. J. (2023) 'Smoothing out the misconceptions of the role of bark roughness in vascular epiphyte attachment', *New Phytologist*, vol. 238, pp. 983–994.

Taylor, A. and Burns, K. (2015) 'Epiphyte community development throughout tree ontogeny: An island ontogeny framework', *Journal of Vegetation Science*, vol. 26, pp. 902–910.

Taylor, A., Weigelt, P., Denelle, P., Cai, L. and Kreft, H. (2023) 'The contribution of plant life and growth forms to global gradients of vascular plant diversity', *New Phytologist*, vol. 240, https://doi.org/10.1111/nph.19011.

Taylor, A., Zotz, G., Weigelt, P., Cai, L., Karger, D. N., König, C. and Kreft, H. (2022) 'Vascular epiphytes contribute disproportionately to global centres of plant diversity', *Global Ecology and Biogeography*, vol. 31, pp. 62–74.

Tryon, R. M. (1964) 'Evolution in the leaf of living ferns', *Bulletin of the Torrey Botanical Club*, vol. 21, pp. 73–85.

Tryon, R. M. (1986) 'The biogeography of species, with special reference to ferns', *The Botanical Review*, vol. 52, pp. 117–156.

Vieira, E. M. and Izar, P. (1999) 'Interactions between aroids and arboreal mammals in the Brazilian Atlantic rainforest', *Plant Ecology*, vol. 145, pp. 75–82.

Wagner, K., Mendieta-Leiva, G., and Zotz, G. (2015) 'Host specificity in vascular epiphytes: A review of methodology, empirical evidence and potential mechanisms', *AoB Plants*, vol. 7, plu092.

Watkins, J. E. and Cardelús, C. L. (2012) 'Ferns in an angiosperm world: cretaceous radiation into the epiphytic niche and diversification on the forest floor', *International Journal of Plant Sciences*, vol. 173, pp. 695–710.

Wolf, J. H. D. (2005) 'The response of epiphytes to anthropogenic disturbance of pine-oak forests in the highlands of Chiapas, Mexico', *Forest Ecology and Management*, vol. 212, pp. 376–393.

Woods, C. L. (2017) 'Primary ecological succession in vascular epiphytes: The species accumulation model', *Biotropica*, vol. 49, pp. 452–460.

Yang, S.-J., Sun, M., Yang, Q.-Y., Ma, R.-Y., Zhang, J.-L. and Zhang, S.-B. (2016) 'Two strategies by epiphytic orchids for maintaining water balance: Thick cuticles in leaves and water storage in pseudobulbs', *Aob Plants*, vol. 8, plw046.

Zotz, G. (2016) *Plants on Plants—The Biology of Vascular Epiphytes*, Springer International Publishing, Cham.

Zotz, G. and Bader, M. Y. (2009) 'Epiphytic plants in a changing world-global: change effects on vascular and non-vascular epiphytes', in U. Lüttge, W. Beyschlag, B. Büdel, and D. Francis (eds) *Progress in Botany*, Springer Berlin, Heidelberg.

Zotz, G. and Thomas, V. (1999) 'How much water is in the tank? Model calculations for two epiphytic bromeliads', *Annals of Botany*, vol. 83, pp. 183–192.

Zotz, G., Weigelt, P., Kessler, M., Kreft, H. and Taylor, A. (2021) 'EpiList 1.0: A global checklist of vascular epiphytes', *Ecology*, vol. 102, e03326.

15

INSECTS IN FOREST ECOSYSTEMS

Andrea Battisti

Introduction

The vast majority of the terrestrial planet is covered by forest biomes since a long time and over a large variety of climatic conditions. This has created enormous possibilities for the animal community to diversify and adapt to local conditions at different spatial scales. Arthropods, and in particular insects, play a major role in the community because of the great potential to adapt and build up trophic networks linking below-ground and above-ground communities. Latitude and elevation are the main factors affecting the distribution and abundance of animal communities in terrestrial habitats, being related by specific lapse rate of temperature (0.6°C every 100 m of elevation and 1° of latitude) (Hodkinson, 2005). The distribution and abundance of insects are mainly driven by temperature gradients associated with latitude and elevation, although topography may play a decisive role at a local scale (Figure 15.1). Consequently, communities associated with mountain forest ecosystems are generally more fragmented than those associated with forests of low elevation, even if they grow across a wide latitudinal span.

The human intensification of farming activities has caused a progressive reduction in forest areas with significant impacts on both plant and animal communities. In addition, the anthropogenic climate change is a further disturbance in the relationships established in the trophic webs (Raven and Wagner, 2021). In the temperate zone, the transformation of forest into agricultural land and intensive silviculture has changed the landscape dramatically, resulting in fragmentation of the distribution range that is eventually responsible for the present structure of insect communities. Furthermore, the plantation of tree species outside their native range, either intentional or unintentional, has promoted the spread and abundance of particular species, which found very favourable conditions in the new stands, ultimately resulting as pests with great economic impact. Human-induced changes in forest ecosystems have also affected the communities of arthropods associated with the soil and responsible for the turnover of nutrients, with different types of feedback on tree growth and ecosystem persistence. This is the typical case of forest ants, a kind of 'ecosystem engineers'

DOI: 10.4324/9781003324072-18

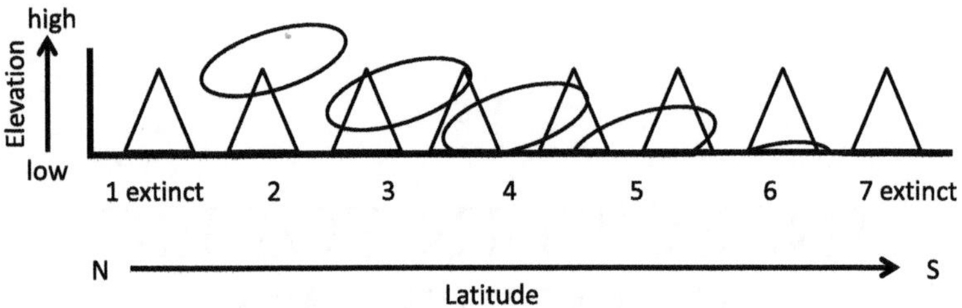

Figure 15.1 Effects of latitude and elevation on the distribution of organisms in the Northern Hemisphere (modified from Gorodkov, 1985). The horizontal axis represents the latitude and the vertical axis the elevation. The triangles along the horizontal line identify mountains that occur at different latitudes. The oblique lenses represent the potential range of the organism as determined by climate (both inclination and width of the lens may vary depending on the reaction norms of the individual species). The potential range may shift from south to north, and from low to high elevation, and vice versa, depending on warming and cooling of the climate. Gorodkov exemplified different types of distribution, which can be observed in relation to the position of the lens along the gradient, that is, 2: montane local (endemic), 3: montane wide, 4: disjuncted plain-montane, 5: continuous plain-montane, 6: plain local (endemic). Types 1 and 7 drive to extinction. Types 4 and 5 are the most common and associated with wider geographic distribution.

whose activity is directly affected by human activity and climate, becoming less abundant in simplified ecosystems and in colder habitats (Hölldobler and Wilson, 1990; Underwood and Fisher, 2006).

Biodiversity

There is little doubt that insects represent the bulk of the biodiversity of forest ecosystems, occupying them with a huge number of species all the available ecological niches associated with trees (Figure 15.2). It should be noted that estimates on the total number of species inhabiting the planet are made by censuses in especially rich tropical forests (Basset et al., 2012), where many taxa are still undescribed. The functional role of biodiversity and the ecosystem services associated with it are gaining more importance, together with the awareness that human activities and climate change threaten the survival of many important species. Fragmentation and ecotonization of forest habitats are thought to be the main reason for the reduction of biodiversity, especially for species characterized by narrow niches and reduced mobility (Muys et al., 2022).

The intrinsic value of biodiversity is easy to understand; however, its study requires large efforts and the availability of expertise, which is becoming rare. The census of species and their abundance, the knowledge of the relationships with the other organisms are essential points for monitoring the status of forest ecosystems in space and time. The phylum of arthropods is by far the most diverse, with more than 1.2 million species described, mainly from terrestrial ecosystems, with an estimated number of species between 2.9 and 12.7 million (Hamilton et al., 2013). The success of this group relies on the small size (mm to few cm), high access to oxygen, fast reproduction both sexual and asexual, high adaptive potential to micro-niches and to the specialization in relation to the feeding

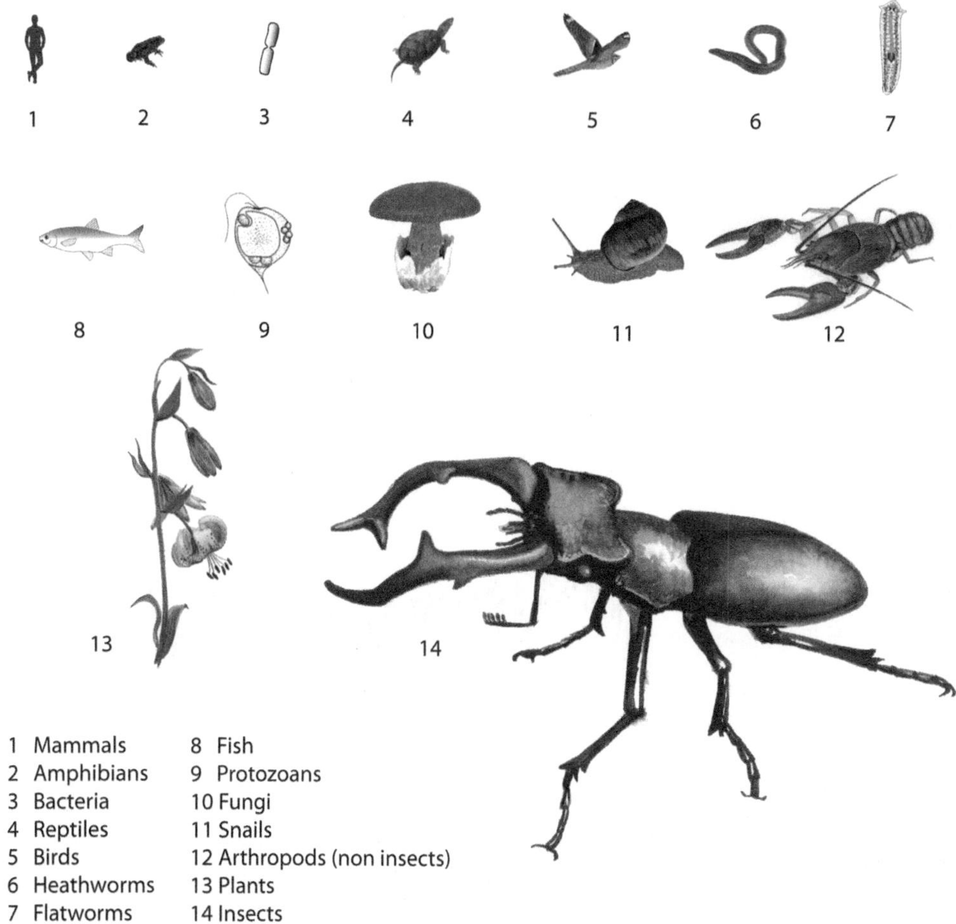

Figure 15.2 Distribution of the number of species among classes of the animal kingdom represented by the relative size of a component of each group (redrawn from Speight et al., 2008).

habit; its origin goes back in time to the beginning of diversification of plants in the primary era. Insects are the most species-rich group among arthropods, followed by mites, and both are very important for the functioning of the forest ecosystems and in general for the survival of many species of plants and animals, which are totally dependent on them, including man (Wilson, 1992). Biodiversity assessment and appreciation is becoming more accessible with the development of digital technologies (artificial intelligence and machine learning) as those available in the Global Biodiversity Information Facility (https://www.gbif.org/), an international network and data infrastructure funded by the world's governments and aimed at providing anyone, anywhere, open access to data about all types of life on Earth.

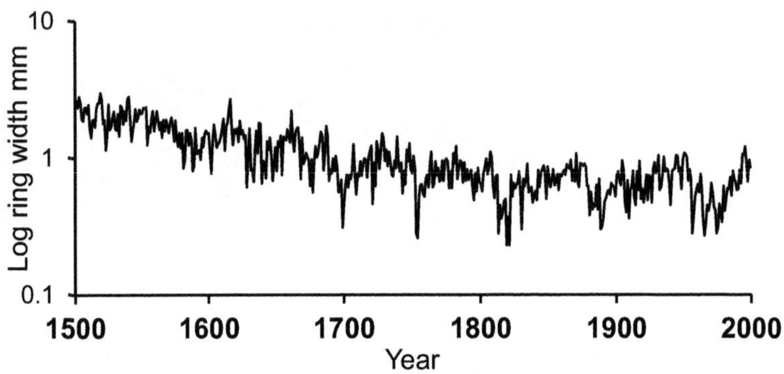

Figure 15.3 Section of a larch trunk from the Italian Central Alps with the indication of the growth rings periodically reduced, as shown by the graph. Reductions correspond to outbreaks of the larch bud moth *Zeiraphera griseana* (data kindly provided by Marco Carrer, University of Padova).

Insects and forests

The easiest and more functional way to classify herbivorous insects associated with forest trees is their grouping by feeding guilds (Table 15.1) and then by the level of specialization. In general, monophagous species are associated exclusively with one genus of host plants, oligophagous with one family, and polyphagous with more than one family. A similar type of specialization and classification can be used for the higher trophic level, that is, that of predators and parasitoids of herbivorous insects, although the specialization is weaker with reaching higher levels of the trophic network.

The specialization and niche adaptation of forest insects are generally explained by the avoidance of competition among organisms interested in exploiting the same part of the plant. This can be seen in space as well as in time, for example, with shifting of the distribution range or the phenological window of activity, respectively. The specialization of insect herbivores is particularly driven by the relationship between the preference expressed by the

Table 15.1 Major feeding guilds of main insect orders

Feeding guilds	*Orthoptera*	*Heteroptera*	*Homoptera*	*Lepidoptera*	*Diptera*	*Coleoptera*	*Hymenoptera*
Sap suckers		X	X				
Gall makers			X		X		X
Leaf feeders	X			X	X	X	X
Shoot feeders				X		X	
Wood borers				X		X	X
Cone and seed		X	X	X	X	X	X
Parasitoids					X		X
Predators	X	X			X	X	X

mother at the oviposition for a given host plant and the performance of the progeny (Jones, 2022). Specialization and niche adaptation may also depend on the relationships between insect herbivores and their natural enemies, which often act as major regulating factors (regulation from above or top-down). The specialization ultimately ends in speciation and this explains why the trophic pyramids of biomass and number of species are inverted in forest ecosystems: in the lower trophic levels, a few species represent a large biomass while many species with smaller biomass are present in the upper trophic levels. In general, however, the first hurdle to being an insect herbivore in a forest is related to the scarce nutrient content of plant tissues, or to the occurrence of compound which may deter insect feeding (bottom-up regulation). Lignin and cellulose are the most abundant components of plant tissues and are not exploited by the large majority of forest insects, with a few exceptions observed when complex symbiotic interactions with microorganisms occur. Green tissues such as leaves and shoots are generally more palatable, but they occur for a short period of time and often become unpalatable because of hardening or replenishing with secondary metabolites such as resins and phenols. Liquids that occur inside cells or in the vascular systems are generally rich in nutrients, although these are difficult to access and highly specialized mouth parts are required. Other plant tissues produced for the sake of reproduction such as flowers, seed, and fruits are also exploited by herbivorous insects, which have to face a highly irregular occurrence of this resource in space and in time, often becoming very limiting for their success.

As plants are unpromising food for insects, because they largely comprise indigestible structural compounds, insects have evolved many different strategies to feed on plants including associations with mutualistic symbionts, which can be important mediators of direct and indirect interactions between herbivorous insects and their host plants (Schoonhoven et al., 1998). How microorganisms can tilt the balance in favour of the insect are provided by recent advances in our mechanistic understanding of the signalling pathways and biochemical responses involved in plant defence and insect counterattack, and how these responses can be modulated by both pathogenic or nonpathogenic microorganisms. Finally, how insect mutualists interact with plants is potentially of major applied significance. The invasion of new pests has often been facilitated by their mutualists, and some novel interactions have resulted in new and more virulent insect pests. The advent of modern technologies exploring the microbiota associated with insects, such as DNA metabarcoding, has opened new perspectives in research (Diehl et al., 2022).

In spite of these limitations, insects may become sometimes so abundant that their host plants can be completely deprived of their functional tissues such as the leaves and may come to death, or the insects may attack directly the vascular system and cause a quick dieback. In general, however, the consumption of plant tissues by insects is limited and not clearly perceived by a non-expert until a threshold which is between 5 and 10 per cent of the biomass, depending on the tree species. It is thought that this amount of herbivory is functional to the nutrient cycling in the ecosystem and in this way can be considered as beneficial for tree growth (Mattson and Addy, 1975). Consumption of roots certainly exists, but it is unfortunately little documented, with very few species ranked as pests (Wainhouse, 2005). Arthropods, however, are extremely abundant in soil and play an essential role in the decomposition of soil organic matter and dead wood. They may contribute, together with earthworms (Annelida), to the breakdown of the litter and its incorporation in the soil, where the process will be completed by mites (Acarina) and springtails (Collembola), in cooperation with a large number of soil microorganisms (Speight et al., 2008). The study of invertebrate communities of forest soils is facilitated by the availability of environmental DNA analyses and related databases (Kirse et al., 2021). In the well-studied experimental basin of Hubbard Brook (New Hampshire, United States), it has been shown that recurrent insect defoliation interannual causes a variance in the distribution of biomass across trophic levels, implying a variation in plant nutritional quality. Synchronous fluctuations of a main consumer guild (such as foliage-feeding Lepidoptera) involved interannual variation in soil mineral nutrient availability (Stange et al., 2011), which in turn is able to modify leaf chemistry and thus the susceptibility to herbivores (Frost and Hunter, 2004, 2008). Variation in the amount and quality of litter is likely to affect the structure and functionality of soil fauna, although this is an aspect of forest ecology, which has not been thoroughly considered and deserves attention in the future. In general, there is a need to characterize better the links between above-ground and below-ground communities of insects and associated organisms, which are mediated by the host plant responses (Wardle et al., 2004).

Ecosystem services are also affected by forest insect pests (Boyd et al., 2013) in different ways. A reduction in tree health affects the value placed on forest resources by the society. This may happen in a number of ways and can have very different and unexpected impacts. For example, iconic tree species such as chestnuts (*Castanea* spp.) and elms (*Ulmus* spp.) can be swept away by combinations of endemic and exotic pests and pathogens (Santini and Battisti, 2019). The non-monetary nature of cultural services means that it is difficult to track their value, and this can lead to unexpected and apparently rapid shifts in the public value placed on them. Social amplification is a term used to describe dramatic changes in public concern, often caused by positive feedback through new and old media. Such events can greatly increase the pressure on policymakers to respond to threats to cultural services (Boyd et al., 2013). International networks such as the International Union of Forest Research Organizations (IUFRO) tree mortality network may illustrate how the role of insects in causing tree mortality is widespread and become of interest to a large audience.

Population dynamics

The outbreaks of forest insects are common in all types of forests and may cause dramatic consequences for the conservation of the habitat and human economy (Barbosa et al., 2012). Insect abundance can generally be explained by both endogenous and exogenous factors. As endogenous factors are generally indicated those related to the biology of the species (e.g., fecundity, voltinism, sex ratio, dispersal), the host plant (e.g., susceptibility, tolerance, resistance), the competitors and facilitators (e.g., other herbivores of the same guild, associated microorganisms), and natural enemies (e.g., parasitoids and predators), while the only exogenous but very important factor is the weather. Of course, interactions between endogenous factors are numerous and sometimes very complex; in addition, all of them are affected by the weather. The idea that the weather is a major driver of the upsurge in population density over large and different areas was developed by Moran (1953) and is still in need of study (Liebhold *et al.*, 2004).

Endogenous factors may generate predictable responses if they are density-dependent, and for this reason are of great value to managers for taking decisions about possible actions to contrast the population growth and to limit the impacts on the ecosystems. Density-dependence may occur over a different time (e.g., months or years) in relation to the nature of the regulation mechanism involved. There are models that attempt to combine general density-dependent regulation with variations of the weather, especially in relation to climate-change predictions (Barbosa et al., 2012).

Outbreaks of forest insects have been studied for several species and various explanations have been attempted. In some cases, it has been possible to link the increase of the population density with a change in some environmental variables that were considered to positively affect the growth, or to negatively affect the action of regulation factors. In other cases, a periodicity of the outbreak event has been observed and explained with complex mechanisms, with a period variable between a few years up to decades. Density dependence has been shown in a few cases to be responsible for such a periodicity, although the precise mechanisms are often unknown. Cyclic population behaviour may be manifested as synchronized or non-synchronized fluctuations between disjunct populations. A third behaviour that has recently received much attention is travelling waves in populations of periodic species. Travelling waves mean that the population peak densities move in one main direction across space. In the winter moth (*Operophtera brumata*), a continental-scale wave has been shown in Europe, where outbreaks occur every 9–10 years over a distance of nearly 4,000 km (Tenow *et al.*, 2013). Similarly, the waves of the larch budmoth (*Zeiraphera diniana*) population travel across the European Alps with a wave speed of 220–250 km per year and cycle length of 8–9 years (Bjørnstad et al., 2002). Although periodic travelling waves in ecology have become increasingly recognized in several fields of study, their origin is unknown. The theory concerns 'reaction-diffusion' models, and in the insect case trophic interactions (e.g., predator–prey or herbivore–plant) and dispersal (spatial movement of reactants or individuals) are involved (Tenow et al., 2013). In the larch budmoth model, an interaction with parasitoids predicts directional waves if dispersal is unidirectional or habitat productivity varies across the landscape (Bjørnstad et al., 2002). For the same system, a tri-trophic model demonstrated that landscape gradients in favourable habitats alone could induce waves from epicentres (Johnson, Bjørnstad & Liebhold 2004). In the case of the larch budmoth, a thorough dendrochronological analysis of larch in the French Alps has allowed deciphering the effect of the climate on population dynamics (Saulnier et al. 2017) (Figure 15.3). The analyses revealed cycle periodicities of 4, 8, and 16 years throughout the entire 1500–2003 time series, except from 1690 to 1790 and since the early 1980s. These abrupt changes could reflect a physiological response to climatic variations, and future dynamics appear difficult to predict based on the expected climate change.

Sometimes, the outbreak may become destructive for the ecosystem and start a succession phase, which may last for several decades. In these cases, the insects generally kill the trees over large areas and often become so numerous that they overcome the resistance mechanisms and spread over large territories. The largest event of this type ever recorded is the outbreak of the mountain pine beetle *Dendroctonus ponderosae* in Northwestern America, which has affected more than 15 million ha of coniferous forest (Kurz *et al.*, 2008), and it is further expanding. In Europe, conifer trees and especially Norway spruce are being massively killed by bark beetles in the last decades (Hlásny et al., 2021).

Forest insects may also participate in outbreaks of plant pathogens, especially when they act as vectors (Santini and Battisti, 2019). The association can be purely neutral or may benefit one component of the system, or both (mutualistic symbiosis). Numerous examples are associated with diebacks of tree species over large areas, as in the case of elms (*Ulmus* spp.) and cypress (*Cupressus* spp.). The relationships seem to be very important, especially for pathogens outside of their native range, as in the recent case of the pine wood nematode *Bursaphelenchus xylophilus*.

Natural enemies are very important for the regulation of populations at both low and high density, and are often taken into consideration for pest management programmes. Insect parasitoids are extremely common and diverse; they differ from the true parasites because they always bring the host to death, and in this way closer to predators. As they need the host for the development of their progeny, they have developed a sophisticated mechanism of host location and finding, which makes them rather specific and reliable for biological control actions. Because of the life history, they are characterised by a delay in the regulation, which is largely predictable in population models. Parasitoids can be specialized on every developmental stage of the host, adjusting their size accordingly. Most species are included in only two insect orders, that is, Diptera and Hymenoptera, where there are families that are entirely specialized in this behaviour. On the contrary, insect predators are less specialized and occur in very many taxonomic groups. Some common predators, such as ants, may behave as such only occasionally, so that they are less reliable than parasitoids for biological control. Other predators of insect herbivores are vertebrates, especially birds. These animals seem to play an important role in keeping populations at low density, while they are unable to reduce the numbers in case of outbreak. Interestingly, a global experiment with model caterpillars indicated an increasing predation towards the equator, with a parallel pattern of increasing predation towards lower elevations (Roslin et al., 2017). Finally, insect pathogens such as viruses, bacteria, fungi, and nematodes may affect the population dynamics and can also be used for direct control at high host density. Viruses and bacteria need to be ingested with food and are generally specific, especially for viruses. They may cause spectacular collapses in host populations when densities are high. Entomopathogenic fungi are common agents of mortality in the soil or under the bark, and more generally when insects are protected under high moisture conditions. The mode of action is based on the penetration of the insect cuticle, and the mechanism is generic. Entomopathogenic nematodes are associated with the water film in the soil or in the wood, where they actively search for their hosts. These are killed by bacteria carried by the nematodes that ultimately thrive as saprophagous. They are quite generic in their action, so they can be multiplied on laboratory hosts and then released against target insects in the field.

Insects in a changing environment

In addition to the intrinsic vulnerability of trees (see Chapters 8 and 18 of this volume), forest insect communities can be strongly affected by several environmental factors, part of which are related to forest management. Accumulating evidence of climate change over the past decades has fuelled debate and research on its ecological consequences. Meteorological data confirm general increases in ambient temperatures and changes in precipitation patterns, which have been connected to changes in species ranges (Parmesan et al., 1999; Battisti et al., 2005), advanced phenology (Walther et al., 2002), and phenological asynchrony among interacting species (Visser and Both, 2006). Much less is known about consequences for population dynamics, but it seems likely from first principles that there would be effects for some species on reproductive success, life histories, mean abundance, and variation in abundance.

Insect herbivores, generally poikilothermic, are highly sensitive to changes in their thermal environment (Uvarov, 1931). Furthermore, responses of insect herbivores are likely based on their life cycle in combination with the life cycle of their host plant (Bale et al., 2002). The development rate generally increases with increasing temperature to some maximum, above which there are frequently sharp increases in mortality associated with the decreasing development rate (Uvarov, 1931). The development rate of insects in mid- to high-latitude systems should increase with climate warming as ambient temperatures are well below that which permit maximum rates of feeding and growth (Hodkinson, 2005).

An increased development rate could lead to increased voltinism in facultative multivoltine species, as predicted for several bark beetle species. Recent research has found increased voltinism in macrolepidopteran species thought to be strictly uni- or bivoltine, as a consequence of climate warming (Altermatt, 2010). Winter mortality is likely to decrease under increasing temperatures (Ayres and Lombardero, 2000), although decreased snow cover (and therefore decreased insulation of overwintering sites) can reverse the pattern. Warmer winters may permit some non-diapausing species to continue feeding and development during months that were previously too cold. For example, larvae of the pine processionary moth (*Thaumetopoea pityocampa*) have a higher probability of survival if winter temperatures do not often fall below specific feeding thresholds (Battisti et al., 2005).

The first case of indirect effects goes through the host plant, with the lengthening of the growing season (especially in spring) for plants at mid to high latitudes. As many insect species evolved to match their feeding activity with certain developmental stages in the host plant, it is likely that insect herbivores are quite efficient at matching the start of their feeding to the time of budburst (Visser and Both, 2006). How well this synchrony is maintained under climate warming will depend on how well the physiological controls on insect development match that of their host plants (Ayres, 1993). The primary and secondary metabolites of woody plants can be highly sensitive to environmental factors (Herms and Mattson, 1992). Climate change will produce situations where trees are not in a good fit with the new conditions, predicting an increase in the frequency of stressful conditions in plants. Stressed plants frequently undergo biochemical changes (Koricheva et al., 1998) that may affect survival and reproduction of some forest insects. That plant stress can trigger insect outbreaks is a long-standing hypothesis in forest entomology, although there are variable consequences for insects because of partly nonlinear responses of plant defences to water and nutrient availability (Herms and Mattson, 1992) and to differential effects of environmental conditions on constitutive versus inducible plant defences (Lombardero et al., 2000). Finally, increasing atmospheric carbon (CO_2) promotes photosynthesis and leads to an increase in the C/N ratio, which makes the plant tissues less attractive to pests. In addition, increasing CO_2 suppresses the production of jasmonates and ethylene and increases the production of salicylic acid, and these differential responses of plant hormones affect specific secondary chemical pathways. In addition to changes in secondary chemistry, elevated CO_2 decreases rates of water loss from leaves, increases feeding rates, and alters nutritional content (Zavala et al., 2013). The completion of the cycle of some insects may therefore take longer than under a milder climate, and the time during which the pests remain exposed to natural enemies is longer.

The second case of indirect effects goes through natural enemies, which can exert powerful forces on the population dynamics of herbivorous insects. Assessing their response to climate warming and understanding the consequences for density-dependent feedback systems is crucial to understanding possible shifts in outbreak frequencies in the future. The distribution and abundance of natural enemies, like those of their prey, can be influenced by direct climatic effects, but the temperature sensitivity might increase (Berggren et al., 2009) or decrease (Furlong and Zalucki, 2017) with trophic level, in which case natural enemies would be more or less affected than the herbivores on which they feed. For arthropod parasitoids and predators, phenological synchrony can affect host/prey availability and host/prey size at the time of attack. Specialist enemies should be under strong selection to track phenological changes in their prey, which might make them less likely to become temporally uncoupled from their prey (Klapwijk et al., 2012). Increases or decreases in the phenological match between parasitoids and hosts could change the population dynamics of enemies through effects on prey attack rates and conversion of attacks into fecundity. Insect pathogens, for example, fungi, bacteria, and viruses, play an important role in the dynamics of many insect populations. Temperature can be important in both infection rate and defence responses within the host, with different thermal optima for host and pathogen. Fungal pathogens tend to be favoured by increases in humidity, especially when temperatures are high, but not too high. Outside of their host, the resting stages of nuclear polyhedral viruses are vulnerable to UV radiation, so increases in UV radiation could influence viral controls on some insect populations. As with specialist predators and parasitoids, the most pronounced consequences of climatic effects on herbivore–pathogen interactions may be in the decline phase of outbreaks.

In conclusion, the direct effects of climatic change on the herbivorous insect will be generally positive as a result of increased winter survival, faster development rates, and sometimes an increased number of generations per year. Some insect herbivores will benefit from an increase in the frequency of stressful events for plants, but others will not, depending on how tree defences will be affected by the changing climate. Insects that exploit young, nutritionally favourable foliage may be sensitive to climatically driven changes at the start of the growing season and to temperatures that influence the relative rates of leaf maturation and insect development after budburst. The nature of physiological controls on plant versus insect phenology will influence the extent to which plant and insect phenology remains coupled under climate change. Changes in arthropod natural enemies could lead to an increased or decreased strength of top-down control on the herbivorous insect. The magnitude of the effect, however, will depend on the attack strategy of predators and host usage in parasitoids. Pathogens are likely to increase their infection rate under increasing humidity and temperature. An important step to consider is how climatic effects on insect individuals can be scaled up to effects on species range and population dynamics. Predictions regarding the dynamics of populations under climate change require integration of the direction and strength of direct and indirect effects.

The geographic range of herbivorous insects can change over years to decades based on changes in climate, among other things (Gaston, 2003). Although adequate data are only available for some species in some regions, there are quite clear signals of recent pole-ward range expansion in numerous insect species (Parmesan et al., 1999), including outbreaking forest insects, such as the pine processionary moth and bark beetles in Northern American (Battisti and Larsson, 2023). Whether new outbreaks occur within or beyond the current range, candidate explanations are the same: release from adverse climatic conditions, natural enemies, competition with other herbivores, and host plant resistance.

The overall consequences of climate change on the vulnerability of forests to attack by insect pests are still under investigation and difficult to predict. There is still considerable uncertainty, as it is difficult to connect the individual to the population responses, which is a crucial step for predicting the extent of possible attacks. According to the most reliable theoretical models, climate change may lead to increases in both frequency and magnitude of insect outbreaks, depending on the characteristics of each species and the type of forest, while it is evident that climate change is leading to considerable changes in the distribution of all organisms, including trees.

Climate change has also aspects that seem to be going in the opposite direction. The increase in atmospheric carbon promotes photosynthesis and leads to an increase of the C/N ratio, which makes the plant tissues less attractive to pests. The completion of the cycle of some insects may therefore take longer than under a milder climate, and the time during which the pests remain exposed to natural enemies is longer. Natural enemies have also been predicted to benefit from climate change more than herbivorous insects, leading to better control. In addition, it is therefore appropriate to consider thoroughly the actions related to the choice or selection of tree species in the forests of temperate zones in the near future. For the same reasons, it is necessary to intensify the attention on plant health and increase the knowledge about both native and invasive pest species.

References

Altermatt, F. (2010) 'Climatic warming increases voltinism in European butterflies and moths', *Proceedings of the Royal Society B — Biological Sciences*, vol 277, pp. 1281–1287.

Ayres, M. P. (1993) 'Plant defense, herbivory and climate change', in P. Kareiva, J. G. Kingsolver and R. B. Huey (eds) *Biotic Interactions and Global Change*, Sinauer, Sunderland MA.

Ayres, M. P. and Lombardero, M. J. (2000) 'Assessing the consequences of global change for forest disturbance from herbivores and pathogens', *Science of the Total Environment*, vol. 262, pp. 263–286.

Barbosa, P., Letourneau, D. K. and Agrawal, A. A. (2012) *Insect Outbreaks Revisited*, Academic Press, New York.

Bale, J. S., Masters, G. J., Hodkinson, I. D., Awmack, C., Bezemer, T. M., Brown, V. K., Butterfield, J., Buse, A., Coulson, J. C., Farrar, J., Good, J. E. G., Harrington, R., Hartley, S., Jones, T. H., Lindroth, R. L., Press, M. C., Symrnioudis, I., Watt, A. D. and Whittaker, J. B. (2002) 'Herbivory in global climate change research: direct effects of rising temperature on insect herbivores', *Global Change Biology*, vol. 8, pp. 1–16.

Basset, Y., Cizek, L., Cuenoud, P. (2012) 'Arthropod diversity in a tropical forest', *Science*, vol. 338, pp. 1481–1484.

Battisti, A. and Larsson, S. (2023) 'Climate change and forest insect pests', in J. Allison et al. (eds) *Forest Entomology and Pathology. Vol. 1 Entomology*, Springer, Berlin.

Battisti, A., Stastny, M., Netherer, S., Robinet, C., Schopf, A., Roques, A. and Larsson, S. (2005) 'Expansion of geographic range in the pine processionary moth caused by increased winter temperatures', *Ecological Applications*, vol. 15, pp. 2084–2096.

Berggren, Å., Björkman, C., Bylund, H. and Ayres, M. P. (2009) 'The distribution and abundance of animal populations in a climate of uncertainty', *Oikos*, vol. 118, pp. 1121–1126.

Bjørnstad, O. N., Peltonen, M., Liebhold, A. M. and Baltensweiler, W. (2002) 'Waves of larch budmoth outbreaks in the European Alps', *Science*, vol. 298, pp. 1020–1023.

Boyd, I. L., Freer-Smith, P. H., Gilligan, C. A. and Godfray H. C. J. (2013) 'The consequence of tree pests and diseases for ecosystem services', *Science*, vol. 342, p. 823.

Diehl, J.M.C., Kowallik, V., Keller, A. and Biedermann, P.H.W. (2022) 'First experimental evidence for active farming in ambrosia beetles and strong heredity of garden microbiomes', *Proceedings of the Royal Society B*, vol. 289, 20221458.

Frost, C. J. and Hunter, M. D. (2004) 'Insect canopy herbivory and frass deposition affect soil nutrient dynamics and export in oak mesocosms', *Ecology*, vol 85, pp. 3335–3347.

Furlong, M.J. and Zalucki, M.P. (2017) 'Climate change and biological control: the consequences of increasing temperatures on host-parasitoid interactions' *Current Opinion in Insect Science*, vol. 20, pp. 39–44.

Gaston, K. J. (2003) *The Structure and Dynamics of Geographic Ranges*, Oxford University Press, New York.

Gorodkov, K. B. (1985) 'Three-dimensional climatic model of potential range and some of its characteristics. I.', *Entomologicheskoye Obozrenyie* (translated: Entomological Review, Scripta Technica 1986), vol. 64, pp. 295–310.

Hamilton, A.J., Novotny, V., Waters, E. K., Basset, Y., Benke, K.K., Grimbacher, P.S., Miller, S.E., Samuelson, G.A., Weiblen, G.D., Yen, J.D.L. and Stork, N.E. (2013) 'Estimating global arthropod species richness: refining probabilistic models using probability bounds analysis', *Oecologia*, vol. 171, pp. 357–365.

Herms, D. A. and Mattson, W. J. (1992) 'The dilemma of plants: to grow or defend', *Quarterly Review of Biology*, vol. 67, pp.283–335.

Hlásny, T., König, L., Krokene, P., Lindner, M., Montagné-Huck, C., Müller, J., and Seidl, R. (2021) 'Bark beetle outbreaks in Europe: State of knowledge and ways forward for management', *Current Forestry Reports*, vol. 7, pp 138–165.

Hodkinson, I. D. (2005) 'Terrestrial insects along elevation gradients: species and community responses to altitude', *Biological Reviews of the Cambridge Philosophical Society*, vol. 80, pp. 489–513.

Hölldobler, B. and Wilson, E.O. (1990) *The Ants*, Harvard University Press, Cambridge MA.

Johnson, D.M., Bjørnstad, O.N. and Liebhold, A.M. (2004) 'Landscape geometry and travelling waves in the larch budmoth', *Ecology Letters*, vol. 7, pp. 967–974.

Jones, L.C. (2022) 'Insects allocate eggs adaptively according to plant age, stress, disease or damage', *Proceedings of the Royal Society B*, vol. 289, 20220831.

Kirse, A., Bourlat, S.J., Langen, K. and Fonseca, V.G. (2021) 'Unearthing the potential of soil eDNA metabarcoding—towards best practice advice for invertebrate biodiversity assessment', *Frontiers in Ecology and Evolution*, vol. 9, 630560.

Klapwijk, M. J., Ayres, M. P., Battisti, A. and Larsson, S. (2012) 'Assessing the impact of climate change on outbreak potential', In P. Barbosa, D. K. Letourneau and A. A. Agrawal (eds) *Insect Outbreaks Revisited*, Academic Press, New York.

Koricheva, J., Larsson, S. and Haukioja, E. (1998) 'Insect performance on experimentally stressed woody plants: a meta-analysis', *Annual Review of Entomology*, vol. 43, pp. 195–216.

Kurz, W. A., Dymond, C. C., Stinson, G., Rampley, G. J., Neilson, E. T., Carroll, A. L., Ebata, T. and Safranyik, L. (2008) 'Mountain pine beetle and forest carbon feedback to climate change', *Nature*, vol. 452, pp. 987–990.

Liebhold, A., Koenig, W. D. and Bjørnstad, O. A. (2004) 'Spatial synchrony in population dynamics', *Annual Review Ecology Evolution Systematics*, vol. 35, pp. 467–490.

Lombardero, M. J., Ayres, M. P., Lorio, P. L. Jr. and Ruel, J. J. (2000) 'Environmental effects on constitutive and inducible resin defences of *Pinus taeda*', *Ecology Letters*, vol. 3, pp. 329–339.

Mattson, W. J. and Addy, N. D. (1975) 'Phytophagous insects as regulators of forest primary production', *Science*, vol. 190, pp. 515–522.

Moran, P. A. P. (1953) 'The statistical analysis of the Canadian lynx cycle', *Australian Journal of Zoology*, vol. 1, pp. 163–173.

Muys, B., Angelstam, P., Bauhus, J., Bouriaud, L., Jactel, H., Kraigher, H., Müller, J., Pettorelli, N., Pötzelsberger, E., Primmer, E., Svoboda, M., Thorsen, B.J. and Van Meerbeek, K. (2022) *Forest Biodiversity in Europe*. From Science to Policy 13. European Forest Institute.

Parmesan, C., Ryrholm, N., Stefanescu, C., Hill, J. K., Thomas, C. D., Descimon, H., Huntley, B., Kaila, L., Kullberg, J., Tammaru, T., Tennent, W. J., Thomas, J. A. and Warren, M. (1999) 'Poleward shifts in geographical ranges of butterfly species associated with regional warming', *Nature*, vol. 399, pp. 579–583.

Raven, P.H. and Wagner, D.L. (2021) 'Agricultural intensification and climate change are rapidly decreasing insect biodiversity', *Proceedings of the National Academy of Sciences*, vol. 118, e2002548117.

Roslin, T. et al. (2017) 'Higher predation risk for insect prey at low latitudes and elevations', *Science*, vol. 356, pp. 742–744.

Santini, A., Battisti, A. (2019) 'Complex insect–pathogen interactions in tree pandemics', *Frontiers in Physiology*, vol. 10, p. 550.

Saulnier, M., Roques, A., Guibal, F., Rozenberg, P., Saracco, G., Corona, C. and Edouard, J.-L. (2017) 'Spatiotemporal heterogeneity of larch budmoth outbreaks in the French Alps over the last 500 years', *Canadian Journal of Forest Research*, vol. 47, pp. 667–680.

Schoonhoven, L.M., Jermy, T. and van Loon, J. J. A. (1998) *Insect–Plant Biology*, Chapman & Hall, London.

Speight, M., Hunter, M. and Watt, A. (2008) *Ecology of Insects: Concepts and Applications*, 2nd ed., Wiley-Blackwell, Oxford.

Stange, E. E., Ayres, M. P. and Bess, J. A. (2011) 'Concordant population dynamics of Lepidoptera herbivores in a forest ecosystem', *Ecography*, vol. 34, pp. 772–779.

Tenow, O., Nilssen, A., Bylund, H., Pettersson, R., Battisti, A., Caroulle, F., Ciornei, C., Csoka, G., Delb, H., De Prins, W., Glavendekic, M., Gninenko, Y.I., Hrasovec, B., Matosevic, D., Meshkova, V., Moraal, L., Netoiu, C., Pajares, J., Rubtsov, V., Tomescu, R. and Utkina, I. (2013) 'Geometrid outbreak waves travel across Europe', *Journal of Animal Ecology*, vol. 82, pp. 84–95.

Underwood, E. C. and Fisher, B. L. (2006) 'The role of ants in conservation monitoring: if, when, and how', *Biological Conservation*, vol. 132, pp. 166–182.

Uvarov, B. P. (1931) 'Insects and climate', *Transactions of the Royal Entomological Society of London*, vol. 79, pp. 1–232.

Visser, M. E. and Both, C. (2006) 'Shifts in phenology due to global climate change: the need for a yardstick', *Proceedings of the Royal Society of London Series B—Biological Sciences*, vol. 272, pp. 2561–2569.

Wainhouse, D. (2005) *Ecological Methods in Forest Pest Management*, Oxford University Press, Oxford.

Walther, G. R., Post, E., Convey, P., Menzel, A., Parmesan, C., Beebee, T. J. C., Fromentin, J. M., Hoegh-Guldberg, O. and Bairlein, F. (2002) 'Ecological responses to recent climate change' *Nature*, vol. 416, pp. 389–395.

Wardle, D. A., Bardgett, R. D., Klironomos, J. N., Setälä, H., van der Putten, W. H. and Wall, D. H. (2004) 'Ecological linkages between aboveground and belowground biota', *Science*, vol. 304, pp. 1629–1633.

Wilson, E.O. (1992) *The Diversity of Life*, Harvard University Press, Cambridge, MA.

Zavala, J. A., Nabity, P. D. and DeLucia, E. H. (2013) 'An emerging understanding of mechanisms governing insect herbivory under elevated CO_2', *Annual Review of Entomology*, vol. 58, pp. 79–97.

16

BRYOPHYTES IN FOREST ECOSYSTEMS

Diversity, function and management

Nicole J. Fenton, Kristoffer Hylander, Emma Pharo and Charles E. Zartman

Introduction

Bryophytes are a polyphyletic group that have existed for millions of years, emerging in the Cambrian-Ordovician 450 million years ago, with approximately 20,000 species found globally (Bechteler et al. 2023). This includes species in three divisions that differ from one another morphologically and reproductively. The recent suggestion is that hornworts split from mosses and liverworts, which are sister groups that developed in parallel (Bechteler et al. 2023). There are approximately 7,500 species of liverworts, 12,000 species of mosses, the most diverse and abundant group, which also includes the peat mosses, and 300 species of hornworts (Bechteler et al. 2023). Despite their separation into different taxonomic groups, together they differ from vascular plants in a few fundamental ways that make a significant difference to their distribution and diversity. Specifically, bryophytes

- have no reinforced vascular tissue, which means they are rarely taller than a few centimetres;
- like all plants, alternate between haploid and diploid generations. However, in contrast to vascular plants, in bryophytes the haploid gametophyte stage is the dominant;
- have spores as the means of sexual reproduction, most of which are about 10–50 μm and are easily borne by the wind;
- have various means of vegetative reproduction including specialized structures and plant fragments; and
- have a poikilohydric habit, which means that many bryophytes can fully dehydrate and suspend function in times of drought.

In this chapter, which is an updated version of a previous text, we look at patterns of bryophyte diversity at different spatial and temporal scales, variation in bryophyte function across major forest biomes and the influence of human disturbance on forest bryophyte diversity.

DOI: 10.4324/9781003324072-19

Species diversity

Global and continental scales

The presence of a global latitudinal diversity gradient in bryophytes has long been debated, as several hotspots of bryophyte diversity are not in the tropics, for example New Zealand. However, recent evidence has shown that there is a latitudinal diversity gradient in liverworts and hornworts, with higher richness in the tropics. It is currently unknown whether this is also true for mosses, as phylogeny and species distributions have not been adequately collated to complete these analyses (Patino and Vanderpoorten 2018). However, for pleurocarpous mosses (certain families with a branching growth form, often growing hanging or in carpets), it has been shown that species turnover among similar-sized tropical areas is higher than among the same-sized temperate areas, which leads to higher total diversity in the tropics (Hedenäs 2007).

Bryophytes are found in most forests around the world, from sub-arctic to tropical, even in mangroves (Salazar Allen et al. 2022). In general, bryophytes seem to have the highest biomass and diversity in relatively cool and moist forests, such as tropical montane cloud forests and temperate rain forests (e.g. Frahm and Gradstein 1991). However, boreal forests can also be species rich, especially in areas with high humidity (Dynesius et al. 2009, Fenton and Bergeron 2008). Compared to montane cloud forests, tropical lowland forests have been assumed to be species-poor (Frahm and Gradstein 1991), but more recent studies have shown high bryophyte diversity in tropical lowland forests around the world (Shen et al. 2023, Berdugo et al. 2022). Bryophytes do not seem to play an important role in tropical deciduous forests, where they can be very scarce, both in terms of cover and species richness (KH, personal observation). Climatic gradients across continents clearly influence the composition and richness of forest bryophytes (Möls et al. 2013). Along the continental to oceanic gradient, it is apparent that the number of liverworts increases with higher oceanity (Vanderpooten and Goffinet 2009).

Interestingly, a number of species in forests in such diverse systems as moist Afromontane forests in southwest Ethiopia, temperate forests embedded in Mediterranean-type vegetation in South Africa and boreal forests along streams in Sweden, all show rather high levels of species richness or alpha diversity at the scale of 0.02 ha, even if the tropical example has the highest numbers (Figure 16.1). The relative order of species richness among different forest ecosystems at the scale of 0.02 ha is also mirrored in the order when pooling the cumulative number of species over 10–20 sites (compare the curves in Figure 16.1). Interestingly, the diversity is quite similar across three continents if the curves for the datasets from Australia, non-stream boreal Sweden and boreal Canada are compared.

Landscape scale

There are many gradients in environmental variables that regulate forest types as well as bryophyte communities across landscapes. Typically, the topography of the landscape determines the variation in solar radiation and can create contrasting moisture conditions. These gradients will directly influence the distribution of bryophytes across landscapes in all biomes, with typically higher species richness in landscape positions with a cool and moist microclimate, even if some species are confined to sunny and dry conditions (Dynesius et al.

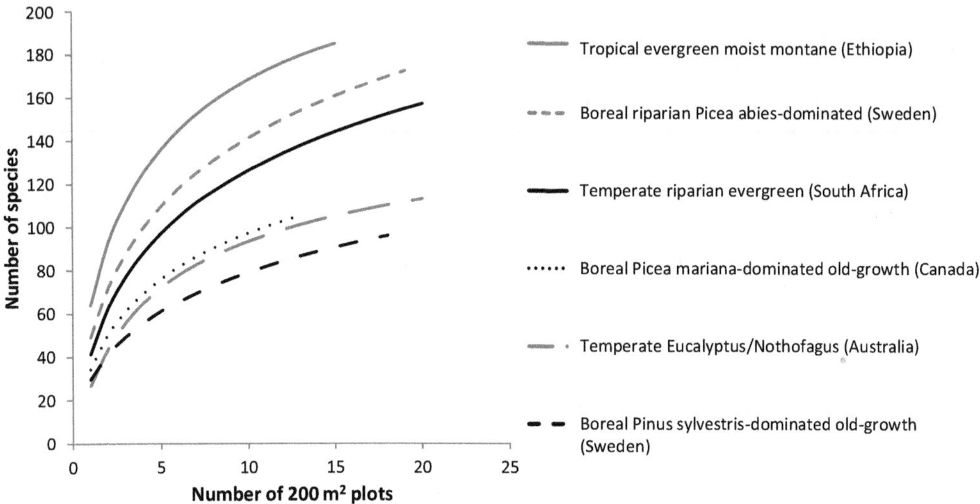

Figure 16.1 Species accumulation curves for bryophytes in scattered 0.02 ha plots (in most cases 10 x 20 m) in different biomes and forest environments across the world. Data unpublished or reanalysed from: Hylander, K. and Nemomissa, S. (2009) 'Complementary roles of home gardens and exotic tree plantations as alternative habitats for Ethiopian montane rainforest plant biodiversity', *Conservation Biology*, vol. 23, pp 400–409; Hylander, K. and Hedderson, T.A. (2007) 'Does the width of isolated ravine forests influence moss and liverwort diversity and composition?—a study of temperate forests in South Africa', *Biodiversity and Conservation*, vol. 16, pp. 1441–1458; Hylander, K. and Dynesius, M. (2006) 'Causes of the large variation in bryophyte species richness and composition among boreal streamside forests', *Journal of Vegetation Science*, vol. 17, pp. 333–346; Pharo, E.J., Meagher, D.A. and Lindenmayer, D.B. (2013) 'Bryophyte persistence following major fire in eucalypt forest of southern Australia', *Forest Ecology and Management*, vol. 296, pp. 24–32.

2009, Tng et al. 2009). Topography also affects bryophyte communities indirectly through its regulation of soil conditions, humidity and the dominant vegetation in boreal, tropical and southern Nothofagus forests (Barbé et al. 2020, Pinheiro Costa et al. 2020, Toro Manrìquez et al. 2020).

Differences in environmental conditions and forest types across landscapes also determine disturbance regimes, which in turn affect bryophyte community composition and diversity. Forests regenerated after different types of disturbances, for example, fires and insect disturbances, can have different community compositions depending on the amount of deadwood or the light regime (Schmalholz et al. 2011). Another illustration of the importance of disturbances for bryophytes at the landscape level is the change in composition and diversity among sites differing in time since stand replacing wildfire in eastern Canada, which determines the abundance and richness of peat mosses (*Sphagnum* spp.) and liverworts on the forest floor (Fenton and Bergeron 2008). In Fennoscandia, wet forests can be species-rich and/or species-poor depending on hydrology, pH and forest type (Hylander and Dynesius 2006).

Stand scale

The importance of the variability in type and quality of microhabitats within the forest as a controlling factor for bryophyte diversity at the stand scale is difficult to overstate. In dry forest types, bryophytes tend to be confined to certain microhabitats, for example, seepage areas along streams unless they are dry habitat specialist epiphytes. A relative scarcity of forest floor bryophytes can also be typical in certain temperate deciduous forests, but here the reason is that it is difficult for bryophytes to establish and survive the leaf litter that accumulates on the forest floor. Similarly, in lowland and montane tropical moist to wet forests, bryophytes are generally rather scarce on the forest floor. In certain other forest types, bryophytes cover the ground (Hylander et al. 2005).

Overall, different bryophyte species utilize different sets of microhabitats such as dead wood, shaded rock surfaces, mineral soil and seepage areas in boreal forests (Hylander et al. 2005), while in moist temperate and tropical forests, epiphytic species generally dominate, even if many other microhabitats are also utilized (Hylander & Nemomissa 2009; Shen et al. 2023; Berdugo et al 2022). Even leaves have their own bryophyte flora in moist tropical forests that could be species-rich (epiphylls; Zartman et al. 2012; Mežaka et al. 2022). Consequently, the presence of microhabitats of various kinds generally increases bryophyte species richness. However, bryophyte richness can also depend on the size of the regional species pool for certain substrate types. For example, there seem to be more species that can potentially colonize moderate to high-pH riparian forests than low-pH riparian forests in Fennoscandia (Hylander and Dynesius 2006).

On an even smaller scale, gradients in pH, surface orientation and moisture holding capacity can correlate with a turnover in species composition within pieces of dead wood, rock surfaces or tree bark. Leachate from deciduous trees with higher pH leaves could influence the underlying bryophyte vegetation on, for example, boulders in a profound way (Weibull and Rydin 2005). Similarly, the bark characteristics (e.g. pH, bark structure) of tree species can result in quite different levels of bryophyte diversity (Löbel et al. 2006). In lowland tropical environments across the globe, the diverse epiphytic community varies with height on the host tree (de Oliveira and der Steege et al. 2015; Berdugo et al. 2021; Shen et al. 2023). While the highest canopy community is similar across large and small landscapes, trunk bryophyte communities seem to display higher beta diversity (de Oliveira and der Steege et al. 2015, Berdugo et al. 2021), although there does not seem to be strong evidence for changes in community drivers from canopy to tree base (Shen et al. 2023).

Temporal scale

Since many bryophytes can expand clonally and die off in some parts while growing in other parts, bryophyte colonies could, theoretically, be very old. Therefore, bryophyte population dynamics in forests are tightly bound to the dynamics of their substrates and display patch-tracking meta-population dynamics (Snäll et al. 2005). One could thus assume that bryophytes on dead wood are more dynamic than epiphytes, which are still more dynamic than on the forest floor or on boulders. Epiphyllous communities are probably the most dynamic, with reported generations of liverworts within a year in lowland Amazonian forests (Zartman et al. 2012). Another rather temporary substrate in forests is newly exposed mineral soil, for example in tree uprooting. This microhabitat can house a quite distinct bryophyte community, in which at least some components are quite short-lived (Jonsson and

Esseen 1990). Several of those species are also adapted to brighter conditions and can dominate early successional stages after large-scale disturbances.

Dispersal by bryophytes is primarily leptokurtic, with wind-deposited propagule density decreasing rapidly with distance from the source colony (Patino and Vanderpoorten 2018). However, it is now clear that the vast majority of undeposited spores disperse in a distance-independent manner (Barbé et al. 2016; Vanderpoorten et al. 2019). However, not all bryophytes produce small diaspores in large quantities and significant dispersal limitation has been found at local to regional scales in most biomes (Vanderpoorten et al. 2019, Ledent et al. 2020). Several studies have found that species' reproductive traits including frequency of reproduction and propagule size and habitat connectivity influence dispersal (Patino and Vanderpoorten 2018).

Variation in bryophyte function across forest biomes

Bryophytes influence a variety of ecosystem functions in different biomes, particularly in many forest ecosystems where bryophytes accumulate significant biomass (e.g. over 20 kg/m^2 in paludified boreal forests, Lecomte et al. 2006, see also Vanderpoorten and Goffinet 2009). For example, superficially different systems such as boreal forest floors (Turetsky et al. 2012), epiphytic bryophyte mats on tree branches in humid temperate forests (Pypker et al. 2006) and epiphytic and epiphyllous bryophyte mats in humid tropical forests can all be seen as analogous (Zartman et al. 2012). These different systems provide a variety of different ecosystem functions including water retention, nutrient cycling and habitat for other organisms.

Water retention

All plants are dependent on water for the turgor of their tissues and for photosynthesis (Glime 2007). Bryophytes, similar to lichens (Chapter 17), are poikilohydric plants that must absorb and retain the water available to them in the above-ground environment, either within a shoot or within a colony. Therefore, when a significant mass of bryophytes are present within an ecosystem, they have a significant impact on the water balance of this system (Turetsky et al. 2012, Pypker et al. 2006; Figure 16.2).

The water retained within bryophyte colonies has a multitude of effects on ecosystem function, which vary with the position of the bryophyte colony. In forests where bryophyte colonies cover the mineral soil, the presence and the composition of the bryophytes significantly change water availability in the mineral soil (Oishi 2018). An extreme example of this is seen during paludification, where *Sphagnum* spp. mosses replace *Pleurozium schreberi* and other feather mosses in some boreal forests. As a consequence of this replacement, the mineral soil is eventually isolated from the surface of the organic layer (generated by the mosses) by a perpetually flooded anoxic zone (Lecomte et al. 2006).

The role of bryophyte colonies growing over coarse woody debris, branches or leaves, while functionally similar, results in different consequences. In these environments, the water retained by bryophytes facilitates the development of other plants. On coarse woody debris and branches, bryophytes promote vascular plant establishment and growth (Clark et al. 2005). For example, epiphytic bryophytes in cloud forests can harvest water droplets (Clark et al. 2005). Vascular plants are then able to access this water source, which with their waxy cuticle they normally would not be able to do (Clark et al. 2005, Pypker et al. 2006). Similarly, epiphytes in the upper canopy transfer water (and nutrients) to epiphytes

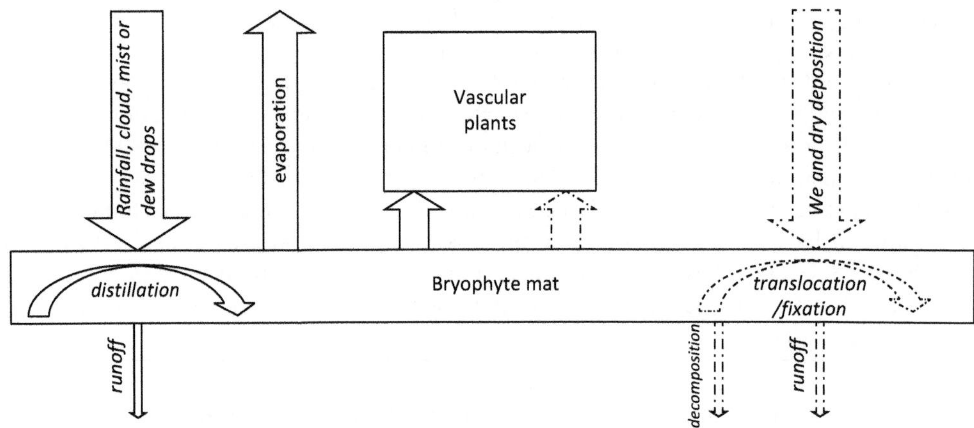

Figure 16.2 Summary of effects of bryophyte mats in various positions (over mineral soil, on branches or on leaves) on water (solid lines) and nutrients (dashed lines) fluxes.

and epiphylls lower in the canopy (Clark et al. 2005). Furthermore, epiphytes and epiphylls in forest canopies store and then evapo-transpire significant amounts of water, influencing the global water cycle (Porada et al. 2018)

Nutrient cycling

Bryophytes influence nutrient cycles in a variety of ways, but most can be mediated through one of three processes: retention, decomposition and nitrogen fixation (Figure 16.2). One of the simplest ways in which bryophytes influence nutrient cycles is simply by the retention of nutrients. Bryophytes obtain a significant portion of their mineral requirements from either dry (dust) or wet (rain and leachate) deposition (Glime 2007). Nutrient availability is typically less concentrated in these sources than in the mineral soil, and consequently bryophytes are well adapted to retain the nutrients that are available to them (Glime 2007). Once adsorbed or absorbed, the nutrients can remain within their tissues for significant periods, as several studies have indicated that bryophytes, like lichens, are able to recycle nutrients within their tissues (i.e. from old towards young; Glime 2007). For example, in one study where marked nitrogen was applied to a bryophyte mat, nitrogen was moved from the senescent segments to the growing segments, but none was released to the environment (Eckstein and Karlsson 1999 in Glime 2007). Similarly, epiphytic bryophytes have been shown to retain 50 per cent of atmospheric deposition in tropical montane forests (Clark et al. 2005).

The decomposition rate of bryophytes, both on the forest floor and in the canopy (Lang et al. 2009), is notoriously low. Many studies have compared the decomposition rates of bryophytes to vascular plants and lichens (e.g. Lang et al. 2009), and all come to the conclusion that bryophytes decompose much slower than these other groups. The cause for this slower rate is less clear, however. Bryophytes lack lignin, the structural component that slows decomposition in woody plants; however, their structural carbohydrates (pentosans) and aromatic compounds do seem to resist decomposition (Lang et al. 2009). In addition, bryophytes have a variety of secondary metabolites that seem to inhibit microbial growth, fungal growth (Opelt et al. 2007) and grazing by invertebrates (Glime 2007), three pathways that lead to vascular plant decomposition. Consequently, bryophytes seem to be primarily decomposed by a subset of tolerant fungi (Davey et al. 2012). Little is known about bryophyte–fungi interactions; however, both mycorhizal and saprophytic fungi have been identified on and within green and senescent bryophytes (Davey et al. 2012).

In addition to their retention capacity and low decomposition rate, bryophytes also host a variety of bacteria species, including cyanobacteria (DeLuca et al. 2002). Cyanobacteria, some of which have the capacity to fix nitrogen via heterocysts (specialized cells), have been found in symbiosis (Bay et al. 2013) with a wide variety of bryophyte species, including forest floor species (e.g. *Pleurozium schreberi*), *Sphagnum* spp., epiphylls and epiphytes (Opelt et al. 2007, Lindo and Whiteley 2011). Long considered inconsequential in the nitrogen budget of ecosystems, a variety of studies have now indicated that the moss-cyanobacterial unit contributes significantly to the overall budget of forests (DeLuca et al. 2002). This is particularly true in nitrogen-poor areas such as boreal forests, as nitrogen fixation rate appears to increase with a decrease in available nitrogen, either naturally (via succession) or anthropogenically (Lindo et al. 2013).

When these three mechanisms are taken together (retention, slow decomposition and nitrogen fixation), bryophytes clearly can have a significant impact on the ecosystem nutrient cycles, particularly in nutrient-poor environments. The consequences for ecosystem functions are numerous. Bryophyte colonies (terrestrial, aerial, or epiphyllous) retain nutrients that otherwise would have been lost to the ecosystem via leaching (Clark et al. 2005, Lindo and Whiteley 2011, Turetsky et al. 2012). These colonies are nutrient storehouses, with different nutrients either adsorbed to their surfaces, or converted into complex compounds and stored within bryophyte cells.

The fate of nutrients captured by bryophytes is less clear. Several studies have demonstrated that the nutrients retained by the bryophytes or fixed by cyanobacterial epiphylls are subsequently used by vascular plants. This occurs either through absorption via mycorrhizal hyphae, which are particularly abundant around senescent mosses, through leaching as cellular contents of the bryophytes are lost during wet/dry cycles, or via a more direct transfer as was found for nitrogen fixed by cyanobacteria living in epiphyllous liverworts (Glime 2007). However, most studies examining the residence time of nutrients within bryophytes find that nutrients are retained for significant periods, for years in nearly all cases (Turetsky et al. 2012). In these situations, the nutrients contained within the bryophyte would be functionally unavailable, and some have argued that bryophytes contribute to decreasing nutrient availability in many ecosystems. However, while the rate of nutrient release from bryophyte mats is quite likely slow, and although the relationship with mycorrhizal fungi remains unresolved, these nutrients do remain in the system and are eventually released. Without bryophyte mats, these nutrients may well have been definitively lost from the system via leaching, and the amount of nitrogen would have been significantly less. This is particularly true of aerial bryophyte mats that retain nutrient-rich layers on branches, effectively generating a soil within the canopy environment (Lindo and Whitely 2011).

Habitat

While bryophytes may seem diminutive, when observed at the appropriate scale, they become a forest unto themselves and thus represent a three-dimensional habitat that is inhabited and used by a variety of organisms. Glime (2007) reviews the variety of different organisms that live in bryophyte mats; these include cyanobacteria, protists, fungi and mesofauna. In general, these groups are very poorly studied. However, the few studies that have looked at the different inhabitants of bryophytes indicate significant diversity at different spatial scales, and some levels of specialization for different bryophyte species. Fungi from a wide variety of groups are associated with all types of bryophytes (Döbbeler and Hertel 2013), and Davey et al. (2012) found that different species of fungi were associated with different bryophyte species.

Forest bryophytes and habitat change

Most of the studies of human use and our effect on forests examine the impacts of fire, timber harvesting and fragmentation. We focus here on aspects of forest management that are particular to bryophytes. For many issues that have been important in the forest management literature over the last 20–30 years, such as the question of remnant size and edge effects, bryophytes follow a similar pattern to other species, with results depending on the interaction between the particular bryophyte community, their environment and the characteristics of the disturbance (Fischer and Lindenmayer 2007).

Bryophyte response to human disturbance of forests needs to be understood in the context of the earlier discussion about the importance of suitable microhabitats. The importance of microhabitats cannot be overstated in seeking to understand bryophyte diversity in both natural and human-altered forests. While microhabitat and macro-habitat are often closely correlated, the history of events within any one particular area of forest can lead to a rich variety of substrates for bryophytes to inhabit or a relatively poor environment. Because bryophytes are small, suitable microhabitat may be available in a landscape that, on first impression, would appear to be highly unsuitable, such as remnant deadwood from the original forest in exotic pine plantations with thick needle litter (Pharo and Lindenmayer 2009). On the other hand, some substrates take so long time to be created, such as large logs, that they only occur in habitats with a long continuity.

Forest fragmentation

Fragmentation is well understood as a significant landscape change for many taxa. Many bryophytes are widely distributed and therefore, in theory at least, have the potential to disperse between areas of suitable habitat. As discussed above, bryophytes generally have small, light propagules, which suggest that successful dispersal in a fragmented landscape might be possible for many species. There is evidence to suggest that this is indeed the case, with some populations of bryophytes able to spread their propagules into fragmented patches in the landscape (e.g. Hylander 2009). However, in at least some forest landscapes, dispersal appears to be more of a controlling factor than in open biomes (Löbel et al. 2006).

Bryophytes follow a pattern that is common to many other types of organisms in fragmented landscapes. For example, large remnants tend to be superior habitat compared with small remnants (Oishi and Morimoto 2013). As for other taxa, the results are variable, depending on the life history strategy of the species in question. The changes in humidity, light and temperature associated with forest edges is generally negatively associated with bryophyte richness, but this is variable, with some species and systems being much less affected or positively changed (e.g. Barbé et al. 2017). Species that you might expect to be resilient to fragmentation are not necessarily able to adapt to the disturbance. For example, tropical epiphylls might be expected to be resilient to changes since they are adapted to high-turnover environments and regularly produce spores (Zartman et al 2012). However, even among these bryophytes adapted to the ephemeral habitat of canopy leaves, researchers have found evidence for dispersal limitation (Zartman et al 2012).

Timber, pulp and biofuel production in forests

Bryophyte response to timber harvesting is largely determined by the severity of the disturbance, with increasing disruption of the bryophyte community with increasing harvesting intensity (Opoku-Nyame et al. 2021; Hylander et al. 2005). Boreal forests that have had between 30 and 70 years to recover from clear felling have less diverse forest bryophyte communities compared with adjacent old stands, not least because of the lower amount of dead wood per area (cf. Dynesius et al. 2009). However, for both clearfell and selective harvesting, retention of microhabitats to 'lifeboat' some species through the disturbance can soften the effect, especially for species growing on concave surfaces (Schmalholz and Hylander 2011).

The relative contribution of different factors, such as microclimate, dispersal ability and substrate availability, in forest recovery is often not particularly clear. It would be fair to say that given the importance of coarse woody debris for boosting bryophyte diversity in some forest ecosystems, repeated extraction of forest products needs careful management in order to prevent or minimize the ongoing erosion of that resource (Opuko-Nyame et al. 2021). Epiphytic and epiphyllous species are obviously strongly affected by timber harvesting. On the other hand, some of them might be adapted to certain successional stages, which means that a total focus on old-growth stages might put some species in a disadvantage (cf. the species composition in aspen stands regenerated after fires versus the dead wood-rich stands developed after spruce budworm outbreaks; Schmalholz et al. 2011).

The conversion of more natural forests to plantations of exotic trees is an interesting and important issue. Plantations are often found to be hostile to bryophytes, but the severity of the impact can be softened through the retention of microhabitats. The replacement of native trees with exotic 'crop' trees tends to leave little in the way of native biodiversity, although biological legacies such as logs and green tree retention can soften the effect (Pharo and Lindenmayer 2009). However, in southwestern Ethiopia, plantations of *Cupressus lusitanica* did contain quite high bryophyte diversity (Hylander and Nemomissa 2009), and mixed hybrid aspen and spruce plantations in Canada housed a relatively diverse bryophyte flora (Randriamananjara et al. 2023).

Climate change

Climate change is influencing bryophytes in forests in three ways: (1) physiologically; (2) in interaction with their forest habitat; and (3) movement of climate envelopes. First, bryophytes, even tropical bryophytes, are generally adapted to cold temperatures and are able to maintain a positive carbon balance in low temperatures, even below zero (He et al. 2016); therefore, increasing temperatures may be problematic for them. They also tend to have negative carbon balances (photosynthesis is outstripped by respiration) at higher temperatures. This is not generally a problem as species will dry out and become dormant at high temperatures and can then survive much higher temperatures. However, conditions of high humidity and high temperature could have a very negative affect on bryophytes (He et al. 2016). Finally, if climate change results in more frequent cycling of wet and dry periods, or more frequent frost damage, the cost of becoming dormant, and subsequently reactivating the metabolism may ultimately lead to bryophyte death (Zuijlen et al. 2023).

A second aspect is how bryophytes respond within their forest habitat, and how the forest mediates climate change. As a suitable microclimate is essential for bryophytes, one element of this is the preservation of microclimate, which will depend in part on landscape structure and use. For example, edge effects from clear-cuts into forests were more severe in a drought year compared to in normal years (Koelemeijer et al. 2023), and bryophyte communities change less under coniferous forests that maintain a stable microclimate than under mixed forests (Barbé et al. 2020). Thus, climate adaptation of bryophyte conservation needs to consider how to preserve the microclimate or ensure its development in new climate-appropriate regions (Hylander et al. 2021). Species traits and taxonomic groups will vary in their sensitivity to climate change within forests, with studies highlighting the sensitivity of liverworts (Mateo et al. 2016), epixylics (Barbé et al. 2020), small short-lived species and those that don't reproduce asexually (Löbel et al. 2021).

Finally, forest bryophytes will need to ultimately track their forest habitats across the globe as ecosystems shift northward and upward. While there is already some evidence of an increase in altitude of some bryophyte communities (Kiebacher et al. 2023), many species in Europe are not predicted to be able to colonize new climate envelopes fast enough, and the relative ability to migrate is again mediated by species traits, in particular spore size (Zanatta et al. 2020). Taken together, climate change is and will continue to have major impacts on bryophytes in forests into the future.

Conclusion

Bryophyte ecology is labour-intensive and requires specialist knowledge, consequently a thorough inventory of a particular area of forest tends to be highly labour-intensive and difficult to carry out. For this reason, our understanding of the effects of forest harvest is limited to regions where there has been the most work. Canopy species are not well understood, nor are some of the most species-rich environments in tropical countries. However, we do know enough to conclude that some of the same general rules of thumb apply to bryophytes as for other organisms:

- There is a large variation among bryophyte responses to disturbance, depending on their life history strategy;
- Some bryophyte species are resilient to disturbances, such as timber harvesting, within limits on the magnitude of the disturbance and intensity of the change, while others seem very sensitive to even slight changes in microclimate or substrate quality;
- Ongoing landscape scale fragmentation and homogenization of stand ages and substrate availabilities tend to result in the decrease of bryophyte diversity;
- Climate change and increased variability in climate, including extreme events, are an additional challenge over and above the ongoing landscape change that needs to be considered in interaction with land use.

Given that bryophytes are difficult to identify, an important approach for informing managers is to emphasize that a variation in microhabitats generally will favour bryophyte diversity including certain species that are sensitive to alterations of habitats. Important microhabitats for bryophytes in forests could be unaltered springs and small streams, shaded rocky outcrops, dead wood and a variation in tree species with different bark structures and chemistry. A complementary approach based on bryophyte life forms (colony sizes and shapes) can be possible to use in certain forest ecosystems. A manager without expert knowledge could in tropical areas identify stands with high cover of epiphyllous species or plenty of large pendulous or fan-like bryophytes growing on trees and shrubs. In temperate areas, it could be more useful to look out for a variety of life forms growing over woody debris and rocky outcrops and sometimes epiphytic indicating a favourable microclimate, and therefore probable high bryophyte diversity.

References

Allen, N. S., Dauphin, G., Villarreal, J. C., Caswell-Levy, C. A. L. E. B., Cox, E. R., Gudiño, J., ... and Zeballos-Grijalva, K. (2022) 'Bryophytes of mangroves of Bocas del Toro, Panama', *Bryophyte Diversity and Evolution*, vol. 45(1), pp. 133–150.

Barbé, M., Bouchard, M., and Fenton, N. J. (2020) 'Examining boreal forest resilience to temperature variability using bryophytes: forest type matters', *Ecosphere*, vol. 11(8), e03232.

Barbé, M., Fenton, N. J., and Bergeron, Y. (2016) 'So close and yet so far away: long-distance dispersal events govern bryophyte metacommunity reassembly', *Journal of Ecology*, vol. 104, pp. 1707–1719.

Barbé, M., Fenton, N. J., and Bergeron, Y. (2017) 'Are post-fire residual forest patches refugia for boreal bryophyte species? Implications for ecosystem based management and conservation', *Biodiversity and Conservation*, vol. 26, pp. 943–965.

Bay, G., Nahar, N., Oubre, M., Whitehouse, M. J., Wardle, D. A., Zackrisson, O., Nilsson, M. C. and Rasmussen, U. (2013) 'Boreal feather mosses secrete chemical signals to gain nitrogen', *New Phytologist*, vol. 200, pp. 54–60.

Bechteler, J., Peñaloza-Bojacá, G., Bell, D., Gordon Burleigh, J., McDaniel, S. F., Christine Davis, E., ... and Villarreal A, J. C. (2023) 'Comprehensive phylogenomic time tree of bryophytes reveals deep relationships and uncovers gene incongruences in the last 500 million years of diversification', *American Journal of Botany*, vol. 110(11), e16249.

Berdugo, M. B., Gradstein, S. R., Guerot, L., León-Yánez, S., Bendix, J., and Bader, M. Y. (2022) 'Diversity patterns of epiphytic bryophytes across spatial scales: Species-rich crowns and beta-diverse trunks', *Biotropica*, vol. 54(4), pp. 893–905.

Clark, K. L., Nadkarni, N. M. and Gholz, H. L. (2005) 'Retention of inorganic nitrogen by epiphytic bryophytes in a tropical montane forest', *Biotropica* vol. 37, pp. 328–336.

Costa, D. P., Nadal, F. and da Rocha, T. C. (2020) 'The first botanical explorations of bryophyte diversity in the Brazilian Amazon mountains: high species diversity, low endemism, and low similarity', *Biodiversity and Conservation*, vol. 29, pp. 2663–2688.

Davey, M. L., Heegaard, E., Halvorsen, R., Ohlson, M. and Kauserud, H. (2012) 'Seasonal trends in the biomass and structure of bryophyte-associated fungal communities explored by 454 pyrosequencing', *New Phytologist*, vol. 195, pp. 844–856.

DeLuca, T., Zackrisson, O., Nilsson, M. and Sellstedt, A. (2002) 'Quantifying nitrogen-fixation in feather moss carpets of boreal forests', *Nature*, vol. 419, pp. 917–920.

Döbbeler, P. and Hertel, H. (2013) 'Bryophilous ascomycetes everywhere: Distribution maps of selected species on liverworts, mosses and Polytrichaceae', *Herzogia*, vol. 26, pp.361–404.

Dynesius, M., Hylander, K. and Nilsson, C. (2009) 'High resilience of bryophyte assemblages in streamside compared to upland forests', *Ecology*, vol. 90, pp. 1042–1054.

Eckstein, R. L. and Karlsson, P. S. (1999) 'Recycling of nitrogen among segments of Hylocomium splendens as Compared with Polytrichum commune: Implications for clonal integration in an ectohydric bryophyte', *Oikos*, vol. 86, No. 1, pp. 87–96. Published by Wiley on behalf of Nordic Society Oikos. www.jstor.org/stable/3546572

Fenton, N. J. and Bergeron, Y. (2008) 'Does time or habitat make old-growth forests species rich? Bryophyte richness in boreal *Picea mariana* forests', *Biological Conservation*, vol. 141, pp.1389–1399.

Fischer, J. and Lindenmayer, D. B. (2007) 'Landscape modification and habitat fragmentation: a synthesis', *Global Ecology and Biogeography*, vol. 16, pp. 265–280.

Frahm J-P. and Gradstein, S.R. (1991) 'An altitudinal zonation of tropical rain forests using bryophytes' *Journal of Biogeography*, vol. 18, pp. 669–678.

Glime, Janice M. (2007) *Bryophyte Ecology*, Ebook sponsored by Michigan Technological University and the International Association of Bryologists. https://digitalcommons.mtu.edu/bryophyte-ecology1/ Accessed on December 3, 2013.

He, X., He, K. S., and Hyvönen, J. (2016) 'Will bryophytes survive in a warming world? ', *Perspectives in Plant Ecology, Evolution and Systematics*, vol. 19, pp. 49–60.

Hedenäs, L. (2007) 'Global diversity patterns among pleurocarpous mosses', *Bryologist*, vol. 110, pp. 319–331.

Hylander, K. (2009) 'No increase in colonization rate of boreal bryophytes close to propagule sources', *Ecology*, vol. 90, pp. 160–169.

Hylander, K. and Dynesius, M. (2006) 'Causes of the large variation in bryophyte species richness and composition among boreal streamside forests', *Journal of Vegetation Science*, vol. 17, pp. 333–346.

Hylander, K., Dynesius, M., Jonsson, B. G. and Nilsson, C. (2005) 'Substrate form determines the fate of bryophytes in riparian buffer strips', *Ecological Applications*, vol. 15, pp. 674–688.

Hylander, K., Greiser, C., Christiansen, D.M. and Koelemeijer, I.A. 2021, 'Climate adaptation of biodiversity conservation in managed forest landscapes', *Conservation Biology*, vol. 3, e13847.

Hylander, K., and Nemomissa, S. (2009) 'Complementary roles of home gardens and exotic tree plantations as alternative habitats for plants of the Ethiopian montane rainforest', *Conservation Biology*, vol. 23, pp. 400–409.

Jonsson, B. G. and Esseen, P. (1990) 'Treefall disturbance maintains high bryophyte diversity in a boreal spruce forest', *Journal of Ecology*, vol. 78, pp. 924–936.

Kiebacher, T., Meier, M., Kipfer, T., and Roth, T. (2023) 'Thermophilisation of communities differs between land plant lineages, land use types and elevation', *Scientific Reports*, vol. 13(1), 11395.

Koelemeijer, I. A., Ehrlén, J., De Frenne, P., Jönsson, M., Berg, P., and Hylander, K. (2023) 'Forest edge effects on moss growth are amplified by drought', *Ecological Applications*, vol. 33(4), e2851.

Lang, S. I., Cornelissen, J. H. C., Klahn, T., van Logtestijn, R. S. P., Broekman, R., Schweikert, W. and Aerts, R. (2009) 'An experimental comparison of chemical traits and litter decomposition rates in a diverse range of subarctic bryophyte, lichen and vascular plant species', *Journal of Ecology*, vol. 97, pp. 886–900.

Lecomte, N., Simard, M., Fenton, N. and Bergeron, Y. (2006) 'Fire severity and long-term biomass dynamics in coniferous boreal forests of eastern Canada', *Ecosystems*, vol. 9, pp.1215–1230.

Ledent, A., Gauthier, J., Pereira, M., Overson, R., Laenen, B., Mardulyn, P., ... and Vanderpoorten, A. (2020) 'What do tropical cryptogams reveal? Strong genetic structure in Amazonian bryophytes', *New Phytologist*, vol. 228(2), pp. 640–650.

Lindo, Z., Nilsson, M. C. and Gundale, M. J. (2013) 'Bryophyte–cyanobacteria associations as regulators of the northern latitude carbon balance in response to global change', *Global Change Biology*, vol. 19, pp. 2022–2035.

Lindo, Z. and Whiteley, J. (2011) 'Old trees contribute bio-available nitrogen through canopy bryophytes', *Plant and Soil*, vol. 342, pp. 141–148.

Löbel, S., Schroeder, B., & Snäll, T. (2021), 'Projected shifts in deadwood bryophyte communities under national climate and forestry scenarios benefit large competitors and impair small species', *Journal of Biogeography*, vol. 48(12), pp. 3170–3184.

Löbel, S., Snäll, T. and Rydin, H. (2006) 'Species richness patterns and metapopulation processes — evidence from epiphyte communities in boreo-nemoral forests', *Ecography*, vol. 29, pp. 169–182.

Mateo, R. G., Broennimann, O., Normand, S., Petitpierre, B., Araújo, M. B., Svenning, J. C., ... and Vanderpoorten, A. (2016), 'The mossy north: an inverse latitudinal diversity gradient in European bryophytes', *Scientific Reports*, vol. 6(1), 25546.

Mežaka, A., Allen, N.S., Mendieta-Leiva, G., and Bader, M. Y. (2022) 'Life on a leaf: The development of spatial structure in epiphyll communities', *Journal of Ecology*, vol. 110(3), pp. 619–630.

Möls, T., Vellak, K. and Ingerpuu, A. N. (2013) 'Global gradients in moss and vascular plant diversity', *Biodiversity and Conservation*, vol 22, pp. 1537–1551.

Oishi, Y. (2018) 'Evaluation of the water-storage capacity of bryophytes along an altitudinal gradient from temperate forests to the alpine zone forests', *Forests*, vol. 9, p. 433. doi:10.3390/f9070433

Oishi, Y. and Morimoto, Y. (2013) 'Identifying indicator species for bryophyte conservation in fragmented forests', *Landscape and Ecological Engineering*, vol. XX, pp. 1–8.

de Oliveira, S. M. and ter Steege, H. (2015) 'Bryophyte communities in the Amazon forest are regulated by height on the host tree and site elevation', *Journal of Ecology*, vol. 103(2), pp. 441–450.
Opelt, K., Berg, C., Schönmann, S., Eberl, L., and Berg, G. (2007) 'High specificity but contrasting biodiversity of Sphagnum-associated bacterial and plant communities in bog ecosystems independent of the geographical region', *The ISME journal*, vol. 1(6), pp. 502–516.
Opoku-Nyame, J., Leduc, A., and Fenton, N. J. (2021) 'Bryophyte conservation in managed boreal landscapes: Fourteen-year impacts of partial cuts on epixylic bryophytes', *Frontiers in Forests and Global Change*, vol. 4, 674887.
Patiño, J. and Vanderpoorten, A. (2018). 'Bryophyte biogeography', *Critical Reviews in Plant Sciences*, vol. 37(2–3), pp. 175–209.
Pharo, E. J. and Lindenmayer, D.B. (2009), 'Biological legacies soften pine plantation effects for bryophytes', *Biodiversity and Conservation*, vol. 18, pp. 1751–1764.
Porada, P., Van Stan, J. T. and Kleidon, A. (2018) 'Significant contribution of non-vascular vegetation to global rainfall interception', *Nature Geoscience*, vol. 11(8), pp. 563–567.
Pypker, T. G., Unsworth, M. H. and Bond, B. J. (2006) 'The role of epiphytes in rainfall interception by forests in the Pacific Northwest. II. Field measurements at the branch and canopy scale', *Canadian Journal of Forest Research*, vol. 36, pp. 819–832.
Randriamananjara, M. A., Fenton, N. J. and DesRochers, A. (2023) 'How does understory vegetation diversity and composition differ between monocultures and mixed plantations of hybrid poplar and spruce?', *Forest Ecology and Management*, vol. 549, 121434.
Schmalholz, M. and Hylander, K. (2011) 'Microtopography creates small-scale refugia for boreal forest floor bryophytes during clear-cut logging', *Ecography*, vol. 34, No. 4, pp. 637–648.
Schmalholz, M., Hylander, K. and Frego,K. (2011) 'Bryophyte species richness and composition in young forests regenerated after clear-cut logging versus after wildfire and spruce budworm outbreak', *Biodiversity and Conservation*, vol. 20, pp. 2575–2596.
Shen, T., Song, L., Corlett, R. T., Guisan, A., Wang, J., Ma, W. Z., ... and Collart, F. (2023) 'Disentangling the roles of chance, abiotic factors and biotic interactions among epiphytic bryophyte communities in a tropical rainforest (Yunnan, China)', *Plant Biology*, vol. 25(6), pp. 880–891.
Snäll, T., Ehrlén, J. and Rydin, H. (2005) 'Colonization-extinction dynamics of an epiphyte metapopulation in a dynamic landscape', *Ecology*, vol. 86, pp. 106–115.
Tng, D. Y. P., Dalton, P.J. and Jordan, G.J. (2009) 'Does moisture affect the partitioning of bryophytes between terrestrial and epiphytic substrates within cool temperate rain forests?', *Bryologist*, vol. 112, pp. 506–519.
Toro Manríquez, M. D., Ardiles, V., Promis, Á., Huertas Herrera, A., Soler, R., Lencinas, M. V., and Martínez Pastur, G. (2020) 'Forest canopy-cover composition and landscape influence on bryophyte communities in Nothofagus forests of southern Patagonia', *PloS One*, vol. 15(11), e0232922.
Turetsky, M. R., Bond-Lamberty, B., Euskirchen, E., Talbot, J., Frolking, S., McGuire, A. D. and Tuittila, E. S. (2012) 'The resilience and functional role of moss in boreal and arctic ecosystems', *New Phytologist*, vol. 196, pp. 49–67.
Vanderpoorten, A. and Goffinet, B. (2009) *Introduction to Bryophytes.* Cambridge University Press, Cambridge.
Vanderpoorten, A., Patiño, J., Désamoré, A., Laenen, B., Górski, P., Papp, B., ... and Hardy, O. (2019) 'To what extent are bryophytes efficient dispersers?', *Journal of Ecology*, vol. 107(5), pp. 2149–2154.
Weibull, H. and Rydin, H. (2005) 'Bryophyte species richness on boulders: relationship to area, habitat diversity and canopy tree species', *Biological Conservation*, vol. 122, 71–79.
Zanatta, F., Engler, R., Collart, F., Broennimann, O., Mateo, R. G., Papp, B., Muñoz, J., Baurain, D., Guisan, A. and Vanderpoorten, A. (2020) 'Bryophytes are predicted to lag behind future climate change despite their high dispersal capacities', *Nature Communications*, vol. 11, 5601. https://doi.org/10.1038/s41467-020-19410-8; www.nature.com/naturecommunications
Zartman, C. E., Nascimento, H.E., Cangani, K.G., Alvarenga, L.D. and Snäll, T. (2012) 'Fine-scale changes in connectivity affect the metapopulation dynamics of a bryophyte confined to ephemeral patches', *Journal of Ecology*, vol 100, pp. 980–986.
van Zuijlen, K., Kassel, M., Dorrepaal, E., and Lett, S. (2023) 'Frost damage measured by electrolyte leakage in subarctic bryophytes increases with climate warming', *Journal of Ecology*, https://doi.org/10.1111/1365-2745.14236.

17

LICHENS IN FOREST ECOSYSTEMS

Per-Anders Esseen, Göran Thor and Darwyn Coxson

Introduction

Lichens consist of a partnership (symbiosis) between one or more heterotrophic fungi (mostly ascomycetes and basidiomycetes) and an autotrophic green alga (chlorolichens) or cyanobacterium (cyanolichens). Diverse communities of other bacteria are also associated with lichens (Spribille et al., 2016), influencing their function and fitness. The fungi (mycobionts) dominate, and their hyphae envelop the algae or cyanobacteria (photobionts). Photobionts provide carbohydrates by photosynthesis while the mycobionts provides both protection from high light, UV radiation and desiccation and facilitate water storage. This partnership has proved to be extremely successful, and lichens have evolved a large diversity of life forms and adaptations that are widely distributed in forests throughout the world.

Lichens exhibit great variability in colour, growth form and size, ranging from less than 1 cm in many crustose species to over a 1 m in length in some epiphytic species (e.g., *Ramalina menziesii*, *Usnea longissima*). The vegetative body (thallus) is usually classified into three main growth forms (Figure 17.1): crustose (forming a crust), foliose (flattened) and fruticose (erect or pendent, often richly branched), but intermediate types are frequently found.

Unlike plants, lichens lack root systems and therefore lack active mechanisms for regulation of uptake and loss of water (poikilohydry). Nonetheless, many lichens show morphological traits that facilitate regulation of hydration status. These include rhizines on the lower surface of *Peltigera* that wick moisture from the forest-floor surface, development of cushions by reindeer lichens that facilitate retention of water or the thin highly dissected thalli of understory rainforest species such as *Sticta filix and S. lacera* in New Zealand that minimize depression of net photosynthesis at high water contents (Lange et al., 1983). These traits are linked to the main water sources in lichens (rain, humid air, dew) and help to explain lichen distribution and function (Gauslaa, 2014).

The photobiont plays a critical role in lichen growth strategies. Cyanolichens must be wetted by liquid water for activation of photosynthesis while chlorolichens can be activated by humid air. This has major implications for the distribution of these lichens in relation to climate. Epiphytic cyanolichens, for example, dominate in oceanic climates with frequent

DOI: 10.4324/9781003324072-20

rainfall, but in more continental climates are often limited to sites with more humid microclimate.

The functional role of lichens in forest ecosystems depends on the interaction of the traits in each species with forest structure and dynamics. In this chapter, we give a broad overview of lichens in forest ecosystems throughout the world, their roles and interactions with other components in forests, and exemplify their role in selected forest ecosystems. For general information on lichens, see Nash (2008). For more detailed reviews on forest lichens, we recommend Sillett and Antoine (2004), Ellis (2012) and Coxson and Howe (2017).

Distribution of forest lichens

Lichens represent a significant component of biodiversity in forest ecosystems. Although some 20,000 lichen species have presently been described, the total number may be as high as 28,000 species, with an estimated 14,000 species found in the tropics (Lücking et al., 2009). By comparison, 5,000 species may be found in the boreal zone and about 8,000 species in the temperate zone. Microlichens (crustose) dominate by species number and constitute about two-thirds of all lichens, while macrolichens (foliose + fruticose) make up the bulk of lichen biomass in most forests. Many crustose lichens are microhabitat specialists. Lichens can be found on a range of microhabitats within forests, but epiphytes dominate by species numbers. Lücking et al. (2009) estimated that over 60 per cent of global lichen species richness was represented by epiphytic species on bark or on leaves (foliicolous lichens).

Boreal forests

Boreal forests support abundant forest floor and canopy lichens but also hosts species-rich communities on rocks and woody debris. They are less variable in terms of species composition than temperate and tropical forests, containing many widespread circumpolar lichens (Hauck, 2011). Among factors contributing to the lower variability of boreal lichen communities is a similarity in ecological conditions, with cool and/or nutrient-poor soils often limiting plant growth. Boreal lichens vary in response to fire and can be classified as:

- Fire-dependent (e.g., *Carbonicola* spp., *Hertelidea botryosa*) on fire-scarred wood
- Fire-related, associated with postfire successions (e.g., jelly lichens Collemataceae) on deciduous trees)
- Fire-sensitive, preferring fire-refugia with long continuity of forest cover (e.g., *Usnea longissma*).

Terricolous lichens can dominate the forest floor in dry boreal forests (Payette and Delwaide, 2018). Lichen woodlands are especially well-developed on coarse or well-drained soils, often growing under pine (*Pinus*) but also spruce (*Picea*), larch (*Larix*) and birch (*Betula*). The forest floor in lichen woodlands is dominated by chlorolichens, including reindeer lichens (mat-forming species in *Cladonia*) and cup-bearing *Cladonia* lichens.

Epiphytic lichen communities in the boreal forest reflect the dominance of conifers and low tree species diversity. A large number of crustose lichens are associated with bark and decaying wood surfaces (Figure 17.1, top left). They are often host-specific, for example, calicioid species ('pin lichens', crustose species that often have thin stalked fruiting bodies), many of whom are old-growth indicators (Goward and Arsenault, 2018).

Figure 17.1 The crustose lichen *Calicium tigillare* on wood in boreal forest, Norway (top left). The foliose lichens *Lobaria pulmonaria* and *L. scrobiculata* (dark grey, in centre) epiphytic on *Picea glauca* in sub-boreal forest in northern British Columbia, Canada (top middle). The fruticose, pendulous lichens *Alectoria sarmentosa* (light coloured) and *Bryoria* spp. (dark coloured) on *Abies lasiocarpa* in subalpine forest, British Columbia, Canada (top right). The foliicolous lichen *Eremothecella calamicola* on a leaf of *Arenga engleri* in subtropical forest, Iriomote Island, South Japan (middle left). The foliose lichen *Pseudocyphellaria coronata* on *Dacrycarpus dacrydioides* in south temperate New Zealand rainforest (middle right). The foliose lichen *Anzia japonica* on bark of *Abies sachalinensis* in hemiboreal forest, Hokkaido, Japan (bottom left). The fruticose lichen *Cladonia sulphurina* growing in boreal forest floor, Norway (bottom right). Photographs (top left) and (bottom right) by Einar Timdal, (top middle) and (middle right) by Darwyn Coxson, (top right) by Per-Anders Esseen, (middle left) by Kento Miyazawa, and (bottom left) by Yoshihito Ohmura.

A striking feature of old coniferous forests throughout the boreal zone is the abundant growth of epiphytic pendulous (alectorioid) lichens in genera such as *Alectoria*, *Bryoria* (Figure 17.1, top right), *Oropogon* and *Usnea*. These lichens are hairlike and have thin branches, which is an adaptation to the rapid uptake of water from humid air. Alectorioid lichens show a typical vertical zonation, with dark *Bryoria* spp. in the upper canopy and light *Alectoria* spp. in the lower canopy.

In younger boreal forests, foliose chlorolichens, in contrast, constitute most of the epiphyte biomass on boreal conifers, including species of, for example, *Hypogymnia*, *Parmeliopsis*, *Platismatia*, *Tuckermannopsis* and *Vulpicida*. Rich cyanolichen communities can develop in boreal rainforests in Asia (at least in southern Siberia, Russia), Europe (western Norway) and North America (part of western and eastern regions) (DellaSala, 2011). Boreal rainforests often contain *Lobarion* community lichens, a distinct community of lichen epiphytes common also to cool temperate rainforests (see below), including conspicuous foliose cyanolichens such as *Lobaria* (Figure 17.1, top middle), *Nephroma*, *Pseudocyphellaria* and *Sticta*. Rare chlorolichens such as *Anzia* (Figure 17.1, bottom left) also occur in these forests.

Temperate forests

Lichen communities in temperate forests are more variable in species richness than in boreal forests due to higher tree species diversity, warmer climate and greater variation in precipitation and elevation. Major temperate forest ecosystems can be found in Asia (e.g., Russia, China, South and North Korea and Japan), Australasia (Australia, New Zealand), Europe, eastern North America and South America (Chile and Argentina). Temperate forests have experienced significant deforestation from conversion to agricultural and/or urban landscapes (Nascimbene et al., 2013).

Temperate forests in areas with cool and wet climates were historically known for rich epiphytic lichen communities, including members of the *Lobarion* community. *Lobarion* lichens are currently much more restricted in distribution due to the cumulative impacts of air pollution and habitat loss (Seaward, 1977). Many old-growth-associated lichens in European temperate forests are red-listed. Some rare/red-listed species occurring in gaps or other open forest environments are now found mainly on old trees in agricultural landscapes, parks and religious sites.

Temperate rainforests occur in Asia (e.g., China, India, Iran, Japan, Nepal, Russia, South Korea, Turkey), Australasia (Australia, New Zealand), Europe (Ireland, United Kingdom, Portugal, Spain), western North America (from northern California to south-eastern Alaska) and South America (Argentina, Chile) and host diverse lichen communities (DellaSala, 2011). Many of these lichens are associated primarily with old-growth forests, where site continuity and protected microclimates result in the accumulation of species that are otherwise rare in regional landscapes (Sillett and Antoine, 2004). Goward and Arsenault (2018), for instance, found that calicioid species diversity was still increasing five centuries after stand initiation in British Columbia's inland temperate rainforest.

High lichen diversity in foliose *Lobarion* community genera such as *Pseudocyphellaria* (Figure 17.1, middle right) is found in temperate rainforests of New Zealand, Australia and Chile, reflecting their presumed ancestral Gondwanaland origin (Nash, 2008, p. 315). Temperate forests in China and Japan are also very species-rich.

Tropical forests

Tropical forests support high lichen species richness with thousands of undescribed species (Lücking et al., 2009). The rapid conversion of species-rich primary forests to species-poor secondary forests or other land uses makes the documentation of tropical lichen communities an urgent priority (Wolseley and Aguirre-Hudson, 1997; Li et al., 2013). The greatest lichen diversity and biomass within tropical forests occurs in canopy environments. About three-quarters of tropical lichens grow on bark or on leaves; this proportion is higher than in temperate and boreal forests (Lücking et al., 2009). Tropical forests stand out with a higher proportion of foliicolous lichens than in other forests (>1,000 species), representing about 90 per cent of the global diversity in this group. These lichens are most diverse in primary tropical lowland rain forest, including genera like *Eremothecella* (Figure 17.1, middle left). Diversity decreases with increasing elevation and seasonality, whereas host tree species has only a minor influence on the composition of foliicolous lichen assemblages.

In general, lichens are a minor component of the forest-floor vegetation in tropical lowland forests as the dense canopy restricts light penetration. The lichen flora in lowland forests is dominated by crustose epiphytic chlorolichens, with diverse communities in, for example, mangrove forests. Crustose lichens were the dominant lichen growth form in a dry tropical forest in Mexico and had surprisingly high biomass (Miranda-González and McCune, 2020). Diversity of lichens was highest in late succession and increased with tree species richness and decreased canopy cover in lowland rain forests of western Amazonia (Déleg et al., 2021).

The diversity and biomass of macrolichens increase with altitude in tropical forests (Wolf, 1993). Macrolichen communities are best developed in montane forests, particularly in cloud forests, where broad-lobed cyanolichens, for example, *Dendrosticta*, *Lobaria*, *Pseudocyphellaria* and *Sticta*, form a conspicuous epiphyte component. The distribution of lichens on different substrates changes with elevation and climate in tropical montane forests, with lichens on terricolous substrates showing the strongest increase (Rodríguez-Quiel et al., 2019). Water is mainly supplied by rain at lower elevations while lichens are frequently hydrated by fog at high elevations.

Function in forest ecosystems

Lichens interact with forests in numerous ways and support many important ecosystem functions that we are now only beginning to understand. They are involved in biogeochemical cycles, provide food and habitat for many species of animals and interact with forest microclimates (Nash, 2008, p. 274) as discussed below.

Biogeochemical cycling

Lichens and bryophytes (see Chapter 16) represent only a small proportion of the biomass in most forested ecosystems; nonetheless, they can have a disproportionate impact on ecosystem function and nutrient cycling (Asplund and Wardle, 2017; Porada et al., 2014). Lichens, however, by virtue of their active uptake of nutrients from atmospheric sources and potentially high rates of turnover can sequester and release nutrients that might otherwise be unavailable.

Lichens play an especially important role in nitrogen cycling. Lichens with cyanobacterial symbionts are capable of fixing nitrogen from atmospheric sources. This ability to fix atmospheric nitrogen enriches the lichen thallus; fixed nitrogen is then available as further leaching and decomposition occur. Nitrogen fixation in lichens can have significant ecosystem impacts, especially where cyanolichens have a high-standing biomass. In subarctic and boreal ecosystems, terricolous lichens such as *Stereocaulon* are a major source of newly fixed nitrogen (Crittenden, 2000), while in wet temperate rainforests, *Lobarion* lichens provide up to 16 kg of N per hectare annually (Nash, 2008; Sillett and Antoine, 2004).

Lichens are often regarded as slow-growing and long-lived. However, the growth and turnover of forest lichens can be significant. Standing biomass of reindeer lichens can range from 1,000–2,000 kg ha^{-1} in dry continental sites to over 9,000 kg ha^{-1} in wet oceanic sites (Payette and Delwaide, 2018). Epiphytic lichens can likewise achieve substantial biomass accumulation. Stevenson and Coxson (2003) documented 1,200 kg ha^{-1} standing biomass of alectorioid lichens in old-growth subalpine coniferous forests, with annual turnover near 10 per cent.

Mat-forming lichens in boreal forests provide an interesting example of nutrient cycling within individual lichen mats. Mats of *Cladonia stellaris* and *Stereocaulon paschale* often develop a thick layer of necrotic material at the base of the mat and it has been suggested that key elements are recycled internally to support new growth at the apices (Crittenden, 2000).

A more comprehensive review of the impact of lichens on nutrient cycling and fluxes is provided by Asplund and Wardle (2017).

Interactions with animals

Lichens constitute forage, habitat, shelter and nest material for invertebrates and vertebrates in forests but animals, in turn, can be important as dispersal vectors for lichens (Asplund and Wardle, 2017). Herbivory is low in most lichens, probably because of their low-nutrient content, structural characteristics and defence compounds produced by the mycobiont. One example of such an antiherbivore compound is vulpinic acid, a bright yellow pigment that occurs in, for example, the calicioid lichen *Chaenotheca chrysocephala* and the wolf lichen (*Letharia* spp.).

Among invertebrates, insects, mites and molluscs are the main lichen feeders. Cyanolichens with their higher nutrient content are preferred food for gastropods (reviewed by Solhaug and Gauslaa, 2012). One study showed that gastropod grazing on three species of *Lobaria* increased near the ground on tree trunks, suggesting that gastropods may play a role in shaping the vertical distribution of lichens.

Chlorolichens in canopies are also important for invertebrates. Branches in old boreal *Picea abies* forests support much higher lichen biomass than in younger forests (Figure 17.2a). Branches with abundant lichens supported both greater diversity and number of invertebrates (Figure 17.2b) by providing more complex habitat structure than branches with few lichens. Large invertebrates (e.g., flies and spiders) are in turn important as forage for overwintering birds such as tits and the Siberian Jay (*Perisoreus infaustus*). Epiphytic lichens thus play a key role in the function of boreal forest canopies and changes in their biomass may have cascading effects on higher trophic levels.

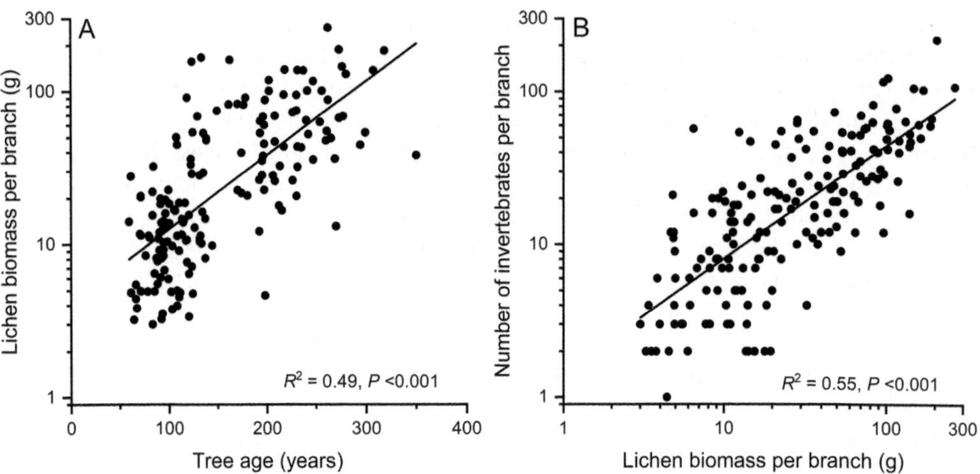

Figure 17.2 Relationship between biomass of epiphytic macrolichens per branch and tree age (A), and between number of invertebrates per branch and lichen biomass (B) in lower canopy of *Picea abies* forests in northern Sweden. Drawn from data in Esseen et al. (1996) and Pettersson et al. (1995).

Among vertebrates, some tree frogs, reptiles and birds are well camouflaged against a background of lichens, but these animals do not eat lichens (Seaward, 1977). Many forest birds use lichens for nest-building. In contrast, both ground and canopy lichens constitute an important part of the diet of some mammals. Reindeer and caribou (subspecies of *Rangifer tarandus*) represent the best examples of a mammal–lichen interaction (see below), but lichens are also preferred food by some species of deer, voles, squirrels and monkeys.

Forage lichens for reindeer and caribou

Subarctic and boreal forests are the main habitats of reindeer and caribou during winter, and lichens are their dominant forage source, often constituting more than 50 per cent of the diet (Heggberget et al., 2002). Reindeer and caribou can locate lichens under snowpack by scent and then use their front hooves to excavate craters. The most important terricolous forage lichens are 'reindeer lichens', particularly *Cladonia arbuscula*, *C. mitis*, *C. rangiferina* and *C. stellaris*, but *Cetraria*, *Stereocaulon* spp. and other species are also eaten. Forage lichens are generally rich in easily digestible carbohydrates, but poor in protein and minerals and the winter diet must be complemented with mushrooms, low shrubs and graminoids to be nutritionally balanced.

High reindeer and caribou densities may lead to overgrazing (Figure 17.3), delaying the recovery of lichen mats. When forest-floor lichens are unavailable, especially during winter, reindeer and caribou often feed on epiphytic lichens. Highly digestible pendulous *Bryoria* spp. are preferred, but other species, such as *Alectoria* and foliose lichens are also consumed. *Bryoria* spp. are an important winter forage for both woodland and mountain caribou ecotypes in western North America. The loss and fragmentation of lichen bearing old forests has become a critical conservation issue for woodland caribou in western North America and reindeer in Fennoscandia.

Figure 17.3 Contrast between heavily grazed (left) and ungrazed forest floor (enclosure from 1949) with reindeer lichens in dry, boreal *Pinus sylvestris* forest in Sweden. Photograph by P.-A. Esseen.

Interactions with climate

Terricolous lichen mats such as *Cladonia* spp. and *Stereocaulon* spp. play a major role in mediating above- and below-ground energy exchange in mid- to late-successional boreal and subarctic forests. Lichens also significantly interact with hydrology in forests through the interception of rain and evaporation and influence diurnal latent heat fluxes by their uptake and loss of water (Porada et al., 2018). Macroclimate, particularly the amount and duration of rainfall, drives the growth of dominant pendulous lichens across a large-scale gradient from continental to oceanic climates (Phinney et al., 2021).

Epiphytic lichens respond strongly to vertical and horizontal gradients in temperature, moisture and light within forest canopies. McCune (1993) noted that as forests age, lichen communities associated with xeric canopy conditions tended to move upwards in the canopy. McCune (1993) proposed the similar gradient hypothesis, namely that: 'epiphyte species are ordered similarly on three distinct spatial and temporal gradients':

1 Vertical differences in species composition within a given stand.
2 Species compositional differences among stands differing in moisture regime but of the same age.
3 Changes in species composition through time in a given stand.

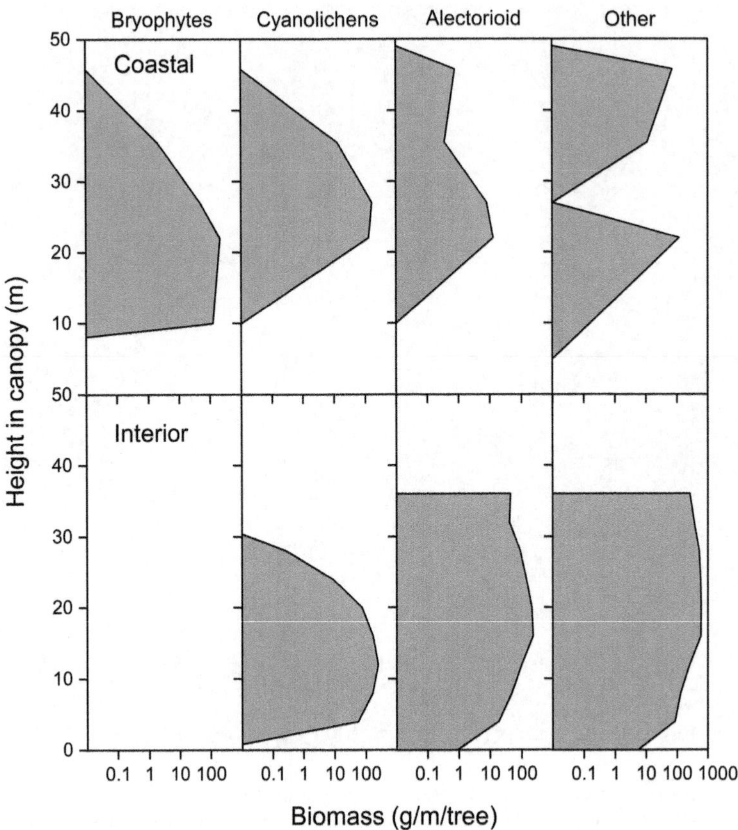

Figure 17.4 A comparison of the vertical distribution of canopy epiphyte functional groups in old-growth (>400 years old) coastal (redrawn from McCune, 1993) and interior (redrawn from Benson and Coxson 2002) wet temperate rainforests in western North America. Interior bryophyte abundance from Darwyn Coxson (unpublished data).

Evidence for this hypothesis can be seen in comparisons between wet coastal and drier interior temperate forests (Figure 17.4) where foliose cyanolichens, which require mesic canopy conditions, occupy only the lower part of the dry interior forests (Benson and Coxson, 2002), but are found in mid- to upper-canopy in wet coastal forests, where bryophytes instead dominate the wet lower canopy (McCune, 1993). The three-dimensional distribution of lichens in canopies is driven by forest dynamics, canopy structure, microclimate and lichen functional traits (Esseen and Ekström, 2023).

Dynamics of lichen communities

The development of lichen communities is tightly linked to forest dynamics, where many lichens can be characterized as 'patch-tracking' organisms that follow tree population dynamics. Changes in biomass and species richness (see 'Distribution of forest lichens' above) are driven by the functional traits (morphology, reproduction, growth, water storage, defence compounds) in local populations as they respond to microclimate, chemical and physical environments, as well as the larger-scale impact of forest disturbance regimes (type, severity, size and interval between disturbances). For example, growth form may change during succession with many crustose lichens being pioneers on bark followed by foliose and fruticose chlorolichens and later by bryophytes, cyanolichens and late-successional crustose lichens.

Epiphytic lichens on bark

Epiphyte communities are controlled by factors that operate at tree-, stand- and landscape level (Ellis, 2012), but mechanisms are complex as factors are often confounded.

Key factors for each level include:

- Tree level: tree species, bark chemistry (bark pH, mineral nutrients), bark structure (texture, stability, water-holding capacity), tree age, diameter, height, inclination and crown structure. The variability at this level is fundamental for species richness during succession as most lichens show some degree of microhabitat specificity. The time scale for epiphyte succession can vary greatly, but generally it is a slow process taking decades to centuries (Figure 17.2a).
- Stand level: tree species composition, horizontal (gaps, clumps of trees) and vertical canopy structure (foliage height profile), age distribution, stand size and shape. Forest structure changes during succession and modifies canopy microclimates, particularly openness and humidity. This, in turn affects lichen biomass accumulation over time (Sillett and Antoine, 2004).
- Landscape level: position in relation to elevation, topography and water bodies, landscape composition and spatial configuration. Landscape heterogeneity affects microclimate, availability of lichen propagules and edge effects.

Epiphytic lichens on leaves

Foliicolous lichens are a ubiquitous feature of tropical rainforests, but these lichens also occur in humid subtropical and temperate forests. This specialized group of leaf epiphytes colonizes a wide range of angiosperms, gymnosperms and ferns (Lücking, 2008). Up to 50 species can occur on a single leave of the size of a hand. Most leaves are short-lived, providing an ephemeral substratum which is difficult to be colonized by organisms. Even in tropical rainforests, the average leaf longevity does not exceed 24–36 months, except for palms and cycads, where leaves may persist for more than 10 years. Foliicolous lichens must therefore complete their life cycle quickly, which implies rapid reproduction and effective propagation (Lücking, 2008).

Lichens on deadwood

The dynamics of lichen communities on deadwood varies with tree species, decay rate, competition with bryophytes, fungi and vascular plants as well as with microclimate (Spribille et al., 2008; Ellis, 2012). The rapid turnover of woody debris in most tropical forests often hinder the development of lichen communities. In contrast, deadwood may persist for decades or centuries in temperate and boreal forests and can host diverse communities on snags, stumps and logs. On treefalls, epiphytes on bark remain for some time and are then followed by wood-inhabiting and finally forest-floor species.

Terricolous lichens

Terricolous lichens are often outcompeted by bryophytes in closed-canopy forests. The successional pattern of terricolous lichens after fire is similar in nutrient-poor lichen woodlands across the boreal zone with five stages (Seaward, 1977, p. 165):

1 Bare soil stage
2 Crustose lichen stage (e.g., *Trapeliopis granulosa*)
3 Cup-lichen stage (e.g., *Cladonia cornuta*, *C. sulphurina;* Figure 17.1, bottom right)
4 First reindeer lichen stage (*C. arbuscula*, *C. mitis*, *C. rangiferina*, *C. uncialis*)
5 A second reindeer lichen stage often develops where *C. stellaris* dominate about 80–120 years after fire in northern lichen woodlands.

In practice, these stages overlap and the sequence is often interrupted by reindeer/caribou grazing or reset by fire. Payette and Delwaide (2018) review factors shaping North American lichen woodlands, emphasizing the role of fire in maintaining successional conditions that favour lichen growth.

Indicators of forest ecosystem health

Epiphytic lichens have long been used as indicators of environmental quality. Their interception of nutrients from atmospheric sources, including entrapment of particles and active uptake of ions, can result in the accumulation of environmental contaminants (Nash, 2008). This combined with their differential sensitivity to pollutants makes epiphytic lichens an important group from which to select indicator species. Standardized methods for lichen indicators have been developed for many regions (Nimis et al., 2002).

Methods of using lichens for conservation surveys have been developed in several countries. Many species are useful as indicators of diversity of epiphytic lichenized and non-lichenized fungi as well as conservation value of forests, including calicioid lichens, *Lobaria pulmonaria* and pendent hair lichens, for example, *Alectoria sarmentosa*. The application of lichen indicators is, however, not without problems. For example, the links between lichens and their habitats are sometimes not fully understood and few studies separate long-term forest continuity from tree age. Ellis (2012) discussed three scenarios that could explain why epiphyte indicators need long continuity of forest cover:

1 They are specialized on certain microhabitats and/or microclimates.
2 They are dispersal-limited (Sillett et al., 2000; Scheidegger and Werth, 2009) and need a continuous supply of suitable habitats.
3 A combination of both.

Conservation biology of forest lichens

Deforestation, landscape fragmentation, forestry, habitat degradation, pollution and climate change have all had dramatic effects on lichens, with many species declining in abundance or regionally extirpated (e.g., Scheidegger and Werth, 2009; Hauck et al., 2013; Allen et al., 2019; Esseen et al., 2022). The boreal felt lichen (*Erioderma pedicellatum*), only found in Alaska, eastern Canada, Kamchatka and Scandinavia, is a prime example of a globally red-listed boreal lichen with highly specific canopy structural requirements (Power et al., 2018).

A major problem in managed forests is the widespread adoption of clear-cutting with short rotations (100 years or less), which hinder the development of species-rich lichen communities (Nascimbene et al., 2013; Esseen et al., 2022). The fundamental issue is that industrial forestry can drive forests well outside of their natural range of variability, shifting age-class distributions towards fewer older forest stands, reduced structural heterogeneity and denser forests.

Anthropogenically caused shifts in abiotic (fire, drought, wind, snow, ice) and biotic (pest insects, pathogens) disturbance regimes may also cause widespread damage to lichen communities. For example, moderate to high-severity wildfires result in long-term losses in species richness and abundance of epiphytic lichens (Miller et al., 2018). Interactions of climate change and large-scale forest pest outbreaks are also having major impacts on epiphytic lichen communities, as many lichens only occur on a few tree species.

In conclusion, key priorities for the conservation of forest lichens and ecosystem functions they support in a global perspective are to:

- Protection of at least 30 per cent of the world's forests (Dinerstein et al., 2019). These forests must be formally protected in a long-term perspective. No forestry measures should be allowed, including removal of dead trees, and invasive species should be actively combatted
- Identify and protect lichen biodiversity hotspots at all spatial scales, including old forests, the remaining boreal and temperate rainforests, swamp forests, riparian forests, particularly spray zones near waterfalls and other humid forests near water bodies
- Protect microrefugia, with suitable microclimates and microhabitats for lichens, throughout biomes, regions, and landscapes, where lichens can find respite from changing climate in future landscapes (Ellis, 2020)
- Maintain tree species diversity and its link to natural disturbance regimes
- Preserve old trees in all kinds of ecosystems.

Priorities in managed forests are to:

- Develop and apply novel forest management methods (e.g., partial cutting, variable thinning, long rotations), as alternatives to clear-cutting
- Preserve large patches of lichen-rich forests that can function as sources of lichen propagules throughout the landscape
- Retain and restore structural elements such as native tree species, deadwood as well as old and large trees
- Non-native tree species should not be allowed, and the use of clonal and breeding plants as well as fertilization should be strictly limited.

Acknowledgements

We thank Kento Miyazawa, Yoshihito Ohmura and Einar Timdal for permission to use their photos.

References

Allen, J. E., McMullin, R. T., Tripp, E. A. and Lendemer, J. C. (2019) 'Lichen conservation in North America: a review of current practices and research in Canada and the United States', *Biodiversity and Conservation,* vol. 28, pp. 3103–3138.

Asplund, J. and Wardle, D. A. (2017) 'How lichens impact on terrestrial community and ecosystem properties', *Biological Reviews,* vol. 92, pp. 1720–1738.

Benson, S. and Coxson, D. S. (2002) 'Lichen colonization and gap structure in wet-temperate rainforests of northern interior British Columbia', *Bryologist,* vol. 105, pp. 673–692.

Coxson, D. and Howe, N. (2017) 'Lichens in natural ecosystems', in J. Dighton, and J. F. White (eds) *The Fungal Community: its Organization and Role in the Ecosystem* (4th ed.), Taylor & Francis, Boca Raton, FL.

Crittenden, P. D. (2000) 'Aspects of the ecology of mat-forming lichens', *Rangifer,* vol. 20, pp. 127–139.

Déleg, J., Gradstein, S. R., Aragón, G., Giordani, P. and Benítez, A. (2021) 'Cryptogamic epiphytes as indicators of successional changes in megadiverse lowland rain forests of western Amazonia', *Ecological Indicators,* vol. 129, 10789.

DellaSala, D. A. (ed) (2011) *Temperate and Boreal Rainforests of the World: Ecology and Conservation,* Island Press, Washington, DC.

Dinerstein, E., Vynne, C., Sala, E., Joshi, A. R., Fernando, S. and others (2019) 'A global deal for nature: Guiding principles, milestones, and targets', *Science Advances* vol. 5, eaaw2869.

Ellis, C. J. (2012) 'Lichen epiphyte diversity: a species, community and trait-based review', *Perspectives in Plant Ecology, Evolution and Systematics,* vol. 14, pp. 131–152.

Ellis, C.J. (2020) 'Microclimatic refugia in riparian woodland: A climate change adaptation strategy', *Forest Ecology and Management,* vol. 462, 118006.

Esseen, P.-A. and Ekström, M. (2023) 'Influence of canopy structure and light on three-dimensional distribution of the iconic lichen *Usnea longissima*', *Forest Ecology and Management,* vol. 529, 120667.

Esseen, P.-A., Ekström, M., Grafström, A., Jonsson, B. G., Palmqvist, K. and others (2022) 'Multiple drivers of large-scale lichen decline in boreal forest canopies', *Global Change Biology,* vol. 28, pp. 3293–3309.

Esseen, P.-A., Renhorn, K.-E. and Pettersson, R. B. (1996) 'Epiphytic lichen biomass in managed and old-growth boreal forests: effect of branch quality', *Ecological Applications,* vol. 6, pp. 228–238.

Gauslaa, Y. (2014) 'Rain, dew, and humid air as drivers of morphology, function and spatial distribution in epiphytic lichens', *Lichenologist,* vol. 46, pp. 1–16.

Goward, T. and Arsenault, A. (2018) 'Calicioid diversity in humid inland British Columbia may increase into the 5th century after stand initiation', *Lichenologist,* vol. 50, pp. 555–569.

Hauck, M. (2011) 'Site factors controlling epiphytic lichen abundance in northern coniferous forests', *Flora,* vol. 206, pp. 81–90.

Hauck, M., de Bruyn, U. and Leuschner, C. (2013) 'Dramatic diversity losses in epiphytic lichens in temperate broad-leaved forests during the last 150 years', *Biological Conservation,* vol. 157, pp. 136–145.

Heggberget, T. M., Gaare, E. and Ball, J. P. (2002) 'Reindeer (*Rangifer tarandus*) and climate change: importance of winter forage', *Rangifer,* vol. 22, pp. 13–31.

Lange, O. L., Büdel, B., Heber, U., Meyer, A., Zellner, H., and Green, T. G. A. (1983) 'Temperate rainforest lichens in New Zealand: high thallus water content can severely limit photosynthetic CO_2 exchange', *Oecologia,* vol. 95, pp. 303–313.

Li, S., Liu, W. Y. and Li, D.-W. (2013) 'Epiphytic lichens in subtropical forest ecosystems in southwest China: species diversity and implications for conservation', *Biological Conservation*, vol. 159, pp. 88–95.

Lücking, R. (2008). 'Foliicolous lichenized fungi', *Flora Neotropica Monograph*, vol. 103, pp. 1–866, The New York Botanical Garden Press, New York.

Lücking, R., Rivas Plata, E., Chaves, J. L., Umaña, L. and Sipman, H. J. M. (2009) 'How many tropical lichens are there... really?', in A. Thell, M. R. D. Seaward and T. Feuerer (eds), *Diversity of Lichenology—Jubilee Volume*, Bibliotheca Lichenologica, vol. 100, pp. 399–418, J. Cramer, Berlin, Germany.

McCune, B. (1993) 'Gradients in epiphyte biomass in three *Pseudotsuga-Tsuga* forests of different ages in western Oregon and Washington', *Bryologist*, vol. 96, pp. 405–411.

Miller, J. E. D., Root, H. T. and Safford, H. D. (2018) 'Altered fire regimes cause long-term lichen diversity losses', *Global Change Biology*, vol. 24, pp. 4909–4918.

Miranda-González, R. and McCune, B. (2020) 'The weight of the crust: Biomass of crustose lichens in tropical dry forest represents more than half of foliar biomass', *Biotropica*, vol. 52, pp. 1298–1308.

Nascimbene, J., Thor, G. and Nimis, P. L. (2013) 'Effects of forest management on epiphytic lichens in temperate deciduous forests of Europe—a review', *Forest Ecology and Management*, vol. 298, pp. 27–38.

Nash, T. H. L. III. (ed) (2008) *Lichen Biology* (2nd ed), Cambridge University Press, Cambridge, UK.

Nimis, P. L., Scheidegger, C. and Wolseley, P. A. (eds) (2002) *Monitoring with Lichens—Monitoring Lichens*, NATO Science Series vol. 7, Kluwer Academic Publishers, Dordrecht, the Netherlands.

Payette, S. and Delwaide, A. (2018) 'Tamm review: The North-American lichen woodland', *Forest Ecology and Management*, vol. 417, pp. 167–183.

Pettersson, R. B., Ball, J. P., Renhorn, K.-E., Esseen, P.-A. and Sjöberg, K. (1995) 'Invertebrate communities in boreal forest canopies as influenced by forestry and lichens with implications for passerine birds', *Biological Conservation*, vol. 74, pp. 57–63.

Phinney, N. H., Gauslaa, Y., Palmqvist, K. and Esseen, P.-A. (2021) 'Macroclimate drives growth of hair lichens in boreal forest canopies', *Journal of Ecology*, vol. 109, pp. 478–490.

Porada, P., Van Stan, J. T. and Kleidon, A. (2018) 'Significant contribution of non-vascular vegetation to global rainfall interception', *Nature Geoscience*, vol. 11, pp. 563–567.

Porada, P., Weber, B., Elbert, W., Pöschl, U. and Kleidon, A. (2014) 'Estimating impacts of lichens and bryophytes on global biogeochemical cycles', *Global Biogeochemical Cycles*, vol. 28, pp. 71–85.

Power, T. D., Cameron, R. P., Neily, T. and Toms, B. (2018) 'Forest structure and site conditions of boreal felt lichen (*Erioderma pedicellatum*) habitat in Cape Breton, Nova Scotia, Canada', *Botany*, vol. 96, pp. 449–459.

Rodríguez-Quiel, E. E., Mendieta-Leiva, G. and Bader, M. Y. (2019) 'Elevational patterns of bryophyte and lichen biomass differ among substrates in the tropical montane forest of Baru Volcano, Panama', *Journal of Bryology*, vol. 41, pp. 95–106.

Scheidegger, C. and Werth, S. (2009) 'Conservation strategies for lichens: insights from population biology', *Fungal Biology Reviews*, vol. 23, pp. 55–66.

Seaward, M. R. D. (ed) (1977) *Lichen Ecology*, Academic Press, London.

Sillett, S. C. and Antoine, M. E. (2004) 'Lichens and bryophytes in forest canopies', in M. D. Lowman and H. B. Rinker (eds), *Forest Canopies* (2nd ed), pp. 151–174, Elsevier Academic Press, Burlington, MA.

Sillett, S. C., McCune, B., Peck, J. E., Rambo, T. R. and Rutchy, A. (2000) 'Dispersal limitations of epiphytic lichens result in species dependent on old-growth forests', *Ecological Applications*, vol. 10, pp. 789–799.

Solhaug, K. A. and Gauslaa, Y. (2012) 'Secondary lichen compounds as protection against excess solar radiation and herbivores', in U. Lüttge, W. Beyschlag, B. Büdel and D. Francis (eds), *Progress in Botany*, vol. 73, pp. 283–304, Springer-Verlag, Berlin, Germany.

Spribille, T., Thor, G., Bunnell, F. L., Goward, T. and Björk, C. R. (2008) 'Lichens on dead wood: species-substrate relationships in the epiphytic lichen floras of the Pacific Northwest and Fennoscandia', *Ecography*, vol. 31, pp. 741–750.

Spribille, T., Tuovinen, V., Resl, P., Vanderpool, D., Wolinski, H. and others (2016) 'Basidiomycete yeasts in the cortex of ascomycete macrolichens', *Science*, vol. 353, pp. 488–492.

Stevenson, S. K. and Coxson, D. S. (2003) 'Litterfall, growth and turnover of arboreal lichens after partial cutting in an Engelmann spruce-subalpine fir forest in north-central British Columbia', *Canadian Journal of Forest Research*, vol. 33, pp. 2306–2320.

Wolf, J. H. D. (1993) 'Diversity patterns and biomass of epiphytic bryophytes and lichens along altitudinal gradients in the northern Andes', *Annals of the Missouri Botanical Garden*, vol. 80, pp. 928–960.

Wolseley, P. A. and Aguirre-Hudson, B. (1997) 'The ecology and distribution of lichens in tropical deciduous and evergreen forests of northern Thailand', *Journal of Biogeography*, vol. 24, pp. 327–343.

18

MAMMALS IN FOREST ECOSYSTEMS

Richard T. Corlett and Alice C. Hughes

Introduction

The 6,526 species of living wild mammals currently recognized (ASM Mammal Diversity Database, April 2023) are divided between 27 orders, all but two of which (Notoryctemorphia and Sirenia) include species that occur in forests. Mammals occupy every substantial area of forest worldwide and play multiple ecological roles (Table 18.1). The diversity of most large orders peaks in the tropics (Rolland et al., 2014). More than 120 species of mammals have been recorded from the richest tropical rainforest sites (e.g. Happold, 1996; Voss and Emmons, 1996) and 63 per cent of all mammal species occur in tropical forests (Pillay et al., 2022). Diversity declines slowly with latitude but rapidly with isolation on oceanic islands, reflecting the poor cross-water dispersal of most taxa. Diversity has also declined since the Pleistocene origin and spread of modern humans (Andermann et al., 2020). The majority of terrestrial ecosystems outside Africa have lost large mammals in the last 125,000 years and declines in vulnerable species have accelerated in recent decades.

Bats are the most widespread mammals and the only native terrestrial species on Hawaii, New Zealand and many other Pacific islands. Rodents now equal this distribution, but no native species occur on the more isolated islands. Artiodactyls (even-toed ungulates) and carnivores (Carnivora) occur in most forests, and primates in most warm regions, but they did not reach New Guinea or Australia, or most small islands, until introduced by humans. Proboscids (elephants and their relatives) were almost as widespread until the late Pleistocene but suffered massively during the megafaunal extinctions mentioned above and are now confined to the tropical and subtropical forests of Asia and Africa. The true insectivores (Eulipotyphla) are also widespread in forests but did not reach New Guinea, Australia or Madagascar, and are recent arrivals in South America, where they are restricted to the northern Andes. The other orders are all geographically more or less more restricted, although many are regionally important (Table 18.2).

The literature on forest mammals is strongly biased taxonomically, but we have tried to provide a comprehensive coverage of all orders of forest mammals, including those small, nocturnal and/or uncharismatic taxa that have attracted only a few specialists. An inevitable corollary of this is that we have had space for only a superficial summary of the vast literature on, for example, forest primates or cats. We have focused on diversity and function (i.e.

DOI: 10.4324/9781003324072-21

Table 18.1 Major ecological roles played by the mammalian orders in forests

Mammalian order	*Ecosystem engineer*	*Consumer of:*										
		Invertebrates	*Social insects*	*Small vertebrates*	*Large vertebrates*	*Leaves*	*Browse*	*Bark/ cambium*	*Fruits*	*Seeds*	*Nectar/ pollen*	*Others*
Monotremata	X	XX	XXX									
Didelphimorphia		XX		X		XX			XX		X	exudates
Paucituberculata		XX		X								
Microbiotheria		X							X		X	
Dasyuromorphia		XX	X	XX	X							
Peramelemorphia	X	XX		X					X			fungi
Diprotodontia	X					XXX	XXX		XX		X	
Tubulidentata			XXX									
Afrosoricida		XXX		X								
Macroscelidea		XX										
Hyracoidea						XX						
Proboscidea	XXX					XXX	XXX	XX	XX			
Cingulata		XXX										
Pilosa			XXX			XXX						
Scandentia		XX		XX					XX			
Dermoptera						XX						
Primates		XX				XXX	X	X	XXX	XX		exudates
Lagomorpha						XX	XXX					
Rodentia	XXX	XX				X		XX	XX	XXX	X	
Eulipotyphla	X	XXX		X								
Chiroptera		XXX		X					XXX		XX	blood
Carnivora		XX	X	XX	XXX				XXX			
Pholidota			XXX									
Perissodactlya	X					XX	XXX		XX			
Artiodactyla	X					XXX	XXX	XX	XX			

Notes: The number of crosses indicates the estimated significance of the order in this role at sites where it occurs: X—present; XX—at least locally significant; XXX—at least locally dominant. Minor roles are omitted. See the text for details. Groups which feed on fruits have important roles in seed dispersal, as do some of those that consume seeds, and many of those which consume nectar or pollen have roles in pollination.

Table 18.2 The recent global distributions of mammalian orders in forests

Mammalian Order	*Common names*	*Tropical America*	*Africa*	*Madagascar*	*Tropical Asia*	*Australia & New Guinea*	*Nearctic*	*Palearctic*
Monotremata	Monotremes					X		
Didelphimorphia	Opossums	X					(X)	
Paucituberculata	Shrew-possums	X						
Microbiotheria	Monito del monte	X						
Dasyuromorphia	Marsupial carnivores					X		
Peramelemorphia	Bandicoots					X		
Diprotodontia	Kangaroos, possums					X		
Tubulidentata	Aardvark		X					
Afrosoricida	Tenrecs		X	X				
Macroscelidea	Sengis		X					
Hyracoidea	Tree hyraxes		X					
Proboscidea	Elephants		X		X			(X)
Cingulata	Armadillos	X					(X)	
Pilosa	Sloths, anteaters	X						
Scandentia	Treeshrews				X			(X)
Dermoptera	Colugos				X			
Primates	Primates	X	X	X	X			X
Lagomorpha	Rabbits, hares	X			X		X	X
Rodentia	Rodents	X	X	X	X	X	X	X
Eulipotyphla	Shrews etc.	X	X		X		X	X
Chiroptera	Bats	X	X	X	X	X	X	X
Carnivora	Cats, dogs, bears	X	X	X	X		X	X
Pholidota	Pangolins	X	X					(X)
Perissodactlya	Tapirs, rhinoceroses	X			X			
Artiodactyla	Cattle, deer, pigs	X	X		X		X	X

Notes: Symbols in parentheses indicate that the order is represented by only one or a few species, usually with ranges that extend from the adjacent tropical region. For the larger orders, the common names are representative but not comprehensive.

ecological roles) in the order-level accounts below, which are structured phylogenetically, and then cover other topics at a more general level.

We have also highlighted forest-dependent species with exceptionally high conservation priority using the EDGE2 rankings (Gumbs et al., 2023). The Evolutionary Distinct and Globally Endangered (EDGE) prioritization metrics combine phylogenetic diversity with extinction risk (from the International Union for Conservation of Nature (IUCN) Red List categories), and thus direct conservation attention at threatened species that embody large amounts of evolutionary history. The EDGE2 protocol differs from the original metric mainly in the consideration of the extinction risk of close relatives when calculating the score, because the threatened species with threatened close relatives have a greater responsibility for their shared evolutionary history than threatened species with secure relatives.

Mammalian orders in forests

Monotremes

Order monotremata

The five species in this order are the only surviving egg-laying mammals and are restricted to Australia and New Guinea. The short-beaked echidna (*Tachyglossus aculeatus*) is found all over Australia and southern New Guinea and is a specialist on ants and termites, making a major contribution to soil turnover (Dundas et al., 2022). The three species of long-beaked echidnas (*Zaglossus*) occur only in New Guinea and have broader diets of earthworms and other subterranean invertebrates. The platypus (*Ornithorhynchus anatinus*) occurs in forest streams in Australia, where it feeds on small aquatic vertebrates and invertebrates. All three long-beaked echidna species are among the top 100 EDGE2 mammal species, reflecting their phylogenetic isolation and the fact that they are all threatened (Gumbs et al., 2023).

Marsupials

Order didelphimorphia

This order includes most of the surviving American marsupials (126 species), known as opossums (Teta et al., 2022). They are almost confined to South and Central America, with only the largest species, the Virginia opossum, extending north into the United States and southern Canada. In Amazonian rainforests, at least 12 species can coexist, a diversity exceeded only by bats and rodents. Most are more or less arboreal and more or less omnivorous, but some focus more on fruit, leaves, exudates or animal prey, and some spend more time on the ground.

Order paucituberculata

This order consists of seven species of poorly known shrew-opossums, six in cold and humid environments in the tropical and subtropical high Andes and one in temperate humid *Nothofagus* forests of Chile and Argentina (Teta et al., 2022). They feed on invertebrates and small vertebrates and spend much of their time in burrows.

Order microbiotheria

The two living species of this order, in the genus *Dromiciops*, are small (20–30 g), nocturnal, arboreal marsupials found in the temperate forests of South America (Fontúrbel et al., 2022). They are omnivorous but consume large amounts of fruit and are important dispersal agents for some plant species.

Order dasyuromorphia

The Dasyuromorphia of Australia and New Guinea includes three recent families, the Dasyuridae, with 77 species, the Mymercobiidae, with one, the termite-specialist numbat (*Myrmecobius fasciatus*), and the Thylacinidae, whose sole survivor, the dog-sized thylacine, became extinct in 1936 (Dickman and Calver, 2023). The numbat is fifth in the global EDGE2 rankings, reflecting its phylogenetic isolation and endangered status (Gumbs et al., 2023). The biggest surviving dasyurid, the 6–14 kg Tasmanian devil, was also extinct on the Australian mainland, but was reintroduced to New South Wales in 2020. In addition to devils, six species of quolls (*Dasyurus* spp.; 0.3–7 kg) feed mainly on other vertebrates and are capable of killing prey several times their own sizes. Devils are also bone-crushing scavengers. Up to four large dasyurid species coexisted in some Australian forests before the introduction of the placental dingo, around 4,000 years ago, and the feral cat and European fox in the last 200 years. The other members of this order are smaller and feed mostly on invertebrates, but even species <50 g kill and eat small vertebrates.

Order paramelemorphia

The forest bandicoots in New Guinea and Australia consume various combinations of plant material, including fruits, invertebrates and macrofungi (Dickman and Calver, 2023). They may be significant dispersers of fungal spores and, possibly, seeds, and they contribute to bioturbation through their digging activities.

Order diprotodontia

This order, with 151 species, includes the most important forest herbivores of Australia and New Guinea (Dickman and Calver, 2023). Up to 20 species can coexist in the same landscape, although with specialization by habitat. The family Macropodidae consists of terrestrial kangaroos and wallabies in addition to arboreal tree-kangaroos. These are browsers and grazers with a ruminant-like digestive system, playing similar ecological roles to the artiodactyls on the ground and colobine primates and sloths in the canopy. The family Phalangeridae includes the largely arboreal brushtail possums and cuscuses, which feed mostly on leaves and fruits. Two species, the bear cuscus (*Ailurops ursinus*) and Sulawesi dwarf cuscus (*Strigocuscus celebensis*), have reached Sulawesi, where the bear cuscus appears to occupy the arboreal folivore niche occupied by colobine monkeys on Borneo (Corlett, 2019). Other families include the Pseudocheiridae (ringtail possums), which are specialized leaf-eaters, and the omnivorous Petauridae, which includes the striped-possums, Leadbeater's possum and several species of gliders. Leadbeater's possum is third in the global EDGE2 rankings, reflecting its phylogenetic isolation and critically endangered status (Gumbs et al., 2023). The order also includes the koala, a specialized folivore, and three species of largely herbivorous, burrowing wombats.

The impacts of native herbivores at natural densities are difficult to assess, but the potential roles of diprotodontid herbivory in Australian forests are illustrated by the effects of overabundant swamp wallabies (*Wallabia bicolor*) on the regeneration of palatable species following exotic predator control (Dexter et al., 2013). Similarly, brushtail possums (*Trichosurus vulpecula*) introduced to New Zealand, where they lack predators, reduce the canopy biomass of preferred tree species and increase mortality (Byrom et al., 2016).

Afrotheria

Order tubulidentata

The only surviving species in this order is the aardvark, which is better adapted to more open habitats but also occurs in forests across much of Africa. It is a specialist feeder on ant and termite nests and shows evolutionary convergence with other myrmecophagous taxa (Delsuc et al., 2014).

Order afrosoricida

The tenrecs (Tenrecidae) are diverse and abundant in Madagascan forests where they co-dominate the small-mammal community with rodents (e.g. Dammhahn et al., 2013). Local species richness varies from 10 to 15 in humid montane forests to two to seven in dry forests (Stanley et al., 2011). The 31 species currently recognized range in size from 5 g to 2.5 kg and include climbing, ground-dwelling, swimming and burrowing species. Little is known about their diets or other aspects of their ecology, although all species appear to feed predominantly on invertebrates and the larger species also take some small vertebrates. The three species of otter shrews (Potamogalidae) are African tenrecs that feed on aquatic invertebrates, fish and amphibians. Southern African forests also support several species of burrowing, insectivorous, golden moles (Chrysochloridae).

Order macroscelidea

The 20 species of elephant shrews (sengis) are confined to Africa and mostly occupy non-forest habitats, but a few species occur in forests (e.g. Rovero et al., 2013), where they feed on the ground, mostly on invertebrates.

Order hyracoidea

The four species of tree hyraxes (*Dendrohyrax*) are small (< 5 kg) arboreal folivores in African forests.

Order proboscidea

Although there are only three species extant today (two African and one Asian), the proboscideans were much more diverse and widespread in the middle to late Pleistocene (Corlett, 2013). Among these recently extinct taxa, at least the many Asian *Stegodon* species and the North American mastodon (*Mammut americanum*) inhabited forests, as probably did at least some gomphotheres of Central and South America. The African and Asian forest elephants selectively consume a variety of plant materials in large amounts and are important—sometimes apparently unique—dispersal agents for trees with very large, large-seeded, fruits (Campos-Arceiz and Blake, 2011). They also share with beavers the ability to re-shape their habitat as ecosystem engineers (Poulsen et al, 2018).

Xenarthra

Order cingulata

This order was formerly much more diverse than the present 22 armadillo species and included several extinct families, such as the glyptodonts and pampatheres, which formed part of the late Pleistocene megafauna of the Neotropics (Corlett, 2013). Diets vary but the forest armadillos feed predominantly on invertebrates. At least four species can coexist in Amazonian rainforests (Voss and Emmons, 1996),

Order pilosa

The two suborders, sloths (Folivora; seven species) and anteaters (Vermilingua; ten species), are mostly confined to tropical forests in South and Central America. In the late Pleistocene, the sloths included terrestrial species the size of modern elephants. Both suborders have specialist adaptations to their diets, with sloths feeding on leaves, and anteaters on ants and termites. All extant sloths are arboreal, but the anteaters include the seven entirely arboreal *Cyclopes*, two semi-arboreal species of *Tamandua* and the terrestrial *Myrmecophaga*.

Euarchontoglires

Order scandentia

This order consists of 23 species of treeshrews found in Asian tropical and subtropical forests, with up to five species coexisting at one site (Emmons, 2000). They range from largely terrestrial to fully arboreal and their diets include various combinations of invertebrates, fruits and small vertebrates (Selig et al., 2019).

Order dermoptera

This order consists of two currently recognized species of colugos, one in the southern Philippines and the other occupying much of Southeast Asia. However, both these species consist of multiple genetically distinct lineages, which could be recognized as species (Mason et al., 2016). They are poorly studied, nocturnal, gliding folivores.

Order primates

The primates are the fourth largest mammalian order, with 517 species, all but a few associated with forests. Although now largely confined to the tropics, gibbons, colobines and macaques were widespread in subtropical Asia into historical times and macaques still extend north into temperate deciduous forests in several places. Maximum diversities are in wet tropical forests, where up to a dozen co-occur in the richest sites (Corlett and Primack 2011). Most primate communities include specialist leaf-eaters, seed predators, frugivores and insectivores, and many include specialists on plant exudates. Where present, primates make up a large proportion of the canopy vertebrate biomass and the frugivorous species are probably third in importance to birds and bats as seed dispersal agents. Many primate species are threatened and four—all Madagascan lemurs—are among the global top 20 EDGE2 species (Gumbs et al., 2023). The bizarre aye-aye—which functions as a sort of primate woodpecker, detecting and excavating insect larvae from wood—is ranked second globally, reflecting its phylogenetic isolation and Endangered status.

Order lagomorpha

Although most species in this order are found in non-forest habitats, there are a few forest specialists and others that occupy open forests with a grassy understory. Snowshoe hares (*Lepus americanus*) occupy boreal and montane forests across North America, where they are active year-round (Wirsing and Murray, 2002). In summer, they eat herbs and new woody growth, but in winter they depend on a few species of woody browse and can leave well-defined browse-lines. These hares in turn are an important food supply for several predators, including lynx, owls and diurnal birds of prey. Other *Lepus* species play similar roles in northern Eurasia (Lyly et al., 2014). Multiple species of cottontail rabbit (*Sylvilagus*) occur in both forest and non-forest habitats in Central and South America (Mora et al., 2021) and there are three localized rabbit species in tropical East Asian forests, but little is known about any of these species.

Order rodentia

Most non-flying mammal species are rodents (2,652 species), but their local diversities are generally lower in forests than those of bats, with a maximum of 20–30 species in some tropical forests (Happold, 1996; Voss and Emmons, 1996). Rodents have chisel-shaped, continually growing, gnawing incisors, coupled with large and complex jaw muscles, which gives them access to mechanically protected plant foods, such as seeds and cambium, although many also eat softer plant foods and invertebrates. There are also specialist folivores and insectivores, particularly on islands, and few-toothed, long-snouted, earthworm-eating species have evolved convergently in the Philippines and on Sulawesi (Rowe et al., 2016). Rodent communities utilize the full volume of forests, with gliding squirrels in Asia and North America and the unrelated gliding anomalures (Anomaluridae) in Central Africa, arboreal members in several families and a variety of ground-dwelling species, some of which burrow. Rodents dominate the diets of numerous medium-sized carnivores, including mammals, birds and snakes, and form an essential link in forest food chains.

In addition to consuming diverse plant materials and invertebrates, forest rodents impact plant reproduction through their consumption of large seeds. Most of this is predation, which may occur before or after seed dispersal, but some terrestrial rodents in several families cope with the fluctuating supply of seeds by storing some of the surplus in scattered caches of one or a few seeds buried in the soil—scatter-hoarding (Lichti et al., 2017). This can be a very effective means of dispersal, even if most cached seeds are eventually retrieved and eaten, and at least one highly successful plant family, the Fagaceae (oaks, chestnuts etc.), depends on it, as do many species in other families.

Both Old and New World porcupines feed on the cambium of preferred tree species and can cause significant damage to plantations. Despite their small sizes, several species of voles can also cause significant damage to seedlings of preferred tree species in boreal forests, with damage most obvious during the peak phase of their population cycles (Lyly et al., 2014). The two species of beaver, found in North America and western Eurasia respectively, are unique in their roles as ecosystem engineers, cutting trees for use as food and in the construction of dams and lodges (Brazier et al., 2021).

Laurasiatheria

Order eulipotyphla

This large order (569 spp.) includes the shrews (Soricidae), the hedgehogs, moonrats and gymnures (Erinaceidae), the moles, desmans and shrew-moles (Talpidae), and the solenodons (Solenodontidae). The order is diverse and abundant in forests across Eurasia and the Americas, south to the northern Andes. They range in size from around 2 g to more than 1 kg, although the few larger species have restricted distributions. The highest shrew diversities are in African tropical forests (8–18 species: Dudu et al., 2005; Stanley et al., 2011), where they are the only members of the order present and eat overlapping combinations of arthropods, earthworms and snails. Diversities are lower in the temperate zone (two to nine species), but with a clearer ecological separation on the basis of body size (Churchfield et al., 1999). Pearson's long-clawed shrew is ranked 11th globally by the EDGE2 prioritization metric because of its taxonomic isolation and endangered status (Gumbs et al., 2023).

The Erinaceidae are mostly larger and much less diverse than the shrews and are absent from the Americas and from most forested areas in Africa. Hedgehogs occur in forests across northern Eurasia, while the non-spiny moonrats and gymnures are confined to moist forests in Southeast Asia (He et al., 2012). Diets, where known, are similar to the larger species of shrew, but they consume more vertebrates and, in some species, plant material. Eurasian and North American forests also often support one or two species of subterranean moles (Talpidae). The two species of *Solenodon* are relatively large (c. 1 kg), venomous, burrowing animals, found only on the Caribbean islands of Cuba and Hispaniola. They have similar diets to the other large members of the order. The Cuban solenodon is ranked fourth globally by the EDGE2 prioritization metric because of its phylogenetic isolation and endangered status (Gumbs et al., 2023).

Order chiroptera

Bats make up nearly a quarter of known mammal species with 1,462 species currently described. Within forests, bats are frequently the most species-rich mammal group, and this is especially marked in the tropics, where they may represent more than half mammal species present. Peak diversities are attained in tropical forests in South America and Southeast Asia, which can support >60 species in one area. Diversity declines with distance from the equator, but bats are found from 70°N to 54°S, and extending beyond tree-covered areas adjacent to polar regions.

Around one-third of bat species (384 species), living largely in tropical forests, feed on fruit and/or nectar, with separate clades involved in the New World (Phyllostomidae) and Old World (Pteropodidae). Most other species are insectivores, but forest-dwelling bats have a wide variety of diets, including blood, fish and small vertebrates, as well as specialists on particular invertebrate groups. Bats play a number of important roles within forest ecosystems, including the regulation of insect populations. They pollinate many plant species and for some are the major or only pollinator. Forest bats also have a major role in seed dispersal, which has a key role in forest regeneration following deforestation. On oceanic islands, large pteropodid fruit bats can be essential mutualists for many plant species. One such species, the little-known Fijian monkey-faced bat, is 18th in the global EDGE2 rankings because it is critically endangered and the only species in its genus (Gumbs et al., 2023). The New Zealand greater short-tailed bat, which feeds on invertebrates in the air and on the ground, is ninth, since there have been no confirmed sightings since 1967 and its only relative, the New Zealand lesser short-tailed bat, is Vulnerable.

Order carnivora

Excluding the semiaquatic pinnipeds, the Carnivora includes 273 species currently placed in 13 families, with the African palm civet, Asian linsangs, Malagasy carnivores, skunks and red panda now each having their own families. All families but the hyenas have representatives in forests. Local diversity of Carnivora communities is highest in the tropics, with 15–25 coexisting species in lowland Southeast Asia, <15 in the Amazon and Central America and < 10 in Africa (Corlett, 2011). Carnivora reached Madagascar once but did not reach New Guinea and Australia until the recent introduction of dogs to both and of cats and foxes to Australia. The largest species in each region, such as tigers in Asia, can kill all but the largest forest herbivore. Most Carnivora, except the hyper-carnivorous cats, also eat some fruit, and the most frugivorous species, including bears, many mustelids, civets (in Asia and Africa) and procyonids (in the Americas), are important dispersal agents for large seeds in large fruits (Corlett and Primack, 2011). The red panda is 12th in the global EDGE2 rankings, reflecting its phylogenetic isolation and endangered status (Gumbs et al., 2023).

Order pholidota

The eight species of pangolins are confined to tropical and subtropical Asia and Africa (Challender et al., 2020). Pangolins are terrestrial or semi-arboreal specialists on ants and termites. Almost all pangolin species are threatened by hunting and trapping, and the four Asian species of *Manis* are ranked within the global top 20 of EDGE2 species because the entire genus is critically endangered (Gumbs et al., 2023).

Order perissodactyla

This order includes five species of tapir, four in South and Central America and one in Southeast Asia, and two forest rhinoceroses in Southeast Asia, plus additional non-forest rhinoceroses and horses. Tapirs have a diverse herbivorous diet, dominated by leaves and fruits. In the Neotropics, where they are the largest surviving seed dispersal agents, they are considered major dispersers of large seeds, while in Southeast Asia they appear to be less important (Corlett, 2017). The forest rhinoceroses are browsers and consumers of fallen fruits. They are large enough to push over saplings and small trees, and at natural densities (which no longer occur anywhere) may have influenced forest structure in a similar way to elephants. All species disperse seeds from fallen fruits (McConkey et al., 2022). Both forest rhinos are critically endangered as a result of hunting and the Sumatran rhinoceros is ranked tenth globally in the EDGE2 list (Gumbs et al., 2023).

Order artiodactyla

Excluding the aquatic cetaceans, even-toed ungulates include 261 species in 10 families and account for half of all terrestrial wild mammal biomass (Greenspoon et al., 2023). Seven families are commonly found in forests: the non-ruminant pigs (Suidae) and peccaries (Tayassuidae), and the ruminant chevrotains (Tragulidae), okapi (Giraffidae), deer (Cervidae), musk deer (Moschidae) and cattle (Bovidae). They did not reach New Guinea, Australia or Madagascar until introduced by humans. They were also absent from remote oceanic islands, although at least pigs are now present on most, and *Bubalus* reached Sulawesi and several Philippine islands naturally. Artiodactyls are most diverse in the moist tropical forests of Central and West Africa, where up to seven species of duikers (small forest antelopes; Bovidae) coexist with several larger species (Corlett and Primack, 2011), followed by tropical Asia, with fewer species in the forests of South and Central America (< 4) and outside the tropics. Being confined to the forest floor limits their ecological roles to browsing plants within reach, trampling plants underfoot, grazing grass in forest gaps and edges, and consuming fruits and other plant material that fall from the canopy (Corlett & Primack, 2011). Browsing impacts can be dramatic when populations are high, with long-term effects on the structure and species composition of the forest. Some species also eat tree bark, and the pigs excavate roots and consume some small animals. Many species disperse small seeds from the fruits they eat in their feces and deer disperse large, hard seeds by regurgitation (Corlett, 2017). Artiodactyls are the major prey of the largest forest carnivores, and their feces support a diversity of specialist beetles and other invertebrates.

Artiodactyls not only dominate terrestrial wild animal biomass but also, as domestic livestock, the total global biomass of mammals (Greenspoon et al., 2023). Cattle are most important, with a greater biomass than even humans, followed by buffaloes, sheep, pigs, goats and camels. Domestic livestock are allowed to graze in many forests and, although their densities are usually much lower than in grasslands, can have significant impacts on structure, function and biodiversity (Li and Jiang, 2021). The creation of pastures for livestock, particularly cattle, is also a major driver of deforestation in many areas.

Mammals versus other animal groups: how important are they in forests?

Most ecological roles played by forest mammals (Table 18.1) are shared with other species, including birds and snakes as predators of vertebrates, numerous small vertebrate and invertebrate species as predators of invertebrates, herbivorous insects as consumers of most plant materials and birds as consumers of fleshy fruits and seeds. Within these broadly defined roles, however, there is usually a part that is exclusive to mammals. At most mainland forest sites with intact faunas, the apex predator is a mammal (e.g., tigers in Asia), although birds or snakes may be apex predators on islands (Corlett, 2011). Mammals also dominate predation on large colonies of social insects, which are well-defended against invertebrate predators, and bats are the major predators on nocturnal flying insects, including insect pests. Total folivory by mammals is probably a small fraction of that by insects but browsing of woody shoots by large mammalian herbivores has an impact on plant growth and form that no other animal equals. Most forests have fruit types that are taken exclusively or mainly by mammals, and even where fruits are shared with birds, mammals may provide complementary seed dispersal services (Corlett, 2017). In the tropics, there are also many plants with flowers specialized for bat pollination, especially in neotropical regions (Ratto et al., 2018).

Management issues

Overexploitation

Although deforestation is more visible and climate change attracts more global attention, the biggest threat today to most mammal species weighing > 1 kg in the tropics and subtropics is hunting, mainly for meat, but also elephants for ivory and, particularly in Asia, a wide range of species for traditional medicine. Other vertebrates are also hunted, but mammals dominate meat offtake almost everywhere, with ungulates, primates, and large rodents generally most important (*see* Chapter 42, 'Wild meat hunting in tropical forests'). Even subsistence hunting by low-density populations using traditional methods depleted populations of preferred species, but these impacts were potentially reversible if the hunters moved to a new area, and remote, non-hunted refugia allowed the persistence of large, slow-breeding species. Over the last 20–30 years, however, hunting pressure on wild mammals has increased almost everywhere, drastically reducing the diversities and densities of large and medium-size mammals in all but the most remote and/or best-protected areas, resulting in local and regional extirpations and threatening global extinctions.

The causes of this explosive growth in hunting pressure are similar across the tropics and much of the subtropics: increased accessibility of forests along new roads and from new settlements; improved hunting technologies, including access to guns in many areas and the use of cheap snares in Asia and Africa; and increased human populations, incomes and demand, particularly in urban areas (*see* Chapter 42, 'Wild meat hunting in tropical forests'). Hunting is initially selective for large-bodied species, but overhunting fuels a demand for progressively smaller species, until small-bodied rodents dominate the catch. Intermediate levels of hunting selectively remove the species responsible for modification of forest structure, seed predation and the dispersal of large seeds, with consequences for the tree community that can be rapid and dramatic, including increased clustering of saplings and a relative decline in the recruitment rates for animal-dispersed species (e.g. Harrison et al., 2013). Conversely, one of the most widespread and conspicuous impacts of carnivore declines has been the massive expansion of deer populations in both North America and Eurasia, with consequent increases in browsing pressure and impacts on both forest structure and the populations of numerous plants and smaller animals.

Unsustainable hunting is bad news both for the mammals and for the diets and livelihoods of the millions of people for whom wild meat provides a major source of protein, micronutrients and income. Sustainable wild meat production for local consumption may be theoretically possible (Fa et al., 2022), but easier forest access, improved hunting technology and the attraction of urban markets are not going away. Conversely, a scattering of success stories across the tropics show that law enforcement, combined with education, community engagement and improved economic opportunities, can largely exclude hunters from protected areas. A combination of these approaches, with urban markets closed, rule-based subsistence hunting in rural areas and well-protected conservation areas acting as refugia may be enough to prevent a new round of mammalian extinctions in forests.

Alien invasive species

Many mammals, from shrews to elephants, have been deliberately or accidentally introduced outside their natural ranges, but most of the literature on impacts refers to ungulates (predominantly deer, goats and pigs), small carnivores (particularly cats, foxes and mongooses), or rodents (particularly rats and mice) (Doherty et al., 2016; Spatz et al., 2022). Alien rats are established on almost all islands and many also have alien ungulates and small carnivores, but introductions are also becoming increasingly widespread in mainland forests. Introductions started millennia ago (e.g., at least 7,000 years ago on Flores; van den Bergh, 2009) and may be hard to recognize without archaeological data. Initial impacts can be dramatic and devastating, including the extinction of prey species (especially flightless bird species) and competitors, but in the longer term, the alien species becomes part of a new equilibrium, and its subsequent elimination may have complex and unpredictable effects (e.g. the Australian dingo; Ripple et al., 2014). The eradication of invasive mammal species from small oceanic islands (up to hundreds of hectares), by hunting, trapping and poisons, has had an increasingly high success rate and is becoming routine (Spatz et al., 2022). Larger islands are now being targeted, as well as exclosures on the mainland.

Restoration and reintroduction

Where local extirpations of mammal populations are the problem, reintroductions are a potential solution. Even when surplus stock is available from wild or captive populations, however, the success rate of reintroductions is low and the process can be complicated, prolonged and expensive (Berger-Tal et al., 2020). In general, translocations from wild populations have a higher rate of success since captivity is associated with an almost inevitable loss of genetic diversity as well as the acquisition of maladaptive behavioral traits. Pre-release behavioral enrichment as well as techniques to ensure fear of humans may be essential for captive-born animals, at least in primates and carnivores, where many behaviors needed for survival in the wild (locomotion, predator avoidance, foraging, social skills etc.) must be learned (Reading et al., 2013).

As noted in the introduction, many large mammal species have become extinct since the middle Pleistocene, so reintroduction is no longer an option. In such cases, it may be worth considering the introduction of a close relative or even an unrelated taxon with a similar ecological role, but such 'taxon substitutions' are highly controversial and should not be undertaken lightly. An additional term, 'rewilding', is increasingly used in the literature to refer to large-scale restoration of fully functioning ecosystems, with the focus often on reintroducing large herbivores and apex predators (Corlett, 2016). The de facto rewilding of parts of Europe over the last 50 years shows what can be achieved in a densely populated region with economic incentives and public support (Ledger et al, 2022).

Conclusions

The Cenozoic is the Age of Mammals, and it is hard to imagine forests without them. High food demand as a result of the calorific requirements of an endotherm lifestyle is behind the major impacts of mammals in forests, making them quantitatively dominant as consumers of most items that they eat. The same demands make invasive mammals particularly destructive. As mammals ourselves, we are biased toward our closest wild relatives, although this has done little to stem the tide of anthropogenic extinctions from the Late Pleistocene onward (Andermann et al., 2020). It is not just species that are threatened, but their evolutionary history since whole clades are under threat. Conservationists are often accused of focusing on charismatic mammals in preference to other deserving taxa, but the future for many species still looks grim. In view of their disproportionate impacts on ecosystem functioning, restoring intact mammal communities would have major benefits for other animals, plants, and ecosystem processes.

References

Andermann, T., Faurby, S., Turvey, S.T., Antonelli, A. and Silvestro, D. (2020) 'The past and future human impact on mammalian diversity', *Science Advances*, vol. 6, eabb2313.

ASM Mammal Diversity Database. (2023). *Mammal Diversity Database* (Version 1.11) www.mammaldiversity.org/

Berger-Tal, O., Blumstein, D.T. and Swaisgood, R.P. (2020) 'Conservation translocations: a review of common difficulties and promising directions', *Animal Conservation*, vol 23, pp. 121–131.

Brazier, R.E., Puttock, A., Graham, H.A., Auster, R.E., Davies, K.H. and Brown, C.M.L. (2021) 'Beaver: Nature's ecosystem engineers', *WIREs Water*, vol. 8, e1494.

Byrom, A.E., Innes, J. and Binny, R.N. (2016) 'A review of biodiversity outcomes from possum-focused pest control in New Zealand', *Wildlife Research*, vol. 43, p. 228.

Campos-Arceiz, A. and Blake, S. (2011) 'Megagardeners of the forest—the role of elephants in seed dispersal', *Acta Oecologica*, vol. 37, pp. 542–553..

Challender, D.W.S., Nash, H.C. and Waterman, C. (2020) *Pangolins: Science, Society and Conservation*. Academic Press, London.

Churchfield, S., Nesterenko, V.A. and Shvarts, E.A. (1999) 'Food niche overlap and ecological separation amongst six species of coexisting forest shrews in the Russian Far East', *Journal of Zoology*, vol. 248, pp. 349–359.

Corlett, R. T. (2011) 'Vertebrate carnivores and predation in the Oriental (Indomalayan) Region', *Raffles Bulletin of Zoology*, vol. 59, pp. 325–360.

Corlett, R. T. (2013) 'The shifted baseline: prehistoric defaunation in the tropics and its consequences for biodiversity conservation', *Biological Conservation*, vol. 163, pp. 13–21.

Corlett, R. T. (2016) 'Restoration, reintroduction, and rewilding in a changing world', *Trends in Ecology & Evolution*, vol 31, pp. 453–462.

Corlett, R. T. and Primack, R. B. (2011) *Tropical Rain Forests: An Ecological and Biogeographical Comparison*, 2nd edition. Wiley-Blackwell, Oxford.

Corlett, R.T. (2017) 'Frugivory and seed dispersal by vertebrates in tropical and subtropical Asia: An update', *Global Ecology and Conservation*, vol. 11, pp. 1–22.

Corlett, R.T. (2019) *The Ecology of Tropical East Asia*, 3rd edition. Oxford University Press, Oxford.

Dammhahn, M., Soarimalala, V. and Goodman, S. M. (2013) 'Trophic niche differentiation and microhabitat utilization in a species-rich montane forest small mammal community of eastern Madagascar', *Biotropica*, vol. 45, pp. 111–118.

Delsuc, F., Metcalf, J.L., Wegener Parfrey, L., Song, S.J., González, A. and Knight, R. (2014) 'Convergence of gut microbiomes in myrmecophagous mammals', *Molecular Ecology*, vol. 23, pp. 1301–1317.

Dexter, N., Hudson, M., James, S., MacGregor, C. and Lindenmayer, D. B. (2013) 'Unintended consequences of invasive predator control in an Australian forest: overabundant wallabies and vegetation change', *PLoS One,* vol. 8, e69087.

Dickman, C.R. and Calver, M.C. (2023) 'Food habits and activity patterns of Australasian marsupials'. In: *American and Australasian Marsupials* (eds. N.C. Cáceres & C.R. Dickman). Springer Nature Switzerland, Cham, pp. 1151–1187.

Doherty, T.S., Glen, A.S., Nimmo, D.G., Ritchie, E.G. and Dickman, C.R. (2016) 'Invasive predators and global biodiversity loss', *Proceedings of the National Academy of Science*, vol 113, pp. 11261–11265.

Dudu, A., Churchfield, S. and Hutterer, R. (2005) 'Community structure and food niche relationships of coexisting rain-forest shrews in the Masako Forest, northeastern Congo'. In: *Advances in the Biology of Shrews II* (eds. J.F. Merritt, S. Churchfield, R. Hutterer & B.I. Sheftel). International Society of Shrew Biologists, New York, pp. 229–239.

Dundas, S.J., Osborne, L., Hopkins, A.J.M., Ruthrof, K.X. and Fleming P.A. (2022) 'Bioturbation by echidna (*Tachyglossus aculeatus*) in a forest habitat, south-western Australia', *Australian Journal of Zoology*, vol. 69, pp. 197–204.

Emmons, L. H. (2000) *Tupai: A Field Study of Bornean Treeshrews.* University of California Press, Berkeley, CA.

Fa, J.E., Funk, S.M. and Nasi, R. (2022) *Hunting Wildlife in the Tropics and Subtropics.* Cambridge University Press, Cambridge.

Fontúrbel, F. E., Franco, L. M., Bozinovic, F., Quintero-Galvis, J. F., Mejías, C., Amico, G. C., Vazquez, M. S., Sabat, P., Sánchez-Hernández, J. C., Watson, D. M., Saenz-Agudelo, P. and Nespolo, R. F. (2022) 'The ecology and evolution of the monito del monte, a relict species from the southern South America temperate forests', *Ecology and Evolution*, vol. 12, e8645.

Greenspoon, L., Krieger, E. Sender, R., Rosenberg, Y., Bar-On, Y. M., Moran, U., Antman, T., Meiri, S., Roll, U., Noor, E. and Milo, R. (2023) The global biomass of wild mammals. *Proceedings of the National Academy of Sciences of the USA*, vol. 120, e2204892120.

Gumbs, R., Gray, C. L., Böhm, M., Burfield, I. J., Couchman, O. R., Faith, D. P., Forest, F., Hoffman, M., Isaac, N. J. B., Jetz, W., Mace, G. M., Mooers, A. O., Safi, K., Scott, O., Steel, M., Tucker, C. M., Pearse, W. D., Owen, N. R. and Rosindell, J. (2023) 'The EDGE2 protocol: Advancing the prioritisation of Evolutionarily Distinct and Globally Endangered species for practical conservation action', *PLoS Biology*, vol. 21, e3001991.

Happold, D. C. D. (1996) 'Mammals of the Guinea–Congo rain forest', *Proceedings of the Royal Society of Edinburgh Section B: Biological Sciences,* vol. 104, pp. 243–284.

Harrison, R. D., Tan, S., Plotkin, J. B., Slik, F., Detto, M., Brenes, T., Itoh, A. and Davies, S. J. (2013) 'Consequences of defaunation for a tropical tree community', *Ecology Letters,* vol. 16, pp. 687–694.

He, K., Chen, J., Gould, G. C.; Yamaguchi, N., Ai, H., Wang, Y., Zhang, Y. and Jiang, X. (2012) 'An estimation of Erinaceidae phylogeny: a combined analysis approach', *PLoS One*, vol. 7, e39304.

Ledger, S.E.H., Rutherford, C.A., Benham, C., Burfield, I.J., Deinet, S., Eaton, M., Freeman, R., Gray, C., Herrando, S., Puleston, H., Scott-Gatty, K., Staneva, A. and McRae, L. (2022) *Wildlife Comeback in Europe: Opportunities and Challenges for Species Recovery.* Zoological Society of London, London.

Li, B.V. and Jiang, B. (2021) 'Responses of forest structure, functions, and biodiversity to livestock disturbances: a global meta-analysis', *Global Change Biology*, vol. 27, pp. 4745–4757.

Lichti, N.I., Steele, M.A. and Swihart, R.K. (2017) 'Seed fate and decision-making processes in scatter-hoarding rodents', *Biological Reviews*, vol. 92, pp. 474–504.

Lyly, M., Klemola, T., Koivisto, E., Huitu, O., Oksanen, L. and Korpimäki, E. (2014) 'Varying impacts of cervid, hare and vole browsing on growth and survival of boreal tree seedlings', *Oecologia,* vol. 174, pp. 271–281.

Mason, V.C., Li, G., Minx, P., Schmitz, J., Churakov, G., Doronina, L., Melin, A.D., Dominy, N.J., Lim, N.T.-L., Springer, M.S., Wilson, R.K., Warren, W.C., Helgen, K.M. and Murphy, W.J. (2016) 'Genomic analysis reveals hidden biodiversity within colugos, the sister group to primates', *Science Advances*, vol. 2, e1600633.

McConkey, K., Campos-Arceiz, A., Corlett, R.T., Sushma, H.S., Ong, L. and Brodie, J. (2022) 'Megafruit and megafauna diversity are positively associated, while megafruit traits are related to abiotic factors, in Tropical Asia', *Global Ecology and Biogeography*, vol. 31, pp. 740–752.

Mora, J.M., Silva, S.M., López, L.I., Burnham-Curtis, M.K., Wostenberg, D.J., French, J.H. and Ruedas, L.A. (2021) 'Systematics, distribution, and conservation status of Dice's cottontail, *Sylvilagus dicei* Harris, 1932 (Mammalia, Lagomorpha, Leporidae), in Central America', *Systematics and Biodiversity*, vol. 19, pp. 74–88.

Pillay, R., Venter, M., Aragon-Osejo, J., González-del-Pliego, P., Hansen, A.J., Watson, J.E.M. and Venter, O. (2022) 'Tropical forests are home to over half of the world's vertebrate species', *Frontiers in Ecology and the Environment*, vol. 20, pp. 10–15.

Poulsen, J.R., Rosin, C., Meier, A., Mills, E., Nuñez, C.L., Koerner, S.E., Blanchard, E., Callejas, J., Moore, S. and Sowers, M. (2018) 'Ecological consequences of forest elephant declines for Afrotropical forests', *Conservation Biology*, vol. 32, pp. 559–567.

Ratto, F., Simmons, B.I, Spake, R., Zamora-Gutierrez, V., MacDonald, M.A., Merriman, J.C., Tremlett, C.J., Poppy, G.M., Peh, K.S.-H. and Dicks, L.V. (2018) 'Global importance of vertebrate pollinators for plant reproductive success: a meta-analysis', *Frontiers in Ecology and the Environment*, vol. 16, pp. 82–90.

Reading, R. P., Miller, B. and Shepherdson, D. (2013) 'The value of enrichment to reintroduction success', *Zoo Biology*, vol. 32, pp. 332–341.

Ripple, W. J., Estes, J. A., Beschta, R. L., Wilmers, C. C., Ritchie, E. G., Hebblewhite, M., Berger, J., Elmhagen, B., Letnic, M., Nelson, M. P., Schmitz, O. J., Smith, D. W., Wallach, A. D. and Wirsing, A. J. (2014) 'Status and ecological effects of the world's largest carnivores', *Science,* vol. 343, 1241484. doi: 10.1126/science.1241484

Rolland, J., Condamine, F. L., Jiguet, F. and Morlon, H. (2014) 'Faster speciation and reduced extinction in the tropics contribute to the mammalian latitudinal diversity gradient', *PLoS Biology,* vol. 12, e1001775.

Rovero, F., Collett, L., Ricci, S., Martin, E. and Spitale, D. (2013) 'Distribution, occupancy, and habitat associations of the gray-faced sengi (*Rhynchocyon udzungwensis*) as revealed by camera traps', *Journal of Mammalogy,* vol. 94, pp. 792–800.

Rowe, K.C., Achmadi, A.S. and Esselstyn, J.A. (2016) 'Repeated evolution of carnivory among Indo-Australian rodents', *Evolution*, vol. 70, pp. 653–665.

Selig, K.R., Sargis, E.J. and Silcox, M.T. (2019) 'Three-dimensional geometric morphometric analysis of treeshrew (Scandentia) lower molars: Insight into dental variation and systematics', *The Anatomical Record*, vol. 302, pp. 1154–1168.

Spatz, D.R., Holmes, N.D., Will, D.J., Hein, S., Carter, Z.T., Fewster, R.M., Keitt, B., Genovesi, P., Samaniego, A., Croll, D.A., Tershy, B.R. and Russell, J.C. (2022) 'The global contribution of invasive vertebrate eradication as a key island restoration tool', *Scientific Reports*, vol 12, 13391.

Stanley, W. T., Goodman, S. M. and Hutterer, R. (2011) 'Small mammal inventories in the East and West Usambara Mountains, Tanzania. 2. Families Soricidae (Shrews) and Macroscelididae (Elephant Shrews)', *Fieldiana Life and Earth Sciences*, vol. 4, pp. 18–33.

Teta, P., Chemisquy, M.A. and Martin, G. (2022) 'Taxonomy and diversity of living American marsupials', In: *American and Australian Marsupials* (eds. N.C. Cáceres & C.R. Dickman). Springer, Switzerland, Cham, pp. 89–113.

van den Berg, G. D., Meijer, H. J., Due Awe, R., Morwood, M. J., Szabó, K., van den Hoek Ostende, L. W., Sutikna, T., Saptomo, E.W., Piper, P. J. and Dobney, K. M. (2009) 'The Liang Bua faunal remains: a 95 k.yr. sequence from Flores, East Indonesia', *Journal of Human Evolution,* vol. 57, pp. 527–537.

Voss, R. S. and Emmons, L. H. (1996) 'Mammalian diversity in Neotropical lowland rainforests: A preliminary assessment', *Bulletin of the American Museum of Natural History*, vol. 230, pp. 3–115.

Wirsing, A. J. and Murray, D. L. (2002) 'Patterns in consumption of woody plants by snowshoe hares in the northwestern United States', *EcoScience,* vol. 9, pp. 440–449.

19

ECOLOGY AND CONSERVATION OF FOREST BIRDS

Malcolm C. K. Soh, Ding Li Yong, Richard T. Corlett and Kelvin S.-H. Peh

Introduction

Birds occur on all continents and all but the smallest islands, inhabit a great variety of habitat types, and some species have adapted to novel anthropogenic environments. The majority of the world's bird species, however, are dependent on forests. All forests have birds for at least part of the year, from the equatorial lowlands to the alpine and arctic treelines, and from the mainland to the most remote oceanic islands. Forest birds fulfil multiple ecological roles and their importance is shown by the selective pressures they have exerted on other forest taxa resulting, for example, in both crypsis and aposematism in insects and in the occurrence of red fruits and flowers in the forest. People also have a peculiar fascination for birds, both scientists—shown by the number of specialist bird journals, as well the proportion of bird studies in more general journals—and the general public. Bird watching is an immensely popular leisure activity in many countries. The online platform, eBird, has enabled amateur birders to contribute high-quality data for science and conservation in many countries (e.g., Cazalis et al., 2020). Bird watching has also been shown to improve health and well-being, with committed bird watchers more likely to recover from stress than casual bird watchers, regardless of their expertise (Randler et al., 2022). Conservationists have leveraged this public interest to develop local, regional, and global bird conservation plans and activities that are the envy of those who work for the conservation of other animals or plants.

In this chapter, we first introduce the world's forests as bird habitat, with a focus on the importance of resource seasonality. We next describe how the post-Cretaceous spread of forest bird lineages has been restricted by ocean barriers and non-forest habitats, resulting in some striking differences between bird communities in similar forests on different continents. Then we discuss the many ecological roles that birds play in forests and their consequences for other species. Bird migrations driven by spatial differences in the seasonal availability of forest resources are then described, showing the importance of forests to migrants as both their breeding and wintering habitats as well as intermediate stopover sites. The last three sections address threats to forest birds, conservation strategies for overcoming these, and the need for long-term studies to improve our understanding of forest birds and our ability to protect them.

DOI: 10.4324/9781003324072-22

Forests as habitats for birds

The complex three-dimensional structure that distinguishes forests from other habitats favours flying as a means of locomotion, but vertebrate flight requires a high-energy, endothermic metabolism, resulting in continuous high food demands. Unlike bats, their nocturnal mammalian counterparts, no birds hibernate, and the demands of flight mean they cannot carry large stores of fat. Small birds, in particular, eat daily or die. This need for a continuous supply of food accounts for many of the general patterns we see in global forest bird communities. Equatorial lowland forests show little seasonality in resource availability, allowing many species of narrow diet specialists to coexist, and explaining their exceptional bird diversity. Moving away from the equator, seasonality increases, reaching a maximum in the boreal forests of the far north (and, to a much smaller extent, their southern counterparts). Birds can adapt to moderate seasonality in forest resources by diet-switching (e.g., from insects to fruits), and many do so, but as seasonality increases, an increasing proportion of species and individuals migrate instead—billions of birds every year. Of course, forests provide more than just food: their structure buffers the climate and gives many opportunities for escape from predators, while their structural complexity provides a greater diversity of nesting sites than any other habitat type.

Forest bird biogeography

Mesozoic forests supported a diversity of birds alongside the non-avian dinosaurs, but fossil and molecular evidence suggests that no arboreal bird lineages survived the end-Cretaceous asteroid impact (Field et al., 2018). Our modern forest avifauna is thus a result of the explosive radiation of birds that occurred in the Palaeocene and Eocene. The late Mesozoic break-up of Gondwanaland meant that these land bird lineages spread into a world that was fragmented by ocean barriers, particularly in the tropics (Corlett & Primack, 2011). Australia was linked through Antarctica to South America until the early Eocene, around 50 million years ago, and there have been intermittent land bridge connections in the Northern Hemisphere, but the terrestrial links between Africa and Eurasia and between South and North America are relatively recent, as is the increasing proximity of Australia to Asia. Madagascar has been an island continuously since the Mesozoic. Seabirds and some migratory land birds cross open oceans by choice, but most forest birds—particularly those that live below the canopy—do not do so willingly and are also unlikely to be blown off-course by storms. Moreover, even where land connections are available, many forest birds are unable or unwilling to cross no-forest habitats. Colonization of new forest areas is therefore most likely for non-specialist bird species.

Today all forests support native birds, and which birds are where is determined partly by environmental factors and partly by which bird lineages colonized successfully and are thus present in the regional species pool. This latter factor means that environmentally comparable forest sites on different continents and islands often share no bird species, few or no genera, and only a minority of families. Except on the remotest islands, the bird lineages that were available have evolved to fill most of the same morphological and ecological space (*see* the section 'Ecological Roles of Forest Birds'), but there are also some striking differences between the avifaunas of different forest regions, which are most easily explained as the consequences of the historical contingencies described above. It is probably not just a matter of which lineages arrived where, but also the order in which lineages arrived, with early arrivals having the advantages both of numbers and of time to evolve local adaptations over later arrivals ('priority effects'). Lineage establishment and diversification may also be limited by competition with mammals. The complementary diversities of muscular-jawed parrots and squirrels in tropical forests, with parrots most diverse in New Guinea, where there are no squirrels, and South America, with few, may be an example of this diversity (Corlett & Primack, 2011).

Some ecological roles of birds, such as foliage-gleaning insectivores, are occupied everywhere by similar-looking species from a diversity of lineages, but there are also examples of apparently unoccupied niches. A single diverse lineage of hole-excavating woodpeckers occurs in most forests worldwide, except in New Zealand, Australia, New Guinea, and Madagascar, with no ecological replacement where they are missing. Both the Old and New Worlds have specialized avian scavengers—vultures—but Old World vultures depend on their acute eyesight to detect carcasses, so they cannot scavenge in forests, where opportunistic mammals and birds dominate, while New World turkey vultures and their relatives detect carcasses below the canopy by smell and dominate scavenging in forests. In other cases, unrelated lineages occupy the same niche in different regions, but with incomplete convergence. Most forests with a long enough flowering season support nectar-specialist birds, but this niche is occupied by hummingbirds in the Neotropics, sunbirds in Africa and Asia, and honeyeaters in New Guinea and Australia, with scattered specialists in other families, including parrots (the lorries and lorikeets from Southeast Asia to Australia), tanagers (Neotropical flowerpiercers), and finches (Hawaiian honeycreepers). Most hummingbirds are small-bodied and can hover for long periods, while most other nectar specialists are larger and prefer to perch when feeding, with consequences for floral evolution in the plants they pollinate.

Ecological roles of forest birds

Birds play diverse ecological roles in forests, functioning as predators of vertebrates and invertebrates, prey, scavengers, seed predators and dispersers, and pollinators. In comparison with forest mammals, which have a similar, energy-demanding, endothermic lifestyle, the major missing role is leaf-eating, which needs a large stomach to process an abundant but low-quality food. The hoatzin in South America appears to be the only specialist avian forest folivore. Most other broadly defined ecological roles are shared with mammals, other vertebrates and invertebrates, but birds dominate in many more narrowly defined roles in seed dispersal or pollination.

In continental forests, the largest predator of other vertebrates is usually a mammal, but large mammals are confined to the ground or lower tree branches. From mid-canopy upward, birds are usually the apex predators, at the top of the food chain with no predators of their own. Most predatory forest birds belong to the families Accipitridae (hawks, eagles etc.) or Falconidae (falcons and their relatives), or the order Strigiformes (owls). Hawks and owls are present in almost all forests worldwide, including some remote oceanic islands, while the falcons are more often open-country birds, although also widespread in forests. Detecting these birds within the forest can be challenging due to their small population sizes and cryptic colourations and behaviours, but larger diurnal species can be more conspicuous. The harpy eagle (*Harpia harpyja*), in Central and South America, can weigh as much as 10 kg, but its relatively short but broad wings allow great manoeuvrability in tropical forests. A study in Panama found that sloths were the main prey, with primates accounting for most of the rest of their diet (Aguiar-Silva et al., 2014). Many birds will scavenge opportunistically on animal carcasses but, as mentioned earlier (*see* the section 'Forest Bird Biogeography'), only the New World turkey vultures and their relatives use their acute sense of smell to locate carcasses below the canopy and can thus dominate the scavenging niche in forests.

The importance of insectivorous birds in forests is demonstrated by the many adaptations forest invertebrates have for avoiding visually hunting predators. It has also been confirmed by exclusion experiments (e.g., in South Africa; Peter et al., 2015). Sudies have demonstrated a spill-over effect whereby forest insectivores reduce pests in crops grown near forest fragments that have been retained. In Costa Rica, forest birds were the main predators of a coffee berry boring beetle (*Hypothenemus hampei*) (Karp et al., 2013). Insectivores usually dominate the mixed-species foraging flocks that are characteristic of forest bird communities, and uncommon in other vegetation types. These flocks are most coherent, structured, and diverse in tropical forests, but they occur at least seasonally in most forests outside the tropics (Mangini et al., 2023). There is still a lot to learn about them, but there is evidence that flock members benefit from increased foraging efficiency and/or reduced predation risk, with the benefits varying between species.

Birds are the dominant consumers of fleshy fruits in forests and the most important dispersers of the seeds they contain. Seed dispersal is essential for most plants and the abundance and mobility of birds makes them excellent dispersal agents (Fleming & John Kress, 2011) (Figure 19.1), particularly for shrubs and small tree growing below the forest canopy, where wind dispersal is impractical. Many bird species eat some fruits, and specialist forest frugivores range in size from 5-g flowerpeckers to 3,000-g (flying) hornbills and 70,000-g (flightless) cassowaries. The maximum sizes of seeds dispersed (< 6 mm to > 50 mm) and the maximum dispersal distances (tens to thousands of metres) roughly scale in proportion to bird size (Corlett, 2021). Most fleshy fruits and their seeds are small, but large birds can swallow small fruits and not vice versa, so the largest frugivores in a forest have irreplaceable roles in seed dispersal. Most birds swallow fruits whole, but Neotropical tanagers and related species crush them in their bills and squeeze out all but the smallest seeds. Seeds that are swallowed are unwanted extra weight, so most frugivores rapidly remove the fruit flesh and dispose of the seeds by defecation or regurgitation. Seeds can also be food, however, and specialist granivores either destroy seeds by breaking them in the bill, as parrots do, or by

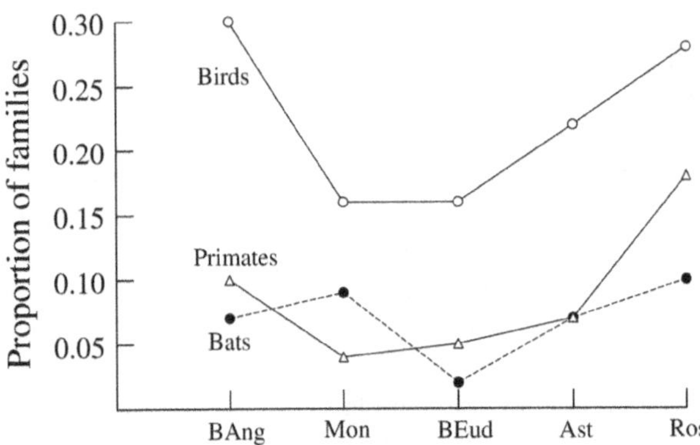

Figure 19.1 The proportion of vertebrate-dispersed angiosperm families dispersed by birds is higher than by mammals (here, primates and bats) (Fleming & John Kress, 2011). BAng, basal angiosperms; Mon, monocots; BEud, basal eudicots; Ast, asterids; Ros, rosids.

grinding them up in the gizzard, as do some forest pigeons and ground-feeding galliforms (such as pheasants and currasows). The dividing line between frugivore and granivore is rarely sharp, however, and all granivores seem to disperse some seeds intact.

Across the northern temperate zone, bird—jays, nutcrackers, and other corvids—are major dispersal agents for large, dry fruits and seeds, including those of oaks (acorns), some pines, and number of other tree species (Pesendorfer et al., 2016). These birds scatter-hoard the fruits in surface litter, with the intention of retrieving them later, but many are not retrieved. Rodents also scatter-hoard many of the same fruits and seeds, but birds can disperse them over much longer distances.

Compared with insects, birds are expensive pollinators, requiring larger flowers with bigger rewards (Corlett & Primack, 2011). In compensation, they can carry larger pollen loads over longer distances and will fly at lower temperatures than most insects. Of the major groups of flower-visiting birds, the Neotropical hummingbirds are the smallest (the Cuban-endemic bee hummingbird weighs only 2 g) and this, coupled with their hovering flight, which removes the need for a perch, probably accounts for the abundance and diversity of hummingbird flowers in Neotropical forests. Plant–bird co-evolution is most visible in the diversity of bill lengths and shapes in co-existing hummingbird species adapted to different flowers. Sunbirds, which are most diverse in Africa, are generally larger and typically visit larger flowers with bigger rewards, as do honeyeaters in Australia. Although relatively few plant species are pollinated by birds in tropical Asian forests, these species are conspicuous to us because bird vision overlaps with ours (with the addition of an extra UV- or violet-sensitive cone pigment). Red 'bird flowers' are found in all tropical forests, although not all flowers pollinated by birds are red. Specialist nectar feeders can only survive if flowers are available year-round, so bird pollination declines in frequency away from the tropics. In more seasonal climates, bird pollination depends on seasonal diet-switching by more generalist birds, or on migrant specialists, and is rare in forests.

Birds also play a role as ecosystem engineers in forests by excavating nest holes for other animals. Globally, woodpeckers dominate in this role, because they occur in most forests worldwide, except in New Zealand, Australia, New Guinea, and Madagascar, and their unique bill structure enables them to excavate holes in timber that is too hard for other species. Some barbets, nuthatches, and parids also make their own holes, but usually only in decayed wood, and they have more restricted global distributions. Holes excavated by these primary cavity nesters are then used by many other species of birds, for nesting or roosting, as well as many mammals and a diversity of other vertebrates and invertebrates.

Migratory birds in forests

Each autumn, billions of birds depart their breeding grounds across the Northern Hemisphere on long migrations towards the subtropics and tropics, where they may spend as many as seven months in wintering habitat. In spring, these birds return north to breed, taking advantage of the summertime resource peak, the long summer days, and the reduced competition from residents, to fuel their energetically demanding breeding activities. For these birds, the risks of migration are clearly outweighed by the benefits. In some species, only the northernmost populations migrate, often not very far, while in others the entire species shuttles seasonally between widely separated areas, not necessarily with the same type of habitat. Tropical forests provide important wintering habitat for species-rich assemblages of migratory landbirds in Asia and South America, while in Africa, most migrants overwinter in more open habitats.

In Southeast Asia, for example, tropical forests provide regular wintering habitat for approximately 90 migratory landbirds breeding in boreal and temperate forests in East Asia (Yong et al., 2015), while at least 60 are Sino-Himalayan and/or short-distance migrants that breed in the mountains of southern China and the Himalayas. Lowland rainforests in the Malay Peninsula support as many as 27 species, with migrants forming 6–15% of total bird abundance at some sites (Yong et al., 2015). By comparison, only five species of Palearctic-breeding migrants have been documented to winter in tropical African forests. In Mexico, 20 species of North American-breeding migrants arrive and winter in tropical rain forest (Rappole and Warner, 1980). Species richness and abundances of Nearctic migrants in the tropical forests of Central America and the Caribbean are typically comparable (Robertson et al., 2011), forming a significant component of the ecological communities they are part of in the winter. At the other end of the migratory flyway, 90 per cent of all the bird species and individuals in the boreal forest region of Canada—three to five billion birds in total—leave their breeding grounds at the end of summer to head south to the United States, Mexico, the Caribbean, or South America (Cheskey et al., 2011).

Satellite tracking has revolutionized recent effort to study migratory routes and habitat usage of species in staging and wintering grounds with increasing levels of resolution. Since the mid-2000s, the advent of increasing small transmitters such as geolocators has enabled the migration ecology and routes of smaller landbird species to be better documented, while offering insights on stopover habitat usage, especially Nearctic-breeding migrants in the Neotropics (McKinnon et al., 2013). One of the best-studied forest-wintering migrants in tropical Asia, the Oriental honey buzzard, *Pernis ptilorhynchus*, has been tracked extensively using satellite transmitters from their breeding areas in Japan and found to use stopover and wintering sites in a consistent manner (Figure 19.2, Sugasawa and Higuchi, 2019). These birds are specialist feeders on bee and wasp nests, which are only seasonally available in northern Asia but occur year-round in the tropics, where some populations do not migrate. Heim et al. (2022) used light-level geolocators to track the blue-and-white flycatcher, *Cyanoptila cyanomelana*, and demonstrated the connectivity of two major stopover sites (in Korea and China) with the breeding grounds in Russia, and the wintering areas in the lowland forests of the central Philippines.

Threats to forest birds

Deforestation in the tropics (see Chapter 27, 'Tropical deforestation, forest degradation and the role of REDD+ in climate change mitigation') is the foremost threat to forest birds given the high rates of forest loss, mainly for agriculture. Selective logging (see Chapter 30, 'The ecology of logged forests') also degrades habitat by altering forest structure, resource availability, and microclimate (Peh et al., 2005). While more sustainable than clear-cutting, selective logging can reduce nesting and foraging opportunities for specialized birds (Miranda et al., 2020). Tropical birds are particularly susceptible as they have limited dispersal abilities, lower reproductive output, lower population densities and specialized habitat needs, compared to their temperate counterparts (Tobias et al., 2013). Among tropical birds, larger, insectivorous, and frugivorous species that nest in the understory are most vulnerable (e.g., Newbold et al., 2014). Insectivores are particularly threatened due to their poor dispersal abilities, loss of suitable microhabitat and preference for nesting on in the understorey or ground, which carries a higher risk of predation (Sherry, 2021).

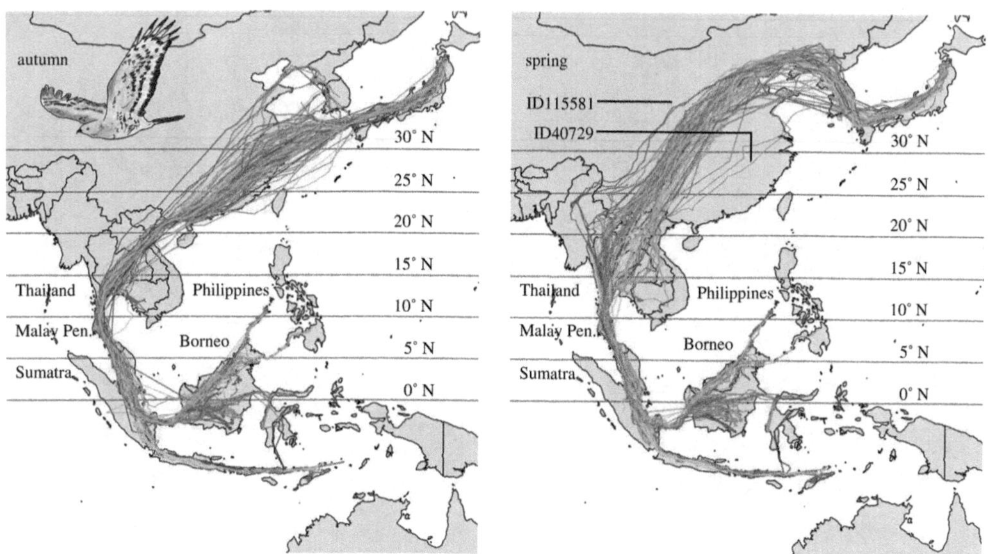

Figure 19.2 Oriental honey buzzards' migratory routes in autumn and spring (Sugasawa and Higuchi, 2019). The varying shaded lines denote different individual routes; ID115581 and ID40729 showed more divergent spring migration routes. Overall, the Oriental honey buzzards made substantial stopovers during the spring migration with high location fidelity, presumably preferring stopovers with a high abundance of bees.

Forest loss and degradation are known to affect migratory landbirds in their tropical wintering grounds (Ko et al., 2022; Yong et al., 2015) and have been widely presumed to be a major cause of decline detected in the breeding grounds (Higuchi and Morishita, 1999). Declines of wintering bird communities in tropical forests have been widely reported, although this is probably better documented in the New World tropics (Rappole, 1996; Robbins et al., 1989; Wunderle and Waide, 1994). In Puerto Rico's dry forests, wintering bird communities have exhibited strong declines with mist-netting catch-rates of wintering species about one-third of that 20 years ago (Faaborg et al., 2013). Tropical Asia has suffered the highest rates of forest loss globally and it is expected that the extensive deforestation taking place throughout the lowlands can impact several globally threatened and forest-dependant winterers (Yong et al., 2015). Wells (2010) reported that densities of wintering Siberian blue robin, *Larvivora cyane,* were substantially lower in hill-slope forests and in disturbed or agricultural habitats compared to old-growth lowland (dipterocarp) forest in the Malay Peninsula, alluding to the species' sensitivity to forest degradation. In Taiwan, Ko et al. (2022) reported an annual decline of 8 per cent for the Fairy Pitta *Pitta nympha* population and found that the drivers of decline are most likely linked to habitat loss and degradation in the non-breeding range.

Species wintering in montane forests have been presumed to face lower risk but montane forests are also under increasing pressure from illegal logging, agricultural expansion, and shifting cultivation (Soh et al., 2019). It is unclear how such extensive land-use change in the mountains of tropical Asia may impact migratory species using upland forest habitat in winter.

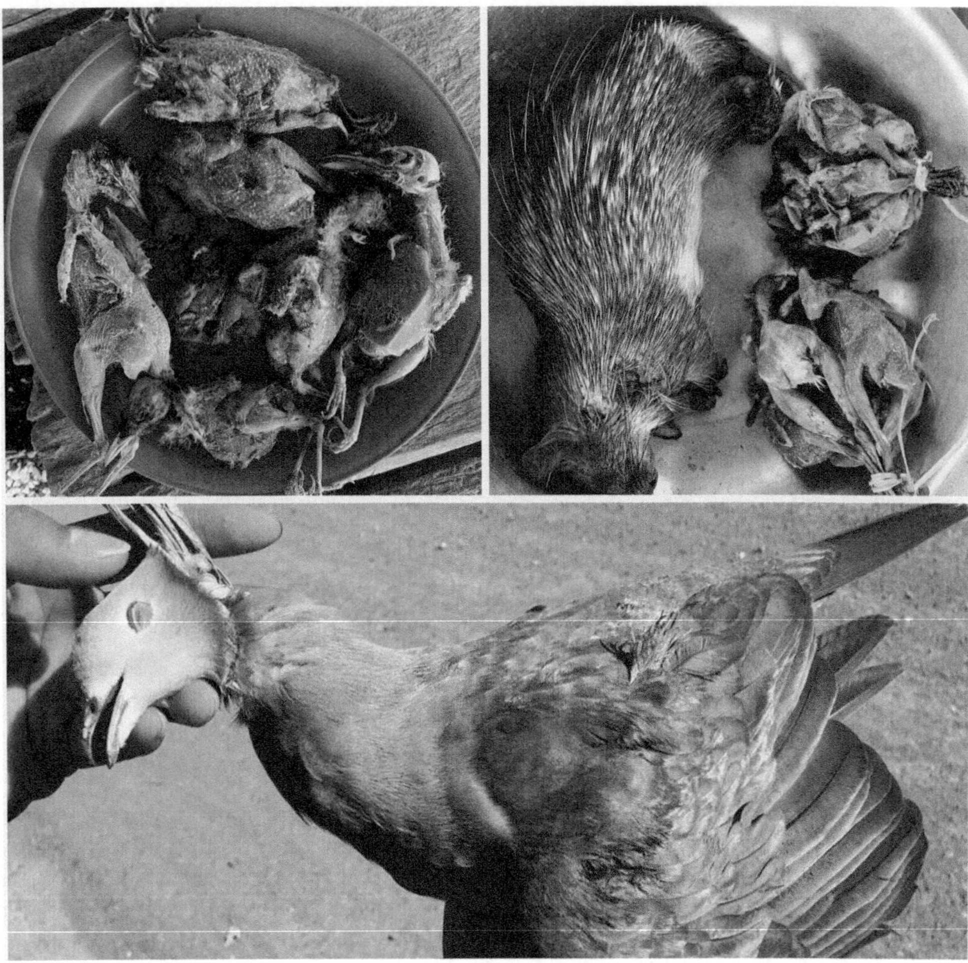

Figure 19.3 Spatial distribution of colour diversity in passerines (Senior et al., 2022).

Forest fragmentation (see Chapter 29, 'Forest fragments and fragmentation') intensifies the effects of habitat loss by isolating populations, impeding gene flow. Small, isolated populations are prone to inbreeding depression and have lower genetic diversity, making them less adaptable to environmental changes. Surrounding land-use affects connectivity; urban areas severely limit dispersal. For example, gene flow for the wrentits, *Chamaea fasciata*, in Santa Monica in the United States was found severely impeded for populations sampled in forest patches surrounded by the oldest and most intensive urban development even though they are fairly common (Delaney et al., 2010). Forest-dependent species like understorey insectivores are more sensitive than vagile species (e.g., Callens et al., 2011). Fragmentation also increases competition and predation (Piper & Catterall, 2003). Sensitive understory birds may avoid edge habitats, move faster, and cover longer distances in a fragmented landscape with high-edge areas (Hansbauer et al., 2008).

In addition to the loss and degradation of their forest habitats, the role of hunting and trapping for food, medicine, cage birds, a variety of cultural and ritual purposes, and for

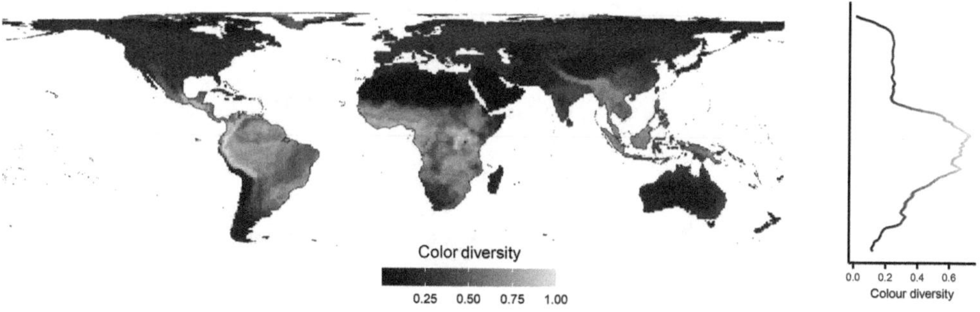

Figure 19.4 Snared birds in Southeast Asian markets. *Top-left*: Broadbill, doves and thrushes sold in rural markets in Laos; *top-right*: a porcupine and bulbuls sold in rural Laos; *bottom*: pink-headed imperial pigeon, *Ducula rosacea*, sold at the roadside in Timor-Leste (Photo credits - *top*: Yong Ding Li, *bottom*: Colin Trainor).

recreation has also had a global impact on forest birds (see Chapter 42, 'Wild meat hunting in tropical forests'). Forest birds are commonly shot in Europe for recreation (Hirschfeld & Heyd, 2005). Larger species are snared in Africa and Asia and shot in South America for domestic consumption and the wild meat trade (Figure 19.3; e.g., Yong et al., 2022). For the pet trade, forest raptors are prized in African markets (Buij et al., 2016), while songbirds are heavily poached in Southeast Asia (Lees and Yuda, 2022). The Asian songbird crisis stems from enormous domestic demand in Indonesia and other regional markets. Aside to their songs, showy and colourful songbirds from the tropics such as the Bali myna, *Leucopsar rothschildi*, and the straw-headed bulbul, *Pycnonotus zeylanicus*, are at greater risk (Figure 19.4; Senior et al., 2022). Exploitation of forest birds can result in broader ecological impacts, affecting organizational levels from individual to ecosystem. For example, the sustained removal of forest birds from their natural habitats can lead to loss of genetic diversity (e.g., Barusan sharma, *Copsychus (malabaricus) melanurus*; Rheindt et al., 2019). The absence of one or two more traded species may cause a breakdown in the structure of mixed-species flocks (Marthy and Farine, 2018). Removing target species from forests can also disrupt ecological roles like seed dispersal. The hunting of large frugivores could subsequently alter the forest plant community composition. For example, unprotected sites in Panama had disproportionately more lianas that were wind dispersed, and seeds that were dispersed by bats, smaller birds and mechanical means than protected sites (Wright et al., 2007).

Invasive bird species (see Chapter 32, 'Biological invasions in forests and forest plantations') also pose a considerable threat to island birds by outcompeting, predating, and spreading disease to forest natives. For example, Japanese white-eyes, *Zosterops japonicus*, were found to have similar dietary requirement and outcompete the honeycreepers native to the Hawaiian archipelago (Hawaiian iiwi, *Drepanis coccinea* and Hawaii elapaio, *Chasiempis sandwichensis*) (Mountainspring and Scott, 1985). These native honeycreepers have also succumbed to avian malaria, a disease that caused by an invasive protozoan parasite, *Plasmodium relictum* and infected southern house mosquitoes (*Culex quinquefasciatus*), likely introduced into the islands via European ships and exotic birds (van Riper III et al., 1986).

To date, climate change has had a relatively modest effect on ecosystems and biodiversity, compared to direct threats such as habitat loss and degradation, overexploitation, and

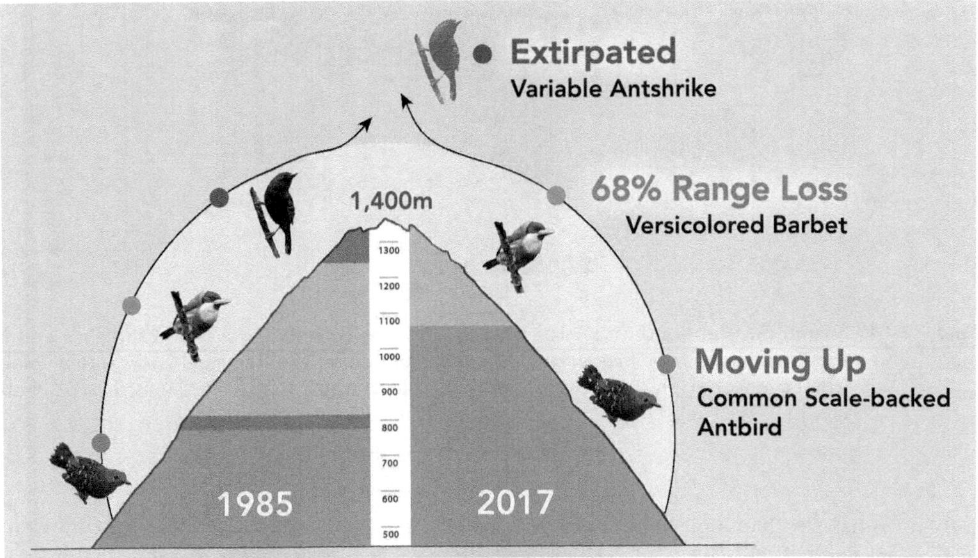

Figure 19.5 Warmer temperatures are driving montane species upslope on the Cerro de Pantiacolla in southern Peru (Freeman et al., 2018). The variable antshrike, a high altitudinal specialist, found in 1985, which was not detected in 2017. The versicolored barbet's range has shrunk by nearly two-thirds while the common scale-backed antbird, a lowland species, has moved to occupy a 17 per cent area further upslope.

Source: Adapted from graphic designed by Jillian Ditner.

invasive species (Malhi et al., 2020). However, climate change already has measurable impacts on forest birds. Morphological changes due to warming climates with less precipitation like smaller size and longer wingspans in an Amazonian primary forest birds reduce energy needs but compromise speed and agility (Jirinec et al., 2023). Tropical montane specialists are extremely vulnerable (Mata-Guel et al., 2023). Increased temperatures reduce population growth of tropical montane species (Neate-Clegg et al., 2021). Upslope distribution shifts expand tropical lowland species' ranges but shrink those of tropical montane species, even driving local extirpations (Figure 19.5; Freeman et al., 2018).

Conservation strategies for preserving forest birds

Human activities over the centuries have greatly altered global forests, with massive consequences for bird populations. Forest bird conservation has taken two main approaches, which correspond to the level of human impact on the landscape. In areas with substantial forest cover still supporting specialized forest birds of high conservation concern, efforts focus on establishing, enforcing, and managing protected areas to maintain habitat quality and integrity. These designated areas provide essential habitats and sanctuaries for forest specialists, ensuring that these are protected from logging, deforestation, and urban development (Cazalis et al., 2020). Protected areas also allow researchers and conservation practitioners to monitor and study bird populations, which is crucial for understanding their needs and developing effective conservation strategies (Vimal et al., 2021). In areas with degraded forests, in contrast, conservation primarily aims to allow natural recovery and/or to enhance bird diversity by improving habitat suitability. For example, the preservation of dead trees—snags—is essential for bird conservation in degraded forests. Snags provide critical nesting sites for cavity-nesting birds like woodpeckers and owls, and foraging opportunities for many bird species (e.g., Nappi et al., 2015). By preserving snags in forested areas, conservation practitioners can ensure that these valuable resources remain available to forest birds. Efforts to preserve dead trees can include leaving them standing in managed forests or actively creating artificial cavities in trees to mimic natural nesting sites.

At the landscape level, enlarging remaining habitat patches to increase the area of suitable habitat, through natural regeneration or habitat restoration, is a critical conservation strategy (Moudrý et al., 2021). Forests can recover from disturbance, although some may take longer to return to their previous state (Cole et al. 2014). Conservation practitioners must carefully monitor the response of forest birds to recovery efforts and adapt strategies as needed to ensure that restored habitats are suitable for the avian species that rely on them (Kamp et al., 2020).

Road construction and both agricultural and urban development can disrupt the connectivity between forested areas for the many bird species that require large and interconnected areas for breeding, foraging, and migration. Therefore, improving functional connectivity between forested patches through habitat corridors, stepping stones, or improving matrix permeability is crucial for preserving forest bird species richness and community structure (Mayhew et al., 2019). Conservation efforts should focus on creating and maintaining green corridors between forested areas to allow birds to move freely and access the resources they need (Beaugeard et al., 2021). But green corridors may not improve the conservation value of small forested areas (Hannon and Schmiegelow, 2002).

Edge effects occur at the boundaries of fragmented forests, where contrasting conditions with forest interior may be unfavourable for forest birds (McWethy et al., 2009). Some key management approaches for dealing with threats to forest birds from edge effects include increasing core habitat area and reducing edge-interior ratio by expanding forest patch size or combining adjacent patches (Slattery and Fenner, 2021) and creating complex surrounding vegetation to provide more resources and shelter (Mesquita et al., 1999). The general goals are reducing hostility of the edges for forest interior species while also enhancing habitat heterogeneity for edge and gap specialists.

Sustainable farming practices that incorporate trees, such as shade coffee plantations, can also contribute to forest bird conservation. Shade-grown coffee provides 'complementary or supplementary habitat' for many forest bird species (Sánchez-Clavijo et al., 2020). Shade coffee farming could benefit both birds and the livelihoods of local communities (Buechley et al., 2015). However, protecting forest remnants is still essential for forest bird conservation in agricultural landscapes (Sánchez-Clavijo et al., 2020).

In conclusion, the conservation of forest birds requires a combination of strategies that will need to be tailored to local conditions. Bird communities dominated by resident specialists, as in the tropical lowlands, will need different strategies from boreal forest bird communities dominated by long-distance migrants. Remote forest areas occupied only by low densities of indigenous people are different from forest fragments in human-dominated landscapes, and selectively logged tropical forests differ from the forestry clearcuts common outside the tropics. Bird conservation has benefited from the unequalled attention that birds have attracted from scientists, skilled amateurs, and a sympathetic general public, and birds are the only group of organisms for which both threats and conservation needs are known for the majority of species.

Knowledge gaps remain regarding the impacts of forestry practices, underlying mechanisms of population decline, habitat needs outside the breeding season, and role of floaters (i.e., juveniles that are not breeding or holding a territory) (Fuller and Robles, 2018). Conservation strategies should also fit the regional context and consider multi-level approaches that account for bird responses to within-patch habitat factors, patch configuration, and landscape composition. For forest birds that are being exploited, the Convention on International Trade in Endangered Species of Wild Fauna and Flora (CITES) regulates international trade, but most trade in living and dead birds, and their parts, is domestic, so domestic regulation and enforcement is crucial (Shepherd et al., 2020). Reducing consumer demand through education and community-based conservation of wild populations can also be effective (Tasirin et al., 2021).

Future research for forest birds

Longitudinal studies of bird populations and communities are crucial in furthering our understanding of how they are affected by anthropogenic impacts over the long term. Most studies are cross-sectional, with inferences drawn from a snapshot in time, and understandably so given logistical, scheduling, and funding constraints. While much knowledge can undoubtedly be gleaned from short-term studies, some biological responses may only be discernible after a considerable length of time. Biological responses to climate change, for instance, are typically not observable over a few years and any measurable effects occur after a few decades. Since long-term studies require sustained funding, space-for-time substitution studies, in which spatial-climatic gradients are used to project climate-driven biotic responses, are sometimes used as a convenient alternative. However, the primary assumption that spatial and temporal gradients are equivalent has rarely been tested (Blois et al., 2013) and should be validated with historical data or simulations (Lovell et al., 2023).

References

Aguiar-Silva, F. H., Sanaiotti, T. M., & Luz, B. B. (2014) 'Food habits of the harpy eagle, a top predator from the Amazonian rainforest canopy', *Journal of Raptor Research*, vol 48, pp. 24–35.

Beaugeard, E., Brischoux, F., & Angelier, F. (2021) 'Green infrastructures and ecological corridors shape avian biodiversity in a small French city', *Urban Ecosystems*, vol. 24, pp. 549–560.

Blois, J. L., Williams, J. W., Fitzpatrick, M. C., Jackson, S. T., & Ferrier, S. (2013) 'Space can substitute for time in predicting climate-change effects on biodiversity', *Proceedings of the National Academy of Sciences*, vol. 110, pp. 9374–9379.

Buechley, E. R., Şekercioğlu, Ç. H., Atickem, A., Gebremichael, G., Ndungu, J. K., Mahamued, B. A., Beyene, T., Mekonnen, T., & Lens, L. (2015) 'Importance of Ethiopian shade coffee farms for forest bird conservation', *Biological Conservation*, vol. 188, pp. 50–60.

Buij, R., Nikolaus, G., Whytock, R., Ingram, D. J., & Ogada, D. (2016) 'Trade of threatened vultures and other raptors for fetish and bushmeat in West and Central Africa', *Oryx*, vol. 50, pp. 606–616.

Callens, T., Galbusera, P., Matthysen, E., Durand, E. Y., Githiru, M., Huyghe, J. R., & Lens, L. (2011) 'Genetic signature of population fragmentation varies with mobility in seven bird species of a fragmented Kenyan cloud forest', *Molecular Ecology*, vol. 20, pp. 1829–1844.

Cazalis, V., Princé, K., Mihoub, J.-B., Kelly, J., Butchart, S. H. M., & Rodrigues, A. S. L. (2020) 'Effectiveness of protected areas in conserving tropical forest birds', *Nature Communications*, vol. 11, 4461.

Cheskey, E., Wells, J., & Casey-Lefkovitz, S. (2011) *Birds at Risk: The Importance of Canada's Boreal Wetlands and Waterways.* Natural Resources Defense Council.

Cole, L. E. S., Bhagwat, S. A., & Willis, K. J. (2014) 'Recovery and resilience of tropical forests after disturbance', *Nature Communications*, vol. 5, 3906.

Corlett, R. T. (2021) 'Frugivory and seed dispersal', In Del-Claro K. & Torezan-Silingardi H. M. (Eds.), *Plant-Animal Interactions* (pp. 175–204). Springer Nature.

Corlett, R. T., & Primack, R. B. (2011) *Tropical Rain Forests: an Ecological and Biogeographical Comparison.* John Wiley & Sons.

Delaney, K. S., Riley, S. P. D., & Fisher, R. N. (2010) 'A rapid, strong, and convergent genetic response to urban habitat fragmentation in four divergent and widespread vertebrates', *PLoS One*, vol. 5, e12767. https://doi.org/10.1371/journal.pone.0012767

Faaborg, J., Arendt, W. J., Toms, J. D., Dugger, K. M., Cox, W. A., & Canals Mora, M. (2013) 'Long-term decline of a winter-resident bird community in Puerto Rico', *Biodiversity and Conservation*, vol. 22, pp. 63–75.

Field, D. J., Bercovici, A., Berv, J. S., Dunn, R., Fastovsky, D. E., Lyson, T. R., Vajda, V., & Gauthier, J. A. (2018) 'Early evolution of modern birds structured by global forest collapse at the end-Cretaceous mass extinction', *Current Biology*, vol. 28, pp. 1825–1831.e2.

Fleming, T. H., & John Kress, W. (2011) 'A brief history of fruits and frugivores', *Acta Oecologica*, vol. 37, pp. 521–530.

Freeman, B. G., Scholer, M. N., Ruiz-Gutierrez, V., & Fitzpatrick, J. W. (2018) 'Climate change causes upslope shifts and mountaintop extirpations in a tropical bird community', *Proceedings of the National Academy of Sciences*, vol. 115, pp. 11982–11987.

Fuller, R. J., & Robles, H. (2018) 'Conservation strategies and habitat management for European forest birds', In G. Mikusinski, J.-M. Roberge, & R.J. Fuller (Eds.), *Ecology and Conservation of Forest Birds* (p. 455) Cambridge University Press.

Hannon, S. J., & Schmiegelow, F. K. A. (2002) 'Corridors may not improve the conservation value of small reserves for most boreal birds', *Ecological Applications*, vol. 12, pp. 1457–1468.

Hansbauer, M. M., Storch, I., Leu, S., Nieto-Holguin, J.-P., Pimentel, R. G., Knauer, F., & Metzger, J. P. W. (2008) 'Movements of neotropical understory passerines affected by anthropogenic forest edges in the Brazilian Atlantic rainforest', *Biological Conservation*, vol. 141, pp. 782–791.

Heim, W., Antonov, A., Beermann, I., Lisovski, S., Sander, M. M., & Hahn, S. (2022) 'Light-level geolocation reveals unexpected migration route from Russia to the Philippines of a Blue-and-white-Flycatcher Cyanoptila cyanomelana', *Ornithological Science*, vol. 21, pp. 121–126.

Higuchi, H., & Morishita, E. (1999) 'Population declines of tropical migratory birds in Japan', *Actinia*, vol. 12, pp. 51–59.

Hirschfeld, A., & Heyd, A. (2005) 'Mortality of migratory birds caused by hunting in Europe: bag statistics and proposals for the conservation of birds and animal welfare', *Berichte Zum Vogelschutz*, vol. 42, pp. 47–74.

Jirinec, V., Burner, R. C., Amaral, B. R., Bierregaard, R. O., Fernández-Arellano, G., Hernández-Palma, A., Johnson, E. I., Lovejoy, T. E., Powell, L. L., Rutt, C. L., Wolfe, J. D., & Stouffer, P. C. (2023) 'Morphological consequences of climate change for resident birds in intact Amazonian rainforest', *Science Advances*, vol. 7, eabk1743.

Kamp, J., Trappe, J., Dübbers, L., & Funke, S. (2020) 'Impacts of windstorm-induced forest loss and variable reforestation on bird communities', *Forest Ecology and Management*, vol. 478, 118504.

Karp, D. S., Mendenhall, C. D., Sandí, R. F., Chaumont, N., Ehrlich, P. R., Hadly, E. A., & Daily, G. C. (2013) 'Forest bolsters bird abundance, pest control and coffee yield', *Ecology Letters*, vol. 16, pp. 1339–1347.

Ko, J. C., Chang, A., Lin, R., & Lee, P. (2022) 'Effects of breeding habitat loss on a threatened East Asia migratory forest bird', *Conservation Science and Practice*, vol. 4, e12772.

Lees, A. C., & Yuda, P. (2022) 'The Asian songbird crisis', *Current Biology*, vol. 32, pp. R1063–R1064.

Lovell, R. S. L., Collins, S., Martin, S. H., Pigot, A. L., & Phillimore, A. B. (2023) 'Space-for-time substitutions in climate change ecology and evolution', *Biological Reviews*, vol. 98, pp. 2243–2270.

Malhi, Y., Franklin, J., Seddon, N., Solan, M., Turner, M. G., Field, C. B., & Knowlton, N. (2020) 'Climate change and ecosystems: threats, opportunities and solutions', *Philosophical Transactions of the Royal Society B: Biological Sciences*, vol. 375, 20190104.

Mangini, G. G., Rutt, C. L., Sridhar, H., Buitron, G., Muñoz, J., Robinson, S. K., Montaño-Centellas, F., Zarco, A., Fanjul, M. E., Fernández-Arellano, G., Xing, S., & Camerlenghi, E. (2023) 'A classification scheme for mixed-species bird flocks', *Philosophical Transactions of the Royal Society B: Biological Sciences*, vol. 378, 20220100.

Marthy, W., & Farine, D. R. (2018) 'The potential impacts of the songbird trade on mixed-species flocking', *Biological Conservation*, vol. 222, pp. 222–231.

Mata-Guel, E. O., Soh, M. C. K., Butler, C. W., Morris, R. J., Razgour, O., & Peh, K. S.-H. (2023) 'Impacts of anthropogenic climate change on tropical montane forests: an appraisal of the evidence', *Biological Reviews*, vol. 98, pp. 1200–1224.

Mayhew, R. J., Tobias, J. A., Bunnefeld, L., & Dent, D. H. (2019) 'Connectivity with primary forest determines the value of secondary tropical forests for bird conservation', *Biotropica*, vol. 51, pp. 219–233.

McKinnon, E. A., Stanley, C. Q., Fraser, K. C., MacPherson, M. M., Casbourn, G., Marra, P. P., Studds, C. E., Diggs, N., & Stutchbury, B. J. M. (2013) 'Estimating geolocator accuracy for a migratory songbird using live ground-truthing in tropical forest', *Animal Migration*, vol. 2, pp. 1–8.

McWethy, D. B., Hansen, A. J., & Verschuyl, J. P. (2009) 'Edge effects for songbirds vary with forest productivity', *Forest Ecology and Management*, vol. 257, pp. 665–678.

Medway, G. G.-H., & Wells, D. R. (1976) *The Birds of the Malay Peninsula: A General Account of the Birds Inhabiting the Region from the Isthmus of Kra to Singapore.* Witherby, London.

Mesquita, R. C. G., Delamônica, P., & Laurance, W. F. (1999) 'Effect of surrounding vegetation on edge-related tree mortality in Amazonian forest fragments', *Biological Conservation*, vol. 91, pp. 129–134.

Miranda, E. B. P., Peres, C. A., Marini, M. Â., & Downs, C. T. (2020) 'Harpy Eagle (*Harpia harpyja*) nest tree selection: Selective logging in Amazon forest threatens Earth's largest eagle', *Biological Conservation*, vol. 250, 108754.

Moudrý, V., Moudrá, L., Barták, V., Bejček, V., Gdulová, K., Hendrychová, M., Moravec, D., Musil, P., Rocchini, D., & Šťastný, K. (2021) 'The role of the vegetation structure, primary productivity and senescence derived from airborne LiDAR and hyperspectral data for birds diversity and rarity on a restored site', *Landscape and Urban Planning*, vol. 210, 104064.

Mountainspring, S., & Scott, J. M. (1985) 'Interspecific competition among hawaiian forest birds', *Ecological Monographs*, vol. 55, pp. 219–239.

Nappi, A., Drapeau, P., & Leduc, A. (2015) 'How important is dead wood for woodpeckers foraging in eastern North American boreal forests?', *Forest Ecology and Management*, vol. 346, pp. 10–21.

Neate-Clegg, M. H. C., Stanley, T. R., Şekercioğlu, Ç. H., & Newmark, W. D. (2021) 'Temperature-associated decreases in demographic rates of Afrotropical bird species over 30 years', *Global Change Biology*, vol. 27, pp. 2254–2268.

Newbold, T., Scharlemann, J. P. W., Butchart, S. H. M., Şekercioğlu, Ç. H., Joppa, L., Alkemade, R., & Purves, D. W. (2014) 'Functional traits, land-use change and the structure of present and future bird communities in tropical forests', *Global Ecology and Biogeography*, vol. 23, pp. 1073–1084.

Peh, K. S.-H., Jong, J. de, Sodhi, N. S., Lim, S. L.-H., & Yap, C. A.-M. (2005) 'Lowland rainforest avifauna and human disturbance: persistence of primary forest birds in selectively logged forests and mixed-rural habitats of southern Peninsular Malaysia', *Biological Conservation*, vol. 123, pp. 489–505.

Pesendorfer, M. B., Sillett, T. S., Koenig, W. D., & Morrison, S. A. (2016) 'Scatter-hoarding corvids as seed dispersers for oaks and pines: A review of a widely distributed mutualism and its utility to habitat restoration', *The Condor*, vol. 118, pp. 215–237.

Peter, F., Berens, D. G., Grieve, G. R., & Farwig, N. (2015) 'Forest fragmentation drives the loss of insectivorous birds and an associated increase in herbivory', *Biotropica*, vol. 47, pp. 626–635.

Piper, S. D., & Catterall, C. P. (2003) 'A particular case and a general pattern: hyperaggressive behaviour by one species may mediate avifaunal decreases in fragmented Australian forests', *Oikos*, vol. 101, pp. 602–614.

Randler, C., Murawiec, S., & Tryjanowski, P. (2022) 'Committed bird-watchers gain greater psychological restorative benefits compared to those less committed regardless of expertise', *Ecopsychology*, vol. 14, pp. 101–110.

Rappole, J. H. (1996) 'The importance of forest for the world's migratory bird species', In *Conservation of Faunal Diversity in Forested Landscapes* (pp. 389–406) Springer.

Rappole, J. H., & Warner, D. W. (1980) 'Ecological aspects of migrant bird behavior in Veracruz, Mexico', *Migrant Birds in the Neotropics: Ecology, Behavior, Distribution, and Conservation.* Smithsonian Institution Press, Washington, DC, pp. 353–393.

Rheindt, F. E., Baveja, P., Ferasyi, T. R., Nurza, A., Rosa, T. S., Haminuddin, R. R., & Gwee, C. Y. (2019) 'The extinction-in-progress in the wild of the barusan shama copsychus (malabaricus) melanurus', *Forktail*, vol. 35, pp. 28–35.

Robbins, C. S., Sauer, J. R., Greenberg, R. S., & Droege, S. (1989) 'Population declines in North American birds that migrate to the Neotropics', *Proceedings of the National Academy of Sciences*, vol. 86, pp. 7658–7662.

Robertson, B. A., MacDonald, R., Wells, J. V, Blancher, P. J., & Bevier, L. (2011) 'Boreal migrants in winter bird communities', *Boreal Birds of North America: A Hemispheric View of Their Conservation Links and Significance* (JV Wells, Editor). Studies in Avian Biology, vol. 41, pp. 85–94.

Sánchez-Clavijo, L. M., Bayly, N. J., & Quintana-Ascencio, P. F. (2020) 'Habitat selection in transformed landscapes and the role of forest remnants and shade coffee in the conservation of resident birds', *Journal of Animal Ecology*, vol. 89, pp. 553–564.

Senior, R. A., Oliveira, B. F., Dale, J., & Scheffers, B. R. (2022) 'Wildlife trade targets colorful birds and threatens the aesthetic value of nature', *Current Biology*, vol. 32, pp. 4299–4305.e4.

Shepherd, C. R., Leupen, B. T. C., Siriwat, P., & Nijman, V. (2020) 'International wildlife trade, avian influenza, organised crime and the effectiveness of CITES: The Chinese hwamei as a case study', *Global Ecology and Conservation*, vol. 23, e01185.

Sherry, T. W. (2021) 'Sensitivity of tropical insectivorous birds to the Anthropocene: A review of multiple mechanisms and conservation implications', *Frontiers in Ecology and Evolution*, vol 9. www.frontiersin.org/articles/10.3389/fevo.2021.662873

Slattery, Z., & Fenner, R. (2021) 'Spatial analysis of the drivers, characteristics, and effects of forest fragmentation', *Sustainability*, vol. 13, 3246.

Soh, M. C. K., Mitchell, N. J., Ridley, A. R., Butler, C. W., Puan, C. L., & Peh, K. S.-H. (2019) 'Impacts of habitat degradation on tropical montane biodiversity and ecosystem services: A systematic map for identifying future research priorities', *Frontiers in Forests and Global Change*, vol. 2. www.frontiersin.org/articles/10.3389/ffgc.2019.00083

Sugasawa, S., & Higuchi, H. (2019) 'Seasonal contrasts in individual consistency of oriental honey buzzards' migration', *Biology Letters*, vol. 15, 20190131.

Tasirin, J. S., Iskandar, D. T., Laya, A., Kresno, P., Suling, N., Oga, V. T., Djano, R., Bawotong, A., Nur, A., Isfanddri, M., Abbas, W., Rihu, N. A., Poli, E., Lanusi, A. A., & Summers, M. (2021) 'Maleo Macrocephalon maleo population recovery at two Sulawesi nesting grounds after community engagement to prevent egg poaching', *Global Ecology and Conservation*, vol. 28, p. e01699.

Tobias, J. A., Şekercioğlu, Ç. H., & Vargas, F. H. (2013) 'Bird conservation in tropical ecosystems', In *Key Topics in Conservation Biology 2* (pp. 258–276).

van Riper III, C., van Riper, S. G., Goff, M. L., & Laird, M. (1986) 'The epizootiology and ecological significance of malaria in Hawaiian land birds', *Ecological Monographs*, vol. 56, pp. 327–344.

Vimal, R., Navarro, L. M., Jones, Y., Wolf, F., Le Moguédec, G., & Réjou-Méchain, M. (2021) 'The global distribution of protected areas management strategies and their complementarity for biodiversity conservation', *Biological Conservation*, vol. 256, 109014.

Wells, D. R. (2010) *The Birds of the Thai-Malay Peninsula* (Vol. 2) Bloomsbury Publishing.

Wright, S. J., Hernandéz, A., & Condit, R. (2007) 'The bushmeat harvest alters seedling banks by favoring lianas, large seeds, and seeds dispersed by bats, birds, and wind', *Biotropica*, vol. 39, pp. 363–371.

Wunderle, J. M., & Waide, R. B. (1994) 'Future prospects for Nearctic migrants wintering in Caribbean forests', *Bird Conservation International*, vol. 4, pp. 191–207.

Yong, D. L., Jain, A., Chowdhury, S. U., Denstedt, E., Khammavong, K., Milavong, P., Aung, T. D. W., Aung, E. T., Jearwattanakanok, A., Limparungpatthanakij, W., Angkaew, R., Sinhaseni, K., Le, T. T., Nguyen, H. B., Tang, P., Taing, P., Jones, V. R., & Vorsak, B. (2022) 'The specter of empty countrysides and wetlands—Impact of hunting take on birds in Indo-Burma', *Conservation Science and Practice*, vol. 4, e212668.

Yong, D. L., Liu, Y., Low, B. W., Española, C. P., Choi, C.-Y., & Kawakami, K. (2015) 'Migratory songbirds in the East Asian-Australasian Flyway: a review from a conservation perspective', *Bird Conservation International*, vol. 25, pp. 1–37.

20

AMPHIBIANS AND REPTILES OF FOREST ECOSYSTEMS

David Bickford, Adrian Garda, Seshadri K.S., Umilaela Arifin, Hiral Naik, Hanyeh Ghaffari and Barbod Safaei-Mahroo

Introduction

Reptiles and amphibians are among the most diverse vertebrate groups with more than 20,000 species (>11,940 reptiles; Uetz et al. 2022, and >8,566 amphibians; Frost 2023). Collectively known as herpetofauna, reptiles and amphibians are ectotherms with strikingly different ecological needs and constraints from other vertebrates (MacNeil et al. 2013). Being ectothermic, herpetofauna can occupy niches that seem extreme or unsuitable (e.g., deserts, torrential rivers, and forest canopies) and because of their life history traits (e.g., biphasic, body size changes, age, and size at maturity) may require the use of multiple habitats (e.g., many amphibians are aquatic or terrestrial at different times in their development). Forests are vital to maintaining species diversity (MacNeil et al. 2013) and promoting species diversification (Acebey et al. 2003; Couvreur et al. 2011). Forests are habitats for over half of all reptile species and the most common habitat for amphibians (Cox et al. 2022). Perhaps surprisingly, herpetofauna affect forest ecosystem processes like decomposition (Wyman 1998), disperse seeds (e.g., Falcón et al. 2020), and even pollinate flowers (e.g., de-Oliveira-Nogueira et al. 2023). Some have even been labeled 'ecosystem engineers' (Flecker et al. 1999).

Many factors explain the distribution and diversity trends displayed by amphibians and reptiles found in forest habitats. Among these are elevation, latitude, biogeography, and features of the climate such as precipitation, relative humidity, and temperature. Interestingly, the evolutionary inertia of lineages found in a particular area are under similar selection pressures. Many of these selection pressures form the basis of parallel trends in evolutionary ecology and functional roles across a diverse array of both amphibian and reptile species around the world.

Like most other groups, herpetofauna diversity is highest in the tropics and decreases toward higher latitudes. There are limited numbers of species of reptiles and amphibians in boreal forests due to short summers and long winters, with spring and summer conditions governing much of their activity, distribution, and behavior. In contrast, tropical forests host higher diversity and abundances of herpetofauna, particularly forests in Latin and South America, central Africa, and Southeast Asia. While these latitudinal gradients are perhaps the most pervasive and recognizable biogeographic patterns, there are important exceptions.

DOI: 10.4324/9781003324072-23

Anurans (frogs), caecilians (worm-like amphibians), squamates (snakes and lizards), and crocodilians are more diverse in tropical regions; however, salamanders and turtles are more species-rich at higher latitudes. This is mainly due to the ancient geographic origins of these lineages and their inability to disperse long distances and/or cross biogeographic barriers. Though studies are limited, climatic (especially temperature for reptiles and humidity for amphibians) constraints might shape the distribution of herpetofauna species within biogeographic regions. Additionally, elevation contributes to physical and physiological barriers that limit species' dispersal and distribution (Moura et al. 2017).

Topographically complex regions seem to promote higher and faster rates of speciation (García-Rodríguez et al. 2021). Mountains can be important refugia and key areas for biodiversity conservation. Climate variation across altitudinal gradients is especially important for ectothermic species like amphibians and reptiles, since they are critically dependent on water availability and temperature. Mountains within drier regions of Brazil, for example, promote orographic rainfall, which supports forest inselbergs in a semiarid area. Historically, higher elevation areas were key to herpetofauna resilience during glacial cycles, acting as refugia for species that could move to more suitable climates as climatic conditions changed.

Biogeographic highlights

With a long history of isolation from other continents, topographic complexity, a series of climate shifts, and mostly low latitudes, South America is home to a tremendous diversity of amphibians and reptiles. Such conditions also promoted the evolution of many endemic lineages, such as poison-dart frogs, tegu lizards, and lineages of vipers and coral snakes, as well as many others. The region is also the cradle of diversification for bufonid toads and hylid tree frogs, which later spread throughout the globe.

Africa, on the other hand, was a cradle of early herpetofauna diversification, but later became much less diverse as other areas diversified more quickly. For example, Neobatrachia (one of the old stem-groups of frogs and toads) are thought to have an African origin and later diversified as they spread. Africa and Madagascar have iconic amphibians including the Goliath and tomato frogs. While salamanders do not occur in Africa, there are a few caecilians and high reptile diversity and endemicity. Perhaps one of the most charismatic reptiles from African and Madagascan forests are chameleons. Completing the ancient Gondwanan areas, the Indian subcontinent has one of the most important amphibian hotspots and is home to many radiations endemic to India and Sri Lanka. Parts of Southeast Asia, Australia, and New Guinea are home to important radiations of direct-developing (no tadpole stage) frogs, venomous elapid snakes, and the largest forest predators in the region, pythons. New Zealand has a few important amphibians and reptiles, but none are unique as the tuatara (*Sphenodon punctatus*), the only extant rhynchocephalian representative.

Moving onto the Laurasian areas, western Asia is now predominantly arid or semiarid, with humid highlands. Its main mountain ranges are covered with temperate broadleaf, mixed, and conifer forests, refuges for a variety of amphibian species of the Middle East. One of its most species-diverse areas is the Irano-Anatolian biodiversity hotspot, which contains a high number of endemic amphibians and reptiles. An endemic group of salamanders, for example (the Salamandridae), has four genera and 16 species. China and Japan have many forest-dependent turtles (e.g., leaf turtles) and salamanders, including the world's largest, the giant salamanders *Andrias davidianus* and *A. japonicus*.

North America has seen an adaptive radiation of lungless salamanders and interesting groups of turtles and snakes. Notably, a group of vipers evolved an acoustic warning signal in the form of rattles at the ends of their tails (rattlesnakes). Europe, on the other hand, is relatively depauperate but has interesting species-rich groups like the lacertids (wall lizards), salamanders, and snakes.

Structural complexity and microhabitats

Forests provide many kinds of microhabitats, areas that are relatively small and found within many different habitats. These microhabitats are complex, three-dimensional, and dynamic and comprise multiple layers of vegetation and landscape elements such as waterways, rock outcrops, soils, and leaf litter, creating structural and microclimatic diversity. Interactions of the structural diversity of habitat, edaphic factors, and uneven distribution of abiotic factors such as sunlight and moisture create unique niches occupied by amphibians and reptiles. The niches available for herpetofauna in forests are broadly governed by the following dimensions: microclimate, precipitation and temperature, refugia/shelter, and sites for oviposition, calling, and display.

Microclimate

Microclimates are a function of temperature and moisture availability in a given portion of the habitat. Forests have numerous microclimatic regimes that are dynamic and drastically different from microhabitats outside forests (de Frenne et al. 2019). Reptiles have a greater tolerance to desiccation than amphibians and are able to exploit a wider range of habitats with thermal maxima other organisms cannot tolerate. Although microclimates vary across different types of forests, fluctuation within a microclimate can be buffered and herpetofauna exploit this relative stability.

Forest canopies, forest floor sun flecks, riparian corridors, shaded areas under large canopy trees, tree falls, and topographic features all provide a variety of microhabitats with very different microclimates inside forest ecosystems. Sun flecks and tree fall gaps, in particular, are used as basking and foraging sites, mainly by reptiles. Structurally complex habitats are so important for reptiles that these features have allowed anole lizards to speciate in response to perch height and perch diameter within forest canopies (Rand 1964; Reagan 1995).

Precipitation and temperature

Forests buffer variation in precipitation and temperature (O'Connor et al. 2021). Amphibians tend to be most affected by precipitation as bioavailable water is a key factor for survival and reproduction for most amphibians (Aragón et al. 2010). In forests, any site that can accumulate water, such as tree holes (standing or fallen), bromeliad and *Pandanus* leaf axils, or root mass cavities of tree falls, are incredibly important breeding microhabitats. Globally, amphibian diversity is affected by rainfall and areas receiving high rainfall often correlate with regions of high species diversity. Typically, amphibians depend on water to breed, but many terrestrial species undergo direct development and do not require free-standing or flowing water. Species that breed terrestrially are found in moist forests, especially cloud forests. For forest reptiles however, precipitation's effect is more indirect on their life history with temperature being a much stronger driver of diversity patterns. Because herpetofauna are ectotherms, their metabolism is directly affected by ambient temperature and reptiles in particular use behavioral thermoregulation to achieve more optimal physiological responses when their environment is suboptimal.

Refugia/shelter

Forests serve as important refugia for herpetofauna both during ecological (recent/short) and geological (ancient/long) time periods. Within a reptile or amphibian life cycle, for example, microhabitat structure provides refugia like physical shelter to maintain moisture, thermoregulate, and/or avoid predators. On the other end of the time spectrum, for example during glacial fluctuations, mountaintop forests served as landscape refugia for populations that later spread, recolonized, and diversified.

Oviposition sites

The structural diversity in forests creates places that serve as oviposition sites and shelter eggs from predators while buffering against environmental extremes. Oviposition sites are ecologically important because they directly influence natality and affect long-term population viability. The choice and use of oviposition sites are different for amphibians and reptiles. Amphibians, having a greater dependency on moisture with their anamniotic eggs, are more restricted in their oviposition site choices than reptiles. Some examples of amphibian oviposition sites are phytotelmata (small cavities in plants where water accumulates apart from connectivity to other aquatic habitats), which reduces predators and buffers eggs against desiccation. Examples are poison-dart frogs from Central and South America, which often breed in leaf axils of bromeliads. Other examples include *Microhyla nepenthicola* from the forests of Borneo laying eggs within pitcher plants in the genus *Nepenthes.*

Reptiles do not appear to have tight associations with specific abiotic aspects of forests. Many reptiles undergo temperature-dependent sex determination (TSD), restricting the choice of oviposition sites. They seek out microhabitats that buffer moisture and temperature to determine the sex of the offspring. The king cobra (*Ophiophagus hannah*), for example, breeds in forests where the female makes a nest from fallen leaves and deposits eggs inside. The leaf mound nest functions as a thermo-regulated incubator and also protects them from torrential rain toward the end of the breeding period. Many forest stream turtles also depend on structural habitat diversity. Species such as the Burmese roofed terrapin, *Batagur trivittata*, lay eggs on sandy riverbanks in forests. Many species of lizards bury their eggs in soil. Some arboreal lizards deposit eggs in nooks created by peeling bark or broken branch stubs.

Calling and display sites

Many herpetofauna rely on acoustic, visual, and chemical cues to communicate. Communication is an integral aspect of life history, functioning to mark and maintain territories, advertise reproductive status, as well as confuse predators and limit competition. Acoustic cues function over relatively long distances whereas visual signals are only effective near other individuals and chemical signals deteriorate over time and distance from the source. Acoustic cues are predominant among anurans and geckos, but there is increasing evidence that most herpetofauna may use multiple cues to communicate (e.g., acoustic and visual cues joined or used sequentially, e.g., see Starnberger et al. 2014). While reptiles may use calling sites, they typically engage in visual displays and the sites of visual display in the forests are often limited by the thermal profiles of the locations and visibility to other individuals.

Forests provide biotic needs

While numerous abiotic factors and their interactions provide potential niches for amphibians and reptiles (see previous section), many species of herpetofauna play important roles in forest ecosystems. Most species of amphibians and reptiles have central roles in forest food webs, being consumers and prey items for larger organisms (see Valencia-Aguilar et al. 2013). Given that amphibians and reptiles inhabit both terrestrial and aquatic aspects of forest ecosystems, they are particularly important because they transfer energy and nutrients between the two habitat types.

Many individual amphibians, with their biphasic life cycle, occupy both aquatic (as larvae) and terrestrial ecosystems (as adults). Consequently, diets vary considerably over the course of their lives. Larvae are often herbivorous, whereas adults are almost exclusively carnivorous. Furthermore, larvae often display plasticity in their trophic level due to food availability, presence of a competitor or predator, and/or water levels and temperature as they undergo metamorphosis (Montaña et al. 2019). Thus amphibian larvae and adults, even of the same species, play distinct ecological roles in their habitats. Reptiles, on the contrary, mainly occupy the same habitat and trophic roles throughout their life. Snakes can be an exception as some species inhabit terrestrial environments and will prey largely on amphibians and other aquatic prey. Understanding the ecological roles of amphibians and reptiles, including their contribution toward ecosystem services, is crucial in shedding light toward current global declines in population numbers. Herein, we follow the designation of the ecological functions of amphibians and reptiles following Valencia-Aguilar et al. (2013) and Cortés-Gomez et al. (2015) by characterizing five categories: bioturbation, energy flow, nutrient cycling, pollination, and seed dispersal.

Bioturbation

Bioturbation is a process in which an organism influences the physical, chemical, and ecological dynamics of the base soil, sediment, or benthic habitat (Cortes-Gomez et al. 2015). Tadpoles are reliable agents for bioturbation in many freshwater forest ponds, lakes, rivers, and streams whereas adult burrowing frogs, snakes, turtles, and lizards change soils where they live with active digging and burrowing. In Panama, for example, tadpoles facilitate primary productivity in stream patches while consuming periphyton and sediments (Ranvestel et al. 2004). Without tadpoles, benthic sedimentation in streams increases and the functional structure of the trophic assemblage is affected (Ranvestel et al. 2004, Connelly et al. 2008, Colon-Gaud et al. 2010). Burrowing species of amphibians (some salamanders, many frogs, and all caecilians) and reptiles (some snakes, crocodilians, turtles, tuatara, and lizards) also create and maintain soil cavities, burrows, and increase soil turnover and nutrient availability in many forests (see Jones et al. 2006).

Energy flow

With their roles as both predator and prey, amphibians and reptiles are critical components in trophic interactions. Through predation or competition, amphibians and reptiles can regulate populations of many other organisms. Spielman and Sullivan (1974) and Rodríguez and González (2000) demonstrated that in the Bahamas and Cuba, tadpoles of *Osteopilus septentrionalis* regulate abundance of the mosquito *Culex quinquefasciatus.* Each tadpole consumes 13–21 mosquito larvae/day. Consequently, the mosquito is never abundant when the frog is present. Similarly, in Puerto Rico, Dial and Roughgarden (1995) suggested that *Anolis* lizards help reduce up to 20 per cent of damage to canopy leaves in the rainforest by preying on arthropods (Blattaria and Orthoptera).

As prey, amphibians and reptiles are high-quality sources of protein and other nutrients for other organisms (Burton & Likens 1975, Wells 2007, Collins & Crump 2009), especially because they often occur in high abundance and convert ingested energy into biomass efficiently (Boyd & Goodyear 1971; Wells 2007).

Both invertebrates (e.g., molluscs, crustaceans, arachnids, coleopterans, hymenopterans, orthopterans, hemipterans, and dipterans) and vertebrates (e.g., caecilians, other anurans, snakes, and tortoises) are predators on amphibians (e.g., see Cortez-Gomez et al. 2015). Many other vertebrates (e.g., fishes, birds, mammals, and even other species of amphibians and reptiles) are also predators for herpetofauna.

Nutrient cycling

A vital facet of supporting ecosystem services (see MEA 2005), nutrient cycling, is critical to forest structure and function. All amphibians and reptiles excrete wastes into their environments and these wastes play a role in forests' nutrient availability. For example, there are measurable decreases of potassium and phosphorus available for leaf litter decomposition when a single frog species (*Eleutherodactylus coqui*) is not present (Beard et al. 2003). Another example is tadpoles in Japan affecting leaf C:N ratios, and hence, both nutrient availability and decomposition (Iwai & Kagaya 2005).

Pollination

Pollination is a critically important process for plants. When animals are involved, it is almost always birds or insects. Although reptiles and amphibians are rarely recognized as pollinators, at least some are capable and important as pollinators. For example, herbivorous lizards can pollinate during foraging activities (Godínez-Álvarez 2004). Sazima et al. (2005, 2009) demonstrated that *Trachylepis atlantica* pollinates the Brazilian tree *Erythrina velutina* while drinking nectar, an important source of water and calories. Only recently have amphibians been added to the list of pollinators with another nectar-driven behavior by the Brazilian treefrog, *Xenohyla truncata* (de-Oliveira-Nogueira et al. 2023), also an active and important seed disperser (see below).

Seed dispersal

Seed dispersal is another crucial process shaping plant populations, forest succession, and heterogeneity and is one of the most important ecosystem services globally (Fall et al. 2023). Again, although the majority of amphibians and reptiles are carnivorous, a handful of species are herbivorous and help disperse the seeds of some plants. For example, Neotropical tortoises and the lava lizard *Tropidurus torquatus* eat many different fruits. As mentioned above, the treefrog *Xenohyla truncata* not only pollinates but also consumes fruits and flower parts of *Anthurium harrisii* and *Erythroxylum ovalifolium.* Studies also suggest that the frog defecates seeds in microhabitats, which increase germination rates (Da Silva et al. 1989, Fialho 1990, and Da Silva & Britto-Pereira 2006). Seed dispersal and germination of several plants in Neotropical arid forests have also been helped by a number of tortoise species (e.g., *Chelonoidis carbonaria, C. chilensis, C. denticulata, Rhinoclemmys funerea,* and *R. annulate*) (Falcón et al. 2020). The lava lizard, *Tropidurus torquatus,* disperses seeds for *Erythroxylum ovalifolium* and *Melocactus violaceus* (Fialho 1990), although it has been suggested that the lizard is highly variable in how effectively it disperses seeds into viable germination microhabitats (Cortes-Figueira et al. 1994).

Drivers of current extinction trends

Despite their clear important roles in forest ecosystems worldwide, herpetofauna are threatened globally with both population declines and extinction. Nearly 41 per cent of amphibians and 21 per cent of reptile species are threatened with extinction (Cox et al. 2022). Existing and emergent threats such as habitat loss, overexploitation, invasive species, climate change, and infectious diseases have synergistic effects on amphibian and reptile populations (Blaustein et al. 1995; Schlaepfer et al. 2005; Sodhi et al. 2008). Global biodiversity declines mainly stem from habitat loss and fragmentation (Haddad et al. 2015; Newbold et al. 2015), and it is also the greatest threat to amphibian and reptile populations (Cushman 2006; Todd et al. 2010).

Forest fragmentation and loss

Global deforestation and forest fragmentation occur mainly because of human activities, including land-use change (conversion to agriculture, orchard, or other land-use forms), fires, urbanization, and dam and road construction. Foraging and reproduction can be adversely affected by habitat loss (Irwin and Irwin 2006) and smaller fragments have fewer species (e.g., Bickford et al. 2010a). Large-scale deforestation destroys important microhabitats, like moist areas under fallen logs and leaf litter.

Logging threatens the Channing's toad (*Sclerophrys channingi*) in the eastern Lower Guinean rainforest in the Congo. Because of the conversion of forest into agricultural land in Morocco, *Testudo graeca* populations have experienced declines (Bayley and Highfield 1996). After a fire in the northern Zagros forest, reptile populations, species diversity, and richness all decreased, and a forest-dependent lizard (*Timon kurdistanica*) disappeared. Several small fossorial reptiles and salamanders cannot reach new suitable forest habitats by crawling through large deforested areas (Gibbons and Buhlmann 2001). The primary effects of roads on reptiles and amphibians include habitat loss, barrier effect, and mortalities caused by attempted road-crossing. For some forest-dwelling species, fragmentation of habitats by roads creates completely impassable barriers. For example, mass road kills in deciduous forests threaten local toads, *Bufo bufo* and *B. eichwaldi*, during the breeding season. Dams and reservoirs are also perilous for reptiles and amphibians in forest ecosystems. For example, the Chinese giant salamander has declined significantly because of dam building (Wang et al. 2004). Dams have also resulted in the isolation of *Neurergus* and *Salamandra* populations.

Invasive species

Globally, invasive alien species are considered the second greatest threat to biodiversity (Bellard et al. 2016). Cargo stowaways and pet trade escapes are primary pathways for introducing invasive reptiles and amphibians (Pitt et al. 2018). In addition, species are purposefully transported as biological control agents (e.g., cane toads in Australia) and sources of food (e.g., the red-eared slider and the American bullfrog). Five species of amphibians and reptiles are among the 30 vertebrate species ranked as the worst invasive species (Global Invasive Species Database, www.iucngisd.org/gisd/). Among the most well-known invasive species is the American bullfrog (*Lithobates catesbeianus*), which preys on native amphibians and reptiles and transmits fatal pathogens (Hulme 2009). The Coqui frog native to Puerto Rico has invaded Hawaii, Florida, the Dominican Republic, and many other areas. With the highest population density of any frog, it has altered ecosystem functions and impacted the invaded countries' economies (Pitt et al. 2018). Introducing the arboreal brown tree snake (*Boiga irregularis*) to Guam resulted in native forest bird and bat extinctions and a severe reduction of native lizard populations (Savidge 1987). Introducing mammalian predators such as Pacific rats (*Rattus exulans*) threatens the tuatara, endemic to New Zealand (Irwin and Irwin 2006). Invasive fire ants (*Solenopsis invicta*) prey on amphibian and reptile species and are capable of suppressing populations of native herpetofauna (Todd et al. 2008).

Overexploitation

Reptiles and amphibians are among the most trafficked groups of wildlife (Herrel and van der Meijden 2014). They are traded for traditional medicine, skin, meat, and as pets (Janssen 2021) both legally and illegally. Many forest-dwelling species are threatened by unsustainable overexploitation, both for local use and international trade. Forest reptiles, such as the Asian forest tortoise (*Manouria emys*) are consumed locally by humans as food. Like all forest turtles in Asia, one of the largest South American river turtles, the Arrau (*Podocnemis expansa*), is consumed unsustainably as eggs, hatchlings, and adults (Pritchard and Trebbau 1984). Amphibians inhabiting forests including Kaiser's mountain newt (*Neurergus kaiseri*), Chinese and Jiangxi giant salamanders (*Andrias davidianus* and *A. jiangxiensis*), Western Nimba toad (*Nimbaphrynoides occidentalis*), Monte Verde toad (*Incilius periglenes*), and Zetek's golden frog (*Atelopus zeteki*) are all listed in CITES Appendix I because they are threatened with extinction by international trade. Overcollection for food is one of the main threats to the Titicaca water frog (*Telmatobius culeus*) and Goliath frog (*Conraua goliath*), as well as most forest-dwelling large river frogs in Asia.

Climate change

Climate change affects both reptiles and amphibians because of their strong ties to water (mostly for amphibians) and their ectothermic life histories. Amphibian extinction is linked to epidemic disease triggered by large-scale warming (Pounds et al. 2019). Approximately one-third of herpetofauna have characteristics that make them very sensitive to climate change. In the Amazon basin, amphibians are highly vulnerable to climate change (Foden et al. 2013) and forest-dwelling lizards like *Gonatodes humeralis* and *Anolis fuscoauratus* may become locally extinct (Diele-Viegas et al. 2019) because of temperature increases. Many forest ectotherms are expected to be vulnerable to climate warming (see Bickford et al. 2010b; Huey et al. 2009).

Conservation

There is no substitute for primary forests when it comes to conserving and maintaining biodiversity (e.g., Gibson et al. 2011), including amphibians and reptiles. In our current global biodiversity crisis, most amphibians and reptiles are threatened and populations almost everywhere are declining. Thus, conserving all forests, especially tropical ones, is one of the highest biodiversity conservation priorities. Likewise, we need to restore habitats that have been degraded or lost, providing connectivity among the remaining forest fragments.

But it is not easy to conserve biodiversity on a planet with more than 8 billion humans. Appropriate planning is needed to successfully design conservation management strategies. This includes determining priority areas for conservation (Azevedo-Ramos and Galatti 2002). This, in turn, relies on reliable and accurate detailed information concerning taxonomy, species diversity, and distribution across particular geographic regions, including species richness, endemism, and similarity of faunal assemblages among sites (Azevedo-Ramos and Galatti 2002).

Amphibians and reptiles, however, also show remarkable resiliency to different drivers of extinction and population decline. Targeted strategies that focus on important habitat characteristics (e.g., breeding habitats, see Bickford et al. 2010 and Moor et al. 2022) could be effective short-term conservation solutions. This would still ensure essential conservation outcomes before more forest habitats need restoration. Likewise, provision and support for existing intact habitats (e.g., protected areas) can enable stable population regulation for reptiles like the iconic Komodo dragon (*Varanus komodoensis*) (Ardiantiono et al. 2018; Purwandana et al. 2020).

Captive breeding and reintroduction can also be potential solutions to avoid extinction, although monitoring the results of reintroduction is important to determine the effectiveness of the program (Canessa et al. 2016). For example, Harlequin toads (*Atelopus* species) bred in captivity in Colombia (Silva 2012), Ecuador (Coloma & Almeida-Reinoso 2012), Germany (Gawor et al. 2012), and the United States (Zippel 2002) have been viable for reintroduction back to their native habitats in Central America.

While determination of priority areas for conservation is crucial, successful conservation efforts will be more effective if corridors between conservation areas are identified and established. Although limited studies have been conducted to assess the effectiveness of newly established corridors, restoring habitat connectivity seems to be an effective approach for amphibian populations. Installing ways that amphibians can cross roads to restore connectivity between wetlands significantly decreases deaths (Scoccianti 2006). Additionally, road closures (e.g., Karthaus 1985), assisted crossing, and barrier fences (e.g., Feldman & Geiger 1989) during seasonal breeding migration of amphibians have successfully prevented mass mortality of amphibians in Germany.

Conclusions

Forest habitats are the most important ecosystems for most amphibians and reptiles. Reciprocally, both reptiles and amphibians are important parts of the communities and ecosystems where they live, central to energy flows and nutrient cycling. Surprisingly, herpetofauna species even provide pollination and seed dispersal services to forest trees. Hence, the forests need amphibians and reptiles just as much as they need forests.

References

Acebey, A., Gradstein, S. R. and Krömer, T. (2003) 'Species richness and habitat diversification of bryophytes in submontane rain forest and fallows of Bolivia', *Journal of Tropical Ecology*, vol. 19, pp. 9–18.

Aragón, P., Rodríguez, M.A., Olalla-Tárraga, M.A. and Lobo, J.M. (2010), 'Predicted impact of climate change on threatened terrestrial vertebrates in central Spain highlights differences between endotherms and ectotherms', *Animal Conservation*, vol. 13, pp. 363–373.

Ardiantiono, A., Jessop, T., Purwandana, D., Ciofi, C., Imansyah, M., Panggur, M., and Ariefiandy, A. (2018), 'Effects of human activities on Komodo dragons in Komodo National Park' *Biodiversity and Conservation*, vol. 27, pp. 3329–3347.

Azevedo-Ramos, C. and Galatti, U. (2002) 'Patterns of amphibian diversity in Brazilian Amazonia: conservation implications', *Biological Conservation*, vol. 103, pp. 103–111.

Bayley, J.R. and Highfield, A. C. (1996) 'Observations on ecological changes threatening a population of *Testudo graeca graeca* in the Souss Valley, Southern Morocco', *Chelonian Conservation and Biology*, vol. 2, pp. 36–42.

Beard, K. H., Eschtruth, A. K., Vogt, K. A., Vogt, D. J., and Scatena, F. N. (2003) 'The effects of the frog *Eleutherodactylus coqui* on invertebrates and ecosystem processes at two scales in the Luquillo Experimental Forest, Puerto Rico' *Journal of Tropical Ecology*, vol. 19, pp. 607–617.

Bellard, C., Cassey P. and Blackburn, T. M. 2016. Alien species as a driver of recent extinctions. *Biology Letters*, vol. 2, 20150623.

Bickford, D., Ng, T. H., Qie, L., Kudavidanage, E. P. and Bradshaw C. J. A. (2010a) 'Forest fragment and breeding habitat characteristics explain frog diversity and abundance in Singapore', *Biotropica*, vol. 42, pp. 119–125.

Bickford, D., Howard, S. D., Ng, D. J. J. and Sheridan, J. A. (2010b) 'Impacts of climate change on the amphibians and reptiles of Southeast Asia', *Biodiversity and Conservation*, vol. 19, pp. 1043–1062.

Blaustein, A. P., Beatty, J. J., Olson, D. H. and Storm, R. H. (1995) 'The biology of amphibians and reptiles in old-growth forests', in *The Pacific Northwest. Gen. Tech. Rep. PNW-GTR-337*, U.S. Department of Agriculture, Forest Service, Pacific Northwest Research Station, Portland.

Boyd, C.E. and Goodyear, C.P. (1971) 'Nutritive quality of food in ecological systems' *Archiv für Hydrobiologie*, vol. 69, pp. 256–270.

Burton T.M. and Likens G.E. (1975) 'Energy flow and nutrient cycling in salamander populations in the Hubbard Brook Experimental Forest, New Hampshire', *Ecology*, vol. 56, pp. 1068–1080.

Canessa, S., Genta, P., Jesu, R., Lamagni, L., Oneto, F., Salvidio, S., and Ottonello, D. (2016) 'Challenges of monitoring reintroduction outcomes: insights from the conservation breeding program of an endangered turtle in Italy', *Biological Conservation*, vol. 204, pp. 128–133.

Collins J.C. and Crump M.L. (2009) *Extinction in Our Times: Global Amphibian Decline*, Oxford University Press, Oxford.

Coloma, L.A. and Almeide-Reinoso, D. (2012) 'Ex situ management of five extant species of *Atelopus* in Ecuador—progress report', *Amphibian Ark Newsletter*, vol. 20, pp. 9–12.

Colon-Gaud, C., Whiles, M. R., Lips, K. R., Pringle, C. M., Kilham, S. S., Connelly, S., Brenes, R. and Peterson, S. D. (2010) 'Stream invertebrate responses to a catastrophic decline in consumer diversity', *Journal of the North American Benthological Society*, vol. 29, pp.1185–1198.

Connelly, S., Pringle, C. M., Bixby, R. J., Brenes, R., Whiles, M. R., Lips, K. R., Kilham, S. and Huryn, A. D. (2008) 'Changes in stream primary producer communities resulting from large-scale catastrophic amphibian declines: can small-scale experiments predict effects of tadpole loss?', *Ecosystems*, vol. 11, pp. 1262–1276.

Cortes-Figueira, L. E., Vasconcellos-Neto, J., Garcia, M. A. and Teixeira de Souza, A. L. (1994) 'Saurocory in *Melocactus violaceus* (Cactaceae)', *Biotropica*, vol. 26, pp. 295–301.

Cortés-Gomez, A. M., Ruiz-Agudelo, C. A., Valencia-Aguilar, A. and Ladle, R. J. (2015) 'Ecological functions of neotropical amphibians and reptiles: a review', *Universitas Scientiarum*, vol. 20, pp. 229–245.

Couvreur, T. L., Forest, F. and Baker, W. J. (2011) 'Origin and global diversification patterns of tropical rain forests: inferences from a complete genus-level phylogeny of palms', *BMC Biology*, vol. 9, pp. 1–12.

Cox, N., Young, B. E., Bowles, P., Fernandez, M., Marin, J., Rapacciuolo, G., Böhm, M., Brooks, T. M., Hedges, S. B., Hilton-Taylor, C. and Hoffmann, M. (2022) 'A global reptile assessment highlights shared conservation needs of tetrapods', *Nature*, vol. 605, pp. 285–290.

Cushman, S. A. (2006) 'Effects of habitat loss and fragmentation on amphibians: a review and prospectus', *Biological Conservation*, vol. 128, pp. 231–240.

Da Silva, H.R. and Britto-Pereira, M. C. (2006) 'How much fruit do fruit-eating frogs eat? An investigation on the diet of *Xenohyla truncata* (Lissamphibia: Anura: Hylidae)', *Journal of Zoology*, vol. 270, pp. 692–698.

Da Silva, H. R., De Britto-Pereira, M. C. and Caramaschi, U. (1989) 'Frugivory and seed dispersal by *Hyla truncata*, a neotropical tree frog', *Copeia*, vol. 1989, pp. 781–783.

de Frenne, P. Zellweger, F., Rodríguez-Sánchez, F., Scheffers, B. R., Hylander, K., Luoto, M., Velland, M., Verheyen, K. and Lenoir, J. (2019) 'Global buffering of temperatures under forest canopies', *Nature Ecology & Evolution*, vol. 3, pp. 744–749.

de-Oliveira-Nogueira, C. H., Souza, U. F., Machado, T. M., Figueiredo-de-Andrade, C. A., Mônico, A. T., Sazima, I., Sazima, M. and Toledo, L. F. (2023) 'Between fruits, flowers and nectar: The extraordinary diet of the frog *Xenohyla truncata*', *Food Webs*, vol. 35, e00281.

Dial, R. and Roughgarden, J. (1995) 'Experimental removal of insectivores from rain forest canopy: direct and indirect effects', *Ecology*, vol. 76, pp.1821–1834.

Diele-Viegas, L. M., Werneck, F. P. and Rocha, C. F. D. (2019) 'Climate change effects on population dynamics of three species of Amazonian lizards', *Comparative Biochemistry and Physiology Part A: Molecular & Integrative Physiology*, vol. 236, 110530.

Falcón, W., Moll, D. and Hansen, D. M. (2020) 'Frugivory and seed dispersal by chelonians: a review and synthesis', *Biological Reviews*, vol. 95, pp. 142–166.

Feldmann, R. and Geiger, A. (1989) 'Protection for amphibians on roads in Nordrhein-Westphalia' Proceedings of the Amphibians and Roads: Toad Tunnel Conference. Rendsburg, Federal Republic of Germany, pp. 51–57.

Fialho, R. F. (1990) 'Seed dispersal by a lizard and a treefrog-effect of dispersal site on seed survivorship', *Biotropica*, vol. 22, pp. 423–424.

Flecker, A. S., Feifarek, B. P. and Taylor, B. W. (1999) 'Ecosystem engineering by a tropical tadpole: density-dependent effects on habitat structure and larval growth rates', *Copeia*, vol. 1999, pp. 495–500.

Foden, W. B., Butchart, S. H. M., Stuart, S. N., Vié, J.-C., Akçakaya, H. R., Angulo, A., DeVantier, L. M., Gutsche, A., Turak, E., Cao, L., Donner, S. D., Katariya, V., Bernard, R., Holland, R. A., Hughes, A. F., O'Hanlon, S. E., Garnett, S. T., Sekercioglu, C. H. and Mace, G. M. (2013) 'Identifying the world's most climate change vulnerable species: a systematic trait-based assessment of all birds, amphibians and corals', *PLoS One*, vol. 8, e65427.

Frost, D. R. (2023) *Amphibian Species of the World: An Online Reference. Version 6.1*, https://amphibiansoftheworld.amnh.org/index.php. American Museum of Natural History, New York. doi.org/10.5531/db.vz.0001, accessed 10 October 2023.

García-Rodríguez, A., Martínez, P. A., Oliveira, B. F., Velasco, J. A., Pyron, R. A., and Costa, G. C. (2021) 'Amphibian speciation rates support a general role of mountains as biodiversity pumps', *The American Naturalist*, vol. 198, pp. E68–E79.

Gawor, A., Rauhaus, A., Karbe, D., VanDerStraeten, K., Lötters, S. and Ziegler, T. (2012) 'Is there a chance for conservation breeding? Ex situ management, reproduction, and early life stages of the harlequin toad *Atelopus flavescens* Duméril & Bibron, 1841 (Amphibia: Anura: Bufonidae)', *Amphibian & Reptile Conservation*, vol. 5, pp. 29–44.

Gibbons, J. W. and Buhlmann, K. A. (2001) 'Reptiles and amphibians', in J. G. Dickson (ed) *Wildlife of Southern Forests: Habitat and Management*, Hancock House Publishers, Surrey, British Columbia and Blaine.

Gibson, L., Lee, T. M., Koh, L. P., Brook, B. W., Gardner, T. A., Barlow, J., Peres, C. A., Bradshaw, C. J. A., Laurance, W. F., Lovejoy, T. E., and Sodhi, N. S. (2011) 'Primary forests are irreplaceable for sustaining tropical biodiversity', *Nature*, vol. 478, pp. 378–381.

Godínez-Álvarez, H. (2004) 'Pollination and seed dispersal by lizards: A review' *Revista Chilena de Historia Natural*, vol. 77, pp. 569–577.

Haddad, N.M., Brudvig, L.A., Clobert, J., Davies, K.F., Gonzalez, A., Holt, R.D., Lovejoy, T.E., Sexton, J.O., Austin, M.P., Collins, C.D. and Cook, W.M. 2015. 'Habitat fragmentation and its lasting impact on Earth's ecosystems', *Science Advances*, vol. 1, e1500052.

Herrel, A. and van der Meijden, A. (2014) 'An analysis of the live reptile and amphibian trade in the USA compared to the global trade in endangered species', *The Herpetological Journal*, vol. 24, pp. 103–110.

Huey, R. B., Deutsch, C. A., Tewksbury, J. J., Vitt, L. J., Hertz, P. E., Álvarez Pérez, H. J. and Garland Jr, T. (2009) 'Why tropical forest lizards are vulnerable to climate warming', *Proceedings of the Royal Society B: Biological Sciences*, vol. 276, pp. 1939–1948.

Hulme, P.E. (ed.). (2009) *Handbook of Alien Species in Europe (Vol. 569)*, Springer, Dordrecht.

Irwin, L. and Irwin, K. 2006. 'Global threats affecting the status of reptile populations', In *Toxicology of Reptiles*, Taylor and Francis, New York.

Iwai, N. and Kagaya, T. (2005) 'Growth of Japanese Toad (*Bufo japonicus formosus*) tadpoles fed different food items', *Current Herpetology*, vol. 24, pp. 85–89.

Janssen, J. (2021) 'A primer to the global trade of reptiles: magnitude, key challenges, and implications for conservation', in S. C. Underkoffler and H. R. Adams (eds) *Wildlife Biodiversity Conservation: Multidisciplinary and Forensic Approaches*, Springer International Publishing, Cham.

Jones, D. T., Loader, S. P. and Gower, D. J. (2006) 'Trophic ecology of east African caecilians (Amphibia: Gymnophiona), and their impact on forest soil invertebrates', *Journal of Zoology*, vol. 268, pp. 117–126.

Karthaus, G. (1985) 'Schutzmaßnahmen für wandernde Amphibien vor einer Gefährdung durch den Staßenverkehr - Beobachtungen und Erfahrungen', *Natur und Landschaft: Zeitschrift fur Naturschutz und Landschaftspflege*, vol. 60, pp. 242–247.

MacNeil, J., MacGowan, B. J., Currylow, A. and Williams, R. N. (2013) 'Forest management for reptiles and amphibians: A technical guide for the midwest', *Purdue Extension*, vol. 10 (2.1), pp. 1977–9842.

Millennium Ecosystem Assessment (MEA) (2005) 'Ecosystems and Human Well-being: Biodiversity Synthesis' World Resources Institute, Washington, DC.

Montaña, C.G., Silva, S. D. G. T. M., Hagyari, D., Wager, J., Tiegs, L., Sadeghian, C., Schriever, T. A. and Schalk, C. M. (2019) 'Revisiting "what do tadpoles really eat?" A 10-year perspective', *Freshwater Biology*, vol. 64, pp. 2269–2282.

Moor, H., Bergamini, A., Vorburger, C., Holderegger, R., Bühler, C., Egger, S. and Schmidt, B. R. (2022) 'Bending the curve: Simple but massive conservation action leads to landscape-scale recovery of amphibians', *Proceedings of the National Academy of Sciences*, vol. 119, e2123070119.

Moura, M. R., Argôlo, A. J. and Costa, H. C. (2017) 'Historical and contemporary correlates of snake biogeographical subregions in the Atlantic Forest hotspot', *Journal of Biogeography*, vol. 44, pp. 640–650.

Newbold, T., Hudson, L. N., Hill, S. L., Contu, S., Lysenko, I., Senior, R. A., Börger, L., Bennett, D. J., Choimes, A., Collen, B. and Day, J. (2015) 'Global effects of land use on local terrestrial biodiversity', *Nature*, vol. 520, pp. 45–50.

O'Conner, J. C., Dekker, S. C., Staal, A., Tuinenburg, O. A., Rebel, K.T., and Santos, M. J. (2021) 'Forests buffer against variations in precipitation', *Global Change Biology*, vol. 27, pp. 4686–4696.

Pitt, W. C., Beasley, J. and Witmer, G. W. (eds) (2018) *Ecology and Management of Terrestrial Vertebrate Invasive Species in the United States*. CRC Press, Boca Raton, FL.

Pounds, A., J., Bustamante, M. R., Coloma, L. A., Consuegra, J. A., Fogden, M. P., Foster, P. N., La Marca, E., Masters, K. L., Merino-Viteri, A., Puschendorf, R. and Ron, S. R. (2006) 'Widespread amphibian extinctions from epidemic disease driven by global warming', *Nature*, vol. 439, pp. 161–167.

Pritchard, P. C. and Trebbau, P. (1984) *The Turtles of Venezuela*, Soc. for the Study of Amphibians and Reptiles, Oxford, Ohio.

Purwandana, D., Imansyah, M. J., Ariefiandy, A., Rudiharto, H., Ciofi, C. and Jessop, T. S. (2020) 'Insights into the nesting ecology and annual hatchling production of the Komodo dragon', *Copeia*, vol. 108, pp. 855–862.

Rand, A. S. (1964) 'Ecological distribution in anoline lizards of Puerto Rico', *Ecology*, vol. 45, pp. 745–752.

Ranvestel, A. W., Lips K. R., Pringle C. M., Whiles M. R. and Bixby R. J. (2004) 'Neotropical tadpoles influence stream benthos: evidence for the ecological consequences of decline in amphibian populations', *Freshwater Biology*, vol. 49, pp. 274–285.

Reagan, D. P. (1995) 'Lizard ecology in the canopy of an island rainforest', in M. D. Lowman and N. M. Nadkarni (eds) *Forest Canopies*, Academic Press, San Diego, CA.

Rodríguez, R. J. and González, B. R. (2000) 'Evaluación de la capacidad depredadora de *Osteopilus septentrionalis* (Anura:Hylidae) sobre larvas de *Culex quinquefasciatus* (Diptera: Culicidae) en condiciones de laboratorio y de semicampo', *Boletín de Malariología y Saneamiento Ambiental*, vol. 11, p. 12.

Savidge, J. A. (1987) 'Extinction of an Island forest avifauna by an introduced snake', *Ecology*, vol. 68, pp. 660–668.

Sazima, I., Sazima, C. and Sazima, M. (2005) 'Little dragons prefer flowers to maidens: a lizard that laps nectar and pollinates trees', *Biota Neotropica*, vol. 5, pp. 185–192.

Sazima, I., Sazima, C. and Sazima, M. (2009) 'A catch-all leguminous tree: *Erythrina velutina* visited and pollinated by vertebrates at an oceanic island', *Australian Journal of Botany*, vol. 7, pp. 26–30.

Schlaepfer, M. A., Hoover, C. and Dodd, C. K. (2005) 'Challenges in evaluating the impact of the trade in amphibians and reptiles on wild populations', *BioScience*, vol. 55, pp. 256–264.

Scoccianti, C. (2006) 'Rehabilitation of habitat connectivity between two important marsh areas divided by a major road with heavy traffic', *Acta Herpetologica*, vol. 1, pp. 77–79.

Silva, C. (2012) 'A conservation program for *Atelopus* species at the Cali Zoo, Colombia', *Amphibian Ark Newsletter*, vol. 19, p. 7.

Sodhi, N. S., Bickford, D., Diesmos, A. C., Lee, T. M., Koh, L. P., Brook, B. W., Sekercioglu, C. H. and Bradshaw, C. J. A. (2008) 'Measuring the meltdown: drivers of global amphibian extinction and decline', *PLoS One*, vol. 3, e1636.

Spielman, A. and Sullivan, J. (1974) 'Predation on peridomestic mosquitoes by hylid tadpoles on grand Bahama island', *The American Journal of Tropical Medicine and Hygiene*, vol. 23, pp. 704–707.

Starnberger, I., Preininger, D. and Hödl, W. (2014) 'The anuran vocal sac: a tool for multimodal signalling' *Animal Behaviour*, vol. 97, pp. 281–288.

Todd, B. D., Rothermel, B. B., Reed, R. N., Luhring, T. M., Schlatter, K., Trenkamp, L. and Gibbons, J. W. (2008) 'Habitat alteration increases invasive fire ant abundance to the detriment of amphibians and reptiles', *Biological Invasions*, vol. 10, pp. 539–546.

Uetz, P., Freed, P, Aguilar, R., Reyes, F. and Hošek, J. (eds, 2022) 'The Reptile Database', www.reptile-database.org, accessed 28 February 2023.

Valencia-Aguilar, A., Cortés-Gómez, A. M. and Ruiz-Agudelo, C. A. (2013) 'Ecosystem services provided by amphibians and reptiles in Neotropical ecosystems', *International Journal of Biodiversity Science, Ecosystem Services & Management*, DOI: 10.1080/21513732.2013.821168

Wang, X. M., Zhang, K. J., Wang, Z. H., Ding, Y. Z., Wu, W. and Huang, S. (2004) 'The decline of the Chinese giant salamander *Andrias davidianus* and implications for its conservation', *Oryx*, vol. 38, pp. 197–202.

Wells, K. D. (2007) *The Ecology and Behavior of Amphibians*, The University of Chicago Press, Chicago, IL.

Wyman, R. L. (1998) 'Experimental assessment of salamanders as predators of detrital food webs: effects on invertebrates, decomposition and the carbon cycle', *Biodiversity and Conservation*, vol. 7, pp. 641–650.

Zippel, K. C. (2002) 'Conserving the Panamanian golden frog: Proyecto Rana Dorada', *Herpetological Review*, vol. 33, pp. 11–12.

21

GLOBAL PATTERNS OF BIODIVERSITY IN FORESTS

Christine B. Schmitt and João de Deus Vidal Jr.

Introduction

Forest ecosystems are of immense importance worldwide due to their high biodiversity and the ecosystem services they provide. They constitute some of the last large and hence unfragmented ecosystems of the world, yet some forest ecosystems are considered severely threatened. For example, although most of the global biodiversity hotspots harbour forest, only small areas of intact natural vegetation remain (Schmitt et al., 2009; Hansen et al., 2020).

According to the latest reports by the Food and Agriculture Organisation of the United Nations (FAO) and the United Nations Environment Programme (UNEP), forests occupy approximately 4.06 billion ha of the global land area, representing approximately 30.8 per cent of global land cover (FAO and UNEP, 2020). Forests also represent 6,630 of the 16,337 IUCN Key Biodiversity Areas, being the most prevalent and largest type of habitat classified under this framework (Birdlife International, 2023). Understanding the patterns of biodiversity in these areas is crucial for strategic conservation planning on a global scale and for monitoring global conservation targets. Furthermore, there is strong international interest in preventing forest degradation and reducing biodiversity loss, with several examples of global development goals focused on addressing this issue, for example, Sustainable Development Goal 15 and Goal A of the Kunming–Montreal Global Biodiversity Framework. Reducing deforestation and forest degradation is also directly related to the mitigation of global climate change due to their role in biosphere carbon fluxes (Schmitt, 2013). However, recent estimates reveal that over the past 30 years, nearly 420 million ha of forest have been converted into other types of land use and more than 80 million ha of primary forest areas have been lost (FAO and UNEP, 2020).

It is well known that for many groups of organisms, the global species richness is highest in the tropics and that tropical rainforests represent some of the most diverse ecosystems on Earth (Mace et al., 2005). The latitudinal diversity gradient is assumed to be positively correlated with area size, long-term environmental stability, functional diversity, kinetics, and productivity and negatively correlated with climatic variability (Kreft and Jetz, 2007; Brown, 2014). Evolutionary processes such as speciation and extinction rates and range

DOI: 10.4324/9781003324072-24

expansion may also play a crucial role (Brown, 2014; Lomolino et al., 2017). For vascular plants, a global analysis showed that the main community assembly processes, such as ecological drift, dispersal limitation, and priority effects, can also vary along the latitudinal gradient (Nishizawa et al., 2022). However, the latitudinal diversity gradient does not apply to all taxa and ecosystems. For instance, the species richness of conifers is high at mid-latitudes, and plant diversity is exceptionally high in the Southern African Cape region and the Mediterranean regions of Chile (Kreft and Jetz, 2007).

This chapter aims to look beyond the latitudinal diversity gradient and identify other vital patterns of biodiversity in forests worldwide. Species are useful units for the assessment of diversity (Mace et al., 2005); however, a comprehensive evaluation of global biodiversity patterns needs to consider more than just simple species numbers (e.g., De Kort et al., 2021; Pollock et al., 2020). Therefore, this chapter presents an overview of the criteria, indicators, and data commonly used to describe biodiversity, followed by a summary of the ecosystem classifications adopted widely in the reporting of global biodiversity patterns. Within this context, the main global patterns of biodiversity in forests are described. It should be noted that both the differences in biodiversity patterns between managed and unmanaged forests, as well as the threats and conservation issues, are beyond the scope of this chapter.

Biodiversity criteria and indicators

Biodiversity refers to the diversity of ecosystems, species, and genes, including functional diversity and phylogenetic diversity. It can be measured with a wide range of criteria and indicators on different taxonomic and spatial scales (Pollock et al., 2020). There is a plethora of available criteria to describe the biodiversity of a given area, but most of it can be summarized by variety, composition, and distribution (Table 21.1).

Biodiversity in terms of composition has generally been assessed in studies with smaller geographic scopes because data on species communities and individual species populations

Table 21.1 Biodiversity criteria and indicators

Criterion	*Indicator (examples)*
Variety	= *the number of different types*
Genetic	• number of alleles and allele combinations (genotypes)
Taxonomic*	• number of species, taxa, families and major lineages (orders)
Range-related*	• number of restricted range (endemic) and migratory species
Functional	• number of life forms, guilds and ecological groups
Composition	= *quantity and quality*
Populations	• population structure and genetic diversity of individual species
Communities	• species lists, presence/absence data, abundance data
Diversity indices	• Shannon, Jaccard, evenness
Distribution	= *spatial pattern*
Genetic	• maps of gene flow, evolutionary history and phylogeny
Variety*	• maps of species richness
Species*	• maps of the distribution of individual species and species groups
Ecological units*	• maps of ecological zones, habitat types, biomes and ecoregions

Source: Modified from Mace et al. (2005).

Note: * Frequently used in global-level analyses.

may be spatially biased or too specific for large-scale analyses (Valdez et al., 2023). Information on genetic diversity, functional variety, or ecological interactions has also been typically used in studies on regional scales, but there is now an increasing number of data available for global-scale analyses (De Kort et al., 2021; Nishizawa et al., 2022; Cornwell et al., 2019).

Analyses at the global level often define variety in terms of the number of species, that is, species richness, or the number of range-restricted species, also called endemics (e.g., Raven et al., 2020). This information is important because it helps to distinguish global areas with unique ecological features. However, it is crucial to note that species richness and endemism are scale-dependent variables, and patterns may vary in relation to the unit of analysis (Sabatini et al., 2022). Furthermore, species richness and distribution maps for particular taxa or species groups (for birds, see Figure 21.1) are common methods for illustrating global biodiversity patterns. Finally, the spatial distribution of ecological units, such as global forest types, is a useful indicator of biodiversity and ecological processes expected at a local level, especially where more comprehensive information on species numbers, identities, and ranges is lacking (Schmitt, 2013).

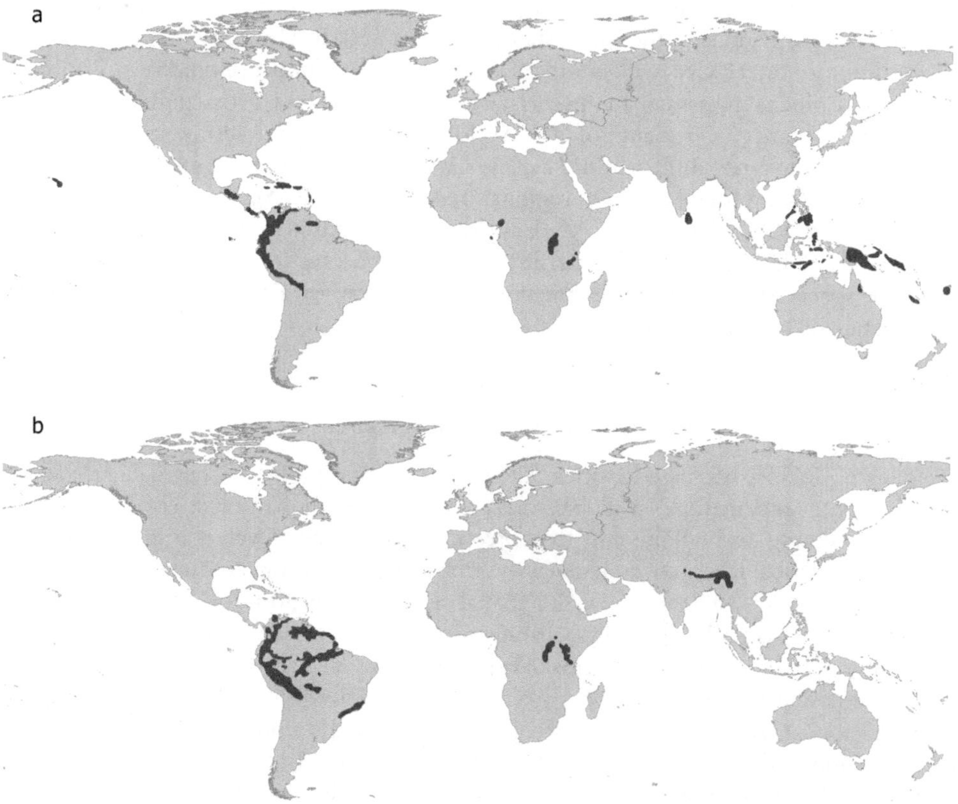

Figure 21.1 Global centres (shown in black) of bird species endemism (a) and richness (b) (adapted from Orme et al., 2005). Detailed methodology available in the original publication.

Global species data

Various organisations and databases make global data sets available for a wide range of individual species or species groups (Cornwell et al., 2019). The most renowned global database is the Red List of Threatened Species by IUCN, which provides free, curated, regular updates of global species data. The latest version of Red List (Version 2022.2) contains assessments of more than 150,300 species. Distribution maps based on known species occurrences and expert opinions are available for approximately 123,600 species, while information on population trends, ecological aspects, and threats to conservation are available for approximately 142,600 species (IUCN, 2023).

Despite the expanding database, the species covered by the IUCN Red List represent only a small fraction of the estimated 2.1 million known species described globally (IUCN, 2023). Furthermore, many species have not yet been scientifically documented (Liu et al., 2022). For example, incomplete inventories and taxonomic revisions are major problems for insects and fungi (García-Robledo et al., 2020; IUCN, 2023). The best available IUCN data for terrestrial vertebrates are for birds, mammals, amphibians, and reptiles, and include most species described globally for these taxa (Table 21.2). For plants, only conifer and cycad species have been fully assessed (IUCN, 2023). Even for well-documented taxa, the present distribution maps may not cover areas where species may be present but have not yet been recorded. Furthermore, data can be biased by significant variations in sampling intensity between different regions of the world (Cornwell et al., 2019; Hughes et al., 2021).

In addition to the IUCN Red List, there is an increasing number of biodiversity initiatives on the world wide web that collate species data for particular taxa, ecosystems or geographic regions. For instance, the Plants of the World database (https://powo.science.kew.org), the Global Biodiversity Information Facility (www.gbif.org/), the Integrated Digitized Biocollections (www.idigbio.org/), regional repositories like SpeciesLink (https://specieslink.net) and RAINBIO (https://gdauby.github.io/rainbio), and even citizen science platforms like iNaturalist (www.inaturalist.org) have become relevant clearing-house for worldwide species data from a wide range of sources. However, data quality control remains problematic (Zizka et al., 2020).

Distribution of the global forest ecosystems

To evaluate global patterns of species diversity in forests, global species data need to be combined with information on global forest distribution. Global forest cover maps based on remote sensing technology aim to represent actual forest cover at a given time. The latest series of global forest cover maps (www.globalforestwatch.org) show the global change in forest cover from 2000 to 2021 at a spatial resolution of 30 m (Hansen et al., 2013). However, these maps do not fully separate natural forests from managed forests, which makes an important difference in terms of expected forest diversity and impacts the ability to access their conservation status properly (Chiarucci and Piovesan, 2020).

In addition, there are global ecosystem templates based on predefined, spatially explicit ecological units. Although they can indicate the geographical location of the major forest ecosystems, they lack information on actual forest cover. For example, the World Wildlife Fund (WWF) developed the global framework of terrestrial ecoregions to prepare a biodiversity map reflecting the distribution of the natural species assemblages and communities of the earth before major changes in land use (Olson et al., 2001). It identifies eight

Table 21.2 Number of species of mammals, birds, amphibians, reptiles and conifers in the IUCN forest habitats

IUCN Habitat	*Mammals*		*Birds*		*Amphibians*		*Reptiles*		*Conifers*	
	(5,973 species)		*(11,188 species)*		*(7,486 species)*		*(10,222 species)*		*(610 species)*	
	No of species	*% of species*	*No of species*	*% of species*	*No of species*	*% of species*	*No of species*	*% of species*	*No of species*	*% of species*
(Sub)tropical moist lowland forest	2,543	43%	5,814	52%	3,515	46%	3,779	37%	114	19%
(Sub)tropical moist montane forest	1,860	31%	4,237	38%	3,667	49%	2,111	21%	189	31%
(Sub)tropical dry forest	1,151	19%	2,177	19%	487	7%	2,032	20%	24	4%
(Sub)tropical swamp forest	338	6%	842	8%	156	2%	200	2%	15	2%
(Sub)tropical mangrove forest*	222	4%	1,085	10%	14	< 1%	222	2%	1	< 1%
Temperate forest	794	13%	993	10%	704	9%	656	6%	321	53%
Boreal forest	116	2%	256	2%	24	< 1%	8	< 1%	16	3%
Subarctic forest	32	< 1%	8	< 1%	7	< 1%	2	< 1%	5	< 1%
Subantarctic forest	10	< 1%	8	< 1%	18	< 1%	2	< 1%	1	< 1%
All forest habitats	4,336	73%	8,667	77%	6,435	86%	6,604	64%	549	90%

Notes: The five taxa have been completely assessed globally (IUCN, 2023); total global species numbers in brackets. Percentage of species is defined as the percentage of all species in this group occurring in that particular habitat. Species can occur in more than one habitat (*IUCN Red List of Threatened Species*, Version 2022-2. www.iucnredlist.org, accessed 5 May 2023).

Subtropical/Tropical abbreviated as (Sub)tropical.

* (Sub)tropical mangrove forest vegetation above high tide level.

realms, 14 major biomes and 825 terrestrial ecoregions. The realms, such as the Neotropic and Afrotropic, represent a division of the landmasses of the world according to large-scale biogeographical differences in species distributions. Based on this biogeographical approach, the framework combines the tropics and subtropics and distinguishes azonal biomes shaped by edaphic and hydrological factors, for example, flooded grasslands and savannas, and mangroves. In contrast, the Global Ecological Zones, defined by the FAO, are based primarily on climatic factors (FAO and UNEP, 2020).

Combining forest cover maps and ecosystem templates can provide useful information on the distribution and status of global forest biodiversity. For example, Schmitt et al. (2009) used the 2005 MODIS Vegetation Continuous Field products, which represent forest cover as continuous tree cover densities from 0 to 100 per cent at a spatial resolution of 500 m. Pixels with more than 10 per cent tree cover were classified as forests to comply with international forest definitions such as those adopted by the FAO and the United Nations Framework Convention on Climate Change (UNFCCC). In the next step, the forest cover map was overlaid with the Global Land Cover 2000 product to exclude forests modified by humans, such as tree plantations and certain types of below agroecosystems (Schmitt et al., 2009). Furthermore, the overlay with the WWF terrestrial biomes framework showed that the tropical and subtropical (hereafter (sub)tropical) moist broadleaf forests comprised the largest forest cover globally, followed by the boreal forests/taiga. However, in 2005, forest cover was estimated only at 60 per cent in the (sub)tropical moist broadleaf forest biome and at 27 per cent in the (sub)tropical dry broadleaf forest biome (Figure 21.2), reflecting the massive and ongoing conversion of forests to other land uses (Potapov et al., 2022). It should be noted that grasslands also contained some forest cover at the selected 10 per cent tree cover threshold (Figure 21.2). This is particularly conspicuous in the (sub)tropical grasslands, savannas and shrublands because some savanna areas, for example, parts of the woody savannas in central and southern Africa and some patches of the Cerrado in South America, have a relatively closed tree cover, while maintaining a significant grass component (Schmitt, 2013).

Finally, the IUCN has been developing its own habitat classification scheme in an ongoing iterative process; it is not spatially explicit and requires expert knowledge in the assignment of species to habitats. The classification scheme differentiates 16 major terrestrial, marine and human-modified habitats based on biogeography and latitudinal zonation (IUCN, 2023). The 'forest habitat' includes nine types of forest (Table 21.2). Due to the ad hoc assignment of species records to different habitats, the IUCN scheme considers the actual habitat status but lacks a global map of the habitat distribution. Although there is relatively good agreement on the definition of boreal and temperate forest types between the WWF and IUCN classification systems, the subdivisions of (sub)tropical forests are not congruent (Figure 21.2 and Table 21.2). For example, the WWF emphasises the difference between

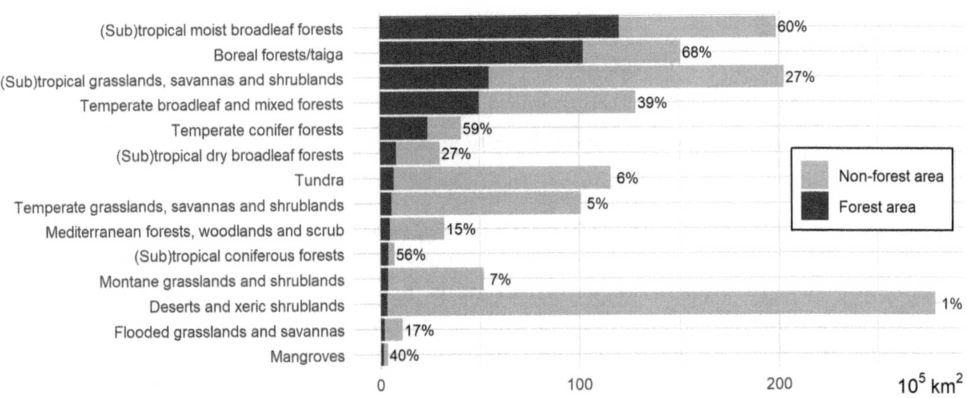

Figure 21.2 Forest and non-forest areas for the 14 WWF biomes (Olson et al., 2001). Forest area is predominantly natural forest at 10 per cent tree cover in the 2005 MODIS Vegetation Continuous Fields dataset; the Global Land Cover 2000 datasets were used to exclude non-natural forest areas such as tree plantations and some types of above agroecosystems (see Schmitt et al., 2009). Subtropical and Tropical is abbreviated as (Sub)tropical. Percentages indicate the proportion of forest cover in each biome.

broadleaf and coniferous forest biomes, whereas the IUCN highlights the difference between lowland and montane forest habitats.

Global patterns of biodiversity in forests

Global maps of species richness for vascular plants, birds, mammals and amphibians clearly indicate the well-known concentration of species numbers in the tropics (Orme et al., 2005; Jenkins et al., 2013; Kreft and Jetz, 2007). Generally, the biome of (sub)tropical moist broadleaf forests has a much higher species richness and endemism than all other biomes. In addition, many tropical rainforest taxa have a long independent evolutionary history, which is a measure of taxonomic uniqueness and irreplaceable genetic diversity (Mace et al., 2005). It could be that the outstanding species diversity of this biome is related to its large area size (see Figure 21.2); however, species richness and area are not statistically related for the 14 terrestrial biomes (Mace et al., 2005).

The enormous diversity in the tropical rainforest is illustrated by the 473 tree species recorded in 1-ha plots in Amazonian Ecuador (diameter at breast height (dbh) ≥ 5 cm) (Valencia et al., 1994). Generally, the number of tree species in 1-ha plots in the Upper Amazon forests can vary between 149 and 307 species (dbh ≥ 10 cm) (Valencia et al., 1994). The lowland Amazon is estimated to harbour at least 16,000 tree species (Ter Steege et al., 2013), which is in stark contrast to the 53 tall tree species that are known to occur in central European broadleaf and mixed forests (Leuschner and Ellenberg, 2017). Conifers are an exception because they have the highest species richness in the temperate forest habitat (Table 21.2).

Tropical tree communities usually harbour some dominant species. For example, ter Steege et al. (2013) estimated that 227 tree species (1.4 per cent of all tree species) account for half of all trees in the Amazon. Similarly, only four woody species (1.5 per cent of all woody species) constitute more than 50 per cent of all woody individuals in the moist montane forest of Ethiopia (Figure 21.3). In addition, there is a much higher proportion of small range or rare species in tropical tree communities than in communities at higher latitudes; a pattern explained to some extent by niche specialisation and dispersal mechanisms (Engel et al., 2022; Brown, 2014). This is illustrated by the long tail of species in the ranked species-abundance distribution (Figure 21.3). For high-latitude assemblages, a similar graph would typically be steep and approximately linear (Brown, 2014).

Similarly to plants, the number of animal species is higher in the biome of (sub)tropical moist broadleaf forests than in all other biomes (Mace et al., 2005). The latest IUCN data for mammals, birds and amphibians illustrate the large number of species in the (sub)tropical forest habitats (Table 21.2). For example, 86 per cent of global amphibian species occur in forest habitats, and amphibians are clearly most diverse in the (sub)tropical moist lowland and montane forests. In contrast, mammals and birds are more diverse in moist lowland forests than in moist montane forests, and a considerable number of these species occur in dry forests (Table 21.2). In fact, the biome of (sub)tropical dry broadleaf forests holds many endemic species (Mace et al., 2005). Individual dry forest areas are highly variable in forest structure and floristic diversity, with up to 121 woody species in a 0.1-ha plot in Colombia (Gillespie et al., 2000). Therefore, considering the relatively small global area of (sub)tropical dry forests which still remains (Figure 21.2), this forest habitat harbours a substantial portion of the world's biodiversity.

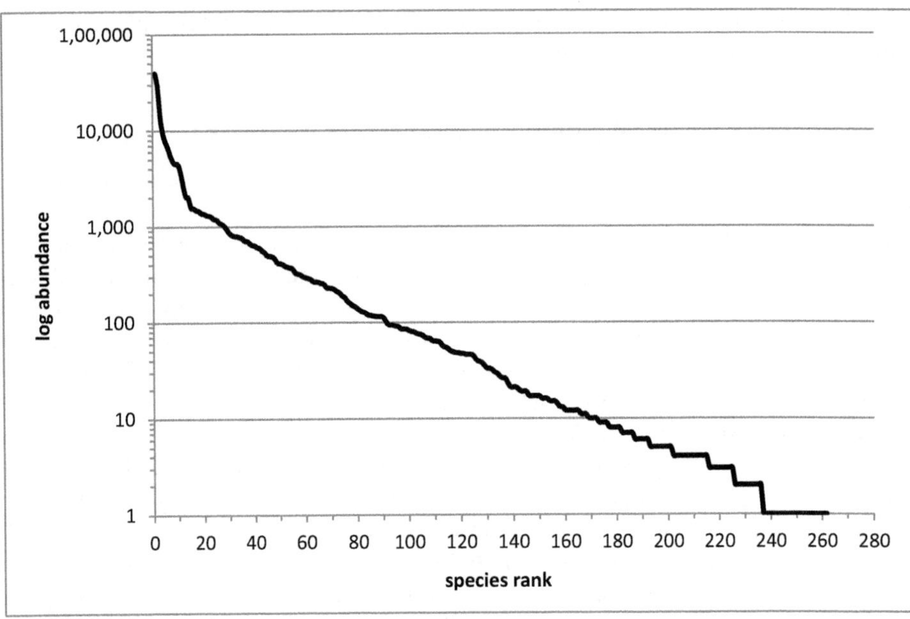

Figure 21.3 Ranked species-abundance distributions for 262 woody species in the Ethiopian moist montane forests (logarithmic scale). Abundance is defined as the number of individuals for trees, shrubs and woody lianas with height $\geq$ 0.5 m in 180 study plots (20m × 20m) (see Schmitt et al., 2013). Four species account for more than 50 per cent of all woody individuals.

Although both species richness and endemism peak in the tropics, their global patterns are not congruent. As a general rule, the global centres of species richness are primarily located on the continents, whereas the centres of endemism are more strongly correlated with large tropical islands and island archipelagos (Orme et al., 2005; Jenkins et al., 2013). For birds, the centres of species richness are concentrated in the moist forest of the Amazon, the Brazilian Atlantic Forest and the continental (sub)tropical mountain areas (Figure 21.1). The same areas are important centres of diversity for vascular plants and mammals; moreover, these taxa are very species rich on some of the southeast Asian islands (Jenkins et al., 2013; Kreft and Jetz, 2007). Although endemism is high on large tropical islands for birds and mammals, continental mountain areas such as Mount Cameroon and the eastern Arc in Africa are also important (Jenkins et al., 2013). On the contrary, amphibian species mostly have small ranges, and thus there are no major centres of endemism globally (Jenkins et al., 2013). The Andes show an outstanding congruence of species richness and endemism for plants and many vertebrate species, which is partly explained by steep altitudinal and climatic gradients that stimulate high species turnover (Perrigo et al., 2020).

Vast differences are apparent in the distribution ranges and endemism patterns of vertebrates and invertebrates. However, the species data for the latter are still extremely sparse (Mace et al., 2005; García-Robledo et al., 2020). There are also striking differences in diversity patterns between vascular plants and bryophytes, especially with regard to the wide distribution ranges and relatively low spatial genetic structure of bryophytes observed on the global scale (Patiño and Vanderpoorten, 2018). Although earlier studies reported an inverse latitudinal gradient of species richness for mosses, with the exception of tropical mountain ranges and cloud forests, this finding needs to be rectified, as taxonomic revisions are ongoing and the sampling effort has been increasing in the tropics (Patiño and Vanderpoorten, 2018).

In addition to global variations in species richness and endemism, there are large differences in species diversity between biogeographic realms for each forest biome. For example, the Neotropical rainforests have a much higher plant species richness (approximately 93,500 species) than rainforests in the Asia-Pacific region (approximately 61,700 species) and the African tropics (approximately 20,000 species) (Corlett and Primack, 2011). The richness of local species also differs with an average of approximately 180, 150 and 95 large tree species per hectare in the Neotropical, Asian, and African rainforests, respectively (Corlett and Primack, 2011). While the Neotropical rainforests are known for their high species richness in birds, bats, amphibians, and bromeliad epiphytes, the Afrotropical rainforests have the highest browser diversity, and the Southeast Asian forests exhibit the highest diversity in gliding animals and dipterocarp tree species (Corlett and Primack, 2011; Jenkins et al., 2013). The underlying reasons for these variations in species diversity include continental drift over the past 200 million years, the glacial cycles during the Pleistocene and current climate patterns (Corlett and Primack, 2011; Lomolino et al., 2017).

In temperate regions, beech forests cover large areas of the Palearctic, Nearctic and Australasia, but differ in the dominant genus, that is, *Fagus* in the northern hemisphere and *Nothofagus* in the south (Lomolino et al., 2017). Furthermore, with a total of around 124 tree species, the temperate broadleaf and mixed forests of eastern North America are more species-rich than those of Central Europe (n = 53) (Leuschner and Ellenberg, 2017). In Europe, mountain ranges such as the Alps blocked the latitudinal shift of species in response to cooler temperatures. This led to multiple species extinction events, whereas in Northern America, species shifts were facilitated by north–south running rivers and mountain ranges (Lomolino et al., 2017). Furthermore, warmer summer temperatures in eastern North America today support a higher species richness than in Europe.

In boreal forests of Europe, Asia and North America, species typically have wider distribution ranges, and species assemblages are relatively homogeneous across larger regions (Nishizawa et al., 2022). In the WWF ecoregions framework, this is reflected by the large size of the northern ecoregions. These areas were delineated with an emphasis on dynamics and processes, such as variations in climate, fire disturbance regimes and large vertebrate migrations (Olson et al., 2001). Conifers are the dominant species in northern boreal forests, where they cover vast areas with only five to ten species co-occurring in units of 10,000 km^2 (Mutke and Barthlott, 2005). On the contrary, the global centres of conifer species richness are located in temperate and (sub)tropical forest areas (Table 21.2), especially in southeast Asia, western North America and Mexico (Mutke and Barthlott, 2005). Temperate conifer forests are recognised as an individual biome by the WWF due to their unique species assemblages (see Figure 21.2). In particular, temperate rainforests along the coast of western North America, New Zealand and Chile are dominated by huge, slow-growing trees. These forests also have high epiphyte diversity and many endemic species (Lomolino et al., 2017). Similarly, the Mediterranean forests, woodlands and shrubs cover only a small area globally (Figure 21.2) but are recognised for their distinct plant species diversity.

There are notable variations in forest diversity along the elevation gradient. In temperate forests, broadleaf tree species are replaced by conifers at higher elevations (Olson et al., 2001), while temperate coniferous forests can extend as far south as southern Mexico in mountain areas (Lomolino et al., 2017). Individual (sub)tropical moist montane forests are global centres of species richness and can harbour a substantial number of endemic species along altitudinal gradients (Perrigo et al., 2020). They often form part of biodiversity hotspots, for example, in the Western Ghats and Sri Lanka, the tropical Andes, and the mountains of southwest China (Schmitt et al., 2009). The forests in the eastern Arc Mountains of Tanzania and Kenya support at least 96 species of endemic vertebrate and 800 species of endemic vascular plants (Burgess et al., 2007). This exceptional diversity cannot be explained by a single mechanism; more recent speciation due to heterogeneous habitats, a minimal influence of Pleistocene climatic changes and long evolutionary histories in isolation seem to play a role. Ethiopian moist montane forests (Figure 21.3) are floristically related to the eastern Arc forests but have fewer endemic species (Schmitt et al., 2013). Overall, global (sub)tropical moist montane forests support large proportions of the worldwide vertebrate diversity (Table 21.2), although their area is much smaller than that of global (sub)tropical moist lowland forests.

Finally, soil type can strongly influence species diversity in forests worldwide. For example, tree diversity in the Amazon is partly determined by soil type (Ter Steege et al., 2013). Furthermore, the highly diverse (sub)tropical moist broadleaf forests biome comprises areas of relatively low plant species richness, such as (sub)tropical freshwater and peat swamp forests. An assessment of tree species diversity (dbh ≥ 10cm) in 1-ha plots revealed that the mean species richness in Ecuador's swamp forest was 74 as opposed to 239 in adjacent non-flooded forest (Pitman et al., 2014). Southeast Asian peat swamp forests and well-drained forests comprised between 30 and 122 species and between 100 and 290 species per hectare, respectively (Posa et al., 2011). Fewer tree species are presumably related to problems of dispersal, germination, establishment and growth caused by seasonal extremes in water levels (Corlett and Primack, 2011). Mangrove forests also contain relatively low plant species richness and are formed by a highly specialised group of trees and shrubs that thrive in saline soils subject to tidal flooding. However, (sub)tropical mangrove and swamp forests are crucial habitats for a diverse range of bird species worldwide (Table 21.2). In temperate broadleaf and mixed forests, the diversity of species, especially understorey and herbaceous plants, can be highly variable depending on local climate, soil type and fire frequency (Lomolino et al., 2017).

Conclusions

In recent years, we have seen a significant increase in novel databases for global forest and biodiversity information ranging from species geographic records and life history traits to open access satellite data. However, integrating and analysing these datasets with different scope and resolution still presents a major challenge today (Pollock et al., 2020). Furthermore, there are still vast gaps in our knowledge of particular taxa, especially invertebrates and fungi that are important components of forest biodiversity (IUCN, 2023).

There has been much emphasis internationally on the outstanding biodiversity of tropical rainforests, which are included within the (sub)tropical moist broadleaf forests biome in biogeographical terms. This review shows that not all tropical rainforests are equally rich in species and that there are significant variations in species diversity between rainforests in the Neotropics, the Afrotropics and the Asian-Pacific region. Moreover, a focus on total species numbers is bound to overlook globally unique forest ecosystems other than the (sub-)tropical moist forests. For instance, locally, the (sub)tropical dry broadleaf forests, Mediterranean forests, woodlands and scrub, and the temperate coniferous forest biomes can host a large number of species, including many endemics. Although boreal forest ecosystems are relatively poor in species, they are functional in a harsh environment and support exceptional ecological processes such as large vertebrate migrations.

Most studies on global diversity patterns of particular species or species groups do not consider the spatial distribution of forest cover and forest ecosystems. Even if forest ecosystems are considered, the prevalence of different global classification systems complicates comparisons of biodiversity patterns, especially for the (sub)tropical forests. Therefore, the clear documentation of the source and spatial resolution of species data, forest cover data and forest classification systems are crucial when comparing studies on forest biodiversity. Finally, considering the ongoing high rates of global deforestation and forest degradation, as well as the prevalent impacts of climate change, it remains more important than ever to strengthen global efforts to protect forest ecosystems and their diversity throughout the world.

References

Birdlife International (2023) 'The World Database of Key Biodiversity Areas. Developed by the KBA Partnership', www.keybiodiversityareas.org, accessed 16 May 2023.

Brown, J.H. (2014) 'Why are there so many species in the tropics?', *Journal of Biogeography*, vol. 41, pp. 8–22.

Burgess, N.D., Butynski, T.M., Cordeiro, N.J., Doggart, N.H., Fjeldsa, J., Howell, K.M., Kilahama, F.B., Loader, S.P., Lovett, J.C., Mbilinyi, B., ... Stuart, S.N. (2007) 'The biological importance of the Eastern Arc Mountains of Tanzania and Kenya', *Biological Conservation*, vol. 134, pp. 209–231.

Chiarucci, A. and Piovesan, G. (2020) 'Need for a global map of forest naturalness for a sustainable future', *Conservation Biology: The Journal of the Society for Conservation Biology*, vol. 34, pp. 368–372.

Corlett, R.T. and Primack, R.B. (2011) *Tropical Rain Forests. An Ecological and Biogeographical Comparison*, Wiley-Blackwell, Oxford, UK.

Cornwell, W.K., Pearse, W.D., Dalrymple, R.L. and Zanne, A.E. (2019) 'What we (don't) know about global plant diversity', *Ecography*, vol. 42, pp. 1819–1831.

De Kort, H., Prunier, J.G., Ducatez, S., Honnay, O., Baguette, M., Stevens, V.M. and Blanchet, S. (2021) 'Life history, climate and biogeography interactively affect worldwide genetic diversity of plant and animal populations', *Nature Communications*, vol. 12, p. 516.

Engel, T., Blowes, S.A., McGlinn, D.J., Gotelli, N.J., McGill, B.J. and Chase, J.M. (2022) 'How does variation in total and relative abundance contribute to gradients of species diversity?', *Ecology and Evolution*, vol. 12, e9196.

FAO and UNEP (2020) *The State of the World's Forests 2020*, FAO and UNEP.

García-Robledo, C., Kuprewicz, E.K., Baer, C.S., Clifton, E., Hernández, G.G. and Wagner, D.L. (2020) 'The Erwin equation of biodiversity: From little steps to quantum leaps in the discovery of tropical insect diversity', *Biotropica*, vol. 52, pp. 590–597.

Gillespie, T.W., Grijalva, A. and Farris, C.N. (2000) 'Diversity, composition, and structure of tropical dry forests in Central America', *Plant Ecology*, vol. 147, pp. 37–47.

Hansen, M.C., Potapov, P.V., Moore, R., Hancher, M., Turubanova, S.A., Tyukavina, A., Thau, D., Stehman, S.V., Goetz, S.J., Loveland, T.R., ... J. R. G. Townshend (2013) 'High-resolution global maps of 21st-century forest cover change', *Science*, vol. 342, pp. 850–853.

Hansen, M.C., Wang, L., Song, X.-P., Tyukavina, A., Turubanova, S., Potapov, P.V. and Stehman, S.V. (2020) 'The fate of tropical forest fragments', *Science Advances*, pp. 1–9.

Hughes, A.C., Orr, M.C., Ma, K., Costello, M.J., Waller, J., Provoost, P., Yang, Q., Zhu, C. and Qiao, H. (2021) 'Sampling biases shape our view of the natural world', *Ecography*, vol. 44, pp. 1259–1269.

IUCN (2023) *The IUCN Red List of Threatened Species*. Version 2022-2, www.iucnredlist.org/, accessed 5 May 2023.

Jenkins, C.N., Pimm, S.L. and Joppa, L.N. (2013) 'Global patterns of terrestrial vertebrate diversity and conservation', *Proceedings of the National Academy of Science USA*, vol. 110, pp. e2602–e2610. doi:10.1073/pnas.1302251110

Kreft, H. and Jetz, W. (2007) 'Global patterns and determinants of vascular plant diversity', *Proceedings of the National Academy of Science USA*, vol 104, pp. 5925–5930.

Leuschner, C. and Ellenberg, H. (2017) *Ecology of Central European Forests*, Revised and extended version of the 6th German edition, Springer International, Cham.

Liu, J., Slik, F., Zheng, S. and David B. Lindenmayer, D. B. (2022) Undescribed species have higher extinction risk than known species. *Conservation Letters*, vol. 15, e12876.

Lomolino, M.V., Riddle, B.R. and Whittaker, R.J. (2017) *Biogeography. Biological Diversity Across Space and Time*, Fifth edition, Sinauer Associates Inc. Publishers, Sunderland, MA.

Mace, G., Masundire, H., Baillie, J., Ricketts, T., Brooks, T., Hoffmann, M., Stuart, S., Balmford, A., Purvis, A., Reyers, B., ... Williams, P. (2005) *Biodiversity, Ecosystems and Human Well-being: Current State and Trends*, Volume 1, Millennium Ecosystem Assessment, Island Press, Washington, DC.

Mutke, J. and Barthlott, W. (2005) 'Patterns of vascular plant diversity at continental to global scales', *Biologiske Skrifter*, vol. 55, pp. 521–531.

Nishizawa, K., Shinohara, N., Cadotte, M.W. and Mori, A.S. (2022) 'The latitudinal gradient in plant community assembly processes: A meta-analysis', *Ecology Letters*, vol. 25, pp. 1711–1724.

Olson, D.M., Dinerstein, E., Wikramanayake, E.D., Burgess, N.D., Powell, G.V.N., Underwood, E.C., D'Amico, J.A., Itoua, I., Strand, H.E., Morrison, J.C., ... Kassem, K.R. (2001) 'Terrestrial ecoregions of the world. A new map of life on earth', *BioScience*, vol. 51, pp. 933–938.

Orme, D.L., Davies, R.G., Burgess, M., Eigenbrod, F., Pickup, N., Olson, V.A., Webster, A.J., Ding, T.-S., Rasmussen, P.C., Ridgely, R.S., ... Owens, I.P.F. (2005) 'Global hotspots of species richness are not congruent with endemism or threat', *Nature*, vol. 436, pp. 1016–1019.

Patiño, J. and Vanderpoorten, A. (2018) 'Bryophyte biogeography', *Critical Reviews in Plant Sciences*, vol. 37, pp. 175–209.

Perrigo, A., Hoorn, C. and Antonelli, A. (2020) 'Why mountains matter for biodiversity', *Journal of Biogeography*, vol. 47, pp. 315–325.

Pitman, N.C.A., Andino, J.E.G., Aulestia, M., Cerón, C.E., Neill, D.A., Palacios, W., Rivas-Torres, G., Silman, M.R. and Terborgh, J.W. (2014) 'Distribution and abundance of tree species in swamp forests of Amazonian Ecuador', *Ecography*, vol. 37, pp. 902–915.

Pollock, L.J., O'Connor, L.M.J., Mokany, K., Rosauer, D.F., Talluto, M.V. and Thuiller, W. (2020) 'Protecting biodiversity (in all its complexity): new models and methods', *Trends in Ecology & Evolution*, vol. 35, pp. 1119–1128.

Posa, M.R.C., Wijedasa, L.S. and Corlett, R.T. (2011) 'Biodiversity and conservation of tropical peat swamp forests', *BioScience*, vol. 61, pp. 49–57.

Potapov, P., Hansen, M.C., Pickens, A., Hernandez-Serna, A., Tyukavina, A., Turubanova, S., Zalles, V., Li, X., Khan, A., Stolle, F., ... Kommareddy, A. (2022) 'The Global 2000–2020 Land Cover and Land Use Change Dataset derived from the Landsat Archive: first results', *Frontiers in Remote Sensing*, vol. 3, 856903.

Raven, P.H., Gereau, R.E., Phillipson, P.B., Chatelain, C., Jenkins, C.N. and Ulloa Ulloa, C. (2020) 'The distribution of biodiversity richness in the tropics', *Science Advances*, vol. 6, eabc6228.

Sabatini, F.M., Jiménez-Alfaro, B., Jandt, U., Chytrý, M., Field, R., Kessler, M., Lenoir, J., Schrodt, F., Wiser, S.K., Arfin Khan, M.A.S., ... Bruelheide, H. (2022) 'Global patterns of vascular plant alpha diversity', *Nature Communications*, vol. 13, 4683.

Schmitt, C.B. (2013) 'Global tropical forest types as support for the consideration of biodiversity under REDD+', *Carbon Management*, vol. 4, pp. 501–517.

Schmitt, C.B., Burgess, N.D., Coad, L., Belokurov, A., Besançon, C., Boisrobert, L., Campbell, A., Fish, L., Gliddon, D., Humphries, K., ... Winkel, G. (2009) 'Global analysis of the protection status of the world's forests', *Biological Conservation*, vol. 142, pp. 2122–2130.

Schmitt, C.B., Senbeta, F., Woldemariam, T., Rudner, M. and Denich, M. (2013) 'Importance of regional climates for plant species distribution patterns in moist Afromontane forest', *Journal of Vegetation Science*, vol. 24, pp. 553–568.

Ter Steege, H., Pitman, N.C.A., Sabatier, D., Baraloto, C., Salomao, R.P., Guevara, J.E., Phillips, O.L., Castilho, C.V., Magnusson, W.E., Molino, J.-F., ... Silman, M.R. (2013) 'Hyperdominance in the Amazonian Tree Flora', *Science*, vol. 342. doi:10.1126/science.1243092

Valdez, J.W., Callaghan, C.T., Junker, J., Purvis, A., Hill, S.L.L. and Pereira, H.M. (2023) 'The undetectability of global biodiversity trends using local species richness', *Ecography*, vol. 2023, e06604.

Valencia, R., Balslev, H. and Paz y Miño C, G. (1994) 'High tree alpha-diversity in Amazonian Ecuador', *Biodiversity and Conservation*, vol. 3, pp. 21–28.

Zizka, A., Antunes Carvalho, F., Calvente, A., Rocio Baez-Lizarazo, M., Cabral, A., Coelho, J.F.R., Colli-Silva, M., Fantinati, M.R., Fernandes, M.F., Ferreira-Araújo, T., … Antonelli, A. (2020) 'No one-size-fits-all solution to clean GBIF', *PeerJ*, vol. 8, e9916.

PART IV

ENERGY AND NUTRIENTS

22
MYCORRHIZAL SYMBIOSIS IN FOREST ECOSYSTEMS

Leho Tedersoo

Introduction

Mycorrhiza is a term for the symbiotic interaction between plant roots and fungal mycelium in which the tissues of both partners are specifically differentiated for improved exchange of carbon and nutrients. Plants provide photosynthetically fixed carbon to fungi in exchange for water-dissolved nutrients (Smith and Read, 2008). Mycorrhizal fungi are more efficient than plant roots and root hairs in capturing dissolved mineral nutrients due to their 10-fold smaller diameter compared to root hairs and more plasticity in growth and branching. Mycorrhizal symbiosis is not always beneficial for growth, because both partners optimize their individual benefits, and there are other benefits not related to growth per se. In natural conditions, 85–92 per cent of vascular plant species are typically colonized by mycorrhizal fungi (Brundrett and Tedersoo, 2018), and this association is obligatory for most mycorrhizal plants and fungi, depending on the type of mycorrhiza.

The benefits of mycorrhizal symbiosis are many-fold for both plants and fungi (Smith and Read, 2008). In addition to enhanced nutrition, mycorrhizal fungi protect plants from hazardous chemicals (Polycyclic Aromatic Hydrocarbons (PAHs), allelochemicals, etc.) and heavy metals by immobilizing these molecules. Mycorrhizal fungi may also protect plants from various pathogens and herbivores by alerting neighbouring plants via hyphal connections, inducing systemic resistance, producing antibiotics in the soil environment, and occupying the physical niche for pathogens (Whipps, 2004). Deep-rooted trees and mycorrhizal fungi collectively retain critical water potentials in soil to ameliorate drought stress (Gehring et al., 2017). In addition, the roots of perennial plants and trees provide a relatively stable habitat compared to decomposing litter. Plants and fungi collectively defend the mycorrhizal root system from grazers and pathogens by the use of chemical weapons.

Types of mycorrhiza

Based on the anatomy of symbiosis, four principal types of mycorrhiza are distinguished, viz. arbuscular mycorrhiza (AM), ectomycorrhiza (EcM), ericoid mycorrhiza (ErM) and orchid mycorrhiza (OM) (Figure 22.1). Taxonomically distinct groups of plants and fungi are associated with different types of mycorrhiza. These types are widely distributed in various

DOI: 10.4324/9781003324072-26

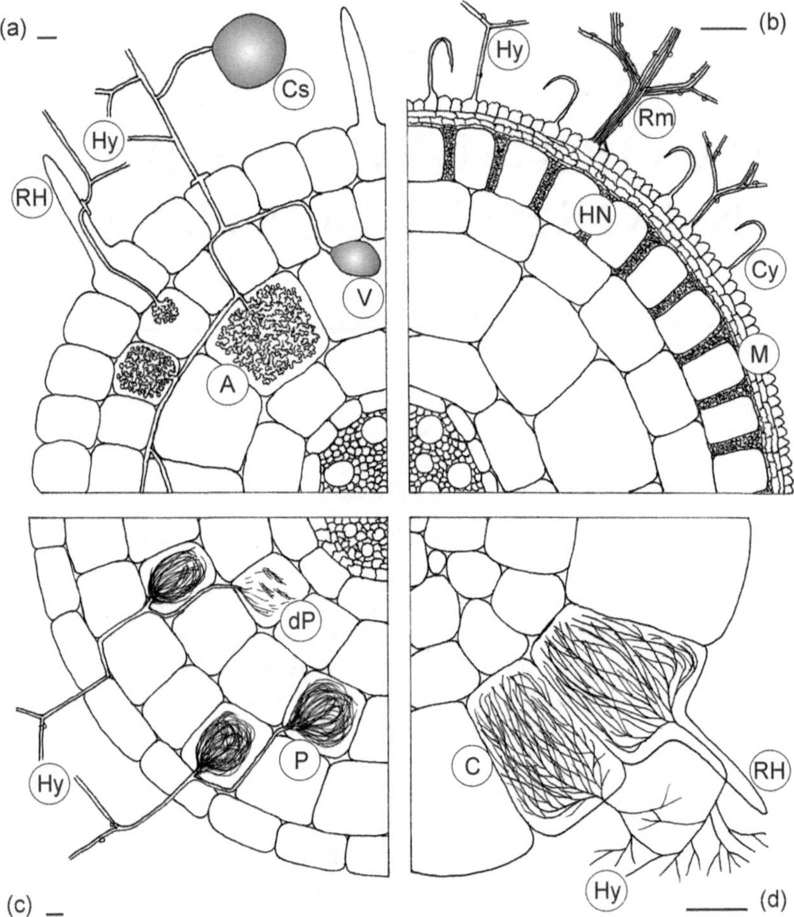

Figure 22.1 Structural differences among four major types of mycorrhiza: (a) arbuscular mycorrhiza; (b) ectomycorrhiza; (c) orchid mycorrhiza; (d) ericoid mycorrhiza. Abbreviations: A, arbuscules; V, vesicles; RH, root hair; Hy, extraradical hyphae; Cs, chlamydospores; M, mantle; HN, Hartig net; Cy, cystida; Rm, rhizomorphs; P, pelotons; dP, digested pelotons; C, coils. Bar, 20 μm.

ecosystems and continents, except Antarctica. The associated fungi differ in their access to organic nutrients (Figure 22.2).

In arbuscular mycorrhiza, fungal colonization of plant roots is mostly intracellular. The bush-like intracellular structures termed ***arbuscules*** constitute the interface for nutrient transfer between the symbionts. In fungi, AM associations have at least twice in the ancestors of the fungal phylum Glomeromycota and class Endogonomycetes of Mucoromycota (Soudzilovskaia et al., 2020). In vascular plants, the AM condition is ancestral, and currently around 72 per cent of all plant species are obligately AM and 7 per cent are facultatively AM. Arbuscular mycorrhiza is predominant in herbs and in tropical plants, including most tropical trees.

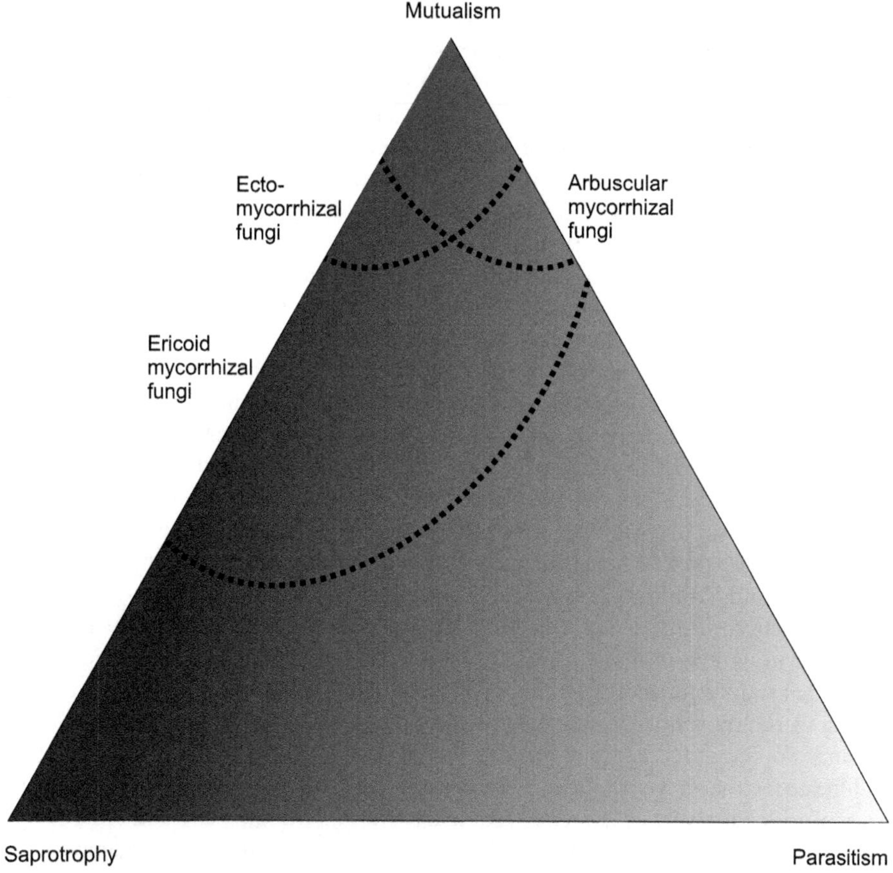

Figure 22.2 Relative placement of mycorrhizal fungi in the parasitism–mutualism–saprotrophy continuum (above dashed line). Orchid mycorrhizal fungi may belong to all types of trophic, depending on the plant species.

Ectomycorrhiza is characterized by the presence of a *fungal mantle*, which covers the root tips, and a *Hartig net* that constitutes plant root epidermal cells densely surrounded by fine, highly branched hyphae. Hartig net plays a key role in nutrient exchange between the partners. Ectomycorrhiza has evolved >20 times repeatedly in AM or secondarily non-mycorrhizal (NM) plants in the Carboniferous (Pinaceae) to Tertiary. In fungi, EcM root associations have independently evolved in approximately 80 lineages in the same time frame (Tedersoo, 2017; Sheikh et al., 2022). EcM fungi commonly produce *rhizomorphs*—hyphal strands that are efficient in soil exploration and nutrient transportation. EcM fungi include most soil-inhabiting macrofungi and truffles. Ectomycorrhiza occurs in 2 per cent of all vascular plant species, but these plants include ecologically and economically important timber species in the northern temperate (Pinaceae, Fagaceae, Salicaceae), tropical (Dipterocarpaceae, Caesalpinioideae of Fagaceae) and southern temperate (Myrtaceae, Nothofagaceae) ecosystems (Tedersoo, 2017). EcM associations are obligatory to all fungal partners involved and to a vast majority of plant partners. Certain plants (e.g., Salicaceae, Myrtaceae and various Australian shrubs) associate with both AM and EcM fungi (Teste et al., 2020).

In ericoid mycorrhiza, fungal colonization occurs exclusively inside root cells, where fungi form clew-like structures termed *coils*. Ericoid mycorrhiza has evolved in Ericaceae and the associated fungi mostly belong to Helotiales, Chaetothyriales and Sebacinales that also act as root endophytes (asymptomatic root-associated fungi) in non-Ericaceae plants and as peat and humus saprobes. Whereas ErM fungi are facultative symbiotrophs, the ericaceous host plants have obligatory associations. ErM accounts for 1.5 per cent of plant species due to extensive speciation in the Ericaceae family.

Orchid mycorrhizal symbiosis is characterized by dense intracellular colonization of root cells by fungal hyphae that form clew-like structures termed *pelotons*. Upon ageing, the pelotons are digested by the plant cells. The OrM association is restricted to orchids (10 per cent of all vascular plants), most of which associate with saprotrophic fungi from the Ceratobasidiaceae, Tulasnellaceae and Serendipitaceae families of the Basidiomycota. Certain terrestrial orchids have shifted to associate with typical EcM fungi, leading to partial or full loss of photosynthesis capacity in many unrelated orchid groups that parasitize mycorrhizal networks. Besides orchids, partly and fully mycoheterotrphic plants include several minor taxonomic groups that associate with EcM fungi (Pyroleae and Monotropeae tribes of Ericaceae) or AM fungi (mostly tropical groups, e.g., certain members of the Burmanniaceae and Triuridaceae families). Because myco-heterotrophic associations are certainly non-beneficial to fungi and it is difficult to reliably measure costs and benefits in associations involving green orchids, consideration of OrM and mycoheterotrophic associations as mycorrhizal is questionable (Merckx, 2013).

Because of the morphoanatomical features of the mycorrhiza structures as well as evolutionary and trophic differences among the associated fungi, mycorrhiza types exhibit functional differences that have important implications for their physiology and functioning at the scale of individuals to ecosystems. Compared to AM fungi, EcM and ErM fungi have limited enzymatic capacities to release micro- and macronutrients from organic matter and to take up simple organic compounds such as amino acids and oligopeptides. EcM fungi acidify the soil, slow down nutrient cycling and reduce the activities of soil-borne pathogens compared to AM symbiosis (Tedersoo & Bahram, 2019). The EcM and ErM symbioses are also associated with greater soil carbon sequestration compared to the AM symbiosis (Soudzilovskaia et al., 2019; Ward et al., 2021).

Distribution of mycorrhizal plants and fungi

The distribution of mycorrhizal types has conspicuous patterns on a global scale and across local nutrient gradients (Read et al., 2004; Tedersoo, 2017). Arctic grasslands and heathlands are dominated by ErM shrubs (*Empetrum, Cassiope, Vaccinium*), EcM shrubs (*Salix*, *Betula*, *Dryas)* and herbs (*Bistorta vivipara*) and NM sedges (*Carex*). Very similar plant communities occur above the treeline in the Alps, the Himalayas, the Rocky Mountains and many other northern mountain chains as relics of historical glacial connections. The subartic and subalpine forest tundra is dominated by small EcM trees and their bush-like forms (Betulaceae, Salicaceae and Pinaceae) with an understorey of ErM plants. The same EcM and ErM groups also dominate in the vast belt of boreal forests. The AM understorey plants become increasingly abundant southwards and in more productive habitats.

Temperate forests are dominated by both EcM (Fagaceae, Pinaceae, Betulaceae, Salicaceae, Tiliaceae, etc.) and AM (Oleaceae, Ulmaceae, Rosaceae, Sapindaceae, Lauraceae, etc.) trees with nearly exclusively AM understorey in deciduous forests and AM-ErM understorey in coniferous forests. The same plant families with many additional groups of tropical origin are also common in subtropical forests, where AM usually gains dominance over EcM in lowland habitats.

In tropical lowland and montane rainforests, most of the trees and understorey are AM. The relative abundance of various AM plant families and particularly EcM families strongly differs among continents mainly because of historical biogeographic reasons rather than soil properties. Of tropical EcM plants, Dipterocarpaceae are the most important plant family in South Asia, where this group often forms monodominant stands in lowland habitats in both moist and dry tropical forest biomes. EcM Fagaceae are common in tropical montane forests in both Southeast Asia and Central America. Throughout tropical lowland rainforests of Central and South America, EcM *Coccoloba* (Polygonaceae), *Neea* and *Guapira* (Nyctaginaceae) are widely distributed but usually occur in low numbers as understorey trees. In coastal areas, *Coccoloba* species may form monodominant stands on sandy soils. The South American EcM tree legume genera *Dicymbe* and *Aldina* are mostly distributed in seasonally flooded areas on white sands, where they may gain monodominance. African woodlands receiving <600 mm annual rainfall are composed exclusively of AM vegetation, with a diverse ErM flora in Cape province and montane forests. In semi-open miombo woodlands, EcM vegetation is clearly dominant and is represented by the *Berlinia* and *Afzelia* groups of the Fabaceae and *Uapaca* (Uapacaceae). Both groups may form monodominant stands in miombo woodlands, but also in rainforests of the Congo basin, which is mostly dominated by AM vegetation. In tropical and subtropical Australia, vast semiarid areas are dominated by EcM–AM dual mycorrhizal Myrtaceae (*Eucalyptus*, *Melaleuca*), Fabaceae (*Acacia*) and Casuarinaceae. These taxa are present as a minor component in Australian remnant rainforest patches, which are dominated by AM. Due to its geological history and position on the migration routes, the island of New Guinea has floristic elements of both Australia and Southeast Asia including the EcM Dipterocarpaceae, Fagaceae, Nothofagaceae and Myrtaceae, and diverse ErM shrubs. Temperate forests of the Southern Hemisphere are dominated by the EcM Nothofagaceae and *Eucalyptus* in Australia, whereas Nothofagaceae dominate in temperate montane forests of New Zealand and southern South America, but the Drankensberg mountains in southernmost Africa are EcM-free. However, the vast majority of tree genera including Podocarpaceae and Araucariaceae gymnosperms and the forest understorey is AM in southern temperate forests. ErM Ericaceae are also common in the understorey of southern temperate forests, tropical montane rain forests and above the tree line. In both temperate and tropical montane forests, Ericaceae and Orchidaceae are common as epiphytes, and their mycorrhizal symbionts acquire nutrients from bark and accumulating debris. Subantarctic islands are covered by AM and NM herbs and a few ErM shrubs. The lack of EcM vegetation in many remote islands can partly be explained by a stronger dispersal limitation.

At the landscape scale, soil properties drive the relative competitive abilities of mycorrhizal types and their prevalence (Read et al., 2004). In the most nutrient-poor habitats, ErM and NM plants dominate on organic and mineral soils, respectively. Non-mycorrhizal cluster roots developed by Cyperaceae and Proteaceae may enable sufficient nutrient uptake in ultrapoor soils (Lambers et al., 2008). At less severe nutrient limitation, EcM associations appear to be more beneficial compared to AM due to the acquisition of simple organic nutrients and the formation of rhizomorphs for efficient nutrient transportation. These benefits are, however, metabolically costly to plants and therefore AM plants tend to dominate in more productive and phosphorus-rich soils where competition for light becomes increasingly important (Read et al., 2004; Lambers et al., 2008).

Throughout their biogeographic range, EcM trees tend to form monodominant stands more commonly than AM trees (Torti et al., 2001). Such a clumped distribution of conspecific individuals poses certain benefits and risks. Monodominance enhances inbreeding because pollen and seed dispersal among patches is limited due to geographic distances. Furthermore, plant species accumulate taxon-specific pathogens that decrease the establishment of conspecific seedlings (Liu et al., 2012). However, monodominance has also certain benefits, such as more efficient signalling to trigger mast fruiting at irregular time intervals to reduce the load of pathogens. Many monodominant EcM plants exhibit recalcitrant litter, and by positive feedback from mycorrhizal fungi, monodominant plants may secure internal cycling of nutrients (Newbery et al., 1997). The presence of an abundant mycorrhizal inoculum may outweigh the negative feedback from soil pathogens in EcM plants.

Functional role of mycorrhizal fungi in forest ecosystems

Mycorrhizal fungi play a fundamental role in nutrient cycling in forest ecosystems. Forest ecosystems are limited by nitrogen, phosphorus or both of these key elements. To acquire these nutrients, one quarter to two-thirds of the primary production is allocated belowground. Up to 20 per cent of the fixed carbon is supplied to EcM fungi (Hobbie, 2006), which account for up to 40 per cent of soil microbial biomass, especially in low-nutrient soils (Högberg et al., 2010). These large amounts of carbon are required for the development and maintenance of extensive extraradical mycelium and fruit-bodies, and production of extracellular exudates that are composed of degradation enzymes, organic acids and simple sugars. A wide variety of mycorrhizosphere protists and prokaryotes relies on these compounds for their nutrition and in return these organisms fix dinitrogen, synthesize vitamins, signal molecules, enzymes and allelochemicals that suppress pathogens (Bueé et al., 2009). These heterotrophic microbes, EcM fungi and roots are estimated to account for approximately 60, 25 and 15 per cent of soil respiration, respectively (Heinemeyer et al., 2007). After death, the components of the recalcitrant cell walls of EcM and ErM fungal hyphae and glycoproteins produced by AM fungi account for much of the soil humus (Clemmensen et al., 2013).

Mycorrhizal fungi and plants are integrated by multiple links to the metanetworks of ecosystems (Figure 22.3; Simard et al., 2013). In addition to providing exudates to mycorrhizosphere microbial consortia, both plants and fungi provide a rich food source for direct consumers. Plants are grazed by a variety of herbivorous animals above- and belowground and degraded by pathogenic and saprotrophic fungi. Hyphae of mycorrhizal fungi are attacked by enzymes of mycoparasitic bacteria and other fungi. Fungal mycelium is actively grazed by soil mesofauna such as nematodes, mites and springtails. In addition to these organisms, fungal fruit-bodies are consumed by larvae of gnats and flies, beetles, snails, rodents and large ungulates such as elk and wild boar. Invertebrates and mammals provide an active means of dispersal for the fungi as many spores remain intact or their germination is triggered by passing through the digestive tract. Moreover, ornamented spores are easily adhered to the fur, appendices or slime of soil animals.

Mycorrhizal fungi connect multiple conspecific and heterospecific tree individuals of various ages via hyphal networks, which has inspired the *Avatar* movie. These networks

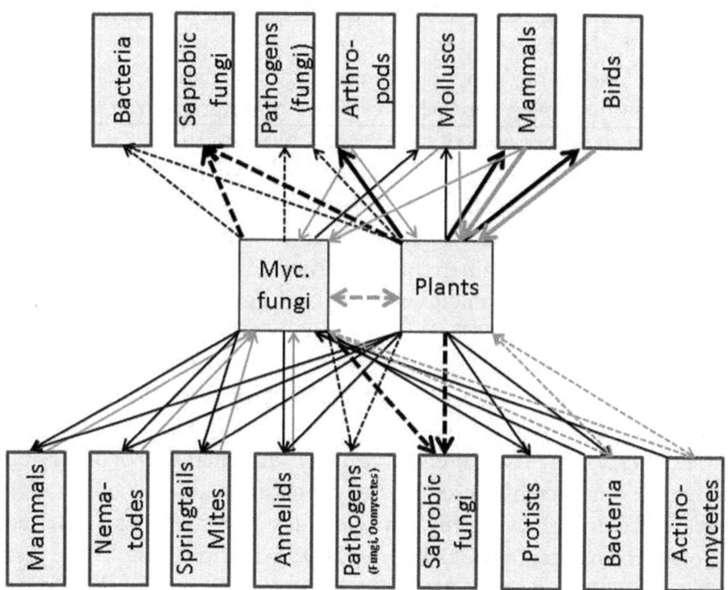

Figure 22.3 Simplified structure of metanetworks in forest ecosystems emphasizing on the relationships of mycorrhizal fungi and plants with other major organism groups aboveground (top) and belowground (bottom). Interactions among other organisms are excluded for clarity. Arrowheads indicate the benefit of the relationship. Thick lines indicate the ecologically most important relationships. Black solid lines, food web interactions; black dashed lines, decomposition; grey solid lines, dispersal; grey dashed lines with double arrows, mutualistic nutrient exchange.

facilitate carbon-to-nutrient trade between plants and fungi and ensure the stability of symbiosis (Kiers et al., 2011). Besides carbon transfer from plants to fungi, a substantial proportion of carbon is re-transferred to other trees (Klein et al. 2016). Tree-derived carbon may also move from parent trees to their offspring (kin) rather than genetically distantly related seedlings, suggesting kin selection (Simard et al., 2018). Although the ecological relevance of tree-to-tree carbon transfer remains a matter of debate, mycorrhizal networks sustain various mycoheterotrophs, proving that such reversed carbon flow may occur.

Forest disturbance

The effects of natural disturbance and forest management on mycorrhizal associations strongly depend on the type and severity of disturbance, mycorrhiza types, particular ecosystem, and interaction of these factors. Forests are naturally affected by various disturbance events such as wildfire, herbivore attack and pathogen outbreak. The magnitude of these effects depends on ecosystem resilience, that is, adaptation of organisms to these disturbance events by self-protection and regeneration ability. Disturbance usually results in the death or weakening of plants that limits photosynthesis and thus alters nutritional balance with mycorrhizal fungi belowground. Typically, loss of aboveground biomass results in restructuring carbon flow to rapid regeneration aboveground, which causes a reduction of mycorrhizal biomass and activity (Štursova et al., 2014).

Intensive fires are one of the most destructive types of natural disturbance that destroy most or all aboveground biomass and the forest floor, and significantly increase soil pH. Heat itself probably kills up to 90 per cent of mycorrhizal biomass and the rest is subject to be lost within one to two years, depending on the resprouting ability of vegetation, a widespread adaptation in fire-prone ecosystems. Fires induce seedling establishment from the dormant seed bank that rapidly develop mycorrhizal associations with mycelium that has persisted in deeper soil or simultaneously germinated from spores. Similarly to plants, many fungi adapt to periodic fires by evolving thickened cell walls and germination triggered by heat or an increase in soil pH (Peay et al., 2009). Wildfires may result in a nearly complete turnover of the resident fungal communities, depending on ecosystem adaptation, intensity, and interval of fires. Low-intensity periodical fires often have a negligible effect on soil fungal communities in fire-adapted ecosystems.

Pathogens and herbivores may alter the quality and quantity of litter input. Reduced photosynthesis results in decrease in plant and fungal reproductive effort and a loss in soil fungal biomass, disfavouring mycorrhizal species that produce large amounts of mycelium and are thus costly to maintain (Kuikka et al., 2003). Harmful soil-borne pathogens, such as *Heterobasidion parviporum* and *Phytophthora cinnamomi* or bark beetles may cause the death of entire patches of trees. Because the root systems of adult trees extend >20 m, soils under such patches probably retain some of the mycorrhizal biomass, but relative proportion compared to saprotrophs and opportunistic pathogens is reduced (Štursova et al. 2014).

Waterlogging of soil may result from annual periodic rain events, snow melt, construction work, etc. Excess water reduces the oxygen content and periodically causes the soil habitat to become anaerobic. Mycorrhizal fungi seem to be relatively tolerant of waterlogging because they can withstand periodic inundation. Many wetland plants have mechanisms to provide oxygen belowground to ameliorate anaerobiosis. Persistent waterlogging of mesic forests usually results in high mortality of vegetation that clearly has a detrimental effect on symbionts belowground.

Harvesting of trees has severe impacts on mycorrhizal fungi due to the disruption of a carbon source, exposition of the forest floor to solar radiation and disturbance to the soil. The extent of disturbance depends on the time and machinery used for harvesting. In clear-cuts, the vast majority of mycorrhizal organisms usually die within one year of harvesting. The remaining seedlings are unable to sustain the energy-demanding mycorrhizal networks. Thus, they retain the depauperate mycota comprised of fungi that require low amounts of carbon (Durall et al., 2005). After harvesting, soil water circulation is substantially altered due to soil exposure to the sun, highly reduced transpiration flow and disrupted deep water uptake. This usually results in standing water in humid sites and excessive soil desiccation in dry sites. The altered water regime may have severe consequences for natural and assisted forest regeneration. The deposition of harvesting residues, especially large-diameter dead wood on the forest floor, is useful for maintaining soil moisture and continuation of soil development. However, piles of debris may hamper the regeneration of shade-intolerant trees and seedling recruitment of small-seeded plants. The preservation of retention trees is very important for securing seed source of selected tree species and for maintaining both the soil water balance and mycorrhizosphere organisms in the soil (Durall et al., 2005). Tropical rainforests are especially intolerant to clear-cutting due to rapid loss of soil to erosion and reduction of humidity in tree canopies that is required for optimal functioning of the trees and the retention of epiphyte and animal communities. Highly valued EcM tropical timber trees such as Dipterocarpaceae are particularly sensitive to heavy cutting and disturbance due to their patchy occurrence and rapid loss of vitality of EcM mycelium in soil. For AM trees, the consequences of harvesting are usually less severe because the understorey vegetation usually provides an abundant mycorrhizal inoculum. Terrestrial species of Ericaceae also regenerate rapidly, because they obtain their suitable mycorrhizal fungi from the soil and roots of neighbouring plants.

Global change

Industrialization has led to environmental pollution with greenhouse gases, plastic litter, excess nutrients and various hazardous chemicals including heavy metals. Altered air and water temperature affects precipitation patterns that cause shifting of biomes and result in the extinction of specialist organisms that are unable to migrate across fragmented landscapes at the pace of climate change. Changes in both temperature and precipitation patterns are predicted to result in further aridification of relatively dry southern temperate and subtropical areas that will become more susceptible to fire and lose tree cover. Due to warming and humidification, boreal forests will expand greatly into northern arctic and tundra ecosystems. Changes in climate and land use have resulted in shifts in dominant mycorrhizal types (Soudzilovskaia et al., 2019). For example, aridification has led to the contraction of EcM-dominated oak savannas in North Africa and miombo woodlands in North-Central and South-Central Africa, and their replacement with dry AM savanna in the Holocene (Linder, 2001), with these processes currently accelerating. Similarly, in Mediterranean biomes globally, ErM Ericaceae and AM plants of semideserts will probably expand at the expense of EcM-dominated tree communities (Tedersoo, 2017). Conversely, EcM plants appear to gain nutritional benefits over AM plants in northern temperate forests (Terrer et al., 2016).

To physiologically cope with changing climate, mycorrhizal systems optimize their nutrient allocation, whereas plants select the most tolerant or efficient fungal genotypes. In most ecosystems, global warming is expected to enhance primary production, which renders soil nutrients more limiting to growth. Experiments manipulating global change on a small scale have revealed a speeding up of carbon and nitrogen cycling belowground partly due to activities of mycorrhizal fungi and greater allocation to mycorrhizal biomass (Johnson et al., 2013). However, multiple global change drivers may synergistically hamper soil functioning (Rillig et al., 2019).

Mycorrhiza applications in forestry

The application of mycorrhizal fungi has received great attention in both agriculture and forestry. Inoculation with mycorrhizal fungi promotes the growth and survival of seedlings, especially when planted in harsh site conditions or in non-native soil (Hoeksema et al., 2018; Averill et al., 2022). In practice, the greatest growth benefits of inoculation occur with EcM trees, because their natural inoculum in the soil is patchy and degrades rapidly after severe disturbance such as clear-felling. Plantations of several economically important EcM tree families such as Pinaceae and Dipterocarpaceae may completely fail without pre-inoculation. Inoculation with EcM fungi is usually performed by adding spore slurry or pieces of axenically grown mycelium to sown seeds or young containerized seedlings. Compressed beads consisting of dried spores are commercially available for inoculation programmes. However, only a few fungal taxa are commonly used for afforestation / reforestation. The selected strains may not be well suited to a particular host or environment. For example, the same fungal inoculum has been frequently used to produce EcM seedlings of both pines, eucalypts, and dipterocarps in southern Asia. These tree species do not co-occur naturally and neither do their fungi. On non-native hosts, mycorrhiza development is usually poor, and the association may be only partly functional, resulting in no or limited benefits (Brundrett et al., 2005). Therefore, it is recommended to use multiple strains for inoculation and to let both tree genotypes and the environment select the best suitable fungi. Long-term monitoring programmes have revealed that the originally inoculated fungi may be replaced by indigenous species within a few years after planting. Nurseries considering fungal inoculation have to compromise between fertilization and aboveground biomass on one hand and mycorrhization and normal root development on the other hand. Reduced fertilization results in longer rotation period, but healthier seedlings that are better prepared for re-planting and have thus greater potential for survival. Inoculation with forest soil not only results in good mycorrhiza development, but also co-introduces soil-borne pathogens.

Throughout the 20th century, vast areas of tropical and subtropical land have been replanted with exotic EcM forestry trees, mostly fast-growing pine, eucalypt or wattle species for pulp and construction wood. As trees were usually imported as containerized seedlings, they retained their mycorrhizal fungi from the potting soil. In some regions of the world, these co-introduced mycorrhizal fungi are surprisingly diverse because of multiple introductions from various sources. Co-introduced mycorrhizal fungi probably play a key role in the invasiveness of EcM trees in several regions such as New Zealand, South Africa, and Central South America (Traveset and Richardson, 2014). Invasive EcM trees and their symbiotic fungi alter soil nutrient cycling and function by reducing the rate of litter decomposition, which also results in greater flammability. Deep-rooted eucalypts, oaks and their mycorrhizal fungi may cause desiccation of topsoil due to more efficient water uptake and greater transpiration rates compared with native vegetation. The altered soil conditions, allelochemicals and dramatically increased fire frequency reduce the establishment of native vegetation and fauna that are not adapted to intensive fires. Invasive AM vegetation has usually lower impact on the ecosystem services because these plants link to native AM fungal networks. While most mycorrhiza inoculation applications aim to improve plant productivity, mushroom production of mycorrhizal fungi is becoming increasingly important. Edible mushrooms, in particular truffles, represent a high-revenue crop that rewards long-term investments into forestry plantations. Natural crops of highly prized edible forest mushrooms are declining because of increasing loads of pollution, non-optimal management, changes in land use, and climate that all degrade natural productive habitats. Truffles are inoculated by use of spores and the seedlings are transplanted into sites that preferably lack pre-existing EcM vegetation. Native EcM fungi may have superior competitive abilities, and therefore, tree plantations with European trees and truffles are commonly established in Southern Hemisphere. For many highly prized edible forest mushrooms such as matsutake (*Tricholoma matsutake*), porcini (*Boletus edulis*) and the golden cantharelle (*Cantharellus cibarius*), cost-effective inoculation and seedling propagation methods are yet to be developed. Therefore, proper management of productive stands is essential to ensure the harvest. In rural areas, sustainable harvesting and trade of wild mushrooms offer revenues to people in all parts of the world (Härkönen et al., 2003). Unfortunately, forest management is typically carried out by governmental or large international logging companies that are not aware or do not care about the alternative forest revenues such as wild mushrooms, fruits and medicinal plants. Habitat conservation and careful management planning, along with the establishment of plantations are promising for attaining the sustainable use of these forest products in the long term. So far, there is no evidence that intensive collection of fruit-bodies reduces the reproductive output in fungi, because always a few fruit-bodies not found by mushroom hunters remain for sporulation. Transportation of collected mature fruit-bodies within the forest may facilitate the distribution of spores. The mycelium of these prized mushrooms can live in soil for decades.

Mycorrhizal associations have received negligible attention in the conservation of forests. Both rare and common plants are associated with the same mycorrhizal networks, except for certain orchids. Because we know too little about the autecology of mycorrhizal fungal species, habitat conservation offers the best possible means for protecting fungi and plants with specific mycorrhizal associations.

Conclusions

Mycorrhizal associations play an integral role in the mineral nutrition of plants and nutrient cycling in forest ecosystems. Disruption of mycorrhizal connections reduces primary production, soil microbial biomass and biodiversity. The fungi involved in the types of mycorrhiza differ in their capacity to degrade soil organic compounds and transfer nutrients. Climate change, particularly in temperature and precipitation, causes shifts in vegetation and the prevailing types of mycorrhiza, which may dramatically alter ecosystem nutrient cycling. These changes are also important to consider in the management of exotic forest plantations and understanding the invasion biology of plants. In addition to improving timber and fruit production, mycorrhizal systems offer alternative benefits for the cultivation and commercial harvesting of highly prized edible mushrooms.

References

Averill, C., Anthony, M.A., Baldrian, P., Finkbeiner, F., van den Hoogen, J., Kiers, T., Kohout, P., Hirt, E., Smith, G.R. and Crowther, T.W. 'Defending Earth's terrestrial microbiome', *Nature Microbiology*, vol. 7, pp. 1717–1725.

Brundrett, M.C. and Tedersoo, L. (2018) 'Evolutionary history of mycorrhizal symbioses and global host plant diversity', *New Phytologist,* vol. 220, pp. 1108–1115.

Brundrett, M.C., Malajczuk, N., Mingqin, G., Daping, X., Snelling, S. and Dell, B. (2005) 'Nursery inoculation of *Eucalyptus* seedlings in western Australia and Southern China using spores and mycelial inoculum of diverse ectomycorrhizal fungi from different climatic regions', *Forest Ecology and Management,* vol. 209, pp. 193–205.

Bueé, M., De Boer, W., Martin, F., van Overbeek, L. and Jurkevich, E. (2009) 'The rhizosphere zoo: an overview of plant-associated communities of microorganisms, including phages, bacteria, archaea, and fungi, and some of their structuring factors', *Plant and Soil,* vol. 321, pp. 189–212.

Clemmensen, K.E., Bahr, A., Ovaskainen, O., Dahlberg, A., Ekblad, A., Wallander, H., Stenlid, J., Finlay, R.D., Wardle, D.A. and Lindahl, B.D. (2013) 'Roots and associated fungi drive long-term carbon sequestration in boreal forest', *Science*, vol. 339, pp. 1615–1618.

Durall, D.M., Jones, M.D. and Lewis, K.J. (2005) 'Effects of forest management on fungal communities', in J. Dighton, J.F. White and P. Oudemans (eds) *The Fungal Community. Its Organization and Role in the Ecosystem*, CRC Press, Boca Raton, FL.

Gehring, C.A., Swaty, R.L. and Deckert, R.J. (2017) 'Mycorrhizas, drought, and host-plant mortality', in Johnson NC, Gehring CA, Jansa J. (eds). *Mycorrhizal Mediation of Soil*, Elsevier, Oxford.

Härkönen, M., Niemelä, T. and Mwasumbi, L. (2003) 'Tanzanian mushrooms: Edible, harmful and other fungi', *Norrlinia*, vol. 10, pp. 1–200.

Heinemeyer, A., Hartley, I.P., Evans, S.P., Carreira de la Fuente, J.A. and Ineson, P. (2007) 'Forest soil CO_2 flux: uncovering the contribution and environmental responses of ectomycorrhizas', *Global Change Biology*, vol. 13, pp. 1786–1797.

Hobbie, E.A. (2006) 'Carbon allocation to ectomycorrhizal fungi correlates with belowground allocation in culture studies', *Ecology*, vol. 87, pp. 563–569.

Hoeksema, J.D., Bever, J.D., Chakraborty, S., Chaudhary, V.B., Gardes, M., Gehring, C.A., Hart, M.M., Housworth, E.A., Kaonongbua, W., Klironomos, J.N. and Lajeunesse, M.J. (2018) 'Evolutionary history of plant hosts and fungal symbionts predicts the strength of mycorrhizal mutualism', *Communications Biology*, vol. 1, 116.

Högberg, M.N., Briones, M.J.I., Keel, S.G., Metcalfe, D.B., Campbell, C., Midwood, A.J., Thornton, B., Hurry, V., Linder, S., Näsholm, T. and Högberg, P. (2010) 'Quantification of effects of season and nitrogen supply on tree below-ground carbon transfer to ectomycorrhizal fungi and other soil organisms in a boreal pine forest', *New Phytologist*, vol. 187, pp. 485–493.

Johnson, N.C., Angelard, C., Sanders, I.R. and Kiers, E.T. (2013) 'Predicting community and ecosystem outcomes of mycorrhizal responses to global change', *Ecology Letters*, vol. 16, pp. 140–153.

Kiers, R.T., Duhamel, M., Beesetty, Y., Mensah, J.A., Franken, O., Verbruggen, E., Fellbaum, C.R., Kowalchuk, G.A., Hart, M.M., Bago, A., Palmer, T.M., West, S.A., Vandenkoornhuyse, P., Jansa, J. and Bücking, H. (2011) 'Reciprocal rewards stabilize cooperation in the mycorrhizal symbiosis', *Science*, vol. 333, pp. 880–882.

Klein, T., Siegwolf, R.T.L. and Körner, C. (2016) 'Belowground carbon trade among tall trees in a temperate forest', *Science*, vol 352, pp. 342–344.

Kuikka, K., Härmä, E., Markkola, A., Rautio, P., Roitto, M., Saikkonen, K., Ahonen-Jonnarth, U., Finlay, R.D. and Tuomi, J. (2003) 'Severe defoliation of Scots pine reduces reproductive investment by ectomycorrhizal symbionts', *Ecology*, vol. 84, pp. 2051–2061.

Lambers, H., Raven, J.A., Shaver, G.A. and Smith, S.E. (2008) 'Plant nutrient-acquisition strategies change with soil age', *Trends in Ecology and Evolution*, vol. 23, pp. 95–103.

Linder, H.P. (2001) 'Plant diversity and endemism in sub-Saharan tropical Africa', *Journal of Biogeography*, vol. 28, pp. 169–182.

Liu, X., Liang, M., Etienne, R.S., Wang, Y., Staehelin, C. and Yu, S. (2012) 'Experimental evidence for a phylogenetic Janzen-Connell effect in a subtropical forest', *Ecology Letters*, vol. 15, pp. 111–118.

Merckx, V. (2013) *Mycoheterotrophy: The Biology of Plants Living on Fungi*. Springer, Berlin.

Newbery, D.M., Alexander, I.J. and Rother, J.A. (1997) 'Phosphorus dynamics in a lowland African rain forest: the influence of ectomycorrhizal trees', *Ecological Monographs*, vol. 67, pp. 367–409.

Peay, K., Garbelotto, M. and Bruns, T.D. (2009) 'Spore heat resistance plays an important role in disturbance-mediated assemblage shift of ectomycorrhizal fungi colonizing *Pinus muricata* seedlings', *Journal of Ecology*, vol. 97, pp. 537–547.

Read, D., Leake, J.R. and Perez-Moreno, J. (2004). 'Mycorrhizal fungi as drivers of ecosystem processes in heathland and boreal forest biomes', *Canadian Journal of Botany*, vol. 82, pp. 1243–1263.

Rillig, M.C., Ryo, M., Lehmann, A., Aguilar-Trigueros, C.A., Buchert, S., Wulf, A., Iwasaki, A., Roy, J. and Yang, G. (2019) 'The role of multiple global change factors in driving soil functions and microbial biodiversity', *Science*, vol. 366, pp. 886–890.

Sheikh, S., Khan, F.K., Bahram, M. and Ryberg, M. (2022) 'Impact of model assumptions on the inference of the evolution of ectomycorrhizal symbiosis in fungi', *Scientific Reports*, vol. 12, 22043.

Simard, S.W. (2018) 'Mycorrhizal networks facilitate tree communication, learning, and memory', in: Baluska, F. (ed.). *Memory and Learning in Plants*, Springer, Berlin.

Simard, S.W., Martin, K., Vyse, A. and Larson, B. (2013) 'Meta-networks of fungi, fauna and flora as agents of complex adaptive systems', in C. Messier, K.J. Puettman and K.D. Coates (eds) *Managing Forests as Complex Adaptive Systems*, Routledge, Oxon, UK.

Smith, S.E. and Read, D.J. (2008) *Mycorrhizal Symbiosis*, Academic Press, London.

Soudzilovskaia, N.A., van Bodegom, P.M., Terrer, C., van't Zelfde, M., McCallum, I., McCormack, M.L., Fisher, J.B., Brundrett, M.C., de Sá, N.C. and Tedersoo, L. (2019) 'Global mycorrhizal plant distribution linked to terrestrial carbon stocks', *Nature Communications*, vol 10, 5077.

Soudzilovskaia, N.A., Vaessen, S., Barcelo, M., He, J., Rahimlou, S., Abarenkov, K., Brundrett, M., Gomes, S., Merckx, V. and Tedersoo, L. (2020) 'FungalRoot: global online database of plant mycorrhizal associations', *New Phytologist*, vol. 227, pp. 955–966.

Štursova, M., Šnajdr, J., Cajthaml, T., Barta, J., Šantručkova, H. and Baldrian, P. (2014) 'When the forest dies: the response of forest soil fungi to a bark beetle-induced tree dieback', *The ISME Journal*, vol. 8, pp. 1920–1931.

Tedersoo, L. (ed). (2017) *Biogeography of Mycorrhizal Symbiosis*. Springer, Berlin.

Tedersoo, L. and Bahram, M. (2019) 'Mycorrhizal types differ in ecophysiology and alter plant nutrition and soil processes', *Biological Reviews*, vol. 94, pp. 1857–1880.

Terrer, C., Vicca, S., Hungate, B.A., Phillips, R.P. and Prentice, I.C. (2016) 'Mycorrhizal association as a primary control of the CO2 fertilization effect', *Science*, vol. 353, pp. 72–74.

Teste, F.P., Jones, M.D. and Dickie, I. (2020). 'Dual-mycorrhizal plants: their ecology and relevance', *New Phytologist*, vol 225, pp. 1835–1851.

Torti, S.D., Coley, P.D. and Kursar, T. (2001) 'Causes and consequences of monodominance in tropical lowland forests', *American Naturalist*, vol. 157, pp. 141–153.

Traveset, A. and Richardson, D.M. (2014) 'Mutualistic interactions and biological invasions', *Annual Reviews in Ecology, Evolution and Systematics*, vol. 45, pp. 89–113.

Ward, E.B., Duguid, M.C., Kuebbing, S.E., Lendemer, J.C., Warren, R.J. II and Bradford, M.A. (2021) 'Ericoid mycorrhizal shrubs alter the relationship between tree mycorrhizal dominance and soil carbon and nitrogen', *Journal of Ecology*, vol. 109, pp. 3524–3540.

Whipps, J.M. (2004) 'Prospects and limitations of mycorrhizas in biocontrol of root pathogens', *Canadian Journal of Botany*, vol 82, pp. 1198–1227.

23
BIOGEOCHEMICAL CYCLING

David Paré, Daniel Markewitz and Håkan Wallander

This chapter will review the three following topics:

(1) Nutrient cycling at three scales and integration. Specifically, what organisms are involved? What are the main drivers of the cycles of nutrients at the three scales and how these cycles interact?
(2) Resilience to disturbance and ecosystem retrogression
(3) Drivers of site productivity including ecosystem conditions and effects of management.

It will take a global perspective and include the most recent findings especially the role of plants and microbial symbioses into acquiring nutrients from dead organic matter and from non-readily available forms.

Forest nutrient cycles

Of the 34 elements known or believed to be essential to plants, animals and microorganisms, six are known as plant macro nutrients: nitrogen (N), phosphorus (P), potassium (K), calcium (Ca), magnesium (Mg) and sulfur (S). N and P are the ones that are the most studied and that most often limit plant growth (Sterner and Elser 2002). These elements are found in relatively small amounts in the soil in a form that is available to plants and are usually efficiently recycled and conserved within the ecosystem. The study of forest nutrient cycling uses three interlinked cycles as proposed by Switzer and Nelson (1972):

The biochemical cycle (or cycling within organisms)

This cycle includes fluxes of elements that are recovered from dying tissues. Nutrient resorption is a fundamental process through which plants withdraw nutrients from leaves before abscission. A large portion of mobile elements such as N and P are recovered prior to leaf shedding, but other elements such as K could also be substantially recycled within the plant (Vergutz et al. 2012). Typically, resorption rates are around 50 per cent but can be as high as 80 per cent (Aerts 1996). Resorption from roots may be important but is difficult to quantify because little information is available. Nutrients that play a structural role in the plant such as

 DOI: 10.4324/9781003324072-27

Ca are not recycled biochemically by plants. In infertile soil conditions, a consistent strategy for conserving elements is through long-leaf lifespan. This allows for a greater photosynthetic C return per unit of nutrient used (high nutrient use efficiency) (Chapin 1980). For example, black spruce (*Picea mariana* Mill.) can maintain its needles for up to 13 years.

The biogeochemical cycle (or cycling within the ecosystem)

The biogeochemical cycle involves fluxes of nutrients from the trees to the soil in solid form (leaves, dead roots, wood, other plants such as mosses) or in soluble form (stemflow, leaching). It also includes the transformation of nutrients that accumulate in soil organic matter and litter into soluble forms through the processes of decomposition and mineralization. Finally, the uptake of nutrients from the soil completes this cycle. Quantifying or describing these fluxes necessitate a definition of system boundaries. It can be as large as a biome, but usually a small area is used, such as a forest stand. The plant rooting depth is often used as the lower boundary for biogeochemical nutrient cycling. There is also a chemical boundary as nutrients that are present in recalcitrant organic matter or in rock or clay minerals that have limited solubility can be considered as being outside of the ecosystem boundaries while still being in the rooting zone. The solubilization of these elements can be considered as geochemical input. The term critical zone (National Research Council 2021) is used to identify the boundary for biogeochemical cycles. It is defined as the heterogeneous, near-surface environment in which complex interactions involving rock, soil, water, air and living organisms regulate the natural habitat and determine the availability of life-sustaining resources.

Biogeochemical fluxes can be affected by numerous factors at various temporal and spatial scales. When comparing across biomes, climate strongly controls nutrient fluxes and these fluxes closely correlate with ecosystem productivity. Cold temperatures and excessive humidity or dryness can reduce plant productivity and nutrient uptake, the flow of nutrients in organic forms to the soil and decomposition or mineralization of organic nutrients in the soil. Suboptimal temperatures and water content limit organic matter decomposition and release of nutrients. The decomposition process follows an exponential relationship with temperature, which is often represented by a Q_{10} relationship where the rate of the process increases every 10°C rise by a given factor. A Q_{10} of 2, indicating a doubling of the rate for every 10°C rise, is widely used (Davidson and Janssens 2006). The relationship between soil water availability and decomposition (and concomitant nutrient mineralization) follows a bell-shaped curve with a sharp rise and decline at both ends and a large plateau in between. Apart from microclimatic conditions, major factors affecting biogeochemical cycles are soil type and forest community composition (see the section 'Plant–soil interactions'), and all these factors are to a certain degree interrelated.

Several soil properties affect nutrient conservation and facilitate efficient nutrient cycling. The accumulation of soil organic matter enhances nutrient retention in two ways: first, it contains nutrients in proportions that are relatively similar to that of plants, thus promoting efficient uptake upon decay, and second, organic matter possesses both negatively and positively charged functional groups that provide cation and anion exchange sites where nutrients that are in the form of cations (K^+, Ca^{2+}, Mg^{2+}, , NH_4^+) or anions (NO_3^-, $H_2PO_4^-$) can be maintained in a form easily accessible to plant roots. Soil properties such as drainage and the abundance of small particles (clay and silt) have a great influence on the accumulation of organic matter. Other soil properties such as the presence and type of clay also strongly influence the soil exchange capacity. Very acidic and very basic soils have the potential to sequester P anions into recalcitrant forms that are poorly available to plants or microbes. Nutrients that are not retained may be lost from the soil by leaching. Trees develop a root system that together with mycorrhizal fungi is extremely efficient in preventing these losses. While tree roots can extend to great depths, the bulk of fine roots are located within the top 30 cm of soil, especially in boreal forests (Jackson et al. 1997). A very high density of fine roots in the top-soil horizon is often observed in tropical, boreal and temperate forests.

The geochemical cycle (or ecosystem gains and losses)

This cycle involves both inputs of nutrients to ecosystems of atmospheric, geological and biological origin as well as outputs or losses from ecosystems through solid, gaseous or liquid form. The same issues of ecosystem boundaries apply as for other cycles. Gains and losses of elements can vary rapidly over short distances as they are affected by soil types, the biota (tree composition for example) or forest stand development stage. While plots of a few square meters are often considered, the boundaries of a watershed (i.e., drainage basin) are often used to constrain the limit of these studies, especially when the focus is on the ecosystem-scale nutrient balance. A watershed is beneficial in defining rainfall inputs and stream water outputs but may include a mosaic of conditions within the basin with various rates of geochemical cycles.

The main ecosystem input fluxes (often defined in $kg.ha^{-1}.yr^{-1}$) are from mineral weathering, particularly for base cations and P, and from atmospheric inputs in rainfall, fine particles and aerosols and N fixation. The highest N-fixing rates are observed for the few species of plants that carry symbiotic N-fixing bacteria with rates that can reach over 100 $kg.ha^{-1}$ of N annually. Non-obligatory symbiotic organisms such as the blue-green algae Nostoc are present in the soil on decaying wood and in certain mosses and have rates that are generally lower than 3 $kg.ha^{-1}.yr^{-1}$. As N fixation is a highly energy-demanding process, there is some evidence that N-fixing activity is regulated by the amount of N available in the soil in a negative feedback loop. When levels of available N are high, N fixation rates are often low. This is observed in N-fixing algae in moss communities (Gundale et al. 2011) as well as in natural successional pattern from early N-fixing plants that performed well in N-poor soil to non-nitrogen-fixing communities. Ecosystems lose small amounts of nutrients continuously through the processes of leaching and denitrification but episodic disturbances can cause high losses through biomass removal, fire (mostly through oxidation when the material is combusted) and erosion.

Overview of the three cycles

The nutrients circulating in a forest ecosystem originate from geochemical inputs. However, on a yearly basis, biochemical and biogeochemical cycles strongly dominate the fluxes of nutrients (Table 23.1). The bulk of the amounts of nutrients required by plants for their growth come from these two fluxes. The pool of nutrients that is cycling quickly in trees and within the tree–soil system makes it possible to produce high amounts of biomass in relatively poor soils. Because the maintenance of a foliar cover creates a major nutrient requirement within forest ecosystems on a yearly basis, having a large portion of the requirement satisfied by foliar resorption alleviates the need to prospect the soil for additional nutrients and

Table 23.1 Nutrient fluxes in a tropical rain forest, a deciduous temperate forest and a boreal forest

	N	*P*	*K*	*Ca*	*Mg*
Tropical (rain forest, Brazil) (from Markewitz et al. 2004)					
Uptake	149	3.3	65	115	27
Precipitation inputs	4	0.03	5	3	1
Leaching	12	0.1	1.4	9.9	5.2
Leaching loss as % of uptake	**8**	**3**	**2**	**9**	**19**
Biogeochemical recycling efficiency (%)	**91**	**97**	**98**	**91**	**81**
Temperate (Northern hardwood, USA) (from Whittaker et al. 1979)					
Uptake	114	13	47	50.6	8.5
Precipitation inputs	6	0.1	1.1	2.6	0.7
Stream leaching	2.3	0.02	1.7	11.7	2.8
Leaching loss as % of uptake	**2**	**0**	**4**	**23**	**33**
Biogeochemical recycling efficiency (%)	**98**	**100**	**96**	**77**	**67**
Boreal (mixed, Alaska) (from Van Cleve et al. 1983)					
Uptake	42.9	4.7	23.9	30.1	7.3
Precipitation inputs (n.a.: not available)	n.a.	n.a	n.a.	n.a.	n.a.
Stream leaching	0.7	0.2	0.3	3.1	0.6
Leaching loss as % of uptake	**2**	**4**	**1**	**10**	**8**
Biogeochemical recycling efficiency (%)	**98**	**96**	**99**	**90**	**92**

Notes: Fluxes are in $kg.ha^{-1}.yr^{-1}$. Numbers in bold are expressed in % of uptake. Leaching losses as a percentage of uptake give an approximation of the intensity of recycling within the ecosystems: 100 minus this estimate gives as a percentage the proportion of the plant uptake that is met by biogeochemical recycling.

thus compete with other plants or microbes. Greater geochemical inputs such as fertilization or soil richness can reduce the rates of resorption, but this is not always observed (Aerts 1996). Plants adapted to nutrient-poor environments generally have slow growth rates, slow growth response to nutrient addition and greater leaf longevity, which is associated with a greater production of anti-herbivore products (Chapin 1980). These properties are conducive to a positive retroaction between soil fertility and nutrient cycling. Table 23.1 illustrates some nutrient fluxes in a tropical, a temperate and a boreal forest ecosystem. A universal property of forest ecosystems is that the quantity of nutrients lost in leachates, in the absence of severe disturbance, represents a small fraction of the amounts cycling within these systems on an annual basis, especially for N and P. Climate, topography, soil drainage and parent material define potential limits to nutrient cycling rates. Sites with low geochemical inputs are highly dependent on biochemical and biogeochemical cycles. This greatly reduces the difference in forest productivity between rich and poor sites (Legout et al. 2020), in contrast to agricultural sites. Nevertheless, these rates are very dynamic in space and time and are dependent on time since disturbance (see the following section) and on the biological composition of the ecosystem (see the section 'Plant–soil interactions'). Forest management activities influence nutrient cycling at the scale of all three interlinked cycles. They have direct impacts by enhancing nutrient inputs or outputs and indirect impacts by influencing the biological composition of the ecosystem or the soil properties that control the rates of nutrient cycling.

Nutrient cycling: change over time

Stand development

Biogeochemical cycling of nutrients through forests is defined by the balance of inputs and outputs from the ecosystem and drives changes in ecosystem storage (Figure 23.1). Nutrient cycling in forest ecosystems changes during stand development and species succession. These changes in nutrient cycles differ depending on the initial conditions of the site, ideas that developed originally during studies of primary or secondary forest succession. In primary succession, forest development occurs on recently exposed or deposited parent materials (e.g., deglaciation, sand dunes, lava flows). In these scenarios, the parent material is often nutrient-rich (but see Olson 1958 for the consideration of sand dunes) and can contribute substantially to the nutrients accumulated by plant biomass. In these cases, however, there is limited organic matter or actively cycling nutrients in the soil because the sites have not been previously vegetated. Studies have been conducted to examine primary succession using a chronosequence approach at the edge of retreating glaciers in Alaska (Chapin et al. 1994). In

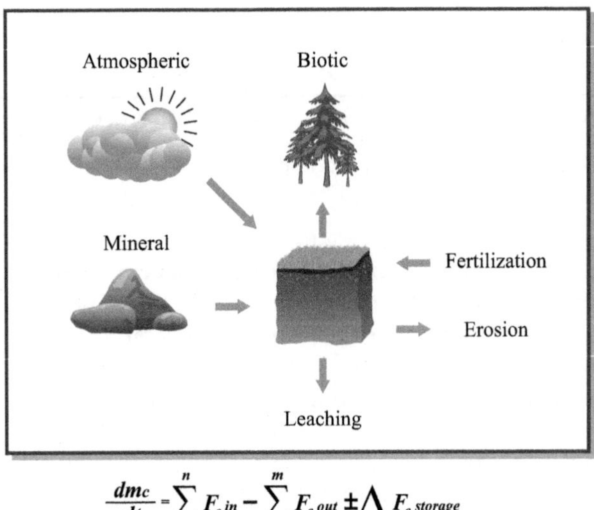

Figure 23.1 Generalized model of biogeochemical cycling of elements in ecosystems demonstrating a balance of inputs, outputs and change in storage. The equation represents the change in the mass of an element *c* with time $\left(\frac{dm_c}{dt}\right)$ as a sum of all the fluxes in from 1 to *n* of element *c* $(\sum_{i=1}^{n} F_c in)$ minus all the fluxes out from 1 to *m* of element *c* $\left(\sum_{o=1}^{m} F_c out\right)$ plus or minus any change in storage of element c $(\Delta F_c\ storage)$.

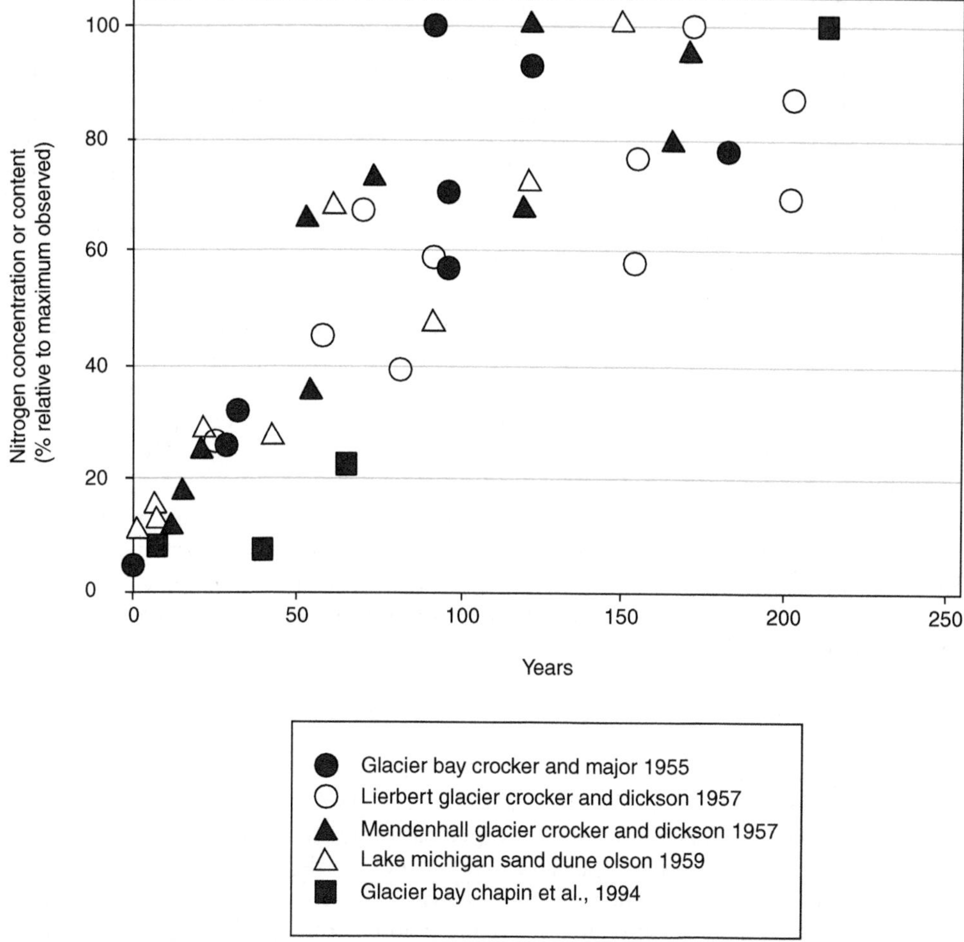

Figure 23.2 Patterns of soil N accumulation in primary successional sequences. Data at each location are presented as relative to the maximum observed at that location.

these ecosystems that extended over 200 years, a successional sequence of species from blue-green algae to Sitka spruce (*Picea sitchensis*) was described and the accumulation of plant biomass was associated with a pattern of decreasing soil pH, increasing contents of soil C and N, and an increase in soil exchangeable cations. In a chronosequence of primary succession on sand dunes along the shores of Lake Michigan, Olson (1958) described a plant succession over ~300 years from bunch grasses (*Andropogen* sp.) to jack pine (i.e., *Pinus banksiana*) and black oak (*Quercus velutina*), but a similar pattern of accumulation of soil C and N with an increase in exchange capacity and exchangeable acidity. The development of nutrient cycles in these ecosystems includes an accumulation of nutrients in living biomass and soil organic matter pools (see Figure 23.2 for relative rates of soil N accumulation) as well as an increase in the intrasystem cycling of nutrients (see biogeochemical cycles defined above). The case of N is particularly interesting as N content in primary soil substrates is often quite limited such that N fixation, as noted above, can be a critical process for N inputs. Thus, whether N fixers are part of the plant succession can dramatically alter N accumulation and dynamics.

These nutrient accumulations during stand development on newly exposed substrates contrast with secondary succession where sites have previously been vegetated but have been reset to an earlier stage of plant succession through some form of forest disturbance (e.g., harvest, fire, landslide, or hurricane). An important difference in stand development after disturbance is that there may be substantial organic matter and nutrients already circulating in the ecosystem. In studies of stand development in old agricultural fields, previous fertilizer inputs that have accumulated in soil may also be used by plants and thus transfer soil stocks to plant biomass (Richter et al. 2000). Furthermore, rates of plant regeneration may be slower in primary succession that requires the development of soil and the import of plant stock (i.e., seeds) while in secondary succession, rates may be quite rapid due to a large seed bank and prolific root or stump sprouting. There have been a substantial number of studies that have investigated secondary succession after a range of disturbances: agriculture (Richter and Markewitz 2001), fire (Nave et al. 2011) and harvest (Nave et al. 2010). In many of these studies, there is a period shortly after disturbance where nutrient availability exceeds plant demand such that losses of elements from the ecosystem in solution (groundwater or streams) are apparent. These processes were amply demonstrated by Bormann and Likens (1979) in the watersheds of the Hubbard Brook ecosystems of New Hampshire. As plants begin to once again re-vegetate the site and increase plant demand, losses of elements from the ecosystem can decline.

Supply and demand

During ecosystem development, both the supply of and demand for nutrients change. This balance was described by Vitousek and Reiners (1975) as a critical framework for understanding when ecosystems should be expected to tightly retain nutrients as compared with periods when the ecosystem appears leaky; in other words, outputs from the ecosystem in drainage waters or gases would be large and nearly equal to inputs (see Figure 23.3 for conceptual model). Very shortly after substrate exposure or disturbance, the demand for nutrients is low due to an absence of plant demand. Nutrients coming into the ecosystem with wet deposition, for example, should go through the ecosystem rapidly. In the case of forest disturbance, we might observe an additional release of nutrients that had accumulated in forest floor or soil organic matter such that losses might exceed that being contributed through rainfall.

In a young and vigorously growing forest, the demand by plants for essential elements will be high and thus retention of these elements in growing biomass can be great (Figure 23.3). During this early phase, supply can be high for some elements (for example, rock-derived elements such as P or Ca) or low (i.e. N). The situation for N is dynamic and can be highly affected by plant succession or the previous history of the site (Richter et al. 2000). N mineralization of organic matter on site can increase N availability after disturbance and such increases have been demonstrated in many studies after forest harvest (Nave et al. 2010). In contrast, newly exposed substrates can have a limited N supply (Chadwick et al. 1999) as can sites after disturbance from severe fire. As stands continue to develop and plant growth slows, so will plant demand and thus there is increased potential for loss of nutrients outside the defined ecosystem boundaries. However, there is little evidence for a greater leakiness for old or mature forests. Recent studies suggest that, rather than being lost, nutrients accumulate

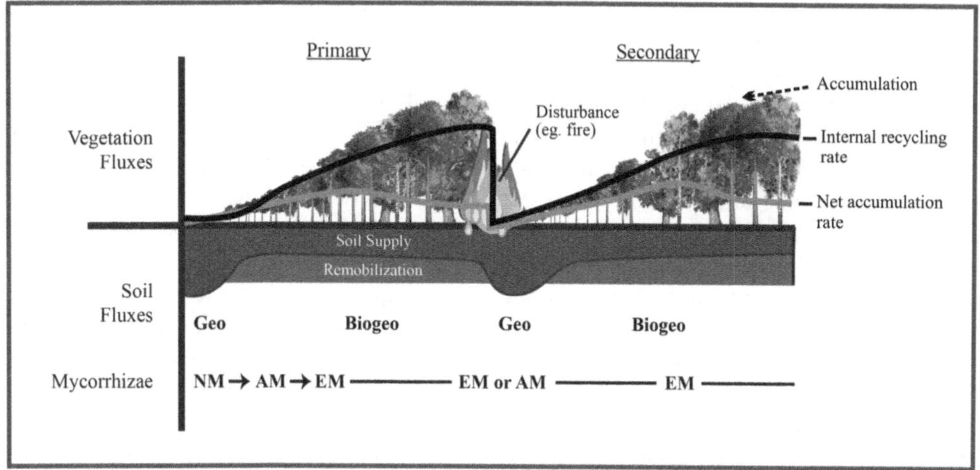

Figure 23.3 Conceptualized patterns of changes with time in rates of nutrient fluxes along primary and secondary successions. The height of stands represents the accumulating stocks of carbon and nutrients in the vegetation. Lines represent the yearly fluxes of nutrients accumulating in the vegetation and the requirement that is met by internal recycling. Soil supplies and the portion that comes from remobilization are shown in the opposite direction. Geochemical N fixation is frequent in early primary succession while disturbance can enhance nutrient availability in secondary succession. The dominance of biogeochemical versus geochemical cycles is indicated as well as that of non-, arbuscular- and ectomycorrhizal fungi (NM, AM and EM, respectively) as observed in temperate and boreal climates.

in the soil (Lovett et al. 2018) or in dead organic matter (Lajtha 2020) building a reserve that may be useful to maintain tree nutrition in the period of greater needs. This pattern of supply versus demand during ecosystem development is not observed for non-essential plant elements since there is no plant demand (e.g., Na), and such elements demonstrate a relatively constant balance of inputs and outputs. For those elements that are plant essential but are not perceived as limiting to growth, an intermediate pattern is expected to develop (i.e., some decline in outputs relative to inputs but some output flux is still measurable). The dichotomy of tight versus leaky relates most to the geochemical cycle of ecosystem inputs and outputs as described above and should not overshadow the efficient and persistent internal biogeochemical cycles that are sustained even in well-developed forest ecosystems. These efficient cycles are particularly relevant for plant-limiting nutrients of N and P where individual atoms of the element may cycle through the ecosystem many times despite a relative geochemical balance of inputs and outputs.

Change in limiting element

During any phase of forest growth, there is presumed to be some element that limits the growth of the forest or species of interest (i.e., Liebig's law of the minimum). Over time, there has been an increasing recognition of co-limitations to forest growth that has been demonstrated by a range of irrigation × fertilizer studies (Albaugh et al. 1998) and CO_2 × nutrient studies (McCarthy et al. 2010). Despite this recognition, much work has adopted the paradigm of the limiting element with a particular focus on N and P. In a millennial-scale chronosequence study in Hawaii, Chadwick et al. (1999) proposed a developmental sequence in which ecosystems are most limited by N early in development because P is available from rock weathering at that time. Over time, P is slowly converted from available to unavailable forms as originally hypothesized by Walker and Syers (1976). Changes in P availability are described as progressing from a condition of high availability due to weathering of the primary mineral apatite to a state of low availability as weathered and biogeochemical cycling P is slowly adsorbed on Fe and Al oxides and hydroxides, as predominate in highly weathered soils, such that P is no longer plant available (Peltzer et al. 2010). During this developmental sequence in the Hawaii chronosequence, P became increasingly more limiting to forest growth such that atmospheric dust inputs of P displaced soil-derived inputs as the predominant ecosystem supply. During this same millennial-scale development period, N accumulated in the ecosystem and became less of a constraint to growth. Similar patterns in nutrient limitation from N to P have also been proposed during secondary stand development (see Figure 23.4 for data comparison). Davidson et al. (2007) measured increasing N capital in a chronosequence of secondary tropical forests while P availability and capital remained constant, suggesting increasing P limitation.

In all these cases of biogeochemical cycling through time, both inputs and outputs can be dynamic and other macro (i.e., K) and micro (i.e., B, Mo) elements can constrain productivity and cycling in the ecosystem. These changes can have effects on vegetation composition with feedbacks on nutrient cycling (see case study in the next section).

Plant–soil interactions

Plant and soil organisms play important roles in regulating biogeochemical cycling. As indicated above, forest stand composition is influenced by succession, disturbance, management activities and the presence of invasive species. Forest tree species composition, together with associated organisms including plants of different strata, soil fauna and microorganisms, influence nutrient capture by the ecosystem as well as nutrient retention and biogeochemical nutrient cycling rates. Of particular importance are plant associations with mycorrhizal fungi. These associations are almost ubiquitous among trees, as well as other plants, and large amounts of carbon are invested in exchange for nutrients, especially under nutrient-poor conditions. The fine mycelia formed by mycorrhizal fungi have an outstanding capacity to explore the soil substrate in the search for nutrients. They extend beyond the depletion zones around roots and enter micropores and cavities in the soil where roots have no access (Smith and Read 2008). Arbuscular mycorrhiza, the most ancient form of mycorrhizal association, evolved with the first land plants that colonized the planet around 400 million years ago. This was considered a necessary step in order to obtain enough P when moving from the water to the riparian zones where these plants flourished. Arbuscular mycorrhizal fungi colonize around 80 per cent of all plant species and are the most common association in tropical forest trees where P is usually the growth-limiting element (Smith and Read 2008). Ectomycorrhizal (EM) associations evolved much later and are the dominant associations in temperate and boreal forests. In these ecosystems, N is usually the growth-limiting element because of slow decomposition rates, which results in N retention in the organic material. This effect is expressed more fully further north since lower temperatures retard decomposition. For this reason, forest trees are usually more strongly dependent on their EM symbionts in northern ecosystems (Smith and Read 2008).

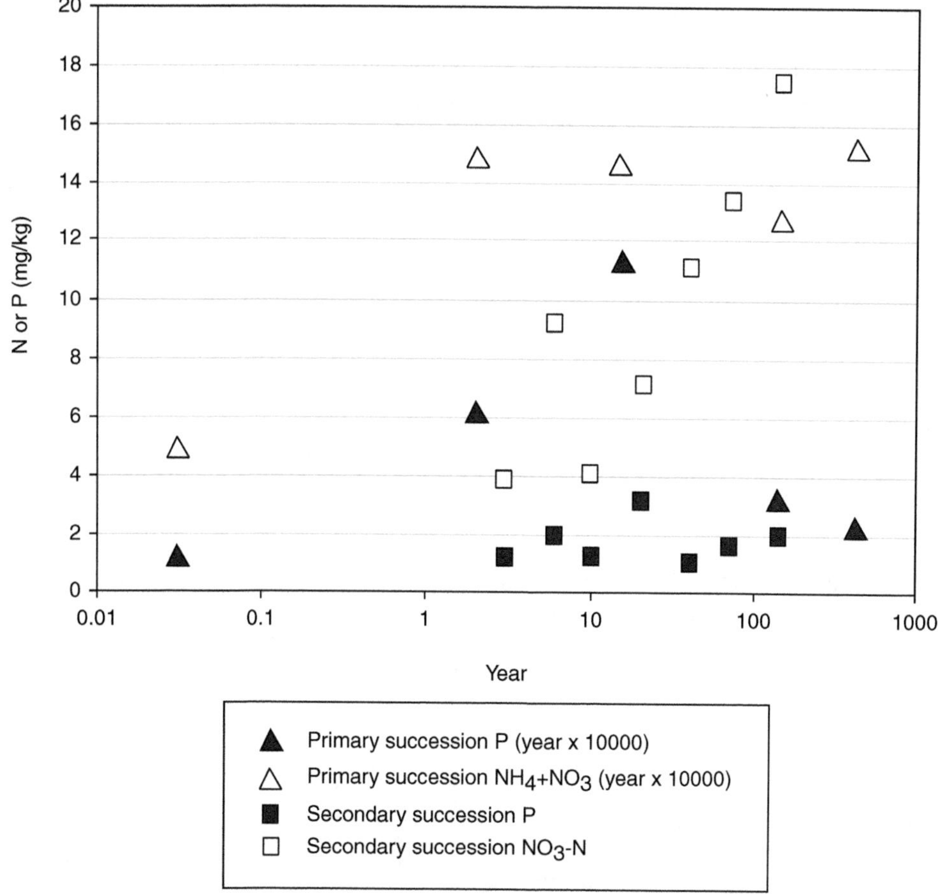

Figure 23.4 Patterns of soil N and P availability throughout a primary chronosequence over a millennial timescale from Hawaii (Chadwick et al. 1999) and a secondary chronosequence over a decadal timescale from the Brazilian Amazon (Davidson et al. 2007).

Plant–soil interaction and biogeochemical cycling

The chemical and physical characteristics of plant litter, which varies with species and plant components, affect rates of decay and the accumulation of soil organic matter. In boreal forests, ericaceous shrubs and mosses, typically associated with conifers, slow down the rates of cycling of N through the production of recalcitrant organic molecules and through the cooling of the soil surface (Thiffault et al. 2013). The impact of plant litter can be complex since the physical and chemical properties will influence the community of decomposers, which in turn will affect the decomposition rates. While the decomposition process is largely conducted by soil fungi and bacteria, larger organisms, such as earthworms, can influence the process dramatically both by physically reducing the size of organic matter and by altering its chemical composition. In a controlled experiment, Laganière et al. (2010) observed that deciduous litter decayed much faster than conifer litter only when lying on its own soil, which hosted earthworms. This example shows that litter properties were only part of the explanation. Tree species influenced the litter decay rates in great part by harboring a distinct soil fauna.

It is generally considered that EM fungi evolved from saprophytic fungi (Wolfe et al. 2012) and many species have retained the oxidative or enzymatic machineries to attack organic matter while mining for nutrients, especially N (Rineau et al. 2013; Lindahl and Tunlid 2015). EM fungi produce large amounts of external mycelium when exploring the soil for nutrients, and residues of this mycelium are important for soil formation and carbon sequestration. Recent research revealed that EM associations contributed to more than 50 per cent of the soil organic matter in northern boreal forests (Clemmensen et al. 2013), probably an effect of the large production and slow decomposition rates of roots and external EM mycelium in nutrient-poor forest soils. When carbon is sequestrated through this mechanism, N will be retained in N-rich fungal residues, which form complexes with plant-derived condensed tannins (Adamczyk et al. 2019). Ectomycorrhizal fungi may thus enhance C sequestration and N retention through their residues but at the same time reduce C sequestration and N retention when they mine the organic matter for N (Lindahl et al. 2021). The outcome of these opposing processes (Figure 23.5) is critical for the C and N balance of forest ecosystems. At the global scale, EM ecosystems have been shown to enhance C sequestration by 1.7 times per unit of N relative to AM ecosystems (Averill et al. 2014).

The production of recalcitrant N-rich litter by EM fungi has been suggested as a way to reduce competition from more N-demanding species that lack the N mining capacity of EM

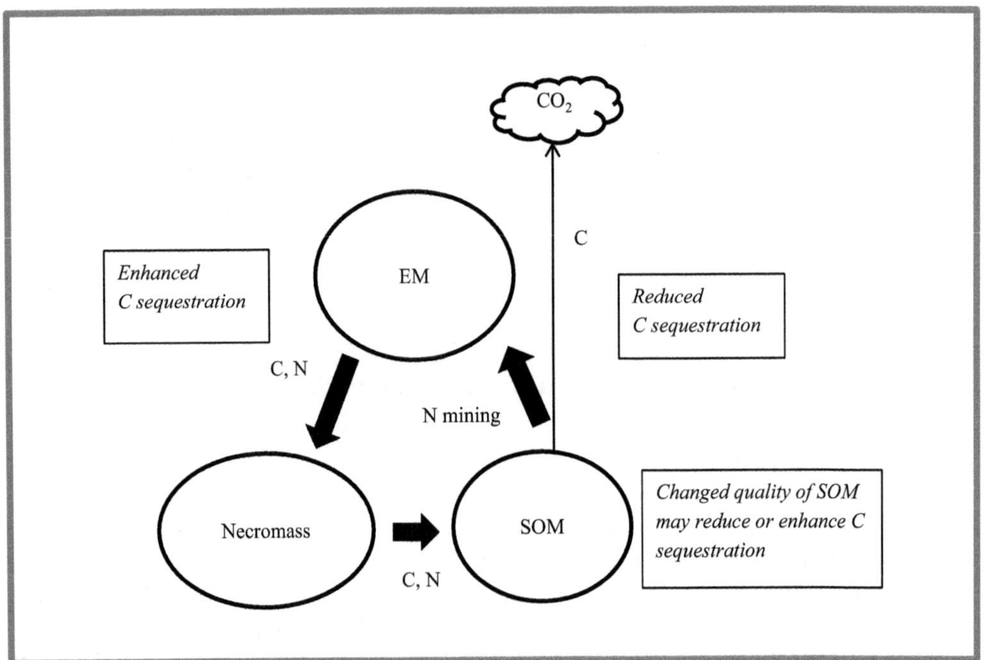

Figure 23.5 Ectomycorrhizal fungi (EM) effects on C and N cycling in forest soil. Carbon is released when EM mine for N in soil organic matter. But C and N are also sequestered when recalcitrant residues of EM accumulate in the soil. The quality of soil organic matter also changes after fungal colonization which will affect its decomposability.

associations. In support for this hypothesis, Näsholm et al. (2013) demonstrated strong N retention in EM mycelia in forests in northern Sweden, which could explain the exclusion of N-demanding understory, which proliferate on clear cuts or after forest fires when the dominance of EM symbionts has diminished (Figure 23.3).

Biotic influence on geochemical inputs

Symbiotic N fixation activities contribute to building the ecosystem N stocks especially in primary successions (see the section, 'Nutrient cycling: change over time'). However, such a process is found only in the root systems of a few tree and plant species. The architectural structure of trees influences the capture of nutrients. Atmospheric depositions will accumulate to a larger extent in conifers, which have foliage the whole year around, compared with deciduous trees, which drop their foliage for winter, especially under conditions of high atmospheric pollution (Berger et al. 2009). Overall, this will lead to a higher atmospheric input in coniferous compared with deciduous forests and potentially to greater soil acidification and leaching losses. Mycorrhizal fungi also play a role in nutrient capture by accelerating mineral weathering. Many EM species together with associated bacteria produce acids that can release nutrients such as P from primary minerals (Finlay et al. 2020;).

Regulation of nutrient cycling by EM fungi

N deposition case study

The growth of EM mycelia depends on below-ground allocation of carbon, which is regulated by the nutrient status of the trees (Ericsson 1995). Deficiency of N and P enhances, while deficiency of K and Mg reduces, below-ground carbon allocation and EM growth (Ericsson 1995). Fertilization or elevated N deposition usually results in retarded EM growth since carbon allocated below ground will be reduced and a large portion will be used instead to incorporate inorganic N into amino acids to enhance above-ground shoot growth (Wallander 1995). The composition of the EM communities is also influenced by N input and this will have consequences for uptake and leaching of nutrients from the soil. For instance, EM species with a high capacity to mine organic matter for N usually decline when the availability of inorganic N increases (Taylor et al. 2000), which may lead to the accumulation of SOM since N mining species degrade SOM (Lindahl et al 2021). These species are carbon-demanding, produce large external mycelium and are classified as long-distance or medium-distance types, according to Agerer (2001). In contrast, species that proliferate in N-rich areas are less carbon-demanding, produce low amounts of mycelia and are classified as short-distance or contact types (Agerer 2001). Kjøller et al. (2012) found the dominance of contact types, low EM growth and high N leaching at a forest edge exposed to elevated N levels from a nearby poultry farm, while EM growth increased and N leaching declined when moving toward the forest interior where N deposition declined. Similar changes in EM communities were found along a N deposition gradient in Alaska (Lilleskov et al. 2002).

It can be assumed that reduced EM mycelial production at elevated N levels reduces the overall capacity to take up other nutrients such as P, K and Mg from the soil. But shifts in EM communities may also enhance species that are more efficient in uptake of nutrients other than N. For instance, mycorrhizal roots formed by *Paxillus involutus* increased in abundance along the N gradient studied in Alaska by Lilleskov et al. (2002) and this species is known to be efficient in P uptake (Colpaert et al. 1999). Although most species that are enhanced by N deposition are contact types with low mycelial production and low capacity to degrade organic matter, *P. involutus* is an exception. It has a high capacity to degrade organic matter but in contrast to other species, this capacity does not decline under conditions of high inorganic N levels (Rineau et al. 2013).

In contrast to N and P, deficiency of K and Mg results in reduced below-ground carbon allocation due to a shortage of carbon in the leaves and needles, and impaired carbon loading into the phloem when Mg is in short supply (Ericsson 1995). For this reason, Mg deficiency was suggested to be one of the major causes for the forest decline in central Europe that occurred during the second half of the last century (Schulze et al. 2005). The poor Mg status in these sites was a result of low bedrock Mg levels, low atmospheric inputs and severe soil acidification due to extensive SO_2 emissions.

The differential response of carbon allocation and EM growth to different nutrients may stem from the fact that K and Mg are relatively abundant in most ecosystems and specific mechanisms to increase the availability of these nutrients have thus not evolved. On the other hand, N and P deficiency is common in many ecosystems and plants have evolved a number of traits to cope with these stresses. For these reasons, directions for sustainable forest management should make sure that the removal of nutrients through harvesting and leaching does not result in K and Mg deficiencies unless these nutrients are added as fertilizers.

Succession–retrogression case study

Chronosequence studies often show continued weathering of soil minerals during progression (Peltzer et al. 2010) and a peak in ecosystem biomass will eventually be reached, followed by a retrogressive phase when biomass declines due to deficiency of P. The reason for this is that P-containing minerals are relatively rare and P released during weathering is often transformed into occluded forms that are unavailable for plant uptake. The community of plants will change when the ecosystem undergoes these changes from a progressive to a retrogressive stage. Plants producing dauciform roots, with a high capacity to release P through exudation of citric acid, will become more common, and plants forming strong associations with mycorrhizal fungi, especially arbuscular mycorrhizal fungi, which are efficient in taking up P at low concentrations, will increase in abundance during the retrogressive stage (Peltzer et al. 2010). Turpault et al. (2009) found a small but significant enhancement of apatite weathering at 20 cm soil depth when exposed to EM fungi, and Wallander and Thelin (2009) found stimulated colonization of apatite by EM fungi in forests with low P status. It can thus be expected that EM-induced weathering of P-containing minerals increases when forests undergo a transition from N to P deficiency, but a similar mechanism is much more uncertain when forests approach Mg or K deficiency.

Biogeochemical services

The biogeochemical cycling of nutrients through forest ecosystems provides a critical service in sustaining productivity. The biochemical, biogeochemical and geochemical cycles of nutrients facilitate the recycling of elements that allows for the continued photosynthetic capture of CO_2. As ecosystems progress through primary or secondary succession, the opened or closed nature of the cycle will change relative to plant demand and one element might develop predominance over another as the limiting element in the ecosystem changes. Forest ecosystems have evolved plant–microbe interactions to facilitate the capture of these critical elements that over the long sweep of time have sustained these ecosystems over cycles of disturbance and regeneration. Managing and maintaining these fundamental biogeochemical cycles will require continued research and enlightened application of this knowledge.

References

Adamczyk, B., Sietiö, O. M., Straková, P. et al. (2019) 'Plant roots increase both decomposition and stable organic matter formation in boreal forest soil', *Nature Communication*, vol. 10, 3982.

Aerts, R. (1996) 'Nutrient resorption from senescing leaves of perennials: Are there general patterns?', *Journal of Ecology*, vol. 84, pp. 597–608.

Agerer, R. (2001) 'Exploration types of ectomycorrhizae—A proposal to classify ectomycorrhizal mycelial systems according to their patterns of differentiation and putative ecological importance', *Mycorrhiza*, vol. 11, pp. 107–114.

Albaugh, T., Allen, H. L., Dougherty, P. M., Kress, L. and King, J. S. (1998) 'Leaf area and above- and belowground responses of loblolly pine to nutrient and water additions', *Forest Science*, vol. 44, pp. 317–327.

Averill, C., Turner, B. L. and Finzi, A. C. (2014) 'Mycorrhiza-mediated competition between plants and decomposers drives soil carbon storage', *Nature*, vol. 505, pp. 543–545.

Berger, T. W., Inselbacher, E., Mutsch, F. and Pfeffer, M. (2009) 'Nutrient cycling and soil leaching in eighteen pure and mixed stands of beech (*Fagus sylvatica*) and spruce (*Picea abies*)', *Forest Ecology and Management*, vol 258, pp. 2578–2592.

Bormann, F. H. and Likens, G. E. (1979) *Pattern and Process in a Forested Ecosystem*. Springer-Verlag, New York.

Chadwick, O. A., Derry, L., Vitousek, P. M., Huebert, B. J. and Hedin, L. O. (1999) 'Changing sources of nutrients during four million years of ecosystem development', *Nature*, vol. 397, pp. 491–497.

Chapin, F. S. III, Walker, L. R., Fastie, C. L. and Sharman, L. C. (1994) 'Mechanisms of primary succession following deglaciation at Glacier Bay, Alaska', *Ecological Monographs*, vol. 64, pp. 149–175.

Chapin, F.S. III (1980) 'The mineral nutrition of wild plants', *Annual Review of Ecology and Systematics*, vol. 11, pp. 233–260.

Clemmensen, K. E., Bahr, A., Ovaskainen, O., Dahlberg, A., Ekblad, A. and others (2013) 'Roots and associated fungi drive long-term carbon sequestration in boreal forest', *Science*, vol. 339, pp. 1615–1618.

Colpaert, J. V., van Tichelen, K. K., van Assche, J. A. and Van Laere, A. (1999) 'Short-term phosphorus uptake rates in mycorrhizal and non-mycorrhizal roots of intact *Pinus sylvestris* seedlings', *New Phytologist*, vol. 143, pp. 589–597.

Davidson, E. A. and Janssens, I. A. (2006) 'Temperature sensitivity of soil carbon decomposition and feedbacks to climate change', *Nature*, vol. 440, pp. 165–173.

Davidson, E. A., de Carvalho, C. J. R., Figueira, A. M., Ishida, F. Y., Ometto, J. P. B. and others (2007) 'Recuperation of nitrogen cycling in Amazonian forests following agricultural abandonment', *Nature*, vol. 447, pp. 995–998.

Ericsson, T. (1995) 'Growth and shoot: Root ratio of seedlings in relation to nutrient availability', *Plant and Soil*, vol. 168–169, pp. 205–214.

Finlay, R. D., Mahmood, S., Rosenstock, N., Bolou-Bi, E. B., Köhler, S. J., Fahad, Z., Rosling, A., Wallander, H., Belyazid, S., Bishop, K. and Lian, B. (2020) 'Reviews and syntheses: Biological weathering and its consequences at different spatial levels—from nanoscale to global scale', *Biogeosciences*, vol. 17, pp. 1507–1533.

Gundale, M. J., Deluca, T. H. and Nordin, A. (2011) 'Bryophytes attenuate anthropogenic nitrogen inputs in boreal forests', *Global Change Biology*, vol. 17, pp. 2743–2753.

Hoffland, E., Kuyper, T. W., Wallander, H., Plassard, C., Gorbushina, A. and others (2004) 'The role of fungi in weathering', *Frontiers of Ecology and the Environment*, vol. 2, pp. 258–264.

Jackson, R. B., Mooney, H. A. and Sculze, E. D. (1997) 'A global budget for fine root biomass, surface area and nutrient contents', *Proceedings of the National Academy of Sciences*, vol. 94, pp. 7362–7366.

Kjøller, R., Nilsson, L.-O., Hansen, K., Schmidt, I. K., Vesterdal, L. and Gundersen, P. (2012) 'Dramatic changes in ectomycorrhizal community composition, root tip abundance and mycelial production along a stand-scale nitrogen deposition gradient', *New Phytologist*, vol. 194, pp. 278–286.

Laganière, J., Paré, D. and Bradley, R. L. (2010) 'How does a tree species influence litter decomposition? Separating the relative contribution of litter quality, litter mixing, and forest floor conditions', *Canadian Journal of Forest Research*, vol. 40, pp. 465–475.

Lajtha, K. (2020) 'Nutrient retention and loss during ecosystem succession: revisiting a classic model', *Ecology*, vol. 101, e02896.

Legout, A., Hansson, K., van der Heijden, G., Laclau, J., Mareschal, L., Nys, C., Nicolas, M., Saint-André, L. and Ranger, J. (2020) 'Chemical fertility of forest ecosystems. Part 2: Towards redefining the concept by untangling the role of the different components of biogeochemical cycling', *Forest Ecology and Management*, vol. 461, 117844.

Lilleskov, E. A., Fahey, T. J., Horton, T. R. and Lovett, G. M. (2002) 'Belowground ectomycorrhizal fungal community change over a nitrogen deposition gradient in Alaska', *Ecology*, vol. 83, pp. 104–115.

Lindahl, B. D. and Tunlid, A. (2015) 'Ectomycorrhizal fungi—potential organic matter decomposers, yet not saprotrophs', *New Phytologist*, vol. 205, pp. 1443–1447.

Lindahl, B. D., Kyaschenko, J., Varenius, K., Clemmensen, K. E., Dahlberg, A., Karltun, E. and Stendahl, J. (2021) 'A group of ectomycorrhizal fungi restricts organic matter accumulation in boreal forest', *Ecology Letter*, vol. 24, pp. 1341–1351.

Lovett, G. M., Goodale, C. L., Ollinger, S. V., Fuss, C. B., Ouimette, A. P., Likens, G. E. (2018) 'Nutrient retention during ecosystem succession: a revised conceptual model', *Frontiers in Ecology and Environment*, vol. 16, pp. 532–538.

Markewitz, D., Davidson, E., Moutinho, P. and Nepstad, D. (2004) 'Nutrient loss and redistribution after forest clearing on a highly weathered soil in Amazonia', *Ecological Applications*, vol. 14, pp. S177–S199.

McCarthy, H. R., Oren, R., Johnsen, K. H., Gallet-Budynek, A., Pritchard, S. G. and others (2010) 'Re- assessment of plant carbon dynamics at the Duke free-air CO_2 enrichment site: Interactions of atmospheric $[CO_2]$ with nitrogen and water availability over stand development', *New Phytologist*, vol. 185, pp. 514–528.

Näsholm, T., Högberg, P., Franklin, O., Metcalfe, D., Keel, S. G. and others (2013) 'Are ectomycorrhizal fungi alleviating or aggravating nitrogen limitation of tree growth in boreal forests?', *New Phytologist*, vol. 198, pp. 213–221.

National Research Council (2021) *Basic Research Opportunities in Earth Science*, The National Academies Press, Washington, DC.

Nave, L. E., Vance, E. D., Swanston, C. W. and Curtis, P. S. (2010) 'Harvest impacts on soil carbon storage in temperate forests', *Forest Ecology and Management*, vol. 259, pp. 857–866.

Nave, L. E., Vance, E. D., Swanston, C. W. and Curtis, P. S. (2011) 'Fire effects on temperate forest soil C and N storage', *Ecological Applications*, vol. 21, pp. 1189–1201.

Olson, J. S. (1958) 'Rates of succession and soil changes on southern Lake Michigan sand dunes', *Botanical Gazette*, vol. 119, pp. 125–170.

Peltzer, D. A., Wardle, D. A., Allison, V. J., Baisden, W. T., Bardgett, R. D. and others (2010) 'Understanding ecosystem retrogression', *Ecological Monographs*, vol. 80, pp. 509–529.

Richter, D. D. and Markewitz, D. (2001) *Understanding Soil Change*, Cambridge University Press, Cambridge.

Richter, D. D., Markewitz, D., Heine, P. R., Jin, V., Raikes, J. and others (2000) 'Legacies of agriculture and forest regrowth in the nitrogen of old-field soils', *Forest Ecology and Management*, vol. 138, pp. 233–248.

Rineau, F., Shah, F., Smits, M. M., Persson, P., Johansson, T. and others (2013) 'Carbon availability triggers the decomposition of plant litter and assimilation of nitrogen by an ectomycorrhizal fungus', *ISME Journal*, vol. 7, pp. 2010–2022.

Schulze, E. D., Beck, E. and Muller-Hohenstein, K. (2005) *Plant Ecology*, Springer, Berlin.

Smith, S. E. and Read, D. J. (2008) *Mycorrhizal Symbiosis*, third edition. Academic Press, San Diego, CA.

Sterner, R. W. and Elser, J. J. (2002) *Ecological Stoichiometry, The Biology of Elements from Molecules to the Biosphere*, Princeton University Press, Princeton, NJ.

Switzer, G. L. and Nelson, L. E. (1972) 'Nutrient accumulation and cycling in loblolly pine (*Pinus taeda* L.) plantation ecosystems: The first twenty years', *Soil Science Society of America Journal Proceedings*, vol. 36, pp. 143–147.

Taylor, A. F. S., Martin, F. and Read, D. J. (2000) 'Fungal diversity in ectomycorrhizal communities of Norway spruce (*Picea abies* (L.) Karst.) and beech (*Fagus sylvatica* L.) along north-south transects in Europe', *Ecological Studies*, vol. 142, pp. 343–365.

Thiffault, N., Fenton, N. J., Munson, A. D., Hébert, F., Fournier, R. A. and others (2013) 'Managing understory vegetation for maintaining productivity in black spruce forests: A synthesis within a multi- scale research model', *Forests*, vol. 4, pp. 613–631.

Turpault, M.-P., Nys, C. and Calvarusco, C. (2009) 'Rhizophere impact on the dissolution of test minerals in a forest ecosystem', *Geoderma*, vol. 153, pp. 147–154.

Van Cleve, K., Oliver, L., Schlentner, R., Viereck, L. A. and Dyrness, C. T. (1983) 'Productivity and nutrient cycling in taiga forest ecosystems', *Canadian Journal of Forest Research*, vol. 13, pp. 747–766.

Vergutz, L., Manzoni, S., Porporato, A., Novais, R. F. and Jackson, R. B. (2012) 'Global resorption efficiencies and concentrations of carbon and nutrients in leaves of terrestrial plants', *Ecological Monographs*, vol. 82, pp. 205–220.

Vitousek, P. M. and Reiners, W. A. (1975) 'Ecosystem succession and nutrient retentions: A hypothesis', *BioScience*, vol. 25, pp. 376–381.

Walker, T. W. and Syers, J. K. (1976) 'The fate of phosphorus during pedogenesis', *Geoderma*, vol. 15, pp. 1–19.

Wallander, H. (1995) A new hypothesis to explain allocation of dry matter between ectomycorrhizal fungi and pine seedlings. *Plant and Soil*, vol. 168–169, pp. 243–248.

Wallander, H. and Thelin, G. (2009) 'Influence of phosphorus and potassium fertilization on weathering of apatite in Norway spruce forest with low phosphorus status', *Soil Biology and Biochemistry*, vol. 40, pp. 2517–2522.

Whittaker, R. H., Likens, G. E., Bormann, F. H., Easton, J. S. and Siccama, T. G. (1979) 'The Hubbard Brook ecosystem study: Forest nutrient cycling and element behavior', *Ecology*, vol. 60, pp. 203–220.

Wolfe, B. E., Tulloss, R. E. and Pringle, A. (2012) 'The irreversible loss of a decomposition pathway marks the single origin of an ectomycorrhizal symbiosis', *PLoS One*, vol. 7, pp. 1–9.

24
FOREST HYDROLOGY

André St-Hilaire

Introduction

Forest ecosystems are the largest providers of water on Earth, encompassing 70 percent of the freshwater supply on the planet (Sun et al., 2023). Forest ecozones are characterized by a latitudinal gradient, starting with tropical forests near the equator, temperate forests in mid-latitudes, to boreal forests in latitudes approximately between 50° and 65°. This latitudinal gradient is one of the main causes of the variability of key hydrological processes such as precipitation, interception, water storage, evapotranspiration, and run-off. The objective of this chapter is to summarize some of the recent (2019 or later in most cases) advances in our current knowledge related to the hydrology of forested watersheds and how some anthropogenic impacts may alter hydrological processes and water quality.

Main hydrological processes

Although forested watersheds at low latitudes are pluvial, those located in the boreal forest are generally either nivo-pluvial or fully nival, i.e., their hydrogram variability is dominated by snowmelt in the spring, which generates peak flows and volumes during this season.

In general, forests are seen by hydrologists as water interceptors and evapotranspirators, acting as regulators of run-off and groundwater availability. Many studies have indicated that forested areas have a lower groundwater recharge than elsewhere, but some relatively recent studies indicated that there are some exceptions to this rule. Ouyang et al. (2019) have found that an increase in forested lands was linked to an increase in groundwater recharge in the humid-subtropical ecozone. This may be simply due to the higher availability of water in this ecozone than those located away from the Equator. Costa et al. (2023) made an important distinction between forests with shallow (< 5 m) and deep water tables, the former being underrepresented in recent studies even though such forests encompass 50 percent of the Amazon. Species composition in the Amazon forest depends in part on the depth of the water table. Higher mortality and lower productivity were observed in forests with shallow water table.

Large forested areas are associated with increased cloud cover and precipitation. This is particularly true at low latitude and is associated with high variability of sensible heat fluxes (Li et al., 2022). Interception of rainfall and/or snow is a major redistributor of water. The stem flow also redistributes water both spatially and temporally. It has been stated to be the

DOI: 10.4324/9781003324072-28

most important source of rainwater for trees because it enters the soil near the tree in a relatively small area (Liang, 2020). This author describes the so-called 'double funnelling' effect of stem flow, as it concentrates flow from the canopy along the stem, but also concentrates percolation in the soil near the trees.

In cold regions, hydrological processes become more complex because precipitations vary in phase (liquid vs. solid) according to seasons. In boreal systems, 20–40 percent of the total annual precipitation is solid (He and Pomeroy, 2023). When snow falls in the forested area, a significant portion of solid precipitation is intercepted by the canopy. Snow interception is influenced by canopy closure and structure, and by forest maturity, with older trees intercepting more snow due to larger and stronger branches. In addition, snow is often redistributed by wind, according to the spatial repartition of open and forested areas. Snow storage was shown to decrease as a function of increasing forest area in the watershed and increasing crown or canopy closure. The complex combined impacts of wind and canopy on snow distribution have been summarized by Dickerson-Lange et al. (2021) and are synthesized in Table 24.1.

In most cases, this relatively simplistic view of forest as a water sink via interception, evapotranspiration/sublimation, and stemflow has been validated by a number of paired catchment studies where the control site is forested and the impacted site is fully or partially logged (e.g., Van Loon et al., 2019). This approach allows for comparison of watersheds with similar forest structure and vegetation productivity prior to logging. Peel (2009) noted two important constraints on paired catchment studies: (1) they tend to be completed in small (< 10 km^2) drainage basins and (2) their geographical distribution does not cover all the different climates. However, more general regional studies indicated that the key factor that impacts evapotranspiration and run-off in forested catchments is vegetation productivity, especially in tropical forests (Onuchin et al. 2018). In boreal forests, local studies have shown that canopy interception can vary between 5 percent and 16 percent of total rainfall in the summer and the fraction of intercepted water that is evaporated can represent up to 29 percent of total evapotranspiration (Hadiwijaya et al., 2020).

The hydro-forest nexus (higher water availability → increase in vegetation biomass → higher water demand) is obviously a key determinant of forest ecosystems. Roebroek et al. (2020) exemplify this role for both oases and gallery forests that can tap into groundwater supplies to compensate for possible insufficient precipitation.

Onuchin et al. (2018) also indicated that there are several conflicting studies on the hydrological role of forests. They state that some of the possible confusion stems from the dominance of studies conducted at latitudes lower than those where boreal forests exist. The latter represent nearly 30 percent of all forested ecosystems on the planet. Therefore, more

Table 24.1 Combined impact of canopy and wind on snow storage, according to Dickerson-Lange et al. (2021)

Wind	*Canopy closure*	*Snow storage*
High	High	Medium-high
High	Low	Low
Low	High	Medium
Low	Low	Medium-high

work is still required to fully understand the variability of hydrological process in all forest ecosystems.

Harvesting

The impact of harvesting on the water budget was the focus of many paired-catchment studies in the past century, and some are still ongoing. In fact, Kuraji (2022) states that the number of long-term monitoring sites has been on the rise in recent years. Picchio et al. (2021) reviewed 155 of these studies and concluded that increases in run-off are of the order of 45 percent after harvest. The persistence of increased water yield will depend on harvesting intensity (i.e., percentage of drainage basin area that was logged), but in the case of large clear-cuts, the impact can last for several decades. This is generally true when run-off is compared annually between the harvested and control catchments. However, when studies focus on the low flow periods, a decrease in flow is often associated with harvesting. Coble et al. (2020) reviewed 25 studies conducted in the United States and Canada and indicated that for 16 of the 25 catchments, a decrease in low flows was still observed many years after harvesting.

Li et al. (2022) found that in contrast to the general perception that harvesting and the associated loss of evapotranspiration decrease the net flux of water vapour towards the atmosphere, some small-scale deforestation increases rather than decreases clouds and precipitation in the Amazon. They concluded that the heterogeneity of the land surface (i.e., loss of smoothness at the earth surface) is one of the main causes of this phenomenon.

Groundwater recharge is also affected by harvesting, with notable increases associated with clearing. In their review of 25 studies in semi-arid environment, Sanz et al. (2005) found between 7.8 percent and 12.6 percent post-harvest increase in groundwater recharge.

The impact of thinning on hydrological processes was recently reviewed by del Campo et al. (2022). The researchers conducted a meta-analysis using information from studies covering a large range of climate, forest ecosystems, and time scales. They reasserted that throughfall, net precipitation, and soil moisture generally increase after thinning 50 percent or more of the tree biomass, with percentages of moisture increase varying between 14 percent and over 50 percent. Stemflow and transpiration decreased in most cases, and those impacts can last more than eight years. In their review of the impact of thinning in drought-prone Mediterranean forests, Tague et al. (2019) indicated that declines in leaf and branch area index alter the below-canopy microclimate with greater wind and higher temperature. They state that these changes may offset, in part, the decrease in evapotranspiration associated with thinning. The same authors also noted that in semi-arid environments, increases in bare soil surfaces via thinning can sometimes reduce infiltration because of the formation of a crust at the surface.

The impact of harvesting on water quality has also been investigated for many decades. Most studies in the 20th century focused on two variables of water quality: temperature and sediments. Studies completed during this period and beyond focused in part on the impact of riparian vegetation (or a lack thereof) on stream temperature. St-Hilaire et al. (2023) have reviewed research on lotic thermal regimes in Canada during the last 60 years and concluded from this review that clear-cutting in the riparian zone resulted in sustained increases in river temperature, while thinning had a less consistent impact, depending on its intensity and location along the river system. Several recent studies have advocated for riparian zone maintenance and reforestation as a mitigating factor to maintain cool streams (e.g., Jackson et al., 2021).

The rate of soil erosion is inversely correlated with forest cover, and when the latter is below 60 percent, erosion increases significantly (Kumar et al., 2022). Globally, over the last 100 years, rates of erosion, and therefore sediment yields in rivers, have increased. Deforestation (and the associated road infrastructure) has been identified as one of the main drivers of increased turbidity, along with the subsequent conversion of land use to agriculture (Ervard et al. 2020). For instance, Netzer et al. (2019) focused on the Lower Mekong watersheds and found that from 2001 to 2013, deforestation in the area caused an increase in sediment yields in all of the 22 sub-watersheds included in the study. Maintaining riparian vegetation has also been known to mitigate sediment yield in lotic systems by stabilizing banks (which can supply more than 50 percent of sediments in stream) and filtering run-off.

Other water quality variables are affected by deforestation. Several studies have shown that riparian vegetation acts as a sink for nutrients, with removal rates exceeding 80 percent in some cases (Singh et al., 2021). Vegetation also acts as a global sink in mercury biogeochemical cycling, which means that deforestation can result in mobilization of this element (Sonke et al., 2023).

Fires

While rivers that drain natural forested areas generally have good water quality (Klimas et al. 2020), fires, like harvesting, can alter the water quantity and quality in many ways. Some of the differences between pre- and post-fire hydrology are similar to those identified for harvesting, that is, loss of interception, evapotranspiration, and faster snowmelt in northern regions, thereby altering the timing and amplitude of spring floods. In the western United States, a review of 78 sites where fires occur showed that the maximum storage of water in the solid phase decreased by more than 30 mm/year, and the date at which snow has completely melted advanced by 9 days on average. Faster snowmelt, by as much as 3 mm/day, was also observed in high-severity burnt forests (Smoot and Gleason, 2021).

Although forest fires are typically associated with loss of evapotranspiration, in some cases, such as in central Chile where 500,000 ha burnt in 2017, evapotranspiration actually increased in the following years because of the high demand from resprouted forest (Pacheo and Fernandes, 2021). The variability in post-fire hydrological response of watersheds was also highlighted by Robinne et al. (2019) in their review of 82 Canadian/American studies on wildfires, but changes were observed in mean annual or seasonal flows, peak flow timing, and stream flashiness in 23 percent of the cases. The review of Agbeshie et al. (2022) on the impacts of forest fires concluded that high intensity fires change the soil properties and typically increase bulk density and hydrophobicity, leading to less water infiltration and more erosion.

Bare soils and increased run-off can increase sediment loads significantly after forest fires. Sun et al. (2023) stated in their literature review that forest fire-hydrology research has largely focused on sediment export over the last few decades and that predictive tools are lacking. Despite this, Robinne et al. (2020) indicated that numerous Canadian and American studies observed post-fire increases in suspended sediment concentrations and bedload transport. Chen and Chang (2023) also reviewed 62 studies and concluded that post-fire increases in turbidity are positively correlated with fire intensity, and also with the degree of anthropogenic post-fire activities such as salvage logging.

Forest fires also lead to larger exposure of streams to incident solar radiation and, hence, to increases in water temperature. Chen and Chang (2023) indicated that these increases tend to be site specific and vary with topography. In their study in the western United States, Alizadeh et al. (2021), noted that post-fire stream temperature increases along the burnt areas can be 2 to 3 times higher than in unburnt portions of the watershed.

Other water quality variables can also be affected by fires. For instance, Spence et al. (2020) found that the concentrations of dissolved organic carbon and nutrient were higher in a partially burnt catchment than in an unburnt paired catchment in the Canadian boreal forest. This is corroborated by many other studies that indicate an increase in post-fire nitrogen input through combustion, oxidization, and ash deposition (Gustine et al. 2022). Robinne et al. (2020) also concluded that severe fires in the Canadian Rocky Mountains result in increased concentrations of nitrogen and phosphorus in boreal streams.

Diseases

The impact of parasitic outbreaks on watershed hydrology, chiefly those of Mountain Pine Beetles (MPB) in mid- to northern latitudes, has been studied for over 20 years. Recent studies reiterated the well-known fact that canopy loss, which is one of the results of MPB infection, decreases evapotranspiration and therefore increases water yields, especially in wet years (Ren et al. 2021). In dry years, however, if the areal coverage of MPB-caused mortality is high, the water yield can actually decrease, possibly due to high ground evaporation. Another recent study focused on root drainage in stands with and without MPB attacks in Alberta (Canada). Root drainage did not vary significantly between stands. Recent studies have all underlined the complexity of the impact of MPB on hydrology, with high temporal and spatial variability (Goodbrand, 2021).

Modelling

Forest watersheds are increasingly subject to multiple stressors and management increases in complexity with the multiplication of objectives that include ecosystem health, water conservation, carbon sequestration, economic considerations, etc. (Sun et al., 2023). In this context, modelling tools have become more important and more complex over the past few decades. Sun et al. (2023) critically reviewed a considerable number (47) of forest hydrological models, with a view to examine how they can be of assistance in addressing issues of ecosystem services trade-offs, climate change adaptation, cumulative effects, urban forestry, and watershed restoration. They noted that approximately 15 percent of the publications related to distributed hydrological models emphasized forest ecosystems. During the last couple of decades, an increase in the effort to couple carbon and water cycles in models has been observed.

Given the strong feedback loop between forests and water, a relatively large modelling effort has targeted possible mitigation/adaptation to climate change through forest management. One of the challenges associated with this effort is the relatively coarse resolution of climate models, which limits their ability to capture adequately land surface–atmosphere interactions. Dynamical or statistical downscaling of the global model outputs is required to provide greater insight in local interactions (e.g., at the watershed scale). Some of the scenarios modelled indicate that future vegetation changes in forested watersheds can mitigate some of the expected changes in snow accumulation and evapotranspiration. For instance, Rasouli et al. (2019) used the Cold Regions Hydrological Modelling platform in northwestern Canada and found that the projected increase in vegetation in some watersheds can compensate for the predicted decrease in snow accumulation.

Models are also slowly making their way into the operational management of harvesting operations. For instance, Salmivaara et al. (2020) used the Spatial Forest Hydrology Model in combination with a k-nearest neighbour (machine learning) algorithm to predict forest trafficability, which is defined by two key characteristics: rut depths and rolling resistance.

Some hydrological models have been used to simulate the impacts of forest fires. Zema et al. (2020) adjusted the Morgan–Morgan–Finney (MMF) model to better predict those impacts. Changes in evapotranspiration and soil hydrological depths were required to better predict post-fire seasonal run-off volumes and sediment yields in Mediterranean forest watersheds. Although routine application of complex forest-fire-hydrological models is not yet widespread, the increase in data availability, data storage capacity, and computing power will enable forest managers to include such tools in future management.

Coupling hydrological models with vegetation models can also be useful in economic analyses. For instance, in their recent analyses of the Huanjiang karst mountainous region of China, Zhan et al. (2020) used a hydrological model to investigate the impact of afforestation and found that an increase in forest cover by 10–30 percent would increase the sustainability of the system and would have a positive impact in maintaining the regions' gross domestic product (GDP).

Some attempts have also been made to couple forest hydrological models with simple water quality models to generate scenarios. An example is from Southern Portugal, where Rodrigues et al. (2020) used the Revised Universal Soil Loss Equation, which accounts for forest composition via a cover management factor, to investigate future soil erosion.

Conclusion

The recent literature briefly summarized in this chapter shows that our understanding of basic hydrological processes has increased in the last few decades. The forest–water interaction has garnered increased interest in view of the pending climate crisis. Forests are seen as important mitigators of local and regional shifts in river/stream hydrograms caused by anthropogenic impacts and in water quality. The modelling community has been highly active and new tools are being developed to assist managers in critical decision-making.

References

Agbeshie, A. A., Abugre, S., Atta-Darkwa, T. and Awuah, R. (2022) 'A review of the effects of forest fire on soil properties', *Journal of Forestry Research*, vol 33, pp. 1419–1441.

Alizadeh, M. R., Abatzoglou, J. T., Luce, C. H., Adamowski, J. F., Farid, A. and Sadegh, M. (2021) 'Warming enabled upslope advance in western US forest fires', *Proceedings of the National Academy of Sciences*, vol 118, e2009717118.

Chen, J. and Chang, H. (2023) 'A review of wildfire impacts on stream temperature and turbidity across scales', *Progress in Physical Geography: Earth and Environment*, vol 47, pp. 369–394.

Coble, A. A., Barnard, H., Du, E., Johnson, S., Jones, J., Keppeler, E. and Wagenbrenner, J. (2020) 'Long-term hydrological response to forest harvest during seasonal low flow: Potential implications for current forest practices', *Science of the Total Environment*, vol 730, 138926.

Costa, F. R., Schietti, J., Stark, S. C. and Smith, M. N. (2023) 'The other side of tropical forest drought: do shallow water table regions of Amazonia act as large-scale hydrological refugia from drought?', *New Phytologist*, vol 237, pp. 714–733.

del Campo, A. D., Otsuki, K., Serengil, Y., Blanco, J. A., Yousefpour, R. and Wei, X. (2022) 'A global synthesis on the effects of thinning on hydrological processes: Implications for forest management', *Forest Ecology and Management*, vol 519, 120324.

Dickerson-Lange, S. E., Vano, J. A., Gersonde, R. and Lundquist, J. D. (2021) 'Ranking forest effects on snow storage: A decision tool for forest management', *Water Resources Research*, vol 57, e2020WR027926.

Evrard, O., Chaboche, P. A., Ramon, R., Foucher, A. and Laceby, J. P. (2020) 'A global review of sediment source fingerprinting research incorporating fallout radiocesium (137Cs)', *Geomorphology*, vol 362, 107103.

Goodbrand, A. R. (2021) *Mountain pine beetle and forest harvest effects on hydrologic processes and streamflow in the Alberta Foothills.* PhD thesis, University of Alberta.

Gustine, R. N., Hanan, E. J., Robichaud, P. R. and Elliot, W. J. (2022) 'From burned slopes to streams: How wildfire affects nitrogen cycling and retention in forests and fire-prone watersheds', *Biogeochemistry*, vol 157, pp. 51–68.

Hadiwijay, B., Isabelle, P. E., Nadeau, D. F. and Pepin, S. (2021) 'Can a physically-based land surface model accurately represent evapotranspiration partitioning? A case study in a humid boreal forest', *Agricultural and Forest Meteorology*, vol 304–305, 108410.

He, Z. and Pomeroy, J.W. (2023) 'Assessing hydrological sensitivity to future climate change in the Canadian southern boreal forest', *Journal of Hydrology*, vol 624, 129897.

Jackson, F. L., Hannah, D. M., Ouellet, V. and Malcolm, I. A. (2021) 'A deterministic river temperature model to prioritize management of riparian woodlands to reduce summer maximum river temperatures', *Hydrological Processes*, vol 35, e14314.

Klimas, K., Hiesl, P., Hagan, D. and Park, D. (2020) ' Prescribed fire effects on sediment and nutrient exports in forested environments: A review', *Journal of Environmental Quality*, vol 49, pp. 793–811.

Kumar, R., Kumar, A. and Saikia, P. (2022) 'Deforestation and forests degradation impacts on the environment', in V. P. Singh, S. Yadav, K. K. Yadav and R. N. Yadava (eds.), *Environmental Degradation: Challenges and Strategies for Mitigation.* Water Science and Technology Library, vol 104, pp. 19–46.

Kuraji, K. (2022) 'Long-term monitoring and research in forest hydrology: Towards integrated watershed management', *Water*, vol 14, 2556.

Li, W., Franssen, H. J. H., Brunner, P., Li, Z., Wang, Z., Wang, Y. and Wang, W. (2022) The role of soil texture on diurnal and seasonal cycles of potential evaporation over saturated bare soils–Lysimeter studies. *Journal of Hydrology*, vol. 613, 128194.

Liang, W. L. (2020) 'Effects of stemflow on soil water dynamics in forest stands' in D. F. Levia, D. E. Carlyle-Moses, S. Lida, B. Michalzik, K. Nanko and A. Tischer (eds.), *Forest-Water Interactions*, Ecological Studies, vol 240, pp. 349–370.

Netzer, M. S., Sidman, G., Pearson, T. R., Walker, S. M. and Srinivasan, R. (2019) 'Combining global remote sensing products with hydrological modeling to measure the impact of tropical forest loss on water-based ecosystem services', *Forests*, vol 10, 413.

Onuchin, A. A., Burenina, T. A., Balzter, H. and Tsykalov A. G. (2018) 'New look at understanding hydrological role of forest', *Siberian Journal of Forest Science*, vol 5, pp. 3–18.

Ouyang, Y., Jin, W., Grace, J. M., Obalum, S. E., Zipperer, W. C. and Huang, X. (2019) 'Estimating impact of forest land on groundwater recharge in a humid subtropical watershed of the Lower Mississippi River Alluvial Valley', *Journal of Hydrology: Regional Studies*, vol 26, 100631.

Owuor, S. O., Butterbach-Bahl, K., Guzha, A. C., Rufino, M. C., Pelster, D. E., Díaz-Pinés, E. and Breuer, L. (2016) 'Groundwater recharge rates and surface runoff response to land use and land cover changes in semi-arid environments', *Ecological Processes*, vol 5, 16.

Pacheco, F. A. and Fernandes, L. F. S. (2021) 'Hydrology and stream water quality of fire-prone watersheds', *Current Opinion in Environmental Science and Health*, vol 21, 100243.

Peel, M. C. (2009) 'Hydrology: Catchment vegetation and runoff', *Progress in Physical Geography*, vol 33, pp. 837–844.

Picchio, R., Jourgholami, M. and Zenner, E. K. (2021) 'Effects of forest harvesting on water and sediment yields: A review toward better mitigation and rehabilitation strategies', *Current Forestry Reports*, vol 7, pp. 214–229.

Rasouli, K., Pomeroy, J. W. and Whitfield, P. H. (2019) 'Are the effects of vegetation and soil changes as important as climate change impacts on hydrological processes?', *Hydrology and Earth System Sciences*, vol 23, pp. 4933–4954.

Ren, J., Adam, J. C., Hicke, J. A., Hanan, E. J., Tague, C. L., Liu, M. and Abatzoglou, J. T. (2021) 'How does water yield respond to mountain pine beetle infestation in a semiarid forest?', *Hydrology and Earth System Sciences*, vol 25, pp. 4681–4699.

Robinne, F. N., Hallema, D. W., Bladon, K. D. and Buttle, J. M. (2020) 'Wildfire impacts on hydrologic ecosystem services in North American high-latitude forests: A scoping review', *Journal of Hydrology*, vol 581, 124360.

Rodrigues, A. R., Botequim, B., Tavares, C., Pécurto, P. and Borges, J. G. (2020) 'Addressing soil protection concerns in forest ecosystem management under climate change', *Forest Ecosystems*, vol 7, 34.

Roebroek, C. T., Melsen, L. A., Hoek van Dijke, A. J., Fan, Y. and Teuling, A. J. (2020) 'Global distribution of hydrologic controls on forest growth', *Hydrology and Earth System Sciences*, vol 24, pp. 4625–4639.

Salmivaara, A., Launiainen, S., Perttunen, J., Nevalainen, P., Pohjankukka, J., Ala-Ilomäki, J. and Finér, L. (2020) 'Towards dynamic forest trafficability prediction using open spatial data, hydrological modelling and sensor technology', *Forestry*, vol 93, pp. 662–674.

Sanz, D. B., del Jalón, D. G., Teira, B. G. and Martínez, P. V. (2005) 'Basin influence on natural variability of rivers in semi-arid environments', *International journal of river basin management*, vol 3, pp. 247–259.

Singh, R., Tiwari, A. K. and Singh, G. S. (2021) 'Managing riparian zones for river health improvement: An integrated approach', *Landscape and ecological engineering*, vol 17, pp. 195–223.

Smoot, E. E. and Gleason, K. E. (2021) 'Forest fires reduce snow-water storage and advance the timing of snowmelt across the Western U.S.', *Water*, vol 13, 3533.

Sonke, J. E., Angot, H., Zhang, Y., Poulain, A., Björn, E. and Schartup, A. (2023) 'Global change effects on biogeochemical mercury cycling', *Ambio*, vol 52, pp. 853–876.

Spence, C., Hedstrom, N., Tank, S. E., Quinton, W. L., Olefeldt, D., Goodman, S. and Dion, N. (2020) 'Hydrological resilience to forest fire in the subarctic Canadian shield', *Hydrological Processes*, vol 34, pp. 4940–4958.

St-Hilaire, A., Oyinlola, M., E. Rincon, E. and Ferchichi, H. (2023) 'Six decades of research on river temperature in Canada', *Canadian Water Resources Journal*, vol 48, pp. 450–474.

Sun, G., Wei, X., Hao, L., Sanchis, M. G., Hou, Y., Yousefpour, R. and Zhang, Z. (2023) 'Forest hydrology modeling tools for watershed management: A review', *Forest Ecology and Management*, vol 530, 120755.

Tague, C. L., Moritz, M. and Hanan, E. (2019) 'The changing water cycle: The eco-hydrologic impacts of forest density reduction in Mediterranean (seasonally dry) regions', *WIREs Water*, vol 6, e1350.

Van Loon, A. F., Rangecroft, S., Coxon, G., Breña Naranjo, J. A., Van Ogtrop, F. and Van Lanen, H. A. 2019. Using paired catchments to quantify the human influence on hydrological droughts. *Hydrology and Earth System Sciences*, vol 23, pp. 1725–1739.

Xu, R., Li, Y., Teuling, A. J., Zhao, L., Spracklen, D. V., Garcia-Carreras, L. and Fu, B. (2022) 'Contrasting impacts of forests on cloud cover based on satellite observations', *Nature. Communications*, vol 13, 670.

Zema, D. A., Nunes, J. P. and Lucas-Borja, M. E. (2020) 'Improvement of seasonal runoff and soil loss predictions by the MMF (Morgan-Morgan-Finney) model after wildfire and soil treatment in Mediterranean forest ecosystems', *Catena*, vol 188, 104415.

Zhan, C., Zhao, R. and Hu, S. (2020) 'Energy-based sustainability assessment of forest ecosystem with the aid of mountain eco-hydrological model in Huanjiang County, China', *Journal of Cleaner Production*, vol 251, 119638.

25

PRIMARY PRODUCTION AND ALLOCATION IN FOREST ECOSYSTEMS

Frank Berninger and Kelvin S.-H. Peh

Introduction

Photosynthesis produces almost all dry biomass in plants, which is a key factor in primary global production. Increased photosynthesis is of interest to farmers and foresters. However, the relationship between photosynthesis and dry biomass production is complicated in some systems, such as tropical forests (see Chapter 5, 'Tropical forests'), which have high photosynthesis but low wood biomass production. In addition, there is a moderate correlation between the annual variations in photosynthetic production and the growth of stem biomass, highlighting the uncertain relationship between these two interconnected traits and suggesting that the percentage of photosynthetic production used for stem growth varies between years and locations (Rocha *et al.* 2006; Gea-Izquierdo *et al.* 2014). This chapter examines the complex relationships between photosynthesis and wood growth, the factors affecting photosynthetic production and the way plants allocate photosynthesis products for biomass production.

Primary production

Primary production is the synthesis of organic compounds from atmospheric carbon dioxide, mainly through photosynthesis. Primary producers, including plants and algae, are responsible for this process. This chapter focuses on plants. Gross Primary Production (GPP) is the total amount of carbon that plants fix as carbohydrates in photosynthesis, but not all contribute to plant growth due to loss of plant respiration, also known as autotrophic respiration (H_A). After autotrophic respiration, the remaining portion of the GPP is the *Net Primary Production* (NPP), the rate of plant new organic matter production. However, not all NPPs contribute to the increase in net biomass, as a portion is lost due to senescence and plant mortality. The relationship can be explained as follows:

Biomass production = GPP – H_A – senescence – mortality
= NPP – senescence – mortality

DOI: 10.4324/9781003324072-29

When analysing the biomass accumulation in tropical forests, we only consider wood biomass, but non-woody tissue, such as plant leaves and roots, also uses some resources available for biomass production. Therefore, wood production is lower than total (biomass) production. Another factor to consider for forest-level production is the carbon balance, where a large pool of carbon in the soil can be lost by heterotrophic respiration (H_R). Carbon loss can also be caused by dissolved organic carbon in flow, but usually not significant. Although senescence and mortality are important for calculating GPP, the carbon balance in forests is not affected immediately, as dead biomass remains in the soil until it decomposes. However, harvesting entire trees or parts of them immediately removes carbon from the ecosystem. Therefore, the carbon balance of the ecosystem is the difference between GPP and heterotrophic and autotrophic respiration, minus carbon losses in run-off.

Ecosystem carbon balance = GPP – H_A – H_R – carbon lost through run-off

Figure 25.1 describes the various components of photosynthetic production, growth, and carbon balances for three forest types, and Table 25.1 provides definitions of the terminology used. All three types depicted in Figure 25.1 exhibit a high rate of photosynthetic production, which would result in a high growth rate if all photosynthetic products were used for stem growth. For information on typical GPP and NPP values of different biomes, see Chapter 38 entitled 'Forest carbon budgets and climate change'.

Environmental effects on photosynthetic production

Land availability is a key factor that limits plant productivity at a site scale (Running 2012). However, primary production is also influenced by various other environmental factors, including temperature, solar radiation, water, nutrients, and plant respiration, which can have both positive and negative effects. For example, plants grown in warm and wet environments generally exhibit higher productivity levels compared to those grown in colder regions of the world. This chapter will provide an overview of how some of these environmental factors affect photosynthetic production.

Light and temperature

The first step in photosynthesis is the absorption of light by leaf chlorophyll, and unabsorbed photons do not contribute to photosynthesis. As a result, low leaf areas lead to a reduction in light absorption and a decrease in photosynthetic production. In addition to the leaf area index (LAI), which represents the total leaf area projected per ground area, the spatial layout of leaf plays an important role in light absorption and photosynthesis. Empirical measurements and models provide several, albeit imperfect, ways to estimate light interception in forest stands. For example, hemispheric photographs captured with fish-eye lenses help to calculate how much the free sky is visible and the amount of light in each direction. These parameters help determine the light interception of the canopy above the sampling point. Direct measurements of light reflection using satellite images are another technique. Nevertheless, each method is based on non-trivial assumptions regarding the distribution of leaves and light interception by non-photosynthetic organs such as tree stems. Consequently, different techniques tend to produce different estimates of intercepted radiance. This discrepancy was caused by different assumptions in different methods (Chen *et al.* 2006).

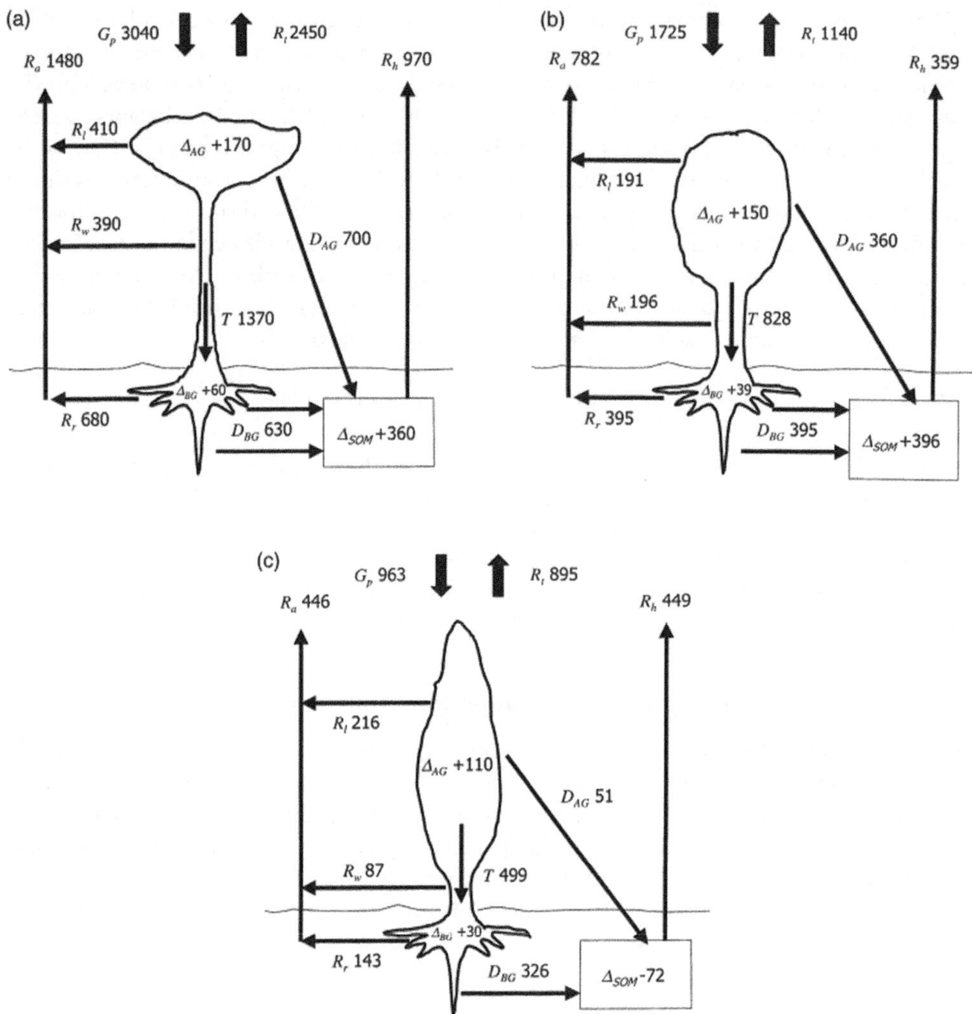

Figure 25.1 Productivity of a boreal, a temperate, and a humid tropical ecosystem (all in g C m^{-2} yr^{-1}). Forests are (a) tropical evergreen rainforest from the Amazon (Brazil), (b) a deciduous broadleaf forest from Tennessee (USA), and (c) a northern boreal black spruce forest from Canada. G_p = gross primary productivity (GPP), R_t = ecosystem respiration, R_a = autotrophic respiration, R_w = aboveground wood respiration, DAG = aboveground senescence and mortality, DBG = belowground senescence and mortality, ΔAG = aboveground net biomass carbon increment; ΔBG = belowground net biomass carbon increment; ΔSOM = net increment in soil organic carbon. (Reproduced from Malhi et al., 1999.)

Photosynthetic production in a leaf in response to the light it intercepts is a curvilinear relationship (Monteith 1994)—it saturates at high light intensities, causing the leaf's instantaneous light use efficiency ($LUE_{L,}$ the rate of photosynthesis divided by the quantity of light absorbed by the leaf) to decrease. However, there are different definitions of light use efficiency, and readers should understand which definition is being used in each scientific paper.

Table 25.1 Definition of different productivity terms

GPP	Gross primary production	Photosynthetic production of all organisms in an ecosystem excluding losses by respiration
H_A	Autotrophic respiration	Respiration of photosynthesizing organisms
H_R	Heterotrophic respiration	Respiration of non-photosynthesizing organisms
NPP	Net primary production	$NPP = GPP - H_A$ Growth of biomass of all plants excluding losses of biomass through litter production and herbivory
NEE	Net ecosystem exchange	$NEE = H_R + H_A - GPP = H_R - NPP$ Difference of respiration and the gross primary production. Net carbon balance of an ecosystem. Negative numbers usually indicate a carbon sink and positive numbers a carbon source to the atmosphere (see NEP below)
NEP	Net ecosystem production	$NEP = -NEE = GPP - H_A - H_R$ Usually -1 × NEE.
NBP	Net biome productivity	Carbon balance for large regions. Average NEP minus losses from disturbance.
I_B	Net Biomass growth	$NPP - L - M$ Biomass growth in an ecosystem including losses from litter and tree mortality. Can be measured as the different of biomass at two points in time.
L	Litter production	Death of biomass. Usually excluding sapwood turnover and litter production due to tree mortality.
M	Tree mortality	Biomass loss due to the mortality of whole trees.

At low light levels, initial light use efficiency—or the light use efficiency of photosynthesis—is relatively constant across plant species and slightly higher than the theoretical rate of photosynthesis, derived from the stoichiometry of photosynthesis (Long *et al.* 1993). Canopies have a higher light use efficiency than single leaves at medium and high light intensities because shaded leaves in the canopies operate at higher light use efficiencies, while sunlit leaves operate at lower ones (Long *et al.* 1993).

Figure 25.2 illustrates the relationship between photosynthetic production and light levels in leaves and canopies of Scots pine at Hyytiälä, both in the immediate and long term. While the relationship between intercepted irradiance and photosynthesis is nonlinear for a single leaf at a moment, a linear relationship emerges for canopy production over longer periods, which was first described by John Monteith about 20 years ago (Monteith 1994; Turner *et al.* 2005) and is still widely used in the remote sensing-based products to estimate plant production (see Running *et al.* 2004 for a description of the MODIS GPP algorithm). However, this linear relationship is only an approximation, and other environmental factors can affect plant light use efficiency and productivity, which must be considered. This is discussed in more detail below.

Temperature is a key environmental factor that affects the relationship between absorbed irradiance and productivity, and limits photosynthetic production in various ways. Although photosynthetic rates decline at very high and low temperatures, photosynthetic production has a broad temperature optimum, and its temperature response can adapt to changing environmental conditions (Hikosaka *et al.* 2006). However, long-lasting effects due to low temperatures can be observed in boreal forests, where low rates of photosynthesis during winter months are partially due to suboptimal temperatures and deactivation of the photosynthetic apparatus. The recovery process of these photosynthetic systems is driven by

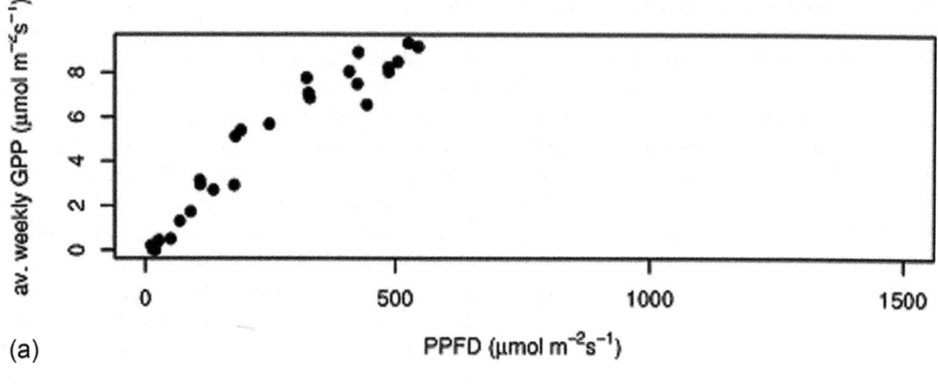

(a)

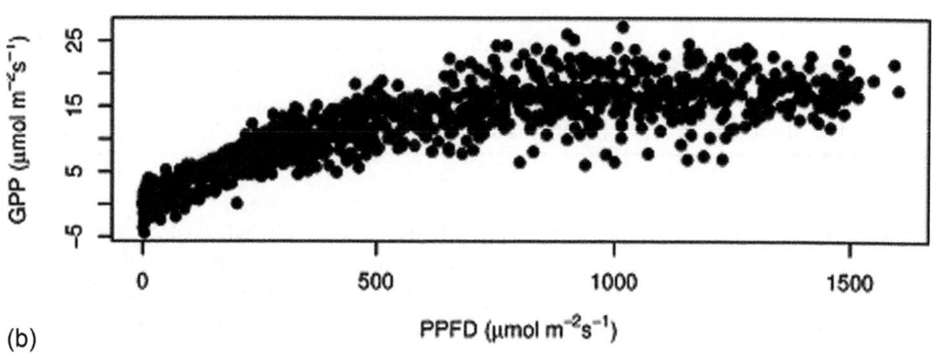

(b)

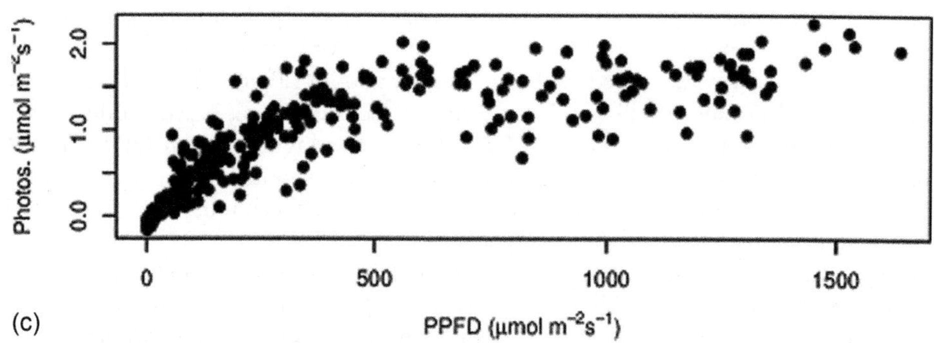

(c)

Figure 25.2 Rates of photosynthesis of Scots pine at different light levels. (a) Long-term weekly responses of gross primary production to incoming photosynthetically active radiation. (July–November 2010), (b) the half hourly response of gross primary production to incoming photosynthetically active radiation (July 2010), and (c) the response of a single Scots pine shoot to photosynthetically active radiation (July 2008). Note that the relationship becomes more linear from the bottom to the top of the graphs. Data is from Hyytiälä SMEARII.

temperature (Mäkelä *et al.* 2004; Suni *et al.* 2003) and is relatively slow, taking several weeks to complete. The recovery process is slower in more northern boreal ecosystems than in southern ecosystems (Gea *et al.* 2010). Douglas fir systems in maritime environments have minimal differences between winter and summer states of photosynthesis due to the relatively high winter temperatures. On the other hand, continental boreal coniferous forests show a slow recovery process that lasts several weeks (Gea *et al.* 2010).

Water

Plants rely on water availability for survival and growth, making it the most limiting environmental factor for plant productivity. Water is essential for photosynthesis and other functions, including transpiration, which transports water from the roots to the rest of the plant and in turn facilitates the movement of nutrients from the soil to the growing tissues of the plant. Transpiration is driven by diffusion through stomatal pores, which also controls CO_2 uptake (Chaves *et al.* 2003). During drought, stomatal aperture is reduced, limiting nutrient transport and CO_2 uptake, leading to reduced productivity. Experimental evidence shows that mean annual precipitation strongly affects net primary productivity (NPP) across ecosystems (Sala *et al.* 2012).

Studies linking productivity to multiple environmental factors have shown that soil water and nitrogen levels are not independent in Patagonian steppe systems, as soil water controls nitrogen levels (Austin and Sala 2002) because many well-drained drier soils have more open nitrogen cycles. With the predicted increase in drought frequency and severity, productivity in many regions is expected to decrease in the future.

Nutrients

Nitrogen is essential for photosynthetic production and its concentration in leaves affects the GPP of trees through multiple mechanisms. About half of the nitrogen in leaves is found in the main photosynthetic enzyme Ribulose-1,5-bisphosphate carboxylase/oxygenase (RuBisCO), which strongly correlates with the photosynthetic capacity of leaves (Reich *et al.* 1995), especially for sun leaves and species with low LAI (where there is no effect on light use efficiency at low light intensities). Additionally, leaves with higher nitrogen concentrations tend to have a larger leaf area per dry leaf mass (Coll *et al.* 2011), which increases the interception of light per unit of leaf mass. Nitrogen has a greater impact on photosynthetic capacity in deciduous broad-leaf trees than in conifers (Reich *et al.* 1995; Figure 25.3), possibly due to the dual role of foliage in nitrogen storage and metabolism, which is more pronounced in conifers (Vapaavuori *et al.* 1995).

Nitrogen is typically considered the primary limiting nutrient for plant productivity in temperate and boreal forests, while in tropical forests, phosphorus availability is the main factor affecting productivity (Quesada *et al.* 2012). Phosphorus is crucial for adenosine triphosphate and sugar phosphate metabolism, and its deficiency may limit photosynthesis and primary production at the community level. However, tropical montane forests and some lowland tropical forests may still have limited productivity due to nitrogen availability (Quesada *et al.* 2012).

In temperate and boreal forests, base cations can enter the system through atmospheric deposition. High levels of air pollution causing a drastic increase in base cations, especially extractable calcium, and magnesium, can negatively affect nutrient cycling and tree

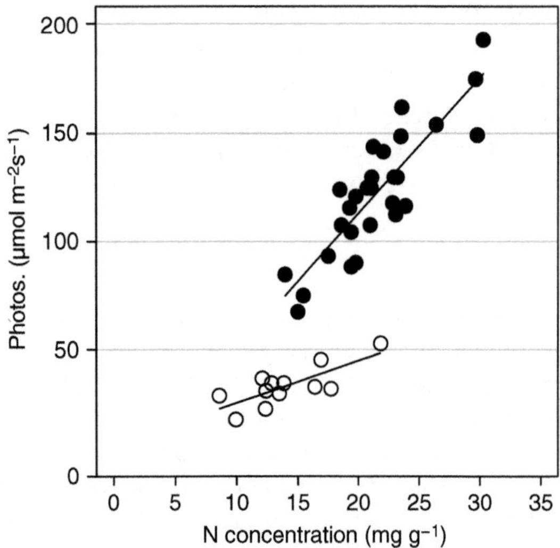

Figure 25.3 Rates of photosynthesis of different coniferous (closed symbols) and deciduous (open symbols) species in Wisconsin as a function of the leaf nitrogen concentration. (Redrawn from Reich et al., 1995.)

productivity (see Chapter 23, 'Biogeochemical cycling'). Deficiencies in other nutrients may also limit tree productivity, especially in anthropologically modified habitats like old field and drained peatland.

Soil resources are critical for plant productivity and are responsible for a significant portion of wood and leaf NPP when calcium, nitrogen, and water availability are considered (Baribault *et al.* 2010). These factors are likely to act in concert with other environmental factors to jointly regulate productivity (Baribault *et al.* 2010).

Respiration

About half of the carbon produced through photosynthesis is lost through respiration, and respiration rates increase with temperature. Trees in warm regions may have higher carbon losses through respiration than those in cold regions (Berninger and Nikinmaa 1997). Respiration can be divided into three components (Warren-Wilson 1967): maintenance respiration, growth respiration, and substrate-induced respiration. Maintenance respiration, which relies on metabolic enzyme turnover, is temperature-dependent, while growth respiration and substrate-induced respiration depend on growth rates and the chemical composition of new biomass. Growth respiration is determined by chemical composition of the tissues, directly reflecting the metabolic costs of producing new biomass, which are higher for energy-rich tissues like protein-rich leaves than for cellulose-rich plant stems (Amthor 2000).

Although respiration in trees is dependent on temperature, its rate does not increase from high to low latitudes at a global scale. Trees are capable of acclimating to higher temperatures, and their respiration rate only declines if they were exposed to such temperatures for extended periods of time. As a result, Waring *et al.* (1998) proposed that, when considering large geographical and environmental ranges, respiration approximates half of all photosynthetic production. The ratio of GPP to respiration may decrease with tree size, but the effect is likely insignificant (Mäkelä and Valentine 2001). Therefore, it can be concluded that under current climatic conditions, respiration accounts for approximately half of GPP. However, this ratio is not constant and may be influenced by climate variability.

Species diversity and composition

The production of plant biomass provides energy for heterotrophs. With the high rate of species loss in tree-dominated ecosystems due to human disturbance, it is increasingly important to understand the consequences of declining species diversity on plant production. 'Natural' experiments in Mediterranean (Vilà *et al.* 2007), temperate (Caspersen and Pacala 2001), and tropical forests (Peh 2009), as well as tropical tree plantations (Erskine *et al.* 2006), have shown a positive relationship between species richness and wood production at the landscape scale. The association between species diversity and NPP is thought to occur through two mechanisms: (1) areas with more species may have increased productivity due to resource-use complementarity and facilitation, or (2) production may depend heavily on the traits of individual species and their presence or absence.

Species composition can affect productivity as certain species may grow faster than others under specific resource availability conditions. In northern hardwood forests, *Quercus rubra* has been found to grow faster than *Quercus alba* under similar resource conditions (Comas and Eissenstat 2004). The species composition of a stand can therefore affect the growth response to resource availability. Additionally, species composition can affect resource levels through resource use (Fujinuma *et al.* 2005) and litter cycling rates (Dijkstra 2003). For example, abundance of *Tilia americana* and *Acer saccharum* is strongly correlated with calcium availability, and the latter is also associated with increased nitrate levels (Baribault *et al.* 2010). However, species abundance in the northern hardwood forest is not correlated with NH_4+ availability.

Plant allocation

Plants must distribute resources obtained from the environment and those produced by the plant to various tissues and important functions—such as growth, reproduction, and defence—in a balanced manner. This allocation is regulated by several factors. When considering allocation from a productivity perspective, it is important to keep the intended outcome in mind. For example, in agriculture, the harvest index is a critical factor that measures productivity, and allocation to reproduction is crucial for fruit production. On the other hand, in tree plantations, stem growth is considered the most important allocation aspect since it is the site of wood production.

Productivity and allocation are closely linked, as growth is determined by the amount of NPP allocated to different plant tissues. In practice, increasing crop and tree yields is often achieved by altering allocation patterns rather than by increasing NPP itself. For example, although grain yield has increased significantly in agriculture over the last century, genetic improvements in NPP have been negligible, and the gains have instead come from changes in allocation (Richards 2000). Improved agricultural practices, such as the use of fertilizers, have also contributed to the increase in wheat productivity. However, it is not clear how productivity and allocation are related in woody plant species (Pulkkinen *et al.* 1989).

Allocation is based on the transport of carbohydrates from the leaves to the sites of growth. While water is transported in the xylem down a water potential gradient that runs from the roots to the leaves, carbohydrates in the phloem are transported via an active process called translocation. In translocation, the solutes move from mature leaves (sources) to sinks, which can be any other organ, such as roots, stems, young leaves, or seeds. The direction of translocation is influenced by plant allocation and can change throughout the season. For example, early in the season translocation can primarily occur downwards from the leaves to the roots, before shifting later in the season when allocation redirects to the flowers and seeds of the plant. Hormonal regulation also plays a role, with auxins and gibberellins increasing shoot growth and abscisic acid promoting root allocation (for details, see Taiz and Zeiger 2010).

The regulation of translocation of carbohydrates and the related plant growth in trees is not well understood quantitatively. Mathematical models are often used to analyse the complex process of translocation. Water relations are important for translocation (Nikinmaa *et al.* 2013), but these models have not yet been applied to long-term tree development. A simpler approach by Thornley (1991) modelled trees as interconnected labile carbon and labile nitrogen reservoirs. Transport of carbon and nitrogen depended on the concentration differences of these elements between these reservoirs, with growth proportional to the product of carbon and nitrogen concentrations. Although this extremely simplified approach simulates a realistic forest stand development, it has not been widely used.

Allocation between fine roots and leaves

The balance between fine root and leaf allocation is assumed to exist in plants (Pearsall 1927), with the root-shoot ratio depending on nutrient supply, particularly nitrogen. The shoot fraction of birch trees increases as they approach optimal nutrition (Ingestadt and Ågren 1991; Figure 25.4), allowing for vegetative growth, thereby achieving a higher net growth rate and competitive success for aboveground resources, such as light. This shift in allocation from fine roots to leaves maximizes plant productivity under variable nutrient availability. Low nutrient levels necessitate a shift in allocation to root production for basic plant maintenance, resulting in limited foliar investment compared to high nutrient conditions. This is consistent with the 'mechanistic' Thornley model which is described above. The optimal models of allocation give similar results to the functional balance of allocation between roots and shoots (Mäkelä and Sievänen 1987). The role of nutrition in determining allocation between roots and shoots, however, has been challenged by Coyle and Coleman (2005), suggesting that higher root–shoot ratios with fertilizer application in young *Populus deltoides* are an ontogenetic effect because those ratios were dependent on tree size and not on nutrient availability experimentally.

Interestingly, tropical forests differ in their allocation of NPP between wood and fine roots, while allocation to foliage remains relatively constant across different sites (Malhi *et al.* 2011). Thus, the allocation trade-off appears to exist between wood and fine roots. However, tropical forests on fertile soils may have higher NPP and allocate more of it to fine roots, which challenges the prediction of resource allocation theory (Doughty *et al.* 2013).

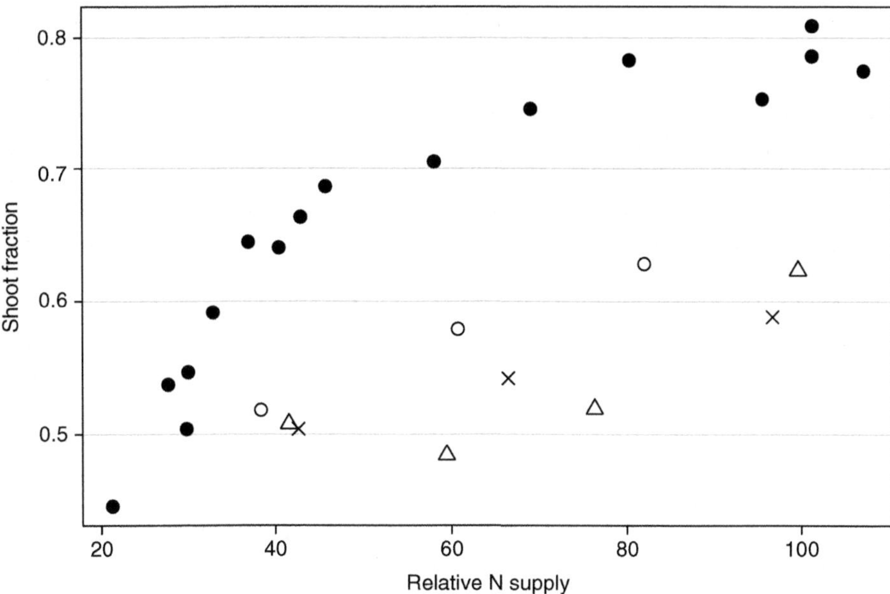

Figure 25.4 Shoot fraction for pendolous birch (*Betula pendula*) (closed circles), Scots pine (*Pinus sylvestris*) (triangles), Contorta pine (*Pinus contorta*) (open circles), and Norway spruce (*Picea abies*) (crosses) as a function of relative nitrogen supply. (Redrawn from Ingestadt and Ågren, 1991.)

Drought

Compared to the effects of nutrients, we know less about how drought affects the distribution of resources between roots and shoots. Moderate droughts reduce the uptake of water and nutrients from the soil, which should increase root allocation based on the functional balance theory. However, low soil water potentials can cause fine root mortality, and severe droughts can directly limit root growth. Despite this, several experimental studies have shown that root–shoot ratios increase under drought conditions (Zhang *et al.* 2004). Meier and Leuschner (2008) studied the allocation of fine roots in European beech along a precipitation gradient and found that although fine root biomass did not differ between stands with different precipitation levels, drier stands had thinner roots and higher root turnover, resulting in more resources being used belowground.

Water availability varies in time and space, but plants in even arid regions can employ different water use strategies to access sufficient water for survival. In a greenhouse experiment, Li *et al.* (2000) found that the water use strategy of *Eucalyptus microtheca* F. Muell. seedlings depended on the distribution of annual rainfall in their native region. Seedlings from regions with even rainfall distribution were less affected by drought in terms of root–foliage ratio than those from regions with pronounced dry seasons. However, the root–foliage ratio was always higher in the drought treatment than in the control. Li *et al.* (2000) linked these changes in allocation to plant strategies, which allowed them to withstand periods of drought. Plants from regions with a diffuse dry season and low average rainfall use a conservative water use strategy with high allocation to roots, enabling steady growth when water is scarce. In contrast, plants from regions with pronounced drought periods use an aggressive water use strategy with high water use and comparatively low allocation to roots, allowing them to exploit short pulses of water and achieve high growth rates when water is available.

Schwinning and Ehleringer (2001) used a simple model to simulate optimal allocation, water use, and rooting depth for desert plants, finding two optimal evolutionary stable allocation strategies for plant growth and resource use. The first strategy involves growing deep root systems with low stomatal conductance and allocating more resources to the roots. This allows for a long-lasting water supply and helps maintain a positive carbon balance during drought (see O'Grady *et al.* 1999). The second strategy involves growing shallow root systems that can harvest rainfall water efficiently. Water is rapidly consumed before it evaporates from the surface or percolates below the ground. This means that when water supply is high, the root–shoot ratios are low and the stomatal conductance is elevated. Stomatal conductance, water use, and photosynthetic production decline rapidly in the latter strategy as the surface soil water store is exhausted. But this strategy is less successful for Mediterranean trees as their mortality rates increase with decreasing root–shoot ratios (Padilla and Pugnaire 2007). Different plant adaptive traits, such as drought deciduousness, rooting depth, or sclerophyllous leaves, can improve drought tolerance and resistance. Allocation strategies and drought tolerance and resistance are interconnected, and many viable strategies exist. Even if plants shed their leaves or reduce their physiological activity during drought, allocation may not change significantly due to the primary allocation occurring during wet periods.

Recent meta-analyses of plant allocation show that plants in drier regions allocate a higher proportion of their production to growth. Based on 164 studies, Eziz et al. (2017) reported that under controlled drought the biomass fraction of roots increased, while the biomass fraction of all other plant organs decreased (see Figure 25.5). Qi et al. (2019) analysed published data of root–shoot ratios across different biomes. Climate variation and plant characteristics explained an equal proportion of the root–shoot-ratio. Allocation to roots, as indicated by the root–shoot ratio, was usually higher for vegetation with a lower annual mean precipitation.

In summary, the distribution of resources between fine roots and foliage is affected by environmental factors such as nutrient and water availability. When resource uptake from the roots is restricted, the allocation to the roots generally increases.

Allocation to woody plant tissue

Although the allocation of biomass to different plant tissues, such as foliage and roots, has been extensively studied, woody plant tissue also plays a crucial role in tree biomass. While woody stems do not perform photosynthesis or nutrient uptake, they primarily act as support structures for foliage in the canopy. Two main theories of the importance of stems in trees have emerged since the late 19th century. Pressler (1864) suggested that stems provide mechanical support for branches and leaves, while Jaccard (1913) proposed that stems serve as a water pipeline from roots to foliage. Both theories indicate that to grow new foliage, trees must also allocate resources towards the necessary stem, coarse roots, and branches to support the foliage mechanically and hydraulically.

To understand the mechanical function of tree stems, researchers have analysed how tree diameter changes with height up the stem. This tapering shape helps to maintain a constant load-bearing capacity, preventing stem breakage (Domec and Gartner 2002). Young trees

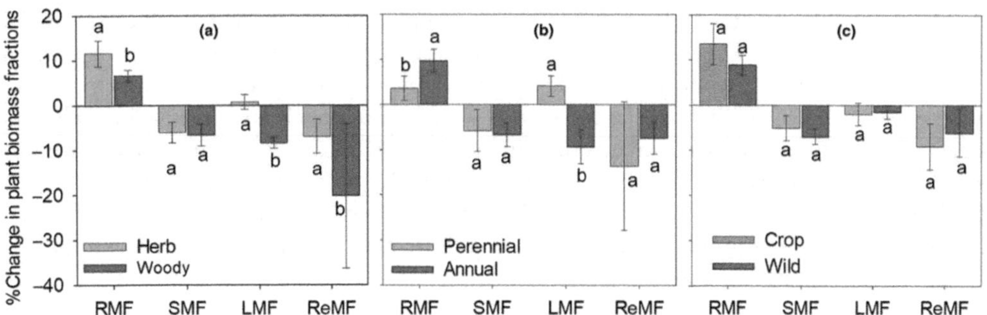

Figure 25.5 Biomass fraction of root (RMF), stem (SMF), leaf (LMF), and reproductive (ReMF) as affected by drought in different plant forms. Data is derived from controlled drought experiments published in the literature. Comparisons are: (a) herbaceous and woody plants, (b) perennial and annual herbaceous plants, and (c) cultivated and wild plants. Error bars describe the 95% confidence intervals. Different letters indicate significant difference of the response ratios between plant groups based on heterogeneity test. Note that the effects of drought on all biomass ratios are significant (i.e., all ratios deviate from 1). Effects of drought are expressed as percentage change relative to the control (%). (Reproduced from Eziz et al., 2017.)

also maintain safety margins against breakage over time and with different stand densities (King 1990). Additionally, tree diameter growth responds to mechanical stimuli, indicating that plants can adapt their allocation to woody tissues (Valinger 1992).

To model plant allocation, hydraulic theories of stem form are more commonly used than mechanical theories. This is because tree water transport functions near its operational limit and increases in transpiration during dry and hot periods can lead to xylem failure. Thus, during drier conditions, more carbohydrates may be allocated to woody parts of the tree that conduct water from the roots to the foliage to reduce the risk of cavitation and xylem failure. The risk of cavitation and xylem failure could be increased by a longer hydraulic pathway, and water transport might effectively limit productivity and height growth in old trees (Yoder *et al.* 1994). Mencuccini and Grace (1996) have shown that tall *Pinus sylvestris* (Scots pine) faces more difficulty in transporting water from roots to foliage compared to shorter counterparts. A theoretical equation proposed by Whitehead *et al.* (1984) links hydraulic conductance, climate, stomatal conductance, and foliage mass, suggesting that stomatal conductance, transport resistance of wood, and evaporative demand are closely linked and contribute to consistent water potential in the leaves.

The 'pipe model theory' proposed by Shinozaki *et al.* (1964) is the basis for most quantitative studies on tree allocation. This theory suggests that the foliage is connected to the roots through pipes, leading to a linear relationship between the stem's cross-sectional area and the foliage mass (or foliage area). This relationship is interpreted as a hydraulic model of tree form, where the mass of foliage and sapwood area are related. The production of new foliage requires sufficient support from the pipe system in the branches and stem. Modelling studies (e.g. Berninger and Nikinmaa 1997) have indicated that allocation to the stem increases with tree height, as longer pipes are required to support the foliage. Allocation to the stem is also high when foliage biomass increases rapidly. Processes such as sapwood turnover, height growth, and the foliage to sapwood area ratio are important determinants of tree growth and stand dynamics.

More recently, West *et al.* (1997, 1999) attempted to combine the mechanical and hydraulic theories of tree growth using geometric scaling laws, which consider the scaling of tubes between organs, to explain the size of the branching network (i.e. trunks, branches, and petioles) and vessel diameter in the wood. While these models are effective in explaining large-scale differences between plants of vastly different sizes, they may not work as well for small-scale differences between trees of similar sizes (Nygren and Pallardy 2008). In conclusion, the approach of West *et al.* (1997, 1999) is useful for understanding differences between trees of different sizes and ecosystems, but may not be adequate for explaining allocation differences at finer scales.

What determines stem growth of trees: changes in allocation or changes in production?

Both tree production per unit of leaf area and the amount of foliage increase with higher resource availability. When trees are irrigated or fertilized, a higher allocation of carbohydrates to the stem, which could be due to decreased root growth or different pipe model ratios, can contribute to higher growth. The main drivers of changes in stemwood production in trees from sites with different levels of fertility are still uncertain, but there are emerging ideas. For example, Coll *et al.* (2011) used a process-based model to analyse the response of *Populus* spp. (poplar) to nitrogen fertilization and found that changes in photosynthetic capacity and increases in GPP dominate the response to nitrogen fertilization. In conifers, Mäkelä (1997) focused on allocation responses and reported that increased foliage biomass leads to increased stand productivity when less carbon is invested in root growth.

In summary, the growth of stemwood and other biomass in forests is determined by the combination of primary production and allocation. Although photosynthetic production in forest ecosystems is well understood, less is known about plant allocation, which is intricately linked with production through feedback cycles. Therefore, gaining a better understanding of the short- and long-term dynamics of these processes is crucial.

References

Amthor, J. S. (2000) 'The McCree-de Wit-Penning de Vries-Thornley respiration paradigms: 30 years later', *Annals of Botany*, vol 86, pp. 1–20.

Attwell, B. J., Kriedeman, P. E. and Turnbull, C. G. J. (1999) *Plants in Action, Adaptation in Nature, Performance in Cultivation.* Macmillan Education, Melbourne.

Austin, A. T. and Sala, O. E. (2002) 'Carbon and nitrogen dynamics across a natural precipitation gradient in Patagonia, Argentina', *Journal of Vegetation Science*, vol 13, pp. 351–360.

Baribault, T. W., Kobe, R. K. and Rothstein, D. E. (2010) 'Soil calcium, nitrogen, and water are correlated with aboveground net primary production in northern hardwood forests', *Forest Ecology and Management*, vol 260, pp. 723–733.

Berninger, F., Coll, L., Vanninen, P., Mäkelä, A., Palmroth, S. and Nikinmaa, E. (2005) 'Effects of tree size and position on pipe model ratios in Scots pine', *Canadian Journal of Forest Research*, vol 35, pp. 1294–1304.

Berninger, F. and Nikinmaa, E. (1997) 'Differences in pipe model parameters determine differences in growth and affect response of Scots pine to climate change', *Functional Ecology*, vol 11, pp. 146–156.

Caspersen, J. P. and Pacala, S. W. (2001) 'Successional diversity and forest ecosystem function', *Ecological Research*, vol 16, pp. 895–903.

Chaves, M. M., Maroco, J. P. and Pereira, J. S. (2003) 'Understanding plant responses to drought — from genes to the whole plant', *Functional Plant Biology*, vol 30, pp. 239–264.

Chen, J. M., Govind, A., Sonnentag, O., Zhang, Y., Barr, A. and Amiro, B. (2006) 'Leaf area index measurements at fluxnet-canada forest sites', *Agricultural and Forest Meteorology*, vol 140, pp. 257–268.

Coll, L., Schneider, R. Domenicano, S, Messier, C. and Berninger, F. (2011) 'Quantifying the effect of nitrogen induced physiological and structural changes on poplar growth using a carbon-balance model', *Tree Physiology*, vol 31, pp. 381–390.

Comas, L. H. and Eissenstat, D. M. (2004) 'Linking fine root traits to maximum potential growth rate among 11 mature temperate tree species', *Functional Ecology*, vol 18, pp. 388–397.

Coyle, D. R. and Coleman, M. D. (2005) 'Forest production responses to irrigation and fertilization are not explained by shifts in allocation', *Forest Ecology and Management*, vol 208, pp. 137–152.

Dijkstra, F. A. (2003) 'Calcium mineralization in the forest floor and surface soil beneath different tree species in the northeastern US', *Forest Ecology and Management*, vol 175, pp. 185–194.

Domec, J. C. and Gartner, B. (2002) 'Age- and position-related changes in hydraulic versus mechanical dysfunction of xylem: inferring the design criteria for Douglas-fir wood structure', *Tree Physiology*, vol 22, pp. 91–104.

Doughty, C. E., Metcalfe, D. B., da Costa, M. C., de Oliveira, A. A. R., Neto, G. F. C., Silva, J. A., Aragao, L. E. O. C. and 6 co-authors (2013) 'The production, allocation and cycling of carbon in a forest on fértile terra preta soil in eastern Amazonia compared with a forest on adjacent infertile soil', *Plant Ecology and Diversity*, vol 7, pp. 41–53.

Erskine, P. D., Lamb, D. and Bristow, M. (2006) 'Tree species diversity and ecosystem function: can tropical multi-species plantations generate greater productivity?', *Forest Ecology and Management*, vol 233, pp. 205–210.

Eziz, A, Yan Z., Tian, D., Han, W., Tang, Z., Frang, J., (2019) 'Drought effect on plant biomass allocation: a meta-analysis', *Ecology and Evolution*, vol 7, pp. 11002–11010.

Fujinuma, R., Bockheim, J. and Balster, N. (2005) 'Base-cation cycling by individual tree species in old-growth forests of Upper Michigan, USA', *Biogeochemistry*, vol 74, pp. 357–376.

Gea-Izquierdo, G., Bergeron, Y., Huang, J. G., Lapointe-Garant, M. P., Grace, J. and Berninger, F. (2014) 'The relationship between productivity and tree-ring growth in boreal coniferous forests', *Boreal Environmental Research*, vol 19, pp. 363–378.

Gea-Izquierdo, G., Mäkelä, A., Margolis, H., Bergeron, Y., Black, A. T., Dunn, A., Hadley, J., Tha Paw, U., Falk, M., Wharton, S., Monson, R., Hollinger, D. Y., Laurila, T., Aurela, M., McCaughey, H., Bourque, C., Vesala, T. and Berninger, F. (2010) Modelling acclimation of photosynthesis to temperature in evergreen boreal conifer forests. *New Phytologist*, vol 188, pp. 175–186.

Hikosaka, K., Ishikawa, K., Borjigidai, A., Muller, O. and Onoda Y. (2006) 'Temperature acclimation of photosynthesis: mechanisms involved in the changes in temperature dependence of photosynthetic rate', *Journal of Experimental Botany*, vol 57, pp. 291–302.

Ingestadt, T. and Ågren, G. I. (1991) 'The influence on nutrition on biomass allocation', *Ecological Applications*, vol 1, pp. 169–174.

Jaccard, P. (1913) 'Eine neue Auffassung über die Ursachen des Dickenwachstums. Naturwissentschaftliche Zeitung Forst- und Landwirtschaft', vol 11, pp. 241–279.

King, D. A. (1990) 'The adaptive significance of tree height', *American Naturalist*, vol 135, pp. 809–828.

Li, C., Berninger, F., Koskela, J. and Sonninen, E. (2000) 'Different origins of *Eucalyptus microtheca* differ in their ability to acclimate to drought', *Australian Journal of Plant Physiology*, vol 27, pp. 231–238.

Long, S. P., Postl, W. F. and Bolhár-Nordenkampf, H. R. (1993) 'Quantum yields for uptake of carbon dioxide in C3 vascular plants of contrasting habitats and taxonomic groupings', *Planta*, vol 189, pp. 226–234.

Mäkelä, A. (1997) 'A carbon balance model of growth and self-pruning in trees based on structural relationships', *Forest Science*, vol 43, pp. 7–24.

Mäkelä, A., Hari, P., Berninger, F., Hänninen, H., Nikinmaa, E. (2004) 'Acclimation of photosynthetic capacity in Scots pine to the annual cycle of temperature', *Tree Physiology*, vol 24, pp. 369–376.

Mäkelä, A. and Mäkinen, H. (2003) 'Generating 3D sawlogs with a process-based growth model', *Forest Ecology and Management*, vol 184, pp. 337–354.

Mäkelä, A. and Sievänen, R. (1987) 'Comparison of two shoot-root partitioning models with respect to substrate utilization and functional balance', *Annals of Botany*, vol 59, pp. 129–140.

Mäkelä, A. and Valentine, H. T. (2001) 'The ratio of NPP to GPP: evidence of change over the course of stand development', *Tree Physiology*, vol 21, pp. 1015–1030.

Malhi, Y., Baldocchi, D. D. and Jarvis, P. G. (1999) 'The carbon balance of tropical, temperate and boreal forests', *Plant, Cell and Environment*, vol 22, pp. 715–740.

Malhi, Y., Dougthy, C. and Galbraith, D. (2011) 'The allocation of ecosystem net primary productivity in tropical forests', *Philosophical Transactions of the Royal Society B*, vol 366, pp. 3225–3245.

Meier, I. C. and Leuschner, C. (2008) 'Below-ground drought response of European beech: fine root biomass and carbon partitioning in 14 mature stands across a precipitation gradient', *Global Change Biology*, vol 14, pp. 2081–2095.

Mencuccini, M. and Grace, J. (1996) 'Developmental patterns of above-ground hydraulic conductance in a scots pine (*Pinus sylvestris* L.) age sequence', *Plant, Cell and Environment*, vol 19, pp. 939–948.

Monteith, J. L. (1994) 'Validity of the correlation between intercepted radiation and biomass', *Agricultural and Forest Meteorology*, vol 68, pp. 213–220.

Nikinmaa, E., Hölttä, T., Hari, P., Kolari, P., Mäkelä, A., Sevanto, S. and Vesala, T. (2013) 'Assimilate transport in phloem sets conditions for leaf gas exchange', *Plant, Cell and Environment*, vol 36, pp. 655–669.

Nygren, P. and Pallardy, S. G. (2008) 'Applying a universal scaling model to vascular allometry in a single-stemmed, monopodially branching deciduous tree (attim's model)', *Tree Physiology*, vol 28, pp. 1–10.

O'Grady, A. P., Eamus, D. and Hutley, L. B. (1999) 'Transpiration increases during the dry season: patterns of tree water use in eucalypt open-forests of Northern Australia', *Tree Physiology*, vol 19, pp. 591–597.

Padilla, F. M. and Pugnaire, F. I. (2007) 'Rooting depth and soil moisture control Mediterranean woody seedling survival during drought', *Functional Ecology*, vol 21, pp. 489–495.

Pearsall, W. H. (1927) 'Growth studies VI, On the relative size of organs in plants', *Annals of Botany*, vol 41, pp. 549–556.

Peh, K. S.-H. (2009) *The relationship between species diversity and ecosystem function in low- and high-diversity tropical African forest*. PhD thesis, University of Leeds, UK.

Pressler, M. R. (1864) 'Das Gesetz der Stammbildung. Arnoldische Buchhandlung', Leipzig, no 153.

Pulkkinen, P., Pöykkö, T., Tigerstedt, P. M. A. and Velling, P. (1989) 'Harvest index in northern temperate cultivated conifers', *Tree Physiology*, vol 5, pp. 83–98.

Qi, Y., Wei, W., Chen, C. and Chen, L. (2019) 'Plant root-shoot biomass allocation over diverse biomes: a global synthesis', *Global Ecology and Conservation*, vol 18, e00606.

Quesada, C. A., Phillips, O. L., Schwarz, M., Czimczik, C. I., Baker, T. R., Patiño S., Fylla, N. M. and 40 co-authors. (2012) 'Basin-wide variations in Amazon forest structure and function are mediated by both soils and climate', *Biogeosciences*, vol 9, pp. 2203–2246.

Reich, P. B., Kloeppel, B. D., Ellsworth, D. S. and Waiters, M. S. (1995) 'Different photosynthesis-nitrogen relations in deciduous hardwood and evergreen coniferous tree species', *Oecologia*, vol 104, pp. 24–30.

Richards, R. A. (2000) 'Selectable traits to increase crop photosynthesis and yield of grain crops', *Journal of Experimental Botany*, vol 51, pp. 447–458.

Rocha, A. V., Goulden, M. L., Dunn, A. L. and Wofsy, S. C. (2006) 'On linking interannual tree ring variability with observations of whole-forest CO_2 flux', *Global Change Biology*, vol 12, pp. 1378–1389.

Running, S. W. (2012) 'A measurable planetary boundary for the biosphere', *Science*, vol 337, pp. 1458–1459.

Running, S. W., Nemani, R. R., Heinsch, F. A., Zhao, M., Reeves, M. and Hashimoto, H. (2004) 'A continuous satellite-derived measure of global terrestrial production', *BioScience*, vol 54, pp. 547–560.

Ryan, M. G., Hubbard, R. M. Pongracic, S., Raison, R. J. and McMutrie, R. E. (1996) 'Foliage, fine-root, woody-tissue and stand respiration in *Pinus radiata* in relation to nitrogen status', *Tree Physiology*, vol 16, pp. 333–343.

Sala, O. E., Gherardi, L. A., Reichmann, L., Jobbágy, E. and Peters, D. (2012) 'Legacies of precipitation fluctuations on primary production: theory and data synthesis', *Philosophical Transactions of the Royal Society of Biological Sciences*, vol 367, pp. 3135–3144.

Schwinning, S. and Ehleringer, J. D. (2001) 'Water use trade-offs and optimal adaptations to pulse-driven arid ecosystems', *Journal of Ecology*, vol 89, pp. 464–480.
Shinozaki, K., Yoda, K., Hozumi, K. and Kira T. (1964) A quantitative analysis of plant form–the pipe model theory. I. Basic analyses', *Japanese Journal of Ecology*, vol 14, pp. 133–139.
Suni, T., Berninger, F., Vesala, T., Markkanen, T., Hari, P., Mäkelä, A., Ilvesniemi, H., Hänninen, H., Nikinmaa, E., Huttula, T., Laurila, T., Aurela, M., Grelle, A., Lindroth, A., Arneth, A., Shibistova, O., and Lloyd, J. (2003) 'Air temperature triggers the commencement of evergreen boreal forest photosynthesis in spring', *Global Change Biology*, vol 9, pp. 1410–1426.
Taiz, L. and Zeiger, E. (2010) *Plant Phyiology*. Sinauer Associates, Sunderland, MA.
Thornley, J. H. M. (1991) 'A transport-resistance model of forest growth and partitioning', *Annals of Botany*, vol 68, pp. 211–226.
Turner, D. P., Ritts, W. D., Cohen, W. B., Maeirsperger, T. K., Gower, S. T., Kirschbaum, A. A., Running, S. W., Zhao, M., Wofsy, S. C., Dunn, A. L., Law, B. E., Campbell, J. L., Oechel, W. C., Jung, K. H., Meyers, T. P., Small, E. E., Kurc, S. A. and Gamon, J. A. (2005) 'Site-level evaluation of satellite-based global terrestrial gross primary production and net primary production monitoring', *Global Change Biology*, vol 11, pp. 666–684.
Valinger, E. (1992) 'Effects of wind sway on stem form and crown development of Scots pine (*Pinus sylvestris* L.)', *Australian Forestry*, vol 55, pp. 15–21.
Vapaavuori, E. M., Vuorinen, A. H., Aphalo, P. J. and Smolander, H. (1995) 'Relationship between net photosynthesis and nitrogen in Acots pine: seasonal variation in seedlings and shoots', *Plant and Soil*, vol 168–169, pp. 263–270.
Vilà, M., Vayreda, J., Comas, L., Ibáñez, J. J., Mata, T. and Obón, B. (2007) 'Species richness and wood production: a positive association in Mediterranean forests', *Ecology Letters*, vol 10, pp. 241–250.
Waring, R. H., Landsberg, J. J. and Williams, M. (1998) 'Net primary production of forests: A constant fraction of gross primary production?', *Tree Physiology*, vol 18, no 2, pp. 129–134.
Warren, W. J. (1967) 'Ecological data on dry-matter production by plants and plant communities', in E. F. Bradley and O. T. Denmead (eds.), *The collection and processing of Field data*. John Wiley & Sons, New York.
West, G. B., Brown, J. H. and Enquist, B. J. (1997) 'A general model for the origin of allometric scaling laws in biology', *Science*, vol 276, pp. 122–126.
West, G. B., Brown, J. H. and Enquist, B. J. (1999) 'A general model for the structure and allometry of plant vascular systems', *Nature*, vol 400, pp. 664–667.
Whitehead, D., Edwards, W. R. N. and Jarvis, P. G. (1984) 'Conducting sapwood area, foliage area, and permeability in mature trees of *Picea sitchensis* and *Pinus contorta*', *Canadian Journal of Forest Research*, vol 14, pp. 940–947.
Yoder, B. J., Ryan, M. G., Waring, R. H., Schoettle, A. W. and Kaufmann, M. R. (1994) 'Evidence of reduced photosynthetic rates in old trees', *Forest Science*, vol 40, pp. 513–527.
Zhang, Y., Zang, R. and Li, C. (2004) 'Population differences in physiological and morphological adaptations of *Populus davidiana* seedlings in response to progressive drought stress', *Plant Science*, vol 166, pp. 791–797.

PART V

FOREST CONSERVATION AND MANAGEMENT

26
NATURAL REGENERATION AFTER HARVESTING

Nelson Thiffault, Lluís Coll, Douglass F. Jacobs, Marie Ange Ngo Bieng and Eliott Maurent

Introduction

Natural regeneration of forest stands replaces dead or removed mature trees with recruits of sexual or vegetative origin, many of which will eventually form the new canopy. Natural and anthropogenic disturbances modify stand structure and resource availability, releasing growing space for other plants. Harvesting-related disturbances interact with stand characteristics, influencing available ecological niches for new individuals. Thus, forest regeneration is intimately linked to stand disturbances.

Natural regeneration differs from artificial regeneration, which involves planting or direct seeding for stand establishment. Natural regeneration is often the most cost-effective method for reestablishing desirable species after harvesting (Chudy *et al.* 2022). However, the outcomes of natural regeneration are less predictable due to complex interactions among factors such as seed availability, management history, stand and soil characteristics, light, water, nutrients, species autecology, plant interactions, predation, and browsing. A successful natural regeneration after harvesting can lead to overstocked stands, requiring thinning to optimize productivity.

In this chapter, we summarize the key concepts of natural forest regeneration dynamics after harvesting. We examine the impact of harvesting on ecosystems and its interactions with stand characteristics, soil, and environmental properties, affecting the success of regeneration. We illustrate these concepts by investigating the effects of harvesting and its interaction with ecosystem properties on natural forest regeneration in four distinct ecosystems: boreal, temperate deciduous, dry Mediterranean, and wet tropical forests.

Harvesting as a disturbance

Forest harvesting serves many purposes in silviculture. The main objective is usually to extract commercially valuable wood products and alter the characteristics and density of forest stands. Timber harvesting can also be used to purposefully modify stand structure, develop specific visual characteristics, manage wildlife habitats, or promote or dampen various understory responses (Nyland 2016). The wide range of harvesting objectives result in varying

DOI: 10.4324/9781003324072-31

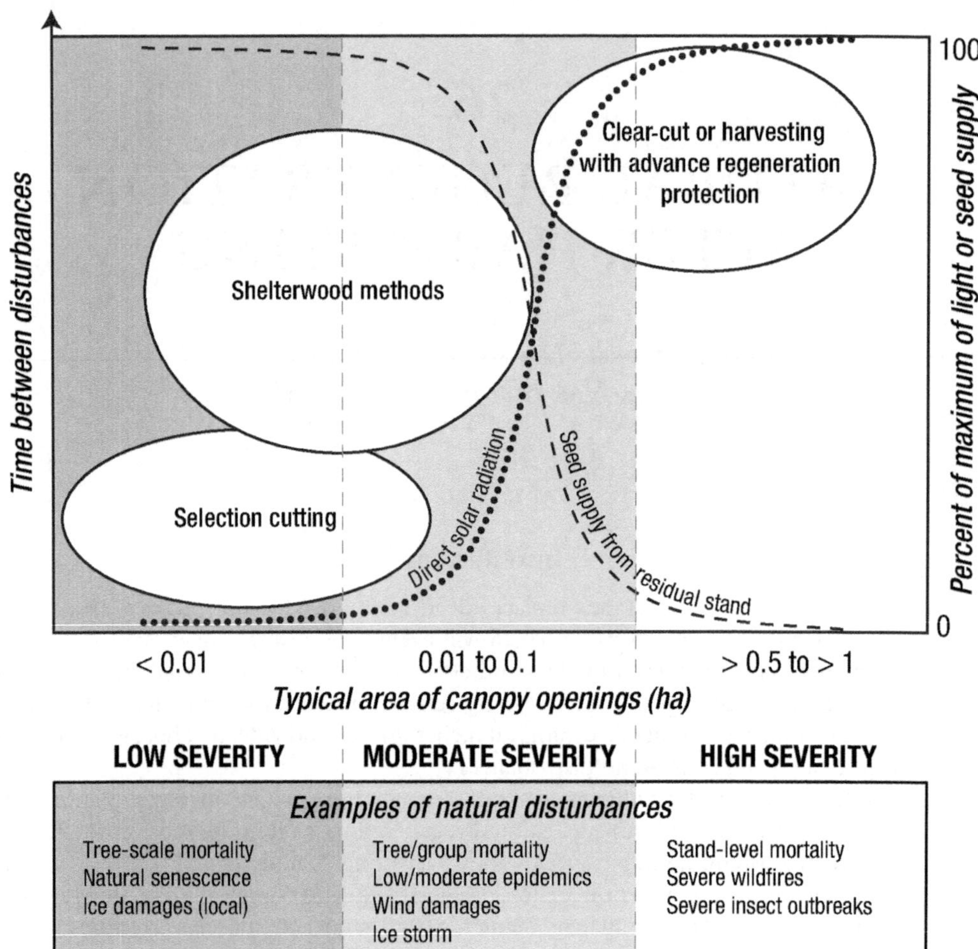

Figure 26.1 Conceptual representation of regeneration methods within a gradient of severity, clearing size, and frequency. (Adapted from Raymond et al. 2013, Coates and Burton 1997, and Smith 1962). Severity is defined by the degree to which the canopy is removed, microclimate and soils are altered, the extent of organic layer consumption or removal, and the degree to which living organisms are killed (Haeussler and Kneeshaw 2003).

approaches that differ in severity, spatial scale, and frequency. These approaches interact differently with ecosystem characteristics and functions. Figure 26.1 illustrates the distribution of harvesting regeneration methods based on severity, clearing size, and time between disturbances. It also shows relative light availability and seed supply after harvesting. Natural disturbances of low, moderate, or high severity are also illustrated. Selection cutting involves frequent, low-intensity cuttings, mimicking small gap natural mortality caused by tree senescence. It creates small gaps with established regeneration, sometimes using partial mechanical soil scarification to enhance seed germination. Light levels remain low due to partial canopy retention. Clear-cut harvesting mimics severe natural disturbances like wildfires to a certain extent, removing the entire canopy and altering environmental conditions. Complete canopy removal favours light-demanding species with wind-dispersed seeds or persistent seedbanks. Some variants of clear-cutting protect advance regeneration by limiting machinery to specific trails.

The severity of harvesting determines its impact on key environmental resources, affecting seed germination, seedling emergence and survival, and advance regeneration. Canopy removal has various impacts: increased understory light levels, altered seedbeds, and changes in soil temperature and moisture. For instance, canopy transmittance exhibited a negative exponential relationship with basal area, density, and stand density index in trembling aspen (*Populus tremuloides*)- and paper birch (*Betula papyrifera*)-dominated residual stands of Ontario, Canada (Parker and Sharma 2018). Careful logging around advanced growth was characterized by higher soil temperatures than control, unharvested plots in forested peatlands of northwestern Quebec, Canada (Botroh *et al.* 2023). Harvesting can affect soil moisture, with reports of drying in the surface layer up to 3.5 cm in boreal forest areas of Newfoundland, Canada (Moroni *et al.* 2009). Organic layer disturbance or removal often resulting from harvest operations promotes germination of yellow birch (*Betula alleghaniensis*) in eastern Canadian mixedwood forests (Elie *et al.* 2009) and black pine (*Pinus nigra*) in Mediterranean forests (Lucas-Borja *et al.* 2012). Canopy removal also influences the aerial seed bank, encompassing seeds in fruits or ovulate cones on standing trees.

The process of regeneration

Natural regeneration of forest stands relies on the presence of ample, viable seed supply or perennating organs and suitable substrate for seed establishment (Kozlowski 2002). Figure 26.2 provides a summary of this process. The original stand usually serves as the primary seed or propagule source, though some forests may have long-lived soil seed banks, and trees at the cut edges can contribute a significant seed fraction if the cut size allows dispersal. For instance, red maple (*Acer rubrum*) seedlings abundantly recruit from persistent seed banks in temperate forests (Hille Ris Lambers and Clark 2005). Wind-dispersed seeds generally have limited dispersal capacity inversely related to seed size, while animal-dispersed seeds do not show a clear relationship with propagule mass. Small animals tend to avoid recent clearings, restricting the dispersal range of relatively large walnut fruits by rodents to within 50 m seed trees (Hewitt and Kellman 2002). In contrast, birch species (*Betula* spp.) produce small seeds that can travel long distances, aided by wind movement along the snow surface (Houle 1998). Certain boreal species like black spruce (*Picea mariana*) and jack pine (*Pinus banksiana*) maintain aerial seed banks on branches, serving as reliable seed sources until natural disturbances occur. Additionally, tree regeneration can occur through vegetative reproduction via stump sprouting, root suckering, rhizomes, or layering from parent trees. Such capacity is more common among angiosperms than gymnosperms.

Seed and vegetative sources are affected by various abiotic and biotic factors like predation, desiccation, flooding, and diseases, occurring before or after harvest. Germination, after dispersal, is influenced by habitat (e.g., seedbed characteristics) and biotic (e.g., aboveground biomass of dominant plants) filters. The severity and timing of harvesting can either promote or hinder these filters (Figure 26.1). For asexual recruits, the auxin-to-cytokinin ratio plays a crucial role in promoting shoot development from dormant buds.

Survival and growth of newly established seedlings depend on environmental conditions (e.g., light, temperature, moisture, nutrients, and compaction) in the harvested area, as well

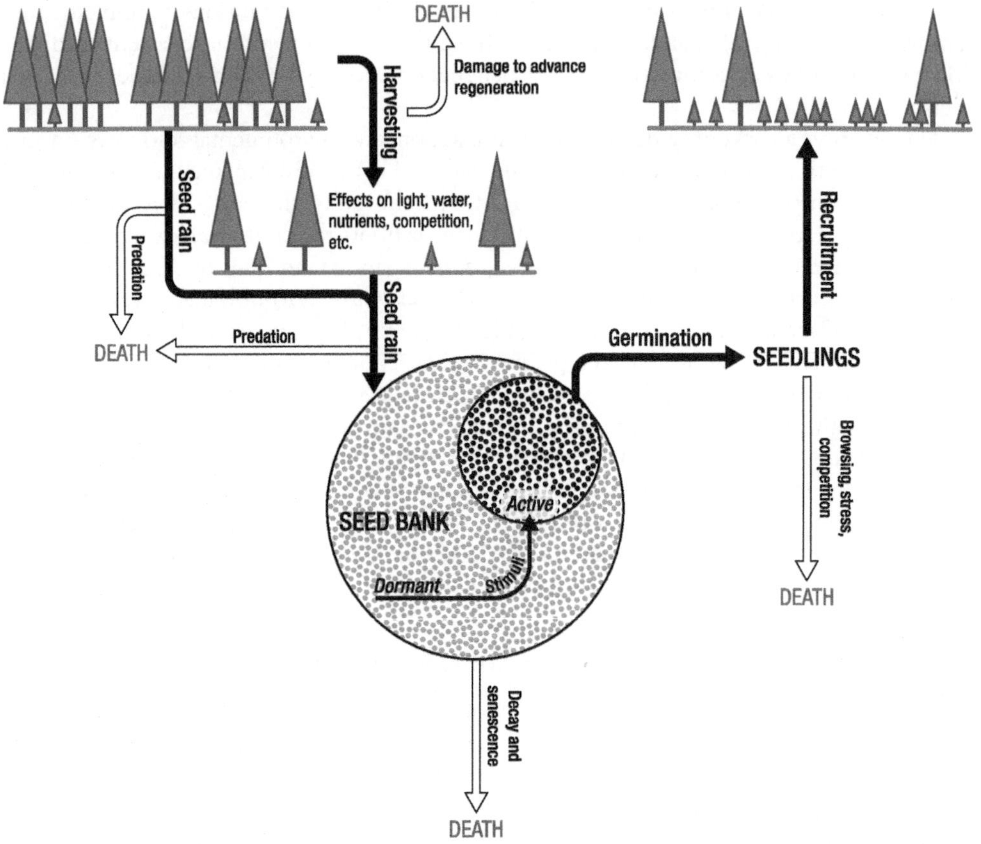

Figure 26.2 Main factors affecting natural regeneration following harvesting. (Adapted from Nyland, 2016.)

as competitive and facilitation interactions, biotic factors like insects, diseases, and browsing, and damages from logging operations. Species vary in their sensitivity to these factors, shaping regeneration dynamics. The severity of the cut directly impacts these factors (Picchio *et al.* 2020), and seedlings that survive and show adequate growth are recruited into the regenerating cohort. The following sections illustrate how the general process of natural regeneration following harvesting takes place in contrasting North American, European, and tropical ecosystems. Transposition to other contexts and ecosystems can be made using, for example, key functional traits characterizing the vegetative and regeneration stages of the tree life cycle.

Natural regeneration after harvesting in boreal ecosystems

Boreal ecosystems have a subarctic climate that affects growing conditions through factors like precipitation, air temperature, soils, and the length of the growing season. The boreal forest experiences severe winters, short growing seasons, cold soils, and slow decomposition rates, limiting plant productivity. Dominated by conifers, it undergoes natural disturbances primarily driven by insect outbreaks and wildfires (Shorohova *et al.* 2023). The understory of nutrient-poor, moist to dry boreal forest sites comprises mainly mosses (e.g., *Pleurozium* and *Ptilium*), lichens (e.g., *Cladina*), and ericaceous shrubs (e.g., *Kalmia, Rhododendron,* and *Vaccinium*).

Some conifer tree species in the North American boreal forest require high-intensity fires or increased heat for cone opening and seed dispersal (Greene *et al.* 1999). Non-serotinous species can also recolonize sites through in situ aerial seed banks (Michaletz *et al.* 2013). Natural seeding for regeneration typically occurs within 5 years post-harvest in these northern ecosystems, as seedbed quality declines due to leaf litter accumulation. Some species also reproduce vegetatively through layering, a common trait among conifers. However, this occurs long after harvesting when moss growth encourages adventitious rooting.

Clear-cutting, along with its variations that protect advance regeneration, is the dominant harvesting approach used in boreal stands globally. Clear-cutting removes most canopy trees, creating open sky conditions with abundant light (Figure 26.1). It can promote the growth of competing species, necessitating mechanical soil preparation to expose patches of mineral soil and humus, creating suitable seedbeds and reducing competitor cover (Montoro Girona *et al.* 2018).

The southern part of the North American boreal forest is characterized by a few more tree species not present in its northern part. Most dominant tree species in this region do not exhibit seed dormancy in the soil for longer than a year (Greene *et al.* 1999). Balsam fir (*Abies balsamea*), which is often a dominant conifer, forms dense understory seedling banks (Figure 26.3a) that can persist for decades under low light conditions and positively respond to canopy gaps (Parent and Ruel 2002). Norway spruce (*Picea abies*), a prominent conifer in the Scandinavian boreal forest, establishes well on moss substrates in small clear-cuts, with higher establishment close to the surrounding stand (Hanssen 2003). Trembling aspen, a northern hardwood species, reproduces through aggressive root suckering when aboveground parts are removed (Frey *et al.* 2003). Clear-cutting is the preferred method to regenerate this species, but it often lacks substantial coarse woody debris, which can affect regeneration for certain species such as white spruce (*Picea glauca*).

In some cases, shade-tolerant boreal forest stands can be harvested and naturally regenerated using shelterwood techniques that expose the mineral substrate to promote germination (Montoro Girona *et al.* 2023). This involves selective removal of the canopy, leaving seed trees for regeneration. The increased understory light levels support germination, establishment of new regeneration, or growth stimulation of existing advance regeneration.

Natural regeneration after harvesting in temperate deciduous ecosystems

Temperate deciduous forests span vast areas in America, Asia, Europe, and the southern hemisphere (South America, Africa, and Australia). These ecosystems undergo seasonal leaf shedding by dominant species, which impacts regeneration. Temperate deciduous forests are typically harvested using variants of the selection cut system, which maintains a permanent cover and is favourable to the establishment and growth of shade-tolerant species such as sugar maple (*Acer saccharum*) and yellow birch (Figure 26.3b).

Due to their fertile soil conditions, deciduous forests have faced extensive logging, land clearing for agriculture, and urbanization historically (Bergmeier *et al.* 2010). In the Central Hardwood Forest Region in North America, covering over 89 million ha, with approximately 50% forested (Johnson *et al.* 2009), forests mainly developed as secondary forests from the

Figure 26.3 (a) Natural seedling of balsam fir established in the understory of a mature stand in northern Québec (Canada). (Photograph by N. Thiffault.) (b) Harvest trail from a selection cut in a sugar maple-dominated stand of Maine (USA). (Photograph by N. Thiffault.) (c) Experimental group selection cuttings of 0.2 and 0.4 ha applied in a Supra-Mediterranean Scots pine stand near Solsona, Lleida (Spain). (Photograph by Santiago Martín Alcón.) (d) Damage to soil in a trail used for harvesting in a tropical stand in Costa Rica. (Photograph by E. Maurent.)

late 19th to the mid-20th century due to land abandonment amid competition with new prairie farms. Anthropogenic disturbances like clear-cutting, fire, and grazing have influenced these forests, leading to dominance by shade-intolerant species like oaks (*Quercus* spp.) and yellow poplar (*Liriodendron tulipifera*). However, recent management practices have shifted toward fire suppression and selective cutting, favouring single-tree selection (Abrams 2003). Simultaneously, the region has seen an increase in white-tailed deer (*Odocoileus virginianus*) populations, resulting in greater herbivory pressures and selective browsing. This favours browse-tolerant species, leading to a shift in forest composition (Beguin *et al.* 2016).

The shift in the natural dominance of temperate forests is an important issue worldwide, as naturally regenerating stands are considered of low quality and biodiversity (Dudley *et al.* 1986). Northeastern North America, for instance, has witnessed a decline in natural oak regeneration, a matter of great importance for forest managers. Oaks hold economic value for timber, and their presence is crucial for wildlife habitat, aesthetics, and culture (Johnson *et al.* 2009). Even-aged silviculture, specifically clear-cutting (see Figure 26.1), has been advocated as the most reliable approach to establish oak regeneration in the region. However, successful oak regeneration in even-aged systems largely depends on the presence of advanced regeneration or stump sprouting, with seed contribution being minimal (Morrissey *et al.* 2008). Site characteristics also play a role, with higher-quality sites typically dominated by sugar maple, red maple, yellow poplar, and white ash (*Fraxinus americana*), while oak species thrive better on poorer sites. In southern Indiana, United States, oak regeneration success after 21–35 years depends on site characteristics and the pre-harvest abundance of oaks (Morrissey *et al.* 2008), with nearly half of the dominant oak stems originating from stump sprouts. Although oak abundance declines during the early stages of regeneration, dominant oak stems increase during the subsequent stem exclusion stage due to their relatively greater drought tolerance, enabling competitiveness against other species during consecutive extreme drought years. Intermittent silvicultural manipulation, such as selective thinning of less desirable competitors, appears necessary to ensure oak regeneration success in even-aged systems.

Natural regeneration after harvesting in dry ecosystems

Challenges arise in natural regeneration after harvesting dry ecosystems exposed to highly seasonal rainfall. This climatic pattern strongly affects tree phenology, seed production, germination, and seedling establishment (Ray and Brown 1994; Matías *et al.* 2012). In the northern Mediterranean Basin, characterized by pronounced bi-seasonality, forests exhibit complex structures and compositions shaped by environmental heterogeneity and human influence (Nocentini and Coll 2013). Forest types vary along elevational and drought gradients, each posing unique constraints on natural regeneration. Thermo- and Meso-Mediterranean forests, found at lower altitudes, are dominated by drought-tolerant thermophile *Pinus* species and evergreen sclerophyllous oak species. These species reproduce using various mechanisms, such as serotinous cones, resprouting, and long-lived seed banks, to respond to large disturbances like fire or overgrazing. Limited information is available on their response to different harvesting methods (Rodríguez-García *et al.* 2010). Generally, Thermo- and Meso-Mediterranean pines require ample light to regenerate effectively (Calama *et al.* 2017), necessitating substantial openings consistent with the disturbance regimes in their distributional area.

Adequate natural regeneration of Aleppo pine (*Pinus halepensis*), the most widespread coniferous species in the Mediterranean, is uncommon in fire-free systems due to various microsite factors (Nathan and Ne'eman 2004). Applying soil and ground vegetation treatments like mechanical scarification or high-intensity burning of slash after cutting can enhance recruitment (Osem *et al.* 2013). These treatments reduce competition from herbaceous species by diminishing the soil seed bank and perennating tissues. The success of post-harvest natural regeneration in Maritime pine (*Pinus pinaster*) is highly dependent on seedbed characteristics, such as soil texture and litter, affecting soil water-holding capacity (Rodríguez-García *et al.* 2011). In dry environments, the success of natural regeneration for these pine species also relies on post-harvest microclimate conditions, particularly the amount of spring and summer rainfall (Rodríguez-García *et al.* 2010).

Evergreen oak species in the Mediterranean have traditionally been managed as coppice systems, involving total or partial removal of standing biomass to promote asexual regeneration through sprouting (Retana *et al.* 1992). However, the decline in firewood and charcoal use has reduced the frequency of basal coppicing, leading to increased stand density and stagnation, with signs of decay in some stands (Serrada *et al.* 2017). To transition these stands into conventional forest systems with seed-based regeneration, a progressive cutting approach, preserving the best stems for acorn production, has been proposed. Yet, this process faces challenges from high seed predation rates before and after dispersal (Espelta *et al.* 2009). Moreover, predicted drought increases due to global change are expected to reduce acorn production and seedling survival (García-Barreda *et al.* 2023), adding further difficulty to seedling establishment. There is also a recruitment bottleneck for evergreen oaks at the sapling stage: seedlings require cover for establishment (protection from high evaporative demand), but increasing light levels are necessary for their growth (Martín-Alcón *et al.* 2015).

As elevation increases, vegetation communities transition from Thermo- and Meso-Mediterranean to Supra- and Montane-Mediterranean. These forests primarily consist of conifers like *P. nigra* and *P. sylvestris*, along with deciduous species such as *Quercus faginea* and *Quercus pubescens*. Natural regeneration of these species after harvesting necessitates protection against increased evaporation and moisture stress for survival. This has led to the widespread use of shelterwood systems or progressive cuttings in the region (Barbeito *et al.* 2011). In pine-dominated forests, group selection cuttings can be considered for stands with uneven-age structures (Piqué *et al.* 2001) (Figure 26.3c). Additionally, soil preparation may be required to mitigate herbaceous competition and break thick organic layers in some cases (del Cerro Barja *et al.* 2009).

Natural regeneration after harvesting in wet tropical ecosystems

Tropical forests cover 18 million km^2 and constitute 45% of the world's forests. Their climates are influenced by the North-South oscillations of the Intertropical Convergence Zone and their topography, resulting in varying precipitation patterns from low annual but highly seasonal rainfall to consistent rain. Tropical ecosystems exhibit significant variation across climatic zones (Ghazoul and Sheil 2010). Wet tropical forests are renowned for their high biodiversity, hosting approximately half of the world's species richness on just 7% of the land. This rich biodiversity is crucial for providing ecosystem services, including carbon storage and climate stabilization (Gamfeldt *et al.* 2013). However, tropical forests are highly vulnerable and face numerous threats (Hosonuma *et al.* 2012), primarily from agricultural activities (Pendrill *et al.* 2022), as well as timber logging, road construction, climate change, and mining.

Selective logging practices are commonly used in old-growth tropical forests, potentially associated with pre- and post-harvest operations to promote tree regeneration, such as liana removal, thinning, or poison girdling. However, logging activities can have adverse effects on forest soil (Figure 26.3d), causing compaction due to heavy machinery (DeArmond *et al.* 2019). Moreover, tree mortality increases in the remaining stand, resulting in approximately 17 trees lost for each tree harvested (Naves *et al.* 2020). Selective cuts in dense and liana-filled areas can lead to large canopy gaps, which provide increased resources, particularly light and space. These gaps are quickly colonized by pioneer tree species, such as *Cecropia* sp. and *Musanga* sp., known for their short lifespan and fast-growing strategies, and characterized by high foliar area and low wood density. Within the tropical forest understory, there are seedlings of slow-growing, late-successional species, which benefit from the available light resulting from gap openings. The coexistence of pioneer and late-successional species can lead to peak diversity during succession, which slightly diminishes compared to that observed in old-growth forests, where only late-successional species persist (Mirabel *et al.* 2020).

Regeneration of tropical forests after harvesting relies on a combination of endogenous (linked to the pre-disturbed forest stand) and exogenous (associated with the surrounding environment and disturbance) drivers. Among these, disturbance intensity and cutting cycle duration play significant roles (e.g., Sist *et al.* 2021). Of these drivers, those related to disturbances are the most manageable and crucial for achieving successful regeneration and sustainable harvesting practices. To reduce disturbance intensity, it is necessary to minimize the number of trees harvested and the resulting damage inflicted upon the forest stand for each harvested tree through reduced-impact logging (e.g., Putz *et al.* 2008).

Conclusion

A successful regeneration after harvesting faces significant barriers, particularly in the context of climate change. It is crucial to maintain seed and propagule sources on sites during harvest operations to ensure natural stand renewal. However, the impacts of increasing temperature and drought can lead to irregular seed crops for many species, as illustrated by examples in managing hardwood-softwood mixtures (Kabrick *et al.* 2017). Additionally, local genotypes may not be adapted to future climate conditions, potentially affecting the ability of new forests to provide ecosystem services in the future (Boisvert-Marsh *et al.* 2022).

The limited availability of suitable germination substrates due to the absence of appropriate disturbance can further impede regeneration success (Aubin *et al.* 2016). Other constraints include damage to advance regeneration during harvest (Thiffault *et al.* 2023), the presence of woody debris (Fraver *et al.* 2002) and their procurement for biomass (Gouge *et al.* 2021), rapid colonization by alien invasive species (Lanzer *et al.* 2017) or native competing species (Barrette *et al.* 2022; Ménard *et al.* 2019), soil compaction (Naghdi *et al.* 2023), and other characteristics of skid trails (Harvey and Brais 2002). Furthermore, seedling survival under increased temperature extremes and drought conditions, as well as high herbivore populations, poses additional challenges (Dey *et al.* 2019), among others.

Forest ecosystems are dynamic, undergoing changes in species composition, distribution, spatial patterns, leaf area, volume, and other physical characteristics over time. These changes result from the growth of individual plants and their interactions within the ecosystem. Energy and nutrient fluxes also undergo modifications. Natural regeneration of forest stands following harvesting is a crucial part of this dynamic process. Harvesting acts as a disturbance in the ecosystem, initiating secondary succession. The extent of the disturbance and the characteristics of the ecosystems (including macroscale climatic and biogeographic factors) influence the trajectory of succession by impacting seed and propagule sources, germination and survival conditions, and environmental resource availability. The examples we have presented (boreal, temperate deciduous, and European dry and wet tropical ecosystems) demonstrate how specific abiotic and biotic factors can shape forest regeneration. The success of natural regeneration after harvesting is contingent upon intricate interactions among these factors.

Acknowledgements

We thank Dr. Patricia Raymond for insightful discussions regarding the gradient of harvesting severity illustrated in Figure 26.1. Mario Beltrán and Pau Vericat provided helpful comments on the section devoted to Mediterranean forests. We are indebted to the anonymous reviewers as well as to Dr. Alain Leduc for commenting on an earlier version of this chapter.

References

Abrams M. D. (2003) 'Where has all the white oak gone?', *Bioscience*, vol 53, no 10, pp. 927–938.

Aubin I., Munson A. D., Cardou F., Burton P. J., Isabel N., Pedlar J. H., Paquette A., Taylor A. R., Delagrange S., Kebli H., Messier C., Shipley B., Valladares F., Kattge J., Boisvert-Marsh L. and McKenney, D. (2016) 'Traits to stay, traits to move: a review of functional traits to assess sensitivity and adaptive capacity of temperate and boreal trees to climate change', *Environmental Reviews*, vol 24, no 2, pp. 164–186.

Barbeito I., Lemay V., Calama Sainz R., Argimiro R. and Cañellas I. (2011) 'Regeneration of Mediterranean *Pinus sylvestris* under two alternative shelterwood systems within a multiscale framework', *Canadian Journal of Forest Research*, vol 41, pp. 341–351.

Barrette M., Boucher Y., Dumais D. and Auger, I. (2022) 'Clear-cutting without additional regeneration treatments can trigger successional setbacks prolonging the expected time to compositional recovery in boreal forests', *European Journal of Forest Research*, vol 141, pp. 629–639.

Beguin J., Tremblay J.-P., Thiffault N., Pothier D. and Côté, S. D. (2016) 'Management of forest regeneration in boreal and temperate deer–forest systems: challenges, guidelines, and research gaps', *Ecosphere*, vol 7, no 10, article e01488.

Bergmeier E., Petermann J. and Schröder E. (2010) 'Geobotanical survey of wood-pasture habitats in Europe: diversity, threats and conservation', *Biodiversity and Conservation*, vol 19, no 11, pp. 2995–3014.

Boisvert-Marsh L., Pedlar J. H., de Blois S., Le Squin A., Lawrence K., McKenney D. W., Williams C. and Aubin, I. (2022) 'Migration-based simulations for Canadian trees show limited tracking of suitable climate under climate change', *Diversity and Distributions*, vol 28, no 11, pp. 2330–2348.

Botroh, A.-M., Paré, D., Cavard, X., Fenton, N., Valeria, O., Marchand, P. and Bergeron, Y. (2023) 'Nine-years effect of harvesting and mechanical site preparation on bryophyte decomposition and carbon stocks in a boreal forested peatland', *Forest Ecology and Management*, vol 540, article 121020.

Calama R., Manso R., Lucas-Borja M. E., Espelta J. M., Piqué M., Bravo F., del Peso C. and Pardos M. (2017) 'Natural regeneration in Iberian pines: a review of dynamic processes and proposals for management', *Forest Systems*, vol 26, no 2, article eR02S.

Chudy R., Cubbage F., Siry, J. and Chudy J. (2022) 'The profitability of artificial and natural regeneration: a forest investment comparison of Poland and the U.S. South', *Journal of Forest Business Research*, vol 1, no 1, pp. 1–20.

Coates K. D. and Burton P. J. (1997) 'A gap-based approach for development of silvicultural systems to address ecosystem management objectives', *Forest Ecology and Management*, vol 99, no 3, pp. 337–354.

DeArmond D., Emmert F., Lima A. J. N. and Higuchi N. (2019) 'Impacts of soil compaction persist 30 years after logging operations in the Amazon Basin', *Soil and Tillage Research*, vol 189, pp. 207–216.

del Cerro Barja A., Lucas Borja M. E., Martínez García E., López Serrano, F. R., Andrés Abellán M., García Morote, F. A. and Navarro López, R. (2009) 'Influence of stand density and soil treatment on the Spanish Black Pine (*Pinus nigra* Arn. ssp. Salzmannii) regeneration in Spain', *INIA: Sistemas y Recursos Forestales*, vol 18, no 2, pp. 167–180.

Dey D. C., Knapp B. O., Battaglia M. A., Deal R. L., Hart J. L., O'Hara K. L., Schweitzer C. J. and Schuler T. M. (2019) 'Barriers to natural regeneration in temperate forests across the USA', *New Forests*, vol 50, no 1, pp. 11–40.

Dudley N., Mansourian S. and Vallauri D. (1986) 'Forest landscape reforestation in context', in S. Mansourian, D. Vallauri and N. Dudley (eds.), *Forest Restoration in Landscapes: Beyond Planting Trees*. Springer, New York, NY.

Elie J.-G., Ruel J.-C. and Lussier J.-M. (2009) 'Effect of browsing, seedbed, and competition on the development of yellow birch seedlings in high-graded stands', *Northern Journal of Applied Forestry*, vol 26, no 3, pp. 99–105.

Espelta J. M., Cortés P., Molowny-Horas R., Cortés P. and Retana J. (2009) 'Acorn crop size and pre-dispersal predation determine inter-specific differences in the recruitment of co-occurring oaks', *Oecologia*, vol 161, no 3, pp. 559–568.

Fraver S., Wagner R. G. and Day M. (2002) 'Dynamics of coarse woody debris following gap harvesting in the Acadian forest of central Maine, USA', *Canadian Journal of Forest Research*, vol 32, no 12, pp. 2094–2105.

Frey B. R., Lieffers V. J., Landhäusser S. M., Comeau P. G. and Greenway K. J. (2003) 'An analysis of sucker regeneration of trembling aspen', *Canadian Journal of Forest Research*, vol 33, no 7, pp. 1169–1179.

Gamfeldt L., Snäll T., Bagchi R., Jonsson M., Gustafsson L., Kjellander P., Ruiz-Jaen M. C., Fröberg M., Stendahl J., Philipson C. D. et al. (2013) 'Higher levels of multiple ecosystem services are found in forests with more tree species', *Nature Communications*, vol 4, no 1, article 1340.

García-Barreda S., Valeriano C. and Camarero J. J. (2023) 'Drought constrains acord production and tree growth in the Mediterranean holm oak and triggers weak legacy effects', *Agricultural and Forest Meteorology*, vol 334, article 109435.

Ghazoul J. and Sheil, D. (2010) *Tropical Rain Forest Ecology, Diversity, and Conservation*. Oxford University Press, Oxford, UK.

Gouge D., Thiffault E. and Thiffault N. (2021) 'Biomass procurement in boreal forests affected by spruce budworm: effects on regeneration, costs, and carbon balance', *Canadian Journal of Forest Research*, vol 51, no 12, pp. 1939–1952.

Greene D. F., Zasada J. C., Sirois L., Kneeshaw D. D., Morin H., Charron I. and Simard M.-J. (1999) 'A review of the regeneration dynamics of North American boreal forest tree species', *Canadian Journal of Forest Research*, vol 29, no 6, pp. 824–839.

Haeussler S. and Kneeshaw D. D. (2003) 'Comparing forest management to natural processes', in P. J. Burton, C. Messier, D. W. Smith and W. L. Adamowicz (eds.), *Towards Sustainable Management of the Boreal Forest*, NRC Research Press, Ottawa, ON.

Hanssen K. H. (2003) 'Natural regeneration of *Picea abies* on small clear-cuts in SE Norway', *Forest Ecology and Management*, vol 180, no 1–3, pp. 199–213.

Harvey B. and Brais, S. (2002) 'Effects of mechanized careful logging on natural regeneration and vegetation competition in the southeastern Canadian boreal forest', *Canadian Journal of Forest Research*, vol 32, no 4, pp. 653–666.

Hewitt N. and Kellman M. (2002) 'Tree seed dispersal among forest fragments: II. Dispersal abilities and biogeographical controls', *Journal of Biogeography*, vol 29, no 3, pp. 351–363.

Hille Ris Lambers J. and Clark J. S. (2005) 'The benefits of seed banking for red maple (*Acer rubrum*): maximizing seedling recruitment', *Canadian Journal of Forest Research*, vol 35, no 4, pp. 806–813.

Hosonuma N., Herold M., De Sy V., De Fries R. S., Brockhaus M., Verchot L., Angelsen A. and Romijn E. (2012) 'An assessment of deforestation and forest degradation drivers in developing countries', *Environmental Research Letters*, vol 7, no 4, article 044009.

Houle G. (1998) 'Seed dispersal and seedling recruitment of *Betula alleghaniensis*: spatial inconsistency in time', *Ecology*, vol 79, no 3, pp. 807–818.

Johnson P. S., Shifley S. R. and Rogers R. (2009) *The Ecology and Silviculture of Oaks. 2nd Edition*, CABI Publishing, CAB International, Wallingford, UK.

Kabrick J. M., Clark K. L., D'Amato A. W., Dey D. C., Kenefic L. S., Kern C. C., Knapp B. O., MacLean D. A., Raymond P. and Waskiewicz, J. D. (2017) 'Managing hardwood-softwood mixtures for future forests in eastern North America: assessing suitability to projected climate change', *Journal of Forestry*, vol 115, no 3, pp. 190–201.

Kozlowski T. T. (2002) 'Physiological ecology of natural regeneration of harvested and disturbed forest stands: implications for forest management', *Forest Ecology and Management*, vol 158, no 1–3, pp. 195–221.

Lanzer N. B., Lee T. D., Ducey M. J. and Eisenhaure S.E. (2017) 'Sapling white pine (*Pinus strobus* L.) exhibits growth response following selective release from competition with glossy buckthorn (*Frangula alnus* P. Mill) and associated vegetation', *Forest Ecology and Management*, vol 404, pp. 280–288.

Lucas-Borja M. E., Fonseca T. F., Lousada J. L., Silva-Santos P., Garcia E. M. and Andrés Abellán M. A. (2012) 'Natural regeneration of Spanish black pine [*Pinus nigra* Arn. ssp. *salzmannii* (Dunal) Franco] at contrasting altitudes in a Mediterranean mountain area', *Ecological Research*, vol 27, no 5, pp. 913–921.

Martín-Alcón S., Coll L. and Salekin S. (2015) 'Stand-level drivers of tree-species diversification in Mediterranean pine forests after abandonment of traditional practices'. *Forest Ecology and Management*, vol 353, pp. 107–117.

Matías L., Zamora R. and Castro, J. (2012) 'Sporadic rainy events are more critical than increasing of drought intensity for woody species recruitment in a Mediterranean community', *Oecologia*, vol 169, pp. 833–844.

Ménard L.-P., Ruel J.-C. and Thiffault N. (2019) 'Abundance and impacts of competing species on conifer regeneration following careful logging in the eastern Canadian boreal forest', *Forests*, vol 10, no 2, article 177.

Michaletz S. T., Johnson E. A., Mell W. E. and Greene D. F. (2013) 'Timing of fire relative to seed development may enable non-serotinous species to recolonize from the aerial seed banks of fire-killed trees', *Biogeosciences*, vol 10, no 7, pp. 5061–5078.

Mirabel A., Hérault B. and Marcon E. (2020) 'Diverging taxonomic and functional trajectories following disturbance in a Neotropical forest', *Science of the Total Environment*, vol 720, article 137397.

Montoro Girona M., Lussier J.-M., Morin H. and Thiffault N. (2018) 'Conifer regeneration after experimental shelterwood and seed-tree treatments in boreal forests: finding silvicultural alternatives', *Frontiers in Plant Science*, vol 9, article 1145.

Montoro Girona M., Moussaoui L., Morin H., Thiffault N., Leduc A., Raymond P., Bosé A., Bergeron Y. and Lussier, J.-M. (2023) 'Innovative silviculture to achieve sustainable forest management in boreal forests: lessons from two large-scale experiments', in M. Montoro Girona, H. Morin, S. Gauthier and Y. Bergeron (eds.), *Boreal Forests in the Face of Climate Change. Sustainable Management*. Springer, Cham.

Moroni M. T., Carter P. Q. and Ryan D. A. J. (2009) 'Harvesting and slash piling affect soil respiration, soil temperature, and soil moisture regimes in Newfoundland boreal forests', *Canadian Journal of Soil Science*, vol 89, no 3, pp. 343–355.

Morrissey R. C., Jacobs D. F., Seifert J. R., Fischer B. C. and Kershaw J. A. (2008) 'Competitive success of natural oak regeneration in clearcuts during the stem exclusion stage', *Canadian Journal of Forest Research*, vol 38, no 6, pp. 1419–1430.

Naghdi R., Tavankar F., Solgi A., Nikooy M., Marchi E. and Picchio R. (2023) 'Effects on soil physico-chemical properties and seedling growth in mixed high forests caused by cable skidder traffic', *iForest - Biogeosciences and Forestry*, vol 16, no 2, pp. 127–135.

Nathan R. and Ne'eman G. (2004) 'Spatiotemporal dynamics of recruitment in Aleppo pine (*Pinus halepensis* Miller)', *Plant Ecology*, vol 171, no 1–2, pp. 123–127.

Naves R. P., Grøtan V., Prado P. I., Vidal E. and Batista, J. L. F. (2020) 'Tropical forest management altered abundances of individual tree species but not diversity', *Forest Ecology and Management*, vol 475, article 118399.

Nocentini S. and Coll L. (2013) 'Mediterranean forests: human use and complex adaptive systems', in C. Messier, K. J. Puettmann and K. D. Coates (eds.), *Managing Forests as Complex Adaptive Systems. Building Resilience to the Challenge of Global Change*. Routledge, New York, NY.

Nyland R. D. (2016) *Silviculture: Concepts and Applications. Third Edition*. Waveland Press, Inc., Long Grove, IL.

Osem Y., Yavlovich H., Zecharia N., Atzmon N., Moshe Y. and Schiller G. (2013) 'Fire-free natural regeneration in water limited *Pinus halepensis* forests: a silvicultural approach', *European Journal of Forest Resesarch*, vol 132, pp. 679–690.

Parent S. and Ruel J.-C. (2002) 'Chronologie de la croissance chez des semis de sapin baumier (*Abies balsamea* (L.) Mill.) après une coupe à blanc avec protection de la régénération', *The Forestry Chronicle*, vol 78, no 6, pp. 876–885.

Parker W. C. and Sharma, M. (2018) 'Influence of post-harvesting residual stand structure on canopy light transmittance in Ontario's boreal mixedwood forests', *The Forestry Chronicle*, vol 94, no 1, pp. 35–46.

Pendrill F., Gardner T. A., Meyfroidt P., Persson U. M., Adams J., Azevedo T., Bastos Lima M. G., Baumann M., Curtis P. G., De Sy V. et al. (2022) 'Disentangling the numbers behind agriculture-driven tropical deforestation', *Science*, vol 377, article 6611.

Picchio, R., Mederski, P. S. and Tavankar, F. (2020) 'How and how much, do harvesting activities affect forest soil, regeneration and stands?' *Current Forestry Reports*, vol 6, no 2, pp. 115–128.

Piqué M., Beltrán M., Vericat P., Cervera T., Farriol R. and Baiges T. (2001) *Models de gestió per als boscos de pi roig (Pinus sylvestris L.): producció de fusta i prevenció d'incendis forestals. Sèrie: Orientacions de gestió forestal sostenible per a Catalunya (ORGEST)*. Centre de la Propietat Forestal. Departament d'Agricultura, Ramaderia, Pesca, Alimentació i Medi Natural. Generalitat de Catalunya, Spain.

Putz F. E., Sist P., Fredericksen T. and Dykstra D. (2008) 'Reduced-impact logging: challenges and opportunities', *Forest Ecology and Management*, vol 256, no 7, pp. 1427–1433.

Ray G. J. and Brown B. J. (1994) 'Seed ecology of woody species in a Caribbean dry forest', *Restoration Ecology*, vol 2, no 3, pp. 156–163.

Raymond P., Guillemette F. and Larouche C. (2013) 'Les grands types de couvert et les groupements d'essences principales', in C. Larouche, F. Guillemette, P. Raymond and J. -P. Saucier (eds.), *Le guide sylvicole du Québec - Tome 2. Les concepts et l'application de la sylviculture*. Les Publications du Québec, Québec, QC.

Retana J., Riba. M., Castell C. and Espelta J. M. (1992) 'Regeneration by sprouting of holm-oak (*Quercus ilex*) stands exploited by selection thinning', *Vegetatio*, vol 99–100, no 1, pp. 355–364.

Rodríguez-García E., Gratzer G. and Bravo F. (2011) 'Climatic variability and other site factors influences on natural regeneration of *Pinus pinaster* Ait. in Mediterranean forests', *Annals of Forest Science*, vol 68, no 4, pp. 811–823.

Rodríguez-García E., Juez L. and Bravo F. (2010) 'Environmental influences on post-harvest natural regeneration of *Pinus pinaster* Ait. in Mediterranean forest stands submitted to seed-tree selection method', *European Journal of Forest Research*, vol 129, no 6, pp. 1119–1128.

Serrada R., Gómez-Sanz V., Aroca M. J., Otero J., Bravo-Fernández A. and Roig S. (2017) 'Decline in holm oak coppices (*Quercus ilex* L. subsp. *ballota* (Desf.) Samp.): biometric and physiological interpretations', *Forest Systems*, vol 26, no 2, article e06S.

Shorohova E., Aakala T., Gauthier S., Kneeshaw D., Koivula M., Ruel J.-C. and Ulanova N. (2023) 'Natural disturbances from the perspective of forest ecosystem-based management', in M. Montoro Girona, H. Morin, S. Gauthier and Y. Bergeron (eds.), *Boreal Forests in the Face of Climate Change. Sustainable Management*. Springer, Cham.

Sist P., Piponiot C., Kanashiro M., Pena-Claros M., Putz F. E., Schulze M., Verissomo A. and Vidal E. (2021) 'Sustainability of Brazilian forest concessions', *Forest Ecology and Management*, vol 496, article 119440.

Smith D. M. (1962) *The Practice of Silviculture. Seventh Edition*. John Wiley & Sons, Inc., New York, NY.

Thiffault N., Hoepting M., Marchand M., Moulin M. and Deighton H. (2023) 'Eastern white pine regeneration abundance, stocking, and damage along a gradient of harvest intensity', *The Forestry Chronicle*, vol 99, no 1, pp. 92–102.

27

TROPICAL DEFORESTATION, FOREST DEGRADATION, AND THE ROLE OF REDD+ IN CLIMATE CHANGE MITIGATION

John A. Parrotta

Introduction: Tropical forest loss and its climate change implications

Forests today cover an estimated 31% of the earth's land surface (4.06 billion hectares), of which 93% are natural forest and 7% are planted (FAO, 2020). They contain a substantial proportion of the world's terrestrial biodiversity and play a major role in the global carbon cycle, absorbing approximately 11 Gt[1] of carbon dioxide (CO_2) annually from the atmosphere as they grow (IPCC, 2014), and storing carbon for extended periods of time in biomass, dead organic matter, and soil carbon pools. Carbon sequestration by forests represents up to 29% of annual anthropogenic CO_2 emissions between 2011 and 2020 (Friedlingstein et al., 2022; Donegan et al., 2022). Of the global forest carbon stocks (including soils to 1 m depth), an estimated 55% (471 Gt C) is stored in tropical and subtropical forests, of which more than half is stored in biomass (Pan et al., 2011).

However, deforestation and forest degradation—resulting from agricultural expansion and conversion to pastureland, infrastructure development, destructive logging, mining, and climate change impacts such as increased extent and severity of drought, forest fires, and pest and disease outbreaks—is also a significant source of carbon emissions. These emissions are estimated at approximately 4.1 $GtCO_2$/year between 2011 and 2020, equivalent to 10% of the total anthropogenic CO_2 emissions of 39.8 $GtCO_2$/year (Friedlingstein et al., 2022).

Given the important roles that tropical and subtropical forests play in the global carbon cycle and for biodiversity conservation, reversing trends in deforestation and forest degradation in Latin America, Africa, and South and Southeast Asia has become a global priority. The potential for enhancing the capacity of forests and forested landscapes to sequester carbon and help mitigate climate change has attracted increasing interest in recent decades. Current estimates of the mitigation potential from reducing deforestation vary widely (between 0.4 and 5.8 $GtCO_2$/year). However, recent research suggests that the capacity of existing forests

DOI: 10.4324/9781003324072-32

to continue acting as a carbon sink is declining, and that they could soon become a net carbon source due to the impacts of climate change, including extreme weather events, fire, drought, and forest pest and disease outbreaks (Donegan et al., 2022).

What is REDD+?

REDD+[2] is a mechanism for climate change mitigation developed in the United Nations Framework Convention on Climate Change (UNFCCC) and being implemented through a growing number of international organizations, governments, and a variety of financial and other institutions. The aim of REDD+ is to reduce CO_2 and other greenhouse gas emissions from deforestation and forest degradation, and to enhance forest carbon stocks in 'developing' countries (principally in tropical and subtropical regions) by providing financial and other incentives to governments, landholders, and/or communities for managing their forest lands towards this end.

REDD+ can be seen as complementary to other global and regional programmes and initiatives aiming to halt and reverse deforestation, biodiversity loss, and associated land degradation. At the global level, these include the Bonn Challenge for forest landscape restoration launched in 2011, the Paris Agreement (adopted in 2015), and the 2021 Glasgow Leaders' Declaration on Forests and Land Use (2021) arising from the UNFCCC, the Convention on Biological Diversity's Global Biodiversity Framework (GBF) adopted in 2022, and the UN Decade on Ecosystem Restoration (2021–2030).

The topic of reducing emissions from deforestation and forest degradation was first introduced in the 2005 meeting of the UNFCCC's Conference of the Parties, and further elaborated two years later in the UNFCCC's 'Bali Action Plan'. In 2010, the UNFCCC Conference of the Parties in Cancún, Mexico, reached an agreement on policy approaches and positive incentives for reducing greenhouse gas emissions from forests. The Cancún decision on REDD+ specifically encourages developing countries to pursue climate change mitigation actions in the forest sector by (1) reducing emissions from deforestation, (2) reducing emissions from forest degradation, (3) conservation of forest carbon stocks, (4) sustainable management of forests, and (5) enhancement of forest carbon stocks. The Cancún decision also specified that these five REDD+ activities, among other things, should be country-driven, consistent with the objective of environmental integrity and take into account the multiple functions of forests and other ecosystems, be implemented in the context of sustainable development and reducing poverty, be consistent with the climate change adaptation needs of the country, be results-based, and promote sustainable management of forests. The Warsaw REDD+ agreements (2013) finalized the details for funding, implementation, and monitoring of REDD+ activities under the UN-REDD programme.

The development of REDD+ raised hopes and expectations in many quarters, particularly among those who see the potential for significant environmental and socio-economic 'co-benefits' beyond greenhouse gas emissions. These anticipated co-benefits include conservation of forest biodiversity, water regulation, soil conservation, timber, forest foods and other non-timber forest products, and direct social benefits—jobs, livelihoods, land tenure clarification, carbon payments, enhanced participation in decision-making, and improved governance.

However, there have been concerns that the delivery of these co-benefits could be impaired for a number of reasons, including the complex governance arrangements involved in REDD+ programmes and activities, and the international financial mechanisms that underpin them; the environmental and social risks and inequity associated with various aspects of REDD+ policy development, planning, and implementation, for example, issues of sovereignty, insecure land tenure and rights; the high likelihood of 'leakage' (displacement of deforestation and forest degradation from a REDD+ project area to another location); and long-standing difficulties in addressing the underlying causes of deforestation and forest degradation (summarized in Parrotta et al., 2012, 2022).

The following sections examine deforestation and forest degradation in tropical and subtropical regions of the world where REDD+ activities are either underway or may be carried out in the future. The proximal and underlying causes of forest loss and degradation are discussed, as well as their impacts on the provision of ecosystem services,[3] with a particular focus on forest carbon sequestration. Forest and landscape management activities that may be implemented in countries to meet REDD+ objectives will be considered, as will the likely environmental impacts and their social and economic consequences (both positive and negative) on people most affected by REDD-related changes in agricultural and forest management policies, incentives, and governance structures.

Deforestation and forest degradation in tropical and subtropical regions

Deforestation

Deforestation is the conversion of forest land to another land use, generally the result of transformation of forested lands to other land uses. Deforestation includes areas of forest converted to agriculture, pasture, water reservoirs, and urban areas, and excludes areas where trees have been removed as a result of harvesting or logging, where the forest is expected to regenerate naturally or with the aid of silvicultural measures. Deforestation also includes areas where, for example, the impact of disturbance, overuse, or changing environmental conditions affects the forest to an extent that it cannot sustain a tree cover.

At the global level, deforestation rates remain high, with an estimated 420 million hectares (Mha) of forests lost between 1990 and 2020 (FAO, 2020), more than 90% of this occurring in the tropics. In recent decades, however, large-scale forest planting efforts, natural expansion of forests, and successes in slowing deforestation rates in some countries have reduced the rate of net forest loss, which declined from 7.8 Mha per year between 1990 and 2000 to 5.2 Mha per year in 2000–2010 and 4.7 Mha per year in 2010–2020 (FAO, 2020). At the regional level, Africa had the largest annual rate of net forest loss in 2010–2020, at 3.9 Mha, followed by South America, at 2.6 Mha. The rate of net forest loss has increased in Africa in each of the three decades since 1990, while it has declined substantially in South America, however, to about half the rate in 2010–2020 compared to the 2000–2010 period. Asia had the highest net gain of forest area in 2010–2020, followed by Oceania and Europe (FAO, 2020).

Although deforestation rates have fallen in some countries, continued pressure on forests would suggest that rates of forest loss in tropical and subtropical countries are likely to remain high in the foreseeable future. Recent human impacts across global forest ecosystems have not been equal, with some forest types now under severe threat. Both tropical and subtropical dry and montane forests have been converted to a large extent because they are located in climates highly suitable for agriculture and grazing. Mangrove forest area declined by 19% from 1980 to 2005 (FAO, 2007), a trend that continues due to ongoing land clearing, coastal development, conversion for aquaculture, and changes in hydrological regimes. Extensive areas of freshwater swamp and peat forests in Southeast Asia, which store vast amounts of carbon in their soils, have been lost or severely degraded in recent decades by unsustainable logging and agricultural expansion, including oil palm plantations.

Agricultural expansion has been the most important direct cause of global forest loss, accounting for approximately 80% of deforestation worldwide, with the majority occurring during the 1980s and the 1990s through conversion of tropical forests (Gibbs et al., 2010). The conversion of forest lands for commercial agriculture and pasture has been responsible for approximately two-thirds of deforestation in Latin America, while in Africa and tropical and subtropical Asia forest clearing for subsistence farming is the major driver of land use change (Kissinger et al., 2012).

The underlying drivers of forest loss include increasing global natural resource consumption, rapid population growth, and the often over-riding effects of economic globalization and global land scarcity. These are exacerbated by problems of weak governance, inadequate policies, lack of cross-sectoral coordination, 'perverse' incentives that encourage forest conversion, and illegal activities (Kissinger et al., 2012; Pandit et al., 2018; Mansourian et al., 2019).

Forest degradation

Forest degradation may be broadly defined as a reduction in the capacity of a forest to produce ecosystem services, such as carbon storage and wood products, due to anthropogenic pressures and environmental changes. Such degradation typically results in the reduction in affected forests' structural complexity and habitat diversity, changes in the diversity of tree species, and loss of productivity. Forests may be degraded from several perspectives depending on the cause, particular goods or services of interest, and the temporal and spatial scales considered. Forest degradation is widespread and has become an important consideration in global policy processes that deal with biodiversity, climate change, and forest management. The International Tropical Timber Organization estimated that up to 850 Mha of tropical forest could already be characterized as degraded (ITTO, 2002).

The proximate drivers of forest degradation include unsustainable logging, over-harvest of biomass fuels and non-timber forest products (NTFPs), over-grazing, human-induced fires (or fire suppression in dry forests), and poor management of shifting cultivation (ITTO, 2002; Pandit et al., 2018). For example, destructive or otherwise unsustainable timber extraction accounts for more than 70% of tropical forest degradation in Latin America and Asia (Kissinger et al., 2012). Fuelwood collection and charcoal production, which account for an estimated 40% of global removal of wood from forests (FAO 2006), as well as forest grazing, are major causes of forest degradation, particularly across Africa (Kissinger et al., 2012).

Although fire is a natural element in many forest ecosystems, humans have altered fire regimes across an estimated 60% of terrestrial habitats (Shlisky et al., 2009). In recent decades, fires have increased in their frequency and extent into tropical rainforests with the expansion of agriculture, forest fragmentation, unsustainable shifting cultivation, and logging. In addition to direct human-induced and disturbance-related forest degradation, climate change poses an additional and growing threat to global forest ecosystems, increasing the frequency and intensity of severe drought, fires, and forest pest and disease outbreaks that can result in a long-term reduction of forest cover.

Deforestation and forest degradation can often act synergistically. Deforestation fragments forest landscapes, which often results in degradation of remaining forests due to edge effects (e.g., drying of the forest floor, increased fire frequency, increased tree mortality, and shifts in tree species composition). Degradation can also lead to subsequent deforestation. For example, poorly planned logging activities typically increase road access to remaining forest interiors, facilitating shifting cultivation and other land clearing activities, hunting, illegal logging and mining, and fire. Degraded forests can also remain in a degraded state for long periods of time if degradation drivers (e.g., fire, human, and livestock pressures) remain, or if ecological thresholds have been passed beyond which forests cannot recover their structure and composition through natural successional processes.

Impacts of deforestation and forest degradation on carbon sequestration and storage

CO_2 emissions associated with forest conversion

Tropical and subtropical forests store an estimated 247 Gt C in biomass (both above- and belowground) (Saatchi et al., 2011). When forests are converted to croplands, often through burning, a large portion of carbon stored in aboveground vegetation is immediately released to the atmosphere as CO_2 (and other greenhouse gases), or over time, through the decomposition of debris (Figure 27.1). Carbon in soils following deforestation can also become a large source of emissions because of increased soil respiration with warmer ambient temperatures. There is also increased soil loss with higher flooding and erosion rates, with carbon being transported downstream where a large fraction of the decayed organic matter is released as CO_2.

CO_2 emissions associated with forest degradation

Carbon emissions from forest degradation are often pooled with deforestation to estimate emissions from land-use change (e.g., Houghton, 2003). They are more difficult to assess than those from deforestation due to a lack of consistent data resulting from different analytical methodologies, as well as inconsistencies in definitions of degraded forests. Nonetheless, forest degradation is estimated to generate a very significant proportion, between 25% and >65%, of total forest-related carbon emissions (Donegan et al., 2022).

Unsustainable forest management is a significant cause of forest degradation in many tropical regions, as noted above. Poor logging practices create large canopy openings and cause significant damage to remaining trees, subcanopy vegetation, and soils, and in some cases facilitate the introduction of invasive plants, pests, and diseases. During timber harvest, a substantial portion of biomass carbon (approximately 50%) can be left as logging residues, and about 20% of harvested wood biomass is further lost in the process of manufacturing wood products (Pan et al., 2011). There is a further loss of carbon from decomposition or combustion of harvest residues, and from harvested wood that is used for fuelwood and paper production.

CO_2 emissions associated with forest fires

The impact of forest fires on carbon emissions is particularly significant (van der Werf et al., 2009). In extreme drought years, carbon emissions from tropical forest fires can exceed those from deforestation (Houghton, 2003). For example, total estimated carbon emissions

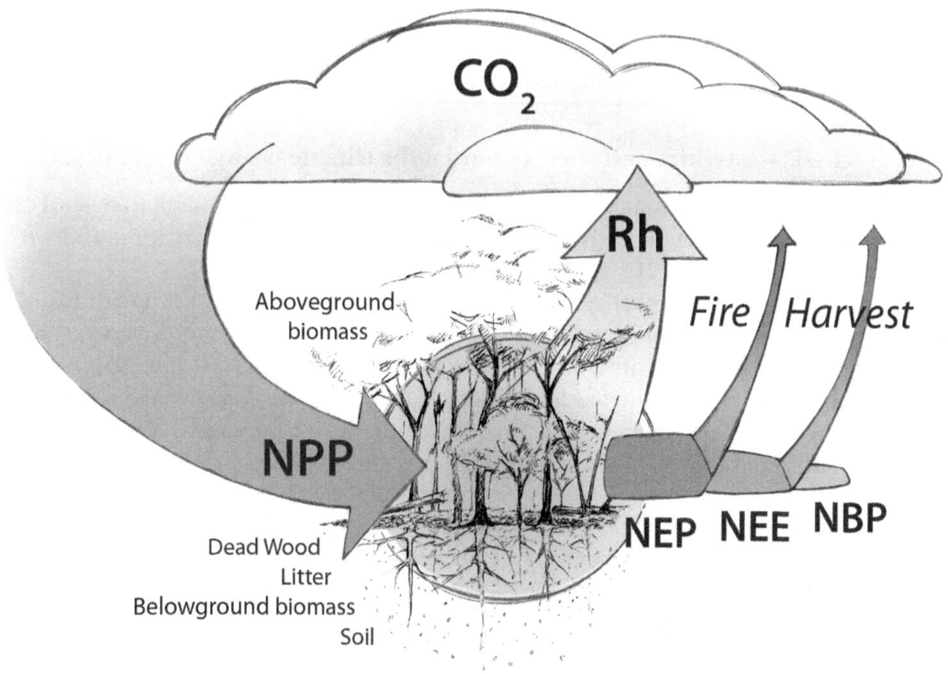

Figure 27.1 The major carbon fluxes in forest ecosystems. Net primary production (NPP) quantifies the amount of organic matter produced annually. Most of this carbon uptake is offset through losses from the decomposition of litter, dead wood, and soil C pools (Rh = heterotrophic respiration). The net balance (net ecosystem production, NEP) is further reduced through direct fire emissions to yield net ecosystem exchange (NEE), from which harvest losses are subtracted to estimate the annual C stock change in forest ecosystems (net biome production, NBP). Positive NBP indicates increasing forest carbon stocks, a sink from the atmosphere, while negative NBP indicates a carbon source. NEE is reported from the perspective of the atmosphere and has the opposite sign convention. (Source: Figure by Avril Goodall [Canadian Forest Service], reprinted from Parrotta et al., 2012).

from tropical forest fires during the 1997–1998 El Niño event, which reduced rainfall and increased susceptibility to fire over large forest areas in the tropics, were 0.83– 2.8 Gt C yr^{-1} (Cochrane, 2003).

In recent decades, the frequency and size of forest fires have increased in many tropical regions, including in areas where fires have not been known to occur historically, and where forest ecosystems are not well adapted to fire. In 2015 alone, fires burned about 4% of the total forest area in the tropical domain, and approximately two-thirds of the 98 Mha of forest affected by fire worldwide were in Africa and South America (FAO, 2020). Such fires are often associated with deforestation, destructive timber harvesting, and other land-use practices associated with human encroachment that increase the frequency, intensity, and impacts of fire in these ecosystems.

Shifting cultivation (also known as swidden or 'slash-and-burn' agriculture), practised throughout the tropics and subtropics, also contributes to overall greenhouse gas emissions. However, carbon sequestration by secondary forest regeneration during the fallow, or abandonment, phase may offset a significant proportion of these losses.

REDD+ activities and their potential for climate change mitigation

A number of forest and associated land management options offer potential to meet REDD+ climate change mitigation objectives while serving biodiversity conservation and broader societal needs (Kapos et al., 2012, summarized below).

These management actions may be applied at varying spatial scales both within and outside of the forest sector to address the proximate as well as underlying drivers of deforestation and forest degradation in tropical and subtropical regions as well as approaches for increasing forest cover on degraded forest lands (Table 27.1). Actions aimed at maintaining existing carbon and biodiversity by effectively reducing deforestation and forest degradation are more likely to have the greatest and most immediate benefits compared to those that aim to restore them.

Actions that address drivers of deforestation and forest degradation are essential to reduce greenhouse gas emissions from forest landscapes. Improvements in agricultural practices on lands currently in use for food and livestock production have a particularly important role to play in REDD+ strategies. Sustainable agricultural intensification and other improvements in existing production systems, including agroforestry and shifting cultivation, may help both to limit the increase in demand for new land (and associated deforestation and forest degradation) and reduce the direct impacts such as those from unsustainable shifting cultivation, the use of fire in land preparation and management, and the application of agrochemicals. Such improvements can also help to enhance carbon stocks across agro-forest landscapes (Kapos et al., 2012).

One of the most effective means for protecting forests from degradation and loss—and thereby conserve forest carbon stocks and reduce greenhouse gas emissions—is the establishment of protected areas or other conservation units. Management approaches for such areas may range from strict protection of biodiversity to allowing multiple uses, including limited extractive activities. Protected area governance may also vary, with some managed by government authorities and others by private landowners or by local communities. Establishing protected areas may not always result in a net reduction of carbon emissions, particularly if deforestation and forest degradation pressures shift to *unprotected* areas elsewhere.

Since the establishment of REDD+ mechanisms for payments for reductions in carbon emissions (i.e., results-based payments), most payments (>95%) made through the Green Climate Fund have been awarded to countries—particularly in Latin America—for emission-reducing activities, including protected area establishment and management. This has resulted in a total avoided 11.4 $GtCO_2eq$ emissions estimated over the period 2006–2020 for the 17 countries involved (Donegan et al., 2022).

In many regions, reducing emissions from forest degradation will also require actions to reduce the impacts of extraction of forest products including timber, fuelwood, and other NTFPs. Such actions may include improved timber harvesting practices such as reduced impact logging, as well as promoting the sustainable harvesting and use of NTFPs. Depending on national and local circumstances, such actions and the policies used to promote them can

Table 27.1 Relevance of management interventions to the five REDD+ activities

Forest management type & management actions likely to be used in REDD	*Relevance to REDD+ activities*				
	Reduce emissions from deforestation	*Reduce emissions from forest degradation*	*Carbon stock enhancement*	*Sustainable management of forests*	*Conservation of carbon stocks*
Improving agricultural practice					
Sustainable agricultural intensification	••	•	•		•
Agroforestry		•	•		
Sustainable shifting cultivation	•	••	•		•
Fire management	•	••	••	••	•
Protection measures	••	••	•	•	••
Reducing impacts of extractive use					
Reduced impact logging		••		••	
Efficiencies, alternative production, or substitution of fuelwood and NTFPs		••		••	
Hunting regulation		••		•	
Restoration/Reforestation					
Assisted natural regeneration	•	•	••	•	
Afforestation & reforestation primarily for wood/fibre production			••		
Reforestation primarily for biodiversity and ecosystem services		•	••		
Landscape scale planning & coordination	••	••	•	••	•

Source: Kapos et al. (2012).

Note: Some interventions have a strong and direct role to play in a given REDD+ activity (••), while others may have less immediate relevance but may still play a role (•).

play an important role in REDD+ through their contribution to reducing forest degradation and promoting sustainable management of forests.

The enhancement of forest carbon stocks may include a broad set of management actions involving various forms of forest restoration, reforestation, and afforestation. Assisted natural regeneration in deforested or degraded areas can be used in many locales to accelerate natural secondary forest development by removing stressors or barriers to natural regeneration (such as recurrent fire, grazing, or dominant invasive vegetation). Reforestation may also be achieved through establishment of planted forests, using either plantation monocultures or mixed species plantings (Lamb et al., 2005). In some cases, enrichment planting may be used to enhance the composition of existing degraded forests (ITTO, 2002; Stanturf et al., 2019). All of these actions have the potential to yield positive results in terms of carbon sequestration as well as the conservation of biodiversity and provision of forest ecosystem services, particularly if designed in ways that will enhance resilience to current and anticipated future environmental conditions.

Although not included among the actions eligible for payments under REDD+, the storage of carbon in long-lived wood-based products from sustainably managed forests can also contribute significantly to overall greenhouse gas emission reductions, particularly when such products can substitute for others (such as concrete, steel, or plastics) whose production results in high CO_2 emissions.

Factors influencing the environmental impacts of REDD+ activities

The forest and land management activities outlined above can have highly variable impacts on carbon, biodiversity, and provision of forest ecosystem services depending on the location, scale of implementation, initial conditions, historical impacts, forest type, and the wider landscape context. While one or more of these actions may be included in REDD+ programmes and strategies, coordination and planning at landscape and broader scales are key to minimizing the negative impacts and ensuring positive outcomes for both carbon and biodiversity. Where REDD+ interventions have been integrated with national development strategies and plans, this appears to have allowed for greater convergence of development finance and other resources towards addressing the ultimate drivers of deforestation and forest degradation (Kapos et al., 2022; Mansourian et al., 2022). Such integration makes it more likely to secure lasting changes, avoid risks of leakage, and secure additionality of REDD+ activities. However, from the limited evidence on REDD+ outcomes to date (summarized in Parrotta et al., 2022), it is not yet possible to reach any firm conclusions regarding questions related to 'permanence' and 'leakage' of carbon gains reported from recent and current REDD+ activities. This is due to the limited information available from national reporting on REDD+ results, the relatively early stage of REDD+ implementation at subnational levels, and the short period of time during which REDD+ activities have been implemented.

Different management actions require different time periods to deliver benefits for carbon, biodiversity, and other ecosystem services (Kapos et al., 2012). In some cases, actions that yield positive carbon sequestration benefits in the short term may fail to deliver biodiversity conservation benefits and in some cases may cause negative impacts. Trade-offs between carbon and biodiversity outcomes can occur both locally and at wider spatial scales. For example, plantations of introduced species may provide large and rapid—though short-lived—carbon benefits but contribute little to local biodiversity and provision of valued ecosystem services. At landscape scales, efforts to alleviate deforestation pressure on natural forests through agricultural intensification and associated inputs of agrochemicals can have detrimental impacts on biodiversity and increase greenhouse gas emissions (Kapos et al., 2012). Not all impacts on carbon and biodiversity are easily anticipated or measured. Impacts can occur outside the area of management and/or in the future (Kapos et al., 2012). Indirect impacts resulting from displacement of land use pressures or extractive activities (e.g., following the creation of protected areas) are particularly problematic. Unintended increases in net greenhouse gas emissions may result if, for example, constraints on timber harvesting lead to the replacement of wood products with more emissions-intensive alternatives such as concrete, steel, or plastics.

Both the magnitude and the direction of the environmental impacts can change over time. For example, fire suppression in naturally fire-dependent forest ecosystems can lead to increased carbon stocks in the short term, but can be severely detrimental in the long term for both carbon and biodiversity if the accumulation of fuel (flammable biomass) leads to catastrophic fires.

It is generally accepted that of the five REDD+ activities, reducing deforestation and forest degradation has by far the greatest potential to yield positive results in terms of carbon sequestration and biodiversity conservation. As a means to enhance forest carbon stocks, restoration of critical habitats such as riparian forests in degraded watersheds, or creation of forest corridors to improve forest connectivity in fragmented landscapes, can provide significant benefits for biodiversity while at the same time can enhance the provision of many forest ecosystem services valued by people (Kapos et al., 2012).

There is still much uncertainty, however, about the potential impacts on biodiversity of other activities to enhance forest carbon stocks and those related to the sustainable management of forests. Further, there is both uncertainty and concern about how all REDD+ activities may directly and indirectly affect the well-being of people, especially indigenous and local communities who may depend on a variety of ecosystem services other than carbon (Mansourian et al., 2022).

Social and economic impacts of REDD+ activities

While REDD+ has the potential to generate positive impacts for climate mitigation and biodiversity conservation, the way in which it is implemented will determine its social and economic effects (Strassburg et al. 2012; Kapos et al., 2022), as summarized in Figure 27.2. The primary objectives of REDD+, that is, avoiding deforestation and forest degradation, can greatly benefit landless or otherwise disadvantaged people in rural areas who typically have a greater dependence on products derived from forests for their subsistence and livelihood needs, and are often disproportionately impacted by the loss of forests and the environmental services they provide.

REDD+ success and permanence of results with respect to carbon sequestration are contingent on the participation and support of local stakeholders who may be directly or indirectly affected, as well as on ensuring positive biodiversity and other environmental impacts of REDD+ activities. Evidence from social evaluations of REDD+ interventions indicates that, where direct and indirect benefits are clearly visible to local stakeholders, and have been delivered, community engagement is strong and projects have achieved positive carbon and social outcomes, at least in the short term. However, there is as yet limited direct evidence of the degree to which longer-term social and environmental outcomes are being achieved in current REDD+ activities (Kapos et al., 2022).

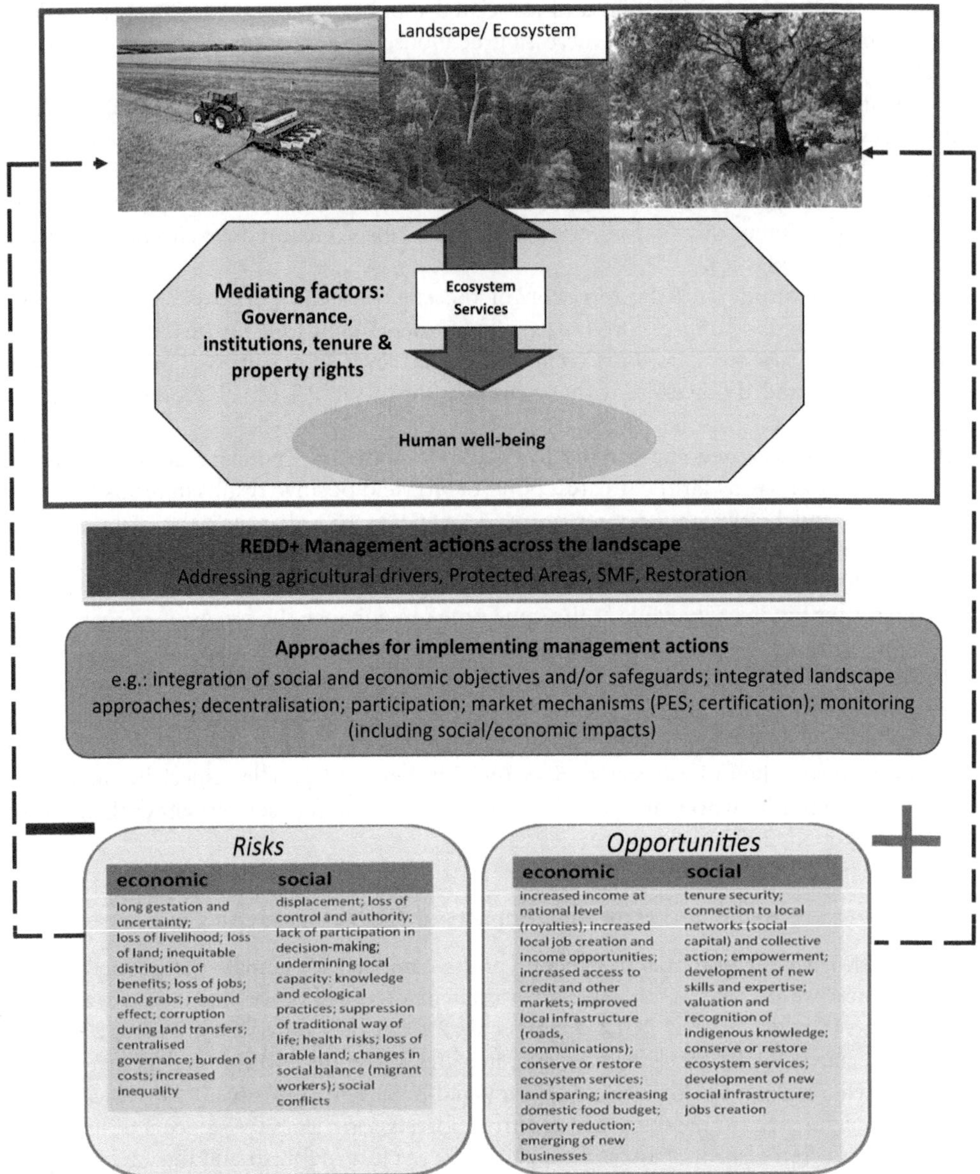

Figure 27.2 Economic and social impacts of REDD+ management actions on different stakeholders within a landscape. (Source: Reprinted from Strassburg et al., 2012, in Parrotta et al., 2012.)

It is generally accepted that tenure security and associated authority for local decision-making support better environmental management, as well as the realization of livelihood benefits in the context of REDD+ and comparable forest landscape restoration initiatives (Strassburg et al., 2012, Mansourian et al., 2019; ITTO, 2020; Mansourian et al., 2022).

Evidence from past experience, including other payment for environmental services (PES) programmes, strongly suggests that pursuing these social objectives in REDD+ planning and implementation is likely to not only make the process more equitable but also increase the likelihood of achieving carbon and biodiversity goals (Strassburg et al., 2012). To date, however, measuring implementation of safeguards in REDD+ activities has been limited, with no requirements concerning performance and outcomes. Complexity surrounding safeguards and proliferation of diverse standards currently affect not only their implementation but also accountability in their application (Kapos et al., 2022).

Integrated landscape planning and management is a powerful tool for addressing and reconciling the many environmental, social, and economic aspects relevant to REDD+ inside and outside forests (IPBES, 2018; Mansourian et al., 2019). Careful and inclusive (participatory) spatial planning can positively influence the distribution of winners and losers across the landscape so that REDD+ could serve the interests of the most vulnerable groups, thereby increasing the likelihood of positive impacts on both equity and environmental effectiveness (Strassburg et al., 2012).

Conclusion

REDD+ was conceived as a climate change mitigation strategy, focusing almost exclusively on measures to enhance the role of forests in sequestering carbon and reducing CO_2 and other greenhouse gas emissions associated with deforestation and forest degradation in tropical and subtropical countries.

Conservation, sustainable management, and restoration of forests are critical not only to efforts to help mitigate climate change, but also to stem the loss of biodiversity which underpins the capacity of forests to provide the ecosystem services that sustain the livelihoods of people worldwide. The success of REDD+ initiatives will depend on the degree to which REDD+ planning and implementation of specific activities recognize—and address—the complexity of historical and contemporary environmental, social, economic, and political factors that drive deforestation and forest degradation. It can also learn from decades of experience as to why efforts to date to slow these processes have often failed and recognize the importance (and more immediate value) of the broad range of other ecosystem services (beyond carbon sequestration and storage) that forests provide to people, particularly to those segments of rural society that depend most directly on forest goods and services to sustain their livelihoods.

Notes

1 $GtCO_2$: 10^9 tonnes of carbon dioxide equivalent.

2 Reducing emissions from deforestation and forest degradation, plus the sustainable management of forests, and the conservation and enhancement of forest carbon stocks.

3 Forest ecosystem services include ***supporting services*** such as nutrient cycling, soil formation, and primary productivity; ***provisioning services*** such as food, water, timber, and medicine; ***regulating services*** such as erosion control, climate regulation, flood mitigation, purification of water and air, pollination, and pest and disease control; and ***cultural services*** such as recreation, ecotourism, educational and spiritual values (MA, 2005).

References

Cochrane, M.A. (2003) 'Fire science for rainforests', *Nature,* vol. 421, pp. 913–919.

Donegan, E., Sandker, M., Achard, F., Vancutsem, C., Pekkarinen, A., Johsson, Ö., and Fox, J. (2022) 'Outcomes and influences of REDD+ implementation on carbon', in J. Parrotta, S. Mansourian, N. Grima and C. Wildburger (eds.), *Forests, Climate, Biodiversity and People: Assessing a Decade of REDD+*. IUFRO World Series vol. 40, International Union of Forest Research Organizations, Vienna, pp. 61–83. www.iufro.org/fileadmin/material/publications/iufro-series/ws40/ws40.pdf

FAO. (2006) *Global Forest Resources Assessment 2005*, Forestry Paper 147, Food and Agriculture Organization of the United Nations, Rome. www.fao.org/forest-resources-assessment/past-assessments/fra-2010/en/

FAO. (2007) *The World's Mangroves 1980-2005: A Thematic Study Prepared in the Framework of the Global Forest Resources Assessment 2005*, FAO Forestry Paper 163, Food and Agriculture Organization of the United Nations, Rome.

FAO. (2020) *Global Forest Resources Assessment 2020*, Forestry Paper 163, Food and Agriculture Organization of the United Nations, Rome. www.fao.org/3/CA8753EN/CA8753EN.pdf

Friedlingstein, P., Jones, M.W., O'Sullivan, M., Andrew, R.M., Bakker, D.C., Hauck, J., Le Quéré, C., Peters, G.P., Peters, W., Pongratz, J. and Sitch, S. (2022) 'Global carbon budget 2021', *Earth System Science Data*, vol. 14, pp. 1917–2005.

Gibbs, H.K., Ruesch, A.S., Achard, F., Clayton, M.K., Holmgren, P., Ramankutty, N. and Foley, J.A. (2010) 'Tropical forests were the primary sources of new agricultural land in the 1980s and 1990s', *Proceedings of the National Academy of Sciences,* vol. 107, pp. 16732–16737.

Houghton, R.A. (2003) 'Revised estimates of the annual net flux of carbon to the atmosphere from changes in land use and land management 1850-2000', *Tellus*, vol. 55B, pp. 378–390.

IPBES. (2018) *The IPBES Assessment Report on Land Degradation and Restoration.* L. Montanarella, R. Scholes, and A. Brainich (eds.). Secretariat of the Intergovernmental Science-Policy Platform on Biodiversity and Ecosystem Services, Bonn, Germany, 744 pp. https://ipbes.net/assessment-reports/ldr

IPCC. (2014) *Climate Change 2014: Mitigation of Climate Change. Contribution of Working Group III to the Fifth Assessment Report of the Intergovernmental Panel on Climate Change.* O. Edenhofer, R. Pichs-Madruga, Y. Sokona, E. Farahani, S. Kadner, K. Seyboth, A. Adler, I. Baum, S. Brunner, P. Eickemeier, B. Kriemann, J. Savolainen, S. Schlömer, C. von Stechow, T. Zwickel, and J.C. Minx (eds.). Cambridge University Press, Cambridge, United Kingdom and New York, NY. www.ipcc.ch/report/ar5/wg3/

ITTO. (2002) *Guidelines for the Restoration, Management and Rehabilitation of Degraded and Secondary Tropical Forests*, ITTO Policy Development Series No. 13. International Tropical Timber Organization, Yokohama, Japan, 138 pp. www.itto.int/guidelines/

ITTO. (2020) *Guidelines for Forest Landscape Restoration in the Tropics.* ITTO Policy Development Series No. 24. International Tropical Timber Organization (ITTO), Yokohama, Japan, 140 pp. www.itto.int/guidelines/

Kapos, V., Kurz, W.A., Gardner, T., Ferreira, J., Guariguata, M., Koh, L.P., Mansourian, S., Parrotta, J.A., Sasaki, N., Schmitt, C.B., Barlow, J., Kanninen, M., Okabe, K., Pan, Y., Thompson, I.D., and van Vliet, N (2012) 'Impacts of forest and land management on biodiversity and carbon', in J.A. Parrotta, C. Wildburger, and S. Mansourian (eds.), *Understanding Relationships between Biodiversity, Carbon, Forests and People: The Key to Achieving REDD+ Objectives,* IUFRO World Series vol. 31. International Union of Forest Research Organizations, Vienna, pp. 53–80.

Kapos, V., Vira, B., Harris, M., O'Leary, A., and Wilson, R. (2022) 'Influence of REDD+ implementation on biodiversity, livelihoods and well-being', in J. Parrotta, S. Mansourian, N. Grima and C. Wildburger (eds.), *Forests, Climate, Biodiversity and People: Assessing a Decade of REDD+*. IUFRO World Series vol. 40. International Union of Forest Research Organizations, Vienna. www.iufro.org/fileadmin/material/publications/iufro-series/ws40/ws40.pdf

Kissinger, G., Herold, M., and De Sy, V. (2012) *Drivers of Deforestation and Forest Degradation: A Synthesis Report for REDD+ Policymakers.* Lexeme Consulting, Vancouver.

Lamb, D., Erskine, P.D. and Parrotta, J.A. (2005) 'Restoration of degraded tropical forest landscapes', *Science*, vol. 310, no. 5754, pp. 1628–1632.

MA. (2005) 'Ecosystems and human well-being: synthesis', *Millennium Ecosystem Assessment*. Island Press, Washington DC.

Mansourian, S., Duchelle, A., Sabogal, C. and Vira, B. (2022) 'REDD+ challenges and lessons learnt', in J. Parrotta, S. Mansourian, N. Grima and C. Wildburger (eds.), *Forests, Climate, Biodiversity and People: Assessing a Decade of REDD+*. IUFRO World Series vol. 40. International Union of Forest Research Organizations, Vienna, pp. 111–135. www.iufro.org/fileadmin/material/publications/iufro-series/ws40/ws40.pdf

Mansourian, S., Parrotta, J., Balaji, P, Belwood-Howard, I., Bhasme, S., Bixler, R.P., Boedhihartono, A.K., Carmenta, R., Jedd, T., de Jong, W., Lake, F.K., Latawiec, A., Lippe, M, Rai, N.D., Sayer, J., Van Dexter, K., Vira, B., Visseren-Hamakers, I., Wyborn, C., and Yang, A. (2019) 'Putting' the pieces together: integration for forest landscape restoration implementation', *Land Degradation and Development,* vol. 31, pp. 419–429. https://doi.org/10.1002/ldr.3448

Pan, Y., Birdsey, R.A., Fang, J., Houghton, R., Kauppi, P.E., Kurz, W.A., Phillips, O.L., Shvidenko, A., Lewis, S.L., Canadell, J.G., Ciais, P., Jackson, R.B., Pacala, S., McGuire, A.D., Piao, S., Rautiainen, A., Sitch, S. and Hayes, D. (2011) 'A large and persistent carbon sink in the world's forests', *ScienceExpress,* vol. 333, pp. 988–993.

Pandit, R., Parrotta, J., Anker, Y., Coudel, E., Diaz Morejón, C.F., Harris, J., Karlen, D.L., Kertész, Á., Mariño De Posada J.L., Ntshotsho Simelane, P., Tamin, N.M., and Vieira, D.L.M. (2018) 'Responses to halt land degradation and to restore degraded land', in IPBES (ed.), *The IPBES Assessment Report on Land Degradation and Restoration.* Secretariat of the Intergovernmental Science-Policy Platform on Biodiversity and Ecosystem Services, Bonn, Germany, pp. 435–528.

Parrotta, J., Mansourian, S., Grima, N. and Wildburger, C. (eds.). (2022) *Forests, Climate, Biodiversity and People: Assessing a Decade of REDD+*. IUFRO World Series vol. 40. International Union of Forest Research Organizations, Vienna. 164 pp. www.iufro.org/fileadmin/material/publications/iufro-series/ws40/ws40.pdf

Parrotta, J. A., Wildburger, C and Mansourian, S. (eds.). (2012) *Understanding Relationships between Biodiversity, Carbon, Forests and People: The Key to Achieving REDD+ Objectives*, IUFRO World Series vol. 31. International Union of Forest Research Organizations, Vienna. www.iufro.org/filead min/material/publications/iufro-series/ws31.pdf

Saatchi, S.S., Harris, N.L., Brown, S., Lefsky, M., Mitchard, E.T.A., Salas, W., Zutta, B.R., Buermann, W., Lewis, S.L., Hagen, S., Petrova, S., White, L., Silman, M. and Morel, A. (2011) 'Benchmark map of forest carbon stocks in tropical regions across three continents', *Proceedings of the National Academy of Sciences,* vol. 108, no. 24, pp. 9899–9904.

Shlisky, A., Alencar, A.A.C., Nolasco, M.M. and Curran, L.M. (2009) 'Global fire regime conditions, threats, and opportunities for fire management in the tropics', in M.A. Cochrane (ed.), *Tropical Fire Ecology: Climate Change, Land Use, and Ecosystem Dynamics.* Praxis (ebook).

Stanturf, J., Kleine, M., Mansourian, S., Parrotta, J.A., Madsen, P., Kant, P., Burns, J., Bolte, A. (2019) 'Implementing forest landscape restoration under the Bonn Challenge: a systematic approach', *Annals of Forest Science*, vol. 76, no. 50. https://doi.org/10.1007/s13595-019-0833-z

Strassburg, B.B.N., Vira, B., Mahanty, S. Mansourian, S., Martin, A., Dawson, N.M., Gross-Camp, N., Latawiec, A. and Swainson, L. (2012) 'Social and economic considerations relevant to REDD+ ', in J. A. Parrotta, C. Wildburger, and S. Mansourian (eds.), *Understanding Relationships between Biodiversity, Carbon, Forests and People: The Key to Achieving REDD+ Objectives*, IUFRO World Series vol. 31. International Union of Forest Research Organizations, Vienna, pp. 83–112.

van der Werf, G.R., Morton, D.C., DeFries, R.S., Olivier, J.G.J., Kasibhatla, P.S., Jackson, R.B., Collatz, G.J., and Randerson, J.T. (2009) 'CO_2 emissions from forest loss', *Nature Geoscience,* vol. 2, pp. 737–738.

28

RESTORATION OF FOREST ECOSYSTEMS

John A. Stanturf

Introduction

Forests dominate terrestrial ecosystems, covering about 30% of the terrestrial land mass and accounting for 75% of the gross primary production of the biosphere. Hosting about 80% of the world's biodiversity, forests provide subsistence, employment opportunities, and income to a quarter of the world's population, as well as protection and other ecosystem services for everyone. Forests are in decline, however, through human impact by removal or fragmentation of native vegetation, simplification, reduction in the complexity of managed natural systems, and physical manipulation of soils. Expansion of agriculture, settlements, and infrastructure have contributed to the loss of two billion hectares of forests and woodlands. Forest restoration is needed to reverse deforestation and degradation so that the remaining, relatively untouched forests can be conserved (Lamb 2015), although forest conservation is a critical need along with restoration (Butler and Laurance 2008).

High-profile policy responses to continued threats to forests include the Bonn Challenge (BC), a landscape-scale effort to restore 350 million hectares of forests by 2030. In addition, countries that have not made formal BC commitments have undertaken significant landscape-scale efforts. Forest restoration is a key component of processes under international treaties addressing climate mitigation, landscape degradation, biodiversity conservation, watershed protection, and the UN Sustainable Development Goals, culminating in the UN Decade of Ecosystem Restoration (Stanturf and Mansourian 2020).

Early efforts to restore damaged forests focused on reducing soil erosion and site degradation. The more ambitious ecological restoration seeks to recreate, initiate, or accelerate the recovery of disturbed ecosystems (SERI 2004). Ecological restoration aiming to replicate "reference" conditions from the past has been implemented only on a relatively small scale. Global warming as well as habitat destruction, fragmentation, and other disturbances, however, means that past conditions may be a poor guide, rendering more degraded forests in need of restoration as novel forests emerge (Figure 28.1). A flexible approach, functional or adaptive restoration, changes the focus from static structural goals to ecosystem dynamics and adaptation to future conditions (Stanturf, Palik, and Dumroese 2014).

Forest and landscape restoration (FLR) underpins the BC, giving equal weight to improving human livelihoods and ecological integrity (Mansourian, Vallauri, and Dudley

DOI: 10.4324/9781003324072-33

Figure 28.1 Forests in Germany degraded by severe drought and bark beetle infestations. (a) 2003 drought effects in mixed forests, Baden-Württemberg. (Photograph by the author.) (b) Multi-year droughts since 2014 caused by elevated temperatures, resulting in severe bark beetle infestation in Norway spruce (*Picea abies*) forests, Sächsische Schweiz. (Photograph by Palle Madsen, InNovaSilva.)

2005). Large-scale restoration is a defining aspect of FLR; the focus on landscapes drives home the social context of restoration. For restoration to be sustainable, human sources of degradation must be eliminated or mitigated, regardless of which interventions are chosen. Most degraded landscapes are not vacant; in fact, they are often inhabited by poor rural populations with insecure tenure (Erbaugh, Pradhan, Adams et al. 2020). Participation and concurrence of local communities, along with other stakeholders, in setting restoration goals is integral to the FLR approach, critical for restoration success, and essential for addressing degradation drivers (Höhl, Ahimbisibwe, Stanturf et al. 2020).

The restoration starting point may be degraded "bare ground," for example, a former agricultural field, pasture, or reclaimed mined land (Holl and Aide 2011). The obstacles to overcome may be biological (e.g., lack of seed sources and dense herbaceous vegetation) or physical (e.g., exposed subsoils and mining spoils). Alternatively, the starting point can be a forest degraded by exploitive harvesting without adequate regeneration, capture of the site by invasive species, suffering from years of fire suppression with buildup of hazardous fuels, or monoculture stands of non-native species (Figure 28.2). Restoration strategies for these conditions utilize silvicultural techniques, including afforestation of non-forest land and reforestation of forest land; natural recovery could be appropriate for either starting point. This chapter discusses the technical approaches and challenges to restoring large landscapes that have been deforested or forested but degraded. Methods for overcoming these limitations are a continuum from natural recovery to three levels of assisted recovery (light, moderate, or intensive; Chazdon, Falk, Banin et al. 2021).

Natural recovery

Natural regeneration and rewilding rely on ecological processes to restore forest vegetation and can be used under appropriate conditions for restoration of deforested or degraded forests.

Natural regeneration

Interventions could be needed to take advantage of natural forest regeneration, including fencing out livestock or wild herbivores, fallowing agricultural land, or removing competing vegetation (Figure 28.3). Natural regeneration may be successful (and permanent) if there are adequate sources of desirable species within the effective dispersal distance and survival and establishment is not limited by herbivores, fires, flooding, or competing vegetation (Borda-Niño, Meli, and Brancalion 2020). Natural regeneration may be ephemeral, however, where human pressure on land is high for other uses like agriculture (Zahawi, Reid, and Holl 2014). Nearby native forest or remnant trees within a degraded stand are sources of regeneration material, or come from seeds dispersed by wind, birds, or mammals (Chazdon and Uriarte 2016). Dispersal by water and even fish is important in wetlands (Gottsberger 1978).

Rewilding

Rewilding is a form of ecological restoration that seeks to restore biodiversity through a network of interconnected reserves and reintroduction of missing keystone species that serve as ecological engineers. Species include apex carnivores (e.g., wolves) and large herbivores (e.g., bison); non-native proxies are allowed for extinct species (Jørgensen 2015). Originally

Figure 28.2 Degraded forest reserve in Ghana restored using the taungya system. (a) The native forest was harvested and planted with teak (*Tectona grandis*) that was destroyed by wildfires. The area was taken over by invasive species. (b) The restored forest ten years after the local community was allowed to plant native trees, interplanted with vegetables. (Photographs by the author.)

Figure 28.3 Natural regeneration development in an exclosure. Secondary broadleaf forests in western Pennsylvania, USA, are under heavy browsing pressure by wild ungulates (white-tailed deer, *Odocoileus virginianus*). Within the fenced exclosure (on the right), natural regeneration proliferates. (Photographs by Susan Stout, USDA Forest Service.)

designed for continental-scale restoration, rewilding is also applied to large ownerships such as rural estates. To date, rewilding has been restricted to temperate and boreal ecosystems, although reversing defaunation in the tropics may be a potential use (Dirzo, Young, Galetti et al. 2014; Galetti, Pires, Brancalion et al. 2017).

Light-assisted recovery

Slightly more effort could be required to protect natural regenerants or increase diversity through underplanting.

Assisted natural regeneration (ANR)

ANR includes fencing out herbivores, installing fire breaks, and controlling weeds, lianas, and vines. Enrichment planting of desirable species can accompany ANR, particularly using fruit or nut trees as well as timber or fodder trees. Farmers allowing native plants to sprout and grow and then pruning them is termed farmer-assisted natural regeneration (FAO 2019; Chomba, Sinclair, Savadogo et al. 2020).

Underplanting

Planting seedlings under a tree canopy can restore degraded forests and increase species diversity, primarily shade-tolerant species. Underplanting large-seeded species in established stands can overcome the lack of suitable seed sources or dispersal limitations and might require herbicides or prescribed fire to control competing vegetation (Dey Gardiner, Schweitzer et al. 2012). Underplanting may be an enrichment option in accelerating forest succession (Soto, Donoso, Vásquez-Grandón et al. 2020).

Moderately assisted recovery

Moderately active measures include direct seeding, planting nursery-raised seedlings or wildlings, or restoration of natural disturbance regimes such as wildfire or inundation. These techniques allow more control over species composition and stand development compared to natural recovery and lightly assisted recovery. Benefits include increased stem density, accelerated canopy closure, increased species richness, and reintroduction of rare species, but the benefits come with higher costs.

A degraded forest could be the starting point following a catastrophic disturbance such as wildfire, windstorm, drought, pest outbreak, or a combination that leaves a partial canopy or a mix of live and dead stems. Severely disturbed forests are often logged to salvage stems and left to regenerate naturally, by direct seeding or planting, or through a combination of methods. Following intense wildfires, for example, the immediate goal may be to reduce soil erosion and protect watersheds, in which case sowing a herbaceous annual cover crop may be needed in addition to planting or sowing trees (Rotich 2022).

The frequent response after a severe disturbance is to restore to predisturbance conditions. Nevertheless, a severe disturbance opens a narrow window in time to convert a forest to a different species composition that may be better adapted to current and future conditions, for example, the conversion of even-aged *Picea abies* in Central Europe to more site-adapted, multilayered, uneven-aged mixed forests (Hilmers, Biber, Knoke et al. 2020).

Direct seeding

Direct seeding avoids some of the risks of relying on natural regeneration; an advantage is better control over species composition. Protection, such as fencing out herbivores, could be needed. Seeds must be collected, stored, and some require pretreatment to maximize germination. Nevertheless, direct seeding is less expensive than planting. Direct seeding is successful only for some species and under certain site conditions. The number of seedlings that successfully establish (i.e., alive after at least one year) is low: 17% for tropical trees, 28% for temperate hardwoods, and 16% for temperate conifers (Grossnickle and Ivetić 2017).

Direct seeding using drones is a new development for restoring large areas following wildfires or other disturbances (Aghai and Manteuffel-Ross 2020). Seed balls typically are used to deliver a mixture of species, possibly incorporating polymer gel antidesiccant, slow-release fertilizer, predator-repellent chemicals, and microbial inocula. Aerial seeding using drones flying close to the ground can more precisely drop seeds than using helicopters or fixed wing aircraft. Drone seeding is an option especially for remote areas, using many of the preferred species and seed pretreatments developed for direct seeding; the lack of effective weed control might be an obstacle (Castro, Morales-Rueda, Navarro et al. 2021).

Planting

Tree planting is a multi-year process that requires more attention and infrastructure than simply digging a hole and planting a seedling. Planting material is usually propagated from locally collected seeds or vegetatively from cuttings and grown in a nursery for one or more years before outplanting. Seeds and cuttings are collected from the wild, seed orchards, or hedging/stooling beds. Obtaining the considerable number of plants necessary to meet current tree planting goals requires efficient nursery methods that produce quality materials

(Haase and Davis 2017). Problems with plant material supply may be a bottleneck for a restoration program, especially for noncommercial species and if insufficient time is allowed for procuring plant materials (e.g., Bannister, Vargas-Gaete, Ovalle et al. 2018).

Common seedling stock types are bareroot, polybags, and hardwall or soft-wall containers. Sometimes bareroot seedlings begin as container seedlings but are then transplanted into nursery beds for additional growth. Wildlings collected from existing forests are an alternative when seed is unavailable or nursery staff lack expertise to grow native plants. Cuttings of some species can be planted directly on site (Stanturf, Palik, and Dumroese 2014). Acceptable material is the stock type and size that yield acceptable results on the outplanting site. Inexpensive flexible polyethylene containers (polybags) filled with local soil are common when technology is limited by cost or availability, but seedlings often suffer from deformed root systems and perform poorly after outplanting (Haase, Bouzza, Emerton et al. 2021). Bareroot or container stocks are preferable; more expensive container seedlings may be a cost-effective alternative for adverse sites or to extend the planting season.

Commercially available planting materials might not have the desired characteristics for restoration. While commercial materials can be genetically improved for fast growth, good form, and desirable wood quality, restoration objectives, however, may prioritize other qualities such as precocious flowering, drought tolerance, or ready sprouting after fire or grazing. Too often, the choice is limited by availability or cost.

The genetics of seeds or seedlings are important but often overlooked. Genetic diversity is critical at all phases, from seed collection to plant materials (Breed, Harrison, Blyth et al. 2019). The key to enhancing ecosystem resilience is maintaining sexual and genetic diversity within species as well as species diversity across the landscape. One advantage of planting over natural recovery is the ability to introduce climate-adapted species or provenances (Dumroese, Williams, Stanturf, et al. 2015).

Overstory manipulation

Alternatively, the starting point could be an intact forest, for example, a plantation monoculture of a non-native species, when the goal is to alter stand structure or species composition. Transforming even-aged forests to a more complex structure, a change in species, or both requires an extended process of partial removal and species replacement (Pommerening and Murphy 2004). The recent trend toward more "natural" forests with complex structure (i.e., multilayered, uneven-aged, or both) is a reaction to the simplification by commercial forestry that has favored even-aged forests of one or a few species (Löf, Sandell Festin, Szydło et al. 2023).

Moderate disturbance regimes that create gaps in forest overstories are common globally and variable density thinning simulates gap-phase dynamics (Brodie and Harrington 2020). The flexible method of "skips and gaps" preserves desirable features of unthinned skips such as underrepresented tree species present in the stand or allows them to be planted in gaps where small groups of trees have been harvested. The matrix area outside of skips and gaps can be thinned to accelerate growth of the remaining trees and increase understory diversity.

Overstory gaps can be filled by releasing saplings or advance regeneration already in the understory, or by supplemental planting (Forbes, Wallace, Buckley et al. 2020). Alternatively, shade-tolerant species can be sown or planted under an overstory and later released by full or partial removal of the overstory (Baumhauer, Madsen, and Stanturf 2005). Gap size and competing vegetation affect success; while large gaps ensure adequate light, they generally are hot and dry conditions that cause moisture stress.

Fire regimes

Restoring natural fire regimes is needed in many fire-adapted forest types with a history of fire suppression. A drier and warmer climate, prolonged warm and dry seasons, extended drought, and invasive grasses and herbs that enhance fire spread increase the frequency and severity of wildfire. Treatments aim to reduce fuels sufficiently so that prescribed burning can be safely conducted (Phillips, Waldrop, Brose et al. 2012). Before reintroducing fire, reducing stem density by thinning, mulching, or herbicides may be necessary. Multiple low-intensity prescribed fires could be required over several years to gradually reduce accumulated fuels. Planting native herbaceous and graminoid understory species and converting the overstory to more fire-adapted species are options (Flatley and Fulé 2016).

Planting patterns and methods

Simple to complex planting patterns are categorized by the number of species, cohorts, and spacing. The simplest patterns use a single species, planted at the same time (single cohort).

Spacing between rows can be uniform; the distance between plants and rows may be the same or different, resulting in an easily managed homogeneous stand with limited diversity. Variations on this design include planting into a cover crop or mini-gaps in a thinned stand. Alternatively, large-seeded species can be planted wide enough apart for infilling by natural regeneration (Gardiner and Oliver 2005). In the tropics, the taungya system is used; edible annual crops are planted between rows of trees for several years until the canopy closes and shade reduces crop yields (Guuroh, Foli, Addo-Danso et al. 2021). The trees benefit from the control of competing vegetation.

Applied nucleation

Scattered individual trees can attract perching birds that disperse seeds into abandoned fields or pastures, but planting single trees or clumps may accelerate recovery. Applied nucleation is more effective (Wilson, Alexandre, Holl et al. 2021); an area is partially planted in a variation on the single or multiple species-single cohort design. This technique has been used to restore tropical pastures (Holl et al. 2020). Small nuclei (islands) of seedlings are planted at some distance from each other; each 100 m^2 nucleus can be a uniformly spaced patch and the area (matrix) between islands is left to regenerate naturally (Benayas, Bullock, and Newton 2008; Zahawi, Holl, Cole et al. 2013). These tree islands could develop in diverse ways: by expanding and eventually coalescing into closed stands or remaining as isolated patches of trees in an agricultural matrix.

Cluster planting

Cluster planting is used to restore late successional species, especially those with large seeds that are not readily dispersed. In Europe, cluster plantings of *Quercus* spp. have been planted either in groups (20–25 seedlings planted 1 m apart) or nests (20–30 seedlings per m^2); spacing between clusters is the expected distance between mature individuals, with enough space for infilling by other species. Group clusters have higher survival, growth, and stand-level tree species diversity than nests (Saha, Kuehne, and Bauhus 2016). Clusters have successfully afforested alpine areas, resulting in mature stands that are resistant to high winds and heavy snow loads (Schönenberger 2001). Usually, a single species is planted in each cluster, but varied species can be used in clusters within a stand or landscape, or several species can be planted together in a cluster. In coastal Brazil, for example, mixed-species clusters of pioneer and non-pioneer species were planted at 30 cm between seedlings and 5 m between clusters (Bertoncello, Oliveira, Holl et al. 2016). Seedlings survived better in clusters than comparable row plantings, especially for non-pioneers, but competition reduced the growth.

Nurse crops

Applied nucleation and cluster planting exemplify facilitation, where a plant in close proximity to another alters the environment and enhances the growth, survival, or reproduction of the neighbor. Nurse crops, either shrubs or trees, are another example of facilitation, especially in challenging environments including drylands, alpine areas, or locations experiencing intense herbivory (Padilla and Pugnaire 2006). Species can be planted together or staggered a few years apart. In the southern United States, cuttings of the fast-growing *Populus deltoides* were planted and two years later, slower growing *Quercus texana* seedlings were planted between the *Populus* rows. After 10 years, the merchantable *Populus* could be harvested, releasing the *Quercus* (Gardiner, Schweitzer, and Stanturf 2001). In a similar system in Denmark using hybrid poplar and *Fagus sylvatica*, the poplar needed 15–20 years to reach merchantable size (Figure 28.4). A different nurse crop system used in Denmark and tested in Uzbekistan and the Kyrgyz Republic uses seedlings planted at the same time, just a few cm apart. The nurse plant, called a sacrificial species because it is culled before reaching maturity, provides

Figure 28.4 Hybrid poplar nurse crop over direct seeded *Fagus sylvatica* on private land in western Denmark. (Photograph by the author.)

protection from ungulates and spring frosts. Both trees and shrubs have been used as sacrificial species (P. Madsen, personal communication).

Nurse trees, often non-native species, are used on severely degraded sites such as mined land to 'catalyze' establishment of native species (Parrotta, Turnbull, and Jones 1997) or on less severely degraded sites to improve site conditions (Lamb 2011). Nurse trees, often pioneer species, grow rapidly and close canopy in a brief time, cooling the soil and inhibiting competing vegetation such as grasses. For example, the non-native *Gmelina arborea* was used in Costa Rica to shade out grasses (Elliott, Blakesley, and Hardwick 2013). After 3–5 years, a diverse understory of native species developed from seeds dispersed from nearby forests by small mammals. Because *G. arborea* is intolerant of shade and produces large seeds, it is non-invasive. The same is untrue of other fast growing exotic species that have been planted in the tropics such as *Acacia mangium* and *Leucaena leucocephala* (Elliott, Blakesley, and Hardwick 2013).

Mixtures

Greater diversity in plantings may be met at the stand level in a number of configurations: in temporary mixtures, such as nurse crops; in simple mixtures, such as single-species in rows or blocks; or in intimate mixtures, where species with complementary growth habits are grown close together (Stanturf, Palik, and Dumroese 2014). The framework species method (FSM) used in the tropics establishes mixtures of species that attract seed-dispersing wildlife (Elliott, Blakesley, and Hardwick 2013; Elliott, Tucker, Shannon et al. 2023). The FSM is a technique of densely planting open sites, close to natural forest seed sources, with indigenous woody species that are able to accelerate ecological succession.

Elliott et al. (2023, p. 3) provided a formal definition of framework species: they are

> woody plants, indigenous to the reference ecosystem, selected for restoration projects because of their tolerance of exposed conditions and their collective ability to inhibit herbaceous weeds and attract seed-dispersing animals, thus accelerating the recovery of forest biomass, structure, biodiversity and ecological functioning.

The FSM is best suited to moderately degraded sites; the minimum number of trees are planted that can close the canopy and shade out weeds within 2–3 years. Density depends on rainfall, with higher density possible in higher rainfall areas. The mixture of framework tree species is planted in the first year and two years of intensive weed control are needed.

The Miyawaki method for establishing high-diversity, high-density plantings has been used primarily for urban afforestation (Miyawaki 1998). Seedlings of 50–100 local native species (main tree species, subspecies, shrubs, and herbs) are planted in a random mix at extremely high density (20,000–30,000 ha^{-1}) following compost addition. Sites are irrigated and weeded for up to three years. "Micro-forest" has been established using this method, primarily in tropical and temperate environments. One trial in a Mediterranean environment was deemed successful even though survival was low (Schirone, Salis, and Vessella 2011). The main advantage of the Miyawaki method is quick development of a diverse overstory but at high cost, thus limiting its usefulness to small areas.

The appropriateness of each restoration method depends on conditions and objectives, which often come down to available material and cost. Planting is the most expensive method and direct seeding costs more than natural regeneration. Natural regeneration guarantees locally adapted genotypes but is limited to local plant materials currently available. Planting is more dependable with the advantages of controlling species composition and the possibility of introducing genetically improved or climatically adapted material. Seed availability limits direct seeding. Direct comparisons of regeneration methods are few and recommendations might risk implicit bias because failures are seldom reported (Reid, Fagan, and Zahawi 2018).

Intensively assisted recovery

Degraded sites have physical barriers to successful restoration that requires expensive interventions to alter soils, landform, or hydrology.

Landform and hydrological modification

Mined land requires stabilization of spoil banks or mitigation of other physical and chemical limitations on plant growth (Macdonald, Landhäusser, Skousen et al. 2015). Often limitations are so severe that only a few tree species can be established; non-native species might ameliorate sites over time, allowing planting or catalyzing recolonization by native species (Parrotta, Turnbull, and Jones 1997). Restoration of steep slopes having shallow or stony soils may require expensive terracing and organic matter addition into planting holes. Building terraces is costly but was necessary in South Korea (Lee, Park, and Park 2015). On a smaller scale, restoring microtopography of former agricultural land creates microsites that favor plant diversity, especially in wetland forests (Simmons, Ben Wu, and Whisenant 2011).

Restoring natural inundation regimes to restore riparian forests can favor adapted species, for example, removing dams, timing water releases, or restoring river flows (Kuz'mina and Treshkin 2012). Plugging drainage channels to raise water levels can restore drained former wetlands and peatlands (Musri, Ainuddin, Hyrul et al. 2020). In uplands, gully restoration, water control structures, and terracing on slopes create sites for trees to flourish (Figure 28.5).

Assisted migration (AM)

AM is the intentional movement of species or populations, especially for rare species with limited adaptation and migration capacities (Williams and Dumroese 2013). Long generation cycles and reduced dispersal ability of trees can be overcome by AM. Different AM techniques are categorized by the distance a species is moved: within a species' current range, beyond a species' range but near the current distribution, or long distances beyond its current range. Besides regulatory barriers, unknowns limit AM, including when to initiate AM, what new provenances or species to move, what effects the new material will have on the receiving ecosystem, and how to manage the novel assemblage present (Breed, Stead, Ottewell et al. 2013).

Landscape designs

Forest restoration must be scaled up to reverse degradation by utilizing the site-level techniques described so far across the landscape and prioritizing areas that conserve biodiversity or protect watersheds, with an eye to adapting to future climate (Lamb, Stanturf, and Madsen 2012). In landscapes where forests are mixed with other land uses such as

Figure 28.5 Mountain afforestation can require physical structures to control runoff and terracing for planting spots. Left photo, water control structures and terraces in South Korea. Right photo, established afforestation on terraces in Uzbekistan. Note that the trees were planted on the lower side of the terrace. (Photographs by the author.)

agriculture, connecting forest remnants by restoring corridors or stepping stones is a biodiversity priority (Saura, Bodin, and Fortin 2014). Restoring forest cover on steep slopes and riparian buffer strips not only reduces soil erosion and sedimentation but also provides habitat diversity and opportunities for species movement (Bentrup, Dosskey, Wells et al. 2012). Restoring groundwater recharge areas in salinity-prone areas increases the depth of the water table (Harper, Dell, Ruprecht et al. 2021).

A resilient landscape will be a mosaic of stands of different ages and structures (Millar, Stephenson, and Stephens 2007). Landscape diversity can be restored by appropriately combining natural regeneration, direct seeding, and planting. As trees grow at different rates and with variable survival, different age cohorts and complex structures develop. Harvesting methods can assist in restoring complexity, for example, by retention harvesting of stands to be converted instead of clear-cutting and planting. Retention harvesting retains a portion of the original stand to maintain structural and compositional diversity (Gustafsson, Baker, Bauhus et al. 2012). Retained patches can be aggregated or dispersed and the cutover areas can be regenerated naturally, by direct seeding, or by planting, depending on the desired species composition.

Restoring forested corridors between existing forests can enhance the movement of animals and other taxa or gene flow. Water courses are natural corridors and restoring forested riparian zones serves multiple restoration objectives. Restoring continental corridors attempts to connect fragments of biologically defined regions or sub-regions. Corridors are a defining element of the three C's approach (i.e., cores, corridors, and carnivores) of rewilding in North America (Soulé and Noss 1998) or the 12 large (i.e., 700–3,000 km) national connectivity initiatives in Australia and New Zealand (Stanturf and Mansourian 2020).

Connectivity need not be continuous to be effective; stepping stones are a series of small habitat patches that connect larger patches. Stepping stones must be large enough to have conservation value, facilitate long-distance dispersal, and connect larger habitats. Individual patches can be established by any method, but applied nucleation and wooded islands are especially appropriate. Stepping stones recommended for restoring montane forest in Sri Lanka provide an example of the process (Gunaratne, Gunatilleke, Gunatilleke et al. 2014). Existing forest fragments were protected from grazing and soils around larger fragments were plowed in a 3–10 m wide buffer strip to reduce competing vegetation and encourage expansion of the fragment by assisted natural regeneration. Stepping stones between fragments were established by planting early successional native tree species, fenced to exclude herbivores until trees were large enough to resist browsing. These patches were protected from fire by a planted buffer of fire-resistant legumes. The maximum distance between stepping stones was 20 m.

Monitoring and long-term management

The need for restoring forest ecosystems is acute and immediate; understandably, much of the current emphasis is on the initial stages of recovery, for example, seedling survival (Banin, Raine, Rowland et al. 2023) and the needs and perceptions of local communities (Höhl, Ahimbisibwe, Stanturf et al. 2020). Nevertheless, long-term success requires management to sustain the restored forest and monitoring to indicate when corrective or further interventions are needed (Stanturf, Mansourian, and Kleine 2017). Restoration is a long-term process and outcomes may not be apparent for decades. If the restoration is successful, new natural assets will be created that may attract unwanted attempts to exploit them. Furthermore, land use or policy changes that occur outside the restored landscape may impact and even threaten its sustainability. A management plan, coupled with a monitoring program, can detect and provide early warning of deviations from the expected development trajectory, possible encroachment, or damaging disturbances.

Monitoring is frequently overlooked because of its perceived cost and complexity (Seidling 2019). With appropriate indicators, monitoring need not be overly complex and local stakeholders can be actively involved, reducing costs, and helping to maintain their interest and support (Evans, Guariguata, and Brancalion 2018; Guariguata and Evans 2020). A monitoring program must focus on indicators that relate to restoration objectives and are meaningful to stakeholders. A cost-effective monitoring program is comprised of a minimum set of indicators, simple and easy to measure (Dey and Schweitzer 2014; Viani et al. 2018). Recent developments in affordable remote monitoring using drones can increase the frequency of monitoring (Pérez, Pilustrelli, Farinaccio et al. 2020).

Monitored variables change over space and time; therefore, indicators must be sensitive enough to detect changes. Some indicators might need to be monitored only for a brief time, while other indicators require long-term monitoring. Long-term indicators initially could require measuring at short time intervals, and then at longer intervals. For example, seedling survival (or mortality) and growth may need measuring annually for 1–5 years and then at 5-year intervals until crown closure and beyond. Different approaches and spatial and temporal intensities of monitoring apply at separate phases of restoration (Hutto and Belote 2013). Surveillance monitoring, for example, continuous forest inventory, can identify areas in need of restoration or potential sources of plant material in planning large-scale restoration, as well as giving warning when restored forests are disturbed by illegal logging. Implementation monitoring is needed to ensure that restoration interventions are conducted as planned and contracted for (e.g., survival of planted material). Effectiveness monitoring is conducted after restoration has been implemented, to evaluate whether interventions were effective in meeting restoration objectives, for example, control of invasive species or livelihood improvements (e.g., household income). Some indicators change slowly, for example, soil organic matter; outcome monitoring might be inappropriate until many years after restoration interventions, for example, in forest landscape restoration, whether ecological integrity was attained and community well-being was improved, or if unintended consequences occurred (Hutto and Belote 2013; Stanturf, Mansourian, and Kleine 2017).

Selecting indicators to monitor should be linked to objectives, suggesting when further interventions are needed (Lamb 2015). Most restoration efforts manipulate vegetation and many indicators have been suggested for monitoring plant development at different scales. Most indicators relate to forest composition, structure, and function (Ruiz-Jaen and Aide 2005; Wortley, Hero, and Howes 2013). Indicators for composition include diversity and abundance of different taxa; structure indicators are easily measured and sensitive to changes, such as height, diameter, and canopy closure; drone-mounted cameras and Light Detection and Ranging (LiDAR) enable repetitive sampling (Camarretta et al. 2020). Functional indicators are more difficult to measure, can take longer time to measurably improve, and they may require laboratory analyses, for example, soil nutrient content and bulk density (Gatica-Saavedra, Echeverría, and Nelson 2017), although the surveillance monitoring protocol for land degradation is field-based (Vâgen and Winowiecki 2020). Monitoring effects on human well-being might involve labor-intensive household surveys (Hicks et al. 2016).

Conclusion

The need to restore forest ecosystems is significant and immediate; degraded forest areas are likely to increase due to global change, including a warmer climate. The current challenges are daunting enough: large areas needing attention, direct costs of interventions as well as opportunity costs to landowners and land users, insufficient quality planting materials, and an uncertain future climate. Forest restoration is a long-term undertaking and just as there are multiple approaches based on the best available knowledge, there are many ways to judge success. The multiplicity of ecological and social conditions means that restoration is sensitive to context and objectives, the starting and ending points. Decisions on where to restore and how actively to intervene are taken under a great deal of uncertainty, but inaction is not an option. The way forward is with humility, collaboratively, with transparency and willingness to change the course based on evidence.

References

Aghai, M., and Manteuffel-Ross, T. (2020), 'Enhancing direct seeding efforts with unmanned aerial vehicle (UAV) "swarms" and seed technology,' *Tree Planters' Notes*, vol 63, no 2, pp. 32–48.

Banin, L. F., Raine, E. H., Rowland, L. M., Chazdon, R. L., Smith, S. W., Rahman, N. E. B., Butler, A., Philipson, C., Applegate, G. G., Axelsson, E. P., Budiharta, S., Chua, S. C., Cutler, M. E. J., Elliott, S., Gemita, E., Godoong, E., Graham, L. L. B., Hayward, R. M., Hector, A., Ilstedt, U., Jensen, J., Kasinathan, S., Kettle, C. J., Lussetti, D., Manohan, B., Maycock, C., Ngo, K. M., O'Brien, M. J.,

Osuri, A. M., Reynolds, G., Sauwai, Y., Scheu, S., Silalahi, M., Slade, E. M., Swinfield, T., Wardle, D. A., Wheeler, C., Yeong, K. L., and Burslem, D. F. R. P. (2023), 'The road to recovery: a synthesis of outcomes from ecosystem restoration in tropical and sub-tropical Asian forests,' *Philosophical Transactions of the Royal Society B: Biological Sciences*, vol 378, no 1867, 20210090. DOI: https://doi.org/10.1098/rstb.2021.0090

Bannister, J. R., Vargas-Gaete, R., Ovalle, J. F., Acevedo, M., Fuentes-Ramirez, A., Donoso, P. J., Promis, A., and Smith-Ramírez, C. (2018), 'Major bottlenecks for the restoration of natural forests in Chile,' *Restoration Ecology*, vol 26, no 6, pp. 1039–1044. DOI: https://doi.org/10.1111/rec.12880

Baumhauer, H., Madsen, P., and Stanturf, J. (2005), 'Regeneration by direct seeding—a way to reduce costs of conversion,' in *Restoration of Boreal and Temperate Forests*, eds. J. Stanturf and P. Madsen, Boca Raton, FL: CRC Press, pp. 349–354.

Benayas, J. M. R., Bullock, J. M., and Newton, A. C. (2008), 'Creating woodland islets to reconcile ecological restoration, conservation, and agricultural land use,' *Frontiers in Ecology and the Environment*, vol 6, no 6, pp. 329–336. DOI: https://doi.org/10.1890/070057

Bentrup, G., Dosskey, M., Wells, G., and Schoeneberger, M. (2012), 'Connecting landscape fragments through riparian zones,' in *Forest Landscape Restoration*, eds. J. Stanturf, D. Lamb and P. Madsen, Dordrecht: Springer, pp. 93–109. DOI: https://doi.org/10.1007/978-94-007-5326-6_5

Bertoncello, R., Oliveira, A. A., Holl, K. D., Pansonato, M. P., and Martini, A. M. (2016), 'Cluster planting facilitates survival but not growth in early development of restored tropical forest,' *Basic and Applied Ecology*, vol 17, no 6, pp. 489–496. DOI: https://doi.org/10.1016/j.baae.2016.04.006

Borda-Niño, M., Meli, P., and Brancalion, P.H.S. (2020), Drivers of tropical forest cover increase: A systematic review. *Land Degradation & Development*, vol 31, no 11, pp. 1366–1379. DOI: https://doi.org/10.1002/ldr.3534

Breed, M., Stead, M., Ottewell, K., Gardner, M., and Lowe, A. (2013), 'Which provenance and where? Seed sourcing strategies for revegetation in a changing environment,' *Conservation Genetics*, vol 14, pp. 1–10. DOI: https://doi.org/10.1007/s10592-012-0425-z

Breed, M. F., Harrison, P. A., Blyth, C., Byrne, M., Gaget, V., Gellie, N. J. C., Groom, S. V. C., Hodgson, R., Mills, J. G., Prowse, T. A. A., Steane, D. A., and Mohr, J. J. (2019), 'The potential of genomics for restoring ecosystems and biodiversity,' *Nature Reviews Genetics*, vol 20, no 10, pp. 615–628. DOI: https://doi.org/10.1038/s41576-019-0152-0.

Brodie, L. C., and Harrington, C. A. (2020), 'Guide to variable-density thinning using skips and gaps,' Gen. Tech. Rep. PNW-GTR-989. Portland, OR: US Department of Agriculture, Forest Service, Pacific Northwest Research Station; 37 p.

Butler, R. A., and Laurance, W. F. (2008), 'New strategies for conserving tropical forests,' *Trends in Ecology & Evolution*, vol 23, no 9, pp. 469–472. DOI: https://doi.org/10.1016/j.tree.2008.05.006

Camarretta, N., Harrison, P. A., Bailey, T., Potts, B., Lucieer, A., Davidson, N., and Hunt, M. (2020), 'Monitoring forest structure to guide adaptive management of forest restoration: A review of remote sensing approaches,' *New Forests*, vol 51, no 4, pp. 573–596. DOI: https://doi.org/10.1007/s11056-019-09754-5

Castro, J., Morales-Rueda, F., Navarro, F. B., Löf, M., Vacchiano, G., and Alcaraz-Segura, D. (2021), 'Precision restoration: a necessary approach to foster forest recovery in the 21st century,' *Restoration Ecology*, vol 29, no 7, e13421. DOI: https://doi.org/10.1111/rec.13421

Chazdon, R. L., Falk, D. A., Banin, L. F., Wagner, M., Wilson, S. J., Grabowski, R. C., and Suding, K. N. (2021), 'The intervention continuum in restoration ecology: rethinking the active–passive dichotomy,' *Restoration Ecology*, e13535. DOI: https://doi.org/10.1111/rec.13535

Chazdon, R. L., and Uriarte, M. (2016) 'Natural regeneration in the context of large-scale forest and landscape restoration in the tropics,' *Biotropica*, vol 48, no 6, pp. 709–715. DOI: https://doi.org/10.1111/btp.12409

Chomba, S., Sinclair, F., Savadogo, P., Bourne, M., and Lohbeck, M. (2020), 'Opportunities and constraints for using farmer managed natural regeneration for land restoration in sub-Saharan Africa,' *Frontiers in Forests and Global Change*, vol 3, 571679. DOI: https://doi.org/10.3389/ffgc.2020.571679

Dey, D. C., Gardiner, E. S., Schweitzer, C. J., Kabrick, J. M., and Jacobs, D. F. (2012), 'Underplanting to sustain future stocking of oak (*Quercus*) in temperate deciduous forests,' *New Forests*, vol 43, no 5–6, pp. 955–978. DOI: https://doi.org/10.1007/s11056-012-9330-z

Dey, D. C., and Schweitzer, C. J. (2014), 'Restoration for the future: endpoints, targets, and indicators of progress and success,' *Journal of Sustainable Forestry*, vol 33, supl, pp. S43–S65. DOI: https://doi.org/10.1080/10549811.2014.883999

Dirzo, R., Young, H. S., Galetti, M., Ceballos, G., Isaac, N. J., and Collen, B. (2014), 'Defaunation in the Anthropocene,' *Science*, vol 345, no 6195, pp. 401–406. DOI: https://doi.org/10.1126/science.1251817

Dumroese, K., Williams, M. I., Stanturf, J. A., and St. Clair, B. (2015), 'Considerations for restoring temperate forests of tomorrow: forest restoration, assisted migration, and bioengineering,' *New Forests*, vol 46, pp. 947–964. DOI: https://doi.org/10.1007/s11056-015-9504-6

Elliott, S., Blakesley, D., and Hardwick, K. (2013), *Restoring Tropical Forests: A Practical Guide*. Kew, UK: Kew Publishing.

Elliott, S., Tucker, N. I., Shannon, D. P., and Tiansawat, P. (2023), 'The framework species method: harnessing natural regeneration to restore tropical forest ecosystems,' *Philosophical Transactions of the Royal Society B*, vol 378, no 1867, 20210073. DOI: https://doi.org/10.1098/rstb.2021.0073

Erbaugh, J. T., Pradhan, N., Adams, J., Oldekop, J. A., Agrawal, A., Brockington, D., Pritchard, R., and Chhatre, A. (2020), 'Global forest restoration and the importance of prioritizing local communities,' *Nature Ecology & Evolution*, vol 4, no 11, pp. 1472–1476. DOI: https://doi.org/10.1038/s41559-020-01282-2

Evans, K., Guariguata, M. R., and Brancalion, P. H. (2018), 'Participatory monitoring to connect local and global priorities for forest restoration,' *Conservation Biology*, vol 2, no 3, pp. 525–534. DOI: https://doi.org/10.1111/cobi.13110.

FAO. (2019), *Restoring Forest Landscapes Through Assisted Natural Regeneration (ANR) — A Practical Manual*. Bangkok: Food and Agriculture Organization of the United Nations.

Flatley, W. T., and Fulé, P. Z. (2016), 'Are historical fire regimes compatible with future climate? Implications for forest restoration,' *Ecosphere*, vol 7, no 10, e01471. DOI: https://doi.org/10.1002/ecs2.1471

Forbes, A. S., Wallace, K. J., Buckley, H. L., Case, B. S., Clarkson, B. D., and Norton, D. A. (2020), 'Restoring mature-phase forest tree species through enrichment planting in New Zealand's lowland landscapes,' *New Zealand Journal of Ecology*, vol 44 no 1, pp. 1–9. DOI: www.jstor.org/stable/26872863

Galetti, M., Pires, A. S., Brancalion, P. H., and Fernandez, F. A. (2017), 'Reversing defaunation by trophic rewilding in empty forests,' *Biotropica*, vol 49, no 1, pp. 5–8. DOI: www.jstor.org/stable/48574991

Gardiner, E. S., and Oliver, J. M. (2005), 'Restoration of bottomland hardwood forests in the Lower Mississippi Alluvial Valley, USA,' in *Restoration of Boreal and Temperate Forests*, eds. J. Stanturf and P. Madsen, Boca Raton, FL: CRC Press, pp. 235–251.

Gardiner, E. S., Schweitzer, C. J., and Stanturf, J. A. (2001), 'Photosynthesis of Nuttall oak (*Quercus nuttallii* Palm.) seedlings interplanted beneath an eastern cottonwood (*Populus deltoides* Bartr. ex Marsh.) nurse crop,' *Forest Ecology and Management*, vol 149, no 1–3, pp. 283–294. DOI: https://doi.org/10.1016/S0378-1127(00)00562-4

Gatica-Saavedra, P., Echeverría, C., and Nelson, C. R. (2017), 'Ecological indicators for assessing ecological success of forest restoration: a world review,' *Restoration Ecology*, vol 25, no 6, pp. 850–857. DOI: https://doi.org/10.1111/rec.12586

Gottsberger, G. (1978), 'Seed dispersal by fish in the inundated regions of Humaitá, Amazonia,' *Biotropica*, vol. 10, no 3, pp. 170–183. DOI: https://doi.org/10.2307/2387903

Grossnickle, S. C., and Ivetić, V. (2017), 'Direct seeding in reforestation–a field performance review,' *Reforesta*, vol 4, pp. 94–142. DOI: https://doi.org/10.21750/REFOR.4.07.46.

Guariguata, M. R., and Evans, K. (2020), 'A diagnostic for collaborative monitoring in forest landscape restoration,' *Restoration Ecology*, vol 28, no 4, pp. 742–749. DOI: https://doi.org/10.1111/rec.13076

Gunaratne, A., Gunatilleke, C., Gunatilleke, I., Madawala, H., and Burslem, D. (2014), 'Overcoming ecological barriers to tropical lower montane forest succession on anthropogenic grasslands: synthesis and future prospects,' *Forest Ecology and Management*, vol 329, pp. 340–350. DOI: https://doi.org/10.1016/j.foreco.2014.03.035

Gustafsson, L., Baker, S. C., Bauhus, J., Beese, W. J., Brodie, A., Kouki, J., Lindenmayer, D. B., Lõhmus, A., Pastur, G. M., and Messier, C. (2012), 'Retention forestry to maintain multifunctional forests: a world perspective,' *BioScience*, vol 62, no 7, pp. 633–645. DOI: https://doi.org/10.1525/bio.2012.62.7.6

Guuroh, R. T., Foli, E. G., Addo-Danso, S. D., Stanturf, J., Kleine, M., and Burns, J. (2021), 'Restoration of degraded forest reserves in Ghana,' *Reforesta*, vol 12, pp. 35–55. DOI: https://doi.org/10.21750/REFOR.12.05.97

Haase, D. L., Bouzza, K., Emerton, L., Friday, J. B., Lieberg, B., Aldrete, A., and Davis, A. S. (2021), 'The high cost of the low-cost polybag system: a review of nursery seedling production systems,' *Land*, vol 10, no 8, pp. 826. DOI: https://doi.org/10.3390/land10080826

Haase, D. L., and Davis, A. S. (2017), 'Developing and supporting quality nursery facilities and staff are necessary to meet global forest and landscape restoration needs,' *Reforesta*, vol 4, pp. 69–93. DOI: https://doi.org/10.21750/REFOR.4.06.45

Harper, R. A., Dell, B., Ruprecht, J. K., Sochacki, S. J., and Smettem, K. R. J. (2021), 'Salinity and the reclamation of salinized lands,' in *Soils and Landscape Restoration*, eds. J. A. Stanturf and M. A. Callaham. New York: Academic Press, pp. 193–208. https://doi.org/10.1016/B978-0-12-813193-0.00007-2

Hicks, C. C., Levine, A., Agrawal, A., Basurto, X., Breslow, S. J., Carothers, C., Charnley, S., Coulthard, S., Dolsak, N., and Donatuto, J. (2016), 'Engage key social concepts for sustainability,' *Science*, vol 352, no 6281, pp. 38–40. DOI: https://doi.org/10.1126/science.aad4977

Hilmers, T., Biber, P., Knoke, T., and Pretzsch, H. (2020), 'Assessing transformation scenarios from pure Norway spruce to mixed uneven-aged forests in mountain areas,' *European Journal of Forest Research*, vol 139, no 4, pp. 567–584. DOI: https://doi.org/10.1007/s10342-020-01270-y

Höhl, M., Ahimbisibwe, V., Stanturf, J. A., Elsasser, P., Kleine, M., and Bolte, A. (2020), 'Forest landscape restoration—what generates failure and success?,' *Forests*, vol 11, no 9, p. 938. DOI: https://doi.org/10.3390/f11090938.

Holl, K., and Aide, T. M. (2011), 'When and where to actively restore ecosystems?,' *Forest Ecology and Management*, vol 261 no 10, pp. 1558–1563. DOI: https://doi.org/10.1016/j.foreco.2010.07.004

Holl, K., Reid, J. L., Cole, R. J., Oviedo, F., Rosales, J. A., and Zahawi, R. (2020), 'Applied nucleation facilitates tropical forest recovery: lessons learned from a 15-year study,' *Journal of Applied Ecology*, vol 57, no 12, pp. 2316–2328. DOI: https://doi.org/10.1111/1365-2664.13684

Hutto, R. L., and Belote, R. (2013), 'Distinguishing four types of monitoring based on the questions they address,' *Forest Ecology and Management*, vol 289, pp. 183–189. DOI: https://doi.org/10.1016/j.foreco.2012.10.005

ITTO. (2020), 'Guidelines for forest landscape restoration in the tropics,' *ITTO Policy Development Series*, No. 24, Yokohama, Japan: International Tropical Timber Organization.

Jørgensen, D. (2015), 'Rethinking rewilding,' *Geoforum*, vol 65, pp. 482–488. DOI: https://doi.org/10.1016/j.geoforum.2014.11.016

Kuz'mina, Z. V., and Treshkin, S. E. (2012), 'Riparian forests and the possibility of their modern restoration,' *Arid Ecosystems*, vol 2, no 3, pp. 165–176. DOI: https://doi.org/10.1134/S2079096112030080

Lamb, D. (2011), *Regreening the Bare Hills: Tropical Forest Restoration in the Asia-Pacific Region*. Dordrecht: Springer. DOI: https://doi.org/10.1007/978-90-481-9870-2

Lamb, D. (2015), 'Restoration of forest ecosystems,' in *Routledge Handbook of Forest Ecology*, eds. K. S. H. Peh, R. T. Corlett and Y. Bergeron. New York: Routledge, pp. 413–426.

Lamb, D., Stanturf, J., and Madsen, P. (2012), 'What is forest landscape restoration?,' in *Forest Landscape Restoration Integrating Natural and Social Sciences*, eds. J. Stanturf, D. Lamb and P. Madsen. Dordrecht: Springer, pp. 3–23. DOI: https://doi.org/10.1007/978-94-007-5326-6_1

Lee, D. K., Park, P. S., and Park, Y. D. (2015), 'Forest restoration and rehabilitation in the Republic of Korea,' in *Restoration of Boreal and Temperate Forests, Second Edition*, ed. J. A. Stanturf. Boca Raton, FL: CRC Press, pp. 217–232.

Löf, M., Sandell Festin, E., Szydło, M., and Brunet, J. (2023), 'Restoring mixed forests through conversion of Norway spruce stands: effects of fencing and mechanical site preparation on performance of planted beech and natural tree regeneration,' *European Journal of Forest Research*, vol 142, pp. 763–772. DOI: https://doi.org/10.1007/s10342-023-01554-z.

Macdonald, S. E., Landhäusser, S. M., Skousen, J., Franklin, J., Frouz, J., Hall, S., Jacobs, D. F., and Quideau, S. (2015), 'Forest restoration following surface mining disturbance: challenges and solutions,' *New Forests*, vol 46, no 5–6, pp. 703–732. DOI: https://doi.org/10.1007/s11056-015-9506-4

Mansourian, S., Vallauri, D., and Dudley, N. (2005), *Forest Restoration in Landscapes: Beyond Planting Trees*. New York: Springer.

Millar, C. I., Stephenson, N. L., and Stephens, S. L. (2007), 'Climate change and forests of the future: managing in the face of uncertainty,' *Ecological Applications*, vol 17, no 8, pp. 2145–2151. DOI: https://doi.org/10.1890/06-1715.1.

Miyawaki, A. (1998), 'Restoration of urban green environments based on the theories of vegetation ecology,' *Ecological Engineering*, vol 11, no 1–4, pp. 157–165. DOI: https://doi.org/10.1016/S0925-8574(98)00033-0

Musri, I., Ainuddin, A., Hyrul, M., Hazandy, A., Azani, A., and Mitra, U. (2020). 'Post forest fire management at tropical peat swamp forest: a review of Malaysian experience on rehabilitation and risk mitigation,' *IOP Conf. Ser.: Earth Environ. Sci.*, vol 504, 012017. DOI: https://doi.org/10.1088/1755-1315/504/1/012017

Padilla, F. M., and Pugnaire, F. I. (2006), 'The role of nurse plants in the restoration of degraded environments,' *Frontiers in Ecology and the Environment*, vol 4, no 4, pp. 196–202. DOI: https://doi.org/10.1890/1540-9295(2006)004[0196:TRONPI]2.0.CO;2

Parrotta, J. A., Turnbull, J. W., and Jones, N. (1997), 'Catalyzing native forest regeneration on degraded tropical lands,' *Forest Ecology and Management*, vol 99, no 1–2, pp. 1–7. DOI: https://doi.org/10.1016/S0378-1127(97)00190-4

Pérez, D. R., Pilustrelli, C., Farinaccio, F. M., Sabino, G., and Aronson, J. (2020), 'Evaluating success of various restorative interventions through drone- and field-collected data, using six putative framework species in Argentinian Patagonia,' *Restoration Ecology*, vol 28, S1, pp. A44–A53. DOI: https://doi.org/10.1111/rec.13025

Phillips, R. J., Waldrop, T. A., Brose, P. H., and Wang, G. G. (2012), 'Restoring fire-adapted forests in eastern North America for biodiversity conservation and hazardous fuels reduction,' in *A Goal-Oriented Approach to Forest Landscape Restoration*, eds. J. Stanturf, P. Madsen and D. Lamb. Dordrecht: Springer, pp. 187–219. DOI: https://doi.org/10.1007/978-94-007-5338-9_9

Pommerening, A., and Murphy, S. (2004), 'A review of the history, definitions and methods of continuous cover forestry with special attention to afforestation and restocking,' *Forestry*, vol 77, no 1, pp. 27–44. DOI: https://doi.org/10.1093/forestry/77.1.27

Reid, J. L., Fagan, M. E., and Zahawi, R. A. (2018), 'Positive site selection bias in meta-analyses comparing natural regeneration to active forest restoration,' *Science Advances*, vol 4, no 5, eaas9143. DOI: https://doi.org/10.1126/sciadv.aas9143.

Rotich, B. (2022), 'A comprehensive framework for forest restoration after forest fires in theory and practice: a systematic review,' *Forests*, vol 13, 1354. DOI: https://doi.org/10.3390/f13091354

Ruiz-Jaen, M. C., and Aide, T. M. (2005), 'Restoration success: how is it being measured?,' *Restoration Ecology*, vol 13, no 3, pp. 569–577. DOI: https://doi.org/10.1111/j.1526-100X.2005.00072.x

Saha, S., Kuehne, C., and Bauhus, J. (2016), 'Lessons learned from oak cluster planting trials in central Europe,' *Canadian Journal of Forest Research*, vol 47, no 2, pp. 139–148. DOI: https://doi.org/10.1139/cjfr-2016-0265

Saura, S., Bodin, Ö., and Fortin, M. J. (2014), 'Stepping stones are crucial for species' long-distance dispersal and range expansion through habitat networks,' *Journal of Applied Ecology*, vol 51, no 1, pp. 171–182. DOI: https://doi.org/10.1111/1365-2664.12179.

Schirone, B., Salis, A., and Vessella, F. (2011), 'Effectiveness of the Miyawaki method in Mediterranean forest restoration programs,' *Landscape and Ecological Engineering*, vol 7, no 1, pp. 81–92. DOI: https://doi.org/0.1007/s11355-010-0117-0

Schönenberger, W. (2001), 'Cluster afforestation for creating diverse mountain forest structures—a review,' *Forest Ecology and Management*, vol 145, no 1–2, pp. 121–128. DOI: https://doi.org/10.1016/S0378-1127(00)00579-X

Seidling, W. (2019), 'Forest monitoring: Substantiating cause-effect relationships,' *Science of the Total Environment*, vol 687, pp. 610–617. DOI: https://doi.org/10.1016/j.scitotenv.2019.06.048

SERI. (2004). *The SER International Primer on Ecological Restoration.* Society for Ecological Restoration International.

Simmons, M. E., Ben Wu, X., and Whisenant, S. G. (2011), 'Plant and soil responses to created microtopography and soil treatments in bottomland hardwood forest restoration,' *Restoration Ecology*, vol 19, no 1, pp. 136–146. DOI: https://doi.org/10.1111/j.1526-100X.2009.00524.x

Soto, D., Donoso, P., Vásquez-Grandón, A. et al. (2020) 'Differential early performance of two underplanted hardwood tree species following restoration treatments in high-graded temperate rainforests,' *Forests*, vol 11, no 4, pp. 401. DOI: https://doi.org/10.3390/f11040401

Soulé, M., and Noss, R. (1998), 'Rewilding and biodiversity: complementary goals for continental conservation,' *Wild Earth*, vol 8, pp. 18–28.

Stanturf, J., and Mansourian, S. (2020), 'Forest landscape restoration: state of play,' *Royal Society Open Science*, vol 7, no 12, 201218. DOI: https://doi.org/10.1098/rsos.201218

Stanturf, J., Mansourian, S., and Kleine, M. (2017), *Implementing Forest Landscape Restoration, A Practitioner's Guide.* Vienna, Austria: International Union of Forest Research Organizations.

Stanturf, J., Palik, B., and Dumroese, R. K. (2014), 'Contemporary forest restoration: a review emphasizing function,' *Forest Ecology and Management*, vol 331, pp. 292–323. DOI: https://doi.org/10.1016/j.foreco.2014.07.029

Vâgen, T., and Winowiecki, L. A. (2020), *Land Degradation Surveillance Framework (LSDF): Field Guide.* Nairobi, Kenya: ICRAF.

Viani, R. A., Barreto, T. E., Farah, F. T., Rodrigues, R. R., and Brancalion, P. H. (2018), 'Monitoring young tropical forest restoration sites: how much to measure?,' *Tropical Conservation Science*, vol 11, pp. 1940082918780916. DOI: https://doi.org/10.1177/1940082918780916

Williams, M. I., and Dumroese, R. K. (2013), 'Preparing for climate change: forestry and assisted migration,' *Journal of Forestry*, vol 114, no 4, pp. 287–297. DOI: https://doi.org/10.5849/jof.13-016

Wilson, S. J., Alexandre, N. S., Holl, K. D., Reid, J. L., Zahawi, R. A., Celentano, D., Sprenkle-Hyppolite, S., and Werden, L. (2021), *Applied Nucleation Restoration Guide for Tropical Forests.* Arlington VA: Conservation International.

Wortley, L., Hero, J.-M., and Howes, M. (2013), 'Evaluating ecological restoration success: a review of the literature,' *Restoration Ecology*, vol 21, no 5, pp. 537–543. DOI: https://doi.org/10.1111/rec.12028

Zahawi, R. A., Holl, K. D., Cole, R. J., and Reid, J. L. (2013), 'Testing applied nucleation as a strategy to facilitate tropical forest recovery,' *Journal of Applied Ecology*, vol 50, no 1, pp. 88–96. DOI: https://doi.org/10.1111/1365-2664.12014

Zahawi, R. A., Reid, J. L., and Holl, K. D. (2014), 'Hidden costs of passive restoration,' *Restoration Ecology*, vol 22, no 3, pp. 284–287. DOI: https://doi.org/10.1111/rec.12098

29

FOREST FRAGMENTS AND FRAGMENTATION

Michael D. Pashkevich, Badrul Azhar, Damayanti Buchori, Robert J. Fletcher, Jr., Jake L. Snaddon and Edgar C. Turner

Introduction

Forest fragmentation is the process of splitting large, continuous areas of forest into increasingly small and separate patches. This process typically reduces the area of continuous forest and creates fragments that can become isolated and impacted by the surrounding environment. Key drivers of forest fragmentation include agricultural expansion, timber extraction, urbanization, and expansion of road networks (Ma et al., 2023). Most forests globally are now degraded and fragmented (Ma et al., 2023), resulting in changes in animal and plant communities and shifts in ecosystem functioning (Haddad et al., 2015).

In this chapter, we review the ecological theory behind forest fragmentation studies and detail how scientists can best study the ecological effects of forest fragmentation around the world. We then discuss the key impacts and associated drivers of forest fragmentation. Next, we investigate the impacts of fragmentation on community interactions and processes, particularly how changes in one component of a community can have cascading impacts on ecosystem functioning. Finally, we consider how tractable restoration strategies can be implemented to mitigate negative impacts of fragmentation, and introduce an ongoing debate on whether forest fragmentation can affect biodiversity both positively and negatively. Throughout this chapter, we consider the effects of fragmentation on forests globally but focus on the tropics, where ongoing rates of forest loss are among the highest globally (Ma et al., 2023), using Southeast Asia as a case study to illustrate our key points.

From the species–area relationship to landscape-scale forest fragmentation studies

Forest fragmentation research is grounded in island biogeography theory. Based largely on the 'Species–Area relationship' (which describes the positive relationship between habitat area and the number of species an ecosystem can support), Island biogeography theory was developed to model the number of species that islands can support (Macarthur and Wilson, 1967). It predicts that larger islands closer to the mainland contain a higher number of species, as larger islands suffer lower rates of extinction and closer islands have higher rates

DOI: 10.4324/9781003324072-34

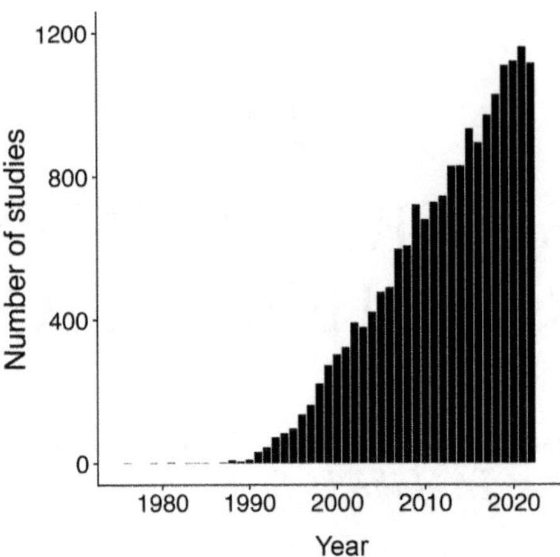

Figure 29.1 Forest fragmentation studies over time, showing that the number of studies has increased dramatically over the last 30 years. We found studies by searching the term 'forest AND fragmentation' on the ISI Web of Science Core Collection. Most fragmentation studies have been observational in nature, with experimental approaches being applied relatively rarely.

of immigration from mainland populations. Island biogeography theory has been supported by numerous observational studies, although whether area and isolation alone drives this relationship is less clear (Ricklefs and Lovette, 1999). In the wake of accelerating forest loss and fragmentation, island biogeography theory was adapted to apply to areas of remaining habitat (Diamond, 1975). In this adaptation, an 'island' is a patch of forest, and the 'mainland' is an area of a larger, similar forest from which the island has been separated.

Drawing on the theory outlined above, increased research attention over the last 40 years has focused on the investigation of the impacts of forest fragmentation on ecosystems worldwide (Figure 29.1). The most scientifically rigorous studies are large-scale forest fragmentation experiments. In contrast to observational studies (which compare ecosystems between areas of intact and fragmented forest and can confound effects of fragmentation with other factors), large-scale experimental studies can partition the direct impacts of fragmentation and isolation from confounding influences of underlying habitat structure or species that are often present in existing fragmented landscapes. For example, agricultural expansion may be more likely to occur in lower-lying areas, leaving fragments on higher ground. Originally designed to answer key questions relating to fragment size alone, the focus of large-scale forest fragmentation experiments has expanded to address a broader range of questions, such as the impact of matrix connectivity and regeneration. Arguably, the most influential of these experiments is the Biological Dynamics of Forest Fragments Project (BDFFP), located in the Brazilian Amazon, which experimentally isolated forest fragments of varying sizes (five 1-ha, four 10-ha, and two 100-ha fragments) (Laurance et al., 2011). Additional large-scale forest fragmentation experiments around the world include the Stability of Altered Forest Ecosystems (SAFE) Project in Sabah, Malaysia (Ewers et al., 2011), the Savannah River Site experiment in South Carolina, USA (Tewksbury et al., 2002), the Wog Wog

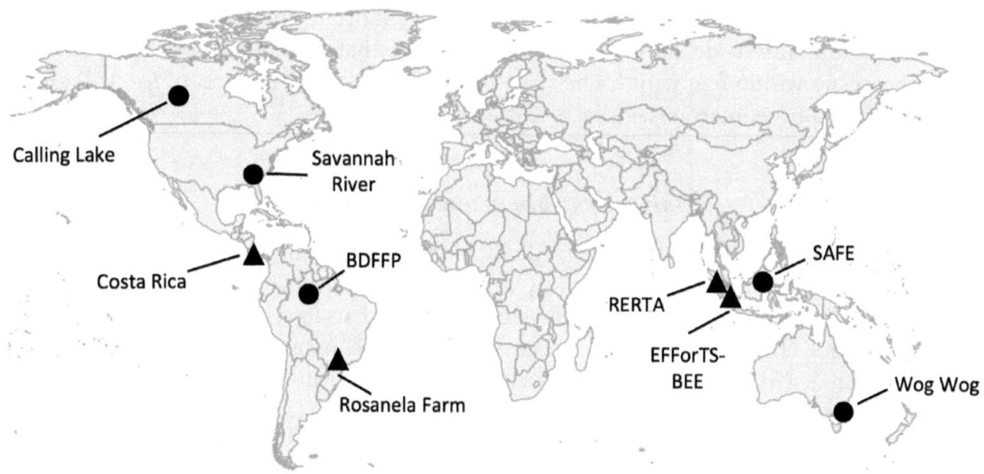

Figure 29.2 Location of some of the most influential large-scale experiments focused on forest fragmentation. Circles indicate studies that are evaluating the ecological impacts of forest fragmentation: Calling Lake (Schmiegelow et al., 1997), Savannah River (Tewksbury et al., 2002), BDFFP (Laurance et al., 2011), SAFE (Ewers et al., 2011), and Wog Wog (Margules, 1992). Triangles indicate studies that are testing strategies to restore ecological complexity in degraded areas, informed by forest fragmentation theory: Costa Rica (Holl et al., 2020), Rosanela Farm (Chazdon et al., 2020), RERTA (Luke et al., 2020), and EFForTS-BEE (Zemp et al., 2023).

Habitat Fragmentation Experiment in New South Wales, Australia (Margules, 1992), and the Calling Lake Fragmentation Experiment in Alberta, Canada (Schmiegelow et al., 1997) (Figure 29.2).

Biological impacts of fragmentation and associated factors

Studies on forest fragmentation have shown that biological communities in forest fragments differ from those in comparable continuous forest in five key ways:

1. Individual fragments only contain a subset of the original community, especially where habitats were varied and species patchily distributed in the original habitat.
2. Population sizes in fragments are lower, increasing the likelihood of stochastic extinction events.
3. Fragments are isolated from other forest areas, reducing immigration and the chance of recolonization once local extinctions have occurred, and potentially increasing levels of inbreeding and loss of genetic diversity.
4. Fragments are impacted by changing conditions as a result of exposure to the surrounding, non-forest area (termed 'edge effects'), potentially further reducing the area for forest specialists.
5. Fragments may be more vulnerable to other perturbations, such as invasion by non-native species, increased hunting pressure, or climate change impacts.

These biological impacts of forest fragmentation often occur simultaneously, precipitating cascading community-wide shifts in biotic interactions that alter functional links and ecosystem processes within fragments. The strength of the biological impacts of forest fragmentation is affected by numerous factors as discussed next.

Habitat patchiness and species rarity

The lower number of species found in individual fragments is, in part, owing to the patchy nature of species distributions, meaning that some species may not be represented in forest fragments simply because they did not occur in that area in the first place (termed 'sample effects'). This may be exacerbated by underlying habitat heterogeneity, which is often correlated with habitat size. For example, in a study on bats in forest fragments in the Amazon, species richness increased with fragment size. However, this effect was also related to vegetation complexity increases, rather than fragment size specifically (Rocha et al., 2017).

A second factor is the intrinsic rarity of many species, which means that, although a species may initially exist in a newly fragmented habitat, it is unlikely to persist in long term. This loss of species over time, which may not occur immediately (therefore sometimes termed 'the extinction debt'), is in line with island biogeography theory, in that losses tend to be more marked and rapid in smaller fragments (Laurance et al., 2011), which contain smaller populations of resident species. For example, in the Amazonian rainforest, Stouffer et al. (2011) found that 44–84% of the original bird fauna had gone extinct in 1-ha fragments 25 years after isolation, but only 31–45% in 10-ha fragments, and 8–16% in 100-ha fragments over this time period. Extinction debt may also take a very long time to be fully paid: continued loss of mammal species in forest fragments in Northeast Australia was still being detected 70 years after the initial fragmentation event (Laurance et al., 2008).

Edge effects

The rapid change in abiotic and biotic factors that occur during fragmentation events in forest fragment edges (called 'edge effects') can precipitate further changes in the fragments and increase species losses (Pfeifer et al., 2017; Ries et al., 2004). Edge effects have been shown to be extensive and important in determining the communities of plants and animals and levels of ecosystem functioning within forest fragments. For example, increased disturbance from higher wind speeds has been measured in forest edges up to 400 m into fragments. Effectively this means that the whole area of fragments less than 50 ha in size may be influenced by edge effects. Other edge effects commonly reported within 100 m of fragments include lower canopy foliage density, increased incidence of tree-fall gaps, higher leaf litter inputs, lower relative humidity, lower soil moisture, and increased air temperature (Laurance et al., 2011; Ries et al., 2004).

Factors associated with edge effects can also interact to exacerbate further changes. Forest trees experience higher mortality closer to forest edges, partly owing to wind damage and partly to physiological stress caused by the rapidly changing abiotic conditions. This changes the gap dynamics of forest edges and can lead to an increase in the prevalence of early successional species and a higher abundance of climbers and lianas (Laurance et al., 2011). Forest edges are also more vulnerable to fire, which can lead to high tree mortality and increase the severity of future burns (Laurance et al., 2011), resulting in even higher mortality of remaining trees and a reduction in the seed bank, hindering forest regeneration.

The matrix matters

The composition or type of surrounding habitat or 'matrix' around fragments can influence a variety of ecological processes (Driscoll et al., 2013), including the effects of forest fragmentation. In real islands, surrounding oceans are often a true barrier to dispersal and provide no resources, but effects can be much more variable for habitat islands, where the matrix represents a potential habitat for terrestrial species in its own right.

The type of matrix surrounding fragments affects local edge effects. For example, some types of matrix, such as plantation forest, may be relatively benign and help to buffer forest edges from the detrimental impacts of wind or changes in microclimate. However, other matrix types, such as pasture or arable areas, provide little buffering capacity and can lead to dramatic and expanding impacts of edge effects into remaining forest fragments. Spillover from the matrix near edges can also be enhanced or minimized. For example, forest areas next to cultivated land may experience drift of pesticides, herbicides, and fertilizers or may trap airborne pollutants resulting in higher rates of deposition in forest edges (Weathers et al., 2001).

The matrix type can also act to support or reduce species migration between fragments. A matrix that is more favourable for forest species can help to reconnect isolated populations, increasing gene flow and reducing the chance of local extinctions. For example, an experimental study on butterflies in meadow patches surrounded by willow or conifer thickets in Colorado found that the degree of matrix permeability for different species was higher in willow than in conifer matrix (Ricketts, 2001).

Habitat isolation

Increasing isolation in forest fragments can leave populations vulnerable to stochastic processes such as birth and death rates, loss of genetic variation, and environmental fluctuations. This can lead to a reduction in species richness in fragments, as once species are lost from isolated fragments there is little chance of recolonization. In a study of dung beetles on forest islands in Venezuela, more isolated islands were found to contain a lower diversity of beetles (Larsen et al., 2008). More isolated and smaller populations are also likely to suffer increased levels of inbreeding, inbreeding depression and loss of genetic diversity, which may also interact with environmental change in fragments to reduce populations still further (e.g. Ismail et al., 2014).

Species-specific characteristics

The exact effects of fragmentation also depend on the characteristics of the species involved (Keinath et al., 2017). Influential species characteristics can include population size and fluctuation, competitive ability, habitat specialization, sensitivity to disturbance, and dispersal ability. These characteristics are often influenced by species-specific traits such as body size, defence and predation structures, and gestation length. As fragmentation often favours species with specific characteristics, fragments tend to contain a non-random subset of the full forest assemblage. For instance, in Venezuela, large-bodied, forest-specialist, and rare dung beetles are more likely to be lost from fragments (Larsen et al., 2008). Species with close biotic interactions may also be particularly vulnerable to fragmentation, as the disappearance of one species will result in the loss of the other. For example, birds that follow foraging army ant colonies to catch fleeing insects are generally lost from forest fragments (Stouffer and Bierregaard, 1995), as the army ants themselves require extensive forest areas to persist.

Region-specific characteristics

Effects of forest fragmentation also depend on regional characteristics. For instance, in the tropics, plant growth rates are amongst the highest globally, and therefore the effects of fragmentation on plant communities may be more quickly reversible than in temperate areas. Human density and local government policy are also highly influential in determining region-specific effects of fragmentation (Box 29.1). Regions with more people are usually under higher environmental pressure, and therefore fragmentation may have more severe and long-lasting impacts. For instance, in 1973, 75.7% (558,060 km^2) of Borneo was forested but, by 2010, extensive human activity had reduced the amount of intact forest to only 28.4% (209,649 km^2) of the island. However, the rate of forest loss was highly variable between geopolitical regions, with Brunei having a greater proportion of intact forest than regions governed by Indonesia and Malaysia (Gaveau et al., 2014).

Synergistic impacts with other environmental change

Other drivers of environmental change, such as invasive species, hunting, pollution and the use of pesticides, and climate change, can all interact with the effects of fragmentation. For example, edge effects can facilitate the colonisation of invasive species into disturbed environments, potentially outcompeting native species (Brook et al., 2008). Fragmented habitats are easier to access for hunting and may be in closer proximity to human populations, leading to greater hunting pressure (Laurance et al., 2011). The ability of species to cope with changes in climate as a result of global warming may be influenced by fragmentation (Ma et al., 2023) and be dependent on the level of structural complexity and habitat fragmentation (Brook et al., 2008). A fragmented habitat may hinder dispersal, effectively isolating species in unfavourable climatic conditions and precipitating higher rates of extinction.

Community interactions and ecosystem functions

Forest fragmentation and the resulting loss of species and changing community interactions can affect a wide range of ecosystem processes, although the severity and direction of these impacts can vary (Brook et al., 2008). The predictability of changes depends on the range of factors that determine particular ecosystem functions (Peh et al., 2014), with functions depending on a few large species being more predictable than those depending on a large number of species or variable environmental conditions. The loss of higher trophic levels can have particularly marked effects, ramifying throughout the food web. For example, the loss of top predators from smaller forest islands in Venezuela resulted in an increased

BOX 29.1 Case Study: Forest degradation and fragmentation in Southeast Asia

The effects of forest degradation and fragmentation are highly region-specific. This is largely because of regional differences in human activity. Southeast Asia, which was almost entirely forested before human activities, has experienced particularly high levels of forest loss and fragmentation over the past 50 years, driven by agricultural expansion, logging, and urbanization (Gaveau et al., 2014). In some regions, very little forest is left. For instance, in Borneo—an island jointly belonging to Malaysia, Indonesia, and Brunei—only 37% of the area that was forested in the 1970s remains intact. This has largely been caused by increased logging activity and the industrial expansion of oil palm agriculture (*Elaeis guineensis*) (Gaveau et al., 2014). Impacts are even greater in more heavily developed areas. For example, Singapore—which is one of the most densely populated countries worldwide—has less than 5% of its remaining natural habitat, and less than 10% of its remaining forest is primary (Corlett, 1992), all of which is fragmented. Large-scale forest fragmentation experiments—such as the Stability of Altered Forest Ecosystems (SAFE) Project in Malaysian Borneo (Ewers et al., 2011)—have shown that the degradation and fragmentation of Southeast Asia's rainforests has caused dramatic declines in biodiversity, changes in ecosystem functioning, and alterations in local human communities. For both Borneo and Singapore, forest degradation and fragmentation has been encouraged by government and local decision-making, which has aimed to improve economic prosperity through the use and commodification of natural resources.

In recent years, increased awareness of the ecological impacts of forest degradation and fragmentation in Southeast Asia has led to enhanced efforts to restore areas of forest that remain. For instance, the Sabah Biodiversity Experiment (in Malaysian Borneo) is testing how enrichment planting of 25 species of native rainforest trees can increase habitat structure, biodiversity, and levels of ecosystem functioning in areas of logged forest (Hector et al., 2011). In Sumatra, the EFForTS-Biodiversity Enrichment Experiment (EFForTS-BEE; Figure 29.2) has established diverse tree islands in areas of oil palm plantation that have been completely denuded of forest. Largely informed by forest fragmentation theory, EFForTS-BEE has shown that diverse tree islands can enhance biodiversity and levels of ecosystem functioning in croplands at no cost to oil palm yields (Zemp et al., 2023). The Riparian Ecosystem Restoration in Tropical Agriculture Project (RERTA; Figure 29.2) is testing strategies to restore degraded areas of land along rivers ('riparian areas'), with implications for re-connecting forest fragments that have become isolated (Luke et al., 2020). These experiments are complemented by government-led programmes, which aim to reverse the impacts of forest fragmentation across vast spatial scales. For instance, the Heart of Borneo initiative—a joint conservation venture between the governments of Malaysia, Indonesia, and Brunei (Sloan et al., 2019)—aims to enhance the coverage of protected areas in Borneo, simultaneously reducing conflict between humans and wildlife and allowing free movement of animal species, particularly charismatic flagship species such as the Bornean orangutan (*Pongo pygmaeus*) and Bornean elephant (*Elephas maximus borneensis*) that are on the verge of extinction. Collectively, these experiments and government programmes show that ambitious conservation action can halt and reverse the ecological impacts of forest degradation and fragmentation, so long as strategies are mindful of and fully involve local communities, with implications for conservation of degraded habitats worldwide.

number of seed predators and herbivores, higher levels of herbivory, and reduced seedling and sapling recruitment (Terborgh et al., 2001). The consequences of changes in function can also depend on characteristics of interacting species. For example, in the Brazilian Amazon, reduced diversity of pollination systems in individual fragments compared to continuous forest led to declines in trees dependent on pollination by birds, flies and non-flying mammals, and self-incompatible species, but increases in hermaphroditic tree species (Girão et al., 2007).

Predicting the impacts of forest fragmentation on ecosystem functioning remains a key challenge for ecologists. Central to this is careful quantification of the exact relationship between factors that predict species loss ('species disassembly rules') and the role that species perform in ecosystems. This depends on the relationship between the response of species to habitat change (response traits) and the effect of species on ecosystem functioning (effect traits). There is evidence that larger bodied dung beetle species may be more sensitive to habitat fragmentation and also more effective in supporting ecosystem functioning (Larsen et al., 2008). Early loss of these species will therefore have a non-random and disproportionately severe impact on ecosystem functioning.

Restoring forest fragments to benefit nature and society

Understanding how ecosystems are affected by forest fragmentation is important for both theoretical and practical reasons. The number of large, intact forest habitats has dwindled globally, and fragmented forest landscapes are now the norm in large parts of the world (Ma et al., 2023). We must protect these forest fragments from further degradation using tailored management and conservation practices based on scientific evidence. Restoration (i.e. enhancing the ecological complexity of degraded areas) is a promising and increasingly commonly used strategy to improve the ecological complexity of fragmented forest landscapes worldwide (Figures 29.2 and 29.3). Large-scale restoration of degraded and fragmented forests is called 'forest-landscape restoration'.

As larger fragments generally support more species and higher levels of ecosystem functioning than smaller fragments, one key way to reverse the impacts of forest fragmentation is to increase the size of existing fragments (Figure 29.3a). Often, this can be done by abandoning managed areas around fragments and allowing the abandoned areas to recover on their own ('natural regeneration'). For instance, in Brazil's heavily fragmented Atlantic Rainforest, it is estimated that natural regeneration could nearly triple the size of existing fragments and, consequently, reduce fragmentation by 44% (Crouzeilles et al., 2020). This expansion is possible owing to the gradual spillover of plants and animals from fragments into the immediate surrounding landscape. However, such strategies could lead to a proliferation of early successional species (Arroyo-Rodríguez et al., 2017), and therefore human intervention (such as pruning of early successional species, or planting of old-growth species) may be needed to enable complete recovery.

As populations in isolated forest fragments are often at greater ecological risk, another key way to reverse the impacts of forest fragmentation is to improve connectivity between existing forest fragments. Such strategies can reconnect isolated populations and—since they are applied in the areas surrounding fragments—improve overall matrix quality, which can have both local and landscape benefits (Brady et al., 2009). One way to improve

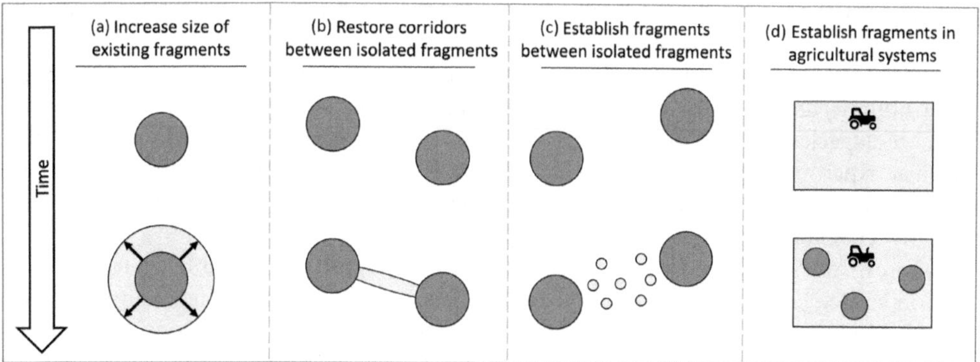

Figure 29.3 Identifying tractable strategies that restore degraded and fragmented forests is needed to conserve forests worldwide. Such restoration strategies are often based on fragmentation theory and seek to reverse the impacts of fragmentation on habitat size (a) and isolation (b–c). Recently, studies have also shown the value of establishing diverse tree islands, which are effectively fragments, in formerly forested areas where forest is no longer present, such as large-scale farmlands (d). Over time, these fragments can benefit biodiversity and ecosystem functioning, and possibly grow in size through seed dispersal processes, with minimal or no impacts on crop yields.

connectivity between fragments and overall matrix quality is to restore or establish corridors, which form continuous highways that link fragments together (Figure 29.3b). In Colombia, fencing-off existing—but degraded—corridors to prevent cattle intrusion and grazing promoted the recovery of native tree species (Calle and Holl, 2019). In Brazil, three million trees from 100 native species were planted to create the largest forest corridor nationwide, linking a small forest fragment, isolated by Rosanela Farm, a large-scale agricultural complex, to a 36,000-hectare forest block (Chazdon et al., 2020). The corridor has promoted movement of endangered species such as the jaguar (*Panthera onca*) and black lion tamarin (*Leontopithecus chrysopygus*) across the landscape (Chazdon et al., 2020). Restoring forested corridors along rivers ('riparian corridors') may be especially helpful, potentially benefitting aquatic and terrestrial communities and also human settlements downstream, by improving river water quality through reduced nutrient run-off from farmlands (Luke et al., 2020). Another way to connect fragments and improve matrix quality is to establish tree islands between larger forest fragments (Figure 29.3c). Unlike corridors, tree islands do not result in continuous highways between fragments, but they can serve as ecological stepping stones, allowing species to 'hop' between fragments and—for small taxa (such as insects)—be an important habitat in their own right. Further, depending on the species that are planted, tree islands can have economic value for local communities, therefore providing a win-win for nature and society. For instance, in Brazil, shade-coffee agroforestry plantings were established as stepping-stones between forest fragments, boosting the biodiversity of birds and butterflies and providing income to local landowners (Chazdon et al., 2020).

Increasingly, restoration strategies are being used to establish forest fragments in areas where forest is no longer present, for instance, in vast agricultural landscapes. Newly established forest fragments can enhance biodiversity and the delivery of ecosystem services, without impacts on or perhaps even increasing crop yields (Zemp et al., 2023). The EFForTS-BEE established fragments in Indonesian oil palm plantations that varied in size (25–1,600 m^2) and tree species number (0–6 species), and found that larger fragments with more tree species most effectively improved ecological complexity (Zemp et al., 2023). A similar experiment in Costa Rica, which established tree islands (ranging from 16 to 144 m^2) in formerly forested areas, found that fragments enhanced multi-taxa biodiversity and a range of ecosystem processes (Holl et al., 2020).

As forests worldwide continue to be degraded and fragmented, identifying tractable restoration strategies that reverse the ecological impacts of degradation and fragmentation is needed urgently. Importantly, any successful strategies will need to be considerate of and fully involve local communities where restoration is occurring. One example of how to do this is to restore fragmented forests by planting non-timber forest product (NTFP) species, which can be used by local people for food, medicine, and cultural purposes.

Does fragmentation always negatively impact ecosystems?

So far in this chapter, we have generally assumed that forest fragmentation has negative impacts on ecosystems. This perspective often arises when forest fragmentation is envisioned as the process of continuous areas of forest being converted into increasingly small and separate patches such that the loss of forest coincides with the change in pattern of remaining forest. Another perspective on fragmentation is to focus specifically on the pattern of how remaining habitat is broken up for a certain amount of habitat loss. In this way, interest is on understanding the contribution of the pattern of 'loss per se' to that of 'fragmentation per se'. The extent to which 'fragmentation per se' is important and whether it tends to have positive or negative effects on ecosystems is currently a major discussion point amongst ecologists (Fahrig, 2017; Fletcher et al., 2018). The debate is focused on habitat fragmentation generally, but its principles are applicable to forest fragmentation specifically.

At the centre of the debate is the concept of scale, as the effects of habitat fragmentation manifest across space and time (Fletcher et al., 2023). For instance, from a spatial perspective, the effects of habitat fragmentation can play out at the landscape scale and also at very localized scales, such as within individual habitat fragments. The perceived impacts of fragmentation can vary depending on the scale that is studied. For instance, at the landscape scale, fragmentation can break up a habitat into many smaller fragments, therefore enhancing landscape-level habitat heterogeneity and potentially increasing overall diversity across the space (Fahrig, 2017). On the other hand, at the individual fragment scale, fragmentation can decrease fragment size, making the patch uninhabitable to species that were previously present, such as wide-ranging species that require large areas of habitat to survive (Haddad et al., 2015).

A comprehensive understanding of how forest fragmentation affects ecosystems requires identifying how effects observed at small scales add up to affect the entire landscape over time. Experiments that assess the impacts of fragmentation over time at scales ranging from the individual patch to landscape levels are therefore needed to identify the underlying mechanisms of fragmentation and quantify the relative proportion of 'good' and 'bad' effects that fragmentation causes (Fletcher et al., 2023).

Conclusion

Fragmentation research has come a long way since the development of the theory of island biogeography. It has moved from a focus on fragment area to a more sophisticated approach that aims to understand the underlying ecological processes determining and resulting from community changes across fragmented landscapes. It is clear that the effects of forest fragmentation are far from straightforward, with multiple factors playing a part in determining the changes observed in biological communities once forests become fragmented. Major gaps still exist in our knowledge of fragmentation effects, including a detailed understanding of the long-term, whole-ecosystem impacts of fragmentation or the interacting impacts of environmental change on communities of plants and animals in fragmented habitats. It is also unclear how these changes will affect earth–atmosphere interactions and how these may add up to affect processes at larger spatial scales. Much of world's forests have already been cleared and fragmented or logged as a result of human activity (Ma et al., 2023). What is now important is to apply understanding of the impacts of fragmentation to minimize negative impacts of ongoing losses, and to develop robust restoration approaches that could restore diverse and functioning forest ecosystems.

References

Arroyo-Rodríguez, V., Melo, F. P. L., Martínez-Ramos, M., Bongers, F., Chazdon, R. L., Meave, J. A., Norden, N., Santos, B. A., Leal, I. R., and Tabarelli, M. (2017). Multiple successional pathways in human-modified tropical landscapes: New insights from forest succession, forest fragmentation and landscape ecology research. *Biological Reviews*, *92*(1), 326–340. https://doi.org/10.1111/brv.12231

Brady, M. J., McAlpine, C. A., Miller, C. J., Possingham, H. P., and Baxter, G. S. (2009). Habitat attributes of landscape mosaics along a gradient of matrix development intensity: Matrix management matters. *Landscape Ecology*, *24*(7), 879–891. https://doi.org/10.1007/s10980-009-9372-6

Brook, B. W., Sodhi, N. S., and Bradshaw, C. J. a. (2008). Synergies among extinction drivers under global change. *Trends in Ecology & Evolution*, *23*(8), 453–460. https://doi.org/10.1016/j.tree.2008.03.011

Calle, A., and Holl, K. D. (2019). Riparian forest recovery following a decade of cattle exclusion in the Colombian Andes. *Forest Ecology and Management*, *452*(August), 117563. https://doi.org/10.1016/j.foreco.2019.117563

Chazdon, R. L., Cullen, L., Padua, S. M., and Padua, C. V. (2020). People, primates and predators in the Pontal: From endangered species conservation to forest and landscape restoration in Brazil's Atlantic Forest. *Royal Society Open Science*, *7*(12), 1–16. https://doi.org/10.1098/rsos.200939

Corlett, R. T. (1992). The ecological transformation of Singapore, 1819-1990. *Journal of Biogeography*, *19*(4), 411. https://doi.org/10.2307/2845569

Crouzeilles, R., Beyer, H. L., Monteiro, L. M., Feltran-Barbieri, R., Pessôa, A. C. M., Barros, F. S. M., Lindenmayer, D. B., Lino, E. D. S. M., Grelle, C. E. V., Chazdon, R. L., Matsumoto, M., Rosa, M., Latawiec, A. E., and Strassburg, B. B. N. (2020). Achieving cost-effective landscape-scale forest restoration through targeted natural regeneration. *Conservation Letters*, *13*(3). https://doi.org/10.1111/conl.12709

Diamond, J. (1975). The island dilemma: Lessons of modern biogeogrpahic studies for the design of natural reserves. *Biological Conservation*, *7*, 129–146.

Driscoll, D. A., Banks, S. C., Barton, P. S., Lindenmayer, D. B., and Smith, A. L. (2013). Conceptual domain of the matrix in fragmented landscapes. *Trends in Ecology & Evolution*, *28*(10), 605–613. https://doi.org/10.1016/j.tree.2013.06.010

Ewers, R. M., Didham, R. K., Fahrig, L., Ferraz, G., Hector, A., Holt, R. D., Kapos, V., Reynolds, G., Sinun, W., Snaddon, J. L., and Turner, E. C. (2011). A large-scale forest fragmentation experiment: The Stability of Altered Forest Ecosystems Project. *Philosophical Transactions of the Royal Society B: Biological Sciences*, *366*(1582), 3292–3302. https://doi.org/10.1098/rstb.2011.0049

Fahrig, L. (2017). Ecological responses to habitat fragmentation per se. *Annual Review of Ecology, Evolution, and Systematics*, *48*(1), 1–23. https://doi.org/10.1146/annurev-ecolsys-110316-022612

Fletcher, R. J., Betts, M. G., Damschen, E. I., Hefley, T. J., Hightower, J., Smith, T. A. H., Fortin, M., and Haddad, N. M. (2023). Addressing the problem of scale that emerges with habitat fragmentation. *Global Ecology and Biogeography*, *32*(6), 828–841. https://doi.org/10.1111/geb.13658

Fletcher, R. J., Didham, R. K., Banks-Leite, C., Barlow, J., Ewers, R. M., Rosindell, J., Holt, R. D., Gonzalez, A., Pardini, R., Damschen, E. I., Melo, F. P. L., Ries, L., Prevedello, J. A., Tscharntke, T., Laurance, W. F., Lovejoy, T., and Haddad, N. M. (2018). Is habitat fragmentation good for biodiversity? *Biological Conservation*, *226*(July), 9–15. https://doi.org/10.1016/j.biocon.2018.07.022

Gaveau, D. L. A., Sloan, S., Molidena, E., Yaen, H., Sheil, D., Abram, N. K., Ancrenaz, M., Nasi, R., Quinones, M., Wielaard, N., and Meijaard, E. (2014). Four decades of forest persistence, clearance and logging on Borneo. *PLoS One*, *9*(7), 1–11. https://doi.org/10.1371/journal.pone.0101654

Girão, L. C., Lopes, A. V., Tabarelli, M., and Bruna, E. M. (2007). Changes in tree reproductive traits reduce functional diversity in a fragmented Atlantic forest landscape. *PloS One*, *2*(9), e908. https://doi.org/10.1371/journal.pone.0000908

Haddad, N. M., Brudvig, L. A., Clobert, J., Davies, K. F., Gonzalez, A., Holt, R. D., Lovejoy, T. E., Sexton, J. O., Austin, M. P., Collins, C. D., Cook, W. M., Damschen, E. I., Ewers, R. M., Foster, B. L., Jenkins, C. N., King, A. J., Laurance, W. F., Levey, D. J., Margules, C. R., … Townshend, J. R. (2015). Habitat fragmentation and its lasting impact on Earth's ecosystems. *Science Advances*, *1*(2). https://doi.org/10.1126/sciadv.1500052

Hector, A., Philipson, C., Saner, P., Chamagne, J., Dzulkifli, D., O'Brien, M., Snaddon, J. L., Ulok, P., Weilenmann, M., Reynolds, G., and Godfray, H. C. J. (2011). The Sabah Biodiversity Experiment: A long-term test of the role of tree diversity in restoring tropical forest structure and functioning. *Philosophical Transactions of the Royal Society B: Biological Sciences*, *366*(1582), 3303–3315. https://doi.org/10.1098/rstb.2011.0094

Holl, K. D., Reid, J. L., Cole, R. J., Oviedo-Brenes, F., Rosales, J. A., and Zahawi, R. A. (2020). Applied nucleation facilitates tropical forest recovery: Lessons learned from a 15-year study. *Journal of Applied Ecology*, *57*(12), 2316–2328. https://doi.org/10.1111/1365-2664.13684

Ismail, S. A., Ghazoul, J., Ravikanth, G., Kushalappa, C. G., Uma Shaanker, R., and Kettle, C. J. (2014). Forest trees in human modified landscapes: Ecological and genetic drivers of recruitment failure in Dysoxylum malabaricum (Meliaceae). *PLoS One*, *9*(2), e89437. https://doi.org/10.1371/journal.pone.0089437

Keinath, D. A., Doak, D. F., Hodges, K. E., Prugh, L. R., Fagan, W., Sekercioglu, C. H., Buchart, S. H. M., and Kauffman, M. (2017). A global analysis of traits predicting species sensitivity to habitat fragmentation: Species sensitivity. *Global Ecology and Biogeography*, *26*(1), 115–127. https://doi.org/10.1111/geb.12509

Larsen, T. H., Lopera, A., and Forsyth, A. (2008). Understanding trait-dependent community disassembly: Dung beetles, density functions, and forest fragmentation. *Conservation Biology*, *22*(5), 1288–1298. https://doi.org/10.1111/j.1523-1739.2008.00969.x

Laurance, W. F., Camargo, J. L. C., Luizão, R. C. C., Laurance, S. G., Pimm, S. L., Bruna, E. M., Stouffer, P. C., Bruce Williamson, G., Benítez-Malvido, J., and Vasconcelos, H. L. (2011). The fate of Amazonian forest fragments: A 32-year investigation. *Biological Conservation*, *144*(1), 56–67. https://doi.org/10.1016/j.biocon.2010.09.021

Laurance, W. F., Laurance, S. G., and Hilbert, D. W. (2008). Long-term dynamics of a fragmented rainforest mammal assemblage. *Conservation Biology*, *22*(5), 1154–1164. https://doi.org/10.1111/j.1523-1739.2008.00981.x

Luke, S. H., Advento, A. D., Aryawan, A. A. K., Adhy, D. N., Ashton-Butt, A., Barclay, H., Dewi, J. P., Drewer, J., Dumbrell, A. J., Edi, Eycott, A. E., Harianja, M. F., Hinsch, J. K., Hood, A. S. C., Kurniawan, C., Kurz, D. J., Mann, D. J., Matthews Nicholass, K. J., Naim, M., ... Turner, E. C. (2020). Managing oil palm plantations more sustainably: Large-scale experiments within the Biodiversity and Ecosystem Function in Tropical Agriculture (BEFTA) programme. *Frontiers in Forests and Global Change*, *2*(January), 1–20. https://doi.org/10.3389/ffgc.2019.00075

Ma, J., Li, J., Wu, W., and Liu, J. (2023). Global forest fragmentation change from 2000 to 2020. *Nature Communications*, *14*(3752). https://doi.org/doi.org/10.1038/s41467-023-39221-x

Macarthur, R. H., and Wilson, E. O. (1967). *The Theory of Island Biogeography* (REV-Revised). Princeton University Press; JSTOR. www.jstor.org/stable/j.ctt19cc1t2.1

Margules, C. R. (1992). The Wog Wog Habitat Fragmentation Experiment. *Environmental Conservation*, *19*(4), 316–325.

Peh, K. S. H., Lin YangChen, L. Y., Luke, S. H., Foster, W. A., and Turner, E. C. (2014). Forest fragmentation and ecosystem function. In C. J. Kettle and L. P. Koh (Eds.), *Global forest fragmentation* (1st ed., pp. 96–114). CABI. https://doi.org/10.1079/9781780642031.0096

Pfeifer, M., Lefebvre, V., Peres, C. A., Banks-Leite, C., Wearn, O. R., Marsh, C. J., Butchart, S. H. M., Arroyo-Rodríguez, V., Barlow, J., Cerezo, A., Cisneros, L., D'Cruze, N., Faria, D., Hadley, A., Harris, S. M., Klingbeil, B. T., Kormann, U., Lens, L., Medina-Rangel, G. F., ... Ewers, R. M. (2017). Creation of forest edges has a global impact on forest vertebrates. *Nature*, *551*(7679), 187–191. https://doi.org/10.1038/nature24457

Ricketts, T. H. (2001). The matrix matters: Effective isolation in fragmented landscapes. *The American Naturalist*, *158*(1), 87–99. https://doi.org/10.1086/320863

Ricklefs, R. E., and Lovette, I. J. (1999). The roles of island area per se and habitat diversity in the species-area relationships of four Lesser Antillean faunal groups. *Journal of Animal Ecology*, *68*(6), 1142–1160. https://doi.org/10.1046/j.1365-2656.1999.00358.x

Ries, L., Fletcher, R. J., Battin, J., and Sisk, T. D. (2004). Ecological responses to habitat edges: Mechanisms, models, and variability explained. *Annual Review of Ecology, Evolution, and Systematics*, *35*(1), 491–522. https://doi.org/10.1146/annurev.ecolsys.35.112202.130148

Rocha, R., López-Baucells, A., Farneda, F. Z., Groenenberg, M., Bobrowiec, P. E. D., Cabeza, M., Palmeirim, J. M., and Meyer, C. F. J. (2017). Consequences of a large-scale fragmentation experiment for Neotropical bats: Disentangling the relative importance of local and landscape-scale effects. *Landscape Ecology*, *32*(1), 31–45. https://doi.org/10.1007/s10980-016-0425-3

Schmiegelow, F. K. A., Machtans, C. S., and Hannon, S. J. (1997). Are boreal birds resilient to forest fragmentation? An experimental study of short-term community responses. *Ecology*, *78*(6), 1914–1932. https://doi.org/10.1890/0012-9658(1997)078[1914:ABBRTF]2.0.CO;2

Sloan, S., Campbell, M. J., Alamgir, M., Lechner, A. M., Engert, J., and Laurance, W. F. (2019). Transnational conservation and infrastructure development in the Heart of Borneo. *PLOS One*, *14*(9), e0221947. https://doi.org/10.1371/journal.pone.0221947

Stouffer, P., and Bierregaard, R. (1995). Use of Amazonian forest fragments by understory insectivorous birds. *Ecology*, 2429–2445.

Stouffer, P. C., Johnson, E. I., Bierregaard, R. O., and Lovejoy, T. E. (2011). Understory bird communities in Amazonian rainforest fragments: Species turnover through 25 years post-isolation in recovering landscapes. *PloS One*, *6*(6), e20543. https://doi.org/10.1371/journal.pone.0020543

Terborgh, J., Lopez, L., Nuñez, P., Rao, M., Shahabuddin, G., Orihuela, G., Riveros, M., Ascanio, R., Adler, G. H., Lambert, T. D., and Balbas, L. (2001). Ecological meltdown in predator-free forest fragments. *Science*, *294*(5548), 1923–1926. https://doi.org/10.1126/science.1064397

Tewksbury, J. J., Levey, D. J., Haddad, N. M., Sargent, S., Orrock, J. L., Weldon, A., Danielson, B. J., Brinkerhoff, J., Damschen, E. I., and Townsend, P. (2002). Corridors affect plants, animals, and their interactions in fragmented landscapes. *Proceedings of the National Academy of Sciences*, *99*(20), 12923–12926. https://doi.org/10.1073/pnas.202242699

Weathers, K. C., Cadenasso, M. L., and Pickett, S. T. A. (2001). Forest edges as nutrient and pollutant concentrators: Potential synergisms between fragmentation, forest canopies, and the atmosphere. *Conservation Biology*, *15*(6), 1506–1514. https://doi.org/10.1046/j.1523-1739.2001.01090.x

Zemp, D. C., Guerrero-Ramirez, N., Brambach, F., Darras, K., Grass, I., Potapov, A., Röll, A., Arimond, I., Ballauff, J., Behling, H., Berkelmann, D., Biagioni, S., Buchori, D., Craven, D., Daniel, R., Gailing, O., Ellsäßer, F., Fardiansah, R., Hennings, N., … Kreft, H. (2023). Tree islands enhance biodiversity and functioning in oil palm landscapes. *Nature*. https://doi.org/10.1038/s41586-023-06086-5

30
THE ECOLOGY OF LOGGED FORESTS

Paul Woodcock, Panu Halme and David P. Edwards

History and significance of logging

Globally, approximately 30% (1.15 billion ha) of forests are primarily managed for wood production, with a further 749 million ha designated for multiple uses (FAO 2020). The historical extent and current trends in logging differ across biomes and countries. In Europe, substantial increases in human influence on temperate forest occurred over 5,000 years ago and from the 16th century forests became increasingly important initially as a source of timber for shipbuilding, with demand for tar and charcoal also driving logging in boreal regions (Wallenius *et al.* 2010). Logging on a large scale tended to occur later in temperate Asia and North America and more remote regions retain important areas of natural forest, but logging across boreal and temperate biomes now often occurs in forest that has already been modified by centuries of human use (Yu *et al.* 2011; Hessburg *et al.* 2019; Aszalos *et al.* 2022). Tropical forests also have a long history of human use (Roberts *et al.* 2017), though the magnitude has expanded rapidly over the past 50–100 years with increases in logging occurring as a result of mechanisation and improved transportation coupled with the opening of global markets.

This chapter is intended to introduce the changes in forest ecology that occur during and after logging, and then describe approaches that are sometimes used to try to support biodiversity and ecosystem functions in logged forest. Note that the effects of logging are variable, and this chapter is not intended to be a systematic and comprehensive assessment of every situation. The focus is also on logging and so should be read in conjunction with other chapters that describe different aspects of forest management. Similarly, the intention is to describe the ecology of logged forest and not to go into detail on socio-economic considerations, legal frameworks, or policy drivers. However, these factors determine when, why, and how logging takes place and therefore are essential to understand. What might seem correct from an ecological perspective will not succeed if it does not consider compatibility with wider societal priorities and practices. Equally, these wider factors can present opportunities where they are aligned with ecological objectives.

DOI: 10.4324/9781003324072-35

Conventional approaches to logging

This section gives a brief overview of logging systems. How each broad system is implemented varies, and this variation *within* systems influences the ecological effects. In boreal and temperate regions, even-aged forestry (clear-cutting, and most forms of shelterwood cutting) and uneven-aged forestry (also named continuous cover forestry or selective cutting) are used, with even-aged systems being the most common in Europe (Kuuluvainen *et al.* 2021; Aszalos *et al.* 2022; see also Chaudhary *et al.* 2016 and Savilaakso *et al.* 2021 for example definitions of terms). Logging in the tropics sometimes uses clear-cutting but more commonly involves selective removal of target tree species above specified diameter limits. The volume of wood removed depends on the density of marketable trees and accessibility, with logging intensities generally highest in Southeast Asia and lower in South America and Africa (Bousfield *et al.* 2020). Although this chapter includes findings from forests that have been planted, it does not consider intensively managed short-rotation plantations in which the ecological effects of logging may be minimal because there is limited value to begin with.

Logging cycles

Forests managed for long-term wood production follow logging cycles, in which timber is harvested and the forest is left to regenerate before relogging. The duration of logging cycles in boreal and temperate systems varies widely, for example, from 60 to 140 years (Brunet *et al.* 2010; Kuuluvainen *et al.* 2021). In the tropics, government-mandated minimum selective logging cycles are often 25–60 years, but these are generally too short to allow for a full recovery of timber stocks. As the extent of unharvested tropical forest declines, relogging is increasingly common —sometimes taking place much earlier than originally intended. When early relogging occurs (whether in the tropics or elsewhere) it typically involves harvesting smaller trees, but despite removing large numbers of stems timber yields are low and long-term potential is severely compromised (Putz *et al.* 2022).

Ecological effects of logging

Investigating logged forest ecology

There are different reasons for understanding logged forest ecology, and these influence the study design and interpretation. One distinction is between understanding the ecological effects of logging and determining the ecological value retained within a logged forest. Studies examining the effects of logging compare the same forest before and after logging, or compare with a similar unlogged forest (or ideally use both comparisons). When a logged forest is compared with an unlogged forest, it is essential that both would have been similar prior to logging. For example, to examine the effects of logging in a secondary forest, the comparator should be a secondary forest of similar age and conditions (e.g. climate, altitude, location, etc.). Studies that examine the effects of logging are most useful where they try to also propose or test solutions based on understanding why changes occur and practical feasibility.

Understanding how logging affects biodiversity and ecosystem functions does not provide a complete picture of the ecological value of logged forest however—particularly given that the alternative is often complete forest clearance for other land uses. One way of determining the ecological value retained within the logged forest is to compare biodiversity, ecosystem functions, etc., with likely alternatives (e.g., agriculture) and with unlogged primary forest. The more closely the logged forest resembles the high conservation value forest versus the alternative, the stronger the case for protecting it from being converted to another land use (from a conservation perspective). Finally, some forestry approaches aim to incorporate processes and structures that would have occurred in primary/natural forest. If these approaches are used, then an understanding of the ecology of unmanaged systems is required.

Heterogeneity in the effects of logging

The effects of different logging approaches vary within systems. This is partly because national legislation often restricts logging in riparian areas and/or on steep slopes and the latter may also be prevented by accessibility. Depending on terrain, large logging concessions will often contain areas that are logged at lower intensities or not at all. Selective logging also varies both within and between regions depending on the distribution and abundance of commercially valuable trees. Using data from six tropical timber producing countries, Putz *et al.* (2019) estimated that on average 69% of the area in selective logging concessions may not be directly affected by logging (e.g. without logging-induced changes in forest structure, but not considering secondary impacts such as increased access associated with road building). This also highlighted variation in the area left intact—ranging from 20% to 97% of the logging blocks studied—and difference within countries. The long-term persistence of intact areas if additional timber harvesting takes place should not be assumed, and neither should their suitability for all species typical of unlogged forest. Equally, this demonstrates that tropical selective logging concessions contain potentially important undisturbed areas and are not homogeneous, so understanding the ecology of logged forest needs to properly encompass this variation.

Forest structure and abiotic conditions

Logging causes changes in abiotic conditions, including increases in light levels, greater variability in temperature, changes in soil chemistry, and soil compaction (Cambi *et al.* 2015; Bowd *et al.* 2019). These changes are most pronounced in clear-cutting, following the removal of almost all tree cover and intensive machine use. The difference in abiotic properties following shelterwood management or selective logging varies. For example, in temperate systems diurnal temperature ranges appear lower in unmanaged than managed forests (Menge *et al.* 2023). By contrast, near-ground midday temperatures and diurnal temperature ranges increase initially in selectively logged tropical forests but may return to a similar microclimate regime within a few years (Molinari *et al.* 2019). In addition to abiotic changes, logging alters the physical structure of forests in several ways. Again, changes are most obvious when clear-cutting removes the whole stand (Figure 30.1). Clear-cutting also reduces the volume of deadwood and the size of remaining material, as it is broken up during logging and then generally removed.

Without mitigation, regenerating stands are homogeneous in composition, age, and structure (Aszalos *et al.* 2022). In systems that are naturally even-aged, over time this regeneration might resemble the pre-logging system whereas in other forest types it will represent a substantial difference (Aszalos *et al.* 2022). In conventional shelterwood management, forest structure also becomes increasingly homogeneous as the canopy is progressively opened up.

Figure 30.1 A five-year-old clear-cut boreal forest in Finland, where energy-wood was harvested, has very little leftover deadwood. Behind is a similar more recently logged area without energy-wood harvesting. (Photograph by Panu Halme.)

In selective logging, non-target vegetation is damaged, with felling, skid trails (corridors in which large vegetation is flattened to extract logs), and logging roads causing damage and mortality (Ellis *et al.* 2019). These result in large gaps and canopy damage and therefore increase light levels at the forest floor. Fast-growing understory plants and/or pioneer tree species are then favoured, and often accompanied by dense tangles of lianas while in the most heavily impacted areas, severe soil compaction restricts regrowth to grassy, scrub-like vegetation or vine tangles (Bousfield *et al.* 2020). However, as described above there is important heterogeneity in effects both within and between selective logging concessions (Putz *et al.* 2019; Figure 30.2). Forest regrowth also closes most canopy gaps within a few years although recovery of mature forest structure can take far longer and will generally not occur in full before relogging takes place.

Biodiversity

Logging impacts on biodiversity are driven by shifts in forest structure, abiotic conditions, and the associated changes in resource availability. Clear-cuts strongly favour species tolerant of open habitats (Demarais *et al.* 2017; Savilaakso *et al.* 2021); therefore, where the conservation of early successional species is needed, regenerating clear-cuts can provide valuable habitat (Schall *et al.* 2018). However, removing large, old trees and woody debris during clear-cutting causes declines amongst many birds, and deadwood-dependent invertebrates, bryophytes, and fungi. The changes in microclimate are also detrimental to some forest specialists. In the absence of mitigation, the structurally homogeneous stand typical of regenerating clear-cuts has limited value for many forest taxa. Over time this recovers, but

Figure 30.2 Examples of recently (<8 years) relogged forest in Borneo and showing variation in vegetation structure in areas more heavily logged (a) compared to less heavily logged (b). (Photograph by Paul Woodcock.)

a meta-analysis by Savilaakso *et al.* (2021) suggests that even-aged forests <80 years old do not support the same level of forest-dependent species as more mature forests (see also Paillet *et al.* 2010; Chaudhary *et al.* 2016). Conventional shelterwood systems can also cause declines across a range of species (Brunet *et al.* 2010), particularly following the final harvest of seed trees. In the case of bryophytes and fungi, adverse effects of shelterwood management may be caused by deadwood removal and changes in soil chemistry. Similarly, many bird species typical of mature forest require structures for nesting and foraging that are rare in conventional shelterwoods (e.g. large, old trees and standing deadwood).

Total species richness in selectively logged forest is often similar to unlogged forest, and sometimes may be higher (Gibson *et al.* 2011; Chaudhary *et al.* 2016; Savilaakso *et al.* 2021). However, this can mask changes in species composition, with generalist and disturbance-tolerant species increasing in abundance following selective logging and old-growth forest specialists declining. In boreal and temperate forests, significant impacts of selective logging result particularly from the removal of large, old trees, which reduces nesting and foraging habitats and limits the accumulation of deadwood. The detrimental effects on some taxa can persist though others may recolonise, indicating that uneven-aged forest management can contribute to biodiversity conservation but natural or near-natural forest is also required (Savilaakso *et al.* 2021).

In tropical forests, the dense post-logging understory regrowth affects insectivorous bird and bat species that require an open understory for effective hunting (Meyer *et al.* 2015; Burivalova *et al.* 2015), whilst changes in microclimate might explain reduced abundance of soft-bodied invertebrates (Ewers *et al.* 2015). This illustrates the vulnerability of many tropical forest species to selective logging disturbance, particularly where relogging has taken place without sufficient recovery time and at higher logging intensities (Burivalova *et al.* 2015). Nonetheless, in large logging concessions the majority of species found in unlogged forest are still present. This is likely in part due to the heterogeneity of selective logging, which means concessions may still contain substantial areas that are less affected and so allow specialists to persist, albeit potentially with smaller total populations (e.g. Montejo-Kovacevich *et al.* 2022). However, the longer-term recovery and maintenance of biodiversity in logged tropical forest is not yet well understood, because there are generally not sufficiently long time series to track changes for 40–50 years or more through multiple logging cycles.

Individual physiology, community structure, ecosystem functioning, and landscape processes

Research on the ecological effects of logging is expanding to examine changes from the physiology of individual animals through to community interactions, ecosystem functioning, and landscape-scale processes. At the individual level, selective logging may alter animal behaviour (e.g. movement), diet, and stress (Franca *et al.* 2016; Bousfield *et al.* 2020), though the consistency of these findings and the extent to which populations are then influenced are not yet clear. Community structure is more consistently affected, with changes in the functional composition of many taxa such as declines in insectivorous birds following selective logging (Burivalova *et al.* 2015), and altered relative abundance of ectomycorrhizal versus arbuscular mycorrhizal fungi after clearcutting (Rodriguez-Ramos *et al.* 2020). These compositional changes can be inferred to indicate shifts in biotic interactions and ecosystem functions, though studies also provide some more direct evidence of changes. Clear-cutting and selective logging both alter decomposition (Ewers *et al.* 2015; Kohout *et al.* 2018), and selective logging changes interspecific interactions (e.g. predator–prey and tree–liana) to potentially make these networks of interactions more susceptible to disturbance or to further species loss (Magrach *et al.* 2016; Hemprich-Bennett *et al.* 2021). Equally, there is evidence of resilience and even amplification of some ecosystem processes in selectively logged forests and so there is a need for additional work to better understand this across biomes (Ewers *et al.* 2015; Malhi *et al.* 2022).

On longer timescales and at landscape scales, logging disturbance differs from natural disturbance dynamics. This difference depends on the system and on the logging approaches used. For example, Aszalos *et al.* (2022) suggested that disturbance size, frequency, and residual structure following clear-cutting and uniform shelterwood management in European boreal and temperate forests differ substantially from the partial periodic mortality that would have occurred in natural forests due to fire, wind, insect outbreaks, etc. These changes in disturbance dynamics may influence longer-term forest resilience, although the relationship is not straightforward (e.g. retaining deadwood can also provide habitat for bark beetles that cause outbreaks). Equally however, other forest types are naturally even-aged as a result of periodic stand-replacing disturbances, and so in these cases even-aged forestry might more closely resemble natural disturbances. There is also a general challenge with understanding how logging alters natural disturbance dynamics because there is often a lack of old-growth forest to determine natural disturbance regimes, though a range of additional data sources can be used (Aszalos *et al.* 2022).

Carbon stocks, emissions, and sequestration

Forests are a major global carbon sink, and so understanding and mitigating the effects of logging on carbon balance is an important part of strategies to combat climate change. This section focuses on changes in forest carbon, although the fate of wood products also affects the final net CO_2 emissions from logging.

Clear-cutting removes much of the aboveground vegetation, with the remainder mostly in deadwood that decays over time. Major reductions in aboveground carbon therefore occur. Meta-analyses suggest that effects on soil carbon stocks are less pronounced, and probably reduced by on average around 8–11% though this is still important given the size of soil carbon pools. There is also variability in estimates between individual studies, a tendency towards greater effects on broadleaf forests than conifers, and risks of large carbon losses if tree planting takes place on peatlands (see Mayer *et al.* 2020 and references therein).

Recovery of aboveground carbon depends on the forest type and management priorities. For example, stands managed primarily for timber production are not intended to reach the carbon stocks found in mature forest. For comparison however, Peichl *et al.* (2023) estimated that clear-cut boreal stands move from being a carbon source to sink within 5–20 years and suggested that an optimum rotation period for carbon sequestration would be 138 years (though carbon is still accumulated beyond this). Recovery of belowground carbon stocks differs between soil types and depths, but may be around 50–100 years (James and Harrison 2016; Mayer *et al.* 2020). For both above and belowground carbon, estimates of long-term recovery are generally based on models rather than field data and vary between sites, so timescales are still subject to uncertainty—particularly for soil carbon (Mayer *et al.* 2020).

In tropical forests, selective logging reduces carbon stocks through the timber directly removed, and due to several associated effects (Ellis *et al.* 2019). These include decomposition of residual material from felled trees and non-target vegetation damaged during felling, as well as vegetation clearance for logging infrastructure (e.g. logging roads and skid trails). Carbon emissions vary depending on logging intensity and harvest methods, but annual emissions from tropical selective logging may be around 6% of the total annual greenhouse gas emissions from tropical timber producing countries (Ellis *et al.* 2019). Logged forests can remain a carbon source for at least a decade as emissions from deadwood decomposition and elevated soil respiration exceed the uptake by forest regrowth (Mills *et al.* 2023). Recovery of aboveground carbon to unlogged forest levels may take approximately 60 years in forest selectively logged at high intensity (Philipson *et al.* 2020). However, carbon emissions and recovery times depend on the logging practices and post-logging measures applied, with recovery time shortened, for example, by restoration treatments and lengthened if forests are relogged without sufficient regeneration periods. Again, there is a lack of longer-term field data on carbon recovery (40–50 years or more), and so most estimates are predictions rather than direct measures.

Secondary effects of logging

Logging operations are associated with several secondary impacts, in particular, arising from road construction, fires, and tree pest and disease outbreaks. Networks of logging roads for transporting timber are increasingly widespread in the tropics. In the Congo Basin, Kleinschroth *et al.* (2019) estimated that total road length in logging concessions doubled between 2003 and 2018, to exceed 100,000 km. Roads make previously remote areas accessible and frequently precede the complete clearance of adjacent forest. Greater accessibility also often results in far higher hunting pressure, with significant declines in medium- and large-bodied mammals in logged African forest attributable more to hunting than to direct effects of logging (Poulsen *et al.* 2013). Closing roads between logging cycles is a key mechanism for limiting the impacts and risks of illegal clearance, but this requires concession owners to be managing forests with an interest in long-term sustainability (Bicknell *et al.* 2015).

Fire risk can be elevated following logging, partly as a consequence of higher volumes of dry, combustible material (e.g. dead and dying non-target vegetation damaged during logging) and partly due to increased exposure to human-caused ignition (Lindenmayer *et al.* 2020; Shvetsov *et al.* 2021). Although fires are an important part of the natural cycle of some forest types, increased frequency and severity can impair the recovery of forest structure and composition and risk crossing a threshold such that the ecosystem instead remains as scrub/savanna. Outbreaks of tree diseases and of herbivorous insects are similarly a part of natural forest dynamics that can be influenced by timber production regimes, particularly in boreal and temperate forests. However, the right balance between different management considerations is not always easy to establish. Whilst stands with a single tree species, homogeneous age structure, and low genetic variation are expected to have a high outbreak risk, management in the form of thinning and deadwood removal can reduce the availability of habitat for some tree pests and diseases (Hlasny *et al.* 2021).

Strategies for maintaining and recovering biodiversity and ecosystem functions in logged forest

Trade-offs in logged forests

Although logging has ecological effects, it is important to see these in comparison with impacts from alternative land uses. For example, even heavily selectively logged forests support considerably higher biodiversity than oil palm plantations. Without the revenue and employment generated by logging there would very likely be increased pressure to convert some forests to other land uses with consequent ecological effects (though note that a forest does not always have to generate high timber profits to be valued sufficiently enough to retain—a combination of national and international policy commitments and mechanisms such as payments for ecosystem services, green finance, etc., mean that timber yields are not the only political, societal, and economic rationale for retaining forest).

Within logged forest, there are then trade-offs over what the forest is managed for. For example, managing forests primarily for timber or carbon or biodiversity will often not be the most optimal for the other two. If forests have multiple objectives, it will be important to understand the trade-offs between different aims. The sections below describe forestry practices that are intended to better maintain or recover biodiversity and ecosystem functions in logged forest, divided into actions that take place before or during logging, and those that occur post-logging. These are generally intended to reduce rather than fully alleviate effects of logging on forest ecology. Decisions on implementation therefore need to consider the ecological benefits and whether these outweigh any associated costs (e.g. potential increased management costs and/or reduced timber yields). Lastly, the effectiveness of different interventions will vary by species—and for some species certain interventions might be detrimental even if biodiversity as a whole is better maintained. As such, implementation for intended biodiversity benefits should take into account how the species of greatest importance for that forest are likely to respond.

Before and during logging

Close-to-nature forestry and related terms

A suite of related management approaches aim to achieve a more natural system, with the intention to help maintain biodiversity and wider ecosystem values. These approaches have several overlapping names, including 'close-to-nature forestry', 'back-to-nature forestry', 'nature-based silviculture', 'natural disturbance-based forest management', 'natural dynamics silviculture', 'ecological forestry', and 'ecological silviculture' (D'Amato *et al.* 2017; Kuuluvainen *et al.* 2021). A range of practices are encompassed within this, and so the ecological effects of close-to-nature forestry and related approaches fall on a continuum between more conventional logging and strict forest protection, depending on what is adopted and how. The diversity in approaches and terminology (including defining what is meant by nature or naturalness) can complicate comparisons and implementation. Examples include maintaining mixed-species stands of native trees, using natural regeneration, retaining deadwood and standing dead trees, and minimising the use of fertilisers and pesticides (Gamborg and Larsen 2003). At larger spatial scales, some forestry approaches aim to emulate processes and disturbance regimes in natural forest (e.g. size and frequency of disturbance). Proponents argue that this can help buffer forests against climate extremes and support the persistence of species typical of old-growth forest (Kuuluvainen *et al.* 2021), and also advise diversifying management rather than switching entirely from one form of forestry to another, for example, considering how uneven-aged forestry and even-aged forestry can complement each other at landscape scales (Aszalos *et al.* 2022).

Retention forestry

Originally developed in North America, retention forestry focuses on maintaining structures that are important for biodiversity and ecological processes but that are typically lost during conventional logging. Retention forestry is applied in boreal and temperate forests managed using clear-cutting, although it is potentially relevant for other systems (Gustaffson *et al.* 2012). The structures retained encompass features that are rare and/or take time to develop, and may include pieces of deadwood, individual trees (often above a minimum diameter), or small patches of unharvested forest. These features can be aggregated or dispersed across the landscape, with a minimum of 5–10% of the total volume of the stand recommended (Gustaffson *et al.* 2012).

Retention forestry can buffer against more extreme changes in microclimate and allow old-growth forest species to persist following clear-cutting (Fedrowitz *et al.* 2014), with areas of intact forest also acting as stepping stones or corridors for dispersal between mature stands. These effects can result in greater species richness and abundance for some forest species in retention forest compared with clear-cuts. However, benefits for other taxa are less apparent: retention forestry does not mitigate impacts on all species adapted to forest interior conditions, whilst open-habitat species can be negatively affected compared with clear-cuts. As such, decisions to use retention forestry on the basis of benefits to biodiversity need to consider suitability given local conservation priorities. Positive effects of retention also increase with retained tree volume and therefore the success of the method depends on

Figure 30.3 A boreal clear-cut in Finland 9 years after harvesting. The tall pine trees are retention trees. Retention level varies in Finland from a few scattered trees to larger patches. (Photograph by Panu Halme.)

the retention level, which can vary widely (2–88% with a mean of 36% in the studies included in a review by Fedrowitz *et al.* 2014—see also Figure 30.3).

The identification of Woodland Key Habitats (WKHs) is related to retention forestry and is applied most prominently in northern Europe. WKHs are selected based on edaphic conditions, hydrology, or geomorphology, or following the identification of key structural features such as late successional forest, riparian forest, aggregations of logs, and forest springs. WKHs are either excluded from logging or logged at lower intensities (Timonen *et al.* 2012). Where WKH or similar patches are incorporated strategically into a reserve network, connectivity between habitats can be enhanced, although small and isolated WKH have more limited biodiversity benefits, particularly for taxa with short dispersal capabilities (Laita *et al.* 2010).

Reduced impact logging (RIL)

RIL is applied in selectively logged forests as an alternative to conventional selective logging and encompasses several practices intended primarily to improve the long-term sustainability of timber harvesting (Putz *et al.* 2008a). The practices employed vary but include one or more of the following: (1) a full inventory and mapping prior to logging, (2) planning and restrictions on skid trails and log decks to avoid unnecessary construction, and (3) pre-harvest liana cutting and directional felling to minimise damage to non-target vegetation and soil (Putz *et al.* 2008b).

By combining planned extraction routes with directional felling, residual damage can decrease by 50% relative to conventional logging (Putz *et al.* 2008a), and a smaller area of forest is affected by soil damage and compaction. Pre-harvest liana cutting further reduces damage to non-target trees and benefits regenerating forest by restricting liana densities post-logging (Alvira *et al.* 2004). When appropriately practised, RIL allows selectively logged forest to maintain a more similar forest structure to unlogged forest, with significantly less canopy damage (Asner *et al.* 2004). Similarly, full implementation of RIL approaches designed to reduce carbon emissions (RIL-C) could decrease the emissions associated with tropical selective logging by over 40%, potentially making an important contribution to reducing carbon emissions globally (Ellis *et al.* 2019).

The extent to which RIL reduces the ecological effects of logging depends on the practices employed, though in general RIL impacts on biodiversity are less than conventional selective logging (Bicknell *et al.* 2014; Burivalova *et al.* 2015). Underlying mechanisms for this difference probably include retaining old hollow trees as nesting habitat, and pre-felling planning to reduce the number of skid trails and logging roads (which limits access by poachers and involves less vegetation clearance). Similarly, practices that maintain a more intact canopy structure may reduce residual damage and fire risk, and in the longer term improved timber yields could provide greater economic and political incentives for forest retention over conversion.

Landscape-scale management

At larger spatial and temporal scales, logging can be implemented in many ways. This creates a spectrum of options for landscape scale management, from using the same approach to logging at the same time across an entire area through to using different logging systems in different parts of the landscape at different times. The most suitable strategy depends on the balance between objectives (e.g. timber production, biodiversity conservation, recreation, etc.). However, for objectives such as timber production there will be an optimum landscape-scale management that is most effective in maintaining biodiversity and other ecosystem functions for a given level of production (Mazziotta *et al.* 2023).

From a spatial perspective, the key planning decision is how to vary timber production intensity across the landscape. This can be framed around the land-sparing versus land-sharing continuum. Assuming that both approaches aim to produce the same amount of timber, land-sparing applies high-intensity production in one part of the forest (potentially including converting to timber plantation) and leaves the remaining forest intact to provide high environmental benefits, whereas land-sharing applies a lower intensity of management across the whole concession (e.g. longer logging cycles, fewer trees harvested). Triad or three-part systems have also been proposed—these divide the landscape into three zones (unlogged, low-intensity logging, and high-intensity logging/plantation) (Betts *et al.* 2021; Duflot *et al.* 2022). There is some evidence that land-sparing logging may have less impact on biodiversity and carbon than land-sharing logging (Edwards *et al.* 2014), but also that improved management plays a key role in determining outcomes (Runting *et al.* 2019). This is an area of active research as part of broader investigation into how effectively land-sparing, land-sharing, and triad approaches can meet multiple objectives (Betts *et al.* 2021).

In addition to logging intensity, at a landscape scale the timing of logging operations can influence the ecological effects. For example, if management results in all forest across the landscape being in a similar successional stage, this will be detrimental to species that require a different stage. Conversely, differences in the timing of logging can give a mosaic of patches in different stages of recovery from disturbance. Planning the timing and type of logging operations can therefore help maintain biodiversity by ensuring that there is spatiotemporal variation in disturbance levels including some relatively intact forest (Duflot *et al.* 2022).

Post-logging

Post-logging management generally seeks to (1) maintain future timber yields and/or (2) assist the recovery of biodiversity and ecosystem functions within logged forest ('ecological restoration'; Gann *et al.* 2019). The objectives overlap but are not necessarily identical. Silvicultural treatments designed to maintain timber yields can include removing vines and understorey vegetation, combined with planting seedlings or saplings of commercial species (enrichment planting). These treatments enhance forest regeneration and carbon sequestration relative to naturally regenerating forest (Philipson *et al.* 2020). Effects on biodiversity depend on the practices employed and vary amongst species. For example, removing vines following logging has negative effects on fruit-eating birds but positive effects on insectivorous birds. Similarly, post-logging interventions that prioritise rapid carbon accumulation or timber could be detrimental to biodiversity if these involve widespread planting of one or a few fast-growing tree species. The examples again highlight the trade-offs in forest management decisions, both within potential objectives (e.g. which aspects of biodiversity to prioritise?) and between management objectives (e.g. how important is rapid carbon sequestration if this is not optimal for biodiversity?) (Cerullo and Edwards 2019).

Ecological restoration can include the silvicultural techniques described above, but may also involve major ecosystem modifications such as reforesting cleared areas and using prescribed fires (Suding 2011; Halme *et al.* 2013). Equally, more subtle measures are sometimes employed, such as creating deadwood and small gaps, and blocking ditches (Halme *et al.* 2013). Ecological restoration can be designed to benefit target groups (e.g. species dependent on fires or on deadwood), though in this case will only be effective if these taxa can colonise the restored sites.

Conclusion

Logging encompasses a wide range of practices and so does not have a consistent set of effects. The picture is further complicated by the lack of long-term (i.e. decadal) data on the impacts of repeated logging, by inherent differences between ecosystems, and by variable impacts within broad logging systems. Nonetheless, some general patterns emerge. The impacts of clear-cutting on habitat structure, carbon stocks, and biodiversity are more severe than the impacts of selective logging, and in both cases forestry practices can reduce effects and/or facilitate recovery to some extent. There is also recognition of the potential conservation value of permanent production forest, with selective logging (in particular) less detrimental to biodiversity than other forms of anthropogenic disturbance (e.g. complete forest clearance for agriculture).

The 20th century saw the economic importance of forests rise as the area managed for production increased globally. In addition to the economic significance, there is now general recognition that forests are a key part of global ambitions on climate change mitigation and biodiversity conservation. These goals will only be met through a combination of protected and production forests, and appreciation that there will often be trade-offs between priorities.

Understanding how and why logging affects forest ecology is important (particularly longer-term changes post-logging), and will be aided by technologies such as Earth Observation, DNA-based monitoring, and bioacoustics (e.g. van Klink *et al.* 2022). This work will be most valuable if it tests and proposes solutions and is clearly communicated. However, despite existing knowledge of what practices can support biodiversity and ecosystem functions in logged forests, these practices are less widespread than conventional logging. Designing and implementing approaches that are ecologically effective therefore also needs to better address the socio-economic and political factors that will ultimately determine uptake. Lastly, it is important to stress that a range of other pressures act on forests, including climate change, fragmentation, and fire. These pressures all interact, and therefore need to be considered when interpreting the effects of logging and when designing future strategies to support biodiversity and ecosystem functions within logged forest.

Acknowledgements

We are very grateful to Zuzana Burivalova and Francis Putz for providing detailed and constructive comments in reviewing this chapter.

References

Alvira, D., Putz, F. E., Fredericksen, T. S. (2004) Liana loads and post-logging liana densities after liana cutting in a lowland forest in Bolivia. *Forest Ecology and Management*, vol. 190, pp. 73–86.

Asner, G. P., Keller, M., Pereira Jr, R., Zweede, J. C., Silva, J. N. M. (2004) Canopy damage and recovery after selective logging in Amazonia: field and satellite studies. *Ecological Applications*, vol 14, pp. S280–S298.

Aszalos, R., Thom, D., Aakala, T., Angelstam, P., Brumels, G., Galhidy, L. *et al.* (2022) Natural disturbance regimes as a guide for sustainable forest management in Europe. *Ecological Applications*, vol 32, e2596. https://doi.org/10.1002/eap.2596

Betts, M. G., Phalan, B. T., Wolf, C., Baker, S. C., Messier, C., Puettmann, K. J. *et al.* (2021) Producing wood at least cost to biodiversity: integrating Triad and sharing-sparing approaches to inform forest landscape management. *Biological Reviews*, vol 96, pp. 1301–1317.

Bicknell, J. E., Gaveau, D. L., Davies, Z. G., Struebig, M. J. (2015) Saving logged tropical forests: closing roads will bring immediate benefits. *Frontiers in Ecology and the Environment*, vol 13, pp. 73–74.

Bicknell, J. E., Struebig, M. J., Edwards, D. P., Davies, Z. G. (2014) Improved timber harvest techniques maintain biodiversity in tropical forests. *Current Biology*, vol 24, pp. R1119–1120.

Bousfield, C. G., Cerullo, G. R., Massam, M. R., Edwards, D. P. (2020) Protecting environmental and socio-economic values of selectively logged tropical forests in the Anthropocene. *Advances in Ecological Research*, vol 62, pp. 1–52.

Bowd, E. J., Banks, S. C., Strong, C. L., Lindenmayer, D. B. (2019) Long-term impacts of wildfire and logging on forest soils. *Nature Geoscience*, vol 12, pp. 113–118.

Brunet, J., Fritz, Ö, Richnau, G. (2010) Biodiversity in European beech forests—a review with recommendations for sustainable forest management. *Ecological Bulletins*, vol 10, pp. 1093–1101.

Burivalova, Z., Lee, T. M., Sekercioglu, C. H., Wilcove, D. S., Koh, L. P. (2015) Avian responses to selective logging shaped by species traits and logging practices. *Proceedings of the Royal Society B*, vol 282, 20150164. http://dx.doi.org/10.1098/rspb.2015.0164

Cambi, M., Certini, G., Neri, F. Marchi, E. (2015) The impact of heavy traffic on forest soils: a review. *Forest Ecology and Management*, vol 338, pp. 124–138.

Cerullo, G. R. and Edwards, D. P. (2019) Actively restoring resilience in selectively logged tropical forests. *Journal of Applied Ecology*, vol 56, pp. 107–118.

Chaudhary, A., Burivalova, Z., Koh, L., Hellweg, S. (2016) Impact of forest management on species richness: global meta-analysis and economic trade-offs. *Scientific Reports*, vol 6, 23954. https://doi.org/10.1038/srep23954

D'Amato, A. W., Palik, B. J., Franklin, J. F., Foster, D. R. (2017). Exploring the origins of ecological forestry in North America. *Journal of Forestry*, vol 115, pp. 126–127.

Demarais, S., Verschuyl, J. P., Roloff, G. J., Miller, D. A., Wigley, T. B. (2017) Tamm review: terrestrial vertebrate biodiversity and intensive forest management in the U.S. *Forest Ecology and Management*, vol 385, pp. 308–330.

Duflot, R., Fahrig, L., Monkkonen, M. (2022) Management diversity begets biodiversity in production forest landscapes. *Biological Conservation*, vol 268, 109514. https://doi.org/10.1016/j.biocon.2022.109514

Edwards, D. P., Gilroy, J. J., Woodcock, P., Edwards, F. A., Larsen, T. H., Andrews, D. J. R. *et al.* (2014) Land-sharing versus land-sparing logging: reconciling timber extraction with biodiversity conservation. *Global Change Biology*, vol 20, pp. 183–191.

Ellis, P. W, Gopalakrishna, T., Goodman, R. C., Putz, F. E., Roopsind, A., Umunay, P. M. (2019) Reduced-impact logging for climate change mitigation (RIL-C) can halve selective logging emissions from tropical forests. *Forest Ecology and Management*, vol 438, pp. 255–266.

Ewers, R., Boyle, M., Gleave, R., Plowman, N. S., Benedick, S., Bernard, H. (2015) Logging cuts the functional importance of invertebrates in tropical rainforest. *Nature Communications*, vol 6, 6836. https://doi.org/10.1038/ncomms7836

FAO. (2020) *Global Forest Resources Assessment—Main Report.* Rome, Italy.

Fedrowitz, K., Koricheva, J., Baker, S. C., Lindenmayer, D. B., Palik, B., Rosenvald, R. *et al.* (2014) Can retention forestry help conserve biodiversity? A meta-analysis. *Journal of Applied Ecology*, vol 51, pp. 1669–1679.

Franca, F., Barlow, J., Araujo, B., Louzada, J. (2016) Does selective logging stress tropical forest invertebrates? Using fat stores to examine sublethal responses in dung beetles. *Ecology and Evolution*, vol 6, pp. 8526–8533.

Gamborg, C. and Larsen, B. J. (2003) 'Back-to-nature'—a sustainable future for forestry. *Forest Ecology and Management*, vol 179, pp. 559–571

Gann, G. D., McDonald, T., Walder, B., Aronson, J., Nelson, C. R., Jonson, J. *et al.* (2019) International principles and standards for the practice of ecological restoration. Second edition. *Restoration Ecology*, vol 27, pp. S1–S46.

Gibson, L., Lee, T. M., Koh, L. P., Brooke, B. W., Gardner, T. A., Barlow, J. *et al.* (2011) Primary forests are irreplaceable for sustaining tropical biodiversity. *Nature*, vol 478, pp. 378–382.

Gustafsson, L., Baker, S. C., Baugus, J., Breese, W. J., Brodie, A., Kouki, J. *et al.* (2012) Retention forestry to maintain multifunctional forests: a world perspective. *BioScience*, vol 62, pp. 633–645.

Halme, P., Allen, K. A., Auninš, A., Bradshaw, R. H. W., Brumelis, G., Cada, W., *et al.* (2013) Challenges of ecological restoration: lessons from forests in Northern Europe. *Biological Conservation*, vol 167, pp. 248–256.

Hemprich-Bennett, D. R., Kemp, V. A., Blackman, J., Struebig, M. J., Lewis, O. T., Rossiter, S. J. (2021) Altered structure of bat-prey interaction networks in logged tropical forests revealed by metabarcoding. *Molecular Ecology*, vol 30, pp. 5844–5857.

Hessburg, P. F., Miller, C. L., Parks, S. A., Povak, N. A., Taylor, A. H., Higuera, P. E. *et al.* (2019) Climate, environment, and disturbance history govern resilience of western North American forests. *Frontiers in Ecology and Evolution*, vol 7. doi: 10.3389/fevo.2019.00239

Hlásny, T., König, L., Krokene, P., Lindner, M., Montagné-Huck, C., Müller, J. *et al.* (2021). Bark beetle outbreaks in Europe: state of knowledge and ways forward for management. *Current Forestry Reports*, vol 7, pp. 138–165.

James, J. and Harrison, R. (2016) The effect of harvest on forest soil carbon: A meta-analysis. *Forests*, vol 7, 308. https://doi.org/10.3390/f7120308

Kleinschroth, F., Laporte, N., Laurance, W. F., Goetz, S. J., Ghazoul, J. (2019) Road expansion and persistence in forests of the Congo Basin. *Nature Sustainability*, vol 2, pp. 628–634.

Kohout, P., Charvatova, M., Stursova, M., Masinova, T., Tomsovsky, M., Baldrian, P. (2018) Clearcutting alters decomposition processes and initiates complex restructuring of fungal communities in soil and tree roots. *ISME Journal*, vol 12, pp. 692–703.

Kuuluvainen, T., Angelstam, P., Frelich, L., Jõgiste, K., Koivula, M., Kubota, Y., *et al.* (2021) Natural disturbance-based forest management: moving beyond retention and continuous-cover forestry. *Frontiers in Forests and Global Change*, vol 4, 629020.

Laita, A., Monkkonen, M., Kotiaho, J. S. (2010) Woodland key habitats evaluated as part of a functional reserve network. *Biological Conservation*, vol 143, pp. 1212–1227.

Lindenmayer, D. B., Kooyman, R. M., Taylor, C., Ward, M., Watson, J. E. M. (2020) Recent Australian wildfires made worse by logging and associated forest management. *Nature Ecology and Evolution*, vol 4, pp. 898–900.

Magrach, A., Senior, R. A., Rogers, A., Nurdin, D., Benedick, S., Laurance, W. F. *et al.* (2016) Selective logging in tropical forests decreases the robustness of liana-tree interaction networks to the loss of host tree species. *Proceedings of the Royal Society B*, vol 283, 20153008. doi: 10.1098/rspb.2015.3008

Malhi, Y., Riutta, T., Wearn, O. R., Deere, N. J., Mitchell, S. L., Bernard, H. *et al.* (2022) Logged tropical forests have amplified and diverse ecosystem energetics. *Nature*, vol 612, pp. 707–713.

Mayer, M., Prescott, C. E., Abaker, W. E. A., Augusto, L., Cecillon, L., Ferreira, G. W. D. *et al.* (2020) Tamm Review: influence of forest management activities on soil organic carbon stocks: A knowledge synthesis. *Forest Ecology and Management*, vol 466, 118127. https://doi.org/10.1016/j.foreco.2020.118127

Mazziotta, A., Borges, P., Kangas, A., Halme, P., Eyvindson, K. (2023) Spatial trade-offs between ecological and economical sustainability in the boreal production forest. *Journal of Environmental Management*, vol 330, 117144.

Menge, J. H., Magdon, P., Wollauer, S., Ehbrecht, M. (2023) Impacts of forest management on stand and landscape-level microclimate heterogeneity of European beech forests. *Landscape Ecology*, vol 38, pp. 903–917.

Meyer, C. F. J., Struebig, M. J., Willig, M. R. (2016) Responses of tropical bats to habitat fragmentation, logging, and deforestation. In C. Voigt and T. Kingston (eds.), *Bats in the Anthropocene: Conservation of Bats in a Changing World*. Springer, Cham. https://doi.org/10.1007/978-3-319-25220-9_4

Mills, M. B., Malhi, Y., Ewers, R. M., Kho, L. K., Teh, Y. A., Both, S. *et al.* (2023) Tropical forests post-logging are a persistent net carbon source. *Proceedings of National Academy of Sciences USA*, vol 120. e2214462120. doi.org/10.1073/pnas.2214462120

Molinari, M. M., Peres, C. A., Edwards, D. P. (2019) Rapid recovery of thermal environment after selective logging in the Amazon. *Agricultural and Forest Meteorology*, vol 278, 107637.

Montejo-Kovacevich G., Marsh C. J., Smith S. H., Peres C. A., Edwards D. P. (2022) Riparian reserves protect butterfly communities in selectively logged tropical forest. *Journal of Applied Ecology*, vol 59, pp. 1524–1535.

Paillet, Y., Bergès, L., Hjältén, J., Odor, P., Avon, C., Bernhardt-Römermann, M. *et al.* (2010) Biodiversity differences between managed and unmanaged forests: meta-analysis of species richness in Europe. *Conservation Biology*, vol 24, pp. 101–112.

Peichl, M., Martinez-Garcia, E., Fransson, J. E. S., Wallerman, J., Laudon, H., Lundmark, T. *et al.* (2023) Landscape-scale variability of the carbon balance across managed boreal forests. *Global Change Biology*, vol 29, pp. 1119–1132.

Philipson, C. D., Cutler, M. E. J., Brodrick, P. G., Asner, G. P., Boyd, D. S., Costa, P. M. *et al.* (2020) Active restoration accelerates the carbon recovery of human-modified tropical forests. *Science*, vol 369, pp. 838–841.

Poulsen, J. R., Clark, C. J., Palmer, T. M. (2013) Ecological erosion of Afrotropical forest and potential consequences for tree recruitment and forest biomass. *Biological Conservation*, vol 163, pp. 122–130.

Putz, F. E., Baker, T., Griscom, B. W., Gopalakrishna, T., Roopsind, A., Umunay, P. M. *et al.* (2019) Intact forest in selective logging landscapes in the tropics. *Frontiers in Forests and Global Change*, vol 2. doi: 10.3389/ffgc.2019.00030

Putz, F. E., Romero, C., Sist, P., Schwartz, G., Thompson, I., Roopsind, A. *et al.* (2022) Sustained timber yield claims, considerations, and tradeoffs for selectively logged forests. *PNAS Nexus*, vol 1, 102. pgac102. https://doi.org/10.1093/pnasnexus/pgac102

Putz, F. E., Sist, P., Fredericksen, T., Dykstra, D. (2008b) Reduced-impact logging: challenges and opportunities. *Forest Ecology and Management*, vol 256, pp. 1427–1433.

Putz, F. E., Zuidema, P. A., Pinard, M. A., Boot, R. G. A, Sayer, J. A., Sheil, D. *et al.* (2008a) Improved tropical forest management for carbon retention. *PLoS Biology*, vol 6(7), e166. doi:10.1371/journal.pbio.0060166

Roberts, P., Hunt, C., Arroyo-Kalin, M., Evans, D., Boivin, N. (2017) The deep human prehistory of global tropical forests and its relevance for modern conservation. *Nature Plans*, vol 3, 17093. https://doi.org/10.1038/nplants.2017.93

Rodriguez-Ramos, J. C., Cale, J. A., Cahill Jr, J. F., Simard, S. W., Karst, J., Erbilgin, N. (2021) Changes in soil fungal community composition depend on functional group and forest disturbance type. *New Phytologist*, vol 229, pp. 1105–1117.

Runting, R. K., Ruslandi, Griscom, B. W., Struebig, M. J., Satar, M., Meijaard, E. *et al.* (2019) Larger gains from improved management over sparing-sharing for tropical forests. *Nature Sustainability*, vol 2, pp. 53–61.

Savilaakso, S., Johansson, A., Hakkila, M., Uusitalo, A., Sandgren, T., Monkkonen, M. (2021) What are the effects of even-aged and uneven-aged forest management on boreal forest biodiversity in Fennoscandia and European Russia? A systematic review. *Environmental Evidence*, vol 10, 1. https://doi.org/10.1186/s13750-020-00215-7

Schall, P., Gossner, M. M., Heinrichs, S., Fischer, M., Boch, S., Prati, D. (2018) The impact of even-aged and uneven-aged forest management on regional biodiversity of multiple taxa in European beech forests. *Journal of Applied Ecology*, vol 55, pp. 267–278.

Shvetsov, E. G., Kukavskaya, E. A., Shestakova, T. A, Laflamme, J., Rogers, B. M. (2021). Increasing fire and logging disturbances in Siberian boreal forests: a case study of the Angara region. *Environmental Research Letters*, vol 16, 115007. DOI 10.1088/1748-9326/ac2e37

Suding, K. N. (2011) Toward an era of restoration in ecology: successes, failures, and opportunities ahead. *Annual Review of Ecology, Evolution, and Systematics*, vol 42, pp. 465–487.

Timonen, J., Gustafsson, L., Kotiaho, J. S., Mönkkönen, M. (2012) Hotspots in cold climate: conservation value of woodland key habitats in boreal forests. *Biological Conservation*, vol 144, pp. 2061–2067.

Van Klink, R., August, T., Bas, Y., Bodesheim, P., Bonn, A., Fossoy, F. *et al.* (2022) Emerging technologies revolutionise insect ecology and monitoring. *Trends in Ecology and Evolution*, vol 37, 872–885.

Wallenius, T., Niskanen, L., Virtanen, T., Hottole, J., Brumelis, G., Angervuori, A. *et al.* (2010) Loss of habitats, naturalness and species diversity in Eurasian forest landscapes. *Ecological Indicators*, vol 10, pp. 1093–1101.

Yu, D., Zhou, L., Zhou, W., Ding, H., Wang, Q., Wang, Y., *et al.* (2011) Forest management in Northeast China: history, problems, and challenges. *Environmental Management*, vol 48, pp. 1122–1135.

31
POLLUTION IN FORESTS

Mikhail V. Kozlov and Elena L. Zvereva

Introduction

The influence of pollution on forests has long been recognised as a serious social and scientific problem (Fowler *et al.*, 2020). Current generations inherited large areas of forests disturbed or even destroyed by pollution, and in some countries the destruction continues to the present, making environmental pollution one of the five main drivers of global biodiversity loss. The extremely slow natural recovery of polluted forests justifies the need to develop preventive and rehabilitation measures. This need, in turn, requires knowledge of resistance, resilience, and tolerance of different types of ecosystems, of disturbance-induced changes in various ecosystems, and of factors affecting their post-disturbance recovery (Kozlov *et al.*, 2009).

The scope of pollution ecology has undergone major changes in the past 10–20 years. Industrial development, environmental legislation, and public attitude have changed its focal research topics from the exploration of acute local damages caused by sulphur dioxide (SO_2), fluorine, and trace elements emitted by large industrial enterprises to studies of the regional effects of ozone (O_3), nitrogen (N), and particulate matter (PM) deposition. The current task is to make pollution ecology an integral part of global change research by merging the gap between studies addressing effects of environmental pollution and studies addressing effects of climate change on ecosystem structure and functions (De Marco *et al.*, 2022).

Pollution, polluters, and pollutants

Any foreign substance, primarily waste from human activities, that renders the air, soil, water, or other natural resources harmful or unsuitable for a specific purpose can be classified as a pollutant. In this chapter, we consider aerial pollution (with the exception of CO_2), leaving out the impacts of contaminated water and solid waste on forests. The pollutants in the ambient air occur as gases (e.g., SO_2, NO_x, and O_3), as aerosol particles (mainly 0.1–3.0 μm in diameter) containing, for example, trace elements, and as cloud droplets (3–20 μm in diameter) containing the ions (e.g., SO_4^{2-}, Cl^-).

DOI: 10.4324/9781003324072-36

Historically, SO_2 was the very first pollutant that caused local but severe forest degradation near large cities and industrial enterprises, especially near non-ferrous smelters and power plants. Fluorine pollution (both gaseous HF and fluorine-containing dust) is primarily associated with the aluminium industry. In high concentrations, both SO_2 and fluorine cause yellowing and chlorosis of leaves and needles, followed by growth retardation and sometimes by the death of trees. Effective control strongly decreased global fluorine emissions by the 1980s and global emissions of SO_2 by the late 1990s.

Sulphur (S) and N compounds can act as either nutrients or stressors for forests. Increased N deposition, associated primarily with NO_x formation during fossil fuel combustion, began to play an important role in European and North American forests several decades ago. Nitrogen deposition does not usually kill trees, but drastically changes the structure and productivity of forests, decreases plant diversity, and predisposes forests to an increased risk of damage by other environmental stressors. Total N emissions stopped increasing in 2013; however, no large-scale response in forest growth or vitality to the reduction of N deposition was observed in Europe yet (De Marco *et al.*, 2022).

Acid rain emerges when SO_2 and NO_x emitted into the atmosphere react with water, oxygen, and other chemicals to form sulfuric and nitric acids. Their deposition causes visible leaf damage (chlorosis or necrosis) and alters plant physiology, hampering photosynthesis, and reducing tree growth. Additionally, soil acidification promotes the leaching of nutrients from upper soil horizons, making nutritional deficiency one of the key factors behind forest degradation in polluted regions (de Vries *et al.*, 2014).

Trace elements (e.g. Pb, Ni, Cu, and Zn) have been very common pollutants since ancient times; they are mostly emitted by smelting industries. They generally do not spread far away from the smelters, but they remain in soil for decades and even centuries and are often responsible for legacy effects, for example, on forest regeneration observed after emission decline. Alkaline dust originating from cement and magnesite factories, as well as from power plants, and organic (mostly hydrocarbon) pollutants emitted by chemical industries generally impose less harmful impacts on ecosystems than SO_2, fluorine, and trace elements. Studies on the effects of radioactive trace elements (^{137}Cs, ^{134}Cs, and ^{131}I) on forests were intensified after the Fukushima Daiichi nuclear power plant accident in 2011 (Yoschenko *et al.*, 2018).

Ozone is formed in sunlight from photochemical reactions of its precursors such as NO_x and volatile organic compounds. Its background concentrations in the Northern Hemisphere have doubled since pre-industrial times, making O_3 the major air pollutant with the potential to affect forest ecosystems worldwide. Ozone causes cellular damage in plants, reduces stomatal control, reduces CO_2 assimilation rates, and induces visible leaf injuries. These effects often accelerate senescence, diminish green leaf area and biomass, and reduce photosynthetic capacity. Hence, O_3 pollution has large impacts on plant functioning and, consequently, on productivity and services of forest ecosystems (Paoletti *et al.*, 2020).

The impacts of contaminants of emerging concern, such as plant hormones, plastic components, flame retardants, surfactants, fragrances, pesticides, etc., on forest ecosystems remain to be explored.

Extent and severity of impacts

Pollution affects the components of an ecosystem both directly and indirectly. Direct effects are caused by the toxicity of pollutants, whereas indirect effects are mediated by habitat loss, changes in microclimate, and alteration in ecosystem structure and species interactions, such as competition, herbivory, facilitation, predation, and mutualism. The effects of pollution can interact with the effects of other stressors, such as climate, pests, and pathogens (Kozlov and Zvereva, 2011; De Marco *et al.*, 2022), making identification of immediate driver(s) awkward.

The acute local effects of pollution on forests that occur near large industrial enterprises are described in detail. The most striking examples of these effects have long been associated with the Canadian smelters (e.g., Sudbury). Today, after the implementation of strong emission controls and rehabilitation programmes in Canada, the biggest industrial polluter is the Norilsk smelter in north-east Siberia. Its emissions have destroyed 24,000 km^2 of boreal forest since the 1960s (Kirdyanov *et al.*, 2020). The consequences of severe pollution impact on forest ecosystems are most comprehensively documented for the Monchegorsk copper–nickel smelter (Murmansk region of Russia) and the Middle Ural copper smelter in Revda (Sverdlovsk region of Russia) (Kozlov *et al.*, 2009; Vorobeichik, 2022).

The regional effects of pollution attracted considerable attention in the 1970s. Although we now understand that the causes of forest decline in those decades were not only air pollution and acid rain, it is evident that the main reason behind the damage to forest health and related ecosystem services in large areas was air pollution (Takahashi *et al.*, 2020). The problem was particularly apparent in central Europe, especially in the 'Black Triangle', an area along the German–Czech–Polish border (Kozlov *et al.*, 2000). In North America, contaminated air masses from the Los Angeles Basin damaged and even killed sensitive ponderosa pines (*Pinus ponderosa*) and Jeffrey pines (*Pinus jeffreyi*) in the San Bernardino Mountains (Bytnerowicz *et al.*, 2008). The recent decline in temperate and subalpine/subboreal forests in Japan and China is tentatively attributed to increased exposure to O_3 (Takahashi *et al.*, 2020), although unequivocal proof of causality is still lacking.

Globally, 8% of forests received in 1985 an S load at which the forests are at risk on acid-sensitive soils; and it is estimated that 17% of the forests will receive this load by 2050. Similarly, by 1990 24% of global forest was exposed to O_3 concentrations >60 ppb (i.e., much higher than the threshold value of 40 ppb, above which direct adverse effects on plants may occur), and this proportion is expected to increase to 50% by 2100 (Fowler *et al.*, 1999). However, predictions on the impact of O_3 on forest vegetation vary greatly between scenarios and regions (Sicard *et al.*, 2017). The increase in forest exposure to pollution will obviously have global consequences for the C, N, S, and P cycles, net primary productivity (NPP), and other characteristics of forest ecosystems.

Effects of pollution on forests

Landscape-level effects

Acid rain and PM deposition can have drastic consequences for forests. Damaged and destroyed forests change the visual perception of landscapes, making them depressive. Extreme deposition of airborne pollutants can transform forests into industrial barrens (Figure 31.1a)—bleak open landscapes, with only small patches of vegetation surrounded by bare land. These landscapes first appeared more than a century ago in North America. In the early 2000s, about 40 industrial barrens had existed worldwide, and most of them were located in boreal and temperate forests next to non-ferrous smelters. The extent of industrial barrens varied from a few hectares (e.g. Harjavalta in Finland) to several hundreds of square kilometres (Norilsk and Monchegorsk in Russia). Despite the overall impression of

(a)

(b)

Figure 31.1 (a) Industrial barren located 4 km south of the nickel–copper smelter in Monchegorsk, north-west Russia. This barren area evolved from dense Norway spruce forest. Note the dead trees, the absence of field layer vegetation, and extensive soil erosion. (b) Modifications of the crown structure of Norway spruce (foreground) and Scots pine (background) in this barren site. Note the dense branches near the ground: they are protected by snow during the winter. (Photographs by V. Zverev.)

'dead land' and the general reduction in biodiversity, industrial barrens still support a variety of life, including regionally rare and endangered species, as well as populations that have evolved specific adaptations to the harsh and toxic environment (Kozlov and Zvereva, 2007). Following emission control, industrial barrens started to recover, but many of them (e.g., in Sudbury and Monchegorsk) still remain extensive.

Ecosystem-level effects

The decrease in NPP is one of the most important ecosystem-level responses to different kinds of disturbance. However, information on the effects of regional pollution on NPP is scarce. The analysis of field data did not reveal the effects of O_3 on forest NPP in Italy (De Marco *et al.*, 2013), although a meta-analysis of experimental data showed that O_3 decreased net tree photosynthesis by 11% and tree biomass production by 7%. However, these estimates are based largely on studies of individual young trees growing in a non-competitive environment, and extrapolation of these results may not be appropriate for predicting the response of mature trees and forests to O_3 (Ainsworth *et al.*, 2012). This uncertainty questions the results of modelling studies reporting, for example, a 14% loss of NPP in China due to O_3 exposure (Yue *et al.*, 2017).

The pollution-induced decrease in NPP affects the nutrient cycle of forest ecosystems (see Chapter 23). Lower density and height of a stand, as well as reduced cover of field layer vegetation, result in a smaller litter input and, consequently, in smaller forest floor thickness and topsoil depth. At the same time, pollution adversely affects soil microbiota and detritophagous arthropods (Zvereva and Kozlov, 2010), thus significantly reducing the litter decomposition rate and leading to an accumulation of non-decomposed plant material, especially near non-ferrous smelters (Kozlov *et al.*, 2009).

The reductions in biomass production and litter decomposition affect the carbon (C) and nutrient budget of forest ecosystems, although the direction of the changes depends on both pollutants and the affected ecosystems. Deposition of SO_2, N, and O_3 and PM can reduce the C storage capacity of forests (Matyssek *et al.*, 2012; Manninen *et al.*, 2015). However, in nutrient-poor soils, N deposition may stimulate forest growth and therefore cause a net C sink (Chapter 37).

Community-level effects

Gordon and Gorham (1963) describe the deterioration of polluted forests as 'peeling off' of the vegetation layers. Trees decline first, followed by tall shrubs, and finally, under the severest pollution, short shrubs and herbs are affected. The generality of this 'downward' pattern of ecosystem destruction is confirmed by a decrease in the magnitude of the effects of pollution on plant abundance, from trees to shrubs and then to dwarf shrubs and herbs (Zvereva and Kozlov, 2012). However, in some circumstances, dwarf shrubs and herbs decline earlier than top canopy trees (Kozlov *et al.*, 2009). The disappearance of some functional groups of plants from polluted forests contributed to the development of the hypothesis on simplification of the ecosystem structure under the impact of pollution.

However, the effects of pollution on the diversity of different groups of forest biota are not uniform (Figure 31.2). While the diversity of soil microfungi, bryophytes, vascular plants, soil/ground living arthropods, and birds generally decreases with industrial pollution, the species richness of insect herbivores often increases (see Kozlov *et al.*, 2022). Nitrogen deposition

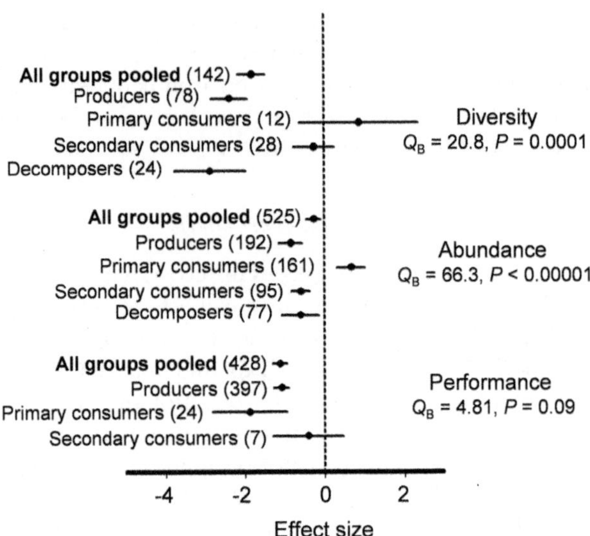

Figure 31.2 A meta-analysis on the effects of industrial pollution on different characteristics and different trophic levels of biota. Producers—vascular plants and bryophytes; primary consumers—herbivorous insects; secondary consumers—predatory and parasitic arthropods; decomposers—soil microfungi and some arthropods. Mean effect sizes (dots), 95 per cent confidence intervals (horizontal lines), and sample sizes (in parentheses). The effect size is the difference between the mean values of the parameter at the polluted and control sites, divided by the pooled standard deviation and adjusted for the sample sizes. Thus, a negative effect size indicates that the parameter under study has lower values in the polluted site(s) than in the control site(s). The effect is significant if the 95% confidence interval does not overlap the zero value. Q_B values indicate the difference among the trophic levels (Kozlov and Zvereva, 2011).

reduces the diversity of forest vegetation (Bobbink *et al.*, 2010), while the presumed adverse impacts of O_3 on biodiversity are mainly projected from fumigation experiments, with limited supporting evidence from polluted forests (Agathokleous *et al.*, 2020).

Differential, and sometimes opposite, pollution effects on organisms that belong to different trophic levels (Figure 31.2) can seriously alter ecosystem structure and function. Pollution generally imposes larger detrimental effects on predatory than on herbivorous arthropods, creating enemy-free space for herbivores, and thus contributing to the increase in their abundance (Zvereva and Kozlov, 2010). Non-trophic interactions can also be affected: the impact of top canopy plants on dwarf shrubs shifted from negative (i.e., competition) in unpolluted forests to generally positive (i.e., facilitation) in heavily polluted barren habitats (Zvereva and Kozlov, 2004). Regrettably, studies on the impacts of pollution on biotic interactions remain scarce.

Population-level effects

Pollution may affect the genetic structure of populations, often leading to the development of tolerance to pollution. These processes are best documented for the microbiota, and have been reported more frequently for plants than for animals. Metal tolerance of plants, especially grasses, provides a well-documented example of rapid evolutionary adaptation. Among long-lived trees, pollution tolerance has been demonstrated for birches and willows. Microevolutionary processes in plant populations could also occur under experimental O_3 exposure, but field evidence is still lacking. Among animals, genetic adaptations to Cu, Cd, and Zn have been described in earthworms, terrestrial isopods, springtails, and insects.

The adverse effects of industrial pollutants on the abundance of organisms generally become stronger with time from the beginning of pollution, whereas the effects on individual fitness become weaker (Figure 31.3). The latter result can be seen as evidence of the development of pollution resistance in the affected populations, while the decline in abundance can be either the mechanism behind this process (survival selection) or the consequence of adaptation, which may involve some costs. Because detected patterns are uniform across the investigated taxa, evolutionary adaptation to the impact of industrial pollution is likely to be a general phenomenon (Kozlov and Zvereva, 2011). However, on a regional scale, 86% of butterfly and moth species in Switzerland were negatively related to N deposition (Roth *et al.*, 2021), indicating that most species did not evolve sufficient tolerance to pollution.

The population densities of insects feeding on birch leaves demonstrated variable, and sometimes opposite, responses to the industrial pollution gradient near the Monchegorsk copper–nickel smelter (Kozlov *et al.*, 2017). Response direction and strength are related to herbivore life history traits: the abundances of oligophagous and polyphagous moth and butterfly species that fed externally on plants and hibernated as larvae generally declined near the smelter, whereas the abundances of monophagous Lepidoptera species that fed inside live plant tissues and hibernated as imagos or pupae were unaffected by pollution (Kozlov *et al.*, 2022).

Pollution effects on a suite of parameters driving population dynamics, such as death rate, life expectancy, survival, and fecundity, are properly documented only for a handful of forest

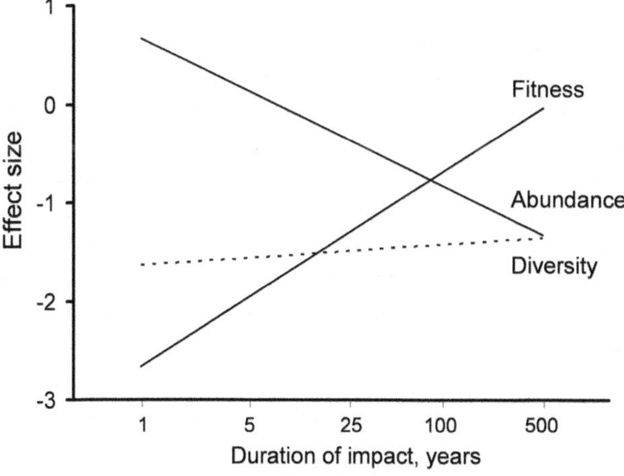

Figure 31.3 Relationships between the duration of pollution impacts and the magnitudes of the pollution effects on individual fitness, abundance, and diversity of organisms. The analysis is based on the data on the following taxa: non-mycorrhizal soil microfungi, vascular plants, bryophytes, and arthropods. The dotted line indicates the non-significant regression model. For an explanation of the effect size, see Figure 31.2 (Kozlov and Zvereva, 2011).

species (Kozlov *et al.*, 2009). These effects are best exemplified in a willow-feeding leaf beetle, *Chrysomela lapponica*, the density of which near a copper–nickel smelter showed a rapid 20-fold decline in the early 2000s, remaining at very low levels thereafter. Time-series analysis and model selection explained this pattern by increases in insect mortality from natural enemies as a consequence of climate warming and pollution decline. Consistent with these results, long-term monitoring of an eruptive leafmining moth, *Phyllonorycter strigulatella*, near a coal-fired power plant revealed a decrease in the strength of rapid (stabilising) density dependence, but an increase in the strength of delayed (destabilising) density dependence with decreasing distance from the polluter. This exciting finding suggests that insect herbivore outbreaks in polluted forests can result from a weakening of rapid density dependence due to predation decline relative to delayed density dependence acting, for example, via changes in host quality (Kozlov, 2022).

Organism-level effects

A great number of publications reported visible plant injury attributable to pollution, such as necrosis, chlorosis, and deformation of leaves. However, the variety of symptoms, subjectivity in estimating damage, and the absence of statistical analysis, especially in older publications, made the interpretation of these data difficult. This is equally true for visual estimates of crown defoliation and discolouration used in forest monitoring programmes. On the contrary, a decrease in needle life span in conifers is a reliable index of the severity of pollution impact (Kozlov *et al.*, 2009).

Pollution frequently modifies the crown structure of trees. Nitrogen and S depositions significantly increase the odds of top damage. At the external borders of industrial barrens, Norway spruces, *Picea abies*, often have vital lower twigs creeping over the ground, whereas the mid-crown is completely dead; the uppermost branches, if alive, have very short needle longevity (Figure 31.1b). Mountain birches, *Betula pubescens* var. *pumila*, are tall, slender trees in unpolluted forests but have a shrubby growth habit in industrial barrens. Highly branching and compact crowns can minimise the impacts of unfavourable environmental conditions on trees and are therefore considered adaptive features (Zverev *et al.*, 2013). At the same time, dense crowns facilitate leaf-tying moth larvae and therefore increase herbivory on shrubby birches relative to slender trees (Zvereva *et al.*, 2014).

The growth and reproduction of vascular plants near industrial polluters are reduced by about 10%, while the growth and reproduction of bryophytes are reduced by approximately 50% relative to unpolluted forests (Kozlov and Zvereva, 2011). The decrease in radial growth of trees is often used for the retrospective analysis of impacts of pollution on forests (Kirdyanov *et al.*, 2020). The adverse effects of O_3 on tree growth, measured in fumigation experiments, are about the same magnitude as the effects of SO_2 and fluorine (Wittig *et al.*, 2009), while field evidence for impacts of ambient O_3 on tree growth ranges from weakly negative (de Vries *et al.*, 2014) to positive (Fenn *et al.*, 2020). Similarly, N depositions can enhance forest growth on low-fertility soils.

Animals living in polluted areas may have a smaller size and a lower fertility. For example, the mean body size (or weight) of terrestrial arthropods decreased significantly near industrial polluters (Zvereva and Kozlov, 2010); decreased clutch sizes have been reported for some bird species breeding in polluted environments (Eeva and Lehikoinen, 2013). Experiments demonstrated that realistic O_3 concentrations (80 ppb and greater) increase mortality and decrease motility of tiny insects. At the same time, responses of herbivorous insects to plants exposed to elevated N or O_3 rank from strongly positive to strongly negative (Kozlov, 2022).

Fluctuating asymmetry, the increase of which presumably reflects an inability of an organism to cope with environmental or genomic stresses during its development, was long ago suggested as a handy index of pollution-induced stress in plants and animals (Graham *et al.*, 1993). However, the rapid accumulation of disconfirming evidence questions the validity of this index and requires rectification of the scope and identification of applicability limits of the underlying theory (Kozlov *et al.*, 2009).

Causality of responses to pollution

Pollution ecology suffers from a generally poor understanding of cause-and-effect relationships. The 'presumption of guilt' is commonly accepted, and changes observed around industrial polluters are often attributed to the toxicity of pollutants without proper justification. On a regional scale, it is even more difficult to identify the immediate cause(s) of forest damage than near industrial polluters, not only because of the generally weaker effects of pollution on forests, but also because other stressors may cause effects that are similar to the effects of pollution. For example, forest damage observed in Finnish Lapland in the late 1980s, which was assumed to develop due to transboundary transfer of emissions from industrial enterprises in north-west Russia, after a thorough examination was attributed to the combination of nutrient deficiency caused by the exceptional weather conditions of winter 1986–1987 with the epidemics of scleroderris canker (Tikkanen and Niemelä, 1995).

The causal links behind the changes observed in polluted forests are far from evident due to both natural variation in the study systems and strong indirect impacts, which may enhance primary disturbance in a positive feedback fashion. This is best exemplified by the restored consequence of events that resulted in the development of industrial barrens near the copper–nickel smelter in Monchegorsk, northwest Russia. Near this smelter, pollution-induced forest thinning and dieback resulted in higher wind speed that increased climatic stress both directly and by changes in snowpack structure, and imposed mechanical damage on plants. A thin and compact snow layer resulted in lower wintertime soil temperatures in industrial barrens compared to unpolluted forests. These changes in microclimate in combination with a pollution-induced decrease in cold hardiness of conifers increased the probability of tree mortality from freezing injury. Forest decline reduced the recovery of topsoil and facilitated soil erosion, with adverse consequences for all components of biota (Kozlov *et al.*, 2009). All these effects are apparent only on the landscape scale and cannot be reproduced experimentally.

The causal relationships between pollution and the changes observed in forest ecosystems could be proved using the criteria of causation (Hill, 1965). These criteria include the strength and consistency of observed association, its specificity and temporality (a logical time sequence of events), existence of the relationship between the dose (the putative cause) and the response, coherence with known biological facts, plausibility, experimental support, and analogy with similar, better known, situations. The more criteria are satisfied, the more confidence we have in our judgement that the association is causal. Nevertheless, controlled experiments remain the main tool for proving causality.

Variation in biotic responses to pollution

The first studies on the effects of pollution on terrestrial biota were driven by economic losses in agriculture and forestry. As a result, the most polluted areas, where forests have been severely damaged or even killed by industrial emissions, have attracted the greatest attention from scientists, who carefully described lifeless industrial barrens around several non-ferrous smelters. The similarity between the first documented examples of extreme forest deterioration created the misleading impression that all types of pollution cause uniform changes in different types of ecosystem.

Further accumulation of observational data, which was especially intensive in the 1980s, demonstrated that biotic responses to pollution are far from being uniform. It soon became clear that the experiences from one system are rarely directly applicable to predict the fate of another system because each impacted area developed in its own way due to a unique history of events. The need to reduce uncertainties in predicting the consequences of the pollution impact called for the quantitative exploration of factors influencing the responses of organisms, communities, and ecosystems to pollution. Accomplishing this task became possible with the development of meta-analysis, which allowed statistical exploration of an unlimited amount of diverse information.

Several meta-analyses (Wittig *et al.*, 2009; Zvereva and Kozlov, 2010, 2012; Li *et al.*, 2017) consistently demonstrated the diversity in responses of terrestrial ecosystems to industrial pollution. The magnitude and even the direction of these responses depend on the characteristics of the polluter (type, amount of emissions, and duration of the impact), the affected organisms (trophic level and life history), the measured character (individual fitness, abundance, and diversity), and the environment (biome and climate). In particular, acidifying pollutants generally cause stronger effects on biota than alkaline pollutants; the effects of non-ferrous smelters emitting both SO_2 and trace elements are generally more detrimental than the effects of polluters emitting only SO_2.

Although the same pollution loads can cause different effects in different ecosystems and/or regions (Figure 31.4), little is known about the properties of the ecosystem that determine its sensitivity to pollution. Differential sensitivities to pollutants are formalised through the concept of critical loads. The critical load for a pollutant defines a deposition level below which even the most sensitive components of ecosystems are not affected. A critical load approach has been widely and successfully used to predict the effects of acid precipitation on ecosystems in Europe (Posch, 2002); it has shaped air pollution control policies since the 1980s.

Another approach to quantifying the sensitivity of ecosystems to disturbances is based on theoretical concepts linking sensitivity with ecosystem structure and functional organisation. In particular, relationships between species diversity and ecosystem stability, which have been and still are widely debated, may have direct implications for understanding how communities respond to pollution and other anthropogenic stressors. However, data collected from polluted ecosystems do not indicate that more diverse ecosystems are more resilient to pollution impacts. The effects of industrial pollution on terrestrial biota were absent in the tundra, lowest in boreal and temperate forests, and highest in desert scrub (Figure 31.4), and this pattern appeared to be related to climate rather than to the diversity of impacted communities (Kozlov and Zvereva, 2011).

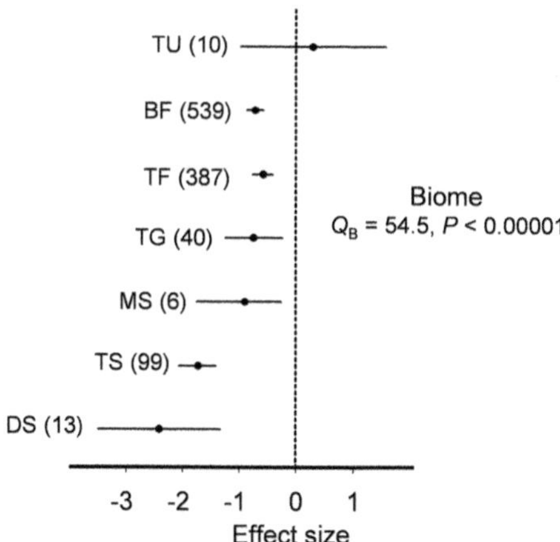

Figure 31.4 A meta-analysis on the effects of industrial pollution on organisms in different biomes. The meta-analysis is based on data on the following taxa: non-mycorrhizal soil microfungi, vascular plants, bryophytes, and arthropods; BF—boreal forest or taiga; DS—desert scrub; MS—Mediterranean scrub; TF—temperate broadleaf deciduous forest; TG—temperate grassland; TS—tropical savannah; TU—tundra. For further explanations, see Figure 31.2 (Kozlov and Zvereva, 2011).

Pollution and climate

Although environmental pollution is an integral part of global change, research addressing the biotic effects of climate change generally does not consider this issue. Similarly, most studies on both the distribution of pollutants and the biotic effects of pollution have neglected the issue of climate change. As a result, experimental studies exploring the combined effects of air pollution and climate change remain uncommon (De Marco *et al.*, 2022).

Stimulation of forest growth by elevating atmospheric CO_2 concentration will increase nutrient demand and, in turn, will result in greater nutrient limitation. At the same time, climate warming will improve N availability for mid- to high-latitude forests through increased biological N fixation and N mineralisation. These changes in nutrient status will likely interact with the effects of S and N deposition, and therefore should be considered simultaneously when projecting future forest dynamics (De Marco *et al.*, 2022).

Contrary to a widespread opinion on the fragility of Arctic ecosystems, these ecosystems appeared to be less sensitive to industrial pollution than low-latitude ecosystems (Figure 31.4). This phenomenon can be explained by several reasons. First, lower temperatures can result in lower stomata opening, decreasing the uptake of gaseous pollutants by plants. Second, the shorter vegetation period decreases plant exposure to pollutants. Third, high-latitude soils generally contain fewer nutrients than low-latitude soils, and, therefore, the adverse effects of toxic pollutants may be counterbalanced to some extent by the fertilising effect of N deposition. Finally, higher temperatures and increased precipitation may enhance mobility and increase the toxicity of pollutants, contributing to the generally higher adverse effects seen in a warmer climate.

The comparison between polluters located at different latitudes suggests that climate warming will modify responses of functional groups of terrestrial biota to pollution. Increases in temperature will enhance the adverse effects of pollution on producers and decomposers,

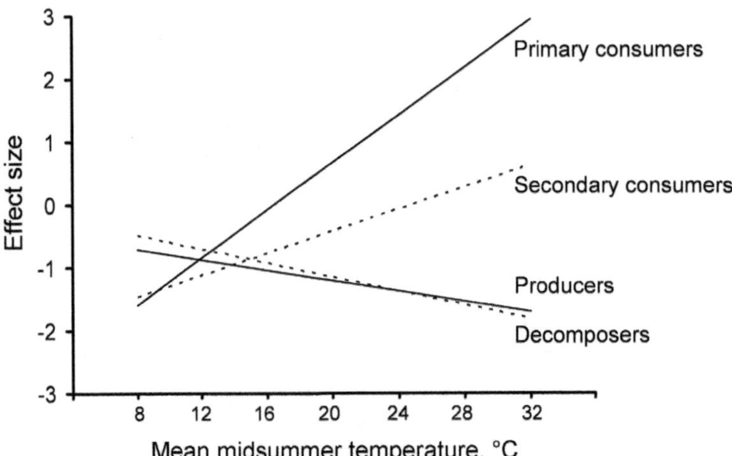

Figure 31.5 Relationships between mean midsummer temperatures in the polluter's locality and the magnitudes of the observed effects of pollution on organisms from four trophic levels. Producers—vascular plants and bryophytes; primary consumers—herbivorous insects; secondary consumers—predatory and parasitic arthropods; decomposers—soil microfungi and some arthropods. Dotted lines indicate non-significant regression models. For an explanation of the effect size, see Figure 31.2 (modified from Kozlov and Zvereva, 2011).

but will mitigate the adverse effects on primary and secondary consumers (Figure 31.5). Thus, warming is likely to increase the harmful impacts of pollution on forests through distortion of ecosystem functioning (Kozlov and Zvereva, 2011).

Forest restoration in polluted habitats

Decommissioning of the polluter or decreasing pollutant deposition due to implementation of the emission control creates prerequisites for natural recovery of forests that have been severely damaged or killed by pollution. This recovery sometimes begins as early as 5–15 years after the emission decline. However, in many situations, the beginning of natural recovery is delayed for several decades due to legacy effects such as slow leaching of toxic elements from contaminated soils (centuries after the cease of emissions), and shortage of nutrients in the thin derelict soil layer (Jonard *et al.*, 2012; Vorobeichik, 2022).

The slow rate or even the absence of natural forest regeneration has prompted many rehabilitation projects. The land reclamation programme in Sudbury, Canada, is the largest and best-known example of a successfully completed restoration of forest ecosystems (Levasseur *et al.*, 2022). The initial motivation for this programme was to improve Sudbury's image as a treeless wasteland, but it gradually turned into the need to restore the landscapes and ecosystems that were destroyed by pollution.

Due to the legacy effects mentioned above, the methods used to restore forests in heavily polluted sites (Kozlov *et al.*, 2000) differ in many details from the restoration measures applied in unpolluted sites (see Chapter 28). Forest restoration around Sudbury started with liming of barren land and seeding of grass, followed by planting tree seedlings, mainly species that were previously common at a site. The greatest success was achieved at heavily polluted sites that received complex, multistage revegetation treatments. Regreening has led to individual trees growing and accumulating nutrients at rates similar to jack and red pine plantations grown in areas not impacted by mining and smelting practices (Levasseur *et al.*, 2022). Twenty-five years after the beginning of restoration, the richness of native understory vascular species at these sites recovered to about the same extent as at naturally recovering, mildly disturbed sites (Rayfield *et al.*, 2005). Thus, it is possible to convert heavily contaminated industrial barrens into forests over a period of about two decades, assuming that there is the social demand for such a conversion and sufficient financial support is provided.

However, restoration of vegetation does not always imply restoration of the full range of fauna and, consequently, of ecosystem structure and functions. In particular, the diversity of the arthropod community lags behind the diversity of the restored plant community (Babin-Fenske and Anand, 2010). Similarly, birds and mammals associated with mature forests can only recolonise the formerly barren areas after the recovery of their habitats. As a result, newly restored sites have sometimes been called 'depauperate imitations', as they superficially resemble undisturbed sites in some ways, but do not have the same diversity or composition of flora and fauna. Therefore, measures that assist natural recovery should have priority over the placement of agricultural soils or soil substitutes rich in nutrients on polluted soils, as had been done at some sites near the copper–nickel smelter in Monchegorsk, Russia (Korotkov *et al.*, 2019).

Future research directions

The future of global forests is a subject of public and political concern due to extensive forest degradation worldwide (Fowler *et al.*, 2020). Addressing this concern requires exploration of the forest ability to cope with the interacting impacts of chronic air pollution and climate change on large (regional to global) spatial scales and the development of management strategies to counteract these impacts (de Vries *et al.*, 2014). The advancement of the understanding of pollution impacts on forest ecosystems requires (1) extension of the network of monitoring sites, especially in the South Hemisphere; (2) further integration of ground and satellite monitoring; (3) generation of flux-based standards and critical levels taking into account the sensitivity of dominant forest tree species; (4) long-term monitoring of N, S, P cycles and base cations deposition together at the global scale; (5) intensification of experimental studies, addressing the combined effects of different abiotic factors on forests ecosystems and on a wide array of tree species; (6) more experimental focus on phenomics and genomics; and (7) development of models integrating air pollution and climate change data from long-term monitoring programmes (De Marco *et al.*, 2022). Especially urgent is the exploration of the impacts of pollution on tropical forests. Among research topics, the most critical knowledge gaps are associated with pollution impacts on belowground processes, spatial structure and demography of populations, development of evolutionary adaptations to combined impacts of pollution and climate warming, functional diversity of communities, and biotic interactions, as well as with links between life history/functional traits of individual species and their tolerance to pollution.

Conclusion

The adverse consequences of the impacts of pollution on forest ecosystems have been under careful investigation for more than a century. Previous studies started from a documentation of forest decline around individual 'super polluters', used experiments to invoke the mechanisms of this decline, and developed rehabilitation measures. Dead forests and lifeless 'moonscapes' near large industrial enterprises for a long time served as textbook examples of the negative effects of human activity on the biota. Now the battle against the 'super polluters' is won in a larger part of the world. However, pollution changed its guise, and forests in many regions remain critically polluted, although the consequences of this pollution are rarely apparent.

The word 'pollution', which for decades was used mainly for SO_2, fluorine, and trace elements, now denotes primarily O_3 and NO_x. Large polluters, whose individual footprints were clearly visible, are replaced by dozens of small polluters. Instead of sharp pollution gradients, which were easy to study because of their limited length and strong correlation between multiple components of aerial emissions, we face patchy and uncorrelated large-scale distribution of different pollutants, which sometimes (but not always) follow shallow regional and global gradients. Consequently, current impacts of regional pollution are rarely documented because even the highest regional levels of O_3, N, and PM usually do not induce detectable ecosystem- or landscape-level changes, such as retardation of tree growth or forest decline. In addition, large-scale variation in environmental factors other than pollutants and their interactions with pollutants greatly complicate the attribution of changes in forest health to pollution.

Thus, the shift in research focus of pollution ecology has led to the change in research methodology. Most importantly, the assessment of forest health around individual polluters is now replaced by the modelling of forest health at regional and global scales based on the distribution of pollutants. However, the precision of these models remains questionable because biotic responses are modelled based on the outcomes of short-term, small-scale exposures of oversimplified study systems to a single pollutant. Only a handful of studies, such as the Aspen-FACE experiment in northern Wisconsin, USA, explored the long-term effects of pollution on the growth of model forest communities in their natural environment. The creation of a network of at least 40–50 experimental sites around the globe, in which natural communities will be exposed to elevated CO_2, N, and O_3 (alone and in combinations), is the only way to substantially improve the quality of the model-based predictions regarding the structure of future forests and services they will provide to humanity.

References

Agathokleous, E., Feng, Z., Oksanen, E., Sicard, P., Wang, Q. and others (2020) 'Ozone affects plant, insect, and soil microbial communities: A threat to terrestrial ecosystems and biodiversity', *Science Advances*, vol 6, eabc1176.

Ainsworth, E. A., Yendrek, C. R., Sitch, S., Collins, W. J. and Emberson, L. D. (2012) 'The effects of tropospheric ozone on net primary productivity and implications for climate change', *Annual Review of Plant Biology*, vol 63, pp. 637–661.

Babin-Fenske, J. and Anand, M. (2010) 'Terrestrial insect communities and the restoration of an industrially perturbed landscape: Assessing success and surrogacy', *Restoration Ecology*, vol 18, suppl 1, pp. 73–84.

Bobbink, R., Hicks, K., Galloway, J., Spranger, T., Alkemade, R. and others (2010) 'Global assessment of nitrogen deposition effects on terrestrial plant diversity: A synthesis', *Ecological Applications*, vol 20, pp. 30–59.

Bytnerowicz, A., Arbaugh, M., Schilling, S., Frączek, W. and Alexander, D. (2008) 'Ozone distribution and phytotoxic potential in mixed conifer forests of the San Bernardino Mountains, southern California', *Environmental Pollution*, vol 155, pp. 398–408.

De Marco, A., Screpanti, A., Attorre, F., Proietti, C. and Vitale, M. (2013) 'Assessing ozone and nitrogen impact on net primary productivity with a generalised non-linear model', *Environmental Pollution*, vol 172, pp. 250–263.

De Marco, A., Sicard, P., Feng, Z., Agathokleous, E., Alonso, R. and others (2022) 'Strategic roadmap to assess forest vulnerability under air pollution and climate change', *Global Change Biology*, vol 28, pp. 5062–5085.

de Vries, W., Dobbertin, M. H., Solberg, S., van Dobben, H. F. and Schaub, M. (2014) 'Impacts of acid deposition, ozone exposure and weather conditions on forest ecosystems in Europe: an overview', *Plant and Soil*, vol 380, pp. 1–45.

Eeva, T. and Lehikoinen, E. (2013) 'Density effects on great tit (*Parus major*) clutch size intensifies in a polluted environment', *Oecologia*, vol 173, pp. 1661–1668.

Fenn, M. E., Preisler, H. K., Fried, J. S., Bytnerowicz, A., Schilling, S. L. and others (2020) 'Evaluating the effects of nitrogen and sulfur deposition and ozone on tree growth and mortality in California using a spatially comprehensive forest inventory', *Forest Ecology and Management*, vol 465, 118084.

Fowler, D., Brimblecombe, P., Burrows, J., Heal, M. R., Grennfelt, P. and others (2020) 'A chronology of global air quality'. *Philosophical Transactions of the Royal Society A*, vol 378, 20190314.

Fowler, D., Cape, J. N., Coyle, M., Flechard, C., Kuylenstierna, J. and others (1999) 'The global exposure of forests to air pollutants', *Water, Air, and Soil Pollution*, vol 116, pp. 5–32.

Gordon, A. G. and Gorham, E. (1963) 'Ecological aspects of air pollution from an iron-sintering plant at Wawa, Ontario', *Canadian Journal of Botany*, vol 41, pp. 1063–1078.

Graham, J. H., Freeman, D. C. and Emlen, J. M. (1993) 'Developmental stability – a sensitive indicator of populations under stress', *Environmental Toxicology and Risk Assessment*, vol 1179, pp. 136–158.

Hill, A. B. (1965) 'Environment and disease: Association or causation', *Proceedings of the Royal Society of Medicine*, vol 58, pp. 295–300.

Jonard, M., Legout, A., Nicolas, M., Dambrine, E., Nys, C. and others (2012) 'Deterioration of Norway spruce vitality despite a sharp decline in acid deposition: A long-term integrated perspective', *Global Change Biology*, vol 18, pp. 711–725.

Kirdyanov, A. V., Krusic, P. J., Shishov, V. V., Vaganov, E. A., Fertikov, A. I. and others (2020) 'Ecological and conceptual consequences of Arctic pollution', *Ecology Letters*, vol 23, pp. 1827–1837.

Korotkov, V. N., Koptsik, G., Smirnova, I. and Koptsik, S. (2019) 'Restoration of vegetation on mine lands near Monchegorsk (Murmansk region, Russia)', *Russian Journal of Ecosystem Ecology*, vol 4, pp. 1–18 (In Russian).

Kozlov, M. V. (2022) 'Population dynamics of herbivorous insects in polluted landscapes', *Current Opinion in Insect Science*, vol 53, 100987.

Kozlov, M. V., Castagneyrol, B., Zverev, V. and Zvereva, E. L. (2022) 'Recovery of moth and butterfly (Lepidoptera) communities in a polluted region following emission decline', *The Science of the Total Environment*, vol 838, 155800.

Kozlov, M. V., Haukioja, E., Niemelä, P., Zvereva, E. and Kytö, M. (2000) 'Revitalization and restoration of boreal and temperate forests damaged by aerial pollution', in J. L. Innes and J. Oleksyn, eds, *Forest Dynamics in Heavily Polluted Regions*, CAB International, Wallingford, UK, pp. 193–218

Kozlov, M. V. and Zvereva, E. L. (2007) 'Industrial barrens: extreme habitats created by non-ferrous metallurgy', *Reviews in Environmental Science and Biotechnology*, vol 6, pp. 231–259.

Kozlov, M. V. and Zvereva, E. L. (2011) 'A second life for old data: Global patterns in pollution ecology revealed from published observational studies', *Environmental Pollution*, vol 159, pp. 1067–1075.

Kozlov, M. V., Zvereva, E. L. and Zverev, V. E. (2009) *Impacts of Point Polluters on Terrestrial Biota: Comparative Analysis of 18 Contaminated Areas*. Springer, Dordrecht, Netherlands.

Kozlov, M. V., Zverev, V. and Zvereva, E. L. (2017) 'Combined effects of environmental disturbance and climate warming on insect herbivory in mountain birch in subarctic forests: Results of 26-year monitoring', *The Science of the Total Environment*, vol 601–602, pp. 802–811.

Levasseur, P. A., Galarza, J. and Watmough, S. A. (2022) 'One of the world's largest regreening programs promotes healthy tree growth and nutrient accumulation up to 40-years post restoration', *Forest Ecology and Management*, vol 207, 120014.

Li, P., Feng, Z. Z., Catalayud, V., Yuan, X. Y., Xu, Y. S. and Paoletti, E. (2017) 'A meta-analysis on growth, physiological, and biochemical responses of woody species to ground-level ozone highlights the role of plant functional types', *Plant, Cell and Environment*, vol 40, pp. 2369–2380.

Manninen, S., Zverev, V., Bergman, I. and Kozlov, M. V. (2015) 'Consequences of long-term severe industrial pollution for aboveground carbon and nitrogen pools in northern taiga forests at local and regional scales', *The Science of the Total Environment*, vol 536, pp. 616–624.

Matyssek, R., Wieser, G., Calfapietra, C., de Vries, W., Dizengremel, P. and others (2012) 'Forests under climate change and air pollution: Gaps in understanding and future directions for research', *Environmental Pollution*, vol 160, pp. 57–65.

Paoletti, E., Feng, Z., De Marco, A., Hoshika, Y., Harmens, H. and others (2020) 'Challenges, gaps and opportunities in investigating the interactions of ozone pollution and plant ecosystems', *The Science of the Total Environment*, vol 709, 136188.

Posch, M. (2002) 'Impacts of climate change on critical loads and exceedances in Europe', *Environmental Science and Policy*, vol 5, pp. 307–317.

Rayfield, B., Anand, M. and Laurence, S. (2005) 'Assessing simple versus complex restoration strategies for industrially disturbed forests', *Restoration Ecology*, vol 13, pp. 639–650.

Roth, T., Kohli, L., Rihm, B., Meier, R. and Amrhein, V. (2021) 'Negative effects of nitrogen deposition on Swiss butterflies', *Conservation Biology*, vol 35, 1766–1776.

Sicard, P., Anav, A., De Marco, A. and Paoletti, E. (2017) 'Projected global ground-level ozone impacts on vegetation under different emission and climate scenarios', *Atmospheric Chemistry and Physics*, vol 17, pp. 12177–12196.

Takahashi, M., Feng, Z., Mikhailova, T. A., Kalugina, O. V., Shergina, O. V. and others (2020) 'Air pollution monitoring and tree and forest decline in East Asia: A review', *The Science of the Total Environment*, vol 742, 140288.

Tikkanen, E. and Niemelä, I. (eds.) (1995) *Kola Peninsula Pollutants and Forest Ecosystems in Lapland.* Finnish Forest Research Institute, Rovaniemi, Finland.

Vorobeichik, E. L. (2022) 'Natural recovery of terrestrial ecosystems after the cessation of industrial pollution: 1. A state-of-the-art review', *Russian Journal of Ecology*, vol 53, pp. 1–39.

Wittig, V. E., Ainsworth, E. A., Naidu, S. L., Karnosky, D. F. and Long, S. P. (2009) 'Quantifying the impact of current and future tropospheric ozone on tree biomass, growth, physiology and biochemistry: A quantitative meta-analysis', *Global Change Biology*, vol 15, pp. 396–424.

Yoschenko, V., Ohkubo, T. and Kashparov, V. (2018) 'Radioactive contaminated forests in Fukushima and Chernobyl', *Journal of Forest Research*, vol 23, pp. 3–14.

Yue, X., Unger, N., Harper, K., Xia, X., Liao, H. and others (2017) 'Ozone and haze pollution weakens net primary productivity in China', *Atmospheric Chemistry and Physics*, vol 17, pp. 6073–6089.

Zverev, V. E., Kozlov, M. V. and Zvereva, E. L. (2013) 'Changes in crown architecture as a strategy of mountain birch for survival in habitats disturbed by pollution', *The Science of the Total Environment*, vol 444, pp. 212–223.

Zvereva, E. L. and Kozlov, M. V. (2004) 'Facilitative effects of top-canopy plants on four dwarf shrub species in habitats severely disturbed by pollution', *Journal of Ecology*, vol 92, pp. 288–296.

Zvereva, E. L. and Kozlov, M. V. (2010) 'Responses of terrestrial arthropods to air pollution: A metaanalysis', *Environmental Science and Pollution Research*, vol 17, pp. 297–311.

Zvereva, E. L. and Kozlov, M. V. (2012) 'Changes in the abundance of vascular plants under the impact of industrial air pollution: a meta-analysis', *Water, Air and Soil Pollution*, vol 223, pp. 2589–2599.

Zvereva, E. L., Roitto, M. and Kozlov, M. V. (2010) 'Growth and reproduction of vascular plants under pollution impact: A synthesis of existing knowledge', *Environmental Reviews*, vol 18, 355–367.

Zvereva, E. L., Toivonen, E. and Kozlov, M. V. (2008) 'Changes in species richness of vascular plants under the impact of air pollution: A global perspective', *Global Ecology and Biogeography*, vol 17, pp. 305–319.

Zvereva, E. L., Zverev, V. and Kozlov, M. V. (2014) 'High densities of leaf-tiers in open habitats are explained by host plant architecture', *Ecological Entomology*, vol 39, pp. 470–479.

32
BIOLOGICAL INVASIONS IN FORESTS AND FOREST PLANTATIONS

Marcel Rejmánek

Introduction

With ever-increasing world trade, biological invasions have become a global phenomenon (Figure 32.1;, Liebhold et al. 2017; Loehle et al. 2023; Mormul et al. 2022). There are two most frequently used definitions of invasive species, also used for other taxa (Simberloff and Rejmánek 2011):

- Non-native species that produce reproductive offspring, often in very large numbers, and spreading over large areas. This definition, usually used by ecologists, does not have any impact connotation.
- Non-native species that spread rapidly, causing environmental or economic damage. This definition (equivalent to "non-native pest species") is often used by managers, particularly in the United States (US).

Obviously, the first definition implies that there is no need to eradicate or control all "invasive" species. The second definition deals with a subset of species encompassed by the first one. Such species, causing environmental or economic damage, will be the subject of this chapter. Among the best-known examples are Africanized bees spreading in the Neotropics, kudzu (*Pueraria montana* var. *lobata*) in the southeastern US, opuntias (Cactaceae) in the Old world, pines in the Southern Hemisphere, Gypsy moth (*Lymantria dispar*) in North America, Red turpentine beetles (*Dendroctonus valens*) in China, feral pigs in Hawaii and many other islands, brushtail possum in New Zealand, beavers in Tierra del Fuego, and brown tree snakes in Guam. Some invasions—e.g., Eurasian earthworms in North America or Argentine ants in southern Africa—are not so conspicuous, but may have profound ecosystem effects.

It is usually expected that unless logged, burnt, or disturbed by storms or floods, forests are, in general, resistant to biological invasions, i.e., penetrations of non-native taxa. Indeed, most natural forests host fewer nonindigenous organisms than do surrounding human-modified habitats. Nevertheless, alien plants and animals have invaded even some old-growth undisturbed forests (e.g., Fridley et al. 2023; Roubik 1988).

DOI: 10.4324/9781003324072-37

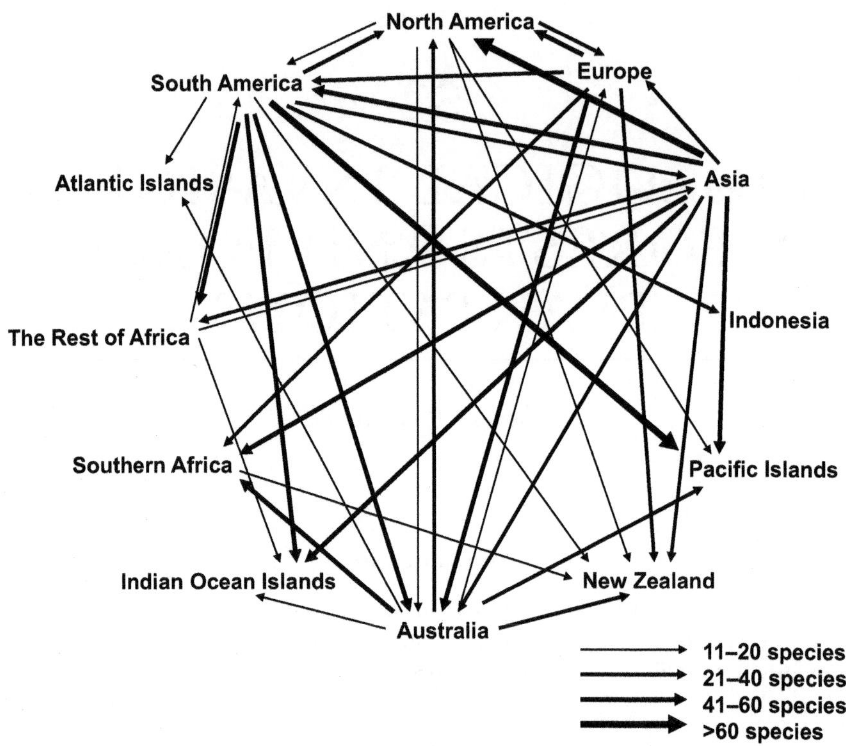

Figure 32.1 The global donor–acceptor network of invasive alien trees and shrubs [derived from Tables 3 and 4 in Rejmanek (2014)]. Arrow thickness corresponds to numbers of species. "South America" also includes Caribbean islands and Central America. Some regions play a major role as donors (e.g., Asia and South America), whereas others are mostly acceptors (e.g., Southern Africa, New Zealand, and Pacific Islands).

This chapter sketches a global overview of biological invasions in forests and forest plantations. Special attention is paid to invasions of alien plants, nematodes, earthworms, snails, slugs, insects, reptiles, birds, and mammals. For each group of organisms, lists of representative invaders are provided, and their impacts in boreal, temperate, and tropical forests are discussed.

Plants

For any non-native species to invade, there must be suitable climatic conditions and resources. Essential resources for plants are light, water, and nutrients. Light and nutrients are more likely available in disturbed ecosystems, and this is the reason why relatively few plant invasions are known from closed, undisturbed forests. Some examples of such species are provided in Table 32.1. More examples may be found in Dawson et al. (2009), Foxcroft et al. (2013), Jayakumar and Nair (2022), Kohli et al. (2009), Martin et al. (2008), Rejmánek (1966), Rejmánek and Richardson (2013), Wagner et al. (2017), Weber (2017), Woodward and Quinn (2011), Polard et al. (2021), and Fridley (2023). Besides straight competition for resources with resident species, it is very often a layer of slow-decomposing

Table 32.1 Some prominent examples of non-native plant species invading forests and forest plantations

Latin name (family)	*Common name*	*Life form*	*Native range*	*Invaded range*	*Habitats and impacts*
Acacia spp. (Fabaceae, Mimosoideae)	Acacia, Wattle	Trees, with nitrogen- fixing rhizobia and ant or bird dispersed seeds	Australia (Africa, Asia)	All temperate and tropical continents, many islands	Open forests Displace native plant species, enrich soil by nitrogen
Acer negundo (Sapindaceae)	Box elder, Ash-leaf maple, Manitoba maple	Dioecious shade-tolerant, fast growing tree	North & Central America	Europe, Australia (New South Wales)	Riparian forests Displaces native plant species
Acer platanoides (Sapindaceae)	Norway maple	Deciduous shade-tolerant wind-dispersed tree	Europe & West Asia	Eastern North America & British Isles	Broad-leaved forests Displaces native plant species
Ageratina adenophora (Asteraceae)	Crofton weed, Eupatory, Snakeroot	Perennial herbaceous shrub	Central America	Asia, Australia, South Africa, many islands	Forests and open areas Forms monospecific understory
Ailanthus altissima (Simaroubaceae)	Tree of heaven	Fast growing, clonal, dioecious tree	China, North Korea	All continents except Antarctica	Open forests, wide range of climatic and soil conditions Forms monospecific stands
Alliaria petiolata (Brassicaceae)	Garlic mustard	Shade-tolerant biennial, not palatable to herbivores	Eurasia	North America	Produces allelochemicals supressing mycorrhizal fungi of native plants
Amorpha fruticosa (Fabaceae)	False indigo-bush	Perennial leguminous nitrogen-fixing shrub	North America, Mexico	Europe, Asia	Riparian habitats Extensive dense stands
Archontophoenix cunninghamiana (Arecaceae)	Bangalow palm, King palm	Shade-tolerant, bird-dispersed palm	Australia	South America	Atlantic rainforest in Brazil Forms compact carpets of seedlings and saplings
Berberis thunbergii (Berberidaceae)	Japanese barberry, Thunberg's barberry	Thorny fleshy-fruiting shrub, veget. reproduction	Japan, East Asia	North America	Forest gaps, floodplains Forms dense thickets

(*Continued*)

Table 32.1 (Continued)

Latin name (family)	*Common name*	*Life form*	*Native range*	*Invaded range*	*Habitats and impacts*
Chromolaena odorata (Asteraceae) (syn.: *Eupatorium odoratum*)	Siam weed, Christmas bush, Devil weed	Evergreen wind-dispersed shrub, climber	North & Central America	Africa, Asia, Australia, Indonesia, many islands	Forest margins and gaps Dense thickets prevent establishment of native species
Cytisus scoparius (Fabaceae)	Scotch broom	Perennial leguminous nitrogen-fixing shrub	Western Europe	North & South America, Australia	Open forests and forest plantations Outcompetes tree seedlings
Fallopia japonica (Polygonaceae)	Japanese knotweed	Tall herbaceous perennial	East Asia	Europe, North America, New Zealand	forests and open wet habitats Outcompetes >80% of native plants
Hedera helix s.l. (Araliaceae)	Ivy, Common ivy, English ivy	Shade tolerant bird-dispersed root climber	Europe & Asia	Australia, USA, New Zealand	Forests and forest plantations Outcompetes native vegetation
Hovenia dulcis (Rhamnaceae)	Japanese raisin tree	Deciduous tree, vertebrate-dispersed	East Asia	South America, East Africa	Subtropical and tropical forests Competes with native plants
Imperata cylindrica (Poaceae)	Cogongrass	Perennial rhizomatous grass	South Asia, Africa, Australia	South & Central America, Southeastern USA	Open forests, disturbed areas Highly flammable pyrophyte
Lantana camara (Verbenaceae)	Lantana, Big sage, Wild sage, Tickberry	Fleshy-fruiting, spiny, shrub	Central & South America	Africa, Asia, Australia, many tropical islands	Open areas, forests and plantations Outcompetes all seedlings
Leucaena leucocephala (Fabaceae)	Jumbay, White popinac	Evergreen shrub or tree, nitrogen-fixing	Central America	All continents except Antarctica	Coastal and riparian habitats Forms dense thickets, preventing regeneration of native trees

Lygodium japonicum (Lygodiaceae)	Japanese climbing fern	Fern with creeping stem and long leaves	East Asia	Southeastern USA	Swamp forests Forms dense mats
Miconia crenata (Melastomataceae) (syn.: *Clidemia hirta*)	Soap bush, Koster's curse	Densely-branching shade-tolerant bird-dispersed shrub	Central & South America	Asia, many islands	Forest edges and gaps Competes with native plants
Microstegium vimineum (Poaceae)	Japanese stiltgrass, Nepalese browntop	Shade-tolerant C4 grass	Asia	Southeastern USA	Broad-leaved forests Displaces native plants
Mikania micrantha (Asteraceae)	Mile-a-minute, Bitter vine	Herbaceous creeping climber, stems >5 m	South & Central America	Asia, Australia	Forest edges, riparian habitats Thick mats prevent any regeneration
Morella (*Myrica*) *faya* (Myricaceae)	Fire tree	Evergreen nitrogen fixing shrub or small tree	Macaronesia	Hawaii	*Metrosideros polymorpha* forests Displaces native plants
Oeceoclades maculate (Orchidaceae)	Monk orchid, African spotted orchid	Terrestrial orchid, self-pollinated	Africa, Madagascar	The Neotropics and Florida	Disturbed and old-growth forests Impact not detected
Pinus radiata (Pinaceae)	Monterey pine, Radiata pine	Fast growing, medium-density softwood	California	Australia, South Africa, Chile	Native eucalyptus orests, fynbos Competes with native plants
Prunus serotina (Rosaceae)	Wild black cherry, Rum cherry	Moderately shade-tolerant, bird-dispersed tree	Eastern North America	Western & Central Europe	Disturbed and undisturbed forests Competes with native plants
Psidium cattleianum (Myrtaceae)	Strawberry guava, Cattley guava	Small fleshy-fruiting, vertebrate-dispersed tree	Tropical South America	Many tropical islands	Disturbed and undisturbed forests Forms dense monocultures
Robinia pseudoacacia (Fabaceae)	Black locust	Tree, 10–30 m tall, strongly suckering, nitrogen-fixing	Eastern USA	All continents except Antarctica	Disturbed forests, riparian habitats Dense stands covering large areas

(*Continued*)

Table 32.1 (Continued)

Latin name (family)	*Common name*	*Life form*	*Native range*	*Invaded range*	*Habitats and impacts*
Rubus armeniuacus (Rosaceae)	Himalayan blackberry (*R. discolor* misapplied)	Fleshy-fruiting shrub, vertebrate-dispersed	Armenia, Iran	North America, Europe	Disturbed and riparian forests Outcompetes native plants
Stenotaphrum secundum (Poaceae)	St. Augustine grass	Perennial, C4, stoloniferous grass	Africa & Americas(?)	Australia	Swamp forests Local extinction of native plants
Syzigium jambos (Myrtaceae)	Rose apple	Fleshy-fruiting tree	Southeast Asia	Central America, Africa, many tropical islands	Forests (all successional stages) Decline in plant species diversity
Triadica sebifera (Euphorbiaceae)	Chinese tallow	Shade-tolerant, fast-growing tree	East Asia, Japan	SE US, India, Pacific Islands	Successional forests Blocking vegetation succession

litter on the ground that prevents establishment of seedlings in exotic-dominated forests. While plants invading disturbed areas are usually fast-growing, light-demanding species that rapidly produce many small wind-dispersed seeds (Rejmánek et al. 2013), plants invading closed forests have very different attributes. They are shade-tolerant, and often vertebrate- (mostly bird-) dispersed. Extended leaf phenology (earlier leaf emergence and/or later leaf senescence) is another attribute that can promote invasiveness of some non-native woody species (Rejmánek 2013a). There are also some unexpected combinations of attributes. For example, Chinese tallow (*Triadica sebifera*) is a fast-growing species with a very short juvenile period, but shade-tolerant. For such species, a model based on whole-plant carbon economics was recently proposed by Fridley et al. (2023). However, even typical disturbed-area invaders can be important in forest dynamics. After any disturbance, some of them can block vegetation succession for a long time, and they can compete with native pioneer species in forest gaps (Dyderski and Jagodzinski 2020; Langmaier and Lapin 2020).

From the healthy forest regeneration point of view, the most important invaders are clonal species, which form dense compact stands, often with high litter biomass (e.g., *Microstegium vimineum, Lantana camara, Fallopia* spp., *Rubus* spp., *Stenotaphrum secundum, Ailanthus altissima,* and Figure 32.2). The second most important group of plant invaders is the climbers (e.g., *Pueraria montana, Hedera* spp., *Lygodium* spp., *Celastrus orbiculatus, Lonicera japonica, Cryptostegia grandiflora,* and *Passiflora* spp.). They can form dense blankets preventing any forest regeneration after disturbances. Fire-promoting grasses (e.g., cogongrass *Imperata cylindrica*, jaragua grass *Hyparrhenia rufa*, molasses grass *Melinis minutiflora*, and Guinea grass *Panicum maximum*) are particularly important in the wet

Figure 32.2 *Ehrharta erecta,* a highly invasive perennial grass native to South Africa, forms a dense cover in the Douglas fir (*Pseudotsuga menziesii*) forests in the Point Reys National Seashore, California. While 15–23 vascular plant species grow in non-invaded 100 m^2 plots, only 4–9 species are present in neighbouring continuously invaded plots of the same size.

tropics where forest trees are not adapted to fires. Invasive woody species associated with nitrogen fixing microorganisms (legumes, *Morella faya*) can completely change successional trajectories (Gaertner et al. 2014). Finally, hybridization with native congeners may cause permanent genetic changes in native populations (e.g., species in the genera *Eucalyptus, Lantana, Platanus, Populus, Rosa, Rubus,* and *Ulmus*).

Some impacts of invasive plants are less obvious (Wohlgemuth et al. 2022). Here are three examples. Invasive *Rhamnus cathartica* changes the vegetation structure of the forest, affecting nest predation of birds. *Berberis thunbergii* provides questing sites for the black-legged ticks that carry the spirochete *Borrelia burgdorferi*, the causative agent of Lyme disease. Dense exotic *Lantana camara* understory in Australian *Eucalyptus saligna* forests seems to be the most critical in the development of forest dieback caused by psyllids (*Glycaspis* spp.). *Lantana* provides cover for the bell miner (*Manorina melanophrys*), a highly territorial bird that drives off other bird species that predate on psyllids. Impacts of invasive plants on forest biodiversity are context dependent and often only careful observations and long-term studies in permanent plots may bring conclusive results (Brown et al. 2006; Norghauer et al. 2011). Also, environmentally safe experiments should be encouraged (Green et al. 2004; Rejmánek 2013b).

Tropical and boreal forests seem to be less invaded by exotic plants than are temperate forests (Rejmánek 1996; Sanderson et al. 2012). However, the degree to which this is a result of differences in propagule pressure is difficult to judge. The quantity of shade and fast postdisturbance recovery of wet tropical forests may be the major reasons that these forests are relatively resistant to plant invasions. Here, the role of different propagule pressure also remains an open question (Chong et al. 2021). Riparian forests are usually more invaded than other types of forests. Frequent disturbances and high input of propagules of invasive plant species are the two underlying causes.

The vast majority of plants invading forests or forest plantations around the world are the result of intentional introductions. Exotic tree plantations, nurseries, gardens and erosion control projects and occasionally also botanical gardens are major local sources of invasive plant species (Reichard 2011; Richardson 2011; Ni and Hulme 2021). Surprisingly, many species, while recognized as influential invaders, are still widely planted. As with many other major groups of invasive organisms, forest fragmentation, roads, and movement of propagules in soil attached to vehicles and forest industry machinery are the major reasons that forests are invaded by non-native taxa. Besides prevention, early detection of harmful invaders is the key to successful eradication. In general, eradication of exotic pest plant infestations in more than 1,000 ha is very unlikely, given a realistic amount of resources (Rejmánek and Pitcairn, 2002). In such situations, besides containment, local eradications, and native seed addition (Moore et al. 2023), species-specific biological control may be an option. A global catalogue of agents for biocontrol of weedy species is being periodically updated (Winston et al. 2014).

Nematodes, earthworms, snails, and slugs

Nematodes (phylum Nematoda) are wormlike in appearance but quite distinct taxonomically from true worms. Plant parasitic nematodes (>1,100 species) are small, less than 2.5 mm in length. They are known to survive periods of desiccation, an ability to survive unintentional transport between countries (Aaders et al. 2017). Because of their size, invasiveness of nematodes is poorly studied and their impacts are not well known (Singh et al. 2013). The most important invasive nematode in forests is the pine wilt or pine wood nematode

(*Bursaphelenchus xylophilus*; Mota and Vieira 2008; De la Fuente and Beck 2019). This species infects pine trees and causes pine wilt disease. It affects, with different severity, more than 30 pine species and several other conifers. It is probably native to North America and is invasive in Japan, eastern Asia, and Portugal. This and related species are vectored by a number of bark beetles and wood borers, most often the cerambycid longhorn beetles in the genus *Monochamus*. Both *Monochamus* and *Bursaphelenchus* may be found in pine chips, unseasoned lumber, and logs. Consequently, they are easily transported. For this reason, the European Plant Protection Organization has banned import of softwood products except kiln-dried lumber from North America.

Semiaquatic and terrestrial earthworms are usually classified into 17 families that belong to the subclass Oligochaeta. Over 3,000 species of earthworms have been described, which may be just about a half of all existing species. Since Darwin's classic *The Formation of Vegetable Mould through the Action of Worms*, published in 1881, there has been increasing appreciation of ecosystem consequences of earthworm activities. So far, about 120 species of earthworms can be classified as invasive species. Representative examples of earthworm species invading forests and forest plantations are presented in Table 32.2. Additional examples can

Table 32.2 Some prominent examples of non-native earthworm species invading forests and forest plantations

Taxa	*Life form*	*Native range*	*Invaded range*
Lumbricideae			
Aporrectodea rosea and *Aporrectodeatuberculata*	Parthenogenetic, endogeic	Western Palearctic[a]	All other realms
Dendrobaena octaedra	Parthenogenetic, epigeic	Western Palearctic	All other realms except Ethiopian
Lumbricus rubellus	Sexually reproducing, epi-endogeic	Western Palearctic	All other realms
Lumbricus terrestris	Cross-fertilization hermaphrodite, aneic[e]	Western Palearctic	All other realms
Octolasion tyrtaeum	Parthenogenetic, endogeic	Palearctic	Nearctic, Neotropical
Megascolecidae			
Amynthas spp., *Metaphire* spp.,	Parthenogenetic, epigeic	Eastern Palearctic[b]	All other realms
Polypheretima spp.			
Glossoscolecidae			
Pontoscolex corethrurus	Sexual and parthenogenetic, endogeic	Neotropical[c]	All other realms
Eudrilidae			
Eudrilus eugeniae	Sexually reproducing, epigeic	Ethiopian[d]	Neotropical

Note: [a]Europe, western Asia, and northern Africa. [b]Eastern Asia. [c]Central and South America. [d]Subsaharan Africa. [e]Litter-feeding.

be found in Chang et al. (2021), Edwards (2004), Hendrix (2006, 2008), James (2011), and Reynolds (2012). The success of invasive earthworms depends to a large extent on a resistant stage (diapause), ability to use diverse food resources, human-mediated transport, and parthenogenesis in many species. Because earthworms are small and their egg capsules are even smaller, they travel with soil and on the roots of living plants without our knowledge. The horticulture industry, topsoil movement by landscapers and other machinery, and use of earthworms as fish bait are major contributors to their long-distance dispersal.

Impacts of earthworm invasions are most pronounced in forests north of the southern limit of the last glaciation in North America and Europe (Frelich et al. 2019; Jang et al. 2022). Such areas have no native earthworm populations due to slow natural range expansions. In the absence of earthworms, a thick, slowly decomposing layer of leaf litter develops on the surface of mineral soil. When earthworms are introduced, they consume the litter and mix the mineral soil with the finely divided organic matter of their faecal material. Reduction of the forest-floor litter layer has many consequences. Nutrient mineralization is accelerated and soil C:N ratio is greatly reduced. Abundance of soil organisms depending on the litter dominated by fungi is probably reduced as well. In some cases, the impacted soil is less cohesive, and herbs and tree seedlings may be uprooted by deer feeding, rather than just deprived of upper parts. This may lead to reductions of forest-floor herb layers and influx of some invasive plant species. In northern hardwood forests, invasions of exotic earthworms may be facilitated by the production of high-quality leaf litter produced by exotic woody invaders like *Rhamnus cathartica* (buckthorn) or *Lonicera* spp. (honeysuckle). Removal of such species may result in reductions of exotic earthworm abundance by 50%. Earthworm burrowing can disrupt mycorrhizae of native species and promote invasive plant species that are either nonobligate or amycorrhizal (*Alliaria petiolata, Berberis thunbergii,* and *Microstegium vimineum*). Invasive plants are conspicuous and are often assumed to be the drivers of environmental changes; however, much of the change may occur belowground, and invasive plants are the beneficiaries rather than drivers of such changes (Nuzzo et al. 2009). With increased human activity and global warming, invasions by earthworms are likely to increase in northern forests in the future (Chang et al. 2021). First projects on biocontrol of invasive earthworms are under way (Nouri-Aiin and Görres 2021).

Snails (gastropods with a coiled shell) and slugs (shell-less gastropods) are also important invasive invertebrates, particularly in forests on islands. For example, about 100 species of land snails are naturalized on the islands of the Pacific (Cowie 2005). Among the best-known examples are *Achatina fulica* (giant African snail), *Arion vulgaris* (Spanish slug, often misidentified as *Arion lusithanicus*; Zemanova et al. 2018), and *Euglandina rosea* (rosy wolfsnail or cannibal snail). *A. fulica* is a macrophytophagous herbivore introduced to Hawaii from Japan. By the 1950s it had become so abundant that 15 predatory snails were introduced as biological control agents; a small number of them became established, the most notable of which is *E. rosea* (Cowie 2005). There is no reliable evidence that *E. rosea* has controlled populations of *A. fulica*, but *E. rosea* was very likely responsible for the decline and possible extinction of Hawaiian tree snails and possibly of other endemic snails. The Hawaiian Islands have no native slugs, but at least 12 introduced slug species are now established. They can reduce seedling survival of endangered plant species by 50%, without significant effects on the survival of invasive plant species (Joe and Daehler 2008). Similarly, the European slug *Deroceras reticulatum* (the grey field slug) significantly damages native herbaceous plants in North American forests, while the invasive *Alliaria petiolata* is avoided (Hahn et al. 2011).

Insects

As far as we know, insects are, besides vascular plants, the most numerous group of invasive organisms. In the continental United States alone, more than 450 exotic herbivorous insect species have colonized forests and urban trees since the European settlement. Only about 15% of these insects have caused notable damage to trees (Aukema et a. 2010). However, some impacts on forests are much more drastic than the impacts of any invasive plant species. Interestingly, the situation in New Zealand is very different: the total number of exotic insect species that affect woody plants is about 200 and only about 5% of them are serious pests, mostly on introduced tree species. Many New Zealand native woody plants appear to be resistant to exotic herbivorous insects, perhaps because of their phylogenetic uniqueness (Brockerhoff et al. 2010). Also, forests in European countries seem to be much less invaded and influenced by invasive herbivorous insects. For example, in the Czech Republic, among 90 exotic forest herbivores, only up to 10 cause any substantial damage (Šefrová and Laštůvka 2005 and personal communication). Several prominent examples of insect species invading forests and forest plantations worldwide are presented in Table 32.3. More examples can be found in Allison et al. (2023), Bentz et al. (2019), Cartwell (2007), Ciesla (2011), FAO (2009), Holway et al. (2002), Liebhold et al. (1995, 2008, 2011), Paine (2008), Paine et al. (2011), and Wylie and Speight (2012).

In contrast to vascular plants, reptiles, birds, and mammals, most exotic insect introductions are unintentional and are typically by-products of economic activities. Most problems from forest insect invasions can be attributed to global trade. Plants imported for nurseries or landscapes have always been an important pathway by which exotic herbivorous insects have arrived. Recently, solid wood packing material has become another important pathway for invasive forest insects. The International Plant Protection Convention recently enacted an international phytosanitary measure (ISPM 15) that limits the movement of raw wood. This should help, but expanding global trade and travel will very likely result in more introductions of exotic forest insects in the near future.

Established exotic insect species usually spread much faster than plants. Species like *Adelges tsugae* (hemlock wooly adelgid), *Cryptococcus fagisuga* (beech scale), or *Lymantria dispar* (gypsy moth) exhibit a radial rate spread of 3 to 30 km/year. Probably the fastest rate of spread ever recorded was that of the Africanized honeybees (resulting from hybridization of *Apis mellifera scutellata* with other subspecific taxa of the same species) in South and Central America: 300 to 500 km/year (Liebhold and Tobin 2008). The spread of exotic species is usually not a continuous process but a combination of more or less continuous spread due to dispersal abilities of individual species and long-distance dispersal events, usually human- or bird-mediated transport.

Most of the invasive forest herbivorous insect species in the continental United States are either sap feeders (192 species) or foliage feeders (158 species). However, phloem and wood borers (71 species) and foliage feeders often exert greater impact than expected (Table 32.4). Detection of insects that feed on phloem or wood has increased markedly in recent years. If species feeding on trees belonging to only one genus are classified as monophagous, species feeding on trees belonging to only one family as oligophagous and species feeding on trees from multiple families as polyphagous, then the majority of invasive forest herbivorous insects in the continental United States are either monophagous (49%) or polyphagous (33%) (Aukema et al. 2010). In contrast, the majority of

Table 32.3 Some prominent examples of non-native insect species invading forests and forest plantations

Latin name (Order)	*Common name*	*Native range*	*Invaded range*	*Host species*	*Tissue attacked/damage*
Adelges tsugae (Hemiptera)	Hemlock wooly adelgid	Eastern Asia	Eastern North America	*Tsuga canadensis* and *T. caroliniana*	Ray parenchyma of twigs/ needle loss, mortality
Agrilus planipennis (Coleoptera)	Emerald ash borer	Asia	North America, eastern Europe	*Fraxinus* spp. (≥16 species)	Xylem, phloem, bark/ dieback, mortality
Anoplophora glabripennis (Coleoptera)	Asian long-horned beetle	Eastern Asia	North America, Europe, Trinidad	*Acer* spp., *Populus* spp., *Ulmus* spp., *Betula* spp., *Aesculus* spp., *Salix* spp.	Xylem, phloem/dieback, breakage, mortality
Apis mellifera hybrid (Hymenoptera)	Africanized honey bees, killer bees	Produced by cross-breeding	South & Central America, southern North America	N/A	Aggressively protective Behaviour
Cameraria ohridella (Lepidoptera)	Horse chestnut leaf miner	Southern Europe	Central and western Europe	*Aesculus hippocastanum*	Leaves/defoliation, reduction of growth and reproduction
Cinara cupressivora (Hemiptera)	Giant cypress aphid	Europe, Near East	Africa, South America	Probably all Cupressaceae	Sap on terminal growth, dessication and dieback
Coptotermes formosanus (Blattodea)	Formosan subterranean Termite	Southern China, Taiwan	Southern US, Hawaii, South Africa	*Quercus* spp., *Fraxinus* spp., *Liquidambar styraciflua*, etc.	Wood and bark/mortality, damage on woody structures
Cryptococcus fagisuga (Hemiptera)	Beech scale	The Caucasus Mts., Greece	Eastern North America	*Fagus grandifolia*	Parenchyma of bark/ fungal infection, mortality
Dendroctonus valens (Coleoptera)	Red turpentine beetle	North America	China	*Pinus* spp., occasionally *Picea* spp.	Phloem/mortality
Dreyfusia nordmannianae (Hemiptera)	Fir woolly adelgid	The Caucasus region	Central and northern Europe	*Abies* spp.	Ray parenchyma of twigs/ needle loss, mortality
Eriococcus orariensis (Hemiptera)	Manuka blight scale	Australia	New Zealand	*Leptospermum scoparium*, *Kunzea ericoides*	Twigs/sap-feeding, dieback, mortality
Gonipterus scutellatus (Coleoptera)	Eucalyptus weevil	Australia	Africa, North & South America, Europe, Asia, New Zealand	*Eucalyptus* spp.	Leaves & soft bark of twigs/ defoliation, dieback of twigs

Heteropsylla cubana (Hemiptera)	Leucaena psyllid	Central & South America	Tropical Asia & Africa, Australia, Pacific islands	*Leucaena** spp., *Albizia* spp.	Leaves, buds/sap-feeding, defoliation, dieback
Hyphantria cunea (Lepidoptera)	Fall webworm	North America	Eurasia	Broadleaf trees (>600 species)	Leaves/defoliation
Linepithema humile (Hymenoptera)	Argentine ant (syn.: *Iridomymex humilis*)	South America	North America, South Africa Europe, Australia, etc.	N/A	Displaces most or all native Ants
Lymantria dispar (Lepidoptera)	Gypsy moth	Eurasia	North America	*Quercus* spp., *Populus* spp., and about 200 other species	Leaves/defoliation
Neodiprion sertifer (Hymenoptera)	European pine sawfly	Europe	North America	*Pinus* spp.	Needles/defoliation
Operophtera brumata (Lepidoptera)	Winter moth	Europe & Near East	North America	Many broadleaf trees	Buds & leaves/defoliation
Pheidole megacephala (Hymenoptera)	Big-headed ant, Coastal brown ant	Africa	North America, Australia, etc.	N/A	Displaces most or all native ants
Phoracantha spp. (Coleoptera)	Eucalyptus borer	Australia	Africa, Europe, South & North America	*Eucalyptus* spp.	Bark and cambium of stressed trees, mortality
Scolytus multistriatus (Coleoptera)	Smaller European elm bark beetle	Europe	North America	*Ulmus* spp., *Zelkova* spp.	Bark/vector of Dutch elm disease (fungus *Ophiostoma*)
Sirex noctilio (Hymenoptera)	Sirex woodwsasp	Eurasia, N. Africa	North & South America, South Africa Australia, New Zealand	*Pinus* spp.	Xylem, phloem/phytotoxic secretion, mortality
Vespula vulgaris (Hymenoptera)	Common wasp, common yellowjacket	Eurasia	Australia, New Zealand, North America, many islands	N/A	Predation on native invertebrates and nesting birds
Xylosandrus crassiusculus (Coleoptera)	Asian ambrosia beetle, granulate ambrosia beetle	Tropical Asia	Africa, North America	>120 host species	Twigs, branches/mortality of some species in plantations

* Because *Leucaena leucocephala* is very often an invasive species itself, *H. cubana* may be considered more of a biological control agent than a forest pest.

Table 32.4 Numbers of non-native forest insect species established in the continental United States

Order	*Number of species*	*Phloem and wood borers*	*Foliage feeders*	*Sap feeders*	*Other*	*Number of high- impact species*	*Percentage of high- impact species*
Blattodea (Isoptera)	1	1	0	0	0	0	0
Orthoptera	1	1	0	0	1	0	0
Thysanoptera	4	0	0	4	0	2	50.0
Diptera	13	0	6	0	7	1	7.7
Hymenoptera	41	2	35	0	4	11	26.8
Lepidoptera	87	5	76	0	6	10	11.5
Coleoptera	119	63	38	0	18	20	16.8
Hemiptera	189	0	0	188	1	18	9.5
Total	455	71	155	192	37	62	13.6

Source: Aukema et al. (2010).

successful exotic herbivorous forest insects in New Zealand seem to be rather polyphagous (Brockerhoff et al. 2010).

One of the recent additions to the most influential non-native insect pests in North America and Europe is the emerald ash borer (EAB) (*Agrilus planipennis*), a beetle native to northeastern Asia. EAB larvae bore through the outer bark of ash trees (genus *Fraxinus*) and feed on the phloem, cambium, and outer xylem, forming serpentine galleries that disrupt the ability of trees to transport nutrients and water. Trees typically die within 2–4 years. Since the first dying ash trees were noticed in the greater Detroit, Michigan, in 2001, tens of millions of ash trees have been killed by this invader in more than 20 US states and two Canadian provinces. It is still spreading, and virtually all of the16 ash species in North America (six of them are commercially important) are seriously endangered. Over the same period, more than $300 million in federal funds have been devoted to batting this invader (Marché 2017). We are still far from completely knowing the consequences of the absence, or substantial reduction of ash trees abundance. Besides the very few positive aspects (increased density of some native insectivorous birds), ash diebacks in North America may substantially change whole ecosystems and may support invasions of non-native shrubs like Amur honeysuckle (*Lonicera maackii*) that in turn may suppress establishment and species diversity of native seedlings (Hoven et al. 2017).

The biology and impact of many forest insects are closely connected with microorganisms such as fungi (Santini and Battisti 2019; Jactel et al. 2020). For example, the very well-known Dutch elm disease is caused by three species of the genus *Ophiostoma* (Ascomycota) that are dispersed in North America by native (*Hylurgopinus rufipes*) and exotic (*Scolytus multistriatus* and *Scolytus schevyrewi*) bark beetles. In an attempt to block the fungus from spreading, the tree reacts by plugging its own xylem tissue with gum and tyloses. As the xylem delivers water and nutrients to the rest of the plant, these plugs prevent them from travelling up the trunk of the tree. As the disease progresses, major limbs die and eventually the entire tree is killed. Another example of insect–fungus association impacts is the considerable damage to pine plantations in the southern hemisphere following invasion of the wood-wasp *Sirex noctilio*, along with its fungus symbiont *Amylostereum areolatum*.

Some exotic insect–fungus relationships are even more complicated. In New Zealand, the Australian scale insect *Eriococcus orariensis* caused significant damage on native manuka (*Leptospermum scoparium*). The insects suck sap and excrete large amounts of excess sap sugars as fine droplets of solution called honeydew. Where the honeydew falls on stems and foliage it provides food for a soot fungus *Capnodium walteri*, which blackens the infested plants. Although the photosynthetic activity of leaves covered with the fungus is reduced, this alone does not cause the death of the plant. Debilitation and death are believed to be the direct result of massive withdrawal of sap nutrients by nymphs and adult females. Initially, heavy attack often resulted in the death of manuka, but later *E. orariensis* suffered from an entomopathogenic fungus *Angatia thwaitesii*, probably also introduced, and due to this spontaneous "biological control" the impact became less serious ((Brockerhoff et al. 2010).

A special group of forest insect invaders are generalist predators. The most prominent cases are wasps. For example, the Eurasian *Vespula germanica* (German wasp) and *Vespula vulgaris* (common yellowjacket) predate on native forest invertebrates and nesting birds in many countries where they have been introduced. Some ant species are particularly important as omnivores and generalist predators (Holway et al. 2002). Out of about 150 ant species that have been introduced by humans beyond their native range, only a few are really widespread and damaging. The best examples are *Linepithema humile* (Argentine ant), *Pheidole megacephala* (big-headed ant), and *Solenopsis invicta* (red fire ant). These species often displace native ants and disrupt existing ant–plant mutualistic relations (Cordero et al. 2023; Warren et al. 2023). Fortunately, invasive ants mostly favour open and disturbed habitats and only rarely invade undisturbed forests.

Preventing the arrival and potential establishment of exotic insects (and any other potential invaders) is ideal. Early detection of newly established exotic species is also important. A localized population of an invader can sometimes be successfully eradicated. The availability of taxonomic experts is an important component of early detection. Containment, or slowing the spread of established species, is a management approach increasingly considered for certain forest insect invasions (Lampert and Liebhold 2023). Suppression of nascent populations along the invasion front prevents their growth and coalescence, ultimately resulting in slower spread into new regions (Liebhold and McCullough 2011). For invasive harmful insect species that are already widespread, biocontrol options should be considered (Bella et al. 2021; Broadley et al. 2022; Kenis et al. 2019).

Reptiles, birds, and mammals

Over 200 species of reptiles are established or spreading outside their natural range, mainly because of inadvertent introductions via the pet trade (Kraus 2009; Meshaka et al. 2022). Two prominent examples of species invading forests are the brown tree snake (*Boiga irregularis*) and Burmese python (*Python molurus bivittatus*). Both are predators, extremely dangerous to native wildlife. The brown tree snake is responsible for extinction of 10 of 13 native bird species, 6 of 12 native lizards, and 2 of 3 bat species on the island of Guam. Some of the lost species were important pollinators or seed dispersers. The Burmese python is a giant constrictor that has invaded swamp forests of southern Florida. It preys mostly on native warm-blooded animals (deer, bobcats, raccoons, herons, etc.), although reptiles including iguanas and crocodilians are also taken (Reed and Rodda 2011). Eradication of snakes is extremely difficult (Guzy et al. 2023).

Almost all established non-native birds are results of intentional introductions. Extreme cases are religious, mainly Buddhist, releases of birds (over 500,000 individuals every year in Hong Kong or Taiwan). Many of the released birds are exotic. Birds seem to be somewhat less important as invaders in forests, particularly on continents. Some examples of non-native bird species invading forests and forest plantations are presented in Table 32.5. More complete lists can be derived from Duncan et al. (2006), Lever (1994, 2005), Long (1981), and Scott et al. (1986).

Native and exotic bird species are usually segregated along gradients related to anthropogenic disturbance (Barnagaud et al. 2014). Most introduced birds are restricted to urban habitats, rural areas, forest edges, and exotic plantations. Even on islands, only a tiny fraction of introduced birds penetrates into undisturbed forests. For example, in Hawaii where more exotic bird species have been established than anywhere else, out of 58 naturalized bird species, only seven have invaded native forests: *Cardinalis cardinalis, Garrulax caerulatus, Garrulax. canorus, Garrulax. pectoralis, Leiothrix lutea, Lophura leucomelana*, and *Zosterops japonicus* (Scott et al. 1986). Similarly, out of 34 established exotic bird species in New Zealand, only two have successfully invaded intact forests: *Fringilla coelebs* and *Turdus merula* (Duncan et al. 2006 and personal communication). In contrast, among 19 bird species introduced by people in Hong Kong, ten invaded secondary forests (Leven and Corlett 2004). In this case, however, over half of these species were very likely present in the area prior to extensive deforestation.

Negative impacts of invasive birds are usually limited to competition with native birds (only rarely conclusively documented) and to dispersal of exotic plant species. However, in some cases the most important impact is the spread of avian pathogens. Via introduced birds infected with *Plasmodium relictum* and the introduced African mosquito *Culex quinquefasciatus*, avian malaria was introduced to and spread throughout the Hawaiian Islands (Neddermeyer et al. 2023). Avian malaria and avian pox together caused the extinction of many native bird species. The only positive impact of exotic birds in some forests is that they can partially substitute for extinct native pollinators and seed dispersers (Aslan et al. 2013; Foster and Robinson 2007). Novel plant–frugivore interactions may vary substantially depending on frugivore species and the seed size of plants (Case et al. 2022).

Across continents and islands, introduced mammals are more successful invaders than birds (Jeschke 2008). Mammals are also more successful as invaders of undisturbed forests, and their direct impacts on native biota are much larger. Some examples of non-native mammal species invading forests and forest plantations are presented in Table 32.6. More examples can be found in Latham et al. (2017), Lever (1994), Long (2003), Nentwig et al. (2009), and Woodward and Quinn (2011). Different groups of mammals are important as soil disturbers (wild boar and feral pigs), drastic habitat modifiers (beaver), competitors (eastern fox squirrel vs. native western gray squirrel, Figure 32.3; North American mink vs. European otter), herbivores (nutria, several deer taxa, rabbits, and brushtail possum), seed predators (rats, wild boar, and feral pigs), and animal predators (mink, feral cats, foxes, and rats). Impacts are particularly strong on islands where animal predators and/or large herbivores were previously absent.

There are many indirect impacts of invasive mammals. For example, non-native wild boar and deer are dispersing non-native ectomycorrhizal fungi that support invasive pine species in northern Patagonia. Moose (*Alces alces*), a non-native herbivore in Gros Monte National Park (Canada), acts as the primary conduit for non-native plant invasions by dispersing propagules

Table 32.5 Some prominent examples of non-native bird species invading forests and forest plantations

Latin name (family)	*Common name*	*Diet*	*Native range*	*Invaded range*	*Impacts*
Acridotheres tristis (Sturnidae)	Common myna (mynah), Indian myna bird	Omnivorous	Asia	Australia, N. America, S. Africa, many islands	Competition with many hollow-nesters and killing of native young birds
Cardinalis cardinalis (Cardinalidae)	Northern cardinal	Mainly granivorous	North America	Hawaii, Atlantic islands	-Not documented
Fringilla coelebs (Fringillidae)	Chaffinch (many subspecies)	Omnivorous, adults mostly granivorous	Eurasia, North Africa in South Africa	New Zealand, rare	-Not documented
Lophura leucomelanos (Phasianidae)	Kalij pheasant	Herbivorous	Asia	Hawaii, Argentina	Dispersal of invasive plant species (e.g., *Passiflora* and *Clidemia*)
Myiopsitta monachus (Psittacidae)	Monk or quaker parakeet	Herbivorous	South America	USA, Mexico, Spain	Eating buds and fruits of native trees
Pycnonotus cafer (Pycnonotidae)	Red-vented bul-bul	Omnivorous	Asia	Hawaii, Fiji, probably eradicated in New Zealand	Dispersal of invasive plant species (*Lantana camara, Miconia clavescens*)
Sturnus vulgaris (Sturnidae)	European starling	Omnivorous	Europe, West Asia, North Africa	Australia, N. America, S. Africa, New Zealand	Seed dispersal, competition with native birds for nesting sites
Turdus philomelos (Turdidae)	Song thrush	Omnivorous	Eurasia	Australia, New Zealand	?
Turdus merula (Turdidae)	Common blackbird, Eurasian blackbird	Omnivorous	Eurasia, North Africa rare in North America	Australia, New Zealand,	Competition with native birds, dispersal of invasive plant species
Tyto alba (Tytonidae)	Barn owl	Predator	All continents except Antarctica	Seychelles, Hawaii	Competes with and predates on native birds
Zosterops japonicus (Zosteropidae)	Japanese white-eye	Omnivorous	East Asia	Hawaii	Competition with native birds,vector for avian parasites

Table 32.6 Some prominent examples of non-native mammal species invading forests and forest plantations

Latin name (Order)	*Common name*	*Life form*	*Native range*	*Invaded range*	*Impacts*
Castor canadensis (Rodentia)	North American beaver, Canadian beaver	Large semi-aquatic rodent	North America	Tierra del Fuego, Finland	Conversion of forests into wetlands
Cervus spp., Odocoileus spp. (Artiodactyla)	Deer, elk	Grazers/browsers	Many continents	Many continents and islands	Prevention of forest regeneration, reduction of vegetation cover and diversity
Felis catus (Carnivora)	Feral cat	Predator	Eurasia	Many continents and islands	Preying on small mammals, birds, reptiles, and amphibians
Myocastor coypus (Rodentia)	Nutria, coypu	Semi-aquatic rodent	South America	North America, Europe, Africa, Japan	Destruction of wetlands, prevention of bottomland forest regeneration
Neovison vison (Carnivora)	America mink	Predator	North America	Continental Europe, Iceland, S. America	Preying on small mammals and birds
Oryctolagus cuniculus (Lagomorpha)	European rabbit	Grazer/bark eater	Europe	Many continents and > 800 islands	Prevention of forest regeneration, reduction of herbaceous vegetation
Rattus spp. (Rodentia)	Rats	Omnivorous rodent	Asia	All continents except Antarctica, many islands	Elimination of native plants, birds, reptiles, insects, and snails
Sciurus carolinensis (Rodentia)	Eastern gray squirrel	Herbivorous and	North America fungivorous rodent	Europe, South Africa	Stripping bark from trees, outcompeting native squirrels*
Sus scrofa (Artiodactyla)	Wild boar and feral pigs	Fast reproducing, omnivorous	Eurasia, North Africa	Americas, North and east Australia, many islands	Extensive rooting, killing plants and ground nesting birds
Vulpes vulpes (Carnivora)	Red fox	Carnivore/frugivore	Eurasia, boreal N. America	North Africa, Australia, eastern North America	Predation on mammals and ground-nesting birds
Trichosurus Vulpecula (Diprotodontia)	Common brushtail possum	Nocturnal, semi-arboreal herbivorous marsupial	Australia	New Zealand	Defoliation of native trees, predation on eggs and chicks

Note: *Also transmitting deadly parapox virus to native squirrels.

Figure 32.3 Introduced eastern fox squirrel (*Sciurus niger*) eating drupes of introduced Chinese pistache (*Pistacia chinensis*) in California. This species is replacing native western grey squirrel (*Sciurus griseus*).

and creating disturbance by trampling and browsing native vegetation. Some introduced mammals have turned out to be vectors of infectious diseases (invasive gray squirrels do not die from the parapox virus but can infect European red squirrels). Invasive rats can cause complete collapse of native small mammal communities in insular forest fragments (Moore et al. 2022).

One of the major problems in European forests is sika deer (*Cervus nippon*) introduced from East Asia. Sika is a serious pest, causing significant damage to forests and forest plantations. It also plays a role in the epidemiology of *Asworthius sidemi*, a nematode affecting native deer (DAISIE 2009). Moreover, hybrids with native deer (*Cervus elaphus*) are fertile and further hybridization or back-crossing is rapidly threatening genetic integrity of the native species.

What does the future hold?

Rates of globalization and almost all human activities are accelerating and so are the rates of intentional and unintentional introductions of non-native biota (Gougherty and Davies 2022; Mormul et al. 2022). The majority of introduced species are not established. However, with global climate change, many new species will be established in areas where they would not have survived before (Dukes et al. 2009; Björkman and Niemelä 2015; Simler-Williamson et al. 2019; Fitnh et al. 2021). At the same time, the impact of biological invasions on forest carbon sequestration is becoming one of the most urgent questions (Peltzer et al. 2010; Nunez et al. 2021). With increasing numbers of introduced species the probability of establishment of some of them increases (Hulme 2012) and with increasing numbers of established non-native species the probability that some of them will have an environmental or economic impact increases as well (Rejmánek and Randall 2004). Therefore, prevention, if we have such an option, is usually cheaper than control (Finnoff et al. 2007).

Early detection of pest species increases the chance of successful eradication (Rejmánek and Pitcairn 2002; Lovett et al. 2016; Suckling et al. 2019). However, once a harmful invader is widespread, eradication is unlikely and several management options, including species-specific biological control, have to be considered (Clout and Williams 2009; Van Driesche et al. 2008; Miller et al. 2010; Wylie and Speight 2012). Ecologically sound forestry practices (e.g., preventing dispersal of alien propagules by machinery, reduced impact of logging, fast reforestation of disturbed areas with native species, or creation of refugia for endangered species) and containment of existing infestations can certainly alleviate problems caused by biological invasions (Lampert and Liebhold 2023; Nahrung et al. 2023; Sitzia et al. 2021). More attention should be given to the role of planted tree species- and age-diversity in enhancing biological resistance of forests to both climate change and biological invasions (Guo et al. 2019; Jactel et al. 2021; Gómez-González et al. 2022; Messier et al. 2022; Ward et al. 2022). Because of the global scale of biological invasions, international collaboration in invasive species research, reporting, and management will be essential (Liebhold et al. 2021).

Acknowledgements

The author greatly acknowledges the support of Clare Aslan, Robert Cowie, Richard Corlett, Richard Duncan, Paul Hendrix, Jonathan Jeschke, Zdeněk Laštůvka, Andrew Liebhold, Dane Panetta, John Reynolds, Chris Stoddart, and Toni Withers who contributed in many ways to the final version of this chapter.

References

Aalders, L. T., McNeill, M. R., Bell, N. L. and Cameron, C. (2017) 'Plant parasitic nematodes survival and detection to inform biosecurity risk assessment', *NeoBiota*, vol. 16, pp. 1–16.

Allison, J., Paine, T. D., Slippers, B. and Wingfield M. J. (eds.) (2023) *Forest Entomology and Pathology, Volume 1: Entomology*. Springer, Dordrecht.

Aslan, C. E., Zavaleta, E. S., Tershy, B., Croll, D. and Robichaux, R. H. (2013) 'Imperfect replacement of native species by non-native species as pollinators of endemic Hawaiian plants', *Conservation Biology*, vol. 28, pp. 478–488.

Aukema, J. E. et al. (2010) 'Historical accumulation of nonindigenous forest pests in the continental United States', *BioScience*, vol. 60, pp. 886–897.

Barnagaud, J-Y., Barbaro, L., Papaix, J., Deconchat, M. and Brockerhoff, E. G. (2014) 'Habitat filtering by landscape and local forest composition in native and exotic New Zealand birds', *Ecology*, vol. 95, pp. 78–87.

Bella, A., Silini, A., Cherif-Silini, H., Bouket, A. C., Moser, W. K. and others (2021) 'The threat of pests and pathogens and the potential for biological control in forest ecosystems', *Forests*, vol. 12, 1579.

Benz, B., Bonello, P., Delb, H., Fettig, C., Puereswaran, D. and Seybold, S. (2019) 'Advances in understanding and managing insec pests of forest trees', in J.A. Stanturf (ed.), *Achiving Sustainable Management of Boreal and Temperate Forests*. Burleigh Dodds Science Publishing, Cambridge, UK.

Björkman, C. and Niemelä, P. (eds.) (2015) *Climate Change and Insect Pests*. CABI, Wallingford, UK.

Broadley, H. J., Boettner, G. H., Schneider, B. and Elkinton, J. S. (2022) 'Native generalist natural enemies and an introduced specialized parasitoid together control an invasive forest insect', *Ecological Applications*, vol. 32, e2697.

Brockerhoff, E. G. et al. (2010) 'Impacts of exotic invertebrates on New Zealand's indigenous species and ecosystems', *New Zealand Journal of Ecology*, vol. 34, pp. 158–174.

Brown, K. A., Scatena, F. N. and Gurevitch, J. (2006) 'Effects of an invasive tree on community structure and diversity in a tropical forest in Puerto Rico', *Forest Ecology and Management*, vol. 226, pp. 145–152.
Cartwell, C. G. (2007) *Invasive Forest Pests.* Nova Science Publishers, New York.
Case, S. B., Postelli, K., Drake, D. R., Vizentin-Burgoni, J., Foster, J. T. and others (2022) 'Introduced galliforms as seed predators and dispersers in Hawaiian forests', *Biological Invasions*, vol. 24, pp. 3083–3097.
Chang, C.-H., Bartz, M. L. C., Brown, G., Callaham, M. A., Cameron, E. K. and others (2021) 'The second wave of earthworm invasions in North America: Biology, environmental impacts, management and control of invasive jumping worms', *Biological Invasions*, vol. 23, pp. 3291–3322.
Chong, K. Y., Corlett, R. T., Nunez, M. A., Chiu, J. H., Courchamp, F. and others (2021) 'Are terrestrial biological invasions different in the tropics?', *Annual Review of Ecology, Evolution, and Systematics*, vol. 52, pp. 291–314.
Ciesla, W. M. (2011) *Forest Entomology: A global Perspective.* Willey-Blackwell, Chichester, UK.
Cloud, M. N. and Williams, P. A. (2009) *Invasive Species Management — A Handbook of Principles and Techniques.* Oxford University Press, Oxford, UK.
Cordero, S., Gálvez, F. and Fontúrbel, F. F. (2023) 'Ecological impacts if exotic species on native seed dispersal systems: A systematic review', *Plants* 2023, 12, 261.
Cowie, R. H. (2005) 'Alien non-marine molluscs in the islands of the tropical and subtropical Pacific: A review', *American Malacological Bulletin*, vol. 20, pp. 95–103.
Craven, D., Thakur, M.P., Cameron, E. K., Frelich, L. E., Beausejour, R. and others (2017) 'The unseen invaders: Introduced earthworms as drivers of change in plant communities in North American forests (a meta-analysis)', *Global Change Biology*, vol. 23, pp. 1065–1074.
DAISIE. (2009). *Handbook of Alien Species in Europe.* Springer, Berlin.
Dawson, W., Burslem, D. F. R. P. and Hulme, P. E. (2009) 'Factors explaining alien plant invasion success in a tropical ecosystem differ at each stage of invasion', *Journal of Ecology*, vol. 97, pp. 657–665.
De la Fuente, B. and Beck, P. S. A. (2019) 'Management measures to control pine wood nematode spread in Europe', *Journal of Applied Ecology*, vol. 56, pp. 2577–2580.
Denslow, J. S. and DeWalt, S. (2008) 'Exotic plant invasions in tropical forests: Patterns and hypotheses', in W.P. Carson and S.A. Schnitzer (eds.), *Tropical Forest Community Ecology*, University of Chicago Press, Chicago, US.
Dukes, J. S. et al. (2009) 'Responses of insect pests, pathogens, and invasive plant species to climate change in the forests of northeastern North America: What can we predict?' *Canadian Journal of Forest Research*, vol. 39, pp. 231–248.
Duncan, R. P., Blackburn, T. M. and Cassey, P. (2006) 'Factors affecting the release, establishment and spread of introduced birds in New Zealand', in R.B. Allen and W.G. Lee (eds.), *Biological Invasions in New Zealand.* Springer-Verlag, Berlin, pp. 137–154.
Dyderski, M. K. and Jagodzinski, A. M. (2020) 'Impact of invasive tree species on natural regeneration species composition, diversity, and density', *Forests* 11, 456. doi:10.3390/f11040456
Edwards, C. A. (ed.) (2004) *Earthworm Ecology*, 2nd ed. CRC Press, Boca Raton, FL.
FAO. (2009) *Global Review of Forest Pests and Diseases.* FAO Forestry Paper 156, FAO, Rome, Italy, pp. 1–222.
Finch, D. M., Butler, J. L., Runion, J. B., Fetting, C. J., Kilkenny, F. F. and others (2021) 'Effects of climate change on invasive species', in T. M. Poland, T. Patel-Weynand, D. M. Finch, C. F. Miniat, D. C. Hayes and V. M. Lopez (eds.) (2021) *Invasive Species in Forests and Rangelands of the United States.* Springer, Dordrecht, pp. 57–83.
Finnoff, D., Shogren, J. F., Leung, B. and Lodge, D. (2007) 'Take a risk: Preferring prevention over control of biological invaders', *Ecological Economics*, vol. 62, pp. 216–222.
Foster, J. T. and Robinson, S. K. (2007) 'Introduced birds and the fate of Hawaiian rainforests', *Conservation Biology*, vol. 21, pp. 1248–1257.
Foxcroft, L. C., Pyšek, P., Richardson, D. M. and Genovesi, P. (eds.) (2013) *Plant Invasions in Protected Areas.* Springer, Dordrecht, Netherlands.
Frelich, L. E., Blossey, B., Cameron, E. K., Dávalos, A., Eisenhauer, N. and others (2019) 'Sideswiped: Ecological cascades emanating from earthworm invasions', *Frontiers in Ecology and the Environment*, vol. 17, pp. 502–510.

Fridley, J. D., Bauerle, T. L., Craddock, A., Ebert, A. R., Frank, S. A. and others (2022) 'Fast but steady: An integrated leaf-stem-root trait syndrome for woody forest invaders', *Ecology Letters*, vol. 25, pp. 900–912.

Fridley, J. D., Bellingham, P. J., Closset-Kopp, D., Daehler, C. C., Dechoum, M. S. and others (2023) 'A general hypothesis of forest invasions by woody plants based on whole-plant carbon economics', *Journal of Ecology*, vol. 111, pp. 4–22.

Gaertner, M., Biggs, R., Te Beest, M., Hui, C., Molofsky, J. and Richardson, D. M. (2014) 'Invasive plants as drivers of regime shifts: Identifying high-priority invaders that alter feedback relationships', *Diversity and Distributions*, vol. 20, pp. 733–744.

Gómez-González, S., Paniw, M., Blanco-Pastor, J. L., García-Cervigón, A. I., Godoy, O. and others (2022) 'Moving towards the ecological intensification of tree plantations', *Trends in Plant Science*, vol. 27, pp. 637–645.

Gougherty, A. V. and Davies, T. J. (2022) 'Most countries are vulnerable to novel pest invasions and under-report the diversity of tree pests', *Global Ecology and Biogeography*, vol. 31, pp. 2314–2322.

Green, P. T., Lake, P. S. and O'Dowd, D. J. (2004) 'Resistance of island rainforest to invasion by alien plants: Influence of microhabitat and herbivory on seedling performance', *Biological Invasions*, vol. 6, pp. 1–9.

Guo, Q., Fei, S., Potter, K.M., Liebhold, A. M. and Wen, J. (2019) 'Tree diversity regulates forest pest invasion', *PNAS*, vol. 116, pp. 7382–7386.

Guzy, J. C., Falk, B. G., Smith, B. J., Wilson, J. D., Reed, R. N. and others (2023) 'Burmese pythons in Florida: A synthesis of biology, impacts, and management tools', *NeoBiota*, vol. 80, pp. 1–119.

Hahn, P. G. et al. (2011) 'Exotic slugs pose previously unrecognized threat to the herbaceous layer in a Midwestern woodland', *Restoration Ecology*, vol. 19, pp. 786–794.

Hendrix, P. F. (ed.) (2006) *Biological Invasions Belowground: Earthworms as Invasive Species*. Springer, Dordrecht, Netherlands.

Hendrix, P. F., Callaham, M. A., Drake, J. M., Huang, C.-Y., James, S. W., Snyder, B. A. and Zhang, W. (2008) 'Pandora's box contained bait: The global problem of introduced earthworms', *Annual Review of Ecology, Evolution, and Systematics*, vol. 39, pp. 593–613.

Holway, D. A., Lach, L., Suarez, A. V., Tsutsui, N. D. and Case, T. J. (2002) 'The causes and consequences of ant invasions', *Annual Review of Ecology and Systematics*, vol. 33, pp. 181–233.

Hoven, B. M., Gorchov, D. L., Knight, K. S. and Peters, V. E. (2017) 'The effect of emerald ash borer-caused tree mortality on the invasive shrub Amur honeysuckle and their combined effects on tree and shrub seedlings', *Biological Invasions*, vol. 19, pp. 2813–2836.

Hulme, P. E. (2012) 'Weed risk assessment: A way forward or a waste of time?' *Journal of Applied Ecology*, vol. 49, pp. 10–19.

Jactel, H., Desprez-Loustau, M.-L., Battisti, A., Brockerhoff, E., Santini, A. and others (2020) 'Pathologists and entomologists must join forces against forest pest and pathogen invasions', *NeoBiota*, vol. 58, pp. 107–127.

Jactel, H., Mororeira, X. and Castagneyrol, B. (2021) 'Tree diversity and forest resistance to insect pests: Patterns, mechanisms and prospects', *Annual Review of Entomology*, vol. 66, pp. 277–296.

James, S. W. (2011) 'Earthworms', in D. Simberloff and M. Rejmánek (eds.) *Encyclopedia of Biological Invasions*, University of California Press, Berkeley, CA, pp. 177–183.

Jang, J., Xiong, X., Liu C., Yoo, K. and Ishii, S. (2022) 'Invasive earthworms alter forest soil microbiomes and nitrogen cycling', *Soil Biology and Biochemistry*, vol. 171, 108724.

Jayakumar, R. and Nair, K. K. N. (2022) 'Invasion and distribution of exotic plants in the tropical forests of Western Ghats, India', *Tropical Ecology*, vol. 63, pp. 522–530.

Jeschke, J. M. (2008) 'Across islands and continents, mammals are more successful invaders than birds', *Diversity and Distributions*, vol. 14, pp. 913–916.

Joe, S. M. and Daehler, C. C. (2008) 'Invasive slugs as under-appreciated obstacles to rare plant restoration: Evidence from Hawaiian Islands', *Biological Invasions*, vol. 10, pp. 245–255.

Kenis, M., Hurley, B., Colombari, F., Lawson, S., Sun, J. and others (2019) *Guide to the Classical Biological Control of Insect Pests in Planted and Natural Forests*. FAO Forestry Paper 182, pp. 1–96. FAO, Rome, Italy.

Kohli, R. K., Jose, S., Singh, H. P. and Batish, D. R. (2009) (eds.) *Invasive Plants and Forest Ecosystems*. CRC Press, Boca Raton, FL.

Kraus, F. (2009) *Alien Reptiles and Amphibians: A Scientific Compendium and Analysis.* Springer, Dordrecht, Netherlands.

Lampert, A. and Liebhold, A. M. (2021) 'Combining multiple tactics over time for cost-effective eradication of invading insect populations', *Ecology Letters*, vol. 24, pp. 279–287.

Lampert, A. and Liebhold, A. M. (2023) 'Optimizing the use of suppression zones for containment of invasive species', *Ecological Applications*, vol. 33, e2797.

Langmaier, M. and Lapin, K. (2020) 'A systematic review of the impact of invasive alien plants on forest regeneration in European temperate forests', *Frontiers in Plant Science*, vol. 11, 524969.

Latham, A. D. M., Warburton, B., Byrom, A. E. and Pech, R. P. (2017) 'The ecology and management of mammal invasions in forests', *Biological Invasions*, vol. 19, pp. 3121–3139.

Leven, M. R. and Corlett, R. T. (2004) 'Invasive birds in Hong Kong, China', *Ornithological Science*, vol. 3, pp. 43–55.

Lever, C. (1994) *Naturalized Animals: The Ecology of Successfully Introduced Species.* T & A D Poyser, London.

Lever, C. (2005) *Naturalized Birds of the World.* T & A D Poyser, London.

Liebhold, A. M., Brockerhoff, E. G., Kalisz, S., Nunez, M. A., Wardle, D. A. and Wingfield, M. J. (2017) 'Biological invasions in forest ecosystems', *Biological Invasions*, vol. 19, pp. 3437–3458.

Liebhold, A. M., Campbell, F. T., Gordon, D. R., Guo, Q., Havill, N. and others (2021) 'The role of international cooperation in invasive species research', in T. M. Poland, T. Patel-Weynand, D. M. Finch, C. F. Miniat, D. C. Hayes and V. M. Lopez (eds.), *Invasive Species in Forests and Rangelands of the United States.* Springer, Dordrecht, pp. 293–303.

Liebhold, A. M., Macdonald, W. L., Bergdahl, D. and Mastro, V. C. (1995) 'Invasion by exotic forest pests: A threat to forest ecosystems', *Forest Science Monographs*, vol. 30, pp. 1–49.

Liebhold, A. M. and McCoullough, D. G. (2011) 'Forest insects', in D. Simberloff and M. Rejmánek (eds.), *Encyclopedia of Biological Invasions*, University of California Press, Berkeley, CA, pp. 238–241.

Liebhold, A. M. and Tobin, P. C. (2008) 'Population ecology of insect invasions and their management', *Annual Review of Entomology*, vol. 53, pp. 387–408.

Loehle, C., Hulcr, J., Smith, J. A., Munro, H. L. and Fox, T. (2023) 'Preventing the perfect storm of forest mortality in the United States caused by invasive species', *Journal of Forestry*, vol. 121, pp. 104–117.

Long, J. L. (1981) *Introduced Birds of the World.* David & Charles, London.

Long, J. L. (2003) *Introduced Mammals of the World.* CSIRO Publishing, Collingwood, Australia.

Lovett, G. M., Weiss, M., Liebhold, A. M., Holmes, T. P., Leung, B. and others (2016) 'Nonnative forest insects and pathogens in the United States: Impacts and policy options' *Ecological Applications*, vol. 26, pp. 1437–1455.

Marché, J. D. (2017) *The Green Menace — Emerald Ash Borer and the Invasive Species Problem*, Oxford University Press, Oxford, UK.

Martin, P. H., Canham, C. D. and Marks, P. L. (2008) 'Why forests appear resistant to exotic plant invasions: Intentional introductions, stand dynamics, and the role of shade tolerance', *Frontiers of Ecology and Environment*, vol. 7, pp. 142–149.

McKenzie, P., Brown, C., Jianghua, S. and Jian, W. (2005) (eds.) *The Unwelcome Guests, Proceedings of the Asia-Pacific Forest Invasive Species Conference.* Food and Agriculture Organization of the United Nations, Regional Office, Bangkok.

Meshaka, W. E., Collins, S. L., Bury, R. B. and McCallum, M. L. (2022) *Exotic Amphibians and Reptiles of the United States.* University Press of Florida, Gainsneville, FL.

Messier, C., Bauhus, J., Sousa-Silva, R., Auge, H., Baeten, L. and others (2022) 'For the sake of resilience and multifunctionlity, let's diversify planted forests!', *Conservation Letters*, vol. 15, e12829.

Miller, J. H., Manning, S. T., and Enloe, S. F. (2010) *A Management Guide for Invasive Plants in Southern Forests*, USDA, Forest Service, General Technical Report SRS-131. Southern Research Station, Asheville, NC.

Moore, E., D'Amico, V. and Trammell, T. L. E. (2023) 'Plant community dynamics following non-native shrub removal depend on invasion intensity and forest site characteristics', *Ecosphere*, vol. 14, e4351.

Moore, J. H., Palmeirim, A .F., Peres, C. A., Ngoprasert, D. and Gibson, L. (2022) 'Invasive rat drives complete collapse of small mammal communities in insular forest fragments', *Current Biology*, vol. 32, pp. 2997–3004.

Mormul, R. P., Vieira, D.S., Bailly, D., Fidanza, K., da Silva, V.F.B. and others (2022) 'Invasive alien species records are exponentially rising across the Earth', *Biological Invasons*, vol. 24, pp. 3249–3261.

Mota, M. M. and Vieira, P. (eds.) (2008) *Pine Wilt Disease: A Worldwide Threat to Forest Ecosystems.* Springer, Dordrecht, Netherlands.

Nahrung, H. F., Liebhold, A. M., Brockerhoff, E. G. and Rassati, D. (2023) 'Forest insect biosecurity. Processes, patterns, predictions, pitfalls'. *Annual Review of Entomology*, vol. 68, pp. 211–229.

Neddermeyer, J. H., Parise, K. L., Dittmar, E., Kilpatrick, A. M. and Foster, J. F. (2023) 'Nowere to fly: Avian malaria is ubiquitous from ocean to summit on a Hawaiian island', *Biological Conservation*, vol. 279, 109943.

Nentwig, W., Kühnel, E. and Bacher, S. (2009) 'A generic impact-scoring system applied to alien mammals of Europe', *Conservation Biology*, vol. 24, pp. 302–311.

Ni, M. and Hulme, P. H. (2021) 'Botanic gardens play key roles in the regional distribution of first records of alien plants in China', *Global Ecology and Biogeography*, vol. 30, pp. 1572–1582.

Norghauer, J. M. et al. (2011) 'Island invasion by a threatened tree species: Evidence for natural enemy release of mahagony (*Swietenia macrophylla*) on Dominica, Less Antiles', *PLos One*, vol. 6, e18790.

Nouri-Aiin, M. and Görres, J. H. (2021) 'Biocontrol of invasive pheretimoid earthworms using *Beauveria bassiana*', *PeerJ*, vol. 9, e11101. DOI 10.7717/peerj.11101

Nunez, M. A., Davis, K. T., Dimarco, R. D., Peltzer, D. A., Paritis, J. and others (2021) 'Should tree invasions be used in treeless ecosystems to mitigate climate change?', *Frontiers in Ecology and the Environment*, vol. 19, pp. 334–341.

Nuzzo, V. A., Maerz, J. C. and Blossey, B. (2009) 'Earthworm invasion as a driving force behind plant invasion and community change in northeastern North American forests', *Conservation Biology*, vol. 23, pp. 966–974.

Paine, T. D. (2008) (ed.) *Invasive Forest Insects, Introduced Forest Trees, and Altered Ecosystems.* Springer, Dordrecht, Netherlands.

Paine, T. D., Steinbauer, M. J. and Lawson. S. A. (2011) 'Native and exotic pests of *Eucalyptus*: A worldwide perspective', *Annual Review of Entomology*, vol. 56, pp. 181–201.

Peltzer, D. A., Allen R. B., Lovett, G. M., Whitehead, D. and Wardle, D. A. (2010) 'Effects of biological invasions on forest carbon sequestration', *Global Change Biology*, vol. 16, pp. 732–746.

Poland, T. M., Patel-Weynand, T., Finch, D. M., Miniat, C. F., Hayes, D. C. and Lopez, V. M. (eds.) (2021) *Invasive Species in Forests and Rangelands of the United States.* Springer, Dordrecht, Netherlands.

Reed, R. N. and Rodda, G. H. (2011) 'Burmese python and other giant constrictors', in D. Simberloff and M. Rejmánek (eds.), *Encyclopedia of Biological Invasions.* University of California Press, Berkeley, CA, pp. 85–91.

Reichard, S. (2011) 'Horticulture', in D. Simberloff and M. Rejmánek (eds.), *Encyclopedia of Biological Invasions.* University of California Press, Berkeley, CA, pp. 336–342.

Rejmánek, M. (1996) 'Species richness and resistance to invasions', in G. Orians, R. Dirzo and J.H. Cushman (eds.), *Ecosystem Functions of Biodiversity in Tropical Forests.* Springer, Berlin, pp. 153–172.

Rejmánek, M. (2013a) 'Extended leaf phenology: A secret of successful invaders?' *Journal of Vegetation Science*, vol. 24, pp. 975–976.

Rejmánek, M. (2013b) 'Assessing the impacts of plant invaders on native plant species diversity', in CONABIO (ed.), *Proceedings of the 2012 Weeds Across Borders Conference.* CONABIO, Mexico City, pp. 63–69. www.weedcenter.org/wab/2012/docs/MemoriaWAB_2012_final.pdf

Rejmánek, M. (2014) 'Invasive trees and shrubs — where do they come from and what we should expect in the future?', *Biological Invasions,* vol. 16, pp. 483–498.

Rejmánek, M. and Pitcairn, M. J. (2002) 'When is eradication of exotic pest plants a realistic goal?', in C.R. Vietch and M.N. Clout (eds.), *Turning the Tide: The Eradication of Island Invasives.* IUCN, Gland, Switzerland, pp. 249–253. www.issg.org/database/species/reference_files/onoaca/Rejmanek.pdf

Rejmánek, M. and Randall, J. M. (2004) 'The total number of naturalized species can be a reliable predictor of the number of alien pest species', *Diversity and Distributions*, vol. 10, pp. 367–369.

Rejmánek, M. and Richardson, D. M. (2013) 'Trees and shrubs as invasive alien species — 2013 update of the global database', *Diversity and Distributions*, vol. 19, pp. 1093–1094.

Rejmánek, M., Richardson, D. M. and Pyšek, P. (2013) 'Plant invasions and invasibility of plant communities', in E. van der Maarel and J. Franklin (eds.), *Vegetation Ecology*, 2nd ed. Wiley-Blackwell, Chichester, UK, pp. 387–424.

Reynolds, J. W. (2012) 'The status of terrestrial earthworm (Oligochaeta) surveys in North America and some Caribbean countries', *Megadrilogica*, vol. 15, pp. 227–247.

Richardson, D. M. (2011) 'Forestry and agroforestry', in D. Simberloff and M. Rejmánek (eds.), *Encyclopedia of Biological Invasions.* University of California Press, Berkeley, CA, pp. 241–248.

Roubik, D. W. (1988) *Ecology and Natural History of Tropical Bees.* Cambridge University Press, Cambridge, UK.

Sanderson, L. A., McLaughlin, J. A. and Antunes, P. M. (2012) 'The last great forest: A review of the status of invasive species in the North American boreal forest', *Forestry,* vol. 85, pp. 329–340.

Santini, A. and Battisti, A. (2019) 'Complex insect-pathogen interactions in tree pandemics', *Frontiers in Physiology*, vol. 10, 550.

Scott, J. M. et al. (1986) *Forest Bird Communities of the Hawaiian Islands.* Allen Press, Lawrence, KS.

Šefrová, H. and Laštůvka, Z. (2005) 'Catalogue of alien animal species in the Czech Republic', *Acta Universitatis Agriculturae et Silviculturae Mendelianae Brunensis*, vol. 53, pp. 151–170.

Simberloff, D. and Rejmánek, M. (eds.) (2011) *Encyclopedia of Biological Invasions.* University of California Press, Berkeley, CA.

Simler-Williamson, A. B., Rizzo, D. M. and Cobb, R. C. (2019) 'Interacting effects of global change on forest pest and pathogen dynamics', *Annual Review of Ecology, Evolution and Systematics*, vol. 50, pp. 381–403.

Singh, S. K., Hodda, M., Ash, G. J. and Banks, N. C. (2013) 'Plant-parasitic nematodes as invasive species: Characteristics, uncertainty and biosecurity implications', *Annals of Applied Biology*, vol. 163, pp. 323–350.

Sitzia, T., Campagnaro, T., Brundu, G., Faccoli, M., Santini, A. and Webber, B. L. (2021) 'Forest ecosystems', in K. Barker and R.A. Francis (eds.), *Routledge Handbook of Biosecurity and Invasive Species.* Routledge, London, pp. 105–127.

Suckling, D. M., Stringer, L. D., Baird, D. B. and Kean, J. M. (2019) 'Will growing invasive arthropod biodiversity outpace our ability for eradication?', *Ecological Applications*, vol. 29, e01992.

Van Driesche, R., Hoddle, M. and Center, T. (2008) *Control of Pests and Weeds by Natural Enemies*, Blackwell, Oxford, UK.

Wagner, V., Chytrý, M., Jiménez-Alfaro, B., Pergl, J., Hennekens, S. And others (2017) 'Alien plant invasions in European woodlands', *Diversity and Distributions*, vol. 23, pp. 969–981.

Ward, S. F., Liebhold, A. M. and Fei. S. (2022) 'Variable effects of forest diversity on invasions by non-native insects and pathogens', *Biodiversity and Conservation*, vol. 32, pp. 2575–2586.

Warren, R. J. and Frankson, P. T. (2023) 'Global change drives synergize with the negative impacts of non-native invasive ants on native seed-dispersing ants', *Biological Invasions*, vol. 25, pp. 773–786.

Weber, E. (2017) *Invasive Plant Species of the World*, 2nd ed. CABI, Wallingford, UK.

Winston, R. L., Schwarzländer, M., Hinz, H. L., Day, M. D., Cock M. J. W. and Julien, M. H. (eds.) (2014) *Biological Control of Weeds: A World Catalogue of Agents and Their Target Weeds, 5th Edition.* USDA Forest Service, Forest Health Technology Enterprise Team, Morgantown, West Virginia. FHTET-2014-04.

Wohlgemuth, T., Gossner, M. M., Campagnaro, T., Marchante, H., van Loo, M. and others (2022) 'Impact of non-native tree species in Europe on soil properties and biodiversity: A review', *NeoBiota*, vol. 78, pp. 45–69.

Woodward, S. L. and Quinn, J. A. (2011) *Encyclopedia of Invasive Species, Vol. 1 & 2.* Greenwood, Santa Barbara, CA.

Wylie, F. R. and Speight, M. R. (2012) *Insect Pests in Tropical Forestry*, 2nd ed. CABI, Wallingford, UK.

Zemanova, M. A., Broennimann, O., Guisan, A., Knop, E. and Heckel, G. (2018) 'Slimi invasion: Climatic niche and current and future biogeography of *Arion* slug invaders', *Diversity and Distributions*, vol. 24, pp. 1627–1640.

Zizka, L. H. and Dukes, J. S. (2011) (eds.) *Weed Biology and Climate Change.* Wiley-Blackwell, Ames, IA.

33

PATHOGENS AND INVERTEBRATE PESTS IN NORTH AMERICAN FOREST ECOSYSTEMS

Sandy M. Smith and Louis Bernier

Introduction

Trees are believed to be the organisms that define forests because of their overriding biomass, structure, and longevity. However, the most significant contributors to forest biodiversity are microbes and invertebrates since they collectively perform a range of functions critically important within forest ecosystems. Bacteria, fungi, and invertebrates are all key players in the development and fertility of forest soils, and living trees form a range of associations with a wide variety of endophytic microbes that ensure their success. Insect pollinators greatly facilitate sexual reproduction in many tree species, and once a tree dies, a succession of wood-boring invertebrates, bacteria, and decay fungi allows for the recycling of its lignocellulosic complex. In so doing, trees are truly the widely recognized basis for biodiversity complexity associated with the world's forests.

Living trees are impacted by a variety of disease-causing microbial pathogens and invertebrates (often referred to collectively as pests) that feed on organs or tissues, either aerial or below ground. Depending on the combination of host, pathogen/pest, and environmental conditions, the outcome of this interaction may vary from no observable effect to decreased growth to tree death. The effect of pathogens and pests is observable at different scales, from individuals to species, and from communities to landscape-level forest stands. Losses caused by pathogens and pests range from the conspicuous to the spectacular, from short-term, catastrophic, and complete tree mortality to long-term successional changes in forest ecosystems. Although this may seem counterintuitive to forest managers, by contributing to stand structure, availability of deadwood, and the creation of suitable habitat for plants and wildlife, many pathogens and pests are important natural drivers of forest composition, succession, and evolution through disturbance. Consequently, the old utilitarian view of healthy forests being characterized by low tree mortality and an absence of forest disease or insects has gradually given way to a broader ecosystem-based concept, one that recognizes disease and pests as natural components of healthy forests, and, in fact, oftentimes define

DOI: 10.4324/9781003324072-38

them (Franklin et al., 1987; Kolb et al., 1994; Castello et al., 1995; Ostry and Laflamme, 2009; Carroll 2023). Today, the diversity of these biological communities, formally termed biodiversity, is quantified and converted to indices that provide a broad measure of forest health.

Unfortunately, pathogens and invertebrate pests may also jeopardize the resilience of forest ecosystems. This has become increasingly apparent with the intensification of management and anthropogenic activities that disturb natural ecosystems. Forestry practices such as silviculture, management, and harvesting have a direct effect on the structure and function of forest ecosystems. Indirect anthropogenic effects are also a cause for concern, particularly in the case of invasive pathogens, pests, and weeds. When these diverse communities are disrupted and replaced with simpler, more uniform complexes, the forest ecosystem is destabilized and its resilience is compromised. Already, a combination of direct and indirect factors has pushed some forest ecosystems over the tipping point, after keystone tree species were destroyed or cascading effects were set in motion (Kenis et al., 2009; Reyer et al., 2015).

In this chapter, we discuss how pathogens and pests shape the structure and functions of forest ecosystems. We address this issue both in the "traditional" context in which natural forests interact with native pathogens and insect pests with which they have co-evolved, and in the current context of intensive forest management and plantation-based forestry, urban forests, biological invasions, and climate change.

Biological features of pathogens and pests

Pathogens are biotic agents that cause disease; they include viroids, viruses, bacteria, phytoplasmas, fungi, oomycetes, nematodes, and parasitic plants. With the exception of nematodes, invertebrates are not considered as pathogens. Disease refers to any malfunctioning of host cells or tissues that results from continuous irritation by a pathogen or environmental factor and leads to the development of symptoms (Agrios, 2005). Fungi are prominently among the most successful plant pathogens and are responsible for well-documented pandemics, such as Dutch elm disease, chestnut blight, and white pine blister rust. However, the last decades have witnessed the emergence of new pandemics caused either by successful interactions between agents (e.g., beech bark disease) or by other agents such as bacteria (e.g., bacterial leaf scorch) and oomycetes (e.g., sudden oak and larch death).

Deleterious herbivores, often referred to more generically as "pests", are the other major group of biotic agents that can have negative impacts on forest ecosystems. A forest pest refers to any animal that interferes with the survival and successful development or reproduction of trees. Invertebrate pests include insects, acari (mites), molluscs (slugs/snails), as well as exotic annelids (earthworms). Well-recognized vertebrate pests, especially in plantations, include rodents (mice, rabbits, squirrels, and porcupines), ungulates (deer and moose), and, infrequently, depending on the context, birds (sapsuckers and cormorants). Most commonly, pests are restricted to specific stages of tree growth, with some attacking only young seedlings, others only feed on early successional stages or mature, closed canopies, while others are found only on older, declining trees (Coulson and Witter, 1984). According to the part of the tree they feed on or the type of localized damage they inflict, invertebrate pests are usually specialized and classified as leaf feeders, tip and shoot feeders, xylem and phloem feeders, root feeders, or seed and cone feeders.

Pests are typically mobile and locate their hosts primarily through chemotaxis (i.e., smell, taste, etc.), initially orienting to the optimal habitat (habitat location), and then refining their search to susceptible targets within that area (host location). Most invertebrates rely on specific chemical cues (kairomones) emitted from their host plants and can recognize healthy or stressed trees from considerable distances, in some cases kilometers. Some species use other cues such as sound for drought-responsive scolytid beetles (Mattson and Haack, 1987) and/or recently burnt areas for fire-responsive species such as cerambycids (long-horned beetles) (McCullough, 1998).

Unlike invertebrates, pathogens that attack aerial parts of plants are unable to actually "locate" their host. Instead, they are dispersed passively by wind, rain splash, running water, or by biological vectors. Many pathogenic fungi produce very large quantities of spores (through asexual reproduction, sexual reproduction, or both) that are released into the air when environmental conditions are favorable, for example, under high moisture conditions. The vast majority of spores land on host plants that are not suitable for infection or colonization. However, when spores land on a suitable host species, the pathogen either establishes itself in the host plant (compatible interaction) or fails to do so (incompatible interaction). Success or failure is determined by a combination of genetically controlled internal plant and microbe factors and is mediated by external, environmentally controlled factors (Asiegbu, 2022).

Pathogens rely heavily on free water movement and air currents for dispersal, and usually produce sporulating structures directly on the surface of infected plants. Pathogens dispersed by biological vectors, however, may produce their sporulating structures inside host tissues. For example, ophiostomatoid fungi produce their sporulating structures in wood galleries made by bark beetles (Figure 33.1). The masses of fungal spores found in these structures are often coated with mucilage, thus promoting their adherence to insects that exit the galleries on their way to colonizing new host trees. Other fungal species emit volatiles that attract insects, and sometimes vertebrates. Some interactions between fungi and vectors are highly specific. For example, female southern pine beetles (*Dendroctonus frontalis*) carry the mutualistic fungi *Ceratocystiopsis ranaculosus* and *Entomocorticium* sp. in specialized structures known as mycangia. These fungi are inoculated by the insect into the inner bark and phloem of the host trees, and their development in galleries has been associated with increased fitness of the developing larvae of *D. frontalis*. Mycangial associates of *D. frontalis* also include actinomycetes that selectively inhibit the sapstaining fungus *Ophiostoma minus*, which is transported phoretically by the insect and has an antagonistic effect on the development of beetle larvae (Scott et al., 2008). Another group of wood-colonizing insects, ambrosia beetles, also have highly specific interactions with fungi and are, in fact, considered to actively grow and maintain fungal gardens (Biedermann et al., 2013; Huler and Skelton, 2023).

Pathogens use various means to penetrate and invade host plant tissues. Specialized structures called appressoria enable some fungal pathogens to physically penetrate host tissue. Furthermore, several pathogens secrete lytic enzymes that weaken host tissues or components thereof, and toxins that allow them to kill host tissue ahead of their progression. The arsenal of pathogens also includes molecules that may block the production of defense molecules by plants or detoxify such molecules if they are produced. Some pathogens also secrete homologs of plant growth hormones, which allow them to alter the physiology of their host (Agrios, 2005).

Fungal pathogens are subdivided into three broad classes based on how they interact with their host. Necrotrophs kill tissues as they grow through them and complete their life cycle (including reproduction) on dead tissue. Several species of canker-causing agents fall

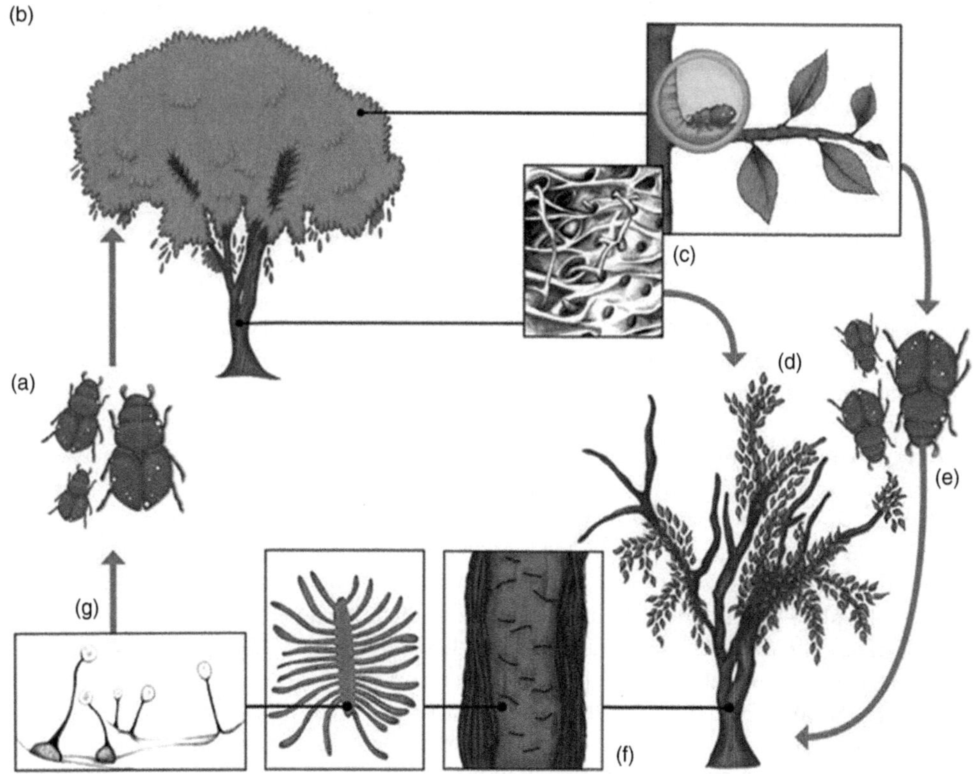

Figure 33.1 Biological cycle of Dutch elm disease (DED), an example of fungal pathogen–insect vector association that has resulted in two devastating pandemics. Young elm bark beetles carrying conidia or ascospores of DED fungi (a) give pathogens access to water-conducting vessels within the xylem when they feed on healthy trees (b). The DED fungi produce yeast-like spores and mycelium when invading the vascular system (c). Infection of susceptible elms causes wilting and eventually death (d). Virgin female bark beetles in search of suitable breeding sites are attracted by elms that were killed or weakened by DED. Many of these beetles carry spores of DED fungi (e) that rapidly colonize galleries in which females have oviposited after mating (f). Within the galleries, the DED fungi produce reproductive structures (g) that include asexual synnemata and sexual perithecia. Both structures produce spores embedded in a drop of sticky mucilage that easily attach to the exoskeleton of young elm bark beetle adults (a) emerging from the galleries. (Figure created by Marilou Desharnais and reprinted from Comeau et al., 2015.)

Note: DED pathogens can also infect healthy trees through root grafts with a diseased individual.

into this category. Biotrophs need host tissue to remain alive for completing their life cycle, and are thus often host-specific, including aggressive pathogens that cause stem and leaf rusts. While the host must remain alive for the pathogen to complete its life cycle, diseases caused by biotrophs are nevertheless highly destructive. The third class of pathogens, termed hemibiotrophs, exhibit an intermediate lifestyle characterized by an initial growth phase in living host tissue, followed by a reproductive phase in dead tissue.

Typically, pests are localized to either the dead or living parts of the tree; the former include saproxylic species such as most wood-boring long-horned beetles or some buprestid beetles as well as carpenter ants. The latter includes the majority of invertebrate species that attack the phloem and/or photosynthetic and meristematic tissues of trees (i.e., leaf-feeding caterpillars, leafminers, shoot and phloem feeders, gallformers, seed and root feeders) as well as vertebrate animals such as ungulates, rodents, and sapsucking bird species (Coulson and Witter, 1984; Allison et al., 2023). Pests can kill trees quickly in a year or less by breaking the vascular transport of the tree (phloem feeders), or more slowly through repeated removal of photosynthetic parts that limit overall tree growth and nutrient storage (e.g., leaf feeders causing defoliation).

Pests play an important role in the death spiral of trees and can act as either the primary or secondary agents in this process, whether in natural forest systems or constrained urban landscapes (Hilbert et al. 2019). Agents that create physical openings for pathogens to invade living trees (e.g., beech bark scale, elm and other scolytid bark beetles) act indirectly while those who play an obligatory role in the pathogen's lifecycle include leaf viruses and phytoplasmas introduced into the tree through the sucking mouthparts of insects such as aphids and scales (Weintraub and Beanland, 2006; Branco et al. 2023).

Pests usually display a preference for feeding on a specific part of the tree, with invertebrates often being solely dependent on this localization for their complete life cycle (Allison et al., 2023). Defoliators are the most common guild and include all pests that feed on living leaf or needle tissues. They can be external feeders such as caterpillars (lepidopteran, coleopteran, and hymenopteran immatures), adult beetles, vertebrates such as ungulates and rodents, and usually consume the photosynthetic material of the tree. Many invertebrates such as leafminers and gallformers also feed internally on these living leaf tissues and are recognized by their characteristic feeding patterns (blotch leafminer, serpentine leafminer, etc.); all have chewing mouthparts and mechanically destroy plant tissue, as do vertebrate pests.

Tip and shoot-feeding pests rarely cause whole tree mortality (Coyle, 2023), however, their impact changes the form and growth of the tree and may cause serious timber defects in plantation forests. Under natural growing conditions, species such as balsam fir adelgid and white pine weevil cause weakened trees to be outcompeted by neighboring trees, thus playing an important role in stand succession. Root-feeding pests spend the majority of their time underground and can lead to overall tree decline (in some cases complete tree mortality). Their impact on natural forest dynamics is unknown but thought to be similar to that in managed plantations, albeit at a lower frequency (Coulson and Witter, 1984; Allison et al., 2023).

Trunk-feeding pests that attack phloem or vascular tissues of the tree stem are usually considered the most damaging of all feeding guilds, leading quickly to whole tree mortality if the trunk or main branches become completely girdled by feeding (Allison et al., 2023). Aggressive *Dendroctonus* bark beetle species such as the mountain pine beetle (*D. ponderosae*) and southern pine beetle (*D. frontalis*) are two well-known pests that have caused consistent, large-scale stand and landscape-level mortality. Recent research has shown that they, along with widespread defoliators such as the spruce and jack pine budworms, are linked to stand breakup and increased fire risk under natural conditions (McCullough, 1998). In all cases, these pests are considered nature's "silviculturalists", as they remove slow-growing, declining trees and stands to allow vigorous understory succession.

Pests that feed on the reproductive parts of trees (flowers, seeds, and cones) can almost completely deplete a tree's annual reproductive potential, often losing more than 90% of a crop (Coulson and Witter, 1984; Strong et al., 2023). In most cases, this is of minor ecological consequence as trees either overproduce reproductive structures or vary their annual output through masting (e.g., family Fagaceae). In so doing, pests are unable to track this unpredictable, yet highly valuable food resource every year.

Ecological effect of pathogens and pests on forest ecosystems

Natural pathogen and invertebrate–host interactions that have co-evolved tend to be at equilibrium (Dinoor and Eshed, 1984). However, this does not imply that the equilibrium persists at all times or over the entire host range. For example, pests that damage or kill individual trees contribute to gap formations in the stand dynamics of temperate forests, while those that impact large, forested areas are often precursors of fire and stand replacement. Natural forest stands can be strongly affected by insect defoliators, such as spruce budworm (*Choristoneura fumiferana*) and jack pine budworm (*C. pinus pinus*) in coniferous forests and forest tent caterpillar (*Malacosoma disstria*) and spongy moth (*Lymantria dispar*, formerly called gypsy moth) in hardwood stands. In most cases, trees do not die immediately from defoliation, but instead become weakened and susceptible to attack by other agents (Coulson and Witter, 1984; Tobin et al., 2023). Whole tree or stand mortality occurs if this is repeated over time. Although these epidemics may be spectacular, they do not compromise forest ecosystems as they are often the disturbance factor needed to transition a mature forest canopy into a new cycle of succession. Studies have shown an increasing number of native pest species that interact with forest systems on such large scales, causing extensive, periodic landscape-level stand mortality and subsequent forest renewal. For example, dendroecological evidence suggests that the larch sawfly (*Pristiphora erichsonii*), an endemic defoliator causing extensive tree mortality, has historically been a major driver in the succession of larch on landscapes across eastern North America (Jardon et al., 1994).

Epidemics caused by introduced pathogens and/or invertebrates may significantly alter the composition of forest ecosystems, impact ecosystem services, and eventually lead to ecosystem meltdown. Chestnut blight, caused by the fungus *Cryphonectria parasitica*, almost eradicated American chestnut (*Castanea dentata*) within a few decades after the pathogen was accidentally introduced from Asia to the United States of America (USA). The impact of the disease has been especially visible in northeastern USA, where chestnuts accounted for more than 20% of the canopy in some areas. The widespread death of these trees allowed other deciduous tree species, such as pignut hickory (*Carya glabra*), red maple (*Acer rubrum*), and sugar maple (*Acer saccharum*), to colonize those areas (McCormick and Platt, 1980). A similar shift in forest composition has occurred following the past century of invasion (from Europe) by spongy moths in oak-dominated forests across eastern North America, with a significant loss of the oak complement (Davidson et al., 1999; Liebhold et al., 2023); concomitant with this change has been responses at multiple trophic levels in native faunal communities. Such a widespread forest restructuring is also being repeated with the loss of ash following the introduction of the emerald ash borer (*Agrilus planipennis*) from Asia into North America.

Diseases, pests, and fire are often investigated separately by pathologists, entomologists, and fire specialists, respectively. However, many important forest and tree health issues result from associations between at least two disturbance agents. For example, the two pandemics of Dutch elm disease (DED) that have devastated native elm populations in Europe and

North America resulted from the acquisition of the exotic fungal pathogens *Ophiostoma ulmi* and *Ophiostoma novo-ulmi* by elm bark beetles, *Hylurgopinus* and *Scolytus* spp. (Figure 33.1). In the absence of the DED fungi, elm bark beetles can hardly be considered as pests, since they cause very little, if any, damage to elm trees. Dutch elm disease illustrates how otherwise benign invertebrates become pests once they serve as vectors for aggressive pathogens. Similar examples include oak wilt in which generalist nitidulid beetles vector the fungal pathogen *Bretziella fagacearum* (Juzwik et al., 2008) and the multi-host Pierce's disease caused by the bacterium *Xylella fastidiosa*, which is vectored by many insects from the leafhopper (Cicadellidae) and spittlebug (Cercopidae) families (Rapicavoli et al., 2019).

Tree diseases may result from successions of attacks by different disturbance agents. This is the case for beech bark disease that occurs when the bark of trees attacked and damaged by the beech scale insect, *Cryptococcus fagisuga*, is colonized by spores of the fungi *Neonectria ditissima* or *Neonectria faginata*. While beech has developed a degree of tolerance to the fungus, the disease has intensified with the arrival of beech bark scale, which feeds by puncturing the bark with sucking mouthparts, thus allowing the pathogen to enter the vascular system of the tree (Houston, 1994). This has led to widespread beech mortality in eastern North America where the two disturbance agents co-occur (Cale et al., 2017).

There are also several cases where insect attacks follow an initial attack by pathogens. For example, trees that have been weakened by root pathogens, such as *Armillaria mellea* and *Heterobasidion* spp., are highly attractive to secondary insects. Stress compounds emitted by weakened trees are the main cues used by all saproxylic beetles to locate potential hosts for colonization. The larger pine shoot beetle, *Tomicus piniperda*, is one recent North American example of the impact of sequential stress factors, where the combination of poor site conditions, root pathogens, and beetle attack on both the shoots and trunks of trees lead to rapid stand mortality (Paine et al., 1997). Some researchers have hypothesized that pathogens that rely on arthropods for dissemination may manipulate their host to make it more attractive to vectors (e.g., McCleod et al., 2005).

A final example of complex interaction is provided by forest declines resulting from interactions among successive predisposing, inciting, and contributing factors (reviewed in Manion, 2003). Predisposing factors, which are usually abiotic and long-term acting, alter the ability of trees to withstand or respond to pathogens or pests. Inciting factors can be abiotic or biotic and are characterized by high intensity and short duration that further weaken the trees and usually result in dieback. These trees may not survive the effect of contributing factors such as canker fungi, decay fungi, or wood-boring insects (Franklin et al., 1987). According to Manion (2003), some forest declines are part of the natural cycling of populations and are not expected to result in the death of ecosystems. Thus, forest declines might, in fact, act as a stabilizing selection agent, whereby competitive dominant trees are selectively killed, whereas stress-tolerant dominant individuals survive and gain the opportunity to contribute to the gene pool.

Effect of changing environments on pathogens and pests

Traditional forest management

So far, we have focused on the impact of pathogens and pests on forest ecosystems. In the following, we address how shifts in the environment, either natural or anthropogenic in nature, affect biotic disturbance agents. The indirect role of abiotic factors in predisposing trees to attack by pathogens and invertebrate pests has already been observed. Environmental factors can also directly affect populations of pathogens and pests. For instance, populations of fungal leaf pathogens fluctuate widely according to climatic conditions. Under conditions of higher moisture, large populations may arise quickly through repeated rapid cycles of asexual reproduction. Additionally, the impact of fires in reducing the inoculum of parasitic dwarf mistletoes (*Arceuthobium* spp.) into coniferous forests of interior western North America is well documented (Parker et al. 2006). Likewise, high-intensity wildfires may reduce populations of pathogens and arthropods that inhabit forest soils, litter, and coarse woody debris (Parker et al. 2006).

Silvicultural practices may increase tree mortality caused by both pathogens and pests. In many localities in the northern hemisphere, hardwood- or mixed stands have been converted into conifer plantations. After a few years of growth, high levels of mortality are sometimes observed in these plantations. Pronos and Patton (1978) reported that mortality caused by the root pathogen *Armillaria mellea* in young red pine plantations in Wisconsin was more noticeable in plantations established on sites that had previously supported oak (*Quercus rubra* and *Quercus alba*). *A. mellea* was present on the roots of living oaks, on which it grew as an epiphyte, without being able to invade the cambial area and cause diseases. However, oaks that were herbicide-killed when the pine plantations were established were heavily colonized by the fungus and became reservoirs for further attacks on red pine saplings. Due to its ability to produce large and long-lived underground rhizomorphs, as well as its capacity to grow saprophytically on dead trees and coarse woody debris, *A. mellea* can persist for decades at suitable sites. In particular, the presence of large stumps provides the fungus with a food base that it may exploit for decades.

Another example of silvicultural prescriptions that can cause unexpected changes in tree growth, form, and structure has been reported when young conifer stands were thinned very early in stand establishment, thereby promoting the survival of pests such as the white pine weevil (*Pissodes strobi*). This major shoot and tip insect kills the tree leader and reduces tree height growth by 1–3 years. Open, sunny microhabitat conditions favor overwintering survival of adult weevils in the ground litter and promote larval feeding and development in the leaders (Coyle, 2023). When tree densities are reduced and stands opened up early in stand development, weevil populations thrive under the increased temperature and attack the remaining tree leaders; they do not decline again until the stand canopy closes (Paine et al., 1997).

Inappropriate logging practices may also promote the spread of pathogens within forest stands. Annosus root and butt rot of coniferous species are caused by the basidiomycete fungi *Heterobasidion* spp., which are efficient colonizers of fresh wounds (Hodges, 1969; Kovalchuk et al., 2022). Although the disease can be expected in unmanaged forest stands, its occurrence in managed stands can increase dramatically if thinning operations are carried out during periods when there are high concentrations of *Heterobasidion* spores in the air. Freshly cut stumps are rapidly colonized by the fungus, which later moves to adjacent, living trees through root contacts. As the mycelium of *Heterobasidion* infects and kills the roots of newly infected trees, several new infection centers around stumps may be created and expand radially over the following years. Similarly, high rates of tree damage, or failure to remove infested material during logging, may result in outbreaks of pest species that can build up in populations large enough to kill healthy living trees in adjacent areas.

In many parts of the world, industrial forestry and the need to protect human dwellings have resulted in aggressive suppression of wildfires. This has led to significant alterations of the landscape, including changes in tree species composition, distribution patterns, and increases in stand density. These changes have, in turn, altered the balance between fire, pathogens, and invertebrate pests. According to Teale and Castello (2011), the rapid expansion of fusiform rust of pines in the southeastern USA in the 20th century is, in part, due to fire suppression. The latter has favored the expansion of slash pine (*Pinus elliottii*) and loblolly pine (*Pinus taeda*), both more sensitive to fire and fusiform rust than longleaf pine (*Pinus palustris*). Fire suppression has has stimulated the regeneration of oak, which is the alternate host required by the fusiform rust fungus (*Cronartium quercuum* f. sp. *fusiforme*) to complete its life cycle, thus allowing pathogen populations to expand.

Fire suppression has also been associated with a suite of diseases and pests in forests of western North America (Parker et al., 2006). Root pathogens and bark beetles appear to have been the main beneficiaries of this situation. The massive outbreak of mountain pine beetle (*Dendroctonus ponderosae*) during the past decade is another example of secondary pest outbreaks believed to be due, at least partially, to fire suppression (Carroll 2023). Although many factors are involved, including warmer winters and drier summers, one of the main drivers is thought to be overmature, declining pine that would have been removed by natural wildfire. Large contiguous stands on the landscape under fire suppression become susceptible to lightning strikes and fuel ignition (Parker et al., 2006). This is also true in eastern North America where mature conifer stands maintained under fire suppression have led to increased susceptibility to spruce budworm (Tobin et al., 2023).

Modern forestry also relies increasingly on plantations. The deployment of monospecific, even-aged plantations poses phytosanitary risks, as it may promote outbreaks that would not otherwise be observed in natural stands characterized by a broader diversity and heterogeneous age structure. For example, the epidemic of *Gremmeniella* canker that started in several US states and Canadian provinces during the 1970s was in large part fueled by the establishment of red pine (*Pinus resinosa*) plantations. The epidemic was worsened by the fact that several tree nurseries that supplied red pine seedlings were unknowingly exporting contaminated, but asymptomatic, material. This led to a rapid expansion of the disease zone before proper quarantine measures were implemented to restrict movement of red pine seedlings (Warren et al., 2011). Similarly, the ongoing outbreak of emerald ash borer (*Agrilus planipennis*) has been fuelled in part by the overabundance of susceptible ash cultivars in eastern North America (Aukema et al., 2010). Urban areas in this region were planting their parks, streets, and new suburbs with nursery stock of green ash (*Fraxinus pennsylvanica)* grafted from similar clonal material highly susceptible to this introduced pest. In turn, this has allowed emerald ash borer populations to establish rapidly and expand spectacularly throughout the region.

Monospecific plantations are not necessarily composed of genetically uniform material since a variety of genotypes are commonly deployed to minimize the risk that an entire plantation is destroyed by a given pathogen or pest. This strategy has been used for decades in Europe for managing poplar leaf rust caused by *Melampsora larici-populina*. Epidemiological studies of European populations of *M. larici-populina* have shown that this pathogen was evolving rapidly in response to the deployment of new varieties of poplars that carried specific resistance genes that had been selected by breeders (Xhaard et al., 2011). Over the years, increasingly complex races of pathogen that had acquired the genes necessary to overcome poplar resistance genes were recovered, thus confirming observations that had been made earlier in agricultural systems, where strong selection pressure for crop resistance is a major driver of pathogen and pest rapid evolution (Stukenbrock and McDonald, 2008).

The introduction of exotic tree species may have unexpected consequences on pathogens and pests, as shown by the rise of *Septoria* canker in North America. Poplar breeding programs initiated in the mid-1900s in the USA and Canada relied extensively on interspecific crosses among native and exotic poplars. Although breeders had selected material with high resistance against the known diseases of aspens and poplars, entire plantations were decimated by a new canker disease caused by a native fungus (*Spherulina musiva*) which, until then, was only known to cause a relatively benign leaf spot disease in native poplars (Feau et al., 2010). Although resistance to *Septoria* canker was successfully integrated into poplar breeding programs, this episode is a potent reminder of the genetic and behavioral plasticity of pathogens and pests. In the case of pests, resistance breeding has been pursued for effective management; however, in most cases, this tactic is still in its infancy. Recent research using molecular tools to identify hybridized lodgepole and jack pine under attack by mountain pine beetle suggests that tree resistance may have some promise (Cullingham et al., 2011); however, tree lifespans and the specialized attack patterns of pests are likely to limit the success of this strategy in the long run.

Invasive species

The early cases of introduced pathogens and pests led to legislation and quarantine measures to prevent the intercontinental spread of these agents. Unfortunately, the efficiency of these measures has been diminished in good part by the explosion in international trade and human movement during the last decades. As a result, new invasive pathogens and pests are being encountered annually in various countries (Aukema et al., 2010; Boyd et al., 2013; Liebhold et al., 2023). The introduction of pathogens and pests into new geographic areas threatens the integrity of global forest ecosystems. In many cases, we have seen epidemic outbreaks of exotic pathogens and pests on native tree species, to the detriment and sometimes extinction of the latter. In others, both deliberate and accidental introductions have led to major concerns over potential invasion meltdown of native ecosystems, those already threatened by shrinkage, intensified management, and urbanization.

Some of our most significant introduced pathogens and pests have resulted in large-scale and costly management programs, including chestnut blight, Dutch elm disease, spongy moth, many species of conifer sawflies, and now hemlock woolly adelgid, emerald ash borer, and the Asian long-horned beetle. Hundreds of exotic forest species are currently regulated in Canada and the USA based on their potential to cause economic loss. Pimentel et al. (2005) reported about 50,000 foreign species in the USA causing losses of up to $120 billion per year, with 42% of endangered or threatened species at risk primarily due to these invaders. Lovett et al. (2006) provide an overview assessing the potential ecological impact of both groups in North American forests based on mode of action, host specificity, virulence, importance of host, uniqueness of host, and phytosociology of the host. Without question, this is the single most important factor affecting the future ecological integrity of many forest ecosystems throughout the world (Jactel et al., 2023), including North American forests. Four distinct phases in the invasion process are broadly recognized: introduction, establishment, spread, and impact (Levine, 2008). To reduce the risks associated with invasive species, the transition from one phase to the next must be reduced as quickly and effectively as possible (Jactel et al., 2023) using a combination of approaches including risk assessment, early detection, rapid response, and control or eradication methods (Raffa et al., 2023).

Although the impact of exotic invasive pests on the landscape can be dramatic, affected tree species may recover if they exhibit genetic diversity for traits such as tolerance and/or resistance to invaders. Since resistant individuals occur at a very low frequency, tree species that include large populations have a better chance of surviving than those in low numbers. Butternut (*Juglans cinerea*) is one example of a tree species put at risk of extinction due to a single pathogen, *Ophiognomonia clavigignenti-juglandacearum*, believed to have been introduced into North America during the 20th century (Woeste et al., 2009). Although the geographic range of butternut is large, encompassing most of eastern temperate North America, it is not an abundant species. The fate of butternut is further impacted by loss of habitat due to anthropic activities and "genetic invasion" from hybridization with exotic congeners (*Juglans ailantifolia* and *Juglans regia*). Interestingly, since these interspecific hybrids usually are highly resistant to butternut canker, controlled interspecific crosses followed by introgression of resistance through backcrosses have been proposed as a strategy against the disease. A similar path had been chosen to save the American chestnut from possible extinction caused by the fungus *C. parasitica* (Anagnostakis, 2012). However, in the case of butternut, the observation of disease resistance within nature has led researchers to advocate resistant germplasm from single-tree intraspecific selection and breeding, rather than developing interspecific hybrids (Michler et al., 2006). This approach is also being implemented in other pathosystems. For example, DED-resistant individuals have been identified in European populations of *Ulmus minor* and North American populations of *Ulmus americana*, leading to renewed interest in restoration planting of native elms in several countries (Martin et al., 2023; Knight et al., 2017). In the case of beech bark disease, the strategy relied on identifying individuals of *Fagus grandifolia* that are resistant to beech scale rather than to the pathogens. Several beech scale-resistant trees have been identified and resistance has been shown to be heritable, allowing for intraspecific resistance breeding (Cale et al., 2017).

Dramatic ecosystem changes may occur as a result of pathogen or pest introductions. For example, whitebark pine (*Pinus albicaulis*), considered a keystone species in subalpine ecosystems of western North America, is under threat across its range as a result of white pine blister rust (caused by the introduced pathogen *Cronartium ribicola*), as well as attacks by native mountain pine beetle, fire exclusion, and climate change. The cascade effects from potential loss of whitebark pine have been predicted to include the provision of high-energy food for wildlife and nurse trees for other species in open terrain, and the retention of snowpack (Smith et al., 2013). Broad-scale ecosystem concerns have also been expressed with respect to other invasive insect species, such as the hemlock woolly adelgid. Although at different spatial and temporal scales, Kizlinski et al. (2002) showed profound changes in stand structure, composition, and ecosystem function in forests dominated by hemlock following invasion by the adelgid.

Similarly, the slow advance of native and introduced earthworm populations has shown significant ecological impacts on forests in northeastern North America. The non-selective feeding, characteristic mucus channels, and soil compaction activities of these newly arrived annelids have been linked to shifts in soil communities, away from native mycorrhizae in favor of bacterial and microbial communities (Hendrix, 2006). As species co-evolved with native forest flora are disrupted and fungi lost from the forest floor, small mammal communities shift from fungal-based voles to omnivorous deer mice, which have been linked to higher tick populations and the spread of human lyme disease in forest stands of eastern North America. These cascading effects are complex and the true impact of such ecological changes on long-term forest vegetation and succession in native forests remains poorly known.

A single invasive species may alter the landscape over a large area without necessarily inducing ecosystem meltdown, as in the case of chestnut blight. However, some areas have suffered regular invasions by different invasive species, which may eventually negatively affect the resilience of forests. This is clearly the case in many urban forests in North America that were first infested by chestnut blight, and then Dutch elm disease in the 20th century. In many of these cities and towns, street chestnuts were replaced with elm and then with ash (*Fraxinus* spp.) or Norway maple (*Acer platanoides*). Ash is now being decimated by the emerald ash borer, an invasive beetle in North America, while maple is the prime target of the Asian long-horned beetle (*Anoplophora glabripennis*), a new species that attacks over ten species of trees and may never be eradicated from North America (Dodds and Orwig, 2011).

Climate change

Several decades ago, plant pathologists began to speculate that climate change might influence the fate of tree-pathogen/pest interactions at the landscape level (Coakley et al., 1999). Evidence is now accumulating that this is happening in various regions of the world (Sturrock, 2012). For instance, shifts to warmer and wetter climate have been associated with outbreaks of *Diplodia* shoot blight of pines in France (Fabre et al., 2011) and *Dothistroma* needle blight in western North America (Welsh et al., 2014). This result is not surprising given that such environmental conditions are known to promote the expansion of plant pathogen populations.

Similar insights have been gleaned by those making predictions about the future impact of climate change on invasive pests in forest systems. Most models point to increasing tree mortality under rising temperatures although the complexity of the models and high degree of uncertainty mean that no specific predictions can be made (Dukes et al., 2009). The general consensus is that the increasing variability and extremes in weather patterns will favor some pest species and not others. The mobility and short generation times of most pests relative to their hosts mean that even though they can disperse and adapt quickly in the short term, shifts in temperature and moisture will have significant effects on survival and reproduction in the long run, especially along range edges where intensive selection pressure occurs. Ultimately, the success of highly specific invertebrate pests will be heavily dependent on the ability of their host itself to survive the effects of rapidly changing, extreme weather patterns (Battisti and Larsson, 2023).

Conclusion

Forest insects and pathogens are the most pervasive and important agents of disturbance in North American forests, historically affecting an area almost 50 times larger than that affected by fire and with an economic impact nearly five times as great (Aukema et al., 2010). Disturbances caused by pathogens and pests are a constant challenge for forest managers but are nonetheless a vital driver of forest composition, succession, and evolution (Tobin et al., 2023). Given these complex ecological processes and natural disturbances, our response to pest and pathogen challenges must be cautious. We need to develop a strong understanding of these processes, our human footprint on them, and the large temporal and spatial scales on which they operate, especially in the remaining intact forests. We must avoid focusing on specific isolated pathogen and pest problems, and instead respond strategically with integrative thinking that supports the development of broad forest stand and landscape resilience.

Many factors are converging that will restrict our ability to retain healthy, resilient forests. Increasing the intensification of forest practices for timber production and land use changes that result in deforestation for agricultural and urban development will dramatically restrict most tree species. This, combined with the rapidly expanding introduction of invasive alien species to most of the world's forests, will exacerbate the effect of pathogens and pests. On top of this is the large unknown of climate change, a new factor that will have major implications for pathogen–insect interactions in our forests. We must be cautious in believing that we can predict, let alone manage, the outcome of disturbances under these complex scenarios.

If we are to successfully retain ecological balance in our forests, our response must become more considered than simply removing every individual (the naturally resistant along with the susceptible) when faced with a pathogen or pest problem. The concept of a healthy forest requires thinking about gene pools and species interactions at the population level in order to achieve successful adaptation under a changing environment. This means working with forest systems to avoid ecological tipping points and supporting natural biodiversity that enables a return to natural states. Moving forward, we need to aim for ecosystem resilience in our forests, where forest health is defined not by the lack of specific pests or pathogens, but by the complex ecological interactions and functioning of diverse forest organisms and communities.

References

Agrios, G. N. (2005) *Plant Pathology*, 5th edition. Elsevier Academic Press, Amsterdam.

Allison, J. D., Paine, T. D., Slippers, B. and Wingfield, M. J. (eds.) (2023) *Forest Entomology and Pathology, vol. 1: Entomology*. Springer Press, Switzerland.

Anagnostakis, S. L. (2012) 'Chestnut breeding in the United States for disease and insect resistance', *Plant Disease*, vol 96, pp. 1392–1403.

Asiegbu, F.O. (2022). 'Basic concepts and principles of forest pathology', in F.O. Asiegbu and A. Kovalchuk (eds.), *Forest Microbiology: Volume 2: Forest Tree Health*. Elsevier Academic Press, Cambridge.

Aukema, J. E., McCullough, D. G., Von Holle, B, Liebhold, A. M., Britton, K, and Frankel, S. J. (2010) 'Historical accumulation of nonindigenous forest pests in the continental United States', *Bioscience*, vol 60, pp. 886–897.

Battista, A. and Larsson, S. (2023) 'Climate change and forest insect pests', in J. D. Allison, T. D. Paine, B. Slippers, and M. J. Wingfield (eds.), *Forest Entomology and Pathology. Vol. 1: Entomology*. Springer Press, Switzerland.

Biedermann, P. H. W., Klepzig, K. D., Taborsky, M. and Six, D. L. (2013) 'Abundance and dynamics of filamentous fungi in the complex ambrosia gardens of the primitively eusocial beetle *Xyleborinus saxesenii* Ratzeburg (Coleoptera: Curculionidae, Scolytinae)', *FEMS Microbiology Ecology*, vol 83, pp. 711–723.

Boyd, I. L, Freer-Smith, P. H., Gilligan, C. A. and Godfray H. C. J. (2013) 'The consequence of tree pests and diseases for ecosystem services', *Science*, vol 342, 1235773.

Branco, M., Franco, J. C., and Mendel, Z. (2023) 'Sap-sucking forest pests', in J. D. Allison, T. D. Paine, B. Slippers, and M. J. Wingfield (eds.), *Forest Entomology and Pathology. Vol. 1: Entomology*. Springer Press, Switzerland.

Cale, J. A., Garrison-Johnston, M. T., Teale, S. A., and Castello, J. D. (2017). 'Beech bark disease in North America: Over a century of research revisited', *Forest Ecology and Management*, vol 394, pp. 86–103.

Carroll, A. L. (2023) 'Forest health in the Anthropocene', in J. D. Allison, T. D. Paine, B. Slippers, and M. J. Wingfield (eds.), *Forest Entomology and Pathology. Vol. 1: Entomology*. Springer Press, Switzerland.

Castello, J. D., Leopold, D. J. and Smallidge, P. J. (1995) 'Patterns and processes in forest ecosystems', *BioScience*, vol 45, pp. 16–24.

Coakley, S. M., Scherm, H. and Chakraborty, S. (1999) 'Climate change and plant disease management', *Annual Review of Phytopathology*, vol 37, pp. 399–426.

Comeau, A.M., Dufour, J., Bouvet, G.F., Nigg, M., Jacobi, V., Henrissat, B., Laroche, J., Levesque, R.C. and Bernier, L. (2015) 'Functional annotation of the *Ophiostoma novo-ulmi* genome: Insights into the phytopathogenicity of the fungal agent of Dutch elm disease', *Genome Biology and Evolution*, vol 7, pp. 410–430.

Coulson, R. N. and Witter, J. A. (1984) *Forest Entomology: Ecology and Management.* Wiley-Interscience Publications, New York.

Coyle, D.R. (2023) 'Tip, shoot, root, and regeneration pests', in J. D. Allison, T. D. Paine, B. Slippers, and M. J. Wingfield (eds.), *Forest Entomology and Pathology. Vol. 1: Entomology*. Springer Press, Switzerland.

Cullingham, C. I., Cooke, J. E., Dang, S., Davis, C. S., Cooke, B. J. and Coltman, D. W. (2011) 'Mountain pine beetle host-range expansion threatens the boreal forest', *Molecular Ecology*, vol 20, pp. 2157–2171.

Davidson, C. B., Gottschalk, K. W. and Johnson, J. E. (1999) 'Tree mortality following defoliation by European gypsy moth (*Lymantria dispar* L.) in the United States: A review', *Forest Science*, vol 45, pp. 74–84.

Dinoor, A. and Eshed, N. (1984) 'The role and importance of pathogens in natural plant communities', *Annual Review of Phytopathology*, vol 22, pp. 443–466.

Dodds, K. J. and Orwig, D. A. (2011) 'An invasive urban forest pest invades natural environments — Asian longhorned beetle in northeastern US hardwood forests', *Canadian Journal of Forest Research*, vol 41, pp. 1729–1742

Dukes, J. S., Pontius, J., Orwig, D., Garnas, J. R., Rodgers, V. L., Brazee, N., Cooke, B., Theoharides, K. A., Stange, E. E., Harrington, R., Ehrenfeld, J., Gurevitch, J. Lerdau, M., Stinson, K., Wick, R. and Ayres, M. (2009) 'Responses of insect pests, pathogens, and invasive plant species to climate change in the forests of northeastern North America: What can we predict?', *Canadian Journal of Forest Research*, vol 39, pp. 231–248.

Fabre, B., Piou, D., Desprez-Loustau, M. L. and Marçais, B. (2011) 'Can the emergence of pine *Diplodia* shoot blight in France be explained by changes in pathogen pressure linked to climate change?', *Global Change Biology*, vol 17, pp. 3218–3227.

Feau, N., Mottet, M.-J. Périnet, P., Hamelin, R. C. and Bernier, L. (2010) 'Recent advances related to poplar leaf spot and canker caused by *Septoria musiva*', *Canadian Journal of Plant Pathology*, vol 32, pp. 122–134.

Franklin, J. F., Shugart, H. H., and Harmon, M. E. (1987) 'Tree death as an ecological process: The causes, consequences, and variability of tree mortanlity', *BioScience*, vol 37, pp. 550–556.

Hendrix, P. F. (2006) 'Biological invasions belowground — earthworms as invasive species', *Biological Invasions*, vol 8, pp. 1201–1204.

Hilbert, D. R, Roman, L. A., Koeser, A. K., Vogt, J. and van Doorn, N. S. (2019) 'Urban tree mortality: A literature review', *Arboriculture and Urban Forestry*, vol 45, pp. 167–200.

Hodges, C. S. (1969) 'Modes of infection and spread of *Fomes annosus*', *Annual Review of Phytopathology*, vol 7, pp. 247–266.

Houston, D. R. (1994) 'Major new tree disease epidemics: Beech bark disease', *Annual Review of Phytopathology*, vol 32, pp. 75–87.

Huler, J. and Skelton, J. (2023) 'Ambrosia beetles', in J. D. Allison, T. D. Paine, B. Slippers, and M. J. Wingfield (eds.), *Forest Entomology and Pathology. Vol. 1: Entomology*. Springer Press, Switzerland.

Jactel, H., Battisti, A., Branco, M., Douma, J. C., Kenis, M., Orazio, C., Robinet, C., Santini, A., Sapundzhieva, A., Seehausen, M.L. and Stoev, P. (2023) 'Management options for non-native forest pests along their invasion pathways', *NeoBiota*, vol 84, pp.1–7.

Jardon, Y., Filion, L. and Cloutier, C. (1994) 'Tree-ring evidence for endemicity of the larch sawfly in North America', *Canadian Journal of Forest Research*, vol 24, pp. 742–747

Juzwik, J., Harrington, T. C., MacDonald, W. L. and Appel, D. N. (2008) 'The origin of *Ceratocystis fagacearum*, the oak wilt fungus', *Annual Review of Phytopathology*, vol 46, pp. 13–26.

Kenis, M., Auger-Rozenberg, M.-A., Roques, A., Timms, L., Pere, C., Cock, M. J. W., Settele, J., Augustin, S. and Lopez-Vaamonde, C. (2009) 'Ecological effects of invasive alien insects', *Biological Invasions*, vol 11, pp. 21–45.

Kizlinski, M. L., Orwig, D. A., Cobb, R. C. and Foster, D. R. (2002) 'Direct and indirect ecosystem consequences of an invasive pest on forests dominated by eastern hemlock', *Journal of Biogeography*, vol 29, pp. 1489–1503.

Knight, K., Haugen, L., Pinchot, C., Schaberg, P. G. and Slavicek, J. M. (2017) 'American elm (*Ulmus americana*) in restoration plantings: A review', in C. Pinchot, K. S. Knight, L. M. Haugen, C. E. Flower, and J. M. Slavicek (eds.), *Proceedings of the American elm restoration workshop 2016*, General Technical Report NRS-P-174, United States Department of Agriculture, Forest Service, Northern Research Station, Newton Square.

Kolb, T. E., Wagner, M. R. and Covington, W. W. (1994) 'Utilitarian and ecosystem perspectives: Concepts of forest health', *Journal of Forestry*, vol 92, pp. 10–15.

Kovalchuk, A., Wen, Z., Sun, H. and Asiegbu, F.O. (2022) '*Heterobasidion annosum* s.l.: Biology, genomics, and pathogenicity factors', in F.O. Asiegbu and A. Kovalchuk (eds.), *Forest Microbiology: Volume 2: Forest Tree Health*. Elsevier Academic Press, Cambridge.

Levine, J. M. (2008) 'Biological invasions', *Current Biology*, vol 18, pp. R57–R60.

Liebhold, A. M., Brockerhoff, E. G., and McCullough, D. G. (2023) 'Forest insect invasions and their management', in J. D. Allison, T. D. Paine, B. Slippers, and M. J. Wingfield (eds.), *Forest Entomology and Pathology. Vol. 1: Entomology*. Springer Press, Switzerland.

Lovett, G. M., Caham, C. D., Arthur, M. A., Weathers, K. C. and Fitzhugh, R. D. (2006) 'Forest ecosystem responses to exotic pests and pathogens in eastern North America', *Bioscience*, vol 56, pp. 395–405.

Manion, P. D. (2003) 'Evolution of concepts in forest pathology', *Phytopathology*, vol 93, no 8, pp. 1052–1055.

Martín, J. A., Domínguez, J., Solla, A., Brasier, C. M., Webber, J. F., Santini, A., Martínez-Arias, C., Bernier, L. and Gil, L. (2023) 'Complexities underlying the breeding and deployment of Dutch elm disease resistant elms', *New Forests*, vol 54, pp. 661–696.

Mattson, W. J. and Haack, R. A. (1987) 'The role of drought in outbreaks of plant-eating insects', *Bioscience*, vol 37, pp. 110–118.

McCormick, J. F. and Platt, R. B. (1980) 'Recovery of an Appalachian forest following the chestnut blight or Catherine Keever — you were right!', *American Midland Naturalist*, vol 104, pp. 264–273.

McCullough, D. G. (1998) 'Fire and insects in northern and boreal forest ecosystems of North America', *Annual Review of Entomology*, vol 43, pp. 107–127.

McLeod, G., Gries, R., von Reuß, S. H., Rahe, J. E., McIntosh, R., König, W. A. and Gries, G. (2005) 'The pathogen causing Dutch elm disease makes host trees attract insect vectors', *Proceedings of the Royal Society B-Biological Sciences*, vol 272, pp. 2499–2503.

Michler, C. H., Pijut, P. M., Jacobs, D. F., Meilan, R., Woeste, K. E. and Ostry, M. E. (2006) 'Improving disease resistance of butternut (*Juglans cinerea*), a threatened fine hardwood: A case for single-tree selection through genetic improvement and deployment', *Tree Physiology*, vol 26, pp. 121–128.

Ostry, M. E. and Laflamme, G. (2009) 'Fungi and diseases — natural components of healthy forests', *Botany*, vol 87, pp. 22–25.

Paine, T. D., Raffa, K. F. and Harrington, T. C. (1997) 'Interactions among scolytid bark beetles, their associated fungi, and live host conifers', *Annual Review of Entomology*, vol 42, pp. 179–206.

Parker, T. J., Clancy, K. M. and Mathiesen, R. L. (2006) 'Interactions among fire, insects, and pathogens in coniferous forests of the interior western United States and Canada', *Agricultural and Forest Entomology*, vol 8, pp. 167–189.

Pimentel, D., Zuniga, R. and Morrison, D. (2005) 'Update on the environmental and economic costs associated with alien-invasive species in the United States', *Ecological Economics*, vol 52, pp. 273–288.

Pronos, J. and Patton, R. F. (1978) 'Penetration and colonization of oak roots by *Armillaria mellea* in Wisconsin', *European Journal of Forest Pathology*, vol 8, pp. 259–267.

Raffa, K. F., Brockerhoff, E. G., Grégoire, J.-C., Hamelin, R. C., Liebhold, A. M., Santini, A., Venette, R. C. and Wingfield, M. J. (2023) 'Approaches to forecasting damage by invasive forest insects and pathogens: A cross-assessment', *BioScience*, vol 73, pp. 85–111.

Rapicavoli, J., Ingel, B., Blanco-Ulate, B., Cantu, D. and Roper, C. (2019) '*Xylella fastidiosa*: An examination of a re-emerging plant pathogen', *Molecular Plant Pathology*, vol 19, pp. 786–800.

Reyer, C. P. O., Brouwers, N., Grant, R. F., Brook, B. W., Rammig, A., Epila, J., Holmgren, M., Langerwisch, F., Leuzinger, S., Lucht, W., Medlyn, B., Pfeifer, M., Steinkamp, J., Vanderwel, M. C., Verbeeck, H. and Villela, D. M. (2015) 'Forest resilience and tipping points at different spatio-temporal scales: approaches and challenges', *Journal of Ecology*, vol 103, no 1, pp. 5–15.

Scott, J. J., Oh, D. C., Yuceer, M. C., Klepzig, K. D., Clardy, J. and Currie, C. R. (2008) 'Bacterial protection of beetle-fungus mutualism', *Science*, vol 322, p. 63.

Smith, C. M., Shepherd, B., Gillies, C. and Stuart-Smith, J. (2013) 'Changes in blister rust infection and mortality in whitebark pine over time', *Canadian Journal of Forest Research*, vol 43, pp. 90–96.

Strong, W. B., Mangini A. C. and Candau, J.-N. (2023) 'Insects of reproductive structures', in J. D. Allison, T. D. Paine, B. Slippers, and M. J. Wingfield (eds.), *Forest Entomology and Pathology. Vol. 1: Entomology*. Springer Press, Switzerland.

Stukenbrock, E. H. and McDonald, B. A. (2008) 'The origins of plant pathogens in agro-ecosystems', *Annual Review of Phytopathology*, vol 46, pp. 75–100.

Sturrock, R. N. (2012) 'Climate change and forest diseases: Using today's knowledge to address future challenges', *Forest Systems*, vol 21, pp. 329–336.

Teale, S. A. and Castello, J. D. (2011) 'Regulators and terminators: The importance of biotic factors to a healthy forest', in J. D. Castello and S. A. Teale (eds.), *Forest Health: An Integrated Perspective*. Cambridge University Press, Cambridge.

Tobin, P. C., Haynes, K. J. and Carroll, A. L. (2023) 'Spatial dynamics of forest insects', in J. D. Allison, T. D. Paine, B. Slippers and M. J. Wingfield (eds.), *Forest Entomology and Pathology. Vol. 1: Entomology*. Springer Press, Switzerland.

Warren, G. R., Harrison, K. J. and Laflamme, G. (2011) 'New and updated information on Scleroderris canker in the Atlantic Provinces', *The Forestry Chronicle*, vol 87, pp. 382–390.

Weintraub, P. G. and Beanland, L. (2006) 'Insect vectors of phytoplasmas', *Annual Review of Entomology*, vol 51, pp. 91–111.

Welsh, C., Lewis, K. J. and Woods, A. J. (2014) 'Regional outbreak dynamics of *Dothistroma* needle blight linked to weather patterns in British Columbia, Canada', *Canadian Journal of Forest Research*, vol 44, pp. 212–219.

Woeste, K., Farlee, L., Ostry, M., McKenna, J. and Weeks S. (2009) 'A forest manager's guide to butternut', *Northern Journal of Applied Forestry*, vol 26, pp. 9–14.

Xhaard, C., Fabre, B., Andrieux, A., Gladieux, P., Barrès, B., Frey, P. and Halkett, F. (2011) 'The genetic structure of the plant pathogenic fungus *Melampsora larici-populina* on its wild host is extensively impacted by host domestication', *Molecular Ecology*, vol 20, pp. 2739–2755.

PART VI

FOREST AND CLIMATE CHANGE

34
FIRE AND CLIMATE
Using the past to predict the future

Marion Lestienne, Justin Waito, Chéïma Barhoumi, Laurent Bremond, Martin P. Girardin, Jacques C. Tardif, Adam A. Ali, Hermann Behling, Julia Unkelbach and Christelle Hély

Introduction

The unique distribution and stand characteristics of global forests are the result of interactions between vegetation and the environment over millennial time spans. The spatial distribution of forests is governed in part by physical processes, such as incoming solar radiation, growth season length, and moisture and nutrient availability, which impose limitations on the establishment and growth of vegetation (Bonan and Shugart, 1989). Additionally, vegetation distribution is influenced by disturbance events such as fire, damage from wind and ice storms, insect outbreaks, and rockfalls and avalanches in mountain regions (Attiwill, 1994). These are important for regeneration and maintenance of the heterogeneous distribution of species and stand structure. Fire has been an important component of ecosystems since the appearance of the first terrestrial plants, with records of fire existing from as far back as 420 million years ago (Bowman et al., 2013). It is present in most of Earth's ecosystems except in areas of sparse vegetation and near the poles (Bowman et al., 2013; Flannigan et al., 2013). It affects hundreds of millions of hectares annually, with the majority being located in grasslands and savannas. Fire is capable of temporarily and rapidly reducing previously vegetated areas to mineral soil. Human manipulation of fire for land-clearing and recreational activities, as well as fire suppression practices, has disrupted the natural pattern of fire activity in many forested regions (Bobek et al., 2017; Bowman et al., 2013; Lestienne et al., 2020b). In addition, human-caused climate change is expected to further alter global fire occurrence, with important consequences for species distribution, ecosystem integrity and function, atmospheric greenhouse gas balance, and human safety (Flannigan et al., 2013). Characterizing the range of fire regime modifications remains a major challenge owing to the large interannual variability in area burned (Figure 34.1) that tends to mask long-term and subtle changes. The importance of fire in Earth's ecosystems necessitates an understanding of historic variability in order to place current observations and future projections in context. In this chapter, we outline the physical processes of fire in relation to the forest environment and climate. We also review the methods used in fire history reconstructions as well as results

DOI: 10.4324/9781003324072-40

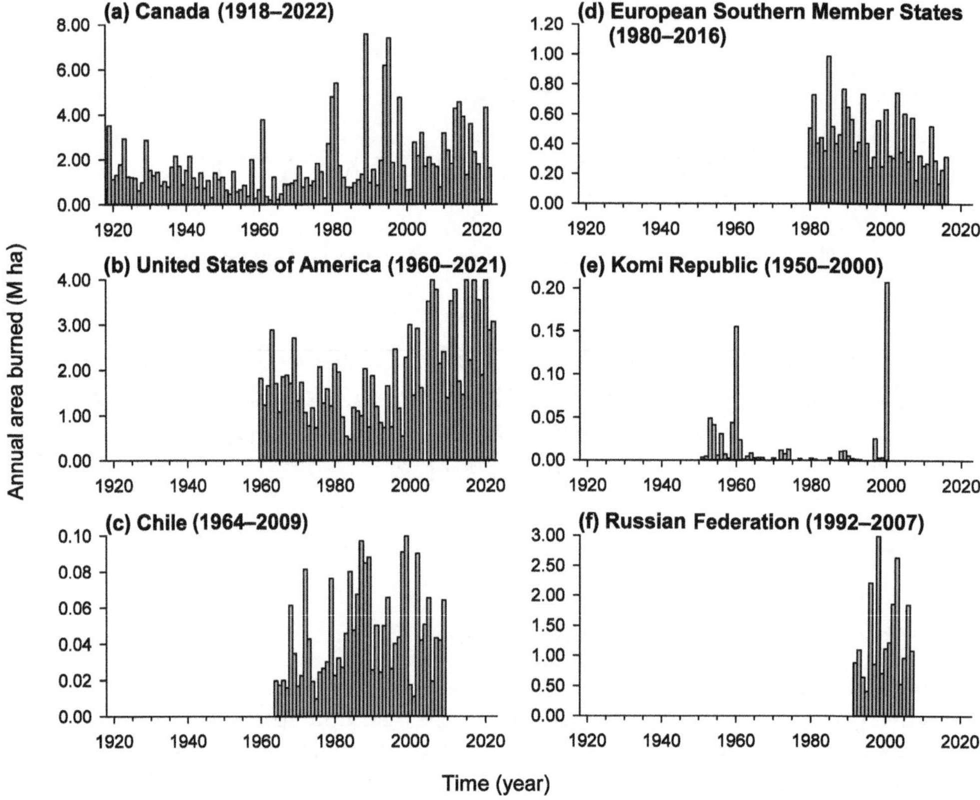

Figure 34.1 Examples of annual area burned time series for various countries: (a) Canada (Van Wagner, 1988, https://cwfis.cfs.nrcan.gc.ca/datamart; last accessed on May 23, 2023); (b) United States of America (www.nifc.gov/fireInfo/fireInfo_stats_totalFires.html; last accessed on February 22, 2022); (c) Chile (https://gfmc.online/inventory/cl_statistic_1964-2004.html; last accessed on February 28, 2022); (d) European Southern Member States (Portugal, Spain, France, Italy and Greece (SCHMUCK et al., 2013); (e) Komi Republic (Drobyshev and Niklasson, 2004); and (f) Russian Federation (Goldammer et al., 2007). Parentheses indicate the periods covered by each fire statistic.

from fire history reconstructions and modeling. The best proxy records for longer term fire reconstructions are tree-ring and lake sediment records. Our discussion will thus focus on fire history originating from these two proxy types in several sites located in boreal, temperate, and tropical regions.

Climate, fuel, and forest fires

To begin, it is worth introducing some concepts and processes associated with forest fires as they relate to landscape processes. Fire in a forested landscape is described through the fire regime concept that encompasses fire ignition (caused by human or lightning), fire type (ground, surface or crown fire), fire behavior (mainly intensity and spread) and severity as well as spatial (size) and temporal (frequency: number of fires per millennium, seasonality) characteristics (Keeley, 2009). Generally speaking, a small number of fires determine the total area burned each year in temperate and boreal forests. For instance, annual area burned in boreal forests of North America and Russia can exceed 7 million and 3 million hectares, respectively. The majority of these are infrequent lightning-caused large crown fires exceeding 200 hectares in size (Goldammer et al., 2007; Stocks et al., 2002). Conversely, in tropical regions (e.g., southern Africa), the cumulative annual area burned by numerous but small anthropogenic fires is equivalent to the lightning-caused area burned (Archibald et al., 2013).

The behavior of individual fires is highly variable and is determined by the interaction between weather, fuels and landscape characteristics. Fires are typically ground fires, surface fires or crown fires, each situated along an increasing burn intensity gradient (with intensity being the energy released during a fire event) (Bowman et al., 2013). Ground fires are slow-moving low-intensity fires that burn through soils with deep organic matter such as in Russian boreal forests (e.g., Kuklina et al., 2022) and Indonesian peatlands (e.g., Jessup et al., 2022; Vetrita and Cochrane, 2019). Despite the low intensity and slow rate of spread, ground fires can burn during several weeks or months and therefore can consume more biomass per hectare than the other fire types. Surface fires represent low-intensity fast-moving fires where thin soil organic matter below the litter layer and vegetation close to the surface (mainly grasses and shrubs) can burn over large areas (e.g., a single fire in Namibia burned 3200 km^2 in Etosha National Park in 2000 (Alleaume et al., 2005)). Crown fires are the most intense and severe of the three types of fires and occur when a fire spreads through the canopy layer (Turner and Romme, 1994; Van Wagner, 1977). Further, the rate of spread through the forest canopy can be described as being passive, active or independent (Van Wagner, 1977). A passive crown fire generally occurs during low wind conditions and requires the release of surface fire energy to be sustained in the crown compartment (Van Wagner, 1977). Active crown fires represent the intermediate class where wind speed is more conducive to spread of fire through the forest canopy. Independent crown fires are the most severe and are characterized by separation of the surface and crown portions of the fire. The type of fire is an important component contributing to the spatial extent of individual fires as well as to the total forested area burned each year.

Large fires are predominantly influenced by climate, often associated with prolonged blocking high-pressure systems in the upper atmosphere over or upstream from the affected regions (Macias Fauria and Johnson, 2008; Pereira et al., 2005). The blocking high-pressure systems cause air subsidence in the upper atmosphere, obstructing the normal west-to-east progress of migratory storms, resulting in typically sunny, warm days that create dry fuel conditions extending over several hundred square kilometers. Such dry fuel conditions facilitate the rapid spread of fires. In North American boreal forests, ignition can occur due to lightning strikes associated with the penetration of short-wave atmospheric troughs or a cold front along the west side of the ridges (Macias Fauria and Johnson, 2008). Additionally, fire spread is enhanced by the strong and gusty surface winds that are commonly associated with the breakdown of these blocking ridges.

In addition to climate, vegetation and fuel types also exert important influences on fire ignition and propagation. For example, fire size and intensity are generally higher in forests dominated by coniferous species compared with those dominated by broadleaf species (Curt et al., 2013; Terrier et al., 2013). This difference is largely due to the quantity and type of organic layer that is available to be burned. The fuel conditions in coniferous forests lead to fire susceptibility because of an abundance of small-diameter, dry organic material (Hély et al., 2000). In contrast, broadleaf forests are less likely to burn because of higher leaf moisture loading and decreased flammability (Campbell and Flannigan, 2000; Päätalo, 1998). In some areas of the world, such as tropical grasslands and savannas, precipitation is an additional indirect factor influencing fire ignition and spread as it is required to initiate plant growth (during the wet season) and subsequent fuel build-up (during the dry season) that allows continuous fire spread on the landscape (Flannigan et al., 2013; Hély et al., 2019). In such situations, the interval between successive fire events depends on the rate of fuel build-up and so ignition is a combination of the annual precipitation amount, time-since-last-fire and favorable climate or anthropic activities.

Landscape characteristics are also important drivers of fire through features such as slope and fire barriers (Cyr et al., 2007). The influence of slope on fire is through preferential fire spread in an up-slope direction. In addition, slope aspect influences fire spread as south-facing slopes are more prone to moisture deficits. However, this influence is limited as severe fires can be relatively impervious to landscape features. Fire occurrence and extent are also affected by natural fire breaks, such as water bodies, that can alter the direction of fire by restricting fire spread in certain directions.

Historical fire records during the instrumental period

Records used in fire history research include written accounts, aerial photographs, and remote sensing (Bowman et al., 2013; Curt et al., 2016; Macias Fauria and Johnson, 2008; Stocks et al., 2002; Van Wagner, 1988). Intuitively, written records provide the longest account of fire data but the resolution is dependent on whether an area was settled, events were recorded and the records survived to the present day (Figure 34.1). Generally speaking, the earliest written records of fire events are sporadic and infrequent. However, records became more spatially explicit and continuous with the advent of systematic record keeping by national institutions (Fritz et al., 1993; Rannie, 2001). Record keeping has further improved with the advent of active fire suppression that allowed collection of first-hand written accounts and boundary maps of fires (Ruffault et al., 2017; Stocks et al., 2002). Since their inception, fire records have been steadily improving as new technologies and monitoring networks have been developed. In many regions of the world, very accurate and reliable records have been available since the 1970s (Figure 34.1).

Following written records, the longest available fire-related records can be found in aerial photography coverage of forested landscapes associated with the development of forest inventories. In the southern boreal forest of North America, records obtained from aerial surveying generally extend back to the early twentieth century (Van Wagner, 1988). These records can be used in fire history reconstructions to locate boundaries between forest types and age classes that are indicative of past disturbance events. A more recent development in fire history mapping is the application of remote sensing techniques involving satellite imagery coupled with geographic information systems analyses (Andela et al., 2019). This method is still being developed through fine resolution satellite products and ongoing classification and validation of interpretations from field studies. For more information on available data sets, visit the Global Fire Monitoring Center website (https://gfmc.online/). However, to understand how forests and other ecosystems respond to fire events and climate change, we need to go further back than the instrumental period.

Fire history reconstruction

Research into past fire activity relies on several indicators, one of the main indicators being the direct analysis of damage caused to trees by fire. Fire as a physical process is the result of a chemical reaction leading to combustion of organic material (Michaletz and Johnson, 2007). Combustion is a fourfold process whereby an endothermic reaction (1) first completely evaporates the fuel water content followed by an exothermic oxidation reaction (2) that breaks down complex molecules (pyrolysis). This process creates volatile organic compounds (Simpson et al., 2011) whose increasing concentration in the surrounding air, combined with increasing temperature, leads to an explosive reaction in the presence of oxygen, which results in flame occurring (3). This third, flaming phase is the most efficient in terms of energy release and proceeds as long as the breakdown of the particle molecules is exothermic enough to produce volatile gas to feed the flame. When the emission of volatile compounds stops, the flame disappears and the exothermic oxidation process, now less efficient, only occurs at the surface of the particle, thereby moving the combustion into the fourth phase (4), known as the glowing phase, which will ultimately cease once the decreasing energy release cannot sustain the high temperatures needed for rapid oxidation. In trees, the exothermic reaction can result in complete combustion or, more often, transitions to a lower energy reaction that leaves some material unconsumed. Burned material left behind after fires (Figure 34.2a) can be used to determine the ages of previous forest cohorts (Heinselman, 1973; Lesieur et al., 2002). In a crown fire ecosystem such as the boreal forest, all standing trees are killed and pioneer species regenerate following fire events. Therein, comparison of stand ages across the landscape reveals the succession of fire events that have occurred on a landscape over time (Figure 34.2). In addition, charred material produced during the flaming and smoldering phases of a fire can become incorporated into soil, lake sediments and peat bogs, thereby preserving a long-term record of fire events.

In a non-lethal surface fire, cambial injury to the stems of living trees can lead to the formation of fire scars on the leeward side (Figure 34.2). Fire scar formation is explained through fluid dynamic processes where the physical behavior of fire in relation to a tree is governed by differential airflow (Gutsell & Johnson, 1996). Airflow patterns around the tree result in the formation of vortices on the leeward side that contribute to fire scar formation by drawing the flame up the stem of the tree, which increases heat exposure time. The size and shape of the scar is dependent on the position of the vortices. Accumulation of organic material around the stem is also a prerequisite for fire scar formation as it will act as the available fuel build-up and fuel bed. The analysis of fire scars is one of the most common methods of inferring past fire activity over the last centuries (Swetnam, 1993). Properly dated through their position along the tree-ring sequence, fire scars can provide information on the year of fire scar formation and seasonality of burn. Comparison of samples with fire scars of the same year provides the spatial extent of individual fires, whereas comparison of multiple fire scar years provides the frequency of fires in a given area (Swetnam, 1993). However, analysis of fire scars is not without limitations, such as limited temporal depth, formation restricted to surface or mixed fire regimes, and only in ecosystems with a presence of trees, and potential confusion between fire scars and scars produced by other means.

The second proxy is charcoal particles that are produced during the fire and deposited as sediment in soil, lakes, wetlands and peatlands, thus producing a record of the events (Higuera et al., 2007; Patterson et al., 1987). Fire events are identified when the charcoal peak component exceeds the background component of the record (Figure 34.3). For example, an analysis of macro-charcoal fragments (particles 0.3–0.5 mg in weight and with

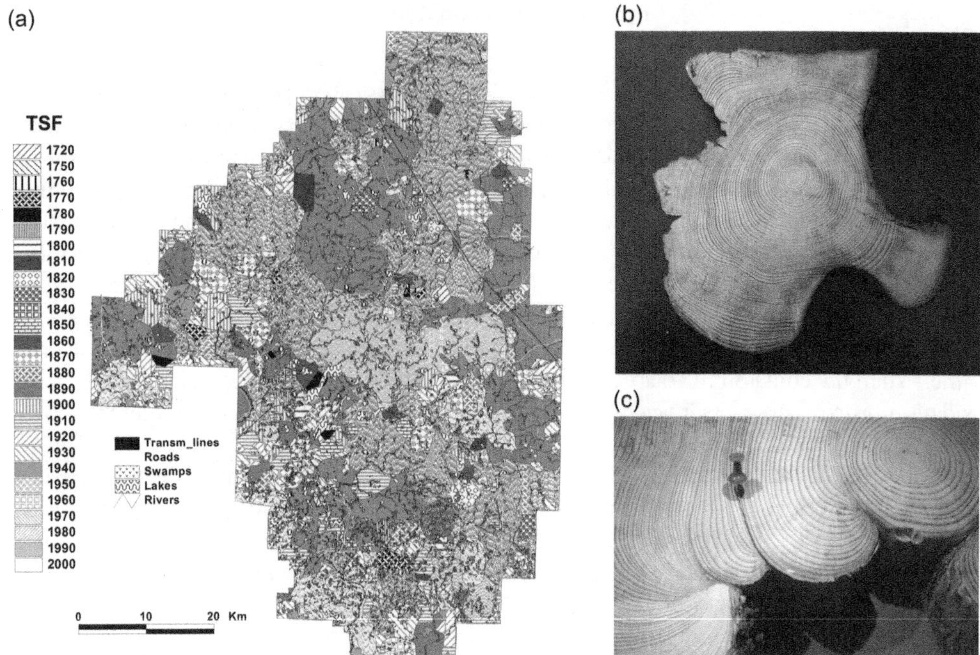

Figure 34.2 (a) Time-since-fire (TSF) map for the Duck Mountain Provincial Forest, Manitoba, Canada, as of 2002. The TSF date (i.e. approximate year of burn) for each of the sites was either provided by the oldest trees in the overstory cohort, by sampling of burned material left on the ground (b), and/or by fire scar date (c). The TSF map indicated that a large portion of the landscape originated from fires between 1880 and 1899. The TSF map also showed that during the early twentieth century, numerous small fires occurred at the southwest periphery. The largest fire of the twentieth century was that of 1961, with about 6% of the study area burned (Gauthier et al., 2009; Tardif, 2004).

Note: Access a colour version of the figure here: https://doi.org/10.6084/m9.figshare.26097829.v1

a diameter >0.2 mm) extracted from soil profiles can yield direct information of past fires in forest stands that underwent incomplete wood combustion (Payette et al., 2012). However, there is a high probability that successive fires burn the charcoal deposited in the soil by a previous fire; thus, numerous events may not be recorded using this method (Kane et al., 2010). The finest particles emitted from fires are air- and runoff-transported over the landscape and can be extracted from lake sediments and used to reconstruct long-term fire history at the landscape and regional scales. Note that it is common practice in such fire history studies to also examine plant macro-remains and pollen assemblages as fire events alter the influx of pollen and macro-remains into sediments by burning the vegetation that produces it (Girardin et al., 2013; Hély and Lézine, 2014; Ledru et al., 2006; Lestienne et al., 2020b; Whitlock and Bartlein, 2003). The careful examination of pollen and macro-remains allows the determination of vegetation assemblages before and after fire events.

The use of charcoal particles for fire history reconstruction requires accurate determination of the age of the fire. In fire reconstructions based on charred particles extracted from soils, age determination is undertaken by the dating of the wood that produced the charcoal

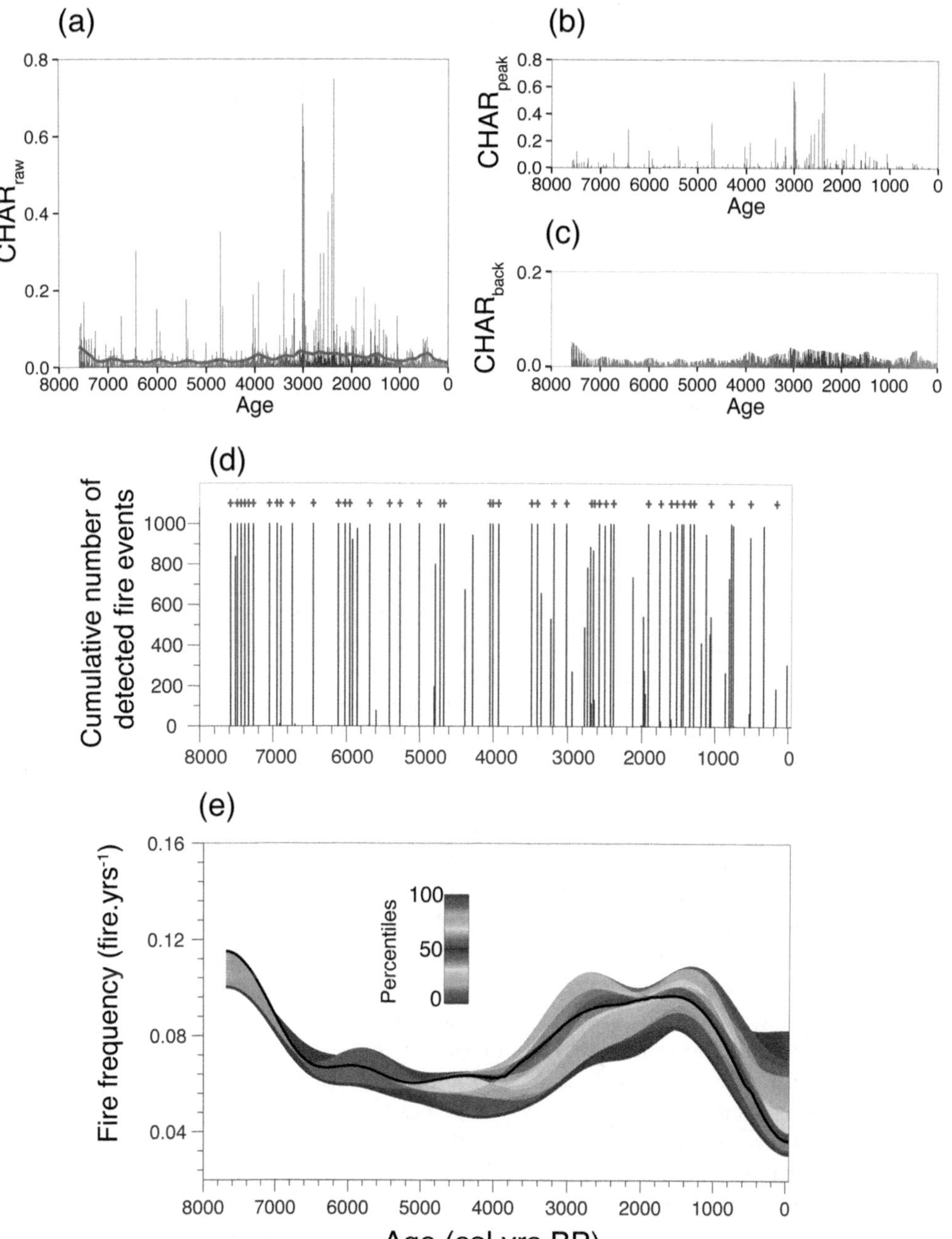

Figure 34.3 Summary of the main analytical steps required for reconstructing past fire frequency from sedimentary charcoal records. (a) Raw charcoal values expressed in Charcoal Accumulation Rate (CHAR $mm^2.cm^{-2}.yr^{-1}$) were smoothed using a locally weighted regression (LOESS) with a 500-year window width, which enabled us to discriminate the CHARpeak (b) and CHARback (c) components. CHARback (c) represented long-distance fires, charcoal redeposition and noise while the CHARpeak (b) represented local

fires. Noise is also present in the remaining CHARpeak component and a Gaussian mixture model is generally used to discriminate the two CHARpeak subpopulations: CHARnoise and CHARfire. The CHARfire component represented the occurrence of one or more fires locally around lakes (<10 km). The Gaussian mixture could be applied globally to the entire records (such as in this example) or locally in order to better fit the centennial to millennial variations in CHAR. Analytical steps (CHARraw filtering technique and window width) that are able to provide well-discriminated fire events are not unique (Higuera et al., 2007) and techniques that aimed at replicating the analysis of a record using ensembles have been proposed (Blarquez et al., 2013). Therein, 1,000 replicates of the steps (a–c) are made using multiple smoothing techniques and window width, and the cumulative number of fires by the ensemble is calculated for each fire event date (d). From this ensemble of fire event dates, it is possible to assess fire frequency (or fire return intervals). In the example illustrated in (e), the tinted bands represent the distribution of the fire frequencies obtained from the 1,000 iterations of the model parameters. From these, one can be confident that the changes from low to high fire frequency ca. 3,000 cal BP are not a consequence of the selection of model parameters but, rather, reflect important changes that have occurred in burning activity (Blarquez et al., 2013)

Note: Access a colour version of the figure here: https://doi.org/10.6084/m9.figshare.26097829.v1

(Payette et al., 2012). Of note with this method is that a lag, which may range from decades to centuries, separates the date of the wood and the date of the actual fire. In contrast to soils, dating of fire events in lake sediment is undertaken through age determination of plant macro-remains or bulk sediments. This procedure ensures continuity of sediments and minimal lag between the effective incorporation of radioactive elements in the organic material and its sedimentation. This methodology also has limitations that mostly involve data analysis methods which require a thorough statistical framework and the source of the deposited charred particles. Scientists have developed several statistical methods to study the charcoal signal (Blarquez et al., 2014; Higuera, 2009), and others have developed methods to compare fire history reconstructions from different sources of information (Aleman et al., 2013; Brossier et al., 2014; Remy et al., 2018). On the other hand, the detection of fire events can only be done when the sedimentation rate allows analysis of sedimentary charcoal over a time step well below the fire return interval. This implies that in a tropical environment, particularly in the savanna where fires are very frequent, it is not possible to identify single fire events. We will then rather speak of "biomass burning".

The process of dating sediments is through analysis of isotopes including lead (^{210}Pb) and carbon (^{14}C). The theoretical underpinning to ^{210}Pb dating is that lead is a product of the uranium decay series, the proportion of which can be used to accurately date the uppermost 150–200 years of sediment (Appleby, 2002). The analysis of ^{210}Pb is made possible because it is continuously added to water bodies from the atmosphere where it becomes incorporated into the sediment. Analysis of the unsupported (atmospheric) ^{210}Pb fraction against the supported (*in situ*) fraction in the sediment is based on its half-life (22.3 years) and provides an approximation of when the isotope was incorporated into the sediment. In order to obtain dates from older sediments, the ^{14}C isotope is often used as it has a half-life of 5,568 years (Björck and Wohlfarth, 2002). ^{14}C is typically used to date material as it is a heavy isotope that is less likely to be preferentially selected by plants (fractionation) and so provides the most accurate date. Accurate dating of material is made possible because known ratios of ^{13}C and ^{12}C have been established, and calibration curves developed, from proxy records such as wood and sedimentary rocks. In addition, the ratio of atmospheric ^{14}C to

^{13}C has been established from wood and so an age determination of material can be made through comparison of the ratio of ^{14}C to the calibration curve developed for other carbon isotopes (Reimer et al., 2020).

Trends in post-glacial fire history worldwide

Areas overview

Fire history has been documented in many regions. We chose to focus on four regions covering four continents: the Mediterranean ecosystems from France (Southern Europe), the continental forest-steppes from Mongolia (Asia), Sub-Saharan Africa and the Canadian boreal forests (North America). All these regions have undergone great climatic changes during the Holocene in response to insolation and anthropic activities.

The Mediterranean region has the largest, most intense and most recurrent fires, whereas the steppes and continental areas have the largest proportion of annual area burnt (Pausas, 2022). Boreal forests are subject to less frequent but more severe megafires due to climate and vegetation composition (Gaboriau et al., 2022; Stephens et al., 2014). The African fire regime is not easy to synthetize for three main reasons. The first reason is that prior to the early 2000s, paleoecological fire reconstructions were done by directly counting charcoal particles included on pollen slides instead of specific charcoal analysis (count and morphology). This implies that the particles counted were very small and could have come from very large regions, thus producing a signal difficult to interpret. The second reason is that the amount of charcoal particles in a sample is often very large. Indeed, many tropical ecosystems have herbaceous components that burn very often, even annually; therefore, a sedimentary sample covering several years may contain charcoal produced by several successive fires. The context is thus very different from that of the other regions. The last reason is that there are very few paleoenvironmental records on a continent very diverse in terms of vegetation, climate and human history.

In this chapter we present a short synthesis of fire histories for the Mediterranean, Asia, Africa and North America. Corsica (France) is an example of the Mediterranean fire history (Figure 34.4). The Corsican vegetation dynamics (Figure 34.4a) and fire history have been obtained with pollen and charcoal analyses (see Lestienne et al., 2020b and Leys et al., 2013 for details). The modeled fire hazard (Figure 34.4b) has been computed for the entire island by calculating the Monthly Drought Code and the Fire Season Length (see Lestienne et al., 2020a, 2022 for details). The fire histories of continental areas and steppes of Mongolia have been documented in two forest-steppe regions: one in the northern-central Khangai mountains for the entire Holocene and the other in the Altai Mountains in western Mongolia for the mid- to late Holocene (Unkelbach et al., 2018, 2019, 2020; Unkelbach and Behling, 2022). Boreal fire history has been reconstructed for mixed-wood forests in eastern Canada (Girardin et al., 2013).

Fire histories

The early Holocene was characterized by an increase in the fire frequency in all regions (Figure 34.4c, d, e, and f). The important climatic changes occurring during this period (due to the higher insolation) is associated with an increase in fuel availability that has directly promoted higher fire frequencies (Whitlock et al., 2010). The modeled fire hazard

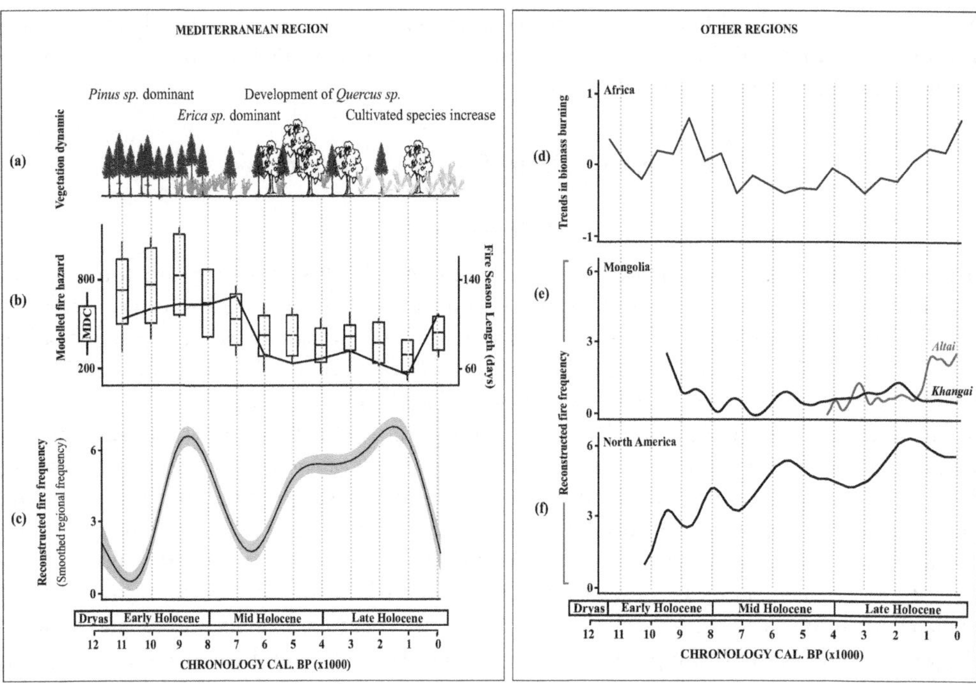

Figure 34.4 (a) General Mediterranean vegetation dynamics and comparison with (b) modeled fire hazard (characterized by the Monthly Drought Code and the Fire Season Length (see Lestienne et al. (2020a) for details) and (c) reconstructed fire frequency. For the modeled fire hazard (b), the boxplots represent the Monthly Drought Code and the curve represents the mean Fire Season Length. To reconstruct the Mediterranean fire frequency (number of fires per millennium) (c), a smooth curve was generated using the Locally Estimated Scatterplot Smoothing (LOESS) method from the fire frequencies of three Corsican lakes: Bastani (Lestienne et al., 2020b), Nino (Lestienne et al., 2020a) and Creno (Leys et al., 2013). These frequencies have been obtained by the analysis of charcoal records with CharAnalysis software (Higuera, 2009). Trends in biomass burning in Africa at the continental scale inferred from charcoal abundance records are represented on panel (d). The composite charcoal series were computed with sites available for Sub-Saharan Africa from the Global Paleofire Database (North Africa excluded, 52 records with good chronologies, time step, continuity and charcoal measurement data; within present forest, savanna and woodland vegetations) with a 500-year bin for the last 12 millennia. Fire frequencies for the other regions are represented in panel (e) for Mongolia (see Unkelbach and Behling, 2022 for details) and (f) for North America in the transition zone of the dense needleleaf and boreal mixedwood forests of eastern Canada (see Girardin et al., 2013 for details).

(Figure 34.4b) confirms this trend for the Mediterranean region (Lestienne et al., 2020a). During that period, the vegetation became more adapted to fire-prone conditions with, for example, the increased dominance of early successional combination of *Erica* sp. and *Pinus* sp. in the Mediterranean region (Figure 34.4a). In the transition zone of the dense needle leaf and boreal mixed wood forests of eastern Canada (Figure 34.4f), the arrival of fire closely follows the northward retreat of the glaciers ca. 13,000–8,000 BP and associated migration of vegetation. In Africa, the increase in fuel availability has been mostly promoted by the increase in precipitation associated with the movements of the intertropical convergence zone. The significant increase in the number of charcoal particles from 11,000 BP to 8,000 BP reflects this increase in precipitation, which caused the development of herbaceous and woody plants that were dense enough to allow thunderstorms, preceding the arrival of the monsoon and signaling the end of the dry season, to start fires.

The decrease in fire frequency in Mongolia around 9,500 BP (Figure 34.4e) was probably due to the presence of denser forest-steppe vegetation compared to later period characterized by a retreat of the forest, in favor of a larger steppe-dominated area (Unkelbach et al., 2021). Fire events remained rare throughout the Holocene. This decrease was also observed in the Mediterranean region and modeling experiments allowed us to infer that this decrease was probably due to climate with wetter conditions and shorter fire season length (Lestienne et al., 2020a). During the Neolithic (~ 6,000 BP) period, human activities became more intense in the Mediterranean region and have favored more frequent fires despite similar climatic conditions as in the previous period (Lestienne et al., 2020b, 2020a).

High fire frequencies have been observed in most regions during the last millennium mostly due to human activities in Africa, Mongolia and the Mediterranean region. For example, in the Altai mountains of Mongolia (Figure 34.4e) this increase coincided with a drying climate but also with the increase in population in the area. Industrialization, changes in herding practices and global warming may also have driven the fire frequency over the last century. In Africa, the last millennium was characterized by an increase in biomass burning with values reaching those from the Holocene Maximum (warm period occurring around 8,000 BP). Given the arid to semi-arid climate inferred at the regional scale of West Africa and the artifacts found at archaeological sites, it can be assumed that at least some of these recent charcoal records indicate intentional ignitions to open landscapes and promotion of the regeneration of pastures for herds.

Modeling of fire behavior

The impact of ongoing climatic changes on forests has become a growing concern, with many studies projecting increasing fire activity worldwide as a result of global warming. Particularly vulnerable to climatic warming are the length of the fire season and the seasonal distribution of fire occurrences (Flannigan et al., 2013; Lestienne et al., 2022). In addition, drought intensity and duration are expected to increase in many regions, which could cause larger, more intense and severe forest fires (Flannigan et al., 2013; Lestienne et al., 2022; Turetsky et al., 2011). However, other ecosystem processes may play a role in determining future fire trajectories. Notably, vegetation dynamics and changes in fuel conditions could modify fire behavior. In the boreal forest, for example, favoring broadleaf species in management planning could provide negative feedback on fire occurrence (Terrier et al., 2013), as shown in pollen and charcoal records during the mid-Holocene in the mixedwood boreal forest and confirmed through modeling experiments focusing on change in fire frequency (Girardin et al., 2013) and fire size (Hély et al., 2020). A high frequency of successive forest fires could also cause regeneration failures in forest stands, thereby reducing fuel loads and also inducing negative feedback on fire (Chaste et al., 2019; Héon et al., 2014). An important challenge in fire science is the quantification of such effects over the short period covered by fire statistics.

Fire history research may allow for the documentation and identification of long-term fire trends that, ultimately, can serve to improve the ability to project future fire impacts.

Modifications of fire activity that serve as a marker between past periods provide a valuable setting for analysis of the sensitivity of fire to various ecosystem and climatic processes. Taking advantage of this opportunity, recent advances include the use of global circulation models (GCMs; also referred to as general circulation models or earth system models) to develop a mechanistic understanding of the evolution of fire in ecosystems at different temporal and spatial scales. A GCM is a three-dimensional time-dependent numerical representation of the atmosphere, oceans, and sea ice Earth compartments, using the equations of motion and including radiation, photochemistry, and the transfer of heat and water vapor. Climatic simulations are available for past millennia and future centuries (Ertugrul, 2019; Gallo et al., 2022; Lestienne et al., 2022; Singarayer and Valdes, 2010). The millennial reconstructions of fire events extracted from lake sediments can be compared with past climatic conditions as simulated using GCMs. One can, for instance, look at whether or not there could be a correlation between variability in fire frequency inferred from lake sediment records during past millennia and drought as simulated by a GCM over the same horizon (Ali et al., 2012; Lestienne et al., 2022). In the case of no correlation, one can test alternative hypotheses such as the influence of human activities or changes in fuel conditions in relation to changes in fire frequency.

More advanced modeling experiments include the coupling of a GCM with an ecosystem simulator to assess the effect of climate change and changing vegetation structure and composition on fire (Lehsten et al., 2009; Smith et al., 2001). Generally, the processes included in the ecosystem simulator cover photosynthesis, respiration, evapotranspiration, organic matter decomposition in soil, post-fire tree mortality, succession and biomass burned (Pfeiffer et al., 2013). In these experiments, vegetation composition is often summarized with the presence and abundance of Plant Functional Types (PFTs) that grow and develop within a given geographical location characterized by its soil type and climate conditions. PFTs are assumed to represent all main species within a plant form (tree or grass) and a given bioclimatic limit (boreal, temperate or tropical) that have the same photosynthetic pathway, as well as ecological affinities and tolerance for light and water. Alternatively, the processes of forest growth may also be simpler and be dictated by yield curves determined from the analyses of forest inventories (de Groot et al., 2013). Through cross-comparison with the fire proxy and vegetation records, these modeling experiments allow testing of different temporal trajectories of fire activity in association with changes in forest types and climatic conditions.

Conclusion

The latitudinal distribution of vegetation on Earth is largely imposed by climatic limitations, but the spatiotemporal distribution of individual species is often the result of physical limitations and of repeated disturbances. In woodlands, for instance, fire leads to species regeneration and therefore contributes to maintaining the health and diversity of the forests. Accordingly, management strategies are now increasingly being focused on nature-based solutions, in which practice impacts should mimic natural disturbance impacts in order to maintain ecosystem integrity and resilience. Additionally, many studies predict that future fire activities will be higher than present ones owing to increasing global temperatures. Yet, fire activity is the result of complex feedback processes between temperature, precipitation, wind, humidity, and vegetation composition and structure; the influence of each of these factors in a given landscape setting is often difficult to tackle when studying short time series such as those provided by fire agencies. The examples presented in this chapter illustrate the need for more work to determine the natural patterns of fire activity and factors contributing to its variability. As outlined here, archival records and fire proxies are suitable for determining past fire, notwithstanding the fact that each type of proxy has its own limitations. Proxy records stored in lake sediments are particularly suitable for regional fire history reconstructions, while information at the stand level can be better obtained from tree-fire scars and charcoal particles preserved in soils.

References

Aleman JC, Blarquez O, Bentaleb I, Bonté P, Brossier B, Carcaillet C, et al. (2013) 'Tracking land-cover changes with sedimentary charcoal in the Afrotropics', *The Holocene*, vol 23, pp. 1853–1862.

Ali AA., Blarquez O, Girardin MP, Hély C, Tinquaut F, El Guellab A, et al. (2012) 'Control of the multimillennial wildfire size in boreal North America by spring climatic conditions', *Proceedings of the National Academy of Sciences*. vol 109, pp. 20966–20970.

Alleaume S, Hely C, Le Roux J, Korontzi S, Swap RJ, Shugart HH, et al. (2005) 'Using MODIS to evaluate heterogeneity of biomass burning in southern African savannahs: A case study in Etosha', *International Journal of Remote Sensing*, vol 26, no 19, pp. 4219–4237.

Andela N, Morton DC, Giglio L, Paugam R, Chen Y, Hantson S, et al. (2019) 'The Global Fire Atlas of individual fire size, duration, speed and direction', *Earth System Science Data*, vol 11, no 2, pp. 529–552.

Appleby PG (2002) 'Chronostratigraphic techniques in recent sediments. Tracking environmental change using lake sediments', *Developments in Paleoenvironmental Research*, vol 1, pp. 171–203.

Archibald S, Lehmann CER, Gómez-Dans JL and Bradstock RA (2013) 'Defining pyromes and global syndromes of fire regimes', *Proceedings of the National Academy of Sciences*, 110(16), pp. 6442–6447. doi:10.1073/pnas.1211466110

Attiwill PM (1994) 'The disturbance of forest ecosystems: The ecological basis for conservative management', *Forest Ecology and Management*, vol 63, no 2–3, pp. 247–300.

Björck S and Wohlfarth B (2002) '^{14}C chronostratigraphic techniques in paleolimnology. Tracking environmental change using lake sediments', *Developments in Paleoenvironmental Research*, vol 1, pp. 205–245.

Blarquez O, Girardin MP, Leys B, Ali AA, Aleman JC, Bergeron Y, et al. (2013) 'Paleofire reconstruction based on an ensemble-member strategy applied to sedimentary charcoal', *Geophysical Research Letters*, vol 40, no 11, pp. 2667–2672.

Blarquez O, Vannière B, Marlon JR, Daniau A-L, Power MJ, Brewer S, et al. (2014) 'Paleofire: An R package to analyse sedimentary charcoal records from the Global Charcoal Database to reconstruct past biomass burning', *Computers & Geosciences*, vol 72, pp. 255–261.

Bobek P, Svobodová HS, Werchan B, Švarcová MG and Kuneš P (2017) 'Human-induced changes in fire regime and subsequent alteration of the sandstone landscape of Northern Bohemia (Czech Republic)', *The Holocene*, vol 28, no 3, pp. 427–443.

Bonan GB and Shugart HH (1989) 'Environmental factors and ecological processes in boreal forests', *Annual Review of Ecology and Systematics*, vol 20, pp. 1–28.

Bowman DM, O'Brien JA and Goldammer JG (2013) 'Pyrogeography and the global quest for sustainable fire management', *Annual Review of Environment and Resources*, vol 38, pp. 57–80.

Brossier B, Oris F, Finsinger W, Asselin H, Bergeron Y and Ali AA (2014) 'Using tree-ring records to calibrate peak detection in fire reconstructions based on sedimentary charcoal records', *The Holocene*, vol 24, no 6, pp. 635–645.

Campbell ID and Flannigan MD (2000) 'Long-term perspectives on fire-climate-vegetation relationships in the North American boreal forest', *Fire, Climate Change, and Carbon Cycling in the Boreal Forest*, vol 138, pp. 151–172.

Chaste E, Girardin MP, Kaplan JO, Bergeron Y and Hély C (2019) 'Increases in heat-induced tree mortality could drive reductions of biomass resources in Canada's managed boreal forest', *Landscape Ecology*, vol 34, pp. 403–426.

Curt T, Borgniet L and Bouillon C (2013) 'Wildfire frequency varies with the size and shape of fuel types in southeastern France: Implications for environmental management', *Journal of Environmental Management*, vol 117, pp. 150–161.

Curt T, Fréjaville T and Lahaye S (2016) 'Modelling the spatial patterns of ignition causes and fire regime features in southern France: Implications for fire prevention policy', *International Journal of Wildland Fire*, vol 25, no 7, pp. 785–796.

Cyr D, Gauthier S and Bergeron Y (2007) 'Scale-dependent determinants of heterogeneity in fire frequency in a coniferous boreal forest of eastern Canada', *Landscape Ecology*, vol 22, pp. 1325–1339.

de Groot WJ, Flannigan MD and Cantin AS (2013) 'Climate change impacts on future boreal fire regimes', *Forest Ecology and Management*, vol 294, pp. 35–44.

Drobyshev I and Niklasson M (2004) 'Linking tree rings, summer aridity, and regional fire data: An example from the boreal forests of the Komi Republic, East European Russia', *Canadian Journal of Forest Research*, vol 34, pp. 2327–2339.

Ertugrul M (2019) 'Future forest fire danger projections using global circulation models (GCM) in Turkey', *Fresenius Environmental Bulletin*, vol 28, no 4, pp. 3261–3269.

Flannigan M, Cantin AS, De Groot WJ, Wotton M, Newbery A and Gowman LM (2013) 'Global wildland fire season severity in the 21st century', *Forest Ecology and Management*, vol 294, pp. 54–61.

Fritz R, Suffling R and Younger TA (1993) 'Influence of fur trade, famine, and forest fires on moose and woodland caribou populations in northwestern Ontario from 1786 to 1911', *Environmental Management*, vol 17, pp. 477–489.

Gaboriau DM, Asselin H, Ali AA, Hély C and Girardin MP (2022) 'Drivers of extreme wildfire years in the 1965–2019 fire regime of the Tłı̨chǫ First Nation territory, Canada', *Écoscience*, vol 29, no 3, pp. 249–265

Gallo C, Eden JM, Dieppois B, Drobyshev I, Fulé PZ, San-Miguel-Ayanz J, et al. (2022) 'Evaluation of CMIP6 model performances in simulating fire weather spatiotemporal variability on global and regional scales', *Geoscientific Model Development Discussions*,Vol 16, pp. 3103–3122. https://doi.org/10.5194/gmd-16-3103-2023

Gauthier S, Vaillancourt M-A, Leduc A, De Grandpré L, Kneeshaw D, Morin H, et al. (2009) ' Ecosystem Management of the Boreal Forest in the Era of Global Change', In: Girona, M.M., Morin, H., Gauthier, S., Bergeron, Y. (eds) Boreal Forests in the Face of Climate Change. Advances in Global Change Research, vol 74

Girardin MP, Ali AA, Carcaillet C, Blarquez O, Hély C, Terrier A, et al. (2013) 'Vegetation limits the impact of a warm climate on boreal wildfires', *New Phytologist*, vol 199, no 4, pp. 1001–1011.

Goldammer JG, Sukhinin A and Davidenko EP (2007) 'Advance publication of wildland fire statistics for Russia 1992-2007', *International Forest Fire News*, vol 37, pp. 85–87.

Gutsell SL and Johnson EA (1996) 'How fire scars are formed: coupling a disturbance process to its ecological effect', *Canadian Journal of Forest Research*, vol 26, no 2, pp. 166–174. https://doi.org/10.1139/x26-020

Heinselman ML (1973) 'Fire in the virgin forests of the Boundary Waters Canoe Area, Minnesota', *Quaternary Research*, vol 3, no 3, pp. 329–382.

Hély C, Alleaume S and Runyan CW (2019) 'Fire regimes in dryland landscapes'. In: D'Odorico P, Porporato A, Wilkinson Runyan C. (eds.), *Dryland Ecohydrology*. Springer, Cham, pp. 367–399.

Hély C, Bergeron Y and Flannigan MD (2000) 'Effects of stand composition on fire hazard in mixed-wood Canadian boreal forest', *Journal of Vegetation Science*, vol 11, no 6, pp. 813–824.

Hély C, Chaste E, Girardin MP, Remy CC, Blarquez O, Bergeron Y, et al. (2020) 'A holocene perspective of vegetation controls on seasonal boreal wildfire sizes using numerical paleo-ecology', *Frontiers in Forests and Global Change*, vol 3.

Hély C and Lézine A-M (2014) 'Holocene changes in African vegetation: Tradeoff between climate and water availability', *Climate of the Past*, vol 10, no 2, pp. 681–686.

Héon J, Arseneault D and Parisien M-A (2014) 'Resistance of the boreal forest to high burn rates', *Proceedings of the National Academy of Sciences*, vol 111, no 38, pp. 13888–13893.

Higuera P (2009) *CharAnalysis 0.9: Diagnostic and Analytical Tools for Sediment-Charcoal Analysis.* User's Guide. Montana State University, Bozeman, MT. Citeseer.

Higuera PE, Peters ME, Brubaker LB and Gavin DG (2007) 'Understanding the origin and analysis of sediment-charcoal records with a simulation model', *Quaternary Science Reviews*, vol 26, no 13–14, pp. 1790–1809.
Jessup TC, Vayda AP, Cochrane MA, Applegate GB, Ryan KC and Saharjo BH (2022) 'Why estimates of the peat burned in fires in Sumatra and Kalimantan are unreliable and why it matters', *Singapore Journal of Tropical Geography*, vol 43, no 1, pp. 7–25. https://doi.org/10.1111/sjtg.12406
Kane ES, Hockaday WC, Turetsky MR, Masiello CA, Valentine DW, Finney BP, et al. (2010). 'Topographic controls on black carbon accumulation in Alaskan black spruce forest soils: Implications for organic matter dynamics', *Biogeochemistry*, vol 100, pp. 39–56.
Keeley JE (2009) 'Fire intensity, fire severity and burn severity: A brief review and suggested usage', *International Journal of Wildland Fire*, vol 18, no, p. 116.
Kuklina V, Sizov O, Rasputina E, Bilichenko I, Krasnoshtanova N, Bogdanov V, et al. (2022) 'Fires on ice: Emerging permafrost peatlands fire regimes in Russia's subarctic taiga', *Land*, vol 11, no 3, p. 322.
Ledru M-P, Ceccantini G, Gouveia SE, López-Sáez JA, Pessenda LC and Ribeiro AS (2006) 'Millenial-scale climatic and vegetation changes in a northern Cerrado (Northeast, Brazil) since the Last Glacial Maximum', *Quaternary Science Reviews*, vol 25, no 9–10, pp. 1110–1126.
Lehsten V, Tansey K, Balzter H, Thonicke K, Spessa A, Weber U, et al. (2009) 'Estimating carbon emissions from African wildfires', *Biogeosciences*, vol 6, no 3, 349–360.
Lesieur D, Gauthier S and Bergeron Y (2002) 'Fire frequency and vegetation dynamics for the south-central boreal forest of Quebec, Canada', *Canadian Journal of Forest Research*, vol 32, no 11, pp. 1996–2009.
Lestienne M, Hély C, Curt T, Jouffroy-Bapicot I and Vannière B (2020a) 'Combining the monthly drought code and paleoecological data to assess Holocene climate impact on Mediterranean fire regime', *Fire*, vol 3, no 2, p. 8.
Lestienne M, Jouffroy-Bapicot I, Leyssenne D, Sabatier P, Debret M, Albertini P-J, et al. (2020b) 'Fires and human activities as key factors in the high diversity of Corsican vegetation', *The Holocene*, vol 30, no 2, pp. 244–257.
Lestienne M, Vannière B, Curt T, Jouffroy-Bapicot I and Hély C (2022) 'Climate-driven Mediterranean fire hazard assessments for 2020–2100 on the light of past millennial variability', *Climatic Change*, vol 170, no 1, pp. 14.
Leys B, Carcaillet C, Dezileau L, Ali AA and Bradshaw RH (2013) 'A comparison of charcoal measurements for reconstruction of Mediterranean paleo-fire frequency in the mountains of Corsica', *Quaternary Research*, vol 79, no 3, pp. 337–349.
Macias Fauria M and Johnson EA (2008) 'Climate and wildfires in the North American boreal forest', *Philosophical Transactions of the Royal Society B: Biological Sciences*, vol 363, no 1501, pp. 2315–2327.
Michaletz ST and Johnson EA (2007) 'How forest fires kill trees: a review of the fundamental biophysical processes', *Scandinavian Journal of Forest Research*, vol 22, no 6, pp. 500–515.
Päätalo M-L (1998) 'Factors influencing occurrence and impacts of fires in northern European forests', *Silva Fennica*, vol 32, pp. 185–202.
Patterson WA, Edwards KJ and Maguire DJ (1987) 'Microscopic charcoal as a fossil indicator of fire', *Quaternary Science Reviews*, vol 6, no 1, pp. 3–23.
Pausas JG (2022) 'Pyrogeography across the western Palaearctic: A diversity of fire regimes', *Global Ecology and Biogeography*, vol 31, no 10, pp. 1923–1932.
Payette S, Delwaide A, Schaffhauser A and Magnan G (2012) 'Calculating long-term fire frequency at the stand scale from charcoal data', *Ecosphere*, vol 3, no 7, pp. 1–16.
Pereira MG, Trigo RM, da Camara CC, Pereira JM and Leite SM (2005) 'Synoptic patterns associated with large summer forest fires in Portugal', *Agricultural and Forest Meteorology*, vol 129, no 1–2, pp. 11–25.
Pfeiffer M, Spessa A and Kaplan JO (2013) 'A model for global biomass burning in preindustrial time: LPJ-LMfire (v1. 0)', *Geoscientific Model Development*, vol 6, no 3, pp 643–685.
Rannie WF (2001) 'The "Grass Fire Era" on the southeastern Canadian Prairies', *Prairie Perspectives: Geographical Essays*. University of North Dakota Grand Forks, vol 4, pp. 1–19.

Reimer PJ, Austin WE, Bard E, Bayliss A, Blackwell PG, Ramsey CB, et al. (2020) 'The IntCal20 Northern Hemisphere radiocarbon age calibration curve (0–55 cal kBP)', *Radiocarbon*, vol 62, no 4, pp. 725–757.

Remy CC, Fouquemberg C, Asselin H, Andrieux B, Magnan G, Brossier B, et al. (2018) 'Guidelines for the use and interpretation of palaeofire reconstructions based on various archives and proxies', *Quaternary Science Reviews*, vol 193, pp. 312–322.

Ruffault J, Moron V, Trigo RM and Curt T (2017) 'Daily synoptic conditions associated with large fire occurrence in Mediterranean France: Evidence for a wind-driven fire regime: Daily synoptic conditions associated with large fire occurrence', *International Journal of Climatology*, vol 37, no 1, pp. 524–533.

Schmuck G, San-Miguel-Ayanz J, Camia A, Durrant T, Boca R and Libertá G (2013) *Forest Fires in Europe, Middle East and North Africa 2012*. Joint Research Centre, European Commission, Institute for Environment and Sustainability, Land Management and Natural Hazards Unit, Luxembourg.

Simpson IJ, Akagi SK, Barletta B, Blake NJ, Choi Y, Diskin GS, et al. (2011) 'Boreal forest fire emissions in fresh Canadian smoke plumes: C 1-C 10 volatile organic compounds (VOCs), CO_2, CO, NO_2, NO, HCN and CH_3 CN', *Atmospheric Chemistry and Physics*, vol 11, no 13, pp. 6445–6463.

Singarayer JS and Valdes PJ (2010) 'High-latitude climate sensitivity to ice-sheet forcing over the last 120 kyr', *Quaternary Science Reviews*, vol 29, no 1–2, pp. 43–55.

Smith JB, Schellnhuber H-J, Mirza MMQ, Fankhauser S, Leemans R, Erda L, et al. (2001) 'Vulnerability to climate change and reasons for concern: A synthesis'. In: McCarthy JJ, Canziani OF, Leary NA, Dokken DJ, and White KS (Eds.), *Climate Change 2001: Impacts, Adaptation, and Vulnerability.* Cambridge University Press, Cambridge, pp. 913–967.

Stephens SL, Burrows N, Buyantuyev A, Gray RW, Keane RE, Kubian R, et al. (2014) 'Temperate and boreal forest mega-fires: Characteristics and challenges', *Frontiers in Ecology and the Environment*, vol 12, no 2, pp. 115–122.

Stocks BJ, Mason JA, Todd JB, Bosch EM, Wotton BM, Amiro BD, et al. (2002) 'Large forest fires in Canada, 1959–1997', *Journal of Geophysical Research: Atmospheres*, vol 107, pp. 5–12.

Swetnam TW (1993) 'Fire history and climate change in giant sequoia groves', *American Association for the Advancement of Science*, vol 262, no 5135, pp. 885–889.

Tardif J (2004) 'Fire history in the duck mountain provincial forest, western Manitoba', *Sustainable Forest Management Network. Project Reports 2003/2004.* University of Alberta, Edmonton, Canada.

Terrier A, Girardin MP, Périé C, Legendre P and Bergeron Y (2013) 'Potential changes in forest composition could reduce impacts of climate change on boreal wildfires', *Ecological Applications,* vol 23, no 1, pp. 21–35.

Turetsky MR, Kane ES, Harden JW, Ottmar RD, Manies KL, Hoy E, et al. (2011) 'Recent acceleration of biomass burning and carbon losses in Alaskan forests and peatlands', *Nature Geoscience* vol 4, no 1, pp. 27–31.

Turner MG and Romme WH (1994) 'Landscape dynamics in crown fire ecosystems', *Landscape Ecology*, vol 9, no 1, pp. 59–77.

Unkelbach J and Behling H (2022) 'The reconstruction of Holocene northwestern Mongolian fire history based on high-resolution multi-site macro-charcoal analyses', *Frontiers in Earth Science*, vol 10.

Unkelbach J, Dulamsuren C, Klinge M and Behling H (2021) 'Holocene high-resolution forest-steppe and environmental dynamics in the Tarvagatai Mountains, north-central Mongolia, over the last 9570 cal yr BP', *Quaternary Science Reviews*, vol 266, pp. 1070–1076.

Unkelbach J, Dulamsuren C, Punsalpaamuu G, Saindovdon D and Behling H (2018) 'Late Holocene vegetation, climate, human and fire history of the forest-steppe-ecosystem inferred from core G2-A in the 'Altai Tavan Bogd'conservation area in Mongolia', *Vegetation History and Archaeobotany*, vol 27, pp. 665–677.

Unkelbach J, Kashima K, Enters D, Dulamsuren C, Punsalpaamuu G and Behling H (2019) 'Late Holocene (Meghalayan) palaeoenvironmental evolution inferred from multi-proxy-studies of lacustrine sediments from the Dayan Nuur region of Mongolia', *Palaeogeography, Palaeoclimatology, Palaeoecology*, vol 530, pp. 1–14.

Unkelbach J, Kashima K, Punsalpaamuu G, Shumilovskikh L and Behling H (2020) 'Decadal high-resolution multi-proxy analysis to reconstruct natural and human-induced environmental changes over the last 1350 cal. yr BP in the Altai Tavan Bogd National Park, western Mongolia', *The Holocene*, vol 30, no 7, pp. 1016–1028.

Van Wagner C (1977) 'Conditions for the start and spread of crown fire', *Canadian Journal of Forest Research*, vol 7, no 1, pp. 23–34.
Van Wagner C (1988) 'The historical pattern of annual burned area in Canada', *The Forestry Chronicle*, vol 64, no 3, pp. 182–185.
Vetrita Y and Cochrane MA (2019) Annual burned area from landsat, mawas, central kalimantan, Indonesia, 1997-2015. *ORNL DAAC*. https://daac.ornl.gov/cgi-bin/dsviewer.pl?ds_id=1708
Whitlock C and Bartlein PJ (2003) 'Holocene fire activity as a record of past environmental change', *Developments in Quaternary Sciences*, vol 1, pp. 479–490.
Whitlock C, Higuera PE, McWethy DB and Briles CE (2010) 'Paleoecological perspectives on fire ecology: Revisiting the fire-regime concept', *The Open Ecology Journal*, vol 3, no 1, pp. 6–23.

35
THE ECOLOGICAL CONSEQUENCES OF DROUGHTS IN FORESTS

Richard T. Corlett

Introduction

Droughts and tree mortality from droughts are natural phenomena, recorded in the historical and paleoecological records (Cook et al., 2018), and the frequencies and intensities of natural droughts have had a major influence on the distributions of tree species globally. In recent decades, however, observations in many parts of the world have shown a significant increase in drought-induced tree mortality (e.g., Fettig et al., 2019; Baumann et al., 2022; McDowell et al., 2022; Ometto et al., 2022; Liu et al., 2023). Global climate models suggest that the weather extremes that have caused this will get worse over future decades in response to anthropogenic climate change (Seneviratne et al., 2021; Hammond et al., 2022; De Luca and Donat, 2023; Tavares et al., 2023). There is therefore a growing concern that forests could be the 'new coral reefs', increasingly vulnerable to global climate change (Brodribb et al., 2020). Forests, like reefs, are dominated by sessile, long-lived, slow-growing organisms, limiting their capacity for rapid adjustment to the changing climates. And, like reefs, they are subject to multiple additional pressures which may exacerbate these impacts and reduce their capacity for resistance and recovery. Moreover, droughts have the potential to convert forests from carbon sinks to sources, further exacerbating climate change.

This chapter therefore aims, first, to clarify how droughts are defined and measured, with a focus on the ecological droughts that result in tree mortality, and to summarize recent progress in modelling and predicting ecological droughts. Our current understanding of how droughts kill trees is then outlined, with a brief discussion of post-drought recovery and legacy effects. Next, observed variation among tree species in drought resistance and resilience is summarized, with a discussion of ecological trade-offs between combinations of drought-related traits. The impacts of droughts on juvenile trees, ground herbs, lianas, and epiphytes are then described, based on, admittedly, incomplete data. Finally, drought impacts on forest structure and species composition are described and discussed.

DOI: 10.4324/9781003324072-41

What are droughts?

The Intergovernmental Panel on Climate Change (IPCC) defines droughts as 'periods with substantially below-average moisture conditions... during which limitations in water availability result in negative impacts for various components of natural systems and economic sectors' (Seneviratne et al., 2021, p. 1570). The vagueness of this definition is unavoidable since different aspects of drought—different variables used to assess the 'below-average moisture conditions'—have different impacts on natural and economic systems. A meteorological drought is characterized by below-average rainfall, a hydrological drought by water shortage in streams and reservoirs, an agricultural drought by reductions in crop yield, and an ecological drought—the focus of this chapter—by plant water stress and the resultant mortality. A single drought event may satisfy all these criteria, if low rainfall leads to a water shortage, reduced crop yields, and plant mortality in natural ecosystems, but the different types of droughts are not always tightly correlated. In this chapter, therefore, droughts are ecological droughts, unless specified otherwise.

Ecological droughts are caused by insufficient soil moisture, often amplified by a concurrent increase in atmospheric evaporative demand (AED). Observational data show that precipitation deficits—meteorological droughts—have occurred in many parts of the world in recent decades and that long-term declines have happened in some parts of Africa and South America, but other regions have got wetter or show no consistent trends (Seneviratne et al., 2021). Increases in AED are more widespread and are driven largely by the exponential increases in vapour pressure deficit (VPD—the difference between the actual and saturated vapour pressure) because of global warming. AED is also influenced by solar radiation and wind speed, but these have been much less important as drivers of recent changes. Long-term observations on soil moisture are rare, but drying trends that have been identified are most often related to increases in AED and thus evapotranspiration, rather than decreases in precipitation (Seneviratne et al., 2021). It is now becoming clear, therefore, that the major driver of the observed increase in tree mortality has not been reduced rainfall. Instead, it has been global warming and the resulting chronic increases in VPD and AED that have made previously non-lethal rainfall deficits lethal.

Droughts vary widely in length, from fast-developing 'flash droughts', lasting only for weeks, to multi-decadal 'megadroughts'. Plant responses to moisture deficits also vary widely in speed, from less than an hour for leaf stomata to several decades for major changes in tree architecture. It is also important to distinguish between rare extreme events and long-term unidirectional changes. Forests may recover from unusual extremes, even if individual trees die, but forest structure and species composition will usually change permanently in response to unidirectional changes in moisture availability.

Although there are many types of droughts, it can still be useful to use standardized drought indices to compare their severity, since this allows for easy comparisons between droughts and between sites (Slette et al., 2020). There are a lot of these indices, with varying properties, but the Standardized Precipitation Evapotranspiration Index (SPEI) and the Palmer Drought Severity Index (PDSI)—both combining precipitation and atmospheric evaporative demand—are most widely reported by ecologists. Also, although droughts are defined by lows in water availability, many observations of severe agricultural and ecological impacts are associated with combinations of low water availability and high temperature, i.e., hot droughts. A global survey of field-documented tree mortality events found a hotter and drier climate signal for events from across the world, from tropical to boreal forests

(Hammond et al., 2022). These compound events can be identified by the concurrence of low precipitation (or low SPEI) and high mean temperature, on an appropriate timescale.

Modelling droughts in a changing climate

The current generation of Earth System Models (i.e., climate models with interactive biogeochemical components) has not been very successful in reproducing observed trends in rainfall, AED, and soil moisture, and their projections for the future have a large spread in many parts of the world (Seneviratne et al., 2021). Modelling studies agree, however, in attributing the recent increase in drought events to human influences on climate, rather than natural variability. Moreover, model agreement, and thus confidence, is high for increasing drought hazard with increasing temperatures, with a substantial increase in risk even from 1.5°C to 2.0°C global warming. Increases in the severity of ecological droughts at 2.0°C are projected with high confidence for several IPCC AR6 regions, including the Mediterranean, the South American Monsoon region (which includes most of the Amazon basin), and southern South America, and with lower confidence in others, with more regions affected as global warming increases above 2°C. The occurrence of compound hot-dry extremes is also projected to increase globally (De Luca and Donat, 2023).

The El Niño/South Oscillation (ENSO) is the most important natural driver of interannual rainfall variation over much of the planet. El Niño events vary in amplitude, frequency, and spatial pattern, and no two are the same. This high internal variability makes ENSO very difficult to model, and the current generation of models shows as wide a spread of projections for the 21st century as did earlier versions, not even agreeing on the sign of possible changes in amplitude (Beobide-Arsuaga et al., 2021). However, even if El Niño events are not strengthened by climate change, higher temperatures and VPD will worsen the impacts of ENSO-driven declines in rainfall.

Rising CO_2 concentrations are both the biggest single driver of anthropogenic climate change and a potentially large, but still poorly understood, influence on plant physiology and thus plant–soil feedbacks. In both theory and experiments, rising CO_2 can increase water use efficiency (WUE, the ratio of photosynthesis to transpiration) at the leaf level by reducing stomatal conductance. Current Earth System Models scale this up to the global level, projecting a reduction in the severity of ecological droughts. However, the significance of increased leaf-level WUE in the real world is still uncertain (Vicente-Serrano et al., 2022) and the benefits will be minimal in dry periods when stomata are already closed. Higher CO_2 concentrations may also promote 'structural overshoot' in non-drought periods, with trees and forests increasing aboveground biomass and leaf area above the levels that can be sustained during droughts (McDowell et al., 2022). It is still not clear, therefore, whether rising CO_2 can alleviate ecological droughts and reduce tree mortality under realistic conditions.

Drivers and mechanisms of drought-related tree mortality

Trees die from drought when so many cells—particularly meristematic cells—are killed, that recovery is impossible. This can occur organ by organ—crown dieback and root loss—or at the whole tree level. Our understanding of the mechanisms that connect changes in rainfall and AED with dieback and tree death is still incomplete, but research is gradually filling in the gaps (McDowell et al., 2022). Trees are inherently vulnerable to droughts because they require long-distance transport of water between the roots underground and the photosynthetic cells in the canopy, which may be tens of meters above. A continuous column of water is needed to ensure that water can be pulled up from the soil through the xylem to the transpiring leaves. During droughts, the tension needed to draw water from the soil increases (conventionally described as an increasingly ***negative*** water potential). Plants can regulate water transport by opening and closing the stomata, but closing stomata to avoid excessively negative water potentials also prevents CO_2 uptake. When the water potential becomes sufficiently negative, gas embolisms—air bubbles in the xylem—can break the water column, thus blocking water transport. As each xylem vessel is blocked, the water potential becomes more negative, leading to more embolisms.

Most evidence suggests that it is hydraulic failure—the inability of the tree to transport water from the soil below ground to leaves and other organs above ground—that is the most important mechanism causing dieback and tree mortality. Embolisms appear to be a ubiquitous feature of tree death and plant resistance to hydraulic failure predicts tree species mortality around the world (Anderegg et al., 2016). Carbon starvation because of stomatal closure and leaf shedding may also contribute to mortality, particularly in prolonged droughts, by impairing carbon-dependent hydraulic and other functions. Also, in addition to direct impacts on plant physiological processes, droughts may promote mortality through fungal or insect attacks on trees weakened by hydraulic failure and/or carbon starvation (Fettig et al., 2019; McDowell et al., 2022), and by promoting more frequent and intense fires (see Chapter 8, 'Fire in forest ecosystems'). Moreover, although hydraulic traits are generally good predictors of drought mortality, they are poor predictors of stem growth during severe—but non-lethal—droughts (Smith-Martin et al., 2023), suggesting that additional mechanisms may be involved in influencing stem growth.

Although stomatal closure reduces water loss from leaves, it does not stop it as loss continues through the cuticle and leaky stomata. Indeed, transpiration rates typically increase with rising VPD, despite stomatal closure. Transpiration also cools leaves, which may be particularly important in hot droughts, but the significance of thermal damage in drought-related dieback is currently unclear and the evidence for significant transpirational cooling in forest canopy leaves is limited (Still et al., 2022). Moreover, high temperatures exacerbate cuticular losses (Cochard, 2021), although the significance of this in real-world droughts is unclear. Some trees appear to use a strategy known as vulnerability segmentation, whereby the leaves are more vulnerable to embolism than the stems and are shed first during droughts as a 'safety valve' (Levionnois et al, 2020). Leaf-shedding, whether drought-induced or seasonal, can stop all water loss from leaves, but water is also lost through the bark, with a wide range of variation between species, and this loss may be particularly significant for deciduous species (Wolfe, 2020).

As long as the tree has access to adequate soil water, these losses can be replaced. However, drying soil, caused by a rainfall deficit or VPD-driven increases in evaporation from the soil surface and transpiration by plants, reduces soil conductance. Under severe drought conditions, the soil–root interface can become the largest single resistance in the water supply pathway. Cut off from soil water, the tree is dependent on internal water storage and its survival time will be determined by how quickly continued water loss after stomatal closure results in lethal dehydration of critical tissues. It has been suggested that it is this interruption of the soil–root continuum that is most likely to be the critical initial failure, leading eventually to dieback and mortality, and that the observed xylem embolisms are a result of this failure, not the starting point (Körner, 2019). There is not enough information currently on belowground hydraulics to test this, but the variation in mortality among coexisting tree species and the association of this species-specific mortality with plant hydraulic traits (Anderegg et al., 2016) suggest that the initial failure is in the plant not the soil. The potential role of mycorrhizal fungi in bridging the soil–root interface is another issue that needs to be explored further (Abdalla et al., 2023).

Drought recovery and legacy effects

Post-drought recovery after non-lethal disruptions of water transport caused by xylem embolisms is still not well-understood. At least two mechanisms are potentially involved: embolism refilling, usually within hours of rewetting, and new xylem growth, which takes weeks or months (Gauthey et al., 2022). Springtime refilling of embolisms caused by winter freezing is well documented, but the importance of refilling after drought is still debated, partly because of the lack of a plausible mechanism to explain refilling under tension. Current evidence suggests that it occurs in some species but is relatively rare. The growth of new xylem tissue, in contrast, is easily observed in surviving trees. It is a slow and expensive process, however, and may not be possible in trees suffering from depleted carbohydrate stores after leaf fall or prolonged stomatal closure.

In addition to the immediate effects of drought on trees and forests, lags in recovery—drought legacy effects—may continue for several years afterwards (Kannenberg et al., 2020). Observed legacy effects range from slowed growth to canopy dieback and delayed tree death. Much of the evidence for drought legacies comes from tree ring studies and it is less clear how they influence aspects of tree biology other than diameter growth. The causes also are unclear, with depletion of stores of non-structural carbohydrates, changes in carbon allocation, persistent hydraulic damage, lags in replenishment of the water table, and increased prevalence of pests and diseases, as non-exclusive suggestions. Tree deaths from pest and pathogen infections of drought-weakened trees typically occur 1–4 years after the drought (Fettig et al., 2019).

Legacy effects may also influence the responses of trees and forests to repeated droughts. Most trees will face multiple droughts during their lives, so resilience to drought is necessary for long-term survival. Trees that were less resilient to previous non-lethal droughts are more likely to die in future droughts (DeSoto, 2020). In general, subsequent droughts appear to be more damaging than initial droughts, but the variation between species and ecosystems is large (Anderegg et al., 2020). Conifers and conifer-dominated forests most often show increasing vulnerability, probably as a result of accumulated physiological damage and/or pest and disease infections, while in angiosperms there is evidence in some cases for acclimation, perhaps reflecting their greater anatomical flexibility. However, the responses to recent consecutive severe droughts in the Amazon suggest that this acclimation capacity in angiosperms can be overwhelmed and an experiment in the north-east Amazon found no evidence of acclimation in hydraulic traits (include P_{50} and P_{88}) after 15 years of imposed water deficit (Bittencourt et al., 2020).

Variation among trees in drought resistance and resilience

Although there is evidence for an evolutionary trade-off at the level of individual xylem (or tracheid) conduits between safety (i.e., avoiding dangerously negative water potentials) and efficient water conduction, this trade-off is much weaker at branch level, probably because of the wide variation in the number of conduits per unit area of sapwood (Franklin et al., 2023). At the whole plant level, there is a stronger trade-off between hydraulic safety and drought avoidance, by deep rooting, deciduousness, and/or within-tree water storage (capacitance) (Oliveira et al., 2021; Smith-Martin et al., 2023).

Sensitivity to embolisms at the organ or plant level is usually measured by the water potential at which conductivity is halved, P_{50} (or ψ_{50}), although most plants can survive higher conductivity losses than this, with a range of at least 60–90% across species and methods. The difference between the P_{50} (or P_{88}) and the lowest midday water potential experienced by each species (P_{dry}) is the hydraulic safety margin, HSM_{50} (or HSM_{88}). A tree can potentially increase its HSM, and thus its ability to survive droughts, by producing safer but less efficient conduits, adding more conduits, reducing leaf area and/or stomatal conductance, enhancing within-tree water storage, and/or growing deeper roots. All these strategies have their costs however, and most tree species seem to live very close to their hydraulic limits.

The costs of hydraulic safety mean that the distributions of tree species are broadly constrained by their intrinsic tolerances of water stress. In Australia, increasing mortality risk associated with rising VPD was greatest in tree species near the drier edges of their range (Bauman et al., 2022). However, different species have different combinations of traits and are therefore spread widely along the avoidance–safety axis (Oliveira et al., 2021). Within-community variation in hydraulic traits is striking, particularly in wet forests where P_{50} values may differ more than fivefold and HSM values range from near-zero to more than 5 MPa (Smith-Martin et al., 2023). Even in dry forests, where all species must resist annual meteorological droughts, trees with very different hydraulic strategies may be able to coexist at the landscape scale by occupying sites with different access to groundwater and, at the community level, by hydrological niche partitioning, so that, for example, deep-rooted species with low embolism resistance can coexist with shallow-rooted species with higher resistance.

Oliveira et al. (2021) propose an additional trade-off, between growth and hydraulic safety margins, resulting in a fast-risky to slow-safe axis of variation among species. Indeed, it has been suggested that HSM may help explain the growth-survival trade-off seen in plot data from across the Amazon, with tree species with acquisitive traits prioritizing growth at the expense of hydraulic safety (Tavares et al., 2023). Risky hydraulic strategies are expected to be more common at nutrient-rich sites where fast growth allows rapid recovery from setbacks and supports nutrient-demanding strategies such as deciduousness. In California, forests on sites with more nutrients are prone to structural overshoot in wet years, leading to plant water demand in drought years that cannot be met, resulting in boom-bust cycles of growth and dieback (Callahan et al., 2023). However, as droughts become more frequent—and frequent droughts become hotter and more lethal—there may no longer be enough time between them for forest recovery. At low nutrient sites, in contrast, trees are expected to have higher safety margins, because rapid recovery is not possible, and thus to suffer less damage in a similar drought. Some studies have suggested that large trees are more vulnerable due to height-driven hydraulic vulnerability, but several recent studies found no evidence for this (Bauman et al., 2022). Au et al. (2023) show that younger trees in the upper canopy are more sensitive to drought than older trees—which may have better root systems—but also recover more quickly.

In response to increasing drought risk, trees can potentially migrate to track the climate conditions to which they are adapted, although few species are likely to disperse far or fast enough to keep up with current rates of change (see Chapter 37, 'Plant movements in response to rapid climate change'), or they can adapt or acclimate *in situ*. Genetic adaptation is slow and likely to be constrained by limited genetic diversity in hydraulic traits. Acclimation through phenotypic plasticity is another possibility, but it is difficult to distinguish between 'adaptive plasticity', which increases fitness or survival, and plastic changes with a neutral or negative impact (Rowland et al., 2023). We should not expect plants under severe stress to function optimally, so phenotypic changes during droughts should not be interpreted as acclimation without additional evidence. Facultative shedding of leaves allows trees to rapidly reduce water loss and is common in many species in droughts, but it also reduces photosynthesis, and the shed leaves must be replaced later. Branch shedding and changes in canopy architecture could also be long-term adaptations to sustained drought conditions, but we currently lack robust evidence for this (Rowland et al., 2023). And we know very little about the responses of belowground traits, despite the crucial role of root systems in water uptake.

Seedlings, saplings, lianas and epiphytes in forests

Canopy trees and lianas must go through seedling and sapling stages in which the advantages of reduced exposure to dry air, winds, and solar radiation, and the shorter distance over which water must be transported, are set against the disadvantages of shade, and thus reduced photosynthesis, and a less well-developed root system. Environmental filtering during these juvenile stages can potentially have a long-term influence on future forest floristics and structure, even in the absence of adult tree mortality. Seedling mortality increased by 11% overall in eight plots across a rainfall gradient in Panama during the severe 2015–2016 drought, with the mortality increases largest for wetter forests and for species associated with these forests (Browne et al., 2021).

Both observations and research on drought impacts in forests have focused on trees, particularly canopy trees, since the deaths of these large trees are most conspicuous and have most impact on carbon and water fluxes, and other ecosystem functions. In many forests, however, most plant diversity is in other growth forms, including understory trees, shrubs and herbs, as well as lianas and vascular epiphytes. Impacts on these growth forms have received a lot less attention. Understory plants live in an environment buffered by the canopy above from environmental variation, but are not protected from the most extreme droughts, when leaf-fall, dieback, and mortality may occur. Overall, however, the literature suggests that direct drought impacts on understory plants are usually less than those observed for canopy plants, although the indirect impacts of canopy opening by canopy dieback and tree death can be large.

Lianas are most abundant in seasonally dry forests and there is evidence that they currently have a dry season growth advantage (see Chapter13, 'The ecology of lianas and their increasing influence in tropical forests'). Moreover, lianas are increasing in abundance in some Neotropical forests with long-term drying trends. Despite this, their substantially higher (more than three times on average) hydraulic conductivities than coexisting trees, but similar sensitivities to embolism (P_{50}), suggest that lianas will be highly vulnerable to the continued rise in VPD (Willson et al., 2022). The future of lianas under climate change is therefore still uncertain.

The sharp decline in epiphyte abundance and diversity from everwet to increasingly dry forests in the tropics demonstrates the overriding importance of water relations in epiphyte ecology, but that does not necessarily translate into vulnerability for species that are adapted to a particular position along this gradient. Epiphytes have been considered by different researchers as either particularly vulnerable because they do not have access to ground water or particularly resistant because they are necessarily adapted to periodic dry conditions. The rather sparse observational data does not permit generalization, but a recent study of vascular epiphytes growing on lakeshore trees in the Barro Colorado National Monument saw a decline of 17% in epiphyte numbers after the severe 2015–2016 El Niño event, with wide variation in mortality among species (Einzmann et al., 2022). Epiphyte communities are highly dynamic, so a single event is unlikely to have a lasting impact, but long-term directional changes in drought frequency and severity could lead to substantial changes in the epiphyte community.

Impacts of droughts on forests

The vulnerability of forests to droughts is influenced by multiple abiotic and biotic factors, varying at different spatial scales (Hollunder et al., 2022). Abiotic factors include climate, soils, topography, and water-table depth, while biotic factors include forest structure and species composition. In general—but with some exceptions—forests are less resilient to extreme droughts at sites that are already dry, including ridges and steep slopes within more mesic forests, despite the increased prevalence of drought-related traits at these sites. The vast areas of forest in the Amazon and elsewhere with shallow water tables (< 5 m) appear to be buffered against moderate droughts but may be more vulnerable to severe droughts because of their shallow roots and the lack of drought-tolerant traits (Costa et al., 2023).

In a pan-Amazon study, tree species at everwet sites, as expected, were found to be less resistant to xylem embolisms (less negative P_{50}) than those at drier sites, but they also had higher hydraulic safety margins (HSM_{50}) and thus apparently faced the lowest risk of hydraulic failure (Tavares et al., 2023). Deciduous species had lower HSM_{50}, suggesting that they had hydraulically riskier strategies. Strikingly, in this study, HSM_{50} was the only significant predictor of long-term increases in aboveground biomass, explaining 67% of the variance (Tavares et al., 2023). The authors suggest that this relationship may reflect the impact of hydraulic failure on the residence time of woody biomass through its influence on tree mortality.

The belowground water supply in a forest is shared by all the plant species and individuals in the community. In forestry, thinning of the stand is often recommended to increase resilience to droughts by reducing competition between individual trees, but a meta-analysis found no consistent effects of competition on the growth response to drought (Castagneri et al., 2021) and a study of *Pinus ponderosa* across a wide range of climates and tree sizes concluded that the impact of thinning on growth during droughts is context-dependent (Young et al., 2023). Most of these studies are based on temperate conifer monocultures, but a global review of the impact of tree species diversity on forest response to drought also found it impossible to draw general conclusions (Grossiord, 2020). On the other hand, eddy-covariance measurements in temperate and boreal forests, as well as global satellite data on vegetation water content, suggest that plant hydraulic diversity—particularly community-level variation in HSM—enhances ecosystem resilience during drought, although the mechanism for this is unclear (Anderegg et al., 2018).

When drought kills many large trees of canopy-dominant species, changes in forest structure and composition are immediate, although the long-term consequences will depend on what species dominate the post-drought regeneration (e.g., Fettig et al., 2019). In the contiguous United States, there has been a shift in forest community composition since the year

2000 towards tree communities with increasingly negative community-weighted P_{50} and increasingly large HSM, caused by selective climate-driven tree mortality (Trugman et al., 2020). These changes, which differed in magnitude between forest types, would be expected to make these forests more resilient in the face of increased exposure to extreme droughts. Similar changes in species composition towards increasing dominance by drought-tolerant species have been observed in Ghana (Fauset et al., 2012) and in the Amazon (Esquivel-Muelbert et al., 2019), although in the latter case the change had so far only impacted recruitment. The higher mortality observed in seedlings of species associated with wetter forests during the 2015–2016 droughts in Panama would also result in an increased relative abundance of drought-resistant species if these impacts persisted over time (Browne et al., 2021). Although these changes in forest species composition are expected to increase the resilience of the forest to droughts relative to the baseline composition, the long generation times of forest trees mean that they will inevitably lag behind climate change.

Drought-induced increases in tree mortality are reducing the carbon sink capacity of forests in many parts of the world. In Canada's boreal forests, the average annual tree mortality rate since 1970 has been estimated at 2.7%, with higher mortality in the western region (3.5%) and an acceleration since 2002 (Liu et al., 2023). These losses translate into an annual rate of carbon loss of around 1.5 MgC ha^{-1}. A LiDar-based estimate for the Amazon forests was somewhat higher (2.35 MgC ha^{-1} $year^{-1}$) (Yang et al., 2018). Future projections by Earth System Models depend on how the models represent potential interactions with heat stress and CO_2 fertilization—neither of which are well understood—but an approach using the empirical relationship between climate and vegetation structure predicted a net reduction in aboveground biomass in the tropics over the remainder of this century (Uribe et al., 2023). These projected biomass losses are driven largely by increasing dry season severity and occur mostly in the Americas.

Conclusion

The next few decades are almost certain to see an increase in the frequency and intensity of ecological droughts and, in consequence, a significant increase in tree mortality. How bad tree mortality will get, however, is still uncertain. The uncertainty comes from both the climate projections, for which there is still a large spread between models, and our incomplete understanding of the causal links between droughts and tree death. Research into these links has been plagued by technical problems and measurement artefacts. Although we now have a much better understanding of how the xylem and stomata respond to drought, these are only two sections of the entire soil–plant–atmosphere continuum, while water movements through the soil, across the soil–plant interface, and inside the plant but outside the xylem are still poorly understood. A better understanding of the whole continuum would facilitate the construction of more mechanistic models of tree and forest responses to droughts and thus help make useful predictions for the future. We also need to know how rising CO_2 concentrations will impact plant responses to droughts.

Finally, trees and forests don't simply consume water. They also have a major influence on local, regional, and global water cycles through their ability to lift water up from belowground into the canopy, where it is released into the atmosphere and contributes to local and downwind rainfall. Forests are more efficient water pumps than grasslands and crops because of their deeper roots, larger leaf areas, and lower albedos, so more solar radiation is absorbed. In the Amazon, for example, tree transpiration is a major source of downwind rainfall, with an estimate of almost half of Amazonian transpiration directly raining out over the Amazon (Staal et al., 2018). The contribution of transpiration to rainfall is greatest in the dry season and may increase during droughts. The importance of forests in the water cycle means that deforestation can exacerbate downwind droughts, while reforestation can potentially help alleviate them, particularly in the tropics (Tuinenberg et al., 2023).

References

Abdalla, M., Bitterlich, M., Jansa, J., Püschel, D. and Ahmed, M. (2023) 'The role of arbuscular mycorrhizal symbiosis in improving plant water status under drought', *Journal of Experimental Botany*, vol 74, pp. 4808–4824.

Anderegg, W.R.L., Klein, T, Bartlett, M., Sack, L., Pellegrini, A.F.L., Choat, B. and Jansen, S. (2016) 'Meta-analysis reveals that hydraulic traits explain cross-species patterns of drought-induced tree mortality across the globe', *Proceedings of the National Academy of Sciences of the USA*, vol 113, pp. 5024–5029.

Anderegg, W.R.L., Konings, A.G., Trugman, A.T., Yu, K., Bowling, D.R., Gabbitas, R., Karp, D.S., Pacala, S., Sperry, J.S., Sulman, B.N. and Zenes, N. (2018) 'Hydraulic diversity of forests regulates ecosystem resilience during drought,' *Nature*, vol 561, pp. 531–541.

Anderegg, W.R.L., Trugman, A.T., Badgley, G., Konings, A.G. and Shaw, J. (2020) 'Divergent forest sensitivity to repeated extreme droughts', *Nature Climate Change*, vol 10, pp. 1091–1095.

Au, T.F., Maxwell, J.T., Robeson, F.M., Li, J., Siani, S.M.O., Novick, K.A., Dannenberg, M.P., Phillips, R.P., Li, T., Chen, Z. and Lenoir, J. (2023) 'Younger trees in the upper canopy are more sensitive but also more resilient to drought', *Nature Climate Change*, vol 12, pp. 1168–1174.

Baumann, D. et al. (2022) 'Tropical tree mortality has increased with rising atmospheric water stress', *Nature*, vol 608, pp. 508–533.

Beobide-Arsuaga, G., Bayr, T., Reintges, A. and Latif, M. (2021) 'Uncertainty of ENSO-amplitude projections in CMIP5 and CMIP6 models', *Climate Dynamics*, vol 56, pp. 3875–3888.

Bittencourt, P.R.L., Oliveira, R.S., da Costa, A.C.L., Giles, A.L., Coughlin, I., Costa, P.B., Bartholomew, D.C., Ferreira, L.V., Vasconcelos, S.S., Barros, F.V., Junior, J.A.S., Oliveira, A.A.R., Mencuccini, M., Meir, P. and Rowland, L. (2020) 'Amazonia trees have limited capacity to acclimate plant hydraulic properties in response to long-term drought', *Global Change Biology*, vol 26, pp. 3569–3584.

Brodribb, T.J., Powers, J., Cochard, H. and Choat, B. (2021) 'Hanging by a thread? Forests and drought', *Science*, vol 368, pp. 261–266.

Browne, L., Markesteijn, L., Engelbrecht, B.M.J., Jones, F.A., Lewis, O.T., Manzané-Pinzón, E. and Wright, S.J. (2021) 'Increased mortality of tropical tree seedlings during the extreme 2015–16 El Niño', *Global Change Biology*, vol 27, pp. 5043–5053.

Callahan, R.P., Riebe, C.S., Sklar, L.S., Pasquet, S., Ferrier, K.L., Hahm, W.J., Taylor, N.J., Grana, D., Flinchum, B.A., Hayes, J.L. and Holbrook, W.S. (2023) 'Forest vulnerability to drought controlled by bedrock composition', *Nature Geoscience*, vol 15, pp. 714–719.

Castagneri, D., Vacchiano, G., Hacket-Pain, G., DeRose, R.J., Klein, T. and Bottero, A. (2021) 'Meta-analysis reveals different competition effects on tree growth resistance and resilience to drought', *Ecosystems*, vol 25, pp. 30–43.

Cochard, H. (2021) 'A new mechanism for tree mortality due to drought and heatwaves', *Peer Community Journal*, vol 1, e36.

Cook, B.I., Mankin, J.S. and Anchukaitis, K.J. (2018) 'Climate change and drought: From past to future', *Current Climate Change Reports*, vol 4, pp. 164–179.

Costa, F.R.C., Schietti, J., Stark, S.C. and Smith, M.N. (2023) 'The other side of tropical forest drought: Do shallow water table regions of Amazonia act as large-scale hydrological refugia from drought?', *New Phytologist*, vol 237, pp. 714–733.

De Luca, P.D. and Donat, M.G. (2023) 'Projected changes in hot, dry, and compound hot-dry extremes over global land regions', *Geophysical Research Letters*, vol 50, e2022GL102493.

DeSoto, L. et al. (2020) 'Low growth resilience to drought is related to future mortality risk in trees', *Nature Communications*, vol 11, 545.

Einzmann, H.J.R, Weichgrebe, L. and Zotz, G. (2022) 'The impact of a severe El Niño event on vascular epiphytes in lowland Panama', *Diversity*, vol 14, 325.
Esquivel-Muelbert, A. et al. (2019) 'Compositional response of Amazon forests to climate change', *Global Change Biology*, vol 25, pp. 39–56.
Fauset, S., Baker, T.R., Lewis, S.L., Feldpausch, T.R., Affum-Bafoe, K., Foli, E.G., Hamer, K.C. and Swaine, M.D. (2012) 'Drought-induced shifts in the floristic and functional composition of tropical forests in Ghana', *Ecology Letters*, vol 15, pp. 1120–1129.
Fettig, C.J., Mortenson, L.A., Bulaon, B.A. and Foulk, P.B. (2019) 'Tree mortality following drought in the central and southern Sierra Nevada, California, U.S.', *Forest Ecology and Management*, vol 432, pp. 164–178.
Franklin, O., Fransson, P., Hofhansl, F., Jansen, S. and Joshi, J. (2023) 'Optimal balancing of xylem efficiency and safety explains plant vulnerability to drought'.*Ecology Letters*, vol 26, pp. 1485–1496.
Gauthey, A., Peters, J.M.R., Lòpez, R., Carins-Murphy. M.R., Rodriguez-Dominguez, C.M., Tissue, D.T., Medlyn, B.E., Brodribb, T.J. and Choat, B. (2022) 'Mechanisms of xylem hydraulic recovery after drought in *Eucalyptus saligna*', *Plant, Cell & Environment*, vol 45, pp. 1216–1228.
Grossiord, C. (2020) 'Having the right neighbors: how tree species diversity modulates drought impacts on forest', *New Phytologist*, vol 228, pp. 42-49.
Hammond, W.M., Williams, A.P., Abatzoglou, A.P., Adams, H.D., Klein, T., López, R., Sáenz-Romero, C., Hartmann, H., Breshears, D.D. and Allen, C.D. (2022) 'Global field observations of tree die-off reveal hotter-drought fingerprint for Earth's forests', *Nature Communications*, vol 13, 1761.
Hollunder, R.K., Garbin, M.L., Scarano, F.R. and Mariotte, P. (2022) 'Regional and local determinants of drought resilience in tropical forests', *Ecology and Evolution*, vol 12, e8943.
Kannenberg, S.A., Schwalm, C.R. and Anderegg, W.R.L. (2020) 'Ghosts of the past: How drought legacy effects shape forest functioning and carbon cycling', *Ecology Letters*, vol 23, pp. 891–901.
Körner, C. (2019) 'No need for pipes when the well is dry—a comment on hydraulic failure in trees', *Tree Physiology*, vol 39, pp. 695–700.
Levionnois S., Ziegler, C., Jansen, S., Calvet, E., Coste, S., Stahl, C., Salmon, C., Delzon, S., Guichard, C. and Heuret, P. (2020) 'Vulnerability and hydraulic segmentations at the stem–leaf transition: Coordination across Neotropical trees', *New Phytologist*, vol 228, pp. 512–524.
Liu, L., Peng, C., Schneider, R., Cyr, D., McDowell, N.G. and Kneeshaw, D. (2023) 'Drought-induced increase in tree mortality and corresponding decrease in the carbon sink capacity of Canada's boreal forests from 1970 to 2020', *Global Change Biology*, vol 29, pp. 2274–2285.
McDowell, N.G. et al. (2022) 'Mechanisms of woody-plant mortality under rising drought, CO2 and vapour pressure deficit', *Nature Reviews Earth & Environment*, vol 3, pp. 294–308.
Oliveira, R.S., Eller, C.B., Barros, F., Hirota, M., Brum, M. and Bittencourt, P. (2021) 'Linking plant hydraulics and the fast–slow continuum to understand resilience to drought in tropical ecosystems', *New Phytologist*, vol 230, pp. 904–923.
Ometto, J.P., Kalaba, K., Anshari, G.Z., Chacón, N., Farrell, A., Halim, S.A., Neufeldt, H. and Sukumar, R. (2022) 'Cross-Chapter Paper 7: Tropical Forests', in H.-O. Pörtner et al. (eds.), *Climate Change 2022: Impacts, Adaptation and Vulnerability. Contribution of Working Group II to the Sixth Assessment Report of the Intergovernmental Panel on Climate Change*. Cambridge University Press, CambridgeUK, pp. 2369–2410.
Rowland, L., Ramírez-Valiente, J.-A., Hartley, I.P. and Mencuccini, M. (2023) 'How woody plants adjust above- and below-ground traits in response to sustained drought', *New Phytologist*, vol 239, pp. 1173–1189.
Seneviratne, S.I. et al. (2021) 'Weather and climate extreme events in a changing climate', in V. Masson-Delmotte et al. (eds.), *Climate Change 2021: The Physical Science Basis. Contribution of Working Group I to the Sixth Assessment Report of the Intergovernmental Panel on Climate Change*. Cambridge University Press, UK, pp. 1513–1766.
Slette, I.J., Smith, M.D., Knapp, A.K., Vicente-Serrano, S.M., Camarero, J.J. and Beguería, S. (2020) 'Standardized metrics are key for assessing drought severity', *Global Change Biology*, vol 26, e1–e3.
Smith-Martin, C.M., Muscarella, R., Ankori-Karlinsky, R., Delzon, S., Farrar, S.L., Salva-Sauri, M., Thompson, J., Zimmerman, J.K. and Uriarte, M. (2023) 'Hydraulic traits are not robust predictors of tree species stem growth during a severe drought in a wet tropical forest', *Functional Ecology*, vol 37, pp. 447–460.

Staal, A., Tuinenberg, O.A., Bosmans, J.H.C., Holmgren, M., van Nes, E.H., Scheffer, M., Zemp, D.C. and Dekker, S.C. (2018) 'Forest-rainfall cascades buffer against drought across the Amazon', *Nature Climate Change*, vol 8, 539–543.

Still, C.J. et al (2022) 'No evidence of canopy-scale leaf thermoregulation to cool leaves below air temperature across a range of forest ecosystems', *Proceedings of the National Academy of Sciences of the USA*, vol 119, e2205682119.

Tavares, J.V. et al. (2023) 'Basin-wide variation in tree hydraulic safety margins predicts the carbon balance of Amazon forests', *Nature*, vol 617, pp. 111–117.

Trugman, A.T., Anderegg, L.D.L., Shaw, J.D. and Anderegg, W.R.L. (2020) 'Trait velocities reveal that mortality has driven widespread coordinated shifts in forest hydraulic trait composition', *Proceedings of the National Academy of Sciences of the USA*, vol 117, pp. 8532–8538.

Tuinenberg, O.A., Bosmans, J.H.C. and Staal, A. (2023) 'The global potential of forest restoration for drought mitigation', *Environmental Research Letters*, vol 17, 034045.

Uribe, M.R., Coe, M.T., Castanho, A.D.A., Macedo, M.N., Valle, D. and Brando, P.M. (2023) 'Net loss of biomass predicted for tropical biomes in a changing climate', *Nature Climate Change*, vol 13, pp. 274–281.

Vincente-Serrano, S.M., Miralles, D.G., McDowell, N., Brodribb, T., Domínguez-Castro, F., Leung, R. and Koppa, A. (2022) 'The uncertain role of rising atmospheric CO2 on global plant transpiration', *Earth-Science Reviews*, vol. 230, 104055.

Willson, A.M., Trugman, A.T., Powers, J.S., Smith-Martin, C.M. and Medvigy, D. (2022) 'Climate and hydraulic traits interact to set thresholds for liana viability', *Nature Communications*, vol 13, 3332.

Wolfe. B.T. (2020) 'Bark water vapour conductance is associated with drought performance in tropical trees', *Biology Letters*, vol 16, 20200263.

Yang, Y., Saathchi, S., Xu, L., Choi, S., Phillips, N., Kennedy, R., Keller, M., Knyazikhin, Y. and Myneni, R.B. (2018) 'Post-drought decline of the Amazon carbon sink', *Nature Communications*, vol 9, 3172.

Young, D.J.N., Estes, B.L., Gross, S., Wuenschel, A., Restaino, C. and Meyer, M.D. (2023) 'Effectiveness of forest density reduction treatments for increasing drought resistance of ponderosa pine growth', *Ecological Applications*, vol 33, e2854.

36

FROM WOOD FORMATION TO TREE RING

How tree growth responds to climate change

Roberto Silvestro, Minhui He, Jian-Guo Huang, Hubert Morin and Sergio Rossi

Introduction

Wood is one of the largest carbon pools on Earth (Pan *et al.*, 2011) and a natural and renewable resource of terrestrial ecosystems (Howard *et al.*, 2021). In the last few decades, a new bioeconomy has emerged that encourages and promotes the use of renewable raw materials instead of petrol-based products. We can observe a growing interest in the possibility that wood can substitute fossil fuels and non-biomass materials, especially in the construction industry (Howard *et al.*, 2021). Moreover, the chemical composition of wood allows the creation of alternative biopolymers and textile products (Howard *et al.*, 2021). Wood and all its derived products have a crucial importance in the global bioeconomy and world trade.

Wood results from the gradual accumulation of xylem cells in trees to renew the water transport system, store reserves and ensure mechanical support to stem, branches and leaves (Larson, 1994). The meristems are at the basis of wood formation and, in the broad sense, plant growth (Larson, 1994). Cambium is a lateral meristem designed to produce xylem and phloem cells. Outside the tropics, through a cyclical alternation of periods of activity and rest, the cambium produces the conducting elements during the most climatic favourable periods (Rossi *et al.*, 2016), leaving indelible traces in the form of tree rings. A tree ring is the product of an entire season of cambial activity that, in the boreal and temperate-cold zones of the planet, corresponds to the period between spring and autumn (Rossi *et al.*, 2016). In these environments, temperature is one of the key factors that influence the resumption of meristem activity and water balance in plants during growth (Rossi *et al.*, 2008). For this reason, it is a priority to understand the effects of a changing climate on forest species to predict the potential future evolution and ecological and economic consequences of tree growth modification over time.

Xylogenesis: wood growth at intra-annual scale

Xylogenesis (i.e., wood formation) is the result of the activity of the vascular cambium, a secondary meristem derived from the procambium, which in turn derives from the apical meristem (Larson, 1994). The vascular cambium is usually defined as bidirectional, given

DOI: 10.4324/9781003324072-42

that it is involved in the production of secondary phloem (i.e., inner bark) externally and secondary xylem (i.e., wood) internally. The appearance of vascular cambium in the Middle Devonian (about 360 million years ago) had a huge impact on land colonisation and allowed the development of plants with large bodies by offering mechanical support and a means of efficient water transport.

Wood growth occurs in a well-defined time window. During this period, trees are directly affected by environmental signals at different time scales: long-term general patterns (climate) or punctual events (weather). Investigations at intra-annual scale, on time resolutions shorter than one year, allow the detection of the effects of short-term (or punctual) environmental signals on wood formation and to assess the consequences of environmental changes on radial growth dynamics.

Intra-annual observations of wood formation provide information on the different developmental stages of xylem cells and the timings of their occurrence, thus identifying the phases of xylem phenology. The study of xylem phenology allows several key questions to be answered: (1) When and how does wood production occur? (2) What environmental signals trigger the timings and dynamics of xylogenesis? (3) Does xylogenesis vary between individuals, and how do such differences affect the response of populations to climate change?

Xylem phenology: from sample collection to phenological patterns

Observations of wood formation began in the last century (Knudson, 1913), with histological measurements based on light microscopy. Since then, xylem development has been the subject of investigation and research by several plant physiologists and ecologists. In pioneer studies, cambium was pierced with a thin needle, and the wood samples were analysed at the end of the year to reconstruct the growing season. This technique, nowadays known as pinning, was later largely applied by wood scientists. During the 1970s, several studies regarding wood formation were published, emphasising data analysis and the relationship with climatic factors.

Periodic samplings are based on repeated observations of wood samples of the developing tree ring during the growing season. To be the least invasive possible, wood samples are repeatedly extracted around the stem in the form of blocks or small cores (i.e., microcores). These samples include the dead outer bark, phloem, cambium, developing xylem and recently formed tree rings (Rossi *et al.*, 2006a). Microcoring is performed using instruments such as Trephor, which include the insertion of small cutting tubes into the stem for extracting the sample (Rossi *et al.*, 2006a). The microcores are stored in an ethanol solution at 5 °C. They consist of three parts (wood, cambium and phloem bark) with different consistency and density, requiring specific procedures for sample preparation. Usually, the preparations of microcores involve dehydration in ethanol and D-limonene, and eventually embedding in paraffin or glycol methacrylate. Transverse sections of 10–30 mm thickness are cut from the samples with a rotary or sledge microtome, stained with cresyl violet acetate or safranin and astrablue, and examined under bright-field and polarised light at 400–500x magnifications to differentiate the developing and mature xylem cells.

The cambium annually or periodically renews both the phloem and the xylem. The cambial zone includes the cambium, made up of a single layer of meristematic cells called "initials" and both phloem and xylem mother cells. When an initial divides, it produces a mother cell and another initial (Larson, 1994), thus assuring a perpetual regeneration of the initials. Given that it is impossible to identify unquestionably which cells represent the true initials based on cell morphology, it is common to refer to all of them as cambial cells.

During development, xylem mother cells undergo differentiation by altering both morphologically and physiologically, gradually assuming definite features and differentiating into the specific elements of the stem tissues (Figure 36.1). Cells firstly enlarge by stretching the primary walls, then they produce, thicken and lignify secondary walls and ultimately succumb to programmed cell death (Rossi *et al.*, 2012). During cell enlargement, the cell protoplast is still enclosed in the thin, elastic primary wall. Following positive turgor increased by water movement into the vacuoles, the cell wall stretches, increasing the radial diameter of the tracheid and, consequently, the lumen area. Once their final size has been reached, the cells begin maturing through the formation of a secondary cell wall, its thickening and lignification (Rossi *et al.*, 2012) (Figure 36.1).

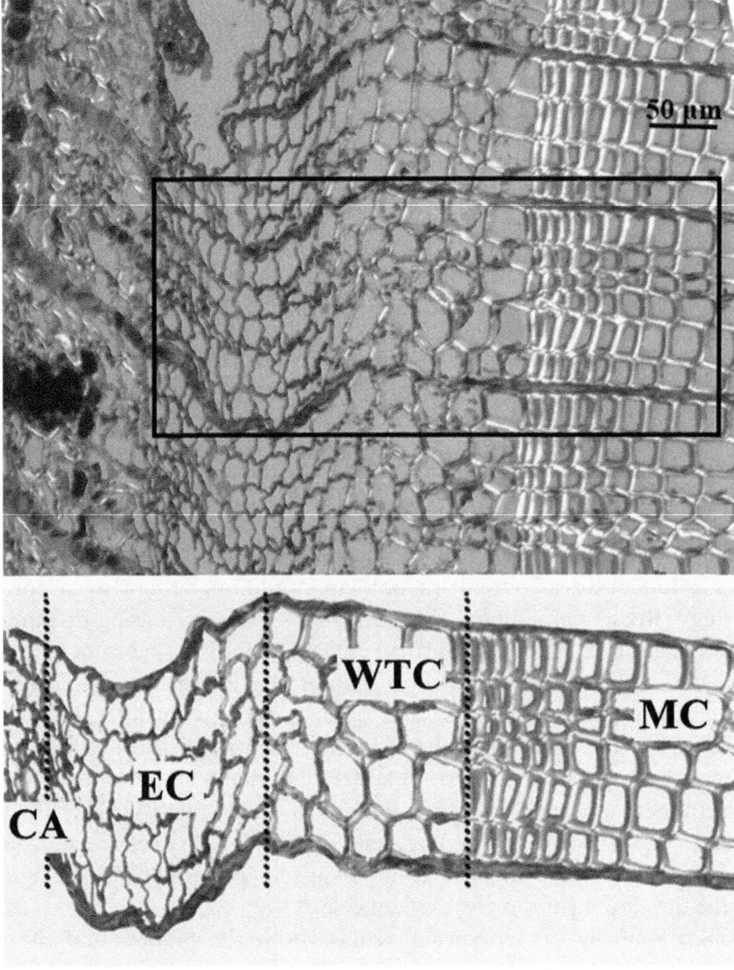

Figure 36.1 Transverse section of a weekly sampled microcore, observed at 400× magnification, for counting the developing tracheids, and classified as (CA) cambium, (EC) enlarging cells, (WTC) wall-thickening and lignifying cells and (MC) mature cells of the previous year (Modified from Silvestro *et al.*, 2022).

Xylem formation can be assessed either by counting cells or by measuring the radial distance covered by the cells (Rossi *et al.*, 2006b; Marion *et al.*, 2007). Counts of cell numbers have been intensively applied in conifers, which have a simple and homogeneous wood structure, mainly composed of tracheids, representing 90% of the total number of xylem cells. On the contrary, because of the presence of vessels and wide variations in cell size, cell count has less frequently been applied in broadleaves, preferring the measurements of the radial distance between developmental zones (Marion *et al.*, 2007).

In cross-section, cambial cells are characterised by thin cell walls and small radial diameters (Rossi *et al.*, 2006b) (Figure 36.1). Observing xylem cells in differentiation, polarised light discriminates between cell enlargement and wall thickening phenological stages (Rossi *et al.*, 2006b). Indeed, due to the cellulose microfibrils arrangement, the developing secondary walls shine under polarised light. On the contrary, there is no glistening in enlargement zones. The progress of lignification is detected with cresyl violet acetate that reacts with lignin and shows a colour change from violet (unlignified secondary cell walls) to blue (lignified cell walls) (Rossi *et al.*, 2006b) (Figure 36.1). The lignification stage is completed when cell walls are entirely blue (Figure 36.1); tracheids attain the mature stage from this moment.

Xylem formation is composed of successive stages (or phases) in which new cells are formed, undergo differentiation and reach the mature stage. Monitoring the timing and observing the spatial distribution of developing xylem cells allow us to identify and describe the phases of xylem phenology and, consequently, the dynamics of xylem growth. In temperate and boreal ecosystems, all species show the same cell differentiation patterns (Rossi *et al.*, 2016). These patterns are connected with the number of cells passing through each differentiation stage and with the gradual accumulation of mature cells in the tree ring (Figure 36.2). Analysis of the dynamics of xylogenesis requires the dates of each stage to be summarised concisely but informatively. The selection and use of descriptive statistics are particularly important with the phenological stages of wood formation because a substantial variability exists within (Lupi *et al.*, 2013) and between (Silvestro *et al.*, 2022, 2023) trees and has to be carefully quantified and analysed.

The growth patterns in Mediterranean climates or in ecosystems with significant summer droughts partially differ from those observed in temperate and boreal regions. In Mediterranean ecosystems, plants show two periods of growth activity within the same growing season: one in spring and a second one in autumn (Camarero *et al.*, 2010). This bimodal growth pattern is associated with the changes in growing conditions. In early summer, water stress can lead to a temporary cessation of wood formation, possibly associated with the formation of tracheids with small lumen areas. Growth resumption in autumn is a response to precipitation. Xylem cell differentiation (enlargement and cell-wall thickening) and, in some cases, cambial activity (Vieira *et al.*, 2014a) can resume. In some Mediterranean species, this bimodal pattern of xylem growth may correspond to a double flushing of buds and branches. However, accurate investigations of xylem anatomy suggest that the autumnal increment period mainly involves rehydration of the stem tissues rather than effective growth (i.e., newly formed cells) (Vieira *et al.*, 2014a).

Environmental factors influencing wood formation

In the last decade, given the economic importance of wood as a raw material and the ecological significance of wood formation in the process of C sequestration in the wood biomass,

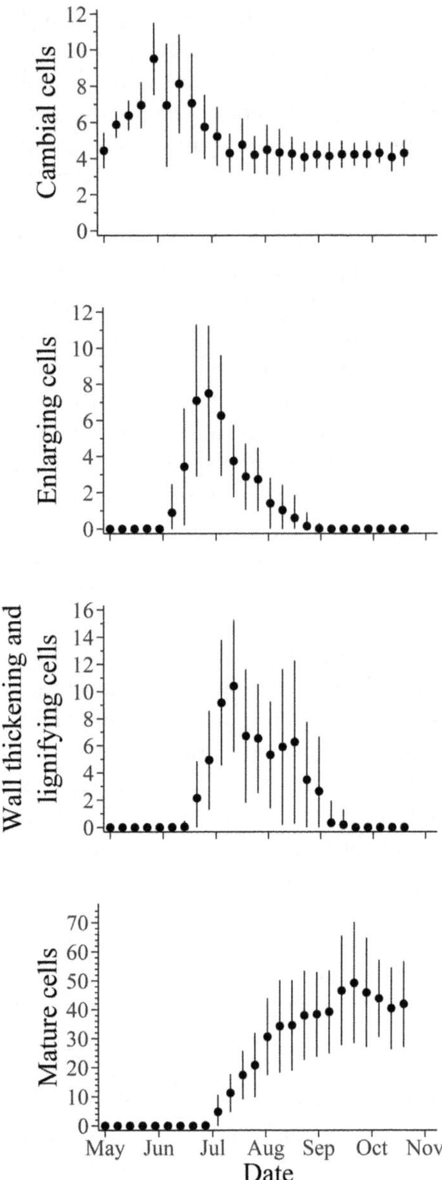

Figure 36.2 Number of cambial, enlarging, wall thickening and lignifying, and mature cells observed in 159 balsam firs during 2018. Dots and bars represent the average and standard deviation, respectively. (Figure from Silvestro et al., 2022.)

it is not surprising that xylogenesis and climate–growth relationships have gained considerable interest worldwide. Several research groups have studied xylem phenology during the growing season or focussed on precise time windows such as the onset or end of growth (Rossi *et al.*, 2003; Vieira *et al.*, 2014a; Silvestro *et al.*, 2022). Investigations have attempted to link xylogenesis with endogenous or environmental factors (Rossi *et al.*, 2012, 2016; Vieira *et al.*, 2014a,b; Huang *et al.*, 2020; Silvestro *et al.*, 2022). Environmental factors affect the wood formation process either directly or indirectly by regulating carbohydrate availability or hormone concentration and sensitivity (Buttò *et al.*, 2020).

Temperature is the driver of biological functions at several levels. Even slight temperature changes can dramatically affect biological processes from cells to populations, with significant ecological consequences. At a small scale, temperature regulates physiological rates and metabolic efficiency, which, in turn, drive cellular reaction rates through kinetic processes. Individuals respond directly to environmental temperature, for example, by modulating their phenology and growth rates (Rossi, 2015).

Monitoring wood formation in natural conditions has historically been used to define the growing season of species and its variability among years and sites according to specific climatic gradients (Rossi *et al.*, 2016; Huang *et al.*, 2020). Nowadays, there is evidence of threshold temperatures controlling the boundaries of the growing season in conifers of cold environments, characterised by an alternation of thermally favourable and unfavourable climate conditions during the year (Deslauriers *et al.*, 2008). These values were demonstrated to be similar among all studied conifer species across a wide geographical range of monitored sites and temperature regimes (Rossi *et al.*, 2008, 2016)

How does the growing season change along a thermal gradient, and how does this change affect cell production? Xylem phenology and cell production were analysed along a latitudinal gradient, from the 48th to 53rd parallels, the range covering the whole closed black-spruce forest in Quebec, Canada (Rossi *et al.*, 2014). As expected, the southern and warmer site showed the highest radial growth, which corresponded to both the highest rates and the longest durations of cell production. More surprisingly, the differences between the other sites in terms of xylem phenology and growth were marginal. Within the range analysed, the relationship between temperature and most phenological phases of xylogenesis was linear. On the contrary, the temperature was related to cell production according to a threshold effect (Figure 36.3). Small increases in the periods of xylogenesis corresponded to substantial increases in cell production, approximately occurring at a May–September average temperature of 14 °C (Figure 36.3). As a consequence, it was deduced that small environmental changes allowing marginal lengthening of the period of cell division could potentially lead to large increases in xylem cell production, with substantial consequences for productivity.

Wood is the result of xylem cell production during a given period (duration) and with a given intensity (rate). Both factors, duration and the rate of growth, play a role in xylem cell production of conifers. However, the importance of these factors has been observed to change according to the species and under different environmental conditions. Several researchers observed that the longer the period of wood formation, the larger the amount of xylem produced (Rathgeber *et al.*, 2011; Vieira *et al.*, 2014b; Silvestro *et al.*, 2022) (Figure 36.4). The timing of onset and the rate of cambial division affect the number of cells in the cambial zone, which, in turn, influences the timing of cell differentiation (Lupi *et al.*, 2010; Rossi *et al.*, 2012). In the case of the occurrence of water deficit during the growing season, individual growth rate assumes a leading control on the annual cell production (Ren *et al.*, 2019). However, if water availability does not represent a constraint, it seems that the larger the number of xylem cells in differentiation, the longer the time needed to complete their maturation.

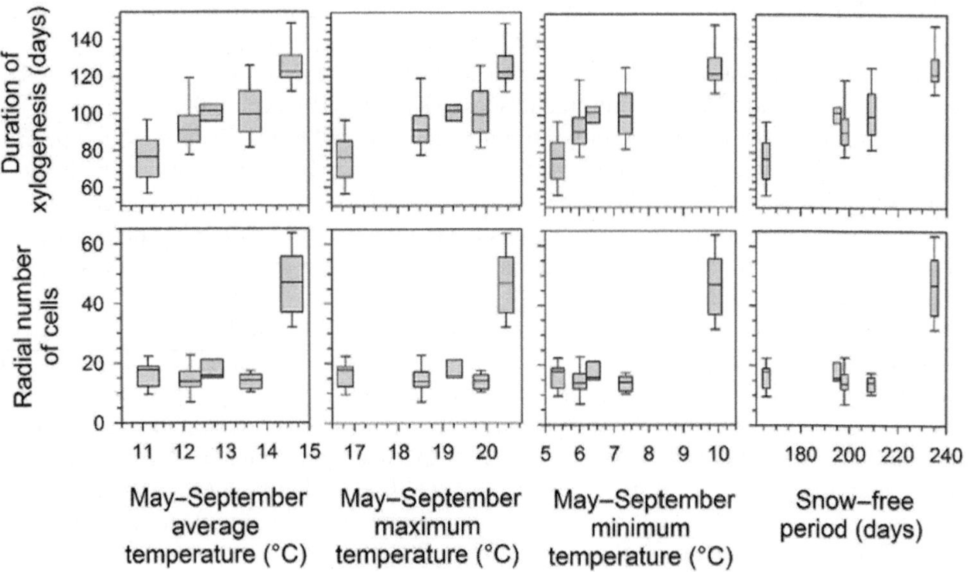

Figure 36.3 Duration of xylogenesis and amount of cell production versus the May–September temperatures and the snow-free period recorded in five study sites in the boreal forest of Quebec, Canada. Boxes represent upper and lower quartiles, whiskers achieve the 10th and 90th percentiles, and the median is drawn as a horizontal solid line. (Modified from Rossi *et al.*, 2014.)

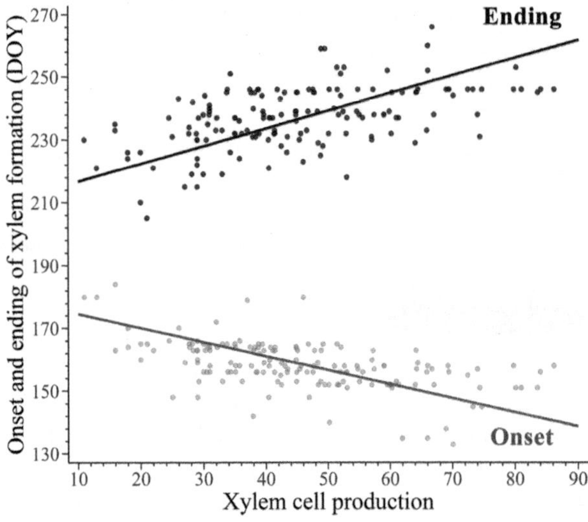

Figure 36.4 Standardized major axis (SMA) regressions among timings of onset (light grey dots) and ending (black dots) of xylem phenology, and the total number of cells in the tree ring at the end of the growing season in 159 balsam firs. (Modified from Silvestro *et al.*, 2022.)

It should not be a surprise that temperature represents a critical ecological factor controlling the timing of the growing season in temperate and boreal climates. However, it emerged that other climatic factors were involved in defining the timings of wood formation. Therefore, we could wonder whether temperature alone can sufficiently explain and describe the xylogenesis patterns. If so, we should observe that xylem cell production follows the seasonal thermal trend, gradually increasing in spring, attaining a maximum around July and finally reducing during late summer and autumn. Rossi *et al.* (2006c) compared the rates of xylem cell production in several North American and European species and found that maximum rate occurs during the summer solstice, when day length, not temperature, culminates. After 21st June, cell production gradually decreases until ceasing. Finally, the influence of temperature on cambium depends on photoperiod, a factor independent of temperature and constant in time (Huang *et al.*, 2020). Trees would have thus evolved by synchronising their growth rates with day length. The advantages of connecting cambial activity with photoperiod are explained by the need to avoid the period thermally unfavourable to growth. So, in ecosystems with a marked seasonality, temperature increases can induce cell division only in a specific period of the year, in spring, when the conditions are and will probably remain favourable for completing the maturation of the newly produced cells.

Water plays several essential roles in plants. At the cell level, water is a solvent for ions and organic molecules, a structuring agent in building proteins and nucleic acids, and a substrate for all enzymatic reactions. More importantly, water and its hydrostatic pressure are essential for xylem cell production and expansion. Indeed, cell division can occur only after a cambial cell has increased in diameter. Similarly, the enlargement phase is a turgor-driven process depending on cellular water uptake and solute accumulation into cellular vacuoles (Deslauriers *et al.*, 2016). In the context of climate change, higher temperatures are expected to produce dramatic consequences on the frequency of summer drought. A water shortage during the growing season could substantially affect wood formation in terms of wood quality (i.e., xylem traits) and quantity (i.e., number of xylem cells produced) and, consequently, wood production (Balducci *et al.*, 2013; Deslauriers *et al.*, 2016). Thus, the role of water in xylem formation is a key step for better understating the drivers of wood growth.

It has been proven that drought can greatly influence the timing of xylogenesis. Although spring temperature remains the main factor for cambial reactivation even in Mediterranean climates, Vieira *et al.* (2014a) found that water deficit played a crucial role in summer by triggering an earlier conclusion of wood formation in maritime pine (*Pinus pinaster*). Moreover, dry conditions during spring and summer strongly limit cambial activity in samplings (Deslauriers *et al.*, 2016). Balducci *et al.* (2013) monitored xylem formation in a controlled environment and showed that a heavy water deficit during June substantially reduced cambial activity, which needed 2–4 weeks to be restored after re-watering, finally producing a narrow tree ring. On the Tibetan Plateau, in wood formation assessment on Qilian junipers (*Juniperus przewalskii*) it has been highlighted that drought can cause both a delayed onset of xylogenesis in spring and an early cessation of xylem differentiation in summer (Ren *et al.*, 2015).

In the next section, we will discuss how the influences of these multiple environmental factors on wood formation at intra-annual scale leave a fingerprint in cell structure and size that, in turn, provides occasions for retrospective studies on past climatic conditions based on tree-ring chronologies.

Dendrochronology: wood growth at interannual scale

Dendrochronology is the science that studies and dates the annual growth layers (i.e., tree-rings) in woody plants, including perennial herbs, shrubs, lianas and trees (Fritts, 1976). Consequently, dendrochronology represents a powerful tool to yield detailed insights into diverse ecological, climatic and earth sciences fields. In various research fields, dendroclimatology has helped to critically advance our understanding of climatology over the past two thousand years and from regional to global spatial scales.

For more than a century, tree-ring climate research has covered all continents, except Antarctica, with most chronologies being developed at the mid- to high-latitudes of the Northern Hemisphere (Figure 36.5). This geographic focus was due to the need to investigate tree-ring samples sensitive to weather. To reconstruct temperature series, we need samples from cold and high-elevation regions; on the contrary, referring to the hydroclimate signal, samples should come from arid and semi-arid areas. Consequently, despite being the most biodiverse, productive and understudied environments, tropical and subtropical regions have usually been neglected. The often-inconsistent growth patterns in trees, sampling difficulties, complex wood anatomy and the lack of physiological knowledge of local wood species are the other reasons for the limited tree-ring sites in the tropics.

According to the International Tree-Ring Data Bank, the most comprehensive repository of tree growth, conifers represent 81% of studied species (Zhao *et al.*, 2019). Sample elevations range from 0 to 4,450 m a.s.l., with a median of 900 m a.s.l. Lengths of the series vary from <50 to >5,000 years (Zhao *et al.*, 2019).

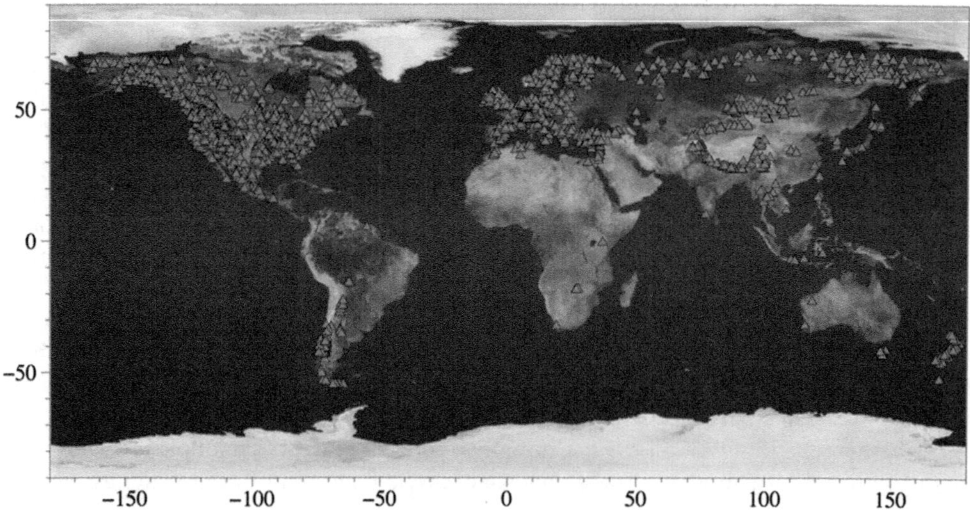

Figure 36.5 Location of the tree-ring sampling sites according to the International Tree-Ring Data Bank (ITRDB), the most comprehensive repository of tree growth.

Tree-ring parameters

Tree-ring width (TRW) measurements from living trees or relict wood (e.g., archaeological, subfossil) are frequently used for the reconstruction of past climate change (Gennaretti *et al.*, 2014; Büntgen *et al.*, 2021). This is the easiest and most direct way to analyse relationships between tree growth and climate data, resulting in several published articles over the past few years. Generally, TRW variation in samples from high-elevation or high-latitude cold sites can reveal changes in growing season temperature (Esper *et al.*, 2016). In lower elevation, temperate and semiarid sites, where plant growth predominantly depends on soil moisture availability, TRW chronologies are more often used to represent hydroclimatic changes (Ljungqvist *et al.*, 2020). In addition to the traditional reconstruction of temperature or hydroclimate index, tree-ring networks are also used to develop reconstructions of other time series events, e.g., seasonal fire dangers or wind speed.

Another approach frequently used in the study of tree rings consists of using stable isotopes. This methodology has several advantages over the traditional TRW approach. Stable oxygen isotopes ($\delta^{18}O$), when used in tree ring studies, have clear fractional distillation mechanisms, can provide robust data with small sample sizes, show insignificant age effects and provide better retention of low-frequency signals than other stable isotopes (Shestakova & Martínez-Sancho, 2021).

Carbon isotopes ($\delta^{13}C$) can serve as proxies for paleoclimatic reconstructions, but they can also be used as tools to deepen our understanding of tree eco-physiology. Based on carbon and oxygen isotopes from tree ring chronologies, results have shown that trends in tree intrinsic water-use efficiency are similar in magnitude to the increase in atmospheric CO_2 over the 20th century (Mathias & Thomas, 2021). In contrast, the role of δ^2H values has yet to be improved, together with the development of new methods for hydrogen isotope analysis (Lu *et al.*, 2023). Overall, the isotope fingerprint is regarded as the most relevant marker of environmental knowledge during tree growth because there is no material transfer or additional fractionation. A multi-proxy strategy based on different isotopes and various components appears to be the most effective method for enhancing climate reconstruction research (Gennaretti *et al.*, 2017). Nevertheless, analysing stable isotope ratios in tree rings remains a time-consuming, labour- and cost-intensive process.

With a large amount of dendroclimatological studies detecting relationships between inter-annual tree-ring series and climate, research is moving on to the intra-annual scale investigation. Indeed, worldwide scientists are interested to know what are the main climatic factors driving the formation of earlywood and latewood, which of these proportions is most sensitive to climate, and how trees adjust their xylem hydraulic function to cope with climate, especially moisture stress. To achieve this goal, one available approach is wood anatomy (Zhirnova *et al.*, 2022). The anatomical features more frequently used include lumen area, anatomical density and wall thickness. In the future, a step forward in dendroclimatology will be to improve interpretations with a more in-depth understanding of tree functioning through quantitative wood anatomy (Gennaretti *et al.*, 2022). Meanwhile, Blue Intensity (BI) is also emerging as a promising alternative tree ring parameter with stronger and more stable climate signals for dendroclimatological research (Jiang *et al.*, 2022). The image-based BI parameter represents a measure of reflected light in the blue wavelengths of the colour spectrum and is widely used for temperature reconstruction (Cao *et al.*, 2022). However, recent studies have demonstrated that records on maximum density frequently result in higher accuracy (Wang *et al.*, 2020, 2022).

Elemental variability (e.g., N, P, K and Ca) in tree rings as an indicator of climate change is sporadically studied, given that fewer elements are available for the uptake into trees and elements are mobile within the tree and between tree rings (Sánchez-Salguero *et al.*, 2019). The results available in the literature help to better understand environmental changes, effects of pollution and long-term droughts, forest dieback processes and, more in general, the biology of trees.

Dendroclimatological reconstruction

Relationships between tree-ring proxy (such as TRW, density or isotope) and climate data are traditionally and generally detected based on a statistical relationship calibrated over the meteorological recorded instrumental period (Anchukaitis *et al.*, 2017). Temperature reconstruction from the linear method over the past decades at various study areas agrees well on several important trends, such as heat waves during medieval times and cooling during the 17th and 19th centuries, but they still exhibit substantial differences during the 13th and 14th centuries. The varying amplitudes and trends of reconstructed temperatures over the last thousand years are the other characteristics of the published series. The most challenging limit to overcome in large-scale hydroclimate reconstruction is the trade-off among highly continuous sample replications, a well-mixed age–class distribution over time and good internal growth coherence. Future studies will likely focus on the accuracy of reconstruction skills and reduce uncertainties in the processes.

A limit of the reconstruction methods is the assumption of linearity and stationarity of the relationship detected from the calibration. This linear relationship also limits dendroclimatological results to the sites where only one principal piece of climate data is supposed to be the driving factor of tree-ring growth. As a complement to statistical methods, process-based models have been used to elucidate relationships between climate and tree-ring proxy (Boucher *et al.*, 2014). Mechanistic models based on tree physiology allow for non-stationarity and non-linearity between tree-ring series and climate data. This approach also enables the extraction of climate signals from tree-rings at sites where multiple climatic factors drive tree growth, thus extending locations for dendroclimatological studies. However, many difficulties limit the use of these models, for example, the parameter adjusting for the complex biological processes usually implies a cautious initialisation and calibration of the model at each particular site of interest, depending on the site environment or tree species.

In this context, Vaganov-Shashkin (VS) and MAIDEN models appear to be promising tools to account for the non-stationarity of tree growth with climate change, especially considering that the MAIDEN model can be linked to investigations on CO_2 variability in atmospheric conditions in paleoclimate reconstructions (Rezsöhazy *et al.*, 2021). The possibility of combining the VS and MAIDEN models for global-scale climate reconstruction is under investigation (Rezsöhazy *et al.*, 2020, 2021)

The recent application of artificial intelligence may open up new avenues for reconstructing past climate conditions and pave the way for future predictions. Artificial neural networks (ANNs) models are proven to be tools more effective for tree-ring series compared to the traditional statistical methods (Khaleghi, 2018). ANN can be considered an improvement of the traditional linear methods. It can be linear or non-linear, and specific training algorithms can solve the overfitting. Complicated relationships between tree-ring parameters and climate or environmental data can be more clearly illustrated if a proper algorithm is found.

Conclusion

Forest productivity is based on wood production. Understanding the mechanisms of xylem production and the environmental factors driving its formation is essential for predicting how trees will respond to climate change in the future. Inter- and intra-annual studies of wood formation allow to analyse how wood production changes in time and according to specific environmental factors from its basic structural units, the tree ring and xylem cell, respectively. A long-term forest management strategy requires knowledge of the natural development dynamics of tree species, which necessarily includes the study of the mechanisms of growth and wood formation and the responses of trees to their environment at several temporal scales. In this sense, tree-based growth models built on the study of tree-ring formation and growth could be valuable decision-making tools in forestry and forest ecology and for the timber industry. If and how tree growth, in terms of tree-ring production, adapts to a changing environment allows scenarios of forest productivity in the near future to be estimated using precise and reliable predictions. At cell level, information on the period and rate of lignification of woody tissues could be useful in tree crops for the production of wood pulp with low lignin content for the paper industry or for the study of the origin of internal tensions in logs, which cause deformation and twisting of timber during seasoning.

In addition to the different temporal scales on which wood formation assessment and dendrochronological studies focus, a crucial difference lies in the temporal resolution of investigation. Intra-annual analyses of wood formation require that samplings, observations or measurements are performed repetitively when growth occurs. Generally, the cost of the analyses and the short time span of research projects allow intra-annual sampling to be performed during a few growing seasons and on a relatively small number of trees. Wood formation assessments are rarely performed for longer than three years and more frequently last just one or two years. On the contrary, dendrochronological assessments can refer to chronologies lasting hundreds to thousands of years, a temporal span defined by the number of overlapped sequences of tree rings found in the sampled individual. The different length of the chronologies, i.e., longer at inter-annual scale and shorter at intra-annual scale, entails two contrasting depths of analysis and different interpretations of the results. Inter-annual chronologies can include large data sets and represent long-term responses of tree growth to climate but fail to distinguish the effects of punctual events during tree-rings formation. On the other hand, dendrochronology can detect the inter-annual influences of climate, such as the effect of the previous year's temperature on current growth. Intra-annual chronologies are mostly very short but provide details of variation in wood production in response to punctual events occurring during the growing season.

The study of intra-annual wood formation has the potential to significantly enhance our understanding of long tree-ring chronologies and improve the accuracy of dendroclimatic reconstructions. By analysing the different phases of xylogenesis, we can gain detailed information on the environmental factors influencing tree growth during a given year. These insights can be used to better understand past climate variability. Climate–growth relationships have opened up interesting fields of exploration and discussion in predicting how a forest and its components, the trees, respond and will evolve under present or future growing conditions. Ecologists and physiologists are now called upon to tackle the complex challenge of combining the results of climate–growth relationships from the different temporal scales of analysis in a convincing, conclusive and inclusive final interpretation of tree growth.

References

Anchukaitis, K. J., Wilson, R., Briffa, K. R., Büntgen, U., Cook, E.R., D'Arrigo, R., Davi, N., Esper, J., Frank, D., Gunnarson, B. E., *et al.* (2017) 'Last millennium Northern Hemisphere summer temperatures from tree rings: Part II, spatially resolved reconstructions', *Quaternary Science Reviews*, vol 163, pp. 1–22.

Balducci, L., Deslauriers, A., Giovannelli, A., Rossi, S. and Rathgeber, C. B. K. (2013) 'Effects of temperature and water deficit on cambial activity and woody ring features in Picea mariana saplings', *Tree Physiology*, vol 33, pp. 1006–1017.

Boucher, E., Guiot, J., Hatté, C., Daux, V., Danis, P. A. and Dussouillez, P. (2014) 'An inverse modeling approach for tree-ring-based climate reconstructions under changing atmospheric CO_2 concentrations', *Biogeosciences*, vol 11, pp. 3245–3258.

Büntgen U., Allen K., Anchukaitis K. J., Arseneault D., Boucher É., Bräuning A., Chatterjee S., Cherubini P., Churakova (Sidorova) O. V., Corona C., *et al.* (2021) 'The influence of decision-making in tree ring-based climate reconstructions', *Nature Communications*, vol 12, 3411.

Buttò, V., Deslauriers, A., Rossi, S., Rozenberg, P., Shishov, V. and Morin, H. (2020) 'The role of plant hormones in tree-ring formation', *Trees - Structure and Function*, vol 34, pp. 315–335.

Camarero, J. J., Olano, J. M. and Parras, A. (2010) 'Plastic bimodal xylogenesis in conifers from continental Mediterranean climates', *New Phytologist*, vol 185, pp. 471–480.

Cao, X., Hu, H., Kao, P.-K., Buckley, B. M., Dong, Z., Chen, X., Zhou, F. and Fang, K. (2022) 'Improved spring temperature reconstruction using earlywood blue intensity in southeastern China', *International Journal of Climatology*, vol 42, pp. 6204–6220.

Deslauriers, A., Huang, J. G., Balducci., L, Beaulieu., M. and Rossi, S. (2016) 'The contribution of carbon and water in modulating wood formation in black spruce saplings', *Plant Physiology*, vol 170, pp. 2072–2084.

Deslauriers, A., Rossi, S., Anfodillo, T. and Saracino, A. (2008) 'Cambial phenology, wood formation and temperature thresholds in two contrasting years at high altitude in southern Italy', *Tree Physiology*, vol 28, pp. 863–871.

Esper, J., Krusic, P. J., Ljungqvist, F. C., Luterbacher, J., Carrer, M., Cook, E., Davi, N. K., Hartl-Meier, C., Kirdyanov, A., Konter O., *et al.* (2016) 'Ranking of tree-ring based temperature reconstructions of the past millennium', *Quaternary Science Reviews*, vol 145, pp. 134–151.

Fritts, H. C. (1976) *Tree rings and climate*. Academic Press.

Gennaretti, F., Arseneault, D., Nicault, A., Perreault, L. and Begin, Y. (2014) 'Volcano-induced regime shifts in millennial tree-ring chronologies from northeastern North America', *Proceedings of the National Academy of Sciences of the United States of America*, vol 111, pp. 10077–10082.

Gennaretti, F., Carrer, M., García-González, I., Rossi, S. and von Arx, G. (2022) 'Quantitative wood anatomy to explore tree responses to global change', *Frontiers in Plant Science*, vol 13, 3204.

Gennaretti, F., Huard, D., Naulier, M., Savard, M., Bégin, C., Arseneault, D. and Guiot, J. (2017) 'Bayesian multiproxy temperature reconstruction with black spruce ring widths and stable isotopes from the northern Quebec taiga', *Climate Dynamics*, vol 49, pp. 4107–4119.

Howard, C., Dymond, C. C., Griess, V. C., Tolkien-Spurr, D. and van Kooten, G. C. (2021) 'Wood product carbon substitution benefits: A critical review of assumptions', *Carbon Balance and Management*, vol 16, pp. 1–11.

Huang, J. G., Ma, Q., Rossi, S., Biondi, F., Deslauriers, A., Fonti, P., Liang, E., Mäkinen, H., Oberhuber, W., Rathgeber, C. B. K., *et al.* (2020) 'Photoperiod and temperature as dominant environmental drivers triggering secondary growth resumption in Northern Hemisphere conifers', *Proceedings of the National Academy of Sciences of the United States of America*, vol 117, pp. 20645–20652.

Jiang, Y., Begović, K., Nogueira, J., Schurman, J., Svoboda, M. and Rydval, M. (2022) 'Impact of disturbance signatures on tree ring width and blue intensity chronology structure and climatic signals in Carpathian Norway Spruce', *Agricultural and Forest Meteorology*, vol 327, 109236.

Khaleghi, M. R. (2018) 'Application of dendroclimatology in evaluation of climatic changes', *Journal of Forest Science*, vol 64, pp. 139–147.

Knudson, L. (1913) 'Observations on the inception, season, and duration of cambium development in the American Larch [Larix laricina (Du Roi) Koch.]', *Bulletin of the Torrey Botanical Club*, vol 40, pp. 271–293.

Larson, P. R. (1994) *The Vascular Cambium: Development and Structure*. Springer Series in Wood Science. Springer

Ljungq vist, F. C., Piermattei, A., Seim, A., Krusic, P. J., Büntgen, U., He, M., Kirdyanov, A. V., Luterbacher, J., Schneider, L., Seftigen, K., *et al.* (2020) 'Ranking of tree-ring based hydroclimate reconstructions of the past millennium', *Quaternary Science Reviews*, vol 230, 106074.

Lu, Q., Liu, X., Treydte, K., Greule, M., Wieland, A., Liu, J., Zhao, L., Zhang, Y., Kang, H., Zhang, L., *et al.* (2023) 'Altitude-specific differences in tree-ring δ2H records of wood lignin methoxy in the Qinling mountains, central China', *Quaternary Science Reviews*, vol 300, 107895.

Lupi, C., Morin, H., Deslauriers, A. and Rossi, S. (2010) 'Xylem phenology and wood production: Resolving the chicken-or-egg dilemma', *Plant, Cell and Environment*, vol 33, pp. 1721–1730.

Lupi, C., Rossi, S., Vieira, J., Morin, H. and Deslauriers, A. (2013) 'Assessment of xylem phenology: A first attempt to verify its accuracy and precision', *Tree Physiology*, vol 34, pp. 87–93.

Marion, L., Gričar, J. and Oven, P. (2007) 'Wood formation in urban Norway maple trees studied by the micro-coring method', *Dendrochronologia*, vol 25, pp. 97–102.

Mathias, J. M. and Thomas, R. B. (2021). 'Global tree intrinsic water use efficiency is enhanced by increased atmospheric CO2 and modulated by climate and plant functional types', *Proceedings of the National Academy of Sciences*, vol 118, no 7, e2014286118.

Pan, Y., Birdsey, R. A., Fang, J., Houghton, R., Kauppi, P. E., Kurz, W. A., Phillips, O. L., Shvidenko, A., Lewis, S. L., Canadell, J. G., *et al.* (2011) 'A large and persistent carbon sink in the world's forests', *Science*, vol 333, pp. 988–993.

Rathgeber, C. B. K., Rossi, S. and Bontemps, J. D. (2011) 'Cambial activity related to tree size in a mature silver-fir plantation', *Annals of Botany*, vol 108, pp. 429–438.

Ren, P., Rossi, S., Gricar, J., Liang, E. and Cufar, K. (2015) 'Is precipitation a trigger for the onset of xylogenesis in Juniperus przewalskii on the north-eastern Tibetan Plateau?', *Annals of Botany*, vol 115, pp. 629–639.

Ren, P., Ziaco, E., Rossi, S., Biondi, F., Prislan, P. and Liang, E. (2019) 'Growth rate rather than growing season length determines wood biomass in dry environments', *Agricultural and Forest Meteorology*, vol 271, pp. 46–53.

Rezsöhazy, J., Gennaretti, F., Goosse, H. and Guiot, J. (2021) 'Testing the performance of dendroclimatic process-based models at global scale with the PAGES2k tree-ring width database', *Climate Dynamics*, vol 57, pp. 2005–2020.

Rezsöhazy, J., Goosse, H., Guiot, J., Gennaretti, F., Boucher, E., André, F. and Jonard, M. (2020) 'Application and evaluation of the dendroclimatic process-based model MAIDEN during the last century in Canada and Europe', *Climate of the Past*, vol 16, pp. 1043–1059.

Rossi, S. (2015) 'Local adaptations and climate change: Converging sensitivity of bud break in black spruce provenances', *International Journal of Biometeorology*, vol 59, pp. 827–835.

Rossi, S., Anfodillo, T., Čufar, K., Cuny, H. E., Deslauriers, A., Fonti, P., Frank, D., Gričar, J., Gruber, A., Huang, J. G., *et al.* (2016) 'Pattern of xylem phenology in conifers of cold ecosystems at the Northern Hemisphere', *Global Change Biology*, vol 22, pp. 3804–3813.

Rossi, S., Anfodillo, T. and Menardi, R. (2006a) 'Trephor: A new tool for sampling microcores from tree stems', *IAWA Journal*, vol 27, pp. 89–97.

Rossi, S., Deslauriers, A. and Anfodillo, T. (2006b) 'Assessment of cambial activity and xylogenesis by microsampling tree species: An example at the Alpine timberline', *IAWA Journal*, vol 27, pp. 383–394.

Rossi, S., Deslauriers, A., Anfodillo, T., Morin, H., Saracino, A., Motta, R. and Borghetti, M. (2006c) 'Conifers in cold environments synchronize maximum growth rate of tree-ring formation with day length', *New Phytologist*, vol 170, pp. 301–310.

Rossi, S., Deslauriers, A., Griçar, J., Seo, J. W., Rathgeber, C. B. K., Anfodillo, T., Morin, H., Levanic, T., Oven, P. and Jalkanen, R.. (2008) 'Critical temperatures for xylogenesis in conifers of cold climates', *Global Ecology and Biogeography*, vol 17, pp. 696–707.

Rossi, S., Deslauriers, A. and Morin, H. (2003) 'Application of the Gompertz equation for the study of xylem cell development', *Dendrochronologia*, vol 21, pp. 33–39.

Rossi, S., Girard, M. J. and Morin, H. (2014) 'Lengthening of the duration of xylogenesis engenders disproportionate increases in xylem production', *Global Change Biology*, vol 20, pp. 2261–2271.

Rossi, S., Morin, H. and Deslauriers, A. (2012) 'Causes and correlations in cambium phenology: Towards an integrated framework of xylogenesis', *Journal of Experimental Botany*, vol 63, 2117–2126.

Sánchez-Salguero, R., Camarero, J. J., Hevia, A, Sangüesa-Barreda, G., Galván, J. D. and Gutiérrez, E. (2019) 'Testing annual tree-ring chemistry by X-ray fluorescence for dendroclimatic studies in high-elevation forests from the Spanish Pyrenees', *Quaternary International*, vol 514, pp. 130–140.

Shestakova, T. A. and Martínez-Sancho, E. (2021) 'Stories hidden in tree rings: A review on the application of stable carbon isotopes to dendrosciences', *Dendrochronologia*, vol 65, 125789.

Silvestro, R., Sylvain, J.-D., Drolet, G., Buttò, V., Auger, I., Mencuccini, M. and Rossi, S. (2022) 'Upscaling xylem phenology: Sample size matters', *Annals of Botany*, vol 130, pp. 811–824.

Silvestro, R., Zeng, Q., Buttò, V., Sylvain, J. D., Drolet, G., Mencuccini, M., Thiffault, N., Yuan, S. and Rossi, S. (2023) 'A longer wood growing season does not lead to higher carbon sequestration', *Scientific Reports*, vol 13, 4059.

Vieira, J., Rossi, S., Campelo, F., Freitas, H. and Nabais, C. (2014a) 'Xylogenesis of Pinus pinaster under a Mediterranean climate', *Annals of Forest Science*, vol 71, pp. 71–80.

Vieira, J., Rossi, S., Campelo, F. and Nabais, C. (2014b) 'Are neighboring trees in tune? Wood formation in Pinus pinaster', *European Journal of Forest Research*, vol 133, pp. 41–50.

Wang, F., Arseneault, D., Boucher, É., Galipaud Gloaguen, G., Deharte, A., Yu, S. and Trou-kechout, N. 2020. 'Temperature sensitivity of blue intensity, maximum latewood density, and ring width data of living black spruce trees in the eastern Canadian taiga', *Dendrochronologia*, vol 64, 125771.

Wang, F., Arseneault, D., Boucher, É., Gennaretti, F., Yu, S. and Zhang, T. (2022) 'Tropical volcanoes synchronize eastern Canada with Northern Hemisphere millennial temperature variability', *Nature Communications*, vol 13, 5042.

Zhao, S., Pederson, N., D'Orangeville, L., HilleRisLambers, J., Boose, E., Penone, C., Bauer, B., Jiang, Y. and Manzanedo, R. D. (2019) 'The International Tree-Ring Data Bank (ITRDB) revisited: Data availability and global ecological representativity', *Journal of Biogeography*, vol 46, pp. 355–368.

Zhirnova, D. F., Belokopytova, L. V., Upadhyay, K. K., Tripathi, S. K., Babushkina, E. A. and Vaganov, E. A. (2022) '495-Year Wood Anatomical Record of Siberian Stone Pine (Pinus sibirica Du Tour) as climatic proxy on the timberline', *Forests*, vol 13, 247.

37

PLANT MOVEMENTS IN RESPONSE TO RAPID CLIMATE CHANGE

Richard T. Corlett

Introduction

The Sixth Assessment Report of the Intergovernmental Panel on Climate Change (IPCC, 2021, 2022) makes for grim reading. Atmospheric concentrations of CO_2 have increased from 278 ppm in 1750, before the industrial revolution, to 421 ppm in 2023: an increase of 50%. The other major greenhouse gases, CH_4 and N_2O, have risen by more than 150% and 20%, respectively. Present-day concentrations of CO_2 are higher than those at any time in the last 2 million years. Because of these atmospheric changes, the Earth has been getting warmer over the last 100 years: an average of 1.1°C warmer globally, and 1.6°C warmer over land. Each of the past three decades has been warmer than the previous one and warmer than all previous decades since records began. Other climatic variables are also changing. Trends in rainfall are less clear and more varied than those in temperature, but heavy rainfall is increasing, as are droughts in some regions.

The global climate models agree well with each other in their predictions for the next 20 years or so, with an additional increase of 0.4–0.5°C in global surface air temperature, giving a total warming of around 1.5°C (2.0°C over land) (IPCC, 2021). Globally, rainfall is expected to increase, but some regions (such as most of the Mediterranean) will get drier and confidence in detailed rainfall predictions in many parts of the world is low. In the longer term, predictions vary greatly even for temperatures, depending on the choice of climate model and the assumptions made about future greenhouse gas emissions and carbon cycle feedbacks. Temperatures over land are expected to increase by 1.4–4.4°C by 2100, compared with 1850–1900, but the range of plausible values is considerably wider (at least 1–6°C).

There is abundant evidence from a variety of sources that the rising temperatures in recent decades, as well as changes in other climatic variables, are impacting wild species and ecosystems (IPCC, 2022). Plant responses to climate change are of particular interest to climate modelers because of the known and potential feedbacks between vegetation—particularly forests—and climate. The global significance of the carbon fluxes through forests is shown by the annual decline in global atmospheric CO_2 concentrations (of 3–9 ppm) at the late-summer peak of photosynthesis in the northern hemisphere, where there is much more forest than in the southern hemisphere. Forests, particularly tropical forests, have also

DOI: 10.4324/9781003324072-43

the major control on the interannual variability in the global CO_2 balance, with the high temperatures and droughts associated with strong El Niño events leading to large increases in the atmospheric CO_2 growth rate. Moreover, in the last two decades global forests have absorbed an estimated 38% of annual global anthropogenic CO_2 emissions, partly in forests recovering from past disturbance but also in mature forests, although large emissions from deforestation and degradation reduce the net forest sink to around 14% of global emissions (Harris et al., 2021). The long-term stability of this crucial sink depends on how plants respond to climate change. Plant responses will also have a major influence on the persistence of animals, particularly the hyperdiverse and relatively plant-dependent invertebrates (Stork and Habel, 2014).

In the face of climate change, plant species must either adapt (genetic change), acclimate (plastic change), or move to stay within their climate envelope (Corlett and Westcott, 2013). Evidence for genetic change is limited so far, but the widely reported changes in plant phenology represent a plastic response that appears in many cases to be adaptive (e.g. Marqués et al., 2023). Other plastic responses are less well understood, but limited studies on tropical trees suggest a useful capacity to acclimate to moderate levels of warming but a limited ability to acclimate to drought (Bittencourt et al., 2020; Ometto et al., 2022). Movement is more difficult for plants than for most animals because they can only move once per generation and plants live much longer than most animals. Movement potential for plants is considered in more detail below.

The paleoecological record

Movement was a near-universal response of plant populations to climate change in the paleoecological record. Species did not necessarily track climate change closely, however, and there is evidence that many plant species in Europe are not in equilibrium with the current climate as a result of post-glacial migration lag (Normand et al., 2011). On the other hand, only one plant species, *Picea critchfieldii*, which was previously widespread and regionally abundant in eastern North America, is known to have become globally extinct due to climate change during the Late Pleistocene (Jackson and Weng, 1999). Overall, the paleoecological record suggests that plant species are more likely to move than acclimate or adapt, even in response to gradual climate change (Wang et al., 2023).

The need for speed

Although climates have changed in the past and have sometimes been warmer than today, there are good reasons to think that 21st-century climate change will be a problem for wild species. A rise of 2.5°C, which currently seems likely to happen before the mid-century, would result in higher sustained surface temperatures than any that we know of in the last 3 million years (IPCC, 2021). Present-day species are unlikely to have retained adaptations to this amount of warming. Moreover, evidence from the paleoecological record suggests that the *rates* of warming and other climate changes over the 21st century will be exceptional since changes in well-studied periods in the past were 10–100 times slower.

A useful concept is the 'velocity of climate change', which is the speed and direction that something needs to move to keep within its current 'climate envelope', i.e. to stay within the range of climate conditions it can tolerate (Brito-Morales et al., 2018). There are two main ways of estimating climate velocity. Local climate velocity is calculated from the rate of change of a climate variable in time and the gradient in space, while climate-analogue velocity is calculated from the distance to the nearest point with an analogous future climate, divided by the time difference. Velocities of temperature change are expected to exceed 1 km year^{-1} over much of the Earth's land surface during the 21st century (Loarie, 2009), although steep temperature gradients will result in much lower local velocities in areas of rugged topography (< 10 m/year on the steepest slopes). Velocities of rainfall change are projected to be similar. Few, if any, plant species have their distributions determined by only one or two climatic factors, however. Multiple variables can be combined using principal components analysis (PCA) (e.g. Rota et al., 2022), but velocities of change in PCA axes are difficult to interpret and it is often more useful to analyze the potential climatic drivers separately (Heikkinen et al., 2020).

For comparison, the estimated global mean velocity of temperature change between the last glacial maximum and the present day was only 5.9 m year^{-1} (Sandel et al., 2011), although more detailed studies show that there were short periods with much higher velocities (400–700 m year^{-1}), between 20,000 and 10,000 years ago (Correa-Metrio et al., 2013). Moreover, the resolution of the paleoecological record is insufficient to rule out even more rapid changes on a century scale. However, these past changes differ from those projected for the 21st century in another important way: the extremes were maintained for relatively short periods, before swinging back. This makes it more likely that plant populations could persist through periods of unfavorable climate as suppressed, non-reproductive, individuals, whereas the forecast monotonic trends for the 21st century and beyond are likely to favor extinction (Correa-Metrio et al., 2013).

It needs to be emphasized that the velocity of climate change is a measure of the exposure of a plant population to climate change, not necessarily the impact of this change. Current ranges are not necessarily limited by climate and a species restricted by non-climatic factors well within its potential climate envelope may not need to move. Moreover, as discussed above, the capacity of plant populations for acclimation and adaptation is still poorly understood. Plant responses exhibited in the paleoecological record suggest that climate niches are, in general, highly conservative (Wang et al., 2023), but the temporal and climatic resolution of these records is not sufficient to make confident predictions on a multidecadal timescale.

The potential for plant movement

Plant individuals are immobile as adults, but plant populations can move if seeds are dispersed and establish beyond the current range at the leading edge, while the plants at the trailing edge fail to regenerate. Most studies have focused on the dispersal phase of the colonization process, but for movement to occur the dispersed propagules must then establish in conditions that are likely to be increasingly dissimilar to the home environment the further a population moves outside its original range (De Frenne et al., 2014). Except at the treeline, forest plants will normally be moving into established forests, which may have been weakened by climate change, but still have the advantages of numbers and local adaptation to non-climatic factors. Currently the role of establishment in promoting or reducing plant movements is poorly understood, but negative impacts are most likely for slow-growing, mature-phase forest taxa moving into existing forests, while pioneer and open-country species are less likely to suffer problems (Corlett and Westcott, 2013).

There have been significant advances in our understanding of the mechanisms of seed dispersal over the last decade or so, both for wind-dispersed taxa (e.g. Trakhtenbrot et al.,

2014) and for dispersal inside the guts of frugivorous animals (e.g. Corlett, 2021; Morales and Morán López, 2022). However, while this mechanistic understanding can help predict median dispersal distances for various combinations of plants and vectors, we are much further from being able to predict the rare long-distance dispersal events—the 99th percentile dispersal distances—which models suggest will have a major influence on plant movement velocities (Zani et al., 2022; Wu et al., 2023). We can say with some confidence that most seeds will be dispersed <1500 m from the parent plant (Corlett and Westcott, 2013), with the upper end of this range occurring mostly outside forests, but we cannot currently rule out a significant long-distance 'tail' to the right of this distribution.

Converting dispersal distances into potential velocities also requires knowledge of the number of dispersal events in the time period under consideration, which depends on the time from dispersed seed to reproductive adult. Assuming most plant species are dispersing 50–1500 m every 1–30 years, the range of plant velocities in unfragmented habitats with no barriers to movement would be 1.7–1500 m per year (Corlett and Westcott, 2013). However, the positive relationship between dispersal distance and plant height (Thomson et al., 2011) and the generally lower dispersal distances and longer times to maturity in forests suggest that most forest plant species will not be capable of moving more than 1000 m per year and many will move less than 100 m. Projected velocities of climate change and potential velocities of plant movement based on observed dispersal distances thus broadly overlap, but most plant velocities are likely to be at the lower end of this range and will be exceeded by climate velocities except in steep topography.

A key question then becomes: how important are occasional long-distance dispersal events on the multidecadal timescales of most concern here? Will the leading edges of plant populations tracking climate change spread slowly at the rates predicted by the simple model above, based on observed dispersal kernels and known dispersal agents, or will population movements depend on unusual long-distance dispersal events—perhaps by atypical vectors—followed by slower infilling of the gaps (Zani et al., 2022)?

Routine dispersal distances >1500 m are most common in species that have small, wind-dispersed seeds, such as epiphytic orchids, and those moved by the largest birds, fruit bats, and terrestrial mammals (Corlett, 2021). At the other extreme, ants, which are major dispersal agents for herb species in temperate deciduous forests, disperse seeds for a mean distance of around 2 m (Gómez and Espadaler, 2013). Forests have been reduced and fragmented worldwide. In general, this is likely to greatly reduce dispersal distances, particularly for plants with vectors that cannot move between fragments. Most forests have also lost some or all of their large-bodied vertebrates, including species that have key roles in the long-distance seed dispersal needed for plants to track climate change. Using trait-based models of seed dispersal for fleshy-fruited plants, Fricke et al. (2022) estimated that losses of the mammals and birds responsible for long-distance dispersal have already reduced the ability of plants to track climate change by 60% globally.

Recent movements

A recent review of studies that reported climate-driven shifts in the ranges of plants and animals in response to contemporary climate change found mixed results (Lenoir et al., 2020). Along gradients in elevation, plant populations generally moved upslope, as expected, but not as fast as needed to keep pace with climate change. This lag may reflect the limited dispersal capacities and long lifespans of many plant species, but it is also possible that the warming experienced so far has simply not been enough to require movement by some species (i.e., thermal safety margins have not been exceeded), or that variations in microclimate over short-distances allow species to compensate without tracking the macroclimate (see climate change refugia, below). Along latitudinal gradients, in contrast, there was no evidence for a general movement of plant populations. In addition to the natural factors suggested above, habitat loss and fragmentation, and the depletion of dispersal agents, are almost certainly impeding latitudinal movements in many parts of the world.

Modeling movements of species and communities

Species distribution models (SDMs) are currently the most widely used method to model the potential impacts of climate change on the distributions of individual plant species. SDMs use correlations between current spatial distributions and current climatic and other (soil, topography, etc.) variables to predict the future distribution from the future climate. Such correlative models are easy to run and are generally fairly accurate in reproducing the current distributions of plants. However, they have several limitations when used for prediction, including the inherent dangers of extrapolation from correlations and their inability to deal with novel future climates (or novel combinations of climate and other factors).

Correlative SDMs typically show that some areas currently occupied will become climatically unsuitable in the future under climate change, while new areas of climatically suitable habitat will become available elsewhere. These results are often interpreted as predicting a shift in overall distribution, but neither losses nor gains are simple processes. At the 'trailing edge' of the distribution, long-lived species may persist for decades or centuries in climatically unsuitable habitats before they are replaced (Svenning and Sandel, 2013), while, at the leading edge, occupation of the newly available habitat will depend on migration ability, which is determined by seed dispersal and generation time, as well as the ability of the species to establish in competition with the existing flora. SDM predictions are usually implemented with two extreme dispersal scenarios, full (unlimited) dispersal and no dispersal, on the assumption that the truth will lie somewhere between these extremes. Recently, there have been many attempts to incorporate realistic dispersal abilities into SDMs, with varying degrees of success (Boisvert-Marsh et al., 2022). However, in most cases, the major limitation is not the modelling process itself, but the availability of the data needed to parameterize these more sophisticated models. This limitation applies even more strongly to fully mechanistic models of plant responses to climate change, although such models are expected to have major advantages, particularly under novel conditions where historical correlations may no longer apply (Urban et al., 2016).

SDMs for individual species can be 'stacked' to predict future plant species assemblages (e.g. Pomoim et al., 2022), although this approach cannot account for interactions between the species projected to newly coexist. Dynamic global vegetation models (DGVMs) take a different approach, simulating the response of a limited range of plant functional types (PFTs), based largely on physiology. DGVMs can explicitly include the processes involved in plant migration, including fecundity, seed dispersal, establishment, growth, and competition with other species, although most applications so far have assumed that every PFT is everywhere, i.e. unlimited dispersal. Recently, Zani et al. (2022) used post-glacial migration rates for dominant European tree taxa estimated from pollen and macrofossil records to test the performance of LPJ-GM, a DGVM with a dynamic migration module. A sensitivity analysis showed that migration rates were most sensitive to seed dispersal parameters and that a 'fat-tailed' dispersal kernel, incorporating long-distance dispersal events, was necessary to generate the observed migration rates for most species. Simulations of migration rates across homogeneous landscapes on millennial timescales are not directly relevant to 21st-century movements across fragmented modern landscapes in response to rapid anthropogenic climate change, but they show the potential to develop more realistic mechanistic models.

Climate change refugia

In all but the flattest topography, local variations in temperature, moisture, and other climate variables create potential 'refugia', where plant populations may persist in favorable conditions outside their broadly defined climate envelope (Finocchiaro et al., 2023). SDMs are usually run at too coarse a spatial resolution to detect them, but the use of finer scale climate projections can capture these topographically controlled variations, revealing refugia that were not apparent at a coarser scale. The ability of 50-m resolution topographic variation in climates to buffer regional climate change appears to be limited, however, although it is possible that variation on a still finer scale (e.g. forested ravines) may provide longer-lasting refugia (Heikkinen et al., 2020). Moreover, microclimate-based SDMs using data from a below-canopy network of climate sensors in complex forested terrain have demonstrated a greater potential for buffering than shown with downscaled regional climate projections (Stark and Fridley, 2022). Collecting climate data on this fine spatial scale is rarely feasible, but measures of topographic diversity are a possible surrogate (Stralberg et al., 2020).

Cryptic refugia north of the main treeline have been invoked to explain improbably high post-glacial tree migration rates in the northern hemisphere and could similarly reduce the movement velocities needed to track future climate change (Corlett and Westcott, 2013). Taken together, the evidence for past and potential future refugia suggests that modeling approaches based on macro (>100 km) or mesoscale (>1 km) climate projections may greatly underestimate the potential for microclimatic buffering of climate change in complex topography and thus overestimate the need for movement. Identifying and protecting future refugia is one of the more plausible suggestions for mitigation of climate change impacts.

The potential consequences of migration lag

Model-based projections of future plant species extinction risks due to climate change can be catastrophic, contrasting strongly with the rarity of extinctions in the paleoecological record and the absence of well-documented recent extinctions (Nic Lughadha et al., 2020). The models may be overestimating extinction risk, because they largely ignore acclimation capacity and are typically run at a spatial resolution too coarse to detect potential refugia, but it is also possible that the combination of unprecedented rates of climate change with massively fragmented landscapes and a suite of other adverse human impacts really will be catastrophic. Extinction, however, is likely to be a slow and untidy process, since plant species can typically persist as individuals well outside the range of conditions that will support a self-sustaining population (Sax et al., 2013). Such individuals may be 'committed to extinction' if conditions do not change, but as long as they persist there is the possibility of rescue by a future reversal in climate change, or by active human intervention.

The impacts of migration lags on the global carbon sink are more easily predicted, since species that fail to track climate change are likely to grow more slowly and fix less carbon as the climate becomes less suitable. Moreover, the forests of the future, whether they replace existing forests or occupy deforested areas, will be dominated by those species that can move fast enough, which are more likely to be relatively short-lived, short-stature pioneers with low-density wood than long-lived, carbon-rich, late-successional species.

Assisted migration

Assisted migration (also called managed translocation and assisted colonization) is the deliberate establishment of a population of organisms *outside* their known historical range to save the species from extinction. First applied to the movement of vulnerable species to predator-free offshore islands, these terms are now most often used in relation to climate change. If a species is vulnerable to extinction because it is isolated on a mountain top or edaphic island, or its habitat has become fragmented, or it simply cannot move fast enough, then why not move it artificially to a suitable site? Opponents highlight the risk of introduced species becoming invasive (Mueller et al., 2008), but the plant traits that make a species likely to need human assistance—including poor dispersal, long life-cycle, low competitive ability—are the opposite of those that favor invasiveness. Moreover, most species translocations are likely to be into similar communities near the edge of the natural range, reducing the risk of unpredictable impacts.

More realistic are concerns for the practicality of assisted migration on anything approaching the scale that is likely to be needed. Successful translocations will require more than simply matching climate variables, since many range limits are likely to be influenced by other factors, including topographic and edaphic variables, and interactions with competitors and natural enemies (Park and Talbot, 2018). Moreover, climate change is a moving target and long-lived species will experience significant changes doing their lifetime, so the risks of present or future maladaptation of translocated plants are high.

Poleward translocations of suitable genotypes *within* the overall native range ('assisted population migration') are likely to be less controversial than translocations outside the range, as is the use of warmth-adapted genotypes of native species in ecological restoration, reforestation, and native plantations ('predictive provenancing'). However, in the longer term these options are unlikely to be enough and we will have to learn how to carry out assisted migration in a way that minimizes the risk: reward ratio. Given the large uncertainties, there is an urgent need for experimental studies in a wide variety of situations, from using translocated species in reforestation of degraded and open habitats to the more challenging task of trying to establish populations in intact natural communities. There are legal barriers to such experiments in many areas and these will need to be modified in a way that does not open the flood gates to uncontrolled plant movements.

Conclusion

Rapid climate change coupled with pervasive fragmentation and a host of other adverse impacts could make the remainder of the 21st century a very bad time for plants. Just how bad is currently unclear, with large uncertainties in both climate projections and climate change impacts on plant persistence. These uncertainties favor uncontroversial, 'no-regrets' adaptation measures, including reducing non-climate impacts, restoring forests, maximizing landscape connectivity, protecting and, where necessary, re-introducing long-distance seed dispersal agents, and protecting refugia where climate change is expected to be less than the regional mean. However, if current climate projections are at all realistic, these measures will not be enough since many plants could not move fast enough even in intact landscapes and refugia cannot hold everything.

Areas of low projected climate change velocity are likely to be the best places for ***in situ*** conservation. Most of these will be in regions of steep and rugged topography, which are also more likely to provide microrefugia for species that cannot track even low-velocity changes. However, sharp declines in area with altitude will limit the long-term potential of many such regions, with well-vegetated foothills perhaps the best long-term prospect. Areas with high projected climate change velocities, which are usually in landscapes with low-relief topography, may be most appropriate for managed translocations and other human interventions. Indeed, the whole concept of 'natural vegetation' in such landscapes is probably meaningless in times of rapid change. Large-scale experiments involving multiple sites and species are urgently needed, but we will not have time to wait for the results, so some degree of risk is unavoidable, even with detailed monitoring and adaptive management.

References

Bittencourt, P. R. L., Oliveira, R. S., da Costa, A. C. L., Giles, A. L., Coughlin, I., Costa, P. B., Bartholomew, D.C., Ferreira, L. V., Vasconcelos, S. S., Barros, F. V., Junior, J. A. S., Oliveira, A. A. R., Mencuccini, M., Meir, P. and Rowland, L. (2020) 'Amazonia trees have limited capacity to acclimate plant hydraulic properties in response to long-term drought', *Global Change Biology*, vol 26, pp. 3569–3584.

Boisvert-Marsh, L., Pedlar, J. H., de Blois, S., Le Squin, A., Lawrence, |K., McKenney, D. W., Williams, C. and Aubin, I. (2022) 'Migration-based simulations for Canadian trees show limited tracking of suitable climate under climate change', *Diversity and Distributions*, vol 28, 2330–2348.

Brito-Morales, I., Molinos, J. G., Shoeman, D. S., Burrows, M. T., Poloczanska, E.S., Brown, C. J., Ferrier, S., Harwood, T. D., Klein, C. J., McDonald-Madden, E., Moore, P. J., Pandolfi, J. M., Watson, J. E. M., Wenger, A. S. and Richardson A. J. (2018) 'Climate velocity can inform conservation in a warming world', *Trends in Ecology & Evolution*, vol 33, pp. 441–457.

Corlett, R. T. (2021) 'Frugivory and seed dispersal'. In: Del-Claro, K. and Torezan-Silingardi, H. M. (eds.), *Plant-Animal Interactions*. Springer Nature, Switzerland, pp. 175–204.

Corlett, R. T. and Westcott, D. A. (2013) 'Will plant movements keep up with climate change?', *Trends in Ecology and Evolution,* vol 28, pp. 482–488.

Correa-Metrio, A., Bush, M., Lozano-García, S. and Sosa-Nájera, S. (2013) 'Millennial-scale temperature change velocity in the continental northern Neotropics', *PLOS ONE,* vol 8, e81958–e81958.

De Frenne, P., Coomes, D. A., De Schrijver, A., Staelens, J., Alexander, J. M., Bernhardt-Römermann, M., Brunet, J., Chabrerie, O., Chiarucci, A., den Ouden, J., Eckstein, R. L., Graae, B. J., Gruwez, R., Hédl, R., Hermy, M., Kolb, A., Mårell, A., Mullender, S. M., Olsen, S. L., Orczewska, A., Peterken, G., Petřík, P., Plue, J., Simonson, W. D., Tomescu, C. V., Vangansbeke, P., Verstraeten, G., Vesterdal, L., Wulf, M. and Verheyen, K. (2014) 'Plant movements and climate warming: intraspecific variation in growth responses to nonlocal soils', *New Phytologist*, vol 202, pp. 431–441.

Finocchiaro, M., Médail, F., Saatkamp, A., Diadema, K., Pavon, D. and Meineri, E. (2023) 'Bridging the gap between microclimate and microrefugia: A bottom-up approach reveals strong climatic and biological offsets', *Global Change Biology*, vol 29, pp. 1024–1036.

Fricke, E., Ordonez, A., Rogers, H. S. and Svenning, J.-C. (2022) 'The effects of defaunation on plants' capacity to track climate change', *Science*, vol 375, pp. 210–214.

Gómez, C. and Espadaler, X. (2013) 'An update of the world survey of myrmecochorous dispersal distances', *Oikos*, vol 36, pp. 1193–1201.

Harris, N. L., Gibbs, D. L., Baccini, A., Birdsey, R. A., de Bruin, S., Farina, M., Fatoyinbo, L., Hansen, M. C., Herold, M., Houghton, R. A., Potapov, P. V., Suarez, D. R., Roman-Cuesta, R. M., Saachi, S. S., Slay, C. M., Turubanova, S. A. and Tyukavina, A. (2021) 'Global maps of twenty-first century forest carbon fluxes', *Nature Climate Change*, vol 11, pp. 234–240.

Heikkinen, R. K., Leikola, N., Aalto, J., Aapala, K., Kuusela, S., Luoto, M. and Virkkala, R. (2020) 'Fine-grained climate velocities reveal vulnerability of protected areas to climate change', *Scientific Reports*, vol 10, 1678.

IPCC. (2021) *Climate Change 2021: The Physical Science Basis. Contribution of Working Group I to the Sixth Assessment Report of the Intergovernmental Panel on Climate Change.* Cambridge University Press, Cambridge and New York.

IPCC. (2022) *Climate Change 2022: Impacts, Adaptation, and Vulnerability. Contribution of Working Group II to the IPCC Sixth Assessment Report.* Cambridge University Press, Cambridge and New York.

Jackson, S. T. and Weng, C. (1999) 'Late Quaternary extinction of a tree species in eastern North America', *Proceedings of the National Academy of Sciences,* vol 96, pp. 13847–13852.

Lenoir, J., Bertrand, R., Comte, L., Bourgeaud, L., Hattab, T., Murienne, J. and Grenouillett, G. (2020) 'Species better track climate warming in the oceans than on land', *Nature Ecology & Evolution*, vol 4, pp. 1044–1059.

Loarie, S. R., Duffy, P. B., Hamilton, H., Asner, G. P., Field, C. B. and Ackerly, D. D. (2009) 'The velocity of climate change', *Nature,* vol 462, pp. 1052–1055.

Marqués, L., Hufkens, K., Bigler, C., Crowther, T. W., Zohner, C. M. and Stocker, B.D. (2023) 'Acclimation of phenology relieves leaf longevity constraints in deciduous forests', *Nature Ecology & Evolution*, vol 7, pp. 198–204.

Morales, J. M. and Morán Lopéz, T. (2022) 'Mechanistic models of seed dispersal by animals', *Oikos*, e08328.

Mueller, J. M. and Hellmann, J. J. (2008) 'An assessment of invasion risk from assisted migration', *Conservation Biology,* vol 22, pp. 562–567.

Nic Lughadha, E. et al. (2020) 'Extinction risk and threats to plants and fungi', *Plants, People, Planet*, vol 2, pp. 389–408.

Normand, S., Ricklefs, R. E., Skov, F., Bladt, J., Tackenberg, O. and Svenning, J.-C. (2011) 'Postglacial migration supplements climate in determining plant species ranges in Europe', *Proceedings of the Royal Society B-Biological Sciences,* vol 278, pp. 3644–3653.

Ometto, J. P., Kalaba, K., Anshari, G. Z., Chacón, N., Farrell, A., Halim, S. A., Neufeldt, H. and Sukumar, R. (2022) 'Cross-chapter paper 7: Tropical forests'. In: H.-O. Pörtner et al. (eds.), *Climate Change 2022: Impacts, Adaptation and Vulnerability. Contribution of Working Group II to IPCC Sixth Assessment Report.* Cambridge University Press, Cambridge and New York, pp. 2369–2410.

Park, A. and Talbot, C. (2018) 'Information underload: ecological complexity, incomplete knowledge, and data deficits create challenges for the assisted migration of forest trees', *BioScience*, vol 68, 251–263.

Pomoim, N., Hughes, A. C., Trisurat, Y. and Corlett, R. T. (2022) 'Vulnerability to climate change of species in protected areas in Thailand', *Scientific Reports*, vol 12, 5705.

Rota, F., Casazza, G., Geneva, G., Midolo, G., Prosser, F., Bertolli, A., Wilhalm, T., Nascimbene, J. and Wellstein, C. (2022) 'Topography of the Dolomites modulates range dynamics of narrow endemic plants under climate change', *Scientific Reports*, vol 12, 1398.

Sandel, B., Arge, L., Dalsgaard, B., Davies, R. G., Gaston, K. J., Sutherland, W. J. and Svenning, J. C. (2011) 'The influence of Late Quaternary climate-change velocity on species endemism', *Science,* vol 334, pp. 660–664.

Sax, D. F., Early, R. and Bellemare, J. (2013) 'Niche syndromes, species extinction risks, and management under climate change', *Trends in Ecology and Evolution,* vol 28, pp. 517–523.

Stark, J. R. and Fridley, J. D. (2022) 'Microclimate-based species distribution models in complex forested terrain indicate widespread cryptic refugia under climate change', *Global Ecology and Biogeography*, vol 21, pp. 562–575.

Stork, N. E. and Habel, J. C. (2014) 'Can biodiversity hotspots protect more than tropical forest plants and vertebrates?', *Journal of Biogeography,* vol 41, pp. 421–428.

Stralberg, D., Carroll, C. and Nielsen, S. E. (2020) 'Toward a climate-informed North American protected areas network: Incorporating climate-change refugia and corridors in conservation planning', *Conservation Letters*, vol 13, e12712.

Svenning, J.-C. and Sandel, B. (2013) 'Disequilibrium vegetation dynamics under future climate change', *American Journal of Botany*, vol 100, pp. 1266–1286.

Thomson, F. J., Moles, A. T., Auld, T. D. and Kingsford, R. T. (2011) 'Seed dispersal distance is more strongly correlated with plant height than with seed mass', *Journal of Ecology*, vol 99, pp. 1299–1307.

Trakhtenbrot, A., Katul, G. G. and Nathan, R. (2014) 'Mechanistic modeling of seed dispersal by wind over hilly terrain', *Ecological Modelling*, vol 274, pp. 29–40.

Urban, M. C., Zarnetske, P. L. and Skelly, D. K. (2013) 'Moving forward: dispersal and species interactions determine biotic responses to climate change', *Annals of the New York Academy of Sciences*, vol 1297, pp. 44–60.

Wang, Y., Pineda-Munoz, S. and McGuire, J. L. (2023) 'Plants maintain climatic fidelity in the face of dynamic climate change', *Proceedings of the National Academy of Sciences*, vol 120, e2201946119.

Wu, Z.-Y., Milne, R. I., Liu, J., Nathan, R., Corlett, R. T. and Li, D.-Z. (2023) 'The establishment of plants following long-distance dispersal', *Trends in Ecology & Evolution*, vol 38, pp. 289–300.

Zani, D., Lebsten, V. and Lischke, H. (2022) Tree migration in the dynamic, global vegetation model LPJ-GM 1.1: Efficient uncertainty assessment and improved dispersal kernels of European trees', *Geoscientific Model Development*, vol 15, 4913–4940.

38

FOREST CARBON BUDGETS AND CLIMATE CHANGE

Yadvinder Malhi, Tina Christmann, Xiongjie Deng, Huanyuan Zhang-Zheng, Sam Moore and Terhi Riutta

Key concepts in forest carbon budgets

The carbon budget of a forest is the product of the balance between processes of carbon gain (photosynthesis, tree growth and carbon accumulation in soils) and processes of carbon loss (respiration of living biomass, tree mortality, microbial decomposition of litter, oxidation of soil carbon, fire, degradation and type of disturbance). These processes are influenced by a number of climatic and environmental variables, such as temperature, atmospheric carbon dioxide (CO_2) concentration, moisture availability and frequency of disturbance. These variables create large differences in carbon allocation and storage patterns between forest types such as the low-latitude humid or dry tropical forests, mid-latitude temperate forests and the high-latitude cold boreal forests, requiring that the major forest biomes need to be treated separately.

Whether in the wet tropics or the boreal regions, old-growth forests are often assumed to be mostly in a state of equilibrium, such that over a period of several years their carbon balance would be neutral. However, increasing evidence from forest plot studies has challenged this view and it is now more commonly accepted that undisturbed areas of forest also sequester carbon (i.e., be a net carbon sink), either because the demographic processes of carbon accumulation can persist for centuries or even millennia or because the ecosystems are responding to global environmental change (see below). This stored carbon in ecosystems is eventually returned to the atmosphere, either naturally, when individual trees or groups of trees die through either natural senescence and/or external environmental disturbance and then decompose, or through human-induced reasons such as logging, burning or forest clearance.

To fully understand forest carbon budgets, a more complete understanding of the processes involved in carbon gain and carbon loss is useful. The carbon budget of a forest is the quantification of carbon fluxes between atmosphere, vegetation and soil, such as photosynthesis, leaves respiration, biomass accumulation in different organs of vegetation, tree mortality, microbial decomposition of litter, etc. (see Figure 38.1). Before diving into the components of forests carbon budget and their response to environmental change, it can be helpful to clarify the terminology of natural forests carbon budget:

DOI: 10.4324/9781003324072-44

The **Gross Primary Production (GPP)** is the total amount of carbon fixed from atmospheric CO_2 into carbohydrates through the process of photosynthesis by plants in an ecosystem, such as a stand of trees in a forest. Globally, total terrestrial GPP is estimated to be about 120 Pg C yr^{-1} (Beer et al. 2010) (1 Pg = 1 Gt = 10^{15} g). GPP represents the total amount of CO_2used during plant photosynthesis.

The **Net Primary Production (NPP)** equals the difference between GPP and plant respiration (autotrophic respiration, R_a), which is carbon used to power the plants' own metabolisms. Hence:

$$NPP = GPP - R_a$$

Global NPP is often estimated to be about half of the GPP, about 60 Pg C yr^{-1} (Waring et al. 1998) though measurements from many old-growth boreal and tropical forests suggest the ratio is more often 0.3–0.4 (DeLucia et al. 2007). Hence, global NPP is in the range 35–60 Pg C yr^{-1} and global plant respiration is about 60–85 Pg C yr^{-1}. NPP could be further partitioned into volatile organic compounds (VOCs), roots exudation and **Biomass Production (BP)** (Vicca et al. 2012).

The **Net Ecosystem Production (NEP)** is the net accumulation of carbon by an ecosystem at a local spatial scale, following the loss of carbon through autotrophic and heterotrophic respiration (R_h). The sum of R_a and R_h is the **Total Ecosystem Respiration (R_{ECO}).**

$$NEP = NPP - R_h \text{ or } NEP = GPP - (R_a + R_h) \text{ or } NEP = GPP - R_{ECO}$$

Heterotrophic respiration is respiration by organisms that gain their carbon by consuming organic matter rather than producing it themselves, including fungi, bacteria and animals, predominantly through the decomposition of dead organic matter. NEP is defined to be negative if an ecosystem is losing carbon over time, for example, respiration is larger than production.

NEP was used interchangeably with **Net Ecosystem Exchange (NEE**) (Kirschbaum et al. 2001; Chapin et al. 2006), and nowadays used synonymously for the land carbon sink or terrestrial carbon sink (SLand).

If a forest stand is enmeshed with a spatial matrix of occasional local disturbance and gradual recovery, the NEP may be positive (a carbon sink) at local scale but at landscape scale this carbon accumulation may be offset by disturbance events that are discrete in space and time, such as fires, pest outbreaks or storm blowdowns. To measure the larger scale ecosystem carbon balance, we need to incorporate the spatial and temporal patterns of the disturbance regime. This concept is captured by the **Net Biome Production (NBP)**, the net carbon balance over larger spatial and temporal scales:

$$NBP = NEP - fFire - fDist - fAnth$$

where fFire is the flux associated with natural or human ignition fire, fDist is the flux associated with semi-natural disturbance processes (e.g., storm blowdowns and pathogens) that do not involve change in land use and fAnth is the flux associated with anthropogenic disturbance processes (e.g., logging, fuelwood harvest) that do not involve change in land use. NBP should equal long-term changes in total carbon reservoirs (e.g., biomass, soil carbon pools), and was referred to as 'net land carbon sink' or 'net land CO2 flux'. NBP is appropriate for describing the net carbon balance of large areas (100–1000 km^2) and longer periods of time (several years and longer). Most natural forests approximately have NBP = NEP, except in fire-prone or logging regions where disturbance creates a mosaic of forest stands of different ages.

Forest carbon fluxes and stocks vary significantly depending on different climates, productivity and disturbance regimes. However, the total flux of carbon in an ecosystem, or in a particular carbon pool of an ecosystem, gives relatively little insight into its importance as a long-term carbon store. The time of retention of carbon in a particular carbon pool of a forest ecosystem has a strong influence on that pool's importance as a carbon store. This concept is captured by the carbon **residence time**, which is defined as the equilibrium carbon pool size divided by carbon input or carbon loss. For example, the annual rate of production of leaves may be much higher than that of wood in a forest, but the short residence times of leaves (around one year) compared to woody biomass (>100 years in many trees) means that wood is a much larger and important carbon store than leaf material.

As another example, an arctic tundra and a tropical forest may store the same amount of carbon as one another (when soil carbon stocks are included), but the carbon turnover in the tropical forest may be 500 times more rapid than it is in the arctic tundra. Carbon isotope methods have been used to show that in the tropics, 85% of the ^{14}C that entered ecosystems during the era of atmospheric nuclear testing in the 1950s had been converted to humus by the early 1990s, whereas this proportion was only 50% in temperate soils and approximately 0% in boreal soils, further indicating a much more rapid turnover of soil organic carbon in the tropics than at high latitudes (Trumbore 1993; Trumbore and Harden 1997).

In assessing the carbon sequestration potential of forest ecosystems now and in the future, we need to consider the different time scales over which carbon gain is measured or estimated. These can vary from daily (solar radiation) to monthly (seasonal) to annual (length of growing season) and longer-term time scales (forest life cycle). Over the long term, the carbon sink capacity of any forest is determined by the size of the above- and belowground woody productivity, rates of input of organic material into soils, and woody and soil carbon residence times. Thus, additional carbon can be stored in an ecosystem only if more carbon is kept for the same period of time or the same amounts of carbon are kept over longer periods of time.

Measuring the forest carbon budget and its fluxes

Measuring the carbon budget of forests requires a variety of scientific methods—which can be grouped into field-based measurements that occur *in situ* and remote sensing and modelling approaches that occur at a distance.

Field-based measurements

Field-based measurements play a vital role in describing the carbon and energy budgets of ecosystems at a plot and landscape scale. Global networks of forest plots have emerged in the 1960s and 1970s for calculating the biomass, soil carbon and forest dynamics, for

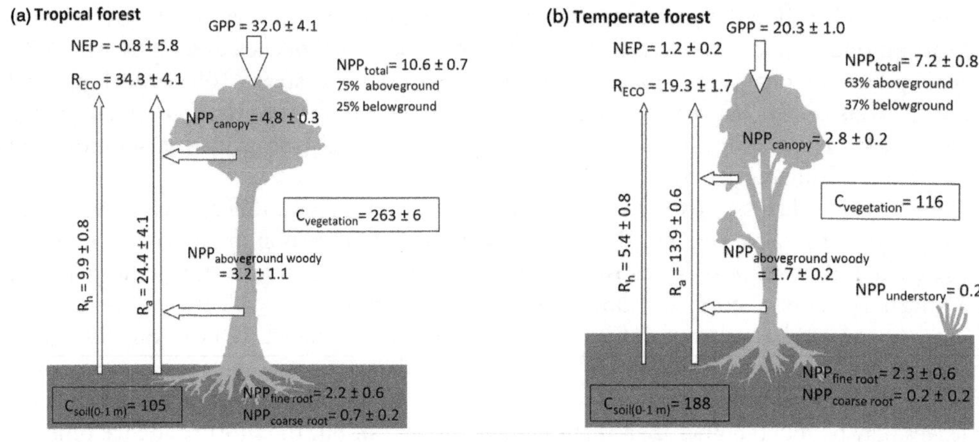

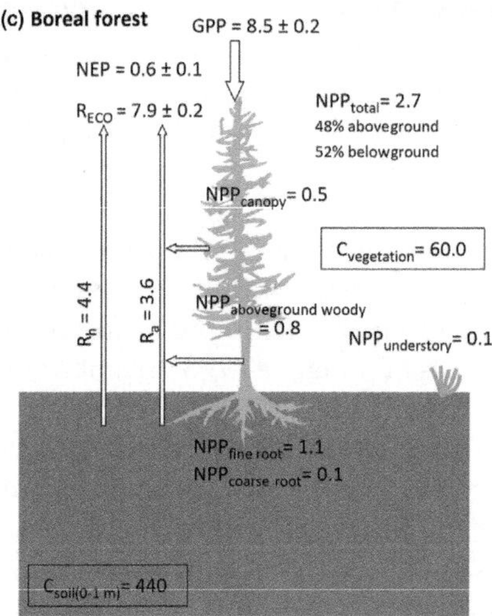

Figure 38.1 Example of vegetation and soil C pools (Mg C ha^{-1}) and C fluxes (Mg C ha^{-1} year-1) in mature, closed canopy forests in tropical, temperate and boreal zones. The values are based on data from the individual sites, not biome means. Upper panel: Humid lowland tropical forest in Amazonia (Caxiuanã, Brazil), evergreen broadleaved forest with ~200 tree species/hectare (Aragão et al. 2009; Malhi et al. 2009; Malhi 2012). Middle panel: Temperate broadleaf deciduous ancient woodland in the UK (Wytham Woods), dominated by broad-leaved Acer pseudoplatanus, Fraxinus excelsior and Quercus robur (Butt et al. 2009: Fenn et al. 2010: Thomas et al. 2011: Fenn et al. 2015). Lower panel: Evergreen, coniferous forest in Saskatchewan, Canada (BOREAS SOB site), dominated by Picea mariana, with a thick, peaty organic layer (Gower et al. 1997; Jarvis et al. 1997; Malhi et al. 1999; O'Connell et al. 2003). In the tropical site, the soil C stock was sampled to 1 m depth, but in the temperate and boreal sites only to 0.3 m (139 Mg C ha^{-1}) and 0.7 m (404 Mg C ha^{-1}) depth, respectively. The additional fraction of C down to 1 m depth was estimated based on Jobbágy and Jackson (2000).

example, the Chinese Ecosystem Research Network (Fu et al. 2010), the CTFS-ForestGEO network (Anderson-Teixeira et al. 2015; Davies et al. 2021), the Forest Plots meta-network (Malhi et al. 2002; Peacock et al. 2007; Blundo et al. 2021) and the FLUXNET network which measures the cycling of carbon, water, and energy based on eddy covariance technique (Baldocchi et al. 2001). Combining these networks provides new knowledge of the assessment of forest structure, biomass and carbon dynamics in space and time (Lewis et al. 2009; Phillips et al. 2009; Hubau et al. 2020). However, it is challenging to link results from these networks with Earth System Science (ESS) macroscopes and remote sensing (Malhi et al. 2021). The Global Ecosystem Monitoring (GEM) network (www.globalecosystemmonitoring.com) aims to fill in the gap between forest inventory data and ESS macroscopes, as well as remote sensing technologies. In late 2020, the GEM network spanned 294 plots with coverage of 178 ha across all tropical continents. GEM sites measure key components of NPP, ecosystem respiration and functional traits, which provide a unique chance to link forest field measurements and remote sensing data.

Remote sensing and modelling

The longevity and spatial coverage of remotely sensed data allows us to monitor variables related to spatiotemporal patterns of carbon stocks (biomass and soil organic carbon) and fluxes (e.g. GPP, NPP, ER, NEP and NBP) (Xiao et al. 2019). Figure 38.2 indicates different remote sensing platforms that are widely applied in carbon cycle assessment. In particular, the distribution and amount of the forest biomass have been mapped and estimated by synthetic aperture radar backscatter (Bouvet et al. 2018; Soja et al. 2021), airborne (Cao et al. 2019; Chan et al. 2021) and spaceborne (Duncanson et al. 2022) light detection and ranging, optical remotely sensed imagery (López-Serrano et al. 2019; Chen et al. 2021), and the combination of multiple sensors (Fatoyinbo et al. 2021; Pötzschner et al. 2022). Vegetation indices calculated from remotely sensed data act as proxies of GPP and NPP. GPP and NPP products derived from the moderate resolution imaging spectroradiometer satellite images have generated spatially continuous GPP and NPP estimations at global scale since 2000 (Running et al. 2004). Recently, solar-induced chlorophyll methods present new potential for assessing GPP (Bai et al. 2022).

Carbon stocks and flows of tropical, temperate and boreal forests

Nature and extent of intact and human-modified forests

Forests cover 30% (3720 million hectares [Mha]) of global land area, with tropical forests being the most extensive (44%), followed by boreal (32%) and temperate (24%) forests (Food and Agriculture Organization 2010). Pre-agricultural Holocene forest area was estimated at 5680 Mha, of which 65% remains today (Goldewijk 2001). Rapid and extensive deforestation is a recent phenomenon driven by modern technology—until 300 years BP, for instance, only 7% of the global forest area was lost (Goldewijk 2001; Williams 2000; Malhi et al. 2002 and references therein). Nowadays, over 50% of the original forest area in the temperate zone has been deforested or fragmented, nearly 25% in the tropics, but only 4% in the boreal zone (Wade et al. 2003). Temperate deforestation peaked in the late 19th and early 20th century, but has stabilised and increased in many areas due to agricultural abandonment and reduced biomass energy use (Food and Agriculture Organization 2010).

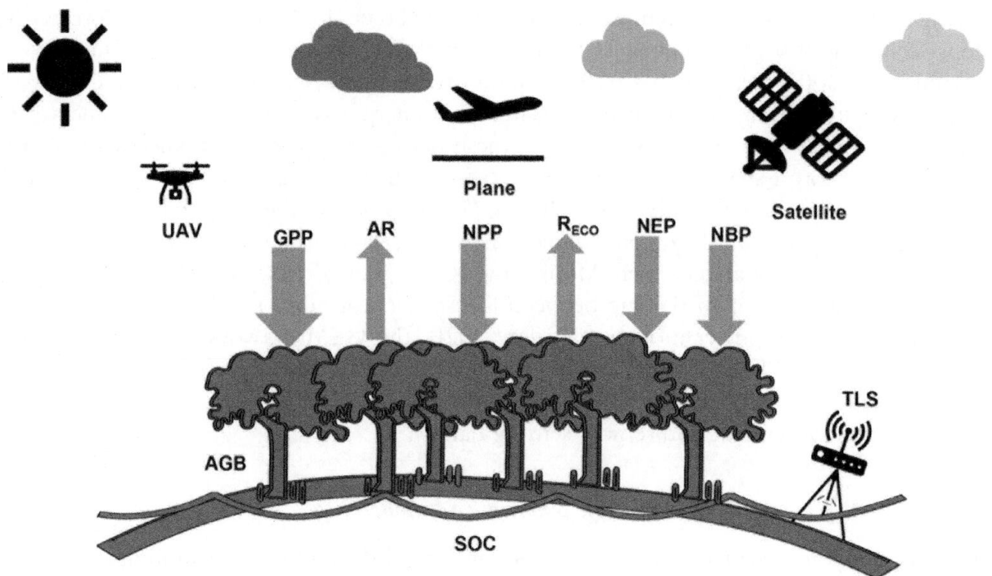

Figure 38.2 Terrestrial laser scanning (TLS), unmanned aerial vehicle (UAV), plane and satellite remote sensing technologies are part of a remote sensing system able to quantify carbon fluxes (gross primary production (GPP), autotrophic respiration (AR), net primary production (NPP), ecosystem respiration (RECO), net ecosystem production (NEP) and net biome production (NBP)) and stocks (aboveground biomass (AGB) and soil organic carbon (SOC)).

Tropical deforestation accelerated in the late 20th century and remains high, with the highest rates in Southeast Asia and Latin America (Waring et al. 1998). Boreal forests have faced less land use pressure, with most of the original area remaining.

Many remaining forests are heavily human-modified, with only 36% of the remaining forest area in a relatively undisturbed state (Food and Agriculture Organization 2010), even less than this (24%) in large, >500 km^2 unfragmented patches (Potapov et al. 2008). In the temperate zone, only 6% of the remaining forest area is relatively undisturbed (Hannah et al. 1995). Almost all northern hemisphere temperate forests were subject to management practices that reduced the forest carbon stocks (harvesting, litter raking and conversion to agricultural land) up to the mid-20th century (Goodale et al. 2002). In the boreal zone, 44% of the forest area is relatively undisturbed, most of this in North Western Canada and Russia (Potapov et al. 2008; Bradshaw et al. 2009). The wood production is sustainable in terms of stable yields and biomass stocks, although not necessarily in terms of biodiversity and ecosystem functioning (Paillet et al. 2010). Approximately 30% of the tropical forests area is relatively intact, most of this in Amazonia or the Congo Basin (Potapov et al. 2008).

Carbon stocks in vegetation and soil

Globally, the main carbon reservoirs are the atmosphere (870 PgC), vegetation (450 PgC), soils (1700 PgC) and permafrost (1200 PgC) (Table 38.1). Tropical forests account for most of the global forest biomass, while more than half of the global soil C is stored in the boreal region (Table 38.1). However, some of the world's most carbon dense (high biomass per area) forests are old-growth forests located in the temperate zone in areas such as the Pacific Northwest, and southern hemisphere temperate zones in Chile, Australia and New Zealand (Keith et al. 2009; Pan et al. 2013). Most of the vegetation C, approximately 60–80% in mature forests, is above-ground (Mokany et al. 2006). For a given shoot biomass, the mean root:shoot ratio is relatively invariant across the forest biomes (Mokany et al. 2006). However, the root:shoot ratio decreases with increasing shoot biomass. Therefore, the root:shoot ratio in mature forests is highest in the boreal zone where the mean shoot biomass is lowest, and the ratio decreases towards the tropics (Mokany et al. 2006; Qi et al. 2019; Ye et al. 2021).

Globally, soils contain more C (in the range of 1800 Pg C) than vegetation, atmosphere or surface oceans. Forest soils constitute approximately half of the soil C pool (about 900 Pg C). The estimates of the global soil C pools are, however, uncertain, due to the high spatial variability and lack of reliable remote sensing techniques (Scharlemann et al. 2014; Jobbágy and Jackson 2000; Scharlemann et al. 2014). In forests, on average, about 25–35% of the carbon is contained in vegetation and 65–75% in soil (Table 38.1). However, there are marked differences among biomes. In tropical forests, typically 42% of the carbon is in vegetation and 58% in the soil, whereas in boreal forests, 10% of the carbon is in vegetation and 90% in the soil. Temperate forests are intermediate. This latitudinal pattern is mainly driven by temperature, which limits the organic matter decomposition and nutrient cycling rates in colder climes more than it limits productivity, resulting in accumulation of soil organic matter. Approximately 30% of the soil C is below the top 1m depth (Jobbágy and Jackson 2000), although this layer is not included in most soil C estimates.

Productivity, carbon fluxes and residence time

Productivity of the forest biomes follows the latitudinal and temperature gradient: both GPP and NPP decrease from the tropics towards the colder regions, particularly so in conifer-dominated boreal forests (Figure 38.1 and Table 38.1). This decline appears more driven by extension of the dormant season (winter) than by temperature effects on plant physiology in the growing season (Malhi 2010). However, as both autotrophic and heterotrophic respirations follow this same pattern of latitudinal decline, NEP in the three forest types is similar (Fig 38.1). NPP in most temperate forests is only slightly lower than in tropical forests, and they are a bigger C sink per unit area (Table 38.1 and Figure 38.1). The current NEP in the temperate zone is probably due to the predominance of young age classes, as the forests are recovering from past disturbance, and due to the positive response to the high nitrogen deposition in large parts of the region (Magnani et al. 2007).

The NPP allocation to above- and below-ground varies with latitude: tropical forests allocate approximately 75% of NPP to aboveground components compared with 63% in temperate forests and 54% in boreal forests (Figure 38.1). The mean residence time of carbon in the vegetation components is similar across biomes, about 1–3 years in leaves and fine roots, and 30–100 years in wood (Malhi et al. 1999). The residence time of soil carbon is similar to that of vegetation in tropical and temperate forests, but much longer (a century) in the boreal zone (Malhi et al. 1999). There are likely large differences between soil layers, with the inactive deep soil carbon having a residence time of more than a millennium and the labile topsoil carbon being circulated with a much higher turnover rate.

Table 38.1 Forest area and forest C stocks and fluxes in tropical, temperate and boreal zones

	Tropical	*Temperate*	*Boreal*	*Global*[a]	*Reference*
Area (Mha)[b]	1620	900	1200	3720	Food and Agriculture Organization (2010)
C density, vegetation (Mg C ha^{-1})	124	57	40	81	Dixon et al. (1994); Saatchi et al. (2011); Thurner et al. (2014); Erb et al. (2018)[e]
C density, soil (Mg C ha^{-1}), 0–1 m depth + 1–3 m depth	107 + 63	128 + 44	276 + 95	167 + 69	See footnote[c]
Biome C stock, vegetation (Pg C)	201	51	48	300	Calculated as vegetation C density × Area
Biome C stock, soil (Pg C), 0–1 m depth + 1–3 m depth	173 + 102	115 + 40	331 + 114	620 + 256	Calculated as soil C density × Area
NPP (Mg C ha^{-1} $year^{-1}$)	8.4	7.4	2.7	6.3	Luyssaert et al. (2007)
NEP (Mg C ha^{-1} $year^{-1}$)	0.6	0.9	0.4	0.6	Calculated as NBP/Area
Biome NPP (Pg C $year^{-1}$)	13.6	6.7	3.2	23.5	Calculated as NPP × Area
NBP (Pg C $year^{-1}$)	1.0±0.5[d]	0.8±0.1	0.5±0.1	2.3±0.5	Pan et al. (2011)
C loss from forest degradation and deforestation (Pg C $year^{-1}$)	1.1±0.8	0	0	1.1±0.8	Ciais et al. (2013)

Notes

[a] Sum or area-weighted mean.

[b] Definition of forest: Land spanning more than 0.5 ha with trees higher than 5 metres and canopy cover >10%, or trees able to reach these thresholds *in situ*.

[c] Calculated by combining global soil organic carbon map (Hiederer and Köchy 2011; Panagos et al. 2012) and FAO maps of global land distribution by land cover type and occurrence of forest, which are part of the Food Insecurity, Poverty and Environment Global GIS database (FGGD). www.fao.org/geonetwork/srv/en/main.home#landcover; www.fao.org/geonetwork/srv/en/metadata.show?id=14066). Estimates of the soil C stock at 1–3 m depth are based on Jobbágy and Jackson (2000).

[d] Estimate for low disturbance forests not subjected to degradation or clearance, equivalent to SLand in Global Carbon Project (Friedlingstein et al. 2022).

[e] Note that there is a lack of consistency in the literature on tropical forests biomass estimation.

Interannual variability in productivity and C fluxes is high in all biomes, largely driven by weather variations. Extreme weather events, such as heat waves and droughts, may switch these ecosystems from C sinks to C sources (Ciais et al. 2005; Gatti et al. 2014), indicating that the forest carbon dynamics are sensitive to changes in climate. Most of the interannual variability in the global terrestrial carbon cycle is driven by tropical vegetation, with warm El Niño years associated with carbon sources in the tropics (caused by a combination of higher temperatures and, in some regions, drought and increased fire activity). The interannual variability of the tropical terrestrial carbon cycle appears to have increased in recent years (Wang et al. 2014).

Boreal forests have an extreme seasonal pattern, with a clear growing season and a dormant season. The seasonality is governed by day length, which in the northernmost parts of the boreal zone ranges from 0 to 24 hours, and temperature, which ranges from sub-zero temperatures in the winter to high twenties in the summer. Medium-term (five day) mean air temperature in the spring is the most important trigger of the onset of photosynthesis in the spring (Suni et al. 2003). Because of the long days in the summer, the diurnal variation in carbon fluxes is smaller than in the lower latitudes. Temperate forests in the mid-latitudes vary in their degree of seasonality, both because of the variation in day length within this large zone and because the maritime regions (particularly on the western edge of the continents where predominant winds come from the oceans) have far less pronounced variation in temperature than the continental regions. In maritime climates, evergreen coniferous trees can photosynthesise throughout the year, although the short day length during the winter limits their productivity. In the more continental regions with larger temperature variation between summer and winter, the seasonal pattern is similar to that in boreal forests, but the growing season is longer, and the day length during the growing season is shorter.

In tropical forests, seasonal variation in productivity and carbon dynamics is linked to rainfall patterns, and the direction and magnitude of the change in productivity between wet and dry seasons depend on the severity of the water stress. Regions with strong dry seasons, such as South-Eastern Amazonia, show a distinct decline in productivity due to water limitations (Restrepo-Coupe et al. 2013). On the other hand, in regions where water does not become a limiting factor during the dry season, such as Equatorial Amazonia and Borneo, seasonal variation in productivity is modest and the difference in cloudiness (and therefore solar radiation) between seasons may be a more important factor behind seasonal patterns than the difference in moisture (Kho et al. 2013; Restrepo-Coupe et al. 2013).

Disturbance regime

Disturbance is the process that distinguishes NEP from NBP and is a source of carbon to the atmosphere. The scale of natural disturbances varies in space and time, which can mean that they are difficult to capture through localised forest census or eddy covariance approaches and are often better described through remote sensing approaches. Typical disturbances include windthrows, natural or human-induced crown and surface fires, insect outbreaks and droughts (Figure 38.3).

Disturbances can have a large effect on the carbon dynamics of the site by reducing stock and altering productivity. Immediately after disturbance, productivity is low due to the diminished photosynthesising biomass, but it can recover quickly and NPP can even exceed pre-disturbance values. Disturbance typically leads to a release of carbon from the vegetation and soil C stocks, and that, coupled with a decrease in GPP, leads to an initial decrease in net carbon uptake, possibly turning the system into a carbon source. NPP and NEP are highest in forests that have started to recover from disturbance, and decline when the forest ages and canopy closes (Magnani et al. 2007).

In regions that lack regular large-scale disturbances such as cyclones or extensive wildfires, the effect of disturbances on the landscape scale carbon balance is dominated by the small-scale disturbances, due to their high frequency. Espírito-Santo et al. (2014) showed that in the Amazon basin, small-scale (<0.1 ha) disturbances account for 88% of the carbon lost due to natural disturbances, while intermediate (0.1–5) and large (>5 ha) scale disturbances account for 13% and 0.2%, respectively. Small gap disturbance rate, resulting from individual tree fall events, is similar in the three biomes, with a return time of approximately 100 years (Hiura et al. 1996; Masek et al. 2008; Espírito-Santo et al. 2014). Intermediate and large-scale disturbances have a return time ranging from a century to greater than 10,000 years, the largest disturbances being extremely infrequent (Hiura et al. 1996; Espírito-Santo et al. 2014).

Montane forests

Approximately 2.4% of the world's forests are montane forests (Schmitt et al. 2009). Forests in high altitudes have different characteristics from lowland forests. Temperature decreases towards higher altitudes, and the change in the forest carbon stocks and dynamics along an elevational gradient shares many features with the change along the latitudinal gradient from tropical to boreal zone, even though the spatial scales are very different, the former variation taking place within a hundred kilometres and the latter across thousands. Temperature is the dominant but not sole environmental variable that varies with elevation, and changes in precipitation, cloud immersion and UV radiation can also be important. Vegetation C stock and productivity typically decrease with elevation (Girardin et al. 2014), while soil C stock in the organic layer (although not in the mineral layer) increases (Leuschner et al. 2007; Girardin et al. 2010). Productivity decreases, and allocation to belowground often increases along with elevation. As an example, in an altitudinal transect in Peruvian Andes, at a lowland humid tropical forest plot at 194 m asl (mean annual temperature 26.4°C), the total NPP is 13.6 Mg C ha^{-1} $year^{-1}$ (Girardin et al. 2010). At the same latitude but at 3,025 m asl in a montane cloud forest (mean annual temperature 12.5°C), the total NPP is 5.09 Mg C ha^{-1} $year^{-1}$. Soil carbon stocks can be very high in tropical montane forests (> 100 Mg C ha^{-1}; (Gibbon et al. 2010)), but are often stored in shallow soils (< 1 m depth) that are vulnerable to loss through erosion and warming. In a warming world, montane forests are likely to be sensitive to climate change, and may turn into net carbon sources as rates of decomposition increase.

Peatland forests

Approximate 104 Mha, or 3%, of the world's forests, are on peat soils, of which about 55% in boreal zone and 45% in the tropics (Zoltai and Martikainen 1996; Page et al. 2011; Pan et al. 2013). The area estimates are, however, scarce and highly uncertain, and new tropical peatlands have recently been described in Amazonia (Lähteenoja et al. 2012) and in the Congo Basin. Excessive moisture and suitable vegetation (high productivity and/or recalcitrant litter quality) are the prerequisites for peat formation. In such environments, litter and coarse woody debris production rate exceeds decomposition rate, which is impeded by anoxic conditions, and the partially decomposed organic matter accumulates as peat. The soil C content of these ecosystems is typically high, with values of about 1810 Mg C ha^{-1} in the northern latitudes (Korhola et al. 1995; Borren et al. 2004; Akumu and McLaughlin 2013) and 2010 Mg C ha^{-1} in the tropics (Page et al. 2011). Tropical peat forests are estimated to hold 70–130 Pg C (Mitchard 2018). Therefore, although peatland forests globally make up a small proportion of the total forest area, they constitute a significant fraction of the total forest soil C pool (at least 22% or 197 Pg; the figures are likely to be even higher given the recent discovery of new forested tropical peatland areas). Vegetation C stock and productivity in peatland forests are similar to the forests on mineral soil in the same region (Bond-Lamberty and Gower 2008; Peregon et al. 2008; Hirano et al. 2012). In contrast to forests on mineral soil, which are small methane (CH_4) sinks, pristine forested peatlands are CH_4 sources to the atmosphere, emitting 0.1–0.2 Mg CH_4-C ha^{-1} $year^{-1}$ (Minkkinen et al. 2002; Couwenberg et al. 2010). However, the CH_4 emissions from forested peatlands are approximately five times lower than emissions from low tree cover and open peatlands, and contribute only about 10% to the total emissions from natural wetlands. The C balance of peatland ecosystems is highly sensitive to human disturbance and changes in climate (Page et al. 2002 Ise et al. 2008; Frolking et al. 2011; Moore et al. 2013), and because of their large soil C pool, they have a potential to turn into substantial C sources. However, these high carbon ecosystems, along with other wetlands, have only recently become a focus of coherent monitoring efforts (IPCC 2014).

Forests carbon dynamics and global change

Forests in the context of the global carbon budget

Since 1750, continuously increasing anthropogenic CO_2 emissions and land use change have perturbed the carbon cycle. Over the period 1750–2011, fossil fuel emissions and cement production have emitted 375±30 Pg C to the atmosphere, with a further 180±80 Pg C emitted by net land use change (mainly deforestation). Of these CO_2 emissions, only 43% (240±10 Pg C) have remained as an observed net increase in atmospheric CO_2 concentrations, with 28% (155±30 Pg C) absorbed by the oceans and 29% (160±90 Pg C) by the terrestrial biosphere (Ciais et al. 2013). In the decade 2002–2011, mean annual fossil fuel and cement production emissions (8.3±0.7 Pg C yr^{-1}) and net land use change emissions dominated by tropical deforestation (0.9±0.8 Pg C yr^{-1}) were apportioned between a 47% (4.3±0.2 Pg C yr^{-1}) atmospheric increase, a 26% (2.4±0.7 Pg C yr^{-1}) ocean sink and a 27% (2.5 ± 1.3 Pg C yr^{-1}) land biosphere sink (Ciais et al. 2013).

Hence, there is a large carbon sink in the land biosphere away from areas of deforestation and degradation. Forests are the major contributor to this sink as the largest components of terrestrial biomass and productivity because trees hold much more carbon per unit area than other types of vegetation. The highest overall rates of total carbon accumulation (considering both belowground and aboveground carbon) are found in the wetter and warmer forests in the tropics, while belowground carbon accumulation does not show a clear pattern between biomes (Cook-Patton et al. 2020). Tropical forests are estimated to gain between 2 and 3.5

tC/ha/year, with the highest carbon gains found in Southeast Asia (Sullivan et al. 2020). If this land biosphere carbon sink would not be not present, and this excess carbon was instead allocated to ocean and atmospheric pools at the current fractions, atmospheric CO_2 would rise at a rate 37% higher than the one currently observed; the global biosphere "brakes" the rate of atmospheric CO_2 rise. Forests are certain to be a large part of this "brake", and one of the major questions in forest carbon cycle ecology is to understand the nature, distribution and stability of this climate change brake.

Pan et al. (2011) presented a synthesis of forest inventory data and long-term carbon cycle studies to estimate a global forest carbon sink (over the period 2000–2007) of 2.30±0.49 Pg C yr^{-1} (i.e. forests account for almost all of the biosphere carbon sink) partitioned between 22% in boreal forests (0.50±0.08 Pg C yr^{-1}), 34% in temperate forests (0.78±0.09 Pg C yr^{-1}) and 44% in intact tropical forests (1.02±0.47 Pg C yr^{-1}). There is a further sink in regrowing tropical forests, but this is closely entwined with gross tropical deforestation and incorporated in the net tropical land use change emission of 0.9±0.8 Pg C yr^{-1}. Between 1984 and 2018, for instance, regrowing degraded and secondary forests in the Amazon, Borneo and Central Africa accumulated between 90 and 130 Tg C yr^{-1} (Heinrich et al. 2023). However, this counterbalanced only 26% of the carbon emissions from tropical forest loss during the period (Heinrich et al. 2023).

Driving factors of forest carbon dynamics

Next, we explore why forests appear to be the cause of such a large net carbon sink. In principle, mature forest stands are close to carbon balance (net ecosystem productivity ~0), with the growth of individual trees being balanced by the death and decomposition of other stands, although there is evidence that some forests can continue to accumulate biomass for millennia. At a landscape scale, the continued growth of some stands is balanced by the degradation of other stands through natural disturbance, leading to expectations of a net biome productivity close to zero for mature landscapes away from anthropogenic disturbance.

In practice, many forest stands are not in carbon balance. Following a disturbance event (whether natural or anthropogenic), forests are short-term carbon sources (for up to several years) as dead organic material decomposition dominates. Eventually, regrowing forests become large carbon sinks as new biomass accumulation and recruitment of trees dominates over mortality and decomposition (Cook-Patton et al. 2020).

There are a variety of stochastic disturbance events, such as insect and pathogen attacks, wildfires, wind throws and extreme weather events which affect forest carbon budgets, both causing carbon release and subsequent carbon gain (Figure 38.3). Further, spatially extensive past disturbance events at broad scale can lead to coherent regrowth and recovery and a large-scale carbon sink. Examples include extensive drought or fire caused by a large-scale weather event such as El Niño, or forest recovery after extensive deforestation followed by agricultural abandonment, as is the case of much of Eastern North America, the Tropical Andes and parts of Europe, or simply a reduction in the intensity of human use of forests caused by population collapse or rural emigration. The latter has been speculated for some tropical regions such as Amazonia following the arrival of European colonists and diseases in the Americas, or West and Central Africa following the peak of the slave trade. Conversely, gradual intensification of human use and degradation of forests for fuelwood or timber can result in long-term carbon sources.

A further agent of change in the carbon balance which can catalyse processes leading to either carbon losses or carbon gains is atmospheric change, encompassing shifts in rainfall and moisture regimes, atmospheric gases and temperatures. Shifts in local rainfall regimes can cause gradual loss or increase in biomass, although sometimes in unexpected directions. Fauset et al. (2012) demonstrated that long-term drying in Ghana caused an increase in forest biomass (i.e. a net biomass carbon sink) because of increasing dominance of fast-growing, high biomass species. Increases in temperature could potentially increase growth rates in temperature-constrained regions such as high-latitude or montane forests, through direct effects on growth rate, through extension of the growing season, or indirectly through increasing litter decomposition and nutrient recycling (Jarvis and Linder 2000). However, warming can also cause soil organic matter decomposition rates to increase, resulting in net carbon emissions from soils and necro mass that could offset or exceed the carbon accumulation in live biomass. Climate change could also cause a shift in disturbance regimes (e.g., the frequency of blowdowns, fires, insect outbreaks or droughts), leading to either net carbon emissions or uptake depending on whether disturbance intensity and frequency increase or decrease. In the Brazilian Amazon, for instance, exacerbated droughts have been shown to cause forest fires which are decoupled from deforestation fires. These "degradation" fires have caused tree mortality and associated carbon losses of up to 1600 Tg CO_2/year in heavy drought years (Aragão et al. 2018).

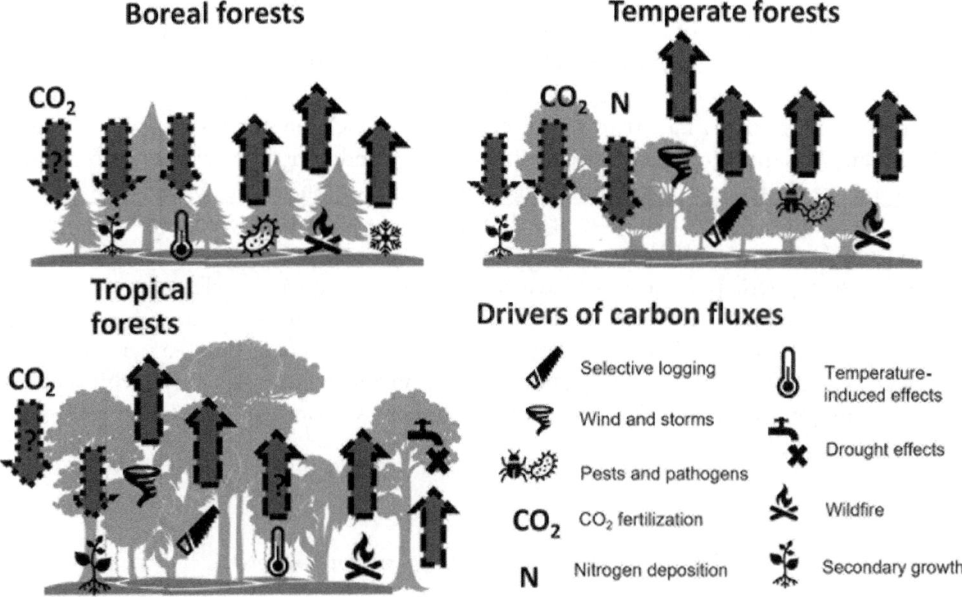

Figure 38.3 General scheme of how disturbances, forest regrowth and atmospheric changes at a forest stand level impact local carbon flux. A carbon gain is indicated by an upward arrow and a carbon loss by a downward arrow. Note that many of these disturbance factors do interact, e.g., dry conditions promote fire which in turn creates gaps in the canopy and promote regrowth. Question marks indicate unclear relationships. Tree shapes from "brgfx / Freepik".

Elevated atmospheric CO_2 can increase rates of photosynthesis, leading to an effect known as CO_2 fertilisation. In the last century, vegetation productivity has increased because of CO_2 fertilisation in all major forest biomes, with the biggest rise expected in tropical forests where GPP is modelled to have risen from 32 PgC in 1900 to 45 PgC in 2020 (Haverd et al. 2020). The evidence for a corresponding rise in biomass productivity over this period is sparse, however. The effect of CO_2 fertilisation is greatest in the tropics because of a greater sensitivity of photosynthesis to CO_2 at high temperatures (Hickler et al. 2008). However, a key uncertainty is whether plants are able to sufficiently utilise such an abundance of CO_2 by accessing other limiting nutrients such as nitrogen and phosphorus. A number of free air carbon dioxide enrichment experiments have been conducted on forests, all of them on temperate secondary forests or plantations. Experiments have shown that most C3 plants (all trees) respond to elevated concentrations of CO_2 with increased rates of photosynthesis, increased productivity and increased biomass (Norby et al. 2005). A range of studies have demonstrated 20–30% increases in biomass in response to elevated CO_2 (Luo et al. 2006). However, this increase is not universal (Korner et al. 2005) and despite the initial increases in productivity observed in trees under elevated concentrations of CO_2, experiments at the ecosystem level and experiments longer than a few years suggest much reduced responses. Furthermore, productivity is not equivalent to carbon storage. If an increase in productivity is in tissues with a rapid turnover (e.g. fine roots and foliage), the enhanced growth may be respired within a year or two, leading to little or no gain in carbon storage. Moreover, higher concentrations of CO_2 enable plants to acquire the same amount of carbon, with a smaller loss of water through their stomata. This increased water-use efficiency reduces the effects of drought. Higher levels of CO_2 may also alleviate other stresses of plants, such as high temperatures. The observation that productivity increases relatively more in low productivity years suggests that the indirect effects of CO_2 in ameliorating stress may be more important than the direct effects of CO_2 on photosynthesis (Luo et al. 1999). While N-deposition, CO_2 fertilisation and increased temperatures may increase plant growth, recent evidence suggests that there may be trade-offs: faster growth could cause higher tree mortality due to shorter tree life spans and reduce or even reverse long-term carbon gains (Brienen et al. 2020).

Finally, atmospheric pollution can have either positive or negative effects on atmospheric carbon balance. In many temperate regions, huge increases in nitrogen deposition (Galloway et al. 2008) are thought to have stimulated forest growth rates and caused a net carbon sink (Magnani et al. 2007). On the other hand, excessive nitrogen is also associated with acid rain, which may inhibit forest growth. Other pollutants such as ozone can also inhibit leaf photosynthesis and forest carbon uptake.

Future trajectories in forest carbon budgets

As the 21st century proceeds the forces of global change will continue to strongly influence the carbon budget and balance of tropical forests, and in return forest ecosystems will have a strong influence on atmospheric CO_2 concentrations and global climate. The future carbon budget of forests will depend on the future trajectories of climate and land use change and associated agents of disturbance, the global economy, and changes in policies and land management.

Deforestation and degradation

Loss of forests through direct deforestation and logging and degradation of forests continue to be substantial sources of carbon, particularly in tropical regions (Malhi 2010; Aragão et al. 2018). South America has seen large increases in deforestation largely due to cutting of the Amazon forest to make space for large-scale agriculture, causing carbon emission of 373 TgC in the period 2001–2019 (Feng et al. 2022). Likewise, in Southeast Asia deforestation has been driven by large-scale agriculture, but for oil palm plantations. In other tropical forests, particularly in Africa and the South American Andes, pastoralism, small-scale agriculture and mining keep encroaching and degrading many remnant natural forests. Whether these deforestation trends prevail will largely depend on the creation of pro-forest policies, engagement with local communities and stakeholders and on enforcement of legal action against forest encroachment.

Fire and other disturbances

Fire regimes have a strong influence on the long-term carbon balance. In many thinly populated tropical forest, boreal forest and semi-arid regions, increasing settlement and population density may increase fire return frequencies, decrease forest recovery between fire events and result in net emissions of carbon. On the other hand, in many long-settled and prosperous regions, such as North America, the Mediterranean and Australia, active fire suppression leads to increases in vegetation biomass, albeit this is vulnerable to extreme fire events because of the accumulated fuel loads. Climate change may also change the frequency and intensity of other forms of disturbance, such as extreme storm events, tropical cyclones and extreme drought or precipitation events, although the details and dynamics of such changes are hard to predict. The Anthropocene is characterised by multiple drivers of change acting in synchrony. The interaction of increasing drought frequency, forest degradation and fire pressure may lead to more sustained degradation and loss of forest, with net carbon emissions (Malhi et al. 2009; Brando et al. 2014).

Atmospheric and climate change

The impacts of atmospheric change on the carbon balance forests are one of the key uncertainties that will determine the CO_2 concentration in the atmosphere at the end of the century. Most ecophysiological models suggest that rising CO_2 will continue to increase photosynthesis rates, tree growth rates and carbon stocks (Ciais et al. 2013). However, this sink is likely to slow down and possibly reverse because of saturation of photosynthesis, feedbacks that increase tree mortality rates, or increased carbon emissions from warming soils. A state where large forested regions enter a positive feedback mode induced by warming and precipitation changes could be regarded as a tipping point, a dangerous threshold that needs to be avoided. Some studies suggest that the Amazon Forest has already passed such a tipping point (Nepstad et al. 2008; Boers et al. 2017).

Second growth

Over the 20th century, many temperate forest regions have passed through a "forest transition" whereby forests have regenerated naturally following a reduction of pressure on forests and increase in forest biomass and area, as marginal and uneconomic agricultural areas are abandoned, forests are no long a source of wood fuel, and as economies move to a more industrial base and intensify agriculture. Over the 21st century, parts of the tropics are also likely to experience such a transition, and there is evidence that it is underway in countries such as Costa Rica and across the Andes. Such "second growth" forests are a substantial carbon sink (Cook-Patton et al. 2020) and will become of increasing importance for regional ecosystem functioning, including carbon dynamics. If degraded and regrowing secondary tropical forests in the Amazon, Borneo and Central Africa are conserved, this could create a future carbon sink of 44–62 Tg C year^{-1} (Heinrich et al., 2023).

The era of forest restoration

The decade between 2020 and 2030 has been proclaimed as the UN Decade on Ecosystem Restoration. As part of this, international efforts across all the major forest biomes have begun to restore forest area, structure and composition. Forest restoration using active restoration methods (such as tree planting, direct seeding, etc.) is experiencing a particular boom due to its potential for carbon sequestration and its appeal for the carbon offsetting sector. If planned and implemented well, such a global wave of forest restoration presents a large opportunity to increase the terrestrial carbon sink along with providing biodiversity and livelihood improvements.

References

Akumu, C. E. and McLaughlin, J. W. (2013) 'Regional variation in peatland carbon stock assessments, northern Ontario, Canada', *Geoderma,* vol 209–210, pp. 161–167.

Anderson-Teixeira, K. J., Davies, S. J., Bennett, A. C., Gonzalez-Akre, E. B., Muller-Landau, H. C., Joseph Wright, S., ... & Zimmerman, J. (2015) 'CTFS-Forest GEO: A worldwide network monitoring forests in an era of global change', *Global Change Biology*, vol 21, pp. 528–549.

Aragão, L. E. O. C., Anderson, L. O., Fonseca, M. G., Rosan, T. M., Vedovato, L. B., Wagner, F. H., Silva, C. V. J., Silva Junior, C. H. L., Arai, E., Aguiar, A. P., Barlow, J., Berenguer, E., Deeter, M. N., Domingues, L. G., Gatti, L., Gloor, M., Malhi, Y., Marengo, J. A., Miller, J. B., Phillips, O. L. and Saatchi, S. (2018). '21st Century drought-related fires counteract the decline of Amazon deforestation carbon emissions', *Nature Communications*, vol 9, 536.

Aragão, L. E. O. C., Malhi, Y., Metcalfe, D. B. and 23 other authors. (2009) 'Above- and below-ground net primary productivity across ten Amazonian forests on contrasting soils', *Biogeosciences*, vol 6, pp. 2759–2778.

Bai, J., Zhang, H., Sun, R., Li, X., Xiao, J. and Wang, Y. (2022) 'Estimation of global GPP from GOME-2 and OCO-2 SIF by considering the dynamic variations of GPP-SIF relationship', *Agricultural and Forest Meteorology*, vol 326, 109180.

Baldocchi, D., Falge, E., Gu, L., Olson, R., Hollinger, D., Running, S., ... and Wofsy, S. (2001). 'FLUXNET: A new tool to study the temporal and spatial variability of ecosystem-scale carbon dioxide, water vapor, and energy flux densities', *Bulletin of the American Meteorological Society*, vol 82, pp. 2415–2434.

Beer, C., Reichstein, M., Tomelleri, E. and 21 other authors. (2010) 'Terrestrial gross carbon dioxide uptake: Global distribution and covariation with climate', *Science,* vol 329, pp. 834–838.

Blundo, C., Carilla, J., Grau, R., Malizia, A., Malizia, L., Osinaga-Acosta, O., ... and De Araujo, R. O. (2021) 'Taking the pulse of Earth's tropical forests using networks of highly distributed plots', *Biological Conservation*, 260, 108849.

Boers, N., Marwan, N., Barbosa, H. M. J. and Kurths, J. (2017) 'A deforestation-induced tipping point for the South American monsoon system', *Scientific Reports*, vol 7, 41489.

Bond-Lamberty, B. and Gower, S. T. (2008) 'Decomposition and fragmentation of coarse woody debris: Re-visiting a boreal black spruce chronosequence', *Ecosystems*, vol 11, pp. 831–840.

Borren, W., Bleuten, W. and Lapshina, E. D. (2004) 'Holocene peat and carbon accumulation rates in the southern taiga of western Siberi', *Quaternary Research*, vol 61, pp. 42–51.

Bouvet, A., Mermoz, S., Le Toan, T., Villard, L., Mathieu, R., Naidoo, L. and Asner, G. P. (2018) 'An above-ground biomass map of African savannahs and woodlands at 25 m resolution derived from ALOS PALSAR', *Remote Sensing of Environment*, vol 206, pp. 156–173.

Bradshaw, C. J. A., Warkentin, I. G. and Sodhi, N. S. (2009) 'Urgent preservation of boreal carbon stocks and biodiversity', *Trends in Ecology & Evolution*, vol 24, pp. 541–548.

Brando, P. M., Balch, J. K., Nepstad, D. C., Morton, D. C., Putz, F. E., Coe, M. T., Silerio, D., Macedo, M. N., Davidson, E. A., Nobrega, C. C., Alencar, A. and Soares-Filho, B. (2014) 'Abrupt increases in Amazonian tree mortality due to drought–fire interactions', *Proceedings of the National Academy of Sciences*, vol 111, pp. 6347–6352.

Brienen, R. J. W, Caldwell, L., Duchesne, L., Voelker, S., Barichivich, J., Baliv, M., Ceccantini, G., Di Filippo, A., Helama, S., Locosselli, G. M., Lopez, L., Piovesan, G., Schongart, J., Villalba, R. and Gloor, E. (2020) 'Forest carbon sink neutralized by pervasive growth-lifespan trade-offs', *Nature Communications*, vol 11, 4241.

Butt, N., Campbell, G., Malhi, Y., Morecroft, M., Fenn, K. and Thomas, M. (2009) *Initial Results from Establishment of a Long-Term Broadleaf Monitoring Plot at Wytham Woods, Oxford, UK*. University of Oxford Report. University of Oxford, Oxford.

Cao, L., Coops, N. C., Sun, Y., Ruan, H., Wang, G., Dai, J. and She, G. (2019) 'Estimating canopy structure and biomass in bamboo forests using airborne LiDAR data', *ISPRS Journal of Photogrammetry and Remote Sensing*, vol 148, pp. 114–129.

Chan, E. P. Y., Fung, T. and Wong, F. K. K. (2021) 'Estimating above-ground biomass of subtropical forest using airborne LiDAR in Hong Kong', *Scientific Reports*, vol 11, 1751.

Chapin, F. S., Woodwell, G. M., Randerson, J. T., Rastetter, E. B., Lovett, G. M., Baldocchi, D. D., … and Schulze, E. D. (2006) 'Reconciling carbon-cycle concepts, terminology, and methods', *Ecosystems*, vol 9, pp. 1041–1050.

Chen, Y., Guerschman, J., Shendryk, Y., Henry, D. and Harrison, M. T. (2021) 'Estimating pasture biomass using Sentinel-2 imagery and machine learning', *Remote Sensing*, vol 13, 603.

Ciais, P., Reichstein, M., Viovy, N. and 30 other authors. (2005) 'Europe-wide reduction in primary productivity caused by the heat and drought in 2003', *Nature*, vol 437, pp. 529–533.

Ciais, P., Sabine, C., Bala, G. and 12 other authors. (2013) 'Carbon and Other Biochemical Cycles', in T. F. Stocker, D. Qin, G.-K. Plattner, M. Tignor, S. K. Allen, J. Boschung, A. Nauels, Y. Xia, V. Bex and P. M. Midgley (eds.), *Climate Change 2013: The Physical Science Basis. Contribution of Working Group I to the Fifth Assessment Report of the Intergovernmental Panel on Climate Change*. Cambridge University Press, Cambridge and New York.

Cook-Patton, S. C., Leavitt, S. M., Gibbs, D. and 29 other authors (2020) 'Mapping carbon accumulation potential from global natural forest regrowth', *Nature*, vol 585, pp. 545–550.

Couwenberg, J., Dommain, R. and Joosten, H. (2010) 'Greenhouse gas fluxes from tropical peatlands in south-east Asia', *Global Change Biology*, vol 16, no 6, pp. 1715–1732.

Davies, S. J., Abiem, I., Salim, K. A., Aguilar, S., Allen, D., Alonso, A., … & Yap, S. L. (2021). 'ForestGEO: Understanding Forest diversity and dynamics through a global observatory network', *Biological Conservation*, 253, 108907.

DeLucia, E. H., Drake, J. E., Thomas, R. B. and Gonzalez-Meler, M. (2007) 'Forest carbon use efficiency: Is respiration a constant fraction of gross primary production?', *Global Change Biology*, vol 13, pp. 1157–1167.

Dixon, R. K., Brown, S., Houghton, R. A., Solomon, A. M., Trexler, M. C. and Wisniewski, J. (1994) 'Carbon pools and flux of global forest ecosystems', *Science*, vol 263, pp. 185–190.

Duncanson, L., Kellner, J. R., Armston, J., Dubayah, R., Minor, D. M., Hancock, S., … and Zgraggen, C. (2022) 'Aboveground biomass density models for NASA's Global Ecosystem Dynamics Investigation (GEDI) lidar mission', *Remote Sensing of Environment*, vol 270, 112845.

Espírito-Santo, F. D. B., Gloor, M., Keller, M. and 18 other authors. (2014) 'Size and frequency of natural forest disturbances and the Amazon Forest carbon balance', *Nature Communications*, vol 5, 3434.

Fatoyinbo, T., Armston, J., Simard, M., Saatchi, S., Denbina, M., Lavalle, M., ... and Hibbard, K. (2021) 'The NASA AfriSAR campaign: Airborne SAR and lidar measurements of tropical forest structure and biomass in support of current and future space missions', *Remote Sensing of Environment*, vol 264, 112533.

Fauset, S., Baker, T. R., Lewis, S. L., and five other authors. (2012) 'Drought-induced shifts in the floristic and functional composition of tropical forests in Ghana', *Ecology Letters*, vol 15, pp. 1120–1129.

Feng, Y., Zeng, Z., Searchinger, T. D., Ziegler, A. D., and 20 others (2022) 'Doubling of annual forest carbon loss over the tropics during the early twenty-first century', *Nature Sustainability*, vol 5, pp. 444–451.

Fenn, K. M., Malhi, Y. and Morecroft, M. D. (2010) 'Soil CO2 efflux in a temperate deciduous forest: Environmental drivers and component contributions', *Soil Biology and Biochemistry*, vol 42, pp. 1685–1693.

Fenn, K. M., Malhi, Y., Morecroft, M. D., Lloyd, C. and Thomas, M. (2015) 'The carbon cycle of a maritime ancient temperate broadleaved woodland at seasonal and annual scales', *Ecosystems*, vol 18, pp. 1–15.

Food and Agriculture Organization. (2010) *Global Forest Resources Assessment 2010*. Food and Agriculture Oraganization, Rome.

Friedlingstein, P., Jones, M. W., O'Sullivan, M., Andrew, R. M., Bakker, D. C., Hauck, J., ... and Zeng, J. (2022) 'Global carbon budget 2021', *Earth System Science Data*, vol 14, pp. 1917–2005.

Frolking, S., Talbot, J., Jones, M. C. and four other authors. (2011) 'Peatlands in the earth's 21st century climate system', *Environmental Reviews*, vol 19, pp. 371–396.

Fu, B., Li, S., Yu, X., Yang, P., Yu, G., Feng, R. and Zhuang, X. (2010) 'Chinese ecosystem research network: Progress and perspectives', *Ecological Complexity*, vol 7, pp. 225–233.

Galloway, J. N., Townsend, A. R., Erisman, J. W. and six other authors. (2008) 'Transformation of the nitrogen cycle: Recent trends, questions, and potential solutions', *Science*, vol 320, pp. 889–892.

Gatti, L., Gloor, M., Miller, J. and seven other authors. (2014) 'Drought sensitivity of Amazonian carbon balance revealed by atmospheric measurements', *Nature*, vol 506, pp. 76–80.

Gibbon, A., Silman, M. R., Malhi, Y. and six other authors. (2010) 'Ecosystem carbon storage across the grassland–forest transition in the high Andes of Manu National Park, Peru', *Ecosystems*, vol 13, pp. 1097–1111.

Girardin, C. A. J., Malhi, Y., Aragão, L. E. O. C. and nine other authors. (2010) 'Net primary productivity allocation and cycling of carbon along a tropical forest elevational transect in the Peruvian Andes', *Global Change Biology*, vol 16, pp. 3176–3192.

Girardin, C. A. J., Silva-Espejo, J. E., Doughty, C. E. and 17 other authors. (2014) 'Productivity and carbon allocation in a tropical montane cloud forest in the Peruvian Andes', *Plant Ecology and Diversity*, vol 7, pp. 107–123.

Goldewijk, K. K. (2001) 'Estimating global land use change over the past 300 years: The HYDE Database', *Global Biogeochemical Cycles*, vol 15, pp. 417–433.

Goodale, C. L., Apps, M. J., Birdsey, R. A. and ten other authors. (2002) 'Forest carbon sinks in the Northern Hemisphere', *Ecological Applications*, vol 12, pp. 891–899.

Gower, S. T., Vogel, J. G., Norman, J. M., Kucharik, C. J., Steele, S. J. and Stow, T. K. (1997) 'Carbon distribution and aboveground net primary production in aspen, jack pine, and black spruce stands in Saskatchewan and Manitoba, Canada', *Journal of Geophysical Research: Atmospheres*, vol 102, pp. 29029–29041.

Hannah, L., Carr, J. L. and Landerani, A. (1995) 'Human disturbance and natural habitat - a biome level analysis of a global data set', *Biodiversity and Conservation*, vol 4, pp. 128–155.

Haverd, V., Smith, B., Canadell, J. G., and six other authors (2020) 'Higher than expected CO2 fertilization inferred from leaf to global observations', *Global Change Biology*, vol 26, pp. 2390–2402.

Heinrich, V., Vancutsem, C, Dalagnol R. and eleven other authors (2023) 'The carbon sink of secondary and degraded humid tropical forests', *Nature*, vol 615, pp. 436–442.

Hickler, T., Smith, B., Prentice, I. C., Mjöfors, J., Miller, P., Arneth, A. and Sykes, M. T. (2008) 'CO_2 fertilization in temperate FACE experiments not representative of boreal and tropical forests', *Global Chang Biology*, vol 14, pp. 1531–1542.

Hiederer, R. and Köchy, M. (2011) *Global Soil Organic Carbon Estimates and the Harmonized World Soil Database, Publications Office of the European Union*, EUR 25225 EN.

Hirano, T., Segah, H., Kusin, K., Limin, S., Takahashi, H. and Osaki, M. (2012) 'Effects of disturbances on the carbon balance of tropical peat swamp forests', *Global Change Biology*, vol 18, pp. 3410–3422.

Hiura, T., Sano, J. and Konno, Y. (1996) Age structure and response to fine-scale disturbances of Abies sachalinensis, Picea jezoensis, Picea glehnii, and Betula ermanii growing under the influence of a dwarf bamboo understory in northern Japan. *Canadian Journal of Forest Research*, vol 26, pp. 289–297.

Hubau, W., Lewis, S. L., Phillips, O. L., Affum-Baffoe, K., Beeckman, H., Cuní-Sanchez, A., ... & Zemagho, L. (2020) 'Asynchronous carbon sink saturation in African and Amazonian tropical forests', *Nature*, vol 579, pp. 80–87.

IGBP. (1998) 'The terrestrial carbon cycle: Implications for the Kyoto Protocol', *Science*, vol 280, pp. 1393–1394.

IPCC. (2000) *Land Use, Land-Use Change, and Forestry: A Special Report of the Intergovernmental Panel on Climate Change*. Cambridge University Press, Cambridge, UK.

IPCC. (2014) *2013 Supplement to the 2006 IPCC Guidelines for National Greenhouse Gas Inventories: Wetlands*. IPCC, Switzerland.

Ise, T., Dunn, A. L., Wofsy, S. C. and Moorcroft, P. R. (2008) 'High sensitivity of peat decomposition to climate change through water-table feedback'. *Nature Geoscience*, vol 1, pp. 763–766.

Jarvis, P. and Linder, S. (2000) 'Constraints to growth of boreal forests', *Nature*, vol 405, pp. 904.

Jarvis, P. G., Massheder, J. M., Hale, S. E., Moncrieff, J. B., Rayment, M. and Scott, S. L. (1997) 'Seasonal variation of carbon dioxide, water vapor, and energy exchanges of a boreal black spruce forest', *Journal of Geophysical Research: Atmospheres*, vol 102, pp. 28953–28966.

Jastrow, J. D., Miller, R. M., Matamala, R. and four other authors. (2005) 'Elevated atmospheric carbon dioxide increases soil carbon', *Global Change Biology*, vol 11, pp. 2057–2064.

Jobbágy, E. G. and Jackson, R. B. (2000) 'The vertical distribution of soil organic carbon and its relation to climate and vegetation', *Ecological Applications*, vol 10, pp. 423–436.

Keith, H., Mackey, B. G. and Lindenmayer, D. B. (2009) 'Re-evaluation of forest biomass carbon stocks and lessons from the world's most carbon-dense forests', *Proceedings of the National Academy of Sciences of the United States of America*, vol 106, pp. 11635–11640.

Kho, L. K., Malhi, Y. and Tan, S. K. S. (2013) 'Annual budget and seasonal variation of aboveground and belowground net primary productivity in a lowland dipterocarp forest in Borneo', *Journal of Geophysical Research-Biogeosciences*, vol 118, pp. 1282–1296.

Kirschbaum, M. U. F., Eamus, D., Gifford, R. M., Roxburgh, S. H. and Sands, P. J. (2001) *Definitions of Some Ecological Terms Commonly Used in Carbon Accounting*. Cooperative Research Centre for Carbon Accounting, Canberra, 2–5.

Korhola, A., Tolonen, K., Turunen, J. and Jungner, H. (1995) 'Estimating long-term carbon accumulation rates in boreal peatlands by radiocarbon dating', *Radiocarbon*, vol 37, pp. 575–584.

Körner, C., Asshoff, R., Bignucolo, O. and six other authors. (2005) 'Carbon flux and growth in mature deciduous forest trees exposed to elevated CO_2', *Science*, vol 309, pp. 1360–1362.

Lähteenoja, O., Reátegui, Y. R., Räsänen, M., Torres, D. D. C., Oinonen, M. and Page, S. (2012) 'The large Amazonian peatland carbon sink in the subsiding Pastaza-Marañón foreland basin, Peru', *Global Change Biology*, vol 18, pp. 164–178.

Leuschner, C., Moser, G., Bertsch, C., Röderstein, M. and Hertel, D. (2007) 'Large altitudinal increase in tree root/shoot ratio in tropical mountain forests of Ecuador', *Basic and Applied Ecology*, vol 8, pp. 219–230.

Lewis, S. L., Lopez-Gonzalez, G., Sonké, B., Affum-Baffoe, K., Baker, T. R., Ojo, L. O., ... and Wöll, H. (2009) 'Increasing carbon storage in intact African tropical forests', *Nature*, vol 457, pp. 1003–1006.

López-Serrano, P. M., Cárdenas Domínguez, J. L., Corral-Rivas, J. J., Jiménez, E., López-Sánchez, C. A. and Vega-Nieva, D. J. (2019). 'Modeling of aboveground biomass with Landsat 8 OLI and machine learning in temperate forests', *Forests*, vol 11, 11.

Luo, Y., Hui, D. and Zhang, D. (2006) 'Elevated CO_2 stimulates net accumulations of carbon and nitrogen in land ecosystems: a meta-analysis', *Ecology*, vol 87, pp. 53–63.

Luo, Y., Reynolds, J., Wang, Y. and Wolfe, D. (1999) 'A search for predictive understanding of plant responses to elevated CO_2', *Global Change Biology*, vol 5, pp. 143–156.

Luyssaert, S., Inglima, I., Jung, M. and 62 other authors. (2007) 'CO_2 balance of boreal, temperate, and tropical forests derived from a global database', *Global Change Biology*, vol 13, pp. 2509–2537.

Magnani, F., Mencuccini, M., Borghetti, M. and 18 other authors. (2007) 'The human footprint in the carbon cycle of temperate and boreal forests', *Nature*, vol 447, pp. 848–850

Malhi, Y. (2010) 'The carbon balance of tropical forest regions, 1990–2005', *Current Opinion in Environmental Sustainability*, vol 2, pp. 237–244.

Malhi, Y. (2012) 'The productivity, metabolism and carbon cycle of tropical forest vegetation', *Journal of Ecology*, vol 100, pp. 65–75.

Malhi, Y., Aragão, L. E. O. C., Metcalfe, D. B. and 13 other authors. (2009) 'Comprehensive assessment of carbon productivity, allocation and storage in three Amazonian forests', *Global Change Biology*, vol 15, pp. 1255–1274.

Malhi, Y., Baldocchi, D. D. and Jarvis, P. G. (1999) 'The carbon balance of tropical, temperate and boreal forests', *Plant Cell and Environment*, vol 22, pp. 715–740.

Malhi, Y., Girardin, C., Metcalfe, D. B., Doughty, C. E., Aragão, L. E., Rifai, S. W., ... and Phillips, O. L. (2021) 'The Global Ecosystems Monitoring network: Monitoring ecosystem productivity and carbon cycling across the tropics', *Biological Conservation*, vol 253, 108889.

Malhi, Y., Meir, P. and Brown, S. (2002) 'Forests, carbon and global climate', *Philosophical Transactions of the Royal Society of London. Series A: Mathematical, Physical and Engineering Sciences*, vol 360, pp. 1567–1591.

Malhi, Y., Phillips, O. L., Lloyd, J., Baker, T., Wright, J., Almeida, S., ... & Vinceti, B. (2002) 'An international network to monitor the structure, composition and dynamics of Amazonian forests (RAINFOR)', *Journal of Vegetation Science*, vol 13, pp. 439–450.

Masek, J. G., Huang, C., Wolfe, R. and four other authors. (2008) 'North American forest disturbance mapped from a decadal Landsat record', *Remote Sensing of Environment*, vol 112, pp. 2914–2926.

Minkkinen, K., Korhonen, R., Savolainen, I. and Laine, J. (2002) 'Carbon balance and radiative forcing of Finnish peatlands 1900–2100 — the impact of forestry drainage', *Global Change Biology*, vol 8, pp. 785–799.

Mitchard, E. T. A. (2018). 'The tropical forest carbon cycle and climate change', *Nature*, vol. 559, pp. 527–534.

Mokany, K., Raison, R. J. and Prokushkin, A. S. (2006) 'Critical analysis of root: Shoot ratios in terrestrial biomes', *Global Change Biology*, vol 12, pp. 84–96.

Moore, S., Evans, C. D., Page, S. E. and seven other authors. (2013) 'Deep instability of deforested tropical peatlands revealed by fluvial organic carbon fluxes', *Nature*, vol 493, pp. 660–663.

Nepstad, D. C., Stickler, C. M., Soares-Filho, B. and Merry, F. (2008) 'Interactions among Amazon land use, forests and climate: Prospects for a near-term forest tipping point', *Philosophical Transactions of the Royal Society B: Biological Sciences*, vol 363, pp. 1737–1746.

Norby, R. J., DeLucia, E. H., Gielen, B. and seven other authors. (2005) 'Forest response to elevated CO2 is conserved across a broad range of productivity', *Proceedings of the National Academy of Sciences of the United States of America*, vol 102, pp. 18052–18056.

O'Connell, K. E. B., Gower, S. T. and Norman, J. M. (2003) 'Net ecosystem production of two contrasting boreal black spruce forest communities', *Ecosystems*, vol 6, pp. 248–260.

Page, S. E., Rieley, J. O. and Banks, C. J. (2011) 'Global and regional importance of the tropical peatland carbon pool', *Global Change Biology*, vol 17, pp. 798–818.

Page, S. E., Siegert, F., Rieley, J. O., Boehm, H.-D. V., Jaya, A. and Limin, S. (2002) 'The amount of carbon released from peat and forest fires in Indonesia during 1997', *Nature*, vol 420, pp. 61–65.

Paillet, Y., Berges, L., Hjälten, J. and 21 other authors. (2010) 'Biodiversity differences between managed and unmanaged forests: Meta-analysis of species richness in Europe', *Conservation Biology*, vol 24, pp. 101–112.

Pan, Y., Birdsey, R. A., Fang, J. and 15 other authors. (2011) 'A large and persistent carbon sink in the world's forests', *Science*, vol 333, pp. 988–993.

Pan, Y., Birdsey, R. A., Phillips, O. L. and Jackson, R. B. (2013) 'The structure, distribution and biomass of the world's forests', *Annual Review of Ecology, Evolution, and Systematics*, vol 44, pp. 593–622.

Panagos, P., Van Liedekerke, M., Jones, A. and Montanarella, L. (2012) 'European Soil Data Centre: Response to European policy support and public data requirements', *Land Use Policy*, vol 29, pp. 329–338.

Peacock, J., Baker, T. R., Lewis, S. L., Lopez-Gonzalez, G., & Phillips, O. L. (2007) 'The RAINFOR database: Monitoring forest biomass and dynamics', *Journal of Vegetation Science*, vol 18, 535–542.

Peregon, A., Maksyutov, S., Kosykh, N. P. and Mironycheva-Tokareva, N. P. (2008) 'Map-based inventory of wetland biomass and net primary production in western Siberia', *Journal of Geophysical Research: Biogeosciences,* vol 113, G01007.

Phillips, O. L., Aragão, L. E., Lewis, S. L., Fisher, J. B., Lloyd, J., López-González, G., … and Torres-Lezama, A. (2009) 'Drought sensitivity of the Amazon rainforest', *Science*, vol 323, pp. 1344–1347.

Potapov, P., Yaroshenko, A., Turubanova, S. and seven other authors. (2008) 'Mapping the world's intact forest landscapes by remote sensing. *Ecology & Society,* vol 13, p. 51.

Pötzschner, F., Baumann, M., Gasparri, N. I., Conti, G., Loto, D., Piquer-Rodríguez, M. and Kuemmerle, T. (2022) 'Ecoregion-wide, multi-sensor biomass mapping highlights a major underestimation of dry forests carbon stocks', *Remote Sensing of Environment*, vol 269, 112849.

Qi, Y., Wei, W., Chen, C. and Chen, L. (2019). 'Plant root-shoot biomass allocation over diverse biomes: A global synthesis', *Global Ecology and Conservation*, vol 18, e00606.

Restrepo-Coupe, N., da Rocha, H. R., Hutyra, L. R. and 22 other authors. (2013) 'What drives the seasonality of photosynthesis across the Amazon basin? A cross-site analysis of eddy flux tower measurements from the Brasil flux network', *Agricultural and Forest Meteorology,* vol 182–183, pp. 128–144.

Running, S. W., Nemani, R. R., Heinsch, F. A., Zhao, M., Reeves, M. and Hashimoto, H. (2004) 'A continuous satellite-derived measure of global terrestrial primary production', *Bioscience*, vol 54, pp. 547–560.

Saatchi, S. S., Harris, N. L., Brown, S. and 11 other authors. (2011) 'Benchmark map of forest carbon stocks in tropical regions across three continents', *Proceedings of the National Academy of Sciences,* vol 108, pp. 9899–9904.

Scharlemann, J. P. W., Tanner, E. V. J., Hiederer, R. and Kapos, V. (2014) 'Global soil carbon: Understanding and managing the largest terrestrial carbon pool', *Carbon Management,* vol 5, pp. 81–91.

Schmitt, C. B., Burgess, N. D., Coad, L. and 17 other authors. (2009) 'Global analysis of the protection status of the world's forests', *Biological Conservation,* vol 142, pp. 2122–2130.

Soja, M. J., Quegan, S., d'Alessandro, M. M., Banda, F., Scipal, K., Tebaldini, S. and Ulander, L. M. (2021) 'Mapping above-ground biomass in tropical forests with ground-cancelled P-band SAR and limited reference data', *Remote Sensing of Environment*, vol 253, 112153.

Sullivan, M. J. P., Lewis, S. L., Affum-Baffoe, and 220 other authors. (2020) 'Long-term thermal sensitivity of Earth's tropical forests', *Science*, vol 368, pp. 869–874.

Suni, T., Berninger, F., Vesala, T. and 14 other authors. (2003) 'Air temperature triggers the recovery of evergreen boreal forest photosynthesis in spring', *Global Change Biology,* vol 9, pp. 1410–1426.

Thomas, M. V., Malhi, Y., Fenn, K. M. and five other authors. (2011) 'Carbon dioxide fluxes over an ancient broadleaved deciduous woodland in southern England', *Biogeosciences,* vol 8, pp. 1595–1613.

Thurner, M., Beer, C., Santoro, M. and eight other authors. (2014) 'Carbon stock and density of northern boreal and temperate forests', *Global Ecology and Biogeography,* vol 23, pp. 297–310.

Trumbore, S. E. (1993) Comparison of carbon dynamics in tropical and temperate soils using radiocarbon measurements. *Global Biogeochemical Cycles,* vol 7, pp. 275--290.

Trumbore, S. E. and Harden, J. W. (1997) 'Accumulation and turnover of carbon in organic and mineral soils of the BOREAS northern study area', *Journal of Geophysical Research: Atmospheres,* vol 102, pp. 28817–28830.

Vicca, S., Luyssaert, S., Peñuelas, J. and 13 other authors. (2012) 'Fertile forests produce biomass more efficiently', *Ecology Letters,* vol 15, pp. 520–526.

Wade, T. G., Riitters, K. H., Wickham, J. D. and Jones, K. B. (2003) 'Distribution and causes of global forest fragmentation', *Conservation Ecology,* vol 7, 7.

Wang, X., Shilong, P., Philippe, C. and nine other authors. (2014) 'A two-fold increase of carbon cycle sensitivity to tropical temperature variations', *Nature,* vol 506, pp. 212–215.

Waring, R. H., Landsberg, J. J. and Williams, M. (1998) 'Net primary production of forests: A constant fraction of gross primary production?', *Tree Physiology,* vol 18, pp. 129–134.

Williams, M. (2000) 'Dark ages and dark areas: global deforestation in the deep past', *Journal of Historical Geography,* vol 26, pp. 28–46.

Xiao, J., Chevallier, F., Gomez, C., Guanter, L., Hicke, J. A., Huete, A. R., … and Zhang, X. (2019) 'Remote sensing of the terrestrial carbon cycle: A review of advances over 50 years', *Remote Sensing of Environment*, vol 233, 111383.

Ye, J., Yue, C., Hu, Y. and Ma, H. (2021) 'Spatial patterns of global-scale forest root-shoot ratio and their controlling factors', *Science of The Total Environment*, vol 800, 149251.

Zoltai, S. C. and Martikainen, P. J. (1996) 'Estimated extent of forested peatlands and their role in the global carbon cycle', in M. J. Apps and D. T. Price (eds.), *Forest Ecosystems, Forest Management and the Global Carbon Cycle*. Springer, Berlin.

PART VII

HUMAN ECOLOGY

39

MULTIPLE ROLES OF NON-TIMBER FOREST PRODUCTS IN ECOLOGIES, ECONOMIES AND LIVELIHOODS

Charlie M. Shackleton

Introduction

During the colonial period, forests were valued by governments and conservators or forest officers mostly for their timber and watershed functions. Although millions of forest-dwelling or adjacent people made extensive uses of "other" forest resources, most of these resources were not acknowledged by colonial foresters or administrators. Other than a few which become global commodities (such as rubber, Brazil nuts and rattan), the majority, at best, attracted interest only as curiosities by anthropologists and ethnographers of the era. As tropical forest deforestation accelerated in the latter half of the 1900s, a peculiar coalition of interests and processes examined the importance and potential of these "other" forest resources through new lenses, positing that they offered solutions or alternatives to the seemingly disparate concerns of (1) tropical forest deforestation, (2) loss of culture and tradition amongst forest peoples and (3) underdevelopment and poor welfare in what were often marginalised and remote communities. These positions were crystallised into a strident call for action on the basis of economic analyses showing that the use of such products could rival the income from tropical timber, whilst simultaneously potentially having far better outcomes in conserving forests and sustaining or improving livelihoods (Peters et al. 1989; De Beer and McDermott 1996). The results were a steadily growing understanding of the various roles played by these resources, an increasingly coherent terminology and the inclusion of these resources in management plans for forests and, in some instances, national or sub-national development policies.

Defining non-timber forest products

Because these "other" products grow in multiple habitats, are used by people for utilitarian and cultural purposes, and may be marketed, they have been examined from various disciplinary perspectives, resulting in an unsatisfying array of names, terms and acronyms.

DOI: 10.4324/9781003324072-46

Fortunately, with time there has been some convergence, and most commentators now refer to them as either non-timber forest products (NTFPs) or non-wood forest products (NWFPs). The latter is an artefact of the structural organisation of the Food and Agriculture Organization (FAO), which had a Wood Division at the time, and is not used as much as the term NTFPs (Belcher 2003), which will be used here. The interdisciplinary nature of the subject also resulted in a range of definitions of the same terms, which over time have been narrowed down. Whilst specific words may differ between definitions, the key elements of the NTFP concept are widely accepted to include (De Beer and McDermott 1996; Belcher 2003; Shackleton et al. 2011): (1) largely wild (i.e. there can be some harvest from fields or cultivated populations; but the bulk of the population of any NTFP species must be in the wild), (2) biological products, (3) from natural, semi-natural and transformed habitats, including urban (i.e. not just forests!), (4) that are used or traded by local communities and (5) the bulk of the benefits (cultural, material or economic) accrue to those local communities that make management decisions pertaining to the NTFPs (Shackleton et al. 2011).

The importance of NTFPs in ecosystem composition and dynamics

The majority of works on NTFPs have focussed on their contribution to provisioning services for human well-being, along with the argument that conserving forests to secure NTFP stocks will help protect natural systems from land transformation and degradation and hence conserve biodiversity. These arguments overlook that NTFPs also play a role in local ecologies and provide supporting and regulating services within forests and other ecosystems (Shackleton et al. 2018), which may at times or in certain situations result in trade-offs between their different roles. They are also components of biodiversity in their own right and need not be relegated to just being a surrogate reason for protecting other species. Indeed, they can constitute a very rich component of forest flora and fauna. For example, approximately 94% of canopy tree species and 77% of sub-canopy tree species encountered in South African indigenous forests have some recorded traditional or commercial use (Geldenhuys 1999). Similarly, Peters et al. (1989) reported that 42% of all the individual trees in one hectare of tropical forest in Peru provided products that were marketed in the closest town, with the proportion being even higher if non-marketed NTFPs were included. In New York City (USA), Hurley and Emery (2018) reported that 72% of street trees in the city offered provisioning services such as fruits or medicines. These three examples show that NTFPs can constitute a significant proportion of local biodiversity in terms of species and individuals.

Many NTFP species are community dominants by either stature or abundance, and thus play a significant role in shaping community composition, dynamics and nutrient cycles, as well as providing resources to other organisms. For example, riparian reedbeds are common sources of fibre for thatching and weaving by rural people around the world. It is also common for reedbed communities to be dominated by only a few species (such as *Phragmites*, *Papyrus*, *Juncus* or *Typha*) and therefore changes in the abundance of such reeds through fire, harvesting or flooding have significant effects on the abundance and performance of other reedbed species, including plants, invertebrates and birds (e.g. Valkama et al. 2008). Reedbeds are also extremely important in riparian nutrient dynamics.

At a species level, the marula tree (*Sclerocarya birrea* subsp. *caffra*) can be used as an example (Hall et al. 2002). It is a widespread tree species in the semi-arid areas of sub-Saharan Africa, with multiple uses by local communities for its fruit (eaten raw or fermented into various liquors; the kernel is also extracted and eaten), bark for medicine, fruit stones for decoration, kindling for fire, truncheons for live fencing and lightweight timber for carving. *S. birrea* is also typically a large tree which plays a significant role in the ecological community as described by Hall et al. (2002). The fruit is a key food source to dozens of vertebrate and invertebrate species. The copious flowering makes it a significant resource for insect pollinators. The species is a preferred host to several mistletoe species (one of which, *Erianthemum dregei*, is also an NTFP to wood carvers) and is the only host to *Agelanthus crassifolius*. The large canopy provides shade, which alters the understorey light, temperature and moisture regimes and promotes a markedly different subcanopy herbaceous composition relative to adjacent areas. The overall productivity in these subcanopy locations is also higher, which in turn alters natural fire intensities. Similar arguments could be made for many other dominant NTFPs, such as bamboos in Southeast Asia, vegetable ivory palms and Brazil nuts in Amazonia, or faunal species such as edible caterpillars. These are all NTFPs whilst also having significant, but largely unstudied, roles in community structure, dynamics or composition.

The ecology of NTFPs

Given that NTFPs span dozens of life forms and functional types (from fish and caterpillars to bulbs and roots, through to leaves, fruits and resins), thousands of species and often many different parts of the same species it is impossible to derive a functional ecology of NTFPs (by functional ecology I mean the functions and roles of specific types of organisms in communities and landscapes and the organismal traits that characterise specific functional groups). But there is one characteristic that unites them all, namely, that they are harvested by humans. The term "harvesting" used here includes a whole range of practices that humans use to obtain the whole or parts of wild organisms for their benefit, such as collection, gathering, trapping, hunting, logging and tapping (Barron et al. 2022). Therefore, the central question within autecological studies (i.e. the ecology of an individual species) of any NTFP species is "how does it respond to harvesting at different levels (frequencies, intensities, seasons and techniques)?" This is typically part of a broader package of applied questions around what is the ecologically sustainable yield for a specific NTFP, how does that yield vary temporally and spatially, and how does harvesting interact with other environmental pressures on a population (such as fire, herbivory, disease, drought and pollution) to undermine or facilitate individual or population persistence and allowable offtake. Interestingly, despite the billions of people involved in harvesting NTFPs for household consumption or trade, and the billions of dollars in value annually that this represents, we have very little autecological knowledge of most NTFP species, even those on high-value markets.

Although it is impossible to derive generalisable biological similarities between all NTFP species, there are some emerging generalities that are potentially useful starting points in helping understand their nature and hence potential management approaches (Ticktin and Shackleton 2011; Fromentin et al. 2022), but one must always be alert to the exceptions:

- As biological organisms all NTFPs can be harvested at ecologically sustainable levels. Whether this can be achieved in practice with a satisfactory "economic" return is a different question.
- Just because an NTFP is harvested, it does not doom the species to inevitable overharvesting and decline as many more conservation-minded observers fear. The

meta-review by Stanley et al. (2012) showed that almost two-thirds of NTFP harvest systems they examined appeared to be ecologically sustainable. Whether or not ecological sustainability is achieved or maintained in the face of changing contexts depends on the governance system, which itself is vested in the cultural, economic and political milieu (Ticktin and Shackleton 2011; Ticktin 2015).

- The supply of NTFPs is never static in time and place, and hence management plans and approaches must be responsive and adaptive.
- The impacts of harvesting NTFPs are evident at various hierarchical scales above and below that of the population (from genetic to ecosystem) and should be examined at all scales for a full understanding and appropriate management (Ticktin 2004).
- The impact of harvesting on organismal vital rates (i.e. rates of survival, growth and reproduction) and, in turn, the sensitivity of the population growth rate to individual vital rates are key attributes in determining ecological persistence and growth (Ticktin and Shackleton 2011).
- It is harder and more complex to manage multipurpose use species and landscapes sustainably than it is for single use species (Herrero-Jáuregui et al. 2013).
- Species with large populations and fast growth rates offer better prospects for ecologically sustainable management than do species with small populations and low growth rates (Gaoue et al. 2016).
- Species with high and continuous recruitment offer better prospects for ecologically sustainable management than do species with low and episodic recruitment (i.e. species with erratic and unpredictable periods of recruitment).

The role of NTFPs in livelihoods

With approximately 5.7 billion users of NTFPs globally (Shackleton and de Vos 2022), their importance to rural and many urban households is undisputed. The nature and magnitude of the contribution vary spatially and temporally (Table 39.1) across rural and urban communities in both the Global North and Global South. Understanding the spatial and temporal variation is important for the management of NTFPs, natural habitats generally and also to optimise the opportunities for human benefits. To be able to do so requires disaggregation of the various roles or contributions that NTFPs make and to whom (Shackleton and Shackleton 2004).

Table 39.1 provides some illustrative figures from a handful of studies. There are many such studies in the international literature. Meta-analyses are compounded by differences in the methods used and discrepancies in what NTFPs are included (or excluded), not to mention whether values are calculated as gross or net. Most early studies were based on once-off questionnaire surveys and did not take costs or seasonal differences into account. As the discipline has progressed so have the methods employed, with the Poverty and Environment Network of CIFOR seeking to inculcate a set of minimum requirements (seasonal or quarterly questionnaires; household and village scale; costs to be captured; analysis by quintiles or quartiles) (Angelsen and Lund 2011; Angelsen et al. 2014). Nonetheless, no single method is without its drawbacks, and questionnaire surveys typically underestimate values of commonly used NTFPs and overestimate the value of less used ones (Gram 2001; Menton et al. 2010). Therefore, several methods should be deployed to foster triangulation and combine quantitative rigour with qualitative depth. Thus, techniques such as household diaries, participant observation, market surveys, key informant interviews and a host of participatory

Table 39.1 Illustrative examples of the range of NTFP contribution to rural household income (cash and non-cash combined)

% Contribution to household income	*Location*	*Reference*
64	Northeast Peru	L'Roe and Naughton-Treves (2014)
59	Cameroon	Angelsen et al. (2014)
44	Northwest Zambia	Kalaba et al. (2013)
40	Northern Pakistan	Hussain et al. (2019)
37	Democratic Republic of Congo	Mendako et al. (2022)
34	Central India	Mitra and Mishra (2011)
32	Southern China	Hogarth et al. (2013)
30	Northern Nigeria	Suleiman et al. (2018)
15	Vietnam	Nguyen and Tran (2018)
15	Southern Malawi	Kamanga et al. (2009)
14	Southeast Bangladesh	Kar and Jacobson (2012)

rural appraisal tools can be used to great success, especially when implemented over extended periods to account for intra- and inter-annual changes in yields, prices and market demand, including times of difficulty or crisis.

What Table 39.1 does not show is the inter- and intra-household variation in the importance of NTFPs. Research findings over the last decade have shown that such variation can be substantial (Angelsen et al. 2014). Commonly reported patterns, with the inevitable exceptions of course, are the following:

- NTFPs contribute significantly to total household income of poorer households than richer ones (e.g. Luswaga and Nuppenau 2022).
- This often also has a gender dimension because often female-headed households are poorer than male-headed ones in most developing countries.
- The absolute quantities consumed may not differ greatly between wealth quintiles, or may even be higher in richer ones, but this does not translate into higher contributions of NTFPs because richer households have greater total household incomes when all sources are considered (e.g. Kalaba et al. 2013; Luswaga and Nuppenau 2022).
- Richer households tend to dominate trade in high-value NTFPs, whereas poor households lead trade in high-volume, low-value NTFPs with low capital requirements (e.g. Cosyns et al. 2011).
- Local trade between households or in local markets can be substantial and provides an income equalising role as poorer households sell common NTFPs to richer ones, who may buy rather than collect their own (e.g. Shackleton and Shackleton 2004).

NTFPs in household provisioning

The most widespread use of NTFPs is the frequent collection of locally accessible resources to meet household needs for food (fruits, seeds, nuts, leafy vegetables, mushrooms, honey and bushmeat), energy (firewood, charcoal and kindling), shelter/protection (wooden poles for housing and fences; thatch grass, reeds or palm fronds for roofing), medicines (bark, roots and leaves), fibres (lianas, palm fronds, rattans, grasses and bark) and agricultural implements (made from local woods). Individual households can use dozens of species during the course of a year, and within communities, hundreds of species are known and used. Shackleton and Shackleton (2004) reported the mean amounts used per household annually for some NTFPs across several sites in South Africa as approximately 58 kg of wild leafy vegetables, 5.2 tons of firewood, 104 kg of wild fruits, and 180 poles for building. Provisioning uses are also widespread in urban and peri-urban settings (e.g. Schlesinger et al. 2015).

NTFPs allow cash saving

A second advantage of NTFP use for household provisioning is that collection of locally available NTFPs allows households, especially the poorest ones, to allocate scarce cash for livelihood needs that cannot be collected from the wild, for example, agricultural inputs such as seeds or fertilisers, school fees or books, buying of stock to trade, or some accumulation of savings against future misfortune, such as medical or transport expenses (Shackleton and Shackleton 2004). Thus, the use of NTFPs for food, energy or any other purposes means a household does not have to purchase that same need.

NTFPs for cash income generation

Historically most NTFPs were used for household provisioning. But there are only very few communities remaining that are totally divorced from the formal economy and monetisation of inter-household transfers and commercial transactions. This presents opportunities for cash generation through the sale of NTFPs on local markets and further afield. Such trade may take the form of selling small amounts at ad-hoc intervals to supplement other livelihood activities, through bulk sales of low-value products on local markets, to value-added products on specialised or niche markets, nationally or internationally. The income earned and its relative contribution to total household income depends on multiple factors, most significantly the nature of the product, the nature of the market and the degree of engagement in trading by the household. Because many households only engage in NTFP trade as a supplementary livelihood activity, the absolute incomes earned per week, month or per season can be low for such households. This reduces the reported mean income earned per household within a community such that external commentators frequently claim that NTFP trade is not a viable rural development or poverty alleviation option. Therefore, it is better to examine the relative cash returns on the basis of per hour worked, which typically reveals that even low-scale NTFP trade is a viable use of time, netting 2–7 times better returns per hour than local wage labour (Shackleton et al. 2007; Mahonya et al. 2019). Thus, if there is sufficient market demand, this may, in some instances, be a better open for low-skilled workers than daily wage labour. For those engaged in NTFP trade on a more full-time basis, incomes can be well above national poverty lines, although it is highly dependent on the nature of the product and market (Shackleton et al. 2008; Cunningham 2011).

Local or wider trade in NTFPs provides other benefits beyond just the cash income. Several researchers have generated substantive lists of non-monetary benefits, such as building of social networks, development of business skills, being one's own boss, building of pride and confidence in making a living against difficult odds, being able to work from home and multitask with other household needs, contributing to maintenance of cultural processes and products based on traditional knowledge by offering cultural products in the public eye, and the like. These cannot be valued in monetary terms, but if deemed valuable by NTFP traders themselves, then they need to be acknowledged in development programmes and policies as constituents of well-being and human capital.

NTFPs as safety-nets

The contribution of NTFPs as safety-nets is receiving increasing attention but it is difficult to place a value upon. The safety-net function is the role that NTFPs play as a fall-back option when households experience misfortune, such as drought, death or retrenchment of a bread-winner, a natural disaster and the like (Wunder et al. 2014). This safety-net option can be mobilised via three means (Shackleton and Shackleton 2004). The first is for a household to increase the use of an NTFP that is already part of their provisioning needs (e.g. to consume more wild foods to replace cultivated or purchased foods). The second is to start using an NTFP that they typically have not used previously (e.g. use of firewood instead of commercial energies). The third is to engage in temporary sale of NTFPs to generate cash to tide them over until they can adapt to the impacts of the misfortune. Once engaged in NTFP trade, many continue trading even after the effects of the misfortunate have passed or the household has adapted. Asking traders why they entered the NTFP trade often elicits a reply that it was in response to some emergency or misfortune for which cash was needed to cope and adjust. If NTFPs were not available as a safety-net, what other coping strategies do rural household have? Several researchers, such as McSweeney (2005), Hunter et al. (2011) and Paumgarten and Shackleton (2011), have explored this, but a great deal more is required to understand what strategies are chosen under different shocks and contexts. The large-scale intercontinental study of Wunder et al. (2014) revealed that 65% of households reported at least one shock over the previous 12 months, many reported more than one shock event. Of these, 44% used NTFPs as a safety-net, although not necessarily as their primary choice. Most studies have shown that most affected households usually first turn to family and friends. However, if the shock is a covariate one, i.e. most or all households in the community are affected (e.g. via drought or a communicable disease epidemic), then this option is limited. Another coping strategy is to decrease consumption rates (of food or energy), which is not viable for extended periods. Alternatives may be to sell household assets or borrow cash from micro-lenders, both of which may offer an immediate coping strategy but often at the expense of longer-term livelihood sustainability. Migration of some family members to seek employment, or a permanent move of the entire household, has also been reported. Along with these other coping strategies, NTFPs offer a variety of safety-nets options which are employed by millions of people around the world every year.

NTFPs in local culture

It is methodologically complicated to place a value on the cultural importance of NTFPs to local communities (but it has been done), but nevertheless its importance should not be overlooked by researchers or policymakers. It may be that the presence of culturally important species or sites underpins a specific community's interest in, or even identity with, a designated natural area and how it is managed or conserved. For example, many traditional communities have sacred or taboo areas where normal extraction, or land transformation activities, is limited or prohibited, with the result that these areas are important for biodiversity and the provision of other ecosystem services (e.g. Bhagwat et al. 2005). At a species level, many rural people make use of a range of species for culturally mandated spiritual needs and beliefs (Cocks and Dold 2004), such as a deterrent against perceived evil influences or bad luck. The cultural values of NTFPs also play a role in cultural tourism opportunities if done in a respectful, responsible and consensual manner. These may include traditional ceremonies at which specific NTFPs are required, promotion of cultural foods and drinks, or the sale of cultural crafts and artefacts made from local NTFPs.

The value of NTFPs in international market chains

Many of the roles described above relate to local use of NTFPs, and hence they are frequently underappreciated by planners and developers. The bigger markets and export opportunities associated with the formal economy preferentially attract their attention. Many NTFPs have certainly "made it" in such markets (such as argan and marula oil, aloe gels, active extracts for medicinals, rattans, mushrooms, narcotics, etc.), and there is significant potential for many more to do so. For example, the over-the-counter value of products containing *Prunus africana* bark is over US$200 million per annum (p.a.), with it all sourced from wild populations. The export of Brazil nuts from Bolivia alone is worth over US$70 million p.a. Shea butter exports from Africa were estimated in 2007 at 150,000 tons of dry shea kernel earning about US$30 million p.a. and growing. Aloe gel exports from South Africa are worth over US$2 million p.a. Such high market values can, over time, prompt different scenarios depending on the governance systems in place. The first scenario is that high prices attract increasing numbers of market players, which may result in conflict over land and the resource and perhaps even overexploitation which jeopardises livelihoods and market supply in the medium- to long term. The second scenario is increasing governance intervention, at either local or higher levels (national and international) to regulate or promote the value chain for sustainable outcomes, ecologically, economically and, increasingly, socially (e.g. fair trade products), as well as to generate tax revenues from the trade. The third is increasing cultivation or domestication of the NTFP in plantation or field settings. This improves the potential surety of supply but drives up financial costs and typically concentrates the value in the hands of fewer operators, unless a conscious effort is made to engage smallholders. It also divorces the NTFP from the wild, undermining its usefulness as a vehicle for promoting habitat and species conservation.

A common feature of international value chains for NTFPs is that the local harvester or producer receives only a small proportion of the total value or cost paid by the final consumer. This varies per product and per country, typically ranging from 10% to 20%. Conscious efforts towards either value addition close to source or equitable sharing can increase this to 40–60%, thereby better fulfilling the requirement (5) of the NTFP concept discussed in Section, Defining non-timber forest products.

Urban NTFPs

Most of the literature and understanding on the uses, roles and values of NTFPs come from studies in rural communities in developing countries. However, there is growing appreciation that NTFPs are also used by urban populations in both developing and developed countries. The recent global estimate by Shackleton and de Vos (2022) indicated that approximately 50% of all NTFP users globally are from such contexts. For example, 27% of urban respondents in eastern Massachusetts (USA) had collected NTFPs in the 12 months prior to the research survey (Short Gianotti and Hurley 2016), whilst 54% had consumed wild berries or mushrooms in Poland (Golos and Kaliszewski 2016). The drivers of use of NTFPs by urban communities are likely to be similar to those in rural communities, ranging from poverty to culture and income generation. However, the ecology of urban NTFP species and the effects of harvesting and other stressors have hardly been researched. Estimates show that the proportion of urban tree species that yield NTFP products (typically fruits, nuts and medicines) is typically high—over 75% in New York (USA) (Hurley and Emery 2018) and approximately 40% in Beijing (China) (Wang et al. 2015). The extensive use of NTFPs in urban settings demands greater consideration of them in urban planning generally and urban greening initiatives specifically.

Conclusion

This chapter has outlined the multiple and significant roles of NTFPs in forest ecology and human well-being. However, the gap between the two is currently large, with ecology usually being the domain of foresters, ecologists and conservationists, whilst the understanding of local livelihoods is perceived as the realm of economists, development planners and anthropologists. If the real "values" of NTFPs are to be manifest for the benefit of natural habitats and people, a shift in mindsets is required. A key shift is for different disciplines to work together to provide an integrated understanding of NTFPs and the need for appropriate policies for their recognition, management and securing of benefit flows. Another shift is for policymakers and landowners and managers to view all landscapes as multi-purpose landscapes that provide various NTFPs alongside other ecosystem services of interest. Thus, NTFPs should be integrated into management plans and recognised in trade-off decisions.

References

Angelsen, A., Jagger, P., Babigumira, R., Belcher, B., Hogarth, N.J., Bauch, S., Börner, J., Smith-Hall, C. and Wunder, S. (2014) 'Environmental income and rural livelihoods: A global comparative analysis', *World Development*, vol 64, pp. S12–S28.

Angelsen, A. and Lund, F. (2011). 'Designing the household questionnaire', in A. Angelsen, H. O. Larsen and F. F. Lund (eds.), *Measuring livelihoods and environmental dependence: methods for research and fieldwork*. Earthscan, London.

Barron, E. S., Chaudhary, R. P., Carvalho Ribeiro, S., Gilman, E., Hess, J., Hilborn, R., Katz, E., Kigonya, R., Masski, H., Mesa Castellanos, L.I., Mograbi, P. J., Nayak, P. K., Queiroz, H., Sidorovich, A., Silvano, R. A. M., Zeng, Y., Djagoun, C., and Danner, M. C. (2022) 'Status of and trends in the use of wild species and its implications for wild species, the environment and people', in J. M. Fromentin, M. R. Emery, J. Donaldson, M. C. Danner, A. Hallosserie and D. Kieling (eds.), *Thematic assessment report on the sustainable use of wild species of the intergovernmental science-policy platform on biodiversity and ecosystem services*. IPBES Secretariat, Bonn.

Belcher, B. M. (2003) 'What isn't an NTFP?', *International Forestry Review*, vol 5, pp. 161–168.

Bhagwat, S. A., Kushalappa, C. G., Williams, P. H. and Brown, N. D. (2005) 'The role of informal protected areas in maintaining biodiversity in the Western Ghats of India', *Ecology and Society*, vol 10, 8.

Cocks, M. L. and Dold, A. P. (2004) 'A new broom sweeps clean: The economic and cultural value of grass brooms in the Eastern Cape province, South Africa', *Forests Trees and Livelihoods*, vol 14, pp. 33–42.

Cosyns, H., Degrande, A., De Wulf, R., Van Damme, P., Tchoundjeu, Z. (2011) 'Can commercialisation of NTFPs alleviate poverty? A case study of *Ricinodendron heudelotii* (Baill.) Pierre ex Pax. kernel marketing in Cameroon', *Journal of Agriculture and Rural Development in the Tropics and Subtropics*, vol 112, pp. 45–56.

Cunningham, A. B. (2011) 'Non-timber products and markets: lessons for export-oriented enterprise development from Africa', in S. E. Shackleton, C. M. Shackleton and P. Shanley (eds.), *Non-timber forest products in the global context*. Springer, Heidelberg.

de Beer, J. H. and McDermott M. J. (1996) *The economic value of non-timber forest products in Southeast Asia* (2nd ed.). IUCN, Amsterdam.

Fromentin, J. M., Emery, M. R., Donaldson, J., Danner, M. C., Hallosserie, A. and Kieling, D. (eds.) (2022) *Thematic assessment report on the sustainable use of wild species of the Intergovernmental Science Policy Platform on Biodiversity and Ecosystem Services*. IPBES Secretariat, Bonn, Germany.

Gaoue, O. G., Jiang, J., Ding, W., Agusto, F. B. and Lenhart, S. (2016) 'Optimal harvesting strategies for timber and non-timber forest products in tropical ecosystems', *Theoretical Ecology*, vol 9, pp. 287–297.

Geldenhuys, C. J. (1999) 'Requirements for improved and sustainable use of forest biodiversity: examples of multiple use forest in South Africa', in J. Poker, I. Stein and U. Werder (eds.), *Proceedings forum biodiversity treasures in the world's forests*. Alfred Toepfer Akademie, Germany.

Golos, P. and Kaliszewski, A. (2016) 'Economic importance of selected non-wood forest products in Poland', *Sylwan*, vol 160, pp. 336–343.

Gram, S. (2001) 'Economic valuation of special forest products: An assessment of methodological shortcomings', *Ecological Economics*, vol 36, pp. 109–117.

Hall, J. B., O. Brien, E. M. and Sinclair, F. L. (2002) *Sclerocarya birrea: A monograph*, School of Agricultural and Forest Science publication no 19. University of Wales, Bangor.

Herrero-Jáuregui, C., Guariguata, M. R., Cárdenas, D., Vilanova, E., Robles, M., Licona, J. C. and Nalvarte, W. (2013) 'Assessing the extent of "conflict of use" in multipurpose tropical forest trees: A regional view', *Journal of Environmental Management*, vol 130, pp. 40–47.

Hogarth, N. J., Belcher, B., Campbell, B. and Stacey, N. (2013) 'The role of forest-related income in household economies and rural livelihoods in the border-region of Southern China', *World Development*, vol 43, pp. 111–123.

Hunter, L. M., Twine, W. and Johnson, A. (2011) 'Adult mortality and natural resource use in rural South Africa: Evidence from the Agincourt health and demographic surveillance site', *Society and Natural Resources*, vol 24, pp. 256–275.

Hurley, P. T. and Emery, M. R. (2018) 'Locating provisioning ecosystem services in urban forests: Forageable woody species in New York City, USA', *Landscape and Urban Planning*, vol 170, pp. 266–275.

Hussain, J., Zhou, K., Akbar, M., Khan, M. K., Raza, G., Ali, S., Hussain, A., Abbas, Q., Khan, G., Khan, M., Abbas, H., Iqbal, S. and Ghulam, A. (2019). 'Dependence of rural livelihoods on forest resources in Naltar Valley, a dry temperate mountainous region, Pakistan', *Global Ecology and Conservation*, vol 20, e00765.

Kalaba, F. K., Quinn, C. H. and Dougill, A. J. (2013) 'Contribution of forest provisioning ecosystem services to rural livelihoods in the Miombo woodlands of Zambia', *Population and Environment*, vol 35, pp. 159–182.

Kamanga, P., Vedeld, P. and Sjaastad, E. (2009) 'Forest incomes and rural livelihoods in Chiradzulu District, Malawi', *Ecological Economics*, vol 68, pp. 613–624.

Kar, S. P. and Jacobson, M. G. (2012) 'NTFP income contribution to household economy and related socio-economic factors: Lessons from Bangladesh', *Forest Policy and Economics*, vol 14, pp. 136–142

L'Roe, J. and Naughton-Treves, L. (2014) 'Effects of a policy-induced income shock on forest-dependent households in the Peruvian Amazon', *Ecological Economics*, vol 97, pp. 1–9.

Luswaga, H. and Nuppenau, E.-A. (2022) 'Non-timber forest products income and inequality status for communities around West Usambara Mountain Forests in Tanzania', *Environment, Development and Sustainability*, vol 24, pp. 11651–11675.

Mahonya, S., Shackleton, C. M. and Schreckenberg, K. (2019) 'Non-timber forest product use and market chains along a deforestation gradient in southwest Malawi', *Frontiers in Forests and Global Change*, vol 2, 71.

Mendako, R. K., Tian, G. and Matata, P. M. (2022) 'Identifying socioeconomic determinants of households' forest dependence in the Rubi-Tele Hunting Domain, DR Congo: A logistic regression analysis', *Forests*, vol 13, 1706.

Menton, M. C., Lawrence, A., Merry, F. and Brown, N. D. (2010) 'Estimating natural resource harvests: Conjectures?', *Ecological Economics*, vol 69, pp. 1330–1335.

Mitra, A. and Mishra, D. K. (2011) 'Environmental resource consumption pattern in rural Arunachal Pradesh', *Forest Policy and Economics*, vol 13, pp. 166–170.

Nguyen, T. V. and Tran, T. Q. (2018) 'Forestland and rural household livelihoods in the North Central Provinces, Vietnam', *Land Use Policy*, vol 79, pp. 10–19.

Paumgarten, F. and Shackleton, C. M. (2011) 'The role of non-timber forest products in household coping strategies in South Africa: The influence of household wealth and gender', *Population and Environment*, vol 33, pp. 108–131.

Peters, C. A., Gentry, A. and Mendelsohn, R. (1989) 'Valuation of an Amazonian rain forest', *Nature*, vol 339, pp. 655–656.

Schlesinger, J., Drescher, A. and Shackleton, C. M. (2015) 'Socio-spatial dynamics in the use of wild natural resources: Evidence from six rapidly growing medium-sized cities in Africa', *Applied Geography*, vol 56, pp. 107–115

Shackleton, C. M. and de Vos, A (2022) 'How many people globally actually use non-timber forest products?', *Forest Policy and Economics*, vol 135, 102659.

Shackleton, C. M., Delang, C., Shackleton, S. E. and Shanley, P. (2011) 'Non-timber forest products: Concept and definition', in S. E. Shackleton, C. M. Shackleton and P. Shanley (eds.), *Non-timber forest products in the global context*. Springer, Heidelberg.

Shackleton, C. M. and Shackleton, S. E. (2004) 'The importance of non-timber forest products in rural livelihood security and as safety-nets: Evidence from South Africa', *South African Journal of Science*, vol 100, pp. 658–664.

Shackleton, C. M., Shackleton, S. E., Buiten, E. and Bird, N. (2007) 'The importance of dry forests and woodlands in rural livelihoods and poverty alleviation in South Africa', *Forest Policy and Economics*, vol 9, pp. 558–577.

Shackleton, C. M., Ticktin, T. and Cunningham, A. B. (2018). 'Nontimber forest products as ecological and biocultural keystone species', *Ecology and Society*, vol 23, 22.

Shackleton, S. E., Campbell, B., Lotz-Sisitka, H. and Shackleton, C. M. (2008) 'Links between the local trade in natural products, livelihoods and poverty alleviation in a semi-arid region of South Africa', *World Development*, vol 36, pp. 505-526

Short Gianotti, A. G. and Hurley, P. (2016) 'Gathering plants and fungi along the urban-rural gradient: Uncovering differences in the attitudes and practices among urban, suburban and rural landowners', *Land Use Policy*, vol 57, pp. 555–563.

Stanley, D., Voeks, R. and Short, L. (2012) 'Is non-timber forest product harvest sustainable in the less developed world? A systematic review of the recent economic and ecological literature', *Ethnobiology and Conservation*, vol 1. www.ethnobioconservation.com/index.php/ebc/article/view/19

Suleiman, M. S., Wasonga, V.O., Mbau, J. S., Suleiman, A. and Elhadi. A. (2018) 'Non-timber forest products and their contribution to households income around Falgore Game Reserve in Kano, Nigeria', *Ecological Processes*, vol 6, 23.

Ticktin, T. (2004) 'The ecological implications of harvesting non-timber forest products', *Journal of Applied Ecology*, vol 41, pp. 11–21.

Ticktin, T. (2015) 'The ecological sustainability of non-timber forest product harvest: Principles and methods', in C. M. Shackleton, A. K. Pandey, A.K. and T. Ticttin (eds.), *Ecological sustainability for non-timber forest products: Dynamics and studies of harvesting*. Routledge/Earthscan, Abingdon.

Ticktin, T. and Shackleton, C. M. (2011) 'Harvesting non-timber forest products sustainably: Opportunities and challenges', in S. E. Shackleton, C. M. Shackleton and P. Shanley (eds.), *Non-timber forest products in the global context*. Springer, Heidelberg.

Wang, H. F., Qureshi, S., Knapp, S., Friedman C. R. and Hubacek, K. (2015) 'A basic assessment of residential plant diversity and its ecosystem services and disservices in Beijing, China', *Applied Geography*, vol 64, pp. 121–131.

Wunder, S., Börner, J., Shively, G. and Wyman, M. (2014) 'Safety nets, gap filling and forests: A global comparative perspective', *World Development*, vol 64, pp. S29–S42.

40

AGRICULTURE IN THE FOREST

Ecology and rationale of shifting cultivation

Olivier Ducourtieux

Introduction

Agriculture is now one of the main types of land use in the world. According to the Food and Agriculture Organization (FAO), forests covered only 31% of the land in the world in 2020, while land used for agriculture—i.e., arable land, permanent crops, and pastures—accounted for 37%. However, in the history of humankind, agriculture is a recent invention (Mazoyer and Roudart, 2005). With the transition from the hunter-gatherer lifestyle to agriculture during the Neolithic Revolution, farmers quickly expanded their activity into forest areas. Here, they developed shifting cultivation, an agricultural practice with temporal rotations in forest ecosystems.

Shifting cultivation still exists today in tropical regions and is the basis of livelihood for probably hundreds of millions of people (Mertz et al., 2009; Cairns, 2017). The range in the population estimate is wide not only because of the intrinsic difficulties in counting smallholders spread out under the tree canopy at the margins of tropical forests, but also because of the confusion that exists in differentiating shifting cultivators from other farmers involved in small-scale agriculture (Thrupp, 1997). We will focus on rotational shifting cultivation, when agriculture takes place for a long time in the forest, but not on its conversion into permanent open fields by shifting and burning the trees. To support the biological, ecological, technical, social, or economic aspects of shifting cultivation, the illustrations and data provided hereafter are based on localized case studies, mostly from Southeast Asia. The resultant generalization may make cultivation appear more concrete and intelligible, but it should be remembered that shifting cultivation systems differ significantly across the globe.

Shifting cultivation principles and fallow ecology

The practices of shifting cultivators vary widely depending on the differences between the cultures, regions, and ecosystems where it is practised. Key researchers in the field, such as Cairns (2015), Palm et al. (2005) and Ramakrishnan (1992), have developed the definitions and typologies of shifting cultivation. Their work helps reduce the diversity of shifting

DOI: 10.4324/9781003324072-47

cultivation practices to a minimal common point: generally, shifting cultivation involves a short phase of cultivation followed by a long phase of fallow development.

The cultivation phase

The cultivation phase starts with selecting a forest plot to be cleared. At the beginning of the dry season, farmers determine the area to clear based on several factors: the available land, household needs and expected yields, the plot's topography, the previous year's harvest and prospects of surplus or shortage, the distance of the new plot from the village and its potential fertility—judged by past crops, soil texture or color, and the standing vegetation, including tree species, density and sizes (Cairns, 2007). The arrangement of the plots can vary as farmers operate at the village community level. Household fields are grouped in a regulated plot allotment with a mandatory rotation system. Elsewhere, farmers may select the plot independently at the household level, with small family fields scattered throughout the forest. However, in both cases, the cultivation is a household activity.

After plot selection, the villagers fell the forest. They begin by clearing the understory with machetes. Then, they cut the bigger trees with axes, large machetes or even chainsaws. To reduce the exhausting nature of the work, trees are often cut above the large base and buttresses of the trunks, sometimes stumps taller than two metres remain (Schmidt-Vogt, 1999). Clear cutting is rare: some trees are left standing in the field because they are too difficult to cut down, or they are too valuable for the villagers, providing other resources: fruit or building materials. Farmers often pollard the standing trees to limit the shading over crops (Rerkasem et al., 2009). The slashed vegetation is left to dry in the field for covering the soil during the dry season.

After a few months (1–4), depending on the local climate and the density of the fallen biomass, the farmers burn the dry biomass. To limit the risk of an uncontrolled fire expanding into the forest, they first clean the border of the plot and then proceed by pushing the burning material against the wind (Ducourtieux, 2006). After the first run, the leftover small branches are grouped and burnt again, while the bigger ones are moved to mark the limits of the fields and are used as a source of firewood for the home. With the first rains, the field is now ready for cultivation. Villagers sow their fields with a dominant source of staple food (carbohydrate)—either cereals such as rice (*Oryza sativa*), maize (*Zea mays*), millets (*Pennisetum glaucum, Setaria italica, Panicum miliaceum, Eleusine coracana*, etc.), sorghum (*Sorghum bicolor*), wheat (*Triticum* spp.), barley (*Hordeum vulgare*), quinoa (*Chenopodium quinoa*) and fonio (*Digitaria* spp.); or tubers such as yams (*Dioscorea* spp.), taro (*Colocasia esculenta*) and cassava (*Manihot esculenta*). Often 30–50 species of vegetables and herbs (e.g., cucurbits, crucifers, peppers, sunflower, groundnut, etc.) are grown in a complex system (Conklin, 1957; Rerkasem et al., 2009). The diversity of crops provides many services to farmers:

- The different species covering the ground ensure (1) the optimal use of sunlight, (2) limited growth of weeds and (3) limited erosion; the varied roots explore the different soil horizons for the optimal use of water and nutrients (Nye and Greenland, 1960; Ramakrishnan, 1992).
- It provides diversified and balanced food for farmers' households (Hladik et al., 1993).

- It minimizes the impact of the farming risks, either from market hazards (price variation, evolution of demand, etc.) or from natural ones (drought or flooding, pests, etc.), which affect only part of the crops (Brookfield, 2001).

In a plot, sowing is neither standardized nor random. Farmers decide how to carry out sowing based on their experience using all the environment's resources, on a very precise scale, per square metre. For example, the sowing density will depend on the slope, with tubers preferentially planted in large heaps of ashes and maize in the wettest part. After sowing, farmers dedicate most of their time to weeding the swidden field, carried out quickly and regularly, to prevent weeds from putting a strain on the yields. For example, weeding done too late lets the weeds bloom and go to seed, complicating the control of weeds (Roder, 2001).

The harvest can last for a few months, beginning with short-cycle crops (e.g., maize); farmers continue to harvest perennial crops (e.g., bananas, cassava, fruit trees, etc.) well into the fallow period. For staple cereal, average yields range from 0.5–2 tonnes ha^{-1} (rice, fonio) up to 4–5 tonnes ha^{-1} (maize) for the first year of cultivation after slashing. The results vary widely according to the species and the pedo-climatic conditions, but the level decreases quickly with successive years of cultivation.[1] Such yields appear low compared to current ones from chemically fertilized agriculture (e.g., 5–12 tonnes ha^{-1}), but are similar (or higher) to manure-fertilized grain farming (Mazoyer and Roudart, 2005). Moreover, we must consider the associated crops: when added to the other services they provide (see above), their value can reach the level of that of the staple crop (Ducourtieux, 2006).

During the cultivation phase, the supply of both the mineral nutrients and organic matter in the soil is depleted; and the soil becomes compacted. Concurrently, the field becomes progressively invaded by weeds, insects and nematodes. This combination of factors contributes to the declining yields observed in the initial crop cycle, requiring intensified efforts for weeding (Roder, 2001; van Keer, 2003). After a span of several years with diminishing returns, farmers ultimately discontinue cultivating the field. This marks the commencement of the second phase in shifting cultivation—the fallow phase—characterized by the natural development of plant life transitioning from herbaceous to forest ecosystems.

Renewal of fertility: rotation and on-site biomass accumulation

In agriculture, the fertility required by the cultivated plants is generally imported to the field, from a wide array of sources: chemical fertilizers, animal manure, alluvial deposits, etc. Renewal of fertility is based on lateral transfer and concentration of nutrients towards the cultivated field (Mazoyer and Roudart, 2005).

Shifting cultivation contrasts with this general rule: soil fertility comes from the biomass accumulated on site during the fallow period. In the fallow period, the organic matter is stored in the above-ground vegetation, but also in the soil. The accumulated biomass is mobilized for the cultivation phase by its mineralization. The process is continuous and spontaneous in the litter as well as in the organic horizons of the soil. Desiccation following the clearing of the plot, and more importantly burning of the slashed vegetation, amplifies and speeds up mineralization. Burning affects the soil by (1) increasing the amount of available nutrients (P, K); (2) alkalinization, which increases the bioavailability of phosphates; and (3) thermal disinfection, which cleans the top ground of pests (insects, nematodes, fungus, etc.) and of dormant weed seed banks (Gerold et al., 2004; Nye and Greenland, 1960; Ramakrishnan, 1992; Roder, 2001).

During the burning, most of the nitrogen is released into the atmosphere. However, the cultivated plants are supplied by the subsequent mineralization of humus, accelerated by both the burning and exposure of the soil to solar light, and by the biological fixation of atmospheric nitrogen by legumes as associated crops (Szott et al, 1999). With the rapid mineralization of the biomass and the accumulation of the crop, soil fertility decreases quickly during the cultivation phase. The labour productivity wanes, inviting farmers to stop cultivating the field, which is then left for regrowth of spontaneous vegetation. A new fallow phase begins, with biomass slowly accumulating above and below the ground: fertility renewal depends on rotation in shifting cultivation (see Figure 40.1).

Fertility is classically presented as proportional to the fallow duration (Ramakrishnan, 1992). The progressive accumulation of biomass has been proven,[2] but more recent research challenges the oversimplified interpretation of linear progression. The staple crop yield is not strictly proportional to the fallow length (Mertz et al., 2008; van Keer, 2003), but overall increases, with diminishing marginal returns (Coomes et al., 2016).

Farmers must find a trade-off between contradictory factors for the duration of the fallow phase. The first one is the available land at the household or village community level, depending mainly on land suitability (e.g., slopes and depth of soil) and demography.

With older fallows, the yields are higher, and the labour required for weeding is reduced; that induces farmers to look for long rotations. However, when dealing with older fallow plots, farmers also find themselves facing a heightened workload due to the necessity of increased slashing, as well as accessing the fields and transporting the harvested crops back.[3] When there is plenty of land available, farmers may decide to limit the area of fallow in rotation and to allocate the remote land to a community reserve. For example, the Phunoy people of Samlang village decided that the difference in yield after a 12- or 15-year fallow (see Figure 40.1) was not sufficient to compensate for the extra access time to the field. They, therefore, decided to follow a 12-year fallow; incidentally, all land further than three-hour walk from the village is left out of the rotation, in reserve (Ducourtieux, 2006).

Ecology of the fallows

As soon as farmers stop weeding the field, spontaneous vegetation tends to develop, beginning with pioneer herbaceous species, whose seeds had arrived[4] since the clearing and that farmers struggled to control by weeding during the cultivation phase.

While the clearing and subsequent cultivation phase are an allogenic succession (human disruption), the secondary succession after the plot is left to fallow corresponds to an autogenic one, with successive seral communities that vary widely according to regional biomes and local ecosystems (Gómez-Pompa et al., 1991; Mukul and Herbohn, 2016). In this chapter, the examples of species pertain to a localized ecological succession in Southeast Asia, around Phongsaly,[5] Northern Laos (see Figure 40.2).

The first seral community comprises pioneer herbaceous plants (e.g., *Pennisetum* spp., *Imperata cylindrica, Ageratum conyzoides* and *Miscanthus* spp.). Scrub species (e.g., *Chromolaena odorata*) quickly grow over the weeds and supplant them in a few years as the second seral community. After five years or more, the first trees appear (i.e., stand initiation), from heliophilous fast-growing species (e.g., *Trema orientalis, Macaranga denticulata* and *Mallotus paniculatus*). When the canopy closes, only shade-tolerant (sciaphilous) plants survive in the understory. The trees are engaged in a competition for light (stem exclusion),

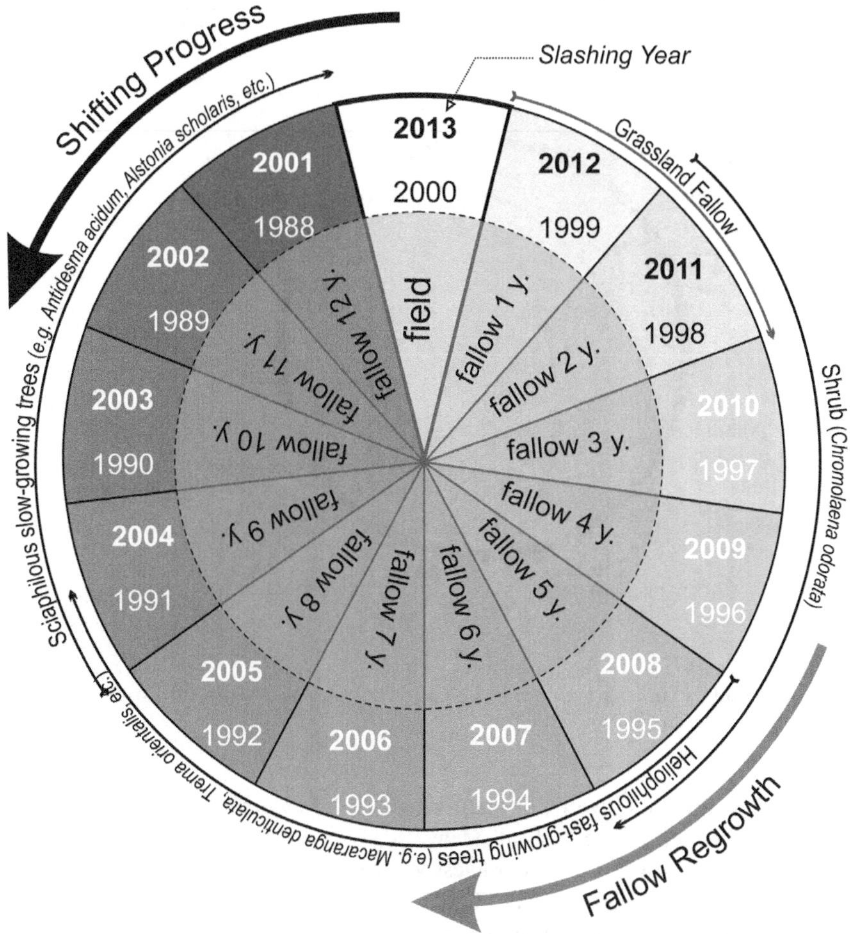

Figure 40.1 Rotation of a long fallow phase and a short cultivation phase in a shifting cultivation. Example from Ban Samlang village, Phongsaly, Northern Laos. (Author's own work.)

which modifies their habits and exposes them to rapid decay in windthrow, leaving room for the last seral community by stand-replacing of sciaphilous slow-growing trees (e.g., *Antidesma acidum*, *Alstonia scholaris* and *Spondias pinnata*). They comprise the old fallow, a secondary forest over ten years old.

Such a sequence is common in forest ecology (Gómez-Pompa et al., 1991), but secondary successions in shifting cultivation fallow differ from spontaneous ones because of four human-based specific and interactive factors (Cairns, 2007; Cairns, 2015; de Jong et al., 2001; Mukul and Herbohn, 2016; Ramakrishnan, 1992; Rerkasem et al., 2009; Schmidt-Vogt, 1999):

- The plot slashing without grubbing benefits the tree species that are able to resprout from the stump. The large scale of the clearing furthers heliophilous species and burning favours fire-tolerant ones for the first steps of the succession. For example, shifting cultivation has

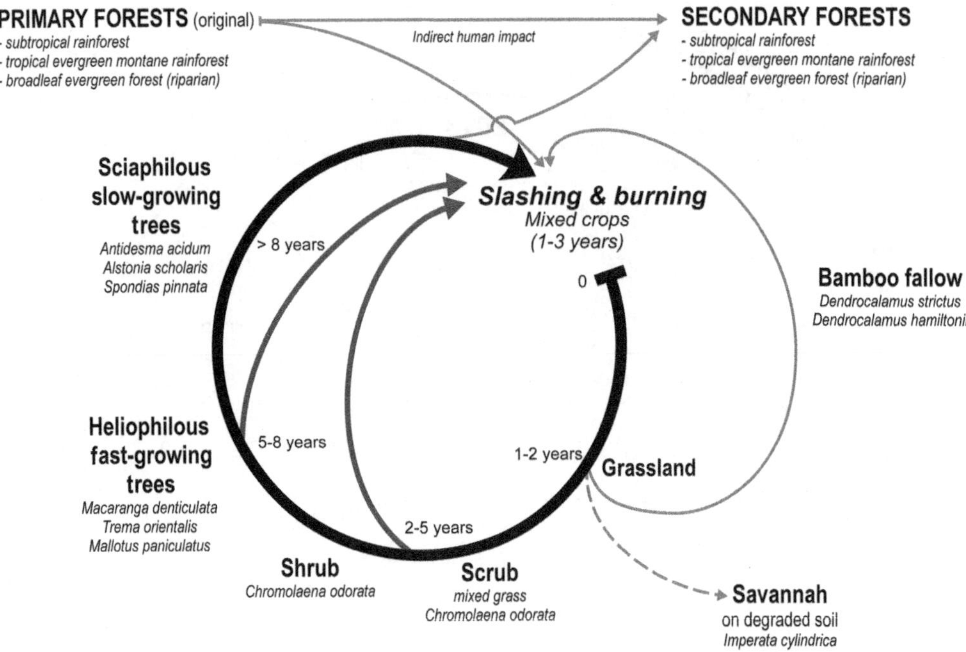

Figure 40.2 Secondary succession in a shifting cultivation fallow. Example from Phongsaly, Northern Laos. (Author's own work.)

contributed to the wide extent of Siam Weed (*Chromolaena odorata*) in Southeast Asia since the end of the 19th century (Roder, 2001).

- Farmers use fallows for grazing their livestock, for hunting and for gathering spontaneous fruits, vegetable, building material (e.g., wood, rattan and weed straw), and medicinal plants, either for self-consumption or for sale.
- Farmers do not cut down all trees when slashing the fallow. They maintain some of them, particularly useful species, such as palms,[6] fruit trees (e.g., *Mangifera* spp., *Carica papaya*, *Musa* spp., etc.), resin trees (e.g., *Styrax benzoin*), etc.. Cycle after cycle, the reproduction of those species is favoured and the plant density is increased in the fallow.
- Farmers may plant perennial plants in the young fallow, either for harvesting[7] (van der Meer Simo et al., 2020; Wood et al., 2016) or for improving fertility renewal.[8]

All these human interventions affect secondary successions and lead to very specific ecosystems (Mukul and Herbohn, 2016). In the context of shifting cultivation, the landscape assumes a mosaic pattern reminiscent of a 'leopard skin' (Heinimann et al., 2017). The size of these mosaic spots[9] is influenced by both the allocation system and the dimensions of the social and productive entities, which can range from village or clan communities to nuclear or extended families. Old fallows are secondary forests, but they differ from old-growth forests. Fallows are not abandoned areas, but productive areas that farmers exploit and manage. In extreme cases, the economic value of products collected from the fallow can reach a level that dissuades the farmers from slashing the fallow, which then becomes a permanent garden, for example, in Indonesia with rubber agroforest in Kalimantan or dammar agroforest gardens in Java (Babin, 2004; Cairns, 2007; Michon, 2005). Therefore, shifting cultivation appears as a variety of agroforestry (Cairns, 2015, 2017).

Forest dynamics and environmental impact of shifting cultivation

Sustainability limits of shifting cultivation: forest degradation and demographic crisis

Agriculture emerged around 10,000 years ago in half a dozen of Neolithic centres, which were mostly in open ecosystems: steppes of the Fertile Crescent, flood plains, peatlands, etc. (Mazoyer and Roudart, 2005)[10]. Rapidly, the expansion of agriculture encountered the forests, the predominant ecosystems which had been expanding since the last Glacial Period (Williams, 2006). To conquer those forest frontiers, farmers had to design new tools, new techniques and new farming systems that lead to the appearance of shifting cultivation. For example, based on artefacts, archaeologists conclude that the influence of agriculture—i.e., shifting cultivation—progressed at an average pace of one kilometre per year in Neolithic Europe.[11] However, should we imagine a pioneer front, inexorably progressing from southeast to northwest, with forest ahead and open field behind? This is unlikely. Clearing a primary dense forest,[12] either a temperate or a tropical one, requires ten times more work than a 15–20 years old fallow. With a population limited to less than 100 million people at the dawn of our era, labour was the scarcest factor of production (Mazoyer and Roudart, 2005). Neolithic farmers were more interested in living in sedentary settlements and in basing their livelihood on rotational shifting cultivation around their village, with long fallows, than in being nomadic along a pioneer deforestation front. Such a rationale still prevails nowadays, when population densities remain low and compatible with long fallows.

The forest space included in the rotation tends to increase with the needs of a growing population.[13] To maintain the yield levels and the duration of the fallow period, the villagers had to expand their territory within the rotational farming system. However, that was not a permanent solution: eventually, they faced impassable obstacles (e.g., large river and mountains) or borders of forests cultivated by other villages; they might be reluctant to increase to an unbearable level the access time to the fields; some might reject the pressure of community rules enforced by more and more distant village or clan authorities. Therefore, some people left the village to find a new community settlement further into the primary forest, while most of the households stayed. The inclusion of primary forests into shifting cultivation rotation combines these two processes of crown expansion and new settlement leap; it ends when the pristine forest resource is no longer available, or too far away. In a mature shifting cultivation system, agroforests of the villages are adjoining, leaving no room for expansion.

To face continued demographic growth, farmers must then rely on speeding up rotation to increase the crop area in limited accessible space. The early stages are barely noticeable,[14] but, with shorter and shorter fallows, (1) biomass accumulation is more and more limited (Ribeiro et al., 2015)—yields drop due to inadequate fertility (see Figure 40.3); (2) the fallow time becomes insufficient for forest secondary successions. The forest is more and more degraded and finally vanishes while people convert to permanent farming (Mazoyer and Roudart, 2005; Williams, 2006).

This classic demographic crisis explains how shifting cultivation contributed to the deforestation of the most fragile forests, close to a Neolithic centre. For example, the Mediterranean forest quickly disappeared, before our era[15], while the Western European

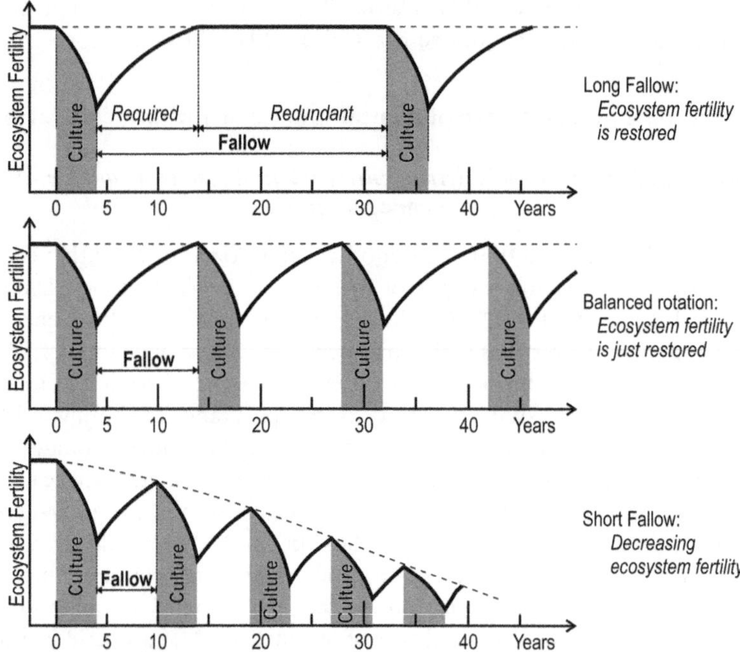

Figure 40.3 Shifting cultivation crisis with accelerating rotation. (Based on Mazoyer and Roudart, 2005, p. 115, from earlier works by Guillemin 1956 and Ruthenberg 1976.)

forest was still relatively extensive at the time of the great medieval clearings ('assarting'), which were not for shifting cultivation, but for permanent agriculture, whose fertility renewal relied on lateral transfer from pastures to fields (Mazoyer and Roudart, 2005). Agriculture was no longer in the forest, but outside the forest instead. Slashing and burning were just two of the techniques used in the pioneer front to convert forests into definitive open fields, in a rationale that cannot be compared with rotational shifting cultivation, which nowadays occurs only in tropical and equatorial forests.

Shifting cultivation and biodiversity: a complex nexus

With deforestation, another facet of the environmental impact of shifting cultivation is biodiversity. At the plot level, the issue is clear: the level of biodiversity in an old-growth forest (either a primary or late seral) exceeds those of a secondary seral community (Cairns, 2015; Fuying et al., 2018; Gerold et al., 2004).

The decrease in biodiversity can be huge when the shifting cultivation faces the demographic crisis, with poor grass and scrub fallows dominating the landscape. However, the comparison becomes more complicated when biodiversity is compared at territorial scales, either village or regional level. The biodiversity of an old-growth forest is nearly isotropic; it increases marginally with the size of the sample plot. In a shifting cultivation landscape, the association of species differs from place to place according to the time since the last cultivation phase (Fuying et al., 2018). The level of biodiversity is the sum of the biodiversity from the different stages of the fallow mosaic plus that of the cultivated field (Rerkasem et al., 2009). The number of species in the fallow varies with the number of seral communities comprising the mosaic of ecosystems (Gerold et al., 2004; Heinimann et al., 2017; Schroth et al., 2004). As these seral communities are successive, the species differ in rotation (see Figure 40.1 and Figure 40.2); the total number of species in a shifting cultivation landscape may exceed those of an old-growth forest (Babin, 2004). However, the biodiversity of an ecosystem cannot be reduced to the number of species it comprises. Specialist species, localized and often rare, may be superseded by generalist species when old-growth forests are converted into secondary forests.

Villagers can also manage the fallow to extract more value from the different ecosystems (Cairns, 2015; Wood et al., 2016), with an ambivalent impact on biodiversity: favouring some valuable species, sometimes by planting them, may increase the total number of species per fallow plot; but if these species become numerous, they can supplant the spontaneous ones, thus reducing biodiversity.[16]

In the shifting cultivation field, the biodiversity combines inter- and intra-specific agrobiodiversity. Up to 30 or 50 different species can be planted in a swidden (Cairns, 2015; Conklin, 1957) and, for each, farmers tend to manage a large set of varieties and cultivars to adapt the production to (1) the micro-specificities of the plot; (2) pest and climate hazards, reducing sensitivity to the hazards by mixing cultivars; (3) the current household needs; and (4) the level of the former harvest[17] (Brookfield, 2001; Wood et al., 2016). The stocks for the different cultivars are managed not only at the household level, but also at the community level with exchanges between households (Conklin, 1957; Rerkasem et al., 2009; Xu Jianchu and Mikesell, 2003). Kept away from the Green Revolution, shifting cultivation has suffered less from the erosion of genetic diversity for the main cereals (rice, maize and wheat). For example, more than half of the 90,000 rice germplasms stored in the International Rice Research Institute (IRRI) world gene-bank (Los Baños, Philippines) come from the uplands of two countries, India and Laos (Xu Jianchu and Mikesell, 2003).

Shifting agriculture, carbon cycles and global climate change

As shifting cultivation involves burning forests, it has been included very early in the causes of global climate change. In one of the first glaring publications about the greenhouse effect in 1978, Wong, an oceanographer, incriminated shifting agriculture for the massive net release of carbon into the atmosphere: 'the atmospheric input of carbon dioxide from burning forests is mainly due to the forest fires from the shifting cultivation in tropical regions'.[18]

Although land preparation (slashing, then burning) causes a massive but one-off emission of CO_2, the fallow phase is a significant carbon sink (McNicol et al., 2015). The net carbon balance is close to zero during the complete cycle of shifting cultivation and cannot be assimilated with the high level of net release when the forest is converted into permanent agricultural land (Achard et al., 2014; Bruun et al., 2009; Tinker et al., 1996; Ziegler et al., 2012). The CO_2 released during the spectacular burning of fallow is part of a short carbon cycle, in which the storage/release process takes a few years (IPCC, 2021). This is quite different from the gaseous emissions of chemical-based farming systems, which come from fossil carbon. However, the long-term balance is not completely neutral because (1) the first conversion of the old-growth forest to shifting cultivation released large amounts of CO_2, since primary forests can store three times more carbon than fallow in long rotation, and (2) if the rotation speeds up, the amount of carbon stored in fallow tends to decrease.

Political and social dimensions of forest dynamics under shifting cultivation

Relying on manual labour and yielding low outputs, shifting cultivation offers restricted land and labor productivity in comparison to input-intensive farming systems.[19] Beyond fulfilling the fundamental needs of families, the surplus labour available for accumulation of capital at the household level remains quite limited. Consequently, the potential for both individual capital accumulation and its appropriation by the broader society is minimal. Unlike other agrarian systems, shifting cultivation proves insufficient to sustain large and densely populated communities (Mazoyer and Roudart, 2005). Historical evidence demonstrates that societies reliant on shifting cultivation struggled to establish robust governing entities. This economic structure has led to asymmetrical power dynamics, resulting in conflicts and power imbalances with neighbouring states that practise open field agriculture. These agricultural systems progressively encroached upon the resources of forest-based communities—encompassing land, forests, water and labour. As a consequence, shifting cultivation populations have faced marginalization, a phenomenon explored by Scott (2009). They were subjected to onerous exactions, including tributes paid in metal (gold or iron), goods (such as grains, animals and wood), or even through forced labour and slavery. Such marginalized communities encountered disdain and severe criticism from dominant powers rooted in open field plains. This approach was further propagated through Western European colonialism, leading to its global dissemination.

Widespread condemnation of shifting cultivation

When the British and French colonial empires expanded during the 19th century, explorers, military, executives and scientists discovered many instances of shifting cultivation in tropical regions. They began rapidly a discourse advocating for the elimination of shifting agriculture. For example, Thorel wrote in 1868 while exploring Indochina, the pearl of the French empire[20]:

> The second mode of rice cultivation, which is practiced in forests, is a barbarian, transitory method that is destined to disappear with the progress of civilization. [...] It is practiced from Saigon to China, but more frequently in Cambodia and Laos, where civilization is still much more backward and where forests are more extensive.

The text encapsulates the colonizers' position towards shifting cultivation, which was considered an archaic, unproductive and degrading practice throughout the time the empires persisted. The condemnation took its origins in the very sources of Western European colonialism: racism, the civilizing mission and greed for land. At the time of independence, the new national powers, built on elites trained by the former colonizers, as well as the United Nations bodies,[21] endorsed this critical position.

In the second half of the 1980s, a new concern rose in the scientific community, civil society, the media and political network: how to prevent the degradation of the environment. The FAO and the World Bank launched in 1985 the Tropical Forestry Action Plan, followed by many bilateral agencies and environmental non-governmental organizations. For the first time, it aimed to stop tropical deforestation for environmental reasons and, of course, shifting cultivation was again presented as the main cause of deforestation. Even the

General Assembly of the United Nations joined the movement at the 1992 Earth Summit in Rio: one of the objectives of Agenda 21 is the recommendation to 'limit and aim to halt destructive shifting cultivation by addressing the underlying social and ecological causes' (articles 11–13).

Drawing from misguided notions regarding shifting cultivation, several critical perspectives emerge. Firstly, the consensus revolves around the belief that shifting cultivation lacks sustainability (see Figure 40.3). Secondly, its perceived low productivity perpetuates a cycle of poverty among forest ethnic minorities, impeding their integration into the broader national economic advancement. Moreover, the adverse externalities stemming from this practice, including deforestation, erosion and biodiversity loss, have repercussions affecting not only the practitioners but also the society at large. This prompts intervention wherein the governments, positioned as the custodians of collective well-being, assume the authority to denounce shifting cultivators, proscribe their methods and endorse alternative agricultural approaches and forest preservation initiatives. In this role, governments often discriminate shifting cultivators, who already grapple with challenges in terms of cultural, linguistic and lifestyle disparities, and without adequate political representation (Ramcilovic-Suominen and Kotilainen, 2020). This governmental stance, notably influenced by urban perspectives and the interests of influential elites, tends to eclipse the concerns of forest minorities (Hecht and Cockburn, 2010; Xu Jianchu and Mikesell, 2003). Moreover, urban-based media, politicians and researchers find it convenient to vilify shifting cultivators by merely highlighting freshly burnt fallow landscapes and framing them as harbingers of tropical forest demise.

Shifting cultivators: forest destroyers or forests gardeners

More and more scientists have disputed this logic by demonstrating that (1) shifting cultivators have proved, on many occasions, their ability to adapt their agricultural practices and their livelihoods to changing environment (Cramb, 2007; Hansen and Mertz, 2006; Michon, 2005) and (2) the quick deforestation that occurred in the Tropics during the 20th century resulted mainly from other causes and drivers than rotational shifting cultivation (Rudel, 2005).

The tropical forest covered around 15% of the earth in the 1950s; this had decreased to 12% by the mid-1970s and is at 6% nowadays.[22] Such a steady pace raises strong and legitimate concerns in terms of biodiversity loss and climate impact. If shifting cultivators are the sole or main perpetrators of forest slaughter as widely accepted, the solution is easy: we just must ban shifting cultivation and provide alternative livelihoods for the populations, with either the promotion of new farming systems or direct compensatory payments for environmental services (Cairns, 2017; Coomes et al., 2016; Gullison et al., 2007; Palm et al, 2005). However, the factors contributing tropical deforestation are not as simple. Many studies (Achard et al., 2002, 2014; Boucher et al., 2011; Curtis et al., 2018; Lambin et al., 2001; Rudel, 2005) have demonstrated that the factors driving rapid deforestation are diverse. Shifting cultivation, as indicated in Figure 40.4, plays a secondary role in this process. The primary drivers include the conversion of forests into permanent agriculture, carried out by both smallholders and large estates. This conversion is mainly for crops like soybean, oil palm, hevea and sugarcane, as well as for activities like ranching. Other significant contributors include logging, the establishment of migration settlements and urban expansion, and encroachments facilitated by roads and extensive infrastructures such as mines and hydropower dams.

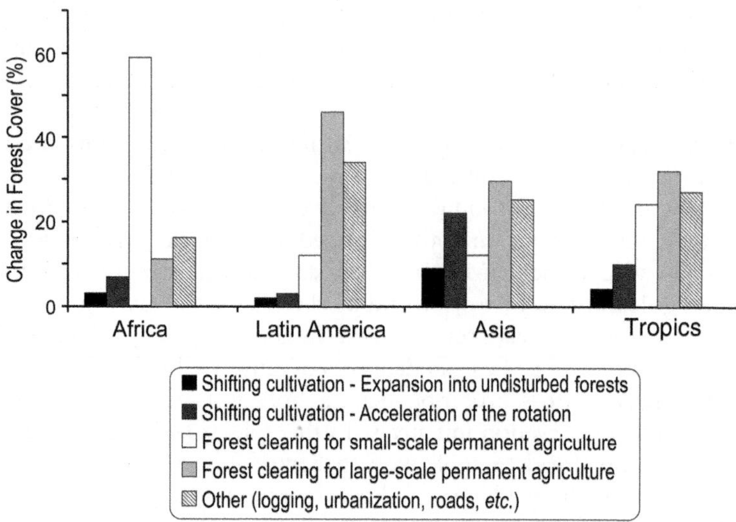

Figure 40.4 Factors of the decline in tropical forest cover. (Based on FAO, 2001, Global Forest Resources Assessment 2000—Main report, FAO, Rome (www.fao.org/documents/show_cdr.asp?url_file=/DOCREP/004/Y1997E/y1997e0t.htm)).

When focusing on shifting cultivation, the environmental policies fail to reduce the progress of deforestation (Achard et al., 2014; Rudel, 2005) and further marginalize the shifting cultivators (Cairns, 2017; Coomes et al., 2016; Dressler et al., 2017; Fox et al., 2009). Furthermore, when a ban on shifting cultivation is enacted and enforced, the fallow forests become accessible to other economic entities. As a result, there is a preference for immediate deforestation over the gradual, long-term degradation associated with shifting cultivation. The misguided approach to tropical forest protection stems from the misconceptions that (1) a demographic crisis is often regarded as a mid- or long-term concern (i.e., timescale confusion) and (2) practices like slashing and burning serve as initial steps in converting forests into open land, similar to the use of bulldozers, for instance. These practices might also overlap with steps in the rotational pattern used for shifting cultivation in forests (Ziegler et al., 2011).

Such misunderstandings lead to overestimations of the population involved in shifting cultivation (Heinimann et al., 2017; Mertz et al., 2009; Thrupp et al., 1997). They base inefficient policies that aim to protect the forest by restricting shifting cultivation, but that may trigger a demographic crisis when the land allowed for fallow is shrunk or when the shifting cultivators' villages are resettled and consolidated on limited land (Lestrelin and Giordano, 2006; Ramcilovic-Suominen and Kotilainen, 2020). Moreover, these policies may serve the interests of powerful economic agents and lobbies, easing their access to forest resources and land (Hecht and Cockburn, 2010; Ziegler et al., 2009).

Conclusion

Shifting cultivation covers a tremendous diversity of farming systems, which originated from the first steps of agriculture. Nowadays, it perdures in tropical forest ecosystems. In principle, shifting cultivation involves soil fertility being renewed by the forest biomass accumulated in situ during the fallow period, resulting in a landscape mosaic of secondary plant formations, including forests. Without (fallow) forest, there is no shifting cultivation: the forest is intrinsic to shifting cultivation.

World maps of deforestation are superimposed on those of shifting cultivation (Hansen et al., 2013; Heinimann et al., 2017; Thrupp et al., 1997). Inferring that shifting cultivation is the main or sole cause of the current tropical deforestation is a too simplistic conclusion, based on confusion, prejudice and collusion of interests (Fox et al., 2009). Another interpretation could be advanced: the tropical forests only last where human societies have the experience to protect them and the interest in doing so. The sole heritage and asset of the shifting cultivators is the forest. The degradation and depletion of the forest only take place over the long term due to the intricate and multifaceted social regulations that the forest farmers have developed. These regulations are designed to safeguard their livelihood, which is far removed from the concept of the tragedy of the commons (Cairns, 2007, 2015, 2017; Cramb, 2007; Forsyth and Walker, 2008; Hecht and Cockburn, 2010). The deforestation and shifting cultivation maps may be understood in a different way; tropical forests remain where shifting cultivators remain (Ziegler et al., 2011). The current eagerness to blame forest farmers overlooks their stewardship and management of forest resources and masks other drivers involved in definitive deforestation (Curtis et al., 2018; Dressler et al., 2017; Rudel, 2005; Williams, 2006), such as permanent agriculture, unsound forest exploitation, urbanization or large-scale infrastructures. Focusing on eradicating shifting cultivation fails to prevent deforestation and even may hasten its trend.

Notes

1 For example, rice yields are often divided by two for the second year, and the field is almost never farmed for more than three years. On the contrary, maize yields decrease too, but less sharply and a field is more often cultivated over four to six successive years in similar soil and climate conditions. The sensitivity to nematodes for aerobic rice roots explains the difference (Roder, 2001).

2 For example, in Northern Thailand, van Keer (2003) found a biomass over 20 tonnes ha^{-1} after a 3-year fallow period, then 30 tonnes ha^{-1} after 7 years, 70 tonnes ha^{-1} after 10 years and 80 tonnes ha^{-1} after 18 years.

3 For example, two labourers that comprise a household can cultivate one hectare; they must carry back home 30 to 50 bags of 50 kg each after harvesting.

4 Either via anemochory (wind dispersal) or zoochory (animal dispersal).

5 500–1,500 m asl, 21°41'N/102°06'E.

6 *Arecaceae* spp., for example, sugar palms (*Borassus flabellifer*, *Arenga pinnata* and *Caryota urens*), oil palm (*Elaeis* spp.), betel palm (*Areca catechu*) and coconut palm (*Cocos nucifera*).

7 For example (Cairns, 2015), fruit trees such as banana (*Musa* spp.); timber species such as teak (*Tectona grandis*); essential oil species such as cardamom (*Amomum* spp.), *Styrax tonkinensis* for benzoin; and other cash crops such as oil palm (*Elaeis* spp.), coffee tree (*coffea* spp.), rubber tree (*Hevea brasiliensis*), rattan (*Calameae*), paper mulberry (*Broussonetia papyrifera*), lacquer tree (*Toxicodendron vernicifluum*) or dammar tree (*Shorea* spp. and *Hopea* spp.), etc.

8 Legume trees, such as *Alnus nepalensis* in Himalaya, *Leucaena leucocephala* in Latin America and Southeast Asia, *Sesbania grandiflora* (edible flowers and fruits) in Southeast Asia, etc. (Cairns, 2015).

9 For example, with plot allotment regulated in a village of 30 nuclear-family households, the surface unit reaches 15–30 ha; if the allotment is dispersed, the unit is limited to 0.5–1 ha.

10 See notably Price T.D. and Gebauer A.B. (eds., 1995), *Last hunters, first farmers: New perspectives on the prehistoric transition to agriculture*, School of American Research, Santa Fe, NM, US.

11 See notably Harris D.R. (ed., 1996), *The origins and spread of agriculture and pastoralism in Eurasia*, Smithsonian Institution Press, Washington, DC.

12 'Primary forest' means here a forest untouched by agriculture, whether hunters and gatherers disturbed it or not.
13 At a demographic growth rate of 0.1% per year, a population doubles in seven centuries; at 0.5% per year, in 140 years; and at 2% per year, in 36 years.
14 For example, the year-to-year yield variability due to climate hazards conceals the average limited reduction when fallow time decreases from 20 to 18 or to 15 years (Mertz et al., 2008).
15 Shifting cultivation is not the sole culprit. The development of harbour cities and the expansion of trade (and wars) by wooden ships required a huge quantity of wood (Williams, 2006).
16 In West Kalimantan, for example, a rubber garden comprises less than 0.02 tree species per square meter, less than half of a woody fallow (Rerkasem et al., 2009).
17 Particularly important for limiting the staple food shortage period the year after a poor harvest.
18 Wong, C.S. (1978), 'Atmospheric input of carbon dioxide from burning wood', *Science*, vol. 200, no 4338, p. 197.
19 For example, with 0.5 ha per worker and a yield of 2 tonnes per hectare (staple and mixed crops, in grain equivalent), the labour productivity reaches only 1 tonne/worker/year, compared to 3 tonnes/worker/year for irrigated rice farming systems, to 10 tonnes/worker/year in three-field rotation with heavy plough farming systems, or more than 500 tonnes/worker/year in mechanized and chemically fertilized agriculture (Mazoyer and Roudart, 2005).
20 See Thorel C. (2001), *Agriculture and ethnobotany of the Mekong Basin*, White Lotus, Bangkok, p. 79 and p. 185.
21 Among others, the Forestry Division of FAO: in 1957, Unasylva published a special issue (vol. 11/1) entitled *Shifting Cultivation: An Appeal by FAO to Governments, Research Centers, Associations and Private Persons Who Are in a Position to Help*, which began with the forewords 'Shifting cultivation, in the humid tropical countries, is the greatest obstacle not only to the immediate increase of agricultural production, but also to the conservation of the production potential for the future, in the form of soils and forests.'
22 FAO and UNEP (2020), *The state of the world's forests 2020: Forests, biodiversity and people*. FAO, Rome.

References

Achard, F., Beuchle, R., Mayaux, P., Stibig, H.-J., Bodart, C., Brink, A., Carboni, S., Desclee, B., Donnay, f., Eva, H. D., Lupi, A., Rasi, R., Seliger, R. and Simonetti, D. (2014) 'Determination of tropical deforestation rates and related carbon losses from 1990 to 2010', *Global Change Biology*, vol 20, pp. 2540–2554.

Achard, F., Eva, H. D., Stibig, H.-J., Mayaux, P., Gallego, J., Richards, T. and Malingreau, J.-P. (2002) 'Determination of deforestation rates of the world's humid tropical forests', *Science*, vol 297, pp. 999–1002.

Babin, D. (ed., 2004) *Beyond tropical deforestation: From tropical deforestation to forest cover dynamics and forest development*, UNESCO/CIRAD, Paris.

Boucher, D., Elias, P., Lininger, K., Roquemore, S. and Saxon, E. (2011) *The root of the problem: What is driving tropical deforestation today?*, Union of Concerned Scientists, Cambridge, MA.

Brookfield, H. C. (2001) *Exploring agrodiversity*, Columbia University Press, New York.

Bruun, T., de Neergaard, A., Lawrence, D. and Ziegler, A. (2009) 'Environmental consequences of the demise in swidden cultivation in Southeast Asia: Carbon storage and soil quality', *Human Ecology*, vol 37, pp. 375–388.

Cairns, M. F. (ed., 2007) *Voices from the forest: Integrating indigenous knowledge into sustainable upland farming*, Earthscan, London.

Cairns, M. F. (ed., 2015) *Shifting cultivation and environmental change: Indigenous people, agriculture and forest conservation*, Earthscan, London.

Cairns, M. F. (ed., 2017) *Shifting cultivation policies: Balancing environmental and social sustainability*, CABI Publishing, Wallingford.

Conklin, H. C. (1957) *Hanunóo agriculture: A report on an integral system of shifting cultivation in the Philippines*, FAO, Rome.

Coomes, O. T., Takasaki, Y. and Rhemtulla, J. M. (2016) 'Forests as landscapes of social inequality: Tropical forest cover and land distribution among shifting cultivators', *Ecology and Society*, vol 21, 20.

Cramb, R. A. (2007) *Land and longhouse: Agrarian transformation in the uplands of Sarawak*, NIAS, Copenhagen.

Curtis, P. G., Slay, C. M., Harris, N. L., Tyukavina, A. and Hansen, M. C. (2018) 'Classifying drivers of global forest loss', *Science*, vol 361, pp. 1108–1111.

De Jong, W., Chokkalingam, U. and Perera, D. (2001) 'The evolution of swidden fallow secondary forests in Asia', *Journal of Tropical Forest Science*, vol 13, pp. 800–815.

Dressler, W. H., Wilson, D., Clendenning, J., Cramb, R., Keenan, R., Mahanty, S., Bruun, T. B., Mertz, O. and Lasco, R. D. (2017), 'The impact of swidden decline on livelihoods and ecosystem services in Southeast Asia: A review of the evidence from 1990 to 2015', *Ambio*, vol 46, pp. 291–310.

Ducourtieux, O. (2006) 'Is the diversity of shifting cultivation held in high enough esteem?', *Moussons*, vol 9 -10, pp. 61–86.

Forsyth, T. and Walker, A. (2008) *Forest guardians, forest destroyers: The politics of environmental knowledge in Northern Thailand*. University of Washington Press, Washington, D.C.

Fox J. M., Fujita Y., Ngidang D., Peluso N., Potter L., Sakuntaladewi, N., Sturgeon, J. and Thomas, D. (2009) 'Policies, political-economy, and swidden in Southeast Asia', *Human Ecology*, vol 37, pp. 305–322.

Fuying D., Yunling, H. and Runguo, Z. (2018) 'Recovery of functional diversity following shifting cultivation in tropical monsoon forests', *Forests*, vol 9, pp. 506–525.

Gerold G., Fremerey, M. and Guhardja, E. (eds., 2004) *Land use, nature conservation and the stability of rainforest margins in Southeast Asia*. Springer, Berlin.

Gómez-Pompa, A., Whitmore, T. C. and Hadley, M. (eds., 1991) *Rain forest regeneration and management*. UNESCO, Paris.

Gullison, R. E., Frumhoff, P. C., Canadell, J. G., Field, C. B., Nepstad, D. C., Hayhoe, K., Avissar, R., Curran, L. M., Friedlingstein, P., Jones, C. D. and Nobre, C. (2007) 'Tropical forests and climate policy', *Science*, vol 316, pp. 985–986.

Hansen, M. C., Potapov, P. V., Moore, R., Hancher, M., Turubanova, S. A., Tyukavina, A., Thau, D., Stehman, S. V., Goetz, S. J., Loveland, T. R., Kommareddy, A., Egorov, A., Chini, L., Justice, C. O. and Townsend, J. R. G. (2013) 'High- resolution global maps of 21st-century forest cover change', *Science*, vol 342, pp. 850–853

Hansen, T. S. and Mertz, O. (2006) 'Extinction or adaptation? Three decades of change in shifting cultivation in Sarawak, Malaysia', *Land Degradation and Development*, vol 17, pp. 135–148.

Hecht, S. B. and Cockburn, A. (2010) *The fate of the forest: Developers, destroyers, and defenders of the Amazon*, 2nd edition. University of Chicago Press, Chicago, IL.

Heinimann, A., Mertz, O., Frolking, S., Egelund Christensen, A., Hurni, K., Sedano, F., Chini, L. P., Sahajpal, R., Hansen, M. and Hurtt, G. (2017) 'A global view of shifting cultivation: Recent, current, and future extent', *PLOS ONE*, vol 12, e0184479.

Hladik C. M., Hladik, A., Linares, O. F., Pagezy, H., Semple, A. and Hadley, M. (eds., 1993) *Tropical forests, people and food: Biocultural interactions and applications to development*. UNESCO, Paris.

IPCC. (2021) Climate Change 2021: The Physical Science Basis. Contribution of Working Group I to the Sixth Assessment Report of the Intergovernmental Panel on Climate Change [Masson-Delmotte, V., P. Zhai, A. Pirani, S.L. Connors, C. Péan, S. Berger, N. Caud, Y. Chen, L. Goldfarb, M.I. Gomis, M. Huang, K. Leitzell, E. Lonnoy, J.B.R. Matthews, T.K. Maycock, T. Waterfield, O. Yelekçi, R. Yu and B. Zhou (eds.)]. Cambridge University Press, Cambridge.

Lambin, E. F. et al. (2001) 'The causes of land-use and land-cover change: Moving beyond the myths', *Global Environmental Change*, vol 11, pp. 261–269.

Lestrelin, G. and Giordano, M. (2006) 'Upland development policy, livelihood change and land degradation: Interactions from a Laotian village', *Land Degradation and Development*, vol 18, pp. 55–76.

Mazoyer, M. and Roudart, L. (2005) *A history of world agriculture: From the Neolithic age to current crisis*. Monthly Review Press, New York.

McNicol, I. M., Berry, N. J., Bruun, T. B., Hergoualc'h, K., Mertz, O. et al. (2015) 'Development of allometric models for above and belowground biomass in swidden cultivation fallows of Northern Laos', *Forest Ecology and Management*, vol 357, pp. 104–116.

Mertz, O., Leisz, S. J., Heinimann, A., Rerkasem, K., Thiha, Dressler, W., Pham, V. C., Vu, K. C., Schmidt-Vogt, D., Colfer, C. J. P., Epprecht, M., Padoch, C. and Potter, L. (2009) 'Who counts? Demography of swidden cultivators in Southeast Asia', *Human Ecology*, vol 37, pp. 281–289.

Mertz, O., Wadley, R. L., Nielsen, U., Bruun, T. B., Colfer, C. J. P., de Neergaard, A., Jepsen, M. R., Martinussen, T., Zhao, Q., Nowig, G. T. and Magid, J. (2008) 'A fresh look at shifting cultivation: Fallow length an uncertain indicator of productivity', *Agricultural Systems*, vol 96, pp. 75–84.

Michon, G. (ed., 2005) *Domesticating forests: How farmers manage forest resources.* IRD/Cifor/Icraf, Bogor.

Mukul, S. A. and Herbohn, J. (2016) 'The impacts of shifting cultivation on secondary forests dynamics in tropics: A synthesis of the key findings and spatio temporal distribution of research', *Environmental Science & Policy*, vol 55, pp. 167–177.

Nye, P.H. and Greenland, D.J. (1960) *The soil under shifting cultivation.* CABI, Wallingford.

Palm, C. A., Vosti, S. A., Sanchez, P. A. and Ericksen, P. J. (eds., 2005) *Slash-and-burn agriculture: The search for alternatives.* Columbia University Press, New York.

Ramakrishnan, P. S. (1992) *Shifting agriculture and sustainable development: An interdisciplinary study from north-eastern India.* UNESCO, Paris.

Ramcilovic-Suominen S. and Kotilainen J. (2020) 'Power relations in community resilience and politics of shifting cultivation in Laos', *Forest Policy and Economics*, vol 115, 102159.

Rerkasem, K., Lawrence, D., Padoch, C., Schmidt-Vogt, D., Ziegler, A. and Bruun, T. B. (2009) 'Consequences of swidden transitions for crop and fallow biodiversity in Southeast Asia', *Human Ecology*, vol 37, pp. 347–360.

Ribeiro Filho, A. A., Adams, C., Manfredini, S., Aguilar, R. and Neves, W. A. (2015) 'Dynamics of soil chemical properties in shifting cultivation systems in the tropics: A meta-analysis', *Soil Use and Management*, vol 31, pp. 474–482.

Roder, W. (ed., 2001) *Slash-and-burn rice systems in the hills of northern Lao PDR: Description, challenges, and opportunities.* IRRI, Los Baños, Philippines.

Rudel, T. K. (2005) *Tropical forests: Regional paths of destruction and regeneration in the late twentieth century.* Columbia University Press, New York.

Schmidt-Vogt, D. (1999) *Swidden farming and fallow vegetation in northern Thailand.* Franz Steiner, Stuttgart.

Schroth, G., Da Fonseca, G. A. B., Harvey, C. A., Gascon, C., Vasconcelos, H. L. and Izac, A.-M. N. (eds., 2004) *Agroforestry and biodiversity conservation in tropical landscapes.* Island Press, Washington, DC.

Scott, J. C. (2009) *The Art of not being governed: An anarchist History of upland Southeast Asia.* Yale University Press, New Haven, CT.

Szott, L. T., Palm, C. A. and Buresh, R. J. (1999) 'Ecosystem fertility and fallow function in the humid and subhumid tropics', *Agroforestry Systems*, vol 47, pp. 163–196.

Thrupp L. A., Hecht S. B. and Browder J. O. (1997) *The diversity and dynamics of shifting cultivation: Myths, realities, and policy implications.* World Resources Institute, Washington, DC.

Tinker, B. P., Ingram, J. S. I. and Struwe, S. (1996) 'Effects of slash-and-burn agriculture and deforestation on climate change', *Agriculture, Ecosystems and Environment*, vol 58, no 1–2, pp. 13–22.

Van Der Meer Simo, A., Kanowski, P. and Barney, K. (2020) 'The role of agroforestry in swidden transitions: A case study in the context of customary land tenure in Central Lao PDR', *Agroforestry Systems*, vol 94, pp. 1929–1944.

Van Keer, K. (2003) 'On-farm agronomic diagnosis of transitional upland rice swidden cropping systems in northern Thailand', PhD thesis, Katholieke Universiteit Leuven, Leuven.

Williams, M. (2006) *Deforesting the Earth: From prehistory to global crisis, an abridgment.* University Of Chicago, Chicago, IL.

Wood, S. L. R., Rhemtulla, J. M. and Coomes, O. T. (2016) 'Intensification of tropical fallow-based agriculture: Trading-off ecosystem services for economic gain in shifting cultivation landscapes?', *Agriculture, Ecosystems & Environment*, vol 215, pp. 47–56.

Xu, J. and Mikesell, S. (eds., 2003) *Landscapes of diversity: Indigenous knowledge, sustainable livelihoods and resource governance in montane mainland Southeast Asia.* Yunnan Science and Technology Press, Kunming.

Ziegler, A. D., Fox, J. M., Webb, E. L., Padoch, C., Leisz, S. J., Cramb, R. A., Mertz, O., Bruun, T. B. and Vien, T. C. (2011) 'Recognizing contemporary roles of swidden agriculture in transforming landscapes of southeast Asia', *Conservation Biology*, vol 25, pp. 846–848.

Ziegler, A. D., Fox, J. M. and Xu, J. (2009) 'The Rubber Juggernaut', *Science*, vol 324, pp. 1024–1025.

Ziegler, A. D., Phelps, J., Yuen, J. Q., Webb, E. L., Lawrence, D., Fox, J. M., Bruun, T. B., Leisz, S. J., Ryan, C. M., Dressler, W., Mertz, O., Pascual, U., Padoch, C. and Koh, L. P. (2012) 'Carbon outcomes of major land-cover transitions in SE Asia: Great uncertainties and REDD+ policy implications'. *Global Change Biology*, vol 18, pp. 3087–3099.

41
INDIGENOUS FOREST KNOWLEDGE

Hugo Asselin

Introduction

There are more than 476 million Indigenous people in the world, in some 90 countries (ILO, 2019). Article 33 of the United Nations Declaration on the Rights of Indigenous Peoples recognizes the principle of self-identification, whereby Indigenous peoples themselves define their own identity based on various criteria, among which are culture, language and the occupation of ancestral lands. Several Indigenous peoples live in forested ecosystems and rely to various extents on ecosystem goods and services to meet their needs. In many countries, Indigenous identities, cultures and practices are closely linked to traditional lands. This intimate connection between cultures and ecosystems is reflected in the vast body of Indigenous forest knowledge. Akin to the traditional ecological knowledge concept, it refers to "a cumulative body of knowledge, practice and belief evolving by adaptive processes and handed down through generations by cultural transmission, about the relationship of living beings (including humans) with one another and with their environment" (Berkes, 2018, p. 8).

Indigenous practices on the land shape forest composition, structure and dynamics into "cultural landscapes" (Cuerrier et al., 2015). These landscapes, which have evolved under the joint influence of natural processes and sustainable cultural practices, tend to have higher biodiversity compared to industrially managed lands, and even compared to protected areas (Pradhan et al., 2019). However, Indigenous knowledge is still largely ignored in forest planning and management (e.g., Sierra-Huelsz et al., 2020), and mere modifications to the prevailing science-and-technology-based forestry regimes will not suffice to achieve "Indigenous forestry". A paradigm shift is required to base forest planning and management on Indigenous values, practices and knowledge, actively involving Indigenous people in forest governance and valuing their own governance systems (Jackson et al., 2021; Teitelbaum et al., 2023). Drawing on examples from various forest ecosystems around the world, this chapter reviews the properties of Indigenous forest knowledge that make it complementary to scientific knowledge, as well as key contributions that Indigenous knowledge can make to forest management.

 DOI: 10.4324/9781003324072-48

Properties of Indigenous forest knowledge

Indigenous and scientific knowledge are not engaged in a "credibility contest" (Trosper and Parrotta, 2012), and they should instead be considered on equal footing. Scientific "validation" of Indigenous knowledge is futile at best, and carries a risk of perpetuating power imbalances (Mazzocchi, 2020). In fact, the intrinsic properties of Indigenous knowledge make it complementary to scientific knowledge in several ways. Indigenous knowledge is cumulative, dynamic, long-term, holistic, local, embedded, moral and spiritual (Menzies and Butler, 2006).

Indigenous forest knowledge is cumulative, as it is the sum of empirical observations acquired through trial-and-error by passing generations. This is somewhat similar to scientific knowledge, also based on empirical evidence. The term "traditional", often used to qualify Indigenous knowledge, should not be understood as meaning "static", "ancient" or "outdated". It rather implies that each generation modifies the accumulated body of knowledge, practices and beliefs according to its own experiences, in a dynamic and adaptive fashion. Indigenous forest knowledge is thus constantly being updated by modifying, adding or deleting information to take into account the evolving context and the apparition of innovations. For example, an Indigenous hunter using a rifle and a global positioning system is using contemporary tools but still can rely on traditional knowledge.

Being rooted in traditions and still mostly transmitted orally, Indigenous forest knowledge provides a long-term perspective, which can prove particularly useful in areas where written archives or instrumental data are recent, discontinuous or unavailable. There is growing concern over the loss of knowledge and cultural erosion that proceeds at an accelerating pace. The United Nations (2022) indeed estimates that up to 50–95% of the world's languages will become extinct or seriously endangered by the end of the 21st century, largely because of acculturation. A study conducted with the Tsimane' people of Bolivia showed a 20% decrease of traditional plant use reports between 2000 and 2009, irrespective of individuals' age, sex, schooling or Spanish fluency (Reyes-Garcia et al., 2013). The decrease was more pronounced for men than for women and for informants living in villages close to market towns than for those in remote settlements. Nevertheless, plant use reports in this study did not significantly differ according to decade of birth (from the 1920s to the 1980s), illustrating the effectiveness of intergenerational knowledge transfer.

Indigenous forest knowledge is holistic, meaning that it considers all elements of the environment (including humans) to be interconnected and influencing each other. For example, in British Columbia (Canada), the Nuu-chah-nulth philosophy of *hishuk'ish tsawalk* (everything is one)

> connects people, animals, plants, and the natural and the supernatural (spiritual) realms in a seamless and interconnected web of life where all life forms are revered and worthy of mutual respect. The land, water, animals and plants are regarded as your kinfolk, not as a commodity that can be exploited.
>
> (Coté, 2016)

Similarly, in Aotearoa New Zealand, the Māori concept of *whakapapa* is a way of understanding the world where everything, living and non-living, is connected and related (McAllister et al., 2023). Such holistic worldviews are in contrast with scientific knowledge, which is reductionist, meaning that it considers the different elements of the environment separately. In addition, scientific knowledge generally assumes that mind and matter are separate, and thus that humans are outside the environment and can control or manage it. Indigenous people rather consider mind and matter on the same level, and that it is not resources that need to be managed, but resources users. In Morocco, Berber communities have developed a forest management system, known as *agdal*, which is based on a holistic worldview (Genin and Simenel, 2011). Following customary laws, they use periodicity of cutting, harvesting quotas, species selection and functional zoning to shape forest landscapes into diversified patches in order to satisfy their material, social and cultural needs, while preserving biodiversity.

Indigenous forest knowledge is local and provides a level of detail that is hard to achieve by other means. For example, in southern India, *amla* trees (*Phyllanthus emblica* and *Phyllanthus indofischeri*) are infested by *Taxillus tomentosus*, a native but invasive mistletoe (Rist et al., 2010). People from the local Soliga Indigenous community get >10% of their cash income from selling the fruit of the *amla* tree. They identified almost three times more secondary host species to mistletoe than scientific surveys (35 vs. 12 species), they pointed to mammals as important mistletoe dispersers (whereas scientific information was restricted to bird dispersers) and they pointed to fire suppression as a possible cause for mistletoe spread in the area. Such information could be key to design effective management strategies.

Although Indigenous forest knowledge is local, most people use extensive territories. Just as scientific studies need several sampling sites over wide areas to be generalizable, it is possible to scale up Indigenous knowledge by interviewing several knowledge holders from different families, communities or peoples, to gather information over large areas (e.g., thousand km^2) and extended periods, something that would be almost impossible—or then very costly—using conventional scientific studies. One such case was documented in the Northern Territory of Australia, where limited and localized surveys suggested major declines of some mammal species, but with inconsistencies in the data. Interviews with Indigenous knowledge holders allowed to substantially expand the spatial and temporal scales of data acquisition (Ziembicki et al., 2013). Indigenous knowledge confirmed decline for several mammal species and pointed towards habitat modifications due to cessation of traditional surface burning practices as a probable cause.

Indigenous forest knowledge is context-dependent, not only environmentally but also culturally. Knowledge is indeed embedded within each culture's idiosyncrasies that shape how people acquire and use knowledge. Hence, knowledge cannot be readily transferred from one context to another or from one community to another. For example, in the Eastern Usambara Mountains of Tanzania, community–forest interactions differ according to altitude (upstream vs. downstream) and distance to forest reserves (Fadhilia et al., 2016). In a study of the botanical knowledge held by 57 Indigenous communities in northwestern South America, only a moderate amount of knowledge was shared, even about shared species (Cámara et al., 2019).

Indigenous knowledge is also moral and spiritual, whereas scientific knowledge generally adopts a mechanistic viewpoint, assuming that phenomena have rational explanations and that events may be connected as cause and effect. Interviews and sharing circles held within 12 Indigenous communities across Canada revealed eight ethical principles inherent to caring for the land: reciprocity, gratitude, humility, agency, responsibility, moderation, mindfulness, and respect (Menzies et al., 2021). These principles translate into seven recommendations for land management and monitoring: identify and listen to community priorities; seek guidance from elders; prioritize youth involvement; foster connections with the land; create opportunities for knowledge transfer; use traditional language; and build a sense of community. Indigenous peoples' practices are based on moral codes, often transmitted orally through storytelling (Fernández-Llamazares and Cabeza, 2018), to ensure that the land is used sustainably. For example, in the Peruvian Amazon, the Kukama people have adapted to declining wildlife populations during five years of extreme flooding by greatly reducing their hunting of white-lipped peccary (*Tayassu pecari*), collared peccary (*Pecari tajacu*), red brocket deer (*Mazama americana*), lowland paca (*Cuniculus paca*), and black agouti (*Dasyprocta fuliginosa*) (Bodmer et al., 2020). They compensated by fishing and buying/trading wild meat with communities from upland sites not affected by flooding.

Key contributions of Indigenous knowledge to forest management

There are several ways in which Indigenous knowledge can contribute to forest management: by providing biological information and ecological insight, as a source of alternative management practices, by favouring biodiversity conservation, by contributing to environmental monitoring and assessment, as a support for social development and as a source of environmental ethics principles (Berkes, 2018).

Indigenous forest knowledge can provide extensive biological and ecological information at levels of detail that would be hard to reach using quantitative scientific research methods. For example, in a study conducted within seven Merap and Punan villages in East Kalimantan (Indonesia), 52 informants were asked if, where and when they had observed eight species of regional conservation interest: rafflesia (*Rafflesia* spp.), black orchid (*Coelogyne pandurata*), sun bear (*Helarctos malayanus*), tarsier (*Tarsius bancanus*), slow loris (*Nycticebus coucang*), proboscis monkey (*Nasalis larvatus*), clouded leopard (*Neofelis diardi/ Neofelis nebulosa*) and orangutan (*Pongo pygmaeus*). The study allowed to gather important information on the elusive rafflesia and clouded leopard, and extended the known distributions of sun bear, tarsier, slow loris and clouded leopard (Padmanaba et al., 2013). It was completed within only six weeks, at a total cost estimated to have been one or two orders of magnitude lower than expert field surveys.

Indigenous forest knowledge sometimes tends to focus on larger or more culturally salient species (Ziembicki et al., 2013), such as cultural keystone species, i.e., species that shape in a major way the cultural identity of a people, as reflected in the fundamental roles they play in diet, materials, medicine, and/or spiritual practices (Garibaldi and Turner, 2004). An example of cultural keystone species is the eastern white pine (*Pinus strobus*), perceived as the "king of the forest" by the Kitcisakik Algonquin community of Quebec, Canada (Uprety et al., 2013). White pine—*cigwâtik* in the local language—is mentioned in traditional stories and myths. It provides habitat for culturally important wildlife species, such as bald eagle (*Haliaeetus leucocephalus*), moose (*Alces americanus*) and marten (*Martes americana*). White pine is also an important medicinal plant species, it is used as timber, and tall individuals towering above the canopy serve as landmarks (Figure 41.1). Shelterwood cuts and understorey plantations are often used to manage white pine stands. According to the Kitcisakik people, association with balsam fir (*Abies balsamea*) should however be avoided, as a legend says the two species are enemies.

One classic example of how forest management practices can be influenced by Indigenous knowledge is the management of traditional forests by the Menominee tribe of Wisconsin (USA) (Mausel et al., 2017). By historically preferring selective cutting of sick, fallen and

Figure 41.1 Two white pines (*Pinus strobus*) towering above the surrounding canopy and used as landmarks by the members of the Kitcisakik Algonquin community. White pine is a cultural keystone species to this community. (Photograph by Pierre Cartier.)

old trees over short-rotation clear-cutting (advised by federal authorities) they were able to produce high-quality wood for which they could get good prices on the market, while maintaining forest cover for cultural activities. The difference between Menominee and conventional forestry in terms of residual forest cover is striking and can easily be seen by typing "Menominee County, Wisconsin" in Google Earth™. Another example of Indigenous forest management involves the monkey-puzzle tree (*Araucaria araucana*), a sacred species to the Mapuche Pewenche of Chile (literally the people—*che*—of the monkey-puzzle tree—*pewen*). The species is listed as endangered by the International Union for Conservation of Nature, following decades of intense exploitation by the timber-extraction industry and substitution by plantations of exotic species (*Pinus radiata*, *Eucalyptus* spp.) or conversion to pastureland. Seeds of the monkey-puzzle tree are an important part of the Mapuche Pewenche diet and people have developed profound knowledge of the species' ecology. Strict cultural norms ensure sustainable seed harvesting, and people make sure to leave some seeds in the trees and on the ground to favour regeneration. People also plant seeds, transplant seedlings and tend to mature trees (protecting them from frost or healing wounded branches). The intimate relationship between the Mapuche Pewenche and *Araucaria araucana* is a mix of stewardship, ritualized exchange and devotion which has been called "mutual nurturing" (Ladio and Reis, 2023).

Non-timber forest products are also used and managed by Indigenous peoples. While edible and medicinal plants have received much attention, Indigenous forest knowledge also extends to more cryptic or lesser-studied species. For example, the California Indian Tribes were reported to traditionally use at least 36 mushroom species as food, medicine or for other uses such as to make a red pigment used in war paints, to start or carry fire, to polish animal skin or to make bowls (Anderson and Lake, 2013). They have developed mushroom management practices involving prescribed burning, based on their precise knowledge of mushroom habitat, seasonality and ecological relationships. The Kaqchikel people in Guatemala also have extensive knowledge of wild edible mushrooms, with more than 40 species and varieties traded in local markets (Mérida Ponce et al., 2019). In Nepal, a study of lichen use among nine communities with three different cultural backgrounds revealed that seven species were used for medicine, food, ritual and spiritual practices, decoration, bedding, and ethno-veterinary purposes (Devkota et al., 2017).

Indigenous forest knowledge can also be used in the management of so-called "nuisance species". For example, beaver (*Castor canadensis*) can cause substantial damage to forest roads in North America by blocking culverts to stop water flow and create ponds where to build huts to raise their young (Figure 41.2). Forestry companies sometimes hire Indigenous trappers to get rid of problematic beavers. However, problems usually arise in summer, when beaver meat is unpalatable and when fur quality is low (Kneeshaw et al., 2010). Hence, trappers are forced to leave the beavers in the forest, feeling guilty for such a waste. It would be more culturally sound to put preventive trapping programs into place, such as the one developed by the Omushkego Cree community in Ontario, Canada (Ahmed et al., 2022). Indeed, as Indigenous trappers possess profound knowledge of habitat use by beavers, they could identify in advance the areas that might be problematic in summer, and harvest the animals at a time of year when fur and meat are usable.

Certain traditional practices are essential to maintain some species in specific areas. For example, controlled fires set by Indigenous people of southern British Columbia (Canada)

Figure 41.2 Beaver (*Castor canadensis*) dam preventing stream water flow through a culvert. The resulting pond flooded the nearby forest road (to the right, not shown on photograph), causing major damage. (Photograph by Claude-Michel Bouchard.)

have contributed to maintain Garry oak (*Quercus garryana*) in the landscape over most of the last 3000 years, despite an unfavourable, wet and cool climate (Pellatt and Gedalof, 2014). Indigenous population decline since the European colonization and the subsequent application of fire suppression policies have caused replacement of Garry oak by conifer species.

Ecological restoration involves management practices to assist the recovery of degraded, damaged or destroyed ecosystems. Uprety et al. (2012) reviewed the main contributions of Indigenous knowledge to ecological restoration. The site-specific nature of Indigenous knowledge makes it particularly applicable to restoration initiatives. Indigenous knowledge can be used to document the characteristics of reference ecosystems, select appropriate species and sites to maximize restoration success, provide expertise through traditional management practices, manage invasive species and conduct post-restoration monitoring.

While protected areas are a key element of biodiversity conservation strategies, they can be problematic from an Indigenous viewpoint. Considering themselves an integral part of forest ecosystems, Indigenous people cannot contemplate "putting a dome" over an area and preventing themselves to use it. Advances have been made in conservation practices, and protected areas increasingly accommodate a range of Indigenous practices (including subsistence hunting and the collection of plants for traditional purposes) (Ens et al., 2021). Moreover, Indigenous-led approaches to the creation and management of protected areas based on traditional governance systems are increasingly put forward in various jurisdictions worldwide (Tran et al., 2020).

Environmental monitoring and assessment are time-consuming and costly. They are thus often limited to short and specific time frames in order to reduce operation costs. Indigenous people are usually present year-round on the land and will be the first ones to notice any change in the environment. They can take measurements on a continuous basis and maintain long-term monitoring, whereas forest managers usually focus on areas where logging is planned in the short term. Governments sometimes maintain permanent sampling plots, but these are often not numerous enough or cannot be sampled frequently enough. Community-based environmental monitoring, such as the Indigenous Guardian Programs in Australia, Aotearoa-New Zealand, Canada and the United States (Artelle et al., 2019), can be implemented to facilitate mapping of Indigenous knowledge, observations, practices and land use in a dynamic and adaptive way.

Participatory geographic information systems (GIS) are used in different settings to inform decision-makers. For example, the Sami people of Sweden and the Cree and Naskapi First Nations of eastern Canada collect field data on the risks and impacts of resource extraction industries on habitat use by reindeer/caribou (*Rangifer tarandus*) and Indigenous livelihoods (Herrmann et al., 2014). The data collected by the Sami, Cree and Naskapi peoples are site-specific, but the accumulation of data over extended areas and during several consecutive years provides a landscape perspective allowing for a thorough understanding of human–environment relationships that would otherwise be difficult to grasp.

The Kumeyaay people in northwestern Mexico is a good example of how participatory GIS can be used to combine Indigenous and scientific knowledge to more accurately monitor plant populations for sustainable management (Andrade-Sánchez *et al.*, 2021). The Kumeyaay use willow (*Salix* sp.) and spiny rush (*Juncus* sp.) to weave baskets used for collecting, preparing and storing food. This cultural activity is central to several families' livelihood. Facing a marked reduction of spiny rush and willow coverage, the Kumeyaay engaged in a participatory mapping project to create a conservation and management plan to ensure that the collection of spiny brush and willow would be both economically viable and ecologically sustainable. Land users and scientists identified cultural sites and measured the abundance and biomass of both species in different zones, which allowed to determine where restoration or conservation was needed, and where sustainable plant collection was possible.

Effective decision-support tools are needed to foster integration of Indigenous knowledge at all steps of the sustainable forest management process, from inventory to monitoring, through planning and management. As social acceptability of forestry operations largely depends on the cohabitation of different forest users seeking different ecosystem goods and services, maps are probably the most important tool to develop. A challenge in this regard is the reluctance of Indigenous people to share culturally sensitive information such as precise locations of sites of interest (Berkes, 2018). Functional zoning offers a working compromise. The territory is divided into zones ascribed to one or more uses (e.g., wood harvesting, berry picking, hunting or spiritual activities). However, such a compartmented view of the land does not correspond to the holistic perception of Indigenous people. A more culturally appropriate alternative is to use participatory mapping to identify the factors affecting Indigenous landscape value, as was done in a study with the Abitibiwinni (Anicinape) and Ouje-Bougoumou (Cree) communities in northern Quebec (Canada) (Bélisle and Asselin, 2021). Landscape state was evaluated on each family hunting ground based on Indigenous landscape value, biophysical variables (e.g., topography, species composition), and disturbance history (e.g., wildfire, logging). Such maps provide guidelines for forest management planning without giving away key information (e.g. precise location of cultural sites), thus granting Indigenous people increased control over the decision-making.

Putting Indigenous knowledge to work increases the social acceptability of forest management practices. Adopting a holistic viewpoint and recognizing the intricate link between Indigenous people and the land can be achieved through local-scale management stemming from a bottom-up decision-making process. Such "human-scale forestry" is more acceptable as it allows more needs to be met through the provision of a wider range of ecosystem goods and services. Not only does this foster social development but also improves community well-being and resilience, as Indigenous knowledge establishes a clear link between healthy land and healthy people. This is illustrated by the *mino pimatisiwin* concept shared by most Indigenous people from the Algic language family in North America, which means "being alive well", and where health is not only defined in terms of individual physiology, but also in terms of social relations, cultural identity and link to the land (Landry et al., 2019). Similar concepts in other parts of the world include *buen vivir* or *sumak kawsay* for the Quechuas in Ecuador, *hauora* for the Māoris in New Zealand and *ubuntu* for the Ngunis in South Africa.

Respect for other beings—living and non-living—is central to several Indigenous peoples' environmental ethics: this principle is enacted by taking only what is needed, not wasting and sharing. O'Flaherty *et al.* (2009) reported four lessons learnt from the Pikangikum Ojibway elders (Ontario, Canada), which nicely summarize Indigenous environmental ethics and complementarity with scientific knowledge. First, humbly accept that management practices can only partly meet the needs of other beings with whom Indigenous people share the land, and who will be affected by management. Second, adopt a precautionary approach and develop a monitoring system based on Indigenous knowledge. Third, establish effective communication channels between Indigenous people and other stakeholders.

Fourth, consider Indigenous and scientific knowledge on equal footing, each equally contributing to an integrated land management strategy.

Conclusion

Several properties of Indigenous forest knowledge make it complementary to scientific knowledge. Perhaps the most striking difference between the two knowledge types is that Indigenous knowledge is based on a holistic perception of the environment, implying that mind and matter are on the same plane. In other words, Indigenous people consider themselves an integral part of forest ecosystems, rather than considering they can control or manage it from the outside. This is far from being trivial, especially considering the current interest for ecosystem-based forest management, an approach that aims to reduce the differences between natural and managed landscapes to maintain ecosystem functions. Considering that Indigenous knowledge and practices have existed for millennia, that Indigenous people are a key component of many forest ecosystems, and that the idea of a pristine, untouched forest is a myth, ecosystem-based forest management provides a good meeting ground for Indigenous and scientific forest knowledge (Asselin et al., 2015).

The various contributions of Indigenous forest knowledge to forest management practices highlighted in this chapter suggest several ways to reduce the gap between natural and managed landscapes, as most Indigenous practices can be considered drivers of the so-called "historic range of variability". Small steps have been made towards integration of Indigenous knowledge and needs in forestry, showing, for example, that it does not severely impact profitability (e.g., Dhital et al., 2013). The next (big) step will need to move from simple integration of Indigenous knowledge within a scientific paradigm to full-fledged Indigenous forestry.

References

Ahmed, F., Liberda, E. N., Solomon, A., Davey, R., Sutherland, B. and Tsuji, L. J. (2022) 'Indigenous land-based approaches to well-being: The Amisk (Beaver) Harvesting Program in subarctic Ontario, Canada', *International Journal of Environmental Research and Public Health*, vol 19, no 12, art 7335. https://doi.org/10.3390/ijerph19127335

Anderson, M. K. and Lake, F. K. (2013) 'California Indian ethnomycology and associated forest management', *Journal of Ethnobiology*, vol 33, no 1, pp. 33–85. https://doi.org/10.2993/0278-0771-33.1.33

Andrade-Sánchez, J., Eaton-Gonzalez, R., Leyva-Aguilera, C. and Wilken-Robertson, M. (2021) 'Indigenous mapping for integrating traditional knowledge to enhance community-based vegetation management and conservation: The Kumeyaay basket weavers of San José de la Zorra, México', *ISPRS International Journal of Geo-Information*, vol 10, no 3, art. 124. https://doi.org/10.3390/ijgi10030124

Artelle, K. A., Zurba, M., Bhattacharyya, J., Chan, D. E., Brown, K., Housty, J. and Moola, F. (2019) 'Supporting resurgent Indigenous-led governance: A nascent mechanism for just and effective conservation', *Biological Conservation*, vol 240, art 108284. https://doi.org/10.1016/j.biocon.2019.108284

Asselin, H., Larouche, M. and Kneeshaw, D. (2015) 'Assessing forest management scenarios on an Aboriginal territory through simulation modeling', *The Forestry Chronicle*, vol 91, no 4, pp. 426–435. https://doi.org/10.5558/tfc2015-072

Bélisle, A. C. and Asselin, H. (2021) 'A collaborative typology of boreal Indigenous landscapes', *Canadian Journal of Forest Research*, vol 51, no 9, pp. 1253–1262. https://doi.org/10.1139/cjfr-2020-0369

Berkes, F. (2018) *Sacred Ecology*, 4th edition, Routledge, New York, NY. www.routledge.com/Sacred-Ecology/Berkes/p/book/9781138071490

Bodmer, R., Mayor, P., Antunez, M., Fang, T., Chota, K., Yuyarima, T. A., Flores, S., Cosgrove, B., López, N., Pizuri, O. and Puertas, P. (2020) 'Wild meat species, climate change, and indigenous Amazonians', *Journal of Ethnobiology*, vol 40, no 2, pp. 218–233. https://doi.org/10.2993/0278-0771-40.2.218

Cámara-Leret, R., Fortuna, M. A. and Bascompte, J. (2019) 'Indigenous knowledge networks in the face of global change', *Proceedings of the National Academy of Sciences of the USA*, vol 116, no 20, pp. 9913–9918. https://doi.org/10.1073/pnas.1821843116

Coté, C. (2016) '"Indigenizing" food sovereignty. Revitalizing Indigenous food practices and ecological knowledges in Canada and the United States', *Humanities*, vol 5, no 3, art. 57. https://doi.org/10.3390/h5030057

Cuerrier, A., Turner, N. J., Gomes, T. C., Garibaldi, A. and Downing, A. (2015) 'Cultural keystone places: conservation and restoration in cultural landscapes', *Journal of Ethnobiology*, vol 35, no 3, pp. 427–448. https://doi.org/10.2993/0278-0771-35.3.427

Devkota, S., Chaudhary, R. P., Werth, S. and Scheidegger, C. (2017) 'Indigenous knowledge and use of lichens by the lichenophilic communities of the Nepal Himalaya', *Journal of Ethnobiology and Ethnomedicine*, vol 13, art. 15. https://doi.org/10.1186/s13002-017-0142-2

Dhital, N., Raulier, F., Asselin, H., Imbeau, L., Valeria, O. and Bergeron, Y. (2013) 'Emulating boreal forest disturbances dynamics: Can we maintain timber supply, aboriginal land use, and woodland caribou habitat?', *Forestry Chronicle*, vol 89, no 1, pp. 54–65. https://doi.org/10.5558/tfc2013-011

Ens, E., Reyes-García, V., Asselin, H., Hsu, M., Reimerson, E., Reihana, K., Sithole, B., Shen, X., Cavanagh, V. and Adams, M. (2021) 'Recognition of Indigenous ecological knowledge systems in conservation and their role to narrow the knowledge-implementation gap', in C. Ferreira and C.F.C. Klütsch (eds.), *Closing the Knowledge-Implementation Gap in Conservation Science*. Springer, New York, NY. https://link.springer.com/chapter/10.1007/978-3-030-81085-6_5

Fadhilia, B., Liwa, E. and Shemdoe, R. (2016) 'Indigenous knowledge of Zigi community and forest management decision-making: a perspective of community forest interaction', *Journal of Natural Resources and Development*, vol 6, pp. 14–21. https://doi.org/10.5027/jnrd.v6i0.03

Fernández-Llamazares, Á. and Cabeza, M. (2018) 'Rediscovering the potential of indigenous storytelling for conservation practice', *Conservation Letters*, vol 11, no 3, art e12398. https://doi.org/10.1111/conl.12398

Garibaldi, A. and Turner, N. (2004) 'Cultural keystone species: Implications for ecological conservation and restoration', *Ecology and Society*, vol 9, no 3, art 1. www.ecologyandsociety.org/vol9/iss3/art1/

Genin, D. and Simenel, R. (2011) 'Endogenous Berber forest management and the functional shaping of rural forests in Southern Morocco: Implications for shared forest management options', *Human Ecology*, vol 39, no 3, pp. 257–269. https://doi.org/10.1007/s10745-011-9390-2

Herrmann, T. M., Sandström, P., Granqvist, K., D'Astous, N., Vannar, J., Asselin, H., Saganash, N., Mameamskum, J., Guanish, G., Loon, J.-B. and Cuciurean, R. (2014) 'Effects of mining on reindeer/caribou populations and indigenous livelihoods: Community-based monitoring by Sami reindeer herders in Sweden and First Nations in Canada', *The Polar Journal*, vol 4, no 1, pp. 28–51. https://doi.org/10.1080/2154896X.2014.913917

ILO. (2019) *Implementing the ILO Indigenous and Tribal Peoples Convention No. 169: Towards an Inclusive, Sustainable and Just Future*. International Labour Organization, Geneva, Switzerland. www.ilo.org/global/publications/books/WCMS_735607/lang--en/index.htm

Jackson, W., Freeman, M., Freeman, B. and Parry-Husbands, H. (2021) 'Reshaping forest management in Australia to provide nature-based solutions to global challenges', *Australian Forestry*, vol 84, no 2, pp. 50–58. https://doi.org/10.1080/00049158.2021.1894383

Kneeshaw, D., Larouche, M., Asselin, H., Adam, M.-C., Saint-Arnaud, M. and Reyes, G. (2010) 'Road rash: Ecological and social impacts of road networks on First Nations', in M.G. Stevenson and D.C. Natcher (eds.), *Planning Co-existence: Aboriginal Considerations and Approaches in Land Use Planning*. Canadian Circumpolar Institute Press, Edmonton, AB, Canada.

Ladio, A. H. and Reis, M. S. (2023). 'Different relational models have shaped the biocultural conservation over time of *Araucaria araucana* forests and their people', in J. A. Whitaker, C. G. Armstrong and G. Odonne (eds.), *Climatic and Ecological Change in the Americas. A Perspective from Historical Ecology*. Routledge, London. https://library.oapen.org/handle/20.500.12657/75346

Landry, V., Asselin, H. and Lévesque, C. (2019) 'Link to the land and mino-pimatisiwin (comprehensive health) of Indigenous people living in urban areas in Eastern Canada', *International Journal of Environmental Research and Public Health*, vol 16, no 23, art. 4782. https://doi.org/10.3390/ijerph16234782

Mausel, D. L., Waupochick Jr, A. and Pecore, M. (2017) 'Menominee forestry: Past, present, future', *Journal of Forestry*, vol 115, no 5, pp. 366–369. https://doi.org/10.5849/jof.16-046

Mazzocchi, F. (2020) 'A deeper meaning of sustainability: Insights from indigenous knowledge', *The Anthropocene Review*, vol 7, no 1, pp. 77–93. https://doi.org/10.1177/2053019619898888

McAllister, T., Hikuroa, D. and Macinnis-Ng, C. (2023) 'Connecting science to Indigenous knowledge: Kaitiakitanga, conservation, and resource management', *New Zealand Journal of Ecology*, vol 47, no 1, art. 3521. https://doi.org/10.20417/nzjecol.47.3521

Menzies, A. K., Bowles, E., Khan, J. S., McGregor, D., Ford, A. and Popp, J. (2021) *Caring for the Land: Indigenous Practices, Community Values, Changes over Time, and Building Values-Based Environmental Initiatives*. University of Guelph, WISE Lab, Guelph, ON, Canada. https://weavingknowledges.ca/app/uploads/2022/06/Menzies-et-al_CaringForTheLandReport_16-03-2022_for-Results-Sharing.pdf

Menzies, C. R. and Butler, C. (2006) 'Introduction. Understanding ecological knowledge', in C. R. Menzies (ed.), *Traditional Ecological Knowledge and Natural Resource Management*. University of Nebraska Press, Lincoln, NE.www.nebraskapress.unl.edu/nebraska/9780803232464/

Mérida Ponce, J. P., Hernández Calderón, M. A., Comandini, O., Rinaldi, A. C. and Flores Arzú, R. (2019) 'Ethnomycological knowledge among Kaqchikel, indigenous Maya people of Guatemalan Highlands', *Journal of Ethnobiology and Ethnomedicine*, vol 15, art 36. https://doi.org/10.1186/s13002-019-0310-7

O'Flaherty, R. M., Davidson-Hunt, I. J. and Miller, A. M. (2009) 'Anishinaabe stewardship values for sustainable forest management of the Whitefeather forest, Pikangikum First Nation, Ontario', in M. G. Stevenson and D. C. Natcher (eds.), *Changing the Culture of Forestry in Canada: Building Effective Institutions for Aboriginal Engagement in Sustainable Forest Management*. University of Alberta Press, Edmonton, Alberta, Canada. www.uap.ualberta.ca/titles/830-9781896445441-changing-the-culture-of-forestry-in-canada

Padmanaba, M., Sheil, D., Basuki, I. and Nining, L. (2013) 'Accessing local knowledge to identify where species of conservation concern occur in a tropical forest landscape', *Environmental Management*, vol 52, no 2, pp. 348–359. https://doi.org/10.1007/s00267-013-0051-7

Pellatt, M. G. and Gedalof, Z. (2014) 'Environmental change in Garry oak (*Quercus garryana*) ecosystems: The evolution of an eco-cultural landscape', *Biodiversity and Conservation*, vol 23, no 8, pp. 2053–2067. https://doi.org/10.1007/s10531-014-0703-9

Pradhan, A., Ormsby, A. A. and Behera, N. (2019) 'A comparative assessment of tree diversity, biomass and biomass carbon stock between a protected area and a sacred forest of Western Odisha, India', *Ecoscience*, vol 26, no 3, pp. 195–204. https://doi.org/10.1080/11956860.2019.1586118

Reyes-Garcia, V., Guèze, M., Luz, A. C., Paneque-Galvez, J., Macia, M. J., Orta-Martinez, M., Pino, J. and Rubio-Campillo, X. (2013) 'Evidence of traditional knowledge loss among a contemporary indigenous society', *Evolution and Human Behavior*, vol 34, no 4, pp. 249–257. https://doi.org/10.1016/j.evolhumbehav.2013.03.002

Rist, L., Shaanker, R. U., Milner-Gulland, E. J. and Ghazoul, J. (2010) 'The use of traditional ecological knowledge in forest management: An example from India', *Ecology and Society*, vol 15, no 1, art 3. http://www.ecologyandsociety.org/vol15/iss1/art3/

Sierra-Huelsz, J. A., Gerez Fernández, P., López Binnqüist, C., Guibrunet, L. and Ellis, E. A. (2020) 'Traditional ecological knowledge in community forest management: Evolution and limitations in Mexican forest law, policy and practice', *Forests*, vol 11, no 4, art. 403. https://doi.org/10.3390/f11040403

Teitelbaum, S., Asselin, H., Bissonnette, J.-F. and Blouin, D. (2023) 'Governance in the boreal forest: what role for local and Indigenous communities?', in M. Montoro Girona, H. Morin, S. Gauthier and Y. Bergeron (eds.), *Boreal forests in the face of climate change*. Springer, New York, NY. https://link.springer.com/chapter/10.1007/978-3-031-15988-6_20

Tran, T. C., Ban, N. C. and Bhattacharyya, J. (2020) 'A review of successes, challenges, and lessons from Indigenous protected and conserved areas', *Biological Conservation*, vol 241, art 108271. https://doi.org/10.1016/j.biocon.2019.108271

Trosper, R. L. and Parrotta, J. A. (2012) 'Introduction: The growing importance of traditional forest-related knowledge', in J. A. Parrotta and R. L. Trosper (eds.), *Traditional Forest-Related Knowledge. Sustaining Communities, Ecosystems and biocultural diversity*. Springer, New York, NY. https://link.springer.com/chapter/10.1007/978-94-007-2144-9_1

United Nations. (2022) 'International Decade of Indigenous Languages 2022–2032', www.un.org/development/desa/indigenouspeoples/indigenous-languages.html

Uprety, Y., Asselin, H. and Bergeron, Y. (2013) 'Cultural importance of white pine (*Pinus strobus* L.) to the Kitcisakik Algonquin community of western Quebec, Canada', *Canadian Journal of Forest Research*, vol 43, no 6, pp. 544–551. https://doi.org/10.1139/cjfr-2012-0514

Uprety, Y., Asselin, H., Bergeron, Y., Doyon, F. and Boucher, J.-F. (2012) 'Contribution of traditional knowledge to ecological restoration: Practices and applications', *Ecoscience*, vol 19, no 3, pp. 225–237. https://doi.org/10.2980/19-3-3530

Ziembicki, M. R., Woinarski, J. C. Z. and Mackey, B. (2013) 'Evaluating the status of species using Indigenous knowledge: Novel evidence for major native mammal declines in northern Australia', *Biological Conservation*, vol 157, pp. 78–92. https://doi.org/10.1016/j.biocon.2012.07.004

42
WILD MEAT HUNTING IN TROPICAL FORESTS

Julia E. Fa and Stephan M. Funk

Introduction

Throughout human history, wild animals, plants and their derivatives have been harvested for a wide range of purposes, from sustenance to medicinal applications. This practice of exploiting the natural world's resources dates back to the beginning of human evolution (Hill 1982). Even today, contemporary Indigenous and local communities continue to rely on natural resources for their daily needs (Wilkie et al. 2005).

Wild meat or "bushmeat", referring to the meat of wild animals, remains an essential component of the diets of millions of people in tropical and subtropical regions. This is primarily because it often represents the most accessible and sustainable source of protein (Abernethy et al. 2013; Fa, Currie, and Meeuwig 2003). Initially, the term "bushmeat" was widely used, encompassing "non-domesticated terrestrial mammals, birds, reptiles, and amphibians harvested for food", excluding insects, crustaceans, grubs, molluscs and fish (Nasi et al. 2008, p. 6). However, there has been a recent shift away from employing the term towards the use of the more inclusive term "wild meat" to avoid geographical associations since bushmeat has been primarily used in an African context. This change in terminology can be traced back to the IUCN-World Conservation Union General Assembly, Resolution 2.64, which defines wild meat as any wildlife terrestrial animal used for food worldwide (IUCN World Conservation Congress, 2000).

In this chapter, we focus primarily on the use of wild meat within country boundaries (for examples of wildlife products available in the Republic of Laos, see Figure 42.1). This is because, despite the increasing international illegal wildlife trade involving wild meat and affecting the same resource base, the predominant use of wild meat remains for domestic consumption within the country where it is harvested (UNODC, 2016).

Hunted and consumed animal groups

Tropical and subtropical landscapes vary in their wildlife communities, human pressures, and vulnerabilities. Unsustainable hunting practices have rapidly driven species loss in Asian forests, with numerous large vertebrate species already extinct in Vietnam due to hunting.

 DOI: 10.4324/9781003324072-49

Figure 42.1 Examples of animal types consumed as wild meat in different parts of the world: (a) hunter in Cameroon having caught a grasscutter (*Thryonomys swinderianus*) (Photograph by Robert Okale); (b) monitor lizards on sale at a market in Vientiane, capital of Lao People's Democratic Republic (Photograph by Jean-Marc Touzet); (c) cockchafer beetles on sale in a Vientiane market, capital of Lao People's Democratic Republic (Photograph by Jean-Marc Touzet); (d) preparing caterpillars for grilling in southern Cameroon (Photograph by Eva Avila Martin); and (e) dressing a tapir (*Tapirus terrestris*) carcass in Mamiraua reserve, Brazilian Amazon (Photograph by Hani El Bizri).

Presently, the most critical issue is in West and Central Africa, but in the next few decades, losses are expected to extend to even the most remote parts of Latin America. These patterns align with the impacts of development and forest loss on the three continents, largely influenced by human population growth. There is substantial evidence highlighting the dire situation of many species primarily due to overhunting, particularly among mammals (Ripple et al. 2016). Over the past few decades, commercial hunting and trading of wildlife have intensified due to population growth, modernized hunting techniques, and increased accessibility to remote forest areas. This transition has raised concerns as it threatens both wild animal populations and human food security in many regions.

About 50,000 wild plants and animals are used for food, energy, medicine, material, and other purposes through fishing, gathering, logging, and terrestrial animal harvesting globally (IPBES, 2022). Over 10,000 species of these (more than 20%) are a source of food for people. In tropical and subtropical regions, a wide array of taxa, ranging from caterpillars to the largest land mammal, the elephant, are consumed. Redmond et al. (2006) reported a staggering total of 2,000 animal species (including insects) hunted globally for wild meat. Among these, approximately 55% were terrestrial vertebrates. Specifically, 638 species of these vertebrates are hunted across tropical and subtropical regions, where nearly 50% are mammals, followed by birds (34.8%), reptiles (13.8%), and amphibians (5.6%). The regional distribution of these taxonomic groups reflects the unique ecological characteristics of each area across the globe.

Hunted mammals include primates, ungulates, and rodents, typically with an average adult body mass equal to or greater than 1 kg (Robinson and Bennett 2004; Robinson and Redford 1991). These species offer a higher return on hunting investment due to their size and greater susceptibility to common hunting techniques such as snares and projectile weapons, especially firearms. Since larger animals tend to yield greater profits, they are often prioritized by hunters. However, as populations of larger animals dwindle, the effort required to capture them eventually outweighs the potential gain. Consequently, hunters shift their focus to mid-sized species, and with continued overexploitation, they may eventually target smaller species (Jerozolimski and Peres 2003). Throughout this process, opportunistic capture of the largest species continues, hindering their recovery even if they are no longer the primary target (Robinson and Bennett 2004). Additionally, snares, which indiscriminately trap various species, are widely utilized in Africa and Asia (Fa, Ryan, and Bell 2005; Harrison et al. 2016; Noss 1998). In regions dominated by African and Asian moist forests, which house relatively more abundant ground-dwelling species, snares are particularly lucrative. Conversely, the trend shifts towards arboreal taxa in Neotropical forests, where ground snares are less profitable. The distribution of hunted mammals in these forest regions underscores the prevalence of smaller prey species in South America compared to Africa and Asia (Corlett 2007; Fa and Peres 2001).

Birds also play a crucial role in the subsistence of rural communities reliant on wildlife for their sustenance, especially cracids and large arboreal galliform birds (chachalacas, guans, and curassows), which are highly regarded as a food source in tropical forests. In semi-arid regions of Brazil, doves, pigeons, and tinamous are commonly hunted for food due to the scarcity or local depletion of larger species (de Albuquerque et al. 2012). In Amazonian forests, up to 47 bird species are consumed for sustenance, with undulated tinamous, anhingas, razor-billed curassows, Muscovy ducks, and olivaceous cormorants being the most frequently hunted. Bird eggs also serve as a significant food source in various regions of the Peruvian Amazon (Gonzalez 2004).

Though birds are less commonly hunted in African forests, many species are still sought after and traded for their meat and traditional medicinal uses. Petrozzi (2018) documented a total of 302 different bird species across 10 West African countries being sold in wild meat markets. Most of these recorded species were categorized as Least Concern, while 23% were classified as Threatened according to the IUCN Red List. In the Ebo Forest of Cameroon, a study reported that birds constituted a significant portion (55%) of hunted species, surpassing mammals (43%) and other taxa (2%). The study also identified bird species rarely recorded elsewhere (Whytock et al. 2016), indicating their importance in the local diet. The higher offtake of larger bird species has notable conservation implications for raptors and hornbills (Trail 2007).

In Asia, the use of wild birds is largely associated with cultural practices and much less with subsistence purposes. Internal and international trade of wildlife including birds, often providing an income for some of the least economically affluent people, is a major conservation challenge in the region (Nijman, 2009; Marshall et al. 2021). In Indonesia, the demand for cage birds is increasing, while wild bird populations decline, primarily driven by aesthetics, social factors, or financial gain (Marshall et al. 2021). In a study involving experts in ornithology to assess the population trends of 38 bird species, including those heavily traded for ornamental purposes (Harris et al. 2015), as many as 14 species had experienced population declines, all of which are commonly traded, while none of the untraded species were classified as declining.

In Papua New Guinean cultures, bird hunting, particularly of cassowaries, is important for their meat, but various parts of the bird, such as bones and feathers, are also used for cultural and ritual purposes. The use of the highly esteemed feathers of birds of paradise for traditional ceremonies, rituals, and festivals is also common in many highland Papua New Guinean tribes. These are incorporated into traditional attire, adorning headdresses, necklaces, and aprons worn during prominent services.

Reptiles and amphibians likewise contribute to the protein intake of human populations. Among reptiles, turtles and tortoises (chelonians) are most heavily exploited for consumption, with the giant Amazon River turtle being a prominent species. Crocodile and alligator meat are considered delicacies in several regions, including Australia, South Africa, Thailand, Ethiopia, Cuba, and parts of the United States (Hoffman and Cawthorn 2012). The consumption of snakes is more opportunistic, but they are important sources of wild meat in Asian countries like China, Taiwan, Thailand, Indonesia, Vietnam, and Cambodia, as well as in West Africa (Brooks et al. 2010; Hoffman and Cawthorn 2012). In East and Southeast Asia, there exists a large and growing market for tortoises and freshwater turtles for pets, meat, and use in traditional medicines (Chen, Chang, and Lue 2009; Cheung and Dudgeon 2006; Nijman and Shepherd 2015).

Amphibians, while less frequently consumed than other vertebrates, still provide a protein source for human populations. Mohneke et al. (2009) identified at least 32 amphibian species, including 3 Urodela and 29 Anura, used as food globally. In South America, frogs of the genus *Telmatobius* have traditionally been consumed by local communities in the Andes of Peru and Bolivia. In Africa, larger frog species, such as the goliath frog, are commonly hunted for food (Gonwouo and Rödel 2008).

Where is wild meat hunted and consumed?

In moist forests in West and Central Africa, Petrozzi et al. (2016) recorded as many as 129 vertebrate species hunted and consumed in a meta-analysis covering a 40-year period across five countries. By class, mammals were the most numerous, accounting for 91 species, followed by reptiles (n = 19), birds (n = 14), and amphibians (n = 2). Mammals also constituted the highest number of individuals and overall biomass traded, with ungulates and large rodents being prevalent. Among trophic animal guilds, herbivores and frugivores dominated. Forest specialists were the most abundant group, and in riverine habitats, reptile biomass nearly equalled that of mammals. Most species and individuals were not threatened according to the IUCN Red List.

Species hunted for wild meat in African savannas have received less attention compared to forests, with ungulates being the most frequently targeted due to their higher abundance in open habitats (Lindsey et al. 2011; Lindsey 2011). For instance, in Zimbabwe, plains zebra, impala, and blue wildebeest are commonly hunted (Lindsey et al. 2011). A nationwide study in Tanzania unveiled that a total of 25 taxa were consumed across ten tribal areas, with antelope being the most frequently mentioned wild meat, particularly dik-dik and duikers. Hare and Guinea fowl were also prominent, while larger species like bushbuck and African buffalo were consumed less frequently (Ceppi and Nielsen 2014).

Hunting and gathering of a variety of vertebrates in Central America and Amazonia is also important for wild meat (Alves and van Vliet 2018). In a review of 78 hunting studies across these regions, a total of 90 mammal species were recorded, encompassing 12 primate genera, 6 ungulate genera, and 8 rodent genera (Stafford et al. 2017). Like Africa, ungulates and rodents constituted most of the wild meat extraction in Neotropical communities. Within the Amazon Basin, the world's largest moist forest block, medium-sized ungulates such as white-lipped peccary, collared peccary, white-tailed deer, and various brocket deer species, along with large rodents like pacas and agoutis, constituted a significant portion of the wild meat offtake (Fa and Peres 2001; Pires Mesquita, Domingo Rodríguez-Teijeiro, and Nascimento Barreto 2018). Additionally, tapirs, the largest mammal in South American tropical forests, were highly sought-after prey species. Primates were also a primary target for hunters in Central and South America, with large cebid monkeys being particularly favoured (Ráez-Luna 1995). In the Colombian Chocó forests, the caiçaras predominantly consumed paca, armadillo, and agouti, while Indigenous communities in the same region primarily hunted primates (Cormier 2006; Ojasti 1996). The Wai Indigenous communities in Guyana harvested pacas, curassow, and black spider monkeys as their main species (Shaffer et al. 2017). In the more open Brazilian cerrado, South American tapir, white-lipped peccary, collared peccary, and various deer species, including marsh deer, pampas deer, grey brocket deer, red brocket deer, and giant anteater, were commonly hunted (Welch 2014).

Asia presents a unique situation compared to other continents, primarily due to its heavy reliance on large-scale wildlife trade characterized by international supply chains spanning long distances. This distinctive circumstance arises from the region's rapid economic growth, which has driven a surge in demand for various natural resources, including land, timber, and non-timber forest products. Furthermore, East and Southeast Asia serve as significant hubs for the consumption of wildlife derivatives, encompassing a wide range of products from traditional medicines using tiger bones to culinary delicacies like shark's fin. This region plays a pivotal role as a major supplier to both legal and illegal segments of the international wildlife market. Information about wild meat use in Asian tropical forests remains limited (Lee et al. 2014). Typically, mammals weighing over 1 kg are targeted, encompassing over 160 species (Corlett 2007). Pigs represent the primary category of large-bodied mammals hunted for personal consumption, posing a significant threat due to hunting for local, regional, and global markets. Even in areas where high-quality forests remain, only a small fraction of the previous vertebrate diversity and abundance is retained (Harrison et al. 2016). In tropical Asia, a mere 1% of the land supports an intact fauna of mammals exceeding 20 kg (Morrison et al. 2007). Local studies corroborate these defaunation effects (Aiyadurai, Singh, and Milner-Gulland 2010; Johnson et al. 2003; Rao et al. 2010).

Who consumes wild meat?

For millions of Indigenous and non-Indigenous communities in tropical and subtropical regions, particularly among the world's rural poor, wild meat often constitutes a vital source of protein. It is also a rich source of micronutrients, contributing to the nutritional well-being of these communities (Golden et al. 2011; Sarti et al. 2015; Sirén and Machoa 2008). For impoverished farmers in tropical forests, livestock husbandry often poses high risks and investment costs, making it an impractical option. In cases where livestock, such as domestic chickens, can be maintained, they typically serve as a form of reserve or are used to fulfil specific cultural needs. In contrast, wild meat is considered an open-access resource, resulting in lower production costs compared to raising livestock.

In addition to its role as a dietary staple, wild meat plays a vital role in the livelihood strategies of impoverished populations, often serving as a significant source of income (Brown and Williams 2003; Milner-Gulland and Bennett 2003). In urban areas, wild meat consumption assumes a different character, becoming more of a commodity product than a dietary necessity. Furthermore, the meat of wild animals is a valuable tradable commodity due to its high value-to-weight ratio and transportability. The growing human population, coupled with the lucrative nature of the wild meat trade, has led to increased demand and extraction rates of wildlife species in various African and South American countries (see below). This commercialization of hunting has expanded the number of hunters who earn or supplement their income through meat sales, leading to higher hunting pressure and reduced sustainability for many wildlife species. As a result, wildlife hunting is regarded as the most widespread form of resource extraction in tropical regions and involves the harvesting of wildlife for various purposes, including food, trophies, traditional uses, medicines, and pets. While the vulnerability of species varies, uncontrolled exploitation can result in significant population declines and even extinction, especially among larger-bodied mammals and birds. This process, known as defaunation (see below), is driven by human activities and contributes to species and population extirpations, altering the dynamics and structure of ecosystems and their associated services.

Levels of wild meat consumption

Per capita wild meat consumption in different tropical regions has been studied, with a focus on the Congo Basin and Central and South America but data for Asia is scarce. These studies vary in methodologies and precision, making direct comparisons challenging. Nonetheless, they can be used to make comparison of wild meat consumption at least for the two main forest blocks in South America and Africa. While many of these studies are relatively old (Ojasti 1996), they still shed light on the patterns of wild meat consumption in different localities, perhaps offering a historical perspective on the utilization of wild animals for food.

In South American tropical forest communities, the average amount of wild meat consumed was approximately 138.9 ± 128.1 g/person/day, providing around 27.0 ± 25.1 g/protein/day. In African communities, wild meat consumption was slightly higher, with an average of 146.9 ± 75.9 g/person/day, contributing approximately 28.5 ± 14.7 g/protein/day. These differences reflect various factors, including landscape productivity, availability of alternatives, consumer wealth, and preferences for wild meat. In South America, consumption varies widely, ranging from 3 to over 500 g/person/day, primarily due to differences in the availability of wild meat. Communities located near productive rivers with a steady fish supply rely less on forest wildlife. Fish can contribute significantly to protein intake, although its availability is often seasonal.

In the Congo Basin, differences in wild meat consumption are more related to lifestyle contrasts, but habitat differences and hunting pressure also play roles. Foragers and farmers may have varying levels of wild meat consumption within the same region. Settlements near productive rivers with consistent fish sources may depend less on forest wildlife for protein. Studies in South America have estimated that 5–8 million people regularly rely on wild meat as a protein source, particularly among the region's poorest. In some areas, such as semi-arid regions, wild mammal meat becomes vital during drought periods when other protein sources are limited. Similarly, in the Yucatan Peninsula of Mexico, wildlife is essential for communities with limited access to surface water bodies and agriculture. Data on the extraction of wild meat in tropical and subtropical regions, particularly Southeast Asia, are limited. Most hunting studies tend to focus on listing hunted species without specifying the number of animals or biomass harvested per unit of hunting area. However, Robinson and Bennett (2004) compiled data on wildlife resources across different ecosystems and estimated the biomass of wild mammals, such as rodents, primates, and ungulates, in various forest types. Their findings indicated that extraction rates were highest in grasslands (744.3 ± 1030.2 kg/km^2), followed by evergreen and moist forests (168.0 ± 193.3 kg/km^2) and deciduous dry forests (126.5 ± 150.2 kg/km^2). In evergreen and moist forest sites, the biomass harvested per person was positively correlated with rainfall.

Difference between Indigenous and non-Indigenous hunting levels

Wildlife, including fish, holds considerable importance for rural and Indigenous communities worldwide, particularly in tropical and subtropical regions. The extent of the use of wildlife consumption for subsistence purposes varies across ecological zones, countries, and continents. Some of these intercontinental variations can be attributed to the productivity of forest ecosystems. For instance, the Central African forests yield substantially more wild meat from mammals compared to those in South America (Fa and Peres 2001). Conversely, the higher ratio of seacoast to land area in Asia and the prevalence of rivers in the Amazon Basin have historically facilitated the accessibility of seafood and inland fisheries, resulting in a higher proportion of fish consumption and a relatively lower contribution of mammalian meat to diets compared to Africa (Robinson and Bennett 1999)

However, the higher levels of wildlife offtake from humid forests compared to savanna ecosystems may seem counterintuitive, given the greater productivity of the latter (Robinson and Bennett 2004) This apparent paradox may be explained by the savanna's capacity to support domesticated fauna and the associated cultural preference for farmed meat over wild meat in such regions (Chardonnet et al. 1995).

Differences in prey species preference and hunting practices among Indigenous peoples and rural communities have been studied in Neotropical and African contexts. Redford and Robinson (1987) and later Redford (1991) employed an index based on the number of animals hunted per consumer year to highlight contrasts in hunting behaviours between Indigenous peoples and colonists in South American tropical and subtropical forests. Among Amazonian Indian communities, mammals were the primary game, followed by birds and reptiles. In contrast, colonists placed mammals first, reptiles second, and birds third. Importantly, Indigenous groups harvested a higher number of animals per consumer year than colonists did. Furthermore, the types of mammals preferred for hunting differed, with primates being the most frequently hunted order by Indigenous communities and rodents favoured by colonists.

A similar study in the Congo Basin by Fa et al. (2016) revealed significant variations in hunted species and extraction rates between Indigenous Pygmy and non-Pygmy groups. Pygmies targeted a narrower range of taxa but captured a higher proportion of prey with greater mean body mass compared to non-Pygmies. Extraction rates, or the number of animals hunted per unit area, were almost double in non-Pygmy sites than in Pygmy sites. However, when converted to extraction per hunter km^2, non-Pygmy groups harvested more per unit area than Pygmies.

Spatial patterns of wild meat extraction

Wild meat extraction data can be sourced from studies of wild meat markets and records of prey taken by hunters in villages or camps. While the bulk of these studies have concentrated on West and Central Africa and South America, there is a scarcity of information about Asian forests. Overall, although there has been an increasing number of publications on wild meat use since the 1980s, many of these studies have focused on small geographical areas over relatively short periods, which can limit the scope of data available for larger-scale analysis. However, even with these limitations, such studies contribute valuable insights into the overall patterns of wild meat extraction across extensive geographical regions.

Creating regional maps that identify hotspots of wild meat extraction has proven to be a valuable tool for understanding and addressing the challenges of conservation and sustainability in these ecosystems. These maps provide a visual representation of the status of wildlife subjected to hunting, aiding decision-makers, protected area managers, and researchers. Notable regional assessments have been carried out, one example being the spatial analysis undertaken in Central Africa by Ziegler et al. (2016). This analysis used data from hunting sites spanning several countries to map the intensity of wild meat extraction. Environmental and anthropogenic variables, including road density and proximity to protected areas, were considered in predicting annual offtake. Such assessments have shed light on the factors influencing extraction patterns in large, forested areas.

Drivers of wild meat use

Urban demand

The impact of subsistence hunting on prey populations can be significant, but the shift towards commercial hunting arguably poses the greatest threat to the vertebrate faunas in the Congo Basin. This is because of the presence of large cities (e.g., Kinshasa and Brazzaville) within the central African forest block (Fa et al. 2019). Commercialization of hunting has increased drastically due to the growing demand for wild meat from urban areas in the region and also because of its consumption as delicacies globally. Subsistence hunting and fishing typically do not pose significant threats to wildlife in areas with low human population densities around rural forest communities. However, as urban centres continue to grow, the commercial trade in wild meat, aimed at supplying these urban populations, poses an escalating danger to both forest animals and the food security of those who rely on hunting.

Urbanization has significantly influenced the demand for wild meat. In urban areas, the choice of several domestic animal protein sources is available, yet many consumers opt for wild meat for reasons beyond its nutritional value. For example, in provincial towns close to wildlife sources, where livestock production is uncommon and market access is limited, wild meat becomes a preferred choice. In large metropolitan cities, far from wildlife sources, the consumption of wild meat is driven more by cultural desires to connect with a rural past or as a luxury item and status symbol. As a luxury commodity, urban consumers often pay higher prices for wild meat compared to rural consumers, encouraging hunters to increase their catch for both income and food. This shift has led to rural people transitioning from traditional subsistence hunting to supplying cities.

The size of cities in Central Africa is rapidly expanding, particularly south of the Sahara. In 1900, only one in ten people lived in urban areas, but now nearly half of all sub-Saharan residents reside in towns and cities. This demographic shift has transformed food production and procurement methods, placing greater pressure on both domestic and wild environments. The expanding urban population has fuelled the growth of wild meat markets in Central Africa, meeting the demand for wild meat, which often complements the scarcity of domestic meat sources. The inability to raise large numbers of livestock, particularly cattle, in the central African forest region due to trypanosomiasis further drives the reliance on wild meat though the availability of imported domestic meats varies in line with each country's economic development. Recent urban migrants who are accustomed to consuming wild meat may choose it for its taste rather than for dietary necessity, and others opt for the cheapest available meat, considering wild meat a normal food item.

This shift towards urbanization has resulted in a lucrative trade in wild animals from rural and protected areas, driven by factors beyond local food security considerations. This commercial trade is the most immediate and significant threat to wildlife in tropical and subtropical regions worldwide. This also has implications for the livelihood strategies of impoverished communities and broader issues of public governance. Studies indicate that wild meat consumption and hunting are positively associated with landscape characteristics such as increasing forest cover (reflecting game availability) and remoteness, which affects market access. People in remote, forested areas depend more on wild meat, while those in more populous, peri-urban regions contribute significantly to the overall hunting effort due to better market access. Market access not only influences wild meat consumption but also provides hunters with opportunities to transit from barter-based to monetary economies, leading to increased income and livelihood diversification.

Wild meat is typically sold as fresh carcasses or smoked in various locations, including markets, roadside stalls, hunters' homes, and restaurants. In regions where wild meat is traded, it passes from the hunter to the consumer through different entry points in the commercial chain. Hunters may sell whole animals to traders or restaurant operators, who then retail them in smaller portions. Alternatively, hunters may process the carcass and sell pieces directly to consumers in their village. In some cases, hunters or intermediaries transport the meat to the point of sale, often the nearest town or city, where traders may buy it for resale.

Wild meat markets, found in almost every sizeable village or town in Africa, serve as major sites for the sale of wild meat. These markets may range from informal gatherings with makeshift counters to more organized structures with purpose-built market buildings. In African wild meat markets, five main actor groups are typically identified: farmer hunters (mainly subsistence hunters), commercial hunters, wholesalers, market traders, and small restaurant operators. Hunters and intermediaries are predominantly men, while sellers are usually women. Commercial hunters rely solely on wild meat for their livelihood, while farmer hunters sell it to supplement their income from agricultural activities. Wholesalers, market traders, and restaurant operators live and work in urban areas, with wholesalers often operating from home. Wholesalers buy meat in bulk from hunters and sell it to retailers like market traders and small restaurants or bars. Market traders operate from market stalls, and small restaurants or bars are dispersed throughout the city. The wildlife trade in most wild meat markets is often poorly regulated by both state and local institutions. In some countries, wildlife species nominally protected from hunting by legislation are still openly consumed as wild meat. These markets are integral to many African countries, especially in West and Central Africa, where the trade has been documented since the 1970s.

Quantitative data on the sale of wild meat in these markets reveal substantial variation in the volume of meat traded per site. This variation may reflect the number of hunters operating in the area, which, in turn, could relate to the population status of prey species. However, open-access hunting is a dynamic system influenced not only by prey abundance but also by individual hunters' responses to changing hunting costs and prices. This complex interplay between ecological and economic factors makes it challenging to assess sustainability reliably.

The global wild meat trade has witnessed a substantial increase in demand, paralleled by the trafficking of pets and animal parts for medicinal purposes. This surge in international trade has dire consequences for several species. Over 50% of Asian tortoises and freshwater turtles are now at risk of extinction due to this trade (van Dijk et al. 2012). Additionally, all eight species of pangolins, found in both Asia and Africa, face the imminent threat of extinction (Aisher 2016; Challender et al. 2020). These unique creatures are hunted extensively for their meat and scales, which are highly sought-after in traditional medicine. China and Vietnam are the primary destinations for trafficked pangolin products, driven by significant demand for these animals' meat and scales, contributing to their endangerment.

What motivates people to consume wild meat?

Several different factors are known to determine the reasons why people consume wild meat.

A. Factors influencing wild meat consumption

Wealth and socio-economic status:

- Wealthy urban households tend to consume more domestic meats, while poorer rural households rely on wild meat as a cheaper and accessible food source.
- The relationship between wealth and wild meat consumption is intricate and depends on various factors.

Urban vs. rural consumption:

- Rural populations generally consume more wild meat due to better access to wildlife, lower hunting opportunity costs, and cheaper wild meat prices.
- Wealthier urban households may still consume wild meat, often purchasing it rather than hunting it.

Urban markets:

- Urban markets in tropical forest regions significantly contribute to wild meat consumption.
- Wealthier urban residents may buy wild meat for prestige, while poorer urban residents may hunt wild meat to save costs.

Role of rural hunters:

- Rural inhabitants are primary suppliers of wild meat to urban markets, hunting for subsistence and income generation.

- Hunting wildlife for urban markets is predominantly a rural activity.

Distance to markets:

- Proximity to wildlife harvest areas affects wild meat consumption, with closer settlements consuming more wild meat.
- As the distance from hunting areas increases, wild meat prices relative to alternative meats also rise.

B. Non-wealth factors influencing wild meat consumption

Taste and tradition:

- Consumer preferences for the taste of wild meat and its role in diversifying diets can influence consumption.
- Wild meat may be reserved for special social events and traditional occasions.

Age and gender:

- Consumption patterns may vary based on age and gender, with younger urban individuals sometimes avoiding wild meat.
- Changing social systems and Westernization trends can impact perceptions of wild meat consumption.

Settlement type and ecology:

- The type of settlement (rural or urban) and the local ecological context can affect wild meat consumption, with significant differences observed between rural and urban areas in various countries.

Social norms and beliefs:

- Local customs, beliefs, attitudes, and social norms influence wild meat consumption. In some regions, wild meat consumption is a long-standing tradition, even if it is not the primary source of animal protein. Men in particular find hunting to be a highly enjoyable and fulfilling activity, driven by tradition, male bonding, and social norms.

Estimates of overextraction

Estimates of wild meat offtake vary widely, encompassing global evaluations as well as more specific calculations within regions. For instance, estimates have been calculated for the Congo and Amazon Basins, each offering a unique perspective on the scale of wild meat extraction. Fa et al. (2003) conducted an estimation and suggested that approximately 4 million tonnes of dressed wild meat are consumed annually in the Congo Basin. In contrast, an estimate by Wilkie and Carpenter (1999) placed this figure at only 1 million tons. Both estimates have their limitations, largely attributed to small sample sizes. Nonetheless, they underscore the substantial impact of wild meat hunting on these ecosystems. To calculate these estimates, Fa et al. (2003) relied on empirical data collected from hunting studies across multiple African countries. Their findings revealed that the quantity of wild meat extracted per unit area in the Congo Basin far exceeded that of the Amazon, signifying the magnitude of the issue in Central Africa.

Extraction rates for species in the Amazon and Congo Basins further illustrate the disparity. Most of the species in the Congo Basin appear to be exploited unsustainably, particularly primates, with a majority exceeding the 20% sustainable threshold. Conversely, the majority of hunted Amazonian species still fall within sustainable extraction levels. These differences in species exploitation between the two continents are predominantly a result of larger human population sizes within a smaller forest area in the Congo Basin, and the fact that a large proportion of what hunters' kill is sold in towns and villages for profit. Therefore, per capita harvest rates (kg/person/yr) in relation to number of consumers show a lower variation for South American settlements than for Africa where they decline significantly from an average of c. 500 kg/person/yr in smaller settlements to 1 kg/person/yr in the largest settlements (Robinson and Bennett 2004). This does not indicate greater consumption rates of wild meat per person but attests to the fact that wild meat is commercialized.

Sustainability of wild meat extraction

In Africa, Fa et al. (2005) gathered information from 36 rainforest sites, reporting annual extractions ranging from 40 to 12,168 carcasses per site, equivalent to 240.3 kg/yr to 84,092.7 kg/yr. This translates to a mean harvest rate per hunter ranging from 100.6 to 164.6 carcasses/yr and biomass from 945.7 kg/yr to 1612.2 kg/yr. These variations are influenced by factors like hunter numbers and habitat conditions.

In the regions where wildlife hunting for meat is common, most harvested animals are mammals, primarily ungulates. A meta-analysis in Afrotropical forests revealed the hunting of 71 mammal species, with ungulates and rodents being the most frequently hunted groups. Ungulates contributed significantly to both the number of carcasses and total biomass extracted. The relationship between body mass and carcass proportion indicated that larger mammals contributed substantially to total bushmeat biomass. The hunting method, especially the use of non-discriminatory snares, played a significant role in shaping the species composition of harvested animals.

Overall, hunting intensity was lower for larger-bodied species, particularly among carnivores and frugivores, with herbivores and insectivores being more heavily targeted. Vulnerability to hunting is influenced by various factors, including size, behaviour, and group living. Large mammals, often frugivores, are vital for forest ecosystem structure and function, making them susceptible to overexploitation. Sustainable hunting practices are essential to maintain wildlife populations and ecosystem health. However, the presence of a high number of hunters, deforestation, and habitat fragmentation can disrupt these dynamics and lead to population declines. Protecting large forest mammals is crucial for preserving tropical forests' long-term ecological stability.

Published studies that have assessed the sustainability of wildlife extraction in tropical forests, as shown in Table 42.1, generally indicate that hunting is often unsustainable. In most cases, more than half of the species considered in each study were found to be unsustainably hunted. Even in situations where the number of species studied was low, more than 50%, and in some cases up to 100%, were found to be unsustainably harvested. These findings suggest that unsustainable wildlife extraction is a common issue in the areas where

Table 42.1 Estimated sustainability and decline in population densities of mammals due to hunting

Country/region—site	*Main reason for hunting*	*Percentage of species hunted unsustainably (number of species studied)*	*Percentage by which densities of target species are lower in moderately to heavily hunted forests than in unhunted forest*	*Reference*
Africa				
Congo Basin		60% (57)		Fa et al. (2002)
CAR—Mossapoula	Subsistence/trade	100% (4)	43.90%	Noss (2000)
Cameroon	Subsistence/trade	100% (2)		Fimbel et al. (1999)
Cameroon	Subsistence/trade	50–100% (6)		Delvingt et al. (2001)
DRC—Ituri I	Subsistence		42.10%	Hart (1999)
DRC —Ituri II	Subsistence		12.90%	Hart (1999)
Gabon—Makokou			43–100%	Lahm (1993)
Equatorial Guinea, Bioko	Subsistence/trade	30.7% (16)		Fa (1999)
Equatorial Guinea, Rio Muni	Trade	36% (14)		Fa and García Yuste (2001)
Equatorial Guinea, Rio Muni	Trade	12% (17)		Fa et al. (1995)
Ghana	Trade	47% (15)		Cowlishaw et al. (2005)
Kenya	Subsistence/trade	42.9% (7)		Fitzgibbon et al. (1995)
Madagascar—Makira Forest	Subsistence	100% (5)		Golden (2009)
Latin America				
Brazil—101 Amazon sites	Subsistence		90%	Peres and Palacios (2007)
Brazil—Mata de Planalto			27–69%	Cullen Jr et al. (2000)
Bolivia	Subsistence	50% (10)		Townsend (2000)
Ecuador—Quehueiri-ono	Subsistence	30% (10)	35.30%	Mena et al. (1999)
Paraguay—Mbaracayu	Subsistence	0% (7)	53%	Hill and Padwe (1999)
Paraguay—Mbaracayu	Subsistence		0–40%	Hill et al. (1997)
Peru—Manu National Park	Subsistence	26% (19)		Ohl-Schacherer et al. (2007)
South/Southeast Asia				
Indonesia—Sulawesi	Subsistence/trade	66.7% (6)		O'Brien and Kinnaird (1999)
Indonesia—Sulawesi	Subsistence/trade	74% (4)		Lee (1999)
India—Nagarahole			75%	Madhusudan and Karanth (2018)

Source: Compiled by Cawthorn and Hoffman (2015).

Abbreviations: CAR = Central African Republic; DRC = Democratic Republic of Congo.

hunting has been examined. Sustainability of hunting practices in these studies has been measured using the Robinson and Redford (1991) index, which has certain limitations that may affect the results. Sustainable hunting may occur in very remote and sparsely populated locations or in areas beyond the influence of external markets. For example, mature markets like the one in Ghana may still have some sustainably hunted species, particularly smaller ones, as larger species have already been overhunted.

Additionally, assessing the impact of hunting on animal populations through population density estimates of target species has been suggested as an indicator of sustainability. However, this approach can be equivocal because hunted areas are expected to have lower population densities compared to non-hunted areas, and declining stocks may not necessarily indicate unsustainable use. Peres's estimates of standing stocks of mammals in various Amazonian locations that have been subject to hunting to varying degrees clearly show the impact of hunting pressure and forest type on these populations. Therefore, it is challenging to determine definitively whether species assemblages have been hunted sustainably or not, as sustainability depends on the balance between production and extraction, which can vary greatly in different contexts.

Defaunation

Evidence from archaeology and palaeontology suggests that pre-modern societies may have driven animal species to extinction. There have been documented mass extinction events of large-bodied vertebrates in various regions, including Europe, parts of Asia, North and South Americas, Madagascar, and several archipelagos (Young et al. 2016). These extinctions are the subject of debate, with discussions around whether they resulted from human overhunting after the Pleistocene or from climatic and environmental changes. Recent analyses, such as Andermann et al., (2020), increasingly point to the role of increasing human population size in past extinctions.

In more recent history, extinction events induced by overexploitation have been common. European settlers, armed with superior technology, expanded their frontiers, leading to uncontrolled market hunting and subsistence killing in North America during the late 19th century. This unregulated hunting caused severe declines, as seen in the near-extinction of once-vast bison herds and the loss of the passenger pigeon, once the most numerous birds globally (Bucher 1992).

While deforestation, habitat degradation, and climate change are highly visible threats to biodiversity, overhunting is equally serious, often resulting in environments that appear pristine but lack wildlife, especially large-bodied species (Peres et al. 2006). Meta-analyses, such as Benítez-López et al. (2017), show significant reductions in bird and mammal abundances in hunted areas compared to unhunted areas, with declines more pronounced near roads and settlements. Accessibility to major towns with wild meat markets exacerbates depletion.

Remote sensing data indicates that only 23.5% of current forest ecosystems was considered intact in 2008, as defined by an unbroken expanse of natural ecosystems without significant human activity. However, this does not account for "empty forests" resulting from hunting (Redford 1992). Benítez-López et al. (2019) mapped mammal defaunation in tropical forests, revealing an average abundance decline of 13% across species, with larger mammals experiencing greater declines. Defaunation (declines of 10% or more) affected large areas of pantropical forests, intact forests, wilderness areas, and protected areas, particularly in West and Central Africa and Southeast Asia.

For instance, the Atlantic Forest biome in eastern South America has witnessed unprecedented local extinctions of medium- to large-bodied mammals. Although only 10.8% of the original forest cover has been converted to other land uses, a mere fraction of mammal populations now persists in the remaining forest fragments. These patches are highly accessible to hunters, resulting in severe population declines and geographic range contractions. In the Serra do Mar bioregion of the Atlantic rainforest, mammalian biomass plummeted by up to 98% in intensively hunted areas (Galetti et al. 2017). This level of overhunting is confirmed by mapping the fate of surrogate Neotropical large mammal species, revealing high depletion rates (Jorge et al. 2013).

Ecological consequences of defaunation

Defaunation, the decline or extinction of animal populations due to various factors including hunting and habitat destruction, has far-reaching consequences that extend beyond the affected species. These cascading effects result in altered ecosystems, diminished ecosystem services, and challenges to human food security.

In predator–prey systems, prey species may initially benefit from the removal of their predators, which can trigger a chain reaction affecting various ecosystem services. For instance, the overhunting of sea otters in the northern Pacific coast of North America led to a decrease in their population. Sea otters feed on sea urchins, which, in turn, consume kelp. With fewer sea otters, sea urchin populations surged, decimating kelp forests. Researchers estimated that the loss of otters led to a significant decline in carbon storage in kelp forests, with an estimated value of $205 million to $408 million in the European Carbon Exchange.

Ecosystem services such as pest control and, to a lesser extent, pollination services by birds and bats are also affected by defaunation. Experiments in Indonesian cacao agroforestry fields revealed that insect herbivore abundance increased significantly when wildlife was excluded, resulting in a 31% decrease in crop yield in this billion-dollar industry.

Defaunation significantly impacts plant regeneration and, consequently, carbon storage. Exclusion experiments have shown that reduced seed predation and herbivory result in increased seedling density, survival, recruitment, and understory vegetation cover. However, these experiments cannot fully replicate real-world defaunation, as they may exclude some animal species while sparing others. Real-world defaunation often leads to a decline in local tree diversity over time because it directly affects tree species that rely on animals for seed dispersal. Moreover, defaunation alters the spatial structure and dynamics of tree populations, leading to changes in local tree diversity. The absence of large-seeded animal-dispersed species shifts plant composition towards species dispersed by abiotic factors or smaller animals. This results in reduced carbon storage, as large-seeded animal-dispersed trees play a crucial role in carbon sequestration. The impact of defaunation on carbon storage is more significant in forests with high proportions of animal-dispersed species. Defaunation also causes genetic changes in tree populations, affecting dispersal distances of mammal-dispersed tree species and increasing genetic similarity among tree communities. These changes reduce the forest's ability to store carbon.

Zoonotic disease

Defaunation can also influence the prevalence and transmission of zoonotic diseases. Experiments in East Africa that excluded large wildlife led to an increase in rodent populations and their flea vectors carrying *Bartonella* spp., a disease that affects humans (Young et al. 2014). Similar scenarios have been observed for Hantavirus and Lyme disease. Consequently, defaunation can elevate zoonotic disease risks for human populations.

In recent decades, new infectious diseases like SARS, Ebola, Nipah, West Nile fever, and HIV/AIDS, all linked to wildlife, have emerged, impacting human health globally (Fa et al. 2022; Milbank and Vira 2022)). Post the emergence of coronaviruses SARS and MERS after virus spillover from bats via other wild and farmed animals, Afelt et al. (2018) anticipated another +. from bats in Southeast Asia, which materialized as COVID-19, likely transmitted from horseshoe bats through intermediary mammals to humans (Boni et al. 2020; Zhou et al. 2020). The area, suffering extensive deforestation and human demographic growth, creates a conducive environment for zoonotic epidemics, and land-use alterations have increased the proximity and diversity of bat-borne viruses to human habitats, elevating zoonotic risks. Nipah virus, appearing in 1998, illustrates the effects of habitat change on disease spillover. Originating from fruit bats to pigs and subsequently humans, it manifested as encephalitis and respiratory illness, with deforestation and climate change as its primary catalysts (Chua et al. 2000; Chua et al. 2002). Habitat change resulted in bats infiltrating cultivated orchards and coming in close contact with pig livestock, with the virus eventually infecting humans. The onset of zoonotic diseases is tied to intricate dynamics involving land-use modifications, movements of wild disease reservoirs, the interaction between wildlife and domestic animal farming, and government policies.

Civets, which were instrumental in SARS transmission in China, were farmed, highlighting the role of both wild and farmed animals in disease emergence and propagation (Shi et al. 2020). Mink is highly susceptible to infection with several viruses that also infect humans. In late 2020, European and North American authorities and researchers discovered SARS-CoV-2 infections in farmed mink, raising concerns about biosecurity standards in the industry (Larsen et al. 2021; Oude Munnink et al. 2021). In response, some countries, like the Netherlands, ceased mink production. Fortunately, the mink-adapted variants in 2020 were not more contagious than human-adapted strains and did not spread extensively (Zhou et al. 2020). However, despite these concerns, many countries continued mink farming during the pandemic, with some even resuming operations after temporarily halting them, as seen in Denmark.

Due to significant repercussions of the COVID-19 pandemic, the newly implemented bans on wildlife trade in China have received substantial local backing (Rizzolo et al. 2023).

Need for solutions

Addressing the issue of wild meat consumption, particularly in urban areas, is of utmost importance due to its impact on wildlife populations and ecosystem health. Wild animals are hunted for various purposes, including local consumption, sale in urban markets, and cultural or personal preferences. The extent to which hunters sell wild meat varies based on factors such as their economic needs, access to markets, and cultural values. For example, Indigenous groups may sell a smaller proportion of the wild game they hunt compared to non-Indigenous communities.

Wild meat hunting plays a crucial role in the livelihoods and food security of rural and Indigenous communities. However, in some regions, unsustainable hunting practices are causing wildlife populations to decline, affecting ecosystem services like nutrient cycling and carbon capture. Unsustainable hunting is often driven by the demand for wild meat in rapidly growing urban centres, where it is considered a luxury item, fetching higher prices than in rural areas. This demand leads to overhunting and a classic "Tragedy of the Commons" scenario.

The uncontrolled wild meat trade, coupled with habitat loss, poses a significant threat to wildlife in tropical and subtropical regions worldwide. Even in areas like the Amazon, which were previously thought to be less affected by urban demand, there is evidence of a shift from subsistence hunting to supplying city markets. As urbanization continues to rise globally, the demand for wild meat is increasing. Fortunately, reducing or eliminating wild meat consumption in cities is unlikely to affect access to other protein sources, as domestically produced and imported animal products like chicken and fish typically provide most of the dietary protein in urban areas. To address this issue, it is crucial to focus on social science research alongside ecological monitoring. Understanding why urban consumers choose wild meat is essential to developing policies that promote alternative, more abundant, and affordable protein sources.

One proposed solution is wildlife farming, which has been suggested since the 1950s. The idea is to raise wild animals for their meat in controlled environments, reducing the pressure on wild populations. However, the success of wildlife farming depends on various factors, including the species' biology, cost-effectiveness, and cultural acceptance. While some success stories exist, wildlife farming is unlikely to fully satisfy consumer demands and poses conservation risks. Given this, efforts to create hardier domestic livestock breeds and expand domestic animal farming should also be considered as part of a sustainable landscape planning strategy. In some cases, wildlife farming in peri-urban areas may help meet urban demand for wild meat and reduce pressure on wild populations, but it is not a food security or conservation solution.

In conclusion, addressing the demand for wild meat in urban areas requires a multifaceted approach, including social science research, policy development, and sustainable alternatives to wild meat consumption. Wildlife farming, while a potential solution, must be carefully managed to avoid negative impacts on wildlife populations and ecosystem health.

References

Abernethy, K.A., Coad, L., Taylor, G., Lee, M.E., Maisels, F., 2013. Extent and ecological consequences of hunting in Central African rainforests in the twenty-first century. *Philosophical Transactions of the Royal Society B: Biological Sciences* 368, 20120303. https://doi.org/10.1098/rstb.2012.0303

Afelt, A., Frutos, R., Devaux, C., 2018. Bats, coronaviruses, and deforestation: Toward the emergence of novel infectious diseases? *Frontiers in Microbiology* 9, 702. https://doi.org/10.3389/fmicb.2018.00702

Aisher, A., 2016. Scarcity, alterity and value: Decline of the pangolin, the world's most trafficked mammal. *Conservation and Society* 14, 317. https://doi.org/10.4103/0972-4923.197610

Aiyadurai, A., Singh, N.J., Milner-Gulland, E.J., 2010. Wildlife hunting by indigenous tribes: A case study from Arunachal Pradesh, north-east India. *Oryx* 44, 564–572. https://doi.org/10.1017/S0030605309990937

Alves, Rômulo Romeu Nóbrega, van Vliet, N., 2018. Wild fauna on the menu, in: Alves, R. R. N., Albuquerque, U.P. (Eds.), *Ethnozoology*. Elsevier, Oxford, UK, pp. 167–194.

Andermann, T., Faurby, S., Turvey, S.T., Antonelli, A., Silvestro, D., 2020. The past and future human impact on mammalian diversity. *Science Advances 6*, eabb2313. https://doi.org/10.1126/sciadv.abb2313

Benítez-López, A., Alkemade, R., Schipper, A.M., Ingram, D.J., Verweij, P.A., Eikelboom, J.A.J., Huijbregts, M.A.J., 2017. The impact of hunting on tropical mammal and bird populations. *Science* 356, 180–183. https://doi.org/10.1126/science.aaj1891

Benítez-López, A., Santini, L., Schipper, A.M., Busana, M., Huijbregts, M.A.J., 2019. Intact but empty forests? Patterns of hunting-induced mammal defaunation in the tropics. *PLoS Biology* 17, e3000247. https://doi.org/10.1371/journal.pbio.3000247

Boni, M.F., Lemey, P., Jiang, X., Lam, T.T.-Y., Perry, B.W., Castoe, T.A., Rambaut, A., Robertson, D.L., 2020. Evolutionary origins of the SARS-CoV-2 sarbecovirus lineage responsible for the COVID-19 pandemic. *Nature Microbiology* 5, 1417–1418. https://doi.org/10.1038/s41564-020-0771-4

Brooks, S.E., Allison, E.H., Gill, J.A., Reynolds, J.D., 2010. Snake prices and crocodile appetites: Aquatic wildlife supply and demand on Tonle Sap Lake, Cambodia. *Biological Conservation* 143, 2127–2135. https://doi.org/10.1016/j.biocon.2010.05.023

Brown, D., Williams, A., 2003. The case for bushmeat as a component of development policy: Issues and challenges. *International Forestry Review* 5, 148–155. https://doi.org/10.1505/IFOR.5.2.148.17414

Bucher, E.H., 1992. The causes of extinction of the passenger pigeon, in: Power, D.M. (Ed.), *Current Ornithology*. Plenum Press, New York, pp. 1–36. https://doi.org/10.1007/978-1-4757-9921-7_1

Cawthorn, D.-M., Hoffman, L.C., 2015. The bushmeat and food security nexus: A global account of the contributions, conundrums and ethical collisions. *Food Research International* 76, 906–925. https://doi.org/10.1016/j.foodres.2015.03.025

Ceppi, S.L., Nielsen, M.R., 2014. A comparative study on bushmeat consumption patterns in ten tribes in Tanzania. *Tropical Conservation Science* 7, 272–287. https://doi.org/10.1177/194008291400700208

Challender, D.W.S., Heinrich, S., Shepherd, C.R., Katsis, L.K.D., 2020. International trade and trafficking in pangolins, 1900–2019, in: Challender, D.W.S., Nash, H.C., Waterman, C. (Eds.), *Pangolins; Science, Society and Conservation*. Academic Press, London, pp. 259–276. https://doi.org/10.1016/B978-0-12-815507-3.00016-2

Chardonnet, P., Fritz, H., Zorzi, N., Feron, E., 1995. Current importance of traditional hunting and major contrasts in wild meat consumption in sub-Saharan Africa, in: Bissonette, J.A., Krausman, P.R. (Eds.), *Integrating People and Wildlife for a Sustainable Future. Proceedings of the First International Wildlife Management Conference*. The Wildlife Society, Bethesda, Maryland, pp. 304–307.

Chen, T.-H., Chang, H.-C., Lue, K.-Y., 2009. Unregulated trade in turtle shells for Chinese traditional medicine in East and Southeast Asia: The case of Taiwan. *Chelonian Conservation and Biology* 8, 11–18. https://doi.org/10.2744/CCB-0747.1

Cheung, S.M., Dudgeon, D., 2006. Quantifying the Asian turtle crisis: Market surveys in southern China, 2000–2003. *Aquatic Conservation* 16, 751–770. https://doi.org/10.1002/aqc.803

Chua, K.B., Bellini, W.J., Rota, P.A., Harcourt, B.H., Tamin, A., Lam, S.K., Ksiazek, T.G., Rollin, P.E., Zaki, S.R., Shieh, W.-J., Goldsmith, C.S., Gubler, D.J., Roehrig, J.T., Eaton, B., Gould, A.R., Olson, J., Field, H., Daniels, P., Ling, A.E., Peters, C.J., Anderson, L.J., Mahy, B.W.J., 2000. Nipah virus: A recently emergent deadly paramyxovirus. *Science* 288, 1432–1435. https://doi.org/10.1126/science.288.5470.1432

Chua, K.B., Chua, B.H., Wang, C.W., 2002. Anthropogenic deforestation, El Niño and the emergence of Nipah virus in Malaysia. *Malaysian Journal of Pathology* 24, 15–21.

Corlett, R.T., 2007. The impact of hunting on the mammalian fauna of tropical Asian forests. *Biotropica* 39, 292–303. https://doi.org/10.1111/j.1744-7429.2007.00271.x

Cormier, L., 2006. A preliminary review of neotropical primates in the subsistence and symbolism of Indigenous Lowland South American peoples. *Ecological and Environmental Anthropology* 2, 14–32.

Cowlishaw, G., Mendelson, S., Rowcliffe, J.M., 2005. Evidence for post-depletion sustainability in a mature bushmeat market: Sustainability of bushmeat markets. *Journal of Applied Ecology* 42, 460–468. https://doi.org/10.1111/j.1365-2664.2005.01046.x

Cullen Jr, L., Bodmer, R.E., Pádua, C.V., 2000. Effects of hunting in habitat fragments of the Atlantic forests, Brazil. *Biological Conservation* 95, 49–56. https://doi.org/10.1016/S0006-3207(00)00011-2

de Albuquerque, U.P., de Lima Araújo, E., El-Deir, A.C.A., de Lima, A.L.A., Souto, A., Bezerra, B.M., Ferraz, E.M.N., Maria Xavier Freire, E., Sampaio, E.V. de S.B., Las-Casas, F.M.G., de Moura, G.J.B., Pereira, G.A., de Melo, J.G., Alves Ramos, M., Rodal, M.J.N., Schiel, N., de Lyra-Neves, R.M., Alves, R.R.N., de Azevedo-Júnior, S.M., Telino Júnior, W.R., Severi, W., 2012. Caatinga revisited: Ecology and conservation of an important seasonal dry forest. *The Scientific World Journal* 2012, 1–18. https://doi.org/10.1100/2012/205182

Delvingt, W., Dethier, M., Auzel, P., Jeanmart, P., 2001. La chasse villageoise Badjoué, gestion coutumière durable ou pillage de la ressource gibier, in: W. Delvingt (ed.), *La Forêt des Hommes: Terroirs Villageois En Forêt Tropicale Africaine.* es Presses Agronomiques de Gembloux, Gembloux, Belgium, pp. 65–92.

Fa, J.E., 1999. Hunted animals in Bioko Island, West Africa: Sustainability and future, in: Robinson, J.G., Bennett, E.L. (Eds.), *Hunting for Sustainability in Tropical Forests, Biology and Resource Management Series.* Columbia University Press, New York, pp. 168–198.

Fa, J.E., Currie, D., Meeuwig, J., 2003. Bushmeat and food security in the Congo Basin: Linkages between wildlife and people's future. *Environmental Conservation* 30, 71–78.

Fa, J.E., García Yuste, J.E., 2001. Commercial bushmeat hunting in the Monte Mitra forests, Equatorial Guinea: Extent and impact. *Animal Biodiversity and Conservation* 22, 31–52.

Fa, J.E., Juste, J., Del Val, J.P., Castroviejo, J., 1995. Impact of market hunting on mammal species in Equatorial Guinea. Conservation biology 9, 1107–1115. https://doi.org/10.1046/j.1523-1739.1995.951107.x

Fa, J.E., Nasi, R., Funk, S.M., 2022. *Hunting Wildlife in the Tropics and Subtropics, Ecology, Biodiversity and Conservation.* Cambridge University Press, Cambridge.

Fa, J.E., Olivero, J., Farfán, M.A., Lewis, J., Yasuoka, H., Noss, A., Hattori, S., Hirai, M., Kamgaing, T.O.W., Carpaneto, G., Germi, F., Márquez, A.L., Duarte, J., Duda, R., Gallois, S., Riddell, M., Nasi, R., 2016. Differences between pygmy and non-pygmy hunting in Congo Basin forests. *PLoS One* 11, e0161703. https://doi.org/10.1371/journal.pone.0161703

Fa, J.E., Peres, C.A., 2001. Game vertebrate extraction in African and Neotropical forests: An intercontinental comparison, in: Reynolds, J.D., Mace, G.M., Redford, K.H., Robinson, J.G. (Eds.), *Conservation of Exploited Species.* Cambridge University Press, Cambridge, pp. 203–241.

Fa, J.E., Peres, C.A., Meeuwig, J., 2002. Bushmeat Exploitation in Tropical Forests: an Intercontinental Comparison. *Conservation Biology* 16, 6. https://doi.org/10.1046/j.1523-1739.2002.00275.x

Fa, J.E., Ryan, S.F., Bell, D.J., 2005. Hunting vulnerability, ecological characteristics and harvest rates of bushmeat species in afrotropical forests. *Biological Conservation* 121, 167–176. https://doi.org/10.1016/j.biocon.2004.04.016

Fa, J.E., Wright, J.H., Funk, S.M., Márquez, A.L., Olivero, J., Farfán, M.Á., Guio, F., Mayet, L., Malekani, D., Holo Louzolo, C., Mwinyihali, R., Wilkie, D.S., Wieland, M., 2019. Mapping the availability of bushmeat for consumption in Central African cities. *Environmental Research Letters* 14, 094002. https://doi.org/10.1088/1748-9326/ab36fa

Fimbel, C., Curran, B., Usongo, L., 1999. Enhancing the sustainability of duiker hunting through community participation and controlled access in the Lobéké region of southeastern Cameroon, in: Robinson, J.G., Bennet, E.L. (Eds.), *Hunting for Sustainability in Tropical Forests.* Columbia University Press, New York, pp. 356–374.

Fitzgibbon, C.D., Mogaka, H., Fanshawe, J.H., 1995. Subsistence hunting in ArabukoSokoke Forest, Kenya, and its effects on mammal populations. *Conservation Biology* 9, 1116–1126.

Galetti, M., Pires, A.S., Brancalion, P.H.S., Fernandez, F.A.S., 2017. Reversing defaunation by trophic rewilding in empty forests. *Biotropica* 49, 5–8. https://doi.org/10.1111/btp.12407

Golden, C.D., 2009. Bushmeat hunting and use in the Makira Forest, north-eastern Madagascar: A conservation and livelihoods issue. *Oryx* 43, 386. https://doi.org/10.1017/S0030605309000131

Golden, C.D., Fernald, L.C.H., Brashares, J.S., Rasolofoniaina, B.J.R., Kremen, C., 2011. Benefits of wildlife consumption to child nutrition in a biodiversity hotspot. *Proceedings of the National Academy of Sciences* 108, 19653–19656. https://doi.org/10.1073/pnas.1112586108

Gonwouo, L.N., Rödel, M.-O., 2008. The importance of frogs to the livelihood of the Bakossi people around Mount Manengouba, Cameroon, with special consideration of the Hairy Frog, Trichobatrachus robustus. *Salamandra* 44, 23–34.

Gonzalez, J. A., 2004. Human use and conservation of economically important birds in seasonally flooded forests of the Northeastern Peruvian Amazon., in: Silvius, K., Bodmer, R., Fragoso, J. (Eds.), *People in Nature: Wildlife Conservation in South and Central America.* Columbia University Press, New York, pp. 344–361.

Harris, J. B. C., Green, J. M. H., Prawiradilaga, D. M., Giam, X., Giyanto, Hikmatullah, D., Putra, C. A., Wilcove, D. D. 2015. Using market data and expert opinion to identify overexploited species in the wild bird trade. *Biological Conservation* 187, 51–60.

Harrison, R.D., Sreekar, R., Brodie, J.F., Brook, S., Luskin, M., O'Kelly, H., Rao, M., Scheffers, B., Velho, N., 2016. Impacts of hunting on tropical forests in Southeast Asia: Hunting in Tropical forests. *Conservation Biology* 30, 972–981. https://doi.org/10.1111/cobi.12785

Hart, J.A., 1999. Impact and sustainability of indigenous hunting in the Ituri Forest, Congo-Zaire: A comparison of unhunted and hunted duiker populations, in: Robinson, J.G., Bennett, E.L. (Eds.), *Hunting for Sustainability in Tropical Forests*. Columbia University Press, New York, pp. 106–153.

Hill, K., 1982. Hunting and human evolution. *Journal of Human Evolution* 11, 521–544. https://doi.org/10.1016/S0047-2484(82)80107-3

Hill, K., Padwe, J., 1999. Sustainability of Aché hunting in the Mbaracayu reserve, Paraguay, in: Robinson, J., Bennett, E.L. (Eds.), *Hunting for Sustainability in Tropical Forests*. Columbia University Press, New York, pp. 79–105.

Hill, K., Padwe, J., Bejyvagi, C., Bepurangi, A., Jakugi, F., Tykuarangi, R., Tykuarangi, T., 1997. Impact of hunting on large vertebrates in the Mbaracayu Reserve, Paraguay. *Conservation Biology* 11, 1339–1353. https://doi.org/10.1046/j.1523-1739.1997.96048.x

Hoffman, L.C., Cawthorn, D.-M., 2012. What is the role and contribution of meat from wildlife in providing high quality protein for consumption? *Animal Frontiers* 2, 40–53. https://doi.org/10.2527/af.2012-0061

IPBES, 2022. *Thematic assessment of the sustainable use of wild species of the Intergovernmental Science-Policy Platform on Biodiversity and Ecosystem Services*. Zenodo. https://doi.org/10.5281/ZENODO.6448567

IUCN World Conservation Congress, 2000. Resolution 2.64, 2–4.

Jerozolimski, A., Peres, C.A., 2003. Bringing home the biggest bacon: A cross-site analysis of the structure of hunter-kill profiles in Neotropical forests. *Biological Conservation* 111, 415–425. https://doi.org/10.1016/S0006-3207(02)00310-5

Johnson, A., Singh, S., Dongdala, M., Vongsa, O., 2003. *Wildlife hunting and use in the Nam Ha National Protected Area: Implications for rural livelihoods and biodiversity conservation. December 2003*. Wildlife Conservation Society, Vientiane, Lao PDR.

Jorge, M.L.S.P., Galetti, M., Ribeiro, M.C., Ferraz, K.M.P.M.B., 2013. Mammal defaunation as surrogate of trophic cascades in a biodiversity hotspot. *Biological Conservation* 163, 49–57. https://doi.org/10.1016/j.biocon.2013.04.018

Lahm, S.A., 1993. Utilization of forest resources and local variation of wildlife populations in Northeastern Gabon, in: Hladik, C., Hladik, A., Pagezy, H., Linares, O., Koppert, G., Froment, A. (Eds.), *Tropical Forests, People and Food, Man and the Biosphere Series*. UNESCO, Paris, pp. 213–226.

Larsen, H.D., Fonager, J., Lomholt, F.K., Dalby, T., Benedetti, G., Kristensen, B., Urth, T.R., Rasmussen, M., Lassaunière, R., Rasmussen, T.B., Strandbygaard, B., Lohse, L., Chaine, M., Møller, K.L., Berthelsen, A.-S.N., Nørgaard, S.K., Sönksen, U.W., Boklund, A.E., Hammer, A.S., Belsham, G.J., Krause, T.G., Mortensen, S., Bøtner, A., Fomsgaard, A., Mølbak, K., 2021. Preliminary report of an outbreak of SARS-CoV-2 in mink and mink farmers associated with community spread, Denmark, June to November 2020. *Eurosurveillance* 26, 1–6. https://doi.org/10.2807/1560-7917.ES.2021.26.5.210009

Lee, R.J., 1999. Impact of subsistence hunting in North Sulawesi, Indonesia, and conservation options, in: Robinson, J.G., Bennett, L.E. (Eds.), *Hunting for Sustainability in Tropical Forests*. Columbia University Press, New York, pp. 455–472.

Lee, T.M., Sigouin, A., Pinedo-Vasquez, M., Nasi, R., 2014. *The harvest of wildlife for bushmeat and traditional medicine in East, South and Southeast Asia: Current knowledge base, challenges, opportunities and areas for future research*, Occasional Paper. CIFOR, Bogor, Indonesia.

Lindsey, P., 2011. *An analysis of game meat production and wildlife-based land uses on freehold land in Namibia: Links with food security*. TRAFFIC East/Southern Africa, Harare, Zimbabwe.

Lindsey, P.A., Romañach, S.S., Matema, S., Matema, C., Mupamhadzi, I., Muvengwi, J., 2011. Dynamics and underlying causes of illegal bushmeat trade in Zimbabwe. *Oryx* 45, 84–95. https://doi.org/10.1017/S0030605310001274

Madhusudan, M.D., Karanth, K.U., 2018. Hunting for an answer: Is local hunting compatible with large mammal conservation in India?, in: Robinson, J.G., Bennett, L.E. (Eds.), *Hunting for Sustainability in Tropical Forests*. Columbia University Press, New York, pp. 455–472.

Marshall, H., Glorizky, G.A., Collar, N.J., Lees, A.C., Moss, A., Yuda, P., Marsden, S.J., 2021. Understanding motivations and attitudes among songbird-keepers to identify best approaches to demand reduction. *Conservation Science and Practice* 3, e507. https://doi.org/10.1111/csp2.507

Mena, V.P., Stallings, J.R., Regalado, J.B., Cueva, R.L., 1999. The sustainability of current hunting practices by the Huaorani, in: Robinson, J.G., Bennett, L.E. (Eds.), *Hunting for Sustainability in Tropical Forests.* Columbia University Press, New York, pp. 57–78.

Milbank, C., Vira, B., 2022. Wildmeat consumption and zoonotic spillover: Contextualising disease emergence and policy responses. *The Lancet Planetary Health* 6, e439–e448. https://doi.org/10.1016/S2542-5196(22)00064-X

Milner-Gulland, E.J., Bennett, E.L., 2003. Wild meat: The bigger picture. *Trends in Ecology & Evolution* 18, 351–357. https://doi.org/10.1016/S0169-5347(03)00123-X

Mohneke, M., Onadeko, A.B., Rödel, M.O., 2009. Exploitation of frogs–a review with a focus on West Africa. *Salamandra* 45, 193–202.

Morrison, J.C., Sechrest, W., Dinerstein, E., Wilcove, D.S., Lamoreux, J.F., 2007. Persistence of large mammal faunas as indicators of global human impacts. *Journal of Mammalogy* 88, 1363–1380. https://doi.org/10.1644/06-MAMM-A-124R2.1

Nasi, R., Brown, D., Wilkie, D., Bennett, E., Tutin, C., van Tol, G. and Christopherson, T, 2008. *Conservation and Use of Wildlife-Based Resources: The Bushmeat Crisis, CBD Technical Series.* Secretariat of the Convention on Biological Diversity, Montreal.

Nijman, V., 2009. An overview of international wildlife trade from Southeast Asia. *Biodiversity and Conservation* 19, 1101–1114.

Nijman, V., Shepherd, C.R., 2015. Analysis of a decade of trade of tortoises and freshwater turtles in Bangkok, Thailand. *Biodiversity and Conservation* 24, 309–318. https://doi.org/10.1007/s10531-014-0809-0

Noss, A., 2000. Cable snares and nets in the Central African Republic, in: Robinson, J., EL Bennett (Eds.), *Hunting for Sustainability in Tropical Forest.* Colombia University Press, New York, pp. 282–305.

Noss, A.J., 1998. The impacts of cable snare hunting on wildlife populations in the forests of the Central African Republic. *Conservation Biology* 12, 9.

O'Brien, T.G., Kinnaird, M.F., 1999. Differential vulnerability of large birds and mammals to hunting in North Sulawesi, Indonesia, and the outlook for the future, in: Robinson, J.G., Bennett, L.E. (Eds.), *Hunting for Sustainability in Tropical Forests.* Columbia University Press, New York, pp. 199–213.

Ohl-Schacherer, J., Shepard, G.H., Kaplan, H., Peres, C.A., Levi, T., Yu, D.W., 2007. The sustainability of subsistence hunting by Matsigenka Native communities in Manu National Park, Peru. *Conservation Biology* 21, 1174–1185. https://doi.org/10.1111/j.1523-1739.2007.00759.x

Ojasti, J., 1996. *Wildlife Utilization in Latin America: Current Situation and Prospects for Sustainable Management, FAO Conservation Guide.* Food & Agriculture Org., Rome, Italy.

Oude Munnink, B.B., Sikkema, R.S., Nieuwenhuijse, D.F., Molenaar, R.J., Munger, E., Molenkamp, R., Van Der Spek, A., Tolsma, P., Rietveld, A., Brouwer, M., Bouwmeester-Vincken, N., Harders, F., Hakze-van Der Honing, R., Wegdam-Blans, M.C.A., Bouwstra, R.J., GeurtsvanKessel, C., Van Der Eijk, A.A., Velkers, F.C., Smit, L.A.M., Stegeman, A., Van Der Poel, W.H.M., Koopmans, M.P.G., 2021. Transmission of SARS-CoV-2 on mink farms between humans and mink and back to humans. *Science* 371, 172–177. https://doi.org/10.1126/science.abe5901

Peres, C.A., Barlow, J., Laurance, W.F., 2006. Detecting anthropogenic disturbance in tropical forests. *Trends in Ecology & Evolution* 21, 227–229. https://doi.org/10.1016/j.tree.2006.03.007

Peres, C.A., Palacios, E., 2007. Basin-wide effects of game harvest on vertebrate population densities in Amazonian forests: Implications for animal-mediated seed dispersal. *Biotropica* 39, 304–315. https://doi.org/10.1111/j.1744-7429.2007.00272.x

Petrozzi, F., 2018. Bushmeat and fetish trade of birds in West Africa: A review. *Vie et Milieu* 15, 58–64.

Petrozzi, F., Amori, G., Franco, D., Gaubert, P., Pacini, N., Eniang, E.A., Akani, G.C., Politano, E., Luiselli, L., 2016. Ecology of the bushmeat trade in West and Central Africa. *Tropical Ecology* 57, 545–557.

Pires Mesquita, G., Domingo Rodríguez-Teijeiro, J., Nascimento Barreto, L., 2018. Patterns of mammal subsistence hunting in eastern Amazon, Brazil: Mammal Hunting Patterns. *Wildlife Society Bulletin* 42, 272–283. https://doi.org/10.1002/wsb.873

Ráez-Luna, E.F., 1995. Hunting large primates and conservation of the Neotropical rain forests. *Oryx* 29, 43–48. https://doi.org/10.1017/S003060530002086X
Rao, M., Htun, S., Zaw, T., Myint, T., 2010. Hunting, livelihoods and declining wildlife in the Hponkanrazi Wildlife Sanctuary, North Myanmar. *Environmental Management* 46, 143–153. https://doi.org/10.1007/s00267-010-9519-x
Redford, K.H., 1991. The ecologically noble savage. *Cultural Survival Quarterly* 15, 46–48.
Redford, K.H., 1992. The empty forest. *BioScience* 42, 412–422. https://doi.org/10.2307/1311860
Redford, K.H., Robinson, J.G., 1987. The game of choice: Patterns of Indian and colonist hunting in the Neotropics. *American Anthropologist* 89, 650–667. https://doi.org/10.1525/aa.1987.89.3.02a00070
Redmond, I., Aldred, T., Jedamzik, K., Westwood, M., 2006. *Recipes for Survival: Controlling the Bushmeat Trade* (Report). Ape Alliance; WPSA, London.
Ripple, W.J., Abernethy, K., Betts, M.G., Chapron, G., Dirzo, R., Galetti, M., Levi, T., Lindsey, P.A., Macdonald, D.W., Machovina, B., Newsome, T.M., Peres, C.A., Wallach, A.D., Wolf, C., Young, H., 2016. Bushmeat hunting and extinction risk to the world's mammals. *Royal Society Open Science* 3, 160498. https://doi.org/10.1098/rsos.160498
Rizzolo, J.B., Zhu, A.L., Chen, R., 2023. Support for wildlife consumption bans and policies in China post-Covid-19. *Oryx* 1–10. https://doi.org/10.1017/S00306053220015933
Robinson, J.G., Bennett, E.L., 1999. Carrying capacity limits to sustainable hunting in tropical forests, in: Robinson, J.G., Bennett, E.L. (Eds.), *Hunting for Sustainability in Tropical Forests, Biology and Resource Management Series.* Columbia University Press, New York, pp. 13–30.
Robinson, J.G., Bennett, E.L., 2004. Having your wildlife and eating it too: An analysis of hunting sustainability across tropical ecosystems. *Animal Conservation* 7, 397–408. https://doi.org/10.1017/S1367943004001532
Robinson, J.G., Redford, K.H., 1991. Sustainable harvest of neotropical forest mammals, in: Robinson, J.G., Redford, K.H. (Eds.), *Neotropical Wildlife Use and Conservation.* University of Chicago Press, Chicago, IL, pp. 415–429.
Sarti, F.M., Adams, C., Morsello, C., van Vliet, N., Schor, T., Yagüe, B., Tellez, L., Quiceno-Mesa, M.P., Cruz, D., 2015. Beyond protein intake: Bushmeat as source of micronutrients in the Amazon. *Ecology and Society* 20, https://doi.org/10.5751/ES-07934-200422
Shaffer, C.A., Milstein, M.S., Yukuma, C., Marawanaru, E., Suse, P., 2017. Sustainability and comanagement of subsistence hunting in an indigenous reserve in Guyana: Sustainability and comanagement. *Conservation Biology* 31, 1119–1131. https://doi.org/10.1111/cobi.12891
Shi, J., Wen, Z., Zhong, G., Yang, H., Wang, C., Huang, B., Liu, R., He, X., Shuai, L., Sun, Z., Zhao, Y., Liu, P., Liang, L., Cui, P., Wang, J., Zhang, X., Guan, Y., Tan, W., Wu, G., Chen, H., Bu, Z., 2020. Susceptibility of ferrets, cats, dogs, and other domesticated animals to SARS–coronavirus 2. *Science* eabb7015. https://doi.org/10.1126/science.abb7015
Sirén, A., Machoa, J., 2008. Fish, wildlife, and human nutrition in tropical forests: A fat gap? *Interciencia* 33, 186–193.
Stafford, C.A., Preziosi, R.F., Sellers, W.I., 2017. A pan-neotropical analysis of hunting preferences. *Biodiversity and Conservation* 26, 1877–1897. https://doi.org/10.1007/s10531-017-1334-8
Townsend, W.R., 2000. The sustainability of subsistence hunting by the Sirionó Indians of Bolivia, in: Robinson, J.G., Bennett, E.L. (Eds.), *Hunting for Sustainability in Tropical Forest.* Columbia University Press, New York, pp. 267–281.
Trail, P.W., 2007. African hornbills: Keystone species threatened by habitat loss, hunting and international trade. *Ostrich* 78, 609–613. https://doi.org/10.2989/OSTRICH.2007.78.3.7.318
UNODC, 2016. *World Wildlife Crime Report 2016: Trafficking in Protected Species.* United Nations Office on Drugs and Crime (UNODC).
van Dijk, P.P., Iverson, J.B., Shaffer, H.B., Bour, R., Rhodin, A.G.J., 2012. *Turtles of the World, 2012 Update: Annotated Checklist of Taxonomy, Synonymy, Distribution, and Conservation Status (No. 5)*, Chelonian Research Monographs. Chelonian Research Foundation, Lunenberg, MA.
Welch, J.R., 2014. Xavante ritual hunting: Anthropogenic fire, reciprocity, and collective landscape management in the Brazilian cerrado. *Human Ecology* 42, 47–59. https://doi.org/10.1007/s10745-013-9637-1

Whytock, R.C., Buij, R., Virani, M.Z., Morgan, B.J., 2016. Do large birds experience previously undetected levels of hunting pressure in the forests of Central and West Africa? *Oryx* 50, 76–83. https://doi.org/10.1017/S0030605314000064

Wilkie, D.S., Carpenter, J.F., 1999. Bushmeat hunting in the Congo Basin: An assessment of impacts and options for mitigation. *Biodiversity and Conservation* 8, 927–955.

Wilkie, D.S., Starkey, M., Abernethy, K., Effa, E.N., Telfer, P., Godoy, R., 2005. Role of prices and wealth in consumer demand for bushmeat in Gabon, Central Africa. *Conservation Biology* 19, 268–274. https://doi.org/10.1111/j.1523-1739.2005.00372.x

Young, H.S., Dirzo, R., Helgen, K.M., McCauley, D.J., Billeter, S.A., Kosoy, M.Y., Osikowicz, L.M., Salkeld, D.J., Young, T.P., Dittmar, K., 2014. Declines in large wildlife increase landscape-level prevalence of rodent-borne disease in Africa. *Proceedings of the National Academy of Sciences* 111, 7036–7041. https://doi.org/10.1073/pnas.1404958111

Young, H.S., McCauley, D.J., Galetti, M., Dirzo, R., 2016. Patterns, causes, and consequences of anthropocene defaunation. *Annual Review of Ecology, Evolution and Systematics* 47, 333–358. https://doi.org/10.1146/annurev-ecolsys-112414-054142

Zhou, P., Yang, X.-L., Wang, X.-G., Hu, B., Zhang, L., Zhang, W., Si, H.-R., Zhu, Y., Li, B., Huang, C.-L., Chen, H.-D., Chen, J., Luo, Y., Guo, H., Jiang, R.-D., Liu, M.-Q., Chen, Y., Shen, X.-R., Wang, X., Zheng, X.-S., Zhao, K., Chen, Q.-J., Deng, F., Liu, L.-L., Yan, B., Zhan, F.-X., Wang, Y.-Y., Xiao, G.-F., Shi, Z.-L., 2020. A pneumonia outbreak associated with a new coronavirus of probable bat origin. *Nature* 579, 270–273. https://doi.org/10.1038/s41586-020-2012-7

Ziegler, S., Fa, J.E., Wohlfart, C., Streit, B., Jacob, S., Wegmann, M., 2016. Mapping bushmeat hunting pressure in Central Africa. *Biotropica* 48, 405–412. https://doi.org/10.1111/btp.12286

INDEX

Note: Endnotes are indicated by the page number followed by "n" and the note number e.g., 637n8 on page 637.

aardvarks 288

Abies spp. (fir): *A. balsam a* 14, 19, 20, 22, 23, 34, 109, 403, 645; *A. lasiocarpa* 13, 14; *A. nephrolepsis* 17, 18; *A. nordmannia* 36; *A. sachalinensis* 17, 18; *A. sibirica* 15–16; auto-cycling 23; in boreal forest composition 7, 9, 12; cold temperatures, adaptation to 19; in mountain forests 82; in subtropical forests 52, 53

Abitibiwinni community 649

Acacia spp. 347, 435

Acer spp. (maple): *A. mono* 18; *A. macrophyllum* 30; *A. pictum susp. mono* 29, 34, 40; *A. platanoides* 29–30, 34, 527; *A. rubrum* 401, 405, 521; *A. saccharum* 29, 34, 35, 37, 124, 159, 387, 403, 405, 521; *A. ukurunduense* 18; evergreen-deciduous mosaics 39; mesic sites with loamy soils 35; in mountain forests 77; pollination by bees 41; swamp and riparian areas 35; temperature tolerance 34

adaptive management 46, 586

Adelges spp. 44, 159, 501

Aextoxicon punctatum 30

Afzelia spp. 347

Agelanthus crassifolius 615

Ageratum conyzoides 628

agoutis 645, 658

agriculture in forests (shifting cultivation): agrobiodiversity 633; biodiversity of, vs. old-growth forest 632–633; biomass accumulation in fallow period 627–628; burning of dry biomass 626, 627–628; carbon cycle 633–634; colonial disdain for 634; contemporary vilification of 635, 636, 637; crop diversity 626–627; crop rotation 627–628; deforestation, minor contribution to 635–636, 637; demographic crisis 631–632, 636; ecology of fallow-period growth 628–631; global diversity of 625; greenhouse gas emissions 418, 633; harvest 627; historical marginalization of practitioners 634–635, 636; livestock grazing 630; low outputs vs. input-intensive systems 634; Neolithic origins 625, 631; planting in the fallow 630; plot selection and clearance 626, 630; products harvested from fallows 631; sowing 626, 627; staple crops 626; as stewardship of forests 637; weeding 627

Alaska *see* United States

alder *see Alnus* spp.

Aldina spp. 347

Alectoria spp. 271, 274, 278

Alliaria petiolata 44, 500

alligators 657

allocation: definition 387; drought 389–390; fine root–leaf balance 388–389; productivity, closely linked to 387; stem growth as most important aspect 388; translocation 388; in tropical forests 388; woody plant tissue 390–392; *see also* primary production

Alnus spp. (alder) 7, 12, 17–18, 20, 29, 35, 52, 53, 55, 637n8

Alstonia scholaris 629
ambrosia beetles 518
American bullfrog 321
amphibians and reptiles: in Africa 315; biogeography 315–316; bioturbation 318; biphasic life cycle 318; caecilians 315, 318–319; calling and display sites 317–318; chameleons 315; climate change effects on 321–322; climatic constraints 315; conservation strategies 322; crocodilians 315, 318–319; dams and reservoirs as threat to 320–321; deforestation and fragmentation as threat 320–321; ectothermic lifestyles 314, 315, 317, 321–322; hunting and trafficking 321, 657; as invasive species 321; latitudinal gradients 314–315; microclimates 316; nutrient cycling 318; oviposition sites 317; pollination 318; precipitation and temperature 316–317; as predators and prey 319; refugia 315, 317; seed dispersal 320; small ranges of amphibians 334; South American diversity 315; tadpoles 318, 319; tropical diversity 314, 322, 333; *see also individual genera and species*
Amylostereum areolatum 504
Andropogon spp. 361
Angatia thwaitesii 505
Anodopetalum biglandulosum 31
Anolis spp. 316, 319, 321–322
anomalures 290
anteaters 289, 658
Anthurium harrisii 320
Antidesma acidum 629
ants: anthropogenic impacts on 241–242; as ecosystem engineers 241; epiphytes and 235; as invasive species 321, 491, 505; lianas and 216; as predators 248; seed dispersal 41, 582; *see also individual genera and species*
Anzia spp. 271
Apanteles fumiferana 102
Araucaria spp. 30, 43, 51, 54, 55, 82, 88, 347, 646
Arceuthobium spp. 523
Argentina 29, 30, 42, 43, 49, 54, 55, 271, 286
Argentine ant 491, 505
Arillastrum gummiferum 193
armadillos 289, 658
Armillaria spp. 84, 522, 523
Arrau turtle 321
Athrotaxis selaginoides 31
ash *see Fraxinus* spp.
Asian forest tortoise 321
aspen *see Populus* spp.
Aspen-FACE experiment 488
assisted migration *see* plant movement response to climate change
Asworthius sidemi 509
Atelopus spp. 322
Atherosperma moschatum 31, 32
Aucoumea klaineana 174, 201
Australia: amphibians and reptiles 315, 321; birds 301, 303; bryophytes 256; continental drift 300; deer, overgrazing by 43; distinctive tree flora 30, 31–32; drought 177, 557; epiphytes 232, 233; forest connectivity initiatives 437–438; forest tourism 36; gap dynamics 38; Indigenous forest knowledge 644, 648; intact primary forest 42; invasive animal species 44; lichens 271; mammals 286, 287, 288, 292, 293, 295; *Pinus radiata* monocultures 36; mountain forests 74; mycorrhizal symbiosis 347; rainforest climates 34; saprophytic plants 39; soil 35; subtropical forest 49, 51, 55; temperate forest 403; timber use 35; wildfires 89, 120, 127, 129, 130, 603; wild meat hunting 657; wind events 136, 140, 141
autumnal moth (*Epirrita autumnata*) 101
aye-ayes 290

Bahamas 53
bald eagle 645
Bali myna 307
balsam fir adelgid 520
bamboo 38, 40, 615
bandicoots 287
bark beetle outbreaks: climate change effects on 45, 96, 109; conifer defence mechanisms 103; critical host patch size 104; *Dendroctonus frontalis* (southern pine beetle) 103, 104, 518, 520; *Dendroctonus ponderosae* (mountain pine beetle) 85, 96, 97, 104, 106, 108, 109, 248, 376, 520; drought stress 110; extreme damage in North America 97, 103, 110; fire suppression, forest vulnerability increased by 108; girdling of trees 103; forest management, effects of 108; fungi, symbiosis with 103; *Ips typographus* 96, 106; mass-attack strategy 100, 103; in mountain forests 86; phloem as non-renewable plant resource 100; pole-ward range expansion 251; predators and parasites 103–104; spatial pattern 104; symbiotic fungal associates 103; transition from endemic to epidemic 103, 104
Bartonella spp. 669
basswood *see Tilia* spp.
bats: diet 292; diversity 290, 291; forest fragmentation, impact of 448; logging, effects of 463; regulation of insect populations 292; seed dispersal 292, 307, 582; widespread distribution 283; zoonotic diseases 669
bears 25, 40, 86, 292, 645

bees 41, 304, 491, 501
beavers 25, 289, 291, 491, 506, 647
beech *see Fagus* spp.
beech scale insect 501, 520, 522
Beilschmiedia spp. 32, 52
bell miner 498
Berber communities 644
Berberis thunbergii 498, 500
Berlinia spp. 347
Betula spp. (birch): *B. alleghaniensis* 34, 43, 401, 403; *B. costata* 18; *B. ermanii* 17, 18; *B. grossa* 37; *B. kenaica* 13; *B. nana* 14; *B. neoalaskana* 12, 13, 14; *B. papyrifera* 13, 14, 20, 23, 37, 106, 401, 637n7, 401; *B. pendula* 15, 16, 23, 34; *B. platyphylla* 17; *B. pubescens* 15, 482; cold temperatures, adaptation to 19; at continental treeline 8; evergreen-deciduous mosaics 39; forest tent caterpillar 109; frequent fire regimes 38; gap dynamics 38–39; lichens and 269 in oceanic boreal forests 7; overgrazing by bison 43; pollution tolerance 481; root–shoot allocation ratio 388; seed dispersal by wind 20, 23, 401; swamp and riparian areas 35
Bhutan 52
big-headed ant 505
Biological Dynamics of Forest Fragments Project 446
biological invasions: biological control 498, 500, 505, 510; birds 506; as by-product of economic activities 501; carbon cycles affected by 509; climate change effects on 509; climbing plant invaders 497; clonal plant invaders 497; closed-forest plant invaders 497; containment 505; definition 491; early detection, importance of 498, 505, 510; earthworms 44, 159, 491, 492, 499–500, 517, 526; fire-promoting grasses 497–498; forest gap plant invaders 497; forestry practices and 510; hybridization with native congeners 498, 509, 526; insect–fungus relationships 504–505; insects 501–505; intentional introductions 498, 506; mammals 506, 509; nematodes 498–499, 509; propagule pressure 498; reptiles 505; riparian forest susceptibility to 498; snails and slugs 500; tropical forest resistance to 498; *see also individual genera and species*
birch *see Betula* spp.
birds *see* forest birds
bison 43, 125, 293, 428, 667
black lion tamarin 453
blotch leafminer 520
blue-and-white fly-catcher 304
Bolivia 222, 620, 643, 657
Bonn Challenge 413, 426
boreal forests: Alaska–Yukon boreal 12–13; Altai–Sayan boreal 16; animal species 24, 25, 300, 304; auto-cyclic stand replacement after major disturbance 21, 22–23; canopy interception of rain and snow 373; carbon cycle 595, 597; Central Siberian boreal 16; clear-cutting 403; climate 7, 10; climate change effects on 24–25, 178; cold temperatures, adaptation to 19, 25; conifer forests outside boreal zone 9, 335; conifers, dominated by 7, 10, 12, 335; conservation efforts 24; continental boreal climes 7, 10, 13, 14, 15, 16, 17, 18; continental treeline 8, 10, 13, 14; Eastern North American boreal 14, 23; ericaceous shrublands 7, 8, 10, 403; failure of regeneration after major disturbance 22, 23; floristic subdivision, systems of 10–12; forest management practices 24; genetic diversity 168, 173–174, 175, 177; geographical extent 7–8; hemiarctic transition 9; hemiboreal transition 9, 14, 19; Holocene origins of 7; homogeneous species assemblages 335; human exploitation and degradation 23–24; insect outbreaks, adaptation to 20; lichens 8, 12, 13, 14, 15, 18, 25, 269, 271, 274, 278, 403; low numbers of tree species 164, 177; mixedwood forest 9, 12, 14, 16, 21, 23; mixture maintenance through cyclic disturbance 21, 23; mosses 7, 8, 12, 13, 14, 18, 21, 403; mycorrhizal symbiosis 364, 366–367; natural regeneration after harvesting 402–403; nitrogen limitation 386; North Atlantic Maritime boreal 18; Northeastern Siberian boreal 17; North European boreal 15; North Pacific Maritime boreal 18; nutrient cycling 7, 365; oceanic boreal climes 7, 10, 12, 14, 15, 17, 18; Okhotsk–Kamchatka boreal 17; paludification 15, 16, 21, 259; peatlands 7, 9, 21; permafrost 7, 13, 15, 16, 17, 18, 21, 24, 25; photosynthesis rates 385; succession of tree species in absence of major disturbance 21, 23; Transbaikalian boreal 16–17; understory vegetation 7, 12, 13, 14, 15, 16, 19, 20, 22, 23; West–Central North American boreal 13–14, 23; Western Siberian boreal 15–16; wide distribution ranges 335; wildfires 12, 13, 14, 16, 19–20, 21, 22, 23, 120, 129, 537, 543, 544; wind dispersal of pollen 168; wind events, adaptation to 20
Borneo 65, 450, 451, 597, 604
Borrelia burgdorferi 498
Brachystegia laurentii 198
Brazil: birds 334; cluster planting 434; drought 602; forest inselbergs 315; improving connections between forest fragments 453; logging, effects on gene diversity 178; natural

regeneration of fragmented forests 452; plot-based studies 67; subtropical forest 49, 54; tropical forest 61, 67, 213, 222; wildfires 602; wild meat hunting 656, 658
Bretziella fagacearum 522
brocket deer 645, 658
Brosimum rubescens 198, 199, 200
brown tree snakes 321, 491, 505
brushtail possums 40, 43, 287, 288, 491, 506
bryophytes: basic characteristics 255; climate change effects on 263–264; dispersal of spores 259; evolution of 255; forest disturbance, effects on distribution 257, 264; forest fragmentation, effects of 262–263, 264; forest management for preservation of microhabitats 264–265; global distribution 256, 334; as habitat 261–262; logging, effects of 263, 264, 463; microhabitat distribution 258; nutrient cycling 260–261; population dynamics 258–259; taxonomy 255; topographic distribution 256–257; water retention 259–260; *see also* mosses
Bryoria spp. 271, 274
Bufo spp. 320–321
Bupestris spp. 130
Burmese python 505
Burmese roofed terrapin 317

Caesalpinia eriostachys 199
Calluna spp. 82
Cambodia 657
camels 293
Cameroon 656
Canada: balsam fir regeneration 109; birds 304; boreal forest 12, 13, 14, 18, 304, 560; bryophytes 256, 257, 263; droughts 560; fragmentation of forested landscape 24; harvesting disturbances 401; hydrology, effects of climate change on 377; Indigenous forest knowledge 644–645, 647–648, 649; industrial pollution 477, 479; insect disturbances 95, 97; invasive insect species 504; lichens 271, 279; mammals 286; mountain pine beetle outbreaks 376; regulation of exotic forest species 525; Sudbury land reclamation programme 486–487; temperate forest 29; wildfires 122, 153, 376, 543, 544
Canary Islands 49, 51, 52, 53
cane toads 321
Capnodium walteri 505
Carapa guianensis 178
carbon budgets *see* forest carbon budgets
Carbonicola spp. 269
caribou 24, 25, 274, 278, 648
Carica papaya 630
Carpinus spp. 29, 42
Carya spp. 29, 40, 53, 521
cassowaries 302, 657
Castanea spp. 29, 246, 291, 521
Castanopsis spp. 51, 52
cats 287, 292, 506
cattle 42, 43, 293, 453, 662
Cecropia spp. 151, 407
Cedrela spp. 53
Cedrus spp. 88, 89
Celtis spp. 29
Ceratocystiopsis ranaculosus 518
Cerberiopsis candelabra 193
Cecropia spp. 193
Cetraria spp. 274
Chaenotheca chrysocephala 273
Chamerion angustifolia 130
Channing's toad 320–321
chestnut *see Castanea spp.*
Chile: conifer forests 335; deforestation 88; distinctive tree flora 30; epiphytes 232, 233; forest tourism 36; high plant diversity 328; Indigenous forest knowledge 646; intact primary forest 42; lichens 271; major disturbance and early successional species 37; mammals 286; montane vegetation 30, 35; overgrazing 43; *Pinus radiata* monocultures 36; soil 35; 36; wildfires 376
China: afforestation 377; amphibians and reptiles 315; birds 304; drought 571; emerald ash borer 105; industrial pollution 477; moist montane forest 336; non-timber forest products 620; subtropical forest 49, 50, 52, 55, 56, 57, 154; temperate forest 29, 34, 271; wildlife trade bans 669; wild meat hunting 657, 663
Chinese Ecosystem Research Network 592–593
Chinese giant salamander 315, 320–321
Chromolaena odorata 628, 630
Chrysomela lapponica 482
Chusquea spp. *see* bamboo
Cinnamomum spp. 52
Cirsium mexicanum 199, 200
civets 292, 669
Cladina spp. 403
Cladonia spp. 269, 273, 274, 275
classical monodominance: ectomycorrhizal association 199–200; endogenous species characteristics and exogenous forest system characteristics 201; escape from herbivory via leaf defence 201; lack of exogenous disturbance 197; large seed in deep leaf litter 199; masting leading to predation satiation 200; poorly understood biology of 196; poor seed dispersal leading to gregarious habit 201; probabilistic conceptual framework 201–203; shade tolerance under closed canopy 198;

slow decomposition rate 198–199; *see also* monodominance in tropical lowland forests
Clements, Frederick 148, 151
climate change: biological invasion, effects on 509; birds, effects on 307–308; in boreal forests 24–25, 178; bryophytes, effects on 264; composite provenancing 180; deer, effects on 164; epiphytes, effects on 241, 558–559; forest fragmentation and 450; genetic diversity and 178–180; hydrology management and 376; increasing impact on forest ecosystems 1; insects, effects on 241, 249–251; insect disturbances and 45, 96, 100, 102, 109–110, 111, 248; in mountain forests 82, 89; mycorrhizal symbiosis, effects on 351; natural regeneration after harvesting, effects on 407; permafrost thaw 24; pests and pathogens, effects on 527; in subtropical forests 56, 57; succession and 158–160; in temperate forests 44–45; temperature rise 1, 579; tropical cyclones 20, 142; uncertainty of predictive models 579, 585; wildfires 88, 89, 415; wind events 142–143, 603; *see also* droughts; fire and climate; forest carbon budgets; greenhouse gas emissions; plant movement response to climate change; tree growth response to climate change
Coccoloba spp. 347
Codia mackeeana 193
Coelogyne pandurata 645
coffee berry borer (*Hypothenemus hampei*) 97, 302
Colombia 333, 453, 658
colugos 289
common yellowjacket wasp 505
Convention on International Trade in Endangered Species of Wild Fauna and Flora (CITES) 310, 321
Coqui frog 321
Cowles, Henry Chandler 148
Corylus spp. 43
Costa Rica 178, 222–223, 302, 454
COVID-19 669
Cree First Nation 648, 649
crocodiles 657
Cronartium ribicola (white pine blister rust) 167, 517, 526
Cryphonectria parasitica (chestnut blight) 44, 517, 521, 525, 526, 527
Cryptocarya spp. 52
CTFS- ForestGEO network 592–593
Cuba 657
Cupressus spp. 248, 263
curassow 658
cuscuses 287
Cynometra alexandri 200
Cytisus scoparius 44
Czech Republic 501

Dacrycarpus dacrydioides 32, 35
Dacrydium cupressinum 32, 35
Declaration on the Rights of Indigenous Peoples (United Nations) 642
deer: climate change effects on 45; as invasive species 43, 506, 509; in mountain forests 86; overgrazing 43, 294; plant community composition, regulation of 41, 405; predators, decline of 294; seed dispersal 293; *see also individual species*
defoliators: foliage as renewable resource 100; grazers vs. wasteful feeders 102–103; parasitoids 102; periodicity and synchronicity of outbreaks 101, 102; predator–prey cycles 101; photosynthetic capacity reduced by 101; plant secondary metabolites as deterrents 101; principal damaging species 101; viral diseases of 102; *see also* insect disturbances; spruce budworm
deforestation: agricultural expansion as leading cause 415; in boreal forests 594; current trends 603; definition 414; global rates of 414; greenhouse gas emissions 416; recent reduction in net loss rate 414; in subtropical forests 49, 52, 53, 55, 57, 332; synergy with forest degradation 415–416; in temperate and boreal mountain forests 87–88; in temperate forests 594; in tropical forests 61, 65, 66, 67, 69–70, 165, 178, 332, 412, 414, 415, 594, 600, 613; *see also* REDD+
degradation of forest: in boreal forests 23–24, 595; definition 415; greenhouse gas emissions 416; in subtropical forests 49, 52; Synergy with deforestation 415–416; in temperate forests 595; in tropical forests 415, 595; unsustainable management 416; *see also* REDD+
Dendrobium cunninghamii 236
dendrochronology: Blue Intensity parameter 573; definition 572; elemental variability 574; hydroclimatic data from temperate and semiarid sites 573; importance to climatology 572; inter-annual influences of climate 575; stable isotope analysis 573; tree-ring width (TRW) variation as tracker of climate change 573; temperature data from cold sites 573; tropics understudied 572; wood anatomy 573; *see also* tree growth response to climate change; xylogenesis
Dendrosticta spp. 272
Denmark 434, 669
Dicymbe spp. 199, 200, 347
dingos 287, 295

Diprion pini (common sawfly) 41
Distylium spp. 52
dogs 292
Dominican Republic 321
Douglas-fir *see Pseudotsuga menziesii*
Drimys spp. 31, 54
droughts: already dry forests less resilient 559; angiosperm acclimation to 556; atmospheric evaporative demand (AED) 553, 554; carbon sink capacity of forests reduced by 560; carbon starvation 555; climate change effects on 552, 553, 554, 560, 579, 602; conifer vulnerability 556; definition of ecological drought 553; drought-tolerant species, selection pressures for 559–560; El Niño 554, 559; embolism refilling 556; epiphytes 241, 558–559; extreme events vs. unidirectional change 553; fire 555; fungal and insect attacks on weakened trees 555, 556; genetic acclimation 558; hot droughts 553–554; hydraulic failure 554–555; hydraulic safety margin (HSM) 556–557, 559, 560; leaf-level water use efficiency 554; leaf-shedding 555, 558; legacy effects 556; lianas 558; as major influence on tree species distribution 552, 557; mycorrhizal fungi 556; migration of tree species 558; new xylem growth 556; Palmer Drought Severity Index 553; phenotypic plasticity as acclimation 558; recent increase in tree mortality 552; rising atmospheric carbon dioxide levels, influence on plant physiology 554; seedling mortality 558, 560; soil water 555–556; Standardized Precipitation Evapotranspiration Index 553–554; stand thinning as drought management technique 559; structural overshoot 554, 557; thermal damage 555; types of 553; vapour pressure deficit (VPD) 553, 554, 555, 557, 558; variable length 553; water cycle 560–561
Dryobalanops aromatica 188, 199
duikers 293
dung beetles 452

earthworms: ecological roles of 41, 246, 365, 499; genetic adaptation to pollution 481; as invasive species 44, 159, 491, 492, 499–500, 517, 526; mammalian predators of 286, 290, 291; taxonomy 499
Easter Island 55
Ebola 669
echidnas 286
ecosystem-based management 23–24, 516–517, 650
Ecuador 237, 333, 336, 649
EFForTS- Biodiversity Enrichment Experiment 451, 454
Elaeocarpus spp. 52
elephants 56, 283, 289, 293, 294, 295, 451, 655
elephant shrews 288
Eleutherodactylus coqui 319
elk 41, 125, 348
El Niño 120, 194, 554, 559, 580, 597, 600
emerald ash borer (*Agrilus planipennis*) 44, 104, 105, 504, 521, 524, 525
Empetrum spp. 82, 346
Entomocorticium spp. 518
Eperua spp. 195
epiphytes *see* vascular epiphytes
Eremothecella 272
Erianthemum dregei 615
Erica spp. 82, 544
Ericaceae: boreal understory dominated by 7, 403; ericoid mycorrhiza 346, 347, 351; fire and 543; in hemiarctic transition 8; in mountain forests 82, 84, 88; nitrogen cycling slowed by 365; in oceanic boreal climes 7, 10; tree regrowth inhibited by, after insect disturbances 106
Eriococcus orariensis 505
Erioderma pedicellatum 279
Erythrina velutina 319
Erythroxylum ovalifolium 320
Ethiopia 263, 333, 336, 657
Eucalyptus spp.: *E. baxteri* 32; *E. cypellocarpa* 32; *E. melliodora* 167; *E. microtheca* 389; *E. obliqua* 32; *E. regnans* 36; *E. saligna* 498; *E. tricarpa* 177; *E. viminalis* 32; frequent fire regimes 32, 37, 38; monoculture plantations 646; in mountain forests 82; mycorrhizal symbiosis 347, 352; in subtropical forests 55
Eucryphia spp. 30, 32
Evolutionary Distinct and Globally Endangered (EDGE) rankings 286, 287, 290, 291, 292, 293
extinction debt 43, 448

Fagus spp. (beech): beech bark disease 522, 526; C:N ratio 36; *F. crenata* 29, 37, 40; *F. grandifolia* 29, 34, 43, 124, 526; *F. sylvatica* 30, 34, 434; global distribution 335; masting 40; mesic sites with loamy soils 35; in mountain forests 77–78
fairy pitta 305
falcons 301
Faroe Islands 8, 18
Finland 8, 41, 477, 483
fir *see Abies* spp.
fire: animal mortality 125; animals reliant on 130; asexual recolonization 126, 129, 130; boreal forests 12, 13, 14, 16, 19–20, 21, 22, 23, 25, 120, 129, 537, 543, 544; burn

intensity gradient 537; coniferous vs. broadleaf fuel 537–538; crown damage to plants 123–124; droughts and 555; drying of fuel 120, 121; El Niño events 120; elongation 122; fire-adaptive traits 125; fire regime concept 536–537; flaming vs. smouldering combustion 119, 121–122; forest management, effects of 545; greenhouse gas emissions 416–417; ground, surface, and crown fires 119–120, 121–122, 537; human-caused fires 37, 120, 121, 160, 415, 417, 535, 537, 545; indigenous people, burning regimes of 121, 160, 545; lightning ignition 120, 121, 537; logging, fire risk heightened by 466; mycorrhizal symbiosis affected by 350; natural fire breaks 538; pests and pathogens affected by 523, 524; physical phases of fire 539; rapid salvage, destructive effects of 130; recruitment from seeds 126–130; remote sensing fire records 538; residual stands 122; restoring natural fire regimes 433, 546; root damage to plants 122, 124–125; serotiny 19, 22, 82, 124, 129, 158, 180, 403, 405; slope 538; stem damage to plants 122–123; temperate forests 37–38, 45, 537; in tropical forests 121, 416–417, 537; whole-plant function, effects of damage on 124; wind 120, 122; *see also* fire and climate

fire and climate: aerial survey fire records 538; blocking high-pressure systems 537; Canadian boreal forests 543, 544; charcoal particle analysis 539–542, 543, 544, 545, 546–547; climate change 88, 89, 415, 535, 545, 546; drying climate of last millennium 545; fire scar analysis 539, 547; Holocene period 543–545; Mediterranean 543, 545; modeling long-term fire trends 545–546; Mongolia 543, 545; Neolithic period 545; precipitation as cause of fuel growth 538, 545; sub-Saharan Africa 543, 544, 545; written fire records 538, 546

Fitzroya cupressoides 30, 36

FLUXNET 593

forest birds: avian pathogens, spread of 506; biogeography 300–301; bird watchers 299, 309; in boreal forests 300; centres of species richness 334; climate change effects on 307–308; competition with mammals 301; conservation strategies 308–309; continuous food supply, need for 300; deforestation and logging as threat to 304–305; diet-switching, as response to seasonality 300, 303; disease 307; as ecosystem engineers 303; evolution of 300; foraging flocks 302; fragmentation as threat to 307; frugivores and granivores 302–303, 304, 307; habitat, forest as 300; hunting and trapping of 306–307, 656–657; insectivores 302, 304, 306; invasive bird species 307, 506; long-term studies, need for 310; migration, as response to seasonality 300, 303–304; nectar specialists 301, 303; parrots 302–303; pollination 301, 303; predators 301; scavengers 301–302; seed dispersal 302, 303, 306, 582; in tropical forests 300; ubiquity in forests 299, 300; vision 303; woodpeckers 301, 303; *see also individual genera and species*

forest carbon budgets: atmospheric change 601, 603; boreal forests 595, 597, 603; carbon dioxide fertilisation 554, 560, 603; definition 589; deforestation trends 602–603; disturbances 597–598, 600–601, 602, 603; field-based measurements 591, 593, 598; forest restoration efforts 604; global forest carbon sink 599–600; Gross Primary Production (GPP) 380, 590, 593, 595, 597–598, 602; human use and degradation as carbon source 600; interannual variability 597; latitudinal and temperature gradient 595; montane forests 598, 601; Net Biome Production (NBP) 590–591; Net Ecosystem Exchange (NEE) 590; Net Ecosystem Production (NEP) 590, 591, 593, 595, 597–598; Net Primary Production (NPP) 380, 590, 593, 595, 597–598; old-growth forest as carbon sink 589; pollution effects 602; regrowing forest as carbon sink 600–601, 604; remote sensing and modelling 593, 598; residence time 591, 597; soil vs. vegetation carbon stores 595, 597–598; subtropical forests 55–56; temperate forests 595, 597, 603–604; tipping point between carbon sink and carbon source 604; Total Ecosystem Respiration (R_{ECO}) 590; tropical forests 595, 597–598, 604; *see also* primary production

Forest Plots meta-network 593

forest tent caterpillar (*Malacosoma disstria*) 101, 102, 105, 108, 109, 521

foxes 25, 44, 287, 292, 295, 506

fox squirrel 506

fragmentation of forest: climate change and 450; community interaction changes, effects on ecosystem functions 450, 452; definition 445; edge effects 448–449, 450; global extent of 452, 454; 'good' and 'bad' effects at different scales 454; habitat isolation 449; improving connectivity between fragments 452–454; island biogeography theory 445–446, 448; human regional population density 450; involving local communities in restoration 454; large-scale fragmentation experiments 446–447; matrix types 449, 453; natural regeneration strategy 452; region-specific characteristics 450, 451; seed dispersal distance

reduced by 582; species rarity 448; species-specific characteristics 449–450; synergy with other environmental changes 449–450
Framework Convention on Climate Change (United Nations) 332, 413
France: mountain reforestation 88, 89, 90
Fraxinus spp. (ash): emerald ash borer 44, 104, 105, 504, 521, 524, 525, 527; *F. americana* 405; *F. mandshurica* 18; *F. nigra* 29; *F. pennsylvatica* 524; swamp and riparian areas 35; urban planting of 527
French Guiana 222
fungi: bark beetle symbiosis with 103; biotrophs 519; bryophyte–fungi interactions 261; decaying logs as hosts 39; earthworms and 41; endophytes 84, 346; entomopathogenic fungi 249, 250; fire damage and recovery 125, 126; gap-partitioning hypothesis 153; hemibiotrophs 519; human consumption of 35; incomplete inventory 330; logging, effects of 463; mushroom farming 352–353; necrotrophs 518–519; saprophytic plants and 39; spore dispersal 518; *see also individual genera and species*; lichens; mycorrhizal symbiosis

gap dynamics: in boreal forests 14, 23, 152, 153; diversity maintained by 153; early-successional forest, patches of created by 153; ecosystem management goals and 152–153; in mountain forests 82–83; niche-based models 153, 155–156; pests, effects on 521; stochastic processes 153; in temperate forests 37, 38–39, 152; in tropical forests 152; variable density thinning 432
genetic diversity of trees: Baker–Fedorov hypothesis 168; in boreal species 168, 173–174, 175, 177; climate change effects on 178–180; clinal variation 175–177; constraints on genetic study 164; forest management strategies and 179–180; gene flow between populations 168, 178–179; genetic improvement programs 164; genomes, sequencing of 180; genomic selection 181; Hardy–Weinberg model 165, 166; harvesting, effects on 177–178; high level of, in forest trees 164, 178; intraspecific diversity 165, 175, 177, 180, 526; migration 166; mitotic rate hypothesis 167; mutation 167; natural selection 167; neutral vs. adaptive genetic markers 165; phylogeography 173–174; random genetic drift 165–166; tree density, minimal effect of 173; in tropical species 168–173, 174, 177–178
Geopyxis spp. 126
German wasp 505
Ghana 218, 560, 601, 667
giant African snail 500
Gilbertiodendron dewevrei 196, 197, 198, 200, 201
Glasgow Leaders' Declaration on Forests and Land Use 413
Gleason, Henry A. 149
gliding squirrel 290
Global Biodiversity Framework 413
global circulation models / earth system models 545–546, 554, 560
Global Ecosystem Monitoring network 593
Global Land Cover 2000 332
global patterns of biodiversity in forests: criteria and indicators for biodiversity 328–329; distribution of forest ecosystems 330–333; elevation gradient 335–336; endemism, island and mountain hotspots of 334, 336; evolution 328; global forest cover maps 330, 331–332; global species data 330, 336; international interest in conservation 327; latitudinal diversity gradient 327–328, 333, 334–335; range expansion 327–328; soil type 336; species richness, continental hotspots of 334; tropics, high biodiversity of 327, 333, 336; tropics, regional variation in biodiversity of 335, 336; vascular plants vs. bryophytes 334; vertebrates vs. invertebrates 334
Glycaspis spp. 498
Gmelina arborea 435
goats 87, 293, 295
golden cantharelle 352
Goliath frog 315, 321
Gonatodes humeralis 321–322
gray squirrel 40, 44, 506, 509
Green Climate Fund 418
Greenland 8, 18
greenhouse gas emissions: carbon dioxide fertilisation of plants 554, 560, 602; continued rise since Industrial Revolution 1, 80, 579, 600; deforestation 416; degradation of forest 416; logging 465; permafrost thaw 21, 25; reduction programs 413, 418, 420, 423; shifting (swidden) agriculture 418, 633; unintended increases 421
Greenwayodendron suaveolens 174
Gremmeniella spp. 524
grey field slug 500
Guam 491, 505
Guapira spp. 347
Guatemala 647
Guibourtia demeusei 193
Guyana 658
gymnures 291
gypsy moths 491, 501b

hares 290, 657
harpy eagle 301
harvesting timber: clear-cutting 177, 264, 279, 304, 350, 374, 400, 401, 403, 405, 437, 460, 461, 462, 464, 465, 467, 470, 646; compaction from heavy machinery 407; coppicing 406; greenhouse gas emissions 416; retention harvesting 437; selective techniques 304, 403, 406, 407, 460, 461, 462, 463, 465, 468; shelterwood techniques 403, 406, 460, 461, 463; *see also* logged forests, ecology of
Hawaii 283, 307, 321, 364, 491, 500, 506
hawks 86, 301
heath hen 125
hedgehogs 291
hemlock looper (*Lambdina fiscellaria*) 14, 101, 102
herpetofauna *see* amphibians and reptiles
Hertelidia botryosa 269
Heterobasidion spp. 84, 350, 522, 523
heterotrophic respiration 39, 268, 346, 348, 349, 381, 387, 590, 595
HIV/AIDS 669
hoatzin 301
hornbills 302
horses 292
hurricanes *see* tropical cyclones
hydrology of forests: in boreal forests 373; canopy interception of snow and rain 373, 375; climate change mitigation 376; cloud cover and precipitation 372; deep vs. shallow water tables 372; erosion 375, 377; evapotranspiration 372, 373, 374, 375, 376, 377; fire, effects of 375–376, 377; groundwater recharge 372, 374; harvesting, effects of 374–375; hydro-forest nexus 373; latitudinal gradient 372; management of 376–377; modelling 376–377; mountain pine beetle outbreaks, effects of 376; nival watersheds 372; nivo-pluvial watersheds 372; pluvial watersheds 372; riparian zone maintenance 374–375; run-off 372, 373, 374, 375, 377; snowmelt 372, 375; stemflow 372–373, 374; in tropical forests 373; water quality 374–376
Hymenaea courbaril 178
Hypogamnia spp. 271

Iceland 8, 18
Ilex spp. (holly) 52, 53
Imperata cylindrica 628
India 52, 271, 315, 336, 633, 644
Indigenous forest knowledge: biodiversity promoted by 642; burning regimes 121, 160, 545, 647–648; complementary to scientific knowledge 643, 650; context dependence of 644; culture and identity linked to land 642, 649; cumulative and dynamic nature of 642, 643; ecological restoration 648; erosion of 643; environmental monitoring and assessment 648; focus on culturally salient species 645; forest management practices made socially acceptable by 649; functional zoning 649; as holistic 643–644, 649, 650; ignored in planning and management 642; Indigenous Guardian Programs 648; as local 644, 648; as moral and spiritual 644–645, 649; non-timber forest products 646–647; nuisance species management 647; oral transmission 643; participatory geographic information systems 648–649; protected areas 648; shaping of forest ecologies 642, 650; sustainable practices 645–646, 649; *see also individual peoples*
Indonesia: birds 307, 656; forest fragmentation 450, 451, 453–454; hunting 656; Indigenous forest knowledge 645; mammals 290; tropical forest 61; wildfires 537; wild meat hunting 657
insects: agriculture and silviculture, effects on distribution of 241–242; biodiversity of 242–243; climate change effects on 241, 249–251; feeding guilds 244; geographic range, changes in 251; incomplete inventory 330; insect pathogens 248–249, 250; mutualist symbionts 245–246, 248; natural enemies 248, 250, 251; nutrient cycling 246; parasitoids 248; as plant pathogen vectors 248; specialization and niche adaptation 244–245; temperature gradients, distribution driven by 241; trophic networks 241, 244; *see also* ants; insect disturbances; moths
insect disturbances: bark beetles vs. defoliators 95, 100; in boreal forests 20, 97; climate change effects on 96, 100, 102, 109–110, 111, 248, 251; congregation to avoid mating failure 100; density-dependence 247, 250; endemic to epidemic transition patterns 97–99; endemic phase research, need for 110; endogenous factors, complex interactions of 247; fast and slow variables 99, 100; forest collapse 102; forest dynamics affected by 95, 105–106; forest management effects 106–109, 110–111; host specificity 105; increase of 96; invasive species 104–105; markets flooded by damaged-tree harvesting 95; monophagous, polyphagous, and oligophagous species 95; Moran effect 102, 247; in mountain forests 97; in natural monocultures 97; nitrogen spikes caused by 106; periodicity 247–248; plant stress as trigger of 250, 251; potential for economic devastation 95, 247; public

concern 246; reciprocal feedback systems 99–100; spatial scales 99–100; temperature extremes, effects on insect life cycles 96; in tropical forests 97; *see also* bark beetles; defoliators; spruce budworm
interacting anthropogenic threats 1
Intergovernmental Panel on Climate Change 553, 579
intermediate disturbance theory 139, 154
International Union for Conservation of Nature (IUCN): habitat classification scheme 332–333; Red List of Threatened Species (IUCN) 271, 286, 330, 646, 656, 657
Intertropical Convergence Zone 406
invasive species *see* biological invasions
invertebrate pests: chemotaxis 517–518; classification by feeding type 517, 520–521; climate change effects on 527; forest dynamics, contribution to 521; genetic diversity as source of tolerance 526; invasive species 525–527; native pest species 521; pathogen attacks followed by insect attacks 522; pest–host equilibrium 521; as precursors of fire and stand replacement 521; predisposing, inciting, and contributory factors 522; primary vs. secondary agents of plant death 520; silvicultural practices as enabler 523–524; wildfires 523, 524; *see also* insect disturbances
Iran 271
Irvingia gabonensis 174
Italy 479

jack pine budworm (*Choristoneura pinus pinus*) 101, 520, 521
jaguars 453
Jamaica 138, 140
Japan: amphibians and reptiles 315, 319; birds 304; gap dynamics 37, 40; industrial pollution 477; pine wood nematode 499; 'snow forests' 34; spring ephemerals 39; subtropical forest 52, 53; temperate forest 29, 34, 271
Japanese giant salamander 315
Japanese white-eyes 307
jays 40, 86, 273, 303
Juglans spp. 40, 526
Julbernardia seretii 198, 200, 201
Juncus spp. 648–649
Juniperus przewalskii 571

Kaiser's mountain newt 321
Kalmia spp. 403
kangaroos 287
Kaqchikel people 647
Kenya 336
king cobra 317
Kitcisakik Algonquin people 645
koalas 86, 287
Komodo dragon 322
Korea 29, 34, 271, 304, 436
krummholz vegetation 9, 10, 35
Kukama people 645
Kumeyaay people 648–649
Kunming–Montreal Global Biodiversity Framework 327
Kunzea spp. 130

Lagarostrobos franklinii 31, 35
Lantana camara 498
Laos 633
larch *see Larix* spp.
larch budmoth (*Zeiraphera diniana*) 101, 247
larch sawfly 521
Larix spp. (larch): in boreal forests 7, 12; cold temperatures, adaptation to 19; *L. cajanderi* 17; *L. decidua* 88, 90; *L. gmelinii* 16, 17; *L. laricina* 13; *L. sibirica* 15, 6, 19; lichens and 269; in mountain forests 77–78
Laureliopsis philippiana 30
lemurs 290
leopards 24, 645
Leptospermum scoparium 505
Letharia spp. 273
Leucaena leucocephala 435
lianas: abundance negatively correlated with mean precipitation 217–218; basic characteristics 211; cold, vulnerability to 221–222; droughts 558–559; dry season growth advantage 219–221, 558; habitat complexity increased by 211, 216; as human resource 216–217; increasing abundance in neotropical forests 222–223; latitudinal distribution 219; local distribution 217, 221; net forest carbon storage reduced by 212, 215; as nutrition source 211, 216; rapid recruitment after disturbance 212, 217; soil depleted by 212, 214; species diversity 211, 216; tree mortality not increased by 213; tree recruitment, growth, and fecundity reduced by 212, 213–215; vascular systems, large and efficient 219, 221–222
Libocedrus bidwillii 32, 35
lichens: animal–lichen interactions 273–274; anthropogenic impacts on 279; on bark 277; basic characteristics 268; in boreal forests 269, 271, 274, 278, 403; chlorolichens 268, 269, 271, 272, 273, 277; cyanolichens 268–269, 271, 272, 273, 276, 277; on deadwood 277–278; dynamics of lichen communities 276–278; epiphytic 268, 269, 271, 272, 273, 274, 275, 277, 278, 279; forest management for preservation of 279; as indicators of forest ecosystem health 278; on leaves 277;

macroclimatic distribution 275; microclimatic distribution 275–276; as non-timber forest product 647; nutrient cycling 272–273; photosynthesis 268; in temperate forests 271, 278; terricolous 269, 272, 273, 274, 275, 278; in tropical forests 272, 277; water regulation 268, 275; *see also individual genera and species*
linden *see Tilia* spp.
Lindera spp. 52
Liriodendron spp. 29, 405
Lithocarpus spp. 52
Lobaria spp. 271, 272, 273, 278
logged forests, ecology of: abiotic changes 461; alternative land uses vs. logging 466; biodiversity 462–464; canopy gaps 461–462; carbon cycles 464–465, 469; close-to-nature forestry 467; disease and insect outbreaks 466; ecosystem changes 464; fire risk heightened 466; global scale of logging 459; heterogeneous effects of logging 461, 462, 463, 470; landscape-scale management 469–470; logging cycles 460; natural disturbance dynamics vs. 464; post-logging management 470; reduced impact logging 468–469; retention forestry 467–468; roads for logging, effects of 465–466; stand composition 461; study types 460–461; timber vs. carbon vs. biodiversity trade-offs 466, 470; *see also* harvesting timber
longhorn beetles 499, 518, 520, 525, 527
long-tailed Abert's squirrel 86
Lonicera spp. 44, 497, 500, 504
loris 645
Lupinus spp. 44
Lyme disease 498, 669
lynx 25, 86, 290

Macaranga spp. 151, 628
Machilus spp. 52
Macrolobium acaciifolium 194
Madagascar: amphibians 315; birds 301, 303; as island habitat 300; mammals 288, 290, 292, 293; mass extinction 667; subtropical forest 49
Magnolia spp. 52, 53
Malaysia 450, 451
Mallotus paniculatus 628
mammals: Afrotheria 288–289; as apex predators 293; artiodactyls 293; bioturbation 287; birds, competition with 301; browsing, effects of 288, 293, 294; carnivores 292, 293, 294; deforestation for pastureland 293; distribution 283; diversity, decline of 283; domestic livestock grazing in forests 293; endotherm lifestyle 296; Euarchontoglires 289–291; feces 293; hunting 294–295, 656; insect populations, regulation of 292, 293–294; as invasive species 295, 296; Laurasiatheria 291–293; marsupials 286–288; monotremes 286; primates 289–290, 301, 656, 658; reintroduction 295; rodents 290–291, 656; seed dispersal 287, 289, 290–291, 292, 294, 582; in tropical forests 283; Xenarthra 289; *see also individual genera and species*
Mangifera spp. 630
mangrove forests: deforestation 415; floristic simplicity of 65; monodominance 194–195, 336; wind event damage 136, 141
Manilkara multifida 178
Māori people 643, 649
maple *see Acer* spp.
Mapuche Pewenche people 646
marsh deer 658
martens 645
matsutake 352
Melaleuca spp. 347
Melampsora larici-populina 524–525
Melocactus violaceus 320
Menominee tribe 645–646
Metrosideros polymorpha 175, 193
Mexico 49, 52, 53, 104, 304, 335, 648, 660
Michelia spp. 52
Microberlinia bisulcata 197
Microhyla nepenthicola 317
Microstegium vimineum 500
migration *see* plant movement response to climate change
mink 44, 506, 669
Miscanthus spp. 628
Mitragyna stipulosa 193
mixedwood forest: in Alaska–Yukon boreal 12; in hemiboreal transition 9; stand dynamics 21
MODIS Vegetation Continuous Fields 331–332
moles 291
Mongolia 17, 543, 545
monkeys 287, 289–290, 645, 658
Monochamus spp. 130
monodominance in tropical lowland forests: as challenge to conventional perceptions 188; ectomycorrhizal association 188, 192, 195; in low-nutrient forests 195–196; non-persistent dominance 192, 193–194; persistent dominance 192–193; in seasonal floodplains 193, 194; in successional forests 193–194; in waterlogged forests 194–195; *see also* classical monodominance
Monte Verde toad 321
moonrats 291
moose 24, 25, 40, 41, 506, 517, 645
Mora spp. 194, 200
Morchella spp. 126

Morocco 320, 644
mosses: in boreal forests 7, 8, 12, 13, 14, 18, 21, 403; *Ceratodon purpureus* 129; feathermosses 7, 13, 14, 125, 259; fire damage and regeneration 125, 129; latitudinal gradient of species richness 334–335; pleurocarpous mosses 256; *Polytrichum* spp. 129; *Sphagnum* spp. 7, 21, 125, 257, 259, 261; *see also individual genera and species*
moths: *Hemerocampa pseudotsugata* 84; *Heterocampa guttivitta* 41; *Malacosoma americanum* 41; *Zeiraphera diniana* 84
mountain forests: tropical 56, 57, 66, 69, 230, 234, 256, 260, 272, 308, 333, 336, 337, 347, 387, 599; *see also* temperate and boreal mountain forests
Musa spp. 630
Musanga spp. 407
Myanmar 52
mycorrhizal symbiosis: arbuscular mycorrhiza (AM) 84, 343, 344, 346, 364, 368, 464; in Arctic and alpine regions 346; benefits of 343; in boreal forests 364, 366–367; climate change effects on 351; co-introduced mycorrhizal fungi 352; definition 343; in drought conditions 556; ectomycorrhiza (EcM) 84, 345–346, 364, 366–368, 464, 506; ericoid mycorrhiza (ErM) 84, 343, 346; forest disturbance, effects of 349–351; habitat conservation 353; herbivore disturbance 350; inoculation 351–352; logging disturbance 350–351; in monodominant tree stands 348; nutrient cycling 346, 348–349, 364; in nutrient-poor soils 347; in nutrient-rich soils 347–348; orchid mycorrhiza (OM) 343, 346; pathogen disturbance 350; soil acidification 346; soil carbon sequestration 346; in subantarctic islands 347; in temperate forests 347, 364; in tropical forests 347, 364; waterlogging disturbance 350; *see also individual genera and species*

Naskapi First Nation 648
natural regeneration after harvesting: in boreal forests 402–403; climate change effects on 407; cost-effectiveness 399; in dry ecosystems 405–406; as dynamic process 407–408; harvesting as disturbance 399–401, 407, 408; seed supply 401, 407, 428; survival and growth of seedlings 401–402; in temperate deciduous forests 403–405; in tropical forests 406–407; unpredictability 399; vegetative reproduction 401; *see also* restoration of forest ecosystems
Neea spp. 347
nematodes 248, 249, 348, 492, 498–499, 509, 517, 627, 637n1
Nepal 52, 271, 647
Neolitsia spp. 52
Neonectria spp. 522
Nepenthes spp. 317
Nephroma spp. 271
Netherlands 669
Neurergus spp. 321
New Caledonia 193
New Guinea 286, 287, 292, 293, 301, 303, 315, 347
New Zealand: amphibians and reptiles 315, 321; anthropogenic fires 38; bats 283, 292; birds 301, 303; bryophytes 256; conifer forests 335; deer, overgrazing by 43; distinctive tree flora 30; epiphytes 232, 233; evergreen coastal rainforests 32, 34; forest connectivity initiatives 437–438; gap dynamics 38; Indigenous forest knowledge 648; intact primary forest 42; invasive bird species 506; invasive insect species 501, 504, 505; lichens 268, 271; major disturbance and early successional species 37; Māori *whakapapa* 643; masting 40; mycorrhizal symbiosis 352; montane vegetation 30, 35; mountain forests 74; *Pinus radiata* monocultures 36; possums 40, 43, 288, 491; saprophytic plants 39; soil 35
Nguni people 649
Nimba toad 321
Nipah virus 669
non-timber forest products (NTFPs): aloe 620; as alternative to deforestation and degradation 613, 614; argan 620; bamboo 615; as biodiversity components 614; Brazil nuts 613, 615, 620; cash income generation from 618–619; cash saving enabled by 618; caterpillars, edible 615; colonial lack of interest in 613; as community dominants 614; definition 614; domestication and cultivation 620; economic analysis, methods of 616, 620; in ecosystem composition 614–615; fuel 618; honey 617–618; household provisioning 617–618; household wealth differences in production and consumption of 617–618; Indigenous use of 646–647; international market chains 620; in local culture 613, 619–620; management and policy 616, 621, 622; marula tree 614–615, 620; mushrooms 352–353, 617–618, 620, 646–647; non-monetary social benefits 619; *Prunus africana* 620; rattan 613, 620; reeds 614; resource conflict 620; rubber 613; as safety-nets 619; shea butter 620; sustainable harvesting 615–616, 620; twentieth-century revival of interest in

613; urban populations 621; vegetable ivory palms 615
Norway: boreal forest 8, 15; wind events 141
Nostoc spp.
Nothofagus spp. (southern beech): early succession species after major disturbance 37; global distribution 335; masting 40; in mountain forests 82; *N. alpina* 30; *N antarctica* 35; *N. betuloides* 31; *N. cunninghamii* 31; *N. dombeyi* 30; *N. fusca* 32; *N. menziesii* 32, 35; *N. moorei* 31; *N. nitida* 30; *N. obliqua* 30; *N. pumilio* 31; *N. truncata* 32; non-persistent dominance 193
numbats 287
nun moth (*Lymantria monacha*) 101
nutcracker birds 86, 303
nutrient cycling: atmospheric depositions 358, 364, 367; biochemical cycle 356–357, 358–359; biogeochemical cycle 357–358, 359; climate effects on nutrient fluxes 357; critical zone 357; ecosystem boundaries, definition of 357, 358; forest management activities 359, 368; geochemical cycle 358, 359; harvesting 362, 368; leaching 357, 358, 359, 368; limiting elements 363–364, 368; long leaf lifespan 357, 359; mineral weathering 358, 364, 367, 368; mycorrhizal symbiosis 364, 366–368; nitrogen fixation 358, 362; plant litter characteristics 365–366; primary succession 360–362, 367, 368; progression–retrogression 368; resorption 356, 358–359; secondary succession 362, 368; soil properties 357–358; supply and demand 362–363; tree species composition 364, 366

oak *see Quercus* spp.
Omushkego Cree community 647
Ophiognomonia clavigignenti-juglandacearum 526
Ophiostoma minus 518
Ophiostoma ulmi (Dutch elm disease) 44, 504, 517, 521–522, 525, 527
opossums 286
orangutan 451, 645
Oriental honey buzzard 304
Oropogon spp. 271
otters 506, 668
owls 290, 301

pacas 645, 658
Pachira quinata 178
paleo-Antarctic rainforest 51
paludification 15, 16, 21, 259
pampas deer 658
Panama 151, 221, 222, 223, 301, 307, 318, 558, 560
pangolins 292, 663
Papua New Guinea 657
Paraguay 54
Parashorea chinensis 198
Paris Agreement 413
Parmeliopsis spp. 271
pathogens: basic characteristics 517; biological vectors 518, 522; climate change effects on 527; climatic conditions 523; forest dynamics, contribution to 521; genetic diversity as source of tolerance 526; insect attacks after pathogen attacks 522; invasive species 525–527; passive dispersal 518; pathogen–host equilibrium 521; penetration of plant tissues 518; predisposing, inciting, and contributory factors 522; resistance breeding 524–525; silvicultural practices as enabler 523–524; wildfires 523, 524
Paxillus involutus 367–368
peatlands: biodiversity 336; in boreal forests 7, 9, 21; carbon cycle 599; formation of 598–599; tree-planting on 465; in tropical forests 62, 65, 336; wildfires 537
peccaries 645, 658
Peltigera spp. 268
Peltogyne gracilipes 196
Pennisetum spp. 628
permafrost: boreal adaptation to 7, 21; greenhouse gas emissions from thaw of 21, 25; stability affected by climate change 24
Peru 614, 657
Philippines 290, 304
Phoebe spp. 52
Phyllanthus spp. 644
Phyllocladus aspleniifolius 32
Phyllonorycter strigulatella 482
Phytophthora cinnamomi 350
Picea spp. (spruce): in boreal forest composition 7, 9, 12; C:N ratio 36; cold temperatures, adaptation to 19; evergreen-deciduous mosaics 39; lichens and 269, 273; in mountain forests 82; *P. Abies* 15, 34, 36, 45, 89, 101, 106, 179, 180, 273, 403, 431, 482; *P. critchfieldii* 580; *P. engelmanii* 89; *P. glauca* 12, 13, 14, 19, 20, 23, 34, 124, 168, 173, 177, 180, 403; *P. jezoensis* 17, 18, 24; *P. mariana* 13, 14, 19, 20, 22, 158, 168, 173, 175, 180, 357, 401; *P. obovata* 15; *P. rubens* 29; *P. sitchensis* 30, 36, 89, 181, 361; *P. smithiana* 86; in subtropical forests 52, 53; *see also* spruce budworm
Pierce's disease 522
pigs 44, 293, 295, 491, 506, 658, 669
Pikangikum Ojibway people 649
Pilgerodendron uviferum 30, 31
pine *see Pinus* spp.

pine processionary moth (*Thaumetopoea pityocampa*) 101, 249, 251
pine shoot beetle 522
pine wood nematode (*Bursaphelenchus xylophilus*) 248, 499
Pinus spp. (pines): in boreal forests 7, 9, 12; C:N ratio 36; cold temperatures, adaptation to 19; dry sites with sandy soils 35; evergreen-deciduous mosaics 39; fire and 37, 38, 543; as invasive 491 lichens and 269; in mountain forests 77–78, 82; *P. albicaulis* 86, 526; *P. banksiana* 13, 14, 19, 22, 361, 401; *P. cembra* 86; *P. contorta* 13, 19, 177, 180; *P. elliottii* 524; *P. flexilis* 86; *P. geradiana* 86; *P. halepensis* 406; *P. jeffreyi* 477; *P. koraiensis* 34; *P. massoniana* 52; *P. nigra* 88, 90, 401, 406; *P. palustris* 124, 524; *P. pinaster* 180, 406, 571; *P. pinea* 179; *P. ponderosa* 86, 124, 168, 477, 559; *P. pumila* 17, 18, 86; *P. radiata* 36, 646; *P. resinosa* 29, 523, 524; *P. rubra* 166; *P. sibirica* 15, 16, 17, 22, 86; *P. strobus* 35, 156, 167, 177, 645; *P. sylvestris* 15, 16, 17, 19, 34, 35, 36, 179, 389, 391, 406; *P. taeda* 524; *P. yunnanensis* 52; pine wilt disease 499; seed dispersal 303; in subtropical forests 52, 53
plant movement response to climate change: assisted migration 45, 89, 179–180, 436, 585; dynamic global vegetation models (DGVMs) 583–584; extinction risk 584–585; generation time 580, 583; movement as near-universal response in paleoecological record 580; post-glacial migration lag 580; recent movements 582–583; refugia 584, 586; seed dispersal ranges 581–582; species distribution models (SDMs) 583, 584; velocity of climate change 580–581, 586
Platanus spp. 53, 498
Platismatia spp. 271
platypuses 286
Pleurozium spp. 259, 261, 403
Podocarpus spp. 32, 51, 54, 55
podzols 62–63, 66, 195
poison-dart frogs 315, 317
Poland 621
pollution: acid rain 476, 477, 602; in Arctic ecosystems 485; carbon cycle, effects on 479, 602; cause and effect, difficulty of establishing 483; climate change, interaction with 485–486, 487; community-level effects 479–480; crown structure of trees, damage to 482; definition 475; direct vs. indirect effects 476–477; ecosystem-level effects 479; fluctuating asymmetry 483; fluorine 476; industrial barrens 477–479, 483, 484; interaction with other stressors 477, 483; landscape-level effects 477–479; local effects near large industrial enterprises 477, 488; net primary productivity (NPP) decrease 477, 479; nitrogen deposition 476, 479–480, 482, 485, 488, 602; nutrient cycle, effects on 479; organism-level effects 482–483; ozone (O_3) 476, 477, 479, 480, 488, 602; particulate matter (PM) 475, 477, 488; 'peeling off' effect 479; population-level effects 480–482; regional effects 477; research directions 487, 488; restoration 486–487; slow natural recovery from 475, 486; sulphur deposition 476, 477, 482, 484, 485, 488; tolerance, development of 480–481; trace elements 476, 480–481; variation in biotic responses to 484
poplar *see Populus* spp.
Populus spp. (poplar): asexual reproduction from suckers 126; in boreal forest composition 12; cold temperatures, adaptation to 19; forest tent caterpillar 108, 109; frequent fire regimes 38; insect disturbances 102; in mountain forests 82; nitrogen fertilization, response to 392; *P. balsamifera* 13, 174; *P. deltoides* 388, 434; *P. tremula* 15, 16, 17; *P. tremuloides* 12, 13–14, 20, 22, 23, 37, 106, 168, 174, 401, 403; *P. trichocarpa* 13, 180; seed dispersal by wind 20, 23, 151; *Septoria* canker 525; swamp and riparian areas 35
porcini 352
porcupines 25, 291, 517
Portugal 271, 377, 499
possums *see* brushtail possums
primary production: allocation, closely linked to 387; autotrophic respiration 380, 381, 590; carbon balance 381; definition 380; Gross Primary Production (GPP) 380, 590, 595, 597–598, 602; harvesting 381; land availability 381; light 381, 383–385; Net Primary Production (NPP) 380, 590, 593, 595, 597–598; nutrients 385–386; respiration 386–387; senescence and mortality 380, 381; species diversity and composition 387; temperature 384–385; in tropical forests 381; water 385; wood vs. non-woody tissue 381; *see also* allocation; forest carbon budgets
Prosopis africana 177
Prumnopitys spp. 32
Prunus spp. 29, 40, 41, 44, 620
Pseudocyphellaria spp. 271, 272
Pseudotsuga menziesii 29, 30, 36, 53, 89, 90, 126, 175
Pterophylla racemosa 32, 35
Ptilium spp. 403
Pueraria montana 491
Puerto Rico 137, 222, 305, 321
Pygmy groups 660

Pyrrosia eleagnifolia 236
pythons 315

Quechua people 649
Quercus spp. (oak): C:N ratio 36; cluster planting 434; coppicing 406; dry sites with sandy soils 35; evergreen-deciduous mosaics 39; evergreen Mediterranean species 406; frequent fire regimes 37, 38; gap dynamics 39; masting 40; in mountain forests 77–78, 82; mycorrhizal symbiosis 352; natural regeneration, decline in 405; oak wilt 522; *Q. alba* 387, 523; *Q. faginea* 406; *Q. garryana* 648; *Q. mongolica* 29, 34; *Q. oleoides* 200; *Q. pubescens* 406; *Q. robur* 30, 34; *Q. rubra* 29, 40, 43, 44, 387, 523; *Q. taxana* 434; *Q. velutina* 361; seed dispersal 291, 303; shade intolerance 405; in subtropical forests 52, 53; temperature tolerance 34

rabbits 44, 290, 506, 517
raccoon dogs 44
Rafflesia spp. 645
Ramalina menziesii 268
rats 295, 321, 506, 509
REDD+: addressing drivers of deforestation and degradation 418; aims of 413; carbon sequestration vs. biodiversity trade-offs 420–421; co-benefits 413; displacement of deforestation from REDD+ projects 414, 420; establishment of 413; improvement of harvesting practices 418, 420; indigenous and local communities, effects on 421; integration of REDD+ projects with national development strategies 420; local stakeholders, engagement of and benefits to 421–423; payments for carbon emission reductions 418; protected forest areas 418 forest restoration, reforestation, and afforestation 420; sustainable agriculture 418; unintended consequences 420–421; *see also* deforestation; degradation of forest
red deer 43
red-eared slider 321
red fire ant 505
red pandas 292
red squirrel 44, 509
red turpentine beetles 491
reindeer 25, 87, 274, 278, 648
reptiles *see* amphibians and reptiles
restoration of forest ecosystems: adaptation to future conditions 426, 431, 436; applied nucleation 433, 438; assisted migration 45, 89, 179–180, 436; assisted natural regeneration 430; bare ground as starting point 428; carbon storage 604; cluster planting 433–434; degraded forest as starting point 428; direct seeding 431; forest corridors, restoration of 436–438; forest and landscape restoration 427–428; framework species method 435; genetic diversity 432; groundwater restoration 437; international agreements 426; landform and hydrological modification 436; landscape resilience 437; local communities 428, 438; long-term management plans 438–439; Miyawaki method 435; monitoring 438–439; natural fire regimes, restoration of 433; natural regeneration 428, 435–436; non-native plantation monoculture as starting point 432; nurse crops 434–435; overstory manipulation 432–433; planting 431–432, 433–436; planting patterns 433; reference conditions from the past 426; rewilding 295, 428, 430; social context 428; stepping stone habitat patches 438; underplanting 430
Rhamnus cathartica 44, 498, 500
rhinoceroses 56, 292–293
Rhizophora spp. *see* mangrove forests
Rhododendron spp. 82, 403
Riparian Ecosystem Restoration in Tropical Agriculture 451
Robinia pseudoacacia 44
Romania 97
rosy wolfsnail 500
Rubus idaeus 40
Russia: birds 304; boreal forest 8, 15, 16–17, 18; industrial pollution 477, 479, 481, 483, 487; insect disturbances 97, 101; lichens 279; mountain forest 74; temperate forest 271; wildfires 536

Sabah Biodiversity Experiment 451
Salamandra spp. 321
Salix spp. (willow) 8, 9, 12, 20, 35, 53, 193, 346, 449, 481, 482, 648–649
Sami people 648
saprophytic plants 39
SARS disease 669
Sasa spp. *see* bamboo
savanna 29, 30, 34, 37, 45, 62, 65, 121, 129, 177, 331, 332, 351, 466, 535, 538, 542, 657, 660
Schima spp. 52
Sclerocarya birrea 614–615
Scolotyneae: *see* bark beetles
Scotland 89
Senna reticulata 193
Septoria spp. 525
serpentine leafminer 520
Sequoiadendron giganteum 88
sheep 42, 87, 293

shifting cultivation *see* agriculture in forests
Shorea spp. 52, 196, 637n7
shrews 291
Siberian blue robin 305
Siberian silk moth (*Dendrolimus sibiricus*) 101, 102
sika deer 43, 509
Singapore 451
Sirex noctilio 504
sloths 289, 301
solenodons 291
Soliga people 644
Sorbus spp. 18
South Africa 256, 347, 352, 618, 620, 649, 657
Spain 271
Spanish slug 500
Spherulina musiva 525
Spondias spp. 173, 629
spongy moth (*Lymantria dispar*) 101, 105, 521, 525
spruce *see Picea* spp.
spruce budworm (*Choristoneura fumiferana*): *A. balsamea* and 14, 20, 105–106; birds as predators of 101; climate change effects on 96, 102, 109; extreme damage in North America 97; forest dynamics affected by 105–106, 520; as grazer 102; in mature vs. immature stands 101; natural forest stands 521; parasitoids 102; *see also* defoliators; insect disturbances
squirrels: acorn distribution 40; complementary diversity with parrots 301; as disease vectors 509; as forest pests 517; as invasive species 44, 506, 509; lichens eaten by 274; *see also individual species*
Sri Lanka 315, 336, 438
steppe 9, 16, 17, 385, 543, 545, 631
Stereocaulon spp. 273, 274, 275
Sticta spp. 268, 271, 272
Straw-headed bulbul 307
Styrax spp. 53, 630, 637n7
subtropical forests: Asian subtropics 51, 52–53, 55, 56, 57; Australia 51, 55; biogeography 50–52; Canary Islands 53; carbon economy 56, 57, 412; climate 49, 55, 56–57; climate change effects on 56, 57; conifers 335; cyclones 49–50; definition by latitude 49, 56–57; deforestation 49, 52, 53, 55, 57, 332; Fagaceae, abundance and diversity in northern forests 52; fauna 56, 57; fragmentation 57; Madagascar 54; ; North American subtropics 53; phosphorous limitation 50; secondary and degraded forest 49, 52; soil 50; South America 51–52, 54–55; southern Africa 51, 54; transitional entity, often regarded as 56, 57; tropical origins of southern forests 54
succession: arrested succession 159; in boreal forests 21, 23, 150, 158; climate change effects on 158–160; compositional change 150–151; definition 148; dispersal limitation 155–156; disturbance interactions 156; divergent successional pathways 156; diversity 154; forest management and 160; history of concept 148–149; initial floristic composition succession model 83, 151; intermediate disturbance hypothesis 154; invasive species 159; legacy structures 150; mechanisms of 151–152; in mountain forests 148; precocious succession model 156; primary vs. secondary 148; recruitment limitation 155; relay floristic succession model 82–83, 151; reorganization stage 158; stochastic processes 155–156; structural change, stages of 149–150; in temperate forests 37, 150, 151, 160; in tropical forests 150, 151; wind events, effects of 137–138; *see also* gap dynamics
Sumatra 65, 451
Sustainable Development Goals 327, 426
Sweden 8, 40, 89, 141, 256, 367, 648
Swietenia macrophylla 174, 177
Symphonia globulifera 178
Symplocos spp. 52

Taiwan 134, 305, 657
Tanzania 336, 644, 657
tapirs 292, 658
tarsiers 645
Tasmanian devils 287
Taxillus tomentosus 644
Telmatobius frogs 657
temperate and boreal mountain forests: assisted migration 89; carbon cycle 599; climate change effects on 82, 89; communities of species 81–82; deforestation 87–88; disturbances and succession 76, 82–83; ecosystem services 87; ecosystem structure 82; elevation 77, 79, 80, 88; Ericaceae as keystone group 82, 84; geographical extent 74; invasive species 90; land use abandonment in Europe 88; litter, humus, and nutrients 83–84; partial pressure of carbon dioxide and oxygen 80; plant–animal interactions 84–86; plant–fungi interactions 84; precipitation 79–80, 89; reforestation 88; season length 81, 89; soil 81, 90; solar irradiance 77, 79; wildfires 89; wind events 89
temperate forests: biodiversity 335; biogeography 29–32; boreal forest, transition to 34; canopy structure and gap dynamics 37, 38–39; carbon cycle 595, 597; climate change effects on 44–45; coniferous 335, 337; ecosystem services 35; extinction debt 42–43; folivorous

insects 41; fragmentation 42–43, 45; frequent fire regimes 37–38, 45; grazing animals 41; human settlement and land clearance 35, 42; indigenous inhabitants 42; invasive species 44, 45; lichens 271, 278; lianas 219; logging and planted regeneration 42; major disturbance and succession 37; mycorrhizal symbiosis 347, 364; natural regeneration after harvesting 403–405; nitrogen limitation 386; non-timber forest products 35; nutrient cycling and decay processes 36; overgrazing 43; plant–animal interactions 40–41; plant species interactions 39–40; pollination by bees 41; precipitation 34; reforestation with coniferous monoculture 36, 42; seasonality 34; selection cut system 403; seed dispersal by animals 40–41; soil animals 41; soil and site conditions 35; temperature 32–34; timber use 35; tourism 36; wind dispersal of pollen 168
tenrecs 288
Testudo graeca 320
Thailand 657
Thuja spp. 19, 30, 43
Tilia spp. (linden/basswood) 18, 29, 35, 39, 41, 53, 347, 387
tigers 24, 292, 293, 658
Timon kurdistanica 320
Titicaca water frog 321
tomato frogs 315
tortoises 319, 320, 321, 657, 663
Trachylepis atlantica 319
tree growth response to climate change: artificial neural network methods 574–575; forest management informed by 575; increasing interest in climate–growth relationships 567–569; intra-annual scale investigation 573; large xylogenetic effects of small environmental changes 569; limitations of reconstruction methods 574; MAIDEN model 574; temperature record for last millennium 574; tree-ring width (TRW) as tracker of climate change 573; Vaganov-Shashkin model 574; water deficit, effects on xylogenesis 569, 571; *see also* dendrochronology; xylogenesis
tree hyraxes 288
treeshrews 289
Trema orientalis 628
Triadica sebifera 497
tropical cyclones (hurricanes): biodiversity, effects on 138; climate change effects on 20, 142, 602; ecosystem succession, effects on 138; extra-tropical cyclones 133; geographical range of 133; large areas of forest damaged by 134, 141; in mangrove forests 136; nutrient cycling, effects on 137; recovery of forest from 137, 139, 140; 141, 142
Tropical Forestry Action Plan 635
tropical forests: amphibians and reptiles 314; animal dispersal of pollen 168; biodiversity 61, 333, 406; birds 300, 303, 304; biogeography 63; carbon cycle 61, 64, 67, 215, 406, 412, 416, 595, 599, 600; classification of types 63–64; clear-cutting, intolerance to 350–351; climate 61, 62, 63; conifers 335; definition by latitude 61; deforestation 61, 65, 66, 67, 69–70, 165, 178, 332, 412, 600, 613; degradation 66–67, 193, 415; fire 67, 69; dominant species 333; forested wetlands 65; fragmentation 69, 450, 451; genetic diversity 168, 173, 174, 177–178; high numbers of tree species 164–165; human habitation 61–62, 65; hydrology 372, 373; Late Cretaceous tree lineages 63; lichens 272, 277; montane 56, 57, 66, 69, 230, 234, 256, 260, 272, 308, 333, 336, 337, 347, 385, 598; mycorrhizal symbiosis 347, 364; natural regeneration after harvesting 406–407; phosphorus limitation 62, 385; plot-based and other field-based studies 67–69; secondary and structurally altered forests 66–67, 193; selective logging 407; soil 62–63, 64, 66, 67, 195–196; tropical dry forests 64–65; tropical rainforests 64; tropical montane forests 66, 69; *see also* agriculture in forests; lianas; monodominance in tropical lowland forests; vascular epiphytes; wild meat hunting in tropical forests
Tropidurus torquatus 320
Tsimane' people 643
Tsuga spp. 29, 30, 34, 35, 37, 39, 43, 52, 159
tuataras 315, 318, 321
Tuckermannopsis spp. 271
tundra 8, 21, 87, 346, 351, 484, 591
Turkey 271
turtles 315, 316, 317, 318, 321, 657, 663

Ulmus spp. (elm): pests and pathogens 44, 246, 248, 504, 517, 521–522, 525526; in temperate forests 29; in subtropical forests 53
United Kingdom 29, 271
United Nations: Decade on Ecosystem Restoration (United Nations) 413, 426, 604; Declaration on the Rights of Indigenous Peoples; Food and Agriculture Organisation (FAO) 10, 331, 332, 614, 625, 635; Framework Convention on Climate Change (UNFCCC) 332, 413
United States: Alaskan boreal forest 8, 12, 18; amphibians and reptiles 321; birds 304, 306; chestnut blight 521; deer 41; droughts 557, 560; drought epiphytes 232; Indigenous forest knowledge 645–646, 646–647, 648,

649; industrial pollution 477, 488; insect disturbances 95, 97, 103, 104, 105; invasive insect species 501, 504; kudzu invasion 491; lichens 279; mammals 286; non-timber forest products 614, 620; oak regeneration 405, 434; regulation of exotic forest species 525; 'snow forests' 34; snowmelt 375; spring ephemerals 39; subtropical forest 49, 53; temperate forest 29; wildfires 120, 127, 156, 158, 375, 376; wild meat hunting 657; wind events 136, 137, 140, 141, 156; wolves 41; *see also* Hawaii
Usnea spp. 268, 269, 271

Vaccinium spp. 15, 82, 346, 403
vascular epiphytes: anthropogenic impacts on 235; cold and aridity, intolerance of 232; dispersal and establishment 235–236; droughts 558–559; evolution of 230–231; habitat complexity increased by 235; heterogeneous biogeographical distribution 231, 232–233; host-tree characteristics 236–237; interception and storage of water and decomposing organic matter 234–235; niche partitioning 234; taxonomy 230–231; scale dependencies 237–238; seed plants vs. pteridophytes 233; Southern Hemisphere abundance of 232; successional patterns 236; temperate rainforests, species diversity in 335; tropical latitudes, preference for 232; water and nutrient constraints, adaptations to 232, 236
Venezuela 55, 195, 449–451
Vietnam 654, 657, 663
voles 291
Vulpicida spp. 271
vultures 301, 302

Wai communities 658
wallabies 287, 288
Weinmannia spp. 30, 54
white pine weevil 520, 523
white-tailed deer 40, 43, 86, 405, 658
wild meat hunting in tropical forests: birds 656–657; commercial hunting 655, 659, 661, 662, 665; consumer motivations 663–664; consumption rates 659–660; cultural uses of animal products 656, 657, 658, 661; definition of wild meat 654; extinction and endangerment 654, 659, 663, 667–668; extraction rates 664–665; as income source 659, 661, 662, 663; Indigenous and rural communities 660, 669; international trade 656, 658; mammals 656, 657, 658, 665; regional mapping of 661; reptiles and amphibians 657; sustainability 654–656, 665–667, 669–670; urban demand 661–662, 663, 669–670; variety of species 655–656, 657; as vital and affordable protein source 654, 658, 669; wildlife farming as solution 670; wild meat markets 662–663; zoonotic disease 668–669
willow *see Salix* spp.
wind events: biodiversity impacts 138–139; in boreal forests 20, 137, 140, 142; climate change effects on 142–143; chronic wind stress 134; ecosystem services affected by 135; fauna, post-event mortality of 137; forest blowdowns 133; fragmented forests, vulnerability of 136; individual tree-level damage 135; invasive species enabled by 138, 140; in mountain forests 89; nutrient cycling, effects on 137; productivity increases in recovering forests 137; recovery timescales 139–140; residual trees, productivity reduced by damage 137; soil damage 137; soil characteristics, effects on root system strength 141–142; spatial scales 134; species susceptibility 136, 141; stand-level effects 135; in subtropical forests 49–50; succession effects 137–138; supercell thunderstorms 133; in temperate forests 133, 137, 140; topographic susceptibility 141; tornadoes 134; tree and forest attributes for damage susceptibility 135–136; in tropical forests 20, 133, 137, 140, 142; velocity 140; *see also* tropical cyclones
winter moth (*Operophtera brumata*) 101, 108, 247
Wollemia nobilis 88
wolves 25, 41, 86, 428
wombats 287
World Bank 635
World Wildlife Fund (WWF) ecoregions framework 330–331, 332–333, 335
wrentits 306

Xenohyla truncata 319, 320
Xylella fastidiosa 522
xylogenesis: assessment of xylem formation 567; bimodal growth pattern 567; cambial cell development 565–567; evolution of 565; hydrostatic pressure 571; periodic sampling of microcores 565, 575; photoperiod 571; seasonal thermal trends 571; time window of wood growth 564, 565; *see also* dendrochronology; tree growth response to climate change

Zetek's golden frog 321
Zimbabwe 657
zoonotic disease 668–669

For Product Safety Concerns and Information please contact our EU representative GPSR@taylorandfrancis.com Taylor & Francis Verlag GmbH, Kaufingerstraße 24, 80331 München, Germany

Batch number: 10399613

Printed by Printforce, the Netherlands